STRENGTH OF METALS AND ALLOYS

(ICSMA 7)

Proceedings of the 7th International Conference
on the Strength of Metals and Alloys
Montreal, Canada, 12–16 August 1985

Comptes rendus de la 7e conférence internationale
sur la résistance des métaux et des alliages
Montréal, Canada, 12–16 août 1985

Volume 2

INTERNATIONAL SERIES ON THE STRENGTH AND FRACTURE OF MATERIALS AND STRUCTURES

General Editor: D. M. R. TAPLIN, D.Sc., D.Phil., F.I.M.

Other titles in the series

CARLSSON & OHLSON	Mechanical Behaviour of Materials (ICM 4) (2 volumes)
EASTERLING	Mechanisms of Deformation and Fracture
FRANCOIS	Advances in Fracture Research (ICF 5) (6 volumes)
GARRETT & MARRIOTT	Engineering Applications of Fracture Analysis
GIFKINS	Strength of Metals and Alloys (ICSMA 6) (3 volumes)
HAASEN, GEROLD & KOSTORZ	Strength of Metals and Alloys (ICSMA 5) (3 volumes)
MILLER & SMITH	Mechanical Behaviour of Materials (ICM 3)
OSGOOD	Fatigue Design, 2nd edition
PIGGOTT	Load Bearing Fibre Composites
RADON	Fracture and Fatigue: Elasto-Plasticity, Thin Sheet and Micromechanisms
SIH & FRANCOIS	Progress in Fracture Mechanics
SMITH	Fracture Mechanics: Current Status, Future Prospects
TAIT & GARRETT	Fracture and Fracture Mechanics: Case Studies
TAPLIN	Advances in Research on the Strength and Fracture of Materials (ICF 4) (6 volumes)
VALLURI	Advances in Fracture Research (ICF 6) (6 volumes)

STRENGTH OF METALS AND ALLOYS

(ICSMA 7)

Proceedings of the 7th International Conference
on the Strength of Metals and Alloys
Montreal, Canada, 12–16 August 1985

Edited by

H. J. McQUEEN, J.–P. BAILON
J. I. DICKSON, J. J. JONAS
M. G. AKBEN

Volume 2

PERGAMON PRESS

OXFORD · NEW YORK · BEIJING · FRANKFURT
SÃO PAULO · SYDNEY · TOKYO · TORONTO

U.K.	Pergamon Press, Headington Hill Hall, Oxford OX3 0BW, England
U.S.A.	Pergamon Press, Maxwell House, Fairview Park, Elmsford, New York 10523, U.S.A.
PEOPLE'S REPUBLIC OF CHINA	Pergamon Press, Qianmen Hotel, Beijing, People's Republic of China
FEDERAL REPUBLIC OF GERMANY	Pergamon Press, Hammerweg 6, D-6242 Kronberg, Federal Republic of Germany
BRAZIL	Pergamon Editora, Rua Eça de Queiros, 346, CEP 04011, São Paulo, Brazil
AUSTRALIA	Pergamon Press Australia, P.O. Box 544, Potts Point, N.S.W. 2011, Australia
JAPAN	Pergamon Press, 8th Floor, Matsuoka Central Building, 1-7-1 Nishishinjuku, Shinjuku-ku, Tokyo 160, Japan
CANADA	Pergamon Press Canada, Suite 104, 150 Consumers Road, Willowdale, Ontario M2J 1P9, Canada

First edition 1986

Library of Congress Cataloging in Publication Data

International Conference on the Strength of Metals
and Alloys (7th : 1985 : Montréal, Québec)
Strength of metals and alloys (ICSMA 7)
(International series on the strength and
fracture of materials and structures)
English and French.
1. Metals — Congresses. 2. Alloys — Congresses.
3. Physical metallurgy — Congresses. I. McQueen, H. J.
II. Title. III. Series.
TA460.I532 1985 620.1'63 85–9585

British Library Cataloguing in Publication Data

International Conference on the Strength of Metals
and Alloys (*7th: 1985: Montreal*)
Strength of metals and alloys (ICSMA7) —
(International series on the strength and fracture
of materials and structures)
1. Strength of materials 2. Metals 3. Alloys
I. Title II. McQueen, H. J. III. Series
620.1'63 TA405

ISBN 0–08–031640–9

In order to make this volume available as economically and as rapidly as possible the authors' typescripts have been reproduced in their original forms. This method unfortunately has its typographical limitations but it is hoped that they in no way distract the reader.

Printed in Great Britain by A. Wheaton & Co. Ltd., Exeter

International Committee
ICSMA Conferences

ICSMA 7 Organizing Committee

Comité d'Organization de CIRMA 7

President	H.J. McQueen	Concordia University
Co-president	J.-P. Baïlon	Ecole Polytechnique
Treasurers	J.I. Dickson	Ecole Polytechnique
Trésoriers	S. Turenne	Ecole Polytechnique
Keynote Lecturers *Conférenciers invités*	J.J. Jonas	McGill University
Technical program	M.G. Akben	McGill University
Comité éditorial	W.R. Tyson	Phys. Met. Res. Lab., CANMET
	G.E. Ruddle	Phys. Met. Res. Lab., CANMET
	J. Masounave	Inst. Génie Matériaux, CNRC
	S. Lalonde	Ecole Polytechnique
	P. Lequeu	McGill University
Industrial sponsors	T.E. Dancy	SIDBEC
Soutien industriel	T.R. Leduc	Steel Co. of Canada
Arrangements	A.E. Blach	Concordia University
Technique	K. Conrod	Concordia University
Accommodation	S.V. Hoa	Concordia University
Logement	D.C. Doan	Concordia University
Exhibition	C. Forget	Leco Instruments Ltd
Exposition	G. Hamel	Techmet Canada Ltée
Social	R. McCallum	Eastern Steel Castings
Divertissements	H. Jonas	
	N.D. Ryan	Concordia University
	J. Bowles	Concordia University
	P. McQueen	
	L. Dickson	
Secrétariat	C. Labelle	Ecole Polytechnique
	I. Crawford	Concordia University
International	D.M.R. Taplin	Trinity College, Dublin

Sponsoring Organisations

Organismes de soutien

Ecole Polytechnique, Génie métallurgique
McGill University, Metallurgical Engineering
Concordia University, Mechanical Engineering
Institut de Génie des Matériaux, CNCR
Physical Metallurgy Research Laboratory, CANMET
Canadian Committee for Research on the Strength
and Fracture of Materials
Canadian Council and Montreal Chapter of
the American Society for Metals
Metallurgical Society of Canadian Institute
of Mining and Metallurgy
Ordre des Ingénieurs du Quebec
Canadian Society of Mechanical Engineers
Institut Canadien d'Ingénieurs
Canadian Steel Industry Research Association
Aluminum Company of Canada
ELKEM Metal Canada
STELCO Inc. (Steel Co. of Canada)
Hawker-Siddeley Canada Inc.
Lavalin Inc.

Contents of Volume 2

9. TOUGHNESS AND MICROSTRUCTURES — TENACITE ET MICROSTRUCTURES

Contents of Volume 1

2. ANISOTROPY AND TEXTURE – ANISOTROPIE ET TEXTURE

6. CREEP RESISTANCE — FLUAGE

Contents of Volume 3

INVITED PAPERS

SUBMITTED PAPERS

SECTION 7

Superplasticity

Superplasticité

Quantitative Discussion on Superplastic Deformation Mechanisms of Mg-4.3Al-1.0Zn-0.4Mn Alloy

C. Puquan and Z. Min

Department of Metals and Technology, Harbin Institute of Technology, People's Republic of China

ABSTRACT

Sheet specimens of Mg-4.3Al-1.0Zn-0.4Zr alloy have been elongated in the range from $\dot{\varepsilon}=3.33\times10^{-6}s^{-1}$ to~$1.67\times10^{-1}s^{-1}$, and the strain rate sensitivity indexes m were measured by means of the sudden change of strain rate which was suggested by W.A. Backofen. The contribution of deformation mechanisms, these are grain boundary sliding (GBS). Diffusion creep(DC) and dislocation slip(DS) to total elongation were estimated respectively in region I, II and III of $\log\sigma$-$\log\dot{\varepsilon}$ and m-$\log\dot{\varepsilon}$ curve with the help of the replicas of transmission electron micrograph. The results calculated are as follows:

region I : $r_{GBS}=35\%$, $r_{DC}=37.8\%$, $r_{DS}=19.6\%$.
region II : $r_{GBS}=58\%$, $r_{DC}=28.5\%$, $r_{DS}=12.3\%$.
region III: $r_{GBS}=19\%$, $r_{DC}=6.3\ \%$, $r_{DS}=69\ \%$.

According to above deat, the mutual effects and accommodation relationships of GBS. DC and DS during superplastic flow are discussed.

INTRODUCTION

Up to now, superplasticity has been considered that there are three deformation mechanisms during superplastic flow, there are GBS. DC and DS [1.2.3.4]. The purpose of this paper are to discuss and analyze the present quantitative methods, to estimate the contribution of GBS. DC and DS to total elongation for Mg-4.3Al-1.0Zn-0.4Mn alloy at superplastic and nonsuperplastic deformation condition, to analyze further the relationships between the main and accommodation deformation mechanisms.

EXPERIMENTAL PROCEDURE

SPECIMEN: test specimens were machined with a gage section 20mm wide, 2mm thick and 54mm long. A number of the sheet specimens were polished on one side and the maker lines were then scribed on this surface using the machine with a fine diamond needle. Normally, there 20 lines paralled to the tension axis and 20 lines in the transverse direction, with the avarage spacing between two lines being 15μm.

TEST: using simasu test machine, the sheet specimens were elongated respectivly at 375°C with different strain rates, the strain rate sensitivity indexes m of this material were also measured. the crosshead speed changed from $\dot{\varepsilon}=3.33\times10^{-5}s^{-1}$ to $1.67\times10^{-1}s^{-1}$ for establishing the dependences of flow stress and strain rate sensitivity index m on strain rate ($\log\sigma/\dot{\varepsilon}$ and $m/\log\dot{\varepsilon}$ curves). The scribed specimens were strained saparately about 20% with $\dot{\varepsilon}_{I}=1.67\times10^{-5}s^{-1}$, $\dot{\varepsilon}_{II}=8.33\times10^{-4}s^{-1}$ & $\dot{\varepsilon}_{III}=1.67\times10^{-1}s^{-1}$ selected at 375°C for quantitative metallography.

METALLOGRAPHY: with the help of the replicas of transmission electron micrograph, the surfaces of deformed specimens on which the marker lines were scribed were observed and the marker lines offsets, the diffusion band widthes and the variations of spacing between two markers in a same grain were measured. Using the formulars below, the contribution of GBS. DC and DS mechanisms to elongation were quantitative estimated one by one.

EXPERIMENTAL RESULTS

The best deformation temperature of this material was defined as 375°C. The maximum flow stress and strain-rate sensitivities m were measured at with different strain-rate. Fig.1 show the $\log\sigma/\log\dot{\varepsilon}$ and $m/\log\dot{\varepsilon}$ curves of Mg-4.3Al-1.0Zn-0.4Mn alloy.

MEASUREMENT OF GBS: to evaluate the strain supplied by GBS, some measure the offsets of transverse marker(Ut) perpendicular to tension and bring the value of Ut in to formular(1a), others measure the offsets of longitudinal marker(We) paralled to tension and then bring it in to the formular(1b), see Fig.2. The former is easy to use, but it neglected the effect of DC, so the results according to formular(1a) are not exact. The latter considered the effect of DC, but it brings the difficult of measuring accurately angle β.

This paper improve formular(1a) and propose the formular (1c). The formular (1c) as well solve the effect of DC, as also keep formular easy to use.

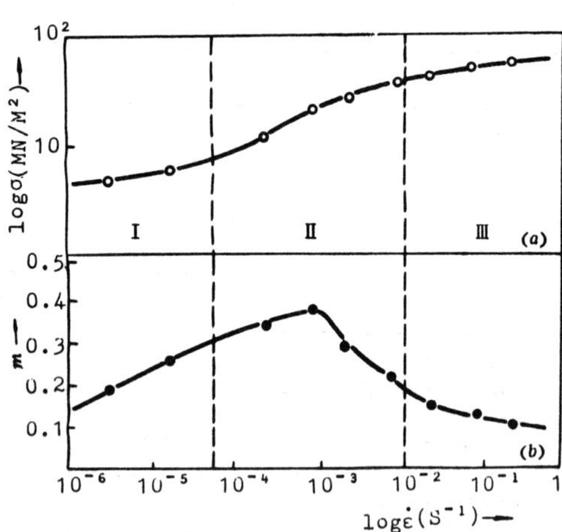

Fig.1 Mg-4.3Al-1.0Zn-0.4Mn alloy dependences of (a) flow stress σ and (b) strain rate sensitivities m on strain rate in logarithm (d=13.36μm, T=375°C)

(a)

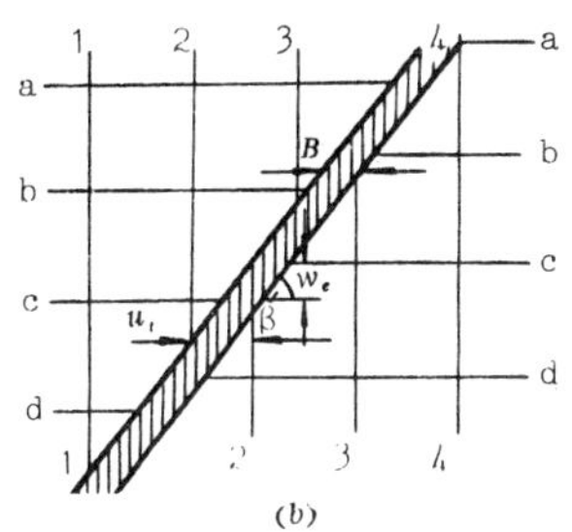

(b)

Fig.2 (a) the replica of specimen deformed at 375°C with $\dot{\varepsilon}_I=1.67\times10^{-5}s^{-1}$ to 18% (b) the scheme of estimating GBS in the state DC accompaning GBS.

$$\varepsilon_{GBS}=kn\bar{u}_t \qquad (1a)$$
$$\varepsilon_{GBS}=k'n\bar{w}_e/tg\beta \qquad (1b)$$
$$\varepsilon_{GBS}=kn(u_t-B) \qquad (1c)$$

where k, k' are experimental constant, n is the number of grain in longitudinal, $\bar{U}_t$ is the offset of transverse marker at boundary $\bar{w}_e$ is the offset of longitudinal marker at boundary, B is the width of diffusion band in longitudinal, β is the angle of grain boundary with the tension axis. The formular (1a) is the special form of formular 1c without diffusion band (B=0). This paper estimate the contribution of GBS to elongation in region I, II and III, using formular (1b) and (1c) respectively. The values obtained list in Table 1.

Table 1. The contribution values of GBS in the region I. II. III to elongation calculated with the formulars (1b) and (1c).

formular		$\dot{\varepsilon}_I=1.67\times10^{-5}s^{-1}$	$\dot{\varepsilon}_{II}=8.33\times10^{-4}s^{-1}$	$\dot{\varepsilon}_{III}=1.67\times10^{-1}s^{-1}$
(1b)	ε_{GBS} %	5.4	9.7	4.2
	r_{GBS} %	30	46	14
(1c)	ε_{GBS} %	6.3	11.6	5.7
	r_{GBS} %	35	58	19

MEASUREMENT OF DC AND DS: the diffusion bands result from substances transfering from longitudinal grain boundary to transverse ones under the action of tension stress. Grain boundary migration make the equaxial grains change to ellipsoidal grains, so that DC can provide a part of extensibility, see Fig.3. This paper measured the widthes of diffusion bands and estimated the strain provided by DC in three regions using formular (2) [8]. The results calculated list in table 2.

$$\varepsilon_{DC}=nB \qquad (2)$$

where n is the number of grain in longitudinal after deformation and B is the width of diffusion band in longitudinal. If two marker lines were scribed in a same grain, the spacing between two marker lines would be increase with elongation of

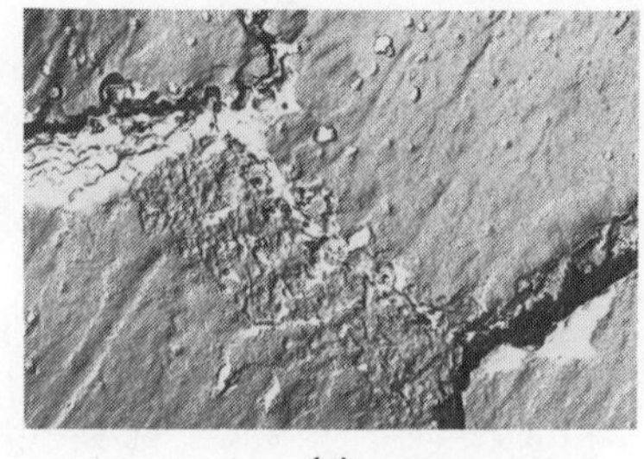

(a)

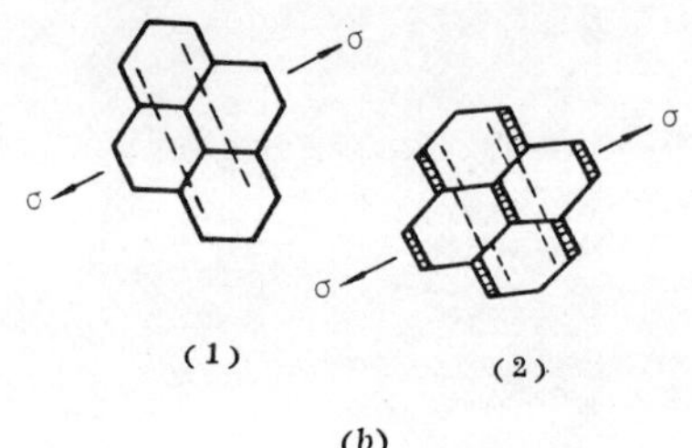

(b)

Fig.3 (a) the replica of the specimen deformed with $\dot{\varepsilon}=8.33\times10^{-4}s^{-1}$ at 375°C to 20%, (b) the scheme of estimating ε_{DC}. (1)— befor deformation, (2)— after deformation.

Table 2. The contribution of DC in three regions to elongation

	$1.67\times10^{-5}s^{-1}$	$8.33\times10^{-4}s^{-1}$	$1.67\times10^{-1}s^{-1}$
$\bar{B}$ μm	1.38	0.99	0.38
ε_{DC} %	7.2	5.7	1.9
r_{DC} %	37.8	28.5	6.3

grain during extension. The auther [2] suggested the formular (3) and the strain provided by DS were calculated in this paper. Fig.4 are the replica and scheme of estimating the contribution of DS to elongation. The results list in table 3.

(a)

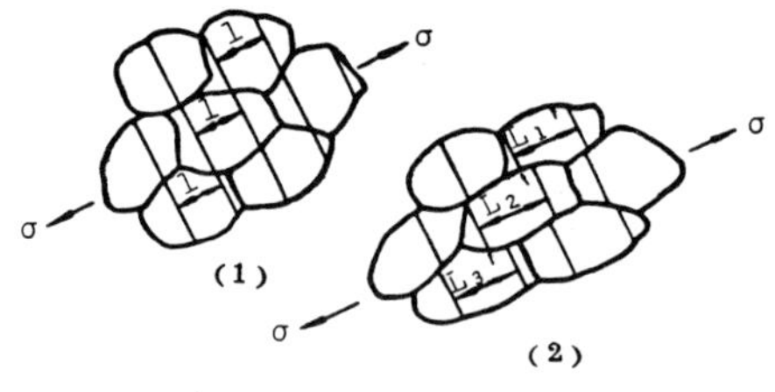

(b)

Fig.4 (a) the replica of the specimen deformed at 375°C with $\dot{\varepsilon}=1.67\times10^{-5}s^{-1}$ to 18% (b) the scheme of estimating the contribution of DS to elongation. (1)—befor deformation, (2)—after deformation.

$$\varepsilon_{DS}=(L'-l)/l \qquad (3)$$

Table 3. The contribution of DS in three regions to elongation

	$1.67\times10^{-5}s^{-1}$	$8.33\times10^{-4}s^{-1}$	$1.67\times10^{-1}s^{-1}$
(L'-l) μm	0.53	0.37	3.10
ε_{DS} %	3.5	2.5	20.7
r_{DS} %	19.6	12.3	69.0

DISCUSSION

For convenientely, the proportion and the sum of r_{GBS}, r_{DC} and r_{DS} calculated above are ploted in Fig.5. It show the three deformation mechanisms all exist in the region I, II and III. Nothing but, each proportion to total elongation is different from other.

In the region II, the GBS accounts for 58 per cent of the total elongation, but DC and DS account for 29 and 12 per cent of the total elongation respecttively. The quantitative results confirm that GBS play a main role in superplastic flow, DS and DC accompany GBS simultaneously to accommodate GBS. At 375^{0}C of superplastic flow, the critical strength of grain boundary is lower than one of grain, with the result that the load stress first reaches the strength of grain boundary and make the grain slide along boundary. But GBS is not possible to be movement without resistances, because grain boundary itself is not smooth flat and there are block grains during GBS (e.g. at triple point). When GBS develop difficulty, stress concentration is produced in local. With the stress concentration increasing, the dislocation in grain mantle [5] is possible to be operated. The result of DS can reliefs the stress concentration produced by GBS and make GBS take continiously. In the same time, DC also play a role in deformation process as DS, it can be helpfull to retain combination between grains. Precisely because of mutual accommodation of three mechanisms so that uniform deformation process can endure.

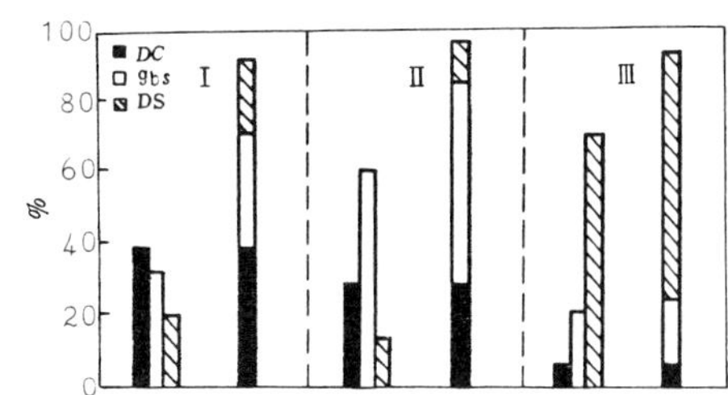

Fig.5 the proportion and the sum of r_{GBS}, r_{DC} and r_{DS} in region I, II, III.

In the region I, DC amount to 38 per cent of total elongation and the proportion of GBS decrease, these are because that with the lower strain rate, substances have enough time to transfer and the load stress almost be reliefed by DC. In the region III, the strain rate belong to the range of original plastic deformation, so that the DS amount to 69 per cent of total elongation. The extensibility contributed by GBS. DC and DS is not equal to the total elongation of specimen. It is possible as the result follow events: grain switching their neighbours [6], new surface emerging [7].

CONCLUSIONS

1. Quantitative study confirm that superplastic deformation consists of three deformation mechanisms: GBS. DC and DS.
2. GBS plays a main role in optimum condition of superplastic flow, With the strain rate decreasing or increasing, GBS in proportions to elongation all decrease.
3. Having suggested a formular (1c) through summing up the formular (1a) and (1b) to estimate simply GBS contribution to elongation.
4. In the region I, the strain supplied by DC is maximum, with

strain rate increase, this contribution to elongation decrease. In region III DC plays adominant role in deformation.

REFERENCES

1. I.I.Novikov, V.K.Portnov and T.E.Terentiva: Acta Met., V.25, 1139 (1977)
2. R.Z.Valiev and O.A.Kaibyshev: Physica Status Solidicas, V.44, 65 (1977)
3. D.Lee: Acta Met., V.17, 1057 (1969)
4. T.G.Langdon: J. Mater. Sci.,: V.16, 2613 (1981)
5. R.C.Gifkins: J. Mater. Sci., V.13, 1926 (1978)
6. M.F.Ashby and R.A.Verall: Acta Met., V.21, 149 (1973)
7. A.E.Geckinli: Met. Sci., V.17, 12 (1983)
8. I.I.Novikov, Svelhplasgutchnosge splavof se wulitramelkim zeronom. (1982)

Cavitation in Superplastic Aluminium Alloys

J. Pilling, B. Geary and N. Ridley

Department of Metallurgy and Materials Science, University of Manchester/UMIST, Grosvenor Street, Manchester M1 7HS, UK

ABSTRACT

The accumulation of cavitation damage in an aluminium alloy, Supral 220, during superplastic flow has been examined as a function of strain, strain rate, temperature and stress state. It has been found that the volume fraction of cavitation is primarily controlled by strain, but that the evolving grain size results in continuous cavity nucleation. Cavitation damage can be substantially reduced by carrying out deformation under a superimposed hydrostatic pressure of approximately half of the flow stress.

KEYWORDS

Superplasticity; aluminium alloy 2004; Supral 220; cavitation; cavity nucleation; cavity growth; hydrostatic pressure.

INTRODUCTION

Aluminium alloys including 2004 (Supral) and several in the 7000 series can be thermomechanically processed to develop fine grain structures which show considerable potential for superplastic deformation. In the 7000 series alloys the optimum strain rates for superplastic flow tend to be relatively slow ($\sim 10^{-4} s^{-1}$) while for Supral superplasticity is maintained to much higher strain rates (10^{-3} to $10^{-2} s^{-1}$) so allowing shorter forming times [1]. However, superplastic aluminium alloys tend to be prone to cavitation damage during forming and whilst this will limit the strain to failure, the presence of cavities in the formed component may adversely affect its service behaviour. In the present paper factors which influence cavitation in Supral 220 (Al-6%Cu-0.4%Zr-0.12%Si-0.1%Ge) will be identified and methods of minimising such damage will be examined.

EXPERIMENTAL

Specimens of Supral 220 have been deformed in uniaxial and biaxial tension at constant strain rates at temperatures between 430°C and 490°C, with and without the superposition of hydrostatic pressure. The level and distribution of cavitation damage which occurred during deformation was measured as a function of strain using precision densitometry and quantitative metallography [2].

RESULTS AND DISCUSSION

Effect of Strain, Strain Rate and Temperature

The variation of the volume fraction of voids (plotted logarithmically) with strain, is shown in Fig. 1. for specimens deformed in uniaxial tension at a strain rate of $4.7 \times 10^{-4} s^{-1}$ over a range of temperatures, and for samples deformed at 460°C at a range of strain rates. It can be seen that the rate of increase of void volume fraction with strain is independent of both temperature and strain rate, a feature consistent with void growth controlled by the plastic deformation of the matrix [3]. However, if the variation of the void volume fraction at a given strain is plotted against temperature or strain rate, Fig. 2, then it is apparent that the level of cavitation damage at a fixed strain increases with increasing deformation temperature or decreasing strain rate.

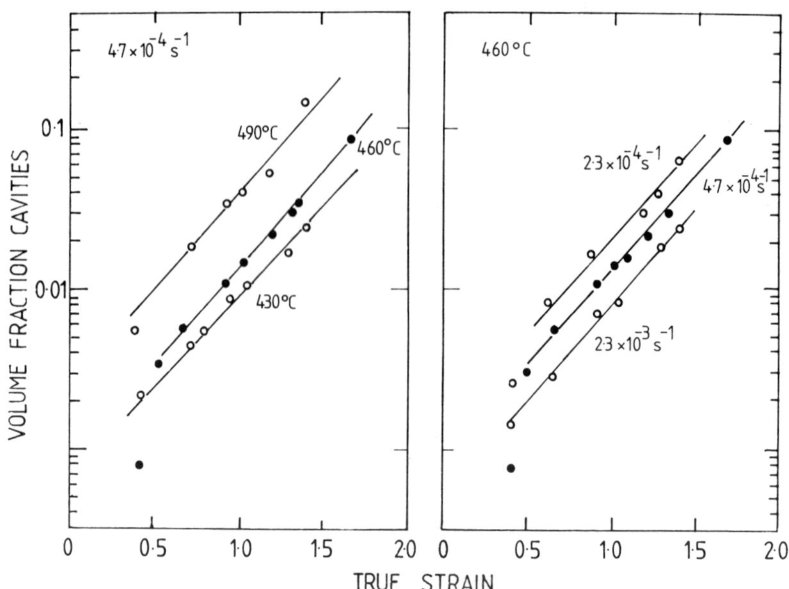

Fig. 1. Variation of the volume fraction of voids with strain at different temperatures and strain rates.

It has been suggested that cavities develop because the local stresses generated by the sliding and rotation of grains during superplastic deformation cannot be relaxed sufficiently quickly by accommodation processes such as diffusion and dislocation activity in, or adjacent to, the grain boundaries [4,5]. Once formed the cavities themselves contribute to the accommodation of deformation and also grow in response to that deformation. It would then be expected, that as the strain rate is decreased or the temperature increased, that the stresses generated by sliding and rotation would be reduced and could be accommodated more quickly. Thus it might be inferred that the level of cavitation damage would decrease, as has been observed in 7475 aluminium alloy [6], whereas in this study the opposite has been found to be true. The difference in behaviour between the two alloy types can be accounted for by examining the microstructural stability of each alloy.

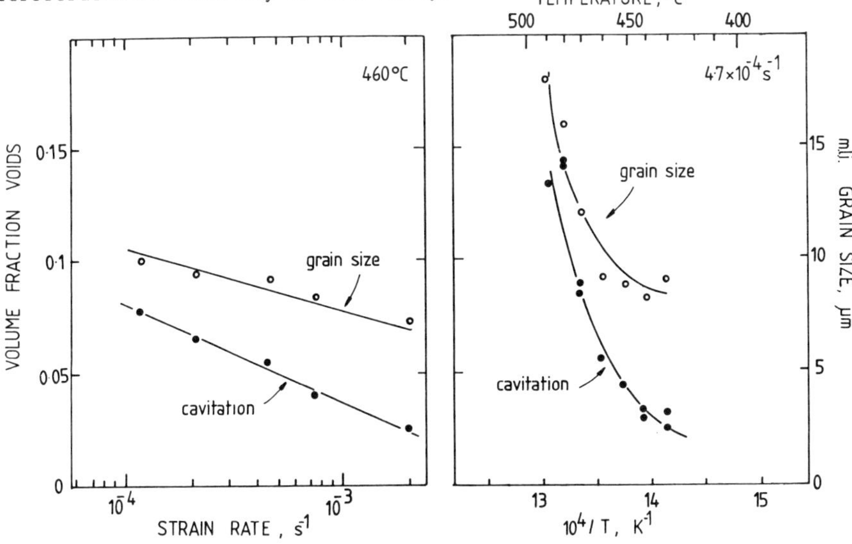

Fig. 2. Effect of strain rate and temperature on the volume fraction voids and grain size (m.l.i) at 300% elongation.

The grain size, d, of Supral 220 for a given superplastic strain increases with increasing temperature or decreasing strain rate, Fig. 2, whilst recent studies have shown the grain size of 7475 remains almost constant [6,10]. The variation in the number of small voids per unit volume with strain in Supral 220 is compared with that in 7475 in Fig. 3 after deformation at the same strain rate. Although the cavity growth rates, $dr/d\epsilon$, are similar in the two alloys [7], the effect of the continually increasing grain size in Supral 220 is to promote substantial cavity nucleation. Thus it would appear that the normally beneficial effects of temperature and slower strain rates are offset by the increased grain size which limits the accommodating processes through increased transport distances [8].

Biaxial deformation

The extent of cavitation damage developed during biaxial deformation is compared to that accumulated at the same constant effective plastic strain rate in uniaxial tension in Fig. 4. It can be seen that the rate of increase of the volume fraction cavities, when plotted against equivalent plastic strain, $\bar{\epsilon}$, (eqn. 1) is similar in both samples though the level of cavitation in the biaxial specimen is greater at any given strain. This would suggest that cavity nucleation is easier in biaxial deformation than

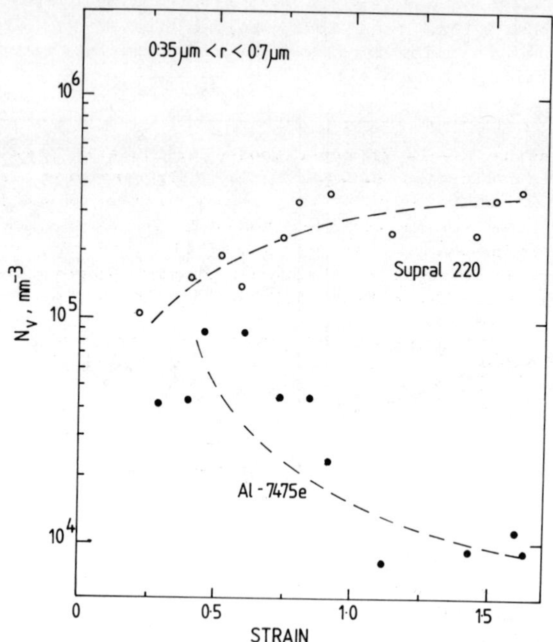

Fig. 3. Effect of strain on the population density of voids with radii 0.35μm <r< 0.7μm in a stable (Al-7475E) and unstable (Supral 220) microstructure.

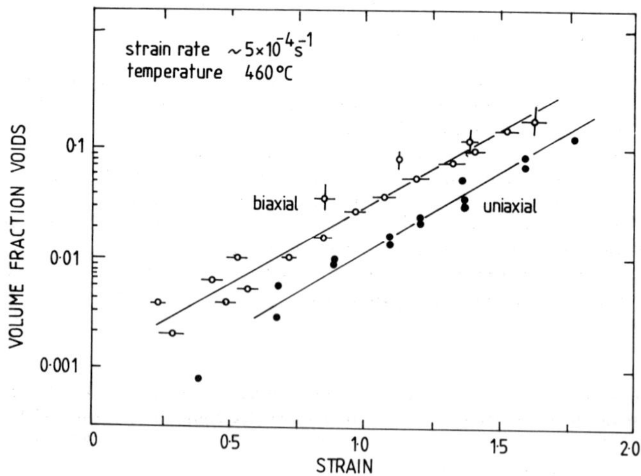

Fig. 4. Comparison of cavitation levels in biaxial and uniaxial deformation.

in the uniaxial case.

$$\bar{\epsilon} = \frac{\sqrt{2}}{3} \left[\left[\epsilon_1 - \epsilon_2\right]^2 + \left[\epsilon_2 - \epsilon_3\right]^2 + \left[\epsilon_3 - \epsilon_1\right]^2 \right]^{1/2} \tag{1}$$

Hydrostatic pressure

The effect of increasing levels of superimposed hydrostatic pressure, P, on the accumulation of cavitation damage during bulge forming is shown in Fig. 5. It is evident that the rate of increase of the volume fraction of voids decreases and the strain required to nucleate voids increases with increasing hydrostatic pressure. The effect of the confining pressure can be explained in terms of changes in the mean and maximum principal stresses and their effects on cavity nucleation and growth.

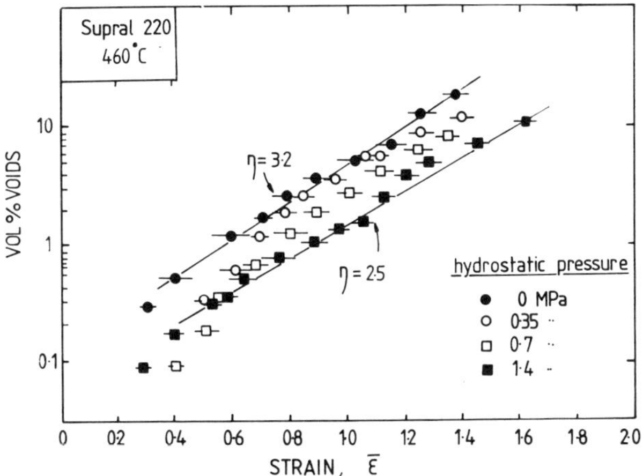

Fig. 5. Effect of hydrostatic pressure on the accumulation of cavitation damage during bulge forming.

Cavity nucleation is most likely to be a fracture dominated process whilst the initial stages of cavity growth are governed by diffusion and thus both processes are controlled by the maximum principal stress σ_1 (= σ_e-P), whilst the major part of the growth process, when r>1μm, is controlled by plasticity and the magnitude of the mean stress σ_m (= 2/3σ_e-P), [9]. Because the grain boundaries are subjected to reduced 'tensile' stresses in the presence of the hydrostatic pressure then there is a lower probability of cavity nucleation and the initial stages of growth proceed at a lower rate. In the case of plasticity controlled growth, it is known that the rate of change of cavity volume, v, with strain, ϵ, is given by:

$$dv/d\epsilon = \eta v \tag{2}$$

It has been shown experimentally that the volumetric growth rate factor, η, can be related to the stress state through the following equation [10]:

$$\eta = \eta_o \; [2(\sigma_m/\sigma_e) - 1/3] \tag{3}$$

where η_o is a material parameter related to the microstructure and cavity nucleation. In the case of equi-biaxial tension, this reduces to:

$$n = n_o\ [1-2P/\sigma_e] \tag{4}$$

Thus as the applied hydrostatic pressure becomes more significant with respect to the flow stress (σ_e) then it would be expected that the rate of accumulation of cavity damage with strain would decrease and should cease when the pressure is greater than half the flow stress. Experiments are in hand to examine this. However, because of the effect of stress concentrators in the microstructure, pressures substantially in excess of the flow stress would be required to totally suppress both cavity nucleation and the subsequent growth of voids by diffusive mechanisms. Nonetheless, the effect of increasing the hydrostatic pressure from 0 to $0.15\sigma_e$ is quite marked, with the volume per cent cavitation reduced from 4.69% to 0.76% at a strain of 1.02. Analysis of distribution of void sizes at the same strain in each sample showed that whilst the number of voids per unit volume was only slightly reduced, their mean radius was reduced from 2.6 to 1.4μm.

CONCLUSIONS

The volume fraction of cavitation in Supral 220 is controlled primarily by strain but grain coarsening which occurs during superplastic deformation can have a significant influence through its effect on cavity nucleation. It has been shown that cavitation can be reduced considerably by the application of relatively low hydrostatic pressures during forming.

REFERENCES

1. R. Swale, in 'Superplastic Forming of Structural Alloys' (edited by N.E. Paton & C.H. Hamilton), p. 307. The Metallurgical Society, AIME, Warrendale, Pa., USA. (1982).
2. D.W. Livesey and N. Ridley, Metall.Trans. 9A, 519 (1978).
3. M.J. Stowell, Metal Sci. 17, 92 (1983).
4. W. Beere, Phil.Trans.Roy.Soc. Lond. A. 288, 177 (1978).
5. R.C. Gifkins, in 'Superplastic Forming of Structural Alloys' (edited by N.E. Paton & C.H. Hamilton), p.3. The Metallurgical Society, AIME, Warrendale, Pa., USA. (1982).
6. C.C. Bampton and J.W. Eddington, Metall.Trans., 13A, 1721 (1982).
7. J. Pilling, Submitted to Mat. Sci. and Technol. (1984).
8. J. Spingarn and W.D. Nix, Acta Metall. 26, 1389 (1978).
9. C.C. Bampton and R. Raj., Acta Metall. 30, 2043 (1982).
10. J. Pilling, Unpublished work, University of Manchester, 1984.

Superplastic Deformation and Cavitation Phenomenon in 7475-Aluminium Alloy

M. K. Rao, J. E. Franklin and A. K. Mukherjee

Division of Materials Science and Engineering, Department of Mechanical Engineering, University of California, Davis, CA 95616, USA

ABSTRACT

The rate parameters for the constitutive expression for superplasticity in 7475-T6 aluminum alloy are determined from the analysis of mechanical data. The cavitation phenomenon in this industrially significant alloy is studied and the effect of hydrostatic gas pressure in minimizing the incidence of cavitation is demonstrated.

KEYWORDS

Superplasticity; aluminum alloy; cavitation phenomenon; hydrostatic gas pressure.

INTRODUTION

Fine grained 7475-T6 aluminum alloy has substantial potential for superplastic forming operations [1-3]. It is now possible to produce such fine grain microstructure by thermomechanical processing [2]. The key element in such a processing step is the ability of the Cr-rich intermetallic particles to pin the grain boundaries. The present work was undertaken in order to obtain the rate parameters to elucidate the deformation mechanism for superplasticity in this alloy. A second point of emphasis was to study the cavitation phenomenon. This alloy cavitates easily during superplastic deformation, primarily by decohesion of the particle/grain boundary interface. The application of hydrostatic gas pressure during concurrent superplastic deformation was shown to substantially reduce the incidence of cavitation.

The experimental results were analyzed using the Mukherjee-Bird-Dorn [4] relation for elevated temperature plasticity.

$$(\dot{\varepsilon}kT)/(DGb) = A(b/d)^p (\sigma/G)^n \quad (1)$$

where $\dot{\varepsilon}$ is the strain rate, the diffusivity $D = D_0\exp(-Q/RT)$ and Q is the activation energy, G = shear modulus, σ = stress, d=grain size, p = grain size exponent, n = stress sensitivity parameter = 1/m where m is the strain rate sensitivity, A is a mechanism and structure dependent constant and KT has the usual meaning. The experimentally determined values of n, p, Q and A can assist in identifying the rate controlling deformation mechanisms.

EXPERIMENTAL PROCEDURE AND RESULTS

The 7475-T 6 alloy has an average grain size of 9 μm. The grain size was measured by taking the cube-root of the product of grain sizes in three mutually perpendicular directions. Grain sizes between 10 and 35 μm were obtained by statically annealing the specimens at 530°C for various lengths of time. Mechanical tests were conducted in an argon atmosphere using a computer-interfaced MTS machine and a constant strain rate test program. The tests were performed primarily at 457°C at strain rates of 10^{-5} to 5×10^{-3} per sec to a maximum true strain level of 1.4. The cavitation experiments were conducted in an Instron machine at constant crosshead speed using a specially designed apparatus that enabled us to maintain a hydrostatic pressure of argon gas during the tensile tests.

Mechanical Properties

The true stress-true strain curves were used to obtain the double logarithmic plot of the true stress vs strain rate at various strain levels as shown in Fig. 1. The curves show the typical sigmoidal shape. At low strain rate range there is clear evidence of strain hardening. The tendency for this behavior decreases with increase in strain rate. The values for strain rate sensitivity parameter, m = $(\partial\log\sigma/\partial\log\dot{\varepsilon})$ was determined at various strain level from the slope of the $\log\sigma$ vs $\log\dot{\varepsilon}$ curves. The grain size sensitivity parameter parameter, p was determined from the slope of double logarithmic plot of stress vs initial grain size at various strain levels. The activation energy for superplastic flow in the temperature range 437 to 517°C was obtained from a plot of $\log(\sigma^{-n}\, G^{n-1}T)$ versus 1/T as shown in Fig. 2. The average value of the activation energy Q is 162 KJ/mole and it was not a function of strain or grain size. The structure parameter A was obtained from a normalized plot of $[(\dot{\varepsilon}kT)/(DGb)](d/b)^p$ versus σ/G. The A-values at various strains were obtained by extrapolating the straight line plots for different strain levels to intersect the y-axis at $\sigma/G = 10^{\circ}$. The estimated A-value decreases as the imposed strain level is increased. The results for these rate parameters are summarized in Table 1.

TABLE 1 The Variation of Rate Parameters with Superplastic Strain

Parameter	ε=0.5	ε=0.7	ε=1.0
m	0.549	0.531	0.520
n	1.82	1.88	1.92
p	2.16	1.56	1.30
Q(KJ/mole)	146	173	166
A	5.30×10^6	9.3×10^3	2.2×10^2

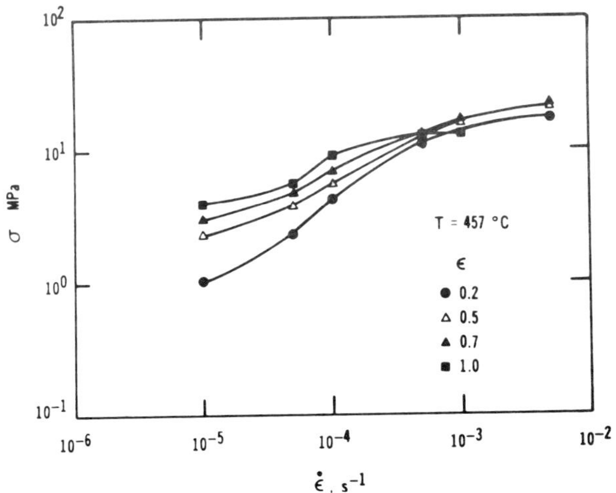

Fig. 1. True stress vs true strain rate behavior.

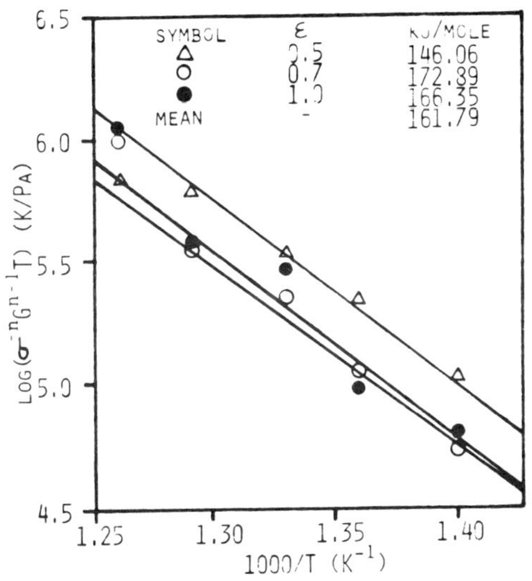

Fig. 2. Activation energy plot for various strains.

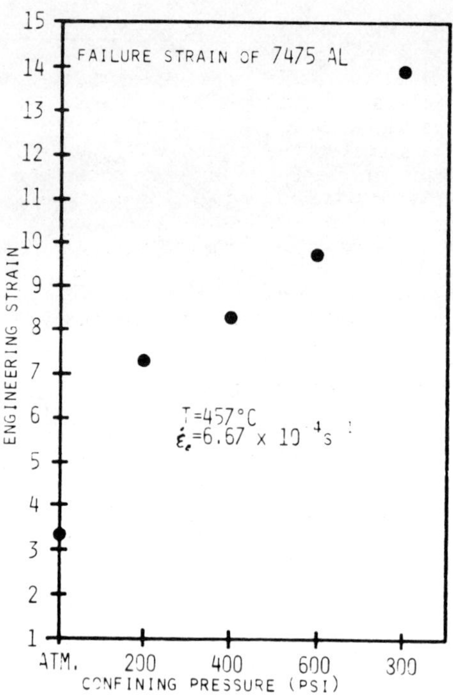

Fig. 3. Engineering failure strain vs confining pressure.

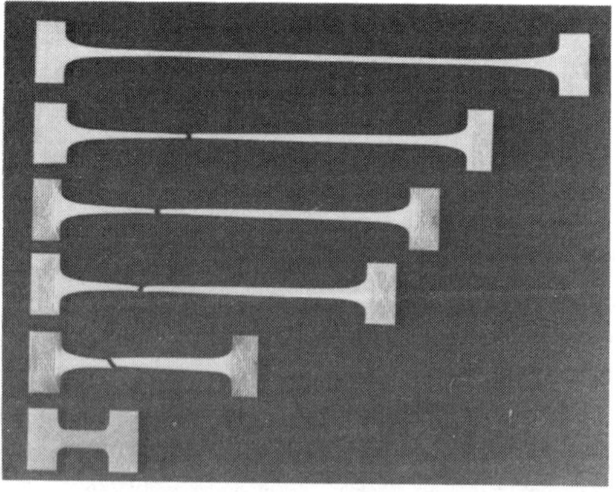

Fig. 4. 7475 Al failure strain at various pressures. (top to bottom) p=800 psi, 1330% w/o failure; p=600 psi, 980%; P=400 psi, 830%; P=200 psi, 720%; P=atm, 330%; undeformed sample. (T=457°C, $\dot{\varepsilon}_E$=6.67x10^{-4}s^{-1})

Cavitation Aspects

Preliminary studies indicated that the constituent particles present in this alloy are intimately involved in the nucleation of cavities during superplastic deformation. TEM investigation revealed grain boundary cavities at an early strain associated with such particles. The extent of cavitation was definitely a function of strain. The effect of hydrostatic gas pressure on the stress-strain curves was dramatic. One observed significant increase in terminal ductility at 200 to 600 psi gas pressure over atmospheric condition and yet bigger increase at 800 psi gas pressure. This increase in terminal ductility is demonstrated in Fig. 3. Using 800 psi confining pressure and an engineering strain rate of 6.7×10^{-4} s^{-1} at 457°C, the specimen did not fail even at 1330 percent engineering strain. A pictorial depiction of specimen elongation at various level of confining gas pressure is shown in Fig. 4.

DISCUSSION AND CONCLUSIONS

The values of the rate parameters shown in Table 1 are in agreement with the general trend and predictions of the common models for superplasticity. The stress sensitivity has a value close to two which is the usual prediction of models. The measured activation energy (162 KJ/mole) for superplasticity is close to that for lattice diffusion [5] in pure aluminum, i.e., 142 KJ/mole. The value of the grain size exponent p is approximately equal to two but it decreases with increase in strain. It is likely that this change in the value for p is associated with change in grain shape, i.e., some extent of grain elongation during superplastic flow. Optical microscopy revealed that under all testing condition, negligible changes in the grain size occurred although deformation induced grain elongation did take place to some extent. The discrepancy between this observation and that of Hamilton et al. [3] may be due to the fact that this was a different "heat", i.e., type B category of ingot. Additionally, it maybe due to the fact that in the present work grain size was quoted after making measurement in three mutually perpendicular directions. The decrease in the estimated value of the parameter A with increasing strain is very likely associated with increase in dislocation density as well as with changes in the value of grain size sensitivity parameter with strain. TEM studies clearly revealed increased dislocation density with increase in superplastic strain and significant level of dislocation-particle interaction. The activation energy for superplasticity was closer to that for volume diffusion. In Al-Zn eutectoid, which does not contain any precipitates, the activation energy for superplasticity is equal to that for grain boundary diffusion. It may be that in this alloy the dislocation has to climb over the blocking intermetallic particles in the grain interior before they can arrive at their annhilation sites, possibly at grain boundaries. In a sequential event, the slower step, i.e., the lattice diffusion controlled climb of the dislocations over the particles will control the rate. This may explain the higher value for the activation energy for superplasticity that is observed here. A recent investigation by Belzunce, Reese and Sherby [6] on the same alloy suggests the possibility of two independent creep contributions. These are, grain boundary sliding accomodated by slip, representing "mantle" deformation and slip creep presenting "core" deformation. Our electron microscopic observation is in agreement with the above suggestion.

Both optical microscopy and TEM study revealed a process of interlikage of cavities at atmospheric pressure as a function of strain. The cavities link up and elongate along the direction of tensile axis. This suggests that the cavities probably grow by continuum hole growth process [8]. The superplastic ductility in this alloy is obviously increased [7] very substantially by superimposing argon gas pressure. The most remarkable aspect of this extraordinary ductility is the fact that it has been achieved by virtually eliminating internal grain boundary cavitation and hence internal necking. The ongoing investigation will now try to ascertain if the effect of superimposed pressure is to reduce the rate of nucleation of cavities, the rate of growth of cavities or both. It will also address the question whether the decrease in or sometimes the lack of cavitation is because they just got sintered away or if they are simply being crushed by the imposed pressure and only appear to be eliminated. Such questions need be answered before the beneficial effect of hydrostatic pressure on cavitation can be fully assessed.

ACKNOWLEDGEMENTS

This work was supported in part by the U.S. Department of Energy, Grant number DE-AT03-79ER10508.

REFERENCES

1. J. A. Wert, N. E. Paton, C. H. Hamilton and M. W. Mahoney, Met. Trans., 12A, 1265 (1981).
2. N. E. Paton, J. A. Wert, C. H. Hamilton and M. W. Mahoney, J. of Metals, 34, 21 (1982).
3. C. H. Hamilton, C. C. Bampton and N. E. Paton in Superplastic Forming of Structural Alloys, (edited by N. E. Paton and C. H. Hamilton), P. 173, TMS-AIME, Warrendale, PA (1982).
4. A. K. Mukherjee, Annual Reviews of Materials Science, 9, 191 (1979).
5. T. S. Lundy and J. F. Murdock, J. Appl. Phy., 33, 1671 (1962).
6. J. M. Belzunce, K. Reese and O. D. Sherby, to be published.
7. C. C. Bampton, M. W. Mahoney, C. H. Hamilton, A. K. Ghosh and R. Raj., Met. Trans., 14A, 1583 (1983).
8. J. W. Hancock, Met. Sci., 10, 319 (1976).

Correlation Between Mechanical Properties and Microstructure in a Ni-modified Superplastic Ti-6Al-4V Alloy

B. Hidalgo-Prada and A. K. Mukherjee

Division of Materials Science and Engineering, Department of Mechanical Engineering, University of California, Davis, CA 95616, USA

ABSTRACT

The superplastic deformation (SPD) properties of Ti-6Al-4V modified by the addition of 2 percent Ni (Ti-6Al-4V-2Ni) have been investigated in the temperature range between 750-870°C and strain rates from 10^{-5} - 10^{-2} s^{-1}. It was found that the effect of microstructural changes occuring during SPD produced strain hardening and strain softening. Metallographic evidence is presented to show that the observed strain hardening is due to deformation-enhanced grain growth of both α and β phases. The strain softening was primarily due to grain size refinement. Maximum attainable super-plastic ductility was found to be associated with a dynamic balance between strain hardening and softening. Wedging and pinching off of the α-grains by the more diffusable β-phase through grain boundary diffusion seems to be an important mechanism in the deformation process of the Ni-modified Ti-alloy in the superplastic temperature and strain-rate range studied.

KEYWORDS

Titanium alloys, superplasticity, deformation mechanisms, strain hardening, strain softening, dynamic balance.

INTRODUCTION

Many titanium alloys exhibit superplastic deformation behavior in the temperature range where the α and β phases co-exist. Ti-6Al-4V, the most commonly used Ti alloy, is superplastic at temperatures between 850°C and 950°C, with optimum superplastic properties attained near 925°C [1], the temperature frequently used for superplastic forming (SPF) operations with this alloy. However, lower SPF temperatures would be desirable to reduce oxidation problems, shorten forming cycle times, and reduce die costs. Thus, there is technological interest in lowering the forming temperature of Ti-6Al-4V by addition of a suitable β-stabilizer. Furthermore, alloy additions with high diffusivities can accelerate creep rates, allowing the creep process essential to superplastic deformation to proceed at reasonable rates below the conventional SPF temperatures for the base alloy. Alloy additions of β-stabilizers such as Ni, Fe, Co, that diffuse rapidly in Ti, have been found to lower the optimum SPF temperature of Ti-6Al-4V alloy [2,3]. Among these, Ti-6Al-4V alloy modified by the addition of 2 percent Ni (Ti-6Al-4V-2Ni) appears to possess the best potential in lowering the SPF temperature without sacrificing the ease of its formability. This paper describes the work carried out to analize the superplastic deformation properties of the Ni-modified Ti-6Al-4V alloy and to correlate them to the microstructure. Furthermore, the mechanisms responsible for the observed stress-strain behavior in the Ti-6Al-4V-2Ni alloy and their correlation with the microstructural evolution produced by the deformation, are detailed.

EXPERIMENTAL PROCEDURES

The Ti-6Al-4V-2Ni alloy used in this investigation had a nominal composition (wt percent) of 5.78 Al, 0.08 Fe, 0.01 Co, 2.10 Ni and balance Ti. Original 7 kg ingots of the Ni-modified alloy (cast by TIMET) were broken down in the conventional manner, and finish-rolled in the α+β phase field to produce a fine mixture of equiaxed α and β grains. The final material was received as a sheet with thickness of approx. 1.3 mm, from which test specimens with tensile axis parallel to the rolling direction were machined. To optimize the initial microsturcture, tensile samples were annealed in purified argon atmosphere for 1 hour at 815°C.

Uniaxial tensile tests were conducted using an MTS servohydraulic machine interfaced with a PDP/11 computer and fast digital data acquisition system. A Quad Elliptical Radiant heating furnace provided a heating rate of 200°C/min, with a phaser power controller which maintained stable temperatures to within ±1°C. All tests were conducted in a purified argon atmosphere. Upon completion of the tests, specimens were quenched under load in pre-chilled argon. Samples were held at the test temperature for 20 minutes before testing.

To determine the SPD behavior pertinent to this investigation, two types of high temperature tests were conducted: 1) Continuous tensile tests to ascertain the effect of rate of deformation on the microstructural evolution, at constant temperature and for different strain levels up to fracture. 2) To minimize the effect of microstructural changes observed during SPD of the Ni-modified alloy, tensile specimens were pre-strained at temperature of 815°C and strain rate of $2 \times 10^{-4}\ s^{-1}$ up to a total true strain of 0.20.

Samples from deformed specimens were investigated by scanning (SEM) as well as by transmission (TEM) electron microscopy.

RESULTS AND DISCUSSION

It has been determined that a large volume fraction of the β-phase in α-β titanium alloys, is essential to SPD [1]. Furthermore, it has been found that equal volume fraction of the two phases in a duplex alloy, produces optimum superplastic properties [4]. The static annealing of the Ni-modified α+β Ti at 815°C for 1 hour, (SA-1-815) produced volume fractions of the α-phase (f_V^{α}) and β-phase (f_V^{α}) approximately 1:1 (49 percent β). In addition, this heat treatment also yielded similar phase sizes (λ^{α} = 6.3 μm, λ^{β} = 5.7 μm). The main purpose of the annealing program was to obtain a suitable and similar initial microsturcture for the tensile tests. Furthermore, contrary to earlier belief, it has been determined in previous work [5] as well as in the present investigation, that the microstructures of the base and Ni-modified alloys undergo substantial evolution during SPD. Thus, the concept of constant microstructure during deformation for this alloy becomes irrelevant. Instead, optimization of the initial microstructure for superplasticity is more desirable.

The temperature and strain rates selected for deformation in this investigation were such that the specimens deformed in region II of superplasticity. The results illustrated in Fig. 1, show the effect of the rate of deformation on the SPD behavior on the Ni-modified alloy. These test were conducted at high, intermediate and low strain rates (5×10^{-3}, 2×10^{-4}, 5×10^{-5} s^{-1}) for the optimum temperature of 815°C and up to true strain of 1.0. Fig. 1 also illustrates the processes of strain hardening and strain softening occurring during deformation. The strain softening observed at high strain rate

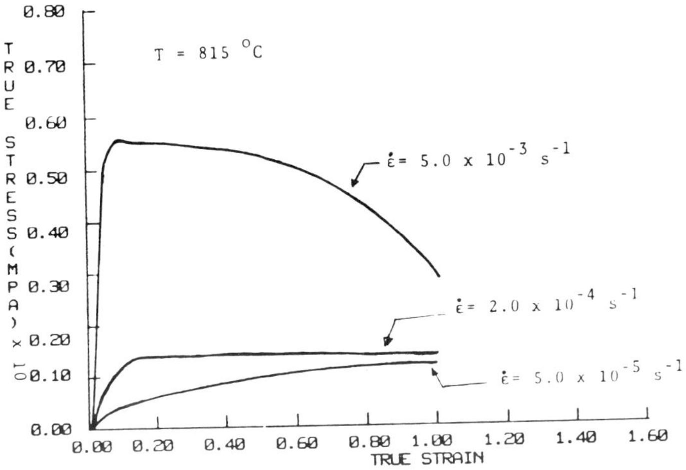

Fig. 1. Effect of strain rate on SPD behavior in Ti-6Al-4V-2Ni at optimum SPF temperature of 815°C and ε_T = 1.0.

(5.0×10^{-3} s^{-1}), was primarily due to grain refinement of both α and β phases, as illustrated in Fig. 2. This process of grain refinement during SPD is due to dynamic recrystallization, which produces fine equiaxed new grains replacing the large grains. At low strain rates (5.0×10^{-5} s^{-1}), significant strain hardening effect was observed, as shown in Fig. 1.

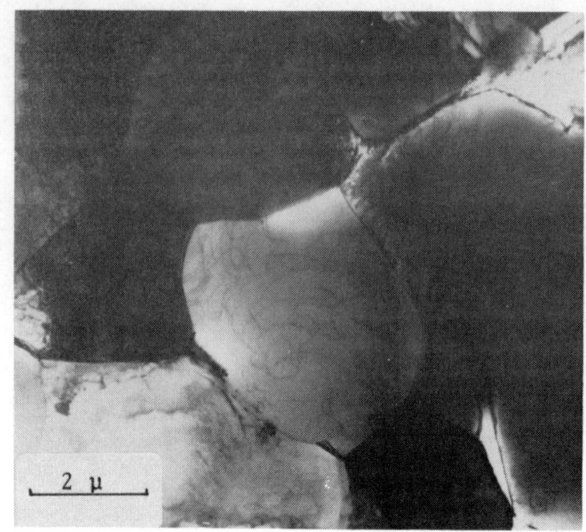

Fig. 2. Dynamic Recrystallization process associated with strain softening during SPD in Ti-6Al-4V-2Ni, at T = 815°C, $\varepsilon = 5.0 \times 10^{-3}\ s^{-1}$, $\varepsilon_T = 1.0$.

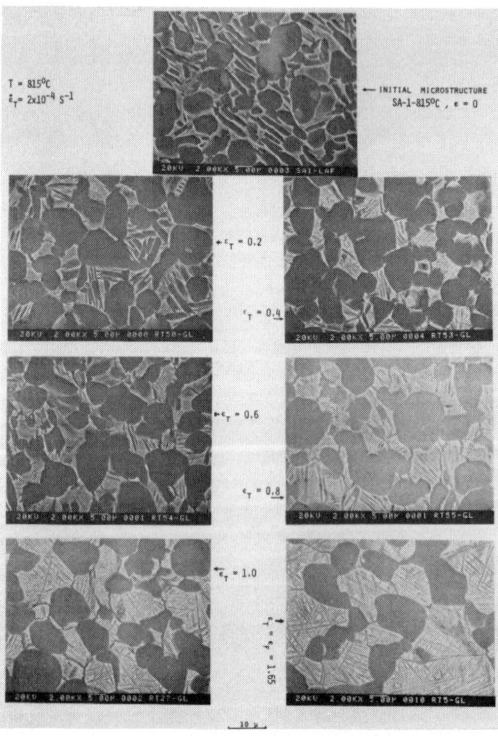

Fig. 3. Microstructural evolution as a function of strain showing the effect of deformation enhanced grain growth in Ti-6Al-4V-2Ni at constant, temperature and strain-rate.

Microstructural correlation is presented in Fig. 3. This micrograph shows that for test conditions leading to strain hardening, there is a deformation-enhanced growth of both α (dark) and β (light) phases. In the absence of overt cavitation, the maximum attainable superplastic ductility was associated with a dynamic balance between strain hardening (due to grain growth) and strain softening (due to recrystallization). This behavior was observed at intermediate strain rate (2.0×10^{-4} s^{-1}) as illustrated in Fig. 1.

For further characterization of the stress dependence of superplastic strain-rate, the true activation energy was calculated from pre-strained test samples deformed at constant strain rates of 2.0×10^{-4} and 5.0×10^{-4} s^{-1} and variable temperatures. Narrow temperature increments between 780°C (1053K) and 880°C (1153K) were used, to allow some degree of microstructural stability at each temperature and flow stress level. Furthermore, pre-strained specimens were used to obtain an activation energy value consistent with some degree of steady-state condition, with regard to the observed microstructural evolution. The average value of the true activation energy, Q_T, was found to be 145 KJ/mole. The activation energy for lattice self-diffusion Q_V, for the β-phase in Ti, obtained from radioactive ^{44}Ti tracer, have been found to be [6] about 248 KJ/mole. Thus, considering the activation energy for grain boundary diffusion, $Q_{GB} \simeq 0.6\ Q_V$, we obtain $Q_{GB}(\beta\text{-Ti}) \simeq 149$ KJ/mole, which is in good agreement with the $Q_T \approx 145$ KJ/mole determined. Hence, it appears that the true activation energy for SPD in the Ni-modified Ti alloy, is similar to that of grain boundary diffusion.

Previous investigations [2] on the mechanism for SPD in Ti-6Al-4V-2Ni, have suggested that the flow stress-strain rate behavior of each phase

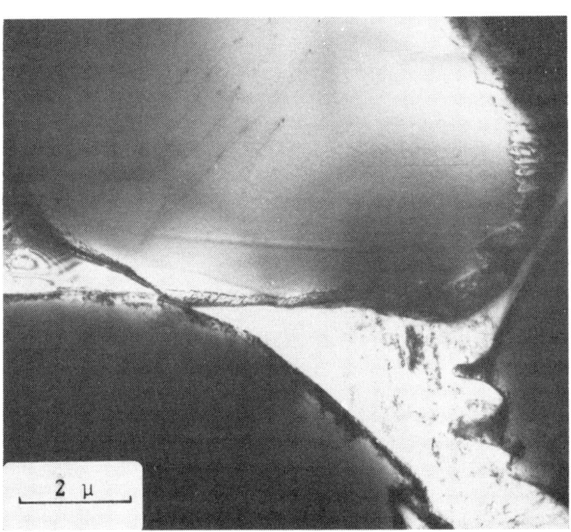

Fig. 4 Accomodation process during SPD in Ti-6Al-4V-2Ni, showing β-phase injection between α-phase grains. SA-1-815, T = 815°C, $\varepsilon = 2.0 \times 10^{-4}$ s^{-1} $\varepsilon_T = 0.6$.

could be represented by the Ashby-Verrall model [7]. Our microstructural correlations of SPD behavior in the range of temperatures and strain-rates tested suggest otherwise. It appears that the controlling deformation process is that of interphase accomodation in agreement with the Spingarn-Nix model for deformation in two phase materials by diffusion creep [8]. As shown in Fig. 4, there is an apparent wedging and pinching off of the harder phase α by the more diffusible β-phase during the accomodation process. Furthermore, as predicted by the model, the interphase boundaries -- appear to be curved -- concave in the more diffusive β-phase, and convex in the harder phase α.

CONCLUSIONS

The superplastic deformation of Ti-6Al-4V-2Ni alloy deformed in region II showed strain hardening at low strain rates and strain softening at higher strain rates. The strain softening is primarily due to dynamic grain refinement. The strain hardening is due to deformation enhanced grain growth. Maximum attainable superplastic ductility was associated with a dynamic balance between hardening and softening behavior.

ACKNOWLEDGEMENT

The authors would like to acknowledge the encouragement of Dr. Alan Rosenstein of the Air Force Office for Scientific Research and the support from AFOSR 82-0081.

REFERENCES

1. N. E. Paton and C. H. Hamilton, Met. Trans., 10A, 241 (1979).
2. J. A. Wert and N. E. Paton, Met. Trans., 14A, 2535 (1983).
3. J. R. Leader, D. F. Neal and C. Hammond, Submitted for publication (1984).
4. D.M.R. Taplin and T. Chandra, J. Mat. Sci., 10, 1642 (1975).
5. G. Gurewitz, Ph.D. Thesis, University of California at Davis (1983).
6. J. F. Murdock, et al., Acta Metall., 12, 1033 (1964).
7. M. F. Ashby and R. A. Verrall, Acta Met., 21, 149 (1973).
8. J. R. Spingarn and W. D. Nix, Acta Met., 26, 13 (1978).

Regularities of Mechanical Behaviour and Evolution of Structure of Aluminium and Its Alloys Under Superplasticity

M. M. Myshlyaev*, O. N. Senkov* and V. A. Likhachev**

*Institute of Solid State Physics, USSR Academy of Sciences, Chernogolovka, Moscow district 142432. USSR
**Institute of Mathematics and Mechanics, Leningrad University, Leningrad 198904, USSR

ABSTRACT

The specific features of the mechanical behaviour, the evolution of the microstructure and the fracture kinetics of aluminium have been studied under the superplastic conditions. The superplastic behaviour of aluminium is found to be due to the "in-situ" dynamic recrystallization, which impedes the fracture development and provides for the practically unrestricted deformation without hardening.

KEYWORDS

Aluminium alloys; superplasticity; mechanical behaviours; substructure; fracture; dynamic recovery; dynamic recrystallization.

INTRODUCTION

The purpose of this experiment is to investigate the nature of high-temperature ductility peak observed in aluminium and its alloys near 450°C.

EXPERIMENTAL

The compositions of aluminium and aluminium-base alloys studied in this investigation are given in Table 1.

Table 1 Composition of Aluminium Alloys

Materials	Composition, Wt %							
	Fe	Si	Cu	Zn	Ti	Mg	Mn	others	Al
A999	0.000	<0.001	0.000	0.000	0.000	0.000	0.000	<0.001	99.999
A92	0.01	0.05	0.008	0.005	0.002	0.01		0.004	99.92
A5	0.15	0.23	0.01	0.03	0.02	0.04		0.02	99.5
D1	0.34	0.4	4.3	0.2	0.06	0.6	0.6	0.2	base

The torsion testings of specimens were performed using a variable strain-rate torsion machine. The microstructure of the specimens was studied by means of electron microscopy and optical metallography. Details of the technique see in [1].

RESULTS

Poly - and single crystals of aluminium and several aluminium alloys exhibit superplasticity (SP) at temperatures near 450°C (Fig. 1). The total strain is here extremely sensitive to the change of temperature and strain rate. The effect of SP falls off with increasing impurity concentration: the maximum in ductility decreases, and the temperature range of the effect's existence gets narrower. Single and polycrystals of equal purity behave qualitatively similar.

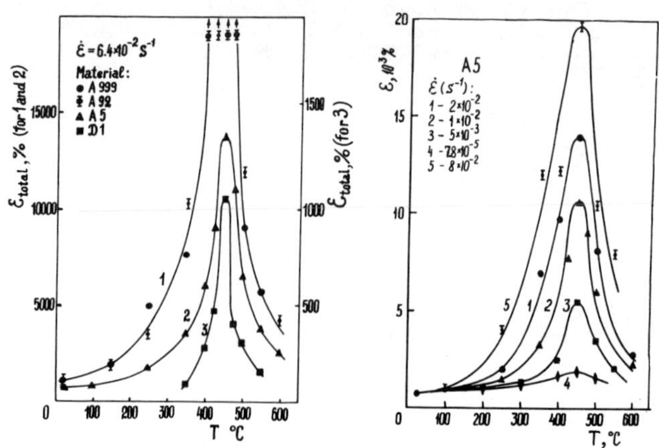

Fig. 1. The dependence of the total strain on temperature for four aluminium alloys (a) and on strain rate for alloy A5 (b).

The stress - strain curves in the SP region demonstrate a multi-stage character [1]. Deformation begins with the hardening stage. When a maximum stress is attained a brief first steady state is observed, then the softening stage ensues to be completed with the prolonged steady-state superplastic flow, where the flow stress remains constant over a large number of twists. The total strain is determined basically by the strain duration at the steady-state SP stage.

At the steady-state SP flow, the stress, σ, the temperature, T, and the strain rate, $\dot{\varepsilon}$, are connected by the relationship (found for A999, A92 and A5)

$$\dot{\varepsilon} = A\,\sigma^{n} \exp\,(-\,Q/RT), \qquad (1)$$

where n=4,4, Q=(33 ± 5) kkal/mol for A999 and A92 and n=5, Q=(55 ± 10) kkal/mol for A5; the parameter $A \sim 10^4 \div 10^5\ s^{-1}/(MPa)^n$

To the steady-state SP flow corresponded a structure consisting of slightly misoriented cells 0,5 - 6 μm in size and strongly misoriented large (30 – 200 μm) subgrains, Fig. 2. The size and misorientation of cells and subgrains substantially depend on the metal purity, temperature and strain rate. Both the subgrains and the cells remain equiaxial and their mean size and misorientation are practically invariable throughout the steady state.

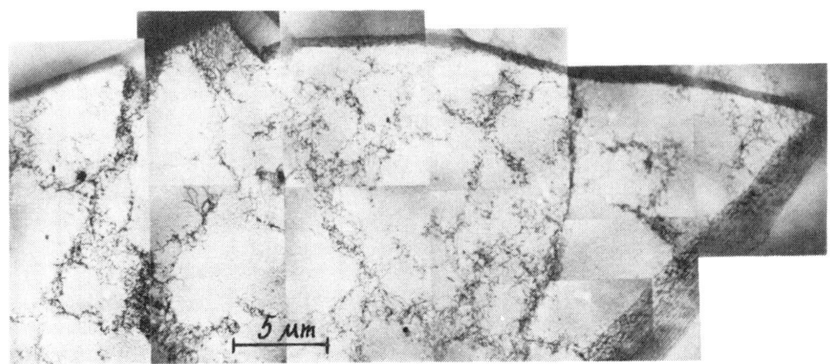

Fig. 2. Substructure of A92 at the superplastic flow. T=450°C, $\dot{\varepsilon}$=0.01 s^{-1}, ε=10000 %.

Cellular substructure with thickened cell walls appeared by the end of the strain hardening stage and persisted throughout the subsequent stages of deformation. With increasing impurity concentration the cells get smaller and their walls become thicker and denser, the dislocation density increases and for the D1 alloy reaches 10^{14} m^{-2}. In either case, the cell wall misorientations do not exceed 1°. Starting from the strains corresponding to the maximum stress on a stress–strain curve, a large number of disintegrating cell walls is observed. At the same time part of the cell walls transforms during deformation into denser subgrain boundaries confining groups of cells with unstable walls.

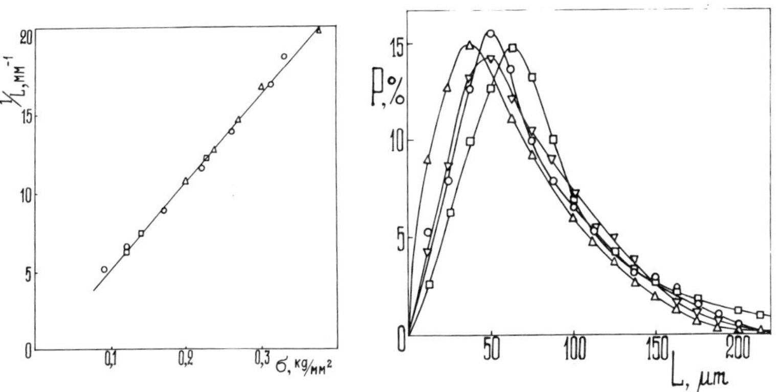

Fig. 3. (a) Subgrain mean size as a function of the SP flow stress: Δ – 420°C; O – 450°; □ – 480°. (b) The distribution of subgrains in size at 450°C: $\dot{\varepsilon}$=0.01 s^{-1}, ε=290 % (Δ), 1000 (∇), 2000 (O) and 15000 (□). A92.

The first subgrains are detected at the strain corresponding to the maximum stress. The formation of subgrains is accompanied by rather intensive disintegration of cell walls and leads to the appearance of the softening stage. The mean size of subgrains determines the flow stress, grows with increasing temperature and decreasing strain rate and only slightly changes during steady-state SP flow, Fig. 3. In spite of very high deformations, large-angle grain boundaries do not appear in pure aluminium, and the subgrains are misoriented by the angles not exceeding $1.5^o - 2^o$ for A999 and $5^o - 7^o$ for A92. In alloy A5 the misorientation of subgrains exceeds 10°; in D1 alloy dynamic recrystallization takes place during SP deformation. The (sub) grain boundaries are often wavy and strongly curved. Indirect data and in-situ studies (the nature of interactions between boundaries and inclusions, boundary meanderings, their ability to turn without destruction, ets.) evidence a high mobility of sub/grain boundaries.

Investigations of the change of specimen densities with strain at various temperatures have shown that the density decreases with increasing strain. Fig. 4 shows a change of the density defect, $\Delta\rho/\rho$, of the alloy A92. The density reduction rate, $d(\Delta\rho/\rho)/d\varepsilon$, is seen to be minimal in the ductility peak at 450^oC under optimal SP conditions. Both increase and decrease of deformation temperature with respect to 450^oC or decrease in the strain rate result in a considerable increase of the density reduction rate.

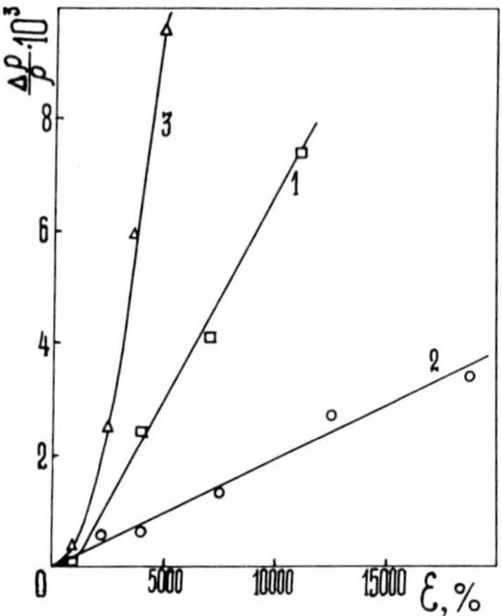

Fig. 4. Change of the density of aluminium A92 during strain at 350^oC (1), 450^o(2) and 630^o(3); $\dot{\varepsilon} = 0.01\ s^{-1}$

The analysis [1] evidences that the observed density reduction is related to the appearance of cracks and voids. In fact, a great number of cracks and elongated voids are observed in the samples failed at temperatures below the ductility peak. The samples fractured at temperatures above the ductility peak demonstrate a great number of voids in the bulk and macroscopic holes 0.5 – 1 mm in size in the sample centre. No voids are actually present in the samples twisted to a considerable extent under optimal conditions and microcracks occuring near the surface penetrate depthward only slightly. A strain rate reduction in the plasticity peak gives rise to the voids formation at the boundaries; by combining, voids form the nerworks penetrating rather deep into the sample interior.

DISCUSSION AND CONCLUSIONS

The superplasticity of aluminium is concluded to differ in nature from that of fine-grained materials [2] since, as it follows from the results and [3], the plasticity peak is not connected with the equiaxial fine grained structure and grain boundary sliding. For example, the ductility of alloy A5 is lower than that of A92 though the first alloy has smaller size of subgrains and higher subgrain misorientations.

The SP of aluminium seems to be governed by the laws of evolution of subgrain structure. The dynamic constancy of the substructure parameters for some thousands of percent strains can be explained as follows. The equiaxity of the cells may be due to a continuous repolygonization, that is the repeated disintegration of the cell walls and subsequent reformation of new cell walls at locations wich keep their average spacing and dislocation density constant [4]. Said process is actually observed throughout the deformation starting with attaining a maximum stress. It would be difficult to account for the preservation of equiaxity of strongly misoriented subgrains by the repolygonization process: the subboundary destructions are never observed, on the contrary, their great stability is noted. Most likely the subgrains remain equiaxial due to the continuous formation and migration of their boundaries. The processes occuring here appear to be similar to dynamic recrystallization but on the subgrain level. We called this process of the continuosly repeated formation and growth of subgrains during deformation the dynamic recrystallization <u>in situ</u> [1,5], by the analogy with the static recrystallization <u>in situ</u> [6] which implies essential growth of subgrains without the large angle boundaries formation. Fresh subgrains form due to coalescence of cells and grow during the course of deformation. In their interior again develops a new cellular structure and the whole process repeats again and again. The continuously repeated formation and growth of subgrains lead to the formation of the dynamically stable substructure and to the appearance of a steady-state SP flow.
Since the dislocations and subboundaries in pure aluminium are very mobile under SP conditions, the forming structure is rather homogeneous and the conditions for the fresh grain boundary nucleation are not realized. However, a decrease of the dislocation mobility due to impurities for example (which consequently gives rise to a more nonhomogeneous strains) has to result in the formation of more misoriented subgrains, and at sufficiently low subboundary mobilities - of even large-angle boundaries, i.e. nuclei of recrystallization [7]. That is what is actually observed.

The proposed mechanism of the substructure development qualitatively explains the shape of the deformation curves of aluminium under SP conditions and other experimental data as well.
It is known that such processes as the dislocation motion in interacting pile-ups [8] and the subgrain migration [9] that make the main contribution to deformation give Eq. (1) with $n=4.5$. This value of n is in good agreement with the experimental ones.
It is also known [4,7] that the activation energy of plastic flow under conditions of a developed dynamic recrystallization may be conside-

rably higher than the energy of selfdiffusion, and this in fact is observed for alloy A5. At the same time, the migration of dislocation boundaries is controlled by selfdiffusion [5,9]. This finds its reflection in the correspondence between the activation energies of the SP flow and the selfdiffusion for A999 and A92 alloys.

The subgrain boundary migration as a variety of dynamic recovery, is an effective restoration mechanism. Its driving into operation naturally increases the plasticity. The more intensive the boundary migration the higher the plasticity. Thus, the plasticity should rise with the temperature and strain rate elevations. This really takes place in the temperature range below the plasticity peak (Fig. 1).
The temperature growth intensifies the processes of diffusuonal void formation. At temperatures above the ductility peak the diffusivity increases so that the voids can easily appear and grow on boundaries and move together with the latter (at a constant strain rate the subboundary migration velocity increases much slower with the temperature [1]). Due to the same reasons a decrease in the velocity of boundary migration should facilitate the formation of voids and promote the sample fracture provided the diffusion rate of atoms remains at the former level. It is just this nature of the metal behaviour that takes plase: the plasticity falls off as the strain rate is decreased, the concentration of impurities increased and the temperature in the range above the plasticity peak enhanced.

It follows from the above data that the superplasticity of aluminium is caused by intensively proceeding dynamic recrystallization in situ, while in A5 and D1 alloys the SP is due to the dynamic recrystallization. Both of them inhibit the development of cracks and voids and make for practically unrestricted deformation without strain hardening.

REFERENCES

1. V. A. Likhachev, M. M. Myshlyaev and O. N. Senkov, in: Problems of Mechanics of Deformed Solids, p. 179 (in Russian). Leningrad Univ. Press, Leningrad (1982).
2. I. I. Novikov and V. K. Portnov, Superplasticity of Fine Grained Alloys (in Russian). Metallurgy Publ., Moscow (1981).
3. S. L. Kuzmin, V. A. Likhachev, M. M. Myshlyaev, Yu. A. Nickonov and O. N. Senkov, Scripta Met., 12, 735 (1978).
4. H. J. McQueen and J. J. Jonas, in: Treatise on Materials Science and Technology, vol.6, Plastic Deformation of Materials (edited by R. J. Arsenault), p. 393. Acad. Press, New York (1975).
5. V. A. Likhachev, M. M. Myshlyaev, O. N. Senkov, S. P. Belyaev, Fiz. Metals and Metallography (USSR), 52, 407, (1981).
6. S. S. Gorelik, Recrystallization of Metals and Alloys (in Russian). Metallurgy Publ., Moscow (1978).
7. V. A. Likhachev and O. N. Senkov, in: Fifth Symposium on Metallography. Proceedings, Part 2, p. 213, Vysoke Tatry, Czechoslovakia (1983).
8. J. Weertman, Trans. ASM, 61, 681 (1968).
9. S. F. Exell and D. H. Warrington, Phil. Mag., 26, 1121 (1972).

Fatigue Deformation of Superplastic Zn-22% Al Eutectoid Alloy

X. Zhu* and B. Ramaswami**

**Fachbereich 12.1 BAU2, Universitaet das Saarlandes, 6600 Saarbrucken, Federal Republic of Germany*
***Department of Metallurgy and Materials Science, University of Toronto, Toronto, Ontario M5S 1A4, Canada*

ABSTRACT

Constant plastic strain amplitude fatigue tests were performed on specimens of Zn-22 percent Al eutectoid alloy at room temperature (298K). A constant plastic strain amplitude of ±0.007 was used in the fatigue tests. The strain rate was varied over the range of 10^{-5} to $10^{-3}s^{-1}$. The tests were interrupted periodically to study the changes in the surface microstructure by TEM using two-stage carbon replicas. The surface microstructure was studied by SEM at the end of the test.

There was very little fatigue hardening under the conditions of fatigue deformation mentioned above. Extensive boundary sliding was observed along Zn/Zn grain boundaries and along Al/Zn interphase boundaries. There was very little boundary sliding along Al/Al grain boundaries.

The difference in the behaviour of the three types of boundaries is discussed in terms of a diffusion-accommodation mechanism of superplastic deformation.

KEYWORDS

Superplasticity; cyclic deformation; boundary sliding, intergranular fracture.

INTRODUCTION

The superplastic behaviour of Zn-22 percent Al eutectoid alloy under tensile deformation has been studied extensively (1,2). The effects of texture and microstructure on the superplastic deformation have been investigated in detail (1, 2). There is agreement among investigators on the importance of grain-boundary sliding for superplastic deformation (1, 2). However, the response of superplastic materials to alternating stress has not received much attention (3-5). The role of grain-boundary sliding on the nucleation of cracks during the fatigue deformation of Zn-0.6 percent Al has been pointed out by Bowden and Ramaswami (5). The present study extends this work to the role of interphase boundaries during cyclic deformation in Zn-22 percent Al eutectoid alloy.

EXPERIMENTAL

The alloy of the eutectoid composition was prepared from aluminium and zinc of 99.999 percent purity. The starting materials were melted in a graphite crucible under a stream of argon in an induction furnace and cast in a graphite mold. The ingots were homogenized for 24 hours at 375^{0}C. The ingots were scalped to remove surface flaws and extruded at 150^{0}C with an extrusion ratio of 21:1. The fatigue test specimens had a rectangular cross-section of 4.0mm x 1.6mm and a gauge length of 5.0mm. The test specimens were solution treated at 375^{0}C for 4 hours and quenched in iced brine at 0^{0}C. The specimens were annealed for 100 minutes at 250^{0}C to produce an average phase size of 1.0μm. They were electropolished in a 15 percent perchloric acid/methanol solution at a potential of 25 V at -40^{0}C using a current density of 200μA/mm^2.

The elongation and m, the strain rate sensitivity, were determined by tensile tests at the strain-rates ranging from 1.87 x 10^{-5} to $10^{-3}s^{-1}$. The fatigue tests were performed at room temperature with a constant plastic strain amplitude of ±0.007 in a floor model Instron testing machine modified for reverse strain. The strain rate was varied from 3.71 x 10^{-5} to 1.54 x $10^{-3}s^{-1}$. The cyclic deformation tests were periodically interrupted to prepare two-stage palladium-carbon replicas which were examined in a Philips EM300 electron microscope operating at 100kV. The surface of the gauge length section of the specimen was examined at the end of the test in a Hitachi S520 scanning electron microscope.

RESULTS

Tensile tests were performed to identify the different regions of superplastic behaviour as a function of the strain rate. The strain-rate sensitivity factor m was in the range 0.20 to 0.23, or region II, for the range of strain rates 3.71 x 10^{-5} to 8.03 x $10^{-4}s^{-1}$ and was 0.175, or in region III, at the highest strain rate 1.54 x $10^{-3}s^{-1}$ used in this series of tests. The elongation was in the range 128 to 153 percent in region II and 73 percent in region III.

Randomly oriented lines were lightly scratched on some of the electropolished tensile test specimens before the test. Grain-boundary offsets and grain rotations, indicating grain-boundary sliding, were observed on these specimens tested in region II of the strain rates. Thus, the Zn-22 percent Al alloy deformed by grain-boundary sliding, a characteristic superplastic mode of deformation, in region II of the strain rates at room temperature.

The fatigue response of the Zn-22 percent Al alloy is shown in Fig. 1 for strain rates in the range 3.71 x 10^{-5} to 1.54 x $10^{-5}s^{-1}$. The average of the peak stresses in the tension and the compression half cycles is plotted against the logarithm of the number of cycles. The average stresses were almost constant for a large number of cycles before decreasing gradually towards the end of the test. None of the specimens had fractured at the end of the test. Intergranular cracking was observed, as shown in Fig. 2, on the surface of the specimens at the end of the test.

The grain size was measured both before the beginning and the end of the fatigue test. There was no significant change in grain size during the test.

The surface microstructure of the specimen after fatigue deformation at a strain rate of 3.33 x $10^{-4}s^{-1}$ is shown in Fig. 3. The relative intensity of the x-ray corresponding to the K_α radiation of aluminum is superimposed on

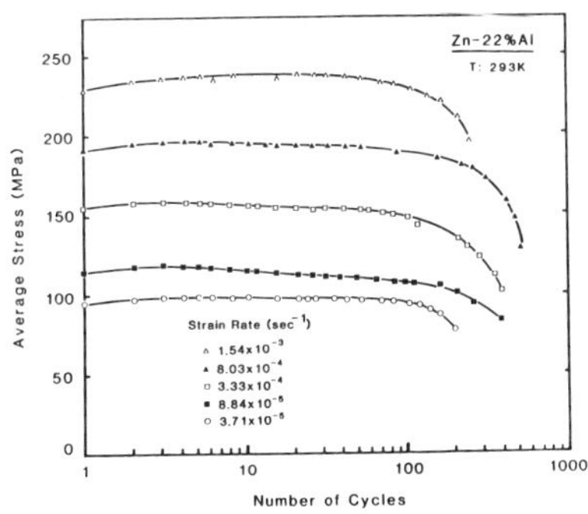

Fig. 1. The cyclic response of the Zn-22 percent Al alloy at various strain rates.

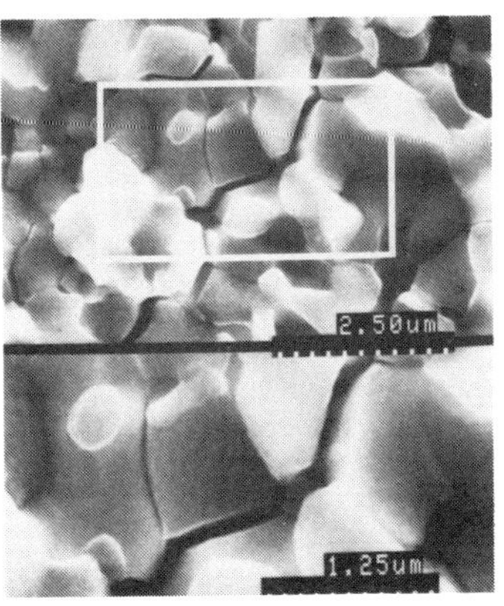

Fig. 2. Surface microstructure after 505 cycles of fatigue deformation at $\dot{\varepsilon}=1.54\times10^{-3}s^{-1}$ showing intergranular cracks.

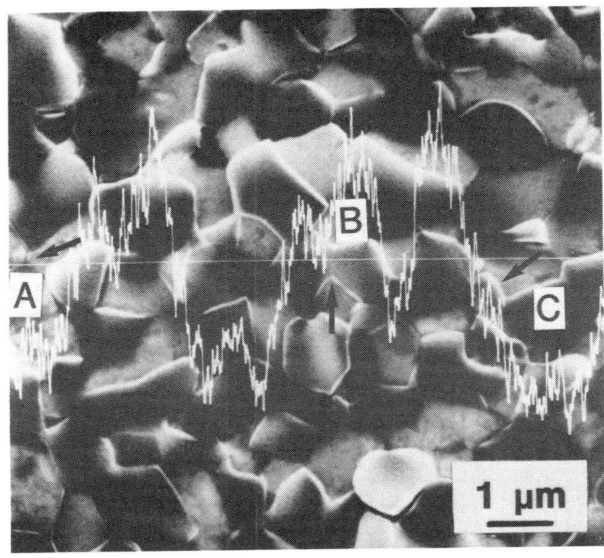

Fig. 3. Surface microstructure after fatigue deformation at $\dot{\varepsilon}=3.33\times10^{-4}s^{-1}$. A:Zn/Zn interface. B:Al/Al interface. C:Zn/Al interface.

the SEM micrograph to identify the Zn-rich and Al-rich phases. The wavy line indicates the composition profile of Al along the white straight line. Grain-boundary sliding results in offsets at the boundaries whenever the sliding displacement has a component normal to the specimen surface. As indicated by arrows at A and C, grain-boundary sliding has occurred along the Zn/Zn grain-boundary and Zn/Al interphase boundary respectively. There is no grain-boundary sliding along the Al/Al grain boundary, as shown at B. Similar results were observed in other SEM micrographs of specimens tested at this strain rate.

The highest m value (0.235) and, hence, the maximum superplastic behaviour were obtained at a strain rate of $8.84 \times 10^{-5} s^{-1}$ in the tensile test. The surface microstructure of the specimen after fatigue deformation at this strain rate is shown in Fig. 4. While extensive grain-boundary sliding occurred along the Zn/Zn grain-boundary and Zn/Al interphase boundary as shown at A and C respectively, very little grain-boundary sliding occurred at Al/Al grain-boundaries as shown at B. The study of two-stage replicas of the surface microstructure by transmission electron microscopy confirmed these observations. The surface microstructure of the specimens subjected to fatigue deformation at other strain rates - viz., $8.03 \times 10^{-4} s^{-1}$ and $3.71 \times 10^{-5} s^{-1}$ - in region II of superplastic behaviour with high m values was similar to those mentioned above.

However, the surface microstructure of the specimen subjected to fatigue deformation at $1.54 \times 10^{-3} s^{-1}$ - in region III - was somewhat different from that described above. The surface microstructure after 200 cycles of fatigue deformation is shown in Fig. 5. There is very little grain-boundary sliding at any of the boundaries.

In summary, during the fatigue deformation of the Zn-22 percent Al alloy at room temperature,

(1) the average peak stress increases with increasing strain rate indicating that the alloy is sensitive to strain rate;
(2) grain boundary sliding occurs predominantly along the Zn/Zn grain boundaries and Zn/Al interphase boundaries with very little or no grain boundary sliding along the Al/Al grain boundary in region II of strain rates characteristic of superplastic behaviour;
(3) there is very little grain-boundary sliding along the Zn/Zn, Al/Al and Zn/Al boundaries in region III of strain rates characteristic of normal behaviour.

DISCUSSION

The three types of interfaces present in the Zn-22 percent Al alloy behaved differently during the cyclic deformation. The extent of boundary sliding was large at Zn/Zn grain boundaries, moderate at Zn/Al interphase boundaries and small or non-existent at Al/Al grain boundaries.

This behaviour is very similar to that observed for the Pb-Sn eutectic (6) and Zn-22 percent Al eutectoid (7) alloys in tensile deformation. Vastava and Langdon (6) have shown that during the tensile deformation of the Pb-Sn eutectic alloy the sliding is maximum at Sn/Sn interfaces, moderate at Sn/Pb interfaces, and very little at Pb/Pb interfaces. Similarly, Shariat et al (7) have shown that during the tensile deformation of the Zn-22 percent Al eutectoid alloy, the sliding is maximum at Zn/Zn interfaces, moderate at Zn/Al interfaces, and very little at Al/Al interfaces. They have explained their re-

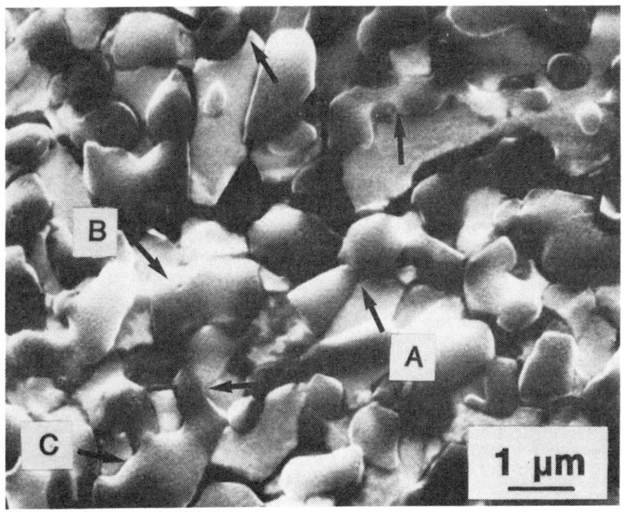

Fig. 4. Surface microstructure after fatigue deformation at $\dot{\varepsilon}=8.84\times10^{-5}s^{-1}$. A:Zn/Zn interface. B:Al/Al interface. C:Zn/Al interface.

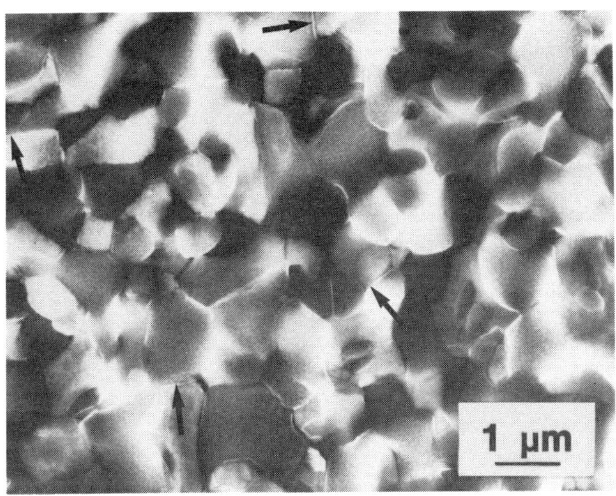

Fig. 5. Surface microstructure after fatigue deformation at $\dot{\varepsilon}=1.54\times10^{-3}s^{-1}$. Arrows indicate different types of interfaces.

sults on the basis of grain-boundary diffusivity (6,7).

The microstructural observations after fatigue deformation can be explained on the basis of the model of grain-boundary sliding developed by Ashby (8). He has shown that the intrinsic velocity of grain-boundary sliding is proportional to δD_{GB}, where δ is the width of the boundary and D_{GB} is the coefficient of grain boundary diffusion. From the coefficients of grain boundary diffusion given in Table 4 of Reference 7 and assuming that the grain boundary width does not depend significantly on the type of interface, it is found that $\delta D_{GB}(Zn)$ is about four orders of magnitude larger than $\delta D_{GB}(Al)$. Therefore, the grain-boundary sliding at Zn-Zn grain boundaries will be larger than at Al/Al grain boundaries. However, the data for diffusion along the Zn/Al interphase boundaries is not available in the literature. It is probable that Zn diffuses faster than Al at Zn/Al interphase boundary, thus contributing to the sliding at the Zn/Al interphase boundary.

CONCLUSIONS

The constant plastic strain amplitude fatigue deformation of Zn-22 percent Al alloy at room temperature did not indicate any hardening during the deformation for many cycles. There was a gradual drop in the peak stress after a large number of cycles. The plastic flow was accommodated by boundary sliding, which was large at Zn/Zn interfaces, moderate at Zn/Al interfaces and negligible at Al/Al interfaces.

ACKNOWLEDGEMENTS

The financial support of the Natural Sciences and Engineering Research Council of Canada (under NSERC Grant A-4933 to B.R.) and the grant of a Fellowship from the Ministry of Education of the People's Republic of China (to X.Z.) are gratefully acknowledged.

REFERENCES

1. T.G. Langdon, Met. Trans. 13A, 689 (1982)
2. J.W. Edington, Met. Trans., 13A, 703 (1982)
3. J. Woodthorpe and R. Pearce, Int. J. Mech. Sci., 16, 699 (1974)
4. J. Woodthorpe and R. Pearce, Metal Sci., 11, 103 (1977)
5. J.W. Bowden and B. Ramaswami, "Strength of Metals and Alloys" ICSMA 6, Ed: R.C. Gifkins, p.683, Pergamon Press (1978)
6. R.B. Vastava and T.G. Langdon, Acta Met., 27, 251 (1979)
7. P. Shariat, R.B. Vastava and T.G. Langdon, Acta Met. 30, 285 (1982)
8. M.F. Ashby, Surface Science, 31, 498 (1972)

Superplastic Deformation Behaviour of Two Microduplex Stainless Steels

N. Ridley and L. B. Duffy

University of Manchester/UMIST, Department of Metallurgy and Materials Science, Grosvenor Street, Manchester M1 7HS, UK

ABSTRACT

The superplastic tensile deformation behaviour of two microduplex stainless steels, AVESTA 3RE60 and IN744, has been compared over the temperature range 850-1000°C, and the structural changes which result from superplastic flow have been examined. While IN744 is superplastic over a wide range of temperatures, significant superplastic behaviour in 3RE60 is limited to temperatures of 950°C or higher. The different behaviours are due to the high critical temperature (~983°C) for σ-phase formation in 3RE60. Hence, σ-phase forms during deformation at temperatures up to 950°C leading to high flow stresses, to an unrealistically high activation energy for Region II superplastic flow and to limited tensile elongations. At high deformation temperatures 3RE60 shows lower levels of cavitation than IN744 primarily because it contains fewer precipitates to assist cavity nucleation.

KEYWORDS

Superplasticity; microduplex alloys; duplex stainless steels; strain rate sensitivity; cavitation; sigma phase.

INTRODUCTION

A number of duplex stainless steels containing approximately equal proportions of α and γ phases are commercially available. These steels can develop fine grain sizes when processed to sheet and because of their microstructural stability they are superplastic at elevated temperatures. Much of the work to date has been concentrated on the earlier developed alloy, IN744, of nominal composition:- Fe-26%Cr-6.5%Ni-0.4%Ti-0.5%Mn-0.5%Si-0.04%C max [1-6]. This steel exhibits marked superplasticity over the temperature range 700°C-1050°C, although it shows some cavitation damage, primarily associated with Ti(CN) inclusions, during deformation [3,7]. Recently several α/γ stainless steels based on the Fe-Cr-Ni-Mo system have been developed and in the present work the superplastic behaviour of one of these alloys, AVESTA 3RE60, of nominal composition:- Fe-18.5%Cr-4.9%Ni-

2.8%Mo-1.5%Mn-1.5%Si-0.03%C, has been examined and compared with IN744.

EXPERIMENTAL

The two steels, AVESTA 3RE60 and IN744, were received in the form of sheet of thickness 1.5mm and 0.75mm, respectively. Tensile specimens of gauge length 10mm and gauge width 5mm were machined with the tensile axis parallel to the sheet rolling direction. Straining was carried out in a high temperature rig attached to an Instron machine using a dry argon atmosphere. Structural changes resulting from superplastic flow were examined using optical microscopy, SEM, X-ray diffraction and densitometry.

RESULTS

Stress-Strain Rate Characteristics

After applying a pre-strain of 50% to remove the initial microstructural banding, values of flow stress, σ, corresponding to a range of strain rates, $\dot{\varepsilon}$, were determined at temperatures varying from 850-1000°C. For both alloys, at each temperature, a sigmoidal log σ versus log $\dot{\varepsilon}$ curve was observed with an increase in deformation temperature leading to a fall in flow stress for a given strain rate, and to an increase in the maximum slope, m, of the curve. However, major differences between the data for the two steels were apparent in that the temperature dependence of flow stress and m value was much more marked in 3RE60 than for IN744. These differences can be clearly seen in Fig. 1 which shows data at 850°C and 1000°C. Although the flow stresses are similar at 1000°C the values for 3RE60 are markedly higher at 850°C. The variation of m with temperature is listed in Table 1 for the two steels. For IN744 the m values in Table 1, are in good agreement with those reported by other workers[2-4, 6].

TABLE 1 Variation of m value with deformation temperature

Temperature	m value	
	3RE60	IN744
850°C	0.41	0.47
900°C	0.51	0.51
950°C	0.65	0.52
1000°C	0.75	0.54

Activation energies for superplastic flow

The flow stress-strain rate data for the two alloys has been used to evaluate activation energies for superplastic flow in Region II of the logarithmic stress-strain rate curve. The relationship:

$$\dot{\varepsilon} = K\,\sigma^{\frac{1}{m}} \exp\left[-\frac{Q}{RT}\right] \tag{1}$$

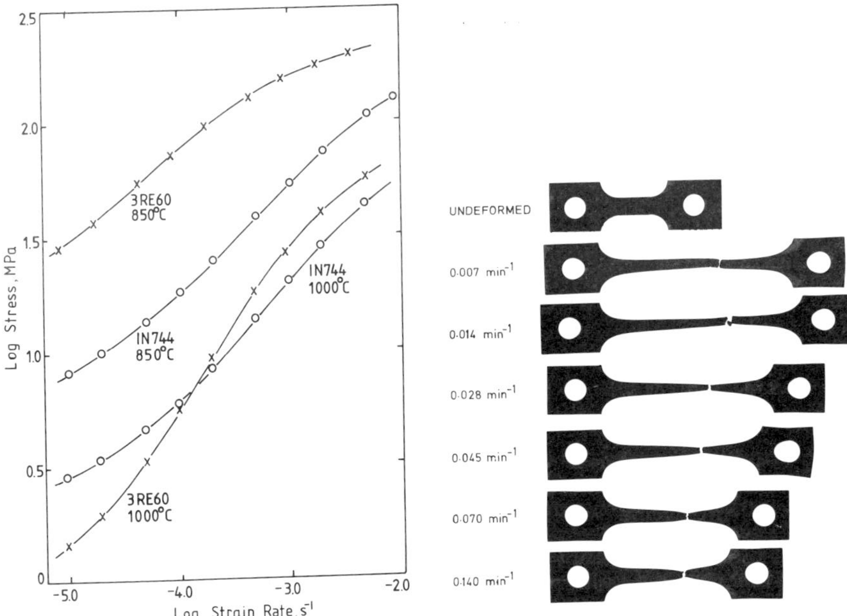

Fig. 1. Stress-strain rate data.

Fig. 2. Fracture profiles of 3RE60 at 900°C.

in which K is a constant and Q is the activation energy, was assumed to describe superplastic flow [8]. At a constant strain rate $Q = Q_{\dot{\Sigma}}$. Hence by plotting $\ell n\ \sigma$ versus 1/T for a constant strain rate gives a straight line of gradient $m\ Q_{\dot{\Sigma}}/R$. Values of $Q_{\dot{\Sigma}}$ obtained for Region II were $161kJmol^{-1}$ for IN744 and $366kJmol^{-1}$ for 3RE60. For IN744 the activation energy obtained is in reasonable agreement with the value of $187kJmol^{-1}$ obtained recently by Kashyap and Mukherjee for similar material [9], although lower than that of previous workers [2, 4]. Published values for the activation energy for grain boundary diffusion of Fe, Cr and Ni in Fe-Ni-Cr alloys are $\sim 180kJmol^{-1}$ [10]. Hence, for IN744 the present observations are consistent with grain boundary diffusion as the rate controlling process for Region II superplastic flow over the temperature range 850-1000°C. The activation energy value obtained for 3RE60 over the same temperature range, although subject to some uncertainty because of the wide variation of m value (Table 1), is difficult to reconcile with diffusional processes.

Elongation to failure tests

Elongation to failure tests were performed on specimens of 3RE60 deformed at constant strain rates ranging from 10^{-4} to $2 \times 10^{-3}s^{-1}$ (0.7-14% min^{-1}). The maximum tensile elongations obtained were lower than those previously reported for IN744 [3, 4], with the differences being increasingly greater with decreasing temperature (Table 2). The results show that the temperature range for which significant superplastic behaviour can be obtained in 3RE60 is much more limited than for IN744.

The profiles of fractured tensile specimens showed two types of failure behaviour. For the IN744 specimens there was litle sign of non-uniform thinning at the fracture surface while similar behaviour was apparent for 3RE60 at 1000°C, 950°C and at the slower strain rates at 900°C (Fig. 2). These specimens undergo cavitation during superplastic flow, as will be seen, and it is the interlinkage of cavities perpendicular to the gauge length which leads to the pseudo-brittle premature fracture observed.

TABLE 2 Maximum elongation to failure, %.

Alloy	Maximum elongation %			
	1000°C	950°C	900°C	850°C
3RE60	700	560	400	225
IN744 (Ref.)	714*	903*	780	781

* interpolated values.

However, specimens of 3RE60 deformed at 850°C and at the faster strain rates at 900°C (Fig. 2) exhibit necking in the vicinity of the fracture face and failure probably occurs due to the combined effects of cavitation and the development of an unstable neck.

The failure behaviour is borne out by the shape of the corresponding true stress-true strain curves. (Fig. 3) For example, at 850°C at the optimum strain rate for superplastic flow in 3RE60, a maximum in the flow stress occurred after only 50% elongation which is untypical of superplastic behaviour. This was followed by a continuing decrease in flow stress and eventual failure involving localised necking. At the higher deformation temperatures, 950°C and 1000°C, after the initial removal of microstructural banding, grain growth occurs and leads to a progressive increase in flow stress throughout most of the test.

Microstructural features

The as-received materials showed microstructural banding with the average grain size, defined as 1.43 times the mean linear intercept, being 8.0μm for 3RE60 and 5.2μm for IN744. During deformation, and following the removal of banding, grain growth occurred and was seen to be a function of strain, strain rate (or time) and temperature. However, grain growth was relatively slow at 850-900°C, and even at higher temperatures the extent of grain growth during the time required to measure flow stress-strain rate relationships, and hence m values, was small. Other microstructural changes involved cavitation and the appearance of an additional phase in 3RE60 during deformation at temperatures of 950°C and below.

The additional constituent in 3RE60 was thought to be σ-phase since calculations based on the composition of the alloy showed that this phase could form below 983°C [11]. Studies were made on the formation of the phase in material which had been superplastically deformed or subjected to isothermal annealing. X-ray diffraction lines corresponding to sigma- phase (~15% volume) were obtained from a specimen annealed at 800°C for 15 hours, while EDAX analysis in an SEM showed the new phase to be enriched in ferrite stabilising elements such as Cr, Mo and Si, and depleted in austenite

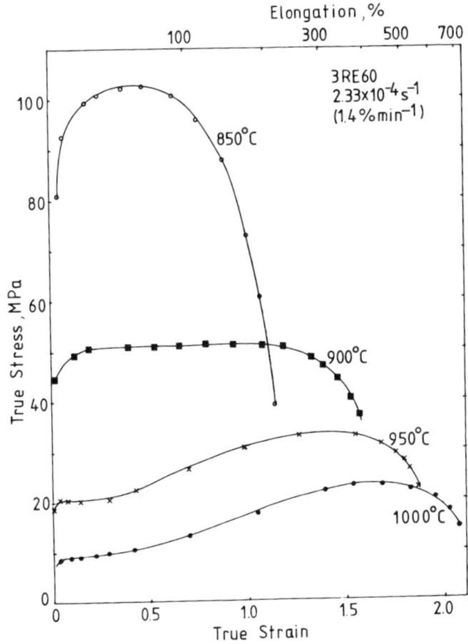

Fig. 3. True stress-true strain data for 3RE60.

Fig. 4. 3RE60 deformed at 900°C. Shows σ-phase and cavitation.

stabilisers such as Mn and Ni. For example after 90 minutes at 900°C the new phase contained: 21.0%Cr-14.5%Mo-3.4%Si-2.4%Ni-0.7%Mn. The distribution of σ-phase in 3RE60 after deformation at 900°C to an elongation of 300% at a strain rate of $2\times10^{-4}s^{-1}$ is shown in Fig. 4. Sigma phase was not detected in IN744 after superplastic deformation or annealing.

Cavitation behaviour

To determine the extent of cavitation in 3RE60 precision density measurements were made on deformed specimens. Density differences between the gauge heads and gauge lengths were attributed to cavitation damage resulting from superplastic deformation. During deformation at 950°C and 1000°C cavitation increased with increasing strain but was relatively independent of strain rate and temperature (Table 3).

For similar deformation conditions 3RE60 cavitates less than IN744. Ridley et al [12] reported > 1% cavities after 200% elongation while Smith et al [3] measured cavitation levels of 2% after 500% elongation. The previous work on IN744 has shown the cavitation level is closely related to the Ti(C,N) content of the steel [7]. Examination of unetched specimens of 3RE60 revealed relatively few inclusions compared to IN744, and this is likely to account for the lower levels of cavitation in the former alloy.

GENERAL DISCUSSION AND SUMMARY

The present work has shown that, despite having similar microduplex structures, the two stainless steels, 3RE60 and IN744, show very different superplastic flow behaviours over the temperature range 850-1000°C. The elongation to failure is appreciably lower for 3RE60 particularly for temperatures of 850-900°C, and the temperature dependence of flow stress is

TABLE 3 Cavitation, %, in 3RE60 on tensile straining

	Cavitation %				
Temperature	100%	200%	300%	400%	500%
950°C	0.12	0.21	0.45	0.66	1.03
1000°C	0.11	0.26	0.47	0.72	1.09

much more marked in 3RE60 than in IN744. This latter behaviour leads to activation energies for Region II superplastic flow which are considerably greater in 3RE60. For IN744 the activation energy is in reasonable agreement with that for grain boundary diffusional control of Region II superplastic flow, whereas the much higher value for 3RE60 is difficult to reconcile with diffusional processes.

These differences between the two alloys are related to different critical temperatures for σ-phase formation. This phase was observed during superplastic flow at temperatures up to 950°C, which is close to the calculated critical temperature of 983°C, but not in IN744 even after prolonged annealing at 800°C. The nucleation of σ-phase at grain boundaries will inhibit the boundary processes associated with superplastic flow thereby raising flow stresses and reducing elongations to failure. The effect on the flow stress will be more marked at lower deformation temperatures where σ-phase forms more rapidly and, as a consequence, this will lead to erroneously high activation energies for superplastic flow.

The alloy 3RE60 shows reasonable superplastic behaviour at ~1000°C, although it undergoes some cavitation damage. However, the alloy contains few precipitates to nucleate cavities so that cavitation tends to be less in extent than for IN744.

REFERENCES

1. R.C. Gibson, H.W. Hayden and J.H. Brophy, Trans. ASM 61, 85 (1968).
2. H.W. Hayden, S. Floreen and P.G. Goodell, Metall. Trans. 3A, 833 (1972).
3. C.I. Smith, B. Norgate and N. Ridley, Metal Sci. 10, 182 (1976).
4. H. Hildebrand, G. Michalzik and B. Simmen, Met. Technol. 4, 32 (1977).
5. B.P. Kashyap and A.K. Mukherjee, J. Mater. Sci. 18, 3299 (1983).
6. B.P. Kashyap and A.K. Mukherjee, Metall. Trans. 14A, 1875 (1983).
7. N. Ridley and C.W. Humphries, in Grain Boundaries, p. E29. Institution of Metallurgists, London (1976).
8. J.W. Edington, K.N. Melton and C.P. Cutler, Prog. in Mater. Sci. 21, 61 (1976).
9. B.P. Kashyap and A.K. Mukherjee, Scripta Met. 16, 541 (1982).
10. A.F. Smith and G.B. Gibbs, Metal Sci., 3, 94 (1969).
11. F.H. Hayes, Unpublished research. University of Manchester, 1984.
12. N. Ridley, C.W. Humphries and D.W. Livesey, in ICSMA-4, p.433. ENSMIM, Nancy (1976).

Strain Anisotropy During Superplastic Flow of an Iron Base Oxide Dispersion Strengthened Alloy

M. J. Luton

Exxon Research and Engineering Company, Annandale, NJ 08801, USA

ABSTRACT

Samples of the oxide dispersion strengthened alloy Incoloy MA 956, with grain sizes of 1.1 and 8.1 μm, were deformed in tension at temperatures between 1050° and 1200°C, at strain rates from 10^{-5} and 10^{-1} s^{-1}. Within this range, the material exhibits a high strain rate sensitivity and specimen elongations of up to 125% were observed. During straining, the cylindrical samples developed an elliptical cross-section indicating marked strain anisotropy. It is shown that the directions of the principal transverse strains are inconsistent dislocation slip in a textured polycrystal. Instead the observations are rationalized in terms of a model for diffusion creep in materials with severe grain shape anisotropy (1).

KEYWORDS

Oxide dispersion strengthened alloys; superplasticity; strain anisotropy; grain shape anisotropy; diffusion creep; threshold stress; grain boundary sliding.

INTRODUCTION

The oxide dispersion strengthened alloy MA 956 is consolidated by extrusion from mechanically alloyed powders. The extruded material inherits the fine grain size that is generated when the heavily cold worked powders recrystallize prior to extrusion. Since the grain boundaries are strongly pinned by the oxide dispersoids, the extruded alloy is highly resistant recrystallization or grain growth at homologous temperatures below 0.85. Accordingly, it is expected that fine grain MA 956 should exhibit superplasticity like the nickel-base ODS alloy MA 6000 (2). The fine grains found in extruded rectangular bars are not generally equiaxed, instead they tend to be elongated and flattened. The diffusion creep of grains with an aspect ratio different from unity was first discussed by Raj and Ashby (3) for two-dimensional grains. More recently, Nix (1) has extended the classical treatments of diffusional creep (4-6) to include three-dimensional grain shape anisotropy. He showed that the strain rate developed is both sensitive to the grain aspect ratio, and different from the predictions based on the two-dimensional model. Implicitly the treatment calls for the material to exhibit a strain anisotropy dictated by the ratio of the grain dimensions. The purpose of the work described in this paper is to explore the influence of grain shape on diffusional creep and superplasticity.

MATERIALS

Two bars of Incoloy MA 956, of nominally different grain size, were purchased from Huntington Alloys. Both were provided in the as-extruded condition. The bars were sectioned on planes normal to the long transverse and extrusion directions, the t and r-directions, respectively. These T and R-sections were examined by optical and transmission electron microscopy, and the grain size determined by automated areal analysis. The grain size, averaged over the two sections was found to be 1.1µm in the first bar and 8.1µm in the second. The grains were elongated in the extrusion direction and flattened along the short transverse (n), as shown in Fig 1. The grain aspect ratio 3.1 on the T-section and 1.5 on the R-

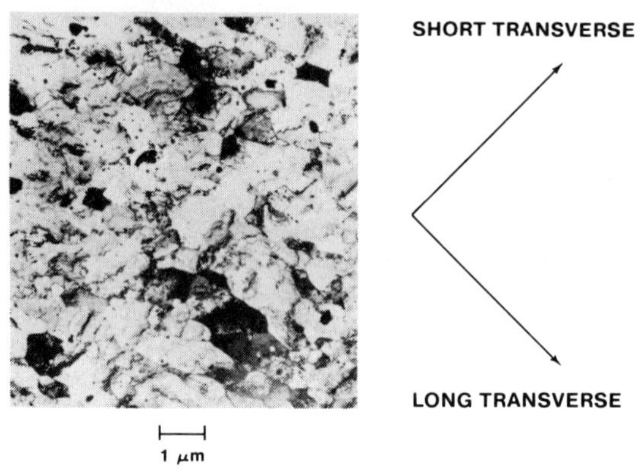

Fig. 1 Transmission electron micrograph of a transverse section of the 1.1µm material.

section for the 1.1µm material and 2.8 and 1.6 respectively for the 8.1µm material. Both bars exhibited a marked $(111)[1\bar{1}0]$ texture, that is typical of recrystallized bcc polycrystals.

EXPERIMENTAL RESULTS AND DISCUSSION

Cylindrical tensile specimens were cut from the two bars with the tensile axis parallel to the extrusion axis. A line was scribed on the end of each sample parallel to the t-direction in the bar. The samples were then subjected to a 3 h anneal at 1200°C in a hot isostatic press at 2 kbars. This treatment was performed to heal the embryonic cracks that were observed on some grain boundaries. No grain growth was detected after this treatment. The test samples were deformed in tension, under an argon atmosphere at temperatures between 1050° and 1200°C. The tests were carried out at constant true strain rates between 10^{-5} and 10^{-1} s^{-1}.

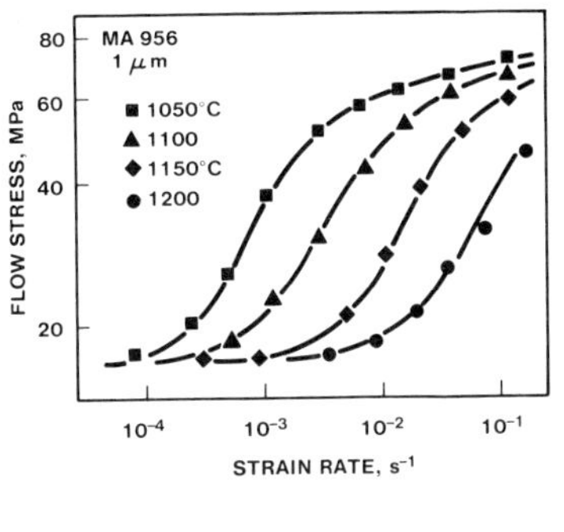

(a)

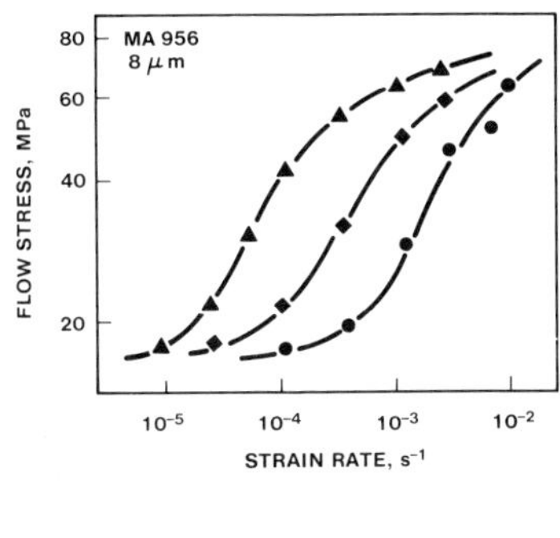

(b)

Fig. 2. Stress plotted versus strain rate for the two materials a) 1.1 μm grain size. b) 8.1 μm grain size.

The results of these experiments are plotted as stress versus strain rate in Fig. 2. It is evident that the two materials exhibit the high strain rate sensitivity that is characteristic of superplastic flow, over a critical range of strain rate that depends on temperature. In the regime of high strain rate sensitivity, the measured elongations were typically greater than 50%, with a maximum observed value of 125%. The temperature dependence of the strain rate for maximum strain rate sensitivity is shown in Fig. 3. The apparent activation enthalpy for the process is 480 kJ/mol,

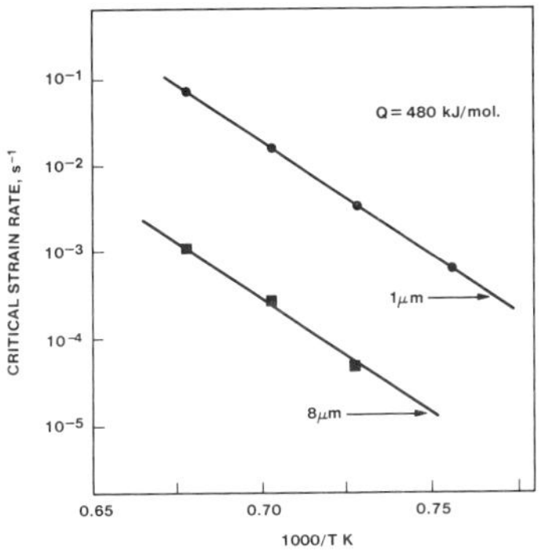

Fig. 3.
An Arhenius plot of the strain rate fo maximum strain rate sensitivity for the two materials.

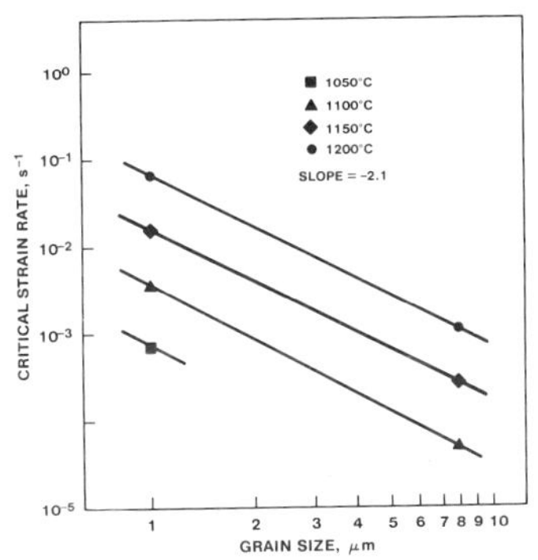

Fig. 4.
A plot of the strain rate for maximum strain rate sensitivity as a function of grain size.

which is comparable to that found for dislocation controlled high temperature deformation in this material (7). The grain size dependence is shown in Fig.4, from which we may conclude that the strain rate for maximum strain rate sensitivity depends on the inverse 2.1 power of the grain size.

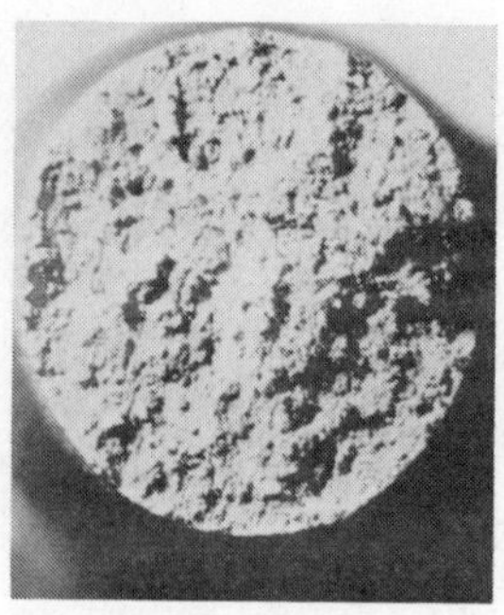

AXIAL STRAIN
0.2

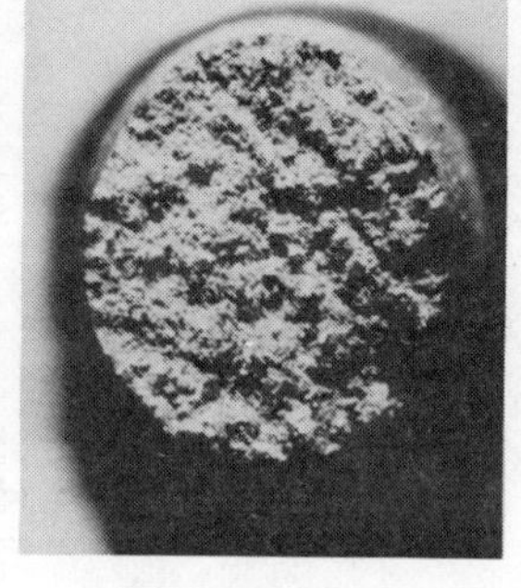

AXIAL STRAIN
0.31

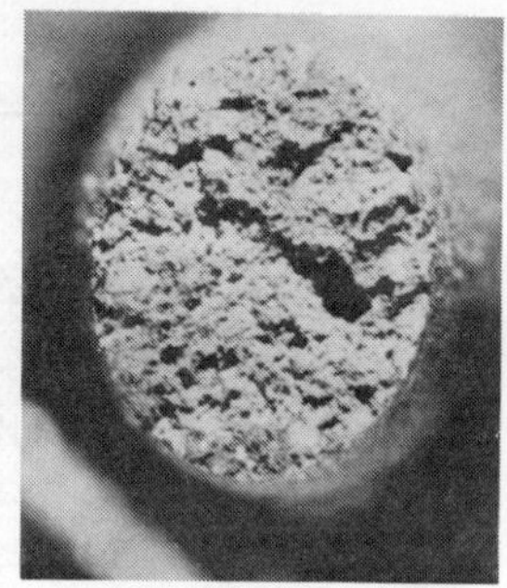

AXIAL STRAIN
0.42

Fig. 5. Micrographs of fracture surfaces of samples of the 1.1μm material, showing the elliptical cross-section.

In all tests performed in the regime of high strain rate sensitivity, the gauge length developed a distinctly elliptical cross-section during straining. This is illustrated in Fig. 5, where it is seen that the eccentricity increases with strain. Furthermore, it was found that the minor axis of the elliptical section was parallel to the fiduciary mark place on the specimen ends indicating the original long transverse, t-direction, in the original bar.

Although the elongations are relatively low, and the apparent activation enthalpy high, the other tensile results have all the characteristics of superplastic flow. The low elongation can be attributed to the persistence of pre-existent cracks. While the high apparent activation enthalpy could be rationalized by invoking the concept of a threshold for diffusional transport (8,9). There is no direct evidence in the present data, however, for this assumption. It is notable that, in all the tests, the maximum flow stress was less than 70 MPa, which is the threshold stress for dislocation creep in single crystals of MA 956 (7). Accordingly, dislocation processes are unlikely to contribute in a major way to the deformation process. So it is reasonable to assume that the dominant mechanism of plasticity is diffusional transport. Indeed, the grain size dependence of the process is comparable with many of the models and experimental results pertaining to diffusional plasticity (3-5,8,10).

The development of anisotropic strains transverse to the tensile axis, however, is inconsistent with the classical models of diffusional creep and superplasticity (3-6). Since the samples used in the present experiments showed a marked preferred orientation, it is necessary to ask whether this phenomenon provides indirect evidence of a contribution from crystal slip. Only small changes in orientation are expected during straining because of of the small strains achieved during the experiments. In the present experiments, the minimum principal strain occurred parallel to the [111] direction in the bar (more correctly; in an equivalent single crystal). It is readily shown (11) that a single crystal strained parallel to the $[1\bar{1}0]$ direction should develop an elliptical section, but the major and minor axes of the ellipse would be rotated 35° from the [111] and the $[11\bar{2}]$ axes. Since this is not the case, we may conclude that the strain anisotropy arises not by crystal slip, but by some anisotropy in the diffusional transport.

The analysis developed by Nix (1) can serve as a basis for an explanation of the present observations. In the analysis, the grain shape is defined in terms of three orthogonal grain diameters. With the "brick-like" grains oriented with their edges parallel to the r, t and n directions, it is convenient to consider diameters measured parallel to these directions. It can then be shown that the three principal strain rates, developed when lattice diffusion is rate controlling, are given by,

$$\dot{\varepsilon}_r = \frac{12D_1\sigma\Omega}{kTd^2} \; \frac{1}{(R_1R_2)^{2/3}} \left|\frac{R_1^2 + R_2^2}{1 + R_1^2 + R_2^2}\right| \tag{1}$$

$$\dot{\varepsilon}_t = \frac{-12D_1\sigma\Omega}{kTd^2} \; \frac{1}{(R_1R_2)^{2/3}} \left|\frac{R_1^2}{1 + R_1^2 + R_2^2}\right| \tag{2}$$

$$\dot{\varepsilon}_n = \frac{-12D_1\sigma\Omega}{kTd^2} \; \frac{1}{(R_1R_2)^{2/3}} \left|\frac{R_2^2}{1 + R_1^2 + R_2^2}\right| \tag{3}$$

where $R_1 = d_r/d_r$, $R_2 = d_r/d_n$ and $d = (d_rd_td_n)^{1/3}$. The grain diameters measured along the r, t and n-directions are d_r, d_t and d_n, respectively. It then follows that,

$$\dot{\varepsilon}_t/\dot{\varepsilon}_n = (R_1/R_2)^2 \tag{4}$$

and

$$\dot{\varepsilon}_t/\dot{\varepsilon}_r = -1/[1 + (R_2/R_1)^2] \tag{5}$$

It can also be shown that the strain rate ratios given in Eqs. 4 and 5 above are the same when grain boundary diffusion is rate controlling.

Now, since the ratio of the transverse strains is proportional to the ratio of the transverse strain rates, it is readily shown that,

$$\varepsilon_t/\varepsilon_n = (d_n/d_t)^2 \tag{6}$$

Evidently the right hand side of Eq. 6 is the inverse square of the grain aspect ratio measured on a plane normal to the r-direction. If the strains, ε_t and ε_n, are expressed in terms of the cross-section diameters, l_t and l_n, respectively the minor and major axes of the ellipse, it can be shown that the ratio

$$l_t/l_n = \exp[-(1-\mu)\varepsilon_r/(1+\mu)] \tag{7}$$

where $\mu = (d_n/d_t)^2$ and ε_r is the axial strain.

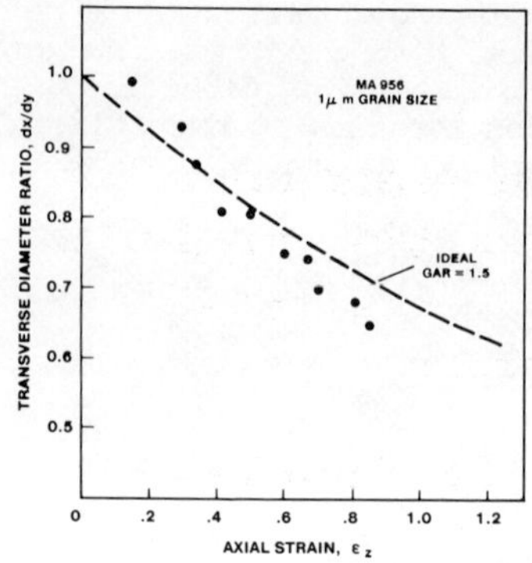

Fig. 6.
A plot of the transverse diameter ratio versus axial strain for the 1.1μm material. The broken line represents Eq. 7.

The deformed samples were measured parallel to the long and short transverse direction and the ratio of these diameters plotted against the axial strain, see Fig. 6. The broken curve in the figure represents Eq. 7 with d_t/d_n set equal to 1.5, the mean value of the transverse aspect ratio, measured from the transmission electron micrographs. The tendency of the experimental points to fall below the curve at high strains can be attributed to the development of non-uniform strain in the gauge section. It is evident that there is quantitative agreement between the predictions of the model and the experimental results.

CONCLUSIONS

The experiments described above show that fine grain MA 956 deforms super-plastically at temperatures above 1050°C. The development of anisotropic strains in the gauge section is attributed to the marked influence of grain shape on the mechanisms of diffusional creep and superplasticity, rather than to a contribution to the strain from crystal slip. It is demonstrated that the magnitude of the strain anisotropy is consistent with the recent model by Nix for diffusion creep in three-dimensional grains.

REFERENCES

1. W.D. Nix, Metals Forum, **4**, 38 (1981)
2. R.F. Singer and G.H. Gessinger, Proc. 2nd Risø Int. Symp. on Metallurgy and Materials Science, N. Hansen et al, eds., Risø National Laboratory, Roskilde, 1981, p. 365.
3. R. Raj and M.F. Ashby, Metall. Trans., **2**, 1113 (1971)
4. C. Herring, J. Appl. Phys.,**21**, 437 (1950)
5. F.R.N. Nabarro, Strength of Solids, The Physical Society, London 1948, p. 75
6. R.L. Coble, J. Appl. Phys. **34**, 1679 (1963)
7. R. Petkovic-Luton and M.J. Luton, Proceedings ICSMA-7 (1985)
8. B. Burton, Met. Sci. J., **5**, 11 (1971)
9. C.M. Sellars and R. Petkovic-Luton, Mater. Sci. Eng. **46**, 75 (1980)
10. B. Burton and G.W. Greenwood, Met. Sci. J., **4**, 215 (1970)
11. W.A. Backofen, Deformation Processing, Addison-Wesley, New York(1972)

Rheological and Microstructural Behaviour of the Low Alloy Steel DIN 34 Cr Ni Mo 6 in Hot Tensile Deformation

A. Sousa Brito and A. P. Loureiro

CEMUL, Centro de Mecânica e Materiais da Universidade Técnica de Lisboa, Av. António José de Almeida, 1000 Lisboa, Portugal

ABSTRACT

Tensile tests at 720 ^{0}C and 750 ^{0}C with strain-rate changes and stress relaxations were made on round specimens of low alloy steel DIN 34 Cr Ni Mo 6 cut along and transverse to the axis of the "as-received" rods, in order to measure the strain rate sensitivity parameter m, the total elongations and the microstructure changes due to the plastic deformation. m-values of around 0.25 were found for strain rates near 5×10^{-3} min^{-1}. Cavitation was observed in the deformed specimens. Total elongation was about 100% in the "transversal" specimens, and about 40% in the "longitudinal" ones.

KEYWORDS

Low alloy steel; hot forming; superplasticity; hot tensile testing; strain rate sensitivity; cavitation; cross-head speed change; load relaxation.

INTRODUCTION

Hot forming of steels is a well known old process implying high temperatures and forces, and so, it is a very expensive method of forming. Furthermore, complexity of shapes of some of the forged parts creates additional problems due to the difficulty of getting a good flow of metal in order to fill completely the space defined by the dies.

Although these problems sometimes induce people to use other cheaper procedures of metal-working, as casting and cutting, the improved properties of forged parts, because of their resulting characteristic structure, has led to develop deeper studies on hot plastic behaviour of steels during forming.

Recently it has been possible to make forging in a better controlled way, in order to obtain the best microstructure in each region of the forged piece according to its particular function. This is due to the increased knowledge of the rheological behaviour of materials (flow stress as a function of temperature and strain rate) and the structural evolution during deformation, linked to an advanced computer aid to design and manufacture (CAD/CAM). In some cases superplastic (SP) behaviour is found (very high elongations at low flow stresses, attained with very fine grain size at high temperatures and low strain rates). Substantial advantages have been shown for the SP forming of certain components. Therefore, the last decade has seen great development in research of SP alloys, namely of the rheological and microstructural behaviour of some carbon and low alloy steels |1,2,3,4|. This paper reports the first results of an investigation in this field for the steel DIN 34 Cr Ni Mo 6, a commercial low alloy steel suitable for application in shafts, gears and mechanical pieces supporting high loads, of which some experience was already got in our laboratory and that had been object of previous characterization work |5|.

EXPERIMENTAL PROCEDURE

The procedure usually practiced includes a preliminar grain refinement by heat or thermomechanical treatments in order to get a SP behaviour. Nevertheless, as this steel was purchased in a quenched and tempered condition (commercial "pre-treated" state), presenting a fine grain size, it was decided to study its behaviour in the "as-received" condition, in order to see if it was already sufficient to exhibit a SP behaviour, thus avoiding more treatments for refinement of the grain.

Characterization of the purchased rods (35 mm and 50 mm diameter) was made by chemical analysis, metallography and dilatometry.

The measured chemical composition is shown in Table 1.

TABLE 1 Chemical Composition of the Steel (%)

C	Cr	Ni	Mo	Mn	Si	S
0.35	1.45	1.37	0.21	0.48	0.28	0.02

Critical temperatures determinations were carried out on a ADAMEL dilatometer: $A_1 = 735^0C$ and $A_3 = 780^0C$. A SEM photomicrograph of an "as-received" sample is presented in Fig. 1, showing an expected tempered martensite fine structure. Subsequent annealings of samples at 700^0C had shown that no significant grain size increase was observed after 4 hours of annealing.

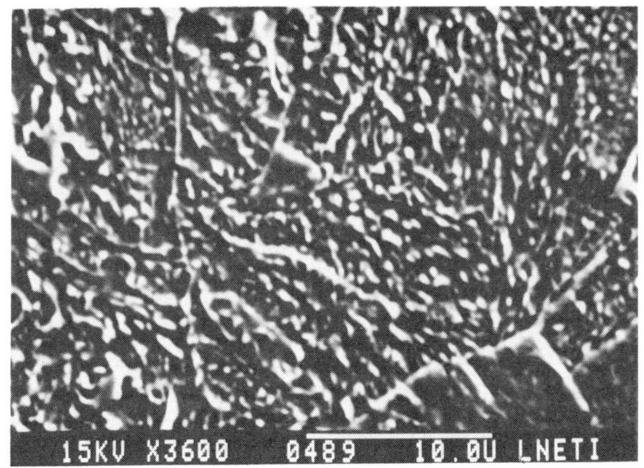

Fig. 1 SEM photomicrograph of an "as-received" sample

A necessary condition for a SP behaviour is a high (usually higher than 0.3) strain rate sensitivity index $m = d\log \sigma / d\log \dot{\varepsilon}$, where σ is the flow stress and $\dot{\varepsilon}$ is the strain rate |6|. So, the mechanical tests were carried out in order to determine the m-values at appropriate temperatures and strain rates.

Another very important aspect is the cavitation that sometimes develops during the plastic deformation |7|. So, the microstructures of the deformed testing samples were also observed in order to check the existence of voids.

Round tensile specimens of 4 mm diameter and 26 mm gauge length were machined, cut both along the rods ("longitudinal" specimens) and perpendicular to their axis ("transversal" specimens), and subjected to a residual stress relaxation treatment by annealing at 400^0C for 1 hour and air cooled.

Using a "INSTRON" tensile machine (model TT-CM-L) with a three zone resistance heating furnace, and a air-tight Inconel capsule which enclosed specimen and grips in argon atmosphere, isothermal tensile tests were carried out at temperatures below and close to A_1 and between A_1 and A_3, as follows: i) conventional tensile tests at constant cross-head speed; ii) tests with sudden cross-head speed change; iii) tests with cross-head arrest, followed by load relaxation, and iv) tests with cross-head speed changes, alternating with load relaxation.

Room temperature tensile properties were also measured using similar samples.

RESULTS AND DISCUSSION

The most significant results of the tensile tests are presented in Tables 2, 3 and 4.

TABLE 2 Room Temperature Tensile Properties

Yeld Strength	98 daN/mm2
Ultimate Tensile Strength	108 daN/mm2
Elongation	11%

TABLE 3 m-values obtained by analysing data from strain rate change tests

T (^{0}C)	$\dot{\varepsilon}$ (min^{-1})	samples	m value from diferent calculating methods |8|	|9|	|10|
720	$5x10^{-3}$	long.	0.21	0.22	0.24
		trans.	0.23	0.23	0.26
750	$5x10^{-3}$	long.	0.22	0.26	0.25
		trans.	0.27	0.28	0.25
	$1.5x10^{-3}$	long.	0.22	0.21	0.23
		trans.	0.30	0.27	0.27

TABLE 4 m-values obtained from load relaxation tests, ploting log σ vs log-$\dot{\sigma}$

T(^{0}C)	long.samples	transv. samples
720	0.18	0.19
750	0.18	0.20

From the results of strain rate change tests a plot log σ vs log $\dot{\varepsilon}$ was made (Fig. 2). This plot permits to measure m-values from the slope of each curve. The largest slopes are found for the lowest strain rate ($\dot{\varepsilon}$ = 10^{-3} min^{-1}) giving a m-value about 0.30 for all of them at that strain rate.

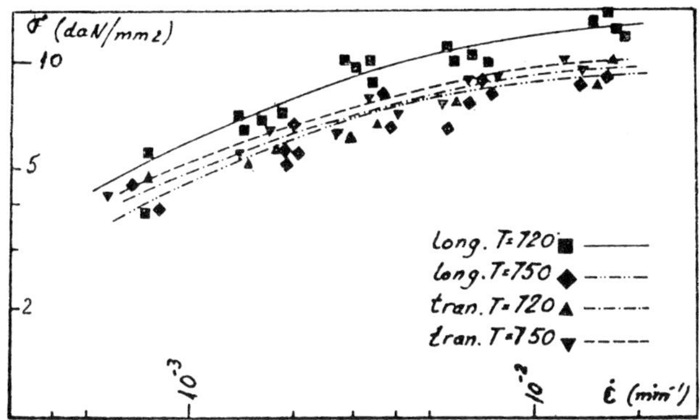

Fig. 2 Relation between σ and $\dot{\varepsilon}$ obtained from strain rate change tests

Most of the m-values obtained were around 0.25,under the value for the already considered superplastic behaviour. Also there is no evidence of relevant variation of m-values obtained both for transversal and longitudinal samples, and at temperatures under and above A_1. The m-values obtained from relaxations tests were much lower than those observed from the other methods, a fact already reported |6|.

Informations on the microstructural changes occuring during the tests on the deformed and undeformed zones were obtained by optical and electronic (SEM) microscopy. Fig. 3 shows a tipical microstructure of the deformed zone. Evident cavitation is presented in all tested specimens, more pronounced near the necking region.

Fig. 3 SEM microstructure of the deformed zone of a testing sample

Conventional tensile tests were made for "longitudinal" and "transversal" samples at temperatures T = 720 and 750 °C and at an initial strain rate which, according to Fig. 2, must correspond to the largest m-values. Results presented on Table 5 show for "transversal" specimens a much larger elongation than for "longitudinal" ones. This behaviour is obviously related to the "banded" structure presented by the "as received" rods, but it was not possible to determine the mechanism that causes shorter elongation for "longitudinal" specimens (with the axis parallel to the bands).

TABLE 5 Ult. Tens. Str. and elongations at high temperature ($\dot{\varepsilon}_0 = 2 \times 10^{-3}\ min^{-1}$)

T (°C)	Samples	σ_{max} (daN/mm2)	elong. (%)
720	long. transv.	7.44 7.42	39.5 100.0
750	long. transv.	6.61 8.40	43.5 113.1

Better m-values may be probably obtained with a grain refinement. Also, influence of the "banded" structure on the total elongation must be studied.

ACKNOWLEDGEMENTS

The authors would like to express their gratitude to the Serviço de Tecnologia de Materiais of LNETI, Lisbon, for facilities in experimental work.

REFERENCES

1. W.B. Morrison, *Trans. ASM*, 61, 424 (1968)
2. B. Walser and O.D. Sherby, *Metall. Trans.* 10 A, 1461 (1979)
3. M.J. Stewart, *Metall. Trans.*, 7 A, 399 (1976)
4. E.W.J. Miller and R. Pearce, *Metalurgia*, 5, 206 (1981)
5. A.V.Seabra, *Thesis*, LNEC, Lisboa, (1968)
6. J.W. Edington, K.N. Melton and C.P. Cutler, *Prog. Mater.Sci.*,21,61 (1976)
7. M.J. Stowell, *Conference Proceedings of Superplastic Forming of Structural Alloys*, AIME (1982)
8. W.A. Backofen, I.R. Turner and D.H. Avery, *Trans. ASM*, 57, 980 (1964)
9. H. Naziri and R. Pearce, *J. Inst. Metals*,97, 326 (1969)
10. J. Hedworth and M.J. Stowell, *J. Mater. Sci.*,6, 1061 (1971)

Strain-Enhanced Grain Growth During Superplastic Deformation

D. S. Wilkinson*, C. H. Caceres and Wu Xin***

**Department of Metallurgy and Materials Science, McMaster University, Hamilton, Ontario, Canada*
***Instituto de Mathematicas, Astronomia y Fisica, Universidad de Cordoba, Cordoba, Argentina*

ABSTRACT

Superplastic flow is generally accompanied by grain growth, at a rate which greatly exceeds that in the absence of deformation. In this paper, experimental data related to this phenomenon are discussed, and some simple models are presented. These account for the behaviour at relatively low strain rates. However, the grain growth behaviour at strain rates corresponding to commercial superplastic forming processes, has yet to be adequately modelled.

KEYWORDS

Superplasticity, grain growth, grain boundary sliding.

INTRODUCTION

Superplastic behaviour has been observed in many fine-grained metals (1-4) and ceramics (5-10), over restricted ranges of temperature and strain rate. The fine grain size is necessary to allow for a high rate of grain boundary sliding, which seems to be an essential component of superplastic flow. As the grain size increases, the superplasticity range becomes smaller and moves to lower strain rates. Thus superplastic forming becomes more difficult.

It is now well established that grain structure is rarely stable during superplastic forming. Both grain shape anisotropy (11-15) and texture (16-18) are destroyed during superplastic deformation. This typicaly requires strain of order 0.5. Once an equiaxed grain structure is established, grain growth is usually observed (13-15, 19-25). All of these processes are driven, at least in part, by the deformation itself. That is, they proceed at a greater rate during deformation than in the absence of deformation. They form a class of phenomena which we call "strain- enhanced microstructure evolution".

EXPERIMENTAL OBSERVATIONS

A recent review of available data (22) shows that strain-enhanced grain growth occurs in a wide range of materials. These include a solid solution Sn - 1% Bi alloy (19), a particle strengthened copper alloy (21), the Cu-P eutectic (14), the Zn-Al eutectoid composition (24), other microduplex alloys (15,20,23) and fine-grained Al_2O_3 (25). It is therefore clear that this phenomenon forms a generic part of the superplastic process. Furthermore, the rate of strain- enhanced grain growth is remarkably similar for a wide range of materials and testing conditions. This is best illustrated in Fig. 1, in which the grain growth rate $\dot{d}$, normalized by the initial grain size d_0, is plotted as a function of strain rate $\dot{\varepsilon}$.

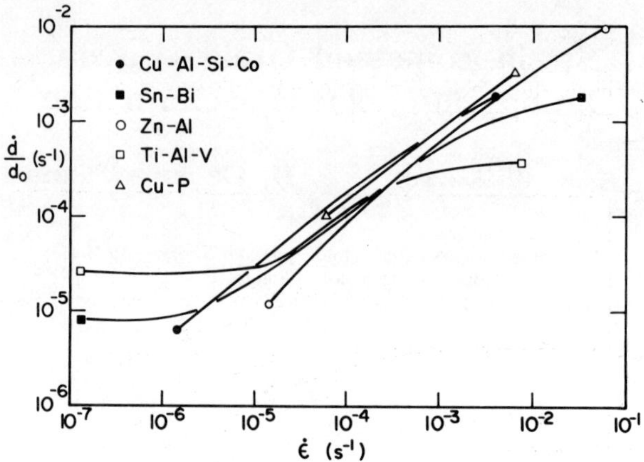

Fig. 1 Normalized grain growth rate as a function of strain rate. Experimental data points have been removed for clarity. The symbols are used purely to identify the materials.

Three regimes are found. At low strain rates for some materials (the Sn and Ti alloys in this case), a lower limit to the grain growth rate is observed. This is of course just the rate of normal grain growth, in the absence of deformation. For the other materials studied, a very stable grain structure was found during static annealing, and no lower limit is observed.

At intermediate strain rates the grain growth rate depends linearly on the strain rate. Moreover, the normalized grain growth rate is essentially the same for each material. This regime can be described by an equation of the form

$$\dot{d} = \lambda\, d_0\, \dot{\varepsilon} \tag{1}$$

where λ is constant of order one.

At higher strain rates, the grain growth behaviour no longer obeys eqn. (1). Instead, the grain growth rate increases more slowly with increasing strain rate. In addition, as indicated by Fig. 1, the rate of grain growth varies for different materials. We might also suspect that the grain growth rate will become temperature dependent. However, there is no experimental evidence to either support or refute this suggestion.

MODELS FOR THE INTERMEDIATE STRAIN RATE REGIME

The uniform response of all materials at intermediate strain rates seems at first remarkable, especially considering the wide range of material microstructures and temperatures used in these studies. Nor can the results be rationalized as corresponding to a single homologous temperature (assuming one can define a suitable value for the microduplex structures). Instead, it appears that in this intermediate strain rate regime, strain-enhanced grain growth is controlled by the morphology of the deformation process itself. Thus there is linear relationship between imposed strain and the resulting increase in grain size. This is clearly illustrated in Fig. 2, where the same data are replotted in terms of the deformation-induced grain growth per unit strain, $\partial d_\varepsilon/\partial\varepsilon$. Here ∂d_ε is the strain-enhanced portion of the grain growth. This is determined by subtracting the grain size reached after static annealing for the same time, from the total grain size. This becomes constant when grain growth is purely strain controlled.

It may still seem surprising that deformation can drive grain growth at the same rate in microduplex and single phase structures. However, we have recently developed (26) models for this process applicable to single phase and microduplex alloys respectively. They are based on different physical concepts. However, both models give essentially the same result, consistent with the observations.

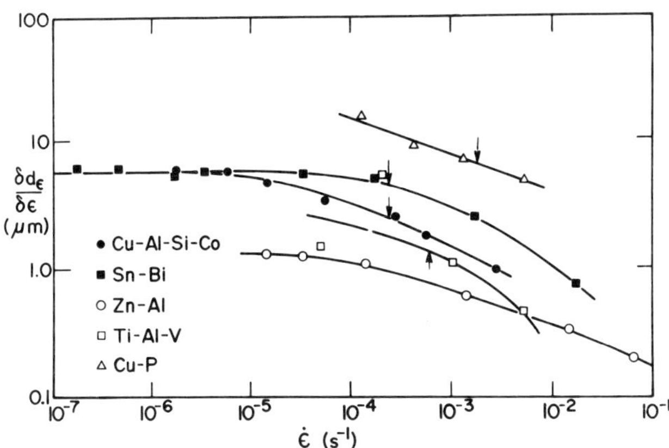

Fig. 2 A plot of the strain-enhanced grain growth per unit strain. The arrows indicate the strain rate at strain rate sensitivity of the flow stress is maximum.

Single Phase Materials

In the case of single phase materials, a model for grain growth based on grain boundary sliding has been developed (26). We assume that grain boundary sliding induces grain boundary migration, consistent with the bicrystal observations of Walter and Cline (27), and the high temperature fatigue observations of Langdon and Gifkins (28). Sliding itself is related to strain rate according to the equation

$$\dot{\varepsilon} = \frac{\alpha \dot{s}}{d} \tag{2}$$

where $\dot{s}$ is the rate of sliding, and α is the fraction of total strain contributed by sliding. The result of this model is an equation for the strain induced grain growth rate, namely

$$\dot{d}_\varepsilon = \alpha b d \dot{\varepsilon} \tag{3}$$

where b is the ratio of grain boundary migration to sliding. Since α and b are both of order one, this reduces to eqn. (1).

Microduplex Materials

This model is not readily applicable to microduplex materials since grain boundary migration must be accompanied by solute redistribution. We have therefore (26) adapted a model proposed earlier by Holm et al. (20). They suggested that the rate of particle coarsening in microduplex alloys will be increased during superplastic deformation. This results from the fact that as grains switch their neighbours particles sitting at grain junctions are agglomerated. The model is therefore based on the Ashby-Verrall model (29) of superplastic flow. This process is illustrated in Fig. 3.

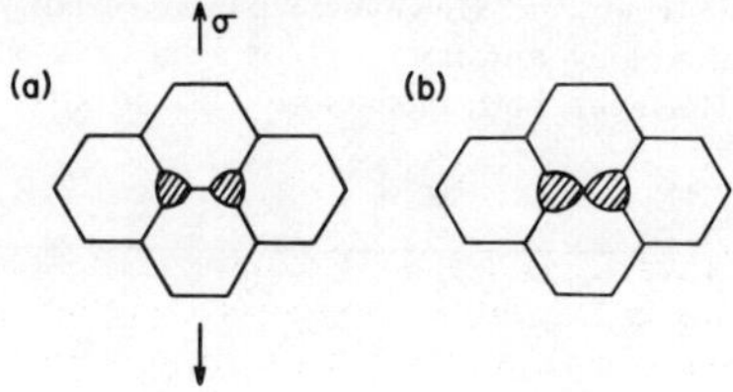

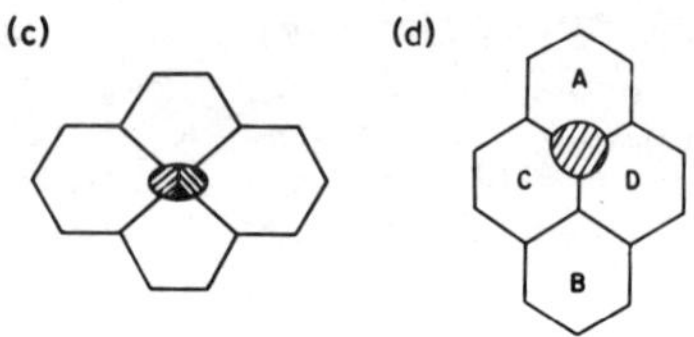

Fig. 3 A model for particle coarsening by grain switching. As grains C and D approach in (b), particles at grain corners agglomerate and coalesce (c). The net result (d), is a decrease in the number of particles by 50% during a strain increment required for one switching event (~ 0.55).

Our model for this process assumes that grain growth keeps up with the Zener limit (30) for microduplex alloys, namely

$$d = \frac{2\,r}{3\,f_v} \tag{4}$$

where r is the average radius and f_v the volume fraction of second phase particles. The result is

$$\dot{d}_\varepsilon = \frac{\ln\left(\frac{1}{1-k}\right)}{3\,\varepsilon_s}\, d\dot{\varepsilon} \tag{5}$$

Here ε_s is the strain required for each switching event (~ 0.55 according to Ashby and Verrall), while k is the fraction of particles removed by this process. Since $\ln(1/(1\text{-}k))/3\varepsilon_s$ is of order one, this result again reduces essentially to eqn. (1).

HIGH STRAIN RATE EFFECTS

As shown in Fig. 2, the simple linear relationship between grain growth and deformation discussed in the previous sections, breaks down at relatively low strain rates. In particular this relationship is not applicable at strain rates for which the strain rate sensitivity of the flow stress is a maximum. Therefore, eqn. (1) cannot be used in the derivation of constitutive laws for superplastic flow applicable to commercial forming conditions. Several workers have used empirical grain growth laws such as (21)

$$\Delta d_\varepsilon = C_1\,\varepsilon\,\dot{\varepsilon}^{-P} \tag{5}$$

or (31)

$$\Delta d = C_2 t^Q \dot{\varepsilon}^{-P} \quad (6)$$

However, these relationships are at best first order approximations, applicable to a limited range of conditions for which data is available. There is therefore a clear need for a greater understanding of the processes which control grain growth at these higher strain rates.

Fig. 2 suggests that an upper limit may exist to the rate of grain growth. This is consistent with the observations of Walter and Cline (27) and Langdon and Gifkins (28), that grain boundary sliding and grain boundary migration are sequential processes. Presumably, grain boundaries slide until they are impeded by obstacles which need to be overcome by a period of migration. Since sequential processes are dominated by the slowest step, it is natural to expect a limit to the grain growth rate, controlled by the rate of grain boundary migration. If this is included in the model for single phase materials, the resulting equation for the grain growth rate (26) is

$$\dot{d} = \dot{d}_a + \frac{\alpha b d \dot{\varepsilon} \dot{d}_m}{\alpha b d \dot{\varepsilon} + \dot{d}_m} \quad (7)$$

where $\dot{d}_a$ is the rate of grain growth in the absence of deformation, and $\dot{d}_m$ the migration limited rate. This equation is shown schematically in Fig. 4, and it is clearly in qualitative agreement with the experimental results in Fig. 1.

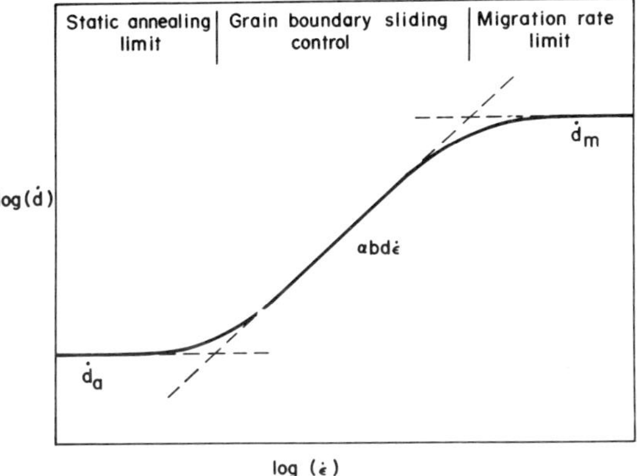

Fig. 4 The model for sliding-induced grain growth predicting three regimes of behaviour.

There are alternative explanations for this phenomenon however which would produce similar results. For example, recrystallization would remove or reduce the driving force for grain boundary migration, thus decreasing the effect of deformation on grain growth. Ghosh (32) has observed recrystallization during superplastic flow. In the microduplex alloys the rate of grain growth may not be sufficient to maintain the grain size at the Zener limit at high strain rates. This would have a similar effect. In all of these cases, we would expect the rate of grain growth at higher strain rates to depend on both temperature and on initial grain size. However, there is as yet insufficient data to determine whether this is indeed the case.

SUMMARY

The rate of strain-enhanced grain growth shows three regimes. At low strain rates, the deformation-free, normal grain growth process dominates. At intermediate strain rates (but below the peak in strain rate sensitivity), grain growth is strain controlled. At still higher strain rates (corresponding to commercial superplastic forming conditions), grain growth is both strain and strain rate dependent, and appears to approach an upper limit. Models have been developed which adequately explain the first two regimes. However the behaviour at higher strain rates is not as well understood. This limits our ability to develop physically based constitutive equations for superplastic flow, applicable to forming operations.

ACKNOWLEDGEMENTS

This work is supported by a grant from the Natural Sciences and Engineering Research Council.

REFERENCES

1. D.M.R. Taplin, G.L. Dunlop and T.G. Langdon, *Ann. Rev. Mater. Sci.* **9** (1979), 151.
2. A.K. Mukherjee, *Ann. Rev. Mater. Sci.* **9** (1979), 191.
3. *Superplastic Forming of Structural Alloys*, Amer. Inst. Metall. Engnrs., New York (1982).
4. S. Reusswig, R. Gleichman, P.G. Zielinski, D.G. Act and R. Raj, *Acta Metall.* **32** (1984), 1533.
5. C. Carry and A. Mocellin, *Proc. Conf. on Deformation of Ceramics* (Plenum, 1984), 391.
6. C. Carry and A. Mocellin, *Proc. Brit. Ceram. Soc.* **33** (1983), 101.
7. J.-G. Wang and R. Raj, *J. Amer. Ceram. Soc.* **67** (1984), 385.
8. J.-G. Wang and R. Raj, *J. Amer. Ceram. Soc.* **67** (1984), 399.
9. T.E. Chung and T.J. Davies, *J. Mater. Sci.* **79** (1979), 143.
10. T.E. Chung and T.J. Davies, *Acta Metall.* **27** (1979), 627.
11. M. Suery and B. Baudelet, *Rev. Phys. Appl.* **13** (1978), 53.
12. B.P. Kashyap and A.K. Mukherjee, *Proc. 2nd Intl. Conf. on Creep and Fracture of Engineering Materials and Structures* (Pineridge, 1978), 185.
13. B.M. Watts and M.J. Stowell, *J. Mater. Sci.* **6** (1972), 657.
14. G. Herriot, M. Suery and B. Baudelet, *Scripta Metall.* **6** (1972), 657.
15. M. Suery and B. Baudelet, *J. Mater. Sci.* **8** (1973), 363.
16. K.N. Melton, C.P. Cutler, J.S. Kallend and J.W. Edington, *Acta Metall.* **22** (1974), 165.
17. C.P. Cutler, J.W. Edington, J.S. Kallend and K.N. Melton, *Acta Metall.* **22** (1974), 665.
18. M. Suery and B. Baudelet, *Proc. Conf. on Superplastic Forming of Structural Alloys* (Met. Soc. AIME, 1982), 105.
19. M.A. Clark and R.T.H. Alden, *Acta Metall.* **21** (1973), 1195.
20. K. Holm, G.R. Purdy and J.D. Embury, *Acta Metall.* **25** (1977), 1191.
21. C.H. Caceres and D.S. Wilkinson, *Acta Metall.* **32** (1984), 415.
22. C.H. Caceres and D.S. Wilkinson, *J. Mater. Sci. Lett.* **3** (1984), 395.
23. M. Suery and B. Baudelet, *Phil. Mag.* **A41** (1980), 41.
24. A.K. Ghosh and C.H. Hamilton, *Metall. Trans.* **10A** (1979), 699.
25. J.D. Fridez, C. Carry and A. Mocellin, *Conf. on Structure-Property Relationships for MgO-Al_2O_3 Ceramics*, MIT (1983).
26. D.S. Wilkinson and C.H. Caceres, *Acta Metall.* **32** (1984), 1335.
27. J.L. Walter and H.E. Cline, *Trans. Metall. Soc. AIME* **242** (1968), 1823.
28. T.G. Langdon and R.C. Gifkins, *Acta Metall.* **31** (1983), 927.
29. M.F. Ashby and R.A. Verrall, *Acta Metall.* **21** (1973), 149.
30. C. Zener, private communication to C.S. Smith, *Trans. Amer. Inst. Min. Engrs.* **175** (1949), 15.
31. A.K. Ghosh, *Proc. Conf. on Superplastic Forming of Structural Alloys* (Met. Soc. AIME, 1982), 85.
32. A.K. Ghosh, private communication, 1984.

Superplasticity and Hardening of Extremely Fine Grained Ultrahigh Carbon-Chromium-Vanadium-Iron Alloys

G. Frommeyer

Max-Planck-Institut für Eisenforschung GmbH, Düsseldorf, Federal Republic of Germany

ABSTRACT

Microcrystalline ultrahigh carbon steels, alloyed with up to 20 wt% vanadium and chromium, have been prepared by melt atomization and powder metallurgical techniques. The resulting microstructure contains a large volume fraction of carbides and a grain size of less than 1 micron. At room temperature, these microcrystalline steels exhibit a high yield stress (2000 MPa) while maintaining adequate toughness for practical applications. Exposure to higher temperatures (> 650 °C) softens the material considerably and it becomes superductile. The large elongation to failure (500 %) and the high value of the strain rate sensitivity, m = 0.5, indicate superior superplastic properties of these extremely fine grained materials.

KEYWORDS

Microstructure; rapid solidification; dispersion hardening; superplasticity.

INTRODUCTION

Ultrafine grained materials exhibit superior mechanical properties compared to materials with a coarser grain structure. The fine grain size increases the flow stress and tensile strength while the room temperature ductility and toughness remains sufficient for practical applications as tool steels. In addition, a fine grained equiaxed two-phase microstructure, which remains thermally stable, provides superplastic properties at higher temperatures (650 °C) (1,2). Fabrication of these microcrystalline alloys is possible using a variety of rapid solidification processes (3). The most promising is melt atomization which produces large quantities of high purity powders that can be subsequently consolidated using powder metallurgy techniques.

In the past few years, there have been many studies characterizing rapidly solidified powders of non-ferrous and ferrous alloys with low and medium carbon contents (4). Recently, Eiselstein and co-workers (5) and Frommeyer et al. (6) reported the microstructural features and extraordinary mechanical properties of rapidly solidified white cast irons with carbon contents up to 4.2 wt%. This material behaves superplastically at 650 °C and has a room temperature tensile strength of 890 MPa.

The present paper describes the microstructure and mechanical properties of rapidly solidified and powder metallurgically processed high alloy steels. Two ultrahigh carbon (C > 2 wt%) steels with different Chromium and Vanadium compositions were selected for this study. These steels contain a large volume fraction of uniformly dispersed carbides within a fine grained alpha iron matrix. A variation in the amount of Cr and V affects the distribution of carbides in the ferritic matrix. Comparison of these two alloys reveals the effect of changing the bimodal size distribution and relative volume fractions of Cr and V carbides on the resulting mechanical properties.

MATERIALS AND EXPERIMENTAL PROCEDURE

Table I lists the compositions of the materials. These alloys were induction melted and atomized by a cold nitrogen gas jet. The rapidly solidified powders experienced cooling rates, on the order of 10^4 - 10^5 K/sec, which were dependent upon the size of the powder particles. Using a sieving procedure, the alloy powders were classified by size. The average particle diameter, dp, was determined to be approximately 80 microns. For the production of bulk specimens, the powders were encapsulated in steel cans and compacted by hot isostatic pressing at 1200 °C. These consolidated bars were subsequently forged to increase their density.

TABLE I

Material	C	Mn	Si	Cr	V	Mo
*X 245 Cr V 5 10	2.45	0.50	0.90	5.25	9.75	1.30
X 220 Cr V 18 6	2.20	0.50	0.50	17.5	5.75	0.50

* 'X' indicates high alloy steel

Tension tests at room temperature and superplastic tests in the temperature range between 650 and 800 °C were performed on the hot isostatic pressed and forged material. The strain-rate-sensitivity parameters were measured using strain-rate-change-tests. The microstructures of the rapid solidification rate (RSR) powders and the compacted specimens were characterized using optical microscopy, scanning electron microscopy (SEM) and x-ray diffraction techniques.

RESULTS AND DISCUSSION

The typical morphologies of the rapidly solidified alloy powder particles are spheroidal and slightly ellipsoidal. Small satellites which have an individual solidification structure are attached to some of the coarser particles. The microstructure

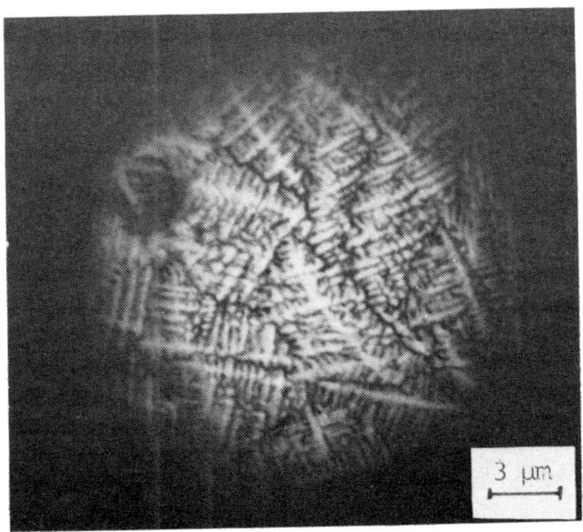

Fig. 1a. SEM photomicrograph of dendritic microstructure of the RSR powder surface (material: X 245 Cr V 5 10)

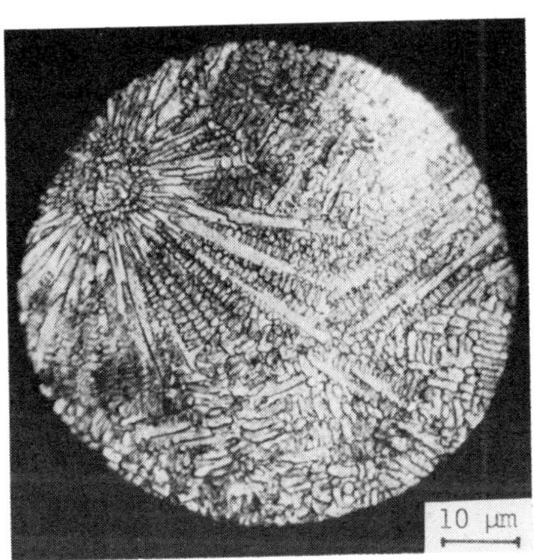

Fig. 1b.
Dendritic microstructure of a RSR powder particle
d_p = 65 micron, lightoptical photomicrograph

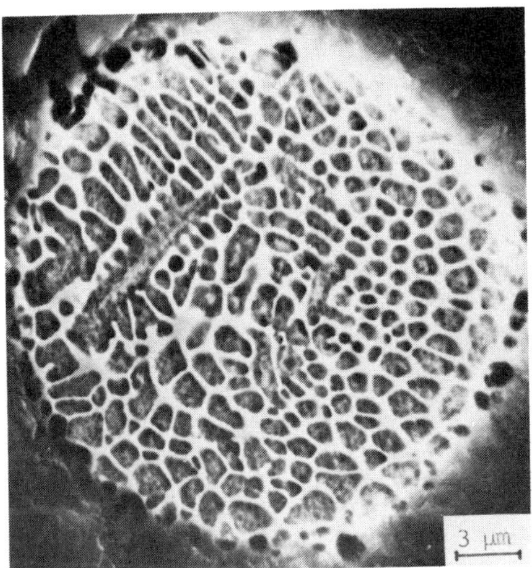

Fig. 1b.
Cellular microstructure of a small RSR powder particle,
$d_p \simeq 25$ micron, SEM photomicrograph

of the larger powder particles (dp > 30 microns) consists of dendrite networks. Figure 1a is a SEM image form the surface of one of the coarser particles. The same dendritic structure observed on the surface in figure 1a is also present in the interior of the larger particles, see figure 1b. Figure 1c reveals the microstructure of one of the smaller particles. In contrast to the larger powders, the small particles have a cellular microstructure. These structural arrangements indicate that the RSR-powders are crystalline (i.e. non amorphous). The majority of the particles are dense and almost free of pores and inclusions.

Measurement of the secondary dendritic arm spacing, λ_p, and the cell size of particles with different diameters allows estimation of the cooling rate, using the empirical relation (7):

$$\lambda \cdot \dot{\eta}^{n} = B \qquad (1)$$

where 'B' is a constant which depends upon the kinetics of the rapid solidification process. The exponent n is about n=1/3 which is in good agreement with other published n values reported for iron-based alloys (8,9). This analysis estimates the cooling rates of these RSR-powders to be on the order of 10^4 to 10^5 K/sec.

X-ray analysis shows that the fcc γ-phase is the major microstructural constituent which forms the dendritic structure of the RSR powders. The γ-phase forms directly from the melt and is maintained upon cooling to room temperature owing to the high cooling rate. Between the dendrites, there is an ultrafine dispersion of chromium and vanadium rich carbides.

During compaction of the powders, the microstructure changes drastically. The γ-phase transforms to bcc α-iron and a large volume fraction of carbides ($>$ 36 vol.%) precipitate out of the former dendrites. Using X-ray powder diffraction of the residue after electrolytic dissolution of the ferritic matrix, the precipitates were identified as predominantly the chromium rich $M_{23}C_6$ carbide and the vanadium rich MC type with a small contribution of the chromium rich M_7C_3 carbide.

Figures 2 and 3 show optical and SEM micrographs of the fine grained equiaxed microstructure of the processed alloys. The

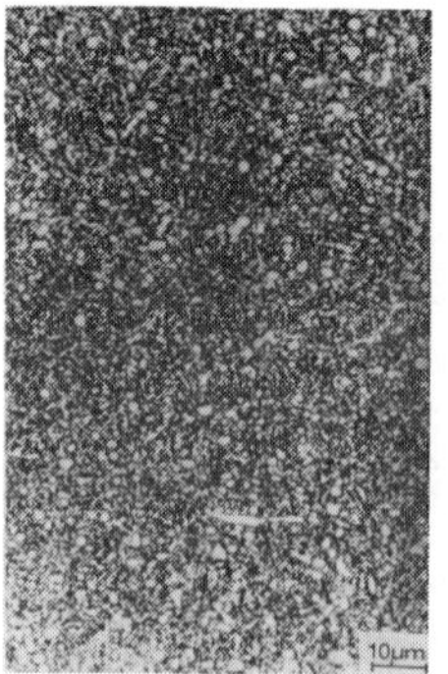

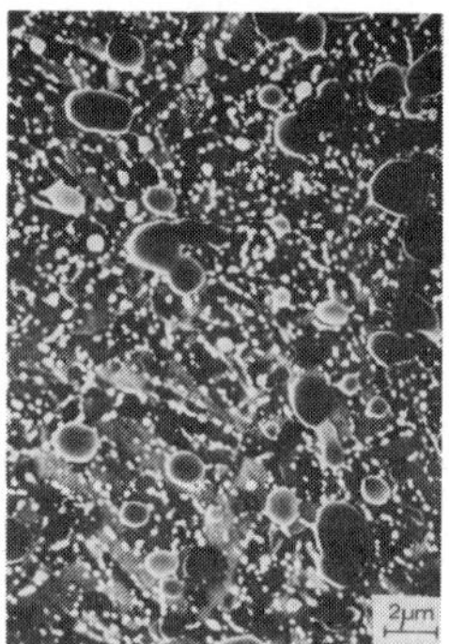

Fig. 2. a) Lightoptical, photomicrographs, b) SEM Microstructure of the consolidated 245 Cr V 5 10 steel. Fig. 2a shows an equiaxed structure, and fig. 2b reveals at higher resolution the bimodal distribution of fine vanadium carbides (0.2 - 0.4 micron size) and larger chromium carbides (1 - 2 micron size)

grain size of the ferrite is similar for both alloys and is on the order of 1 - 2 microns. In the high chromium and the high vanadium content materials there is a bimodal distribution of carbides. These carbides have been identified by energy dispersive spectroscopy of X-rays in the SEM as large chromium rich carbides that range in size from 1 - 2 microns and smaller vanadium rich carbides which are approximately 0.2-0.4 microns. Comparison of figures 2 b and 3 b reveals two distinctly different distributions of carbides. The high vanadium material, in figure 2b, has a smaller average precipitate size and interparticle spacing than the high chromium content material in figure 3b.

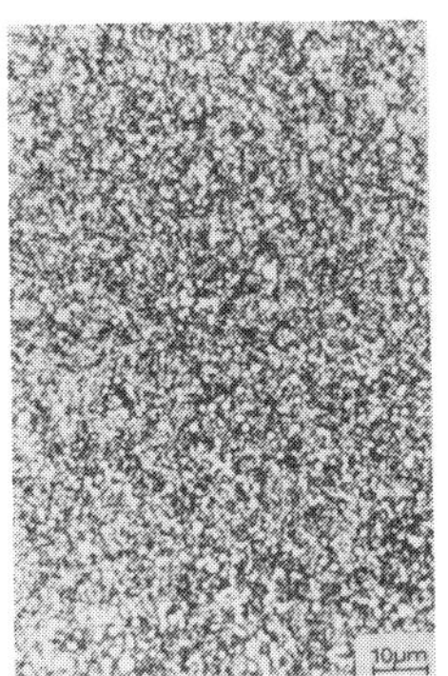

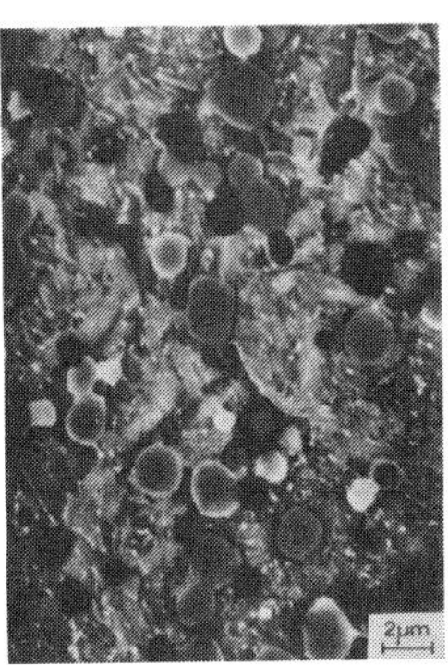

Fig. 3. a) Lightoptical, photomicrographs, b) SEM Microstructure of the consolidated 220 Cr V 18 6 steel. Fig. 3a shows the uniform distribution of chromium carbides within the ferrite matrix. The SEM picture exhibit the darker vanadium rich carbides and the lighter chromium carbides

Figure 4 illustrates the results of room temperature tensile tests. The flow stress ' $_{o}$ ' of different thermally treated samples for obtaining variable interparticle separation ' $_{p}$ ' of the carbides is plotted versus the reciprocal square root of $_{p}$. All measured values fit in a straight line, so that $_{o}$ can be described as a linear function of $_{p}^{-1/2}$. These carbide particle reinforced steels exhibit a remarkable increase in strength from 500 to 2000 MPa. The flow stress of the high vanadium alloy steel is somewhat higher than that of the chromium rich material. Strengthening of paricle - reinforced materials occurs initially when the dispersed particles restrict the deformation of the matrix by mechanical restraint. The amount of the restraint effect is complex and depends on structural factors. In general, it is a function of the ratio of the interparticle separation to the particle diameter $_{p}/d_{p}$ and the elastic sheer moduli of the matrix 'G_m and the second phase 'G_p'.

The strength of the carbide particle-reinforced high alloy steels can be described quantitively by the following relation (10):

$$_{o} = \left(\frac{G_m \, G_p \, |\vec{b}|}{C \; _{p}}\right)^{1/2} \qquad (2)$$

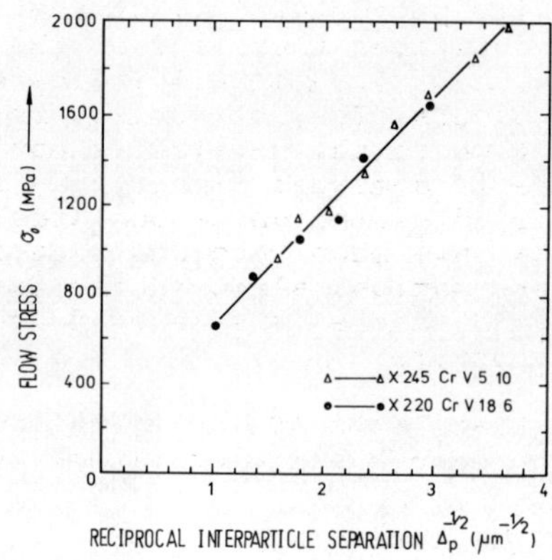

Fig. 4. Influence of the interparticle separation of carbides on the flow stresses of fine grained high alloy steels

Where $\vec{b}$ is the Burgers vector and 'C' defines the theoretical strength of the bcc α-iron matrix. This expression predicts that the yield strength is proportional to the inverse square root of the interparticle spacing of the carbides, when the particles do not yield under load.

Figure 5 illustrates the high temperature deformation results. The data are given on a log-log plot of the flow stress 'σ_o' versus strain rate '$\dot{\varepsilon}$'. Stress-strain rate curves, obtained at two different test temperatures 675 °C and 783 °C, can be divided into three regions. Creep occurs at lower strain rate and relatively low stresses. In the middle region of the stress-strain rate relationship predicts superplastic behaviour. The

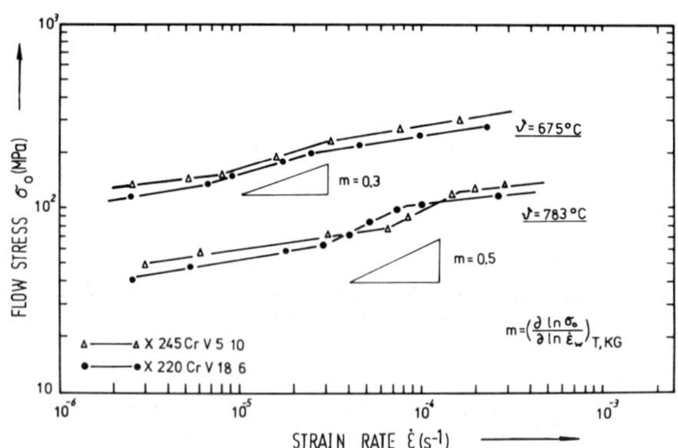

Fig. 5. Superplastic test results show the flow stress as a function of strain rate in a log σ_O - log $\dot{\varepsilon}$ plot for two different test temperatures

strain rate sensitivity parameter 'm', defined as $m = (\partial \ln\sigma_o/\partial \ln\dot{\varepsilon})_T$, describes the slope of the curves. The relatively high m-value of $0.3 \leq m \leq 0.5$ indicates that superplastic deformation occurs. The optimum superplastic properties were found at 783 °C, where the m-value is equal 0.5. The deformation temperature is somewhat higher than for superplastic deformation of unalloyed ultrahigh carbon steels (11). Maximum elongations of > 500 % have been recorded. Fracture occured after a large amount of deformation, although the exact failure mode was not clearly defined, there is some evidence of cavitation revealing a ductile rupture.

Figure 6 shows the strain rate '$\dot{\varepsilon}$' as a function of the reciprocal test temperature in a semi logarithmic plot. The following expression (12):

$$\dot{\varepsilon} = A \left(\frac{\sigma_o}{E}\right)^n \exp(-Q/RT) \qquad (3)$$

allows calculation of the activation energy of superplastically deformed specimens at constant stress from the slope of the curve. 'A' is a constant and depends on structural terms. 'σ_o' is the flow stress, and 'E' defines the elastic modulus.

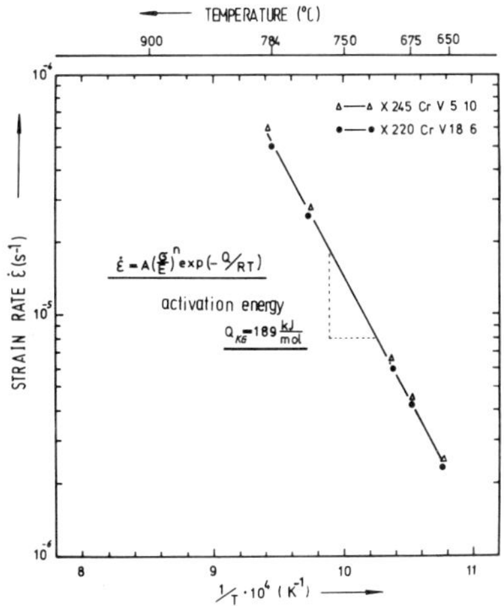

Fig. 6. Semi logarithmic plot of strain rate versus reciprocal temperature for determination the activation energy of superplastic deformation

The Q-value was determined to be 189 KJ/mol. This amount is comparable with activation energies of grain boundary diffusion processes of iron (13).These results lead to the conclusion that diffusion controlled grain boundary sliding could be the dominant mechanism during superplastic deformation of these fine grained ultrahigh carbon vanadium-chromium steels.

CONCLUSION

High strength ultrafine grained vanadium-chromium-iron alloys containing a large volume portion of fine dispersed carbides are applicable as cold work tool steels. Their superplastic properties at higher temperature allows to produce complicated shapes of tool parts by superplastic forming processes.

REFERENCES

1. E. E. Underwood, J. Met. 14 (1962) 914
2. K. A. Padmanabhan and G. Davies Jr., eds., Superplasticity, Springer-Verlag, Berlin - Heidelberg - New York 1980
3. M. Cohen, B. H. Kear and R. Mehrabian, eds., "Rapid Solidification Processing, Principles and Technology I and II", Claitor's Publishing Distribution, Baton Rouge, CA., 1980
4. H. Jones, J. of Mat. Sci., 19 (1984) 1043
5. L. E. Eiselstein, O. A. Ruano and O. D. Sherby, J. Mat. Sci., 18 (1983) 483
6. G. Frommeyer, D. W. Kum and O. D. Sherby, J. Met. Sci., to be published
7. H. Jones, in "Treatise on Materials Science and Technology" Vol. 20, edited by H. Herman (Academic Press New York 1981) p. 1
8. W. E, Brower, R. Strachan and M. C. Flemings, Cast Met. Res. J. 6 (1970) 176
9. P. A. Joly and R. Mehrabian, J. Mat. Sci. 9 (1974) 1446
10. G. Frommeyer, in "Physical Metallurgy", Chap. 29, p. 1854, eds. R. W. Cahn and P. Haasen, Elsevier Science Publishers BV, 1983
11. O. D. Sherby, B. Walser, C. M. Young and E. M. Cady, Scripta Met. 9 (1975) 569
12. A. M. Brow and M. F. Ashby, Scripta Met. 14 (1980) 1297
13. B. Walser and O. D. Sherby, Scripta Met. 16 (1982) 213

SECTION 8

Hot Working and Deformation Processing

Corroyage et traitements thermomécaniques

Elevated Temperature Deformation and Structural Observations in Al-8.4Fe-3.6Ce

D. L. Yaney*, J. C. Gibeling and W. D. Nix***

**Department of Materials Science and Engineering, Stanford University, Stanford, CA 94305, USA*
***Department of Mechanical Engineering, Division of Materials Science and Engineering University of California, Davis, CA 95616, USA*

ABSTRACT

The elevated temperature deformation characteristics of rapidly solidified Al-8.4 wt.% Fe-3.6 wt.% Ce have been investigated. Constant true strain rate compression tests were performed between 523 and 823 K at strain rates ranging from 10^{-6} to 10^{-3} sec^{-1}. At test temperatures below 723 K, the alloy is significantly stronger than oxide dispersion strengthened (ODS) aluminum. However, unlike ODS aluminum, the strength falls rapidly with increasing temperature above 723 K. Since annealing at 773 K for seven hours prior to testing did not produce a similar softening effect, the coarsening of second phase particles was ruled out as a possible explanation for the observed elevated temperature softening. In addition, a change in deformation mechanism does not contribute to the reduction in elevated temperature strength. Hot stage transmission electron microscopy (TEM) was used to show that the second phase particles responsible for the superior low temperature strength of this alloy do not dissolve at temperatures up to 773 K. However, additional TEM studies indicate that particle deformation at the highest test temperatures may be responsible for the softening at elevated temperatures.

KEYWORDS

Rapidly solidified aluminum alloys, compression testing, microstructural coarsening, particle deformation, hot stage microscopy

INTRODUCTION

During the past few years, one of the major goals of aluminum alloy research has been to develop new high strength alloys capable of replacing titanium in a number of elevated temperature applications (1,2). The alloys that have been developed are formed by rapid solidification techniques and depend on a large volume fraction of finely dispersed intermetallic compounds for their elevated temperature strength. In order to insure that the second phase particles formed in these alloys are fine and homogeneously distributed, the alloying elements selected have both high liquid solubility and low solid solubility. Reasonable microstructural stability at elevated temperatures is obtained by requiring that the alloying elements also have low diffusion coefficients in aluminum.

After considering a number of binary and ternary alloys meeting the criteria listed above, several investigators have shown that one of the most promising alloy systems is the Al-Fe-Ce ternary system (1-3). To date however, studies of the elevated temperature deformation characteristics of Al-Fe-Ce alloys have been limited to temperatures less than 650 K. In this paper, the results of compression tests performed on an Al-Fe-Ce alloy at temperatures up to 823 K are described. Structural information obtained from transmission electron microscopy (TEM) of tested as well as untested samples is reported and the factors limiting the elevated temperature strength of Al-Fe-Ce alloys are discussed.

EXPERIMENTAL

The material investigated is a rapidly solidified Al-8.4 wt.% Fe-3.6 wt.% Ce alloy in the as-extruded condition. Constant true strain rate compression tests were conducted in the temperature range of 523 to 823 K at strain rates varying from 10^{-6} to 10^{-3} sec^{-1}. An Instron electromechanical testing machine was used in conjunction with a Hewlett-Packard data acquisition and control system to perform all compression tests.

Untested material as well as selected compression specimens were examined by TEM. The microstructure of tested samples was preserved by rapid quenching from the test temperature. TEM samples were electrolytically thinned in a solution of 20% nitric acid - 80% methanol at approximately 243 K. The majority of thin foils prepared were examined in a Philips EM400ST microscope operated at 120 KeV. High resolution images of particles within the as-extruded material were obtained using this microscope. A Philips EM400 microscope with hot stage attachment was used to study the thermal stablity of the as-extruded material.

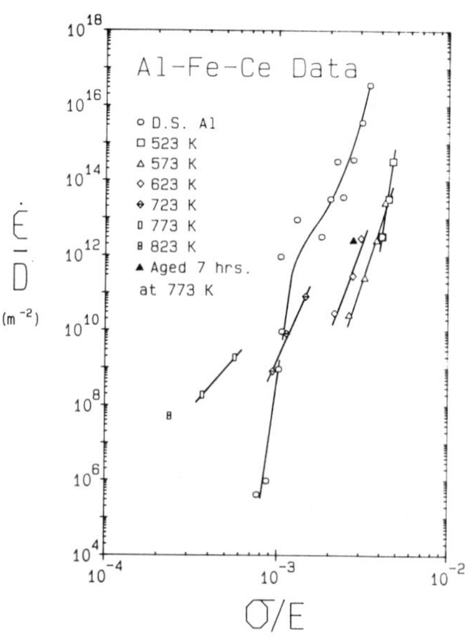

Fig. 1. Temperature compensated strain rate as a function of modulus compensated stress for ODS Al and Al-8.4Fe-3.6Ce. The diffusion coefficient, D is that for lattice diffusion in pure Al. The modulus, E is also for pure Al.

RESULTS

The combined results of all compression tests on the Al-Fe-Ce alloy are shown in Fig. 1. In addition, data for oxide dispersion strengthened (ODS) aluminum (4) have been included for comparison. Strain rate and stress values are normalized by the lattice diffusion coefficient and elastic modulus, respectively, for pure aluminum. At test temperatures below 723 K, the Al-Fe-Ce alloy is significantly stronger than ODS aluminum, and exhibits a stress exponent that decreases from 30 to 12 with increasing test temperature. Above 723 K, the Al-Fe-Ce alloy is considerably weaker than ODS aluminum and the limited data at 773 and 823 K suggest that the stress exponent continues to decrease with increasing temperature.

The results in Fig. 1 show that the Al-Fe-Ce alloy, unlike ODS aluminum, undergoes softening at elevated temperatures that is not related to the temperature dependence of the diffusivity or elastic modulus. In the next section, a number of explanations for the observed softening at elevated temperatures are considered and the validity of each is discussed in terms of additional mechanical tests as well as the TEM results.

DISCUSSION

Normalizing Parameters

In attempting to understand why the Al-Fe-Ce data shown in Fig. 1 do not follow a single power law relationship at all temperatures it is important to consider the validity of the parameters used in the normalization. As mentioned earlier, diffusion coefficient and elastic modulus values for pure aluminum were used to normalize the data in Fig. 1. Although the material being tested contains both iron and cerium, the low solid solubility of these elements in aluminum prevents them from remaining in solution after solidification. Thus, use of the lattice diffusion coefficient for pure aluminum to describe the diffusion controlled climb of dislocations around second phase particles is not unreasonable. With regard to normalization of the stress values, use of the elastic modulus for pure aluminum may not be entirely appropriate. However, since the modulus of the Al-Fe-Ce alloy should not exceed that of pure aluminum by more than 30 percent, any possible changes in Fig. 1 would be small.

Changes in Deformation Mechanism

Another possible explanation for the apparent softening of the Al-Fe-Ce alloy at elevated temperatures is that the deformation mechanism changes as the test temperature is increased. It was noted earlier that the stress exponent appears to decrease significantly with increasing test temperature. Since the Al-Fe-Ce alloy is extremely fine grained ($\simeq 0.5$ µm) and contains a large number of second phase precipitates, this reduction in stress exponent suggests that superplastic deformation may be possible in this material. However, the total elongation for tensile tests conducted at 823 K never exceeded 55 percent. In addition, the tensile data revealed a stress exponent of 6.7 at this temperature. Thus, even though the limited compression data suggest that a change in deformation mechanism may occur at high temperatures, tension tests at 823 K clearly show that a change from climb controlled to superplastic deformation has not occurred.

Microstructural Coarsening

As noted in the Introduction, alloys such as the one being tested rely on a large volume fraction of fine second phase particles for their elevated temperature strength. However, if significant microstructural coarsening were to result from elevated temperature exposure, a loss in strength, such as that seen in Fig. 1, might be expected. In order to gauge the degree of softening due to microstructural coarsening alone, a compression specimen was annealed for seven hours at 773 K prior to testing at 573 K. As shown in Fig. 1., a 27 percent reduction in strength results from this annealing treatment. By comparison, a 94 percent reduction in strength occurs if the alloy is tested at 773 K without prior annealing. Thus, although microstructural coarsening is responsible for some softening, it alone cannot be responsible for the order of magnitude drop in σ/E values seen in Fig. 1.

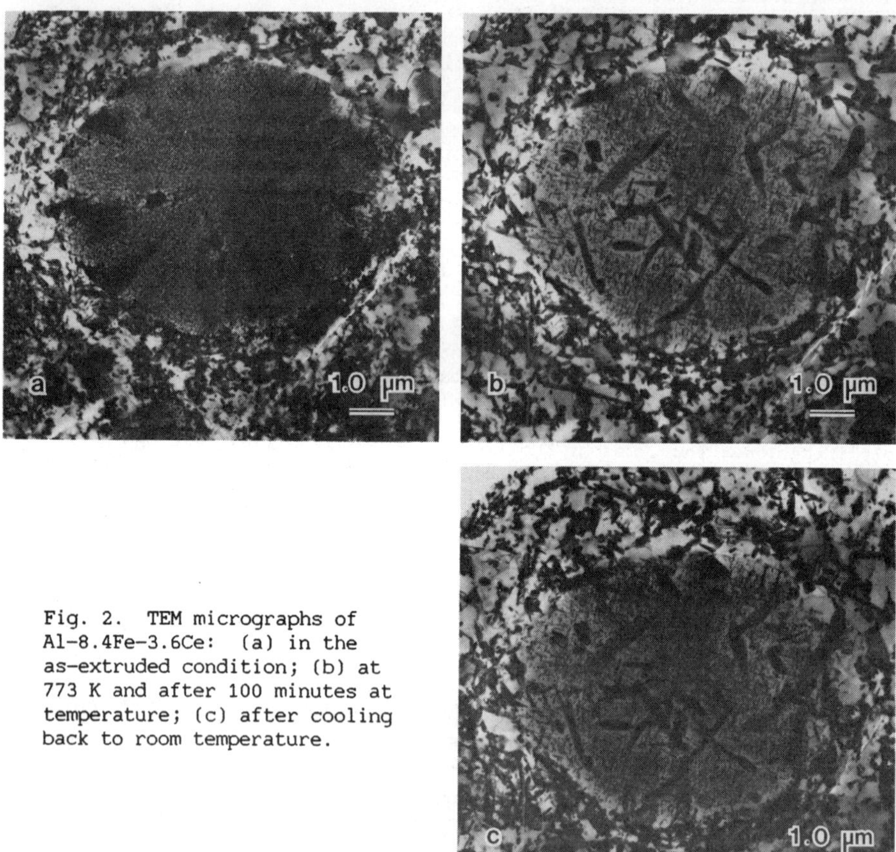

Fig. 2. TEM micrographs of Al-8.4Fe-3.6Ce: (a) in the as-extruded condition; (b) at 773 K and after 100 minutes at temperature; (c) after cooling back to room temperature.

In addition to microstructural coarsening, softening would also occur if the volume fraction of precipitates was reduced by particle dissolution at the highest test temperatures. The results of a TEM hot stage experiment designed to examine this possibility are shown in Fig. 2. A representative view of the as-extruded microstructure can be seen in Fig. 2(a). Note that the particles in the center of Fig. 2(a) do not appear to have been altered by the extrusion process. After 100 minutes at 773 K inside the TEM, Fig. 2(b) shows that noticable microstructural coarsening has taken place. Upon cooling to room temperature, a comparison of Figs. 2(b) and 2(c) reveals that no new particles have precipitated out of solution. Thus, a reduction in the number of second phase particles due to dissolution is not responsible for the loss of strength observed in this Al-Fe-Ce alloy above 723 K. Furthermore, since particle dissolution does not occur, reprecipitation of particles is not responsible for the recovery in strength that takes place upon cooling.

Particle Strength

A number of possible explanations for the softening of an Al-Fe-Ce alloy at elevated temperatures have been considered and rejected. One other possibility is that the second phase particles responsible for the superior low temperature strength of this alloy simply become weak at high temperatures, losing their ability to serve as effective barriers to dislocation motion. Langenbeck et al. (5) have indicated that after extrusion, the majority of precipitates are Al_3Fe type with cerium entering the phase without changing the structure. To avoid confusion, the name 'Al_3Fe' has been retained, although a more accurate chemical formula is actually $Al_{13}Fe_4$ (6). Since Al_3Fe is an equilibrium phase, it has been assumed that the type of particles present in the Al-Fe-Ce alloy were essentially the same in all tests.

A representative TEM micrograph of a specimen deformed 25 percent at a test temperature of 773 K and a strain rate of $1x10^{-5}$ sec^{-1} is shown in Fig. 3. Note that a number of particles (indicated by arrows) exhibit contrast features characteristic of twinned or faulted structures. Previously, Yearim and Shechtman (7) postulated that faults of this type are "grown-in", forming only in Al_3Fe particles that have undergone extensive growth (> 1 μm in length), such as during the 100 hour, 755 K anneal they performed on their Al-Fe-Cr alloys. However, as can be seen in Fig 4(a), faulted particles are also found in the as-extruded microstructure of the Al-Fe-Ce alloy. The complex nature of the faulting is shown in the high resolution micrograph of Fig. 4(b). These particles have not undergone extensive elevated temperature

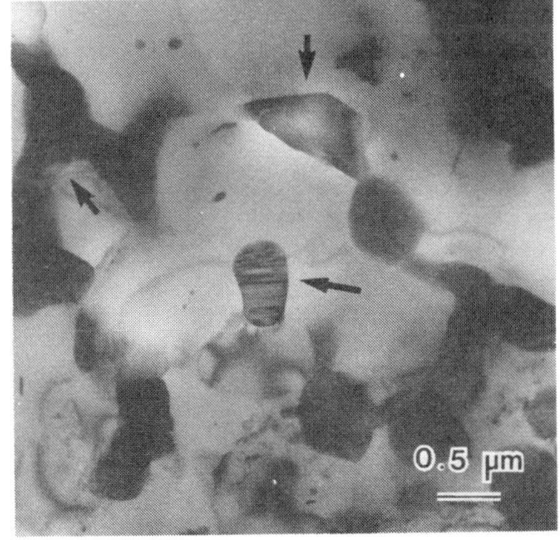

Fig. 3. TEM micrograph of Al-8.4Fe-3.6Ce deformed 25% at 773 K at a strain rate of $1x10^{-5}$ sec^{-1}. Arrows indicate faulted particles.

exposure and in general are quite small. Thus, one cannot rule out the possibility that faulting may be the result of deformation as well as annealing. The presence of particles which are strong and essentially non-deformable at low temperatures but weaken considerably and become deformable at high temperatures would provide a rational explanation for the unusual softening behavior that has been observed. Microstructural coarsening alone is responsible for only a small fraction of the reduction in strength at elevated temperatures. Thus, an additional softening mechanism, such as the reduction in strength of second phase particles with increasing temperature, is needed to explain the observed behavior.

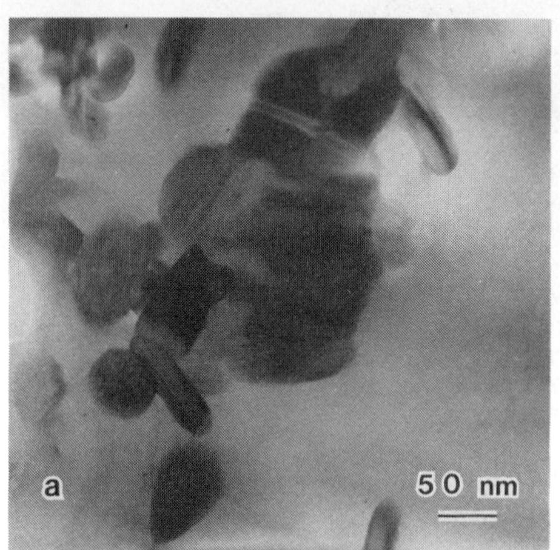

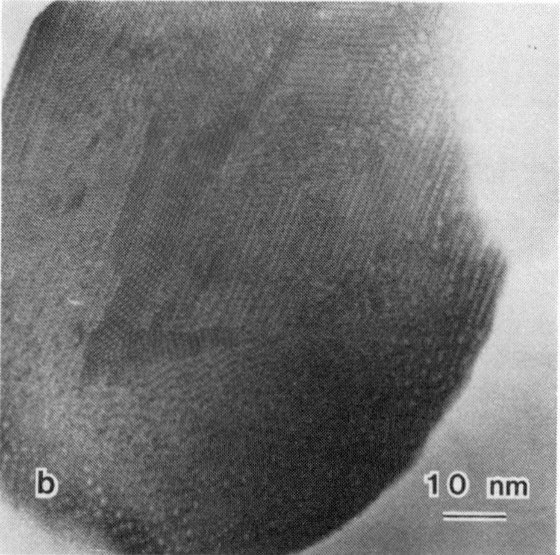

Fig. 4. TEM micrographs of faulted particles in Al-8.4Fe-3.6Ce: (a) Low magnification; (b) high resolution lattice image.

ACKNOWLEDGEMENTS

This work is sponsored by the Air Force Office of Scientific Research under Grant No. 81-0022. The authors wish to thank Dr. S.L. Langenbeck of Lockheed-California for providing the material. The assistance of Prof. J.C. Bravman of Stanford University in obtaining high resolution TEM images is also gratefully acknowledged.

REFERENCES

1. R.E. Sanders, Jr. and G.J. Hildeman, Elevated Temperature Aluminum Alloy Development, Final Technical Report AFWAL-TR-81-4076. Wright-Patterson AFB, OH (1981).
2. S.L. Langenbeck, et al., Elevated Temperature Aluminum Alloy Development, Interim Technical Report, Lockheed Report No. LR 30377. Burbank, CA (1982).
3. W.M. Griffith, R.E. Sanders, Jr. and G.J. Hildeman, in High Strength Powder Metallurgy Aluminum Alloys, Conf. Proc., (edited by M.J. Koczak and J. Hildeman),p. 209. T.M.S.-A.I.M.E., Warrendale, PA (1982).
4. W.C. Oliver and W.D. Nix, Acta metall. **30**, 1335 (1982).
5. S.L. Langenbeck et al., Elevated Temperature Aluminum Alloy Development, Interim Technical Report, Lockheed Report No. LR 30662. Burbank, CA (1984).
6. P.J. Black, Acta Cryst. **8**, 43 (1955).
7. R. Yearim and D. Shechtman, Metall. Trans. **13A**, 1891 (1982).

Strain Rate Sensitivity: Al Versus Al-0.34WT%Cr

M. J. Bull* **and S. Saimoto****

**Alcan International Ltd., Kingston Laboratories, Kingston, Canada*
***Department of Metallurgical Engineering, Queen's University, Kingston, Canada*

ABSTRACT

The yielding behaviour and strain rate sensitivity of polycrystalline Al-0.34 wt% Cr has been examined over the temperature range 78 to 610 K. Comparison of the observed behaviour with that of nominally pure Al and single crystals of Al-0.2 wt% Cr has been made in an attempt to quantify the relative contributions of grain size, solute, precipitates, dislocations and deformation debris to the yielding behaviour and strain rate sensitivity. Using the Haasen plot, at temperatures below 297 K the strain rate sensitivity of Cr solutes is about 5 times larger than that of dislocations.

KEYWORDS

Thermodynamic response; strain rate sensitivity; constitutive relations; solute hardening.

INTRODUCTION

Constitutive relations which relate the microstructural response of engineering materials to plastic flow are benefical in making engineering predictions of material performance during metal deformation processing. Unfortunately, the traditional or conventional determination of mechanical properties such as work hardening response are generally inadequate in sheet formability [1]. For these reasons new phenomenalogical theories [2-5] have been invoked in an attempt to characterize a materials' response as a function of a global structural variable. While benefical in a broad sense, the assumed thermodynamic response of the material to stress is oversimplified, resulting in the inability to clearly differentiate the contributory influences of solutes, precipitates and various deformation products (loops, vacancies) to plastic flow. Therefore, recent precise determinations [6,7] are examined herein.

The method used to differentiate these various contributions to flow stress and strain rate sensitivity has been suggested by Kocks, Ashby and Argon [8] and is designated as the Haasen plot [9], wherein the inverse of the activation volume is expressed as a function of the flow stress. If the flow stress can be approximated by a linear sum of the components due to grain size (gs), solute (s), precipitates (p) and deformation products (D for dislocations and d for debris), and if only the concentration of one species changes during deformation, the activation volume, v, is related to the relative strain rate sensitivities according to $\sigma_o = (\sigma_{gs} + \sigma_p + \sigma_s)$ and

$$\frac{1}{T}\frac{d\sigma}{\partial \ln \dot{\varepsilon}}\bigg|_{T,\Sigma} \equiv \frac{1}{T}\{(\sigma - \sigma_o) M_D + \sigma_s M_s + \sigma_d(M_d - M_D)\} \quad (1)$$

where Σ stands for constant structure. Mulford [9] and Chaturvedi and Lloyd [10] have shown qualitatively using a standard Instron machine that solid solution and precipitate effects can be detected using this technique. Recently MacEwen [11] was able to assess the obstacle strength of interstitial solutes and their spacing.

In the present study, the precise activation volumes of Cr in an aluminum alloy have been measured over the temperature range 78 to 449 K. Attempts are made to quantify the relative contributions of grain size, solute, precipitates, dislocations and deformation debris on the yielding behaviour and strain rate sensitivity of an Al-Cr alloy.

EXPERIMENTAL

The strain rate sensitivity of a super pure aluminum with 0.5 wt % Fe, the same alloy with the addition of 0.34 wt.% Cr and single crystals of super purity Al - 0.2 wt.% Cr have been examined over the temperature range of 78 to 449 K. The material used in this study was especially prepared from super pure aluminum (99.999%) cast in 23 Kg ingots with additions of Fe to control the grain size. After hot rolling to 10 cm thick slab it was cold worked to 1 mm sheet and annealed 2 h at 673 K. The resulting recrystallized grain size was 40 ± 10 μm. Tensile samples for these polycrystalline materials were prepared parallel to the rolling direction with the final dimensions being 2x1x50 mm. The single crystals were grown using a modified Bridgeman technique in a purified graphite mould. The as-grown crystal was homogenized at 873 K for 24 hours and allowed to cool in circulating air. Tensile samples 2.4 x 3.0 x 46 mm were prepared using an EDM parallel to the growth direction. The samples were subsequently etched, and solution heat treated at 848 K and allowed to cool in circulating air.

For constant "structure" determinations, very rapid strain rate changes which eliminate the transient in the load response are required. A specially constructed tensometer similar to that described by Champion, Duesbery and Saimoto [12] which incorporated a magnetic device to eliminate the compliance of the machine and specimen has been used. Strain rate changes (1/10 reduction) were performed at approximately every 1% strain after yielding.

RESULTS AND DISCUSSION

The Additivity Rule for the Flow Stress

The present analysis using the Haasen plot assumes the additivity of the stress contribution from different microstructural obstacles in order to initiate plastic flow. The basic assumption for this rule to hold [13] is that the much weaker species be many times more numerous than the strongest obstacles. This condition should roughly apply to dilute solid solutions, submicroscopic clusters or precipitates, and as previously shown to deformation debris [6,7].

This rule was systematically examined using an analogous procedure to that of MacEwen [11] by assessing the value of M_s at a given temperature from the single crystal data (Fig. 1a) and using it to predict σ_s for the polycrystalline material (Fig. 1b). As expected σ_s is smaller than σ_o whereas M_s is 4 to 5 times M_D. The σ_s contribution of about 10 MN/m² drops precipitously between 273 and 297 K (Table I). However, since this drop is about 1/3 of σ_o, a corresponding drop is difficult to discern especially since the as-grown dislocation portion is continuously decreasing. Considering the σ_o value at 610 K is still about 1/2 that at 78 K, this athermal portion can be attributed to the combined contribution of σ_{gs} and σ_p. In the calculation from the intercept value, σ_s was estimated by subtracting the yield stress value for single crystals of pure Al of similar orientation [14,15]. Considering that this and the imprecise nature of the yield stresses in both single crystal and polycrystalline material could lead to large errors, the deduced values are surprisingly reasonable.

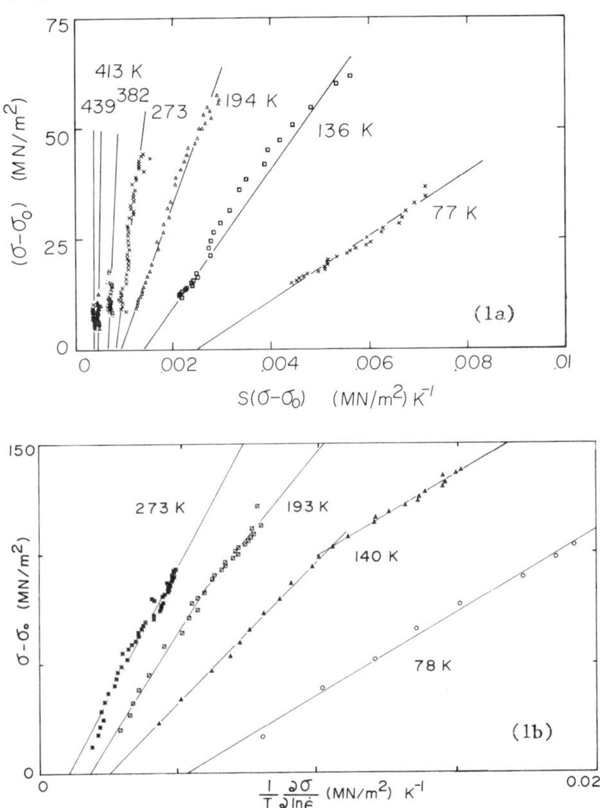

Fig. 1. Plots of flow stress versus inverse of activation volume;

$$\frac{1}{T}\frac{\partial\sigma}{\partial \ln\dot{\varepsilon}} = S(\sigma-\sigma_o) = \frac{k}{v}$$

a) Alloy single crystal where zero was used for σ_o.
b) Alloy polycrystalline material at low temperatures where S is independent of strain rate.

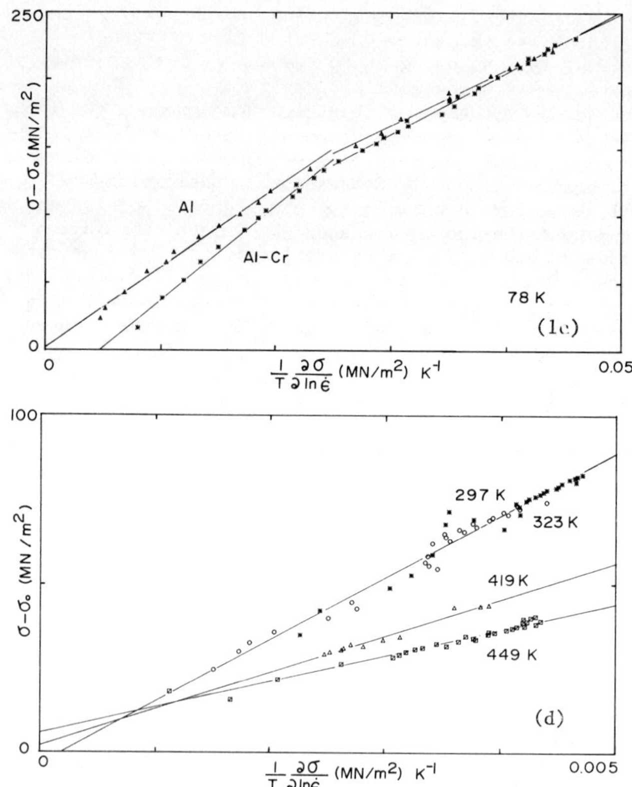

Fig. 1. c) Comparison of pure Al with Al-Cr alloy at 78 K.
d) Alloy polycrystalline material at high temperature where S is dependent on strain rate.

TABLE I Summary of Relative Strain Rate Sensitivities and the Comparison of σ_s to σ_o

Temper-ature	78K		194 K		273		297-450 K		>450K - 610	
	σ yield Contri-bution MN/m².	Rate Sensit-ivity	σ yield Contri-bution	Rate Sensit-ivity	σ yield Contri-bution	Rate Sensit-ivity	σ yield Contri-bution	Rate Sensit-ivity	σ yield Contri-bution	Rate Sensit-ivity
σ_o	40	---	31	---	28	---	26-22	---	21-18	---
Solute (s)	$\sigma_s = 9.7$	$M_s/T = 5.2 \times 10^{-4}$	$\sigma_s = 10.5$	$M_s/T = 2 \times 10^{-4}$	$\sigma_s = 8.4$	$M_s/T = 1.6 \times 10^{-4}$	$\sigma_s(297) \simeq 1.2$	$M_s/T \simeq 1 \times 10^{-4}$	$\sigma_s \simeq 0$	---
Precip-itate (p)	$\sigma_p(A)$	A	$\sigma_p(A)$	A	$\sigma_p(A)$	A	$\sigma_p(A)$	A	σ_p	A to T
Disloc-ations (D)	σ_D	$M_D/T = 1.3 \times 10^{-4}$ $M_{DS}/T = 1.37 \times 10^{-4}$	σ_D	$M_D/T = 3.95 \times 10^{-5}$ $M_{DS}/T = 3.23 \times 10^{-5}$	σ_D	$M_D/T = 3.0 \times 10^{-5}$ $M_{DS}/T = 1.23 \times 10^{-5}$	σ_D	$M_D > M_{DS}$ $M_{DS} \to 0$ Recovery	σ_D	Re-covery
Debris (d)	$\sigma_d(D)$	$M_d > M_D$	$\sigma_d(D)$	$M_d > M_D$	$\sigma_d(D)$	$M_d \sim M_D$	$\sigma_d(D)$ $\sigma_d(p)$	Recovery	$\sigma_d(p)$	Re-covery

M_{DS} from single crystal M_D from polycrystals A - Athermal Contribution T - Thermal Contribution

The intercepts in Fig. 1, were drawn by extrapolating the initial linear part of the $(\sigma-\sigma_o)$ vs $\frac{1}{T}\frac{\partial\sigma}{\partial\ln\dot{\varepsilon}}$ plot, the slope of which is T/M_D. At 78K it was previously shown [6] that the non-linear behaviour at high strains, that is the deviation from Cottrell-Stokes relation [16], is due to debris formation. Comparison of (nominally pure) Al with that of the Al-Cr alloy in Fig. 1c shows that this deviation is also apparent in the latter. Moreover the Al data has an initial intercept on the σ axis with a value near σ_o whereas the Al-Cr alloy shows an $M_s\sigma_s$ value, from which, σ_s was deduced. As the temperature increases to 273 K this deviation due to σ_d disappears since as the defects grow in size, M_d approaches that of M_D.

Constitutive Relationships

At temperatures of 273 K and lower the strain rate sensitivity is independent of the magnitude of the strain rate and the amount of strain rate change, indicating that the pre-expontial to the rate equation is independent of the stress thus meeting the premise of the Haasen plot [6]. In this temperature range, the M_D/T decreases with increasing temperature and the values of M_D and M_{DS} are approximately the same as shown in Table I. As shown elsewhere [17] the constitutive relation in this temperature region follows

$$\sigma-\sigma_o = K\mu T\,[S(\sigma-\sigma_o)]^{\chi}\,\dot{\varepsilon}^{\,M} \qquad (2)$$

where $S = M/T$, $\chi = 0.92$, and σ applies to flow stresses beyond 0.04 strain. Compared to that for pure Al [6] the applicability of this functional relationship has increased from 225 to 273 K indicating that the Cr addition has delayed the recovery effect discussed below.

At temperatures above 273 K, the magnitude of the strain rate and amount of rate change affects the strain rate sensitivity. As summarized in Table I the apparent dislocation rate sensitivity of polycrystalline material becomes considerably larger than that for single crystals. This phenomenon was attributed to multiple slip and possible occurrence of submicroscopic clusters or precipitates. Indeed their existence was confirmed by weak beam microscopy which revealed very fine precipitates of less than 100 Å in the polycrystalline sheet but not in the single crystal. The genesis of these clusters is attributed to the recrystallization anneal at 673 K.

In the temperature range 323 to 449 K, the $(\sigma-\sigma_o)\,d\sigma/d\varepsilon$ vs $S(\sigma-\sigma_o)$ plots showed [17] that a master curve similar to that observed for pure Al [18] exists. This relationship can be approximated by

$$(\sigma-\sigma_o) = K\,S^{\frac{-\alpha}{2+\alpha}}\,\varepsilon^{\frac{1}{2+\alpha}} \quad \text{where } \alpha = 1.42. \qquad (3)$$

Assuming $S = A\,\dot{\varepsilon}^{\,\beta}$, this relation can be simplified to the more familiar relation $(\sigma - \sigma_o) = K\,\dot{\varepsilon}^{\,m}\,\varepsilon^{n}$ where $m = \beta\alpha/(2+\alpha) = \beta(0.5-n)/2$ and $n = 1/(2+\alpha)$. The present relationship shows that m and n are not independent parameters as generally assumed in mechanical testing. These constitutive relationships indicate that their applicability is not general and is highly dependent on the temperature and alloying species.

The approximate correlation between $(\sigma-\sigma_o)\,d\sigma/d\varepsilon$ and $S(\sigma-\sigma_o)$ which results in the work hardening law is an indication that the mean slip distance is proportionally related to the mean interobstacle distance; that is, the activation distance (d) is constant. Under such conditions it was shown by Saimoto and Duesbery [7] that

$$\frac{1}{S} = \frac{2K_0K_2\,N_2\,T}{\dot{\varepsilon}\exp Q/kT - K_0K_2N_2} + \frac{\alpha\,\mu b^2 d}{k} \quad (4)$$

where K_0 and K_2 are constants, N_2 is the density of recoverable species with activation energy Q, and the second RHS term is the strong obstacle constant.

Thus a plot of $(\sigma-\sigma_o)$ vs $(\sigma-\sigma_o)S$ (Fig. 1d) should result in a decreasing slope with increasing temperature as observed. Although the extrapolation of the data to locate the intercept has to be treated with caution, there is an indication that the intercept gives a positive stress value. Since σ_o accounts for the athermal σ_{gs} and σ_p, this effect possibly arises from loop formation around the observed clusters, which inturn act as forest dislocations This effect becomes only detectable where the flow stress has become small.

CONCLUSIONS

Precision strain rate sensitivity measurements have the capacity to differentiate the temperature dependence of the solute contribution to the yield stress from those of grain size and dispersed particles. The strain rate sensitivity below 297 K is independent of the strain rate level and the magnitude of rate change. In this region, the constitutive relation (2) is obeyed whereas at higher temperatures it is dependent on these factors and relation (3) is found.

ACKNOWLEDGEMENT

This study was performed while one of us (M.J.B.) was a recipient of a Stelco Graduate Fellowship. We thank the Natural Sciences and Engineering Council of Canada for generously supporting our studies on plastic flow.

REFERENCES

1. S.S. Hecker, J. Eng. Mater. Tech, 97, 66 (1975).
2. E.W. Hart, Acta Metall. 15, 1545 (1967), Acta Metall. 18, 599 (1970).
3. U.F. Kocks, J. Eng. Mater. Tech. 98, 76 (1976).
4. U.F. Kocks, J.J. Jonas and H. Mecking, Acta Metall. 27, 419 (1979).
5. I.H. Lin, J.P. Hirth and E.W. Hart, Acta Metall. 29, 819 (1981).
6. S. Saimoto and H. Sang, Acta Metall. 31, 1873 (1983).
7. S. Saimoto and M.S. Duesbery, Acta Metall. 32, 147 (1984).
8. U.F. Kocks, A.S. Argon and M.F. Ashby, Prog. Mater. Sci., Vol. 19, (edited by B. Chalmers, J.W. Christian and T.B. Massalski) (1975).
9. R.A. Mulford, Acta Metall. 27, 1115 (1979).
10. M. Chaturvedi and D. Lloyd, Metal Science, 14, 277 (1980).
11. S. MacEwen, Acta Metall. 30, 1431 (1982).
12. H.G. Champion, M.S. Duesbery and S. Saimoto, Scripta Metall. 17, (1983).
13. A.J.E. Foreman and M.J. Makin, Can. J. Phys. 45, Part 2, 511 (1967).
14. W. Staubwasser, Acta Metall. 7, 43 (1959).
15. W.F. Hosford, Jr. R.L. Fleischer and W.A. Backofen, Acta Metall. 8, 187 (1960).
16. A.H. Cottrell, amd R.J. Stokes, Proc. R. Soc. 233A, 17 (1955).
17. M.J. Bull, M.Sc. Thesis, Queen's University at Kingston, Canada (1981).
18. S. Saimoto, H. Sang and L.R. Morris, Acta Metall. 29, 215 (1981).

Effects of Prestrain at High Temperatures on the Strength of Aluminum

S. Kikuchi* and A. Yamaguchi**

**Department of Metal Science and Technology, Kyoto University, Sakyo-ku, 606 Kyoto, Japan*
***Graduate School of Engineering, Kyoto University, Sakyo-ku, 606 Kyoto, Japan*

ABSTRACT

The effects of prestraining at high temperatures on the strength of aluminum have been examined. Polycrystals and single crystals of high purity aluminum were prestrained in tension to a strain of 0.25 above test temperatures. Well-developed subgrain structures were formed by prestraining. The measured subgrain size, λ, is given as a function of the flow stress, τ_s, at the prestrain by $\lambda/b=12(\tau_s/G)^{-1}$, where b is the Burgers vector and G the shear modulus. The flow stress of the prestrained specimen was affected by the initial subgrain size, especially in a small strain region, when it was deformed at a test temperature. The relation between the yield stress, τ_y, and the subgrain size is given by $\tau_y/G= L(\lambda/b)^{-m}$, where L and m are constants. We obtained m=0.81 for aluminum. It was confirmed that the subgrain structures formed by prestraining at high temperatures are thermally stable.

KEYWORDS

High temperature strength; Prestraining; Subgrain size; Aluminum; Yield stress; High temperature deformation.

INTRODUCTION

It is commonly accepted that prestraining causes an improvement of the high temperature strength of metallic materials. Most investigations [1-2] have been undertaken to examine the effect of prestraining at room temperature on the creep strength and rupture elongation at elevated temperatures. However, the dislocation substructures induced by room temperature prestraining are not always stable at test temperatures because recovery and even recrystallization may occur. This instability of substructures is considered unfavorable for the improvement of the high temperature strength of materials.
On the other hand, if prestraining is applied above the test temperature, the stable substructures which are available for strengthening materials at high temperatures can be formed in the crystal. From a metallographic point of view, the possible factors to improving the strength of materials by high

temperature prestraining are the following; (a) subgrain formation, (b) precipitation of second phase particles, and (c) the formation of serrated grain boundaries.

Our main interest is to investigate whether an improvement of the high temperature strength can be achieved in the region of small strains by prestraining above the test temperatures. The reason for this is that the strength at small strains, particularly yield strength, is most important from an engineering point of view.

It is the purpose of the present paper to examine the effects of substructures formed by prestraining at high temperatures on the strength of aluminum at elevated temperatures. Especially, the relationship between the subgrain size and the yield strength is described.

EXPERIMENTAL PROCEDURES

Specimens used here are both polycrystals and <111 > oriented single crystals of high purity aluminum (99.99%). Polycrystalline specimens were machined from a rolled plate and then annealed for an hour at 773 K. The mean grain size is 0.35 mm. Single crystal specimens were cut off by a spark cutter from large single crystals grown by the Bridgman method, and then annealed for 24 hours at 773 K to remove any residual stresses. The dimension of tensile specimens is 5 mm × 3 mm × 30 mm for the polycrystals and 5 mm × 3 mm × 50 mm for the single crystals. All the specimens were prestrained in tension up to a strain of 0.25 in the temperature range from 573 to 773 K and the strain rate range from 10^{-5} to 10^{-3} s^{-1}. After prestraining, the specimens were rapidly cooled to the desired test temperatures and immediately tensile tests were performed. To observe the substructures of prestrained specimens, we used an optical microscope and a transmission electron microscope. Samples for the optical microscopy were electro-polished in a mixture of 20pct perchloric acid in methanol, coated with liquid gallium, held at 333 K for 24 hours, and finally electro-polished again. Samples for the transmission electron microscopy (TEM) were cut from prestrained specimens to thin foils by the spark cutter and those were electro-polished using Tenupole. The subgrain size was measured by a linear intercept method.

RESULTS AND DISCUSSION

Flow stress and Subgrain Size for Prestraining

Figure 1 shows a typical stress-strain curve during prestraining and deformation after it. Specimens were deformed in tension up to a prestrain of 0.25 in the temperature range from 573 to 773 K at various strain rates, cooled rapidly under unloading and then deformed in tension at the desired test temperatures and at a constant cross head speed of 0.1 mm/min. It was found that the steady state deformation is attained at the prestrain and well-developed substructures are formed by prestraining.

The dependence of the average subgrain size, λ, which was measured in both polycrystals and single crystals prestrained, on the applied shear stress, τ_s, at the prestrain was investigated. The results are shown in Fig. 2. It should be noted that the relation between the shear stress, τ, and the normal stress, σ, is $\tau=0.280\sigma$ for <111> oriented single crystals and $\tau=0.327\sigma$ for polycrystals. The average subgrain size is given as a function of the applied shear stress in the following normalized form;

$$\lambda/b = K (\tau_s / G)^{-1} \qquad (1)$$

where b is the Burgers vector, G the shear modulus, K a constant. Equation (1) is applicable to both single crystals and polycrystals. The values of K

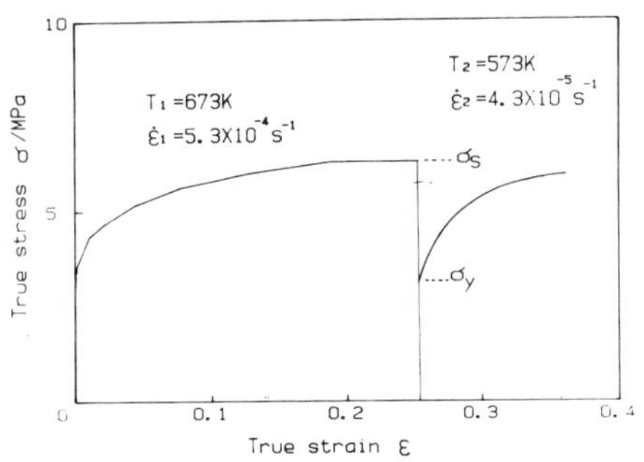

Fig. 1. Example of stress-strain curve for aluminum polycrystal; σ_s and σ_y represent the flow stress at a prestrain of 0.25 and the yield stress at a test condition respectively.

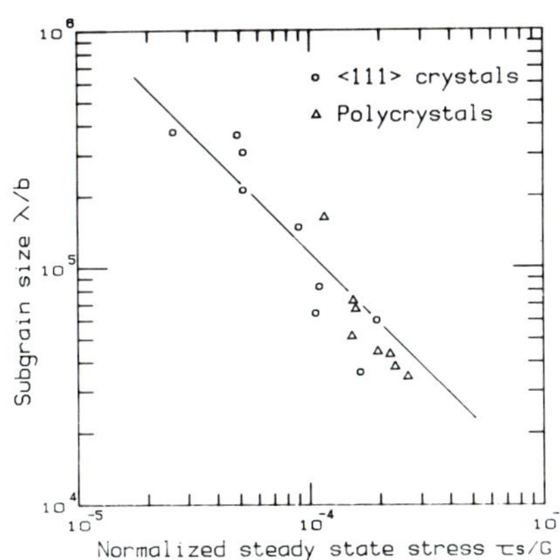

Fig. 2. Log (λ/b) *vs* log (τ_s/G) for aluminum.

were independent of the prestrain temperatures. It was obtained that K=12 from the results in Fig. 2. This value is in good agreement with the value, K=10, reported for a number of materials [3].
In order to confirm that the steady state deformation is attained at the prestrain, the dependence of the strain rate, $\dot{\varepsilon}$, on the applied stress, σ_s, was examined at 573, 673 and 773 K. The strain rate is plotted against the applied stress on a logarithmic scale in Fig. 3. This relation can be expressed in the same form as for the steady state creep rate, namely; $\dot{\varepsilon}=A\sigma_s{}^n$, where A is a constant which depends on the temperature and n is the stress exponent.
It can be seen from Fig. 3 that the straight lines are obtained at each prestrain temperature and the slopes are close to 5. The value of stress exponent, n=5, is in good agreement with the value reported in the case of the high temperature creep for aluminum and other pure metals [4]. The flow stress levels of <111> oriented single crystals were nearly equal to those of polycrystals under the same deformation conditions.

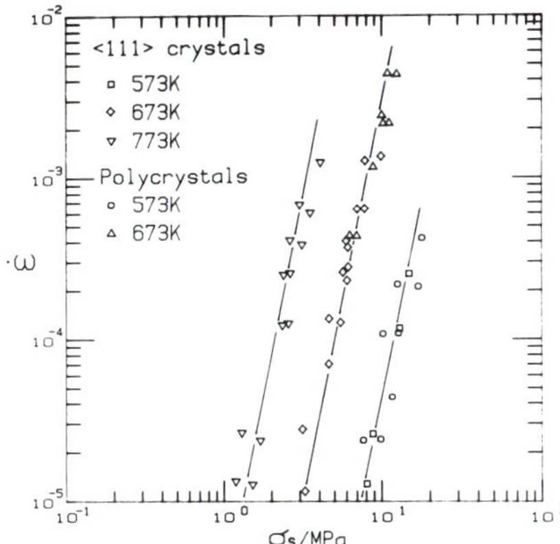

Fig. 3. Log $\dot{\varepsilon}$ *vs* log σ_s for aluminum.

<u>High Temperature Deformation Behavior of Prestrained Specimens</u>

For the prestrained polycrystal specimens with various subgrain sizes, Fig. 4 indicates the typical stress-strain curves at 573 K and $\dot{\varepsilon}=4\times10^{-5}s^{-1}$.
The flow curves of prestrained specimens are divided into two types; the work softening type and the work hardening type. If the initial subgrain size is smaller than that corresponding to the steady state flow stress reached at the test condition, the coarsening of the subgrains occurs during deformation and it leads to work softening. On the other hand, if the initial subgrain size is larger than that of the steady state,

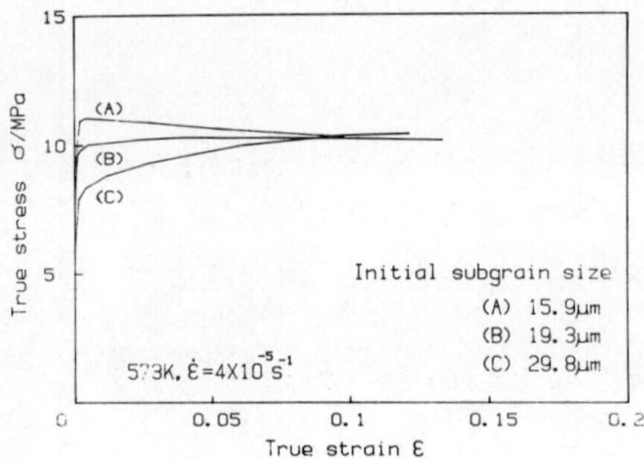

Fig. 4. Stress-strain curves obtained at 573 K for polycrystals with various subgrain sizes.

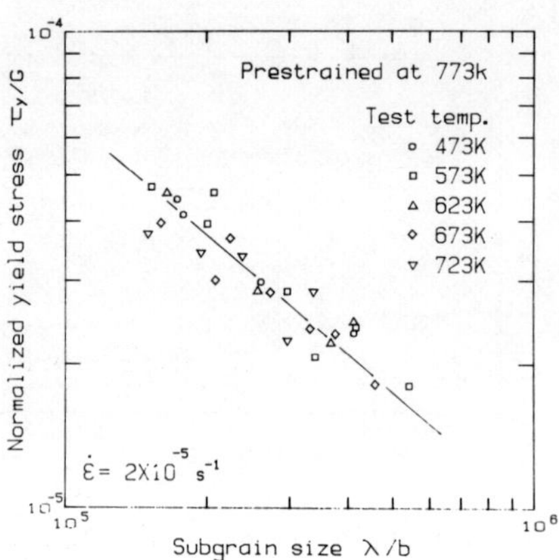

Fig. 5. Log (τ_y/G) *vs* log (λ/b) for aluminum single crystals.

work hardening occurs because the subgrain refinement takes place during deformation. In both cases, the average subgrain sizes of prestrained specimens vary during deformation into the steady state size which is determined by the deforming condition, such as strain rate and temperature. It can be seen that the effect of prestrain on the high temperature strength of aluminum is marked in the small strain region (see Fig. 4). This effect is enhanced as the initial subgrains of the specimens are finer.

Relation between Yield Stress and Subgrain Size

Fig. 5 shows the relation between the yield stress, τ_y, and the initial subgrain size, λ, for <111> oriented single crystals prestrained at 773 K. Here, the normalized yield stress, τ_y/G, is plotted against the normalized initial subgrain size, λ/b, on a logarithmic scale. It is found that the yield stress may be generally expressed as a function of the initial subgrain size in the following form;

$$\tau_y / G = L (\lambda / b)^{-m} \tag{2}$$

where L and m are constants.

We evaluated m=0.81 for aluminum based on the results in Fig. 5. From the results of constant structure creep tests, Sherby *et al.* [5] suggested that the yield stress is proportional to $G(\lambda/b)^{-3/8}$ for pure metals. The value of m (=0.81) which we obtained is not consistent with the value of m (=3/8) derived by Sherby *et al.* from the creep rate equation; our value is rather close to unity obtained by Al-Haidary *et al.* [6].

The yield strength of prestrained specimens is considered to depend strongly on the dislocation link length at subboundaries. Morris *et al.* [7] have proposed that the effective stress, τ_{eff}, necessary to produce dislocation emission from subboundaries is given by

$$\tau_{eff} = Gb / 3\ell \tag{3}$$

where ℓ is the dislocation link length.

Fig. 6 shows the relation between the dislocation link length at subboundaries observed by the TEM and subgrain size. Although the data seem to be somewhat scattering, the linear relation, $\ell=0.003\lambda$, given by Al-Haidary *et al.* [6] is

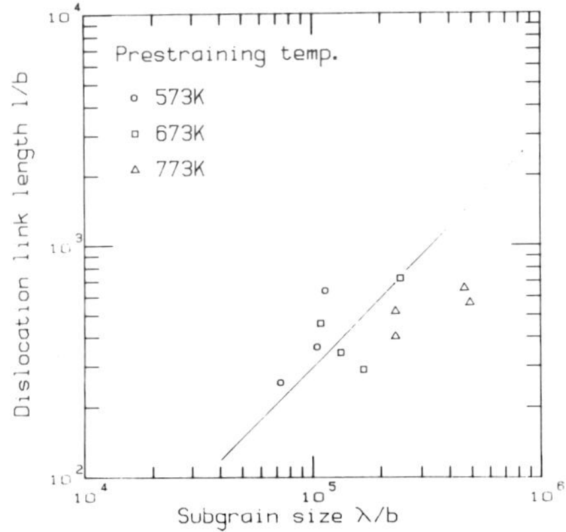

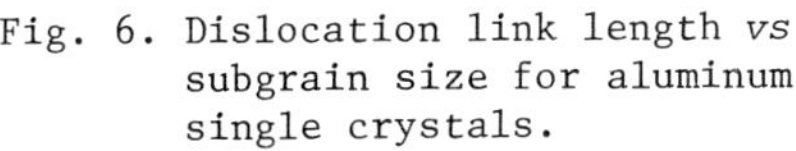

Fig. 6. Dislocation link length *vs* subgrain size for aluminum single crystals.

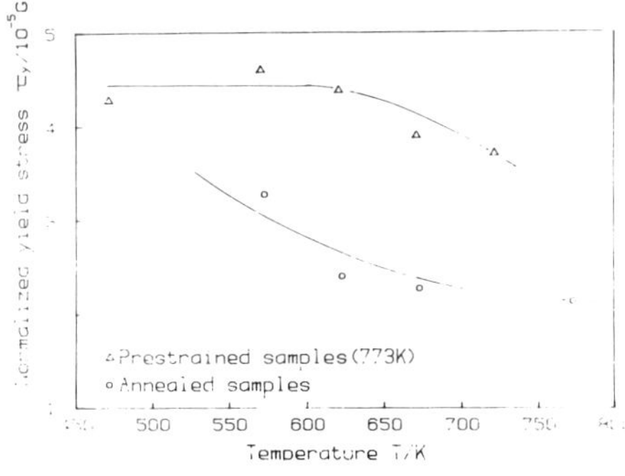

Fig. 7. Effect of temperature on the yield stress for prestrained and annealed specimens of aluminum single crystals.

approximately satisfied. Substituting the above relation between ℓ and λ in Eq. (3) and considering the stress concentration factor, α, at the subboundaries, the yield stress can be expressed as follows;

$$\tau_y / G = \tau_{eff} / \alpha G = (10^2/\alpha) \cdot (\lambda / b)^{-1} \qquad (4)$$

were $\tau_{eff} = \alpha \tau_y$.
We found $\alpha \simeq 10$. This value was obtained by putting the results in Fig. 5 into Eq. (4). Blum *et al.* [8] pointed out that the stress concentration factors at subboundaries range from 2.5 to 10 for aluminum. From this, our value may be considered to be fairly reasonable.

Temperature Dependence of Yield Stress

The relationship of the yield stress with temperature is shown in Fig. 7 for the prestrained single crystals as well as the annealed ones. The yield stress for the annealed specimens decreases monotonously with increasing temperature, while for the prestrained specimens the reduction degree of the yield stress with temperature is very slight. This fact leads to the conclusion that the substructures induced by prestraining are thermally stable.

CONCLUSIONS

Single crystals and polycrystals of high purity aluminum were prestrained in tension up to a strain of 0.25 above the desired test temperatures at various strain rates. As a result, well-developed subgrain structures were formed. The subgrain size is inversely proportional to the flow stress at the prestrain.
The High temperature strength of the prestrained specimens markedly increases in the small strain region compared with that of the annealed ones.
The yield strength of the prestrained specimens depends on the subgrain size, the relation can be expressed by $\tau_y/G = L(\lambda/b)^{-m}$, where m and L are constants. We obtained m=0.81.

The temprature dependence of yield stress for the prestrained specimens is slight. It has been confirmed that the subgain structures induced by prestraining at high temperatures are thermally stable at the test temperatures.

ACKNOWLEDGEMENTS

One of the authors (S. K.) would like to acknowledge Professor B. Ilschner of the Ecole Polytechnique Fédérale de Lausanne (Switzerland) for critical insight that stimulated this investigation.

REFERENCES

1. W. M. Yim and N. J. Grant, Trans. Met. Soc. AIME 227, 868 (1963).
2. B. F. Dyson and M. J. Rodgers, Met. Sci. 8, 261 (1974).
3. T. J. Ginter and F. A. Mohamed, J. Mater. Sci. 17, 2007 (1982).
4. O. D. Sherby and P. M. Burke, Prog. Mater. Sci. 13, 325 (1967).
5. B. Walser and O. D. Sherby, Scripta Met. 16, 213 (1982).
6. J. T. Al-Haidary, N. J. Petch and E. de los Rios, in Yield, Flow and Fracture of Polycrystals (edited by T. N. Baker), p. 33, Applied Science Publishers, London (1983).
7. M. A. Morris and J. L. Martin, Acta Met. 32, 1609 (1984).
8. W. Blum and H. Schmidt, Res Mechanica, 9, 105 (1983).

Restoration of Copper During Stress Relaxation Between Stages of High Strain Rate, Hot Torsion

L. Vasquez* and H. J. McQueen**

**Dept. de Materiales, Universidad Autonoma Metropolitana Unidad, Azcapotzalco, 02200 Mexico DF*
***Mechanical Engineering, Concordia University, Montreal, Canada H3G 1M8*

ABSTRACT

The kinetics of static recovery and recrystallization of ETP Cu were determined while annealing at the deformation temperature under both stress relaxation and unloading conditions after torsion at 450 and 500°C at rates of 0.18 and 1.8 s^{-1}. Restoration was measured by the reduction in flow stress during the interval. Recovery occurs more slowly during stress relaxation than during unloading whereas recrystallization occurs more quickly.

KEY WORDS

Hot deformation, torsion, strain hardening, stress relaxation, static recovery, static recrystallization, kinetics, Avrami constants.

INTRODUCTION

The static softening of metal between stages of either hot or cold working is of considerable importance in reducing forming forces and improving ductility [1-3]. In this the metal is usually free of stress; however, there may be applied stresses such as those in a strip mill from a subsequent stand or coiler, or in a continuous annealing furnace. During stress relaxation after hot deformation, the time for starting recrystallization of γ Fe decreased as the stress increased up to the peak [4]; however in C steel, static restoration was reduced compared to that with stress removal [5]. During annealing after cold work, an applied stress σ increased recovery and retarded recrystallization in Aℓ [6-8], but in Cu, while moderate σ enhanced recovery, high σ accelerated the latter [8]. Relaxation after high strain rate deformation is construed as annealing under an initially high stress.

The progress of static recovery SRV and recrystallization SRX after hot torsion to a fixed strain, which left the metal with a dynamically recovered (DRV) substructure without initiation of dynamic recrystallization DRX [1-3, 9,10], is measured by the strength drop during intervals of holding at the deformation temperature either without releasing the force or with unloading [3,11-13]. The effects on static softening of altering the strain rate $\dot{\varepsilon}$ or the temperature T in the ranges commonly used in industrial hot forming are determined. The usefulness of torsion with its capability for high ε and $\dot{\varepsilon}$ but its gradient of ε and $\dot{\varepsilon}$ is explored.

EXPERIMENTAL TECHNIQUES

The torsion specimens with gage length 25 mm and diameter 6.25 mm were cold twisted 0.5% before vacuum annealing 16 h at 900°C and furnace cooling. This resulted in a grain size of 300 μm and Cu_2O particles (0.019 wt%) in the size range 1 to 2 μm. No other impurities were detected in the ETP copper [14,15]. The specimens were twisted in an argon-purged, double-elliptical radiant furnace on a closed-loop, servo-controlled, hydraulically-powered machine with computer data logging [16]. The equivalent surface stress σ and strain ε were calculated in the usual manner [16]. The deformation conditions were 450°C, 0.18 s^{-1} and 450 and 500°C at 1.8 s^{-1}. The first and last conditions had the same Z ($Z = \dot{\varepsilon}\exp[+220/8.32\,T]$) [15]. While a separate specimen was used for each 0.18 s^{-1} test, only one specimen was used for each of the other conditions by producing complete recrystallization in a 200 s anneal after $\varepsilon = 0.15$ between each test. For each temperature T, this successfully gave reproducible σ-ε curves (Fig. 1) and microstructures, but differing from the annealed one in grain size (60 and 70μm) and texture. It is possible only in torsion if there is adequate ductility and no phase change. The stress relaxation SR tests involved loading the samples at a constant T and $\dot{\varepsilon}$ to $\varepsilon = 0.15$ and relaxing them by holding the position constant. After increasing holding times, the samples were unloaded and immediately reloaded at the previous $\dot{\varepsilon}$. At 1.8 s^{-1}, SR was not achieved due to failure of the program. In unloading UL tests, the samples were completely unloaded during the entire interval. The progress of fractional softening FS or X, was calculated by the expression:

$$FS = X = (\sigma_m - \sigma_r)/(\sigma_m - \sigma_o) \qquad (1)$$

where σ_m is the stress just before relaxation or unloading, and σ_o and σ_r are the initial and reloading stresses measured by 0.1% offset [3,11-13].

RESULTS

The two stage σ-ε curves with SR or UL arrests and the FS for each interval are shown in Fig. 1. In the reloading curves, σ_r diminishes with time until σ_o is reached. Up to FS of about 20%, the reloading curves rejoin the continuation of the first stage curves as observed in SRV [2,3,17,18]. As FS approaches 100%, the reloading curves take on the same shape as the primary curves, confirming that SRX has taken place. At 450°C, 0.18 s^{-1} for UL, the time for 100% FS is 120 s, whereas for SR, it is only 100 s. The higher softening rate in relaxation than in annealing is confirmed by compression at the same $\dot{\varepsilon}$ and T, where the times were 200 s and 100 s respectively [15]. The curves for 1.8 s^{-1} at either 450 or 500°C appear in Fig. 1c. The rise in $\dot{\varepsilon}$ increases σ_m from 91.2 to 100.0 MPa and the strain energy which speeds up 100% FS in UL to about 40 s. With $\dot{\varepsilon}$ constant, the increase to 500°C lowers the strain energy by 20% (σ_m= 83.2 MPa) but raises the thermal activation to halve the softening time to 20 s. Although at the same Z, the tests at 500°C, 1.8 s^{-1} unexpectedly have σ_m 10% lower than tests at 450°C, 0.18 s^{-1} (probably due to textural differences from the altered test procedure). Nevertheless, the increased thermal activation reduces UL softening time from 120 s to about 20 s, which is similar to results in like-Z compression tests with identical σ-ε curves [15].

The strain hardening rate θ derived from the flow curves are presented as a function of stress in Fig. 2. The initial stages shows the commonly observed behavior of declining value with two approximately linear segments [18-

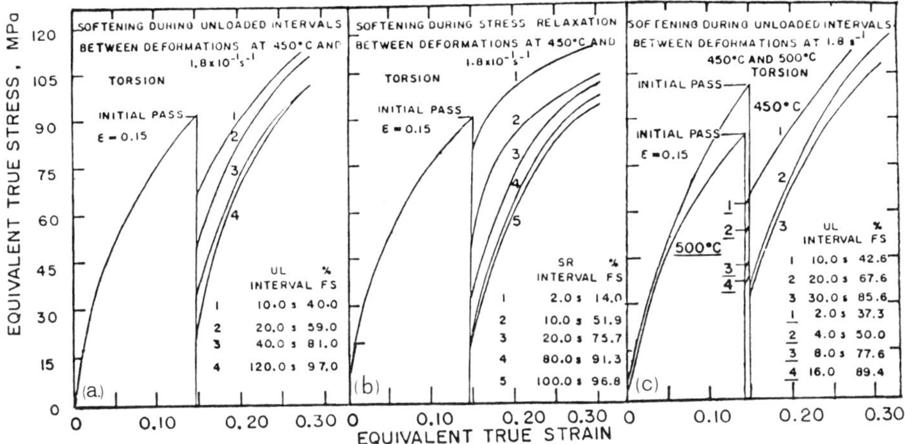

Fig. 1. Initial and reloading flow curves for various delay times and associated FS: 450°C, 0.18 s^{-1} (a) with unloading during interval, (b) with stress relaxation and (c) with unloading at 450°C, 1.8 s^{-1}; 500°C, 1.8 s^{-1}.

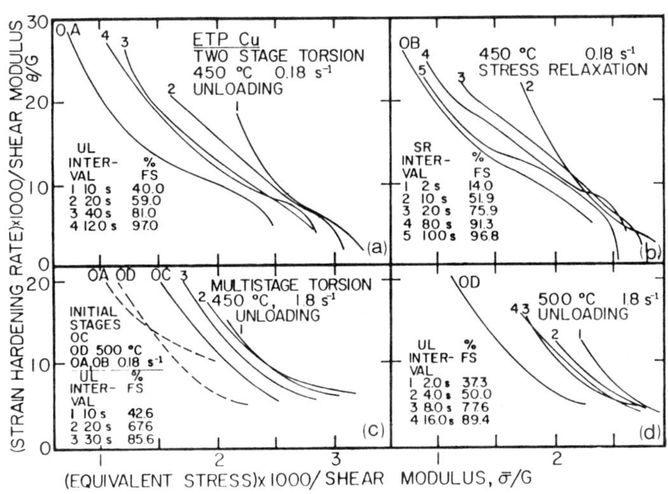

Fig. 2. Relationship between strain hardening rate θ and flow stress after various delay times and associated FS: at 450°C, 0.18 s^{-1} (a) unloading, (b) relaxation, and after unloading (c) 450°C, 1.8 s-1, (d) 500°C, 1.8 s^{-1}.

20]. Their ending before the rapid decline to zero near the σ-ε peak confirms that the arrests came before DRX started. The curves for reloading clearly exhibit an increase in θ with drop in reloading yield stress as a result of either SRV or SRX [18].

Plots of FS or X against log t (Fig. 3) have the traditional S shape usually associated with SRX; however, the shape at short times indicates appreciable SRV. The curves for torsion at 450°C, 0.18 s^{-1} clearly show the advancement of the SR compared to UL at high FS, although at low there appears to be a cross over. The results for compression have a similar relative position of SR and UL but are considerably slowed down except when FS nears 100% . At 1.8 s^{-1} (Fig. 3b), the softening is faster than at the lower $\dot{\varepsilon}$ and is additionally speeded up when T is raised to 500°C. The restoration curves, including compression at very low $\dot{\varepsilon}$ [15], indicate a cross over of the UL and SR curves so that the UL restoration is higher at shorter times. Thus SR, which causes SRX to start sooner than for UL conditions, reduces the rate of static recovery. At short times the S curves for very low $\dot{\varepsilon}$ clearly have linear parts which have been shown to obey recovery kinetics [14,15]. The major part of the S curve was analyzed by the Mehl- Avrami relationship:

$$X = 1 - \exp(-kt^{p}) \qquad (2)$$

The plot of log ln[1/(1-X] vs log t (Fig. 4) exhibits satisfactory straight-lines with slope P and intercept k. For torsion tests at 450°C, 0.18 s^{-1}, the values of p are 1.0 (SR) and 0.8 (UL) and at 1.8 s^{-1} for unloading, p is 1.0 at 450°C and 0.9 at 500°C. For compression p is higher, being 2.1 for SR and 1.6 for UL which are consistent with compression at very low $\dot{\varepsilon}$ [15]. There is no consistent variation in p with T or $\dot{\varepsilon}$, but k increases from 0.08 to 0.6 with $\dot{\varepsilon}$ and is higher for SR than for UL indicating faster growth.

DISCUSSION

The two stage σ-ε curves (Fig. 1) are similar to those determined in previous unloading annealing studies [3,11-13]. In some cases, there has been validation by metallography as there has been here in a preliminary way [3,14,15]. Generally recovery predominates up to 30% softening but at higher degrees, recrystallization assumes a much more significant proportion. Rapid recrystallization in Cu after hot working has been observed previously [9,10]. The usual plateau in the S curve indicating saturation of SRV [3,-11-13] is absent, although observed in both UL and SR tests when initial compression of 0.05 at low $\dot{\varepsilon}$ delayed SRX to long times (~ 10^4 s) [14].

The initial θ-σ curves (Fig. 3) are shifted down to the left as lowered Z reduces θ for a given σ as confirmed by compression at very low $\dot{\varepsilon}$ [15]. The θ-σ curves for 450°C, 0.18 s^{-1} and 500°C, 1.8 s^{-1}, having the same Z are similar, but far from coincident as are the two compression curves with the same Z [15], in agreement with the theory of thermally activated flow. The strain hardening rate on reloading rises as softening increases. Up to 40% FS, the increase in θ is not large, but at higher FS, θ is much higher as progressively more recrystallized grains provide θ values as at the start of straining. The θ-σ curves become parallel to that of the original stage as the σ-ε curves becomes similar to the original. The increases for a given FS are: 1) Similar for UL and for SR, 2) lower for torsion than compression reflecting non-uniform restoration from surface to center, 3) higher for higher $\dot{\varepsilon}$ and lower T consistent with the overall effects on θ. As softening proceeds toward 100% the reloading θ-σ curves become parallel to the initial but remain higher due to unrecrystallized interior regions.

The values of p derived for compression tests in Fig. 4 and at lower $\dot{\varepsilon}$ are 2.1 - 2.8 for SR and 1.6 - 2.0 for UL [15]. These values compare to 2.2 for the recrystallization of Cu determined metallographically in a similar type of hot work and hold experiment [11-13]. For aluminum restoration determined mechanically, p was only 1.4, whereas for recrystallization determined optically it was 3.0 [21]. Thus the high p for SR arises because there is a lower proportion of SRV and a higher of SRX. Preliminary metallographic results indicate that p is in the range 2.4 to 3.2 for both SR and UL. The low p in torsion are likely related to diminishing amounts of SRX with decreasing radius, consistent with similarly tested Al having p ≈ 1.4 [21].

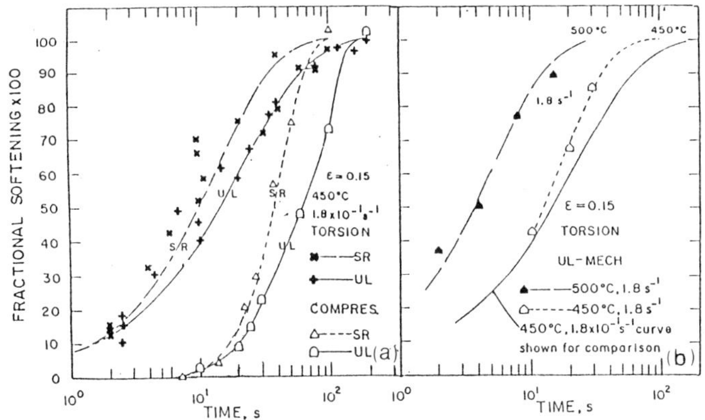

Fig. 3. "S" curves of fractional softening against log time (a) for both UL and SR conditions after torsion and compression prestraining at 450°C, 0.18 s^{-1} and (b) for torsion with UL at 450°C, 1.8 s^{-1} and 500°C, 1.8 s^{-1}.

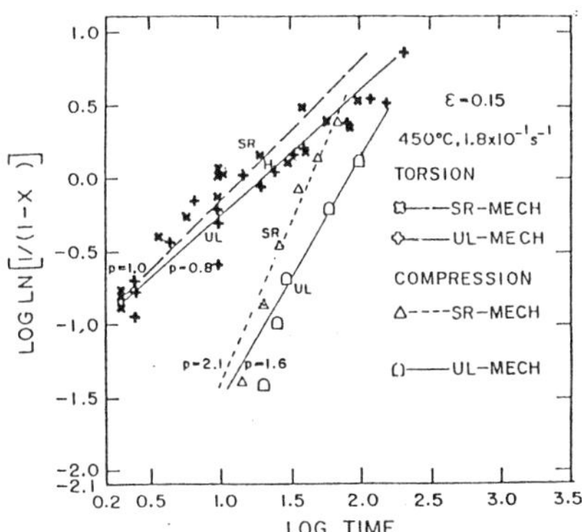

Fig. 4. Avrami-Mehl recrystallization kinetics shown by linear plots according to Eqn. 3 for torsion and compression tests at 450°C, 0.18 s^{-1} for both SR and UL conditions.

The recrystallization is essentially static in both UL and SR since the plastic strain in SR is very small (elastic strain at σ_m). The recrystallization is faster in SR because there is less recovery than in UL, as noted in the cross over of the S curves and corroborated by preliminary metallography [15]. The SR retardation of recovery has been clearly observed where the intial strain was 0.05 [14]. The reduced SRV preserves a high strain energy, i.e. a denser substructure, until SRX starts, which then causes more rapid growth. This agrees with the effects and explanation of the application of a high stress in the annealing of Cu[8]. The retardation of SRX by a small stress [6-8] or after a low strain [5,14] is explained by enhancement of SRV. In simple annealing, the dislocations displace under the influence of an internal backstress within the subgrains and a higher forward stress within the walls [17,22]. A forward stress, up to half the flow stress aids the internal forward stress enhancing recovery,whereas continued application of the flow stress itself causes balancing hardening interactions [17,22]. Alternately the more rapid recrystallization during SR can be considered to be dynamic [4], which usually occurs at a higher critical strain but in a shorter time than does SRX[1,2]. In consideration that the critical strain, or stress, or substructure density, decreases as $\dot{\varepsilon}$ decreases [1,2], one can imagine that as the stress decreases in SR, the critical value is finally reached for the existant substructure (which had not attained the critical value for the initial straining condition). Thus DRX nucleates and proceeds more rapidly in SR than does SRX in UL. This analysis is consistent with that of Wray [4] who found that during SR, the critical stress declined, or the incubation period increased as the hot prestrain decreased below the critical strain for DRX.

CONCLUSIONS

Recovery occurs more slowly and recrystallization more quickly under stress relaxation conditions than under unloading ones. The effect is reduced by increasing the temperature at constant prestrain stress (i.e constant Z) or increasing the strain rate at constant temperature. Static recrystallization is speeded up by the reduced level of recovery which arises because a stress near that for flow induces as many hardening dislocation interactions as softening ones. At prestrains close to the critical strain for dynamic recrystallization the relaxation recrystallization may be dynamic. Torsion results qualitatively agree with compression but differences arise from radial variation in restoration levels due to the linear strain gradient.

ACKNOWLEDGEMENTS

The authors wish to acknowledge the financial support of the Natural Sciences and Engineering Reserch Council of Canada and of the FCAC program in Quebec. L. Vasquez is indebted to the Consejo Nacional de Ciencia y Tecnologia de México for a graduate fellowship and to UAMUA for study leave.

REFERENCES

1. H.J. McQueen and J.J. Jonas, J. Appl. Metalwork, 3, 233-241 (1984).
2. H.J. McQueen and J.J. Jonas, J. Appl. Metalwork, 3, (in press) (1984).
3. H.J. McQueen, M.G. Akben and J.J. Jonas, Microstructural Characterization of Materials by Non-Microscopic Techniques (ed. by J.B. Bilde-Sorenson), pp. 397-404, RISØ Natl. Lab., Roskilde, Denmark (1984).
4. P.J. Wray, Met. Trans., 6A, 1197-1205 (1975).
5. A.H. Ucisik, I. Weiss, H.J. McQueen and J.J. Jonas, Can. Met. Q. 19, 351-358 (1980).
6. J. Hino, P.G. Shewman and P.A. Beck, Trans AIME 194, 873 (1952).
7. J.H. Auld, R.I. Garrod and T.R. Thomson, Acta Met. 5, 74 (1957).
8. P.H. Thornton and R.W. Cahn, J. Inst. Met. 89, 455 (1960-61).
9. H.J. McQueen, Trans. Jap. Inst. Met. 9, Suppl., 170-77 (1968).
10. H.J. McQueen and H. Bergerson, Met. Sci. 6, 25-29, (1972).
11. R.A. Petkovic, M.J. Luton and J.J. Jonas, Met. Sci. 13, 569-72 (1979).
12. R.A. Petkovic, M.J. Luton and J.J. Jonas, Acta Met. 27, 1633-48 (1979).
13. M.J. Luton, R.A. Petkovic and J.J. Jonas, Acta Met., 28, 729-43 (1980).
14. L. Vasquez, Interrupted Hot Compression Testing of Cu (M.Eng. Thesis), McGill University, Montreal, (1981).
15. L. Vasquez, Restoration of Cu During Unloading and Stress Relaxation After Hot Compression (Ph.D. Thesis), Concordia Univ. Montreal (1985).
16. S. Fulop, K. Cadien, M.J. Luton and H.J. McQueen, J. Testing Evaluation, 5, 709-14 (1973).
17. T. Hasegawa, T. Yakou and U.F. Kocks, Acta Met. 30, 235-43 (1982).
18. R.E. Cook, G. Gottstein and U.F. Kocks, J. Mat. Sci., 18, 2650 (1983).
19. U.F. Kocks, J. Eng. Mat. Tech. (ASME H), 98, 76-85 (1976).
20. H. Mecking, Dislocation Modelling of Physical Systems (ed. by M.F. Ashby), pp. 197-211, Pergamon Press, Oxford (1980).
21. H.J. McQueen and N. Ryum, Scand. J. Met., 14, (in press) (1985).
22. W. Blum and H. Schmidt, Res. Mech. 9, 105-20 (1983).

Influence of Grain Size and Strain Rate on the Mechanical Behavior of Polycrystalline Copper at Elevated Temperature

T. Ito and Y. Nakayama

Department of Metallurgical Engineering, College of Engineering, University of Osaka Prefecture, Sakai, Osaka, Japan

ABSTRACT

Polycrystalline copper specimens of 99.99% purity (grain size 80~2000 μm) were deformed in tension with various strain rates at elevated temperatures. Dynamic recrystallization had tendency to occur with decrease in grain size and increase in strain rate. In the fine grained specimens deformed at high strain rate, dynamic recrystallization occurred at the peak stress on the stress-strain curve, and in those deformed at low strain rate, dynamic recrystallization occurred at the first load drop on the stress-strain curve. Dynamic recrystallization would initiate at grain boundaries and grain boundary junctions. In the coarse grained specimens deformed at low strain rate, dynamic recrystallization did not occur. However, in those deformed at high strain rate, dynamic recrystallization occurred by the bulging of original grain boundaries and by nucleating grains in the deformation band within the grain. The orientation relationship between the matrix and the first nucleated grain in the deformation band was analyzed by using the micro-Laue method. It was found that in most cases there existed the rotation relations about the <112> and <111> axes, and that the rotation about the <111> axis would be concerned with the coincidence relation between the matrix and the first nucleated grain.

KEYWORDS

High temperature deformation; polycrystalline copper; grain size; strain rate; stress-strain relationship; dynamic recrystallization; orientation relationship.

INTRODUCTION

When metals are deformed at elevated temperatures, the accumulated strain energy is relieved by dynamic recovery and dynamic recrystallization. Dynamic recrystallization is known to occur in metals of relatively low stacking fault energy [1-6]. In polycrystalline and single crystal copper specimens, if the deformation conditions are satisfied, dynamic recrystallization occurs. Recently, crystallographic studies on dynamic recrystallization in copper single crystals were reported [7-9]. However, on the polycrystalline copper there are few systematic studies on dynamic recrystallization from the view point of the crystallographic aspect. The present study was undertaken to examine the stress-strain relationship, the substructure development, and the initiation of dynamic recrystallization in polycrystalline copper at elevated temperatures in relation to grain size and strain rate, and to understand the mechanism of dynamic recrystallization in the coarse grained specimens by analyzing the orientation relationship between the matrix and the first nucleated grain in the deformation band.

EXPERIMENTAL PROCEDURE

A copper plate of 99.99% purity was supplied in the cold rolled condition. The specimens for tensile deformation were machined with the gauge lengths perpendicular to the rolling direction. Because one of the {111} planes of a few grains would become nearly parallel to side surfaces and the dislocation substructure was able to be observed by etching with the Livingston's solution [10]. The gauge length was 20 mm and the cross section was 2.8X6.0 mm^2. The specimens were annealed between 1073 and 1273 K to obtain various grain size (80~2000 μm). Tensile tests were performed in argon gas atmosphere between 773 and 1073 K with strain rates of 4.2×10^{-5} to $4.2 \times 10^{-2} s^{-1}$. Immediately after turning to unload the machine, the specimens were rapidly cooled by spraying water from a nozzle to preserve the deformed structure. X-ray micro-Laue method (beam diameter 100 μm) was used to determine the orientation relationship between the matrix and the first nucleated grain.

RESULTS AND DISCUSSION

The Stress-strain Relationship and Microstructure

The strain rate dependence of stress-strain curves at 973 K for specimens with grain size of D=90 μm and 600 μm are shown in Fig. 1(a) and (b), respectively. Specimens with D=90 μm show large load drops on the stress-strain curves at low strain rate, and single or two peaks at high strain rate. On the other hand, specimens with D=600 μm have only one peak on the stress-strain curves. The maximum flow stress and the total elongation were increased steeply as the strain rates increased.

Optical micrographs of specimens deformed to the peak stress or to the first load drop on the stress-strain curves are shown in Fig. 2. On the fine grained specimens (D=150 μm), one can observe the mixture of fine (original) and large (recrystallized) grains (Fig. 2(a) and (b)). After the repeated nucleation, the specimen surface is finally occupied by dynamically recrystallized grains regardless of strain rate. As dynamic recrystallization occurs in large areas within a short time, it would be reasonable to consider that dynamic recrystallization initiates at grain boundaries and grain boundary junctions. In the case of coarse grained specimens (D=800 μm) at low strain rate, arrays of voids are formed on the original grain boundaries, which leads to the intergranular fracture (Fig. 2(c)). In the fractured specimens, small oscillations of grain boundaries are frequently observed. However, there is no evidence of dynamic recrystallization. As the strain rate increases, nucleated grains are locally observed around a part of grain boundaries. At the highest strain rate, many nucleated grains surround original grain boundaries (Fig. 2(d)). As grain boundaries become wavy, nucleated grains must be formed by the bulging of original grain boundaries. Nucleated grains are also observed in the deformation band within the grain.

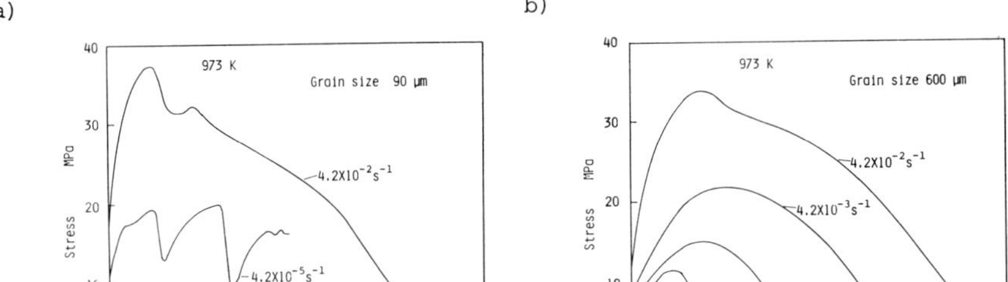

Fig. 1. The strain rate dependence of stress-strain curves at 973 K. a) D=90 μm. B) D=600 μm.

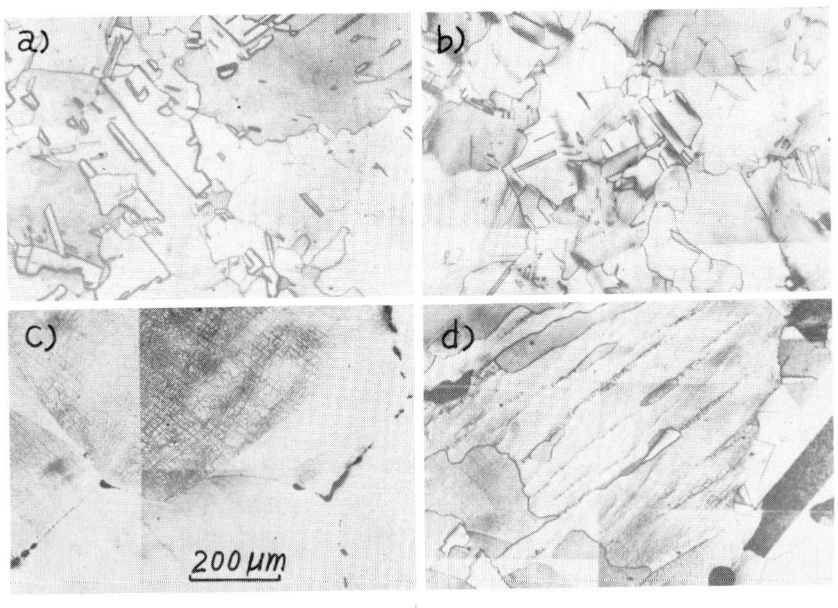

Fig. 2. Optical micrographs. a) Recrystallized. D=150 μm. ε=9%. $\dot{\varepsilon}=4.2\mathrm{X}10^{-5}\mathrm{s}^{-1}$. b) Recrystallized. D=150 μm. ε=10%. $\dot{\varepsilon}=4.2\mathrm{X}10^{-2}\mathrm{s}^{-1}$. c) Not recrystallized. D=800 μm. ε=10%. $\dot{\varepsilon}=4.2\mathrm{X}10^{-5}\mathrm{s}^{-1}$. d) Recrystallized. D=800 μm. ε=15%. $\dot{\varepsilon}=4.2\mathrm{X}10^{-2}\mathrm{s}^{-1}$.

Therefore, the fact that the total elongation strongly depends on strain rate, is concerned with the difficulty in the initiation of dynamic recrystallization. At the strain rate of $4.2\mathrm{X}10^{-2}\mathrm{s}^{-1}$, dynamic recrystallization occurs in all the specimens tested in this experiment. However, at $4.2X10^{-5}s^{-1}$, dynamic recrystallization occurs only in the specimens of D, smaller than 300 μm.

The Orientation Relationship between Matrix and Nucleated Grain

The orientation relationship between the matrix and the nucleated grain was studied in relation to the recrystallization textures in copper single crystals [7-9]. It was found that 1) the nucleation was always due to twinning from subgrain orientations and the subgrain orientation was situated not in the center but within the spread of deformation texture, 2) twinning continued to higher generations, and 3) preferential grain growth took place by a 30° rotation about a common <111> axis of the matrix and the nucleated grains. In the coarse grained Fe-Ni alloy, dynamic recrystallization was studied and it was found that dynamic recrystallization started by the bulging of original grain boundaries, and multiple twins were formed at the growth front [5].

There is little doubt that dynamic recrystallization in coarse grained copper proceeds by forming twins at the growth front. However, the orientation relationship between the matrix and the first nucleated grain is still obscure. Nucleated grains formed by the bulging of original grain boundaries have many twin chains. Therefore, it is not easy to determine which grain is the first nucleated grain. Comparing to them, fortunately, nucleated grains in the deformation band frequently exist as isolated or as a few grains separated with the twin boundaries. Therefore, the orientation relationship between the matrix and the first nucleated grain can be analyzed by the micro-Laue method. Micro-Laue photographs were taken where the X-ray beam covered both the matrix and the nucleated grain. After determining the orientations of the matrix and the nucleated grain separately, the orientation relationship between them was analyzed on twenty two first nucleated grains. It was examined whether there was a simple rotation relation about a common low index axis of the matrix and the nucleated grain such as <100>, <110> , or <111>. However, one could not find such a simple rotation relation.

In polycrystalline copper, the well developed substructure is formed as in Fig. 3. The surface is nearly parallel to the $(1\bar{1}1)$ plane and one can

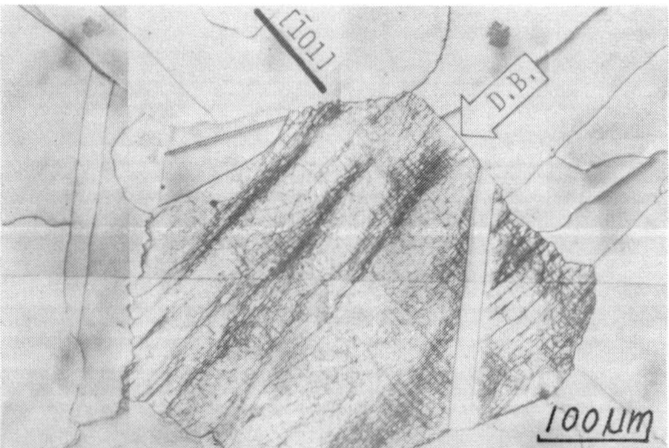

Fig. 3. The formation of deformation band.
T=973 K. ε=15%. $\dot{\varepsilon}=4.2X10^{-2}s^{-1}$.

observe the banded structure composed of small subgrains arranged perpendicular to the primary slip direction. This is the deformation band formed by the activation of the primary slip, (111)[$\bar{1}$01], rotating about the [1$\bar{2}$1] direction. Then, we noticed the rotation about the <112> axis to the matrix and assumed that subgrains which would be the nuclei of dynamic recrystallization rotated about the <112> axis to the matrix. It was found that one pair of the <111> axes in the matrix and the nucleated grain had the rotation about one of the <112> axes. By rotating the matrix about the <112> axis primarily, and the <111> axis secondly, the orientation of the matrix can lie upon that of the nucleated grain.

The above example is shown in Fig. 4. A few nucleated grains with the same orientation are observed (Fig. 4(a)). As in Fig. 4(b), by rotating the matrix 12° about the [1$\bar{2}$1] axis, the (111) plane of the matrix lies upon one of the {111} planes of the nucleated grain. In this case, the [1$\bar{2}$1] axis corresponds to the rotation axis of the deformation band due to the primary slip. Next, as in Fig. 4(c), the orientation of the matrix is consistent with that

a)

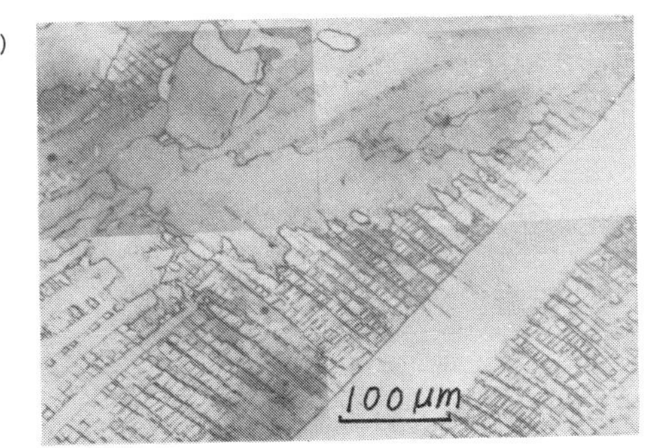

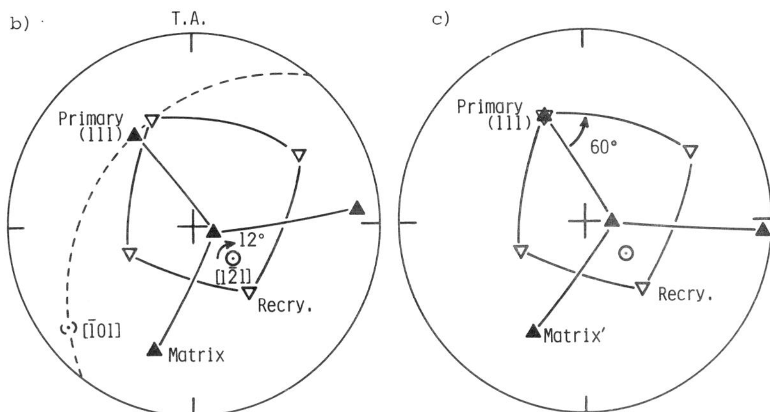

Fig. 4. An example of dynamic recrystallization in the deformation band and the orientation analysis. a) Optical micrograph. b) Rotation about the (111) axis. c) Rotation about the [1$\bar{2}$1] axis. T=1023 K. ε=19%. $\dot{\varepsilon}=4.2\times10^{-2}s^{-1}$.

of the nucleated grain by rotating the matrix 60° about the (111) plane. However, the rotation occurred frequently about the <112> axis besides the [$1\bar{2}1$] axis. The reason will be as follows. In the polycrystalline metals, individual grain is no longer subjected to the uniaxial stress system, even when a specimen is deformed in tension. It is certain that the high density of secondary dislocations are accumulated in the deformation band and give the additional rotation about the <112> axis to the matrix. This additional rotation will contribute for the nuclei of dynamic recrystallization to rotate about various <112> axes.

Secondly, we should consider the rotation about the <111> axis. As an atomic structual model of the grain boundary, the concept of the coincidence lattice was proposed [11, 12]. The boundary between grains with the coincidence relation will be formed easily by the rearrangement of atoms, because it has lower grain boundary energy and needs less migration of atoms to be formed, comparing to the random boundary. By arranging the results which have been obtained on the rotation about the <111> axis, it was found that the rotation angle was not at random, but it seemed to be concerned with the coincidence relation between the matrix and the nucleated grain. Preferentially, we have obtained the rotations about the <111> axis close to $\Sigma 3$ (rotation angle θ=60°), $\Sigma 7$ (θ=21.8°), and $\Sigma 19$ (θ=13.2°) coincidence relations between the matrix and the first nucleated grains.

In accordance with the results, the following mechanism for dynamic recrystallization in the deformation band can be proposed.

1. As the deformation proceeds, the deformation band is formed. In the highly strained region within the deformation band, the nuclei of dynamic recrystallization (subgrains) which rotate about the <112> axis to the matrix will be formed by dynamic recovery.
2. The nucleus which has the coincidence relation about a common <111> axis with its contacting region, will be able to grow to the first nucleated grain observed in the optical microscope.
3. The nucleation will be always due to twinning from the first nucleated grain and multiple twins will be formed at the growth front.

REFERENCES

1. M.J.Luton and C.M.Sellars, Acta Metall. 17, 1033(1969).
2. H.J.McQueen and S.Bergerson, Metal Sci. J. 6, 25(1972).
3. S.Sakui, T.Sakai and K.Takeishi, Tetsu-to-Hagane 62, 856(1976).
4. R.A.Petkovic, M.J.Luton and J.J.Jonas, Acta Metall. 27, 1633(1979).
5. E.Furubayashi and M.Nakamura, Tetsu-to-Hagane 68, 2507(1982).
6. T.Sakai and J.J.Jonas, Acta Metall. 32, 189(1984).
7. G.Gottstein, Proc. 5th Conf. Textures of Materials, Vol. 1, P. 93. Springer, Berlin (1978).
8. G.Gottstein, D.Zabardjadi and H.Mecking, Metal Sci. 13, 223(1979).
9. P.Karduck, G.Gottstein and H.Mecking, Acta Metall. 31, 1525(1983).
10. J.D.Livingston, Direct Observation of Imperfections in Crystals (edited by J.B.Newkirk and J.H.Wernick), P. 115. Interscience, New York (1962).
11. M.L.Kronberg and F.H.Wilson, Trans. AIME 185, 501(1949).
12. Y.Ishida, Bull. Japan Inst.Metal 12, 807(1973).

Static Restoration in the Dynamically Recrystallized Structure of Hot Deformed Nickel

T. Sakai and M. Ohashi

Department of Mechanical Engineering, The University of Electro-Communications, Chofu-shi, Tokyo 182, Japan

ABSTRACT

The static restoration processes after the dynamic recrystallization(DRX) in hot deformed nickel at 1050K ($0.61T_m$) have been studied by means of an interrupted tensile test and the microstructural observations by optical and transmission electron microscopies. The study establishes the existence of three distinct mechanisms of static restoration, namely *metadynamic recrystallization, metadynamic recovery* and classical recrystallization. The metadynamic restoration processes can take place only after the occurrence of prior DRX. Metadynamic recrystallization consists of the continued growth of DRX nuclei formed during deformation. Metadynamic recovery refers to the restoration that occurs only in dynamically growing grains which contain almost no dislocation around the boundaries and have hence no potentiality for static nucleation.

KEY WORDS

Hot deformation, Dynamic recrystallization, Static restoration, Metadynamic recrystallization, Metadynamic recovery, Nickel

INTRODUCTION

The studies on the microstructures developed during hot working and the subsequent structural changes due to static restoration give important information for the microstructural control and the improvement of properties of hot worked products. There have been numerous reports on static restoration process after dynamic recovery operative during hot deformation(1-6). However, few systematic works have been reported the static restoration after dynamic recrystallization(DRX), except the works on the basis of mechanical test by Petkovic et al.(4-6). The present report deals with studies on the static restoration processes after DRX in hot deformed nickel by means of an interrupted tensile test as well as the microstructural observations by optical and transmission electron microscopies(TEM). The effect of dynamic recovery and DRX on static restorations is analysed and mechanisms of the static restorations are discussed in detail.

EXPERIMENTAL

The mechanical tests were carried out on the hot tensile testing machine designed by the authors which was equipped with a H_2 gas quenching apparatus(7). The cooling rate obtained was about 2000K/s. The nickel used in this work had a chemical composition of C; 0.005, Si; <0.002, S; 0.002, Mg; 0.0015 and Ca; <0.002, all in weight per cent. The specimens with a gauge section of 20mm length and 3mm width were prepared from cold rolled 0.6mm thick sheets, the tensile axis being parallel to the rolling direction. The samples were annealed in vacuum at 1230K for 3.6ks. The resulting grain size was 97μm.

The test samples reheated at 1050K($0.61T_m$) were loaded at an initial strain rate of $2 \times 10^{-3}s^{-1}$ to some prescribed strains and unloaded. After a given time interval, the samples were reloaded at the same strain rate and temperature. The degree of softening X_s was determined from equation(1) (4-6).

$$X_s = \sigma_m - \sigma_2 / \sigma_m - \sigma_1 \qquad (1)$$

where σ_m is the flow stress immediately before unloading and σ_1 and σ_2 are the (0.002 offset) yield stress during prestraining and reloading, respectively. A part of the samples was quenched after keeping for various times from 1 ms to 100ks at the deformation temperature. Optical and TEM observations and microhardness measurement were performed on these samples. The specimens for microstrucural observations were prepared by conventional techniques for nickel(8).

RESULTS AND DISCUSSION

Interrupted Tensile Test

Figure 1 shows the effect of prestrain on the fractional softening-holding time (X_s - t) curves at 1050K. The peak flow stress was 103MPa at the strain ε_p of 0.16 at this testing condition (see the insert in Fig. 1). The main results derived are summarized as follows, although the detailed discussion for the results of Fig. 1 will be described in a separate paper (9).
(a) X_s for the prestrain of 0.04 increases with time, approaching a saturation value of about 0.35 after 10ks. This softening can be attributed to static recovery (4-6,9).
(b) For the prestrain of 0.08, the softening proceeds gradually until 1ks, followed by rapid progress to complete softening. The former and the latter are associated with static recovery and static recrystallization, respectively. The as-deformed structures in strains of 0.04 and 0.08 are dynamically recovered, because both strains are considerably less than ε_p.
(c) The deformed microstructure for the prestrain of 0.14 which is 88 per cent of ε_p consists of the mixture of dynamic recovery structure and DRX one (7-10). The softening curve is, therefore, affected by various static restoration processes in both structures  (9).

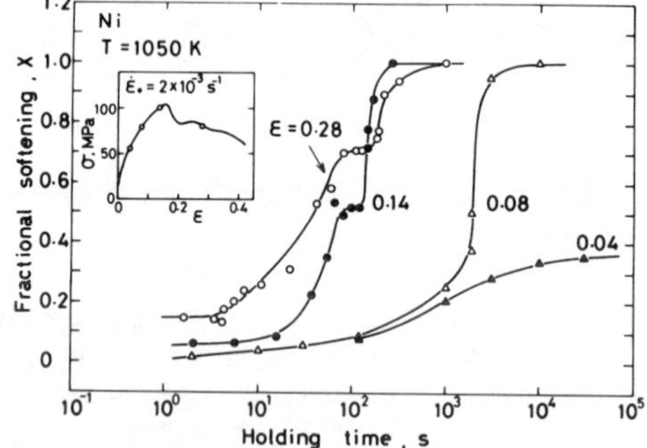

Fig. 1. Effect of strain on the static softening of hot deformed nickel. The insert shows the flow curve at 1050K and $2 \times 10^{-3}s^{-1}$ in which the peak stress appears at strain of 0.16.

(d) The softening curve for the full DRX ($\varepsilon = 0.28$) indicates three distinct softening processes, namely a rapidly initial softening within 1 s and a plateau before 5s (Stage I), a gradual softening from $X_s = 0.15$ to $X_s = 0.72$ and a second plateau at around 100s (Stage II), and complete softening to $X_s = 1$ (Stage III). It is of interest that, if X_s for this strain is compared with that for strain of 0.14, the former is larger than the latter during Stage I and II, and conversely smaller in Stage III.

Microstructural Observation

Optical and TEM observations and microhardness measurement were performed in order to examine the restoration processes operative during each softening stage for the full DRX matrix. The test samples were deformed to strain of 0.28 at 1050K and at $2 \times 10^{-3} s^{-1}$, then gas quenched after having been held for various lengths of time. Mean intercept grain size $\overline{D}$ and number of fine grains below 12μm per unit area N were measured from optical metallographies. Figure 2 shows the changes in $\overline{D}$ (a) and in N (b) with holding time, clearly indicating that two structural changes occur within 5s; namely a rapid grain growth takes place after very short holding time of 0.03s and then grain refinement occurs evidently at around 0.3s, continuing until 10s. These structural changes seem to cause the fractional softening of 0.15 for this period (i.e. Stage I in Fig. 1). $\overline{D}$ - t (or N - t) curve does not show any inflection at holding time of about 100s when the softening curve indicates a second arrest.

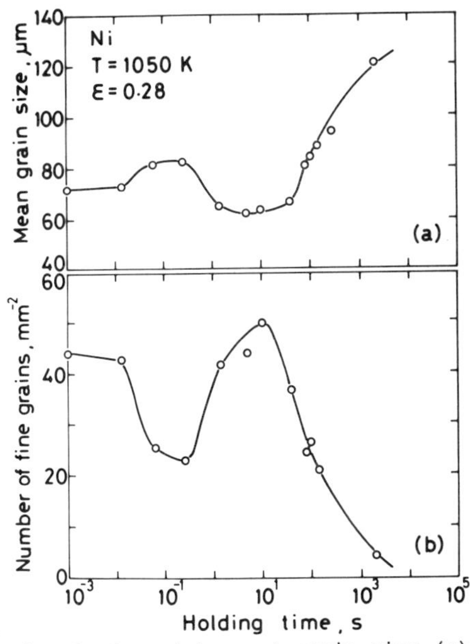

Fig. 2. Mean intercept grain size (a) and number of fine grains below 12μm per unit area (b) during isothermal holding of nickel deformed to strain of 0.28 at 1050K and at $2 \times 10^{-3} s^{-1}$.

The dislocation substructures and new grain formation associated with each softening stage mentioned above were investigated by TEM observations. These examples are presented in Figure 3(a), (b) and (c). Fig. 3(a) shows the microstructure after annealing for 0.03s. Two new grains seem to be nucleated near a DRX grain boundary and to grow rapidly toward the grain interiors. A few dislocation is involved in most part of the new grains, although cell structure is locally formed near the DRX grain boundary. That is, Fig. 3(a) suggests to be microstructural evidence for *metadynamic recrystallization* previously proposed by Petkovic et al.(4-6). The features of substructures seen in Fig. 3(a) did not change significantly except the increase in the number of new recrystallized grains with holding time during Stage I. It is noticeable that the fraction recrystallized at the end of Stage I was evaluated as about 0.1 by TEM observation, and most of new recrystallized grains were isolated and did not grow over 30 to 40μm.

Fig. 3(b) is a typical microstructure at holding time of 10s which corresponds to the begining of Stage II. It can be seen in the grain observed on the left-hand side of Fig. 3(b) that quite a few dislocation exists near the boundary and dislocation substructure is formed progressively toward the grain interior.

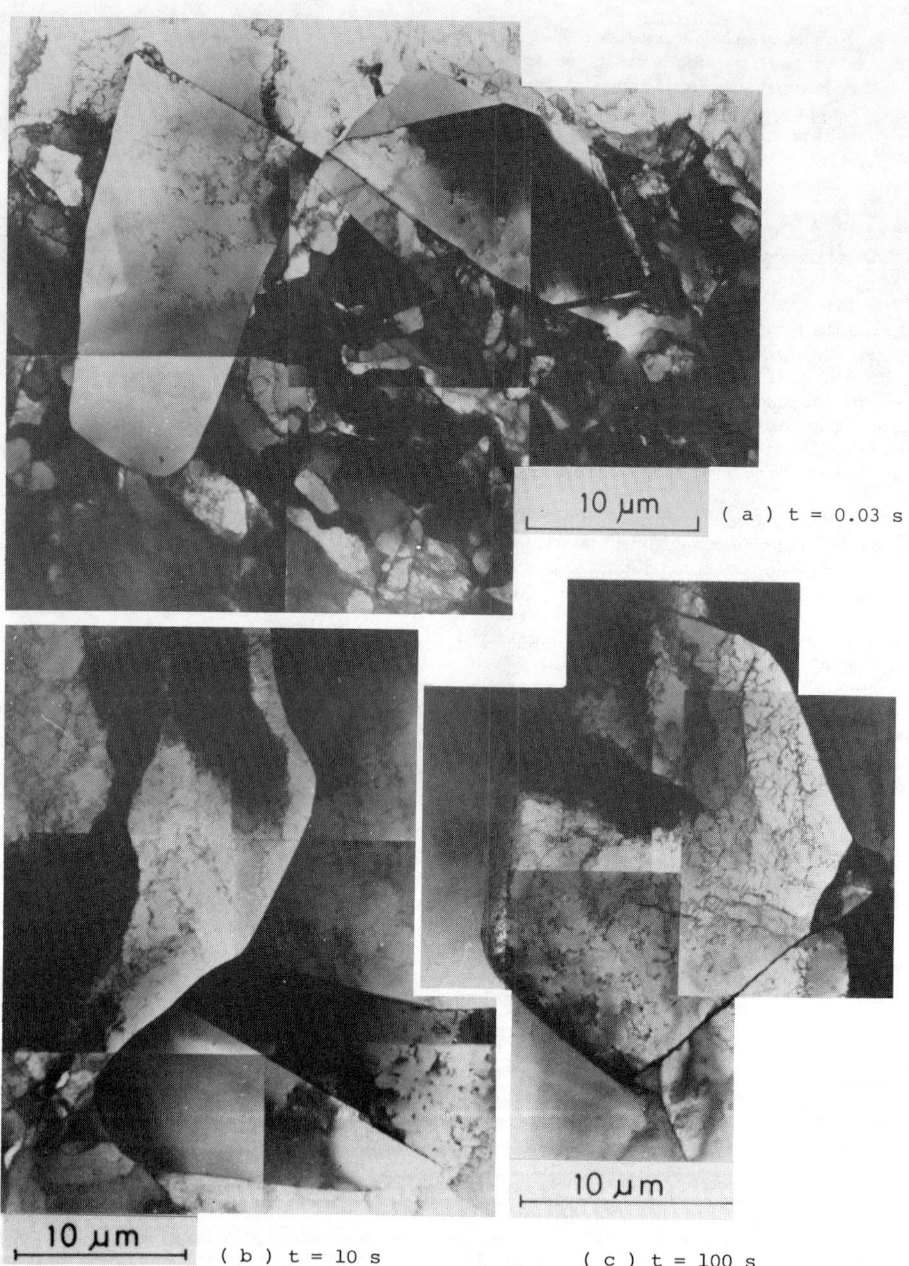

Fig. 3. Typical TEM micrographs of hot deformed nickel after annealing at 1050K for various times as indicated. (a) shows an evidence for metadynamic recrystallization, (b) grain boundary impingement between a growing DRX grain and a recrystallized grain, and (c) a statically recrystallized grain consuming a recovered grain.

This grain appears to be a *growing DRX grain* (9,10). The substructure in this grain is more or less statically recovered as compared with that just after deformation. It is evident also from Fig. 3(b) that grain boundary impingement occurs between a statically recrystallized grain observed on the lower right-hand corner and the growing DRX grain. Under such condition the growth of a new grain could be stopped for a moment or considerably retarded compared with that in homogeneous substructure. The fraction recrystallized at second arrest was about 0.3 under TEM observation, which was much smaller than the fractional softening of 0.72. On the contrary, the proportion of the statically recovered regions which can be seen in Fig. 3(b) or (c) increases with holding time during Stage II. Figure 4 shows the hardness change with annealing time for the prestrain of 0.28. An open symbol and a vertical line show the averaged value of hardness and the scattering, respectively. The hardness at around 100s is extended from the value in full annealed to that in as-hot deformed conditions, evidently indicating the existence of inhomogeneous substructures. This result corresponds to the observation of Fig. 3(b). It can be concluded, therefore, that the softening during Stage II is caused mainly by static recovery in growing DRX grains and the softening associated with the growth of new grains nucleated during Stage I is comparatively small.

Fig. 3(c) shows an example of the typical microstructure after holding for 100s. It is seen that a statically recrystallized grain is consuming a grain containing low dislocation density through recovery during Stage II. Because the static recrystallization rate should be reduced under such a condition as discussed above, even after annealing for 300s there were numerous grains containing substructures (see also Fig. 4). On the contrary, the deformed matrix for strain of 0.14 was fully recrystallized at annealing time of 300s. These differences in static restoration progress in both strains may cause the intersection of both softening curves during Stage III in Fig. 1.

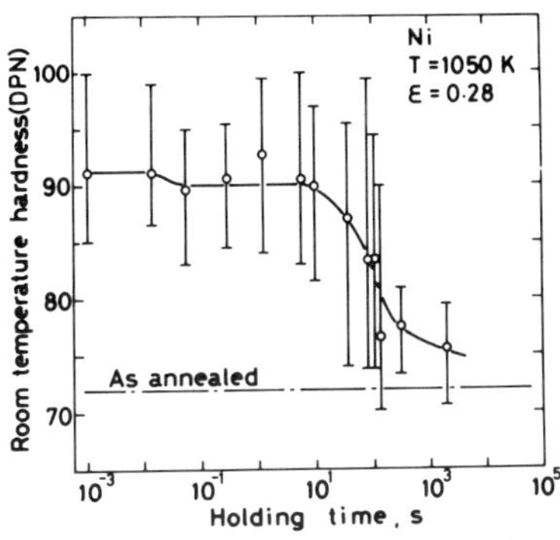

Fig. 4. Vickers hardness number during isothermal annealing of nickel deformed to strain of 0.28 at 1050K and at 2 x $10^{-3}s^{-1}$. An open symbol and a vertical line show the averaged value of hardness and the scattering, respectively.

Static Restoration Mechanisms

Static restoration processes in the DRX structure that occur in a complicated way will be discussed here in connection with the inhomogeneous substructures characteristic of the DRX microstructure. The expected distributions of dislocation density developed during DRX are delineated by full lines in Figure 5, where dislocation density ρ_0 and ρ_c is an initial value for annealed state and a critical value for dynamic nucleation, respectively, and D is a DRX grain size (9,10). They are classified into three types of distributions as shown, which correspond to the

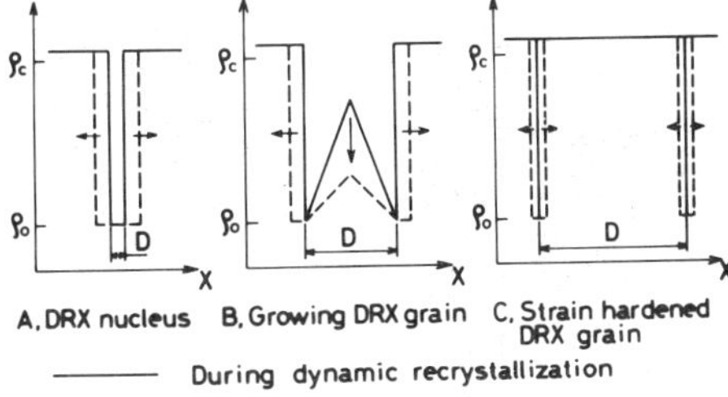

Fig. 5. Expected dislocation density distributions developed in the DRX structure (full lines) and in the restored one after annealing (broken lines). See the text about ρ_0,ρ_c and D.

case of a DRX nucleus A, a growing DRX grain B, an example of which is seen in Fig. 3(b), and a strain hardened DRX grain C after impingement of each DRX grain B, respectively. DRX grain C contains enough dislocations over ρ_c to cause the next cycle dynamic nucleation. These grains A, B and C may be distributed uniformly in the full DRX structure (6, 8-10). On the other hand, their distributions after annealing are assumed as to be shown by broken lines in Fig. 5. Dynamic nuclei (A) can grow continuously immediately after the deformation is ceased. This static restoration has been termed metadynamic recrystallization (6-8). Growing DRX grains (B) contain very few dislocations around their boundaries at which no static nucleation is possible. Under such condition, static recovery takes place mainly within the DRX grains (B), and their static growth occurs also in surrounding region. This type of recovery can occur only after the full DRX and is clearly different from static recovery followed by static nucleation, and is hence termed *metadynamic or post-dynamic recovery*. Classical recrystallization takes place at or near the boundary of DRX grain C with an incubation period.

TABLE 1 Summary of Main Static Restoration in Each Softening Stage.

Stage of softening	Ia	Ib	II	III
Restoration mechanism	Metadynamic recrystallization		Metadynamic recovery	
		Nucleation	Limited growth of new grains	Grain growth
		Classical recrystallization		

Based on the above discussions, the mechanisms of static restoration in each softening stage for the prestrain of 0.28 are summarized as shown in Table 1. The first stage of static softening after the DRX is brought about by metadynamic recrystallization, followed by static nucleation and grain growth near the boundaries of strain hardened DRX grains (C). The degree of fractional softening caused by the above processes is as little as around 0.15 during Stage I. A main restoration process in Stage II is metadynamic recovery within growing DRX grains (B). The growth of metadynamically and statically recrystallized grains is retarded by the existence of the growing DRX grains (B) during Stage II, and thus, they cannot grow over a limited size (for example, 30 to 40μm). The microstructures at the end of Stage II, i.e. around a second arrest, consist of metadynamically and statically recrystallized grains and metadynamically recovered grains originated from DRX grains (B). Both recrystallized grains grow in the recovered regions in Stage III and the growth rate is substaintially slow because of the small difference of dislocation density. It can be concluded, consequently, that the complicated softening processes for the DRX structure are caused mainly by metadynamic recrystallization originated from dynamic nuclei, metadynamic recovery operative within dynamically growing grains and the existence of the latter itself which affect classical recrystallization.

Acknowledgment. The financial support by the Ministry of Education of Japan under a grant of scientific research is gratefully acknowleged.

REFERENCES

1. A.T.Enlish and W.A.Backofen, *Trans. Metall. Soc. AIME*, 230,396(1964).
2. G.Glover and C.M.Sellars, *Metall. Trans.*, 3, 2271(1972).
3. T.Sakai and M. Ohashi, *Tetsu-to-Hagane*, 70, 2160(1984).
4. R.A.Petkovic-Djaic and J.J.Jonas, *J. Iron Steel Inst.*, 210, 256(1972).
5. R.A.Petkovic, M.J.Luton and J.J.Jonas, *Can. Metall. Quart.*, 14, 137(1975).
6. R.A.Petkovic, M.J.Luton and J.J.Jonas, *Acta Metall.*, 27, 1633(1979).
7. S.Sakui, T.Sakai and K.Takeishi, *Trans. Iron Steel Inst. Japan*, 17, 718(1977).
8. M.J.Luton and C.M.Sellars, *Acta Metall.*, 17, 1033(1969).
9. T.Sakai and M.Ohashi, The University of Electro-Communications, Tokyo (to be published).
10. T.Sakai and J.J.Jonas, *Acta Metall.*, 32, 189(1984).

Formation Mechanism of Dynamically Recrystallized Grains in Austenitic Fe-Ni-C Alloy

T. Maki*, S. Okaguchi and I. Tamura***

**Department of Metal Science and Technology, Kyoto University, Sakyo-ku, Kyoto 606, Japan*
***Graduate School of Kyoto University, Present address: Central Research Laboratories, Sumitomo Metal Industries, Ltd., Amagasaki 660, Japan*

ABSTRACT

A formation process of dynamically recrystallized grains was examined by microstructural observations of the first cycle of recrystallization during deformation under various deformation conditions using an austenitic Fe-31% Ni-0.29%C alloy. The strain range in which the first cycle of dynamic re-crystallization occurs (i.e., between ε_c and ε_s) is spread with an increase in Z (Zener-Hollomon parameter), and the formation mechanism of dynamically recrystallized grains changes with Z. Dynamically recrystallized grains are formed mainly by the bulging mechanism under low Z condition and by the nucleation-growth mechanism under high Z condition. However, under inter-mediate Z condition, the mechanism changes from the bulging type to the nucleation-growth type even during one cycle of recrystallization. The change of formation mechanism from the bulging type to the nucleation-growth type occurs at strain of about 0.2 in the present alloy. The dynamically recrystallized grains are formed in similar manner, at least qualitatively, as the static recrystallization.

KEYWORDS

Dynamic recrystallization; Hot deformation of austenite; Bulging mechanism; Nucleation-growth mechanism.

INTRODUCTION

In the case of static recrystallization, it is known that there are two mechanisms for the formation of recrystallized grains [1]. One is the bulging of pre-existing boundaries due to the strain-induced boundary migration (hereafter referred to as the bulging mechanism), and the other is the nucle-ation and growth of new grains whose orientations are different from the de-formed matrix surrounding them (hereafter referred to as the nucleation-growth mechanism). These two mechanisms have also been observed in a dynamic recrys-tallization which occurs during deformation at elevated temperatures. Most of previous papers indicated that the bulging mechanism is the dominant one for the dynamic recrystallization [2-5]. However, there have been some reports showing that the dynamically recrystallized grains are formed by the nucleation-growth mechanism along grain boundaries or at deformation bands inside grains [6,7]. Although it can be expected that the formation mechanism might depend on the deformation condition (i.e., temperature and strain rate), there seems to be no detailed investigation which makes clear the relation between the formation mechanism of dynamically recrystallized grains and deformation condition.

The purpose of the present study is to make clear the formation mechanism of dynamically recrystallized grains formed under various deformation conditions in austenitic ferrous alloy. When the dynamic recrystallization occurs during hot deformation, the stress-strain curve exhibits the peak stress followed by the steady state flow stress as shown schematically in Figure 1. The first cycle of dynamic recrystallization starts at ε_c which is smaller than ε_p (i.e., the strain at peak stress) and finishes at ε_s. At the steady state deformation beyond ε_s, the dynamic recrystallization occurs repeatedly. In the present investigation, the microstructural change during the first cycle of recrystallization was observed in the specimens deformed to various strains up to ε_s at various combinations of T and $\dot{\varepsilon}$.

EXPERIMENTAL PROCEDURE

An austenitic Fe-31.0 wt%Ni-0.29 wt%C alloy which was prepared by vacuum induction melting was used. Tensile specimens with a reduced central gage section of 10 mm in length and 3.5 mm in diameter were machined from hot rolled bars. These specimens were solution treated at 1573 K for 1.8 ks in a vacuum furnace and oil quenched. Solution treated specimens were austenitic even at room temperature and the mean austenite grain size was 207 μm. Specimens were tensile deformed with an Instron-type tensile machine under various deformation conditions as shown in Table 1. These deformation conditions were selected in order to change the value of Zener-Hollomon parameter Z (= $\dot{\varepsilon}\exp(Q/RT)$) between 10^{10}/s and 10^{14}/s. An activation energy Q of hot deformation which was determined from the peak

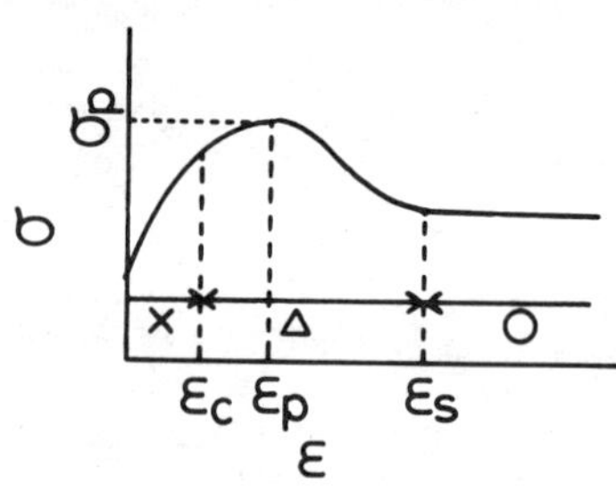

Fig. 1 Schematic illustration of true stress-true strain curve accompanying the dynamic recrystallization.

TABLE 1 Deformation Conditions Performed in the Present Study. Zener-Hollomon Parameter Z (= $\dot{\varepsilon}\exp(Q/RT)$) Was Determined by Using Q = 335 kJ/mol.

Deformation Condition	T / K	$\dot{\varepsilon}$ / s^{-1}	Z / s^{-1}
①	1473	1.7×10^{-2}	1.3×10^{10}
②	1373	1.7×10^{-2}	9.2×10^{10}
③	1273	1.7×10^{-2}	9.3×10^{11}
④	1273	1.7×10^{-1}	9.3×10^{12}
⑤	1173	1.7×10^{-1}	1.4×10^{14}

stress (σ_p) under various deformation conditions was 335 kJ/mol for the present alloy [8]. Specimens were heated by an induction heating in 90% N_2 + 10% H_2 gas atmosphere. For the observation of microstructural change during hot deformation, specimens were rapidly quenched by water spray within 0.2 s after the deformation to various presetting strains less than ε_s. The microstructure was observed by means of optical microscopy on the longitudinal section of specimen.

RESULTS AND DISCUSSION

Figure 2 shows the change in various strains such as ε_c, ε_p and ε_s with the deformation condition Z. The ε_c and ε_s were roughly determined by the microstructural observations in the specimens deformed to various strains. Both ε_c and ε_s increase with an increase in Z. Therefore, the strain range in which the first cycle of dynamic recrystallization occurs markedly changes with Z. For example, the first cycle of dynamic recrystallization starts at strain of about 0.04 and finishes at strain of about 0.2 at $Z = 1.3 \times 10^{10}$/s. On the other hand, when the specimen was deformed at $Z = 1.4 \times 10^{14}$/s, the dynamic recrystallization starts at strain of about 0.25.

Figure 3 shows the progress of the first cycle of dynamic recrystallization under the deformation condition of $Z = 9.3 \times 10^{11}$/s (T = 1273 K, $\dot{\varepsilon} = 1.7 \times 10^{-2}$/s). (a) shows the austenite structure before deformation. When the specimen was slightly deformed ($\varepsilon = 0.08$), the pre-existing grain boundaries become finely serrated and the bulgings of grain boundaries are frequently observed as shown by arrows in (b). This indicates that the dynamic recrystallization already starts by the bulging mechanism at the strain corresponding to about 40% of ε_p. Bulging of grain boundaries becomes more dominant when the specimen was deformed to $\varepsilon = 0.15$ ((c)). However, when the specimen was deformed to $\varepsilon = 0.24$, many small recrystallized grains are observed along austenite grain boundaries as shown in (d). These fine grains have the clear boundaries, indicating that these are formed by the nucleation-growth mechanism. When the specimen was deformed to more than ε_s, the initial austenites are completely replaced with dynamically recrystallized grains as shown in (e). It appears from Fig. 3 that, under this deformation condition, the formation mechanism of dynamically recrystallized grains changes from the bulging mechanism to the nucleation-growth mechanism during the progress of the first cycle of recrystallization.

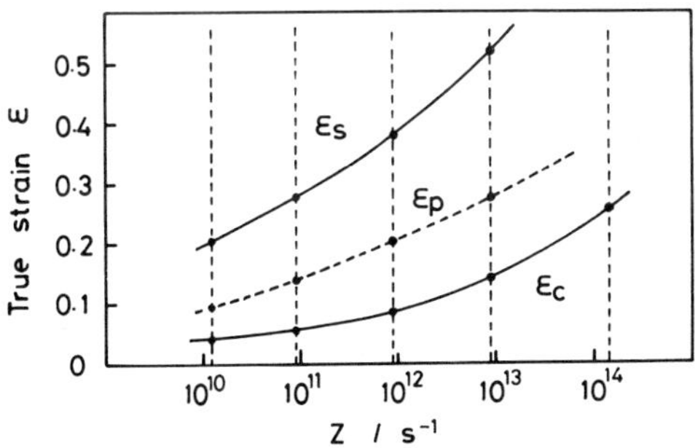

Fig. 2 Change of ε_c, ε_p and ε_s with deformation condition Z.

Figures 4 and 5 show some examples of microstructures in specimens deformed under various conditions of low Z and high Z, respectively. In the case of deformation condition of $Z = 1.3 \times 10^{10}$/s in which the first cycle of recrystallization finishes at small strain (about 0.2), the dynamic recrystallization occurs mainly by the bulging mechanism during the first cycle of recrystallization as shown in Fig. 4. On the other hand, when the specimen was deformed under the condition of $Z = 1.4 \times 10^{14}$/s in which the dynamic recrystallization starts at ε = about 0.25, many fine grains are formed along austenite grain

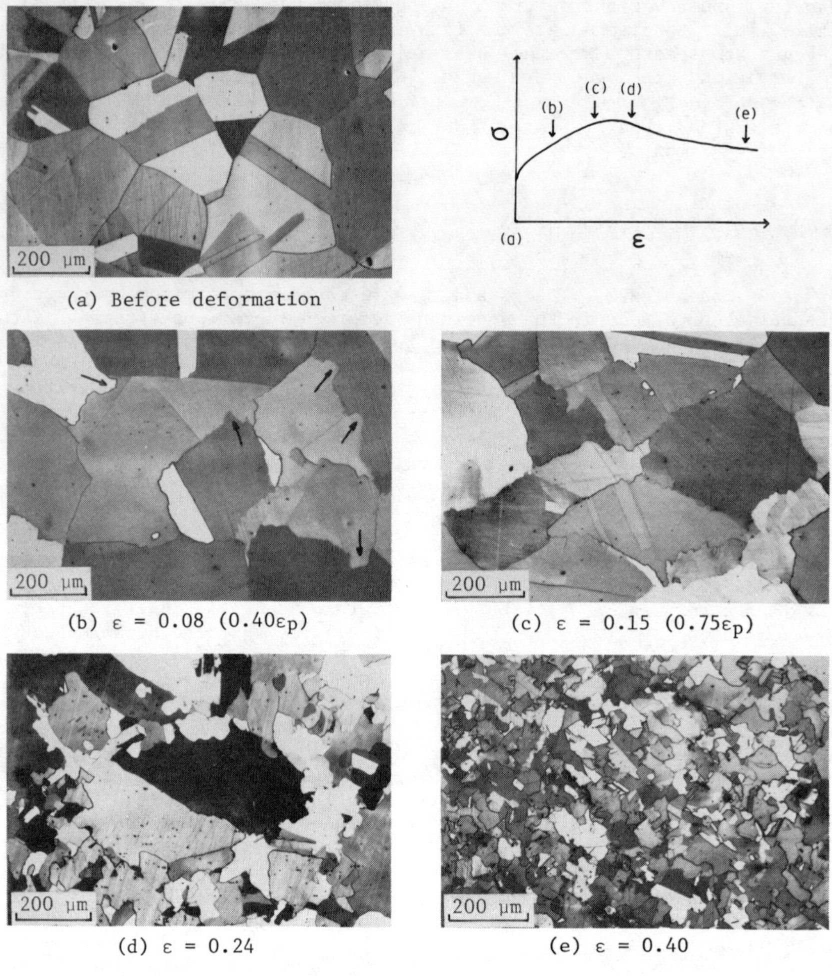

(a) Before deformation

(b) $\varepsilon = 0.08\ (0.40\varepsilon_p)$

(c) $\varepsilon = 0.15\ (0.75\varepsilon_p)$

(d) $\varepsilon = 0.24$

(e) $\varepsilon = 0.40$

Fig. 3 Optical micrographs showing the progress of dynamic recrystallization of austenite under deformation condition of T = 1273 K, $\dot{\varepsilon} = 1.7 \times 10^{-2}/s$ ($Z = 9.3 \times 10^{11}/s$).

boundaries even at very early stage of recrystallization and the bulging of pre-existing grain boundaries is hardly observed as shown in Fig. 5. This indicates that, in the case of Fig. 5, the dynamic recrystallization takes place only by the nucleation-growth mechanism.

Figure 6 is the summary of microstructural observations in the specimens deformed to various strains under various deformation conditions Z. In this figure, the observed formation mechanism of dynamically recrystallized grains at various strains between ε_c and ε_s is indicated on each stress-strain curve. Both the bulging mechanism and the nucleation-growth mechanism were observed during dynamic recrystallization and these mechanisms are closely related to

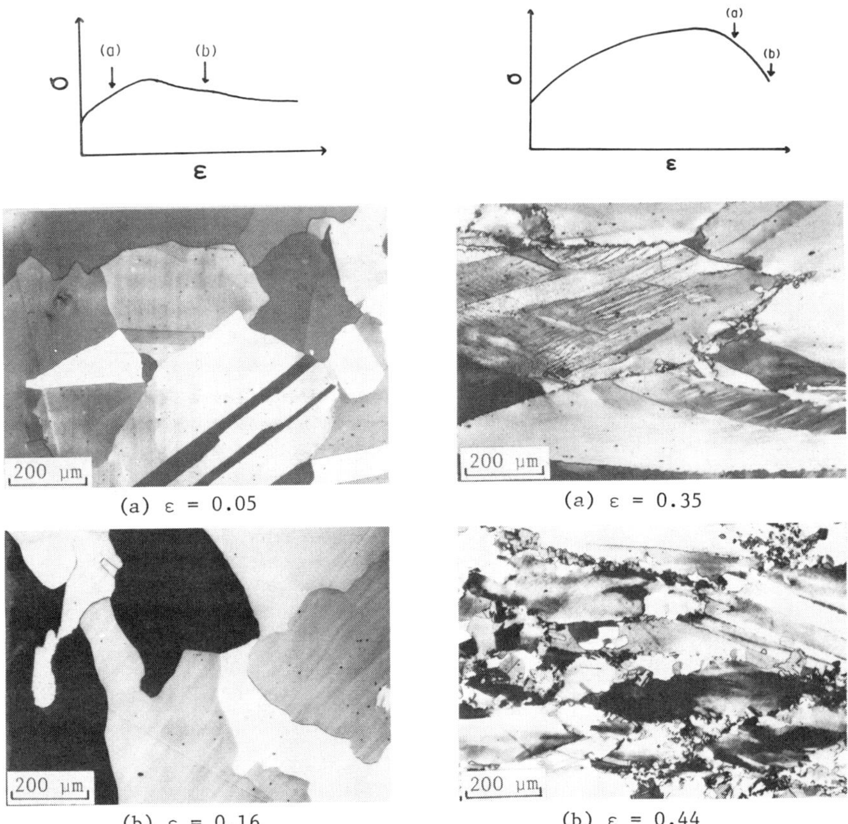

Fig. 4 Optical micrographs showing a partially dynamically recrystallized austenite under condition of T = 1473 K, $\dot{\varepsilon} = 1.7 \times 10^{-2}$/s ($Z = 1.3 \times 10^{10}$/s).

Fig. 5 Optical micrographs showing a partially dynamically recrystallized austenite under condition of T = 1173 K, $\dot{\varepsilon} = 1.7 \times 10^{-1}$/s ($Z = 1.4 \times 10^{14}$/s).

the amount of strain at which the dynamic recrystallization occurs. The formation mechanism changes at the strain of about 0.2 for the present alloy irrespective of deformation condition Z. The bulging mechanism is important when the dynamic recrystallization occurs at strains below about 0.2, and the nucleation-growth mechanism becomes dominant when the dynamic recrystallization takes place at strains above about 0.2.

It is known that, in the case of static recrystallization, the bulging mechanism is important when the amount of deformation is small, e.g., less than 20%, and the nucleation-growth mechanism is dominant when the specimen is heavily deformed [1,9]. Therefore, it can be concluded that the formation mechanism of recrystallized grains is not essentially different between static recrystallization and dynamic recrystallization from the point of view of strain, at least qualitatively.

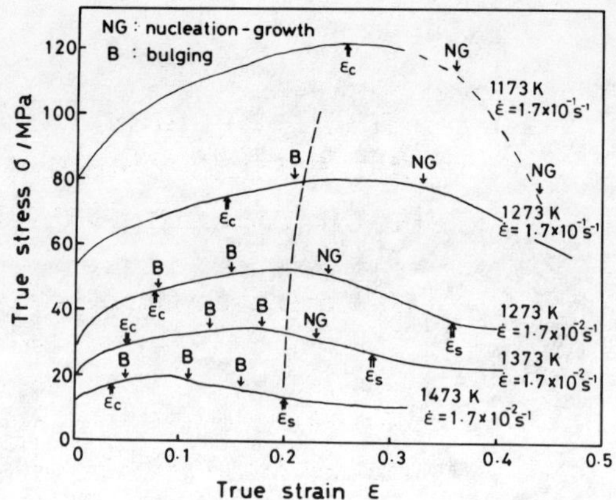

Fig. 6 Summary of the observed formation mechanism of dynamically recrystallized grains during the first cycle of recrystallization under various deformation conditions.

Present results indicate that the formation mechanism of recrystallized grains depends on the strain rather than flow stress during deformation. The flow stress is important for the driving force of dynamic recrystallization because the flow stress is a function of dislocation density. However, the nucleation mechanism might not be governed by the driving force (i.e., the flow stress), but is influenced by the degree of inhomogeneity of deformation, since the local inhomogeneous deformation is necessary for the formation of recrystallized grains by the nucleation-growth mechanism. The inhomogeneity of deformation is a function of strain rather than stress and tends to increase with an increase in strain. Therefore, when the matrix is inhomogeneously deformed (near grain boundaries and at deformation bands inside grains) by the increase in the strain, the nucleation-growth mechanism becomes dominant instead of bulging mechanism.

It has been generally thought that the dynamic recrystallization starts at strain of about $0.7\varepsilon_p$ [10,11]. However, the present observation showed that the dynamic recrystallization already starts at strain of about $0.4\varepsilon_p$ in the case of Fig. 3. Therefore, it can be considered that, when the dynamic recrystallization occurs by the bulging mechanism, it starts at much smaller strains with respect of ε_p than that which has been previously reported. Similar results have been obtained in Ni by Endo et al. [12].

REFERENCES

1. P. A. Beck and P. R. Sperry: J. Appl. Phy. 21, 150 (1950).
2. M. J. Luton and C. M. Sellars: Acta Met. 17, 1033 (1969).
3. G. Glover and C. M. Sellars: Met. Trans. 4, 765 (1973).
4. W. Roberts and B. Ahlblom: Acta Met. 26, 801 (1978).
5. W. Roberts, H. Boden and B. Ahlblom: Met. Sci. 13, 195 (1979).
6. J. P. Sah, G. J. Richardson and C. M. Sellars: Met. Sci. 8, 325 (1974).
7. L. Blaz, T. Sakai and J. J. Jonas: Met. Sci., 17, 609 (1983).
8. T. Maki, K. Akasaka and I. Tamura: in Thermomechanical Processing of Microalloyed Austenite (Ed. by A. J. DeArdo et al.) p.217, AIME (1981).
9. T. Maki, S. Nakagawa and I. Tamura: J. Japan Inst. Met. 44, 1164 (1980).
10. H. J. McQueen and J. J. Jonas: in Plastic Deformation of Metals (Ed. by R. J. Arsenoult) p.393, Academic Press (1975).
11. S. Sakui, T. Sakai and K. Takeishi: Trans. ISIJ, 17, 718 (1979).
12. T. Endo, H. Fukutomi and T. Kishi: Tetsu-to-Hagane 70, 2097 (1984).

Microstructural Evolution of AISI 304 L Grade Stainless Steel During Isothermal Straining by Hot Torsion

G. Carfi*, C. Perdrix, D. Bouleau and C. Donadille**

IRSID - 78105 St-Germain-en-Laye Cedex, France
**Now with Instituto Argentina de Siderurgia, 1104 Buenos-Aires, Argentina*
***Now with USINOR, BP 2-508, 59381 Dunkerque Cedex, France*

ABSTRACT

Isothermal hot torsion tests have been performed on a 304 L stainless steel at temperatures between 900°C and 1200°C in order to study the microstructural evolution up to the onset of dynamic recrystallization. An analysis of the stress strain curves shows that the work hardening range can be decomposed into two substages which can be correlated with changes in the dislocation substructure.

KEYWORDS

Austenitic stainless steel, hot torsion, work hardening, dynamic recovery, dynamic recrystallization, nucleation, microstructure, grain boundary, dislocation cell.

INTRODUCTION

Because of its technological importance, the high temperature deformation behaviour of alloyed and non-alloyed austenite has been widely studied (1,2). Hot torsion testing is one of the more commonly used methods since this type of test enables the mechanical behaviour as well as the microstructural evolution of the material to be investigated as a function of the main parameters governing the deformation (strain $\bar{\varepsilon}$, strain rate $\dot{\bar{\varepsilon}}$, Temperature T, chemical composition and prior history). The case of large strains has been extensively examined and the study and modeling of the associated phenomena of dynamic and static recrystallization are still of current interest (3-5). For high temperatures and strain rates, the microstructural evolution in the early stages of deformation is not so well established. The objective of the present work was to clarify the strain hardening behaviour and the associated structural changes in a 304 L grade austenitic stainless steel. To this end, the stress strain behaviour was analysed in terms of the variation of the derivative of the stress strain curve $\theta = (\partial\bar{\sigma}/\partial\bar{\varepsilon})$ as a function of the flow stress $\bar{\sigma}$ (6) and related to the microstructural observations on specimens deformed by different amounts up to that corresponding to the onset of dynamic recrystallization.

EXPERIMENTAL PROCEDURE

Hot Torsion Testing

The mechanical tests were performed on a commercial 304 L steel (Cr : 18.6; Ni : 11.1; Mo : 0.19; Cu : 0.16; Mn : 1.2; Si : 0.32; C : 0.018; N : 0.006) received in the form of hot rolled bars of 16 mm diameter, using a fully computerized machine developped at IRSID. After a preheating at 1250°C for 30 minutes (resulting grain size $\bar{d}$ = 280 μm) the specimens were rapidly cooled to the test temperature, strained to the preselected amount at the imposed strain rate and finally water quenched. The tests covered the temperature range from 850°C to 1200°C at 50°C intervals and at strain rates of 0.04, 0.4 and 4 s^{-1}. The raw torque-twist data, acquired continuously from the very first stages of straining, were automatically converted and plotted in the form of isothermal equivalent stress - equivalent strain curves taking into account the effects of adiabatic heating as necessary (in particular for $\dot{\bar{\varepsilon}} = 4\ s^{-1}$). Finally, the derived curves $\theta = d\bar{\sigma}/d\bar{\varepsilon} = \theta(\bar{\sigma})$ were calculated and plotted.

Microstructural Observations

The microstructural observations were carried out on selected specimens isothermally deformed at 900°C, 1100°C and 1200°C at strain rates of 0.4 and 4 s^{-1} respectively. The samples were prepared near the periphery of the torsion specimens and observed by optical and electron microscopy in the plane parallel to the specimen axis. The thin foil transmission electron microscopy studies were performed on an EM 400 T instrument equipped with a field emission gun. Convergent beam electron diffraction patterns were used to mesure the local crystallographic misorientation between the observed microstructural units. Some observations were also made by scanning electron microscopy using back scattered electrons at 20 kV to produce crystallographic contrast on electropolished bulk samples.

RESULTS

Fig. 1 shows examples of the experimental $\bar{\sigma} = \bar{\sigma}(\bar{\varepsilon})$ curves and the corresponding derived curves $\theta = \theta(\bar{\sigma})$. On these curves, the points corresponding to the strains $\bar{\varepsilon}$ selected for the metallographical observations are indicated by full circles and arrows.

In the case of the $\bar{\sigma}(\bar{\varepsilon})$ representation, it can be seen that after the initial work hardening stage (here after designated stage I with $\bar{\varepsilon} \leqslant 0,3$), the metal attains a mechanical quasi-steady state (stage II). It is to be noted that for the temperatures and strain rates investigated in these experiments, the curves do not present a sharp maximum.

Examination of the $\theta(\bar{\sigma})$ curves shows that the work hardening stage I can be decomposed into two substages associated with two distinct linear regions in the $\theta(\bar{\sigma})$ relation. These regions are well defined at 1200°C and 1100°C but at 900°C the differences in the slope of the two regions of the curve are only slight.

At 1100°C and for a strain rate of 0.4 s^{-1} the microstructure varies as a function of strain in the following manner :

Stage I

At very low strain (within the first substage) the material deforms by slip simultaneously on several slip planes but the dislocation density remains fairly uniform throughout the grains (Fig. 2 a). The second substage is characterized by the rearrangement of the dislocations into elongated cells which are generally aligned along a single direction within a given grain (Fig. 2 b). During the transition from the first to the second substage, equiaxed and slightly misoriented cells are observed to form on narrow regions on either side of the grain boundaries of the deformed initial grains (Fig. 3).

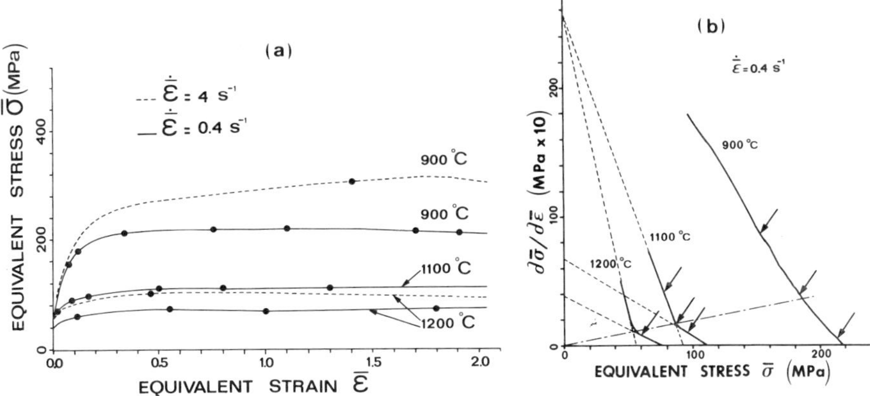

Fig. 1 - Mechanical testing curves : (a) equivalent stress - equivalent strain curves - (b) corresponding derived curves $\theta = \theta\ (\ \overline{\sigma}\ )$.

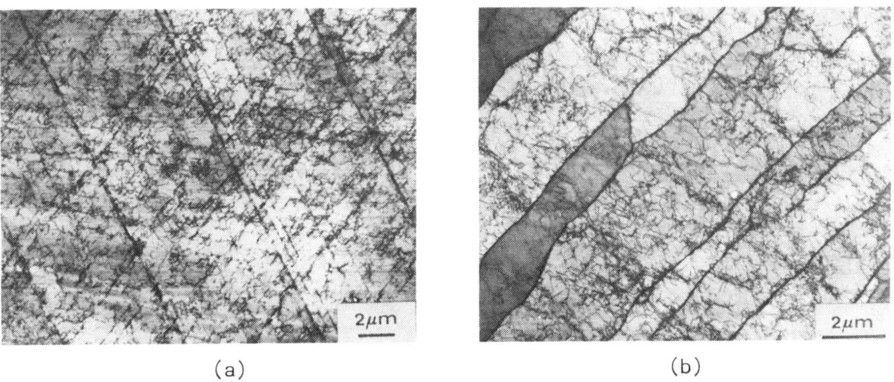

(a) (b)

Fig. 2 - Dislocation substructures devellopped during work hardening at 1100°C at a strain rate of $\dot{\overline{\varepsilon}} = 0.4\ s^{-1}$.
(a) $\overline{\varepsilon} = 0.03$ (first substage) (b) $\overline{\varepsilon} = 0.16$ (second substage).

Stage II

Prior to the onset of dynamic recrystallization which occurs at a strain of about $\overline{\varepsilon} = 0.65$, the elongated cell structure persists and sharpens within the deformed grains. At the same time, new cell walls are formed within the elongated cells or subgrains along directions perpendicular to the major axis (Fig. 4). Little or no evolution is observed in the equiaxed subgrains along the original grain boundaries up to the onset of dynamic recrystallization (Fig. 5). At 1100°C and a strain rate of $0.4\ s^{-1}$ this latter process appears

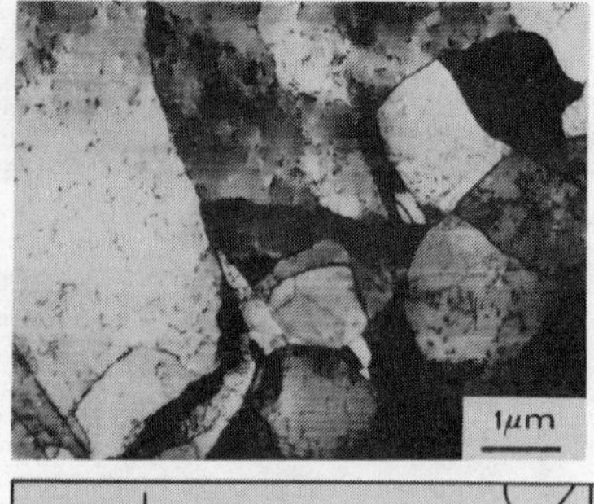

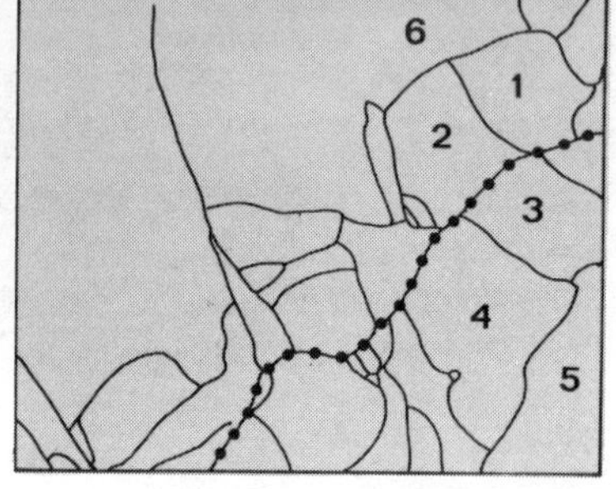

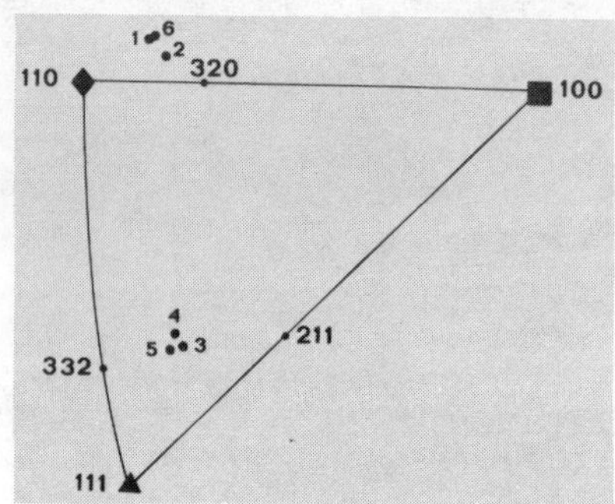

Fig. 3 - Formation of equiaxed cells or subgrains on both sides of the grain boundary of deformed initial grains T = 1100°C, $\dot{\bar{\varepsilon}} = 0.4\ s^{-1}$, $\bar{\varepsilon} = 0.16$

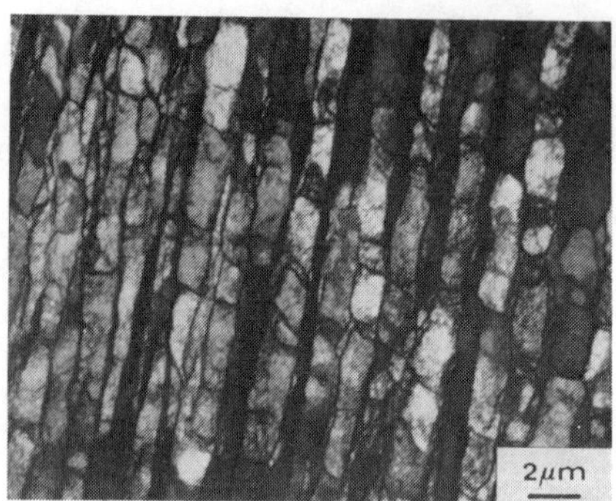

Fig. 4 - Evolution of the elongated cell structure : formation of cell walls perpendicular to the major axis of the cells or subgrains T = 1100°C, $\dot{\bar{\varepsilon}} = 0.4\ s^{-1}$, $\bar{\varepsilon} = 0.5$

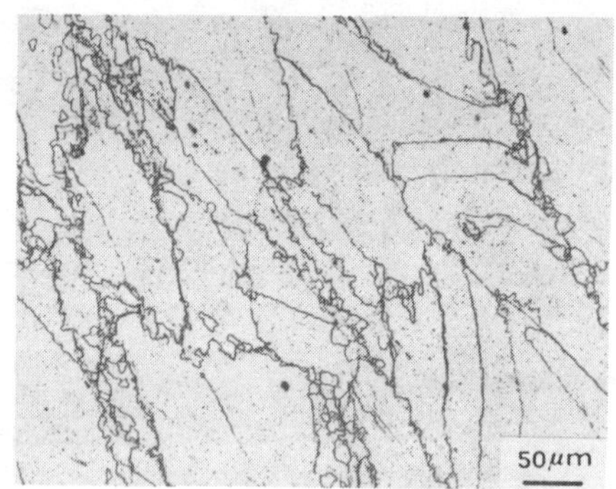

Fig. 5 - Optical micrograph showing the formation of dynamically recristallized grains on the initial grain boundaries T = 1100°C, $\dot{\bar{\varepsilon}} = 0.4\ s^{-1}$, $\bar{\varepsilon} = 0.8$

to occur preferentially by the strain induced boundary migration mechanism at the high angle boundary of the deformed initial grains as indicated by other authors (7) (cf. Fig. 6 b). The thin foil studies, backed up by convergent beam diffraction, confirm that the fine structure observed at the grain boundaries by optical microscopy does indeed correspond to the development of new grains with high angle interfaces relative to the deformed matrix.

The microstructural evolution at the other temperatures and strain rates studied in the present work is in general qualitatively similar to that described above except insofar as the dislocation cell and subgrain sizes vary inversely with the flow stress (8). It is significant, however, to note that in stage II the presence of bands of equiaxed cells within the grains becomes prevalent in the specimen deformed at 900°C (Fig. 6 a). It is thought that these bands may be of similar origin to deformation bands frequently observed in cold worked polycrystals with large grains although in the latter case the dislocation cells are generally elongated in the direction of the band (9).
Once the elongated cell structure is formed throughout the deforming grains, the flow stress remains practically constant and indeed in the conditions of the present experiments it is not strongly affected by the onset of dynamic recrystallization. As in the case of work by other authors, at constant strain rate the critical strain $\bar{\varepsilon}_c$, corresponding to the appearance of the first dynamically recrystallized grains, decreases with increasing temperatures. The values of $\bar{\varepsilon}_c$ estimated at $\dot{\bar{\varepsilon}} = 0.4\ s^{-1}$ are given in Table 1.

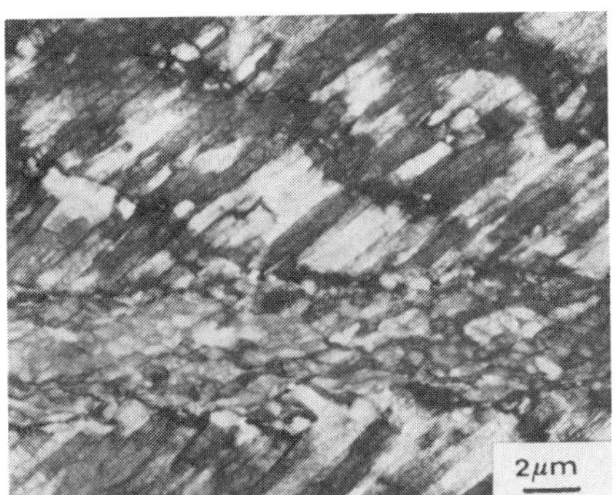

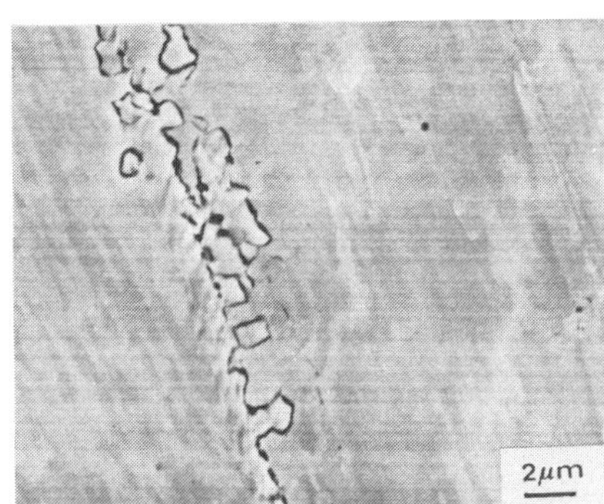

(a) (b)

Fig. 6 - T = 900°C, $\dot{\bar{\varepsilon}} = 4\ s^{-1}$, $\bar{\varepsilon} = 1.4$ a) TEM micrograph : formation of a band of equiaxed cells within a deformed grain. b) SEM micrograph showing evidence for strain induced bulging on a grain boundary.

TABLE 1 - Critical Strain for Dynamic Recrystallization at $\dot{\bar{\varepsilon}} = 0.4\ s^{-1}$.

T°C	900	1100	1200
$\bar{\varepsilon}_c$	1.7	0.65	0.5

DISCUSSION AND CONCLUSIONS

The microstructural evolution observed in the present study up to the onset of dynamic recrystallization may be interpreted within the generally accepted framework of a competition between strain hardening by dislocation glide and dynamic recovery. The importance of recovery is stronger for the higher temperatures and lower strain rates at which the dislocation cells within the grains evolve into weakly misoriented elongated subgrains with a more or less dense dislocation configuration within the subgrains.

The microstructural observations provide evidence of a physical basis to the mathematical treatment of strain hardening at high temperatures in terms of the relation between the slope of the strain hardening curve and the flow stress (3). The decomposition of the strain hardening stage into two distinct substages which has proved useful in the mechanical modeling of hot deformation (10) appears thus to be related to formation of well defined cell structures within the grains and to the development of equiaxed subgrains in the neighbourhood of the initial high angle grain boundaries. Further work is required to verify that the relation between the dislocation substructure and the mechanical testing parameters may be generalized to other deformation modes.

Once the elongated cell structure is well established throughout the grains, the structural modifications are only minor up till the onset of dynamic recrystallization. In particular the width of the equiaxed regions at the initial grain boundaries and the size of the cells in these zones appear to be practically constant. The discontinuous dynamic recrystallization process is initiated preferentially at the initial austenite grain boundaries by a process involving strain induced boundary migration within the zones of equiaxed subgrains. Nucleation of new grains can also occur within the deformed grains particularly at the lower temperatures in the range studied in this work and at high strains. The zones of intragranular nucleation correspond to bands of equiaxed cells crossing the deformed grains which are thought to be analogous to the deformation bands observed in cold deformation of polycrystalline fcc metals.

ACKNOWLEDGEMENTS

The authors are very gratefull to Drs C. ROSSARD and B.J. THOMAS (IRSID - St-Germain-enLaye) for fruitfull discussions and the critical review of the text of this paper.

REFERENCES

1. W.J. Mc G. Tegart and A. Gittins, in *The Hot Deformation of Austenite* (edited by J.B. Ballance), p. 1. Met. Soc. of AIME, New York (1977).
2. H.J. Mc Queen, R. Petkovic, H. Weiss and L.G. Hinton, in *The Hot Deformation of Austenite* (edited by J.B. Ballance), p. 113. Met. Soc. of AIME, New York (1977).
3. B. Ahlblom and R. Sandström, *Intern. Metals Rev.* 1, 1 (1982).
4. W. Roberts and B. Ahlblom, *Acta Metall.* 26, 801 (1978).
5. T. Sakai and J.J. Jonas, *Acta Metall.* 32, 189 (1984).
6. U.F. Kocks, *J. Eng. Mater. Technol.* 97, 76 (1976).
7. P.A. Beck and P.R. Sperry, *J. Appl. Phys.*, 21, 150 (1950).
8. L. Fritzemeier, H.J. Luton and H.J. Mc Queen, *Proceedings of 5th Intern. Conf. Strength of Metals and Alloys* (edited by P. Haasen, V. Gerold and G. Kostorz), p. 95. Pergamon Press, Oxford (1979).
9. R.D. Doherty, in *Recrystallization of Metallic Materials* (edited by F. Haessner), p. 31. Riederer Verlag, Stuttgart (1978).
10. C. Perdrix et al., to be published in *Revue de Métallurgie*, France (1985).

Hot Deformation Characteristics of Worked 301 Austenitic Stainless Steel

N. D. Ryan and H. J. McQueen

Mechanical Engineering, Concordia University Montreal H3G 1M8, Canada

ABSTRACT

High strain, hot torsion tests were performed in the range 900-1200°C at surface strain rates of 0.1 to 5.0 s^{-1} on 301 γ stainless steel. Strain hardening behavior was related to the Kocks-Mecking model in which there is a saturation stress. The critical strain for dynamic recrystallization decreases as temperature rises and strain rate falls. The dependence of flow stress on strain rate was by hyperbolic sine and on temperature by an Arrhenius function. The activation energy was calculated for both isothermal and adiabatic conditions. An acceptable agreement is found between the torsion flow stresses and both compression and extrusion results.

KEY WORDS

Hot torsion, flow stress, dynamic recovery, dynamic recrystallization, strain hardening, substitutional solute, activation energy.

INTRODUCTON

The hot workability of 301 has been addressed in a few previous studies [1-4]. The cam plastometer study of Suzuki et al. [3] is limited in strain and examines a steel with unusually high Al content. Hot extrusion to strains of 0.69 and 1.39 were limited to 700°C, but did show that martensite was not formed above 100°C [4]. The current tests were undertaken to confirm the previous results and to extend them to multistage simulations which will be presented in an additional paper. The torsion test, while having the problem of strain $\bar{\varepsilon}$ and strain rate $\dot{\varepsilon}$ gradient which require some approximation for low strain tests, has the unique capacity for tests of more than three stages with strains greater than 2.3 [5-15].

During hot deformation in similarity to other austenitic strainless steels, Type 301 is expected to undergo softening by both dynamic recovery DRV and dynamic recrystallization DRX [3,10-17]. Substitutional solutes lower stacking fault energy, thereby reducing the dislocation climb and cross slip and hence DRV. While this enhances DRX, solute drag on the migrating boundaries slows it down. The result is that after a broad peak, the flow stress declines to a steady state regime where the grain size is held constant by continual dynamic recrystallization [11-17]. Ductility is greatly increased by DRX as boundary migration isolates the fissures caused by grain boundary sliding, thereby peventing their linkage to induce fracture [11,12,15].

EXPERIMENTAL TECHNIQUES

Type 301 austenitic stainless steel provided by Atlas Steels of Tracy, Que. had a composition of 17.12 Cr, 7.92 Ni, .11 C, 1.12 Mn, .54 Si, .036 P, .002 S, .20 Mo, .019 N, and .0043 O. The specimens, with their axes parallel to the rolling direction, had been machined to close tolerances in the gauge section (l = 25.4 mm, $2r$ = 6.25 mm) to ensure uniform twisting. After annealing (in stainless steel pouches) for thirty minutes at 1050°C followed by quenching in water to avoid carbide precipitation, the grain size was determined by area analysis to be 75 μm. After preheating to deformation temperature for 5 min. in a radiant furnace, the torsion was accomplished on a closed-loop, servo-controlled hydraulic hot testing machine [11,14].

In the calculation of the surface shear stress τ from the torque Γ, the following relationship [8,9,11,14] was used:

$$\tau = \left(\Gamma/2\pi r^3\right)\left(3 + n' + m\right) = \bar{\sigma}/\sqrt{3} \qquad (1)$$

As an approximation, the value of the strain hardening rate n' was taken as zero at the peak stress. The values of the strain rate sensitivity m, which is the reciprocal of the stress exponent n, were determined from slopes of $\log \Gamma$ versus $\log \dot{\bar{\varepsilon}}$ curves as shown in Fig. 1. These values are in good agreement with other similar research [3]. The equivalent stress $\bar{\sigma}$ and surface strain $\bar{\varepsilon}$ were calculated by use of the Von Mises criterion [11,14].

EXPERIMENTAL RESULTS

Representative flow curves for Type 301 steels are presented in Fig. 2. During hot working at temperatures greater than 0.6 Tm, the flow curve exhibits strain hardening at low strains rising to a peak stress, σ_p and strain, ε_p and flow softening which eventually leads to a steady state regime. Such behavior is a result of DRV and DRX [8-15]. Compression curves for 301 [3] are plotted for comparison. As with other types, $\bar{\varepsilon}_p$ decreases with increasing temperature and deceasing strain rate [11,14,15]. Preliminary metallography has shown that in association with the curves tending to steady state, the recrystallized grain size decreases with increase in Z ($= \dot{\bar{\varepsilon}} \exp(Q_{HW}/RT)$, Q_{HW} activation energy for hot working) [6,8-15,18]. Where steady state was not attained at 900°C, some of the original elongated grains still remained. Due to lower Cr, Ni, Mo contents, Type 301 did not fracture at all under the same test conditions as did 304 which implies greater ductility.

The $\bar{\sigma}$-$\bar{\varepsilon}$ curves have been further analyzed by determining the strain hardening rate, $d\bar{\sigma}/d\bar{\varepsilon} = \theta$, which is plotted against $\bar{\sigma}$ in Fig. 3 [19-21]. The work hardening rate decreases sharply with stress, being steeper as the Z value decreases; at constant $\bar{\sigma}$, θ increases as Z increases. Each curve is composed of two linear segments with the bend to lower slope being due to subgrain formation. The extrapolated high-θ segments converge to θ_o (975 MPa) at $\bar{\sigma} = 0$. Because mild steel experiences a faster rate of DRV than the highly alloyed 301, it has lower values of θ and of θ_o (325 MPa). The low-θ segments bend downwards when DRX initiates ($\bar{\sigma}_c, \bar{\varepsilon}_c$) and reach zero at $\bar{\sigma}_p$, $\bar{\varepsilon}_p$. The initiation of DRX occurs at lower $\bar{\sigma}$ as Z decreases. The extrapolation of the low-θ segment of the θ-$\bar{\sigma}$ curve to zero θ gives the saturation stress $\bar{\sigma}_s^*$ which is the stress the flow curve would attain if there were no dynamic

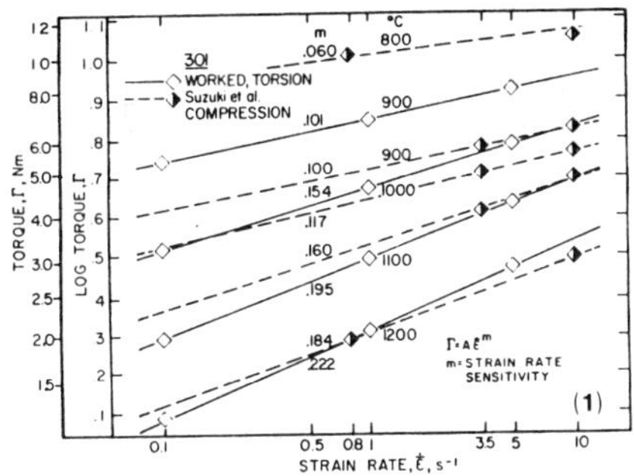

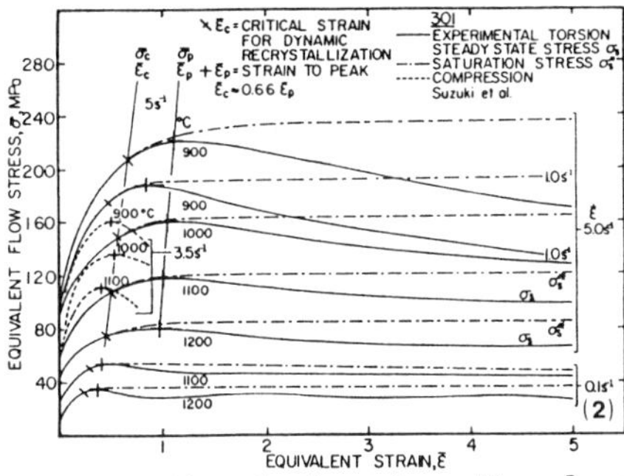

Fig. 1. The logarithmic dependence of torque on the strain rate. The slope m decreases with rising T and is greater than that of the compression alloy due to greater carbide content [3]. The compression data was made comparable by dividing $\bar{\sigma}$ by 27 MPa/Nm, the average factor converting Γ to $\bar{\sigma}$.
Fig. 2. Representative $\bar{\sigma}$-$\bar{\varepsilon}$ curves for Type 301 in torsion and compression. The peak and work softening to steady state are indicative of DRX. $\bar{\varepsilon}_p$ rises with rising $\dot{\varepsilon}$ and falling T and is less in compression [3]. The decrease in σ_s^* with rising T and diminishing $\dot{\varepsilon}$ is the result of DRV.

recrystallization [19,20]. The critical stress $\bar{\sigma}_c$, the peak stress $\bar{\sigma}_p$, and the saturation stress σ_s^* (Fig. 4) all increase at a rising rate as the temperature decreases. However, σ_s^* rises more rapidly in that it changes from 1.02 to 1.07 $\bar{\sigma}_p$ as Z goes from its lowest to highest value.

The dependence of $\bar{\sigma}_p$ on $\dot{\varepsilon}$ is shown in Fig. 5 in accordance with the following equation [6,8-10,14-16,22]:

$$\dot{\varepsilon} = A\left(\sinh \alpha\bar{\sigma}\right)^n \exp\left(-Q_{HW}/RT\right) \qquad (2)$$

The temperature dependence according to Eqn. 2 appears in Fig. 6 for both the steels of Fig. 5. The value of α is taken to be 1.2×10^{-2} MPa^{-1} [9]. The stress exponent n was derived to be 4.4 for the present steel and 4.9 for the compressed steel [3]. The activation energies Q_{HW} were 399 and 437 kJ/mol. When account is taken of deformational heating in the data for 5.0 s^{-1}, the activation energy increases to 458 kJ/mol for the present work. By assuming adiabatic conditions, the temperature increase ($\Delta T = \int\bar{\sigma}d\varepsilon/S\rho$, the specific heat S =.59 J/g°C and the density $\rho = 7.92 \times 10^3$ kg/m^3) is calculated from the deformational work [6,22]. For the extrusion data [4] in Fig. 6b, the stresses were calculated from the extrusion pressures according to the formula of Hughes et al.[7], but with a frictional factor twice as large because of a graphite lubricant instead of glass. The adiabatic correction for the torsion data in Fig. 6a is extrapolated into Fig. 6b and is seen to agree closely with the adiabatic correction of the extrusion data. The activation energies for extrusion are 340 and 508 kJ/mol with correction.

It is possible to rewrite Equation 2 in a constitutive form in which the dependent variable is separated [22]:

$$\bar{\sigma} = 1/\alpha \ \ln\left\{(Z/A)^{1/n} + \left[(Z/A)^{2/n} + 1\right]^{\frac{1}{2}}\right\} \qquad (3)$$

From this equation, the flow stress can be calculated for any Z condition. From Fig. 6, the value of A was determined to be 5.65×10^{14} s^{-1} for the experimental situation and and 5.32×10^{19} s^{-1} for an adiabatic one.

DISCUSSION

To the experimental flow curves in Fig. 2, additions have been made to indicate $\bar{\varepsilon}_c$, $\bar{\sigma}_c$ and σ_s^* derived from Fig. 3. Also appearing is the influence of T and $\dot{\varepsilon}$ on $\bar{\varepsilon}_c$ and $\bar{\varepsilon}_p$ which are related to the variation of $\bar{\sigma}_c$ and $\bar{\sigma}_p$ with Z in Fig. 4. It can also be seen how the experimental flow curve is lowered to $\bar{\sigma}_s$ by DRX compared to one with only DRV which rises to σ_s^* [20]. At high Z the decrease is much greater because DRX removes a higher dislocation density. DRX is initiated at approximately 0.92 $\bar{\sigma}_p$ and 0.67 $\bar{\varepsilon}_p$ for the present work. The Al bearing 301, with a higher work hardening rate, initiates DRX at 0.97 $\bar{\sigma}_p$ and 0.72 $\bar{\varepsilon}_p$ [3]. The compression curves are reasonably similar to the torsion ones, except that the peak is shifted to a lower strain. This arises because the DRX occurs uniformly across the section in compression where the strain is uniform. In torsion, the strain gradient results in recrystallization initiation for rising surface strain at diminishing radius, so that work hardening continues in the interior as the surface recrystallizes [17]. In the range 1 - 5 s^{-1}, the strengths in torsion and compression [3] are substantially the same despite the latter's high Al content; however, the data at 900°C seem much too low so were not used in determining the T dependence. At the greatest Z value, the flow stress of Type 301 was approximately 7% higher than Type 304 with a comparable grain size, due to the precipitation at the low temperature of more carbides which outweigh the effect of substitutional solute [5,8,9,11,15]. When the carbide dissolves at high temperature and because of the lower substitutional solute, the strength of Type 301 decreases below that of Type 304. The carbide forming ability of 301 is due to higher C content than the 304 since both have similar Mo contents.

The work hardening behavior of 301 stainless steel is fairly insensitive to T and $\dot{\varepsilon}$ at low $\bar{\varepsilon}$, but at large $\bar{\varepsilon}$ it is strongly dependent on T and $\dot{\varepsilon}$ (Fig. 3) as a result of their effect on DRV [19-21]. At 140 MPa and 5.0 s^{-1}, θ drops from 306 to 72 MPa as T increases from 900 to 1000°C. Similarly at 900°C, θ drops from 306 to 195 MPa as $\dot{\varepsilon}$ changes from 5.0 to 1.0 s^{-1}. As θ falls, the progression to saturation is slowed i.e., the slope of the $\bar{\sigma}$-$\bar{\varepsilon}$ curve decreases less rapidly so that the curve is higher. This indicates that the DRV is more rapid as tangles are ordering into cell walls than after the subgrains have formed. The low-θ segment of the θ-$\bar{\sigma}$ curves, where subgrain formation occurs, becomes progressively more horizontal as Z increases, thereby increasing σ_s^*. This accounts for the increase of 2 percent over $\bar{\sigma}_p$ at low Z and 7 percent at the highest value of Z shown in Fig. 4. From the demonstrated similarity in values for σ_s^* and $\bar{\sigma}_p$ in Figs. 3 and 4, it is clear that σ_s^* has essentially the same dependence on T and $\dot{\varepsilon}$ as those for $\bar{\sigma}_p$ in Figs. 5 and 6. The Q_{HW} derived by the present analysis for $\bar{\sigma}_p$ or σ_s^* is consistent with the values derived for high temperature by the Kocks-Mecking analysis [20,21].

At 140 MPa and 900°C, the work hardening rate of 301 is approximately five times greater than that of C steel [19]. While both steels have approximately equal peak stresses for similar conditions, σ_s^* for the C steel is 8% greater than the alloy because of the lower magnitude of slope in the low-θ segments in Fig. 3. The higher θ_o for 301 compared to that for C steel is related to the latter's higher stacking fault energy [20,21]. In agreement with Kocks [20], it was found that a plot of log σ_s^* versus T is linear and extrapolates to a finite saturation stress σ_{so}^* at 0°K independent of $\dot{\varepsilon}$, indicating that dynamic recovery can be initiated by stress alone. The value for the present work is close to 14 x 10^3 MPa, while a revised extrapolation for the previously analyzed Type 304 gives 7.5 x 10^3 MPa [20]. This latter value is in good agreement with the authors preliminary results for Type 304. The requirement of almost twice as much stress at 0°K for 301 as for 304 is possibly due to the excess carbides in the former.

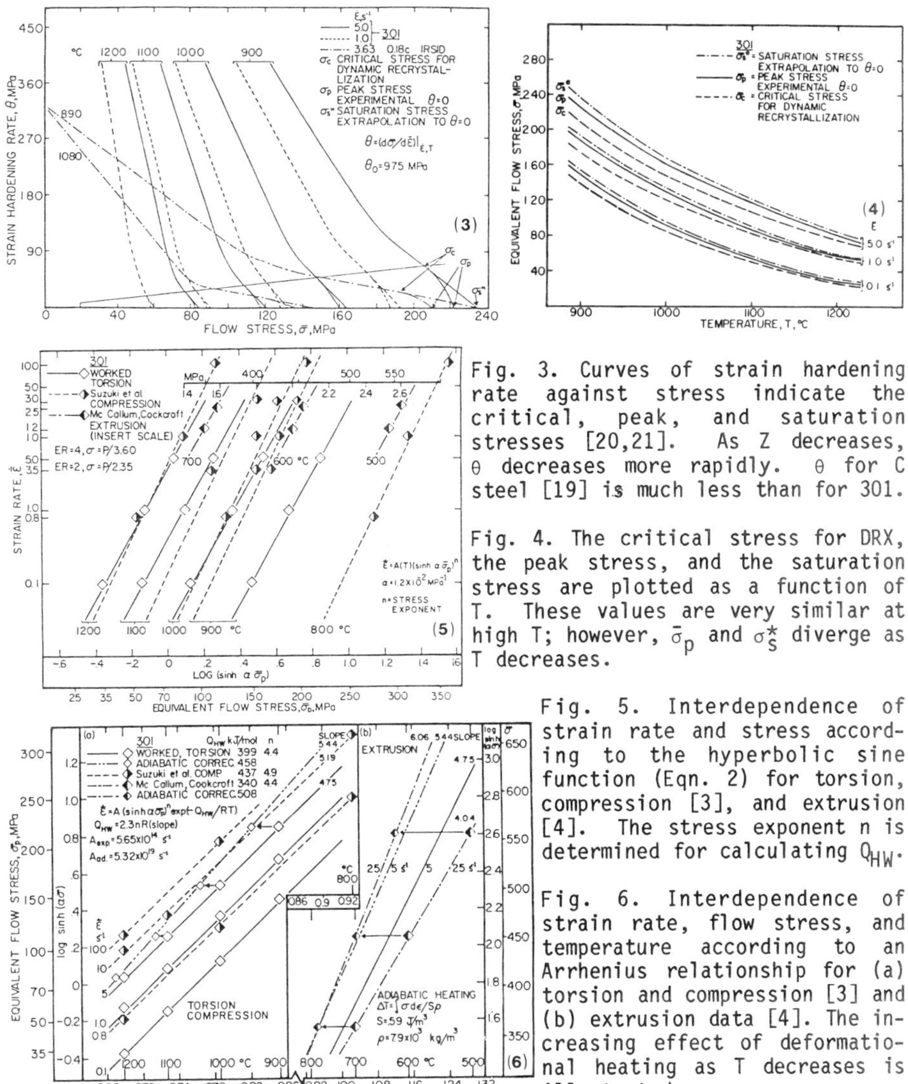

Fig. 3. Curves of strain hardening rate against stress indicate the critical, peak, and saturation stresses [20,21]. As Z decreases, θ decreases more rapidly. θ for C steel [19] is much less than for 301.

Fig. 4. The critical stress for DRX, the peak stress, and the saturation stress are plotted as a function of T. These values are very similar at high T; however, $\bar{\sigma}_p$ and σ_s^* diverge as T decreases.

Fig. 5. Interdependence of strain rate and stress according to the hyperbolic sine function (Eqn. 2) for torsion, compression [3], and extrusion [4]. The stress exponent n is determined for calculating Q_{HW}.

Fig. 6. Interdependence of strain rate, flow stress, and temperature according to an Arrhenius relationship for (a) torsion and compression [3] and (b) extrusion data [4]. The increasing effect of deformational heating as T decreases is illustrated.

The rising effect of deformational heating as the temperature declines is demonstrated in Fig. 6 [6,22,23]. The increase of 92°C at 600°C for extrusion indicates that the real flow stress is 591 instead of the experimentally measured 453 MPa. From Eqn. 3, the flow stress for any condition can be accurately calculated when the values of Q_{HW}, α, A, and n have been determined on the basis of the flow stresses either experimentally measured or corrected for deformational heating. From the stress strain curve, the mean flow stress is readily calculated from which forces can be estimated for rolling or pressures for extrusion [6,8-10,12,15].

CONCLUSION

The following features were revealed regarding the hot deformation characteristics of Type 301 austenitic stainless steel.

(1) During continuous deformation the $\bar{\sigma}$-$\bar{\varepsilon}$ and θ-$\bar{\sigma}$ curves show that both dynamic recovery and dynamic recrystallization take place. The peak stress, peak strain, and strain hardening rate decline as Z decreases.

(2) The critical stress and strain for dynamic recrystallization are 92 percent of the peak stress and 67 percent of the peak strain.

(3) The saturation stress and its slight excess over $\bar{\sigma}_p$ decreases as T rises and $\dot{\varepsilon}$ falls. Both $\bar{\sigma}_p$ and σ_s^* can be adequately related to strain rate by the hyperbolic sine and to temperature by the Arrhenius function.

(4) The activation energy is higher when deformational heating is considered. For different type 301 steels, it differs due to the variations in the carbide and nitride content.

REFERENCES

1. D.W. McDowell, Blast Furnace Steel Plant. 53, 1033 (1965).
2. D.J. Knight and A.R. Palmer, J. Met 18, 578 (1966).
3. H. Suzuki et al, Rep. Inst. Ind. Sci. Univ. Tokyo, 18 [3], 1 1968.
4. R. McCallum and M.G. Cockcroft, Structure Properties of Warm Extruded γ Stainless Steels, NEL Report 517, Dept. Trade and Ind., U.K. (1972).
5. A. Nicholson, Iron and Steel. 37, 290 (1964).
6. C.M. Sellars and W.J. McG. Tegart, Int. Met. Rev. 17, 1 (1972).
7. K.E. Hughes et al, Met. Tech. 1, 161 (1974).
8. W.J. McG. Tegart and A. Gittins, The Hot Deformation of Austenite (edited by J.B. Ballance), p. 1, AIME, New York, NY (1977).
9. H.J. McQueen et al. ibid. p. 113.
10. B. Ahlblom and R. Sandstrom, Int. Met. Rev. 27, 1 (1982).
11. N.D. Ryan et al, Can. Met. Q. 22, 369 (1983).
12. H.J. McQueen and J.J. Jonas, J. Appl. Metalworking, 3, in press (1984).
13. H.J. McQueen and J.J. Jonas, J. Appl. Metalworking, 3, 233 (1984).
14. N.D. Ryan and H.J. McQueen, Proc. Intnl. Conf. New Dev. Stainless Steel Tech., in press, ASM, Metals Park, OH (1985).
15. H.J. McQueen and N.D. Ryan, Stainless Steels '84 in press, Inst. Metals, London (1984).
16. S.L. Semiatin and J.H. Holbrook, Met. Trans. 14A, 1681 (1983).
17. C.M. Sellars, Phil. Trans. R. Soc. Lond. A288 147 (1978).
18. T. Maki et al, Trans. Iron Steel Inst. Japan 22, 253 (1982).
19. Ch. Perdrix, Charact. d'Ecoulement Plastique du Metal T.A.B. à Chaud (CECA 7210-EA 311) IRSID, St. Germaine en-Laye France (1983).
20. U.F. Kocks, J. Eng. Mat. Tech. 98, 76 (1976).
21. B. Nicklas and H. Mecking, Strength of Metals and Alloys (ICSMA5) (edited by P. Haasen et al) vol. 1, p. 351, Pergamon Press, Oxford (1979).
22. H.B. McShane and T. Sheppard, J. Mech Working Tech. 9, 147 (1984).
23. J.J. Jonas and M. Luton, Advances in Deformation Processing (edited by J.J. Burke and V. Weiss) p. 215, Plenum Publ. Co. New York, NY (1978).

ACKNOWLEDGEMENTS

The authors acknowledge the financial support of the Natural Sciences and Engineering Research Council of Canada and of Quebec's FCAC program. They thank P. McQueen for assisting in tests and for drawing the final diagrams.

Effect of Strain Heterogeneity and Adiabatic Heating on the Stress-Strain Behaviour of an Austenitic Stainless Steel

R. Colás* and C. M. Sellars**

*Now at D.I.M.E., Facultad de Ingeniería, U.N.A.M., México, D.F.
**Department of Metallurgy, University of Sheffield, UK, where the research was carried out

ABSTRACT

Stress-strain curves of an AISI type 316 stainless steel obtained from both plane strain and axisymmetric compression tests are compared for a wide range of strain rates. During straining dynamic recrystallisation takes place. A series of divergencies arising in the curves are explained as a function of strain heterogeneity developed in the plane strain tests and of deformational heating. It is found that the heterogeneity is more important at low strain rates and the deformational heating at high strain rates.

KEYWORDS

High temperature tests; high strain rate; hot workability; strain heterogeneity; adiabatic heating.

INTRODUCTION

In hot working testing, much attention is given to the average structure developed during straining. Good correlations between the structure and the material properties are found when the geometry or the testing method are held constant (1,2). Problems due to strain heterogeneity arise when different tests are compared or when the specimen geometry changes (3,4). An additional problem is generated at high strain rates when the deformational heating cannot be dissipated and the specimen is heated adiabatically (5).

This paper discusses differences present in stress-strain curves obtained in plane strain and in axisymmetric compression on an austenitic stainless steel deformed at 1000°C at different strain rates.

EXPERIMENTAL PROCEDURE AND RESULTS

The austenitic steel used in the research was an AISI type 316 (0.024% C, 16.70% Cr, 12.20% Ni, 2.63% Mo, 1.50% Mn, 0.29% Si). The material was received as 25.4 x 50.8 mm bars. Small slabs 100 mm long were cut from the bars and reheated at 1200°C for half an hour before hot-rolling to 10 mm thickness plates that were water quenched and cut into pieces to be used as plane strain specimens (about 55 x 52 x 10 mm). Axisymmetric specimens 10 mm height and 6 mm diameter were machined from the hot-rolled plates with their axes normal to the rolling direction, i.e. parallel to the loading direction of the plane strain specimens. All the specimens were deformed at constant strain rate in a computer controlled servohydraulic machine, with the specimen-tool arrangement shown in Fig. 1, and described in detail elsewhere (1,6). The final equivalent strain was around 1.5 for plane strain tests, but because of experimental limitations the maximum strain in the axisymmetric tests was about 1 (7). The deformation was carried out after 15 min reheating at the testing temperature. Liquid glass was used as lubricant. The initial grain size was 30 μm in all cases. Typical equivalent tensile stress-strain curves are shown in Fig. 2 for a range of equivalent tensile strain rates.

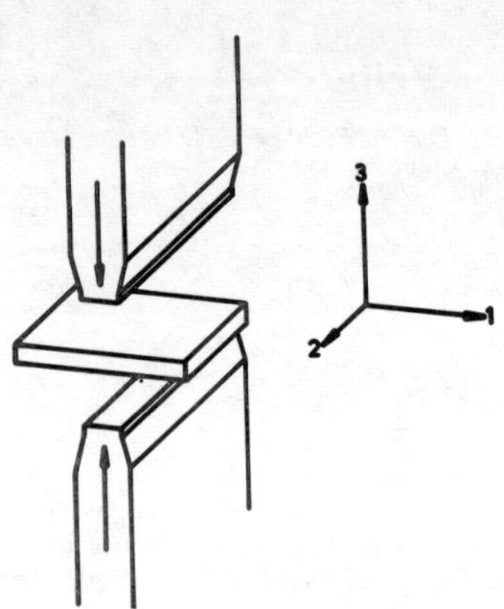

Fig. 1. The specimen-tool arrangement for the plane strain compression test. Directions: 1 - length, 2 - breath, 3 - thickness.

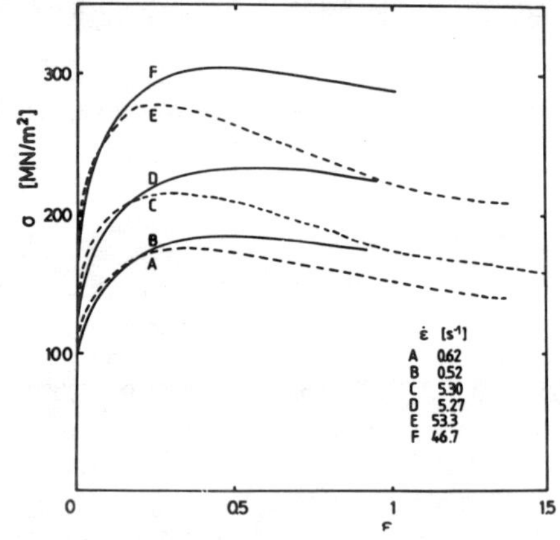

Fig. 2. Stress-strain curves for plane strain compression (broken lines) and axisymmetric compression (full lines).

The strain distribution in plane strain was deduced by deforming specimens with a grid inscribed in the 3-1 plane at the middle of the specimen. Up to six different strain levels at 3 strain rates were tested. After deformation, the distorted grid was photographed and the coordinates of the nodes digitized, then the strain values were deduced in the way described elsewhere (8,9). One of the strain patterns obtained is shown for the upper right hand quadrant of the specimen in Fig. 3. In the lower quadrant the slip line field expected for the final geometry is drawn for comparison. In Fig. 4 the average strain in the slip line field (ε_s) is plotted against the nominal value (ε_n), as deduced from the initial and final dimensions of the specimen, with correction for lateral spreading in the con-

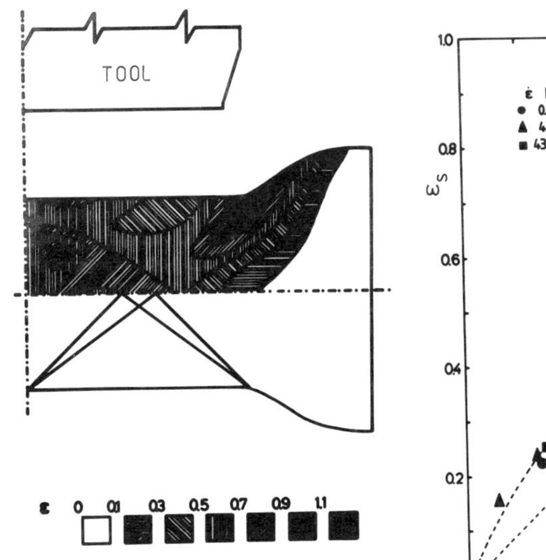

Fig. 3. Experimental strain distribution map and expected slip line field for a specimen deformed at 0.5 s^{-1} to a nominal strain of 0.538. The slip line field includes the area of high shear just outside the tool-specimen interface.

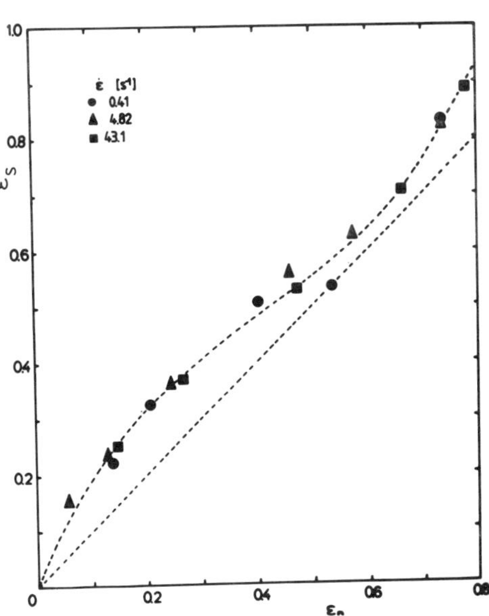

Fig. 4. Relationship between the strain in the slip line field (ε_s) and the nominal strain (ε_n).

version to equivalent tensile strain. A third degree polynomial was fitted to the experimental data, giving:

$$\varepsilon_s = 0.0095 + 2.16\varepsilon_n - 3.64\varepsilon_n^2 + 2.98\varepsilon_n^3 \qquad (1)$$

In order to determine the effect of strain rate on the temperature increase during plane strain testing, some specimens were drilled to the centre and a 1.5 mm diameter inconel sheatted, mineral insulated, chromel-alumel thermocouple was inserted. The emf was recorded by the computer as straining progressed. In order to ensure good thermal contact between the specimen and thermocouple bead, all specimens were deformed to a strain of 0.1 at the lower strain rate, held for two minutes and then deformed to a strain of 1.5 at the different strain rates. The temperature-strain curves obtained are shown in Fig. 5 together with the expected increase of temperature due to adiabatic heating given by (10):

$$\Delta T = \frac{\bar{\sigma}\delta\varepsilon}{c\rho} \qquad (2)$$

where ΔT is the increase of temperature, c the specific heat, ρ the density, $\bar{\sigma}$ the mean stress, ε the interval of strain. $\bar{\sigma}\delta\varepsilon$ represents the increment of work done per unit volume.

DISCUSSION

From Fig. 2 it can be seen that the initial work hardening coefficient is

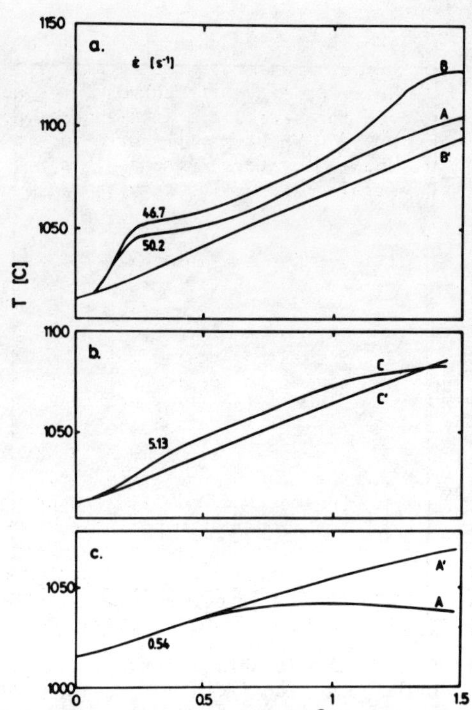

Fig. 5. Temperature-strain curves for specimens deformed in plane strain at the three strain rates. Calculated adiabatic temperature changes are shown by the curves labelled A', B' and C'.

higher for plane strain than for axisymmetric compression, and that the strain to the peak is higher for axisymmetric tests at the same equivalent strain rates. Once the peak in stress is achieved, the curves diverge more markedly. From Fig. 3, it is clear that the strain distribution is heterogeneous in the plane strain specimen. From Fig. 4 the relationship between the different strain values appears to be oscillating with strain. This probably arises from the change of number of "crosses" in the slip line field (9). When the axisymmetric specimens were observed at the end of the tests, no significant barrelling was found, so it is assumed that strain in this test is much more homogeneous.

Three different behaviours can be observed in the temperature-strain curves:

a.) At high strain rates, $\dot{\varepsilon} \simeq 50\ s^{-1}$, the measured curve is always above the calculated one. It can be seen that a bump is produced due to a very high heating rate at low strains, coinciding with the strain range over which the highest divergence between ε_s and ε_n is observed. At intermediate strains the heating rate is about the same for both curves. It is worth mentioning that the bump was previously observed, but it was thought to be due to spurious readings by the thermocouple (10,11).
b.) At intermediate strain rates, $\dot{\varepsilon} \simeq 5\ s^{-1}$, the measured curve is above the calculated one at low strains, but this is inverted towards the end of the test. It can be seen, as in the former case, both curves show equal heating rates at intermediate strains.
c.) At low strain rates, $\dot{\varepsilon} \simeq 0.5\ s^{-1}$, the calculated curve is always above the measured one.

The early parts of the curves drawn in Fig. 2 are reproduced on a larger scale in Fig. 6. The experimental curves are designed as I and V respectively for axisymmetric and plane strain compression. To compare with the curves from plane strain tests, corrections were carried out to the axisymmetric curves in the following order, assuming that strain was homogeneous in these tests:

a.) Correction for strain distribution. The curve was displaced towards the left as result of the ratio of $\varepsilon_n/\varepsilon_s$. Curves are marked as II.
b.) Correction for stress distribution. From the strain distributions it is possible to calculate the effect on the mean stress in the expected slip line fields arising from each component of strain. This is taken from the homogeneous curve, i.e. the axisymmetric curve. The curves corrected for

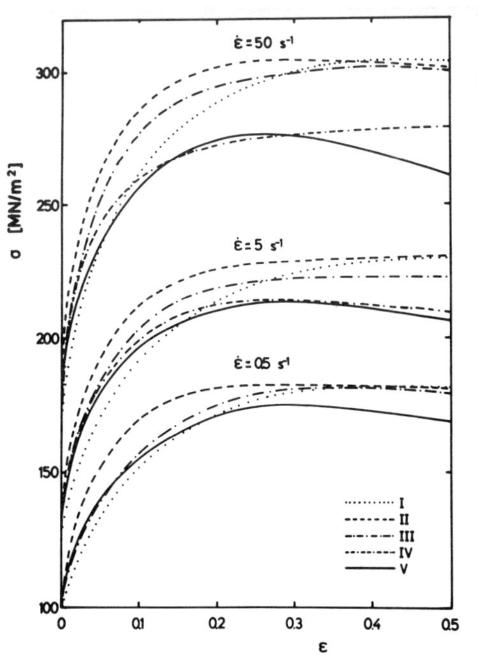

Fig. 6. Stress-strain curves for: I axisymmetric compression, II curves I after correction for strain heterogeneity, III curves I corrected for stress and strain heterogeneity, IV curves III corrected for the temperature increase, V plane strain compression curves.

(a) and (b) are marked as III.

c.) Correction for the increase in temperature. This was done only for the two higher strain rates, it was considered that the axisymmetric specimen was deformed under adiabatic conditions, i.e. following the calculated curves at each strain rate, Fig. 4, and that for plane strain specimens the thermocouple was measuring the temperature in the regions in which the strain was being concentrated. The correction to the stress was made by assuming that the material obeys the relationship given by (7):

$$Z = A\cdot\exp(0.026\sigma) \qquad (3)$$

where $Z=\dot{\varepsilon}\exp(Q/RT)$, $Q=460$ kJ/mol, is the Zener-Hollomon parameter, and A a constant. Corrections (a), (b) and (c) give curves IV.

It can be seen in Fig. 6 that curves IV and V are in excellent agreement for a strain rate of 5 s^{-1}, indicating that the factors taken into account in making the corrections are the major ones leading to the original differences between the axisymmetric and plane strain compression curves. At 0.5 s^{-1}, no temperature correction was attempted because adiabatic deformation cannot be assumed for axisymmetric compression. However, because of the strain locallisation in plane strain compression, a larger temperature rise than in axisymmetric compression would still be expected so that the difference between curve III and V at higher strains is attributable to this. At 50 s^{-1} a difference between curves IV and V develops at higher strains. This is thought to occur because the high temperature rise accelerates dynamic recrystallisation in plane strain compression, leading to accelerated softening which has not been taken into account.

CONCLUSIONS

1. Discrepancies between stress-strain curves, obtained in plane strain compression and in axisymmetric compression under hot working conditions exist even when the former have been corrected to equivalent stresses and strains using the Levy-von Mises relationships.

2. Strain is heterogeneously distributed in plane strain compression tests. The distribution is consistent with the expected pattern of slip line fields and this results in the mean strain in the active fields at any deformation differing from the nominal strain.

3. The differences between the stress-strain curves in plane strain and axisymmetric compression have been shown to arise from the strain distribution in the former, either as a direct result of local strain concentration or indirectly from its effect on local stress and temperature rise.

ACKNOWLEDGEMENTS

The financial support given by the National Council for Science and Technology (CONACyT, México), by Davy-McKee (Sheffield) Ltd and by the Department of Metallurgy of the University of Sheffield to R. Colás is gratefully acknowledged.

REFERENCES

1. J.P. Sah and C.M. Sellars, "Hot Working and Forming Processes", ed. C.M. Sellars and G.J. Davies, Metals Society, London, 1980, 62.
2. W. Roberts, H. Bodén and B. Ahlblom, Met. Sci., 13, 165, 1979.
3. C.M. Sellars, "Les Traitements Thermomécaniques", 24ème Colloque de Metallurgie, Saclay, France, 1981, 111.
4. D.R. Barracloug, H.J. Whittaker, K.D. Nair and C.M. Sellars, J. Test. Eval., 1, 220, 1973.
5. U.S. Lindholm, "Mechanical Properties at High Rates of Strain", Conf. Ser. 21, Inst. Physics, 1974, 3.
6. C.M. Sellars, J.P. Sah, J.H. Beynon and S.R. Foster, "Plane Strain Compression Testing at Elevated Temperatures", Report SRC B/RG/1481, University of Sheffield, 1976.
7. R. Colás, Ph.D. Thesis, University of Sheffield, 1983.
8. J.H. Beynon and C.M. Sellars, to be published.
9. R. Colás and C.M. Sellars, to be published.
10. S.R. Foster, Ph.D. Thesis, University of Sheffield, 1981.
11. R.A. Harding, Ph.D. Thesis, University of Sheffield, 1976.

Effect of Phase Morphology on Restoration Behaviour of Ferrite-Austenite Two-phase Stainless Steel at High Temperatures

T. Chandra, D. P. Dunne and P. Campbell

Department of Metallurgy, University of Wollongong, Australia

ABSTRACT

Dynamic and static restoration processes in a duplex α/γ stainless steel have been investigated in compression for strain rates of 0.004 to $4.0s^{-1}$ at 900^oC. At this temperature, the alloy contained equivolume proportions of α and γ phases, with the γ phase present in one of two different morphologies. In the as-received alloy, γ occurs as stringers (elongated sausage-like shapes) in a matrix of elongated ferrite grains, whereas small Widmanstätten austenite plates are present in very coarse ferrite grains after solution treatment of as-received material at 1350^oC and transforming isothermally at 900^oC. It was found that ferrite recovered dynamically in both types of structures at low strains for either high or low strain rates. On the other hand, the stringer austenite morphology showed profuse static recrystallization within the 3s quench time after straining to a true strain of 0.8 at a strain rate of $4s^{-1}$, whereas Widmanstätten austenite recrystallized to a limited extent under these conditions. The progressive development of the hot worked structures is discussed in relation to the γ morphologies as well as the forms of the flow curves.

KEYWORDS

Duplex stainless steel; recovery; recrystallization; high temperature deformation.

INTRODUCTION

Duplex stainless steels have attracted considerable attention in the last several years because of outstanding engineering properties such as high strength, toughness and resistance to stress-corrosion cracking. These useful properties are related to the microstructure of α/γ steels which are generally produced by hot working processes. It is commonly known that the presence of ferrite in austenitic steels frequently produces hot-shortness and these steels are also susceptible to embrittlement by the presence of sigma phase (1-3). The mechanisms underlying the poor hot workability, which are related directly to restoration processes during and after deformation, are not well understood. Al-Jouni and Sellars (4) have recently reported that dynamic recrystallization of γ phase occurred in 70:30 ferrite/austenite steel with similar kinetics to that in single phase austenite. These authors drew this conclusion from their analysis of stress-strain curves which exhibited the typical form of materials which recrystallize dynamically, but no microstructural evidence was presented to support their conclusion. Chandra et al. (5) on the other hand, found that in a 50:50 ferrite/austenite structure dynamic recovery of α occurred at small strains, but γ showed very limited recovery until high strains ($\varepsilon \simeq 0.5$) were reached. For a strain rate of $\sim 4s^{-1}$ they also obtained a stress-strain curve of the characteristic form associated with dynamic recrystallization, but no recrystallization of α or γ was observed at the strain corresponding to peak stress. At a higher strain of ∿ 0.7 however, limited recrystallization occurred in the γ phase adjacent to σ phase particles.

These observations indicate that the form of the stress-strain is sensitive to the structure of the duplex stainless steel and the present study was undertaken to explore in more detail the effect of the structure of α/γ stainless steels on hot deformation behaviour.

EXPERIMENTAL

The steel used was Sandvik 3 RE60 which has a nominal composition of 0.03C, 1.69 Si, 1.47 Mn, 19.06 Cr, 4.49 Ni and 2.93 Mo. The material was supplied by Sandvik Australia Ltd. in the form of 45mm diameter rod from which cylindrical compression samples 8.5mm dia. x 11.2mm long were machined parallel to the axis of the hot-rolled bar. Two types of microstructure were studied in this investigation. One was the as-received α/γ structure in which axially aligned rods or stringers of γ phase were present in the matrix of elongated ferrite grains. To obtain the second microstructure, the as-received material was solution treated at 1350°C for one hour to dissolve the austenite and then held for one hour at 900°C to produce ∿ 50% γ in the form of Widmanstätten plates in coarse ferrite grains. The stringer of austenite had an average width of 17μm and an average length of 256μm. The heat-treated microstructure, on the other hand, had large ferrite grains of ∿ 660μm diameter with small Widmanstätten plates of austenite typically ∿ 24μm long. Figure 1 shows the two types of microstructures.

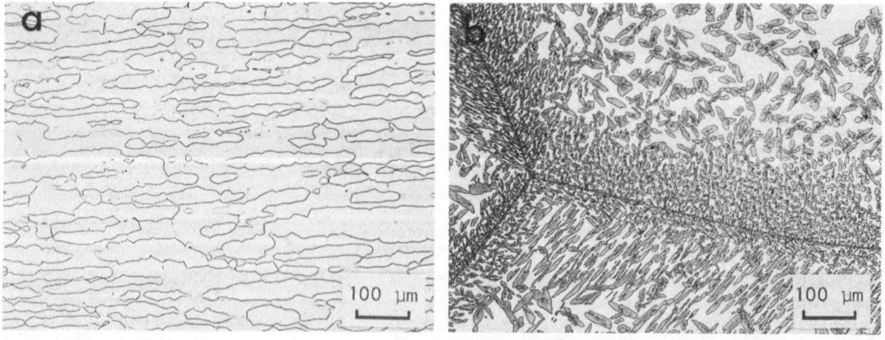

Fig. 1. Optical micrographs of longitudinal section in α/γ alloy, (a) stringer and (b) Widmanstätten structures.

Mechanical testing was carried out in uniaxial compression in an argon atmosphere at constant strain rates in the range 0.004 to $4s^{-1}$ at 900°C. Samples were held at the test temperature for 10 minutes prior to compression to various strains and quenching within 3s of the end of the test. Deformed samples were sectioned longitudinally and transversely for optical metallography. A Joel 120kv transmission electron microscope was used for examination of thin foils.

RESULTS AND DISCUSSION

Stress-Strain Behaviour

True stress-true strain curves at various constant strain rates in uniaxial compression at 900°C for as-received and Widmanstätten samples exhibited similar features, Fig.2. The samples showed the typical form of the curve expected when dynamic recovery occurs, i.e. the flow stress rose sharply to a maximum at relatively low strains and then approached a steady state value. It can be seen from Fig.2 that as the strain rate increased the plateau stress also increased. At all strain rates samples with an austenite

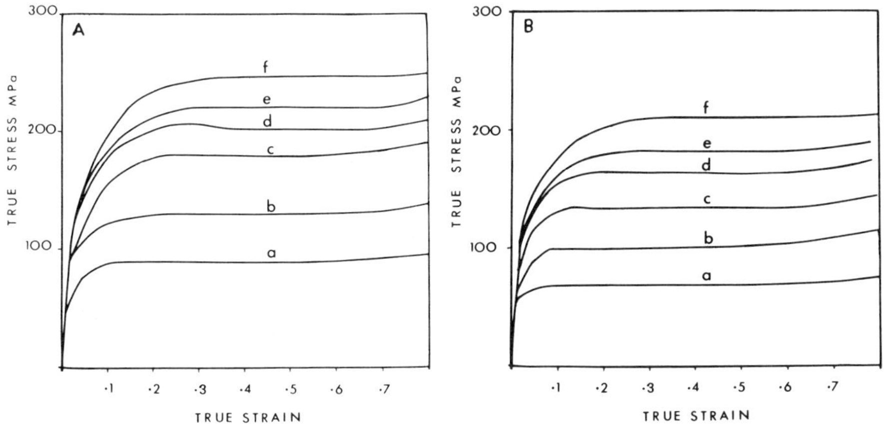

Fig.2. True stress-true strain curves for (A) as-received and (B) heat-treated samples tested at 900°C under uniaxial compression at strain rates (s^{-1}) of (a) .004 (b) .04 (c) 0.4 (d) 0.8 (e) 1.6 (f) 4.0

stringer structure exhibited a higher steady state or plateau stress than the samples with Widmanstätten austenite. It is concluded that the austenite morphology strongly influences the restoration processes occurring in the duplex alloy during and after deformation.

The changes in microstructure at various stages of deformation were also investigated. It is known that the low SFE austenite is the harder phase and that during deformation the softer ferrite flows around austenite (6). It was found on straining the samples with Widmanstätten structure that the plates rotated in ferrite and became aligned normal to the direction of compression when the true strain exceeded $\sim$ 0.4. At low strains alignment was not observed. These observations suggest that at early stages of deformation ferrite tends to flow easily around Widmanstätten γ, thus accounting for the bulk of the total strain.

The effect of deformation on the as-received structure was quite different. The large austenite stringers with banded ferrite are initially parallel to the compression axis, but they tend to break up into smaller stringers at low strains ($\varepsilon \sim 0.3$), and into globular shaped particles at higher strains ($\varepsilon \sim 0.7$). The latter γ morphology allows the softer ferrite to flow more easily between particles as deformation progresses. The results show that the flow behaviour in α/γ alloys is dependent on the alignment and morphology of the phases and it is inferred that these factors influence the relative extents of work hardening and dynamic restoration that occur in the two phase structure.

Strain Distribution

A rough estimate of the level of strain taken by ferrite and austenite in the Widmanstätten structure (with large ferrite grains) was obtained by metallographically measuring the grain strain in the ferrite. For compression to a true strain of 0.8 at a strain rate of $0.4s^{-1}$, $\varepsilon(\alpha)$ was $\sim$ 70% of the total strain, but this percentage was found to be strain rate dependent. The results indicate that the softer ferrite deforms to a larger extent than the harder austenite at low strain rates, but the difference in the strains in the two phases decreases with increase in strain rate. It is suggested that at high strain rates, as less time is available for accommodation processes, the strain is partitioned more equally between the two phases. The lower value of $\varepsilon(\gamma)$ at low and intermediate strain rates, on the other hand, is consistent with interphase boundary sliding in the two phase material which limits stress transference to the γ. It should be noted that the α/γ interfacial surface area in the as-received, banded structure is much lower than for uniformly dispersed Widmanstätten γ in ferrite. Interphase grain boundary sliding would be expected, therefore, to be more prevalent for the Widmanstätten structure with its higher α/γ interfacial area per unit volume (7).

Restoration in Ferrite and Austenite

The microstructural observations in the present work indicated that the rates of recovery and recrystallization of ferrite and austenite were different in the as-received and Widmanstätten structures. Both structures showed ferrite recovery by the presence of well-defined ferrite subgrains at low strain ($\varepsilon \sim 0.15$) at all strain rates. There was some evidence of limited recovery in the stringer γ at low levels of strain, but only at high strains ($\varepsilon \gtrsim 0.8$) did the austenite show marked recovery with a well-defined substructure (Fig.3). No evidence of a similar, recovered substructure was observed in Widmanstätten γ at any level of strain or strain rate (Fig.4). It can be seen in Fig.4 that γ is heavily dislocated while the ferrite has undergone rearrangement of dislocations to a well-defined substructure. These observations indicate that restoration processes in ferrite were far in advance of those in γ, consistent with a greater resistance of austenite to hot working.

In addition to the difference in recovery rate in the two structures, recrystallization of austenite was also found to be structure dependent. Svensson (8) reported that static recrystallization of γ occurred within one

Fig.3. Electron micrograph of as-received alloy showing recovered γ; $\varepsilon = 0.8$, $\dot{\varepsilon} = 0.04s^{-1}$

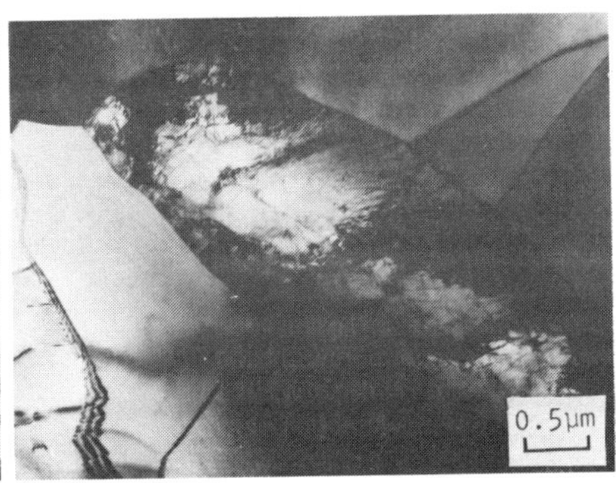

Fig.4. Electron micrograph of Widmanstätten alloy showing a heavily dislocated γ plate surrounded by polygonized α; $\varepsilon = 0.8$, $\dot{\varepsilon} = 4s^{-1}$.

second after deformation at high temperatures in single phase austenitic stainless steel. The time of ∿ 3s involved in quenching the sample immediately after deformation in the present work suggests that austenite is likely to recrystallize statically in the α/γ alloy during the quenching period. In the Widmanstätten microstructure a small number of recrystallized austenite plates was present in a sample strained to a true strain of 0.8 at a high strain rate of $4s^{-1}$, but there was no evidence of recrystallization at low and intermediate strain rates (.004 and $0.4s^{-1}$). It can be seen in the electron micrograph of Fig.5 that recrystallized grains of γ contained twins and few dislocations. It seems likely that these recrystallized grains formed statically, not only because of the 3s quenching time, but also because their features are inconsistent with the irregularly shaped grain boundaries, the absence of twins and the dislocated substructure which characterise dynamic recrystallization of austenite (9-11).

Fig.5. Electron micrograph of Widmanstätten structure showing statically recrystallized γ; $\varepsilon = 0.8$, $\dot{\varepsilon} = 4s^{-1}$.

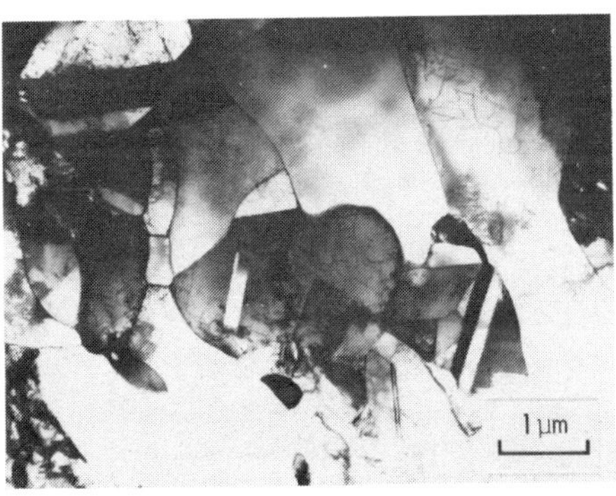

Fig.6. Electron micrograph of as-received alloy showing regions of statically recrystallized γ; $\varepsilon = 0.8$, $\dot{\varepsilon} = 4s^{-1}$.

The stringer austenite structure showed evidence of static recrystallization even at low strain rates and γ was extensively recrystallized when strain rates were high (Fig.6). The difference in the recrystallization response of the two microstructures can be ascribed to different amounts of stored energy which provide the driving force available for recrystallization. Since a critical strain is required for static recrystallization to occur (11), it is concluded that the stringer austenite attains this strain more readily under hot working conditions and undergoes recrystallization more readily than Widmanstätten austenite. Fractional softening experiments (12) also confirmed that static restoration processes were much faster for the stringer structure.

The higher flow stresses in the stringer austenite structure can be attributed to more effective strain transference to the γ (higher ε(γ)) which has an intrinsically higher hot strength than the α. The structural observations of the present work indicate that dislocation density progressively builds up in the austenite, until the strain in the austenite reaches a critical value at which significant dynamic recovery occurs. For the stringer austenite, the retained dislocation density at the end of compression testing was invariably higher than for the Widmanstätten austenite, leading to more rapid static recrystallization.

CONCLUSIONS

The morphology of the γ has a marked effect on the hot flow stress and strain to peak stress; being higher in the stringer structure. Static recrystallization occurred more readily in the stringer, as compared to the Widmanstätten, γ. The hot-worked α showed a well developed subgrain structure for both morphologies.

REFERENCES

1. H.D. Solomon and T.M. Devine, Jr., in *Proc. Conference on Duplex Stainless Steels*, ASM (edited by R.A. Lula), (1983).
2. Y. Maehara, M. Koike, N. Fusino, T. Kunitake, *Trans. Iron & Steel Inst.*, Japan, 23, 240 (1983).
3. Y. Ohmori and Y. Maehara, *ibid.*, 24, 63 (1984).
4. F.E. Al-Jouni and C.M. Sellars, in *4th Riso Int. Symp. on Met. and Mat. Sci.*, Roskilde, Denmark (edited by J.B. Bilde-Sorenson) (1983).
5. T. Chandra, D. Bendeich and D.P. Dunne, in *Proc. Int. Conference on Strength of Metals and Alloys*, Melbourne, Australia (edited R.C. Gifkins) (1982).
6. H. Unkel, *J. Inst. Met.*, 2, 171 (1937).
7. R.N. Stevens, *Met. Rev.*, 11, 129 (1966).
8. U. Svensson, *Hot Deformation of Austenite*, AIME (edited by J.B. Ballance) (1976).
9. G.J. Richardson, C.M. Sellars and W.J. McG. Tegart, *Acta Met.*, 14, 1225 (1966).
10. M.J. Luton and C.M. Sellars, *Acta Met.*, 17, 1033 (1969).
11. P. Sah, G.J. Richardson and C.M. Sellars, *Met. Science*, 8, 935 (1974).
12. P. Campbell, T. Chandra and D.P. Dunne, Unpublished Report, University of Wollongong (1985).

Variation in the Spacing of Dislocations in Subgrain Boundaries with Creep Strain in Type 304 Stainless Steel

M. E. Kassner* **and J. W. Elmer****

**Department of Mechanical Engineering, Naval Postgraduate School, Monterey, CA 93943, USA (on leave from Lawrence Livermore National Laboratory, Livermore, CA 94550, USA)*
***Department of Materials Science and Engineering, Massachusetts Institute of Technology, Cambridge, MA 02139, USA (on leave from Lawrence Livermore National Laboratory, Livermore, CA 94550, USA)*

ABSTRACT

This investigation examined the effect of varying strain upon the spacing of dislocations in subgrain boundaries in Type 304 stainless steel torsionally deformed at 1138 K (0.63 T_m). It was found that the spacing, d, decreased with strain during both transient (primary) and steady-state (secondary) creep. It also appeared that the average subgrain size, λ, decreased with strain during steady state. The results suggested that these two features may not strongly influence creep strength.

KEYWORDS

Transient creep; steady-state creep; subgrain boundary dislocations; misorientation angle; subgrains; forest dislocations; high-temperature strength.

INTRODUCTION

Some general creep theories [1-4] have suggested that the rate-controlling process is associated with subgrain boundaries and that the creep rate is influenced by the spacing of subgrain boundaries (or subgrain size, λ) and/or the spacing, d, of the dislocations that form the boundaries. Recent experimental work [5-8], mostly on Al, seems to be in agreement with this conclusion. In contrast, one of the authors [9,10] recently examined the dependence of transient (stage I) creep stress on λ and the density of dislocations not associated with subgrain boundaries, ρ_I, in 304 stainless steel (d was not examined) and found that ρ_I seems to control the creep rate. This finding seems conclusive at 1023 K and $\dot{\varepsilon} = 5.56 \times 10^{-4}$ s^{-1} where creep conditions are, perhaps, best described as a transition from power law to power-law breakdown ($n \simeq 7$). The analysis of data at 1138 K (0.63 T_m) and $\dot{\varepsilon} = 3.22 \times 10^{-5}$ s^{-1} (power law where $n \simeq 5$) also suggested that dislocation strengthening dominates, but was less conclusive. The aim of the present investigation was to more definitely determine the microstructural feature (λ, ρ_l or d) responsible for creep strength in 304 at 1138 K.

There were two experimental objectives. The primary purpose was to determine the variation of the spacing, d, of dislocations forming subgrain boundaries with transient and steady-state (stage II) creep strain. This would also help complete the analysis of creep microstructures in 304 stainless steel at 1138 K. There was particular interest in learning the dependence of the spacing on the steady-state strain. The material does not experience hardening during steady state. Therefore, if d influences the creep strength, it would not be expected to change during steady state. Second, values of ρ_l and λ over a range of steady-state strain might help determine the important microstructural features that control creep if it is found that d does not influence creep rate. The feature (λ or ρ_l) that is responsible for strength would be invariant during steady state.

Although few studies have determined the dependence of d on strain [11], several studies investigated the dependence of the subgrain misorientation angle, θ, on strain (for Al, Fe, Cu, and Sn [12-17]). The studies provide insight into the d-versus-strain behavior since it is expected that θ and d are related. It appeared from these studies that the misorientation angle increased with increasing strain during transient creep. Therefore, it is possible that the hardening might be related to increasing θ (decreasing d). The results for steady-state creep were inconclusive. While some of the results suggested that the angle across the subgrain boundaries continued to increase with strain (d decreases) during steady-state creep, others suggested that the average misorientation angle remains constant. Part of the difficulty in discerning a trend was that mechanical testing in the investigations was usually performed in tension and failure precluded testing to large strains.

The strain range of steady-state deformation was increased in the present study by performing creep tests in torsion. It was believed, therefore, that a more definite trend in the dependence of d on steady-state strain could be obtained.

EXPERIMENTAL PROCEDURE

Solid torsion specimens of 25.4-mm gage length and 6-mm diameter were machined from 25.4-mm-thick 304 stainless steel plate and annealed in vacuum at 1323 K for one hour. Torsional deformation was performed on the Stanford torsion machine [18]. The torsion specimens were deformed to various transient and steady-state strains and water-quenched. Transmission electron microscopy (TEM) disks were spark-cut from the 3/4-radius position of the specimens and electrochemically polished on a Fischione jet polishing unit. After 0.92 strain (end of stage II creep), voids that presumably formed at grain boundaries generally precluded electrochemical preparation of thin foils. For this reason, foils from one of the two specimens deformed to 0.92 strain required thinning by ion beam milling on a Gatan 600. The foils were then examined on a Jeol 200 CX with a biaxial holder (Jeol BST). Subgrain sizes were determined by a line-intercept method and the dislocation densities were determined by a surface-intersection technique. The procedures are described in detail in Ref. [9].

The average spacing of the dislocations composing subgrain boundaries was determined by examining boundaries at high (100,000X) magnification. Twenty subgrain boundaries were examined in each torsion specimen. It appeared that the boundaries almost always consisted of 3 sets of dislocations with different co-planar Burgers vectors. The dislocations seemed to be arranged in a nodal array (see Fig. 1a). To measure the separation between a set of dislocations (e.g., A dislocations in Fig. 1b), one set of dislocations (e.g., B dislocations) was made invisible by operating the microscope under two beam conditions such that $\vec{g} \cdot \vec{b}_b = 0$. In this study, $\vec{g}$ was always a <111>. About half of the boundaries were photographed from a <110> zone axis and half from a <211> zone axis. Those zones that appeared to maximize the projected separation of the dislocations in contrast were chosen. The image produced under the two-beam conditions consisted of parallel sets of "saw teeth" (A and C dislocations). The average separation, m, of these sets was measured. As shown in Fig. 1b, the projected separation of A (or B) dislocations, d, is 1.15 m if a hexagonal array is assumed. The values of d reported in this study utilized this correction. The values of d are an accurate measure of dislocation separation if the plane of the boundary is parallel to the photographic plane (or if the axis of rotation from the boundary plane to the photographic plane is parallel to m). Since the boundary planes examined were probably misoriented to the photographic plane and the axis of misorientation was not parallel to m, the projected separation of the boundary dislocations was probably less than the actual separation. Since neither the orientations of the boundaries nor m were determined, it was difficult to estimate the systematic underestimation of d. However, other stainless steel studies have assumed that subgrain boundaries lie on {111} planes [19,20]. The <211> and <110> directions make a minimum angle of about 19° and 35°, respectively, with a <111> direction. Therefore, there might be a systematic underestimation of d of about (cos 19° + cos 35°)/2 ≈ 11 percent. The "projection" error is reduced from this value if some component of the <111> to <211> or <110> rotation is parallel to m.

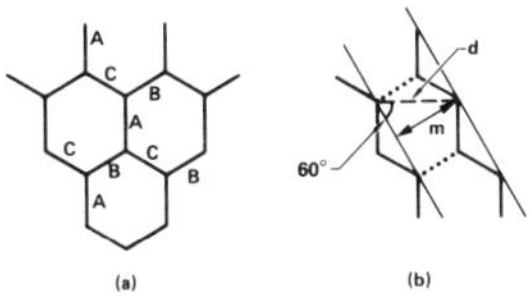

Fig. 1. (a) A nodal array of 3 dislocation sets that form a subgrain boundary; (b) a subgrain boundary image when the TEM is operating under two-beam conditions such that $\vec{g} \cdot \vec{b}_b = 0$.

RESULTS AND DISCUSSION

Specimens were torsionally deformed at 1138 K and an equivalent uniaxial strain rate of 3.22×10^{-5} s^{-1} to strains of 0.009, 0.027, 0.07, 0.15, 0.20, 0.30, 0.38 and 0.92. The equivalent uniaxial flow stress versus equivalent uniaxial strain (based on the usual von Mises conversion) is shown in Fig. 2d. This figure shows that the transient creep range extends to about 0.38 strain. Steady state extends from a strain of 0.38 to about 0.92. The results of the TEM examination of the creep microstructure are shown in Figs. 2a, b, and c. Figure 2a shows the variation in the average spacing of subgrain dislocations with strain. The separation between the subgrain dislocations decreases with strain over the transient range. This trend, as is the trend that the second derivative of d with respect to strain is decreasing, is in agreement with the misorientation angle investigations cited earlier. More important, it appears that d continues to decrease with strain during steady state. Average d values appear to

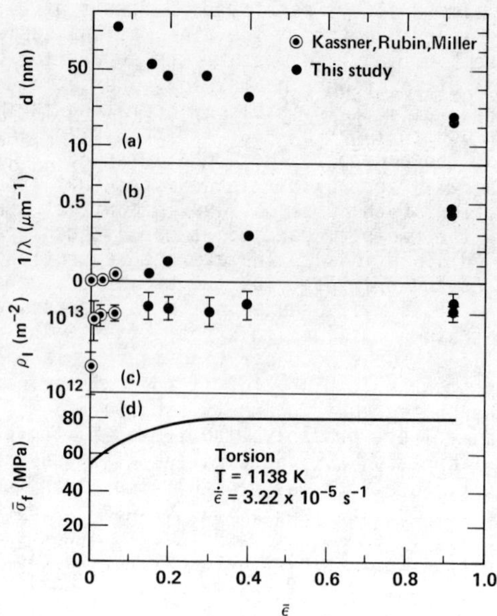

Fig. 2. (a) The variation of the spacing of dislocations in a subgrain boundary; (b) the reciprocal subgrain diameter; (c) the density of dislocations not associated with subgrain boundaries versus strain; (d) the stress versus strain behavior of 304 stainless steel torsionally deformed at 1138 K and an equivalent uniaxial strain rate of 3.22×10^{-5} s^{-1}.

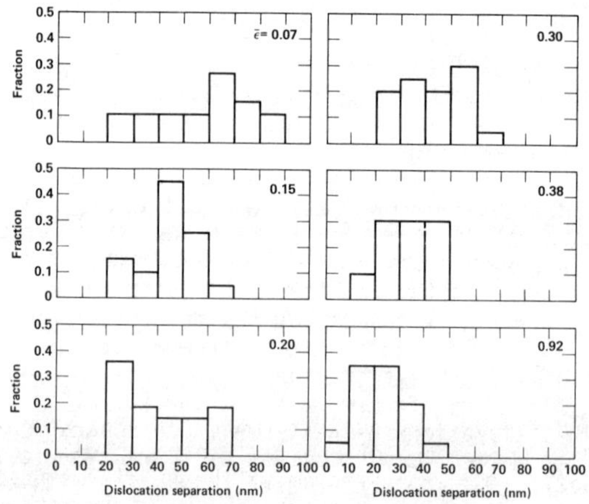

Fig. 3. Histograms illustrating the distribution of observed spacings of dislocations in subgrains, d, at each strain level.

decrease from about 36 nm at 0.38 strain to about 26 nm at 0.92 strain. Although this conclusion is based on limited data, the value for d after 0.38 strain is consistent with the trend of the other transient data and the two d values determined from the specimens deformed to 0.92 strain are in close agreement. If $d \approx \vec{b}/\theta$ is assumed [21], then the misorientation angle for the specimen deformed to 0.38 strain would be about 0.31°. This is in reasonable agreement with the value of 0.40° found by Challenger and Moteff [19] using the Kikuchi shift method for 316 stainless steel that was deformed to steady state at a strain rate/temperature combination comparable to the present study. The dislocation separation values of this study are also in reasonable agreement with the stainless steel creep study by Pahutova et al. [22] that reported separation values in the 20-30 nm range. The variation in dislocation spacing in the boundaries of specimens deformed to a given strain is shown in Fig. 3. The typical distribution illustrated has a standard deviation of about d/3.

As stated earlier, the subgrain boundaries almost always consisted of 3 sets of dislocations, each with, apparently, a different co-planar a/2 <110> Burgers vector. Although the character of the boundaries was not extensively analyzed in this study, a few boundaries were examined. The dislocation-line/Burgers vector orientation relationships could be conveniently examined by assuming that the boundaries lie on {111} planes. It appeared that most areas of the boundaries consisted of dislocations that were nearly screw while in smaller areas the dislocations appeared to have about equal screw-edge character. Analyzing the nature of the subgrain boundaries may be complicated, however, since it was found, from the one boundary plane for which an orientation was determined, that the boundary was oriented about 17° from the {111} plane.

It can be seen in Fig. 2b that $1/\lambda$ not only increases over the transient strain range, as expected, but continues to increase over steady state. In fact, $1/\lambda$ increases by about 50 percent from 0.38 to 0.92 strain. It has been generally found [23] for metals and subgrain-forming alloys that $\sigma_{ss} = k(1/\lambda)$, where σ_{ss} is the steady-state stress and k is a constant. This phenomenological relation predicts that a 50 percent increase in $1/\lambda$ should be accompanied by a 50 percent increase in flow stress. Therefore, based on Figs. 2a and b, it appears that the strength of 304 stainless steel at 1138 K is not strongly related to the spacing of subgrain boundaries (subgrain size) or the spacing of the dislocations that compose subgrain boundaries. It can be seen from Fig. 2c, however, that the density of dislocations not associated with subgrain boundaries, ρ_L, seems to increase slightly over the transient range and remains essentially constant over the steady state. It appears, then, that the density of random (or forest) dislocations might control the creep strength. This conclusion is consistent with the observation from Fig. 2 that the material undergoes a significant fraction (about 25 percent) of the possible hardening before subgrain boundaries are even observed. The conclusion is also consistent with earlier work by one of the authors [9,10].

ACKNOWLEDGEMENT

This work was supported by Lawrence Livermore National Laboratory under the auspices of the U.S. Department of Energy, Contract W-7405-Eng-48, and by the ONR/NPS Foundation Research Program at the Naval Postgraduate School, Monterey, California. The authors appreciate the help with the electron microscopy that was provided by C. J. Echer and W. L. Bell.

REFERENCES

1. J. Weertman, in Proc. 2nd Intern. Conf. on Creep and Fracture of Engineering Materials and Structures, pp. 1-13. Pineridge (1984).
2. W. Blum, Phys. Stat. Sol. 45, 561 (1971).
3. L.I. Ivanov and V.A. Yanushkevich, Phys. Metals and Metallography 17, 561 (1964).
4. K. Maruyama, S. Karashima, and H. Oikawa, Res. Mechanica 7, 21-36 (1983).
5. L. Bendersky, A. Rosen, and A.K. Mukherjee, in Strength of Metals and Alloys, vol. 2, pp. 595-601. Pergamon Press, Oxford (1982).
6. I. Ferriera and R.G. Stang, Acta Metall. 31, 585-590 (1983).
7. M.S. Soliman, T.J. Ginter, and F.A. Mohamed, Phil. Mag. 48, 63-81 (1983).
8. D. Caillard and J.L. Martin, Acta Metall. 31, 813-825 (1983).
9. M.E. Kassner, A.K. Miller, and O.D. Sherby, Metall. Trans. 13A, 1977-1986 (1982).
10. M.E. Kassner, K.A. Rubin, and A.K. Miller, in Strength of Metals and Alloys, vol. 2, pp. 581-587. Pergamon Press, Oxford (1982).
11. W. Blum, A. Absenger, and R. Feilhauer, in Strength of Metals and Alloys, vol. 3, pp. 271-276. Pergamon Press, Oxford (1979).
12. A. Orlova, M. Pahutova, and J. Cadek, Phil Mag. 25, 865-877 (1972).
13. S. Karashima, H. Oikawa, and T. Hasegawa, J. Japan. Inst. Metals 31, 782-787 (1977).
14. A. Orlova, Z. Tobolova, and J. Cadek, Phil. Mag. 26, 1263-1274 (1972).
15. S.H. Suh, J.B. Cohen, and J. Weertman, Scripta Metall. 15, 517-522 (1981).
16. J.J. Jonas, D.R. Axelrad, and J.L. Uvira, Trans. Japan. Inst. Metals 9 suppl., 257-268 (1968).
17. S. Karashima, Iikubo, Wanatabe, and H. Oikawa, Trans. Japan. Inst. Metals 12, 369 (1971).
18. J.L. Robbins, Dissertation, Stanford Univ., Stanford, CA (1964).
19. K.D. Challenger and J. Moteff, Metall. Trans. 4, 749-755 (1973).
20. O. Ajaja and A.J. Ardell, U.S. ERDA Project, Agreement No. 177 under Contract E(04-3) 34 (1977).
21. D. Hull, Introduction to Dislocations, 2nd ed., p. 206. Pergamon Press, Oxford (1975).
22. M. Pahutova, J. Cadek, and V. Cerny, Mater. Sci. Eng. 32, 33-41 (1984).
23. A.K. Mukherjee, in Treatise on Materials Science and Technology, vol. 6, pp. 163-223. Academic Press (1975).

Evolution of Microstructure During Cold Rolling

C. S. Hartley and M. Dehghani

Department of Mechanical Engineering, Louisiana State University, Baton Rouge, LA 70803, USA

ABSTRACT

Microstructural anisotropy developed during cold rolling of mild steel is described in terms of a second rank tensor whose principal axes define the principal axes of the microstructure. The principal values of this tensor are related to the projected grain boundary area on planes normal to the principal axes. The width, rolling and thickness directions are principal directions for the material studied. Changes in the principal values of the material anisotropy tensor in the rolling and thickness directions reflect the progressive elongation and flattening of grains during the process.

KEYWORDS

Microstructure, anisotropy, cold rolling, stereology, grain boundaries

INTRODUCTION

Oriented structures developed during deformation processing are generally described by the number of intersections per unit length of a test line with grain boundaries, P_L, as a function of the orientation of the test line with respect to reference directions in the material. Several methods have been proposed to characterize the results of such measurements. A polar plot of P_L vs. the angle between the test line and a reference direction in the plane of observation, θ, forms the rose of the number of intersections (1,2). The shape of this curve is related to the number and orientation of axes of orientation of the grain boundary network. Saltykov's method of calculating the surface area per unit volume of oriented grains decomposes the structure into components due to isotropic, linear and planar elements. The surface area due to each type of feature is then related to the number of intercepts in the appropriate direction. Hilliard proposes a more general analysis which introduces the concept of a distribution function describing the fraction the total surface area per unit volume oriented normal to a general direction (3).

For an oriented, single phase microstructure the number of grain boundary intercepts per unit length of test line is equal to the grain boundary surface area per unit volume projected onto a plane normal to the test line (1). The reciprocal of this quantity is the mean linear intercept, L, in the direction of the test line. For many oriented structures a polar plot of L *vs*. the orientation of the test line with respect to a reference direction in a plane of observation forms an ellipse with principal axes parallel to the axes of orientation of the grains in the plane (4,5). This suggests that the three-dimensional structure can be described by a second order surface of the form

$$M_{ij}\, x_i\, x_j = 1, \qquad (1)$$

where the coefficients, [M], constitute a symmetric, second rank tensor. The principal directions of this tensor define the principal axes of the microstructure and the principal values correspond to the square of the number of intercepts along principal directions. The surface described by Equation 1 is the representation quadric of the tensor (6). Referred to its principal axes the surface has the equation

$$M_1\, x_1^2 + M_2\, x_2^2 + M_3\, x_3^2 = 1 \qquad (2)$$

which is always a real ellipsoid centered at the origin. The values of the principal semiaxes are equal to the mean linear intercepts in the principal directions and the length of a radius vector from the origin to the surface of the figure represents the mean linear intercept in the direction of the vector.

Plots of L vs. orientation with respect to a reference axis on three mutually perpendicular planes provide sufficient information for the determination of the components of [M]. The intersection of each plane with the representation quadric will be an ellipse centered at the origin. The resulting array of coefficients can then be diagonalized by standard techniques to yield principal values and principal directions for [M]. The following sections describe an investigation of the development of the microstructure of cold-rolled mild steel in terms of the material anisotropy tensor, [M], measured on the undeformed and deformed metal.

PROCEDURE

Specimens of annealed mild steel containing 0.05% C were rolled at room temperature to total reductions in thickness of 31.9%, 46.1% and 59.5% by reducing the specimens 15% per pass. Specimens for measurement of P_L were taken from the center of the rolled sheet in an effort to minimize edge and end effects. Mounted samples were prepared with planes of polish normal to the width (x_1), rolling (x_2) and thickness (x_3) directions for each condition of the metal including the undeformed strip. Samples were polished through one micron alumina abrasive and etched with 2% Nital.

Measurements of P_L were made from photomicrographs taken at 300X for the undeformed metal and the specimen with the lowest total deformation. A magnification of 600X was employed for the more heavily deformed material. Counts of P_L along test lines oriented at 30 degree intervals from the reference directions in each plane were made. Only ferrite/ferrite boundaries were counted. Sufficient counts were taken to provide a coefficient of variance,

$$C.V. = S.D./\bar{P}\sqrt{n}, \quad (3)$$

less than 4%. In equation 3 the superscript bar refers to the average value of P_L, S.D. is the standard deviation and n is the number of measurements.

For each plane of observation data were plotted as L vs. θ and a least squares fit of the form

$$P_L^2 = 1/L^2 = A\cos^2\theta + B\sin^2\theta + C\sin\theta\cos\theta \quad (4)$$

was obtained using a SAS general least squares program. Two corrections were applied to account for inhomogenieties of scale in the microstructures sampled by the photomicrographs. First, diagonal terms of the tensor measured on different faces did not generally coincide. This arises from the fact that the sampling procedure employed precluded any two orthogonal faces from bounding the same volume. Therefore, microscopic inhomogenieties in the scale, i.e. the "average size", of the grain structure, would cause different values to be obtained for the number of intercepts along a particular direction depending on the orientation of the plane of observation. This apparent inconsistency can be removed by using a technique developed by Harrigan and Mann (5) in which scale factors are determined for each plane of measurement such that the product of the scale factor and the value of a diagonal component of the tensor is equal to the product of the corresponding tensor component measured on an orthogonal plane and the scale factor for that plane. This procedure determines the coefficients to within a multiplicative constant, which can be determined by other assumptions or conditions. Harrigan and Mann choose the constant so that the sum of the diagonal elements of [M] is unity. This precludes determination of the absolute values of the components, but preserves the principal directions and the relative values.

In the present work the additional assumption is made that the plane strain conditions which prevail at the center of the strip during rolling result in no deformation of the grain boundary network in the width direction of the strip (x_1). Therefore the scale factors for the rolled specimens are chosen so that the value of M_{11} remains constant with deformation and equal to that for the undeformed material. In addition, to eliminate errors of scale in the undeformed material, the scale factors for the 1-2 and 1-3 faces were chosen so that the value of M_{11} for the undeformed material is the geometric mean of the values measured on the two faces. Values of [M] obtained by this procedure were used to calculate principal values and principal directions for the deformed and undeformed material.

RESULTS

Principal values and principal directions of the microstructures studied are given in Table 1. Principal directions are expressed in terms of the direction cosines of the directions with respect to the coordinate system based on the geometry of the strip. While some microstructural anisotropy is present in the undeformed material, it is not correlated with the direction of subsequent deformation. Microstructures of the deformed specimens exhibit principal directions nearly parallel to the width (x_1), rolling (x_2) and thickness (x_3) directions of the strip. The variation of approximately 5 degrees is considered to be within the experimental error in preparing the specimens.

TABLE 1 Principal Values and Directions for [M]

Specimen Number	Principal Values*	Principal Directions**		
		$\cos\gamma_1$	$\cos\gamma_2$	$\cos\gamma_3$
0	807.7	0.706	-0.616	0.350
	658.9	0.647	0.358	-0.673
	818.2	0.290	0.701	0.651
2	752.7	0.984	-0.168	-0.066
	415.6	0.165	0.985	-0.057
	1256	0.074	0.045	0.996
3	749.1	0.997	0.078	-0.066
	281.8	-0.078	0.997	-0.057
	1276	0.005	-0.003	0.996
4	747.9	0.998	0.056	-0.009
	263.2	-0.056	0.995	-0.087
	1988	0.004	0.087	0.996

* in units of mm^{-2}

** γ_i is the angle between the principal direction and the x_i axis

The properties of the representation quadric require that the mean linear intercepts in the principal directions be equal to the reciprocal of the square root of the corresponding principal value. This provides a method of determining unambiguously the axes of orientation for the structure. The principal mean linear intercepts can then be used in Saltykov's treatment to describe the evolution of the grain boundary network in terms of isotropic, linear and planar elements. Results of these calculations are given in Table 2, where the linear orientation axis has been chosen to coincide with the principal direction corresponding to the largest value of mean linear intercept, i.e. the rolling direction.

TABLE 2 Components of Grain Boundary Area*

Specimen Number	Isotropic	Planar	Linear	Total
0	51.34	0.184	4.321	55.84
2	40.77	8.008	11.07	59.85
3	33.57	8.357	16.62	58.55
4	32.45	17.24	17.47	67.16

* in units of (mm^2/mm^3)

DISCUSSION

Residual microstructural anisotropy present in the undeformed material is due to its prior thermomechanical history, which was unavailable to the authors. No correlation of this residual anisotropy with the geometry of the strip can be expected. Rolling develops a pronounced anisotropy in the microstructure, having characteristic directions related to the geometry of the deformation process. Measurements of the type described in this work are necessary to verify the hypothesis that characteristic directions of the microstructure coincide with principal directions of the deformation process.

Decomposition of the grain boundary area according to Saltykov's method reveals that rolling decreases the isotropic component of the total grain boundary area while increasing the linear and planar components in the rolling direction and the plane of rolling, respectively. Planar components appear to increase relatively more rapidly than linear components, with both reaching comparable values at the highest deformation studied in this work, corresponding to a true plastic strain of -0.90 in the thickness direction. The total area remains substantially unchanged until the highest deformation.

CONCLUSIONS

Microstructural anisotropy developed during cold rolling of mild steel can be described in terms of a second rank tensor whose principal axes correspond to the characteristic directions of deformation. The principal values of this tensor are related to the mean linear intercepts in the principal directions of microstructural anisotropy.

Rolling decreases the isotropic component of the grain boundary area and increases the linear and planar components with the planar component increasing at a relatively more rapid rate with deformation.

ACKNOWLEDGEMENTS

This work is based on research sponsored by the National Science Foundation under Grant No. 84-00991, "Deformation Processing by Hydrostatic Coextrusion".

REFERENCES

1. R. T. DeHoff and F. N. Rhines, Quantitative Metallography, McGraw-Hill, New York, NY (1968).

2. E. E. Underwood, Quantitative Stereology, Addison-Wesley, Reading, MA (1970).

3. J. E. Hilliard, Trans. Am. Inst. Min. Engrs., 224, 1201 (1962).

4. W. J. Whitehouse, J. Microscopy, 101, 153 (1974).

5. T. P. Harrigan and R. W. Mann, J. Materials Sci., 19, 761 (1984).

6. J. F. Nye, Physical Properties of Crystals, Clarendon Press, Oxford (1957).

Obtention d'Alliages d'Aluminium a Microstructure Composite: Ecrouie-Recristallisee par Traitement par Faisceaux Laser ou d'Electrons

J. Merlin, J. Dietz, J. P. Massoud*, D. Brenet et G. Coquerelle*****

**GEMPPM (La 341), INSA, 69621 Villeurbanne Cedex, France*
***Dept Métallurgie, C.E.N.G., 38041 Grenoble, France*
****CMCM, E.T.C.A., Arcueil, France*

ABSTRACT

Interesting directionalized properties may be obtained by realization of "composite alloys", compound of recrystalized and cold-worked strings. This requires a local recrystalization by laser or electron beam treatments. So, at first the localized recrystalization by laser (FL) or electron (FE) beams of an high purity cold-rolled aluminium have been investigated to optimize process parameters. Afterwards some aluminium sheets have been treated with strings at an angle of 0 or 90° with the cold-rolling direction (DL), and tensile samples have been cut at 0,45 and 90° with respect to this direction. The mechanical properties have been investigated and compared to those of homogeneously restaured or partialy recrystalized sheets. To some extent the characteristics are similar to those of partialy recrystalized sheets (high strength, suitable ductility), however anisotropy properties are clearly different and show the composite character of the product. The feasability of these treatments is now proved, but it would be necessary to carry on them on industrial alloys in order to evaluate their actual performances.

MOTS CLES

Aluminium; Recristallisation; Traitement thermique; Faisceau Laser; Faisceau d'Electrons; Composite; Propriétés Mécaniques; Anisotropie.

INTRODUCTION

La modification de la résistance mécanique d'alliages d'aluminium par traitement par Faisceaux Laser (FL) ou d'Electrons (FE) a été envisagée, soit par un durcissement superficiel résultant de l'onde de choc développée par un pulse laser (1-2), soit par une refusion superficielle par FL ou FE d'alliages afin d'affiner les microstructures (3-5). Nous

----

x Ce travail a été entrepris dans le cadre des activités du Centre d'Applications des Lasers de Forte Energie à la Transformation des MATériaux (CALFETMAT)(Lyon).

estimons que des traitements localisés en phase solide sont également susceptibles de modifier le comportement mécanique de certains de ces alliages. En particulier la recristallisation localisée de matériaux écrouis devrait permettre de moduler et surtout de directionnaliser les propriétés en fonction des opérations de formage à froid ou des sollicitations des piéces en cours d'utilisation.
Afin de tester le bien fondé de ces hypothèses nous avons donc choisi un matériau ayant un comportement aussi simple et aussi typique que possible vis-à-vis des phénomènes de restauration-recristallisation, à savoir de l'aluminium de haute pureté (99,99 %) qui nous a été fourni par le C.R.V. (Groupe Cégédur-Péchiney) sous forme de tôles de 1 ou 2 mm d'épaisseur laminées à froid avec un taux de réduction de l'ordre de 10.

OBTENTION D'UNE RECRISTALLISATION LOCALISEE

Pour atteindre l'objectif fixé il est nécessaire de trouver les conditions de traitement susceptibles de conduire à une zone recristallisée de l'ordre du millimètre de large et sur une profondeur la plus importante possible, sans qu'il y ait de fusion superficielle.
Nous avons pu disposer pour effectuer cette étude d'une source laser CO_2 continue de 5 kW (Spectra Physics) à l'ETCA (Arcueil) et d'un canon à électrons de 15 kW (Siacky) au C.E.N. (Grenoble). Quelle que soit la technique (FL ou FE) il apparait que l'on a des caractéristiques de zone recristallisée satisfaisantes avec :
- une densité de puissance communiquée à l'aluminium d'environ 500 W/mm^2
- un temps d'interaction d'environ 5.10^{-2} s (ce qui correspond à des vitesses de déplaçement relatif faisceau-pièce de quelques cm/s si le diamètre du faisceau est de l'ordre de 1 mm).

L'état de surface n'a pas une très grande importance, cependant un dépot préalable de graphite est indispensable dans le cas du traitement laser.

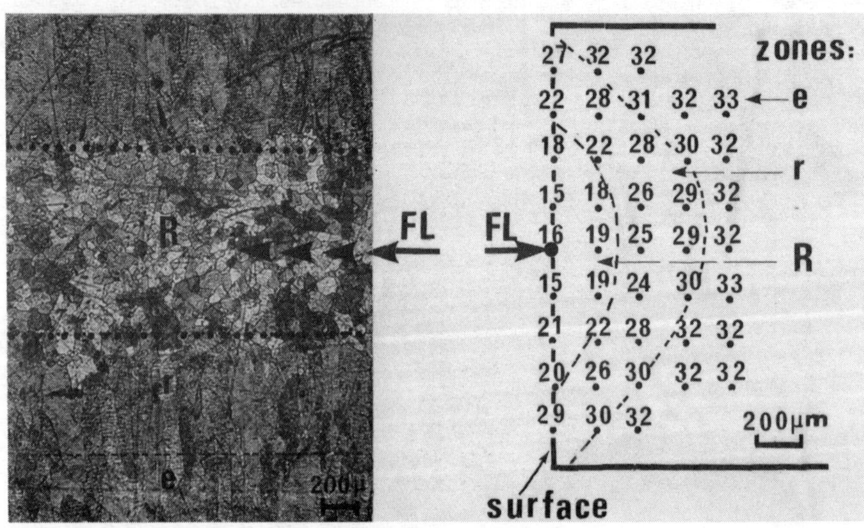

Fig. 1 : Vue de dessus d'un cordon recristallisé par Faisceau Laser (FL). P = 2kW, 0 faisceau 2mm, v = 200mm/s.

Fig. 2 : Evaluation par microdureté de l'étendue du traitement sur une section. (HV_{100}).

Zones : recristalisée "R", restaurée "r", où l'écrouissage persiste "e".

Les zones recristallisées sont alors constituées de grains isotropes de l'ordre de 50 μm (voir fig. 1), à comparer aux grains initiaux très allongés (≃ 1000 x 200 x 40 μm) et aux grains d'environ 100 μm qui peuvent être obtenus sur le même aluminium après traitement "flash" de 30 s à 400°C. L'étendue des domaines recristallisés et restaurés a été évaluée par mesure de micro-dureté Vickers, un exemple est montré fig. 2.
Nous avons en outre observé que la largeur du cordon recristallisé était systématiquement plus élevée lorsque le faisceau se déplaçait perpendiculairement à la direction de laminage de la tôle et cela d'autant plus nettement que le temps d'interaction est faible. Une étude plus détaillée de ce phénomène a été entreprise (6).

MISE EN OEUVRE ET CARACTERISATION D'UNE TOLE à MICROSTRUCTURE COMPOSITE :

Cette étude préliminaire a donc permis le "fibrage" de tôles (de 200 x 150 x 1 mm) par réalisation sur les 2 faces de cordons recristallisés de 1 mm de large et 0,5 mm d'épaisseur avec un pas de 2 mm (voir sur la fig. 3 une coupe selon l'épaisseur d'une tôle ainsi traitée). Cette opération a pu être effectuée avec les 2 techniques FL et FE et les résultats mentionnés ci-après correspondent plus spécifiquement à l'utilisation du FE.
La fig. 4 montre l'aspect d'une tôle après "fibrage". On constate un gonflement dans les zones recristallisées (bien qu'il n'y ait pas eu fusion), gonflement qui s'est accompagné d'une contraction d'environ 6 % de la tôle dans la direction perpendiculaire au sens de déplacement du faisceau. Ce phénomène pourrait avoir son origine dans un relachement des contraintes dans la zone recristallisée.

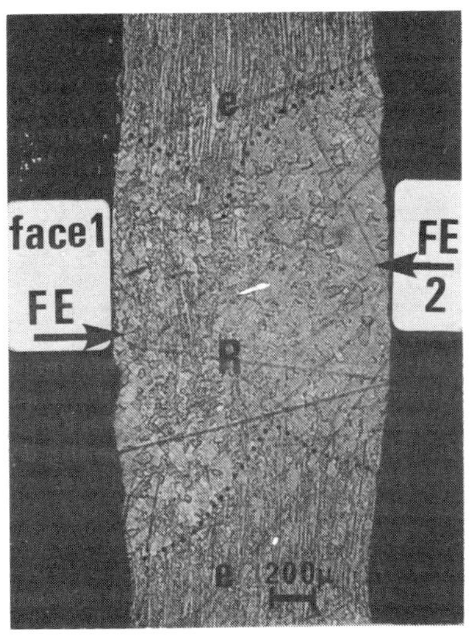

Fig. 3 : Vue en coupe d'un cordon obtenu par recristallisations successives sur les 2 faces.

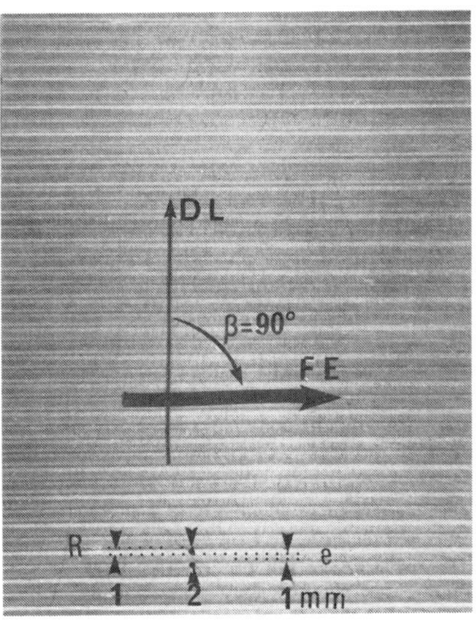

Fig. 4 : Aspect après "fibrage" selon une direction perpendiculaire à la direction de laminage (DL).

La tôle a été fibrée avec un angle β de 0 ou 90° par rapport à la direction de laminage (DL), voir le schéma de la fig. 5 et des éprouvettes de traction de 80 x 8 x 1 mm ont été ensuite prélevées dans la tôle selon un angle α par rapport à DL de 0,45 et 90°; cela a conduit aux 6 situations possibles schématisées fig. 6 selon les orientations respectives des "fibres écrouies" résultant du laminage et du "surfibrage" que constitue en fait les cordons recristallisés.
Après traction à 2 mm/mn les principales caractéristiques mécaniques (limite élastique σ_E, contrainte à la rupture σ_R, allongement à la striction Z et allongement à la rupture A) ont été évaluées selon les différentes directions de fibrage et de prélèvement, ainsi que le coefficient d'anisotropie de déformation r (les valeurs mentionnées étant celles à l'atteinte de la striction). Ces résultats ont été situés par rapport à ceux obtenus sur le même alliage mais traité classiquement et de manière homogène durant des temps variables à 300°C : 4 mn → état restauré (r_o), 10 mn → état 1/4 recristallisé (1/4 R), 20 mn → 1/2 recristallisé (1/2 R). L'ensemble des résultats obtenus est montré sur les fig. 7 a-c et 8 a-c.
En ce qui concerne les niveaux moyens des caractéristiques mécaniques, les tôles "fibrées" avec β = 0 ou 90° se comportent assez sensiblement comme des tôles partiellement recristallisées (caractéristiques de résistance élevées, ductilité acceptable). Cependant l'anisotropie de déformation est sensiblement plus élevée, particulièrement dans la direction de fibrage. Et d'une manière générale ce "fibrage" tend bien à faire disparaître la symétrie d'ordre 4 des propriétés au profit d'une directionnalisation dans le sens de défilement du faisceau ce qui est en accord avec le caractère "composite directionnel" du produit obtenu.
Le comportement des éprouvettes au cours de l'essai a également révélé un certain nombre de singularités qu'il sera bon de préciser :
- la déformation de l'éprouvette n'est pratiquement jamais homogène

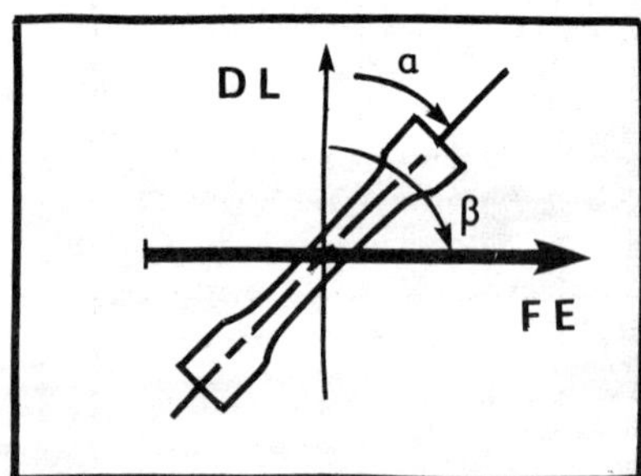

Fig. 5 : Repérage des éprouvettes de traction selon l'angle par rapport à la direction de laminage et selon l'angle de fibrage .

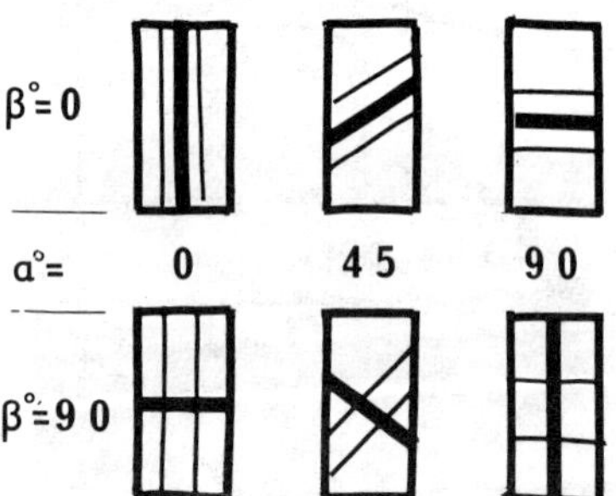

Fig. 6 : Différentes situations possibles pour les orientations respectives des "fibres d'écrouissage" et du " surfibrage" selon les valeurs de et .

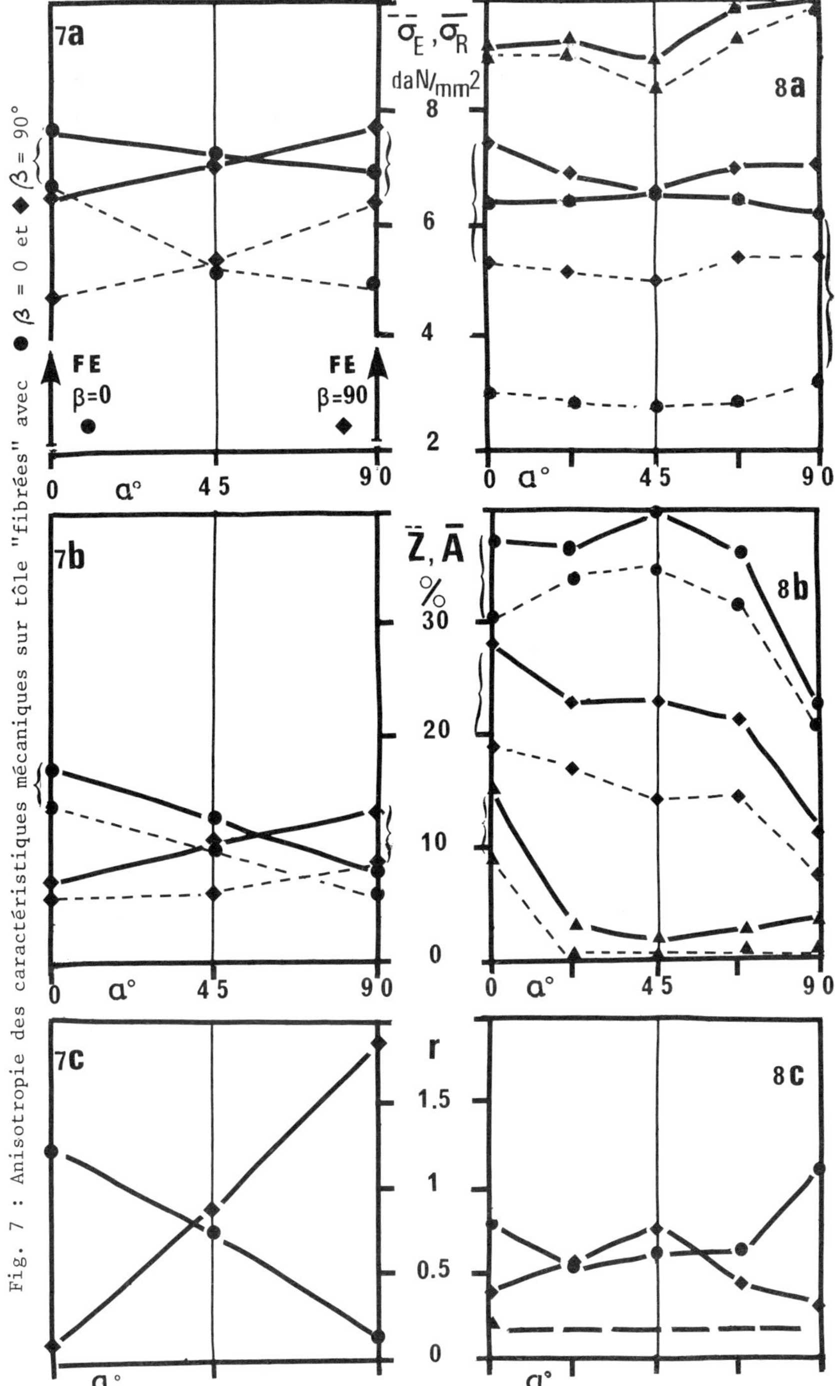

Fig. 7 : Anisotropie des caractéristiques mécaniques sur tôle "fibrées" avec ● β = 0 et ◆ β = 90°

Fig. 8 : Anisotropie des caractéristiques mécaniques sur tôle traitée de manière homogène ▲ r_o, ◆ 1/4 R, ● 1/2 R.

- même lorsque la déformation est franchement localisée la consolidation de l'éprouvette peut se poursuivre ce qui conduit à des strictions assez diffuses.
- lorsque l'axe de traction est confondu avec l'axe du "fibrage" l'allongement à la rupture important observé est associé à un coefficient r élevé ce qui traduit un écoulement dans les cordons recristallisés en quelque sorte "canalisé" par les zones écrouies, ce qui est totalement différent du glissement facile dans les grains (associé à un A élevé et à un r petit) observé en particulier dans les états restaurés dans la direction de laminage
- enfin quelques essais effectués à différentes vitesses de traction semblent montrer une très grande sensibilité du produit obtenu à ce paramètre ce qui n'est pas le cas de l'alliage traité de manière homogène.

CONCLUSION

Les différentes caractéristiques observées sur la tôle d'aluminium pur recristallisée localement soulignent le comportement de type "matériau composite" du produit obtenu. Un certain nombre de paramètres devraient cependant être optimisés et surtout cette étude devrait être étendue à des alliages plus susceptibles d'applications industrielles (alliages Al-Mn et Al-Mg en particulier) pour définir les performances réelles que l'on pourrait attendre de ce type de traitement. La faisabilité d'un tel traitement nous parait cependant d'ores et déjà démontrée, traitement qui pourrait donc en principe être inseré dans la gamme de transformation des alliages avant certaines opérations de formage (emboutissage, pliage) ou sur l'objet fini, afin soit de faciliter la mise en forme, soit d'utiliser une partie de l'énergie mécanique emmagasinée au préalable dans le matériau pour optimiser son comportement ultérieur selon les sollicitations auxquelles il sera soumis.

REFERENCES

1. A.H. Clauer and B.P. Fairant, Appl. of Lasers in Mater. Proc. (A.S.M.) p. 291. (1979).
2. W.F. Bates, Appl. of Lasers in Mater, Proc. (A.S.M.), p. 317. (1979).
3. D.S. Gnanamuthu, Appl. of Lasers in Mater. Proc. (A.S.M.), p. 177. (1979).
4. G. Nicolas, Proc. du 3ème C.I.S.F.E.L. (Lyon), p. 253. (1983).
5. B. Vinet and S. Paidassi, Proc. du 3ème C.I.S.F.E.L. (LYON), p. 245. (1983).
6. J. Dietz and J. Merlin (à paraître).

Theoretical Dependence of Limiting Drawing Ratio on Plastic Strain Ratio

Dong Nyung Lee

Department of Metallurgical Engineering, Seoul National University, Seoul Korea

ABSTRACT

A equation relating the limiting drawing ratio to the plastic strain ratio has been derived based on a generalized form of Hill's quadratic anisotropic yield criterion. The calculated results obtained by the equation agree very well with the experimental data when the exponent is modified to be much larger than 2 which is used in the Hill quadratic yield theory.

KEYWORDS

Limiting drawing ratio ; plastic strain ratio ; anisotropic yield criterion.

INTRODUCTION

The dependence of the limiting drawing ratio (LDR) on the plastic strain ratio or the R value has been well known. Increases of the LDR with the R value are predicted by several theoretical analyses of deep drawing (1,2). Whiteley employed simplifying assumptions of a nonhardening material, and plane strain deformation in the flange to show that

$$\ln(LDR) = \eta\beta \tag{1}$$

where η is a deformation efficiency to account for frictional and bending work and β is the ratio of two plane strain flow stresses corresponding to $\varepsilon_{22} = 0$ and $\varepsilon_{33} = 0$, i.e.,

$$\beta = \sigma_{wall(\varepsilon_{22} = 0,\ \varepsilon_{33} = 0)} / \sigma_{flange(\varepsilon_{33} = 0)} \tag{2}$$

where x_2 and x_3 are the circumferential and thickness directions, and σ_{wall} and σ_{flange} are the flow strengths of the wall and flange. He then used Hill's anisotropic theory (3), which for planar isotropy of the sheet predicts that

$$\beta = \sqrt{(R+1) / 2} \tag{3}$$

so that

$$LDR = \exp\left[\eta\sqrt{(R+1) / 2}\right] \tag{4}$$

There have been more rigorous analyses of deep drawing which allow for both work hardening and thinning or thickening of the flange. These also rely on the Hill theory to characterize the anisotropic behavior. All the theories mentioned above predict more dependence of LDR on R than experimentally observed. Hosford and Kim (2) have suggested that the major problem with these analyses lies in the use of the Hill theory. They calculated the R and β values for sheets of cubic metals based on assumptions of equal strains in all grains, slip restricted to {111}<110> or {110}<111> systems, and textures characterized by a single [hkl] sheet normal with rotational symmetry about that normal. The calculation indicated a much lower variation of β with the R value than that predicted by the Hill theory. Recent works (4,5) have indicated the the shape of yield loci for crystallographically textured fcc and bcc metals could be better represented by a generalization of the Hill's yield criterion of the form of Eq.(5).

$$af(\sigma_{ij}) = F|\sigma_{22}-\sigma_{33}|^a + G|\sigma_{33}-\sigma_{11}|^a + H|\sigma_{11}-\sigma_{22}|^a = 1 \qquad (5)$$

where the exponent a is much larger than the 2 in the Hill's thoery(4) and F, G, and H are constants which characterize the anisotropy. The purpose of this paper is to obtain the expression for the LDR based on Eq.(5)

LIMITING DRAWING RATIO VS. PLASTIC STRAIN RATIO

The constants F, G, and H in Eq.(5) may be evaluated from simple tension tests. Let $\sigma_{y(1)}$ be the tensile yield stress in the x_1 direction. At yielding, $\sigma_{11} = \sigma_{y(1)}$, $\sigma_{22} = \sigma_{33} = 0$, so Eq.(5) becomes $(G+H)\sigma^a_{y(1)} = 1$ or $\sigma^a_{y(1)} = 1/(G+H)$. Similarly, if $\sigma_{y(2)}$ and $\sigma_{y(3)}$ are the tensile yield stresses in the x_2 and x_3 directions, the following relations are obtained.

$$\begin{aligned} \sigma^a_{y(1)} &= 1 / (G+H) \\ \sigma^a_{y(2)} &= 1 / (H+F) \\ \sigma^a_{y(3)} &= 1 / (F+G) \end{aligned} \qquad (6)$$

The flow rules may be developed using the following equation.

$$d\varepsilon_{ij} = d\lambda \frac{\partial f(\sigma_{ij})}{\partial \sigma_{ij}} \qquad (7)$$

Substitution of Eq.(5) into Eq.(7) results in the flow rules ($d|x|^a = \pm a |x|^{a-1}$ according as $|x| \gtrless 0$) : For $\sigma_{11} > \sigma_{22} > \sigma_{33}$ which is the stress state of the wall,

$$\begin{aligned} d\varepsilon_{11} &= d\lambda[G|\sigma_{33}-\sigma_{11}|^{a-1} + H|\sigma_{11}-\sigma_{22}|^{a-1}] \\ d\varepsilon_{22} &= d\lambda[F|\sigma_{22}-\sigma_{33}|^{a-1} - H|\sigma_{11}-\sigma_{22}|^{a-1}] \\ d\varepsilon_{33} &= d\lambda[-F|\sigma_{22}-\sigma_{33}|^{a-1} - G|\sigma_{33}-\sigma_{11}|^{a-1}] \end{aligned} \qquad (8a)$$

For $\sigma_{11} > \sigma_{33} > \sigma_{22}$ which is the stress state of the flange,

$$d\varepsilon_{11} = d\lambda\left[G|\sigma_{33}-\sigma_{11}|^{a-1} + H|\sigma_{11}-\sigma_{22}|^{a-1}\right]$$

$$d\varepsilon_{22} = d\lambda\left[-F|\sigma_{22}-\sigma_{33}|^{a-1} - H|\sigma_{33}-\sigma_{11}|^{a-1}\right] \qquad (8b)$$

$$d\varepsilon_{33} = d\lambda\left[F|\sigma_{22}-\sigma_{33}|^{a-1} - G|\sigma_{33}-\sigma_{11}|^{a-1}\right]$$

Note that for Eqs.(8), $d\varepsilon_{11} + d\varepsilon_{22} + d\varepsilon_{33} = 0$, indicating constant volume. Substitution of $\sigma_{11} = \sigma_{y(1)}, \sigma_{22} = \sigma_{33} = 0$ into Eqs.(8)gives the resulting strains,

$$d\varepsilon_{11} = d\lambda(G+H)|\sigma_{y(1)}|^{a-1}$$

$$d\varepsilon_{22} = -d\lambda H|\sigma_{y(1)}|^{a-1} \qquad (9)$$

$$d\varepsilon_{33} = -d\lambda G|\sigma_{y(1)}|^{a-1}$$

Since the strain ratio for the x_1-direction tension test is defined as $R=R_0 = d\varepsilon_{22} / d\varepsilon_{33}$,

$$R = H / G \qquad (10)$$

Similarly, defining $P = R_{90}$ as the strain ratio in the x_2-direction tension test, $P = d\varepsilon_{11} / d\varepsilon_{33}$ with $\sigma_{22}= \sigma_{y(2)}$ and $\sigma_{11}= \sigma_{33}= 0$, Eqs.(8) result in

$$P = H / F \qquad (11)$$

Equations (10) and (11) allow one to predict the value of the x_3-direction yield stress, $\sigma_{y(3)}$, by conducting x_1-and x_2-direction tension tests and measuring R and P as well as $\sigma_{y(1)}$ and $\sigma_{y(2)}$. From Eq.(6)

$$\left[\sigma_{y(3)} / \sigma_{y(1)}\right]^a = (G+H)/(F+G) = (1/R+1)/(1/R + 1/P)$$

or

$$\sigma_{y(3)} = \sigma_{y(1)}\left[P(1 + R) / (P + R)\right]^{1/a} \qquad (12a)$$

Similarly,

$$\sigma_{y(3)} = \sigma_{y(2)}\left[R(1 + P) / (P + R)\right]^{1/a} \qquad (12b)$$

Substituting $1 = (G+H)\sigma_{y(1)}^a$ from Eq.(6) and dividing by G, Eq.(5) becomes

$$(F/G)|\sigma_{22}-\sigma_{33}|^a + |\sigma_{33}-\sigma_{11}|^a + (H/G)|\sigma_{11}-\sigma_{22}|^a = (1 + H/G)\sigma_{y(1)}^a$$

Substituting $R = H/G$ and $R/P = F/G$ and multiplying by P,

$$R|\sigma_{22}-\sigma_{33}|^a + P|\sigma_{33}-\sigma_{11}|^a + RP|\sigma_{11}-\sigma_{22}|^a = P(R+1)\sigma_{y(1)}^a \qquad (13)$$

Similarly the flow rules, Eqs.(8a), and (8b), reduce to

$$d\varepsilon_{11} : d\varepsilon_{22} : d\varepsilon_{33} = |\sigma_{33}-\sigma_{11}|^{a-1} + R|\sigma_{11}-\sigma_{22}|^{a-1} : \tag{14a}$$
$$(R/P)|\sigma_{22}-\sigma_{33}|^{a-1} - R|\sigma_{11}-\sigma_{22}|^{a-1} : (-R/P)|\sigma_{22}-\sigma_{33}|^{a-1} - |\sigma_{33}-\sigma_{11}|^{a-1}$$

and

$$d\varepsilon_{11} : d\varepsilon_{22} : d\varepsilon_{33} = |\sigma_{33}-\sigma_{11}|^{a-1} + R|\sigma_{11}-\sigma_{22}|^{a-1} : \tag{14b}$$
$$(-R/P)|\sigma_{22}-\sigma_{33}|^{a-1} - R|\sigma_{11}-\sigma_{22}|^{a-1} : (R/P)|\sigma_{22}-\sigma_{33}|^{a-1} - |\sigma_{33}-\sigma_{11}|^{a-1}$$

If the material has rotational symmetry about the x_3-axis (planar isotropy), F=G, L=M, and R=P. In this case, substitution of P=R in Eqs.(13), (14a) and (14b) results in

$$|\sigma_{22}-\sigma_{33}|^a + |\sigma_{33}-\sigma_{11}|^a + R|\sigma_{11}-\sigma_{22}|^a = (R+1)\sigma_{y(1)}^a \tag{15}$$

and

$$d\varepsilon_{11} : d\varepsilon_{22} : d\varepsilon_{33} = |\sigma_{33}-\sigma_{11}|^{a-1} + R|\sigma_{11}-\sigma_{22}|^{a-1} : \tag{16a}$$
$$|\sigma_{22}-\sigma_{33}|^{a-1} - R|\sigma_{11}-\sigma_{22}|^{a-1} : -|\sigma_{22}-\sigma_{33}|^{a-1} - |\sigma_{33}-\sigma_{11}|^{a-1}$$

$$d\varepsilon_{11} : d\varepsilon_{22} : d\varepsilon_{33} = |\sigma_{33}-\sigma_{11}|^{a-1} + R|\sigma_{11}-\sigma_{22}|^{a-1} : \tag{16b}$$
$$-|\sigma_{22}-\sigma_{33}|^{a-1} - R|\sigma_{11}-\sigma_{22}|^{a-1} : |\sigma_{22}-\sigma_{33}|^{a-1} - |\sigma_{33}-\sigma_{11}|^{a-1}$$

For planar isotropy and plane-stress loading ($\sigma_{33} = 0$) Eq.(15) simplifies to

$$|\sigma_{11}|^a + |\sigma_{22}|^a + R|\sigma_{11}\ \sigma_{22}|^a = (R + 1)\sigma_{y(1)}^a \tag{17}$$

The flow strengths, $\sigma_{flange(\varepsilon_{33}=0)}$ and $\sigma_{wall(\varepsilon_{22}=0,\ \varepsilon_{33}=0)}$ in Eq.(2), may be, with reference to Fig. 1, expressed as

$$\sigma_{flange(\varepsilon_{33}=0)} = |\sigma_{11}-\sigma_{22}|_{\varepsilon_{33}=0} \text{ and } \sigma_{wall(\varepsilon_{22}=0,\ \sigma_{33}=0)} = \sigma_{11(\varepsilon_{22}=0,\sigma_{33}=0)}$$

In the flange, where $\sigma_{11} > \sigma_{33} > \sigma_{22}$ and $d\varepsilon_{33}=0$, Eq.(16b) predicts that $\sigma_{33} = (\sigma_{11}+\sigma_{22})/2$. Substitution into Eq.(15) results in

$$2|\sigma_{22}-\sigma_{11}|^a + 2^aR|\sigma_{11}-\sigma_{22}|^a = 2^a(R+1)\sigma_{y(1)}^a$$

Therefore

$$\sigma_{flange} = |\sigma_{11}-\sigma_{22}|_{\varepsilon_{33}=0} = |\sigma_{22}-\sigma_{11}|_{\varepsilon_{33}=0} = 2\sigma_{y(1)}[(1+R)/(2+2^aR)]^{1/a} \tag{18}$$

In the wall, where $\sigma_{11} > \sigma_{22} > \sigma_{33}$, $d\varepsilon_{22}=0$ and $\sigma_{33}=0$, Eq.(16a) predicts that

$$\sigma_{22} = R^{1/(a-1)}\sigma_{11} \ / \ (1+R^{1/(a-1)}) \tag{19}$$

Substitution Eq.(19) into Eq.(17) gives us

$$\sigma_{wall} = \sigma_{11}(\varepsilon_{22}=0,\ \sigma_{33}=0) \tag{20}$$
$$= (R+1)^{1/a}\sigma_{y(1)} \ / \ \left[1 + (R^{a/(a-1)}+R) \ / \ (1+R^{1/(a-1)})^{a}\right]^{1/a}$$

Substituting Eqs.(18) and (20) into Eq.(2) we obtain,

$$\beta = (2+2^{a}R)^{1/a} \ / \ \{2\left[1 + (R^{a/(a-1)} + R) \ / \ (1 + R^{1/(a-1)})^{a}\right]^{1/a}\} \tag{21}$$

Setting a = 2, Eq.(21) reduces to $\beta = \sqrt{(R+1)/2}$, which is equivalent to Eq.(3). Substitution of Eq.(21) into Eq.(1) with the efficiency, η, gives us the relation between the LDR and R-value.

$$\text{In(LDR)} = \eta(2+2^{a}R)^{1/a} \ / \ \{2\left[1 + (R^{a/(a-1)}+R) \ / \ (1+R^{1/(a-1)})^{a}\right]^{1/a}\} \tag{22}$$

Equation (21) was first derived by the present author, but unfortunately it was misprinted (6).

COMPARISON OF CALCULATED AND MEASURED VALUES

Figure 2 shows the calculated relation of the R value and β for various a values. The R value dependence of β decreases with a at the R values above unity. Hosford(2) suggested the $a \approx 6$ for bcc metals and $a \approx 8$ to 10 for fcc metals in the yield locus calculation. Several calculated results of the R -LDR relation are compared with experimental data(7-9)in Fig. 3. The R-LDR relations calculated with a = 8 for cubic metals and a = 4 to 6 for hexagonal metals agree very well with the experimental results.

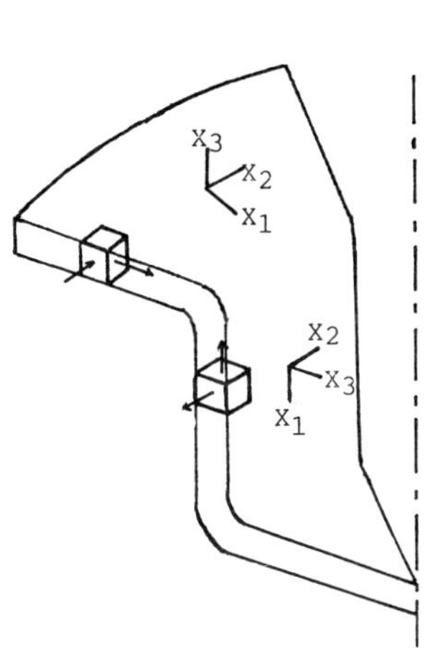

Fig.1. A section from a partially drawn cup showing coordinate system.

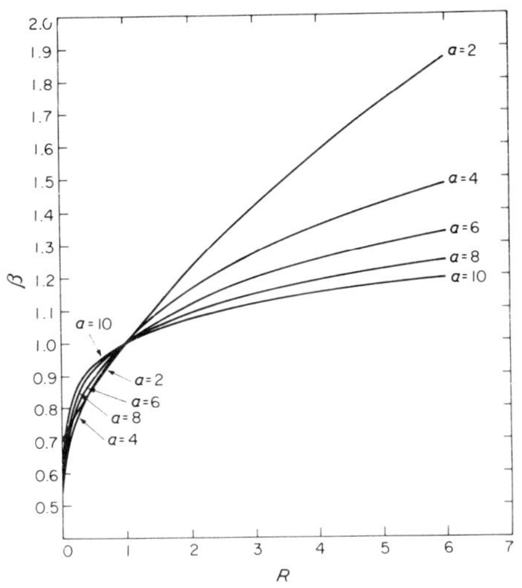

Fig.2. The calculated relation of the R value and β for various a values

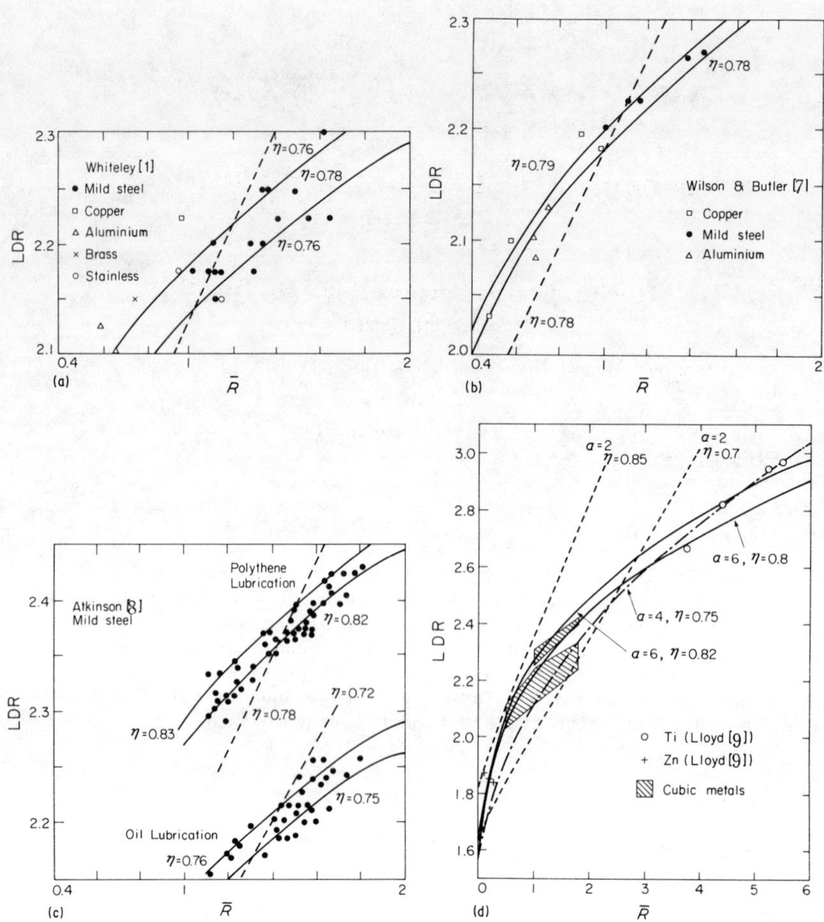

Fig. 3. Limiting drawing ratio vs. average strain ratio $\bar{R}$. The measured data are from Whiteley, Wilson and Butler, Atkinson, and Lloyd. The solid curves and the dashed curves in (a) to (c) are for a=8 and 2, respectively(6).

REFERENCES

1. R.L.Whiteley, Trans. ASM 52, 154(1960).
2. W.F.Hosford and C.Kim, Metall. Trans. 7A, 468(1976).
3. R.Hill, Mathematical Theory of Plasticity, pp.315-40, Oxford University Press(1950).
4. W.F.Hosford and R.M.Caddel, Metal Forming, pp.269-72, Prentice-Hall, Inc. (1983)
5. R.Hill, Math. Proc. Camb. Phil. Soc. 85, 189(1979).
6. D.N.Lee, J.Materials Science Letters, 3, 677(1984).
7. D.V.Wilson and R.D.Butler, J.Inst. Met. 90, 473(1963).
8. M.Atkinson, Sheet Met. Ind. 41, 167(1964).
9. D.H.Lloyd, ibid. 39, 82(1963).

Mechanical Instabilities in Industrial Processes

D. Teirlinck, B. Lelièvre, F. R. Boutin and L. Felgères

Cégédur Péchiney, Centre de Recherches de Voreppe, B.P. 27, 38340 Voreppe, France

ABSTRACT

Flow localization during metal forming processes of aluminum alloys is presented from an industrial point of view. After reporting some examples of localization in Al-Mg alloys, the metallurgical and mechanical approaches are discussed in relation with the experimental results. The need for an engineering criterion for localization of flow, which contains both metallurgical and mechanical parameters, is emphasized.

KEYWORDS

Flow localization ; Aluminum-Magnesium alloys ; shear bands

INTRODUCTION

The localization of flow is commonly observed during the deformation of aluminum alloys (1-3). It can take several forms, from the necking of a tensile specimen to the formation of shear bands. Concerning the latter, some recent reviews (4-5) have presented the main features of shear bands in fcc metals. In particular, very large strains are associated with shear bands, so that fracture can take place readily within the bands, by accumulation of dilational damage (6). From an industrial point of view, these events are redhibitory.

In this paper, some examples of localization during industrial metal forming processes are reported, together with their consequences. Then, the metallurgical and mechanical theories related to these instabilities are reviewed and applied to experimental results.

EXPERIMENTAL RESULTS

The occurrence of shear bands involves plane strain deformation. Many forming processes meet this requirement and therefore are prone to shear band creation.

The following examples are a good illustration of this. The most obvious plane strain process is cold rolling ; fig. 1 shows the shear bands created by hot rolling a 5083 alloy. These bands have been decorated by the preferential precipitation of Mg rich particles during an ageing treatment of one week at

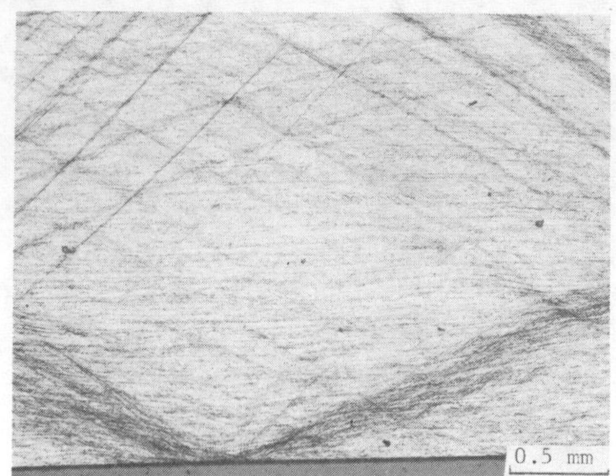

Fig. 1. Deformation microstructure in a hot-rolled Al-4.5 % Mg ($\overline{\varepsilon}$ = 1.6)

125°C (3). They are subsequently revealed by etching with H_3PO_4. The sheet is heavily banded, leading to the formation of cracks near the edges, where the stress state is no longer compressive.

Figure 2 is another example of a shear band leading to crack nucleation : it shows the cold upsetting of a fastener head made from a 5056-H12 : two strong shear bands can be seen, which intersect to form a crack on the specimen free surface.

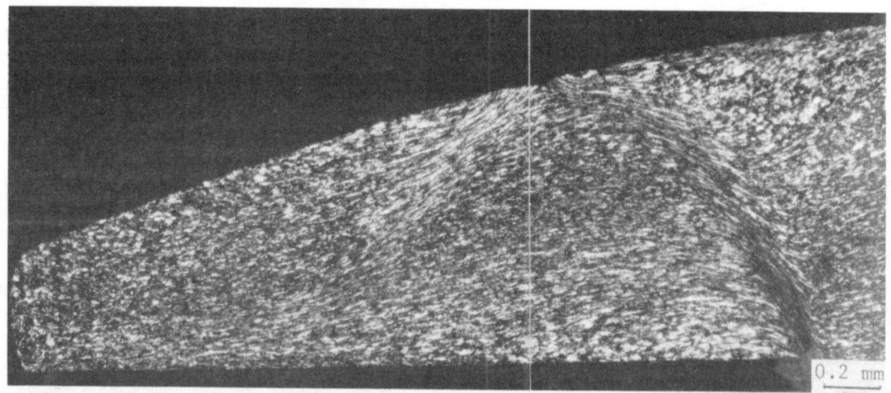

Fig. 2. Microstructure of a cold upsetted fastener head (5056 alloy)

The next example of shear band formation is related to the deep-drawing of cylindrical cups. In fig. 3, such bands are shown in an alloy 5052-H28. Four series of bands are created under the blank holder, at a given angle from the rolling direction, which depends on the anisotropy of the sheet (at about 65° from the rolling direction in the present case). The shear bands cross the

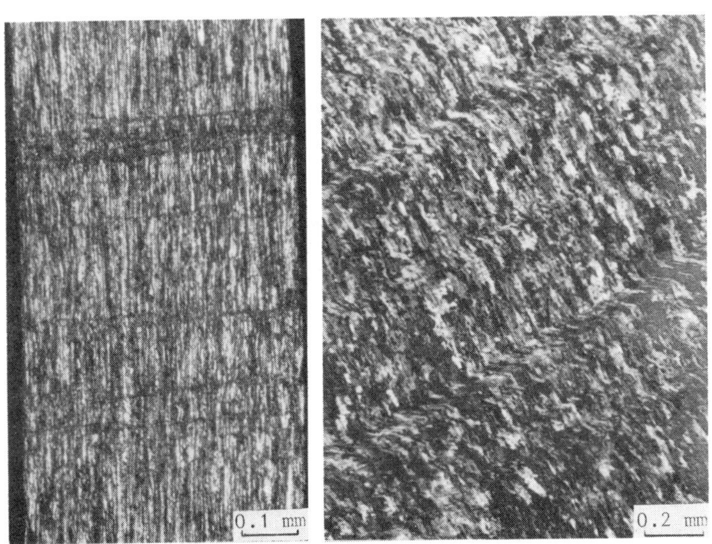

Fig. 3a. Micrograph of the blank thickness

Fig. 3b. Micrograph of the blank surface

Fig. 3. Shear bands created during deep-drawing of a cylindrical cup ; drawing ratio : 2.05 ; cup height : 45 mm

whole thickness of the blank (fig. 3a), and they create a step through the elongated grains (fig. 3b). The bands appear at a critical height of the cup, h = 20 mm in the present case.

DISCUSSION

From the above experimental results, it is obvious that shear localization involves two series of parameters. First of all, the metallurgical state of the material must be considered. Its composition, as well as its thermomechanical history is of prime importance when considering shear band formation. Indeed, the localization is due to some form of structural softening which can have several origins. In the present case, the main phenomenon must be a progressive re-arrangement of the dislocation substructure during deformation ; the Al-Mg alloys presented above are solid solution strengthened. The dislocation substructure at large strains is composed of small cells, with dislocations inside the cells. The Mg atoms interact with the dislocations, thus inhibiting their motion, and the net result is a reduction of dynamic recovery at large strains, i.e. creation of a highly heterogeneous state of deformation : dislocations do not climb, and therefore stay on their glide planes.

As soon as a slip band created within a grain contains enough dislocations, the stress concentration existing ahead of this pile-up will be sufficient to destroy the barrier constituted by the grain, and allow the deformation to continue on this slip plane : the band will propagate.

Following Hatherly (4), the net rate of work-hardening of the material can be written as :

$$\frac{d\sigma}{d\varepsilon} = \frac{\partial\sigma}{\partial\varepsilon} + \frac{\partial\sigma}{\partial\dot{\varepsilon}}\frac{\partial\dot{\varepsilon}}{\partial\varepsilon} + \frac{\partial\sigma}{\partial M}\frac{\partial M}{\partial\varepsilon} - \frac{\partial\sigma}{\partial N}\frac{\partial N}{\partial\varepsilon} \qquad (1)$$

where M is the Taylor factor and N the mobile dislocation density. The condition for flow localization is :

$$\frac{1}{\sigma}\frac{d\sigma}{d\varepsilon} < 0 \qquad (2)$$

As the first two terms on the R.H.S. of equation (1) are either positive or very small, the "softening" may have two origins :
- either a texture or geometrical softening, i.e. a negative value of $\partial M/\partial\varepsilon$ after a sufficient amount of deformation,
- or a structural softening, i.e. an increase of $\partial N/\partial\varepsilon$; this can occur locally in Al-Mg alloys when dislocations are liberated from their solute atoms atmosphere.

If we consider the substructure of these alloys, the localization can be viewed as the breaking of the strengthening due to cell or subgrain walls :

$$\frac{d\sigma}{d\varepsilon} = \frac{d\sigma_o}{d\varepsilon} + \frac{1}{d^n}\frac{dk}{d\varepsilon} \qquad (3)$$

where σ_o is the flow stress of the recrystallized material, d the cell or subgrain size, k a factor expressing the strength of the substructure and n an exponent depending on the nature of the walls.

The condition (2) is realized when $\frac{dk}{d\varepsilon}$ becomes strongly negative ; here again, this occurs when a dislocation pile-up breaks the cell or subgrain wall, thus lowering its strength.

Currently, we are developing a mechanical approach of shear localization, which seems particularly relevant for the deep-drawing of cylindrical cups. It can be summarized as follows :

- First, find a criterion of flow localization in terms of critical strains and/or stresses ; we have used an extension of the Hill's criterion proposed by Cordebois et al. (8).

- Then determine the critical values of strains and stresses for shear localization by pre-deposited grid measurements, and verify the validity of the criterion ; this is also made by a F.E.M. model of deep-drawing of cylindrical cups which includes the Cordebois criterion.

The first results are rather promising and will be reported later.

CONCLUSION

To conclude this presentation of the shear localization phenomena encountered in metal forming processes, we want to emphasize the need for an engineering criterion which could allow a prediction of shear band formation. This criterion should include both metallurgical and mechanical parameters, such as some critical stress or strain, in the same way that the classical Latham-Cockroft or Oyane criteria which are used to determine the formability of alloys in cold upsetting operations.

The consequences of the shear band formation are indeed multiple and rather damaging : edge cracking during cold rolling, cracks at the free surface of cold upsetted products, limited ductility, especially when the material is deformed in tension, destruction of internal protection of deep drawn cups for packaging uses.

REFERENCES

1. K. Brown, J. Inst. Metals, 100, 341 (1972)
2. J.W. Chang and R.J. Asaro, Arch. Mech., 32, 369 (1980)
3. D.J. Lloyd, E.F. Butryn and M. Ryvola, Microstr. Sci., 10, 373 (1982)
4. M. Hatherly, Proc. 6th ICSMA (edited by R.C. Gifkins), p. 1181, Pergamon Press, Oxford (1982)
5. M. Hatherly and A.S. Malin, Scripta Metall., 18, 449 (1984)
6. A. Korbel, V.S. Raghunathan, D. Teirlinck, W. Spitzig, O. Richmond and J.D. Embury, Acta Metall., 32, 511 (1984)
7. J. Gil-Sevillano, P. Van Houtte and E. Aernoudt, Progress in Materials Science, 25, 69 (1980)
8. J.P. Cordebois and P. Ladeveze, Mem. Sci. Rev. Met., 21, 19 (1984)

Effect of a Short Time Austenitization on Mechanical Properties of Steels

P. Braisch*, A. Simon*, G. Beck* and K.-H. Kloos**

**Laboratoire de Métallurgie associé au CNRS LA 159, Ecole des Mines, 54042, Nancy-Cedex, France*
***Institut Für Werkstoffkunde, Technische Hochschule, Darmstadt, Federal Republic of Germany*

ABSTRACT

Rotating-beam and tension tests with small samples of structural steel after a short time austenization by induction throughout heating and subsequently quenched show a characteristic variation of mechanical properties in function of austenitization temperature, different from the variation of hardness. The authors concluded, that this behaviour can be explained with the observed metallurgical structure.

KEYWORDS

Structural steel ; short time austenitization quenching ; rotating-beam test; tension test ; mechanical properties ; metallurgical structure.

INTRODUCTION

As known, the increase of the fatigue resistance of machinery parts from structural steels after a surface strenghtening treatment is due to both, the effect of residual compressive stresses and the improved mechanical properties in the treated surface layer. The achieved increasing ratio depends in an important manner on the chosen process parameters. As very sensitive in function of these parameters it proved to be the induction surface hardening process, which can be performed in a high number of various modes. So investigations on surface induction hardened parts show /1/, that different austenitization conditions lead to different residual stress estates and finally to different fatigue resistances. The variation of the hardness at this can be neglected /2/. But it is of interest to know, in which manner the mechanical properties of steel, i.e. the tensile strength and the fatigue resistance, are influenced by different austenitization conditions. As far as known, investigations in this area were performed only for considerably longer austenitization times than for induction hardening usually used (e.g./3/, 1979 : 10^2 to 10^3 times longer).

In this work the effect of short time austenitization on tensile strength and fatigue resistance of small samples of structural steel, throughout heated, has been investigated.

FACTS

Fatigue tests with induction surface hardened parts show that the fatigue crack can be initiated in each of the different metallurgical layers /4/, which appear in function of the distance from the surface. So for the evaluation of the contribution of mechanical properties to the strength behavior of induction hardened parts, needed is the knowledge of the so-called pure mechanical properties of these different layers. To fabricate test samples with corresponding homogeneous properties becomes a foremost problem in the condition of short time austenitization. This especially because in respect to the needed geometrical design of the test samples a short time austenitization can be done only by induction heating. From Fig. 1 it can be seen, that then the time interval t_{θ_o} corresponding to the needed equalization of

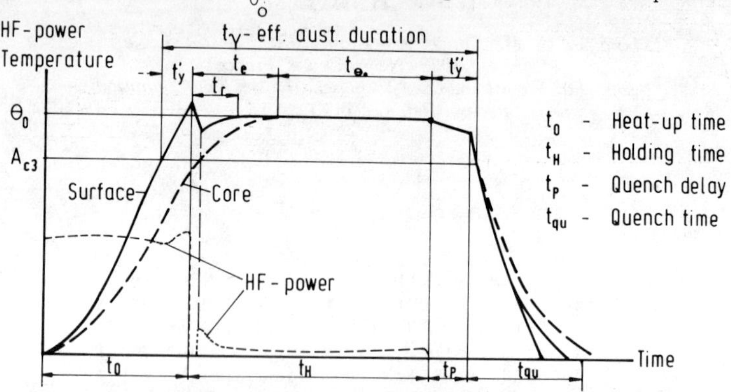

Fig. 1. Temperature and HF power curves used for heating test samples (schematically).

temperature in the whole cross section of the heated sample, is preceded by a time interval $t'_y + t_e$) in wich the temperature is nonuniform and above the nominal austenitization temperature θ_o. The situation is similar during quenching in the time interval t''_y. Therefore, statements concerning mechanical properties of a steel become more and more doubtful as shorter austenitization times are chosen. The quality of such a statement depends on the ratio t_{θ_o}/t_y.

Corresponding with the used experimental equipment, in this work the with a timer controlled holding time t_H and with an optical pyrometer controlled equalization temperature θ_o were chosen as nominal parameters.

EXPERIMENTAL

Cylindrical samples (dia. 4 mm) of 42CrMo4, having a chemistry of C : 0.38, Mn : 0.63, Si 0.28, Cr : 0.97, Mo : 0.23, P+S : 0.014 in percents, have been hardened and tempered for a u.t.s., of about 1000 MPa, then through heated by induction heating using an HF generator of 100 kHz. The in Fig. 1 mentioned time intervals t'_y and t''_y can been evalueted from with termocuples measured heating rate of about 600°C/s and quenching rate of about 200°C/s. For the heating operation a so-called half-shell single-shot inductor and for the quenching operation pressurized oil jets of 60°C at a flow rate of about 2000 m^3/m^2.h were used. The time interval t_e in Fig. 1 was in all cases less then 0.5 s. The quench delay tp was 0.3 s. With the chosen cooling conditions it was possible to obtain at time, the desired metallurgical structures, corresponding to the mentioned different layers in an induction hardened machinery part and specimens nearly free from residual stresses(X-ray measurements). After this heat treatment the samples have been finished, but not annealed.

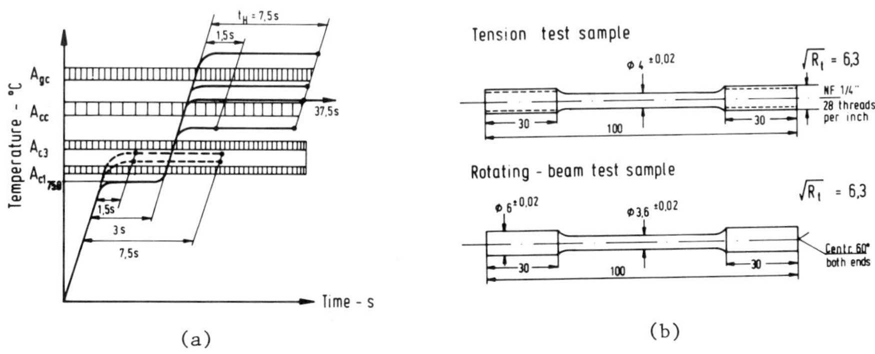

(a) (b)

Fig. 2 a. Representation of the test programme (schematically)
b. Dimensions of the test samples.

The fatigue resistance has been investigated employing a rotating bending machine of 3000 rpm and a uniform bending moment along the tested length of the samples. The statistical analysis of fatigue data have been performed using the staircase method for a preassigned number of cycles $N = 10^7$ and at least 20 specimens for each variant. The tension tests have been performed using fixtures supported by a hydraulic bed /5/ and at least 8 specimens for each variant.

Figure 2 contains schematically the test programme ; the stated figures mean the nominal values of the austenitization durations in s (t_H in Fig. 1). Except the variants in the temperature range Ac_1 - Ac_3 all others have been heated up with an intermediate step below Ac_1 for temperature equalization. The dashed areas mean time depending temperature limits for Ac_1, Ac_3, Acc, Agc. The meaning of the special symbols is : A_{cc} - limit of homogeneous austenite, defined as the temperature limit at which a further increase of hardness ceased /2/, A_{gc} - start of extreme grain coarsing (here without definition).

RESULTS

Figure 3 gives typical stress-strain curves of the tension tests for a martensitic structure. These curves show remarkable rupture strains. Figures 4 and 5 show respectively the variation of the u.t.s. and the fatigue resistance in function of the austenitization conditions. The other specific parameters of the tension test i.e. yield strength, elongation, striction, here not illustrated show the same characteristical configuration. The dependence between grain coarsing and austenitization conditions, determined using optical microscope analysis, is shown likewise in Fig. 5. As for Fig. 2, in Fig. 4 and 5 the in seconds stated figures mean the nominal values of the austenitization durations.

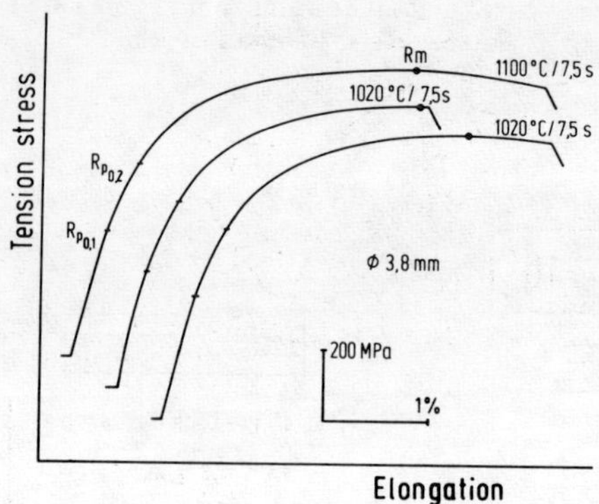

Fig. 3. Typical tension stress - strain curves.

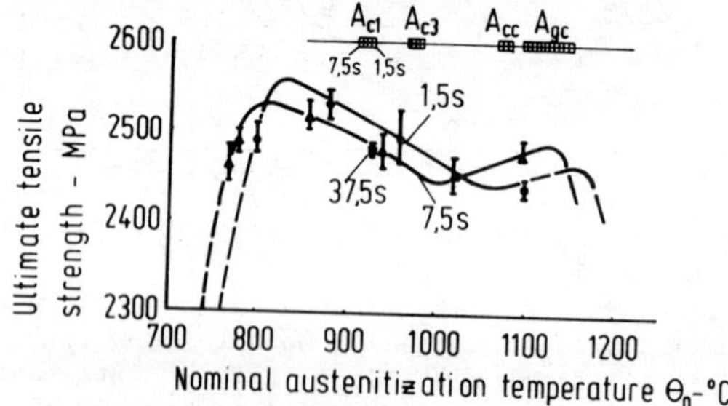

Fig. 4. Variation of u.t.s in function of nominal austenitization conditions.

DISCUSSION AND CONCLUSIONS

The results of both, the tension tests and the fatigue tests with small samples of structural steel after a short time austenitization using an induction throughout heating and subsequently quenched show concordant a characteristic influence of austenitization conditions on the mechanical properties. As Fig. 6 illustrates schematically, there are two maxima and one minimum. This behaviour is essentially different from the behaviour of hardness at which such a minimum has not been appeared, nor in this investigation, nor so far known in others (2). It can be assumed, that the appearance of the mentionned minimum of the mechanical properties is the result of the superposition of the two temperature depending factors : the grain coarsing and the homogenization of metallurgical structure. The presence of the first maximum may be produced by the co-operation of a low, but sufficient homogeneous metallurgical structure and a very fine martensitic structure. An increasing austenitization temperature works towards a growth of grain size, probably without improving first of all the homogeneity, producing in this way the observed minimum. A second maximum appearing with increasing temperature may be explained with a significant higher homogeneization rate than grain growth rate.

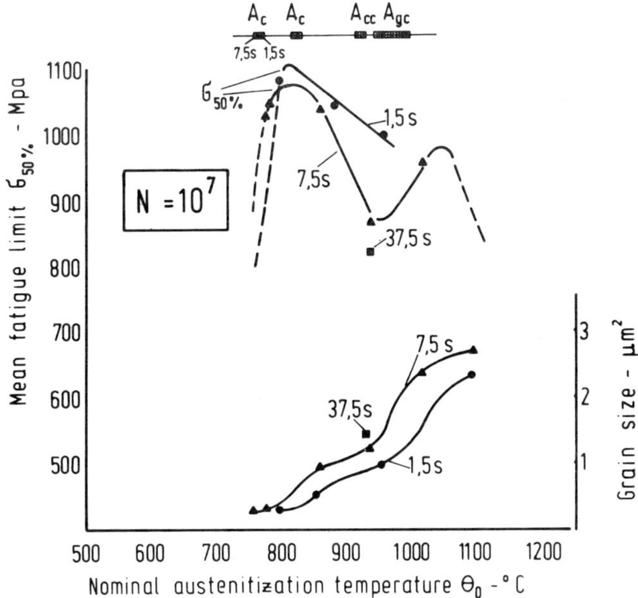

Fig. 5. Variation of fatigue resistance of rotating beam specimens (above) and of grain size (below) in function of nominal austenitization.

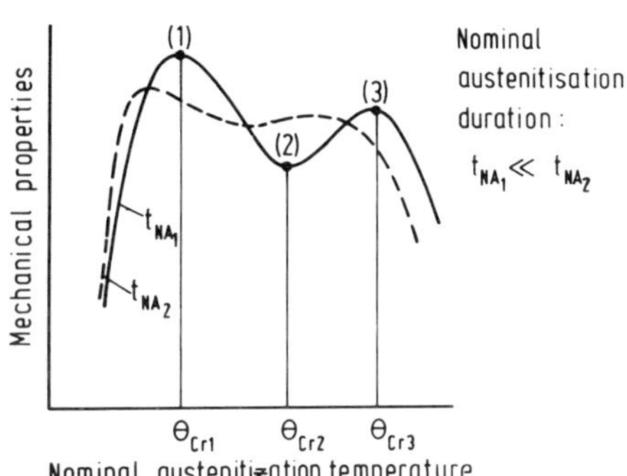

Fig. 6. The characteristic dependence of mechanical properties from austenitization conditions (schematically).

At higher temperatures an extreme grain coarsing and/or the presence of microcracks conduce in a known way towards a rapid degradation of mechanical properties.

It seems, that the temperatures $\theta_{Cr1...3}$ in fig. 6 can be interpreted as critical as to grain coarsing and as to structural homogeneity.

As expected, with increasing austenitisation duration these critical temperatures tend to shift to lower values and the difference between the described maxima and minimum becomes smaller.

This work has been supported especially by the Ministère des Affaires Etrangères of France wich financed a research stage at the Ecole des Mines de Nancy (France).

REFERENCES

1. P. Braisch, VDI-Berichte 506 (1976), p. 47.
2. J. Ohrlich and Others, Atlas zur Wärmebehandlung der Stähle, vol. 3, Stahleisen, Düsseldorf (W.-Germany) 1973.
3. H. Klages, Dr. Ing. Thesis, Technische Universität Berlin (1979).
4. P. Braisch, Dr. Ing. Thesis, Technische Hochschule Darmstadt (1981).
5. O. Zwirlein and Others, Härt. Techn. Mitteilungen 31 (1976) 5, p. 277.

Modified In-line and Accelerated Normalising to Improve Properties and Conserve Time and Energy

D. M. Fegredo and E. F. Connors

PMRL, CANMET, Energy, Mines and Resources, Ottawa K1A OG1, Canada

ABSTRACT

In-line normalising and accelerated normalising experiments were done on Grade 350 WT laboratory hot-rolled plates and the compositions tried had two sulphur (0.005% and 0.03%) and two vanadium (0% and 0.08%) levels. The techniques employed two furnaces; the first rapidly increased the temperature to ~900°C, the second stabilised it at 900°C for any desired time, usually 10 minutes. Air or gentle accelerated cooling (light sprays) was used after heat-treatment. The results show appreciable grain refinement together with improved strength and toughness in comparison with hot-rolled plate. Spray cooling to Ar_1, instead of air cooling, further refines grain size and increases yield strength without loss in toughness.

KEY WORDS

Normalising, in-line, accelerated, sprays, grain-size, strength, toughness.

INTRODUCTION

The incentive of lower costs have recently led to research on processes (1,2) whereby production time is decreased and suitable products obtained for particular applications.

Normalising has been used for decades to refine grain size and is widely accepted by both manufacturer and purchaser. The National Standards of Canada (3) produced under the auspices of the Canadian Standards Association (CSA) states that the availability of any grade of structural quality steel is usually limited by the ability of the specific chemical composition to meet the required mechanical properties for the thickness concerned. Plates are usually delivered as-rolled and may be supplied in the normalised condition when specified by the purchaser. Thus, in cases where normalising is deemed necessary to improve toughness, processes that save time and/or energy merit consideration.

EXPERIMENTAL

Five laboratory-prepared steels, conforming to CSA G40.21 Grade 350WT, were rolled to plate, finishing at ~1050°C, followed by either i) in-line normalising, ii) accelerated normalising, iii) air cooling, or iv) air cooling + conventional normalising. The compositions of the five steels are given in Table 1, along with the composition limits for Grade 350WT. Steels A, D and E were plain C-Mn steels with sulphur concentrations 0.005, 0.03 and 0.017, respectively.

For the in-line normalising treatment, a 250-mm length was cut from 35 and 60 mm plate of steels A-C, following the final pass. After air cooling to Ar_1 (~660°C), the sample was inserted in a muffle furnace (I) set at 1150°C, heated to ~900°C, transferred to a second muffle furnace (II) set at 900°C, and held at 900°C for ten min. Light spray cooling to Ar_1, followed by air cooling, or only air cooling were employed after the normalising treatment. The spray times varied somewhat because different nozzle configurations were used, and by the inherent variability of this process (4,5). This procedure is shown schematically in Fig. 1a.

The accelerated normalising treatment comprized heating 19-64 mm, cold, as-rolled plate (steel D) rapidly to ~900°C in muffle furnace I set at 1100-1200°C, transferring to muffle furnace II set at 900°C, holding ten min. at 900°C, followed by air- or spray-cooling, as before (Fig. 1b).

The conventional normalising treatment was done by putting the 35 mm thick plate in a pre-heated (900°C) furnace for 85 minutes, followed by air cooling. The details of all the heat treatments are summarized in Table 2.

Transverse (T), longitudinal (L) and through-thickness (TT) tensile and Charpy tests were carried out, and the results are summarized in Table 3. Representative microstructures are shown in Fig. 2.

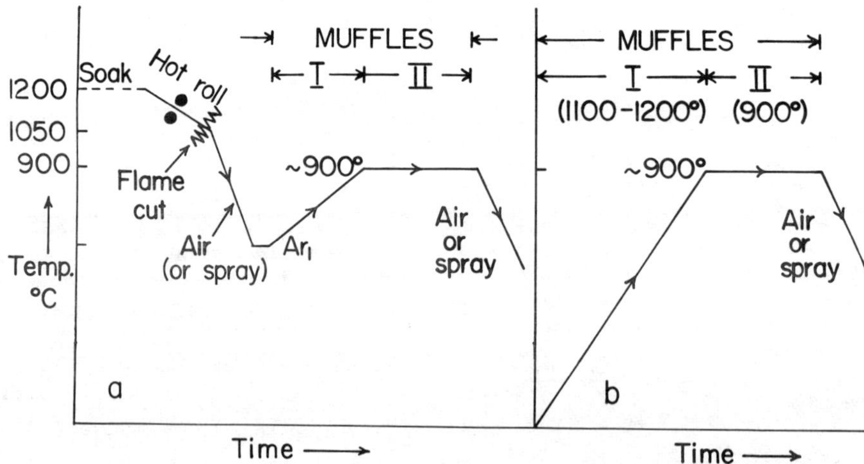

Fig 1(a,b). (a) In-line and (b) accelerated normalising procedures for hot-rolled plate.

TABLE 1 Chemical Analyses, Weight %

Steel	C	Mn	P	S	Si	V	Ti	Al	Other
A	0.18	1.09	0.005	0.005	0.27	-	-	0.035	
B	0.18	1.13	0.005	0.005	0.27	0.08	-	0.04	
C	0.15	1.16	0.005	0.005	0.31	0.035	0.015	0.03	0.008N
D	0.19	1.27	<0.01	0.03	0.26	-	-	0.02	
E	0.19	1.22	0.004	0.017	0.30	-	-	0.04	
350WT	0.22 max	0.8 – 1.5	0.03 max	0.04 max	0.15 – 0.4	*	*	**	

* 0.10 max total grain refining elements
**killed, fine grain practice

DISCUSSION

The specifications for Grade 350WT require a minimum yield strength of 350 MPa and 330 MPa for thicknesses <40 mm and 40-65 mm respectively, and tensile strengths of 480-650 MPa. The minimum elongation is 20%. There are five impact property categories; 1 to 4 define a minimum of 27 J to be absorbed at 0°C, -20°C, -30°C or -45°C respectively. Category 5 is specified by the purchaser.

In the light of the above criteria, the following points may be made from Tables 1-3 and Fig. 2.

1. All in-line and accelerated normalised plates meet the tensile strength requirements except for the 60 mm thick air cooled plate A1. Since the C and Mn contents of the steels are relatively low, particularly steels A-C, the strengths are generally close to the lower bound of the specification. Thus, the air-cooled plates D1 and D4 with higher C and Mn have higher yield strengths of 353 MPa and 356 MPa respectively, than plate A1. Furthermore, normalising substantially increases yield strength over that of hot-rolled plate. For example, the yield strength of A3 (35 mm thick) and D4 (64 mm thick) are ~100 MPa and ~60 MPa higher than those of the 35 mm thick hot-rolled plates A5 and E2 respectively.

2. Both the quantity and morphology of manganese sulphides influence the charpy shelf energies (6,7). This explains the relative shelf energies of the L, T and TT charpy specimens and between those of steels A-C and D and E. Although the hot-rolled plates have sufficiently low transition temperatures to place them in category 4 a marked improvement occurs in both shelf energy and C_V27J(°C) values by normalising, (cf A3 and A5). Furthermore, the 0.005% S content of the in-line normalised plates as well as the equiaxed grain morphology are responsible for the excellent through-thickness impact toughness and tensile ductility.

3. Grain refinement of the hot-rolled plates, (Fig. 2a-c), by the modified normalising procedures, (Fig. 2d-k), is similar, and in some cases superior, to that produced by conventional normalising, Fig. 2(1). This is the only parameter, which when reduced, simultaneously improves strength and toughness (8).

4. The tensile and impact properties of the accelerated normalised plates are comparable to those of the conventionally normalised plate D5.

TABLE 2 Heat Treatment Data

Plate	Thickness mm	Treatment	Muffle I temp °C	Muffle I time min	Mid-* thickness temp °C	Post Muffle II cooling	ASTM No.
A-1	60	in-line	1150	8	893	Air	8 - 8 1/2
A-2	60	"	"	8.6	900	Spray	8 1/2 - 9
A-3**	35	"	"	9.8	890	Spray	9
A-4	60	"	"	9.5	880	Spray	8 1/2 - 9
B-1	60	"	"	8.4	890	Spray	9 1/2 - 10
B-2	60	"	"	7.4	880	Air	8 1/2
C-1	60	"	"	7	880	Spray	9
D-1	19	accelerated	1200	6.7	902	Air	8 1/2 - 9
D-2	35	"	1200	10.6	884	Air	8 1/2
D-3	51	"	1100	16.6	851	Air	8 1/2
D-4	64	"	1200	16	870	Air	8 1/2
A-5	35	hot-rolled					5 1/2 - 6
E-1	19	"					6 1/2 - 7
E-2	35	"					4 1/2
D-5	35	Conventional					8 1/2

* Plate temperature at exit from Muffle I.
**Cooled to 250°C before entry to Muffle I.

TABLE 3 Tensile and Charpy Properties

Plate	UYS/UTS (MPa)* L**	UYS/UTS (MPa)* TT**	El %/RA % L	El %/RA % TT	$C_v 24°(J)/C_v 27J(°C)$* L	T**	TT
A-1	310/476	300/482	38/73	37/67	198/-70	172/-70	126/-58
A-2	340/502	324/498	38/76	38/65	232/-73	170/<-76	111/-37
A-3	361/509	335/492	37/76	35/57	333/<-76	160/<-76	
A-4	328/483	325/483	38/75	34/64	270/-76	176/<-76	128/-44
B-1	403/542	370/548	37/73	32/64	164/-45	136/-45	104/-36
B-2	333/504	302/493	32/69	31/51	168/-40	114/-25	105/-20
C-1	336/480	328/483	36/75	40/66	238/-57	160/-59	87/-30
D-1	356/546		33/68		91/-41	37/-6	
D-4	353/523	367/522	36/69	32/50	77/-53	50/-30	38/-16
A-5	267/480	260/466	35/72	29/49	242/-44	110/-44	
E-1	332/514		34/71		147/-55		
E-2	293/504		33/71		103/-52	58/-20	
D-5	375/534		34/67		102/-54	38/-20	

* UYS/UTS denote upper yield stress and ultimate tensile strength respectively.
C_v24°(J) denotes V-notch Charpy energy absorption (Joules) at 24°C.
C_v27J(°C) denotes temperature corresponding to 27 J energy absorption.
**L, T and TT are longitudinal, transverse and through-thickness respectively.

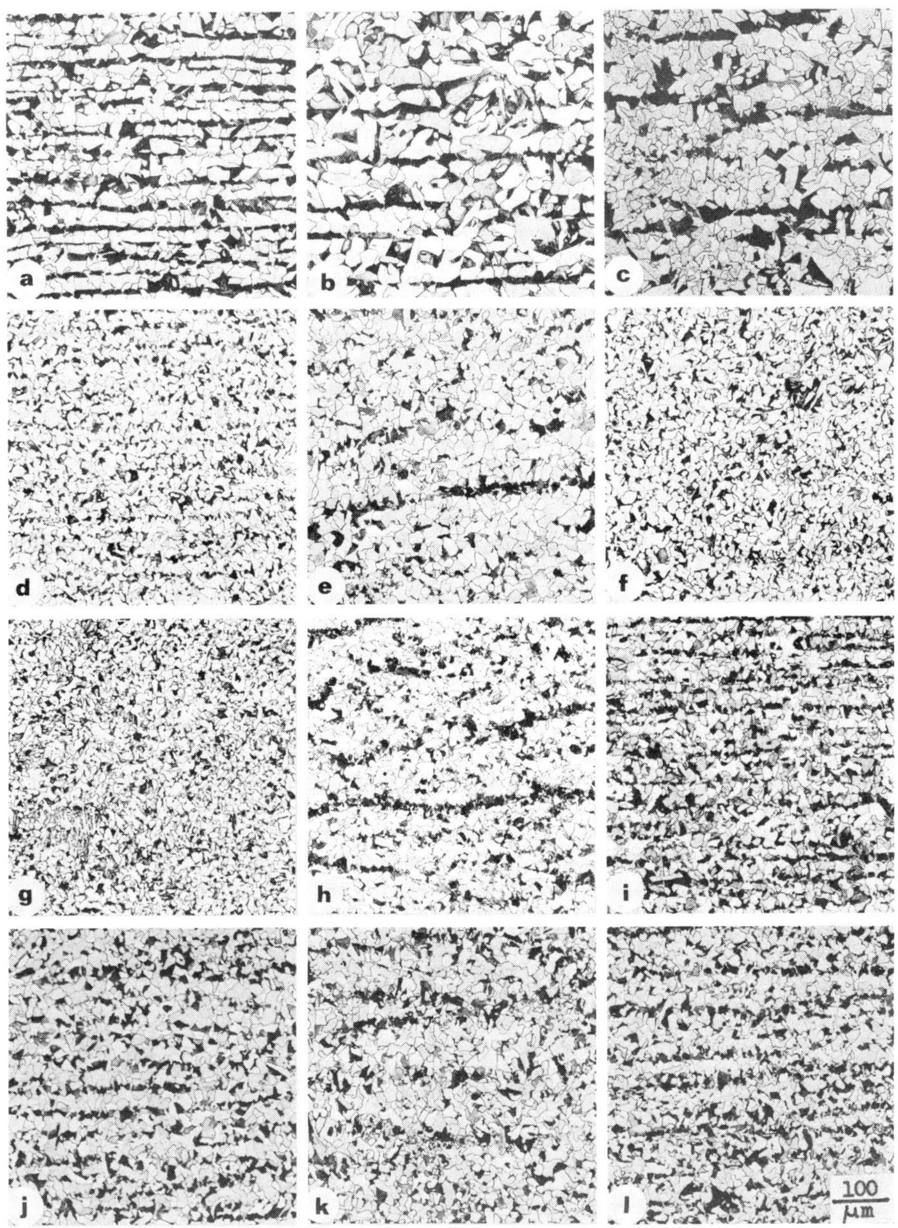

Fig. 2(a-l). Optical micrographs of various plates at mid-thickness. (a-c) as-rolled 19 mm, 35 mm and 60 mm thick plates, steels E, A and B respectively; (d-k) in-line and accelerated normalised plates A-3, A-1, A-2, B-1, B-2, D-1, D-2 and D-4 respectively; (l) conventional normalised 35 mm thick plate, Steel D.

5. Spraying to Ar_1, followed by air-cooling, instead of only air-cooling, in post-muffle II treatment further strengthens the plates without deterioration in toughness by additional grain refinement; (cf Figs. 2e and Fig. 2d or 2f). Banding is also eliminated. The effect of spraying appears most beneficial when V is present as the grains of plate B1, (Fig. 2g), are significantly smaller than those of plate B2, (Fig. 2h). This refinement plus precipitation strengthening results in a marked strength increase without a corresponding decrease in toughness. However, Widmanstatten-type structures occasionally appear in spray-cooled plates, (Figs. 2f,g), indicating incomplete transformation of particular austenite grains to conventional ferrite and pearlite. This type of spray treatment differs from OLAC (9) in that OLAC is directly applied to control-rolled HSLA plates finished a little above Ar_3.

CONCLUSIONS

1. Grain size refinement and improvements in strength and toughness produced by in-line and accelerated normalising procedures are comparable to those obtained by conventional normalising.

2. In-line normalising conserves time and energy. Time is saved by accelerated normalising.

3. The use of accelerated cooling further saves time, refines grain size and improves strength and toughness.

ACKNOWLEDGEMENT

The authors gratefully acknowledge the assistance of Mr. L. Clements and the Rolling Mill Staff.

REFERENCES

1. Fegredo, D.M., (1977). Metals Technol. 4, 417-424.
2. Heedman, P.J., S.P. Rutqvist and J.A. Sjostrom, (1981). Metals Technol. 8, 352-360.
3. National Standards of Canada, (1981). Structural quality steels. Canadian Standards Association CAN 3-G.40.21-M81, Rexdale, Ontario, Canada M9W 1R3.
4. Archambault, P., G. Didier, F. Moreaux and G. Beck, (1984). Metal Progress 126, 67-72.
5. Fegredo, D.M., W.A. Polland, E.F. Connors and D.R. Kiff, (1983). Can. Met. Quart. 22, 453-473.
6. Pickering, F.B., (1971). Towards improved strength and ductility, Climax Molybdenum Co. Symposium, Kyoto, 9-31.
7. Takada, H., K. Kaneko, T. Inoue and S. Kinoshita, (1978). ASTM STP 645, Philadelphia 19103, 335-350.
8. Pickering, F.B., (1978). Physical Metallurgy and the Design of Steels. Applied Science Publishers, London.
9. Tsukada, K., K. Matsumoto, K. Hirabe and K. Takeshige, (1982). Iron and Steelmaker, 9, 21-28.

Effect of Austempering on the Microstructure and Tensile Properties of Ductile Iron

H. Abu-Elfotouh*, O. A. Abu-Zeid**
B. A. Elsarnagawy* and A. M. Eleiche***

**Department of Metallurgy, Military Technical College, Cairo, Egypt*
***Department of Mechanical Design and Production, Cairo University, Cairo, Egypt*
****On leave to Department of Engineering, The American University in Cairo, Egypt*

ABSTRACT

The effects of the transformation temperature and time, as austempering control parameters, on the tensile properties and microstructure of a ductile iron having a carbon equivalent of 3.88 percent have been investigated. Two isothermal transformation temperatures have been used, and were chosen at 270 and 400°C such as to obtain microstructures falling within the zones of lower and upper bainite, respectively. An ultimate tensile strength of 1620 MPa and a corresponding elongation of 5 percent have been obtained when austempering for 10 hours at 270°C, whereas maxima in the ultimate tensile strength of 1087 MPa, the ultimate-to-yield stress ratio of 3 and the elongation percent of 19.7 have been obtained when austempering at 400°C only for one hour. Minimum hardness values of 460 and 265 HV have been obtained after one hour of transformation at 270°C and 400° respectively. The variation of the mechanical properties with the transformation temperature and time has been observed to be directly related to the phase transformations in the microstructure.

KEYWORDS

Austempered nodular iron; microstructure; upper bainite; lower bainite; mechanical properties.

INTRODUCTION

Austempered nodular iron castings can compete favourably with steel forgings in the manufacture of engineering components such as gears and crankshafts (1). This is partly due to the exceptional combination of high strength, ductility and toughness introduced in nodular iron by austempering, in addition to its well-known excellent castability, machinability, wear resistance and damping capacity. Austempered ductile iron has a wide variety of mechanical properties which can be achieved by altering the austempering control parameters such as the austenitization temperature, isothermal transformation temperature and the holding time at each transformation temperature (2). To this end, the present work aims at investigating the effects of selected transformation temperatures and times on the microstructure and tensile properties of a ductile iron having a carbon equivalent of 3.88 percent.

EXPERIMENTAL PROCEDURES

Tensile test specimens were cut from a circular cast ring, of outer diameter 1000mm, wall thickness 23mm and width 65mm, made of a ductile iron of the chemical composition shown in Table 1, and having a ferrite matrix in the as-received state. Specimen dimensions were chosen according to the ASTM standard specifications (3) to be of 8.75 and 35mm gage diameter and length. The austempering treatment processes were performed using commercial neutral salt baths to prevent oxidation,

TABLE 1 Chemical Composition (Wt%) of the Ductile Iron Used (C.E.= 3.88%)

C%	Si%	Mn%	P%	S%	Mg%
3.1	2.31	0.24	0.04	0.017	0.05

carburization or decarburization. All specimens were austenitized at 900°C for 45 minutes. To avoid specimen cracking or warping which might result from rapid heating to this temperature, the specimens were first heated to 400°C inside an air

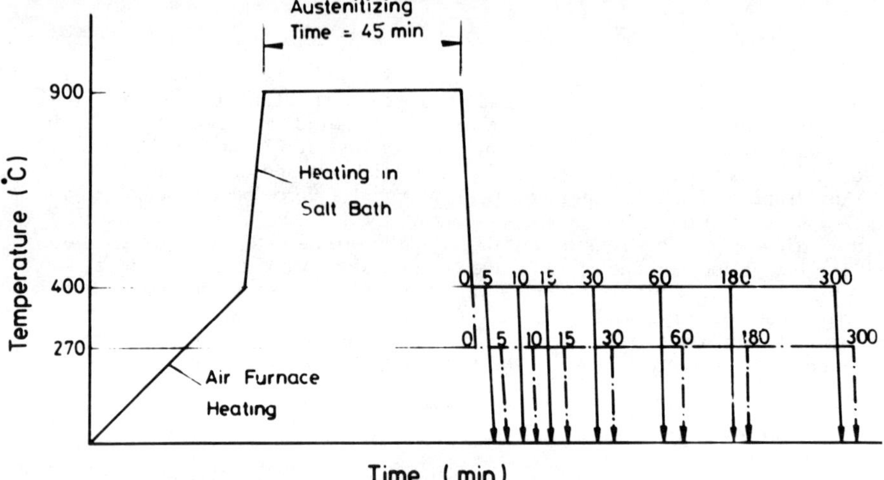

Fig. 1 Representation of the Applied Austempering Thermal Cycles

furnace for 20 minutes, then completely immersed in the salt bath at 900°C. Some of these austenitized specimens were then quenched to 270°C while the remainder quenched to 400°C. Specimens were held at these austempering temperatures for different periods of time ranging from 5 to 600 minutes after which they were further quenched in oil to room temperature. Fig. 1 shows a schematic representation for the different applied austempering thermal cycles. A 100 KN Instron testing machine was used to conduct the tension tests at a constant displacement rate of 0.09 mm/sec, and tensile properties were calculated from the recorded load elongation diagrams. Other samples were simultaneously austempered according to the thermal cycles of Fig. 1. These were used for hardness testing with a load of 4 Kg, and for metallographic examination.

RESULTS AND DISCUSSION

Figure 2 shows the effect of the austempering time on the ultimate tensile strength of the tensile specimens austempered at 270 and 400°C. For these specimens, austempering at 270°C causes the tensile strength to increase at a relative high rate during the first 15 minutes of transformation after which a slight increase in the strength over a longer period of time (about 160 minutes) takes place. Thereafter, the strength is observed to increase again at a higher rate till the 600 minutes transformation time which is the largest period of time used in transformation. These results can be directly related to the microstructure changes that take place during austempering for different times. The photographs of Figs. 3 (a), (b) and (c) show that the rapid increase in tensile strength during the first 15 minutes of transformation at 270°C is associated with a corresponding rapid increase in the amount of lower bainite in the matrix on the expense of martensite. The existence of martensite has been suggested to be associated with severe internal stresses that may lead to a rapid fracture of the quenched specimens (4). The subsequent lower rate of increase in the tensile strength may be associated with a reduction in the transformation rate of austenite into lower bainite as a result of enriching austenite with carbon and its stabilization by the presence of silicon (5). The following rapid increase in strength at longer austempering time intervals is explained by decomposition of austenite to lower bainite, as can be seen from the photograph of Fig. 3 (d).

Le Houillier, Bégin and Dubé (6) examined the effect of austempering time on the amount of bainite. They found a relation similar to that shown for the tensile strength in Fig. 2 for the 270°C austempering temperature. This suggests that the tensile strength is directly related to the amount of lower bainite in the matrix. Increasing the austempering temperature from 270 to 400°C results in a general decrease in the strength except for transformation times less than 10 minutes where martensite is found to be still the dominant phase. The formation of upper bainite instead of lower bainite upon raising the transformation temperature, as can be seen in Figs. 4 (a) and (b) is the main reason for reduction in strength. The peak value in tensile strength observed in Fig. 2 for the tensile specimens austempered for 60 minutes at 400°C may be explained as follows. At first, the strength increases due to the increased amount of transformed upper bainite in the matrix on the expense of austenite and the decrease in the amount of martensite due to enriching of austenite with carbon as can be seen from Fig. 4 (a). For the size of the treated specimens, 60 minutes of austempering is found to be enough for austensite to start decomposing into ferrite and carbides. It has been reported (4) that the mixture of ferrite and carbides found in areas where bainitic ferrite plates have grown into each other suggests that, at this stage, austenite transformation is in the form of eutectoid-type reaction where austenite decomposes into ferrite plates plus carbides on the grain boundaries which lowers the untimate tensile strength.

Ductility of specimens austempered at 270°C has been found to be too small to be measured accurately. On the other hand, Fig. 5 shows that pronounced increased values of ductility as indicated by the elongation percent or the ultimate-to-yield stress ratio have been obtained at 400°C. Both curves in this figure show maxima after 60 minutes of transformation time at this temperature. Before these maxima, ductility increases with increasing the austempering time due to the increase in the amount of retained austenite which is an F.C.C. ductile phase. Maximum value of ductility corresponds to maximum amount of retained austenite in the matrix. The precipitation of carbides on the grain boundaries of ferrite grains is responsible for the decrease in ductility after 60 minutes of transformation.

Figure 6 shows the effect of austempering time on the matrix hardness of the austempered tensile specimens at 270 and 400°C. Minimum values are obtained after 60 minutes of transformation at both temperatures which can be explained by the observation made by Verhoeven and others (5) who found that minimum hardness

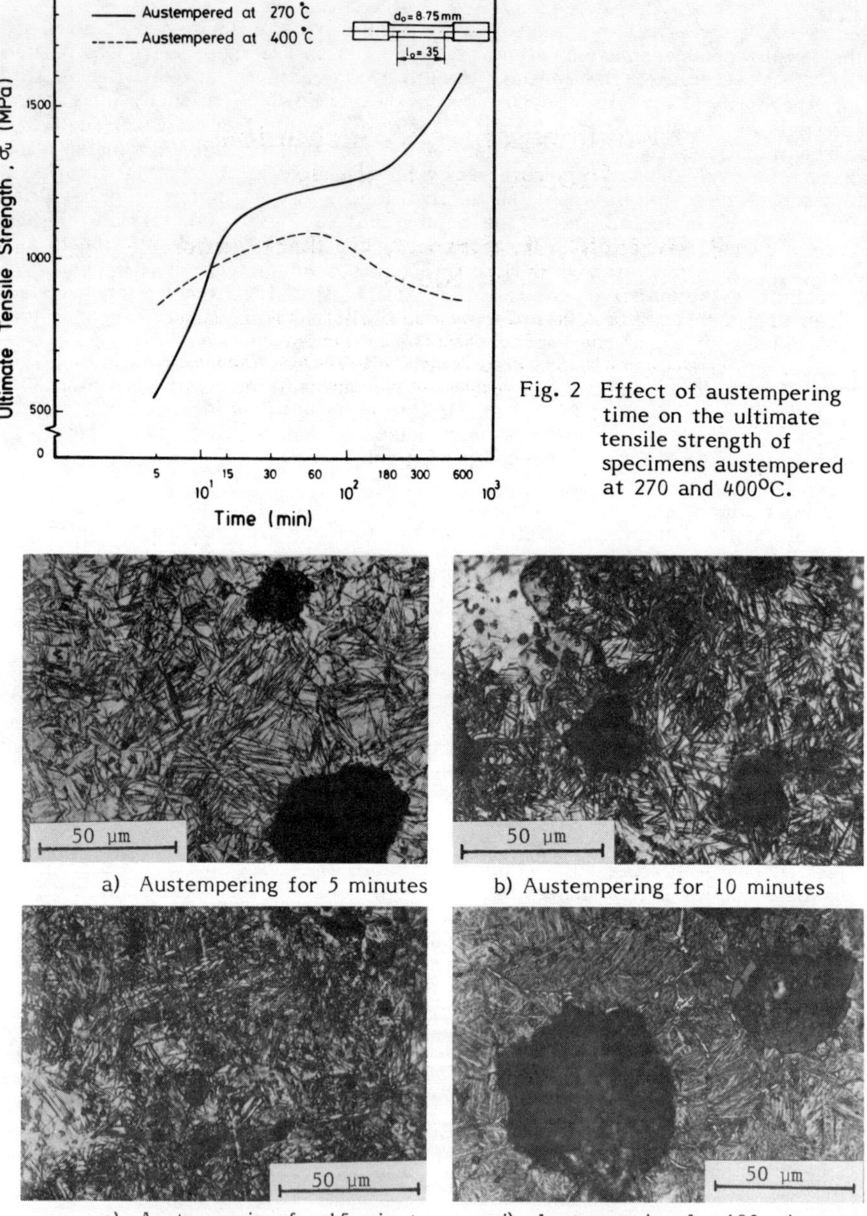

Fig. 2 Effect of austempering time on the ultimate tensile strength of specimens austempered at 270 and 400°C.

a) Austempering for 5 minutes

b) Austempering for 10 minutes

c) Austempering for 15 minutes

d) Austempering for 180 minutes

Fig. 3 Microstructure of lower bainitic ductile iron austempered at 270°C for different times. Etched in nital.

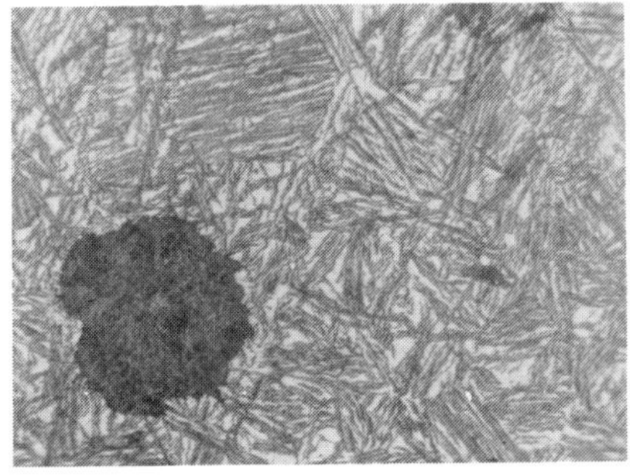

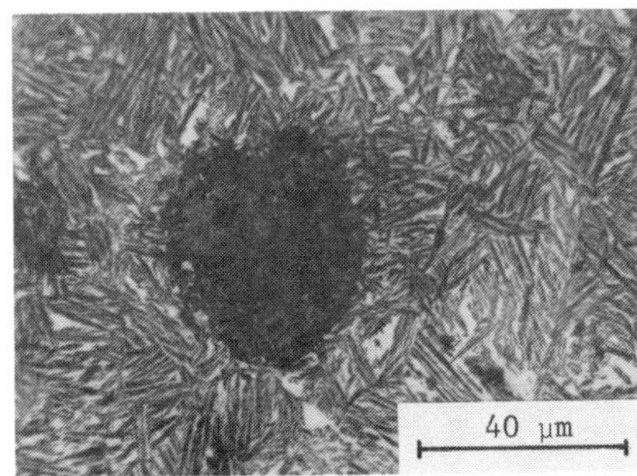

a) Austempering for 30 minutes b) Austempering for 180 minutes

Fig. 4 Microstructures of upper bainite ductile iron austempered at 400°C for 30 and 180 minutes. Etched in nital.

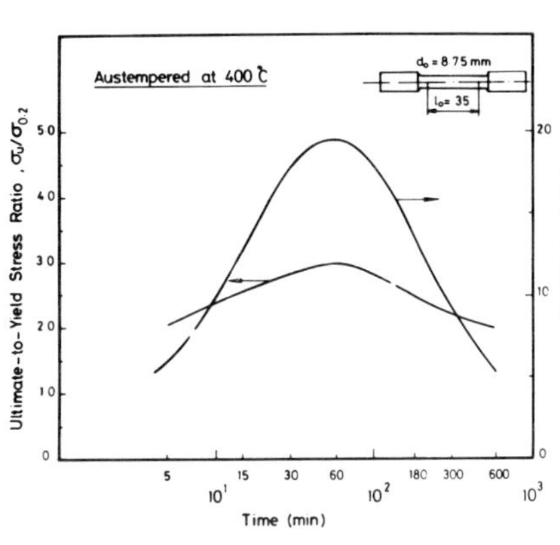

Fig. 5 Effect of austempering time on the elongation percent and the ultimate-to-yield stress ratio of specimens austempered at 400°C.

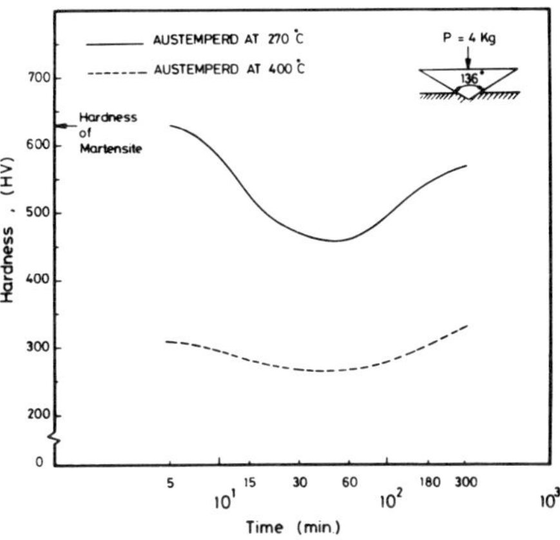

Fig. 6 Effect of austempering time on matrix hardness of specimens austempered at 270 and 400°C.

values correspond to the presence of maximum amounts of stabilized austenite at room temperature.

CONCLUSIONS

The microstructure and hence the mechanical properties of austempered ductile iron are strongly dependent on the austempering isothermal transformation temperatures and time of transformation. High strength, hardness and limited ductility are associated with lower bainitic structure, whereas high ductility with relative lower strength and hardness are associated with upper bainitic structure.

REFERENCES

1. F.S. Rossi and B.K. Gupta. Metal Progress. 25-31, April 1981.
2. J.E. Janowak, R.B. Gundlach, G.T. Eldis and K. Rohrig. Submitted to AFS for publication as a technical report, May 1981.
3. American Society for Testing and Materials. 1976 Annual Book of ASTM Standards. Part 10, E8-69, 120-140, ASTM, Philadelphia (1976).
4. E. Dorazil, E. Mansterova, L. Stransky and A Huvar. AFS International Cast Metal Journal, 53-62, June 1982 .
5. J.D. Verhoeven, A. Elnagar, B. Elsarnagawy, D.P. Cornwell and F. Laabs. Third International Symposium On the Physical Metallurgy of Cast Iron. The Royal Institute of Technology, Stockholm, Sweden. 29-31, August 1984.
6. R. Le Houillier, G. Bégin and A. Dubé. Metall. Trans., Vol. 2, 2645-2653, (1971).

Weld Metal Microstructures in an HSLA Steel

D. N. Hawkins* and M. E. Saggese**

**Department of Metallurgy, Sheffield University, Sheffield S1 3JD, U.K.*
***Materials Department, CNEA, 1429, Buenos Aires, Argentina*

ABSTRACT

Submerged-arc weld runs have been made on a calcium-free and on a calcium-treated HSLA steel using two different commercial wires and six fluxes. The microstructures of the resulting beads have been quantified and their chemical compositions and inclusions analysed, the latter by use of extraction replicas. The object of the work was to gain an understanding of the factors controlling the development of acicular ferrite, (AF), in the weld metal. The major influence was found to be the composition of the inclusions. If they contained $> \sim 30\%$ Al_2O_3 then high proportions of AF were present in the microstructure. However if the oxygen pick-up from the flux is excessive, so causing the inclusion alumina content to fall below this value, then high AF contents will only be maintained if $> \sim \frac{1}{2}\%$ titanium is present in the inclusions.

KEYWORDS

Acicular ferrite; submerged-arc weld microstructures; inclusion compositions.

INTRODUCTION

To achieve high toughness in submerged-arc welds in microalloyed structural steels it is considered necessary for the microstructure of the weld bead to consist mainly of acicular ferrite, (AF). The major factors that control the development of this microstructure are cooling rate, the austenite grain size prior to transformation, the chemical composition of the bead and the inclusions present. Evidence for the effectiveness of inclusions as nucleating agents for acicular ferrite formation has been available for some years (1-3). Further work has shown that their effectiveness is dependent upon their composition, for example that they should be alumina rich (4,5) or that they should contain titanium (6) . The volume fraction of inclusions present in the bead will depend mainly upon the oxygen content of the weld (7), which in turn is dependent upon the type of flux used (5,8,9) and the welding parameters employed (10), and their composition will be determined by the types and levels of deoxidant present. As considerable dilution of the

weld bead by the plate can often occur, variations in plate composition can also be important and it has been suggested that whether or not the plate is calcium-treated can influence the weld microstructure (4).

In the work described in this paper, bead-on-plate welds have been produced, under standardised submerged-arc welding conditions, on two plates of similar chemistry, except that one was calcium-treated. Six different fluxes were used, covering a wide range of basicities, and two different commercial wires. The microstructures of the resulting beads were then quantified, compared, and related to the chemistry of both the matrix and the weld metal inclusions. The object of the work was to gain an understanding of the roles played by flux and wire composition in determining the weld bead microstructure and whether or not it differed when welding calcium-treated plate.

EXPERIMENTAL PROCEDURE

Bead-on-plate welds were deposited on two microalloyed steels, A and B in Table 1, which were in the form of 25 mm thick plate. They had similar

TABLE 1 Chemical Compositions of Plates and Wires, ($^{wt}/_{o}$).

Material	C	Si	Mn	Ni	Mo	Al	Ti
Steel A	0.048	0.33	1.40	0.24	0.006	0.044	0.004
Steel B	0.053	0.33	1.35	0.25	0.005	0.032	0.003
Wire S	0.11	0.20	1.52	0.05	0.46	0.014	< 0.001
Wire T	0.076	0.03	1.39	0.07	0.27	0.020	0.032

chemical compositions, except that steel A was calcium-free whereas steel B was calcium-treated. Each bead was deposited under identical submerged-arc-welding conditions using a single wire feed which carried 600 A d.c., at 32 V, with a travel speed of 460 mm min^{-1}, which gave a heat input of 2.5 kJ mm^{-1} of weld length. On both compositions of steel plate two different wires, S and T, were used, in combination with six different fluxes of basicity indices ranging from 0.82-3.1. The compositions of the wires and fluxes are given in Tables 1 and 2. (In Table 2 B.I. refers to the basicity index of the flux calculated according to the formula suggested by Tuliani et al (11).) There were thus 24 welds in total which for convenience of

TABLE 2 Flux Compositions, ($^{wt}/_{o}$).

Flux	B.I.	SiO_2	Al_2O_3	TiO_2	CaO	MgO	CaF_2	MnO	Fe_2O_3
1	3.1	14.0	17.0	0.69	8.2	28.7	23.0	1.0	1.9
2	1.8	19.5	20.5	0.62	4.8	23.6	19.5	5.2	1.8
3	1.4	44.6	2.2	0.10	38.3	0.6	4.1	5.5	0.2
4	1.0	39.4	14.6	0.78	19.2	10.2	5.3	7.4	0.9
5	0.98	42.1	3.6	2.33	2.8	23.4	3.6	15.6	2.0
6	0.82	38.1	4.5	0.24	5.9	0.4	2.7	45.8	2.5

discussion have been subdivided into two batches, S and T, where each batch consists of 12 welds made with wires S and T respectively. For microstructural assessment each weld was sectioned transverse to the welding direction and, after standard metallographic preparation, the volume fractions of the different ferrite morphologies, (namely acicular ferrite, AF, grain boundary ferrite, GBF, and ferrite with aligned microphases, AC), were quantified by point counting. Similar areas were assessed in each bead and sufficient points were counted to give a 95% confidence limit on the acicular ferrite contents of ± 4%. Within these areas the widths of the prior austenite grains, which were columnar shaped and delineated by grain boundary nucleated ferrite, were also measured. The % AF and grain widths, ($\gamma_{g.w.}$), are given in Table 3. To analyse the inclusions present in each bead

TABLE 3 Weld Bead Microstructures, Oxygen Content and Inclusion Compositions, ($^{wt}/_{o}$).

Weld	$\gamma_{g.w.}$ μm	AF%	O_2ppm	Ti	Al_2O_3	SiO_2	MnO
SA1	123	71	291	1.12	80	8	11
SB1	116	67	286	1.33	65	16	17
SA2	105	73	499	0.51	34	34	31
SB2	120	64	390	0.47	37	33	29
SA3	155	39	703	0.21	25	43	32
SB3	174	21	675	0.15	17	46	36
SA4	147	43	1113	0.29	20	45	35
SB4	148	38	1015	0.26	15	47	38
SA5	119	73	825	1.17	22	41	35
SB5	113	77	843	1.21	11	46	41
SA6	115	37	1256	0.10	12	46	42
SB6	157	35	1166	-	8	47	44
TA1	110	85	318	4.3	76	4	13
TB1	115	87	290	4.9	70	7	15
TA2	109	81	457	3.6	43	23	25
TB2	97	81	426	3.0	39	28	28
TA3	103	79	642	1.86	26	39	31
TB3	105	81	618	2.0	21	39	36
TA4	115	65	1039	1.1	16	46	35
TB4	103	65	1150	1.0	11	47	40
TA5	102	77	840	2.2	16	41	39
TB5	114	73	852	1.9	11	46	40
TA6	108	59	1122	1.1	12	43	42
TB6	141	70	1115	1.1	9	44	48

extraction replicas were taken from the same areas as those used for quantitative metallography. These were then studied in a Philips 400 T electron microscope, which was equipped with STEM and EDAX microanalysis facilities. The average analyses of the inclusions found in each bead are given in Table 3. The oxygen content of each bead is also given in this table. Space precludes the inclusion of the full chemical analysis of each bead. The major variations in their chemical composition will be discussed later.

RESULTS

Weld metal microstructure. The AF contents of the weld beads varied between 59-87% in batch T and from 21-77% in batch S. However there appeared to be no significant difference between the microstructures in the calcium-treated and the calcium-free plates when welded with the same consumables. When using any particular flux, wire T always give higher proportions of AF in the weld bead than wire S. The average prior austenite grain widths varied between 97 and 174 μm, with those in batch S generally being larger than those in batch T when using a similar flux. Within batch S, as the grain width increased, the proportion of AF in the weld bead decreased, whereas in batch T there was no correlation between the two variables.

Weld metal composition. Within each of the two batches the major differences in composition were in Mn (1.28-1.77%), Si (0.26-0.55%) and oxygen (290-1256 ppm) levels, these variations being caused by differences in the flux chemistry. In both batches there were also small variations in the aluminium and nitrogen levels, which were a reflection of the different amounts of these elements in the plate. Welds in batch S had higher molybdenum and lower titanium contents than equivalent welds in batch T, due to the differences in these two elements in the two wires. In batch T it was noticed that, as the oxygen level increased, the proportion of AF decreased, whereas there was no correlation between these two variables in batch S.

Weld metal inclusions. The sizes of the inclusions analysed varied between ¼-3 μm. Whichever flux, wire and plate was used they were found to consist essentially of Al_2O_3, SiO_2 and MnO with small amounts of titanium, with the proportion of the latter being greater in the inclusions in batch T than in those in batch S, presumably because of the titanium in wire T. As the oxygen level of the bead increased the proportions of Al_2O_3 and titanium in the inclusions decreased, while those of SiO_2 and MnO increased. The exception to this was when using flux 5, which contained a higher level of TiO_2 than the other fluxes. The titanium in the inclusions in welds made with this flux increased, notwithstanding the increase in oxygen level. However as these changes in inclusion composition occurred, the ratio of MnO to SiO_2 in the particles remained relatively constant.

DISCUSSION

Previous work has indicated a correlation between the AF content of the weld and the C.E.V., (carbon equivalent value), of the bead composition (4). However, in the work described here, although the range of C.E.V.'s covered was wider than that in the work of Cochrane et al (4), no such correlation was found. This was true also when a more comprehensive C.E.V. formula, that took into account the silicon content of the bead was used. The oxygen level of the weld bead can be predicted from the $^{wt}/_{o}$ (MnO + SiO_2 + TiO_2) of the flux (13), so that variations of these three oxides within the six commercial fluxes used was responsible for the range of oxygen levels observed. Within this range there was no maximum in the AF contents of the welds, as had been reported by other investi-

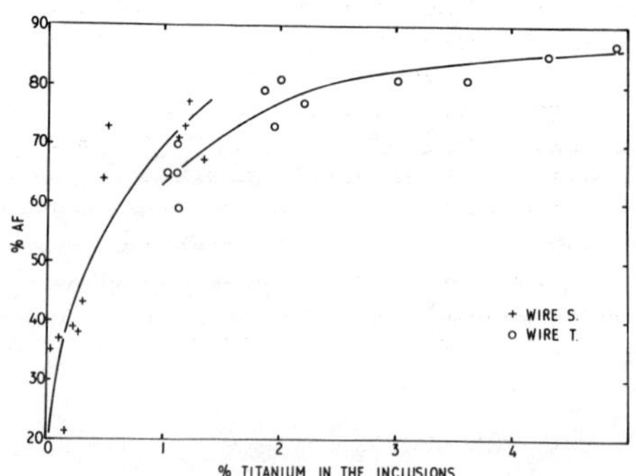

Fig. 1. % Titanium in Inclusions v % AF in the weld metal.

gators (1,5,12). There was only the gradual decrease of AF, with oxygen content, within the T batch of welds. With respect to the observed variations in columnar grain size, it is normally considered that this is controlled by the parent plate grain size, adjacent to the fusion boundary, during solidification. At first sight, therefore, there is no obvious explanation for the large variations that were observed, for each heat affected zone should have experienced the same thermal cycle. Inclusion composition, size and distribution could be a contributory factor. However, the inclusion composition in batch S was very similar to that in batch T, when comparing beads made on the same plate with the same flux, yet the grain sizes were generally larger in batch S. Thus as yet there is no explanation for the differences. Within batch S the trend for AF content to decrease with increase in the average prior austenite grain width is contrary to what might be expected, as AF is an intragranularly nucleated phase (3), whereas the other forms of ferrite which were quantified are nucleated on the austenite boundaries. Further work is therefore being carried out to see whether these effects can be due to differences in inclusion size or distribution and to see if the results are repeatable for wire S using a different range of fluxes. From the chemical analyses of the weld beads it could be seen that even at the highest oxygen level, 1250 ppm, there was sufficient Al, Ti,Si and Mn present to deoxidise the weld pool. At the lowest oxygen levels, ∿ 300 ppm, there was sufficient aluminium present to ensure that the inclusions consisted mainly of Al_2O_3. However as the oxygen level increased, there was insufficient aluminium to fully react with it, so the proportions of Al_2O_3 in the inclusions decreased. The inclusions then became rich with respect to SiO_2 and MnO, as the amount of titanium present was low by comparison. Within each of the two batches it was noticeable that the inclusions in the weld beads on the calcium-treated plate always had lower Al_2O_3 contents than those found in beads made with the same flux on the calcium-free plate; this was an effect of the lower aluminium content of the former plate material. Plates of similar chemistry, except for aluminium, can, if the aluminium difference is large enough, give rise to significant differences in the microstructure of welds in the two materials (4,14). However, in this investigation no significant difference was seen between the microstructures of welds in steels A and B when using similar consumables, probably because the difference in the aluminium contents of the steels was relatively small. The change in inclusion composition with oxygen content is of major importance, for it has been observed that for such particles to act as nucleating agents for AF formation their composition is critical (2-6). This can be seen too from this work, for in batch S as the proportion of Al_2O_3 in the inclusions increases so too does the proportion of AF. (There was no relationship however between these two variables in batch T). Within batch S the exception to this trend was when flux S was used, since beads made using this flux had high levels of AF, although their inclusions had low Al_2O_3 contents. The difference was that the inclusions also contained titanium donated by the flux. Because of the apparent importance of this element, the titanium content of the inclusions of both sets of welds has been plotted against their AF contents in Fig. 1. From this it can be seen that the titanium inclusion content is the one variable in both weld sets that improves their microstructure. Thus it suggests that this element is of prime importance in ensuring that the inclusions act as efficient nucleating agents for AF.

CONCLUSIONS

For the submerged-arc welding conditions studied in this work the major factor controlling the weld bead microstructure was the composition of the inclusions formed within the bead. If they contained $> \sim 30\%$ Al_2O_3 they acted as efficient nucleating agents for the formation of AF. However as the basicity of the flux used was decreased, the oxygen content of the bead increased. There was then insufficient aluminium in the plate and wire to maintain this high proportion of Al_2O_3 in the inclusions, for, unlike titanium, aluminium was not transferred to the weld metal from the flux. Thus at high oxygen levels the proportion of AF in the microstructure decreased markedly unless $> \sim \frac{1}{2}\%$ titanium was present in the inclusions. This level of titanium could be supplied either from the wire or from the flux.

REFERENCES

1. D. J. Abson, R. E. Dolby and P. H. M. Hart, in "Trends in Steels and Consumables for Welding", p.75, Weld. Inst. Cambridge, (1978).
2. R. A. Ricks, P. R. Howell and G. S. Barritte, J. Mater. Sci., 17, 732, (1982).
3. R. E. Dolby, in "Advances in the Physical Metallurgy and Applications of Steels", p.111, Book 284, Met. Soc., London, (1982).
4. R. C. Cochrane, J. L. Ward and B. R. Keville, in "The Effects of Residual, Impurity and Microalloying Elements on Weldability and Weld Properties", paper 16, Weld. Inst. Conf., London, (1983).
5. L. Devillers, D. Kaplan, B. Marandet, A. Ribes and P. V. Riboud, ibid, paper 1.
6. N. Mori, H. Homma, S. Okita and M. Wakabayashi, IIW Doc IX-1196-81, May, (1981).
7. D. J. Widgery, as ref. 1, p.217, (1978).
8. T. W. Eagar, Weld. J., 57, 76-s, (1978).
9. T. H. North, H. B. Bell, A. Koukabi and I. Craig, Weld. J., 58, 343-s, (1979).
10. P. R. Kirkwood, Metal Con., 10, 260, (1978).
11. S. S. Tuliani, T. Boniszewski and N. F. Eaton, Weld. Metal Fab., 37, 327, (1969).
12. R. C. Cochrane and P. R. Kirkwood, as ref. 1, p.103, (1978).
13. M. E. Saggese, A. R. Bhatti, D. N. Hawkins and J. A. Whiteman, as ref. 4, paper 15, (1983).
14. H. Terashima and P. H. M. Hart, as ref. 4, paper 27, (1983).

Effect of Tempering on Tensile Properties of Plain Carbon Dual-phase Steels

T. C. Lei and H. P. Shen

Department of Metals and Technology, Harbin Institute of Technology, Harbin, People's Republic of China

ABSTRACT

Tensile properties and microstructure of dual-phase (M+F) steels with various volume fraction and carbon content of martensite were studied after tempering at 80 to 500 C. The changes in strength and ductility of the steels upon tempering were analyzed with respect to those occured in the two constituents.

KEYWORDS

Dual-phase steels; volume fraction of martensite; carbon content of martensite; ultimate and yield strengths; yield plateau; uniform and total elongations; coefficient of utilization of martensite strength.

INTRODUCTION

The microstructure and property changes upon tempering are very important for practical uses of martensite plus ferrite (M+F) dual-phase steels. Speich(1,2) and Rashid(3,4) have studied the effect of tempering on Mn- and Mn-Si-V dual-phase steels showing that tensile properties cannot be improved by tempering. In our previous studies(5,6) it was found that low temperature tempering may appreciably affect the strength of some dual-phase steels. Davies(7) has reported that tempering has much more effect on water quenched than air-cooled steels. The aim of the present study is to clarify the effect of tempering in a wide range of temperature on the microstructure and tensile properties of plain carbon dual-phase steels with various volume fraction and carbon content of the martensite phase.

EXPERIMENTAL

Materials used were 1 mm thick,cold-rolled sheets of steels

1010(0.09C,0.37Mn,0.17Si) and 1020(0.24C,0.49Mn,0.23Si). Their initial microstructure consists of spheroidized cementite particles and ferrite matrix. Tensile specimens of 35 mm gage length and 12.5 mm width were cut along the rolling direction. Two sets of heat treatment were carried out: (1)Specimens were heated for 15 min to 780 C(steel 1010) or 745 C(steel 1020) and then quenched into 5% NaOH water solution. After that the volume fraction of martensite V_M is 0,18 and 0.35 and the carbon content of martensite C_M is 0.41% and 0.64% respectively for steels 1010 and 1020. Then the specimens underwent a 1 h tempering at 80,120,160,200,350 and 500 C; (2)Specimens of steel 1010 were heated to 760,800,825,840 and 860 C for 15 min and quenched into 5% NaOH water solution with obtaining V_M= 0.14,0.25,0.35,0.45 and 0.70 respectively. The specimens underwent a unique 200 C tempering for 1 h.

Tensile tests were carried out on a 6 T machine with a strain rate of nearly $5x10^{-4} s^{-1}$. The volume fraction of martensite V_M was measured by using the line-cutting method and the carbon content of martensite was estimated with a standard Fe-C phase diagram for various quenching temperatures. Microhardness of the two phases were determined with 5 g load within 10 s. Microstructure of the steels was examined optically and with an electron microscope. The internal friction curves were measured by using a torsion pendulum on specimens 1x2x180 mm with a frequency of about 1 c/s at 20 C in order to find the changes in carbon content of ferrite phase.

RESULTS AND DISCUSSION

1.Strength

The yield(0.2% proof) and ultimate strength values of the steels and the microhardness of the two phases M and F after tempering at various temperatures are shown in Figs 1 and 2. It can be seen that the microhardness of M in steel 1020 is higher than in 1010 because of the higher carbon content,but decreases more rapidly as tempering temperature increases. Microhardness of F in steel 1020 is also higher than in 1010 because of higher V_M and C_M values which result in a higher degree of phase-hardening of the F phase.

Previously(8) we have derived a new expression for evaluating the UTS of dual-phase steels by using the shear lag analysis of the load transfer from M to F. It is

$$\sigma_{bDP} = X_M \, \sigma_{bM} \, V_M + \sigma_{bF}(1-V_M) \qquad (1)$$

with
$$X_M = \frac{1}{K} \left(\frac{\beta}{2\sqrt{3}} + 0.65\right)$$

in which σ_{bDP} is the UTS of steels, σ_{bM} and σ_{bF} are the UTS of the two phases estimated by using the ASTM hardness-to-strength conversion chart on basis of measured microhardness data. V_M is the volume fraction of martensite,while $1-V_M$ is the volume fraction of ferrite. X_M is the coefficient of utilization of the strength of martensite,where K is the strength ratio of M to F. β is the average aspect ratio of the M island

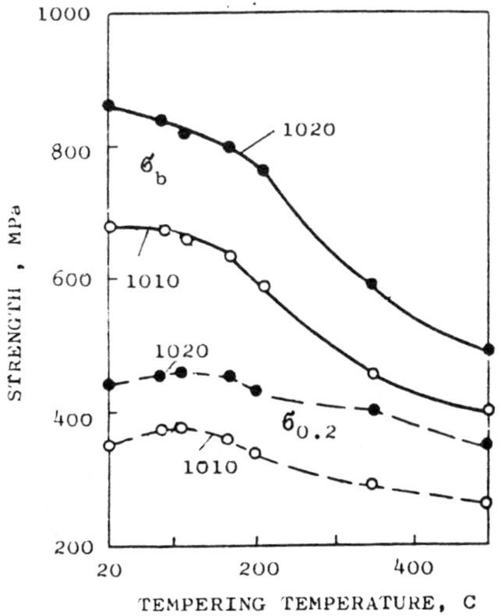

Fig.1 Yield($\sigma_{0.2}$) and tensile (σ_b) strength of dual-phase steels 1010(V_M=0.18,C_M=0.41) and 1020 (V_M=0.35,C_M=0.64) vs tempering temperature.

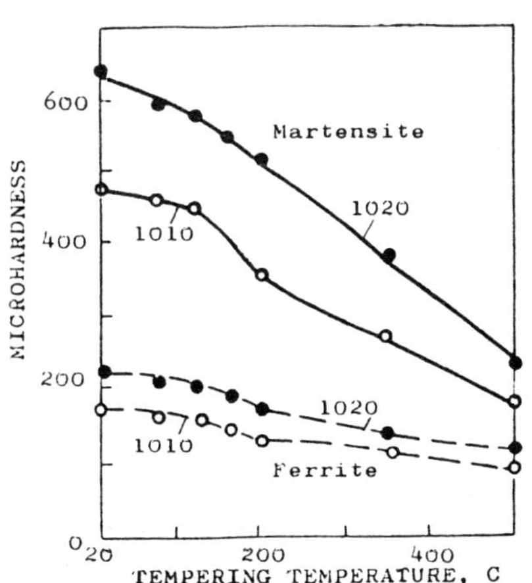

Fig.2 Microhardness of the two phases in dual-phase steels vs tempering temperature.

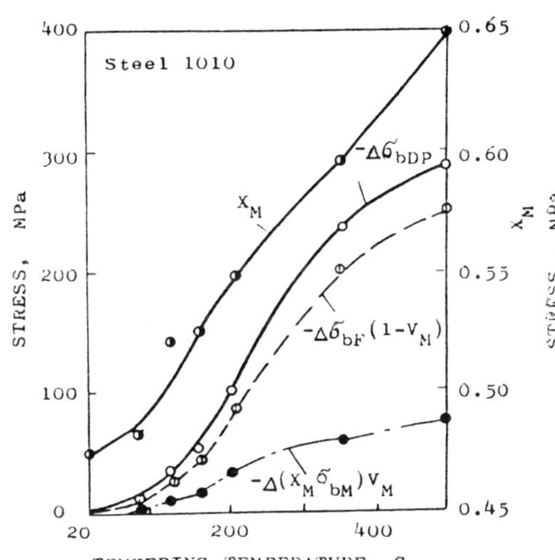

Fig.3 Changes of the terms in Eq.(2) for steel 1010 with tempering temperature.

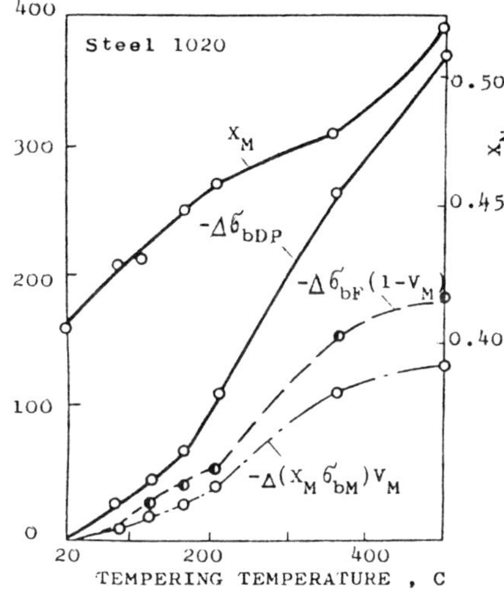

Fig.4 Changes of the terms in Eq.(2) for steel 1020 with tempering temperature.

($\beta = l/d$ in which l is the length and d the width of the island). Therefore, the decrease in UTS of dual-phase steels due to tempering can be expressed as

$$-\Delta\sigma_{bDP} = -\Delta(X_M\,\sigma_{bM})V_M + (-\Delta\sigma_{bF})(1-V_M) \qquad (2)$$

The first righthand term of Eq(2) is the decrease of stress carried by martensite and the second one is the decrease of stress carried by ferrite.

It is clear from Figs 3 and 4 that although the microhardness of M decreases more rapidly than F(see Fig.2), the decrease of stress carried by ferrite matrix plays a more greater role in the decrease of strength of the steels. This is because, firstly, the V_M values are small and, secondly, the coefficient of utilization of martensite strength X_M is less than 1, though it increases obviously with increasing tempering temperature.

Figure 5 shows the changes of strength and ductility of steel 1010 with various V_M after tempering at 200 C. Here, the less the V_M(and hence the more the C_M according to Fe-C phase diagram), the more the decrease in strength values. This tells once more that tempering may give more softening effect for steels with martensite of higher carbon content.

2.Ductility

It can be seen from Fig.6 that the total and uniform elongations of the steels increase with increasing temperature of tempering, especially at temperatures higher than 120 C. Such improvements in ductility of steel 1010 with various V_M have been shown already in Fig.5. The Snoek peak of the internal friction curves shown in Fig.7 was found to decrease quickly after tempering at lower temperatures but slowly at higher temperatures. Decrease in Snoek peak height indicates the escape of interstitial C,N atoms from ferrite causing "purification" of the later. STEM observations, Fig.8, show really the precipitation of $Fe_3C(N)$ particles in ferrite during tempering resulting in an obvious reduction of pinned dislocations and causing the improvement of ductility. The analysis of the two factors, namely, the softening of martensite and the purification of ferrite which can affect the ductility of dual-phase steels during tempering, as shown in Fig.9, indicates that after tempering at lower temperatures the improvement of ductility is essentially due to the purification of ferrite, while after tempering at higher temperatures(higher than 160 C) this improvement is mainly attributed to the softening of martensite.

CONCLUSIONS

1. The YS and UTS values of dual-phase steels can be appreciably affected by tempering due to changes in the microstructure of the two constituents, martensite and ferrite.

2. The ductility of the steels can be improved by tempering due to the purification of ferrite for tempering at lower temperatures and the softening of martensite for higher tem-

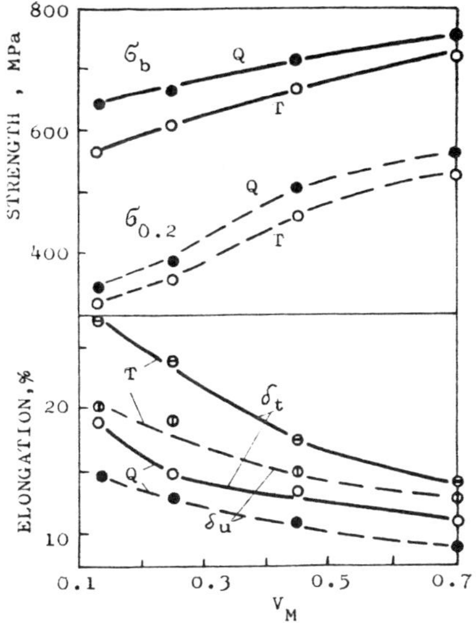

Fig.5 Changes of strength and ductility of quenched(Q) and 200 C tempered(T) dual-phase steels 1010 with volume fraction of martensite. (δ_t and δ_u are total and uniform elongations respectively.)

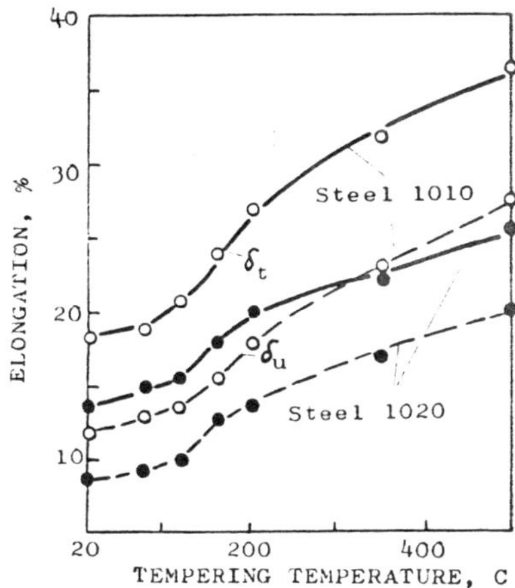

Fig.6 Total and uniform elongations of steels 1010 and 1020 vs tempering temperature.

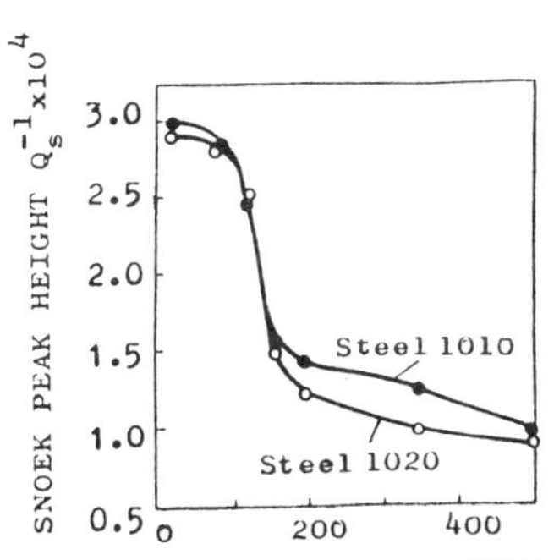

Fig.7 Snoek peak height taken from internal friction curves of the steels vs tempering temperature(with background excluded).

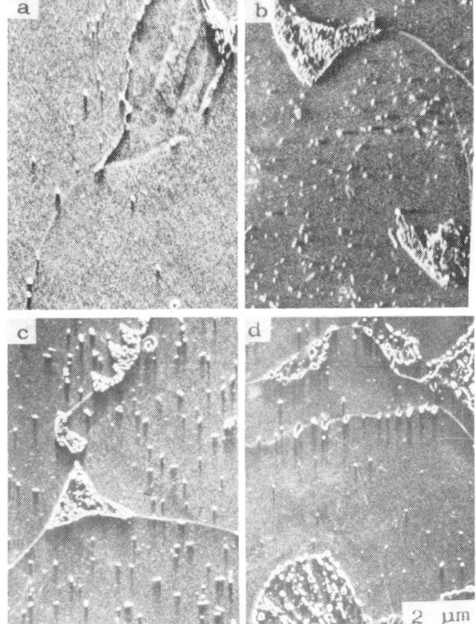

Fig.8 STEM photographs showing the precipitation processes in ferrite with increasing temperature of tempering: a-as-quenched; b-200 C tempered; c-350 C tempered and d-500 C tempered for dual-phase steel 1010 with V_M=0.18 and C_M=0.41.

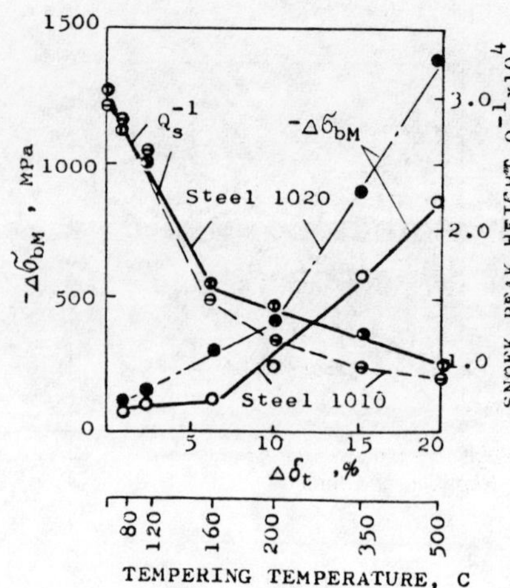

Fig.9 Decrease of martensite strength ($-\Delta\sigma_{bM}$) and Snoek peak height vs increase of total elongation($\Delta\delta_t$) or tempering temperature.

peratures of tempering.

3. Therefore, the strength and ductility of dual-phase steels can be changed in a wide range after tempering even at low temperatures.

REFERENCES

1. G.R.Speich, A.J.Schwoeble and G.P.Huffman, Metall.Trans., 14A, 1079 (1983).
2. G.R.Speich and R.L.Miller, Fundamentals of Dual-Phase Steels, (edited by R.A.Kot and B.L.Bramfitt), p.279, TMS-AIME (1981).
3. M.S.Rashid and B.V.N.Rao, Metall.Trans., 13A, 1697 (1982).
4. M.S.Rashid and B.V.N.Rao, Fundamentals of Dual-Phase Steels, (edited by R.A.Kot and B.L.Bramfitt), p.249, TMS-AIME (1981).
5. H.P.Shen, T.C.Lei, D.J.Li and C.Y.Song, Trans.Metal Heat Treatment, 5, 30 (1984).
6. T.C.Lei, H.P.Shen and J.Pan, Iron and Steel(China), 18, No.7, 46 (1983).
7. R.G.Davies, Fundamentals of Dual-Phase Steels, (edited by R.A.Kot and B.L.Bramfitt), p.265, TMS-AIME (1981).
8. H.P.Shen and T.C.Lei, Metal Science, 18, 257 (1984).

Deformation-aging Strengthening of High Strength Steel 30CrMnSiA

B. D. Hong, T. C. Lei, J. B. Yang and D. M. Jiang

Department of Metals and Technology, Harbin Institute of Technology, Harbin, People's Republic of China

ABSTRACT

The effect of cold deformation and aging after quenching and high temperature tempering on the microsructure and mechanical properties of a high strength commercial steel 30CrMnSiA was studied. The results show that the hardness and strength of the steel increase with increasing degree of cold rolling and an additional strengthening effect can be obtained by a 300°C aging after rolling. Foil TEM observations indicate that the combined strengthening effect of cold rolling and aging lies in the increase of dislocation density and the precipitation of very fine particles of carbides on dislocation networks.

KEYWORDS

Deformation-aging strengthening; quenching; tempering; cold rolling; tensile properties; dislocation density; substructure; transmission electron micrographes; fracture morphology; Strengthening mechenism.

INTRODUCTION

Steel 30CrMnSiA is widely used for manufacturing vessels and other components of various aircraft in aeronautical and space industries, because of its higher strength, good plasticity and practical economy. The steel is often used after quenched and low temperature tempered to obtain high strength. However, the hardness of the steel is too high to form or machine, so it is not suitable for those components which are required with less distortion and accurate dimension. If the steel is treated by quenching and high temperature tempering, though the formability and machinablity are improved, the strength potential of the steel can not be fully used.

However, Many works published in literatures [1,2,3,4] show that the deformation-aging process is an important method for obtaining high strength materials. A higher plasticity and toughness can be possessed in the meantime if the processing parameters are suitable. Moreover, the forming and heat treating of the component parts can be combined ingeniously in the process of deformation-aging, so the energy and time consumption is reduced. To establish a new procedure which gives the steel both formability for manufacturing components and strength as high as posible in practice, the effect of a high temperature tempering-deformation-aging process on structure and properties of the steel 30CrMnSiA was studied.

MATERIAL AND EXPERIMENTAL PROCEDURES

The material used was in sheet form with thickness 2.5mm. It contains 0.28-0.35%C, 0.90-1.20% Si, 0.80-1.10%Mn and 0.80-1.10% Cr. It was rolled at room temperature with deformation degree (ε) 0, 20, 39 and 52% after quenching from 900°C and tempering at 560°C. Part of specimens was in the state normalized from 920°C before rolling. Half of all the specimens was aged at 300°C. The hardness and tensile properties were measured and the microstructure and fracture mode were analyzed with optical microscopy, SEM and TEM (thin foil specimens) in the following states: tempered at 560°C, tempered-deformed and tempered-deformed-aged.

RESULTS AND DISCUSSION

Mechanical Properties

The hardness curves for the specimens quench-tempered and normalized after cold rolling with various deformation degree are shown in Fig. 1a. It can be seen that the hardness values increase obviously with the deformation degree. For example, the HRC value increases 7 and 15 units, respectively, for the specimens quench-tempered and normalized after deformed ε=52%. If an aging process at 300°C is added, the hardness will increase about 5 HRC further.

The tensile strength and yield strength increases in a similar way to the hardness with deformation degree. As shown in Fig.1b, the T.S. and Y.S. of steel 30CrMnSiA increases 36 and 32 kg/mm^2, respectively, after quenching-tempering-deformation (with ε=52%); if an aging process at 300°C is added, the strength of the steel will increase about 10 kg/mm^2 further, which makes the T.S. and Y.S. attain 150 and 140 kg/mm^2, respectively, with elongation δ=9%. These values are near the property levels of the steel in quenched and low temperature tempered state, so it is proved that the process of tempering (at 560°C)-deformation-aging is applicable to meet the requirments of both formability (after high temperature tempering) and high strength (after forming and aging).

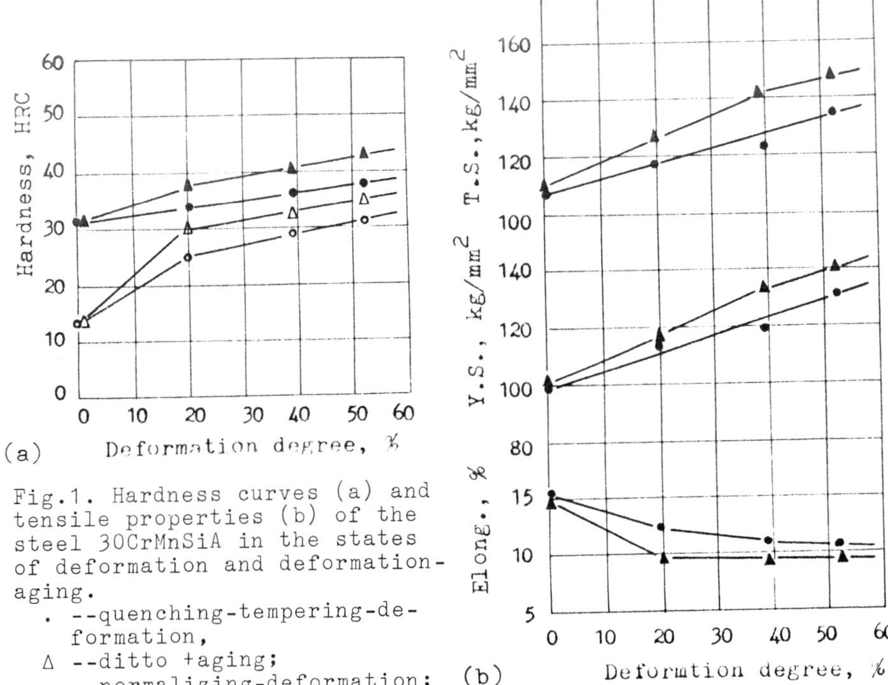

Fig.1. Hardness curves (a) and tensile properties (b) of the steel 30CrMnSiA in the states of deformation and deformation-aging.

- • --quenching-tempering-deformation,
- ▲ --ditto +aging;
- o --normalizing-deformation;
- Δ --ditto +aging.

Microstructures

The effect of deformation-aging on microstructure of the steel 30CrMnSiA in the normalized and quench-tempered states were examined. It can be seen that cold rolling and aging causes the carbides of the steel to grow and the α-phase (matrix) to elongate along the rolling direction (not shown).
The transmission electron micrographes of foil specimens treated by various processes show that under the condition of quench-tempering (at 560°C) the lathes of martensite are ramained partly in the structure, which have the width about 0.2-0.3μm, and the carbides precipitate mainly along the boundaries of lathes (Fig. 2a). In the process of cold rolling the dislocation density in lath is increased and the substructure is refined, it causes the retained carbon in α-phase to precipitate in carbide form along the dislocation networks (Fig. 2b). Aging process (at 300°C) promotes the precipitation and makes it fuller, and the carbides grow further (Fig. 2c). When the deformation degree is increased, the carbides precipitated during aging become finer and closer due to the formation of closer dislocation networks (Fig. 2d).

The Scanning electron micrographes of tensile fracture show that the fractures of the steel are mainly dimple mode (Fig. 3a).

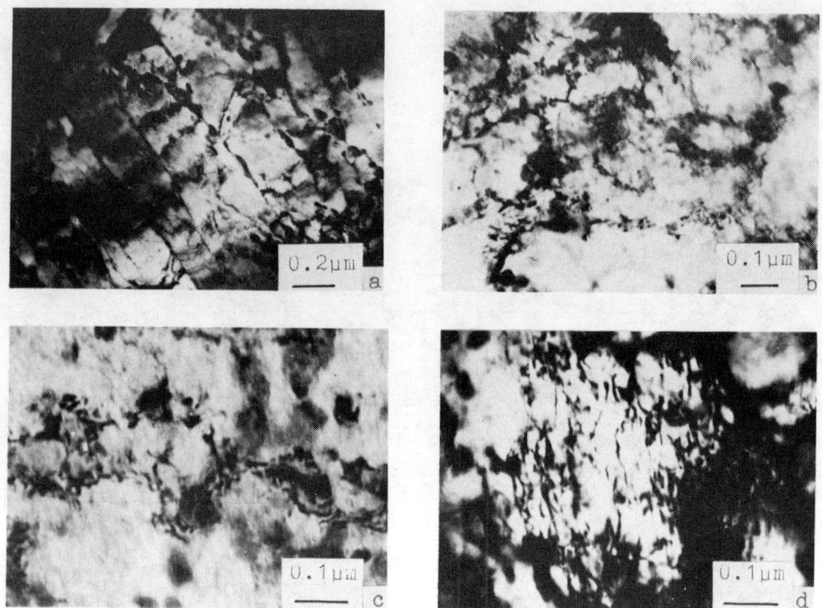

Fig. 2. Transmission electron micrographes showing the effect of deformation-aging on dislocation configuration and substructure of the steel.
a, quenching-tempering (QT);
b, QT+deformation 39% (QTD);
c, QTD+aging;
d, QT+deformation 52% +aging.

After rolling with ε=52% and aging, a layered structure appears on the fracture as shown in Fig. 3b, probably it relates to the elongation of α-phase and banded distribution of carbides due to cold rolling, but the details of the fracture are still dimple mode (Fig. 3c)

Discussion

Strengthening mechanism of deformation-aging. According to the results of this study and previous works, we may come to the conclusion that strengthening of deformation-aging results from many strengthening factors. Therefore, the strength for deformation-aged specimens may be characterized by the sum of various strengthening factors which impede dislocation movement as follows:

$$\sigma = \sigma_1 + \sigma_d + \sigma_\rho + \sigma_\lambda + \sigma_c \qquad (1)$$

where σ_1 is the strength of matrix before deformation, which depends on the resistance of matrix to dislocation moving; σ_d is a factor related to grain size of α-phase, which represents the

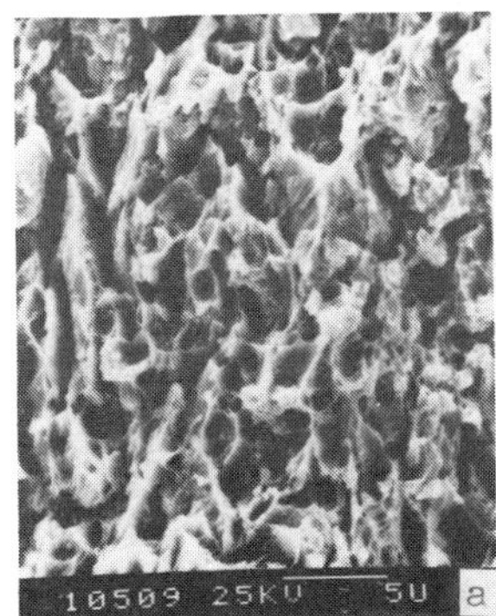

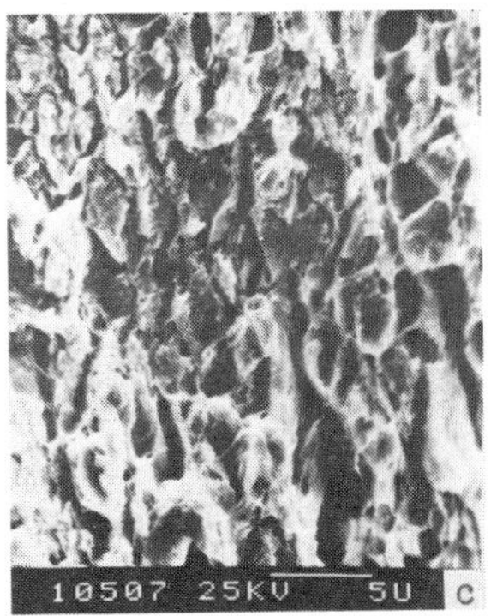

Fig. 3. Scanning electron micrographes showing the effect of deformation-aging on fracture morphology of the steel.
a, quenching-tempering (QT)
b, QT+deformation 52% +aging;
c, ditto, but with larger magnification.

strengthening effect of grain boundary and may be expressed as $k_1 d^{-\frac{1}{2}}$; σ_ρ is a factor related to dislocation density in matrix after deformation, which may be expressed as $k_2 \rho^{\frac{1}{2}}$; σ_λ is a factor related to the spacing and distribution of carbides, which represents the contribution of precipitation strengthening of carbides and may be expressed as $k_3 \lambda^{-1}$; σ_c is a factor related to supersatuation of carbon in matrix, which represents the effect of solid solution strengthening and may be expressed as $k_4 \cdot [C\%]$. Therefore, Eq(1) can be rewritten as follows:

$$\sigma=\sigma_1+k_1 d^{-\frac{1}{2}}+k_2 \rho^{\frac{1}{2}}+k_3 \lambda^{-1}+k_4 [C\%] \qquad (2)$$

In the present study, deformation refines both grain size and substructure (d decreases), increases the dislocation density (ρ increases), so that the strength (T.S. and Y.S.) of the steel increases obviously. Aging after deformation leads to the precipitation of very fine and dispersive carbides (λ decreases), which enhances further the strength of the steel.

Technological process and parameters. From the data mentioned above it can be seen that steel 30CrMnSiA possesses good formability during cold rolling in either quench-high temperature tempered or normalized states for meeting the requirments of manufacturing casings or vessels. However, the strength of the steel in normalize-rolled state is obviously lower than those treated by quench-tempering-rolling (see Fig.1). So, it is a reasonable technological process to treat the steel with quench-tempering before deformation-aging for obtaining the highest strength.

A systematic research [5] has been done to clarify the effect of aging temperature after deformation on the strength of the steel. Results show that no apparent strengthening effect when aging temperature is lower than 200°C, and the strength of the steel decreases after aging when the temperature is higher than 400°C. The best strengthening effect of the steel appears only when aging temperature is between 250-300°C, by which both T.S. and Y.S. can have an increment of about 10 kg/mm^2 (Fig. 1b).

CONCLUSION

1. Higher strengthening effect can be obtained for the steel 30CrMnSiA in the state cold roll-aged after quench-tempered (at 560°C). For instance, tensile strength of the steel can attain 150 kg/mm^2 when deformation degree is 52%, it is 40 kg/mm^2 higher than that in the state quench-tempered (560°C) and remains sufficient plasticity with elong. 9%. It proves that the technological process of quench-tempering (at 560°C)-deformation-aging is able not only to satisfy the requirment of formability, but also to use fully the strength potential of the steel.
2. The strengthening of deformation-aging results from the function of many strengthening factors. The best effect comes from deformation strengthening, which leads to increment of dislocation density in matrix and refinement of substructure. Aging after deformation results in further strengthening due to the precipitation of carbides along dislocation networks.

REFERENCES

1. M. Cohen, J.Iron Steel Inst. (U.K.), No 201, 833 (1963).
2. D. Kalish, J. Metals, No 2, 157 (1965)
3. T. C. Lei et al, Thermomechanical Treatment of Steels, Machine-building Press (Beijing), (1979)
4. E. Tamura, J. Metals Society of Japan, 20, No 12, (1981).
5. T. C. Lei et al, J. Harbin Institute of Technology (supplementary issue), 33 (1981).

Optimisation de l'Analyse Chimique d'un Acier Micro-allie A 0,45% de Carbone pour Obtenir un Niveau de Charge de Rupture de 1000 MPa sans Traitement Thermique

P. Charlier et L. Bäcker

Société des Aciers Fins de l'Est, B.P. 38, 57301 Hagondange, France

RESUME

Les mécanismes métallurgiques permettant l'obtention d'une charge de rupture de 1000MPa sur produits laminés ou forgés sans traitements thermiques sont présentés. La composition analytique en niobium, vanadium, aluminium et azote est optimisée.

MOTS CLES

Aciers micro-alliés, caractéristiques après refroidissement naturel, durcissement par précipitation, niobium, vanadium.

INTRODUCTION

Le développement des aciers micro-alliés pour produits longs destinés à la construction mécanique est relativement récent et la Société des Aciers Fins de l'Est (SAFE) a été une des premières sociétés sidérurgiques à produire une large gamme d'aciers dans ce domaine en lançant en 1979 les aciers METASAFE (1). Leur principale caractéristique est de présenter à l'état brut de laminage ou de forgeage, donc sans traitement thermique, une charge de rupture élevée et constante, quelles que soient les dimensions des produits. La gamme de charges de rupture couverte par cette famille d'aciers est comprise entre 750 et 2000 MPa. L'acier Métasafe 1000 dont il est ici question présente une charge de rupture moyenne de 1000MPa. Son domaine d'application représente une partie importante de l'ensemble des pièces de construction mécanique pour lesquelles une charge de rupture de 1000 MPa est demandée.

L'adoption en 1978 par la SAFE de brevets existants (2) mais non encore appliqués industriellement a permis une mise au point rapide de ces produits. Les études métallurgiques effectuées depuis lors apportent une meilleure compréhension des mécanismes métallurgiques en jeu qui débouche sur l'optimisation des compositions analytiques.

MECANISMES D'OBTENTION D'UNE CHARGE DE RUPTURE CONSTANTE POUR L'ACIER METASAFE 1000

La composition analytique de coulées de Métasafe 1000 apparaît au tableau 1 (coulées C-D-E-F).

	C	Si	Mn	S	Al	V	Nb	N
Coulée A	0,445	0,260	1,570	0,033	0,027	-	-	0,0130
Coulée B	0,530	0,280	1,490	0,030	0,033	0,106	-	0,0120
Coulée C	0,470	0,360	1,550	0,022	0,061	0,120	0,065	0,0300
Coulée D	0,450	0,300	1,520	0,035	0,034	0,120	0,073	0,0220
Coulée E	0,455	0,300	1,500	0,038	0,041	0,116	0,042	0,0185
Coulée F	0,445	0,335	1,475	0,035	0,050	0,110	0,070	0,0185

TABLEAU 1 Composition Analytique des Coulées Etudiées (en %)

Outre la teneur en carbone d'environ 0,45%, on remarque la présence de 1,5% de manganèse, seul élément d'alliage à proprement parler, et des éléments de "micro-alliages" suivants : l'aluminium, le vanadium, le niobium et l'azote.

L'influence des conditions de refroidissement sur les caractéristiques mécaniques en traction de l'acier est examinée de la façon suivante : après maintien 15mn à 1200°C, des lopins de diamètres compris entre 22 et 80mm de la coulée C ont refroidi naturellement à l'air calme.

Les résultats d'essais de traction sur éprouvettes prélevées à coeur de ces lopins, rassemblés au tableau 2, montrent qu'effectivement les charges de rupture et limites d'élasticité sont pratiquement constantes, quel que soit le diamètre des lopins, donc la vitesse de refroidissement.

Diamètre de barre (mm)	V refroidis. entre 800 et 500°C(°C/mn)	Charge de rupture (MPa)	Limite d'élasticité(MPa)	Allongt. (%)	Striction (%)
80	10,5	1010	645	14	14
58	19	1035	680	13	14
40	27	977	635	11	23
22	55	980	675	18	14

TABLEAU 2 Variations des Caractéristiques Mécaniques en Fonction de la Vitesse de Refroidissement, après un Maintien de 15mn à 1200°C ; Acier Métasafe 1000-Coulée C

Les observations en microscopie optique, mettent pourtant en évidence des microstructures très différentes (cf. fig. 1) :

- structure "dure" de bainite pour les barres de plus faibles diamètres
- structure normalement nettement moins dure de ferrite-perlite pour les barres de plus gros diamètres.
- structure mixte, constituée de bandes plus ou moins importantes de bainite dans une matrice ferrito-perlitique pour les barres de diamètres intermédiaires.

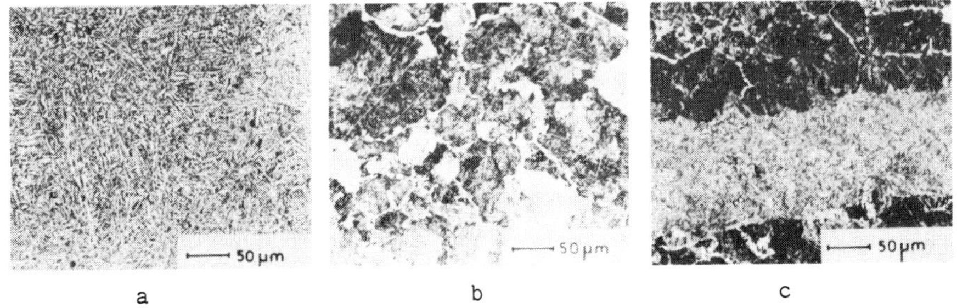

a b c

Fig. 1. Microstructure sur barres de différents diamètres de la coulée C
a : 22mm ; b : 80mm ; c : 40mm

La technique d'isolation des précipités par extraction électrolytique (3) permet de déterminer le taux de précipitation des éléments dispersoïdes pour les différents lopins. Les résultats de ces analyses sur des barreaux de 8mm de diamètre prélevés dans l'axe des lopins montrent que seul le taux de précipitation du vanadium varie (cf. fig. 2) : il est d'autant plus important que la vitesse de refroidissement est faible. Les taux de précipitation du niobium et de l'aluminium sont constants ; la précipitation du niobium au cours du refroidissement naturel est totale, celle de l'aluminium au contraire est nulle.

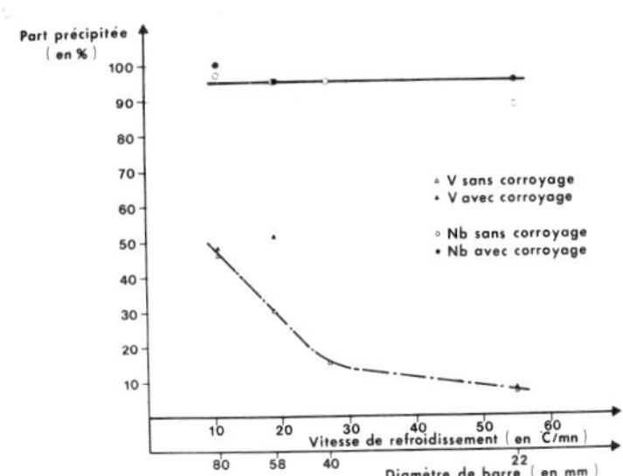

Fig. 2. Pourcentages du vanadium et de niobium précipités par rapport à la quantité totale contenue selon le diamètre des barres

Ainsi il apparaît que l'obtention d'une charge de rupture constante quelle que soit la vitesse de refroidissement des produits est obtenue grâce à un compromis entre trempabilité à l'air et précipitation du vanadium :

- aux plus fortes vitesses de refroidissement, la dureté est obtenue naturellement par effet de "trempe" à l'air (structure bainitique).
- aux plus faibles vitesses de refroidissement, la dureté de la microstructure ferrito-perlitique, normalement plus faible, est relevée grâce aux précipitations fines de vanadium.

OPTIMISATION DE LA COMPOSITION ANALYTIQUE DE L'ACIER METASAFE 1000

L'étude d'un ensemble de caractéristiques sur coulées industrielles dont les analyses sont rassemblées au tableau 1 permet de préciser le rôle des éléments dispersoïdes présents : la coulée A est une coulée de référence sans éléments dispersoïdes ; la coulée B a une analyse similaire à celle de la coulée A avec toutefois une addition de vanadium ; les coulées C-D-E-F sont des coulées de type Métasafe 1000 avec des teneurs en niobium et en azote variables.

Résistance au Grossissement de Grain

Au tableau 3 sont rassemblés les indices de tailles de grains austénitiques (selon ASTM) mesurés après maintien 15mn à 1200°C.

Coulée	A	B	C	D	E	F
Taille de grain austénitique	3,7	3,5	5,7	5,6	6,2	5,4

TABLEAU 3 Taille de Grain Austénitique des 6 coulées A à F après Maintien 15mn à 1200°C

Le vanadium n'apporte pas de résistance particulière au grossissement de grain à la température considérée (comparaison des coulées A et B). Par contre la comparaison entre coulées B d'une part et C à F d'autre part, met en évidence l'effet très favorable du niobium. L'addition d'un excès de niobium par rapport aux possibilités de mise en solution de cet élément n'apporte pas, par contre, de meilleure résistance au grossissement de grain (comparaison des coulées C-D-F d'une part, et E d'autre part).

Mise en Solution des Eléments Dispersoïdes

Le tableau 4 rassemble les résultats de mise en solution des éléments dispersoïdes à 1250°C, température courante de réchauffage avant forgeage ou laminage. Ces valeurs ont été obtenues à partir des analyses des résidus d'extractrions électrolytiques.

Coulée	% V Total	% V Dissous	% Nb Total	% Nb Dissous	% Al Total	% Al Dissous
B	0,106		-	-	0,033	non dosé
C	0,120		0,065	0,030	0,061	0,038
D	0,120		0,073	0,042	0,034	0,021
E	0,116		0,042	0,032	0,041	0,033
F	0,110		0,070	0,030	0,050	0,034

TABLEAU 4 Mise en Solution des Eléments Dispersoïdes Nb, V et Al par Maintien 1H 30mn à 1250°C

La mise en solution du vanadium est totale à la température considérée ; elle n'est que partielle pour l'aluminium et le niobium. La teneur réduite en niobium de la coulée E, proche de la possibilité de dissolution totale à 1250°C, permet de diminuer la quantité de carbonitrures de niobium "grossiers" (fig. 3) présents dans des bandes ségrégées.

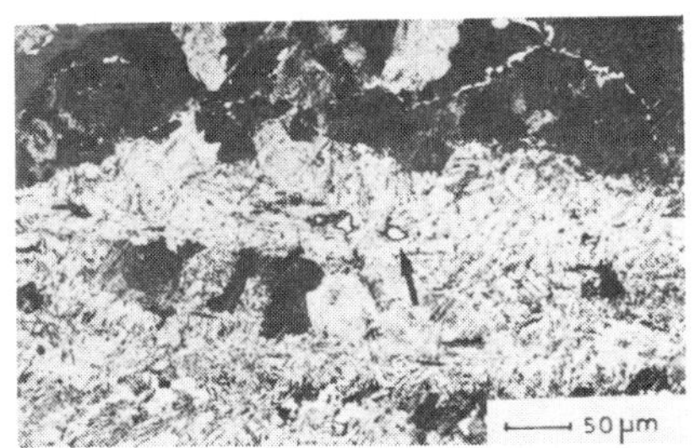

Fig. 3. Alignement de carbonitrures de niobium dans une bande ségrégée (coulée F)

Caractéristiques sur Barres Laminées

Les 5 coulées ont été laminées en barres de "petits" diamètres (Ø=22mm) et de diamètres "moyens" (Ø=55mm) ; les coulées D et F ont de plus été laminées en barres de "gros" diamètres (Ø=115mm).

Le tableau 5 permet de constater que les coulées considérées, sauf celle de référence, présentent des charges de ruptures constantes (mesurées dans l'axe des barres), quelles que soient les dimensions. Les mécanismes métallurgiques explicités précédemment à partir des microstructures et des quantités de vanadium précipitées sont confirmés.

	Coulée A		Coulée B		Coulée C		Coulée D			Coulée E		Coulée F		
	Ø22	Ø55	Ø22	Ø55	Ø22	Ø55	Ø22	Ø55	Ø115	Ø22	Ø55	Ø22	Ø55	Ø115
Rm(MPa)	1008	878	1090	1002	1010	1008	957	953	1005	1002	982	979	1015	961
Re(MPa)	659	506	1002	672	740	687	676	666	708	667	644	667	770	654

TABLEAU 5 Charges de Rupture et Limites d'Elasticité Obtenues par Essais de Traction sur Barres Laminées de 22-55 et 115mm de Diamètres des Coulées A à F (Prélèvements dans l'Axe)

La connaissance de la mise en solution des différents éléments dispersoïdes à 1250°C permet d'évaluer la quantité d'azote "disponible" pour la précipitation de nitrures de vanadium lors du refroidissement naturel que l'on peut comparer aux quantités de vanadium réellement précipitées sur barres de diamètre 55mm (cf. tableau 6).

Coulée	B	C	D	E	F
Teneur en azote "disponible"	0,0120	0,0164	0,0130	0,0130	0,080
Teneur en vanadium précipitée	0,047	0,065	0,013	0,002	0,008

TABLEAU 6 Précipitations du Vanadium sur Barres de Diamètre 55mm (à Coeur) et Teneurs en Azote Disponible pour la Précipitation de Nitrures de Vanadium (en %)

C'est effectivement pour la coulée à teneur en azote "disponible" la plus importante que la précipitation est maximale. La différence entre les coulées B-D et F s'explique par des différences de trempabilité (4).

Les coulées D-E-F n'ont pas des potentiels de précipitation de nitrures de vanadium systématiquement supérieurs à celui de la coulée B, alors que leurs teneurs en azote total le sont très nettement, en raison de la non-redissolution d'une partie des carbonitrures de niobium et de nitrures d'aluminium aux températures de réchauffage pour laminage.

Sur barres de diamètres 115mm, la quantité de vanadium précipitée (0,075%) est en excès par rapport aux possibilités de précipitation sous forme de nitrures, mettant en évidence l'apparition de précipitations de carbures de vanadium pour les plus faibles vitesses de refroidissement. Dans ce cas, les teneurs en azote perdent de leur importance.

CONCLUSION

Un équilibre entre "trempabilité à l'air" et durcissement par précipitations permet d'obtenir pour l'acier Métasafe 1000 une charge de rupture quasiment indépendante des conditions de refroidissement, donc des dimensions des produits forgés ou laminés. Le vanadium est l'agent principal de durcissement par précipitation alors que le niobium permet le contrôle du grain austénitique ; ces deux éléments jouent également un rôle important dans le contrôle de la trempabilité de l'acier dont dépend en grande partie la microstructure des produits.

L'optimisation de la composition chimique de l'acier Métasafe 1000 passe par une réduction de la teneur en niobium initiale à un niveau légèrement en excès par rapport aux possibilités de mise en solution de cet élément à 1250°C (0,04% environ), par une baisse de la teneur en aluminium au niveau minimum nécessaire pour le calmage de l'acier, qui, avec la baisse de la teneur en niobium, permet l'augmentation du "potentiel" de précipitation des nitrures de vanadium pour une teneur en azote donnée de 0,02% environ.

REFERENCES BIBLIOGRAPHIQUES

1. L. Bäcker, P. Charlier, R. Hechelski : Séminaire sur les modifications des exigences de qualité dans la demande d'acier - ONU - TURIN (Italie) Juin 82.
2. Brevets FR 7901333 - 7901819 - 7901820 - 7901821
3. D Henriet, P. de Gelis : Revue de Métallurgie - Août 1984 - p. 703 à 715.
4. P. Charlier : Thèse de Docteur-Ingénieur - Institut National Polytechnique de Lorraine - Novembre 1984.

Mathematical Model for Prediction of Austenite and Ferrite Microstructures in Hot Rolling Processes

P. Choquet*, A. Le Bon* and Ch. Perdrix**

**IRSID, 78105 St Germain-en-Laye Cedex, France*
***Now with USINOR, BP 2-508, 59381 Dunkerque Cedex, France*

ABSTRACT

A model describing the evolution of microstructure during hot working has been developed using hot torsion tests. The derived quantitative relationships predict static recrystallization kinetics, recrystallized grain size, grain growth and ferritic grain size from transformed austenite.

KEYWORDS

Hot rolling, static recrystallization, grain growth, austenite to ferrite transformation, modeling.

INTRODUCTION

Fine grained steels can be obtained by addition of niobium and low rolling temperatures. Considerable work has been done to describe the evolution of austenitic microstructure during hot rolling and the austenite to ferrite transformation (1-4). However steel makers need precise quantitative data in order to adapt compositions and rolling schedules to economic constraints and technological facilities. Moreover reliable predictions of rolling forces are necessary to meet geometrical requirements (thickness, shape and flatness). Simulation by hot torsion testing has proved efficient to improve rolling of plates (5). Simulation of rolling is never perfect because of technological limitations (strain range in compression and strain rate in torsion) but it is a powerful method for modeling and in most cases, results can be extrapolated to industrial processes. Further progress in hot rolling can be made by taking into account the evolution of structure during rolling and its effect on the final structure of the product. IRSID recently developed, by means of the torsion test, a complete model of rolling which describes both flow stress and microstructural evolution (6). This paper only deals with the structural part of the model which calculates static recrystallization kinetics, recrystallized grain sizes, grain growth, austenite to ferrite transformation temperature and ferritic grain size of low carbon-manganese steel (C Mn), niobium microalloyed steels (C Mn Nb) and niobium plus vanadium microalloyed steels (C Mn Nb V).

EXPERIMENTAL PROCEDURE

Material and Experimental Procedure

The hot torsion tests were performed on 6 mm diameter specimens of 50 mm gauge length, machined from hot rolled bars. Specimens were solution treated at 1220°C, deformed under argon protection, and cooled by helium or water quenching at any moment of the process. Metallographical observations were made at the outer part of the specimen after etching in saturated aqueous picric acid containing a wetting reagent. The C Mn steels had a base composition of 0.180 % C - 1.3 % Mn. The C Mn Nb steels had niobium contents varying from 0.025 % to 0.100 %. Deformations were performed at temperatures from 850°C to 1200°C and cooling rates after processing were in the range of 0.25°C/s to 10°C/s.

Double Deformation Method

In order to follow the progress of isothermal recrystallisation, the double deformation method (7-8) has been preferred to long metallographic observations. The principle of this method is illustrated in Fig. 1. Curve (1) is the stress-strain curve during an initial deformation given at a constant strain rate $\dot{\bar{\varepsilon}}$. σI is the value of the flow-stress at the beginning of the test and σ II, that measured at the interruption strain $\bar{\varepsilon}$ 1. After an isothermal holding time t, the specimen is reloaded and the flow-stress is then descibed by curve (2). It can be seen in Fig. 1 that if curve (1) is shifted along the OO' direction and superimposed on curve (2), the two curves are identical after a short transient following reloading which can be associated with static recovery and machine effects. This back extrapolation procedure gives the flow stress level noted σ III in fig. 1. The fractional softening Xa, used in the present study is given by :

$$Xa = (\sigma II - \sigma III)/(\sigma II - \sigma I).$$

The softening kinetics Xa(t) and the recrystallization kinetics $X_R(t)$, determined by metallographic observation, have been compared and were found to be similar (fig. 2). Hence we may write $X_a = X_R$.

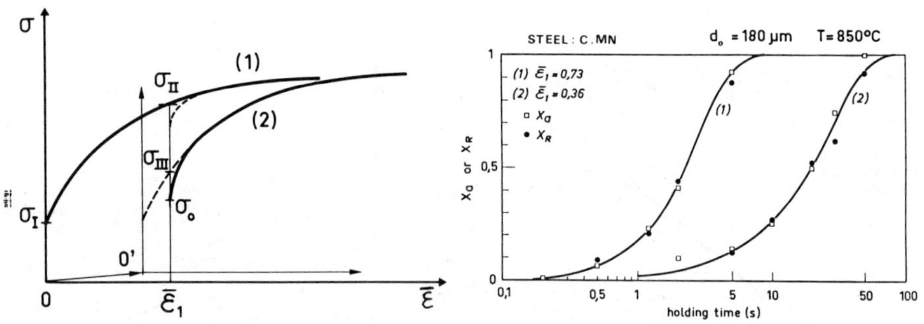

Fig. 1 - Double deformation method

Fig. 2 - Comparison between X_a and X_R

Isothermal Evolution of Austenite

Processing parameters such as strain $\bar{\varepsilon}$, strain rate $\dot{\bar{\varepsilon}}$, temperature T, initial grain size **do** and niobium content have been widely explored to fit mathematical relationships for static recrystallization kinetics, recrystallized grain sizes and grain growth. Additions of vanadium to C Mn Nb steels did not affect the results.

Static recrystallization kinetics have been satisfactorily described by the Avrami equation :

$$X_R = 1 - \exp(\lambda\ t/t_{0.5})^n \qquad (1)$$

$\lambda = \ln 2$

The exponent n and the time for 50 % recrystallization $t_{0.5}$ have been found to be dependant on processing parameters and niobium content according to the following relationships :

$$n = \nu n \exp(-Qn/RT)\, \bar{\varepsilon}^{\,0.5}\, do^{-0.155} \qquad (2)$$

where $\nu n = \nu no + \alpha (Nb)^{\beta}$; $\nu no = 272$; $Qn = Qno + \alpha'(Nb)^{\beta'}$; Qno=9100kcal/mole

$$t_{0.5} = \nu r \exp(Qr/RT)\, \bar{\varepsilon}^{\,p}\, \dot{\bar{\varepsilon}}^{\,-0.28}\, do^{-0.14} \qquad (3)$$

where νr is a function of Nb content and p a function of Nb content and initial grain size

$$Qr = Qro + \gamma \sqrt{Nb} + \Delta Q\,(T) \text{ with } Qro = 70000 \text{ kcal/mole}$$

These relationships apply to the C Mn steel with Nb = 0.
Qr is in good agreement with values obtained in previous work (9-10-11), as can be seen in figure 3. Niobium has a strong effect on the kinetics as shown in Fig. 4. At all temperatures for the C Mn Nb steel (with only 0.010 % Nb) $t_{0.5}$ is ten times larger than for the C Mn steel. An influence of strain rate on the softening kinetics is to be expected since it affects the stress level and hence the driving force for recrystallization. According to relation (3) the time $t_{0.5}$ is divided by a factor of about 2 when strain rate is increased by one order of magnitude.
The exponent n has been found independant of strain rate.

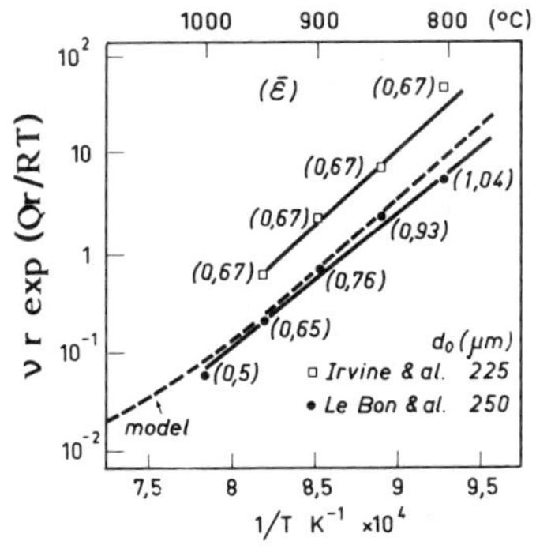

Fig. 3

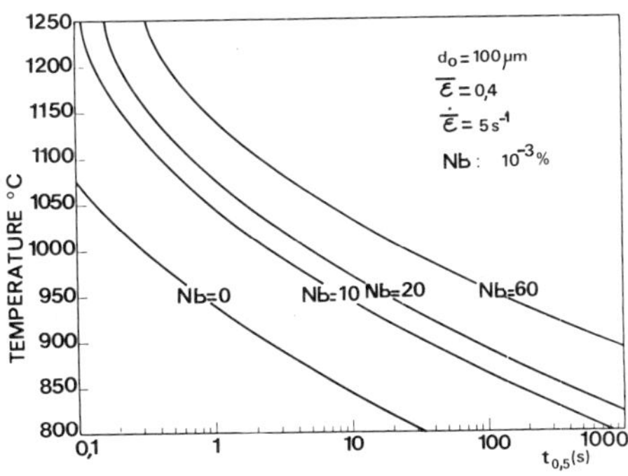

Fig. 4

Recrystallized grain sizes dR have been measured after full completion of recrystallization. The results have been fitted by the two following relations :

$$\text{C Mn steels : } dR = 45 \exp(-5973/RT)\, \bar{\varepsilon}^{-0.6}\, do^{0.374}\, \dot{\varepsilon}^{-0.1} \quad (4)$$

$$\text{C Mn Nb steels : } dR = 472 \exp(-11620/RT)\, \bar{\varepsilon}^{-0.7}\, do^{0.277}\, \dot{\varepsilon}^{-0.1} \quad (5)$$

The influence of strain rate is rather small since when it is increased by an order of magnitude, dR is divided by a factor of about 1.3.

Isothermal grain growth has been followed after deformation and complete recrystallization. The grain size d(t) at time t is related to the recrystallized grain size dR by the equation :

$$d(t)/dR = 1 + \alpha \ln(t/tR) \quad (6)$$

where tR is the time for complete recrystallization (i.e XR = 0.99).
α is the grain growth parameter depending on steel composition :

$$\alpha(\text{CMn}) = 0.195 \text{ and } \alpha(\text{CMnNb}) = 0.098$$

Relation (6) shows that the faster the static recrystallization, the faster the grain growth. It also indicates that both mechanisms have the same activation energy since :

$$\left.\frac{\delta d(t)}{\delta t}\right|_{t=tR} = \alpha \frac{dR}{tR} \quad (7)$$

Evolution of Austenite during Rolling

The effect of cooling rate has been studied since it affects recrystallization and grain growth kinetics. Static recrystallization under cooling conditions can be described by an Avrami equation of the form :

$$X_R = 1 - \exp(\lambda\, W/W_{0.5})^n \quad (8)$$

where : $\lambda = \ln 2$ - W and $W_{0.5}$ are compensated times given by :

$$W = \int_0^t \exp(-Qr/RT)dt \text{ and } W_{0.5} = t_{0.5} \exp(-Qr/RT).$$

Grain growth under cooling conditions can be calculated by the differentiation of relation (6), assuming that tR and $t_{0.5}$ have the same activation energy :

$$d(t)/dR = 1 + \alpha \ln(t/tR) - Qr(1/T - 1/TD)/R \quad (9)$$

where TD is the temperature of deformation and T the temperature at time t.

Deformation of a partially recrystallized structure is followed by static recrystallisation, as shown in fig. 5 where predeformation is $\bar{\varepsilon}1 = 0.36$ and the second deformation, $\bar{\varepsilon}2 = 0.36$, has been given after different holding times to obtain different values of X_R ; the kinetics of recrystallization shown in fig. 5 can still be described by an Avrami equation. In other words the behaviour of a partially recrystallized structure resulting from a predeformation $\bar{\varepsilon}1$, is similar to that of a homogeneous structure having the same grain size as the completly recrystallized structure and uniformly deformed by a ficticious quantity $\bar{\varepsilon}^*$, called residual strain, which has been found to be :

$$\bar{\varepsilon}^* = 1/2\, \bar{\varepsilon}1(1 - X_R) \text{ when } X_R \geq 0.10 \text{ and } \bar{\varepsilon}^* = \bar{\varepsilon}1 \text{ when } X_R < 0.10.$$

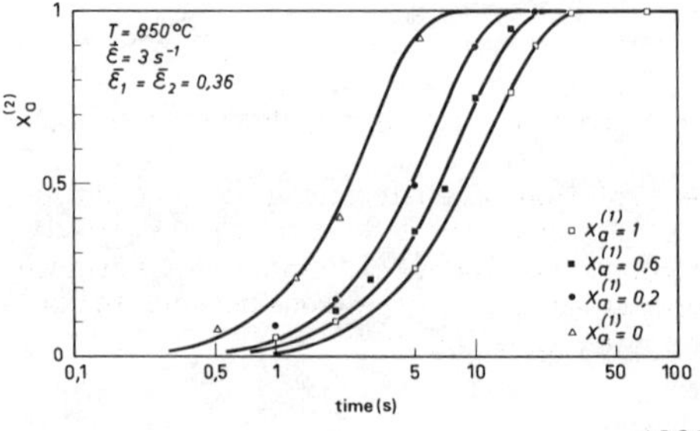

Fig. 5 - Kinetics of static recrystallization ($Xa^{(2)}$) after deformation given on a partially recrystallized structure ($Xa^{(1)}$).

AUSTENITE TO FERRITE TRANSFORMATION

The temperature of Transformation

Ar_3^* is the first parameter which must be known, because it indicates if processing is finished in the austenite range or in the two-phase region. Simulations carried out on the IRSID experimental mill gave the following relation derived from thermal analysis of the specimen after rolling :

$$\mathbf{Ar_3^* = 902 - 527\,C - 62\,Mn + 60\,Si} \quad \text{(compositions in \%)} \qquad (10)$$

Ferritic Grain Size

The ferrite grain size $d\alpha$ has been studied first for a C Mn steel as a function of recrystallized austenitic grain size before transformation $d\gamma$ and cooling rate Cr ; secondly for a C Mn Nb V steel as a function of initial recrystallized grain size $d\gamma$, strain $\bar{\varepsilon}$ and cooling rate Cr.
The results for the C Mn steel are reported on figure 6. A relation has been found between the ratio $d\gamma\ /d\alpha$ and $d\gamma$:

$$d\gamma\ /d\alpha = a + bd\gamma \qquad (11)$$

where $a = aoCr^n$ and $b = boCr^m$.
Figure 7 shows the results for C Mn Nb steel with different strains and constant cooling rate. The derived relationship is :

$$(d\gamma\ /d\alpha\ ) = A\ \ (d\gamma\ /do)^p \qquad (12)$$

where $p = po\,\bar{\varepsilon}^q$, A, po and do are constant.
The exponent q has been found to depend on Cr : $q = qo.\ln(Co/Cr)$
where qo, Co are constant.

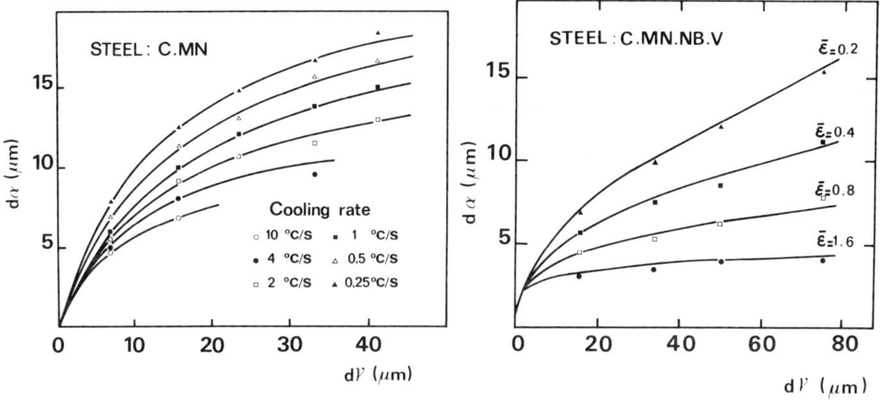

Fig. 6 - $d\alpha$ versus $d\gamma$ and Cr.

Fig. 7 - $d\alpha$ versus $d\gamma$ and $\bar{\varepsilon}$.

PRACTICAL APPLICATIONS

The validity of the model described above has been verified by laboratory simulations of industrial rolling schedules.
Figure 8 shows, for a C Mn Nb steel, the evolution of austenitic grain size during the roughing stage of a hot strip mill. Calculations are compared to the results of hot torsion simulations and a very good agreement is obtained. As shown in fig. 8 the lowering of reheating temperature produces a considerable refinement of austenitic grain size at the entry to the finishing mill.

Two complete rolling schedules have been simulated on the experimental mill at IRSID. After reheating at 1140°C, two slabs with an initial thickness of 70 mm were rolled in the recrystallization range to an intermediate thickness of 42 mm and 24 mm respectively, then finished below the recrystallization temperature to a final thickness of 12 mm, and air cooled. Total deformation of elongated austenite was therefore $\bar{\varepsilon}\,1 = 1.44$ and $\bar{\varepsilon}\,2 = 0.80$. Table 1 shows results of computed and observed ferritic grain sizes for both specimens.

dγ μm before finishing	$\bar{\varepsilon}$	dα μm computed	dα μm measured
56	1.44	4.3	4.4
35	0.80	5.4	5.2

TABLE 1.

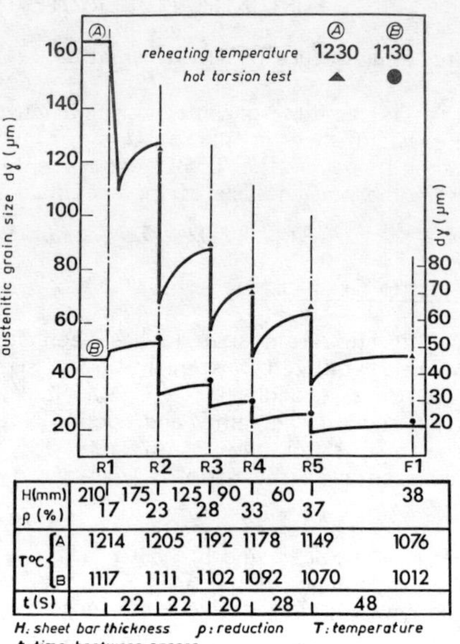

Fig. 8 - Evolution of austenitic grain size in the roughing stage of a hot strip mill for the CMnNb steel.

CONCLUSIONS

A good prediction of the evolution of austenitic microstructure and the resulting ferrite grain size has been obtained.

This model is particularly powerful for industrial application since it takes into account the cooling rate of the product and the succession of deformations; it will therefore help to predict rolling forces and final mechanical properties.

ACKNOWLEDGMENTS

This work was financed by an E.C.S.C. research contract.

REFERENCES

1. M. Korchinsky (Editor), *Microalloying 75*, Union Carbide Corp., New York (1976)
2. J.B. Ballance (Editor), *The Hot Deformation of Austenite*, AIME, New York (1977)
3. C.M. Sellars and GI Davies (Editors), *Hot Working and Forming Processes*, The Metal Society, London (1979)
4. A.J. DeArdo, G. A. Ratz, P.J. Wray (Editors), *Thermomechanical Processing of Austenite*, AIME, New York (1982)
5. A. Le Bon, L.N. de Saint Martin, in ref. (1), p. 72
6. C. Perdrix and al. To be published in the *Revue de Métallurgie, 1985*
7. M. Lamberights and T. Greday in ref. (2), p. 286
8. R.A. Petkovic and J.J. Jonas, *Acta Metall* , **27**, 1633 (1979)
9. A. Le Bon, J. Rofès, C. Rossard, *Met. Sci.*, **9**, 13 (1980)
10. K.J. Irvine, T. Gladman and F.B. Pickering, *J. Iron St. Inst.*,**208**, 717 (1977)
11. C.M. Sellars in ref. (3), p. 3.

Microstructure Evolution During Industrial Rolling of Microalloyed Steel Plates

E. Anelli*, M. Ghersi*, A. Mascanzoni*, M. Paolicchi*, A. Aprile, A. De Vito** and F. De Mitri****

**Centro Sperimentale Metallurgico, Rome, Italy*
***N. Italsider, Taranto, Italy*

ABSTRACT

An analysis of the relationship between microstructure evolution and hot rolling of commercial microalloyed steel plates has been made during a research project sponsored by the European Community (ECSC). The work has involved development of a hot rolling and microstructure integrated model dedicated to plate processing at the Taranto steel works. Low carbon steels of various chemical compositions (Mn, Nb, Nb-Mo, Nb-Ti), used for the production of 12-35 mm thick plates, have been considered. The main metallurgical functions: austenite grain size produced by static and dynamic recrystallization, grain growth, recovery of unrecrystallized volumes, have been used to describe the evolution of the "homogeneous" structures developed during the thermomechanical treatment. The relationship between α and γ grain size determined by air cooling phase-transformation, has also been included. The microstructure model has been trasferred to the plate mill process computer and its forecasting capability verified on specific rolling trials, utilizing process data, supplied on-line by the mill control system.

KEYWORDS

Plate rolling; microalloyed steels; microstructure model; recrystallization.

INTRODUCTION

The control systems presently used in hot-rolling mills were developed essentially to improve productivity and size tolerances; they do not contain models capable of following the evolution of steel structure directly during rolling. Yet the present trends towards improvement in product quality, while minimizing the alloy element content, must involve more stringent control of the thermomechanical process, including control of the corresponding microstructure evolution. This objective can now be attained, both because of the proven possibility of describing microstructure evolution by means of laboratory tests(1,2) and because of the widespread use of automatic control systems in rolling mills. In the last few years microstructure models have been tentatively developed to predict microstructure during hot deformation processes(3,4) extended to optimize practical rolling schedules(5,6). The purpose of the work reported here was the development of an integrated hot rolling model, capable of predicting the evolution of thermal, mechanical and microstructure aspects of C-Mn and microalloyed structural steels, to be used on N. Italsider's Taranto plate-rolling mill.

EXPERIMENTAL PROCEDURES AND RESULTS

Laboratory tests were performed on a C-Mn steel and on three microalloyed steels, the composition of which is given in Table 1.

TABLE 1 Steel Compositions (Wt Pct)

Steel	C	Mn	Si	Al	Nb	Mo	Ni	Ti
1	.150	.80	.27	.013	-	-	-	-
2	.150	1.15	.21	.033	.035	-	-	-
3	.080	1.63	.33	.023	.041	.24	.34	-
4	.062	1.87	.28	.040	.036	-	-	.13

The basic relationships governing the evolution of the microstructure were established by specific tests for each phase of the thermomechanical process, namely: reheating, rolling and cooling. The austenite grain sizes at the beginning of the deformation process, were determined on samples submitted to programmed reheating and water quenching. The hot rolling process was simulated by means of the hot torsion test, using the double torsion technique to evaluate softening and recrystallization kinetics. In order to separate the recovery contribution from softening, a specific graphic procedure was adopted following the technique developed by IRSID(3). The reliability of the method was confirmed by comparison of results with quantitative metallography observations. Specimens water quenched after single torsion tests were used to measure the recrystallized grain size. The equations which describe the effect on microstructure of process variables (time-temperature-deformation) and of initial austenitic grain size were established on the basis of these laboratory tests. Austenite evolution was expressed by means of relevant parameters, namely: recrystallized volume fraction, recrystallized grain size including the effect of grain growth after recrystallization, residual strain and "effective" grain size of unrecrystallized regions (affected by recovery). The recrystallization behaviour of C-Mn and Nb microalloyed steels is compared in Fig. 1.
The curves shown in Fig. 1 were calculated by the model; they refer to two deformation temperatures. At 1000°C (Fig. 1a) the progress of recrystallization in microalloyed steels occurs more slowly than in plain carbon steel. The retarding effect is stronger in the steels with a higher microalloy content.
At lower temperatures the retarding effect on recrystallization kinetics is considerably greater (Fig. 1b) and the onset of recrystallization is delayed.

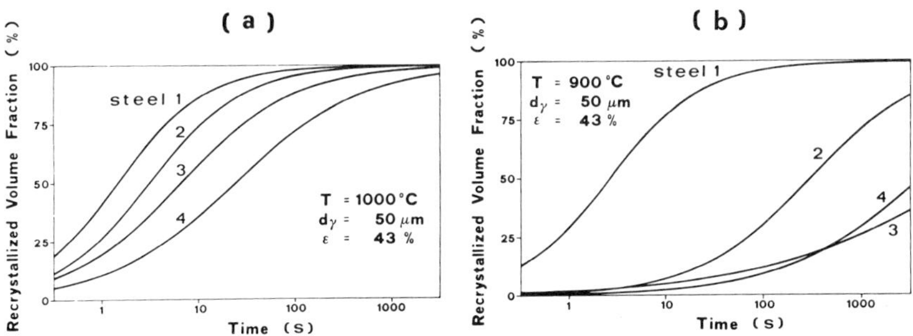

Fig. 1. Comparison of calculated recrystallization behaviour of C-Mn and Nb microalloyed steels at 1000°C and 900°C.

These results confirm that in microalloyed steels, at relatively low temperatures and with times comparable to the interpass times normally used in plate rolling schedules(10-20 s), a large portion of the material volume could be unrecrystallized. In order to extend model capability to prediction of the final microstructure of as-rolled steel plates, the austenite-to-ferrite phase transformation was studied in the range of ferrite-pearlite microstructures, by considering the dependence of ferrite grain size(d_α) on austenite grain size(d_γ) and cooling rate (CR) (Fig. 2).

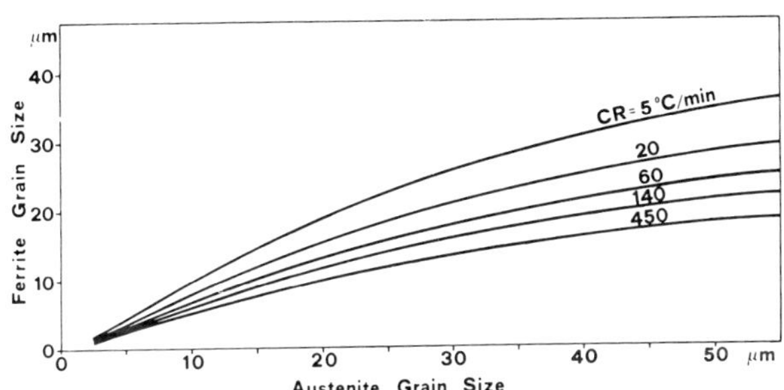

Fig. 2. Dependence of ferrite grain size (d_α) after transformation on austenite grain size (d_γ) and cooling rate (CR).

This analysis was carried out on heat-treated samples, considered as representative of materials with fully recrystallized austenite. Situations corresponding to coarse or inhomogeneous austenite grain sizes, which lead to formation of acicular or Widmanstätten ferrite, were systematically avoided. The relationships between d_α and d_γ expressed by the curves shown in Fig. 2 were used also to describe the structure of steel plate deriving from the transformation of deformed (or partially recrystallized) austenite. Under these conditions the d_γ value was defined as the "effective" grain size, which takes into account the flattening and elongation of austenite grains during deformation, referred to the corresponding volume of deformed austenite. An extensive description of the microstructure model, including details on the basic equations used here, is given in ref.(7).

INTEGRATED HOT-ROLLING MODEL

The integrated model required the assessment of an algorithm which receives, as input, the thermal and mechanical data concerning the plate during the whole rolling schedule and computes the corresponding evolution of the microstructure. The algorithm is thus formed of two basic models: a thermomechanical model and a microstructure model. The thermomechanical model has been adapted to commercial plate-rolling mill operating conditions, having taken into account the thermal balance in the roll bite and temperature evolution during the interpass time. The microstructure model provides the number of homogeneous austenitic structures being formed at any step throughout the rolling process and gives a quantitative description of each of them, while for as-rolled plates with ferrite-pearlite microstructures, it indicates the size of the ferrite grains. The model was checked first on a pilot rolling mill and then verified on the Taranto plate mill. A plain carbon steel and two Nb-microalloyed steels, similar in composition to those investigated in the laboratory tests, were selected for full scale trials, programmed to explore a large spectrum of plate rolling conditions. The mean values of measured and predicted ferrite grain sizes are compared in Fig. 3, where every point is representative of the microstructure of a real steel plate.

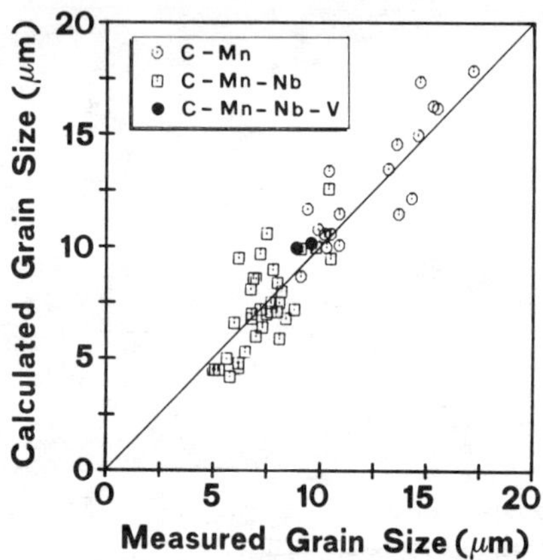

Fig. 3. Comparison of calculated and measured ferrite grain size on as-rolled plates in full-scale trials.

The accuracy of the prediction is acceptable (the standard deviation of the results being of the order of 10%), especially when it is considered that sixty-four plates of three types of steel were rolled to different final thicknesses,

using various rolling schedules, including situations ranging from fully recrystallized to control rolled (highly deformed) austenites.
Even in cases of materials having inhomogeneous grain size distribution in their final ferritic-pearlitic microstructure, the model gives satisfactory detailed predictions. An example of this predictive capability is given in Table 2, whose data refer to the structures reported in Fig. 4.

TABLE 2 Comparison of calculated and measured ferrite grain size distributions. Data referred to the structures of Fig. 4.

PLATE A				PLATE B			
Measured		Calculated		Measured		Calculated	
Vol. %	d_α (μm)	Vol. %	d_α (μm)	Vol. %	d_α (μm)	Vol. %	d_α (μm)
		25	17.0	6.0	34.3	-	-
		34	17.5	18.0	20.6	16.0	22.3
100	17.2	15	18.0	34.0	13.5	29.0	14.8
		10	18.6	34.0	8.3	43.0	9.8
		16	19.0	8.0	4.3	12.0	5.5
$\bar{d}_\alpha$ = 17.2 μm		$\bar{d}_\alpha$ = 17.9 μm		$\bar{d}_\alpha$ = 13.4 μm		$\bar{d}_\alpha$ = 12.7 μm	

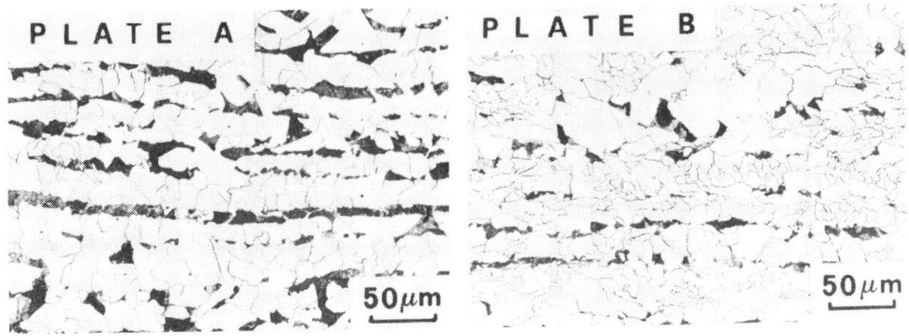

Fig. 4. Ferrite-pearlite microstructures.
Plate A : uniform grain size.
Plate B : heterogeneous grain size.

The measured values of the microstructure parameters were obtained by means of computer-aided image analysis of pictures of metallographic samples, which allows assessment of grain geometry. The good agreement between experimental and calculated results is interesting, because of the prospects of using the

model to prevent formation of inhomogeneous microstructures.
The hot-rolling integrated model is now being used off-line in order to investigate systematically the effects of process variables on the microstructure of as-rolled steel plates. An example of the way the roughing rolling temperatures influence the intermediate and final austenitic grain size, while keeping the finishing rolling conditions unchanged, is presented in Fig. 5. This effect, which is generally underestimated, may become quite important in the case of thick plates, because the smaller the amount of reduction given during finishing, the more effective the roughing temperatures.

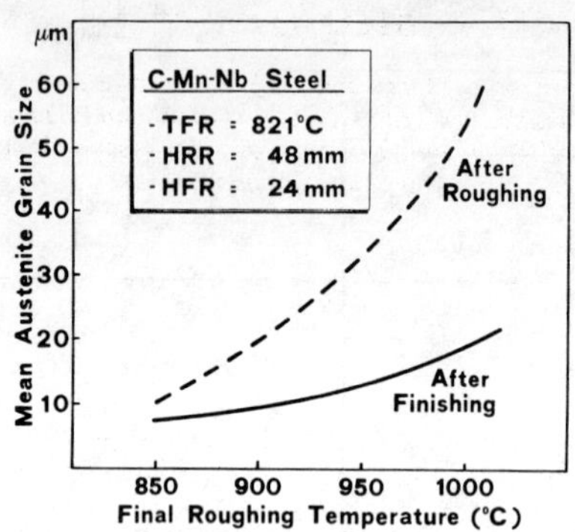

Fig. 5. Relationship between roughing rolling temperature and austenite grain size at the end of roughing and finishing rolling, for constant finishing schedule. (TFR = Final Rolling Temperature; HRR = Transfer Thickness of Plate; HFR = Final Thickness of Plate).

REFERENCES

1. C.M. Sellars, in Hot Working and Forming Processes (edited by C.M. Sellars, G.J. Davies), p.3. The Metals Society, London (1979).
2. R.A.P. Djaic and J.J. Jonas, J. Iron Steel Inst. 210, 256 (1972).
3. C. Perdrix, IRSID Final Report, ECSC Agreement n°7210/EA/311 (1982).
4. W. Roberts, T. Werlefors and A. Sandberg, Intern. Conf. on Tech. and Appl. of HSLA Steels, Philadelphia (1983).
5. Y. Saito et al, International Conference on Steel Rolling, p.1309. The Iron and Steel Institute of Japan, Tokyo (1980).
6. Y. Saito et al, International Conference on HSLA Steels, Wollongong (1984).
7. E. Anelli, M. Ghersi, A. Mascanzoni, M. Paolicchi, CSM Final Report, ECSC Agreement n°7210/EA/410 (1984).

Hot Working Characteristics of Al-Deoxidized Carbon and 0.045% Vanadium Steels

O. Overdal*, H. Gjestland and H. J. McQueen*****

**Research Department, Norsk Jernwerk AS, Mo-i-Rana, Norway*
***Physical Metallurgy, SINTEFF, N7034 Trondheim-NTH, Norway*
****Mechanical Engineering, Concordia University, Montreal H3G 1M8, Canada*

ABSTRACT

A low C and a low C-V steel have been torsion tested in the range 1150-750°C, 0.1-30 s^{-1}. The exponential and Arrhenius functions adequately relate stress, strain rate and temperature. Rolling forces calculated therefrom compare favorably with those measured. The 0.045% V raises the recrystallization temperature by ~50°C, and leads to a ferrite grain refinement of ~20%. Ferrite grain size decreases with falling deformation temperature.

KEYWORDS

Hot torsion, high temperature flow curves, multistage forming, interpass softening, V-steel, HSLA steel,ferrite grain size, dynamic recrystallization

INTRODUCTION

The main objective of controlled rolling is ferrite grain refinement because it enhances both strength and toughness. Largely, this is obtained by austenite grain refinement in the recrystallization temperature range combined with deformation below that. The latter is very effective, because the dislocations introduced enhance ferrite nucleation. Micro-alloying elements like Nb and V, precipitated as carbo-nitride particles during low-temperature hot deformation, stabilize the dislocation substructure and retard recrystallization, thereby facilitating ferrite refinement. To optimize the controlled rolling schedule, it is necessary to have knowledge of the alloy's static recrystallization kinetics and flow stress characteristics to avoid roll separating forces which overload the mill [1-5]. The aim of this torsion testing was to establish the comparative hot working characteristics of Al-killed low C-steels with and without addition of 0.045%V. The effects of deformation temperature T and strain rate $\dot{\varepsilon}$ on rolling forces and ferrite grain size were determined.

EXPERIMENTAL PROGRAM

The steels used had been rolled to universal flats with dimensions 600x25 mm. The rolling forces had been registered by computer with surface temper-

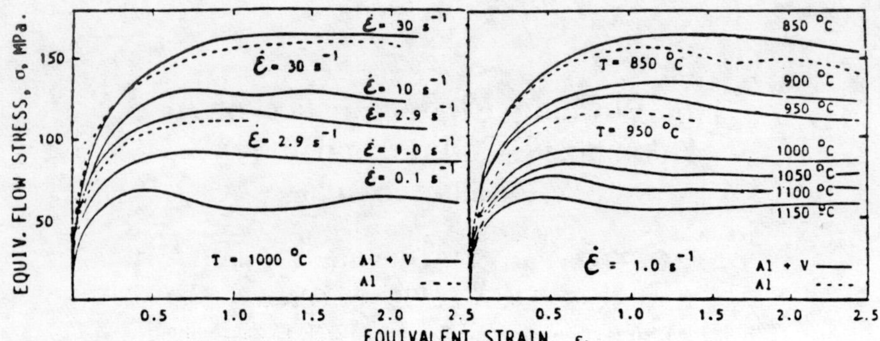

Fig. 1. Representative flow curves for the Al-V and Al steels: a) different $\dot{\varepsilon}$ at 1000°C and b) various T at 1.0 s^{-1}.

TABLE 1: A,Q AND β FOR Al AND Al-V STEELS.

ε	Al A	Q,kJ/mol	β	Al-V A	Q,kJ/mol	β
ε_p	1.84×10^9	296	0.063	1.66×10^9	306	0.068
0.45	1.27×10^9	301	0.080	1.31×10^9	311	0.085
0.40	1.06×10^9	300	0.083	1.41×10^9	314	0.089
0.35	8.76×10^8	300	0.087	1.26×10^9	314	0.093
0.30	5.32×10^8	293	0.090	9.83×10^8	314	0.099
0.25	3.57×10^8	291	0.096	1.44×10^9	323	0.108
0.20	1.33×10^8	279	0.100	5.68×10^8	312	0.114
0.15	7.87×10^7	273	0.108	1.05×10^9	324	0.129

atures measured manually. The test materials from Norsk Jernverk AS had the following composition:

Designation	%C	% Mn	% Si	% P	% S	% N	% Ni	% Cu	% Al	% V
Al	0.09	1.16	0.29	0.006	0.014	0.008	0.06	0.11	0.017	0.00
Al - V	"	"	"	"	"	"	"	"	"	0.045

The torsion specimens, having gage dimensions L=5mm, r=4mm, were heated by induction with temperature being computer controlled. The computer-controlled, hydraulic torsion machine had a maximum velocity $\dot{N}$ of 40 turns/s. The equivalent strain ε and the flowstress σ were computer calculated from the moment M by the following equations:

$$\sigma = 3\sqrt{3}\, M/2\pi r^3 \qquad (1)$$

$$\varepsilon = 2\pi r N/\sqrt{3}\, L \qquad (2)$$

The specimens were preheated at ~1270°C for 5 min. before being deformed continuously at constant T between 1150 and 850°C and at $\dot{\varepsilon}$ between 30 and 0.10 s^{-1}. The Al steel was strained just beyond the peak strain which increased as T decreased. The Al-V steel was given a constant strain of 2.5. The steels were given interrupted tests at 850-1000°C, 0.1 to 2.9 s^{-1} with holding times t_i between the 9 passes (ε_i=0.3) of 1, 10 and 30 sec. After finishing deformation, the steels were cooled at ~3°C/s.

RESULTS AND DISCUSSION: CONTINUOUS DEFORMATION

The flow curves in Fig. 1, exhibiting a peak and decline toward a steady state regime characteristic of dynamic recrystallization [1,2], rise with decreasing temperature and increasing strain rate. The Al-V steel is about 5% stronger than the Al steel only below 1000°C. The stress at the peak

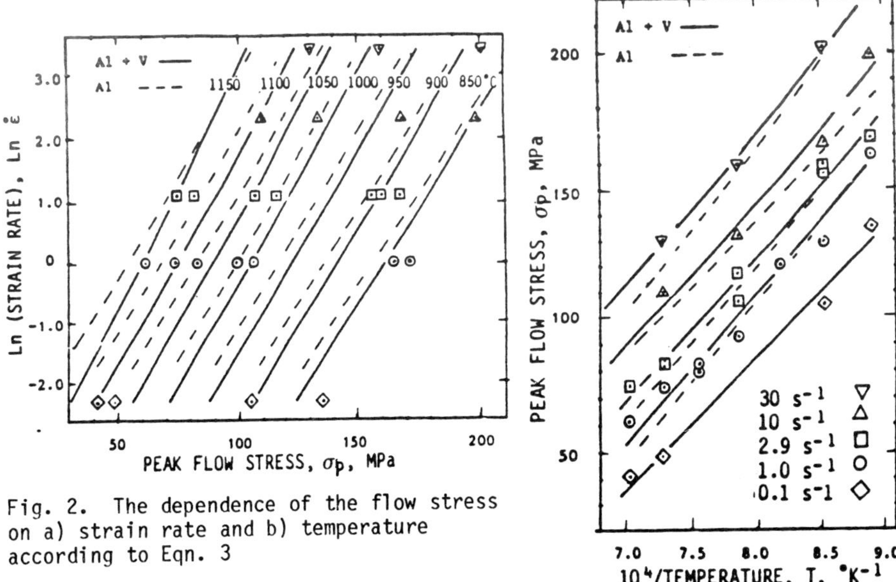

Fig. 2. The dependence of the flow stress on a) strain rate and b) temperature according to Eqn. 3

strain ε_p, or at any given strain, depends on temperature and strain rate according to the expression [4,5]:

$$\dot{\varepsilon} = A \exp(\beta\sigma) \exp(-Q/RT) \qquad (3)$$

The empirical constants A, β and Q found by regression analysis from Fig. 2 are reported in Table 1. In Fig. 2, the lines for the Al steel are consistently at lower stresses than those for the Al-V steel. The constants have been calculated for different strains (Table 1) for the purpose of estimating flow strengths in different rolling passes. As can be seen, the Al-V steel has a somewhat higher activation energy Q (306 kJ/mol) than the Al-steel (296 kJ/mol). This arises from the greater variation in strength with temperature that is caused by precipitation of V-rich particles on dislocations at 950-850°C. These particles slow down grain boundary migration, and hence dynamic recrystallization, which is shown by the increasing strain ε_p to peak-strength at low temperatures for the Al-V steel (Fig. 3) [1,4,6-8]. Sankar et al. [6] found that 0.05% Nb increased Q to 434 from 308 kJ/mol for C steel. The Q for 0.03 Nb steel was reported to be 401 kJ/mol [9]. Generally, Nb has a much stronger effect on the hot workability than V; e.g., at 950°C, 1 s^{-1}, 0.045% V raises the strength from 115 to 125 MPa whereas 0.05% Nb in a 0.12% C steel raises it from 100 to 130 MPa [6]. The raising of ε_p by microalloying elements, has been clearly demonstrated [6-8]; Fig. 3 shows data for C and Nb steels (preheated at 1000°C) with finer grain sizes than the present [6].

After continuous deformation at 1.0 s^{-1}, the Al-V specimens were studied metallographically. The α-grain size decreases linearly as T decreases (Fig. 4) due to increased stored energy and finer dynamically recrystallized γ-grain size which enhanced ferrite nucleation. The Al steel specimen at 850°C, the only one subjected to a continuous strain of 2.5, developed a grain size which is 2 μm larger than the Al-V steel. This is consistent with the higher flow stress and dislocation density in the Al-V steel.

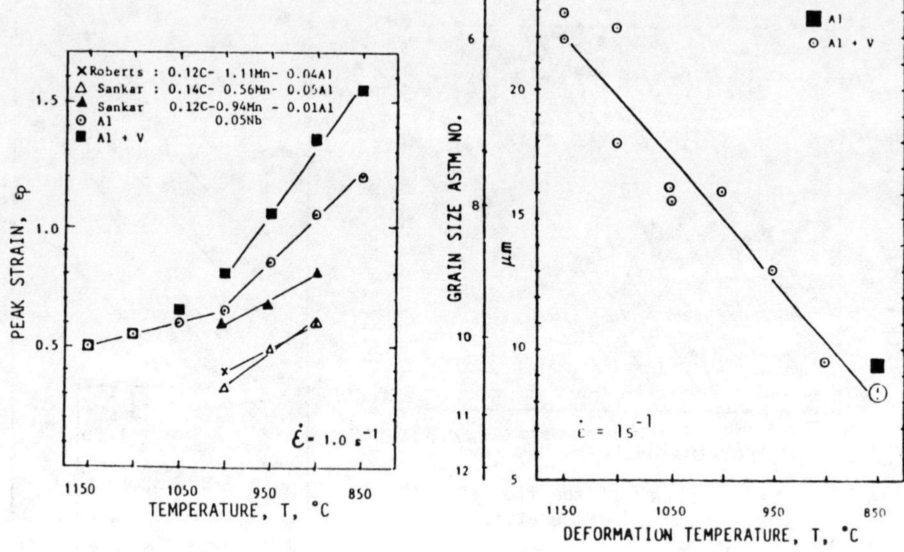

Fig. 3. Strain to the peak stress as a function of temperature for the present and comparative C and Nb steels [6,7].

Fig. 4. Ferrite grain size after a strain of 2.5 at the temperature shown and final cooling at ~3°C/s.

For all stages both steels were rolled with computer registration of the rolling parameters and roll separating forces (Table 2) [10]. The flow stress for each pass was calculated by Eqn. 3 from data in Table 1. This analysis assumes that the structure for each pass is similar to that of material preheated to 1270°C [2-5]. The roll separating forces (Table 2), calculated by several formulas compiled by Sandmark [11], are generally 10-20% lower than the measured values and are even more so in the two finishing passes. The first explanation of this is the formulas' expected accuracy of ±20% [11]. Secondly, the use of data from specimens preheated at 1270°C, whereas the billets recrystallized between passes, could account for the generally low values in the first two calculations [3-5]. Thirdly, the carry over of stronger non-recrystallized material into the two finishing passes is not taken into account. The schedules are successful in limiting force variations to ±10% even though the steels double in strength.

TABLE 2: COMPARISON OF CALCULATED ROLL SEPARATING FORCE WITH VALUES MEASURED AT NORSK JERNWERK AS FOR Al-V STEEL

Rolling parameters						Rolling force (tons)						
							CALCULATED					
Pass no.	$\varepsilon = \ln\frac{H_o}{H}$	$\dot{\varepsilon}$ (s^{-1})	$T_{surf.}$ (°C)	$T_{aver.}$ (°C)	σ_{calc} (N/mm^2)	Measured Fm	Green-Wallace Fc	Fc/Fm	Sims Fc	Fc/Fm	Geleji (μ = 0.35) Fc	Fc/Fm
1	0.25	5.9	1190	1242	58.6	565	536	0.95	492	0.87	605	1.07
2	0.23	6.4	1160	1233	59.4	559	466	0.83	456	0.82	538	0.96
3	0.29	10.2	1130	1215	70.7	697	553	0.79	555	0.80	673	0.97
4	0.29	14.1	1100	1178	80.5	609	559	0.92	590	0.97	684	1.12
5	0.26	18.4	1070	1126	89.0	679	514	0.76	529	0.78	628	0.92
6	0.16	16.8	910	936	110.7	588	445	0.76	481	0.82	544	0.93
7	0.13	13.3	880	901	113	560	376	0.67	392	0.70	460	0.82

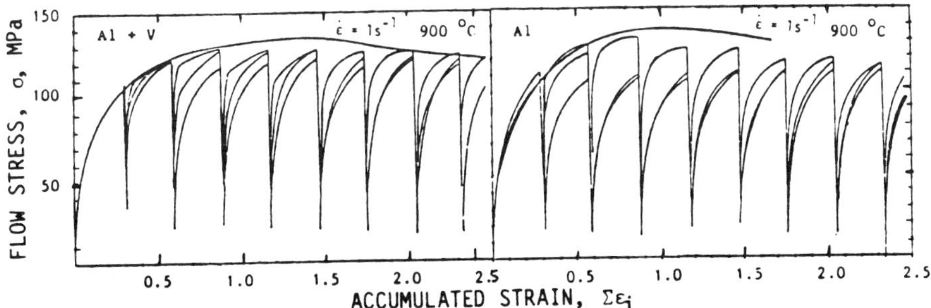

Fig. 5. Multistage isothermal flow curves at 900°C, 1 s^{-1} for passes of 0.3 and intervals of 1, 20 or 30 s. a) Al-V steel and b) Al steel. Continuous deformation curves superimposed.

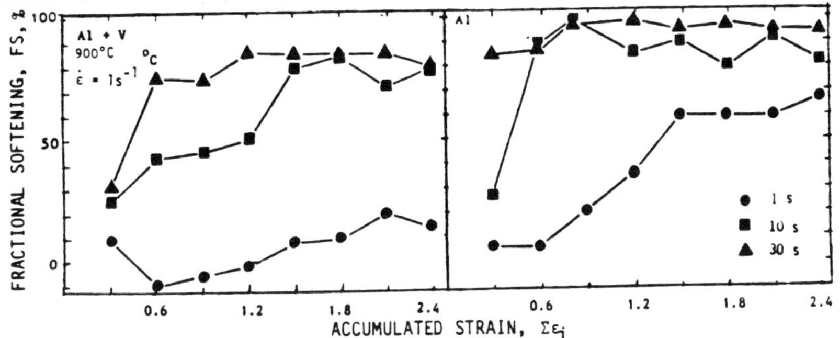

Fig. 6. Fractional softening for 1, 20 and 30 s intervals as a function of accumulated strain at 900°C, 1 s^{-1} for a) Al-V steel and b) Al steel.

RESULTS AND DISCUSSION: MULTIPLE DEFORMATION

The isothermal multistage flow curves are shown in Fig. 5 at 900°C, 1 s^{-1}. For the Al steel, the 30 s intervals, give flow curves almost identical to the first, but with a slight rise possibly due to grain refinement. The 10 s intervals give slightly raised flow curves, possibly due to incomplete recrystallization. The curves for 1 s approximate the continuous curve in the first 3 passes, but gradually separate from it as a result of augmented static or metadynamic recrystallization (growth of dynamic nuclei after deformation ceases) [2]. For the Al-V steels, flow curves for 1 s intervals come fairly close to the continuous curve, indicating that despite some static recovery, the steel strain hardens quickly to the continuous level. The curves for 10 and 30 s are still considerably above the original, indicating incomplete recrystallization or much finer grain size. At 850°C, the Al steel is like the Al-V at 900°C and the Al-V shows even less restoration. At 950°, and more so at 1000°C, the Al-V steel becomes similar to the Al steel at 900°C with much more recrystallization between passes.

From the curves in Fig. 5, fractional softening FS_n in the nth interval was calculated:

$$FS_n = (\sigma_{un}-\sigma_{y(n+1)})/(\sigma_{un}-\sigma_{y1}) \qquad (4)$$

where: σ_{un} is max. stress in the n'th pass and σ_{y1} and $\sigma_{y(n+1)}$ are the yield strengths in the 1st and (n+1)th passes. For 900°C, 1.0 s^{-1}, the fractional softening, plotted against accumulated strain in Fig. 6, increases with rising holding time and strain. The latter arises because the stored energy builds up with each pass until equilibrium between interpass softening and pass strain hardening is reached. The nucleation rate can also be increased by grain refinement during the previous holding period. The very low, even negative softening in the Al-V steel at low cumulative holding times is due to precipitation. Increased T raises the softening, the curves for the Al-V steel at 1000°C being similar to the Al steel at 900°C.

Plots of FS against log t provided traditional S curves for both steels. The Al-V steel recrystallizes slower than the Al steel indicating that the strain induced V-rich particles are pinning the dislocations and grain boundaries [6,8,12]. The maximum difference between the two steels arises on holding 30 s at 900°C, the Al-steel recrystallizes almost completely, whereas the Al-V reaches about 30% FS. Under similar testing conditions, except $\varepsilon = 0.4$ and finer grain size, FS for C steel was about 110% (grain growth) and for 0.05 Nb steel 65%; moreover, these values compared favorably with previous results [6]. To investigate the effect of strain rate on the recrystallization kinetics, FS_1(t=10s) for the Al-V steel at 950°C were determined at 0.1 s^{-1} and 2.9 s^{-1}; FS_1 varies linearly as log $\dot{\varepsilon}$. Thus at industrial rolling rates 10-20 s^{-1}, the stored energy driving recrystallization is sufficiently high to cause the steel to recrystallize totally in 10 s at 950°C.

CONCLUSIONS

The effect of 0.045% V on low C steel's hot working characteristics, as observed by torsion testing, can be stated as follows:

1) The activation energy for dynamic recrystallization is increased from 296 to 306 kJ/mol.

2) The ferrite grain size is refined by 1-2 μm.

3) The derived equivalent flow stresses permit calculation of separating forces with reasonable accuracy.

4) 0.045 V steel has the same recrystallization kinetics as the C steel when its temperature is 50 to 100°C higher.

REFERENCES

1. H.J. McQueen and J.J. Jonas, J. Appl. Metalwork. 3, 233-41 (1984).
2. H.J. McQueen and J.J. Jonas, J. Appl. Metalwork. 3, 410-20 (1984).
3. H.J. McQueen, M.G. Akben and J.J. Jonas, Microstructural Characterization of Materials by Non Microscopic Techniques (ed. by J.B. Bilde-Sorensen), pp. 397-404, RISØ Natl. Lab., Roskilde, Denmark (1984).
4. N.D. Ryan, H.J. McQueen, J. Mech. Metalwork. Tech. 10, in press (1985).
5. N.D. Ryan, H.J. McQueen, J. Mech. Metalwork. Tech. 10, in press (1985).
6. J. Sankar, D. Hawkins and H.J. McQueen, Met. Tech. 6, 325-32 (1979).
7. W. Roberts, Swedish Inst. Metals Res. Report 1858 (1983).
8. I. Weiss and J.J. Jonas, Met. Trans. 10A, 831-40 (1979).
9. C. Ouchi and T. Okita, Trans. Iron Steel Inst. Japan 22, 543-51 (1982).
10. Private Communication, Norsk Jernwerk AS (1983).
11. P.A. Sandmark, Scand. J. Met. 1, 313-18 (1972).
12. R.A. Petkovic, M.J. Luton, J.J. Jonas, Can. Met. Q. 14, 137-45 (1975).

ACKNOWLEDGEMENTS

The authors wish to thank the National Research Council of Norway for financial assistance which made possible the research project and the collaboration. The complete cooperation of Norsk Jernwerk AS is gratefully acknowledged. We are grateful to J. Bowles for rapid preparation of the diagrams.

Thermo-Mechanically Controlled-rolled Low Nickel-Niobium Steel Plate for Cryogenic Service Pipelines

H. Tamehiro, R. Habu, N. Yamada, M. Murata and M. Nagumo

Kimitsu R & D Division, Central R & D Bureau, Nippon Steel Corporation, 1-Kimitsu, Kimitsu City, Chiba Prefecture, 299-11 Japan

ABSTRACT

Low Ni steel plates for cryogenic service pipelines were investigated. In the Thermo-Mechanical Control Process (TMCP) with accelerated cooling, the influence of alloying elements on the properties and microstructures of plates was examined. As a result, through the optimization of chemical composition and process conditions, low-temperature toughness comparable to that of 9% Ni steel was found to be attainable with a 2% Ni-Nb steel.

KEYWORDS

Ni steel; Nb steel; Thermo-Mechanical Control Process (TMCP); Low temperature toughness; Grain-refinement; Heat-affected zone.

INTRODUCTION

Presently, expensive high alloy steels such as 9% Ni steel are used for intra-plant pipes to transport low-temperature fluids, e.g. LNG. It is, however, economically unfeasible to apply such expensive steels to long-distance pipelines for transporting chilled natural gas or dense phase gas.[1),2)] Therefore, cheaper steels having adequate low-temperature toughness will be required if long-distance cryogenic service pipelines are to be realized. This paper approaches this subject from a standpoint of both the alloying design and the thermo-mechanical treatment of steels. In the TMCP (Reheating at a low temperature, thermo-mechanical rolling and accelerated cooling), the effect of alloying elements on the mechanical properties of high purity steels with reduced P, S, N contents were examined, and the relation between the low-temperature toughness and microstructures was discussed.

EXPERIMENTAL PROCEDURE

Laboratory steels reduced in P, S and N contents were vacuum-melted in 300 kg heats and cast as 125 mm thick ingots. The chemical compositions are shown in TABLE 1. These ingots were hot rolled and accelerated cooled to

9 mm thick plates after reheating at 950°C for 1 hr. The finish rolling temperature was in the range from 740° to 760°C after total reduction below 800°C of 80%. The cooling rate of accelerated cooling was about 30°C/s until the cooling stop temperature of around 500°C. The mechanical properties were examined in the transverse direction. Full thickness tensile tests, 2/3 sub-size 2 mm V-notch Charpy tests and Battelle type DWTT were conducted. The heat-affected zone (HAZ) toughness was examined using a welding thermal cycle simulator. In addition, microstructural observations by optical and scanning electron microscopes, observation of precipitates by transmission electron microscope (TEM), etc., were carried out.

TABLE 1 Chemical Composition of Laboratory Steels (wt%)

Steel	C	Si	Mn	P	S	Ni	Nb	Ti	Al	N	Ar_3 (°C) *)
N1	0.019	0.10	0.92	0.002	0.0006	–	0.015	–	0.032	0.0023	800
N2	"	"	"	"	"	1.04	"	–	"	"	763
N3	"	"	"	"	"	2.12	"	–	"	"	724
N4	"	"	"	"	"	3.06	"	–	"	"	690
N5	0.021	0.11	0.96	0.002	0.0004	2.00	–	–	0.032	0.0012	724
N6	"	"	"	"	"	"	0.016	–	"	"	"
N7	"	"	"	"	"	"	"	0.007	"	"	"
N8	0.009	0.10	0.92	0.001	0.0006	1.92	0.020	–	0.030	0.0012	735
N9	0.023	"	"	"	"	"	"	–	"	"	729
N10	0.040	"	"	"	"	"	"	–	"	"	723
N11	0.055	"	"	"	"	"	"	–	"	"	717
N12	0.071	"	"	"	"	"	"	–	"	"	711

*) $Ar_3 = 868 - 396C + 24.6Si - 68.1Mn - 36.1Ni - 20.7Cu - 24.8Cr$ (°C)

EXPERIMENTAL RESULTS

Figure 1 shows the effect of the Ni content on the mechanical properties of Nb steels. Both strength and toughness are improved by the Ni addition. With increasing Ni contents, the microstructure changes from ferrite-pearlite to fine-grained ferrite-bainite. Increase in strength is thought to result from solid-solution hardening, grain-refinement and increased volume fraction of bainite, while improvement in toughness results from the grain-refinement, in addition to the toughening of the ferrite matrix.[3] The grain-refinement comes from the lowered Ar_3 temperature caused by the Ni addition. With a 2% Ni addition, tensile strength (TS) of more than 50 kg/mm^2 and Charpy 50% shear FATT of less than -196°C can be achieved. The shear area at liquified nitrogen temperature is 100%. Separation rarely occurs on the fracture surface. There is no difference in strength between a Ni-free steel and a 1% Ni steel. This is due to the introduction of deformed ferrite(α)-grains in the Ni-free steel because of the high Ar_3 temperature. Figure 2 shows the effect of microalloyed Nb and Ti on 2% Ni steels. Nb improves TS by about 6 kg/mm^2 and Charpy 50% shear FATT by more than 20 deg at the heating temperature of as low as 950°C. However, Ti addition to a Nb steel hardly affects the steel properties. As shown in Photo. 1, α-grains are conspicuously refined by Nb and further refining is not attained by Ti. The increase in strength by the Nb addition is presumably attributable to the grain-refinement and precipitation hardening. Figure 3 shows the effect of C contents and the reheating temperatures on 2% Ni-Nb steels. Increase in C contents slightly enhances strength and deteriorates the low-temperature toughness, because dissolved Nb during heating decreases. Moreover, even if the heating temperature is raised from 950 to 1050°C, the increase in strength in very low C steels is quite small. The reason for this is thought to be that Nb is easily dissolved in very low C steels even at a heating temperature of 950°C. Consequently, a high reheating temperature is not so advantageous because it coarsens

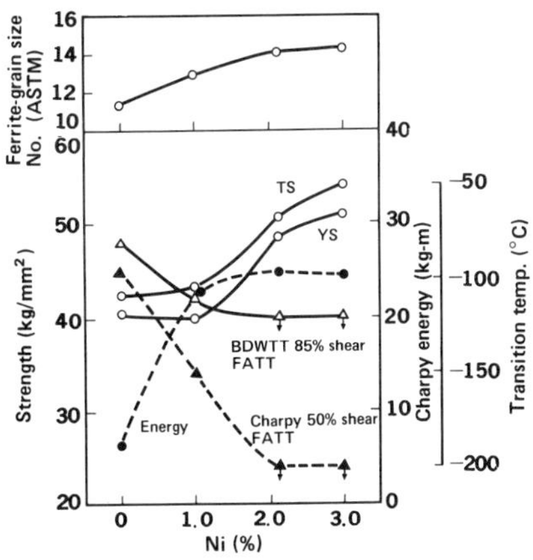

Fig. 1 Effect of Ni content on the mechanical properties of plates (Steel N1 ∿ N4).

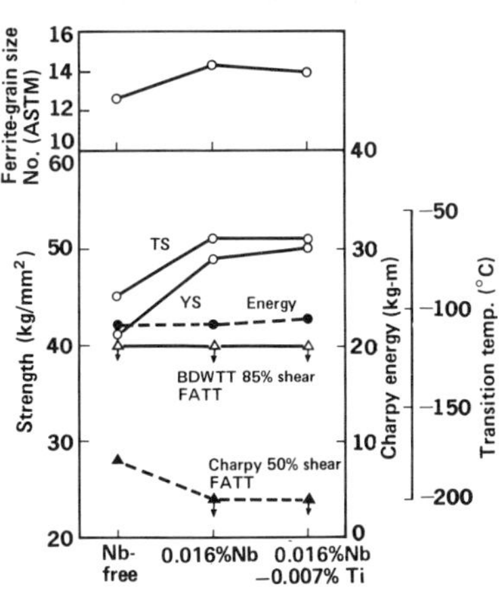

Fig. 2 Effect of Nb and Ti addition on the mechanical properties of plates (Steel N5 ∿ N7).

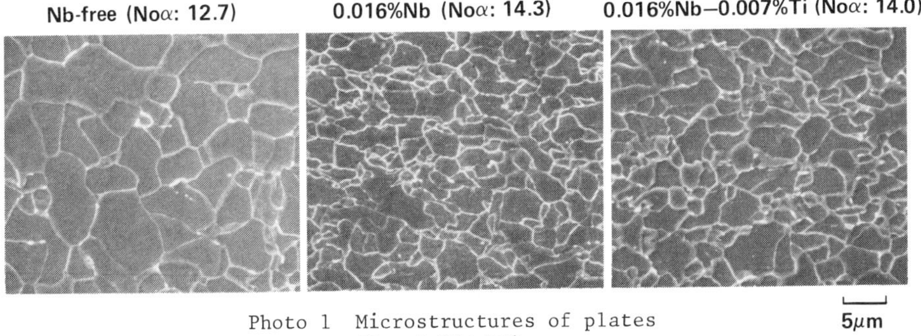

Photo 1 Microstructures of plates (Steel N5 ∿ N7)

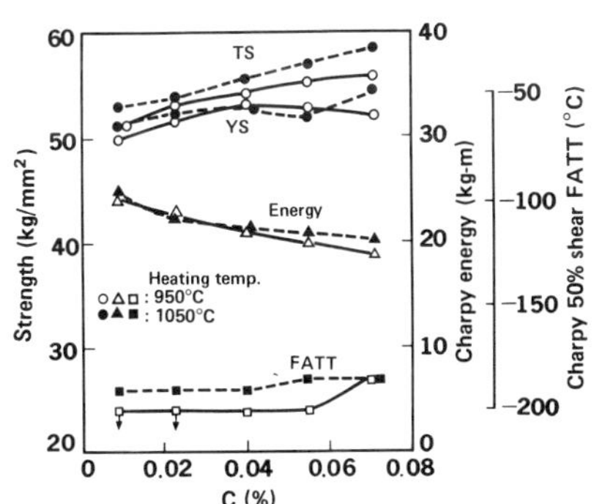

Fig. 3 Effect of C content and reheating temperature on the mechanical properties of plates (Steel N8 ∿ N12).

austenite(γ)-grains during heating and impairs the low-temperature toughness. Figure 4 shows the effect of Ni, Nb and Ti in the simulated HAZ. The transition temperature rapidly drops with increasing Ni contents and Nb further improves it by about 10 deg. Ti addition has no appreciable effect on the HAZ toughness. Ni depressed the nucleation of proeutectoid α at γ-grain boundaries by increasing hardenability and refines the microstructure, besides strengthening the matrix. It is known that microalloyed Ti prevents the grain coarsening during welding and improves HAZ toughness[4]. However, it seems that this effect has been lost when the peak temperature of simulated thermal cycle is as high as 1400°C. Figure 5 shows the effect of C contents on 2% Ni-Nb steels. C contents of 0.02% to 0.04% are suitable from HAZ toughness. Such a C content is also desirable even in the base material properties. In a very low C steel, i.e. about 0.01% or in higher C steel, i.e. more than 0.06%, toughness becomes poor. Increase in the C content increases the volume fraction of high-carbon martensite or bainite which is harmful for low-temperature toughness.

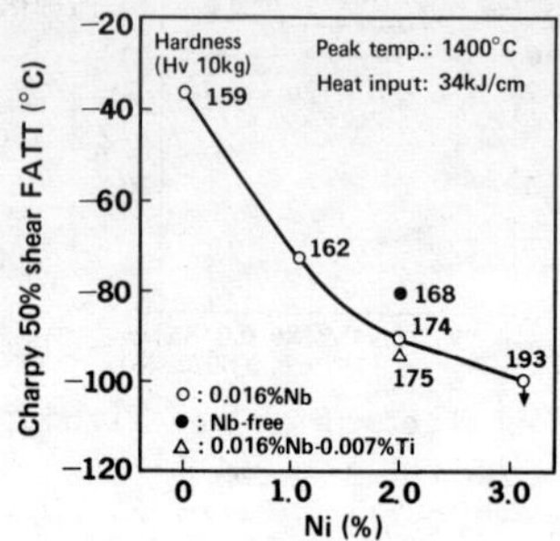

Fig. 4 Effect of Ni content on the low-temperature toughness in the simulated HAZ (Steel N1 ∿ N7).

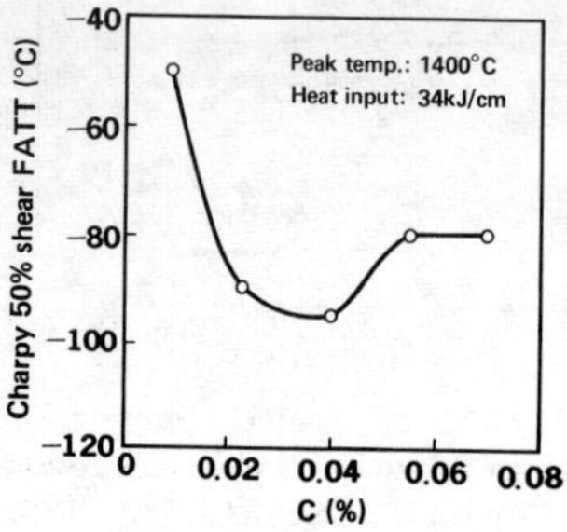

Fig. 5 Effect of C content on the low-temperature toughness in the simulated HAZ (Steel N8 ∿ N12).

DISCUSSION

As mentioned above microalloyed Nb improves not only the base material strength and toughness but also the HAZ toughness. The experiment revealed that Nb dissolved during reheating of ingots plays an important role in improving the properties. From the solubility relationship;

$$\log_{10}[Nb][C + 12/14N] = -6770/T + 2.26$$ [5)]

(T: Absolute temperature °K)

the solubility product [Nb][C + 12/14N] is 5.3×10^{-4} when the heating temperature is 950°C. Therefore, Nb contained in steel N6 is dissolved completely during heating because of a very low C content. Though it is known that Nb in solid solution raises the hardenability of steel and increases the volume fraction of bainite,[6)] it is difficult to find a difference in the microstructures between a Nb-free (Steel N5) and a Nb steels (Steel N6), thus the rise in strength by hardenability seems to be negligible. The increase of strength by grain-refinement and precipitation hardening was analyzed for steels N5 and N6 by using the following Petch equations, and the result is shown in TABLE 2.

$$\Delta YS = K_1(\Delta d^{-1/2}) + \Delta\sigma_{ppt,1}$$

$$\Delta TS = K_2(\Delta d^{-1/2}) + \Delta\sigma_{ppt,2}$$

Where K_1 (= 1.9 kg/mm$^{3/2}$) and K_2 (= 0.8 kg/mm$^{3/2}$) are coefficients depending on grain size,[6)] d is diameter of α-grains and $\Delta\sigma_{ppt,1}$ and $\Delta\sigma_{ppt,2}$ are increases by precipitation hardening. This result shows that the rise in strength by grain-refinement is fairly large, but Nb precipitates of less

TABLE 2 Analysis of Increment of Strength by Nb

	Measured value (kg/mm²)	K (Δd$^{-1/2}$) (kg/mm²)	$\Delta\sigma_{ppt}$ (kg/mm²)
ΔYS	9.8	9.1	0.7
ΔTS	5.8	3.8	2.0

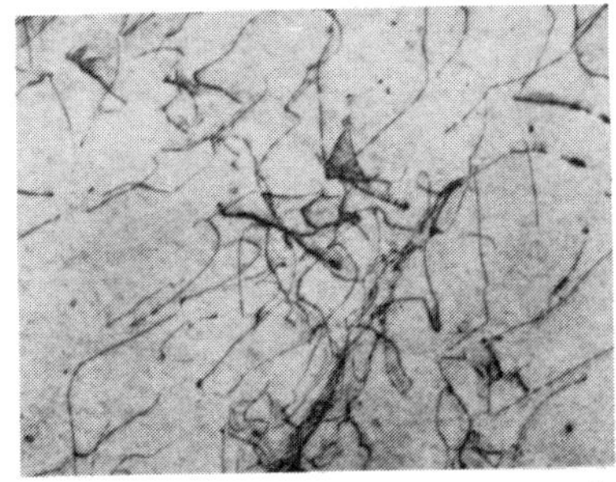

Photo 2 Transmission electron micrograph of Nb-precipitates (Steel N6)

than 100 Å in Photo. 2 undoutedly indicate the increase in strength by precipitation hardening. On the other hand, the improvement in toughness of the base material is considered to be caused by the grain-refinement. For both Nb-free and Nb steels, no separation was observed on the fracture surface. The improvement in transition temperature by grain-refinement (ΔTT) is 120 deg, calculating from the Petch equation;

$$\Delta TT = K\ (\Delta d^{-1/2})$$

where K is 25 deg.mm$^{1/2}$. The grain-refinement by Nb is probably due to the transformation of α from the non-recrystallized γ by a thermo-mechanical rolling. The γ-grains of the Nb steel and Nb-free steels during heating are not uniform with average γ-grain size numbers (ASTM) of 4.7 and 5.2, respectively (Photo 3). γ-grain of the Nb steel is slightly larger than that of the Nb-free steel. Although this γ-grain is refined through the rolling in the recrystallization region, it is hard to think that the recrystallized γ-grains are so refined as to produce an ultra fine grain of 14.3 in α-grain size number in the Nb steel. Figure 6 shows the true stress-true strain curves during hot deformation at 850°C obtained by using a hot deformation simulator. The stress of the Nb steel at the third reduction is remarkably high compared with that of the Nb-free steel. This fact means that the softening of deformed γ-grain through recrystallization is strongly inhibited in the Nb steel. Very fine α-grains may be caused by the transformation from the non-recrystallized γ. Next, let us consider the improvement of HAZ toughness by the Nb addition. Photo 4 shows the microstructure in simulated HAZ. The microstructure of a Nb-free steel consists

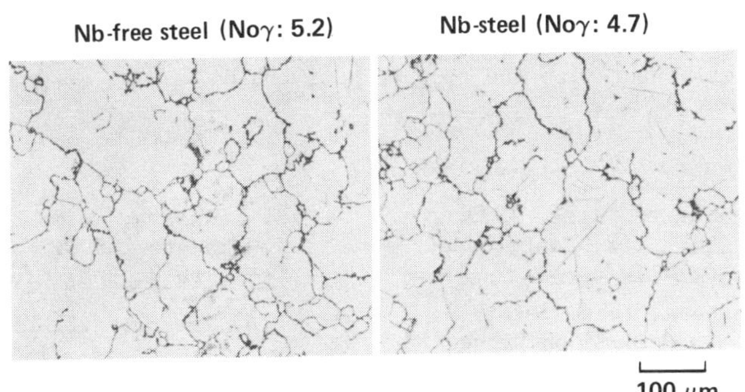

Photo 3 Austenite-grain during heating at 950°C for 1 hr (Steel N5, N6)

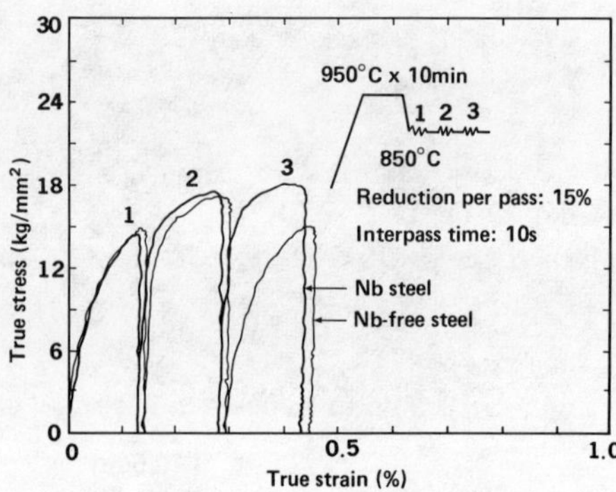

Fig. 6 True stress - true strain curves during hot deformation.

of proeutectoid α formed at the prior γ-grain boundaries and coarse bainite inside γ-grains. The refinement of the coarse γ-grain is insufficient. In a Nb steel, almost uniform acicular or bainitic microstructure is realized with proeutectoid α at γ-grain boundaries being depressed. Such a microstructural change occurs by the effect of increasing hardenability of Nb dissolved during heating (welding).

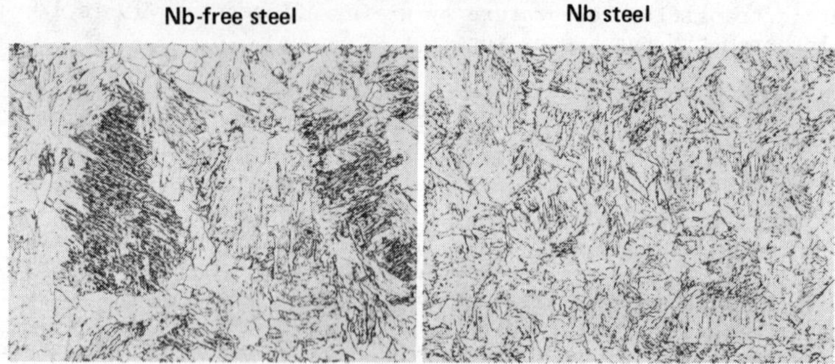

Photo 4 Microstructures in simulated HAZ (Steel N5, N6)

CONCLUSION

Through the optimization of chemical composition (0.03% C - 1% Mn - 0.01% Nb with low P, S and N) and process conditions (950°C reheating, thermomechanical rolling and accelerated cooling), low-temperature toughness comparable to that of 9% Ni steel was found to be attainable with only a 2% Ni addition. Especially, microalloyed Nb plays an important role in this steel. It significantly improves not only the strength and toughness of the base material but also the toughness of the HAZ.

References

1. D.L. Katz and G. King, 1973 National Technical Conference, Calgary, Oct. 1973.
2. M. Dimentberg, Third International Conference on Liquefied Natural Gas, Washington, Sept. 1972.
3. W. Jolly, JISI, 206, 170 (1968).
4. S. Kanazawa, A. Nakashima, K. Okamoto and K. Kanaya, Tetsu-to-Hagané, 61,2589 (1975).
5. K.J. Irvine, F.B. Pickering and T. Gladman, JISI, 205, 161 (1967).
6. C. Ouchi, T. Okita and S. Yamamoto, Tetsu-to-Hagané, 67, 969 (1981).

Influence of Hot Deformation Upon the Transformation Behaviour and the Microstructure of Several Low and Medium Carbon Steel Grades

C. M. Vlad, M. Liphardt and J. Kügler

Stahlwerke Peine-Salzgitter AG, Research Department, Federal Republic of Germany

ABSTRACT

Changes in the solubility and diffusion kinetics of alloying elements as well as of the self diffusion of iron brought about by the "defect structure" of the deformed austenite are altering both nucleation and the growth of the different transformation phases. Whereas the austenite transformation to ferrite and pearlite is accelerated by an increased concentration of vacancies brought about by the hot deformation, the bainitic and the martensitic transformations are shifted to lower temperatures. There is evidence that the decrease in M_s- and B_s-Start temperatures might have been brought about by the development of a strong {123} <412> texture in the austenite.

KEYWORDS

Deformation CTT-diagrammes, defect structure, early ferrite and pearlite nucleation, depression of the M_s- and B_s-start temperatures, {123} <412>-texture of austenite.

INTRODUCTION

The increased tendency in the steel and manufacturing industry to incorporate heat treatment steps directly in a hot rolling or hot forging procedure requires a detailed knowledge of the influence of the deformation upon the transformation behaviour of the austenite. There is a scarcety of data in literature regarding this influence[1) 2)]; in general, the available CTT-transformation diagrams[3) to 7)] are describing the transformation of the austenite without a prior deformation step.
It was the purpose of this work to examine in some details the alteration of the CTT-transformation diagrammes brought about by a prior deformation step of the austenite in some low and medium carbon steel grades used in the manufacture of wire rods, bars[8)] and Dual-Phase-steels[9)].

MATERIAL AND EXPERIMENTAL PROCEDURES

Three different groups of steels were chosen for this investigation: a low carbon-silicon steel grade, 7 Si Mn 64, commonly used as welding wire electrode, three medium carbon steel grades 55 Si 7, 60 Si Mn 5 and 70 Mn Cr 4 containing various amounts of Si-, Mn- and Cr-contents and as well as a D.P. steel grade 10 Mn Si Cr 55 - see Table 1. All the materials investigated were taken from heats previously continuous cast to 165 square billets that subsequently were hot rolled to 7,5 mm dia. wire rods. Cylindrical specimens 4 mm dia and 6,4 mm long were then machined from these wire rods for the dilatometry work. For the D.P. steel 10 Mn Si Cr 55 the specimens for the dilatometry work were machined from 6 mm thick hot bands.

CTT-transformation diagrames without - and with a 50 % deformation were determined on two series of specimens after they were reaustenitized at 1050 °C for 10 minutes by means of a Dilatronic III-Deformation Dilatometer of Theta Industrie Inc. The deformation of the austenite, with the exception of the D.P. steel, was carried out as a single step at temperatures ranging from 980 °C to 850 °C followed by cooling at rates similar to those used with the undeformed specimens. The D.P. steel specimens which were reaustenitized at 1050 °C for 10 minutes were given a 40 % deformation in a single step at 1040 °C. Those specimens reaustenitized at 1150 °C however were deformed in 2 steps to a total deformation of 58 % at 850 °C. The CTT-transformation diagrams of the undeformed series were obtained from the austenitizing temperatures of 1050 ° and 1150 °C respectively.

The microstructures were examined by optical microscopy: standard point counting and mean linear intercept procedures were used for quantitative determination of the volume fractions of each phase. The HV10 hardness measurements were made in accordance with DIN 50133.

Texture determinations on specimens of the steel grade 70 Mn Cr 4 transformed to martensite from an underformed - as well as from a 75 % deformed austenite were made by means of a modified Inverse Polefigure method[10].

RESULTS AND DISCUSSION

The CTT-Diagrammes of the investigated steels are shown in Figs. 1 to 6; It is evident from an inspection of these figures that phase transformations connected with a volume increase will be favoured by an increase concentration of vacancies brought about by the deformation process[11]. As a direct consequence of this, the ferrite-pearlite field regions, as predicted by Nocke et.al.[12] and Samovski[13] have been shifted to higher temperature levels and shorter times. This behaviour has been interpreted to result directly from an increased interfacial area of the austenite-grain boundary[14) in connection with a high frequency of ferrite nucleation[14, 15] and/or with an intergranular formation of ferrite nuclei on the dislocation cells of the deformed austenite[16].

In the high silicon steel grade 7 Si Mn 64 the transformation of the deformed austenite to bainite at the end of the pearlite reaction appears to be retarded; it is possible that this results from the increased nucleation and precipitation of carbon as Fe_3C. A similar-retardation of the bainite reaction has been observed by Smirnov[17]. On alloyed steel grades with retarded transformation characteristics on the other hand, a deformation of the metastable austenite in the lower temperature range of 850 °C will cause a constriction of the bainite field region so that a transformation "free field" is emerging between the bainite and pearlite. This behaviour characteristic for the steel grades 55 Si 7, 60 Si Mn 5 and 70 Mn Cr 4 in which an accelerated pearlite nucleation is taking place, has been observed in the past quite frequently by Coldren et.al.[18] and Bernstein[19] in D.P. steel grades and medium-carbon steels respectively.

In D.P. steel grade, 10 Mn Si Cr 55 the deformation of austenite has shifted the bainite reaction to low formation temperatures of 425 ° to 500 °C, at higher cooling rates of the order at 10 ° to 20 °C/sec. and to high transformation temperatures of 600 ° to 650 °C at slow cooling rates of the order 0,085 to 0,095 °C/sec., depending on the prior austenitic grain size. Such a behaviour is in accordance with a mechanism of increased ferrite nucleation frequency aided by cementite precipitation at slow cooling rates in the deformed austenite and with the impedance of the bainite reaction at high cooling rates by the presence of a deformation texture in the austenite. The depression of the B_s-start-temperature caused by the deformation of the austenite in D.P. steel amounts to about 100 °C, and seems to be independent of the austenite grain size.

The martensitic temperature is also affected by the "defect structure" of the deformed austenite in that the M_s-start-temperature is shifted by as much as 30 ° to 40 °C to lower temperatures. An additional evaluation of the texture of the specimens of 70 Mn Cr 4 steel grade, transformed to martensite after a 75 %-deformation in the lower temperature range of the metastable austenite revealed a 30 % volume fraction of grains to possess a <211> direction //ND. This implies that the deformation of the austenite has resulted in the development of a {123} <412> texture which on quenching to martensite has transformed to a 211 <011> texture. In TM treated steels a similar texture[20] has been observed when the austenite is heavily deformed and subsequently rapidly cooled through the γ/α transformation range. There appears that the presence of a {123} <412> texture in the deformed austenite in combination with a "supersaturation" of the austenite by "lattice defects"[21] are responsible for the decrease of both the M_s- as well as of the B_s-start-temperatures.
More recently Schmidtmann et.al.[22,23] suggested that the decrease in the M_s-start-temperature might have been brought about by the much higher work hardening characteristic of the "deformed" austenite. Such a mechanism has not been observed in this investigation. As a matter of fact the "pure" martensite phase obtained from the deformed austenite on quenching is much softer i. e. it has a lower hardness value HV10 than the martensite obtained from undeformed austenite. This speaks against the Schmidtmann et.al.[22, 23] proposal.

CONCLUSIONS

The deformation of austenite greatly affects the kinetics of the γ/α transformation in steels. While the austenite transformation to ferrite and pearlite is accellerated by a factor 2 to 5 by a 50 % deformation, the bainite field region, depending on the austenite chemistry, may be constricted so that a "free field" may arise between the pearlite and bainite/martensite regions.

Both the M_s- and B_s-start-temperature are lowered by a deformation of the austenite. The lowering of the transformation temperature for bainite and martensite seems to be connected both with the formation of a {123} <412> texture in austenite and the presence of a high concentration of lattice "defects" brought about by the deformation step. No evidence for an increased work hardening characteristic of the deformed austenite to account for the depression of the M_s- and B_s-temperatures was found in this investigation.

ACKNOWLEDGEMENTS

The authors are grateful to Stahlwerke Peine-Salzgitter and to Bundesministerium für Forschung und Technologie for permission to publish.
The work was carried under contract XPS/210679/S194 to BMFT.

Literature

1. Y. E. Smith, C. A. Siebert: Met. Trans. (1971), vol. 2, p. 1711/25.
2. T. L. Capeletti, L. A. Jackman, W. J. Childs: Met. Trans. (1973), p. 1412/24.
3. Atlas zur Wärmebehandlung der Stähle, Bd. 1, Teil 1 und 2, Hrsg. - Max-Planck Inst. und VDEh, Verlag Stahleisen, Düsseldorf 1954/1958.
4. Atlas zur Wärmebehandlung der Stähle, Bd. 2, Hrsg. Max-Planck Inst. und VDEh, Verlag Stahleisen Düsseldorf, 1972.
5. Edelbaustähle, Druckschrift Nr. 500, Ausgabe Juli 1971, Stahlwerke Südwestfalen AG, Hüttental-Geisweid.
6. Handbuch der Baustähle, Röchling'sche Eisen- und Stahlwerke GmbH, Völklingen.
7. Zeit-Temperatur-Umwandlungs-Schaubilder, Mai 1961, DEW AG, Krefeld.
8. C. M. Vlad: Herstellung von Walzdraht und Stabstahl mit der aus der Walzhitze vergüteter Oberfläche, BMFT-Forschungsprogramm XPS/210679/S194, Abschlußbericht, 1984.
9. C. M. Vlad: Stahl und Eisen (1982), Bd. 102, Heft 22, p. 1101/1106.
10. C. M. Vlad: Unpublished Research, UB 755/70, TVW 1 (1970) Stahlwerke Peine-Salzgitter.
11. O. Pawelski, R. Kaspar, L. Reichl, A. Streißelberger: Untersuchung der thermomechanischen Behandlung beim Walzen von hochfesten Baustählen zur Verbesserung ihrer mechanischen und technologischen Eigenschaften, BMFT-Forschungsprogramm, MPI/010580/S014, Abschlußbericht, 1982.
12. G. Nocke, E. Jänsch, P. Lenk: Neue Hütte (1976), vol. 21, p. 468/73.
13. V. A. Samovski: Neue Hütte (1977), vol. 22, p. 628/23.
14. I. Kozasu, C. Ouchi, T. Sampei, T. Okita: Proc. Microalloying 75 - Washington D.C., p. 120/35, Union Carbide Press., New York, 1977.
15. W. Roberts: Scand. J. of Metall. (1980), vol. 9, p. 13/20.
16. D. J. Walker, R. W. K. Honeycombe: Met. Sci. (1978), vol. 10, p. 445/52.
17. M. A. Smirnov et.al.: Fiz. Met. i. Metallov.. (1979), vol. 48, Nr. 4, p. 816/22.
18. A. P. Coldren, G. Tither: J. of Metals (1978), vol. 30, p. 4/7
19. M. L. Bernstein et.al.: Fiz. Met. i. Metallov. (1978), vol. 45, p. 750/61.
20. C. M. Vlad, D. Grzesik: Proc. 4th Int. Conf. on Textures Cambridge (1975), p. 311/25, The Metal Society Press, 1976.
21. E. Schmidtmann, H. Hlawiczka: Arch. Eisenhüttenwes. (1973), vol. 44, p. 520/37.
22. E. Schmidtmann, M. May: Arch. Eisenhüttenwes. (1970), vol. 41, p. 569/75.
23. E. Schmidtmann, M. Grave: Arch. Eisenhüttenwes. (1977), vol. 48, p. 431/35.

Table 1 Chemical analysis of the investigated steels

Steel Quality	Element $^{w}/_{o}$ C	Si	Mn	P	S	N	Al	Cu	Cr	Ni	Mo
7 Si Mn 64	0,04	1,34	1,07	0,010	0,009	0,0143	0,013	0,05	0,02	0,03	≤ 0,01
55 Si 7	0,51	1,68	0,78	0,022	0,014	0,0074	0,015	0,02	0,04	0,05	-
60 Si Mn 5	0,55	1,10	1,08	0,012	0,012	0,0095	0,021	0,04	0,06	0,06	-
70 Mn Cr 4	0,69	0,26	0,86	0,024	0,023	0,0078	0,008	0,04	0,19	0,06	-
10 Mn Si Cr 55	0,09	1,33	1,15	0,007	0,004	0,0133	0,050	≤ 0,02	0,59	≤ 0,02	0,40

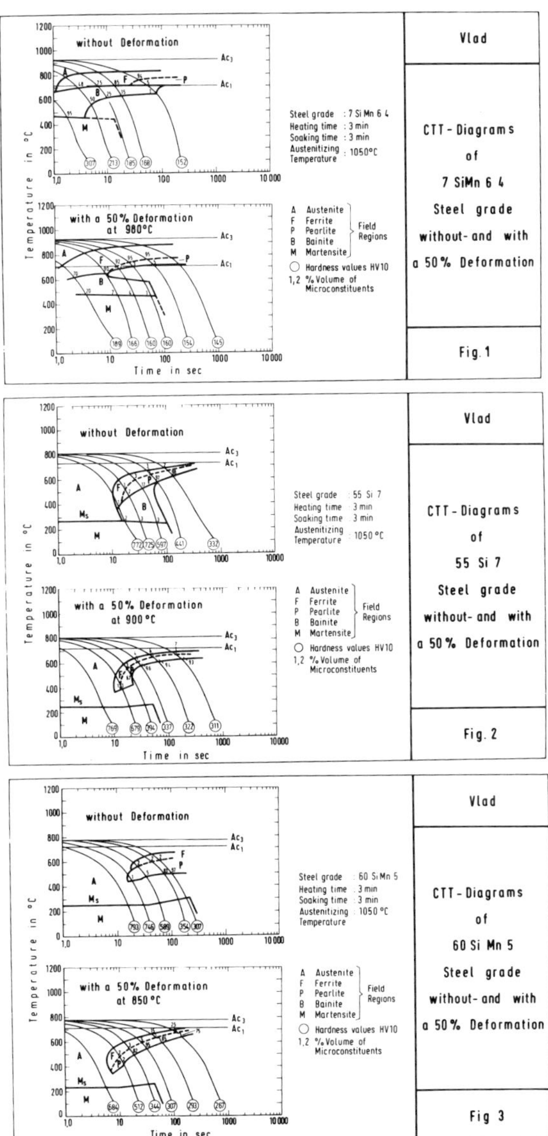

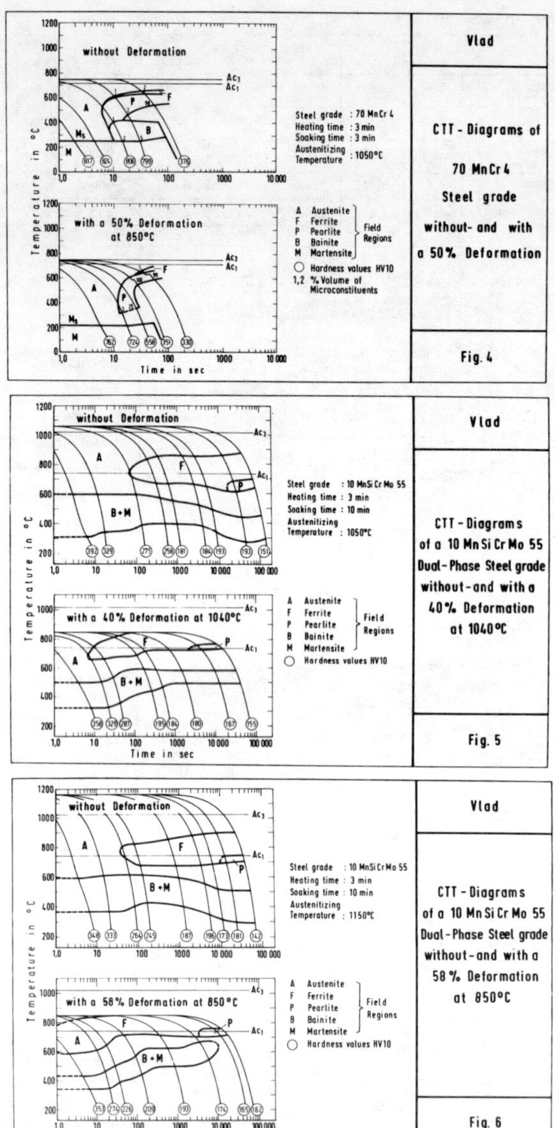

Table 2 The effect of deformation of austenite upon the texture of the martensite of 70 Mn Cr 4 steel grade

Steel Quality	Condition	Pct. volume fraction of grains having a {hkl} direction //ND								
		(110)	(200)	(211)	(310)	(222)	(321)	(411)	(420)	(332)
70 Mn Cr 4	950 °C/30 min 15 °C/sec. to 750 °C/WQ/ 300 °C/30 min AC	14,1	13,1	13,6	12,1	8,4	11,0	7,8	9,0	10,9
	950 °C/30 min deformed 75 % in 3 passes to 750 °C/WQ/ 300 °C/30 min AC	16,5	10,9	30,7	6,0	0,0	10,8	3,9	0,0	21,2

Grain Size Control in the Near Beta Ti-10V-2Fe-3Al Alloy

I. Weiss*, F. H. Froes and P. J. Bania*****

*Wright State University, School of Engineering, Dayton, OH 45435, USA

**Air Force Wright Aeronautical Laboratories, Materials Laboratory, AFWAL/MLLS, Wright-Patterson AFB, OH 45433, USA

***TIMET, Henderson Technical Laboratory, P.O. Box 2128, Henderson, NV 89015, USA

INTRODUCTION

Because of the interplay between recrystallization and grain growth, processing of beta titanium alloys can lead to the formation of a mixed grain size structure containing large and small grains [1]. The mixed grain size structure is undesirable and contributes to poor high temperature flow as well as inferior room temperature properties. Once introduced, it is impossible to remove the mixed grain size structure by heat treatment alone, so that further processing is needed to uniformly refine the microstructure.

Mixed grain size structure was found in hot worked and subsequently annealed Ti-10V-2Fe-3Al alloy [2]. The formation of mixed grain structure is in part a result of the low driving force for growth of statically recrystallized grains. This is due to the presence of a homogeneous and well developed recovered substructure in columnar grains within specimens originally forged 30% at 1225°C (2250°F) or 815°C (1500°F) and annealed at temperatures below 1275°C (2350°F). Specimens containing equiaxed grains also exhibited mixed grain size structure when processed at a lower temperature of 980°C (1800°F) and annealed at temperatures below 1145°C (2100°F) [2]. Mixed grain size structures were also found in the metastable beta titanium alloys, Ti-10Mo-6Cr-2.5Al and Ti-10Mo-8V-2.5Al, following deformation below 855°C (1575°F) and annealing at 980°C (1800°F) [1, 3]. This mixed grain size structure is a result of selective recrystallization in locations with high localized strain (shear bands) [3].

It is the purpose of the present study to investigate the effect of hot deformation and post-deformation heat treatment on the development of mixed grain size structure in specimens containing columnar or columnar+equiaxed grains. That is, to determine whether a true "processing window" exists for which a uniform recrystallized structure results, and which only incorporates deformation and annealing effects.

MATERIAL AND EXPERIMENTAL PROCEDURE

Forging specimens (50mm x 25mm x 25mm) were taken from 305mm cast ingot with a nominal composition of Ti-10V-2Fe-3Al. These were hot die forged to true strain of 36%, 65%, and 105% at temperatures between 925°C (1700°F) and 1370°C (2500°F) as shown in Fig. 1a. A 50 ton hydraulic press was used at a ram speed of 1.26cm/min. Deformed specimens were polished and macro etched in Kroll's solution to groove the deformed grain boundaries. The etched specimens were then vacuum annealed for 1 hour at temperatures between 815°C (1500°F) and 1345°C (2500°F). Following oil quenching the polished surface of the annealed specimen simultaneously shows the prior deformed boundaries ("ghost boundaries") and the position of the newly recrystallized grains (thermally etched) [4].

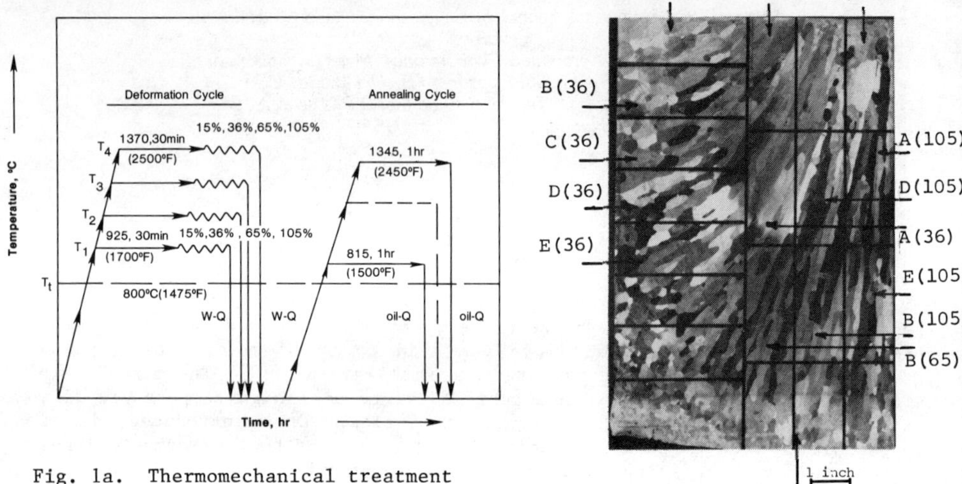

Fig. 1a. Thermomechanical treatment used in this work.

Fig. 1b. Microstructure of the as-cast ingot.

RESULTS AND DISCUSSION

Initial Microstructure

The initial microstructure of the forging blanks are shown in Fig. 1b samples A(36), B(36), C(36), D(36), and E(36) contain equiaxed+columnar grains and were forged to a true strain of 36%. Specimens A(65), B(65), C(65), D(65), and A(105), B(105), C(105), D(105), and E(105) contain columnar grains and were forged to a true strain of 65% and 105%, respectively.

Microstructure of Deformed and Annealed Material

The microstructure of samples forged to a true strain of 36% at 925°C (1700°F) [specimens B(36)], 1090°C (2000°F) [specimens C(36)], 1365°C (2500°F) [specimens D(36)], and subsequently annealed at 1035°C (1900°F) for 1 hour are shown in Fig. 2. Annealing forged samples containing columnar+ equiaxed grains at 1035°C (1900°F) resulted in a mixed grain size structure as shown in Figs. 2a, 2c, and 2d. The recrystallized grains observed exhibit a large radius of curvature indicating a very low driving force for growth into the recovered region [3, 5]. Almost fully recrystallized microstructures are observed when specimens with columnar+equiaxed grains are annealed at 1255°C (2300°F), the result of higher driving force for grain growth in comparison to specimens annealed at 1035°C (1900°F).

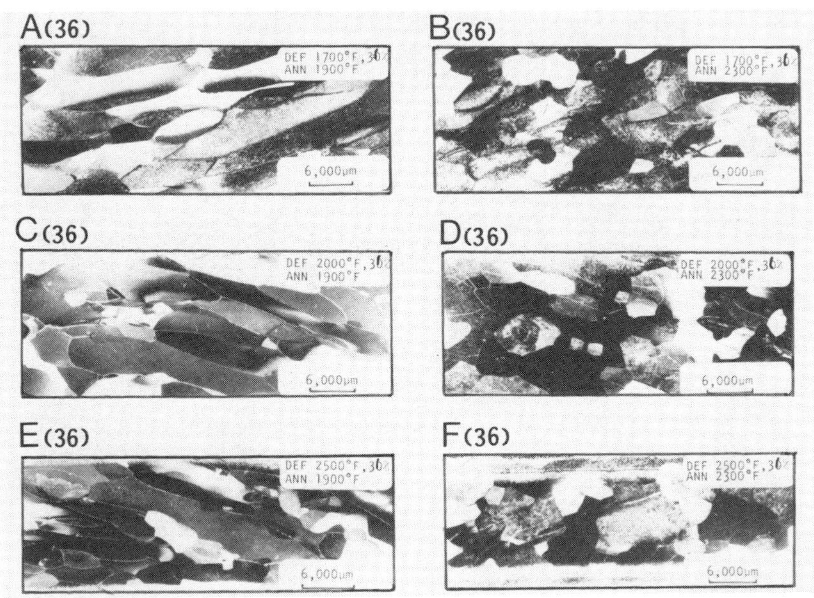

Fig. 2. Microstructure of specimens forged to a true strain of 36% at (A, B) 925°C (1700°F), (C, D) 1090°C (2000°F), (E, F) 1365°C (2500°F), and subsequently annealed at (A, C, E) 1035°C (1900°F) and at (B, D, F) 1255°C (2300°F).

The Processing Window

Effect of processing and annealing temperatures. Figures 3a, 3b, and 3c show the effect of processing and annealing temperatures on the microstructure of specimens forged to a true strain of 36%, 65%, and 105%, respectively. Samples forged to a true strain of 36% at 925°C (1700°F) [specimens B(36)] were found to require the lowest annealing temperature of 1255°C (2400°F) to produce a fully recrystallized microstructure. These specimens have higher driving force for recrystallization and grain growth in comparison with specimens forged above 925°C (1700°F).

However, samples forged below the transus temperature 800°C (1475°F) [samples A(3b)] display fully recrystallized microstructure when annealed at 815°C (1500°F). This is the result of alpha phase precipitation and the formation of alpha/beta interfaces which provide effective nucleation site for recrystallization [2, 6]. Specimens with an initial columnar grain structure forged to true strain of 65% at 1090°C (2000°F) were found to require a lower annealing temperature of 1145°C (200°F) to produce a fully recrystallized structure than samples processed at 925°C (1700°F) [specimens B(65)] and 1365°C (2500°F). Both specimens B(65) and C(65) require higher annealing temperatures of 1310°C (2400°F) and 1200°C (2200°F) respectively. The processing and annealing temperatures of 1090°C (2000°F) and 1145°C

(2100°F) define a "processing window" for which a uniform recrystallied structure results. These observations can be explained in terms of the way columnar grains deform, and the driving force for recrystallization and growth during annealing. The partially recrystallized structure observed following forging at 925°C (1700°F) and annealing at 1200°C (2200°F) is associated with the formation of non-homogeneous substructure in the vicinity of the grain boundaries during deformation, as well as slow migration of the recrystallized grain boundaries during annealing.

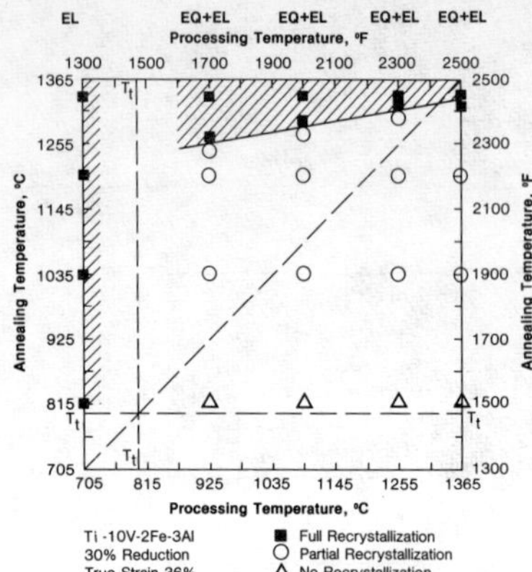

Fig. 3a. The effect of processing and annealing temperatures on the microstructure of specimens forged to a true strain of 36%.

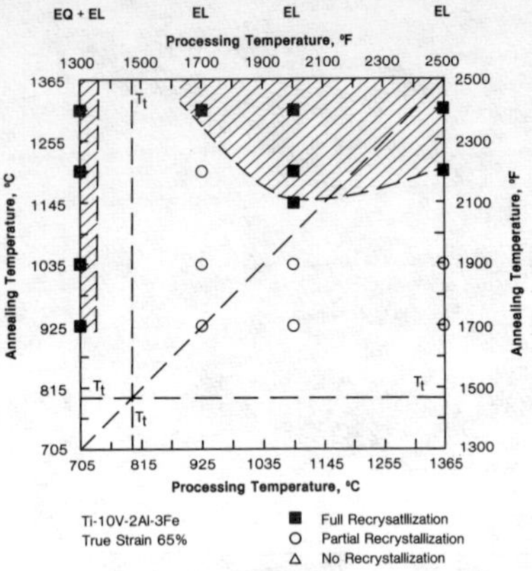

Fig. 3b. The effect of processing and annealing temperatures on the microstructure of specimens forged to a true strain of 65%.

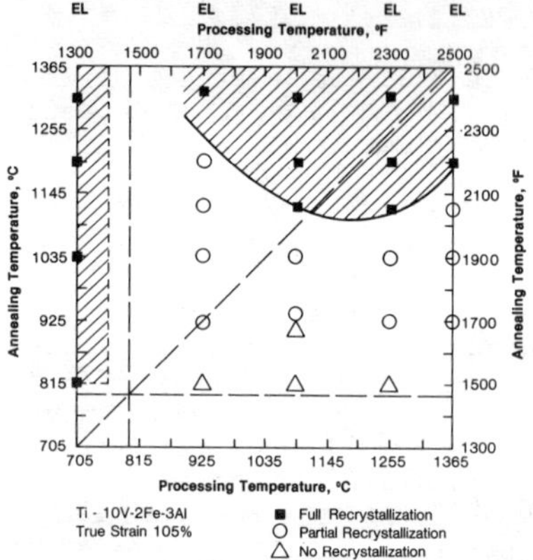

Fig. 3c. The effect of processing and annealing temperatures on the microstructure of specimens forged to a true strain of 105%.

During forging at 925°C (1700°F) deformation tends to localize along the grain boundaries (large as-cast grains) producing finer sub-cells with higher dislocation density in comparison to those formed in the grain interior (Fig. 4). On subsequent annealing, nuclei for static recrystallization form selectively along the grain boundary region where the localized strain is high. These nuclei rapidly produce new recrystallized grains with large radius of curvature (Fig. 5). The well developed and homogeneously distributed recovered substructure present in the grain interior, especially in specimens deformed at 1365°C (2500°F), reduces the stored energy accumulated during deformation to such a low level that the driving force for growth of the recrystallized grain is very low [1]. Under these conditions grain boundary migration of the recrystallized grain is extremely slow and a mixed grain size structure develops.

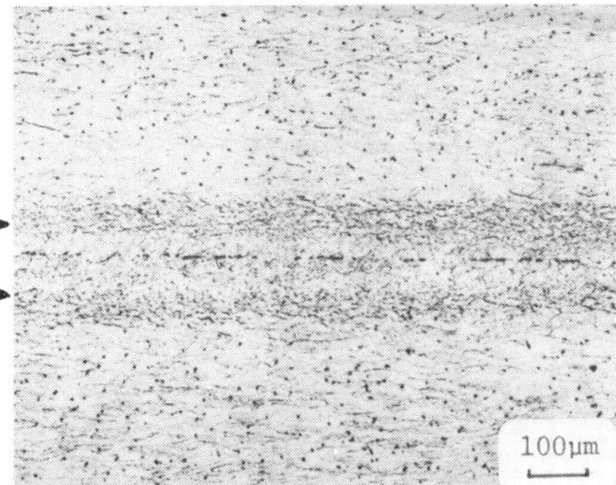

Fig. 4. Non-homogeneous substructure in the vicinity of grain boundary in specimens forged to a true strain of 105% at 925°C (1700°F). (See arrows)

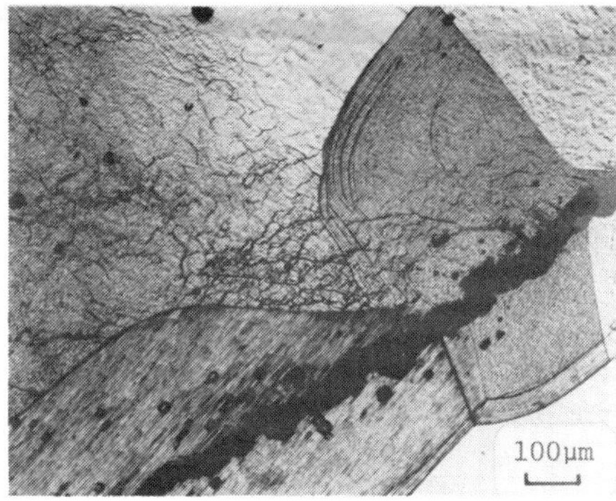

Fig. 5. Nucleation of new recrystallized grains with low radius of curvature along the deformed grain boundary in specimens forged to a true strain of 36% at 1365°C (2500°F) and subsequently annealed at 1200°C (2200°F).

Similar results are obtained for specimens forged to a true strain of 105%. The "processing window" was found to extend to higher forging temperatures, with the minimum occurring at an annealing temperature of 1090°C (2000°F) and a processing temperature of 1200°C (2200°F), compared to the "processing window" for material forged to a true strain of 65% for which the minimum annealing temperature was 1145°C (2100°F) at a processing temperature of 1090°C (2000°C) as shown in Fig. 3c. The shift of the "processing window" is attributed to an increase in driving force for recrystallization and growth when forged to a true strain of 105%. Specimens forged to a true strain of 65% and 105% below the transus temperature were found to require low annealing temperatures of 925°C (1700°F) and 815°C (1500°F) respectively to produce a fully recrystallized microstructure.

Effect of initial grain shape and size. In an earlier investigation it was found that the "processing window" is also associated with the original shape of the ingot grains (equiaxed or columnar), in addition to the particular selection of the forging and annealing temperatures [2]. When both columnar and equiaxed grains are present as typically observed in ingot, it is necessary to anneal this material at a higher temperatuare to achieve fully recrystallized structure than material which contain only equiaxed grains as shown in Fig. 6. For example, samples with original equiaxed grains forged to a true strain of 36% (Fig. 6) require annealing at 1170°C (2150°F) in order to produce a fully recrystallized structure. However, specimens containing equiaxed+columnar grains processed under the same conditions as mentioned earlier, require higher annealing temperature of 1250°C (2300°F) to produce a fully recrystallized structure. This is a direct result of the way columnar and equiaxed grains deform and their response to the annealing treatment. At higher forging strain of 105% both the equiaxed and columnar grains became highly deformed and displayed an elongated morphology with a similar response to the annealing treatment as observed in Fig. 7. The slightly higher annealing temperatures required for material with columnar grains to produce a fully recrystallized structure is again the result of differences in the way equiaxed and columnar grains are deformed (degree of homogeneity).

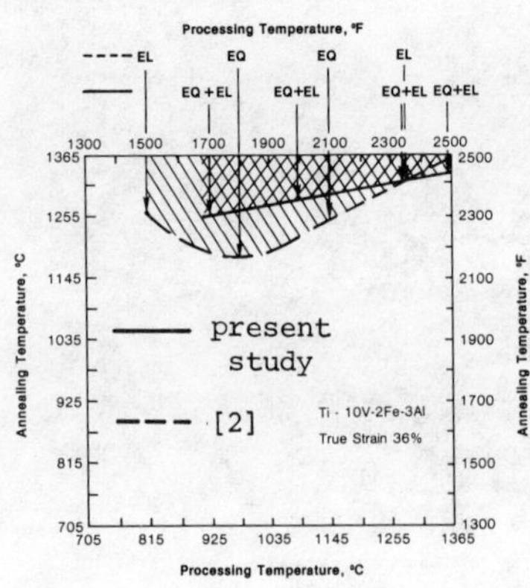

Fig. 6. Effect of initial grain shape on the processing map for material forged to a true strain of 36%.

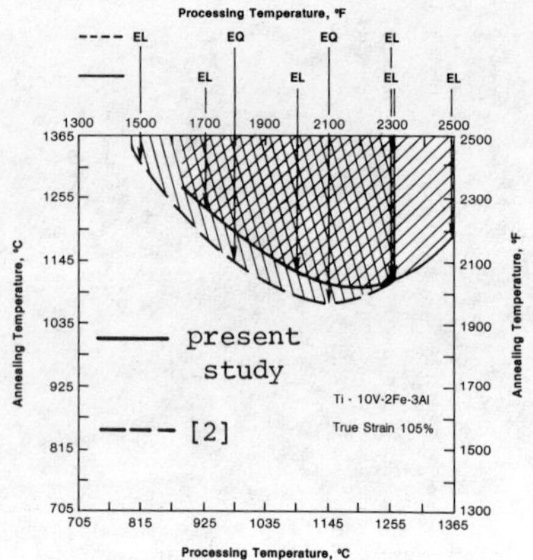

Fig. 7. Effect of initial grain shape on the processing map for material forged to a true strain of 105%.

CONCLUSIONS

1. A recrystallized or mixed grain size structure can develop in Ti-10V-2Fe-3Al alloy depending on forging temperature and reduction, grain shape (columnar or equiaxed), size, and final annealing temperature.

2. A fully recrystallized structure is formed in a specific range of forging temperature reduction and subsequent annealing temperature which defines a "processing window" for the alloy.

3. The minimum annealing temperature for the "processing window" depends on both the forging temperature and forging strain. The lowest minimum annealing temperature of 1145°C (2100°F) is at a forging temperature of 1050°C (2000°F) for a true strain of 65% and is lowered to 1090°C (2000°F) at a forging temperature of 1200°C (2200°F) for a true strain of 105%.

4. The mixed grain size structure is a result of selective recrystallization along grain boundaries due to strain localization during deformation with the driving force for grain growth decreasing rapidly because of competition with recovery.

5. Forging below the transus temperature (705°C [1300°F]) requires a low annealing temperature of 815°C (1500°F) to produce a fine recrystallized microstructure. The result of alpha phase precipitation and the formation of alpha/beta interfaces provides an effective nucleation site for static recrystallization.

REFERENCES

1. F. H. Froes, C. F. Yolton, J. P. Hirth, R. Ondercin, and D. Moracz, Proceedings of the Beta and Near Beta Alloys Symposium, TMS-AIME, Atlanta, GA (March 1983).
2. I. Weiss and F. H. Froes, Proceedings of the 5th International Conference on Titanium, Munich, West Germany (1985), in press.
3. F. H. Froes, C. F. Yolton, and J. P. Hirth, Proceedings of the Beta and Near Beta Alloys Symposium, TMS-AIME, Atlanta, GA (March 1983).
4. I. Weiss, F. H. Froes, and D. Eylon, Met. Trans. A, 15A, 1493 (1984).
5. H. P. Stuwe, Recrystallization of Metallic Material (edited by E. F. Haessner), p. 13. Dr. Reiderer, Verlag GMBH, Stuttgart (1978).
6. H. Margolin and P. Cohen, Titanium '80, TMS-AIME, Warrendale, PA (1980), p. 1555.

Accelerated Spheroidization of Cementite in High Carbon Steel Wire Rod by Drawing at Elevated Temperatures

Seung Eui Nam* and Dong Nyung Lee**

**Department of Metallurgical Engineering, Hong Ik University Seoul, Korea*

***Department of Metallurgical Engineering, Seoul National University Seoul, Korea*

ABSTRACT

Quantitative measurements of the volume fraction of spheroidized cementite particles, the aspect rario and the size distribution of cementite particles showed that the spheroidization of cementite in pearlite of high carbon steel wire rods could be accelerated by wire drawing at elevated temperatures below A_1. A principal mechanism of the spheroidization appeared to be shear fracture of lamellar cementite into smaller particles, whose sharp edges subsequently became blunted by the diffusion of iron and carbon atoms, which was thought to be enhanced by regeneration of dislocations and vacancies by the high temperature deformation. Decreases in the tensile strength and hardness, and an increase and then decrease in the elongation of the wire product with increasing number of drawing passes were attributed to the spheroidization and void formation during the drawing.

KEYWORDS

Spheroidization of cementite ; high carbon steel wire rod ; wire drawing at elevated temperatures.

INTRODUCTION

Formability and machinability of steel can be enhanced by spheroidization of cementite in pearlite of it. The spheroidization of steel can be achieved by annealing slightly below A_1 temperature(1,2). The process can be enhanced by cyclic heating near A_1 temperature(3,4), by cold working prior to annealing (2) and fruther by plastic deformation slightly below A_1 point(5-11). The plastic deformation has been performed mostly by rolling or torsion. Another method of spheroidization is to use quenching and tempering, but the method is hardly applied to the commercial production of high carbon steel wires. The purpose of this work is to investigate if rod or wire drawing slightly below A_1 point can be applied to the accelerated spheroidization and possibly to the production of spheroidized steel wires.

EXPERIMENTAL METHODS

Commercially available high carbon steel rods were used in this experiment. The rods whose diameter ranged from 5.5 to 13mm had chemical compositions given in Table 1.

TABLE 1 Chemical Composition of Materials, %

Wire	C	Si	Mn	P	S	Cu
WR-1	0.85	0.25	0.74	0.022	0.019	-
WR-2	0.79	0.20	0.75	0.020	0.016	-
WR-3	0.74	0.28	0.72	0.020	0.018	-
WR-4	0.68	0.21	0.77	0.019	0.006	-
WR-5	0.65	0.21	0.66	0.017	0.015	0.2

The rods were descaled with emery paper and dipped to a wet lubricant composed of MoS and graphite powders, and then dried in air. The rods coated with the lubricant were heated to predetermined temperatures and drawn at the speed of 10m/min.. Some rods were drawn at room temperature and annealed, to be compared with those drawn at the elevated temperatures. The transmission and scanning electron micrographs of the specimens were examined to understand the spheroidization process. The tensile and micro-Vickers hardness tests (load 500 gr) were also made.

RESULTS AND DISCUSSION

The pearlite lamellar spacings in the wire rods were mostly in the range of 0.12 to 0.15 μm(Fig. 1). Figure 2 shows a wavy pearlite structure of a

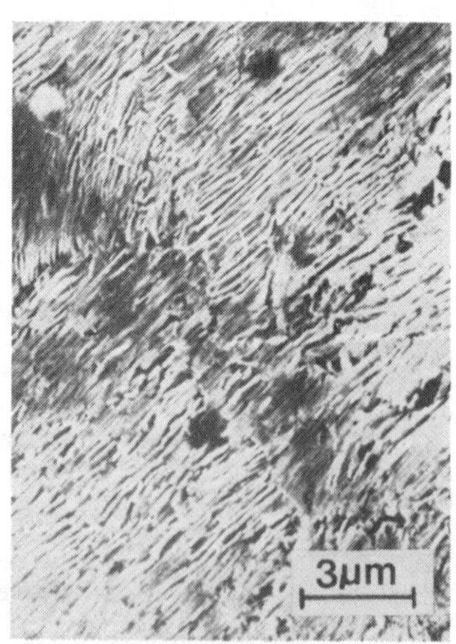

Fig. 1. A scanning electron micrograph of wire rod WR-3.

Fig.2. A scanning electron micrograph of 64% cold drawn RW-3 specimen.

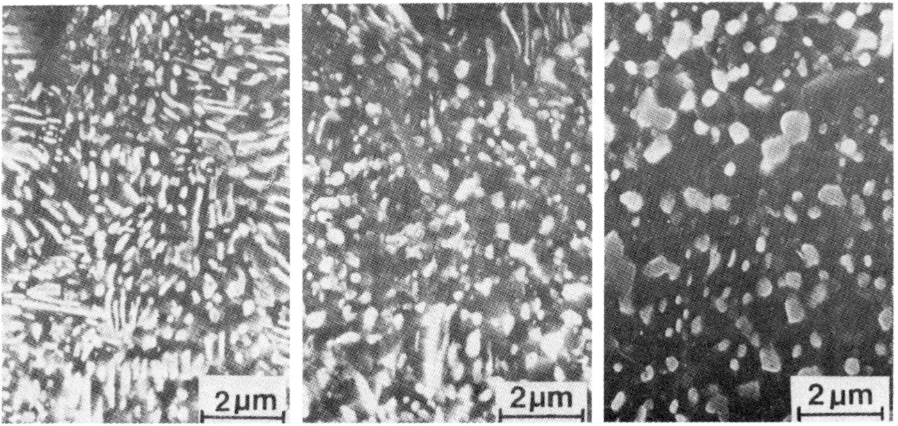

Fig.3. Scanning electron micrographs of WR-3 specimens drawn by (a) 21% (1 pass), (b) 62.8% (3 passes) and (c) 117% (5 passes) at 700°C.

specimen drawn by 64% at room temperature. When the wires were drawn at 700°C rapid spheroidization and particle growth could be achieved as shown in Fig. 3. The wires were reheated to 700°C for about 5 minutes prior to each drawing pass. Even rapid spheroidization cannot be attributed only to an accelerated diffusion of carbon and iron atoms due to generation of dislocations and vacancies during the drawing. The accelerated spheroidization is thought to be achieved mainly by shearing of cementites during the deformation as shown clearly in Fig.4 and subsequent blunting of sharp edges of the sheared particles by diffusion during heating. This easy shearing has not been observed in cold drawing of steel. Figure 5 shows the shape distributions of cementite particles after one pass drawing (21% reduction) at 700°C. The maximum frequency occurred at the aspect ratio of two, and the population of particles having the aspect ratio less than three was over 80%. The particles were assumed to be 'spheroidized' when their aspect ratios were less than three. The volume fraction of spheroidized cementite particles increased with increasing reduction and carbon content as shown in Fig.6. Shearing of

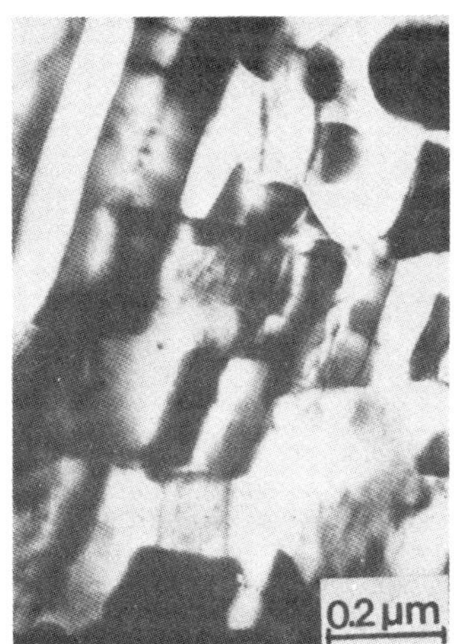

Fig.4. A transmission electron micrograph showing shearing of cementile particles in WR-1 specimen drawn at 700°C

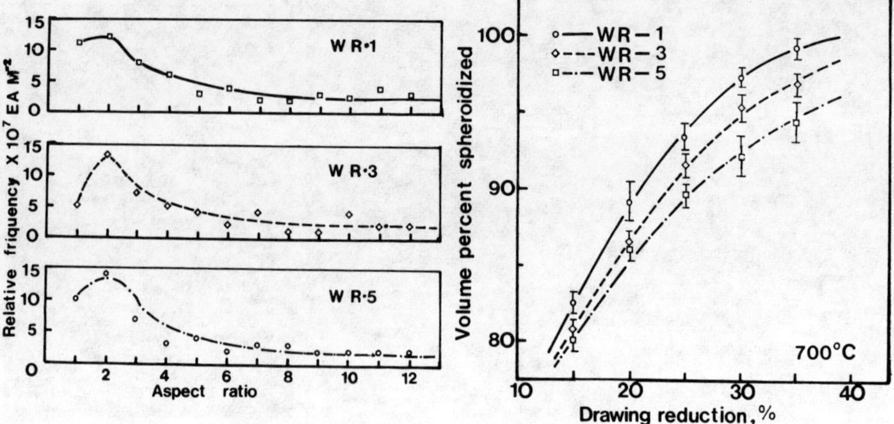

Fig.5. The shape distributions of cementile particles in wire (WR-1,3,5) drawn by 21% at 700°C.

Fig.6. Volume percent of spheroidized cementile particles as a function of drawing reduction per pass at 700°C.

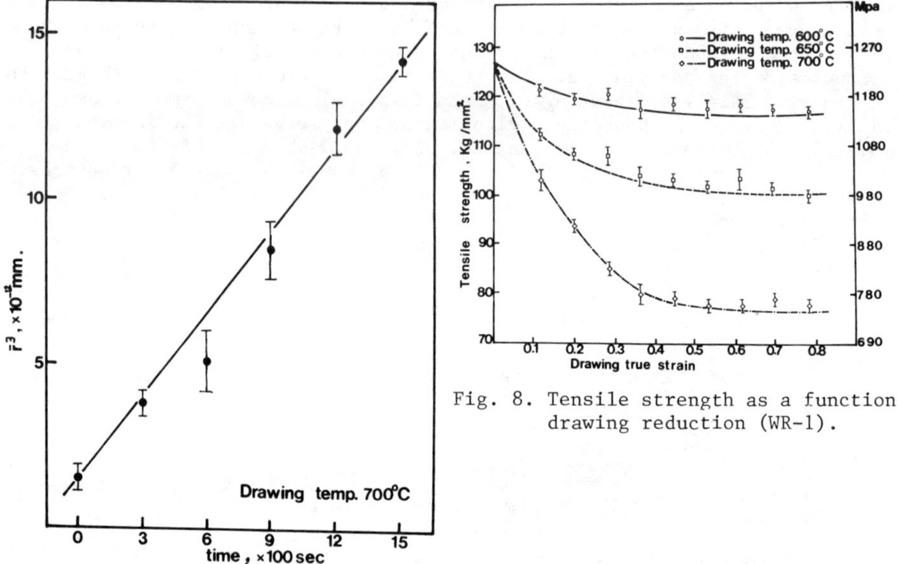

Fig. 8. Tensile strength as a function drawing reduction (WR-1).

Fig.7. Cubic of mean particle radius as a function of heating time at intervals of drawing (WR-1).

cementite particles is expected to increase with increasing reduction. A decrease in carbon content means an increase in soft ferrite phase and a decrease in hard cementite phase, which in turn decrease contribution of the hard phase to the deformation. Therefore, as the soft ferrite increases, the hard cementite particles are expected to be less sheared during deformation. Figure 7 shows the coarsening rate of cementite particles, where time means

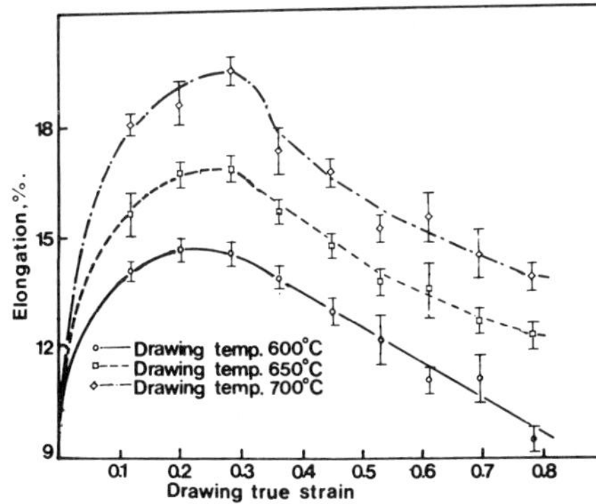

Fig.9. Elongation as a function of drawing reduction (WR-1).

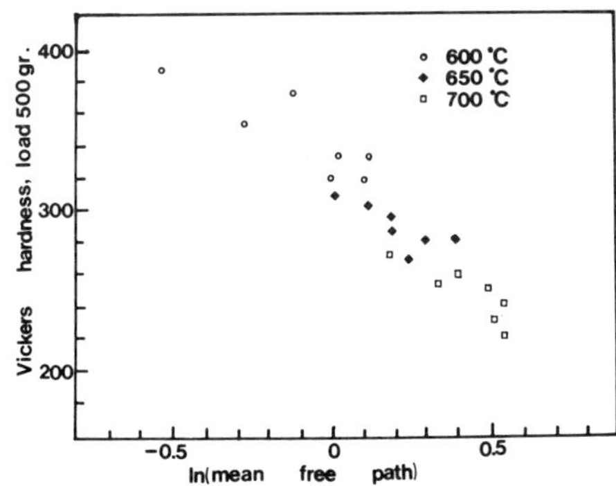

Fig.10. Hardness as a function of mean free path between cementite particles.

the holding time in the furnace between drawing passes. This figure implies a diffusion controlled growth of cementite particles which can be expressed by the following relation(8,12).

$$r^3 - r_0^3 = \alpha\gamma V_{Fe_3C}\, X_c \bar{D}_c t \;/\; (V_{Fe} RT) \qquad (1)$$

where r and r_0 are mean particle radii at time t and t_0, γ is interface energy , X_c is mole fraction of carbon in equilibrium between ferrite and cementite, $\bar{D}_c$ is effective diffusion coefficient of carbon, α is a constant (=8/9)(8) and t is time. Substitution of appropriate quantities [$\gamma = 5 \times 10^{-7}$ J/mm^2, $V_{Fe3C} = 23.97 \times 10^3$ mm^3/mole, $V_{Fe} = 7.3 \times 10^3$mm/mole, $X_c = 7.7 \times 10^{-4}$, R = 8.32 J/(deg. mole), T = (273+700) K] into Eq.(2) and the measured slope of Fig.7 yielded $\bar{D}_c = 2.55 \times 10^{-6}$mm/sec, which is one order of magnitude larger than that obtained from a specimen without any prior deformation. This enhanced diffusion coefficient may be associated with an eccelerated self diffusion of iron in ferrite due to increases in vacancy and dislocation density during deformation at the elevated temperature. The vacancy concentration (mole fraction) $\bar{X}v$ during deformation at temperature T may be expressed as

$$\bar{X}v = C\, \dot{\varepsilon}\, \exp[-Q/(RT)] \qquad (2)$$

where $\dot{\varepsilon}$ is strain rate, C is 5×10^{-3}sec. and Q is 52.7 kJ/mole. The self diffusion coefficient of iron during deformation at T, $\bar{D}_{Fe}$, can be compared with the self-diffusion coefficient of iron at T, D_{Fe}, in the following relation.

$$\bar{D}_{Fe} / D_{Fe} = (1 + \bar{X}v / Xv) \qquad (3)$$

The equilibrium vacancy concentration Xv can be evaluated using the equation of $Xv = \exp[-E/(RT)]$, where vacancy formation energy E is 216 kJ/mole. Substituting the measured strain rate of $\dot{\varepsilon} = 6 \times 10^{-3}$/sec into Eq.(2) and using Eq.(3), we obtain $\bar{D}_{Fe}/D_{Fe} \approx 2 \times 10^4$. The tensile strength of the product decreased as the spheroidization progressed (Fig.8), whereas the elongation increased and then decreased with increasing drawing pass (Fig.9). The behavior can explained by the spheroidization and particle growth, and

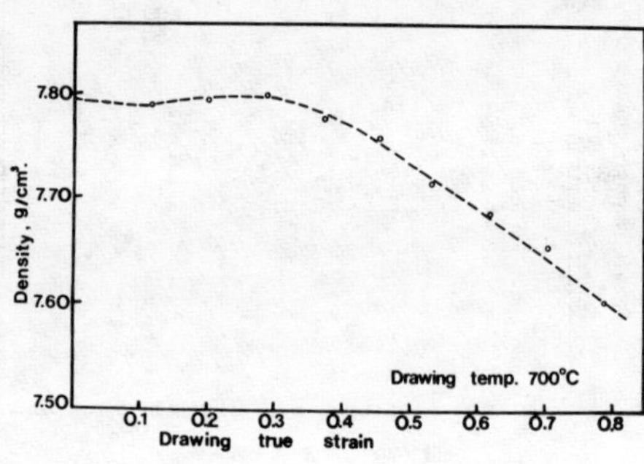

Fig.11. Density of specimen as a function of drawing reduction (WR-5).

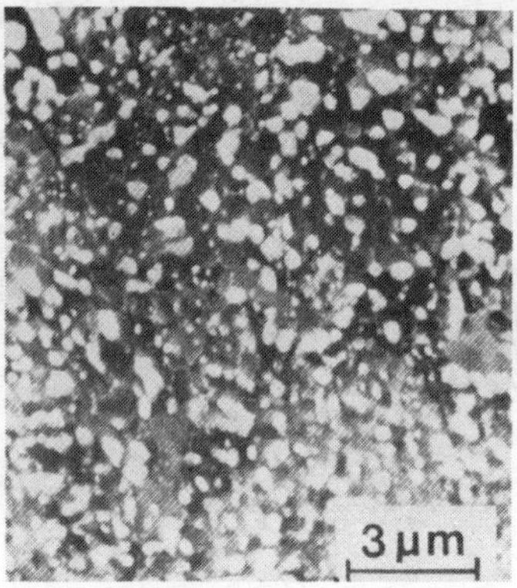

Fig.12. A scanning electron micrograph of 100% sphcroidized WR-3 specimen which was drawn by 65% at room temperature followed by 15% reduction at 700°C

the void formation. The former process is expected to cause the decreases in tensile strenth and hardness(Fig.10) and the increases in elongation, whereas the latter is expected to yield the decreases in both the tensile strength and the elongation. The void formation may be manifested by the density measurement of the products(Fig.11). The drawbacks caused by many drawing passes at elevated temperature could be reduced by many cold drawings plus last single pass of drawing at an elevated temperature. Figure 12 shows an example of such cases, where WR-3 wire was reduced by 65% at room temperature and then drawn by 13% at 700°C, and 100% spheroidization was achieved.

CONCLUSION

The spheroidization of cementite particles in pearlite could be substantially accelerated by drawing at elevated temperatures below A_1 or by a combination of drawings at room temperature and elevated temperature blow A_1.

REFERENCES

1. C.M.Sellar, Metallography 10, 89 (1977).
2. A.H.Holtzman, J.C.Danko, and R.D.Stout, Trans. Met. Soc. AIME 212, 475 (1958).
3. J.Leeder, P.Payson, and W.L. Hodapp, Trans. ASM. 28, 306 (1940).
4. O.E. Cullen, Metal Progress 64, 79 (1953).
5. J.L. Robbins, O.C. Shepard, and O.D. Sherby, JISI. 202, 804 (1964).
6. O.D. Sherby, M.J. Harrigan, L. Shamag and C. Sauve, Trans. ASM 21, 578 (1969).
7. E.A. Chojnowski, W.J. McG. Tegart, Metal Science J. 2, 14 (1968).
8. A.G. Martin and C.M. Sellar, Metallography 3, 259 (1970).
9. D.F. Lupton, and D.H. Warrington, Metal Science J. 6, 200 (1972).
10. S. Chattopadhyay, and C.M. Sellars, Acta Met. 30, 157 (1982).
11. S.E. Nam, and D.N. Lee, J. Korean Inst. of Metals 19, 143 (1981).
12. C.Y. Li, J.H. Blakely, and A.H. Feingold, Acta Met. 14, 1397 (1960).

Forgeage Isotherme de l'Alliage de Titane Ti6242

S. Gautron*, J. P. Immarigeon* et G. l'Esperance**

**Laboratoire de structures et matériaux, Conseil national de recherche du Canada, Etablissement aéronautique national, Ottawa, Ontario K1A 0R6, Canada*

***Ecole Polytechnique de Montréal, Département de génie métallurgique, C.P. 6079, Succ. "A", Montréal, Québec H3C 3A7, Canada*

(S. Gautron, travaille maintenant chez Pratt & Whitney Canada Inc., Longueuil, Québec J4K 4X9, Canada)

RESUME

On a étudié le comportement à chaud du Ti6242 à l'état $\alpha+\beta$, dans le but de générer des données qui pourront servir à la modélisation de la déformation du même alliage à l'état β. On a obtenu les courbes d'écoulement et étudié les transformations microstructurales, à l'aide d'essais de compression uniaxiale, à température et taux de déformation constants. Les températures sont inférieures au transus β et les taux de déformation vont de 0.1 à 3×10^{-5} s^{-1}. Au cours de la déformation, la microstructure initiale évolue vers une microstructure d'équilibre équiaxe et biphasée et la contrainte d'écoulement tend à se stabiliser. La morphologie de la microstructure d'équilibre et le niveau de stabilisation de la contrainte dépendent de la température et du taux de déformation, mais semblent indépendants de la microstructure initiale. Cet état de régime permanent sert de condition limite à une équation constitutive qui, à l'aide de la méthode d'analyse par éléments finis, permet de prédire les changements microstructuraux et les contraintes d'écoulement, lors de la déformation à chaud du Ti6242 à l'état β.

INTRODUCTION

La modélisation de la déformation des alliages en cours de forgeage, afin d'optimiser la microstructure et les propriétés des composantes produites est facilitée par l'application de techniques d'analyse assistées par ordinateur.

Des recherches ont déjà été entreprises pour modéliser le forgeage isotherme (1), un procédé utilisé pour la fabrication d'un disque de compresseur pour turbine à gaz. Les performances d'un tel disque, en cours de service, seront optimales s'il possède une bonne résistance au fluage à la périphérie et une haute résistance mécanique combinée à de bonnes caractéristiques de fatigue oligocyclique au centre (2-4). On peut obtenir ces propriétés multiples dans un même disque grâce à un gradient radial de microstructure. Dans un alliage de titane, tel le Ti6242, ces propriétés seront présentes si la microstructure est de type $\alpha+\beta$ au centre et de type β à la périphérie.

La microstructure β, forgée dans le domaine $\alpha+\beta$, évolue vers la microstructure $\alpha+\beta$ (5-7), et pour une déformation suffisante la transformation devient totale (8). Par le biais de cette transformation, on se propose de produire un disque à propriétés multiples, en utilisant une préforme à microstructure β, et en introduisant une déformation locale au centre du disque pour y obtenir une microstructure de type $\alpha+\beta$, tout en conservant une microstructure β à la périphérie.

En faisant appel à des techniques d'analyse par éléments finis, il est possible de prédire les formes de la préforme et de l'outillage, nécessaires à ce forgeage sélectif. Ce type d'analyse nécessite de l'information concernant le comportement plastique du matériau sous les conditions de mise en forme, qui est généralement donnée par une équation constitutive de plasticité. De façon courante (5), une telle équation relie la contrainte d'écoulement à la température, au taux de déformation et à la quantité de déformation. Bien que satisfaisante à basse température, cette pratique se prête mal à la description du comportement à chaud d'un alliage tel le Ti6242. A haute température, la contrainte d'écoulement ne dépend pas de la quantité de déformation (9), mais de l'histoire de cette déformation, c'est-à-dire la température et le taux de déformation auxquels la déformation a été imposée. On prend cette histoire en considération, en introduisant un ou plusieurs paramètres microstructuraux dans l'équation constitutive:

$$\sigma = f(\dot{\epsilon}, T, S_i)$$

$$\frac{dS_i}{dT} = f(\dot{\epsilon}, T, S_i) \quad \text{pour } i = 1, 2, \ldots, n$$

En tout temps, la contrainte d'écoulement, σ, est fonction du taux de déformation, $\dot{\epsilon}$, de la température, T, et de la valeur instantanée des paramètres microstructuraux, S_i. De plus, la valeur de ces paramètres peut changer avec le temps, suivant les conditions de la déformation. Cette formulation permet de suivre l'évolution des paramètres microstructuraux et ainsi prévoir la microstructure finale, en tout point d'une pièce forgée.

Dans ce travail, on a voulu identifier les paramètres microstructuraux qui contrôlent la déformation dans le Ti6242 et déterminer les effets des conditions de la déformation plastique sur la microstructure. Pour des raisons données ailleurs (10), on peut plus facilement établir ces effets en étudiant le corroyage de la microstructure $\alpha+\beta$ plutôt que celle de type β. Les résultats obtenus (10) pourront servir à la modélisation de la déformation à chaud de la microstructure β.

EXPERIMENTATION

On a simulé le forgeage isotherme du Ti6242 par des essais de compression uniaxiale, à température et taux de déformation constants. Les températures sont inférieures au transus β et les taux de déformation sont compris entre 0.1 et 3×10^{-5} s^{-1}. Une description de l'appareillage et des procédures expérimentales sont données ailleurs **(10)**.

RESULTATS ET DISCUSSION

L'étude microstructurale **(10)** démontre que pour une déformation donnée, la microstructure du Ti6242 est influencée par les conditions de la déformation et en particulier par le taux de déformation. A une température donnée, l'alpha primaire et le bêta transformé de la microstructure initiale (figure 1(a)) sont graduellement redistribués en une microstructure biphasée (figure 1(b)), à grains d'autant plus fins que le taux de déformation est plus élevé, et d'autant plus équiaxes que le taux de déformation est plus faible.

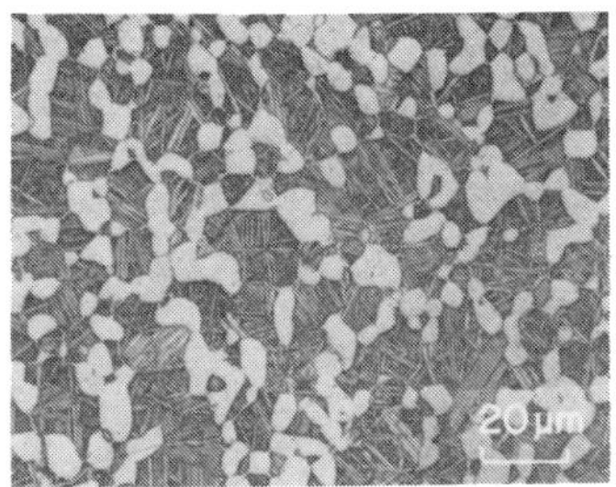

FIGURE 1(a)

Microstructure initiale composée de particules d'alpha primaire et de bêta transformé.

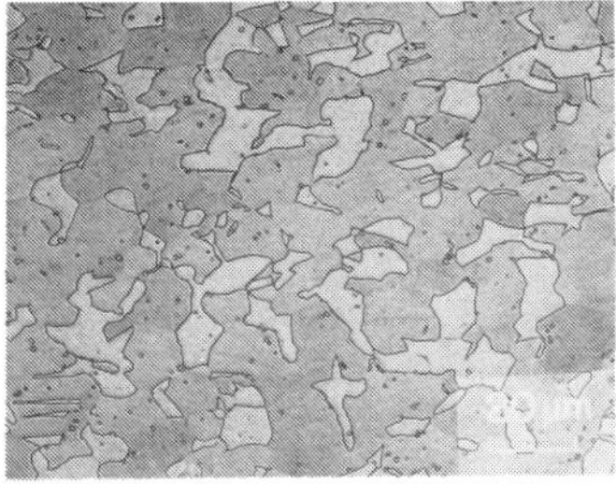

FIGURE 2(b)

Microstructure d'équilibre équi-axe et biphasée, obtenue à 925°C et à 3×10^{-5} s^{-1}, pour une déformation vraie de 1.2.

Cet effet du taux de déformation sur les grains recristallisés se produit aux autres températures d'essai; la proportion des phases en présence varie cependant avec la température. On doit souligner enfin que plus la température est basse, plus le taux de déformation doit être faible, pour que la transformation en microstructure équiaxe et biphasée soit complète (figure 2).

La finesse de la taille des grains recristallisés (<10 µm) et la faible densité de dislocations à l'intérieur de ces grains suggèrent que, dans ces conditions, le glissement aux joints de grains joue un rôle important lors de la déformation de l'alliage. La taille de grain est donc le paramètre microstructural dominant.

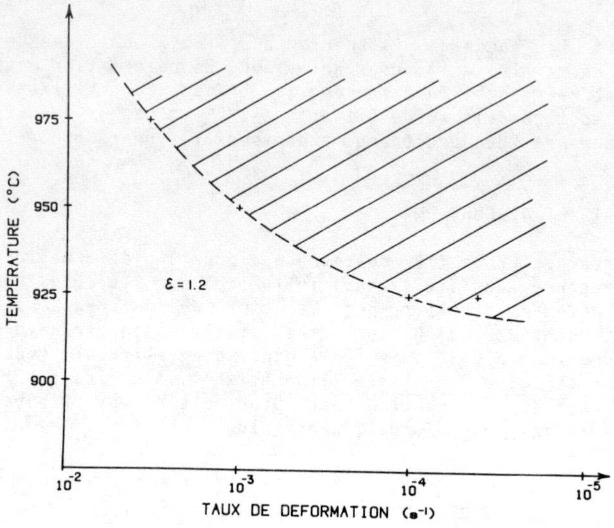

FIGURE 2

Diagramme température-taux de déformation, pour une déformation vraie de 1.2. La zone hachurée représente les conditions d'essai pour parvenir à la microstructure d'équilibre

Les courbes d'écoulement ont pour caractéristique générale d'exhiber un adoucissement plus ou moins prononcé. La principale cause de cet adoucissement est la contribution croissante du glissement aux joints de grains, à la déformation. A déformation élevée, la contrainte d'écoulement tend à se stabiliser à un niveau qui dépend des conditions de la déformation et qui est d'autant plus grand que le taux de déformation est plus élevé ou que la température est plus basse. La stabilisation de la contrainte à déformation élevée semble correspondre à l'obtention de la microstructure équiaxe et biphasée. Dans ces conditions, les résultats suggèrent qu'à une température et un taux de déformation donnés, il existe une contrainte et une microstructure d'équilibre uniques. Ainsi, pour des conditions de déformation fixes, on doit s'attendre à ce que la microstructure β, suffisamment déformée, atteigne le même état d'équilibre que celui obtenu par la microstructure α+β. Sur la base de nos résultats **(10)** et de ceux d'autres chercheurs **(5-7)**, on suggère que la contrainte et la microstructure, à l'état d'équilibre, sont indépendantes de la microstructure initiale.

C'est cet état d'équilibre qui peut servir de base à la modélisation de la déformation dans le matériau initialement à l'état β, en fournissant une condition limite à l'évolution de la contrainte et de la microstructure, à température et taux de déformation constants.

Pour la microstructure β, il a été démontré (7) que la fraction volumique de matériau recristallisé à grains fins augmente avec la déformation, jusqu'à ce que la transformation soit complète. On suggère (10) alors que la contrainte d'écoulement puisse s'exprimer comme étant la réflexion d'une microstructure composite, contenant une proportion de régions molles (régions recristallisées) et une proportion de régions dures (régions non transformées). En supposant que les déformations dans chacune de ces régions se produisent en parallèle (11), on propose une équation constitutive de la forme:

$$\dot{\epsilon} = Y\dot{\epsilon}_m + (1-Y)\dot{\epsilon}_d$$

où $\dot{\epsilon}$ est le taux de déformation imposé, alors que $\dot{\epsilon}_m$ et $\dot{\epsilon}_d$ sont les taux de déformation à l'intérieur des régions molles et dures respectivement. Le paramètre Y proportionne la contribution de chacune de ces régions et varie avec le temps, en fonction du taux de déformation, à une température donnée.

CONCLUSION

Une approche tel que décrit plus haut requiert la connaissance des équations constitutives correspondant aux régions recristallisées et aux régions non transformées, ainsi qu'un modèle microstructural pour décrire l'évolution de la microstructure et donc les changements dans le paramètre Y en fonction des conditions de la déformation. On suggère donc que cela soit l'objet d'études plus détaillées.

REMERCIEMENTS

S. Gautron et G. L'Espérance remercient le Dr. W. Wallace du laboratoire de structures et matériaux de l'EAN, pour l'octroi d'un contrat qui a rendu possible cette recherche et pour l'accès à l'équipement spécialisé de l'EAN. Enfin, les auteurs remercient les Dr. H. Burte et Dr. H. Gegel de ASWAL/MLLM, Wright-Patterson AFB et le Dr. F.L. Semiatin des laboratoires de Battelle, Columbus, pour avoir fourni le matériau utilisé dans ce travail.

REFERENCES

(1) HEWITT, R.L. et al., Isothermal Forging at the National Aeronautical Establishment. DME/NAE Quarterly Bulletin No. 1978 (4), Ottawa (1978), 23 p.

(2) LAHOTI, G.D. et ALTAN, T., Research to Develop Process Models for Producing a Dual Property Titanium Alloy Compressor Disk. Rapport annuel, AFML-TR-79-4156, Battelle's Columbus Laboratories, Columbus, Ohio (1979).

(3) LAHOTI, G.D. et ALTAN, T., Research to Develop Process Models for Producing a Dual Property Titanium Alloy Compressor Disk. Rapport annuel, AFWAL-TR-80-4162, Battelle's Columbus Laboratories, Columbis, Ohio (1980).

(4) LAHOTI, G.D. et ALTAN, T., Research to Develop Models for Producing a Dual Property Titanium Alloy Compressor Disk. Rapport annuel, AFWAL-TR-81-4130, Battelle's Columbus Laboratories, Columbus, Ohio (1981).

(5) DADRAS, P. et THOMAS, J.F. Jr., Met. Trans., 12A, 1867-1876 (1981).

(6) SEMIATIN, S.L. et LAHOTI, G.D., Met. Trans., 12A, 1705-1717 (1981).

(7) SEMIATIN, S.L., THOMAS, J.F. Jr. et DADRAS, P., Met. Trans., 14A, 2363-2374 (1983).

(8) RAUCH, E., CANOVA, G.R. et JONAS, J.J., Communication privée, Université McGill, Montréal (1982).

(9) McQUEEN, H.J. et JONAS, J.J., dans Treatise on Materials Science and Technology, Vol. 6: Plastic Deformation of Materials. Edité par R.J. Arsenault, Academic Press, New York (1975), pp. 393-493.

(10) GAUTRON, S., Forgeage isotherme de l'alliage Ti 6242. Mémoire de maîtrise es sciences appliquées, Département de génie métallurgique, Ecole Polytechnique, Montréal, 229 pages.

(11) GIFKINS, R.C., J. Mat. Sci., 5, 156-165 (1970).

Recovery, Recrystallization and Mechanical Properties in Ti-6Al-4v Alloy

I. Weiss*, G. E. Welsch, F. H. Froes*** and D. Eylon†**

**Wright State University, School of Engineering, Dayton, OH 45435, USA*
***Case Western Reserve University, Dept. of Metallurgy and Materials Science, 10900 Euclid Avenue, Cleveland, OH 44106, USA*
****Air Force Wright Aeronautical Laboratories, Materials Laboratory, AFWAL/MLLS, Wright-Patterson Air Force Base, OH 45433, USA*
†Metcut-Materials Research Group, P.O. Box 33511, Wright-Patterson Air Force Base, OH 45433, USA

KEYWORDS

Hot working; coarse alpha plates; recovery; recrystallization; alpha/alpha sub-boundaries; alpha/beta boundaries; beta phase; extrusion; annealing; Ti-6Al-4V

INTRODUCTION

The alpha and beta phase morphologies strongly influence the mechanical properties of alpha+beta titanium alloys. High aspect ratio alpha phase is associated with low fatigue crack propagation rates, high creep resistance, reduced tensile ductility, and inferior fatigue crack initiation characteristics [1-3]. Tensile and fatigue initiation properties can both be improved by modification of the lenticular alpha plates to a more equiaxed morphology, utilizing hot working [4-9] or thermo-chemical processing [10, 11].

Previous work demonstrated that alpha plates break-up to produce equiaxed alpha grains is strongly dependent on the plate thickness [9, 12]. The break-up of thin alpha plates (defined as having a mean thickness of 3μm) was predominantly by sub-boundary and shear band formation across the alpha plates during deformation. The separation into equiaxed alpha grains was completed by penetration of beta phase along the alpha/alpha interfaces and the sheared zones during subsequent annealing [9, 12]. A similar mechanism is observed for thick alpha plate material (defined as having a mean thickness of ∿6μm) [12]. However here the alpha plates break-up to larger equiaxed alpha grains often containing alpha/alpha boundaries. This latter effect is the result of selective beta penetration of the alpha/alpha boundaries.

The objective of the present work was to study in detail the mechanisms of coarse alpha plates (defined as plates having a thickness of ∿40μm) break-up to a more equiaxed morphology. The coarse alpha plates were obtained by slow cooling from the beta solution treatment temperature. In addition, the influence of microstructural refinement on tensile and fatigue properties was investigated.

MATERIAL AND EXPERIMENTAL PROCEDURES

Mill annealed Ti-6Al-4V was used as the starting stock with the composition given in Table 1.

TABLE 1 Chemical Composition of Ti-6Al-4V (wt %)

Al	V	C	N	Fe	O	H	Ti
6.7	4.1	0.01	0.013	0.18	0.164	0.0044	Balance

Following machining, 76 x 76ϕmm cylinders were beta heat treated at 1010°C (1850°F)/15min followed by 50°C (90°F)/hr slow cooling to produce the coarse alpha plate structure (mean thickness, 30μm) prior to processing. The material was then annealed at 925°C [1700°F]) for 30 minutes to produce the equilibrium amount of lamellar alpha and beta matrix followed by extrusion at the same temperature at a constant speed of 30mm/min, and finally air cooling to room temperature. The 2.3 extrusion deformation was calculated on the basis of 10:1 area reduction. The deformed specimens were then annealed at 925°C (1700°F) for times up to 24 hr (Table 2) followed by slow cooling at 50°C (90°F)/hr. Microstructural analysis was carried out by both optical and transmission electron microscopy. Tensile tests (strain rate of 0.13mm/mm/min [0.005in./in./min]) and load control fatigue tests (R = 0.1) were carried out.

TABLE 2 The Mean Thickness and Volume Fraction of Coarse, Fine, and Recrystallized Alpha Phase

Annealing time	Mean thickness, μm (Coarse lamellar alpha)	Vol%	Mean diameter, μm (Recrystallized alpha)	Vol%
15 minutes	30	70	NA	0
2 hr	30	50	10	30
24 hr	NA	0	20	100

NA = not available.

RESULTS AND DISCUSSION

Starting Microstructure

The thermomechanical treatment conditions used in this work are shown schematically in Fig. 1. The microstructure of material beta solution treated (Cycle I) and subsequently alpha+beta annealed (Cycle II), followed by air cooling is shown in Fig. 2a. Coarse primary alpha plates in a beta matrix containing fine lenticular alpha lamallae are observed. The coarse primary alpha plates are often present along prior beta grain boundaries.

Microstructures of Deformed and Annealed Material

Starting microstructure [8, 9], deformation [3, 9], and post-deformation heat treatment [7, 8] all affect the alpha+beta morphology. The as-extruded microstructure is shown in Fig. 2b. Coarse primary alpha plates are aligned parallel to the extrusion axis. Subsequent annealing for 15 minutes resulted in a microstructure containing a mixture of apparently high (primary) and low (secondary) aspect ratio plates (Fig. 3a). Alpha/alpha boundaries are observed in some coarse plates, and penetration of beta phase along some alpha/alpha interfaces is also detected. Lower aspect ratio primary alpha plates, with more alpha/alpha boundaries were obtained after 2 hr annealing (Fig. 3b). Increasing the annealing time to 24 hr, resulted in an almost entirely equiax primary alpha microstructure (Fig. 3c). Alpha grain growth, from 10μm to about 20μm, also occurs (Table 2). In this condition, the beta phase is located mainly at the triple points of the alpha grains (Fig. 3c).

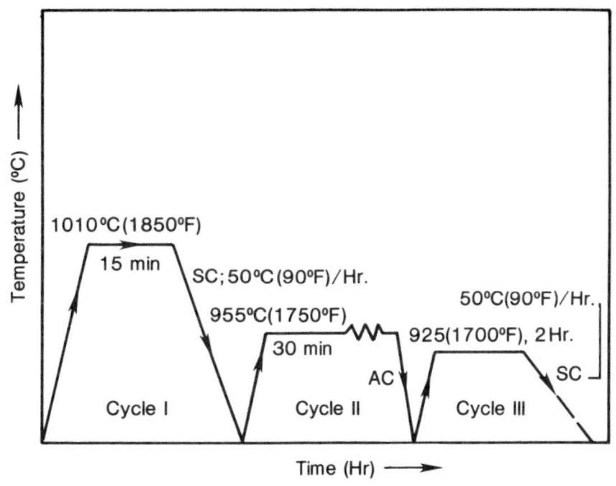

Fig. 1. Thermomechanical treatment used in this work.

Modification of the Coarse Lamellar Alpha Morphology

Extruded material annealed for 15 minutes (Fig. 3a) exhibited an elongated coarse alpha morphology. A TEM examination of the plates revealed elongated sub-cells (Fig. 4) with sub-cell length ranging between 1 and 5μm. The sub-cells were oriented approximately parallel to the extrusion axis with the long side in a <10$\bar{1}$0> direction, the result of dynamic and static recovery during the deformation cycle. Very little beta penetration into these sub-boundaries is observed. A statically recovered and partially recrystallized microstructure is obtained by annealing for 2 hr (Fig. 2b). Some of the low angle alpha/alpha boundaries gradually increase sub-boundary misorientation by coalesence and rotation concurrently with beta phase penetration along these boundaries during annealing [13]. These high angle boundaries bulge and migrate [14] to produce recrystallized grains free of dislocations. These recrystallized grains were first found to nucleate in the vicinity of the alpha/beta interfaces.

Nucleation of the recrystallized grains along the alpha/beta interface is shown in Fig. 5. The recrystallized alpha grains can continue growing at the expense of the adjacent regions containing a recovered sub-structure (Fig. 5). In addition, other regions formed by coalesence and rotation of sub-structure during the early stages of annealing are effectively new recrystallized grains [13]. Thus much of the interior regions of the coarse alpha plates are recrystallized (Fig. 5).

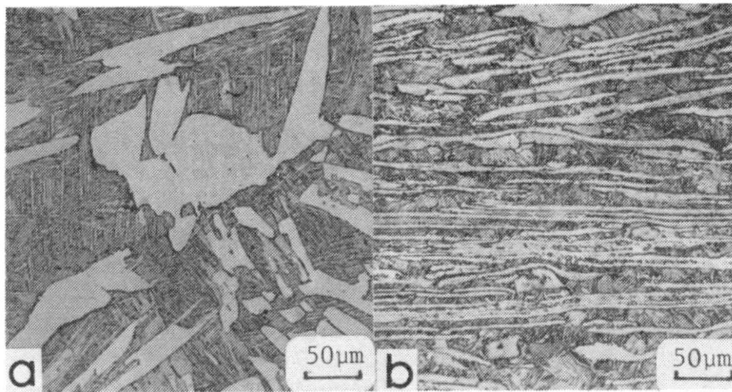

Fig. 2. (a) Microstructure of material beta solution treated (Cycle I) and subsequently alpha+beta annealed (Cycle II), followed by air cooling; (b) microstructure of material hot extruded to a true strain of 2.3 (area reduction 10:1).

Annealing for 24 hr produced a fully recrystallized alpha microstructure with an aspect ratio of 4 or less (Fig. 6). The increased annealing time allows the formation of a more equiaxed alpha, with a larger grain size, by the migration of the high angle alpha/alpha boundaries which formed in the earlier stages of annealing.

The effect of annealing time on the aspect ratio of the primary alpha plates is summarized in Table 2 and in Fig. 7. The mean aspect ratio measured after 15 minutes is approximately 7:1 with occasional plates as high as 14:1

(Fig. 3a). With increased annealing times to 2 and 24 hr lower mean aspect ratio alpha plates of 4 and 2 respectively, are obtained.

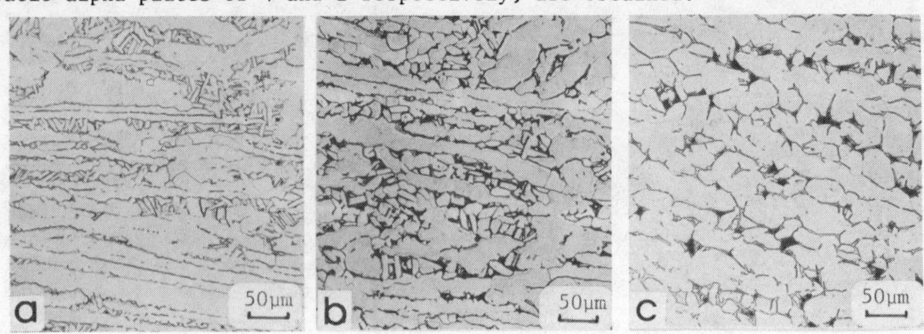

Fig. 3. Microstructures of material hot extruded (10:1) and subsequently annealed for (a) 15 minutes at 925°C (1700°F); (b) 2 hr at 925°C (1700°F); and (c) 24 hr at 925°C (1700°F) and subsequently slow cooled 50°C (90°F)/hr.

Microstructure and Mechanical Properties

The microstructural changes caused by recovery and recrystallization following extrusion and annealing influence the tensile properties (Fig. 8). It was shown earlier [9], that material with a microstructure consisting of lower aspect ratio alpha exhibits higher yield stress than material with higher aspect ratio alpha plates. However, these microstructures (which display high yield stress) are all fully annealed, with very little sub-structure present. In the present work the stable sub-structure (Figs. 4 and 5) in the coarse alpha plates (15 min and 2 hr annealed condition) leads to an increase in yield stress. Maximum yield and tensile stresses of 790MPa and 875MPa, respectively, are observed for material with a mean alpha aspect ratio of 7:1. Lower values of 745MPa and 815MPa are measured in material with an alpha aspect ratio of about 2:1 or less and larger recrystallized grains (Table 2) (Fig. 8). The development of sub-cells in the coarse plates decreases the mean free path from around 30μm (original plate thickness), to about 1-3μm (sub-cell diameter, Fig. 4). This is almost 10 times smaller than the mean free path observed in the fully recrystallized microstructure (Fig. 6), resulting in the highest yield and tensile stresses for the material annealed 15 minutes.

This condition also exhibits the best fatigue life (Fig. 9) while material annealed for 24 hr (20μm grain size) which shows the lowest tensile stress exhibits the lowest fatigue strength.

The ratio of fatigue strength to ultimate tensile strength of all three conditions is calculated in Table 3. No variation in the ratio is observed suggesting that the differences in fatigue strength are the result of differences in tensile strength. However, it should be noted that the higher strength was achieved by microstructure refinement on the grain and sub-grain levels, which can also lead to an improved crack initiation resistance [6].

TABLE 3 Ratio of 2 x 10^6 Fatigue Strength to Ultimate Tensile Strength

Annealing Time	Fatigue Ratio
15 minutes	0.67
2 hr	0.66
24 hr	0.66

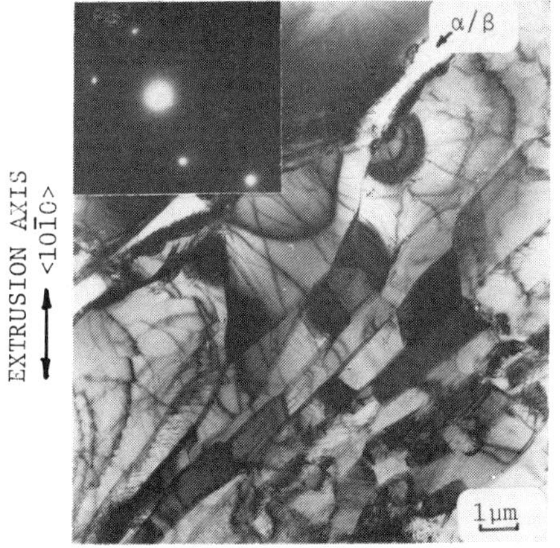

Fig. 4. TEM micrograph of coarse alpha plate interior in material hot extruded (10:1) and subsequently annealed for 15 minutes at 925°C (1700°F) and slow cooled at 50°C (90°F)/hour.

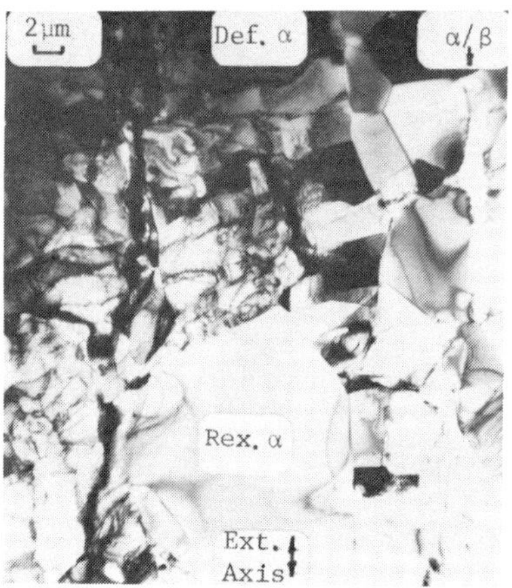

Fig. 5. TEM micrography of material hot extruded (10:1) and subsequently annealed for 2 hr at 925°C (1700°F) and slow cooled at 50°C (90°F)/ hour.

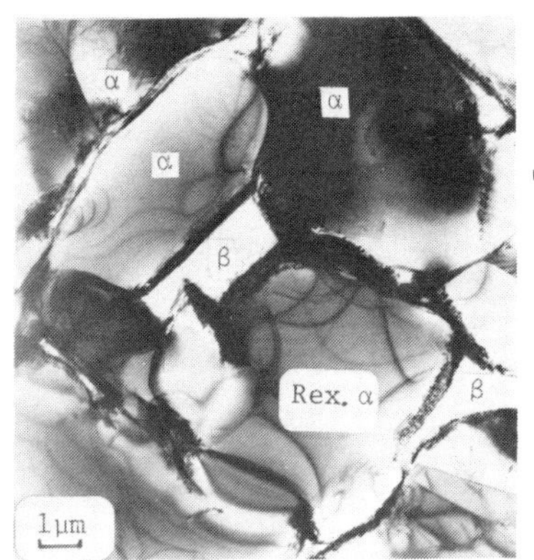

Fig. 6. TEM micrograph of material hot extruded (10:1) and subsequently annealed for 24 hr at 925°C (1700°F) and slow cooled at 50°C (90°F)/hour.

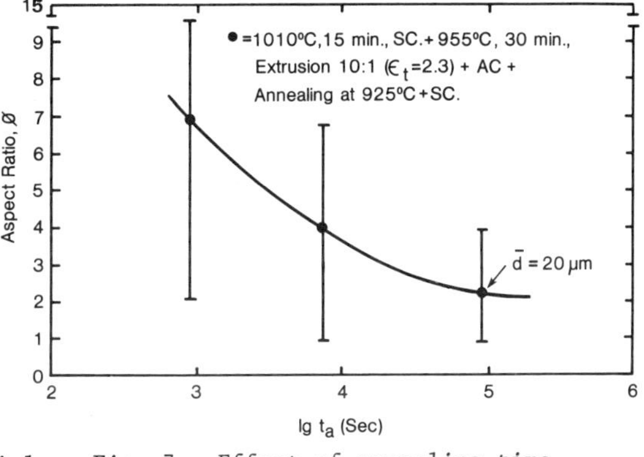

Fig. 7. Effect of annealing time on the mean and range of aspect ratio values of coarse alpha plates.

CONCLUSIONS

1. The mechanism by which a coarse (30-40μm thick) alpha lamellar structure relaxes after extrusion (10:1 deformation ratio) was investigated.
2. A fine sub-structure is developed in the coarse alpha plates after annealing at 925°C (1700°F) for 15 minutes but with little decrease in the aspect ratio of the primary alpha.
3. Annealing for 2 hr resulted in partially recrystallized structure while a 24 hr anneal developed a fully recrystallized structure.

4. The transformation to a more equiaxed morphology occurred initially by formation of alpha/alpha sub-boundaries across the alpha plates during deformation. This is followed during the subsequent annealing by the separation of the alpha grains in the vicinity of the alpha/beta boundary by penetration of the beta phase along the alpha/alpha boundaries. Completion of recrystallization of the lamallae interior occurs by bulging and migration of high angle alpha/alpha boundaries and coalescence of sub-grains to produce a recrystallized structure.

5. The presence of a stable sub-structure in coarse alpha plates results in increased yield and tensile strength (790MPa and 875MPa) compared to fully annealed material (745MPa and 815MPa, respectively).

6. Material annealed for 15 minutes exhibits the best fatigue life while material annealed for 24 hr (20μm grain size) which shows the lowest tensile stress exhibits the lowest fatigue strength.

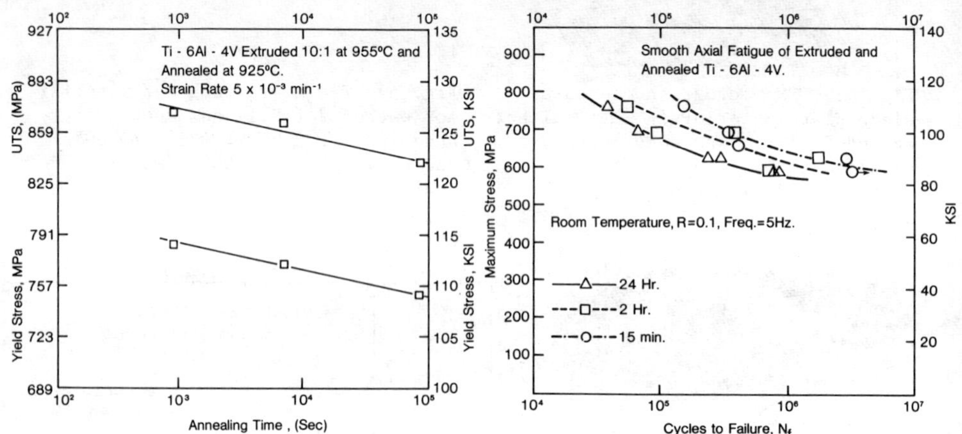

Fig. 8. Effect of annealing time on the yield and tensile stresses.

Fig. 9. Effect of annealing time on fatigue behavior.

REFERENCES

1. G. R. Yoder, L. A. Cooley, and T. W. Crooker, 23rd Structures, Structural Dynamics, and Materials Conference, New Orleans, LA, p. 132 (May 1982).
2. D. Eylon, J. A. Hall, C. M. Pierce, and D. L. Ruckle, Met. Trans. A 7A, 1817 (1976).
3. F. H. Froes and W. T. Highberger, J. of Metals 32, 57 (1980).
4. F. H. Froes, J. C. Chesnutt, R. G. Berryman, G. R. Keller, and W. T. Highberger, 10th National SAMPE Technical Conference, Kiamesha Lake, NY, p. 511 (1978).
5. M. Peters, A. Gysler, and G. Luetjering, in Titanium '80, Science and Technology, Proc. 4th Int. Conference on Titanium, Kyoto, 1980 (edited by H. Kimura and O. Izumi), Vol. 3, p. 1777. AIME, New York (1980).
6. M. Peters and G. Luetjering, in Titanium '80, Science and Technology, Proc. 4th Int. Conference on Titanium, Kyoto, 1980 (edited by H. Kimura and O. Izumi), Vol. 2, p. 925. AIME, New York (1980).
7. H. Margolin and P. Cohen, in Titanium '80, Science and Technology, Proc. 4th Int. Conference on Titanium, Kyoto, 1980 (edited by H. Kimura and O. Izumi), Vol. 2, p. 1555. AIME, New York (1980).
8. M. Peters, G. Luetjering, and A. Gysler, Z. Metallkunde Bd 74, 274 (1983).
9. I. Weiss, G. E. Welsch, F. H. Froes, and D. Eylon, Proc. 5th Int. Conference on Titanium, Munich, West Germany, (1985), in press.
10. W. R. Kerr, P. R. Smith, M. E. Rosenblum, F. J. Gurney, Y. R. Mahajan, and L. R. Bidwell, in Titanium '80, Science and Technology, Proc. 4th Int. Conference on Titanium, Kyoto, 1980 (edited by H. Kimura and O. Izumi), Vol. 4, p. 2477. AIME, New York (1980).
11. L. Levin, R. G. Vogt, D. Eylon, and F. H. Froes, Proc. 5th Int. Conference on Titanium, Munich, West Germany, (1985), in press.
12. I. Weiss, F. H. Froes, D. Eylon, and G. E. Welsch, Met. Trans. A (1985), in press.
13. H. J. McQueen and J. J. Jonas, in Plastic Deformation of Materials (edited by R. J. Arsenaule), Vol. 6, p. 394. Academic Press, New York (1975).
14. J. E. Bailey and P. B. Hirsch, Proc. R. Soc. 267, 11 (1962).

SECTION 9

Toughness and Microstructures

Ténacité et microstructures

Studies of Stress-Corrosion Cracking on Fe-Mn-Al Stainless Steels with Different Carbon Contents

S. C. Chang, T. S. Sheu and C. M. Wan

Department of Materials Science and Engineering, National Tsing Hua University, Hsinchu, Taiwan, Republic of China

ABSTRACT

Fe-Mn-Al stainless steels with different carbon contents and thus different volume fractions of ferrite were studied for their stress-corrosion cracking properties in room temperature 4% NaCl solution. Both constant strain rate test and constant load test were employed in this work.

For the alloys used in this work, the amount of ferrite increases with the aging treatment. The ferrite regions are easier to be corroded than the austenite regions when these alloys were immersed in 4% NaCl solution. The cracks formed on specimens in stress-corrosion cracking test are in general in a direction perpendicular to the applied tensile stress. Stress-corrosion cracks in full austenite alloy are intergranular. In austenite-ferrite dual phase Fe-Mn-Al alloys, stress-corrosion cracks are transgranular in ferrite phase and tend to be blocked by the austenite phase.

KEYWORDS

Fe-Mn-Al Alloys; Stress-Corrosion Cracking; Intergranular; Transgranular; Ferrite; Austenite.

INTRODUCTION

The studies of Fe-Mn-Al alloy systems started in 1930's [1]. Combined with significant cost and density advantages, good mechanical properties, promising corrosion and oxidation resistance, Fe-Mn-Al alloys (Femanal) are good candidates for replacing at least part of the conventional Ni-Cr stainless steels [2-8]. Based on its good marine corrosion resistance and good cavitation erosion resistance, one of the Femanal alloys was selected for a propeller installed on a fishing vessel launched in August 1980 [9].

The resistance to stress-corrosion cracking is one of the necessary properties for safe use of Fe-Mn-Al stainless steels in making stress bearing parts such as propellers. However, little [10,11] data of this category have been reported before.

For conventional Ni-Cr stainless steels, fully ferritic stainless steels are highly resistant to stress-corrosion cracking in the chloride and caustic enviroments that crack the common austeritic stainless steels [12]. Beck et al. [13] and Fontana et al. [14] observed significantly increased resistance to stress-corrosion cracking with increasing ferrite content for cast Ni-Cr stainless steels.

By varying the carbon content, a range of ferrite volume fraction could be obtained in the basically Fe-30%Mn-9%Al-1%Si alloy. It is the purpose of this work to see if the ferrite phase in Femanal stainless steel shows a similar beneficial effect on the resistance to stress-corrosion cracking.

EXPERIMENTAL PROCEDURES

The chemical compositions of the alloys used in this study are listed in Table 1. They are basically 30%Mn, 9%Al, 1%Si, 0-1%C and balance Fe. For alloy B which has a duplex structure, both the compositions of ferrite and austenite phases determined by X-ray microanalyzer are listed with their weighted average. The details of the processing of these alloys were reported elsewhere [11]. The designation adopted for the specimens of different alloy and heat treatment in this work is that the leading letter represents the chemical composition as given in Table 1; the first number of 5 and 7 means the aging temperature of 550°C and 700°C respectively; the following number(s) gives the aging time in hours while ST implies that the specimen is solution heat treated at 1050°C for one hour and then oil quenched.

For stress-corrosion cracking study in room temperature 4% NaCl solution, both constant strain rate and constant load tests were employed. A strain rate of 1.86×10^{-5} sec^{-1} were applied to fracture the specimen in the constant strain rate test. For constant load test, a constant load of 60%, 70% and 80% ultimate tensile load were applied to the specimen through a dead weight loading system designed by the auther. The subsize rectangular tension test specimen with a gauge length of 25 mm and thickness of 2 mm were used in this work.

Before stress-corrosion test, free surface of the specimens were polished with alumina powder to 0.05 micron. They were repolished and etched after the test to examine the crack pathes on them. Optical metallography specimens were etched with 10% Nital solution and the volume fractions of phases were obtained with both plainmeter and linear interception methods.

Table 1 Chemical Compositions of the Alloys (wt%)

Alloy		Al	Mn	Si	C	Fe
A		9.44	30.04	-	0.06	bal.
B	Average	8.94	31.19	0.93	0.57	bal.
	Ferrite	10.05	26.82	1.17		
	Austenite	8.58	32.65	0.85		
C		9.47	30.83	1.01	1.02	bal.

RESULTS

The phases and mechanical properties of the alloys are listed in Table 2. Also listed in table 2 are the maximum stresses of the alloys in constant strain rate stress-corrosion test. For constant strain rate stress-corrosion test, the test duration of a specimen depends on its total elongation. Nevertheless, there is a definite decrease in maximum stress for all the specimens tested.

Alloy A is essentially ferritic with only trace amount of austenite on ferrite grain boundaries. In overaged specimen (A74), brittle βMn phase was observed. Alloy A is susceptible to stress-corrosion cracking in 4% NaCl solution and its mechanical properties is far more inferior to those of alloy B and C. Therefore, most work was then concentrate on alloys B and C.

As shown in Table 2, alloy C is fully austenitic and alloy B has thirty-eight percent volume fraction of ferrite under solution heat treated condition. The volume fractions of ferrite in both alloys increase with aging treatment. Also occurred with aging is the precipitation of FeAlCx within the ferrite grains.

For both constant strain rate and constant load stress-corrosion tests, many surface cracks formed along the entire gauge length of specimens. In general, the cracks are in a direction perpendicular to the applied load. In contrast, there was no surface crack on specimens either tensile tested in air at the same strain rate or immersed in the 4% NaCl solution without stress up to 3 weeks. This indicates that these alloys are susceptible to stress corrosion cracking. When they were immersed in salt water for long

Table 2 Phases, Mechanical Properties in air and the Maximum Stress in Constant Strain Rate Stress-Corrosion Test of the Alloys

Specimen	Phases	Hardness (RA)	Elongation (%)	0.2% YS (MPa)	UTS (MPa)	Sscc (MPa)
AST	α+γ(trace)	58.7	3.5	447	459	395
A54	α+γ(trace)	61.3	4.1	344	364	287
A74	α+βMn+α (trace)	77.1	1.2	380	390	375
BST	γ+α(38%)	63.7	20.3	611	864	652
B54	γ+α*(40%)	68.0	12.3	859	988	830
B74	γ+α*(56%)	65.0	3.0	586	601	524
CST	γ	59.0	56.4	677	863	841
C54	γ+α(11%)	69.3	16.5	1085	1106	1095
C71	γ+α*(trace)	64.2	37.6	778	1003	912
C74	γ+α*(5%)	64.6	38.7	766	1016	993
C78	γ+α*(23%)	64.5	28.2	700	936	841
C716	γ+α*(26%)	64.2	27.2	708	977	870

α* = Fe_3AlC_x precipitated within the ferrite phase

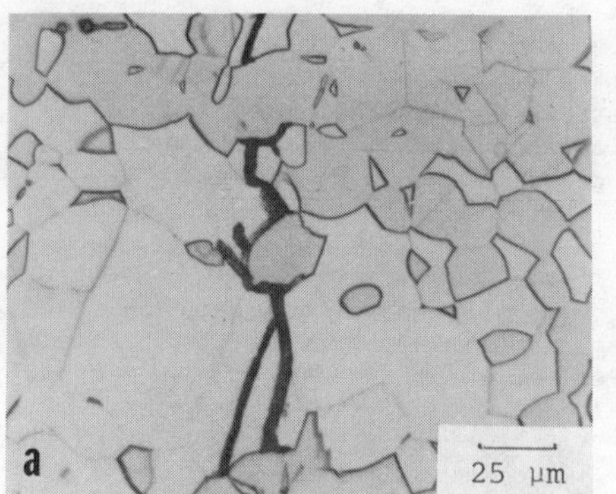

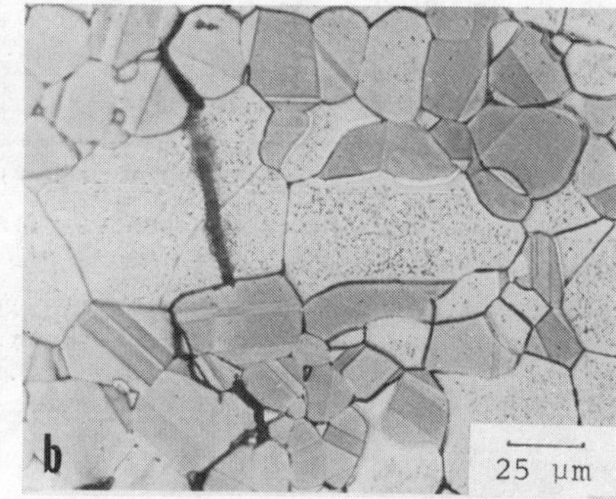

Fig. 1 Stress-corrosion cracks in specimens (a) BST and (b) B54. Cracks are transgranular in ferrite phase and tend to be blocked by the austenite phase.

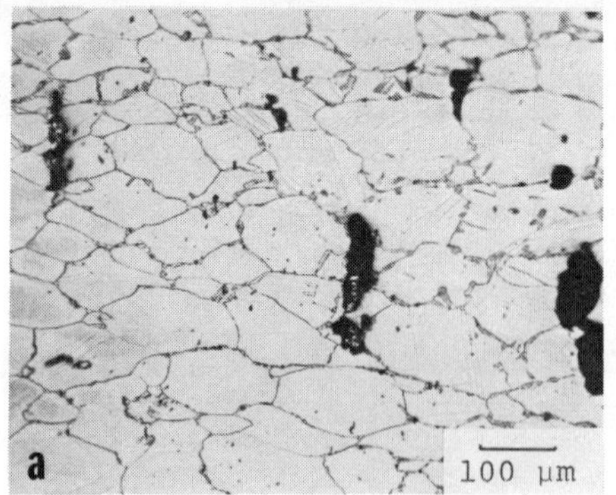

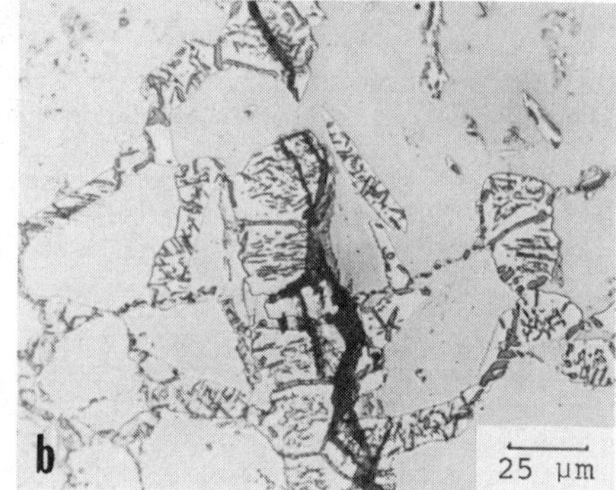

Fig. 2 Stress-corrosion cracks propagate through the ferrite formed along the austenite grain boundaries in specimens (a) C71 and (b) C716.

time, some corrosion did occur. The corrosion is largely occurred on ferrite regions. Identified with an X-ray diffractometer, the corrosion products were mainly Al_2O_3, Mn_5O_8 and Fe_3O_4.

For alloy B and C containing ferrite phase, the stress corrosion cracks were transgranular in the ferrite regions as shown in Fig. 1 and Fig. 2b. The cracks did not pass through the austenite region but propagated along the austenite-ferrite boundary or propagated in the ferrite beneath the austenite and emerged on the other side. In the full austenite specimen (CST) or the specimen with only trace amount of ferrite along austenite grain boundaries (C71), the stress-corrosion cracks propagate intergranularly (Fig. 2a).

Table 3 shows the fracture times in constant load stress-corrosion test of alloy B with different aging treatment and applied load in 4% NaCl solution. Not surprisingly, the time to failure increased with lower applied load. For specimen BST under a tensile stress of 518 MPa, fracture did not occur during the 2000 hour test period. The free surface of specimens were covered with corrosion products, however, cracks similar to those on constant strain rate stress-corrosion tested specimens were observed.

Table 3 Time to Failure of Alloy B in Constant Load Stress-Corrosion Test in 4% NaCl Solution

Specimen	Applied Stress (MPa)	Failure Time (hr)
BST	60%UTS (85%YS) 518	> 2000
	70%UTS (99%YS) 605	216.7
B54	60%UTS (69%YS) 593	301.6
	70%UTS (80%YS) 691	35.5
	80%UTS (92%YS) 790	25.8
B74	60%UTS (62%YS) 360	308.6
	70%UTS (72%YS) 421	127.5

DISCUSSION

In this study, the ferrite phase in Fe-Mn-Al stainless steels is more susceptible to stress-corrosion cracking in room temperature 4% NaCl solution than the austenite phase. The resistance of this system of alloys to stress-corrosion cracking decreases as the amount of ferrite is increased. In the ferrite regions, cracks propagate transgranularly. Cracks do not pass through the austenite region. Whenever a crack propagates to a ferrite-austenite boundary, it goes along the boundary or propagates in the ferrite beneath the austenite and emerged on the other side. These observations are on the contrary to the results of Beck el al. [13] and Fontana et al. [14] on Ni-Cr cast stainless steels.

There are reasons for preferential stress-corrosion cracking of ferrite phase in 4% NaCl solution. Sheu [11] measured the corrosion potential of Fe-Mn-Al alloys in 4% NaCl solution. He found that the alloys became more active when more ferrite phase was formed in aging. It is consistant to the observation in this study that the ferrite regions are easier to be attacked than the austenite regions when immersed in 4% NaCl solution. Wang and Rapp [15] also found that while Fe-30Mn-10Al-Si austenitic alloy was known to posses good marine corrosion resistance [9], a Fe-Al base ferritic alloy with low Mn and Si content, 10% Al may not be sufficient for adequate marine corrosion resistance. Moreover, since both the strength and elongation of ferrite phase were lower than those of austenite phase (Table 2), when loaded, ferrite was the one in the two phase aggregates that sustained a strain and stress of relatively high fraction of its fracture strain and stress.

Pan [10] studied the stress-corrosion cracking of an alloy with chemical composition of 27%Mn, 8.44%Al, 0.77%Si, 0.94%C and balance Fe in which there was no ferrite even after long time aging. The cracks on that alloy were intergranular for solution heat treated and underaged specimens and transgranular for overaged specimens. The alloy C in this study, which contains more Al and Si is also fully austenitic when solution heat treated (CST). The cracks are intergranular under this condition which is similar to the result of Pan's work. However, ferrite formed along the austenite grain boundaries after short time aging which was the easy stress-corrosion cracking path in this alloy. It is reasonable to think that even solution heat treated, there is a very thin layer of ferrite along the austenite grain boundaries and the intergranular cracks in this specimen are actually propagate through the thin layer of ferrite.

CONCLUSION

Fe-Mn-Al stainless steels are susceptible to stress-corrosion cracking in room temperature 4% NaCl solution. The resistance of Fe-Mn-Al alloys to stress-corrosion cracking decreases as the amount of ferrite is increased. Stress-corrosion cracks in full austenite Fe-Mn-Al alloys are intergranular. In Fe-Mn-Al alloys containing both austenite and ferrite phase, the cracks are transgranular in the ferrite grains and tend to be blocked by the austenite grains.

ACKNOWLEDGEMENT

The authors are pleased to acknowledge the financial support of this research by the National Science Council, Republic of China under Grant NSC73-0405-E007-012.

REFERENCES

1. W.Koster and W. Tonn, Archiv fur das Eisenhuttenwesen, 7, 365 (1933-34).
2. W. Justusson, V.F. Zackay, and E. R. Morgan, Trans. ASM, 49, 905 (1957).
3. J. L. Ham and R. E. Cairns, Product Eng. 29, 50 (1958).
4. S. K. Banerji, Met. Prog. 113, No.4, 59 (April 1978).
5. S. K. Banerji, An update on Fe-Mn-Al steels, in Proc. Workshop on Conservation and Substitution Technol. Critical Materials, Vanderbilt University, Nashville, (June 1981).
6. S. K. Banerji, The 1982 status report on Fe-Mn-Al steels, in Proc. Workshop on Trends in Critical Materials Requirements for Steels of the Future-Conservation and Substitution Technology for Chromium, Vanderbilt University, Nashville (Oct. 1982).
7. J. P. Sauer, R. A. Rapp, and J. P. Hirth, Oxid. Met. 18, 285 (1982).
8. R. Wang, M. J. Straszheim and R. A. Rapp, Oxid, Met. 21, 71 (1984).
9. R. Wang and F. H. Beck, Met Prog. 123, No.3, 72 (March 1983).
10. Y. C. Pan, Master Thesis, Tsing Hua University, Hsinchu, Taiwan, Republic of China, (1983).
11. T. S. Sheu, Master Thesis, Tsing Hua University, Hsinchu, Taiwan, Republic of China, (1984).
12. ASM, Metals Handbook, vol. 10, 8th edition, 220.
13. F. H. Beck, J. Juppenlatz and P. F. Wieser, in Stress Corrosion-New Approaches (edited by H. L. Craig, Jr.), p. 381, ASTM STP610, Philadelphia, Pa. (1976)
14. M. G. Fontana, F. H. Beck and J. W. Flowers Met. Prog. 80 , No.6, 99 (December 1961).
15. R. Wang and R. A. Rapp, Proc. Int. Congress on Metallic Corrosion Toronto, Canada, 4, 545 (1984).

Effects of Impurity Segregation on Sustained Load Cracking of a 2¼Cr-1Mo Steel

M. B. D. Ellis, J. J. Lewandowski and J. F. Knott

Department of Metallurgy and Materials Science, University of Cambridge, UK

ABSTRACT

Commercial-purity, high-purity and phosphorus-doped 2¼Cr-1Mo steels have been tested in various as-quenched conditions under load, at 500°C in vacuo. The austenitising treatments were varied to control the amount of sulphur in solid solution available for segregation. It is shown that the amount of segregant available has a large effect on the cracking produced under load, and the reasons for this are discussed.

KEYWORDS

Dynamic impurity segregation, sulphur, phosphorus, sustained load cracking, intergranular failure, sub-notch cracks.

INTRODUCTION

It is well established that the fracture properties of quenched-and-tempered alloy steels are strongly affected by the segregation of trace impurity elements (P, Sn, Sb and S) to microstructural sites, such as prior austenite grain-boundaries or carbide/matrix interfaces. Conventionally, research has been concerned with situations in which the segregation has occurred prior to subsequent fracture at low temperatures, so that the segregant levels have been 'frozen-in' before loading (1,2). Recent work has indicated that, in tests made at moderately elevated temperatures, impurity segregation may occur locally ahead of a crack tip during fatigue crack growth (3) or stress relief (4,5) even though general segregation is not observed. The segregant level therefore appears to be affected by the tensile stress field ahead of the crack tip and it has been proposed that a mechanism of 'stress-assisted segregation' may promote the accumulation of damaging quantities of impurities at critical microstructural sites ahead of a crack or other stress concentrator. Any such effect would mean that specification limits for impurities in commercial alloys based on conventional testing for embrittlement might be over-optimistic.

The present work investigates effects of segregation on cracking at 500°C in 2¼Cr-1Mo steel tested under sustained load conditions. Commercial purity (CP), high-purity (HP) and phosphorus-doped (PD) versions of the steel were tested: detailed compositions are given in Table 1. As-quenched microstructures were produced by one of the following austenitising and quenching treatments: i) 1300°C 1hr/950°C 2hrs OQ.; ii) 950°C 2 hrs OQ; or iii) 1300°C 1 hr→F.B. at 300°C 15 mins OQ; where OQ = oil quench and FB = fluidised bed (see Table 2).

EXPERIMENTAL

The tests were carried out using single-edge notch (SEN) bend specimens, of thickness (B) = 10 mm, depth (W) = 20 mm containing a central 'V' notch of depth (a) = W/3 with a radius of 1 mm and a flank angle of 90°. After quenching, the testpieces were loaded for times up to 24 hours in four-point bend using a Mand Servo-hydraulic testing machine of 30 kN load capacity fitted with a high temperature vacuum chamber operating at a total pressure of 0.3 nbar. The testpieces were held under constant load at 500°C. This temperature was achieved using quartz lamps as a heat source, and could be controlled to within ± 1°C via a thermocouple attached to a microprocessor. Any subsequent crosshead displacement was observed using a chart recorder, plotting displacement vs time and any crack growth was monitored by means of the D.C. potential drop technique. To observe the early stages of crack growth, the load was removed and the temperature was reduced to ambient.

Fractographic observations were made using a Camscan S4 scanning electron microscope operating at 30 kV. Metallographic observations were made on the sectioned and polished notch root regions.

RESULTS

The results for the tests under vacuum are summarised in Table 2. The value of the nominal bending stress σ_{nom}, is calculated from the equation

$$\sigma_{nom} = \frac{6M}{B(W-a)^2} \qquad (1)$$

where B is the SEN testpiece thickness along the notch, (W-a) is the ligament depth below the root of the notch and M is the applied bending moment. For the present series of tests, σ_{nom} = 938 MPa and the two main variables are: i) austenitising treatment (see Table 2) and ii) purity level (see Table 1).

For the C.P. steel, the three different heat treatments produced three markedly different results. The 1300°C/950°C treatment under load at 500°C cracked after 12 hours and failed completely after 18 hours. Figure 1 shows the fracture appearance which exhibits smooth intergranular facets at the notch; slightly further away from the notch root, similar intergranular facets are seen surrounded by ductile failure. The 950°C treatment remained intact after 24 hours under load at 500°C, and no crack growth was observed in the notch root. In contrast, after only 2.5 hours under load at 500°C, the 1300°C/300°C testpiece had cracked considerably with intergranular separation from the notch root. A second 1300°C/300°C specimen was tested under the same conditions, but was unloaded after 1.75 hours. After sectioning and polishing, two isolated cracks were observed; they were both located 0.3 mm to 0.7 mm below the notch root (Fig. 2). The cracks were intergranular with respect to prior austenite grain boundaries.

No cracking was observed in the H.P. or P.D. steels following the 1300°C/300°C treatment for loading times at 500°C of 24 hours and 7 hours.

DISCUSSION

The results can be explained in terms of the austenitising heat-treatments and the effect of these on the solubility of sulphur at the various temperatures. The general effect of austenitising temperature on sulphur solubility may be appreciated by analogy with results obtained by Turkdogan et al.(6). Here, the sulphur solubility in pure iron with 0.43% Mn is found to be ∿ 0.8 ppm at 950°C and 37.0 ppm at 1300°C. At this stage it is important to note that the chromium content also affects the solubility of sulphur in austenite, but its effect is less than that of manganese. On slow cooling from 1300°C to 950°C, sulphur is precipitated from solid solution in the form of MnS, but on quenching from 1300°C to 300°C followed by a 15-minute hold at that temperature the sulphur is more likely to be retained in solid solution and therefore is available for segregation at 500°C.

Previous workers have shown that phosphorus also segregates to austenite grain boundaries (presumably as monolayers) during the austenitising treatment (8,9). With sulphur retained in solid solution in the C.P. steel (following the 1300°C/300°C treatment), cracking occurred under load at 500°C after 2.5 hours, but when the sulphur was precipitated from solid solution (due to the 1300°C/950°C treatment), cracking did not occur until after 12 hours. Sulphur appears to be the major embrittling element when cooling from 1300°C is rapid, because the P.D. steel heat-treated in an identical manner did not show any sign of cracking. The phosphorus level in the P.D. steel is high (310 ppm) and, in other experiments, has been shown to be sufficient to cause reversible temper embrittlement following a tempering treatment, but the sulphur level is a factor of two lower (0.014% versus 0.006%).

When the C.P. 1300°C/300°C steel was held under load at 500°C for 1.75 hours two isolated cracks were found approximately between 0.3 mm to 0.7 mm below the notch (Fig. 2). The cracks were located in prior austenite grain boundaries and did not link with the notch surface.

Using an elastic-plastic, finite-element analysis developed for a similar geometry (10), it is possible to calculate the location of the peak stress below the notch for an applied load such that $\sigma_{nom}/\sigma_Y = 1.61$, which is close to the value employed for the cracked $CP_{(3)}$ specimens (Table 2). The distance below the notch is found to be 0.7 mm, in reasonably close agreement with the metallographic observations (Fig. 2). The HP steel showed no sign of cracking after 24 hours under load at 500°C. This is presumably because the levels of both sulphur and phosphorus are so low that the effect of any segregation on grain boundary cohesion is negligible.

The C.P. 1300°C/950°C steel cracked after 12 hours and finally broke in two after 18 hours. Figure 1 shows the fracture face below the notch. Intergranular patches at the notch appear to have formed after 12 hours, followed by slow crack growth and final failure. It is possible that during the hold at 950°C phosphorus segregates to the austenite boundaries and when held under load at 500°C sufficient further phosphorus segregation to these boundaries occurs to produce the clean intergranular facets at the notch after 12 hours. Clean facets can still be seen at a distance of 1 mm to 1.5 mm below the notch. These facets are surrounded by regions of ductile fracture. Phosphorus is likely to be the main embrittling element in this case due to the fact that sulphur is precipitated from solid solution as MnS at 950°C. Note that the time required to produce cracking is approximately one order of magnitude greater than that for the 1300°C/300°C treatment, which has been attributed to sulphur segregation.

The C.P. steel subjected to the 950°C treatment showed no signs of cracking. The original MnS in the steel remains undissolved even after 2 hours at 950°C and so no deleterious effects due to sulphur segregation are likely to occur. The fine austenite grain size produced at 950°C also enhances resistance to phosphorus-induced intergranular separation compared with that of material with a large austenite grain size, because a fine grain size implies a large grain-boundary area per unit volume and hence a lower degree of coverage by a given amount of impurity.

Attention is now paid to the location of the cracks in the C.P. specimen, subjected to the 1300°C/300°C treatment and unloaded after 1.75 hr at 500°C (see Fig. 2). The intergranular cracks form well below the notch root, roughly at the position of peak stress, suggesting that its position is controlled either by the maximum tensile stress or by the maximum triaxial stress. Two possibilities arise: i) there is general grain boundary embrittlement and the cracks are located where it is because the high tensile stress below the notch is most easily able to produce cracking; ii) the impurity level is highest below the notch, because enhanced segregation has occurred, as a result of the high triaxiality which dilates the lattice. The second situation is analogous to that observed for hydrogen-induced cracking ("static fatigue") in notched bars subjected to steady load (11). Lattice dilation in this case provides extra driving force for hydrogen segregation, such that the local concentration C_L is greater than the uniform concentration C_o by the factor

$$C_L/C_o \propto \exp(\sigma_H \bar{V}/RT) \qquad (2)$$

where $\bar{V}$ is the partial molar volume, R is the gas constant, T is the absolute temperature and σ_H is the hydrostatic stress component = $\sigma_{ii}/3$. It is of interest to note that Troiano (11) observed similar "static fatigue" behaviour for a 5Cr-1Mo steel under constant load at 950°F, but did not report any metallographic observations.

To decide which of the two possibilities is more important in the present experiments, specimens have been simply heated for 2 hrs at 500°C in the absence of applied stress and have then been fractured at -196°C. No intergranular facets below the notch root have been seen, suggesting strongly that "stress-assisted-segregation" is responsible for the cracking at high temperature, but further observations are needed for final confirmation. It is also necessary to carry out Auger analysis of specimens unloaded from 500°C and fractured at -196°C, to confirm that sulphur is, indeed, the element responsible for intergranular embrittlement, as suggested by the heat-treatments. Dynamic segregation of sulphur has been invoked by Bowen et al. (3) to explain fatigue-crack growth behaviour in A533B at 290°C. Further analysis by Hippsley et al. (13) raises the possibility, however, that sulphur may segregate, not to the position of maximum triaxiality ahead of a crack tip (approximately two crack-opening displacements), but to a point somewhere between this position and the crack tip itself. The location of the cracks well below the notch, roughly at the position of maximum tensile stress in the present experiments is therefore intriguing, if the cause of the cracking is stress-assisted-segregation of sulphur.

CONCLUSIONS

i) The prior austenitising heat-treatment affects the susceptibility of a commercial purity 2¼Cr-1Mo steel to sustained load cracking at 500°C in vacuum.

ii) The combination of heat-treatment and impurity levels employed suggest that the cracking is due to the segregation, at 500°C of sulphur remaining in solid solution after quenching.

iii) It is inferred that the segregation is not general, but is "stress-assisted", initially localised to a region of high triaxiality below the notch root.

REFERENCES

1. J. Yu and C. J. McMahon Jr., *Met. Trans. A*, 11A, 277 (1980).
2. J. Yu and C. J. McMahon Jr., *Met. Trans. A*, 11A, 291 (1980).
3. P. Bowen, C. A. Hippsley and J. F. Knott, *Acta Met.*, 32, 637 (1984).
4. C. A. Hippsley, J. F. Knott and B. C. Edwards, *Acta Met.*,28, 869 (1980).
5. C. A. Hippsley, J. F. Knott and B. C. Edwards, *Acta Met.*, 30, 641 (1982).
6. E. T. Turkdogan, S. Ignatowicz and J. Pearson, *J. Iron Steel Inst.*, 180, 349 (1955).
7. B. J. Schulz and C. J. McMahon Jr., *Met. Trans.*, 4, 2485 (1973).
8. J. Q. Clayton and J. F. Knott, *Metal Science*, 16, 145 (1982).
9. G. Clark, R. O. Ritchie and J. F. Knott, *Met. Trans.*, 5, 782 (1974).
10. M. Wall and A. J. Foreman, Harwell Report AERE R11618 (1985).
11. A. R. Troiano, *Trans. ASM*, 52, 154 (1960).
12. R. A. Oriani and P. H. Josephic, *Acta Met.*, 22, 1065 (1974).
13. C. A. Hippsley, H. Rauh and R. Bullough, to be published Acta Met.

ACKNOWLEDGEMENTS

The authors would like to thank Professor R. W. K. Honeycombe, F.R.S. and Professor D. Hull for provision of research facilities. Assistance with mechanical testing was provided by Mr. T. G. Whitworth. Support for two of the authors was provided by a CASE award from SERC and B.P. Ltd. (MBDE) and by a NATO Postdoctoral Fellowship (JJL).

TABLE 1 Bulk analyses of Steels (Wt %)

	C	Mn	Si	S	P	Cr	Mo	Sb	Sn
C.P.	0.15	0.43	0.35	0.014	0.010	2.30	1.10	0.003	0.020
H.P.	0.14	0.63	0.03	0.005	0.001	2.15	1.10	0.001	0.003
P.D.	0.14	0.66	0.01	0.006	0.031	2.16	1.12	0.001	0.001

Fig.1 Notch region in 1300°C/950°C specimen after 18 hrs under load at 500°C

TABLE 2 Summary of results for SEN specimens held at 500°C under a load of 27.8KN

Purity	Hardness (VPN) *	Hardness (VPN) **	0.2% Proof stress at 500°C(MPa)	$\frac{\sigma_{nom}}{\sigma_{y(0.2)}}$	Time at 500°C(hrs)	Comments
$CP_{(1)}$	440	407	842	1.114	18	IG at root of notch. (Fig. 1) IG and ductile below notch. IG Facets clean at notch and contain small MnS away from the notch.
$CP_{(2)}$	459	337	595	1.576	24	No cracks
$CP_{(3)}$	361	340	570	1.646	2½	IG radial cracks from the root of the notch.
$CP_{(3)}$	361	345	570	1.646	1¾	Crack below the notch(Fig. 2)
$PD_{(3)}$	334	338	583	1.609	7	No cracks
$HP_{(3)}$	336	343	642	1.461	21	No cracks

Heat Treatments

(1) 1300°C 1 hr 950°C 2 hrs OQ
(2) 950°C 2 hrs OQ
(3) 1300°C 1 hr F.B. 300°C 15 mins OQ

*As quenched
**After exposure at 500°C

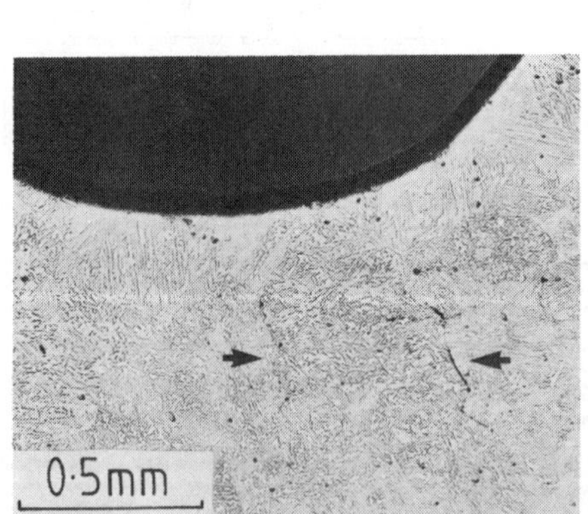

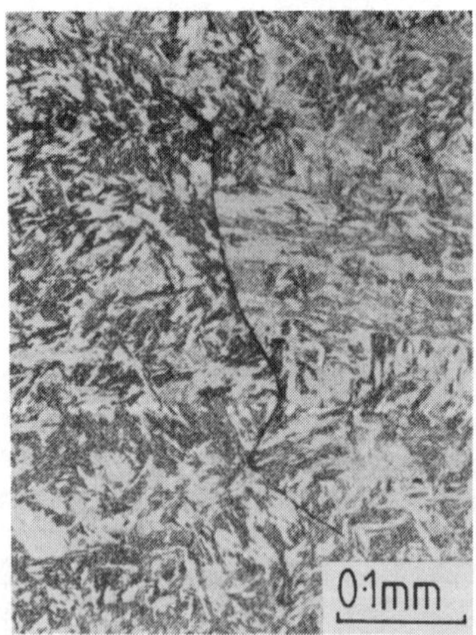

Fig.2 Sub-notch cracks in CP 1300°C/300°C specimen after 1.75 hrs under load at 500°C. Crack on right shown enlarged.

Role of Hydrogen, Sulfur and Other Impurities in Intergranular Fracture in Iron

K. S. Shin* and M. Meshii**

**Department of Mechanical and Aerospace Engineering, Arizona State University, Tempe, AZ 85287, USA*
***Department of Materials Science and Engineering, Northwestern University, Evanston, IL 60201, USA*

ABSTRACT

The intergranular fracture stress was investigated as a function of grain boundary concentrations of sulfur and/or H in polycrystalline MRC-VP iron specimens. The fracture energy was determined as a function of grain boundary sulfur concentration by analyzing the empirical relations with a modified Griffith equation. Two different effects of H, permanent and reversible, were identified. The permanent effect of H was caused by the formation of intergranular microcracks due to the precipitation of supersaturated H and recovered only partially after annealing at high temperatures. The size of microcracks was found to be proportional to the square of H concentration at grain boundaries. The predicted sizes of the microcracks agreed well with those determined metallographically. The present study clearly indicates that there is no synergistic effect between H and sulfur on intergranular fracture of iron regardless of whether the H effect is reversible or permanent. It was found that the primary effect of C is to displace sulfur from grain boundaries and thus to improve the grain boundary strength by reducing the embrittling effect of sulfur. The second effect of C is to increase the frictional stress to dislocation motion and thus to increase the apparent stress for intergranular fracture.

KEYWORDS

Intergranular fracture, H-induced fracture, grain boundary segregation

INTRODUCTION

The grain boundary segregation of impurity elements such as O, P and S is known to cause intergranular fracture in iron (1-9). H is also known to induce intergranular fracture in iron (9-13). A combined effect of H and an impurity element has been the subject of great interest for its practical importance. Two models, predicting additive effect (14,15) and synergistic effect (16), have been proposed for the combined effect on the grain boundary strength.

The presence of C, on the other hand, is known to suppress the intergranular fracture in iron (1-8). The mechanisms suggested for the grain boundary strengthening effect of C are an intrinsic strengthening effect (1-8) and an extrinsic effect either by the displacement of embrittling elements from grain boundaries by C (4,7) or by the scavenging effect due to C-embrittling impurity complexes (3,7).

The objective of the present study is to establish quantitative relations between the intergranular fracture stress and the grain boundary concentrations of sulfur and/or H. The combined effect of sulfur and H and the effect of C on the grain boundary strength of iron are examined from the quantitative analysis of the obtained empirical equations.

EXPERIMENTAL PROCEDURES

Cylindrical tensile specimens and Auger impact specimens were prepared from MRC-VP iron which contained 40 wt ppm sulfur. The interstitial impurity content of the specimens was reduced by purification in a dynamic ZrH_2 purification system. Some of the purified specimens were carburized with 40 wt ppm C (185 at ppm). In order to modify the amount of impurity segregation at grain boundaries, both purified and carburized specimens were quenched from various temperatures into a brine solution at 0° C in an argon atmosphere.

Two effects of H have been reported previously ; they are reversible and permanent effects. The details of reversible effect of H were reported elsewhere (9,17,18). The permanent effect was studied by quenching purified specimens in a H atmosphere. Some of these H_2-quenched specimens were also subsequently annealed at various temperatures for 3 hours in a vacuum of better than 1.3×10^{-4} Pa.

The effect of impurity segregation and/or H on the grain boundary strength was monitored by the fracture stress obtained from tensile tests at 77° K. All the specimens fractured intergranularly without macroscopic plastic deformation. The impurity segregation at grain boundaries was determined by Auger electron spectroscopy. The details of experimental procedures used have been reported elsewhere (9,17).

RESULTS AND DISCUSSION

Effect of Sulfur Segregation

Fig. 1 shows the dependence of the fracture stress of Ar-quenched specimens on sulfur concentration at intergranular surface. The fracture stress σ_F decreased linearly with the sulfur concentration according to the relation

$$\sigma_F \text{ (MPa)} = 573 - 16\, C_s \qquad (1)$$

where C_s is the sulfur concentration at intergranular surface in percent of a monolayer of sulfur. If it is assumed that sulfur distributes equally at both sides of intergranular fracture surface, the fracture stress decreased by 8 MPa per 1% monolayer of sulfur at grain boundaries. The fracture energy, $\gamma = \gamma_s - \gamma_{gb}/2 + \gamma_{pl}/2$, can be calculated as a function of C_s from Eq. (1) assuming a suitable fracture model (17,18) as

$$\gamma = (D/12E)(573 - 16C_s - \sigma_i)^2 \qquad (2)$$

where D, E and σ_i are grain size, Young's modulus and frictional stress for dislocation motion. Fig. 2 illustrates this relation for various σ_i using E = 2.2×10^{11} MPa and D = 200 μm. σ_i is thought to be close to the micro-yield stress which have been reported to be 137 MPa at 77° K in purified Ferrovac E iron (19).

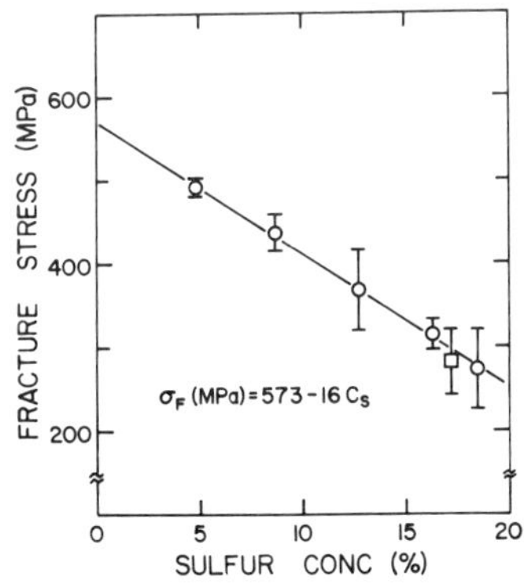

Fig. 1 Dependence of the fracture stress of Ar-quenched specimens on the sulfur concentration at intergranular surface

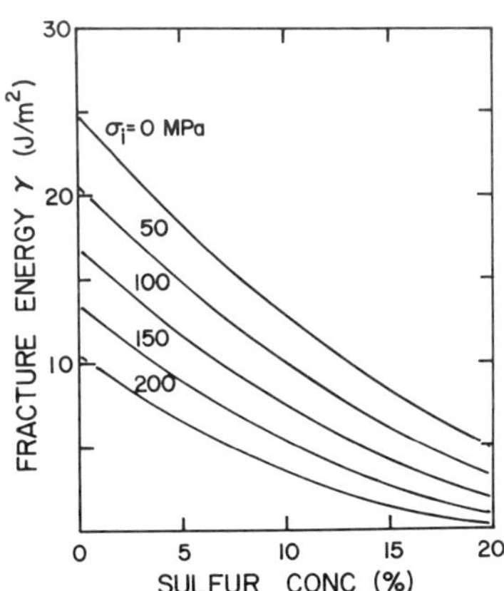

Fig. 2 Dependence of the fracture energy on the sulfur concentration at intergranular surface.

The fracture energy decreases rapidly with sulfur concentration in Fig. 2. The large portion of this decrease comes from a change in plastic work, γ_{pl}. γ_{pl} decreases rapidly with decreasing surface energy (20,21). One possible form of the relation is given by (21)

$$\gamma_{pl} = K\ (\gamma_s - \gamma_{gb}/2)^{m+1} \tag{3}$$

where m was reported to be 6.5 for iron at 77° K (22). The values reported for γ_s and γ_{gb} in the literature are 1.95 and 0.78 J/m², respectively (23). The surface energy for a clean grain boundary, therefore, would be 1.56 J/m². Assuming K = 1, one obtains γ_{pl} = 28 J/m² and thus the fracture energy of 15.6 J/m² for the clean boundary according to Eq. (3). This value agrees with 14 J/m² obtained from the present study with C_s = 0 and σ_i = 137 MPa in Eq. (2).

Effect of Hydrogen

The atomic fraction of H in iron, C_o, in equilibrium with H gas at pressure P (in atm.) is expressed by (24)

$$C_o = 0.00185\ \sqrt{P}\ \exp\ (-3440/T) \tag{4}$$

where T denotes temperature in K. The H concentration at the grain boundary, C_H can be estimated according to the equation

$$C_H/(1-C_H) = \{C_o/(1-C_o)\}\ \exp\ (H_B/RT) \tag{5}$$

When a small amount of H (~ 1 at ppm) was introduced into iron specimens, the effect of reducing the grain boundary strength was reversible (9). No synergistic relation was found between this effect and an effect caused by S at grain boundaries (17). When an iron specimen was quenched from a temperature above 560° C in H_2 gas, a permanent reduction in fracture stress resulted. The fracture stresses of H_2-quenched specimens are plotted against quenching temperature in Fig. 3. In H_2-quenched specimens, the quenching temperature determines not only H concentration but the amount of sulfur segregation at grain boundaries. The combined effect can be expressed as (9)

$$\sigma_F \text{ (MPa)} = 573 - 16\, C_S \text{ (\% monolayer)} - 2.42\, C_H \text{ (at \%)} \tag{6}$$

The permanent effect of H is caused by microcracks at grain boundaries resulting from H precipitation. Metallographic examination of H_2-quenched specimens confirmed intergranular microcracks. The relationship between the size of microcracks and the H concentration at grain boundaries can be obtained with the Griffith criterion (25). The stress, σ_F, required to propagate a crack of length 2C in the plane strain condition is given by

$$\sigma_F = [4G\gamma/(1-\nu)\pi C]^{1/2} \tag{7}$$

where G and ν are shear modulus and Poisson's ratio, respectively. The fracture energy in Eq. (7) depends on the S concentration (Eq. (2) and Fig. 2). In Fig. 4, the size of microcracks, calculated from Eq. (7) with the fracture stress of H_2-quenched specimens, is plotted against the square of the H concentration at grain boundaries using 137 MPa and 200 μm for σ_i and D in Eq. (2). The relation can be represented by a straight line, indicating that the crack size is proportional to the square of the H concentration, hinting that the rate of H precipitation reaction ($H + H \rightarrow H_2$) has the controlling role in the crack formation.

The calculated size of microcracks ranges from 17.8 to 45.5 μm depending upon quenching temperature. In the Table, the average sizes of microcracks determined metallographically are compared with those predicted from the linear relation in Fig. 4.

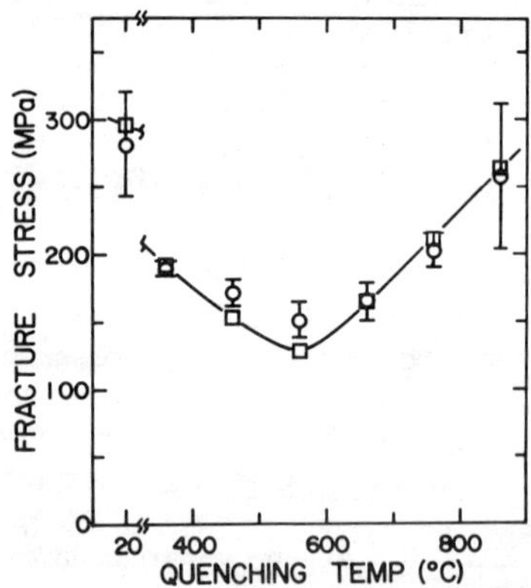

Fig. 3 The fracture stresses of H_2-quenched specimens (circles) are compared with those calculated from Eq. (6) (squares).

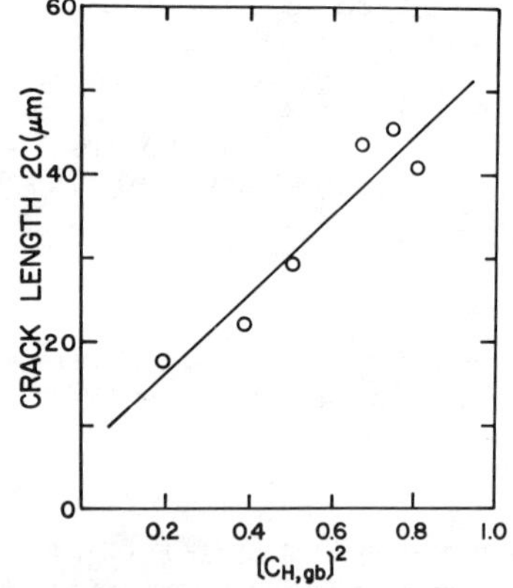

Fig. 4 The microcrack length 2C of H_2-quenched specimens estimated from Eq. (7) is plotted against the square of the grain boundary hydrogen concentration.

Table. Comparison of the sizes of microcrack (μm)

Quenching temperature ($^{\circ}$ C)	560	660	760	860
Predicted	32.9	38.5	42.3	44.9
Measured	25.6	28.4	32.6	35.4

In Fig. 5, the fracture stresses of H_2-quenched specimens which were subsequently annealed in vacuum are plotted against the initial quenching temperature. The vacuum annealing at 200° C hardly affected the fracture stress. All H atoms trapped at grain boundaries must be completely eliminated from the specimens during this treatment. The minimum in the fracture stress around 560° C in Fig. 5 did not disappear even after annealing at 500° C, indicating that the annealing at 500° C for 3 hours was not sufficient to change the amount of S segregation at grain boundaries which was caused by the initial quenching. The fracture stress after vacuum annealing at 800° C decreased monotonically with quenching temperature. The amount of S segregation at grain boundaries is considered to be the same for all specimens. Therefore, the monotonic decrease in the fracture stress is caused by an increase in the microcrack size. The annealing at 800° C apparently did not completely sinter the microcracks. Detailed analysis indicates that the ratio of the crack lengths after and before vacuum annealing at 800° C is ~ 0.4 regardless of the initial quenching temperature (18).

Effect of Carbon

The effect of C on the grain boundary strength of iron was examined from the comparison of the fracture stress-S concentration relations for both purified specimens and carburized specimens (185 at ppm C) quenched from the same temperature in an Ar atmosphere. It has been reported that C can displace sulfur from grain boundaries and improve grain boundary strength (17). In order to examine if C has an intrinsic grain boundary strengthening effect, the fracture stresses of carburized specimens are plotted in Fig. 6 against intergranular sulfur concentration. The result indicates that, at the same sulfur concentration, the fracture stresses of carburized specimens are always higher than those of purified specimens by 40 ~ 55 MPa. However, this does not necessarily indicate that the C has an intrinsic

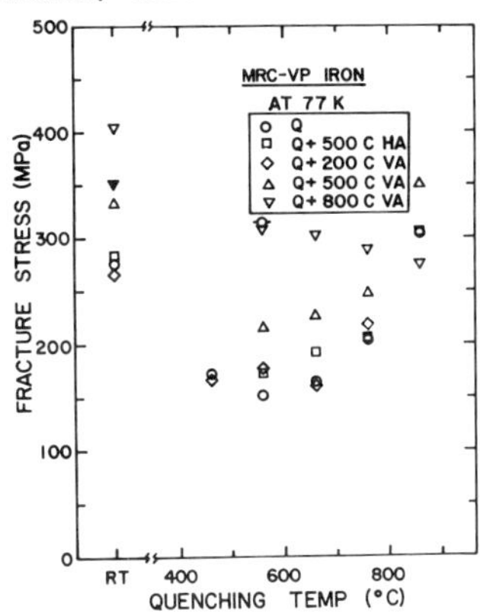

Fig. 5 The fracture stress of H_2-quenched specimens which were subsequently annealed in vacuum is plotted against the previous quenching temperature.

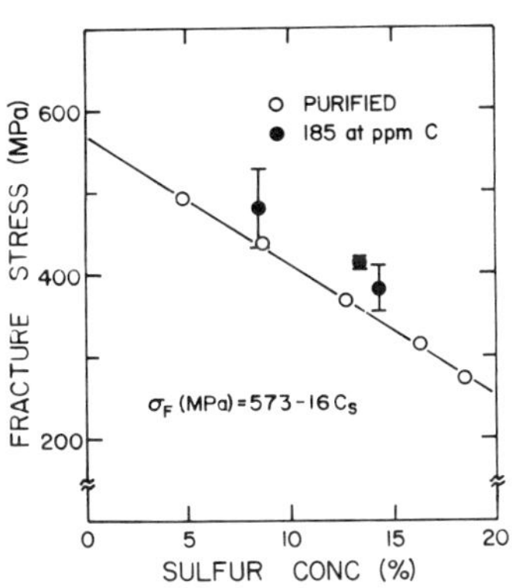

Fig. 6 Dependence of the fracture stress of carburized specimens on the sulfur concentration at intergranular surface.

grain boundary strengthening effect. The C segregation also depended on quenching temperature; the segregation was more pronounced for quenches from 560° C than those from 660° C and 760° C. An increase in frictional stress for edge dislocation motion due to C should influence the dislocation pile-up and therefore the stress concentration (9,17). The microyield stress of iron at 77° K increases by 50 MPa with 50 wt ppm C (19,26). If the micro-yield stress increases linearly with C concentration, the microyield stress of the present carburized specimens containing 40 wt ppm C is expected to be 40 MPa greater than that of purified specimens. This agrees well with the observed difference of 40 ~ 55 MPa in the fracture stress of carburized specimens compared to purified specimens. It is suggested, therefore, that the primary effect of C is to displace sulfur, whereas the second effect of C is to increase the frictional stress to dislocation motion and to reduce the stress concentration for intergranular fracture.

ACKNOWLEDGEMENTS

This work was supported by the Division of Materials Science, U. S. Department of Energy. One of us (K.S.) gratefully acknowledges the assistance of the Faculty Grant-in-Aid Program, Arizona State University.

REFERENCES

1. J. R. Low and R. G. Feustel, Acta Metall. 1 (1953) 185.
2. C. J. McMahon, Jr., Acta Metall. 14 (1966) 839.
3. J. R. Rellick and C. J. McMahon, Jr., Metall. Trans. 1 (1970) 929.
4. H. Erhart and H. J. Grabke, Metal. Sci., 15 (1981) 401.
5. K. Abiko, S. Suzuki and H. Kimura, Trans. Japan Inst. Metals 23 (1982) 43.
6. S. Suzuki, M. Obata, K. Abiko and H. Kimura, Scripta Metall. 17 (1983) 1325.
7. C. Pichard, j. Rieu and C. Gous, Metall. Trans. A 7A (1976) 1811.
8. G. Tauber and H. J. Grabke, Ber. Bunsenges Phys. Chem., 82 (1978) 298.
9. K. S. Shin and M. Meshii, Acta Metall. 31 (1983) 1559; Scripta Metall. 17 (1983) 1121.
10. I. M. Bernstein, Scripta Metall. 8 (1974) 343.
11. M. Cornet and S. Talbot-Besnard, in Environment-Sensitive Fracture of Engineering Materials, Z. A. Foroulis ed., TMS-AIME (1979) p. 411.
12. S. Moriya, H. Matsui and H. Kimura, Mat. Sci. Eng., 40 (1979) 217.
13. R. H. Jones, S. M. Bruemmer, M. T. Thomas and D. R. Baer, Met. Trans. A, 12A (1981) 1621; Scripta Metall. 16 (1982) 615; Metall. Trans. A 13A (1982) 241.
14. K. Yoshino and C. J. McMahon, Jr., Metall. Trans. 5 (1974) 363.
15. J. Kameda and C. J. McMahon, Jr., Metall. Trans. A 14A (1983) 903.
16. R. M. Latanision and H. Opperhauser, Jr., Metall. Trans. 5 (1974) 483; Metall. Trans. A 6A (1975) 233.
17. K. S. Shin and M. Meshii, in Synergism of Microstructure, Mechanisms and Mechanics in Fracture, J. Landes, R. Stoltz and J. M. Wells, ed., TMS-AIME, in print.
18. K. S. Shin, "Effect of Impurity Segregation and Hydrogen on the Grain Boundary Strength of Iron", Ph.D. Dissertation, Northwestern University, 1984.
19. A. Sato and M. Meshii, phys. stat. sol. (a) 20 (1975) 561.
20. M. E. Fine and H. L. Marcus, Metall. Trans. 2 (1971) 1473.
21. C. J. McMahon, Jr. and V. Vitek, Acta Metall. 27 (1979) 507.
22. D. S. Tomalin and C. J. McMahon, Jr., Acta Metall. 21 (1973) 1189.
23. M. C. Inman and H. R. Tipler, Metall. Rev. 8 (1963) 105.
24. N. R. Quick and H. H. Johnson, Acta Metall. 26 (1978) 903.
25. A. A. Griffith, Phil. Trans. Roy. Soc. Lon., 221A (1920) 163.
26. H. D. Solomon and C. J. McMahon, Jr., Acta Metall. 19 (1971) 291.

Proprieties Mecaniques d'Aciers Inoxydables Austenitiques Stable (ZXNCTD 26–15) et Instable (Z2CN 18–10). Role des Traitements Thermiques et de l'Hydrogene Cathodique

P. Huwart*, M. Habashi*, J. P. Fidelle*, P. Garnier** et J. Galland***

**Laboratoire "Corrosion et Fragilisation par l'Hydrôgène",*
***Laboratoire "Chimie Physique du Solide", Ecole Centrale des Arts et Manufactures, 92290 Châtenay-Malabry, France*
****Commissariat à l'Energie Atomique, Service de Métallurgie, B.P. 511, 75752 Paris cedex 15, France*

ABSTRACT

The aim of this study is to decline brittle fracture resistance due to ageing temperature (600°C - 900°C interval) and cathodic hydrogen effects in austenitic stainless steels AISI 304L and AISI 286A modified, using tensile tests and impact energy measurements (Charpy test) at liquid nitrogen temperature.

MOTS CLES

Aciers inoxydables austénitiques, transformation martensitique, traitement de sensibilisation, précipitations, hydrogène.

I INTRODUCTION

Il existe beaucoup d'exemples, dans l'histoire des matériaux industriels, de ruptures catastrophiques dues à la fragilisation par l'hydrogène (FPH). Les aciers inoxydables austénitiques ont longtemps été considérés comme insensibles à l'hydrogène (Blanchard, 1960) ; cependant, leurs conditions d'emploi de plus en plus sévères ont rendu indispensable une analyse plus fine de leur comportement vis-à-vis de l'hydrogène. Aujourd'hui la fragilisation par l'hydrogène des aciers inoxydables austénitiques est un phénomène reconnu (Briant, 1980 ; Odegard, 1980 ; Andriamiharisoa, 1981 ; Shehu, 1981 ; Habashi, 1982), bien que le mécanisme en demeure obscur dans bien des cas.

II MATERIAUX ET METHODES EXPERIMENTALES

Les compositions chimiques des aciers étudiés figurent au tableau 1 ; il s'agit dans tous les cas de coulées industrielles.

TABLEAU 1. Composition chimique des aciers inoxydables austénitiques étudiés

élément % / acier	C	Ni	Cr	Mn	Si	S	P	Mo	V	Ti	Al	B
Z2CN 18-10	0,026	10,38	18,52	1,79	0,345	0,011	0,023	0,11	—	0,005	—	—
X4NCTD 26-15	0,038	26,2	15,2	1,27	0,71	0,003	0,018	1,11	0,25	2,1	0,025	0,006
X1NCTD 26-15	0,007	26,1	15,1	0,010	0,008	0,003	0,003	1,13	0,217	2,0	0,025	0,006

Les essais de traction ont été effectués sur des éprouvettes plates d'épaisseur 2 mm et de longueur utile 60 mm, avec une vitesse de déformation de l'ordre de $10^{-4}.s^{-1}$, et les essais de résilience sur des éprouvettes Charpy V normalisées (10x10x55 mm). Les éprouvettes ont subi les traitements thermiques, sous argon, suivants :

Pour l'acier inoxydable Z2CN 18-10 :
- hypertrempe à l'eau après mise en solution une heure à 1050°C ;
- traitement thermique de sensibilisation avec montée en température à 300°C/h jusqu'à une température T_S comprise entre 650°C et 850°C, maintien à cette température pendant 3 heures puis refroidissement au four.

Pour l'acier ZXNCTD 26-15 :
- hypertrempe (980°C/1h, trempe à l'huile)
- hypertrempe + traitement thermique de vieillissement avec maintien 16 heures à une température T_V comprise entre 500°C et 980°C puis refroidissement au four.

Les éprouvettes de traction et de résilience ont été chargées cathodiquement avant essai dans un bain de sels fondus à un potentiel de -2 Volts/Ag, à 300°C et pendant 20 heures (Andriamiharisoa, 1981). Le rapport C/Co entre les concentrations en hydrogène "à coeur" et à la surface de l'éprouvette de traction, dans ces conditions de chargement, est de l'ordre de 0,1 (Crank, 1975).

III.RESULTATS EXPERIMENTAUX

III.1 ESSAIS DE TRACTION EN PRESENCE OU NON D'HYDROGENE INTERNE

III.1.a ACIER Z2CN 18-10

Les essais de traction effectués à la température ambiante et en absence d'hydrogène montrent que le phénomène de sensibilisation n'apparaît pas nettement ; par contre il existe un domaine de températures de sensibilisation (700°C à 800°C) pour lesquelles le métal est particulièrement sensible à l'hydrogène, fig. 1. On enregistre au niveau de la contrainte réelle maximale R, de l'allongement rationnel maximal ${}_m$ et de la striction à la rupture Z_r, des variations notables en fonction de la température de sensibilisation, et aussi selon que le métal a été ou non chargé en hydrogène. Nous présentons ici la variation de R avec T_S à fig. 2. Il s'avère que le phénomène de sensibilisation est à l'origine d'une baisse des caractéristiques mécanique de cet acier. Dans le cas du métal exempt d'hydrogène, cette baisse intervient dans un domaine de température relativement peu étendu, 50°C, de part et d'autre d'une température dite de

sensibilisation optimale (T_S 800°C), qui correspond à la présence aux joints de grains d'une précipitation continue de carbures de chrome.
Dans le cas du métal chargé en hydrogène le phénomène s'étend sur un domaine de températures beaucoup large, allant de 600°C à 870°C environ, fig. 2. Dans ce domaine de températures, R et $_m$ sont nettement abaissés. Il semble donc que la présence de la martensite formée autour des joints de grains, conséquence de la précipitation, joue le rôle de court-circuit de diffusion d'hydrogène produisant ainsi la fragilisation des joints. Les interactions du phosphore ségrégé aux joints de grains avec l'hydrogène peuvent encore rendre ces joints moins résistant (Huwart, 1984).

III.1.b ACIERS ZXNCTD 26-15

TABLEAU 2. Propriétés mécaniques des nuances d'acier ZXNCTD 26-15 rôles de l'état structural et de l'hydrogène interne. Essais de traction à 20°C et à -196°C.

Paramètre / Acier	Re MPa	R MPa	m	F%	Etat structural
Z1NCTD 26-15	250	700	0,28	15	Austénitique
	700	1100	0,13	45	Vieilli à 720C/16h
Z4NCTD 26-15	320	700	0,28	10	Austénitique
	750	1100	0,13	28	Vieilli à 720C/16h

TABLEAU 2 b Propriétés mécaniques des nuances d'acier ZXNCTD 26-15 vieillies à 720°C/16h : rôle de l'hydrogène interne. Essais de traction à -196°C.

Paramètre / Acier	Re MPa	R MPa	m	F%	Etat structural
Z1NCTD 26-15	850	1150	0,21	11	Vieilli à 720°C/16h
Z4NCTD 26-15	800	1500	0,18	10	Vieilli à 720°C/16h

Les tableaux 2a et 2b montrent la variation des propriétés mécaniques des aciers ZXNCTD 26-15, en présence et en absence d'hydrogène. Il faut noter que la limite d'élasticité de l'acier Z4NCTD26-15 austénitique ou vieilli est supérieure à celle ce l'acier Z1NCD 26-15 dans les mêmes conditions, vraisemblablement en raison de la teneur plus élevée en carbone dans la première nuance. La ductilité des deux nuances vieillies et tractionnées à -196°C est légèrement supérieure à celle mesurée à la température ambiante. Il semble que cette augmentation soit due au mode de déformation plastique par mâclage et que ce mode soit favorisé par l'abaissement de la teneur en carbone. La fragilisation par l'hydrogène de l'acier ZXNCTD 26-15 est plus grande à la température ambiante qu'à -196°C (Holbrook, 1980). Par contre, la nuance d'acier vieilli ayant une teneur élevée en carbone, possède un indice de fragilisation F% plus faible que celui de l'acier Z1NCTD 26-15 vieilli.

L'énergie de défauts d'empilement est élevée lorsque l'austénite contient des teneurs élevées en nickel et en carbone. Cette forte énergie favorise le glissement déviés des dislocations (JAOUL, 1969). Les dislocations peuvent repartir de façon uniforme l'hydrogène dans la matrice austénitique, la concentration d'hydrogène C_H est alors faible et la fragilisation par l'hydrogène n'est pas favorisée. Par ailleurs le traitement de vieillissement (720°C/16h) provoque la formation de précipités ' [Ni_3(Al, Ti)] qui gênent le mouvement de dislocations et appauvrissent la matrice en nickel. L'énergie de défauts d'empilement est alors diminuée et le glissement dévié devient difficile favorisant ainsi l'augmentation de C_H aux obstacles pour atteindre C_K nécessaire à fragiliser ces obstacles (PRESSOUYRE, 1982).

III.2 ESSAIS DE RESILIENCE A -180°C EN PRESENCE OU NON D'HYDROGENE INTERNE

III.2.a ACIER Z2CN 18-10

Dans des travaux antérieurs, par des essais de résilience à basse température, (Habashi, 1982 ; Huwart, 1984) nous avons montré que l'acier Z2CN 18-10 hypertrempé à 1050°C/1h, puis soumis à un revenu pendant 3 heures à une température T_S comprise entre 600°C et 850°C, est le siège d'un phénomène de précipitation de carbures de chrome aux joints de grains.

L'importance du phénomène (sensibilisation) est optimale à T_S 750°C, fig. 3. La présence d'hydrogène interne dans l'austénite sensibilisée abaisse l'énergie de rupture par choc et élargit le domaine de températures dans lequel cette baisse intervient (650°C à 870°C), fig. 3. Ces résultats présentent de grandes analogies avec ceux obtenus lors des essais de traction à -196°C (fig. 2). Afin de pouvoir comparer correctement ces deux types d'essais, nous avons à partir des courbes de traction, calculé la densité d'énergie de déformation $W(T_S)$, c'est-à-dire, l'aire située sous la courbe (force-déplacement) rapportée au volume utile de l'éprouvette de traction correspondante. Les courbes obtenues pour le métal chargé ou non en hydrogène, fig. 4, sont tout-à-fait en accord avec les courbes de résilience.
Par conséquent les interactions hydrogène-dislocations mobiles ne jouent pas un rôle primordial dans la fragilisation par l'hydrogène de l'acier Z2CN 18-10 sensibilisé puisque celle-ci peut être mise en évidence même par l'essai de résilience. L'hydrogène provoque donc un endommagement lors de sa diffusion dans la martensite.

III.2.b ACIERS ZXNCTD 26-16

La figure 5 montre que l'énergie de rupture par choc varie en fonction de la température de vieillissement de la même façon pour les deux nuances d'acier non chargées en hydrogène : on enregistre une baisse brutale de 500°C à 700°C puis une diminution plus faible de 700°C à 800°C. La baisse brutale est plus accentuée pour l'acier moins chargé en carbone ; à partir de 800°C on enregistre à nouveau une augmentation de l'énergie de rupture jusqu'à 980°C. Ajoutons que les courbes d'évolution de l'énergie de rupture par choc de ces aciers avec la température de vieillissement présentent une grande analogie avec les courbes de dilatométrie obtenues par ailleurs.
Le chargement en hydrogène interne d'éprouvettes Charpy V normalisées ne modifie pas leur énergie de rupture, fig. 5, contrairement au cas de l'acier Z2CN 18-10.
La diffusion de l'hydrogène, au cours du chargement cathodique, dans l'austénite stable vieillie entre 500°C et 980°C, c'est-à-dire contenant les phases et ne provoque donc aucun endommagement dans l'acier ZXNCTD 26-15.
Les valeurs de la densité d'énergie de déformation calculées à partir des courbes de traction sont représentées sur la fig. 6. La fragilisation par

l'hydrogène de l'acier ZXNCTD 26-15 est maximale à 20°C pour le métal vieilli ; la chute de cette énergie due au chargement en hydrogène est la même pour les deux aciers.
Il semble donc que les interactions hydrogène-dislocations mobiles, dans la structure austénitique stable et au niveau des interfaces précipités-matrice, jouent un rôle important dans la fragilisation par l'hydrogène interne de l'acier type ZXNCTD 26-15.

IV CONCLUSIONS

D'après les résultats obtenus à l'issue de cette étude nous pouvons dégager les conclusions suivantes :
1- l'essai de résilience à basse température est un essai suffisamment sensible pour représenter la variation de la microstructure des aciers inoxydables austénitiques stables ou non avec les traitements de vieillissement ou de sensibilisation. De plus, compte tenu que la vitesse de déformation est très élevée dans ces essais, les interactions hydrogène-dislocations sont inhibées dans l'acier chargé en hydrogène interne. Par contre, lorsque l'hydrogène provoque un endommagement dans l'acier lors du chargement, l'essai de résilience peut mettre en évidence l'ampleur de cet endommagement : c'est le cas de l'acier Z2CN 18-10 sensibilisé.

2- La densité d'énergie de déformation plastique (aire sous toute la courbe de traction) W(Ts) donne une valeur exacte de la fragilisation par l'hydrogène. Les variations de la résilience et de $W(T_S)$ avec la température de sensibilisation de l'acier Z2CN 18-10, présentent une grande analogie.

REFERENCES

Andriamiharisoa, H. (1981). Thèse de Docteur-Ingénieur, Ecole Centrale, France.
Blanchard, P., A.R. Troiano, (1960). Mémoires scientifiques de la Revue de Métallurgie LVII, n°6, p. 409.
Briant, C.L. (1980). "Hydrogen effectc in metals", AIME publ. p. 527.
Crank, J. (1975). The mathematics of diffusion. Clarendon Press, Oxford.
Habashi, M., J. Galland, (1982). Mémoires et études scientifiques de la Revue de métallurgie, juin, p. 311.
Habashi, M., I. Nedbal, J. Galland, (1982). Congrès "Hydrogène et Matériaux". Ed. Azou, P., Châtenay-Malabry, France. Réf. F10.
Holbrook, J.H., and A.J. West, (1980). Third Int. Conf. "Effect of hydrogen on behavior of materials", Pergamon Press.
Huwart, P., M. Habashi, M. Tvrdy J. Galland, (1984). Sixth Int. Conf. on fracture, New Delhi, Inde, Vol. 4, p. 2435.
Huwart, P. (1984). Thèse de Docteur-Ingénieur, Ecole Centrale des Arts et Manufactures, Châtenay-Malabry, France.
Jaoul, B. (1965). Etude de la plasticité et application aux métaux. Ed. Dunod, Paris.
Odegard, B.C., A.J. West, (1980). Third Int. Conf. on effect of hydrogen on behavior of materials, Pergamon Press.
Shehu, Y. (1981). Thèse de Docteur Ingénieur, Université Paris XI, France.
Pressouyre, G.M., Dollet, J., Vieillard-Baron, B, (1982). Memoires et études scientifiques de la revue de métallurgie, avril p. 161.

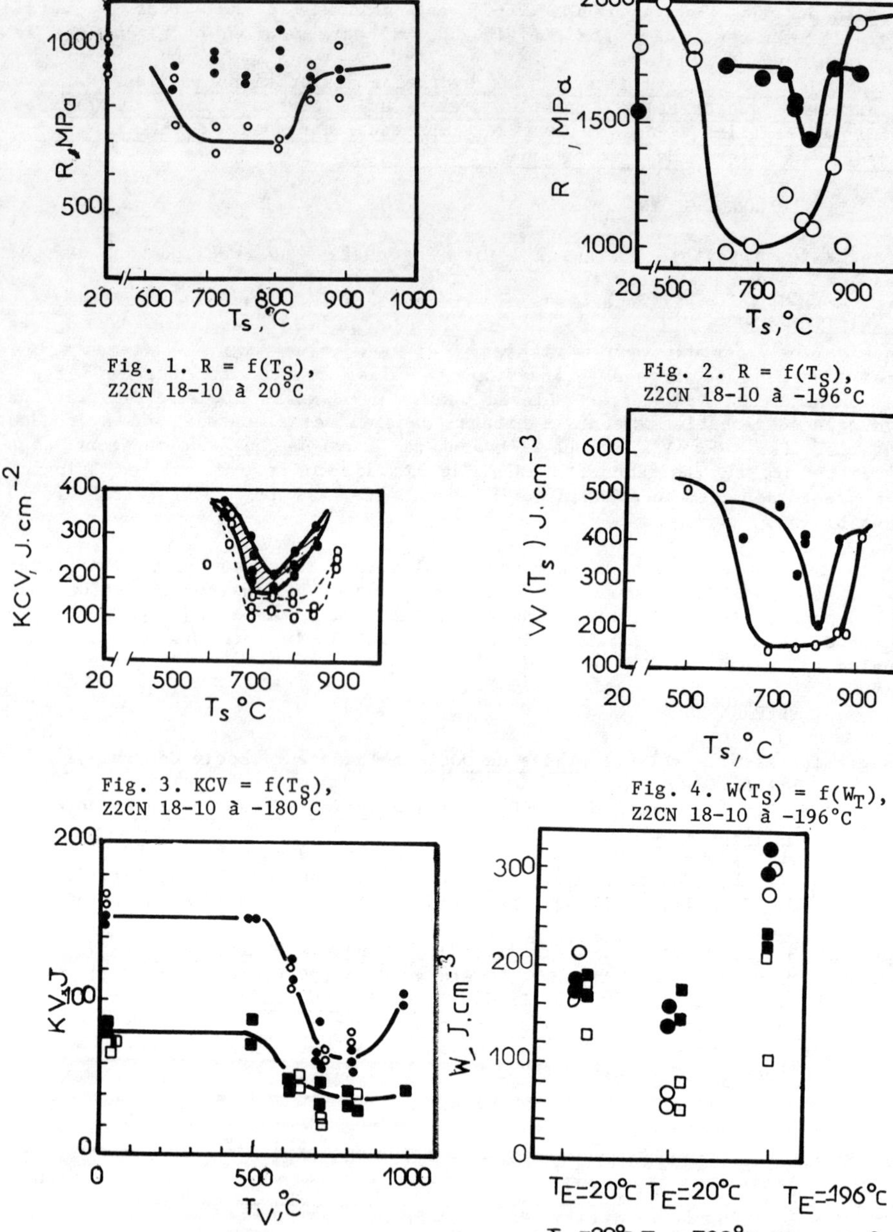

Fig. 1. $R = f(T_S)$, Z2CN 18-10 à 20°C

Fig. 2. $R = f(T_S)$, Z2CN 18-10 à -196°C

Fig. 3. $KCV = f(T_S)$, Z2CN 18-10 à -180°C

Fig. 4. $W(T_S) = f(W_T)$, Z2CN 18-10 à -196°C

Fig. 5. $KCV = f(T_V)$, ZXNCTD 26-15 à -180°C

■,● sans H; □, o avec H

●, o aciers Z2CN 18-10 et Z1NCTD 26-15

■, □ acier Z4NCTD 26-15

Fig. 6. $W(T_S) = f(T_V)$, ZXNCTD 26-15 à -196°C

Influence de l'Allongement des Inclusions au Cours du Corroyage sur l'Anisotropie de Ductilite et l'Amorcage de Defauts dans des Barres d'Aciers de Construction Mecanique

F. Moussy

Institut de Recherches de la Sidérurgie Française (IRSID) 185 rue du Président Roosevelt 78105 Saint Germain en Laye, France

RESUME

La rupture ductile des aciers en sollicitation statique est le résultat final du développement d'un endommagement constitué par des microcavités amorcées sur des inclusions. De nombreux aciers présentent une anisotropie de ductilité liée à l'anisotropie de forme des inclusions.

- Une approche qualitative constituée d'une observation des mécanismes de déformation en Microscopie Electronique à Balayage (MEB) montre que l'anisotropie macroscopique de ductilité dépend essentiellement des hétérogénéités de déformation autour des inclusions. Les mécanismes en compression parallèlement ou perpendiculairement à la direction de laminage sont analogues à ceux en traction perpendiculairement à la direction de laminage.
- Une approche quantitative consiste à caractériser la ductilité en traction et en compression parallèlement ou perpendiculairement à la direction de laminage. Les résultats montrent que les ductilités en compression long et travers sont égales et corrélées à la ductilité en traction travers, et que l'anisotropie de ductilité est acquise dès les premiers stades de corroyage et n'évolue plus avec un corroyage ultérieur.

L'ensemble de ces résultats permet de mieux comprendre les causes de l'anisotropie des aciers en ductilité, ténacité et fatigue. La directionalité de toutes ces propriétés dépend de l'orientation relative des inclusions par rapport aux directions principales du tenseur des contraintes ou des déformations.

MOTS CLE

Fissuration, ductilité, inclusions, anisotropie, endommagement.

INTRODUCTION

La rupture ductile d'un acier est toujours le résultat de l'apparition et de la propagation d'une macrofissure interne ou externe, que la pièce étudiée soit initialement entaillée ou non. L'apparition de cette macrofissure est le résultat final du développement d'un endommagement hétérogène non localisé en début de déformation et qui devient brutalement instable.

Cet endommagement est constitué par l'amorçage, la croissance et la coalescence de microcavités autour de particules de seconde phase, particules qui, en début de déformation, sont essentiellement des inclusions non métalliques. L'anisotropie de ductilité, de ténacité et de résistance à la fatigue est donc gouvernée principalement par la forme des inclusions dans le produit fini. Alors que les tôles fortes qui possèdent trois directions principales d'anisotropie ont fait l'objet de nombreuses études essentiellement en sollicitation de traction, les produits longs ne possédant que deux directions principales d'anisotropie n'ont fait l'objet que de peu de travaux. De plus les barres sont souvent destinées à être forgées ou filées à froid et sont donc soumises à des états de contrainte et de déformation triaxiaux très sévères.
Aussi l'aptitude à la déformation de ces produits en sollicitation de traction ou de compression est déterminante tant du point de vue de la mise en forme que du point de vue des propriétés d'emploi de la pièce formée. La connaissance d'une part qualitative des mécanismes de déformation à l'échelle microscopique, d'autre part de l'évolution de la ductilité en fonction de paramètres microscopiques permet d'obtenir une analyse complète et générale des phénomènes, cette analyse pouvant être transposée à d'autres aciers et modes de sollicitations que ceux étudiés dans cette recherche.

MECANISMES DE DEFORMATION

Les aciers étudiés sont des nuances contenant 0,3 à 0,4% de Carbone.

- Déformation en traction

La méthode d'observation discontinue en MEB d'éprouvettes polies a été décrite dans une publication antérieure [1] ; elle permet de suivre le voisinage d'une même inclusion à différentes étapes de la déformation.

* <u>Traction parallèlement à la direction longitudinale</u>. En traction long, les inclusions sont parallèles à la surface de l'éprouvette. Seule une matrice à perlite globulisée a été observée. Dès le début de la déformation, les inclusions se fragmentent perpendiculairement à la direction de traction. Les cavités ainsi formées s'allongent mais la croissance latérale n'intervient que tardivement (fig. 1) ; c'est cette dernière croissance qui conduit à la coalescence des cavités et à la rupture de l'acier. Autour d'une inclusion, la déformation est sensiblement homogène. Les bandes de glissement sont sinueuses ; ceci est dû probablement à la morphologie de la cémentite sous forme de globules finement dispersés. Le rôle de ces globules est de freiner le développement de bandes de glissement rectilignes et espacées, ce qui favorise une meilleure homogénéité des déformations à l'échelle microstructurale. La coalescence s'effectue par déchirure perpendiculairement à la direction de traction, par arrachement et non par cisaillement.

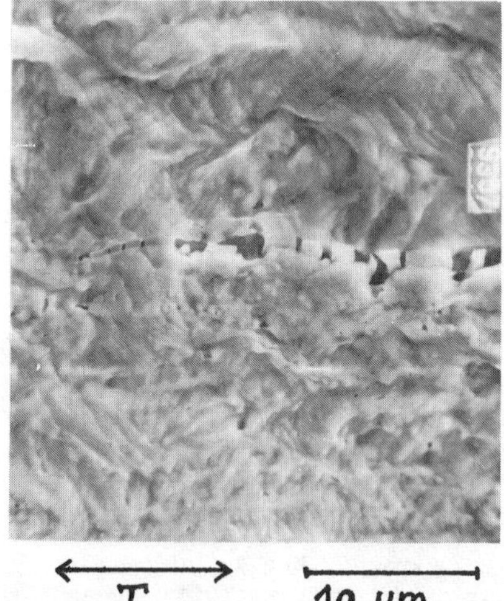

Fig. 1 : Mécanisme de déformation en traction long.

* <u>Traction perpendiculairement à la direction longitudinale</u>. Dans ce cas, les inclusions peuvent être soit parallèles, soit perpendiculaires au plan de l'éprouvette. C'est ce dernier cas qui a été retenu.
Trois états microstructuraux ont été étudiés : perlite lamellaire, perlite globulisée et martensite revenue. Pour les deux premiers, un mécanisme de déformation très caractéristique apparaît (fig. 2). Les microcavités amorcées sur les inclusions sont toujours accompagnées du développement de une, deux, trois ou quatre bandes de glissement formant un angle de 45° avec la direction de traction.

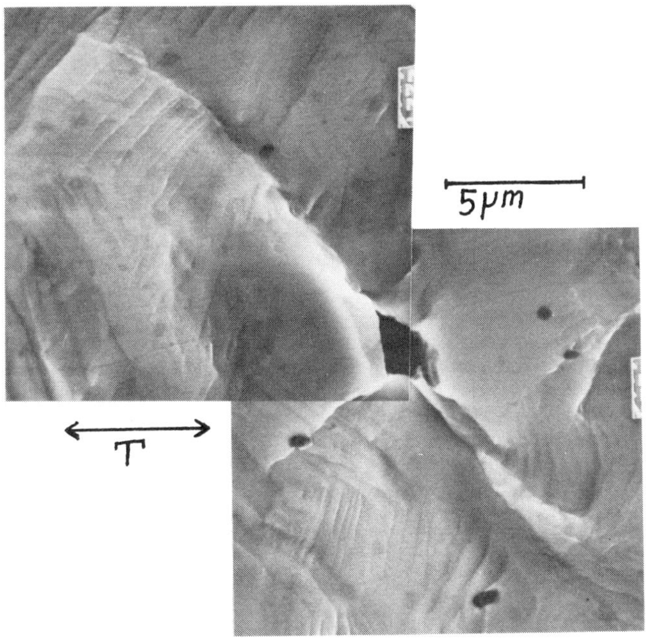

Fig. 2 : Mécanisme de déformation en traction travers.

Ces quatre bandes, cas le plus fréquent, délimitent quatre cadrans dont deux sont très déformés alors que les deux autres restent rigides et très peu déformés. La microcavité se comporte comme une microfissure sollicitée en mode I. La rupture s'effectue par cisaillement entre deux cavités reliées par une bande de glissement ; il n'y a pas dans ce cas d'arrachement. On retrouve que les bandes de glissement sont plus rectilignes pour la perlite lamellaire que pour la perlite globulisée. L'état martensite revenue ne présente que très peu de bandes de glissement. La coalescence intervient brutalement après une faible croissance des microcavités ; la forte densité de dislocations en est probablement la cause.

- Déformation en compression

Certains auteurs [2] ont étudié les mécanismes de fissuration à l'échelle macroscopique. Nous avons ici utilisé la même méthode que celle utilisée en traction en déformant des cylindres par compression, ces cylindres étant préalablement polis diamant afin de révéler les inclusions.

* Compression parallèlement à la direction longitudinale. Les inclusions sont parallèles à la surface latérale polie de l'éprouvette (fig. 3.a). Les cavités s'amorcent sur les inclusions plus tard qu'en traction. Les inclusions présentent une certaine plasticité. Ces deux phénomènes sont dus à la valeur négative de la contrainte moyenne σ_m. La croissance des cavités s'effectue suivant la direction θ (repère cylindrique r, θ, z). Les inclusions se fissurent à 45° par rapport aux directions z et θ c'est-à-dire suivant les directions de cission maximale. La croissance des cavités est analogue à une sollicitation de traction travers. La coalescence des cavités conduit à une macrofissure dont l'orientation sera discutée.

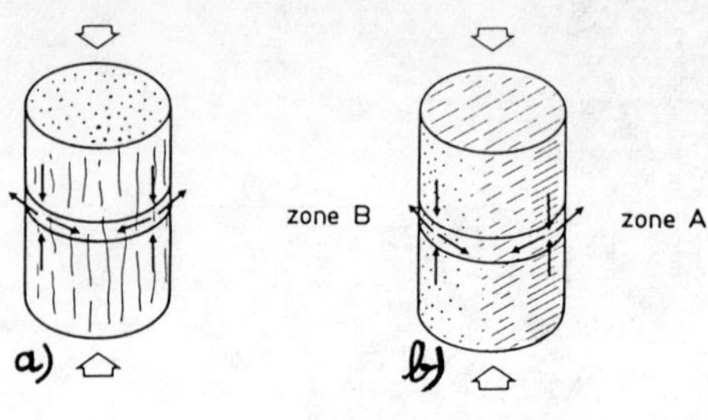

Fig. 3 : Position relative du système de déformation par rapport à l'état inclusionnaire.

* Compression perpendiculairement à la direction longitudinale. Sur un même échantillon les inclusions sont soit parallèles, soit perpendiculaires à la surface polie (fig. 3.b). La rupture survenant dans les zones B, seules ces plages ont été observées.

L'amorçage, la croissance et la coalescence des cavités sont tout à fait analogues à celles rencontrées en traction travers. Néanmoins, les quelques différences sont dues essentiellement au fait que σ_m est négatif : absence du phénomène des bandes de glissement en croix, meilleure homogénéité microscopique des déformations, amorçage plus tardif, rupture des inclusions parallèlement à la direction de compression z indiquant l'équivalence avec une traction suivant θ.

CARACTERISATION QUANTITATIVE DE LA DUCTILITE POUR DIFFERENTES MORPHOLOGIES INCLUSIONNAIRES

Le paramètre étudié est le facteur de forme des inclusions λ.
Afin de le faire varier sans faire varier d'autres paramètres, le produit a été étudié après différents taux de corroyage (τ : de 1 à 100), tous les échantillons étant traités thermiquement ensuite.
On a mesuré la striction à rupture long et travers et les taux d'écrasement limite en compression long et travers. Les principaux résultats sont les suivants :

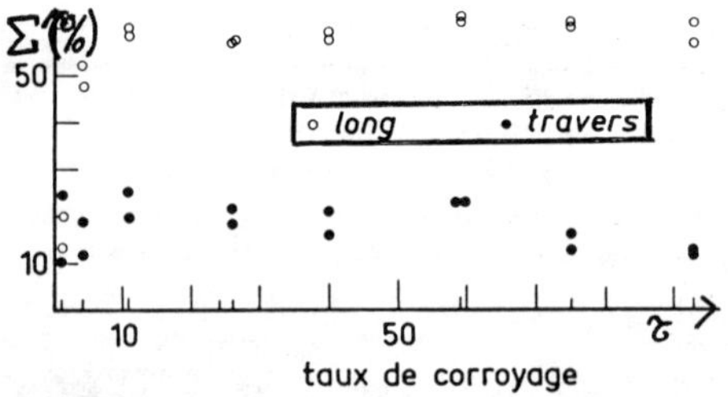

Fig. 4 : Evolution de la striction en fonction du taux de corroyage.

1 - Dès que τ atteint une valeur voisine de 4, les ductilités long et travers en traction et en compression n'évoluent plus avec τ (de 4 à 100) (Fig. 4). Nous avons vérifié par métallographie quantitative que les inclusions se déforment continuement avec τ. Ceci montre que dans un domaine étendu de λ, dès que les inclusions se sont allongées, la ductilité ne dépend plus de la forme des inclusions.

2 - La ductilité en traction travers est toujours inférieure à celle mesurée en long.

3 - Les taux limites d'écrasement en long et en travers sont identiques, ce qui ne signifie pas pour autant que le métal est isotrope.

4 - Il existe une très bonne corrélation entre la striction travers d'une part et les taux limites d'écrasement en long et travers d'autre part. Les points issus de cette étude sont sur la droite déterminée par d'autres auteurs [3].

5 - La déformabilité des inclusions en fonction de τ est continue. Une méthode d'analyse mise au point à l'IRSID a montré qu'il était possible de mesurer l'évolution de λ par classe de taille. On peut ainsi dans un acier contenant plusieurs types d'inclusions, analyser quantitativement la déformabilité de chaque type [4].

DISCUSSION

La connaissance des mécanismes de déformation permet d'interpréter les résultats relatifs à la caractérisation de la ductilité des aciers et les liens existant entre la traction et la compression. Elle permet en outre de comprendre qualitativement l'anisotropie de résistance à la propagation d'une fissure en ténacité ou en fatigue.

1 - Les mécanismes de déformation en traction travers et en compression long et travers sont identiques. σ_m en compression long et travers est identique ce qui explique l'égalité des taux limites d'écrasement en long et travers ; par contre, la traction travers bien qu'ayant des mécanismes voisins a une valeur de σ_m plus grande, donc la rupture surviendra plus tôt ; la déformation à rupture en traction bien que corrélée à la déformation à rupture en compression long et travers est donc plus faible.

2 - En compression cylindrique, il est couramment admis que la direction de fissuration est la direction de cission maximale [5].
Dans les essais présents, cette direction de cission maximale est à 45° des directions θ et z. En compression travers, la fissure est effectivement à 45° par rapport à θ ou z. Par contre en compression long (fig. 5) quand τ augmente, l'orientation est d'abord à 45° puis en escalier formé de "marches" orientées alternativement à 45° et parallèles à z pour finalement être entièrement parallèles à z quand τ est élevé . Ce phénomène vient d'une compétition entre les aspects mécaniques et métallurgiques : la mécanique des milieux continus indique un critère de cission maximale alors que la métallurgie montre que la rupture est conditionnée par la morphologie inclusionnaire. Pour les compressions travers (plage B figure 3) et long (à faible τ), les plans tangents à la surface de l'éprouvette sont métallurgiquement isotropes ; la condition mécanique suffit. Par contre pour les compressions long (fig. 3.a) à τ élevé (inclusions allongées), les plans tangents à l'éprouvette sont métallurgiquement orientés, fibrés et c'est la microstructure qui impose la direction de fissuration. Dans la réalité, on passe progressivement de l'un à l'autre quand les inclusions s'allongent.

3 - Lorsque l'on est en présence d'une fissure ou d'une entaille, l'état de contraintes développé présente une forte triaxialité. L'analyse de la position relative des directions principales et du signe des contraintes par rapport à la direction des inclusions permet, en connaissant les mécanismes de déformation, de déterminer quels sont les sens de prélèvements et de sollicitation des éprouvettes qui conduisent à la ténacité ou à la résistance à la fatigue la plus faible.

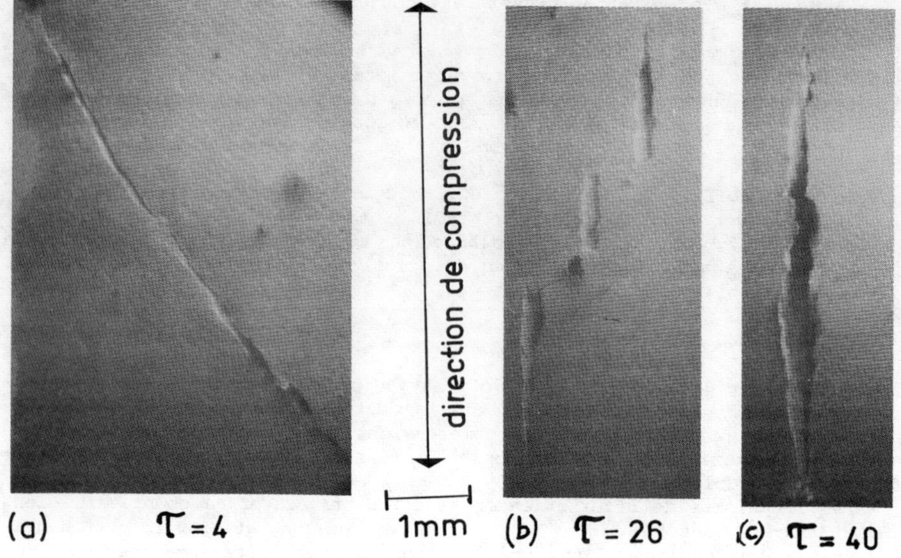

Fig. 5 : Mode de fissuration en compression long en fonction du taux de corroyage.

CONCLUSIONS

L'examen des mécanismes de déformation couplé à une caractérisation de la ductilité en traction et compression long et travers sur des barres d'acier a permis de préciser les causes microstructurales de l'anisotropie de ductilité et les relations existant entre les ductilités mesurées pour différentes sollicitations. Les résultats sont transposables à des sollicitations à triaxialité élevée, par exemple en présence d'une fissure ou d'une entaille. Enfin cette analyse a montré qu'une analyse purement mécanique macroscopique ne suffit parfois pas à prévoir la direction de fissuration dans un matériau métallique, la fissuration étant le résultat d'un état de contraintes mécaniques qui agit sur un milieu dont les caractéristiques microscopiques sont déterminantes.

REMERCIEMENTS

Nous remercions la CECA pour le support financier de cette recherche ainsi que Monsieur Claude QUENNEVAT pour la réalisation délicate de la partie expérimentale.

BIBLIOGRAPHIE

1. Y. KURITA, L. ROESCH et C. SAUZAY, Mem. Sci. Rev. Met., Vol. 74, n°12, 731 (1977).
2. T. ERTÜRK, W.L. OTTO et H.A. KUHN, Met. Trans., 5, 1883 (1974).
3. F. LEROY, D. LOURDELET, P. PORTEVIN, L. ROESCH et G. SANZ, Rapport Interne IRSID. RE 752 (1980).
4. M. BLANC et F. MOUSSY, Mem. Et. Sci., Vol. 81, n°12, 635 (1984).
5. H. KUDO et K. AOI, J. Japan Soc. Tech. Plasticity, 8, 17 (1967).

Microstructure-Toughness Relationships in the Cleavage Fracture of Pressure-Vessel Steel

P. Bowen and J. F. Knott

Department of Metallurgy and Materials Science, University of Cambridge, UK

ABSTRACT

By testing a wide range of heat-treated microstructures in pressure vessel steel it has been possible to show that the local (microscopic) cleavage fracture stress, σ_F^*, is controlled by the size distribution of carbides, particularly by the coarsest carbides in the distribution. The variation in σ_F^* with carbide size may be interpreted by means of a modified Griffith equation. The observed temperature independence of σ_F^*, for the wide range of microstructures considered, provides strong support for a tensile-stress-controlled fracture criterion.

KEYWORDS

microstructure-toughness relationships, pressure vessel steel, cleavage fracture, microscopic cleavage fracture stress, critical tensile stress.

INTRODUCTION

The cleavage fracture of lath microstructures, such as martensites and bainites, has been the subject of much investigation in recent years. In general, it is not easy to extend micromechanistical models of cleavage fracture, established for simple ferrite-carbide distributions, to these more complex microstructures. Indeed, there is little agreement on the principal microstructural features that control failure. Some workers have suggested that the packet size controls toughness in bainitic steels [1,2], others have suggested that the carbide size distribution is critical particularly in quenched and tempered steels [3,4,5], while further work has indicated the involvement of non-metallic inclusions [6]. A dislocation mechanism (essentially controlled by the matrix flow properties) has also been proposed [7]. Only in a few cases have the mechanisms been supported by detailed microstructural studies e.g. [1] and [3].

Further complications can arise because different measurements of "toughness" are often compared e.g. the ductile-brittle transition temperature DBTT, the fracture toughness K_{IC} and the microscopic cleavage fracture stress σ_F^*. For the purposes of relating the measured parameter to microstructural

SMA2-K*

features, the DBTT does not involve appropriate dimensions and the K_{IC} value is ambiguous, because it involves a combination of stress and distance and is also a function of yield stress. In the present paper the variation of 6_F* with microstructural parameters is examined for a wide range of lath martensites, lower and upper bainites and ferritic-pearlitic microstructures, possessing widely different matrix strengths. Microstructural details are given elsewhere [8]. The parameter 6_F* is studied, because it is perhaps the easiest "toughness" parameter to interpret. Through a modified Griffith equation it can be related directly to the size of microcrack, d, that causes failure using

$$6_F^* = (4E\gamma p/\pi(1-\nu^2)d)^{\frac{1}{2}} \quad (1)$$

where γp is the effective surface energy, E is Young's modulus and ν is Poisson's ratio. One inherent difficulty in the use of equation (1), is that the value of γp is not accurately known even for simple ferrite-carbide microstructures. Equation (1) is still of great value, however, if a consistent linear relationship can be established between the measured value of 6_F* and the reciprocal square root of the measured microcrack size. A previous investigation [9] showed that the packet size does not influence 6_F* (or K_{IC}) for autotempered martemsites i.e. even when the carbide size is very small ($\leqslant$ 30 nm) and also suggested tentatively that manganese sulphide inclusions might conceivably control the cleavage fracture process. Subsequent work did not support this suggestion [8,10]. In the present study attention has therefore been directed towards the possible influence of carbides on cleavage fracture.

EXPERIMENTAL

All tests were performed on a commercial grade of A533B of the composition given in Table 1. Details of heat-treatments are reported elsewhere [8].

TABLE 1 Composition of A533B plate, wt.%

C	Mn	Ni	Mo	P	S	Sn
0.25	1.51	0.63	0.53	0.005	0.005	0.005

Tensile tests were performed over the temperature range from -196 to 20°C, and the 0.2% proof stresses are shown in fig. 1. Slow notched bend tests were carried out between the temperatures of -196 and -100°C, such that fracture occurred at or before general yield. The testpiece design was similar to that used by Griffiths and Owen [11] so that their finite element analysis could be used to calculate the maximum tensile stress present ahead of the notch at failure from the fracture load. This stress was then equated to 6_F*. Carbide sizes were measured where possible <u>both</u> by thin-foil TEM and high resolution SEM, and complete size distributions were obtained for all conditions studied - see ref. 8.

RESULTS

The measured values of 6_F* are given as a function of test temperature in fig. 2. These 6_F* values appear to fall into two distinct groups: values in the range from 1200-2200 MPa, obtained from bainitic and pearlitic microstructures; values in the range from 3100-3800 MPa obtained from as-transformed and lightly tempered martensitic microstructures. An important observation is that little variation in 6_F* with test temperature is seen for any single microstructural condition i.e. <u>6_F* is essentially temperature</u>

independent.

The coarsest observed carbide widths are shown in Table 2 for some of the microstructures considered. It is apparent from Table 2 that the coarsest observed carbide width obtained by thin foil TEM is typically one half of that obtained by high resolution SEM. Average values of σ_F^* are also shown in Table 2, and it can be seen in general that as the carbide size decreases, σ_F^* increases. In fig. 3, measured σ_F^* values are plotted versus the reciprocal square root of the coarsest observed carbide widths, obtained by SEM (where possible). Also shown in fig. 3 are 8 points taken from ref.[4] and 15 points taken from ref.[3](for spheroidised microstructures).

TABLE 2 Coarsest observed carbide widths and average σ_F^* values

		Coarsest observed carbide width/ diameter (nm)		σ_F^* (MPa)
		TEM	SEM	
As transformed autotempered martensite	(▽)	36	-	3700
Martensite tempered 1h at 290°C	(▽')	110	-	3300
Martensite tempered 8h at 615°C	(●)	260	600	3300
Lower bainite/upper bainite	(◇)	350	720	1900
Upper bainite/divorced pearlite	(□)	290	1000	1800
Upper bainite tempered 120h at 670°C	(□†)	500	1100	1200
Ferrite/divorced pearlite	(■)	-	340	2200

The carbide widths in these previous studies were measured by optical microscopy [4] and from TEM extraction replicas [3] respectively. For many of the conditions tested in this present study a reasonable straight line fit through the origin can be obtained, see fig. 3. From the slope of this straight line, and using equation 1 a γp value of 6 Jm^{-2} is derived.

DISCUSSION

The temperature independence of σ_F^* for a given microstructure in a temperature range where variations in yield stress are observed, strongly suggests a fracture criterion which is essentially controlled by tensile stress. This is consistent with equating σ_F^* to the maximum tensile stress present ahead of a notch at failure. It is also consistent with the use of equation 1. This predicts an essentially temperature-independent σ_F^*, similar to that obtained for mild steel. The modulus E is virtually independent of temperature over the range examined and γp is also essentially athermal if it represents work done in the crack-tip region of a microcrack of constant size.

It is also encouraging to note that a reasonable linear relationship is observed between measured values of σ_F^* and the reciprocal square root of the coarsest observed carbide widths (diameters). It is, however, also clear that some microstructures do not conform well to this linear relationship. As transformed martensites in particular give lower σ_F^* values than would be predicted from their carbide size distributions. Note that these discrepancies are accentuated in fig. 3 because these carbide sizes were obtained by thin foil TEM and are compared with carbide sizes obtained by high resolution SEM, although from Table 2 it can be deduced that these differences are unlikely to correct the observed discrepancies completely.

A more complete explanation for the "low" 6_F* values obtained from such martensitic microstructures is given if the most potent microcracked carbides are present at embrittled prior austenite grain boundaries [8,10]. This plausibly produces a reduction in γp which results in decreased 6_F* values. In contrast, the martensites tempered at 615°C, give higher 6_F* values than would be predicted from their carbide sizes. Metallographic observations suggest for such conditions, that considerable spheroidisation of carbides has occurred. It is therefore possible that the shape of the carbide particles is also of importance. Some support for this statement can be inferred from the 6_F* values obtained in previous studies on spheroidised steels [3,4] also shown in fig. 3. In general, the 6_F* values obtained from spheroidised microstructures are higher than predicted from equation 1. It should be noted that equation 1 considers the propagation of a through-thickness edge crack and it is possible that the propagation of a microcrack from a spheroidised carbide should be considered as the propagation of a penny-shaped crack [4].

In summary, the cleavage fracture of a wide range of lath-type microstructures appears to be controlled primarily by the sizes of the coarsest carbides present.

CONCLUSIONS

1. The value of 6_F* is found to be independent of temperature for a wide range of heat-treated lath microstructures.

2. A linear relationship is obtained between the measured value of 6_F* and the reciprocal square root of the coarsest observed carbide width for a number of heat-treated microstructures. This relationship can be modelled by the use of a modified Griffith energy balance. Refinements to this simple model appear to be necessary to take full account of the effects of changes in carbide shape i.e. spheroidisation, and of the effects of locally different environments i. e. if the carbide is present on an embrittled grain boundary.

REFERENCES

1. P. Brozzo, G. Buzzichelli, A. Mascanzoni and M. Mirabile, *Met. Sci.* 11, pp. 123-129, 1977.
2. M. J. Roberts, *Metall. Trans.*, 1, pp. 3287-3294, 1970.
3. D. E. Hodgson and A. S. Tetelman, "Fracture", pp. 266-277, 1969 (ed. P. L. Pratt), Chapman and Hall, London.
4. D. A. Curry and J. F. Knott, *Met. Sci.*, 12, pp. 511-514, 1978.
5. K. Wallin, T. Saario and K. Torroneu, *Met. Sci.*, 18, pp. 13-16, 1984.
6. M. J. Khan, T. Shoji and H. Takahashi, *Met. Sci.*, 16, pp. 118-126, 1982.
7. D. A. Curry, *Met. Sci.*, 18, pp. 67-76, 1984.
8. P. Bowen, Ph.D. Thesis, "The effects of microstructure on toughness in pressure vessel steel", Univ. of Cambridge, October 1984.
9. P. Bowen and J. F. Knott, *Met. Sci.*, 18, pp. 225-235, 1984.
10. P. Bowen and J. F. Knott, *Acta metall.*, 32, pp. 637-647, 1984.
11. J. R. Griffiths and D. R. J. Owen, *J. Mech. Phys. Solids*, 19, pp. 419-431, 1971.

ACKNOWLEDGEMENTS

One of the authors (PB) was supported in the course of this work by a CASE award from AERE, Harwell, and latterly by a Goldsmiths Junior Research

Fellowship at Churchill College, Cambridge. Useful and detailed discussions with Drs. S. G. Druce, C. A. Hippsley and J. A. Hudson at Harwell are gratefully acknowledged. Thanks are also due to Professor R. W. K. Honeycombe, F.R.S. and Professor D. Hull for provision of research facilities.

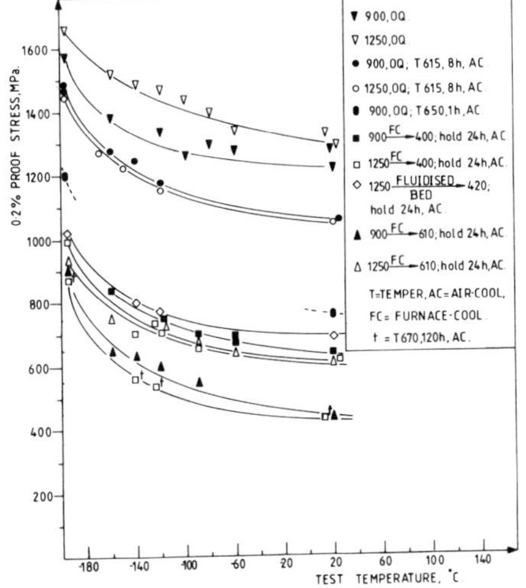

Fig. 1. 0.2% proof stress plotted versus test temperature.

Fig. 2. Values of σ_F^* plotted versus test temperature.

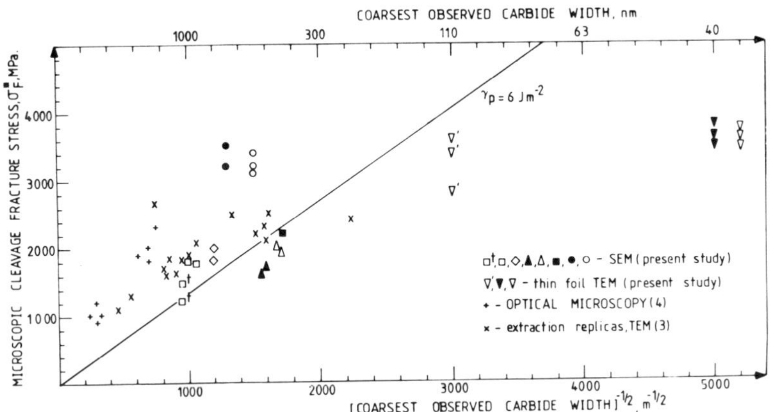

Fig. 3. Values of σ_F^* plotted versus the reciprocal square root of the coarsest observed carbide width.

Microstructural Effects on Fracture Toughness (J-integral) of a Medium-carbon Low Alloy Steel SAE 4130

P. Nguyên-Duy

Institut de Recherche d'Hydro-Québec, IREQ, Varennes, Québec, Canada

ABSTRACT

Quenched and tempered SAE 4130 steel is used intensively in the high voltage transmission line towers and line hardwares in Hydro-Québec; it is necessary to evaluate the fracture toughness of this steel of different microstructures resulting from different heat treating conditions such as quenching and tempering, normalizing, annealing and spheroidizing. The fracture toughness quantified by a computerized elastic partial unloading method using a single four-point bend specimen. The variation of critical value of J-integral, J_{Ic} in terms of heat treating conditions is very remarkable (0.037 to 0.160 $MJ\text{-}m^{-2}$). The variation of fracture toughness was interpreted in terms of microstructural change resulting from different heat treating conditions.

KEYWORDS

J-integral, fracture toughness, steel, heat treating conditions.

INTRODUCTION

Quenched and tempered SAE 4130 steel is used intensively in the high voltage transmission line towers and line hardwares in Hydro-Québec; it is necessary to evaluate the fracture toughness of this steel of different microstructures resulting from different heat treatments.

The fracture toughness quantified by J-integral was measured in using the elastic partial unloading J_R curve method (1). In this method, J-integral value is evaluated according to the equation

$$J_{i+1} = \left[J_i + \left[\frac{2A_{i,i+1}}{b_i B} \right] \right] \left[1 - \left[\frac{1 + 0.76 \left[\frac{b_i}{W} \right]}{b_i} \right] \left[a_{i+1} - a_i \right] \right] \quad (1)$$

The second term of this equation takes into account the crack extension effect on the J-integral value; when Δa is equal to 0, the equation (1) reduces to the equation

$$J_{i+1} = J_i + \left[\frac{2A_{i,i+1}}{b_i B} \right] \tag{2}$$

A computer program was written to monitor the test, to control the servo-hydraulic machine and to calculate in real time the intermediate value of the J-integral and, finally, evaluate the critical value of the J-integral (2).

In this test method, which uses four-point-bend specimens, the compliance function developped elsewhere (3) for pure bending enables evaluation of the crack length

$$a_i = W(0.998265 - 3.81662U_x - 1.80596U_x^2 + 32.3104U_x^3 - 44.1566U_x^4 - 52.6788U_x^5) \tag{3}$$

where U_x is a function of material properties and specimen dimensions

$$U_x = \frac{1}{\left[\frac{4WBE}{(BS-SS)(1-\nu^2)} \right]^{1/2} \left[\frac{COD}{P} \right]^{1/2} + 1} \tag{4}$$

MATERIALS

The chemical composition of the quenched and tempered (1 hour tempering) SAE 4130 steel is shown in Table 1

TABLE 1 Chemical Composition of SAE 4130 Steel

C	Mn	P	S	Si	Cu	Ni	Cr	Mo	Al
0.36	0.63	0.013	0.029	0.21	0.18	0.06	0.93	0.16	0.026

All specimens were originally in this condition (Q&T 1 hr) but were subsequently heat treated under various regimes as reported in Table 2.

After these heat treatments, hardness measurements on the different specimens varied from $38R_C$ to $84R_B$. Very different microstructures were also obtained from these treatments as illustrated in Fig. 1. In the case of the quenched and tempered specimens (specimens no 1, 2, 3, 4) a tempered martensite structure is seen with martensite needles of different sharpness. In the case of

TABLE 2 Different Heat Treating Conditions

Specimen no	Heat Treating Conditions	Hardness (R_C)	Microstructure
1	Q&T at 600^oC, 1hr (1)		Tempered martensite
2	Q&T at 500^oC, 1hr (2)	38	Tempered martensite
3	Q&T at 500^oC, 2hr	35	Tempered martensite
4	Q&T at 500^oC, 3hr	31	Tempered martensite
5	Normalized,925^oC, 0.5hr air cool	20	Acicular ferrite & sup. bainite
6	Normalized,925^oC, 1hr air cool	22	Acicular ferrite & sup. bainite
7	Annealed,850^oC, 0.5hr furnace cool	85R_B	Gross lamellar perlite & matrix of ferrite
8	Annealed,850^oC, 1hr furnace cool	86R_B	-id-
9	Spheroidized,770^oC, 6hr slow cool	81R_B	Partly spheroidized perlite in a ferrite matrix
10	Spheroidized,770^oC, 10hr slow cool	81R_B	-id-

(1) results from previous work (4)
(2) initial conditions of all specimens

normalized specimens (no 5 and 6), a mixte structure of acicular ferrite and bainite is observed whereas specimens no 7 and 8 present a normal ferrito-perlite structure; finally in the case of specimens 9 and 10 fine colonies of perlite can be seen and it is possible to identify the beginning of the spherodizing process of perlite.

The final dimensions of specimens were
BS = 128 mm; SS = 64 mm; W = 32 mm; B = 16 mm
where BS, SS, W and B are respectively large span, small span, width and thickness . Nine of the four-point-bend bars identified from 2 to 10 derived from the present program while specimen no 1 came from a previous study and was a quenched and tempered (600^oC for 1 hr) specimen (4).

TESTING

All specimens were fatigue precracked according to a prescribed method, with a precrack length (a_o) ranging from a_o/W 0.520 to a_o/W 0.627.

The elastic partial unloading test is started with two independant loadings up to a load level corresponding to 10% and 15% of bending load calculated from the following formula (5)

$$P_L = \frac{4B(W-a_o)^2 \sigma_y}{3(BS-SS)} \qquad (5)$$

These two preliminary loadings allow verification of the accuracy of the compliance function; if crack length values a_o obtained from these two loadings are not very different from each other and close to the value measured previously on a precracked specimen, the test is started.

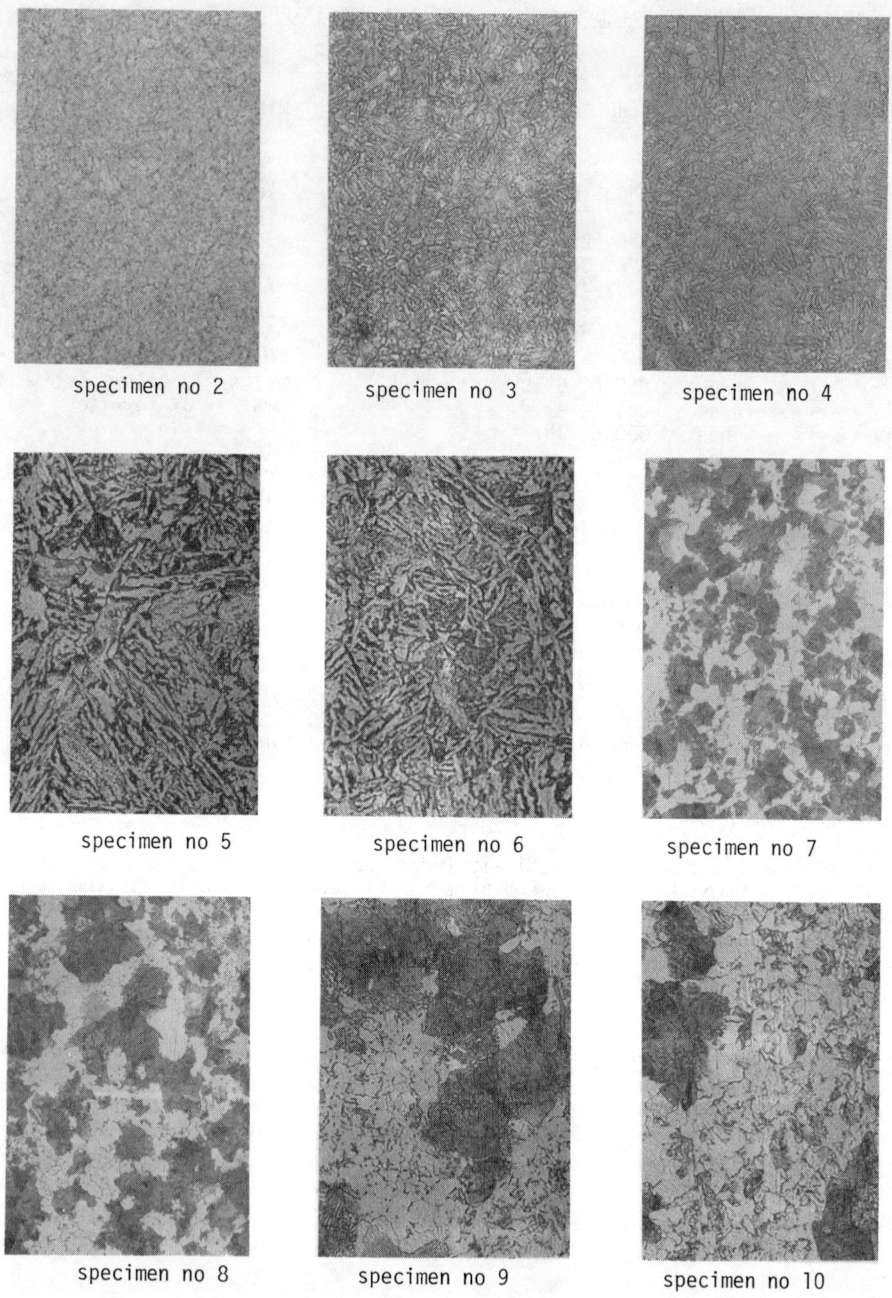

Fig. 1. Different microstructures from various heat treating conditions, magnification 500X.

The testing machine was controlled by the stroke-control mode. Unloading was performed at each 5% of the total gauge length of the COD gauge mounted at notch mouth for a amount of 15% of the actual load P_i.

For each unloading, 100 pairs of points (P-COD) were registered, but only 60 central pairs were saved for a regression analysis performed on line, enabling the evaluation of specimen compliance (COD/P) and the area under the (P-δ) curve. By replacing these two quantities in equations (4), (3) and (1), the J-integral can be calculated.

At the end of each test, the J_R curve method was used to evaluate critical value of J-integral, J_{Ic}. It is important to mention that all crack length values previous to minimum crack length were considered equal to this minimum value. Specimens used for the elastic partial unloading tests were heat tinted and then broken in liquid nitrogen. Initial crack length a_o and total crack extension Δa_{tot} were measured using the following formula

$$a_o \, , \Delta a_{tot} = \frac{A}{BG_xG_y} \tag{6}$$

where A is the cracked ared measured on photograph of broken specimen halves; B is specimen thickness, G_x and G_y are respectively the magnification factors of the photograph in x and y directions (6) .

A typical set of results obtained from this computerized elastic partial unloading method is illustrated in Fig. 2. All results are given in Table 3.

INTERPRETATION OF RESULTS

Interpretation of the results was divided into two distinct parts; the first part is related to the testing method itself while the second correlates material fracture toughness to heat treating conditions.

Testing Method Interpretation

Using only a single specimen, the computerized elastic partial unloading J_R curve method allows the determination of the critical value of the J-integral, J_{Ic} with reliability. The compliance function used for pure bending (equation 3) permits accurate evaluation of crack extension Δa, the total crack extension Δa_{tot} being very comparable with physical measurements on broken specimen halves. Some particular points relative to the method should be mentionned; for instance in general, the crack length decreases for some first unloadings up to nearly the end of elastic region and then begins to increase; this negative crack extension disappears if a light preloading was applied to the specimen. Up to now, negative crack extension was not taken in account in our calculation; minimum crack length is considered to be the initial crack length

$$\Delta a_i = a_i - a_{min} \tag{7}$$

The method allows also determination of what is called the instability threshold (7) of the crack; This corresponds to the moment on the J- a_i curve where the J-integral decreases for a finite crack extension.

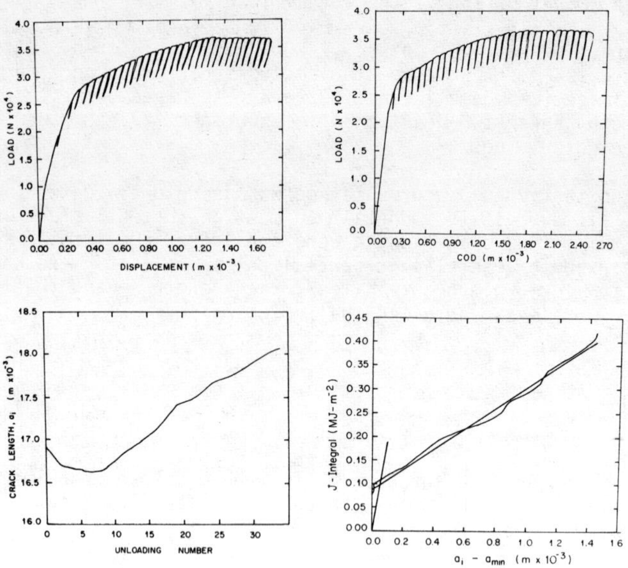

Fig. 2 Typical set of results from computerized elastic partial unloading method.

TABLE 3 Test Results

Specimen	a_o (mm)	a_{tot} (mm)		$CMOD_{tot}$ (mm)	Instability		J_{Ic} ($MJ\text{-}m^{-2}$)
					CMOD (mm)	a(mm)	
1							0.148
2	18.58	3.89[a]	4.08[b]	1.275	1.05	2.53	0.156
3	17.52	4.25	4.48	1.875	1.35	3.39	0.160
4	16.74	5.90	6.12	1.875	1.50	4.28	0.155
5	17.84	-	-	0.225	0.225	-	0.037
6	17.59	-	-	0.225	0.225	-	0.039
7	17.06	0.78	0.90	1.875	-	-	0.093
8	17.28	0.77	0.88	1.875	-	-	0.097
9	20.06	0.50	0.60	2.250	-	-	0.093
10	16.61	1.44	1.56	2.550	-	-	0.101

(a) by the compliance function
(b) by physical measurements

Correlations between Fracture Toughness and Heat Treating Conditions

As shown in Table 3, heat treating conditions in this study,

- the nature of heat treatment
- the residence time at a specified temperature
- the cooling rate

affect strongly the numerical value of critical value of J-integral, J_{Ic}.

For the evaluation of J_{Ic}, only results prior to the crack extension instability were considered. This instability threshold occurs for all quenched and tempered and normalized specimens. It is important to note that for the same or smaller value of COD, the crack extension was much larger in the case of quenched and tempered and normalized specimens.

Quenched and tempered specimens, independantly of the tempering time (1hr, 2hr, 3hr) possess closely similar J_{Ic} values, 0.156, 0.160 and 0.155 $MJ\text{-}m^{-2}$ respectively, but crack extension and COD corresponding to the instability threshold increase with tempering time. This is an indication of the capability of materials to accomodate crack extension.

Annealed and spheroidized specimens do not exhibit for the same or larger COD (up to 2.55 mm), the instability threshold. J_{Ic}, however, is lower than the value obtained for quenched and tempered specimens. Spheroidizing and slow cooling increase considerably the capability of the material (specimen no 10) to accomodate crack extension. Results are summarized in Fig. 3.

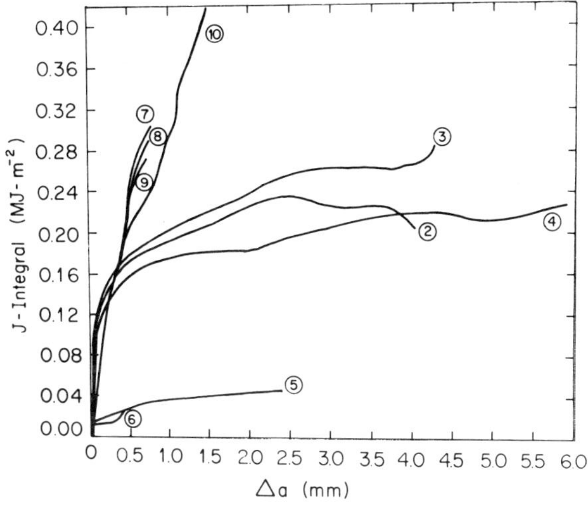

Fig. 3. Comparative J-Δa curves obtained from different specimens.

Fracture surfaces of the different specimens were studied by means of scanning electron microscopy. Fracture surface morphology depends strongly on heat treating conditions, varying from a very brittle aspect (cleavage, normalized specimens) to a very ductile fracture surface (quenched and tempered, spheroidized specimens), passing through intermediate morphology (annealed specimens); these fractures are illustrated in Fig. 4.

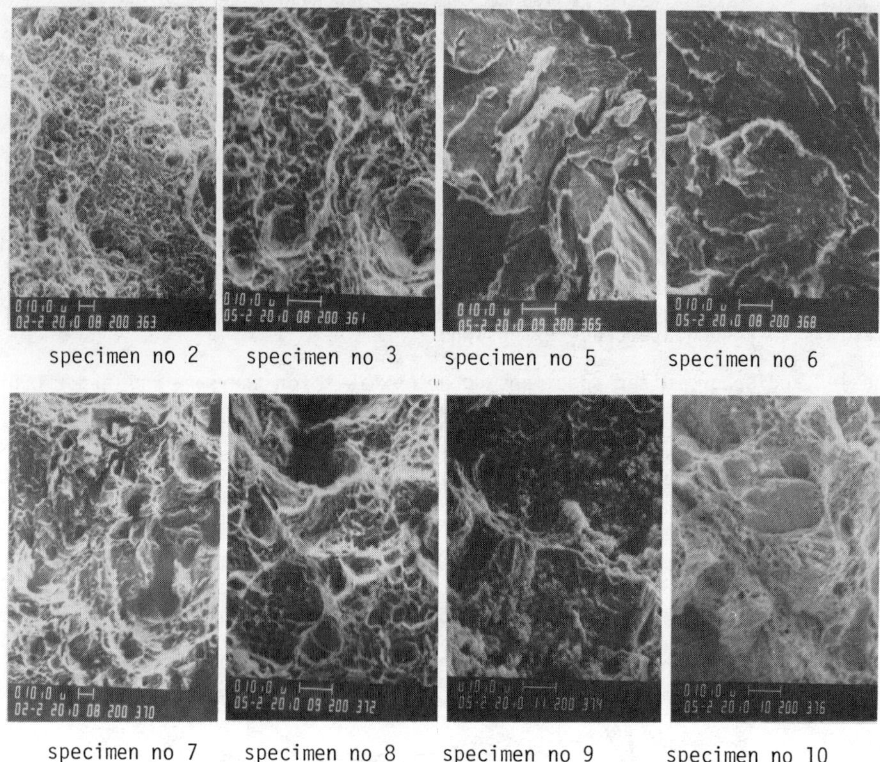

specimen no 2 specimen no 3 specimen no 5 specimen no 6

specimen no 7 specimen no 8 specimen no 9 specimen no 10

Fig. 4. Fracture surfaces of different specimens.

CONCLUSIONS

1.The computerized elastic partial unloading method is a powerful tool for the reliable evaluation of the fracture toughness of material using a single specimen. This method allows the determination of the instability point when it occurs.

2.The fracture toughness quantified by J-integral is considerably affected by different heat treating conditions which are clearly reflected on fracture surface morphology.

ACKNOWLEDGEMENTS

The author would like to express his thanks to Dr. E.A. Dancy for the critical reading of the manuscript, A. Lapointe for the establishement of the computer program, J.G. Millette for all heat treatments and microstructural works and C. Lanouette for the diagrams.

REFERENCES

1. P.C. Paris, H. Tada, A. Zahoor and H.A. Ernst, Elastic-Plastic Fracture, ASTM STP 668, J.D. Landes, J.A. Begley and G.A. Clarke, Eds., American Society for Testing and Materials, p. 5-36, (1979).
2. P. Nguyên-Duy and A. Lapointe, ASME Pressure Vessel and Piping Conference and Exhibition, New-Orleans, (1985).
3. H.A. Ernst, P.C. Paris and J.D. Landes, Fracture Mechanics Thirteenth Conference, ASTM STP 743, R. Roberts, Ed., American Society for Testing and Materials, p. 476-502, (1981).
4. P. Nguyên-Duy and S. Lalonde, The Eleventh Canadian Fracture Conference (CFC11), Ottawa, (1984).
5. P. Albrecht, W.R. Andrews, J.P. Gudas, J. A. Joyce, J.F. Loss, D.E. McCabe, D.W. Schmidt and W.A. VanDerSluys, J. of Test. and Evaluation, JTEVA, 10, 245,(1982).
6. P.Nguyên-Duy and J.Flamand, Annual Winter Meeting of the ASME, Boston, (1983).
7. J.W.Hutchinson and P.C.Paris, Elastic-Plastic Fracture, ASTM STP 668, J.D. Landes, J.A. Begley and G.A. Clarke, Eds., American Society for Testing and Materials, p. 37-64, (1979).

Influence of Inclusion Content, Texture and Microstructure on the Toughness Anisotropy of Low Carbon Steels

D. M. Fegredo, B. Faucher and M. T. Shehata

**CANMET, Energy, Mines and Resources Canada, Ottawa K1A 0G1, Canada*

ABSTRACT

The fracture toughness anisotropy of steels containing two levels of sulphur, with and without rare earth additions to modify the sulphides, has been studied as a function of rolling temperature. Pancaked microstructures and crystallographic textures develop at low rolling temperatures, and sulphides are significantly deformed in the non rare-earth treated steels. It is shown that S-L/T-L fracture toughness anisotropy increases when the rolling temperature decreases, as a result of decreasing S-L toughness. The main contributors to this decrease are pancaked microstructure and flattened sulphides. Pancaked microstructure appears to be the more damaging component, however, the two behave synergistically when both are present.

KEYWORDS

Inclusion, texture, anisotropy, toughness, low carbon steels, microstructure, pancaked grains and modification.

INTRODUCTION

Many papers in the literature have pointed out the deleterious effect of sulphide inclusions on fracture toughness of steels (1-5), and several parameters have been proposed to quantify the relative effects of volume fraction, amount of flattening or aspect ratio, and inclusion size and distribution (6-9). Type II sulphides are found in fully killed steels (10) and are highly plastic, particularly at low rolling temperatures. The weak inclusion-steel interface parts readily at low normal stresses so that flattened sulphide inclusions produce fracture toughness anisotropy. Low rolling temperatures also result in a finer equiaxed grain size which increases both strength and toughness (11). However, if rolling temperatures are lowered enough, recrystallization becomes difficult or ceases altogether, giving a pancaked microstructure. Strong textures are generated at low rolling temperatures (11-15), and these two factors of structure and texture can, besides inclusions, also influence fracture toughness anisotropy.

TABLE 1 Chemical Compositions of Various Steels Used (wt %)

Steel	C	Mn	Si	S	P	Cr	Ni	Mo	Al	Ce+La
A	0.08	1.43	0.20	0.007	0.009	0.04	0.02	0.006	0.08	-
B	0.09	1.30	0.24	0.007	0.01	0.05	0.02	0.006	0.06	0.02
C	0.07	1.11	0.18	0.02	0.01	0.04	0.02	0.006	0.05	0.034
D	0.06	1.20	0.09	0.022	0.008	0.03	0.02	0.006	0.025	-

It is the purpose of this work to determine how the fracture toughness anisotropy of steels containing two levels of sulphur (~0.007% and ~0.02%), with and without added RE to modify the sulphides, is affected by soaking and rolling temperature.

EXPERIMENTAL METHODS

Four different steel heats, A-D (Table 1) made in a basic electric furnace, were cast as 227 kg, hexagonal cross-section, tapered ingots. The ingots were sliced and forged at 1250°C, then the forgings were cut to billets which were rolled after being soaked at temperatures ranging from 680° to 1200°C. Rolling to one-third of the thickness (19.1 mm) was done in 8 passes for plates soaked below 800°C and in 5 passes for plates soaked at 800°C or higher. Embedded thermocouples measured finishing temperatures. All plates were air-cooled after rolling, and particular plates are identified in the text by the steel and soaking temperature, e.g., A-680.

Small scale Wedge Open Loading (WOL) specimens were made according to the design shown in Fig. 1. The sides were slotted after fatigue pre-cracking. S-L (fracture on the rolling plane in the rolling direction) and T-L (fracture on the longitudinal section in the rolling direction) specimens were loaded to failure at different temperatures with a cross-head speed of 0.5 mm/min. Texture determinations were made by X-ray diffraction using the energy dispersion method (11) employing a Li-drifted Si detector and a multichannel analyzer. Fractographic observations were made on a Cambridge Instruments Stereoscan MK 11A scanning electron microscope equipped with an energy dispersive X-ray analyzer. A Quantimet 900 was employed to determine volume fraction, size, and aspect ratio of sulphide inclusions present in the steels. The as-polished specimens were etched in 10% oxalic acid to darken the sulphide inclusions and achieve better contrast for the measurements.

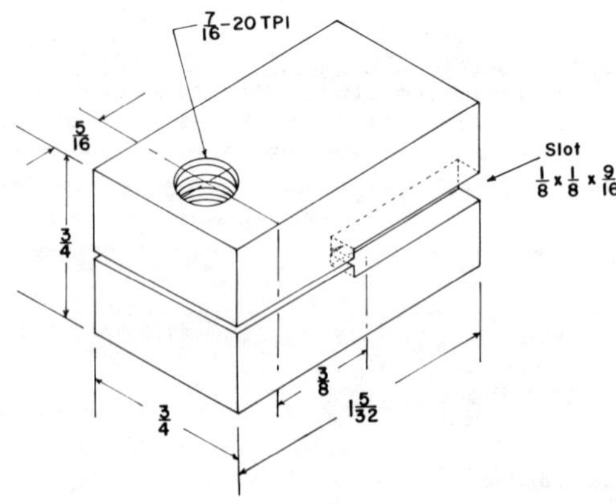

Fig. 1 - Small-scale Wedge Open Loading (WOL) specimen which is slotted on the sides subsequent to fatigue pre-cracking. All dimensions in inches (1 in. = 25.4 mm).

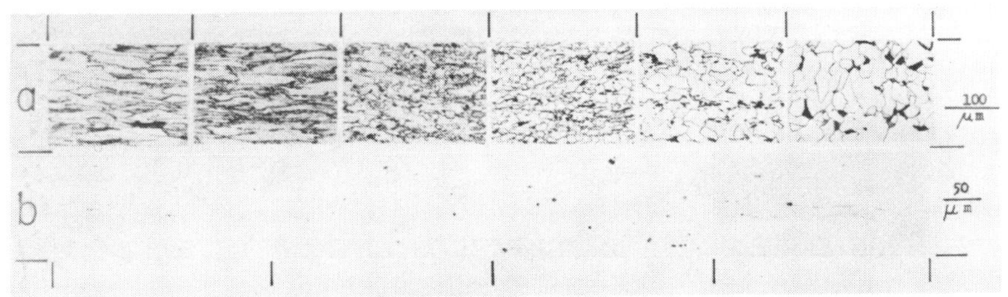

Fig. 2 - Longitudinal sections showing from left to right
(a) Microstructures of B-680, B-740, B-800, B-860, B-920 and B-1200 plates. Finishing temperatures were respectively: 680, 720, 755, 810, 860 and 980°C.
(b) Types of inclusions occurring in A-740, B-740, C-740, D-740 plates.

EXPERIMENTAL RESULTS

The Ac_3 transformation temperatures of the steels is ~880°C. Therefore, the various soaking temperatures of 680°-1200°C put them in the α, α-γ and γ conditions prior to rolling. These conditions are reflected in the plate microstructures. Figure 2(a) shows longitudinal sections of steels B for the various soaking (and associated finishing) temperatures. Pancaked grains and equiaxed grains form at low and high rolling temperatures respectively, with mixed structures occurring in between. Figure 2(b) shows the types of inclusions obtained in longitudinal sections. Table 2 presents sulphide inclusion data measured on the rolling and longitudinal planes. The inclusion anisotropy indexes are the ratio of the parameters (7,8) characterizing inclusion effects in the T-L and S-L orientations.

Figure 3 schematically describes the types of load vs. extension curves obtained in WOL tests. Plastic absorption energies were calculated as the areas under the curves to the ultimate load L_u when ultimate and maximum loads coincide, or to the maximum load L_m when (i) no ultimate forms, e.g., curve "a", "bI", or (ii) a lower load ultimate forms after a load drop, e.g., curve "c". The curve types and experimental energies are given in Table 3 for various test temperatures.

TABLE 2 Inclusion characterization: volume per cent (f); average dimension in rolling (a), transverse (b) and short-transverse (c) directions; average aspect ratios; inclusion parameters ($P_{T-L} = f\ a^{1/2}/b$; $P_{S-L} = f\ a^{1/2}/c$) (7); inclusion anisotropy index ($\eta_W = P_{T-L}/P_{S-L}$; $\eta_M = (1 + 2a/b)/(1 + 2a/c)$ (8))

	f(%)	a(μm)	b(μm)	c(μm)	a/b	a/c	P_{T-L}	P_{S-L}	η_W	η_M
A-680	0.080	3.34	2.55	1.05	1.35	3.74	0.57	1.39	0.41	0.44
A-920	0.080	2.87	1.80	1.33	1.64	2.44	0.75	1.02	0.74	0.73
B-680	0.106	2.57	1.66	1.50	1.73	2.05	1.02	1.13	0.90	0.87
B-740	0.106	2.93	1.85	1.71	1.83	1.96	0.98	1.06	0.92	0.95
B-800	0.106	2.75	1.69	1.68	1.76	1.92	1.04	1.05	0.99	0.93
B-860	0.106	2.82	1.81	1.75	1.69	1.88	0.98	1.02	0.97	0.92
C-800	0.220	3.08	1.97	1.44	1.77	2.53	1.96	2.68	0.73	0.75
D-800	0.202	4.06	2.66	1.12	1.39	3.96	1.53	3.63	0.42	0.42
D-920	0.202	3.78	2.31	1.26	1.71	3.17	1.70	3.12	0.55	0.60

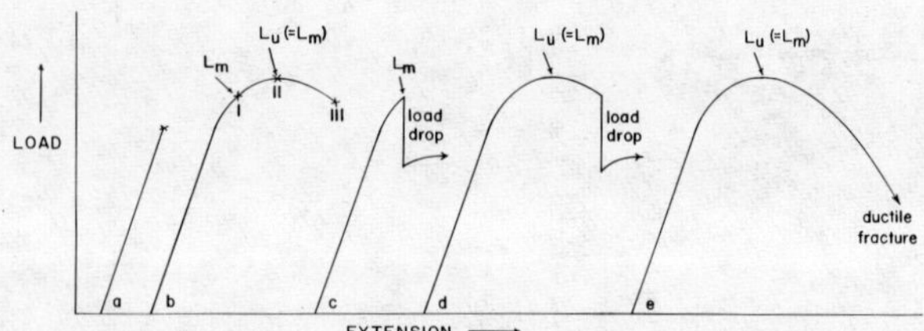

Fig. 3 - Types of load vs. extension curve obtained in tests of WOL specimens from various plates.

TABLE 3 Curve Type and Plastic Energies in Joules for Some WOL Specimens Tested at Various Temperatures

Steel plate	T-L specimens			S-L specimens					
	-60	-40	-20	-60	-40	-20	0	25	50
A-680	e:20	c:10	d:16		a:0	a:0	a:0		c:1
740	e:20	c:13	e:16			a:0			
800	e:34	e:27				bI:2			
860	bI:31	e:39			bI:29	bIII:32			
920					e:22	e:21			
1200				e:34		e:55			
B-680	e:46	c:16			bI:0.3	bI:24	bII:24	e:19	
740	c:23	e:20		bI:28	bIII:33				
800	e:46	e:42		e:50	e:53				
860	e:34			bIII:50	e:48				
920				bIII:52					
B-680N*					bIII:52				
C-680	e:35	e:19	e:42					a:0	
740		e:24	d:28					bIII:11	
800	bII:35	c:22				bI:3		bIII:26	
860	e:35					bIII:41			
1200				bIII:30		e:20			
D-680		e:22	e:10					a:0	a:0
740		e:6	e:5					a:0	
800			e:11					bIII:4	
860			e:11					e:7	
920				e:8	e:3				
1200					e:8				

B680N*: Normalized B-680 specimen.

TABLE 4 Relative Intensities of Texture Components for Various Soaking Temperatures of Steel B

Plane analyzed	Axis	Rolling plane texture	Pole HKL	Soaking temperature, °C					
				680	740	800	860	920	1200
Rolling plane	R.N.	(001) <110>	(001)	6.0	5.9	4.8	3.7	0.6	0.2
		(112) <110>	(112)	6.1	7.1	3.9	5.1	1.6	0.6
		(111) <110>	(111)	10.3	10.6	4.3	4.2	1.7	0.65
		Random Factor $\hat{\sigma}$		0.11	0.13	0.35	0.37	0.80	0.90
Longitudinal section	T.A.	(001) <110>	(110)	2.2	2.7	1.1	0.7	-	-
		(112) <110>	(111)	0.5	0.8	-	0.4	0.3	-
		(111) <110>	(323)	1.3	2.3	-	1.0	-	-
		Random Factor $\hat{\sigma}$		0.87	0.80	0.95	0.91	0.95	0.99

Three well-resolved, strong rolling plane texture components form at lower soaking and finishing temperatures (Table 4). An abrupt transition to a random orientation occurs between the 860°C and 920°C soaking temperatures.

DISCUSSION

Toughness of S-L specimens, as represented by plastic energy absorbed and type of curve, Table 3, increases with increase in soaking temperature especially in the range 680° to 860°C. Since the toughness of T-L specimens in any particular A-D steel does not vary as much with soaking temperature, fracture toughness anisotropy between S-L and T-L specimens increases markedly with decrease in soaking temperature. Observation of this effect for steel B implies a definite influence of grain shape or texture, or both, on toughness anisotropy, since the effects of sulphide volume per cent and morphology are small for this steel. Essentially no plastic energy (0.3 J) is absorbed in the type "bI" fracture of a B(680) S-L specimen tested at -40°C. SEM Examination revealed blunting at the tip of the fatigue pre-crack and no crack extension prior to brittle fracture. Some plastic energy is absorbed in B(680) S-L specimens at -20°, 0° and 25°C but total ductile failure, curve type "e", only occurs at 25°C. Normalizing a B(680) S-L specimen to produce random texture and equiaxed grains alters the curve type to "bIII" in a test at -40°C and a very substantial amount of plastic energy is absorbed, 52 J. The B(740) S-L specimens contain some polygonal grains and are more ductile than B(680) S-L specimens but less ductile than B(800, 860, 920, 1200) S-L specimens which group together at high plastic absorption energies in Table 3 at a temperature of -60°C. The latter result is significant in that (001) texture intensity abruptly drops for soaking temperatures >860°C (Table 4) without any correspondingly significant change in fracture behaviour. This suggests that the pancaked microstructure and not texture is the potent embrittling agent here and leaves the question open whether the (001) texture intensity is insufficient or would have a minor effect even if it was the sole strong component. T-L specimen values are scattered in Table 3 but considerable isotropy exists with S-L specimens for the higher soaking temperatures.

Comparison of S-L specimens A-920 and D-920 at -40°C (Table 3) confirms the decrease of steel toughness which usually accompanies an increase of inclusion content. The effect of inclusion shape (a/c ratio in Table 2) on fracture toughness can be deduced, for example, from S-L specimens C-800 and D-800 at +25°C. Both exhibit ductile fracture, but the energy absorbed by the partially modified steel C with globular inclusions is almost one order of magnitude larger than the energy absorbed by the untreated steel D with elongated inclusions. However, the effect of inclusion shape is more drastic for steels that have been rolled at low temperatures. For example, S-L specimens of steel A-680 with elongated inclusions exhibit a brittle behaviour at test temperatures well above those where the modified steel B-680 is ductile. Furthermore, A-680 and D-800 S-L specimens have about the same inclusion aspect ratio, and D-800 with a higher inclusion content (higher P_{S-L}) should be of lower toughness. Nevertheless, the D-800 S-L specimen tested at 25°C is ductile, while A-680 is brittle at the same temperature. Since both steels must exhibit a strong texture because of their soaking temperature less than 920°C, it is concluded that brittle behaviour is promoted by the pancaked microstructure. However, the B-680 S-L specimen which is also heavily pancaked but has less damaging inclusions, has a behaviour even more ductile at 25°C than the D-800 steel. It thus appears that there is a synergistic effect of inclusion shape and pancaked microstructure on fracture toughness.

CONCLUSIONS

A very marked toughness anisotropy exists between S-L and T-L specimens when pancaked grains are formed on the rolling plane. S-L specimens from these plates are far more susceptible to brittle fracture than T-L specimens. S-L toughness improves and S-L/T-L anisotropy consequently decreases with increasing soaking and finishing temperature, i.e., with increasing amounts of polygonal grains in the microstructure.

There is a synergistic effect of inclusion shape and pancaked microstructure on toughness. Thus, it is particularly important to produce a low sulphide aspect ratio by RE-modification and to reduce the sulphur content when low finishing temperatures are unavoidable.

ACKNOWLEDGEMENTS

Sincere thanks are due to Messrs. W.W. Koch, D. Kiff and E. Cousineau for technical assistance, to Drs. K. Pickwick and C.M. Mitchell for SEM and texture examinations respectively, and to the Foundry, Metal Forming and Mechanical Testing Sections.

REFERENCES

1. J.M. Hodge, R.H. Frazier and F.W. Boulger, Trans AIME 215:745; 1959.
2. K.-H. Schwalbe, Eng. Fract. Mech. 9:795; 1977.
3. G. Bernard, T. Hersant, F. Molièxe and F. Moussy, Mém. Scient. Rev. Métall. 667; 1979.
4. G.R. Speich and W.A. Spitzig, US Steel Corp.; Rep. 76-H-044 (091) - Sect. 1, AD-A100 694; Monroeville, PA; 1981; W.A. Spitzig, Metall. Trans. 14A:471; 1983.
5. V.P. Raghupathy, V. Srinivasan, H. Krishnan and J. Chandrasekharaiah, J. Mater. Sci. 17:2112; 1982.
6. T.J. Baker, K.B. Gove and J.A. Charles, Metals Techn. 183; 1976.
7. A.A. Willoughby, P.L. Pratt and T.J. Baker, "Advances in fracture research (Fracture 81)"; ed. D. François; Pergamon Press; 179; 1982.
8. J. Ménigault, J.Y. Dauphin, J. Foct and G. Mesmacque, Mém. Etudes Scient. Rev. Métall. 80:17; 1983.
9. T.J. Baker and J.A. Charles, J. Iron Steel Inst. 210:702; 1972.
10. D.M. Fegredo, Can. Metall. Quart. 15:21; 1976.
11. C.M. Mitchell and D.M. Fegredo, Can. Metall. Quart. 14:265; 1975.
12. C.M. Mitchell and J.D. Boyd, CANMET Report ERP/PMRL 76-30(J); Ottawa, Ontario; 1976.
13. D.L. Bourell, Metall. Trans. 14A:2487; 1983.
14. D.L. Bourell and O.D. Sherby, Metall. Trans. 14A:2563; 1983.
15. D.M. Fegredo, Can. Metall. Quart. 14:243; 1975.

Relationship Between the Properties and Microstructure of 4% Cr and 2% Cr-2%Mn Steels

R. L. Reuben*, C. S. Wright and T. N. Baker****

**Department of Offshore Engineering, Heriot-Watt University, Edinburgh, UK*
***Department of Metallurgy, University of Strathclyde, Glasgow, UK*

ABSTRACT

Four vacuum melted low carbon steels, two based on 2%Cr-2%Mn and one each on 4%Mn and 4%Cr were solution treated in the range 1150°C to 1300°C and controlled rolled. The steels were finish-rolled between 950°C and 800°C and finally air cooled to room temperature. The mechanical and toughness properties were studied as a function of chemical composition, solution temperature and finish rolling temperature. Measurements were made of the mean free distance, which was correlated with the strength data. The best combination of properties was found for the niobium containing chromium steels with 0.2% proof stresses >900Nmm^{-2} and DBTT <-20°C. NbC particles were observed to outline a boundary network in the 2%Cr steel, but a more uniform finer dispersion was present in the 4%Cr steel. When the mfd was plotted as a function of strength the relationship σ_{yp} -15.4 (2.1Mn%-2.0Cr%) = 135 + 24d$^{-\frac{1}{2}}$ was found, with r equal to 0.83. This is regarded as a first attempt in understanding the factors controlling the strength of acicular steels.

INTRODUCTION

Low alloy steels having a non-polygonal structure have been considered for structural applications. A range of structures including acicular ferrite, bainite and low carbon martensite have been studied and several attempts have been made to quantify the structure/property relationships. These have been discussed by Brozzo et al [1] and by Naylor [2] for structures produced by water quenching after hot rolling. Prior austenite grain size, bainite packet size and lath width have all been suggested as the structural unit determining the mechanical and toughness properties.

The materials studied have included manganese irons [3] and steels [4-9], Cr steels [1,2,10-13] and Ni steels [5]. In particular, lath structures have been quantified in several of these studies, but structures consisting of acicular ferrite have received much less attention. The present work investigated the structure/property relationships in Cr-Mn and Cr steels after controlled-rolling and air cooling to give acicular structures. The influence of both soaking temperature and finishing rolling temperature was studied.

EXPERIMENTAL

The composition of the vacuum melted alloys is given in Table I. Prior to rolling, the material was solution treated (ST) for 1 hr at temperatures in the range 1150°C to 1300°C and then reduced from 30mm square bar to 16mm dia. rod in five passes of ~30% per pass, finishing rolling at temperatures (FRT) between 950°C and 800°C. The alloys were air cooled to room temperature. Mechanical and impact properties were determined and optical and electron microscope structural studies were carried out [14].

RESULTS

A summary of the results is given in Table II.

Table I Compositions of the Alloys in Weight Percent

Alloy	C	Mn	Cr	Nb
A	0.076	3.9	<0.01	0.083
B	0.082	2.09	1.97	0.094
C	0.073	2.06	1.99	<0.005
D	0.12	0.08	4.76	0.093

Si <0.06 Al <0.007 N, Mo, V, P, S <0.004

Table II Mechanical, Impact and Structural Properties

Specimen	0.2% Proof Stress Nmm^{-2}	Ultimate Tensile Stress Nmm^{-2}	55J DBTT, T°C	$d^{-\frac{1}{2}}$ $mm^{-\frac{1}{2}}$
A 1200/950	790	990	-40	24.9
A 1200/870	838	1013	-25	25.2
A 1200/800	929	1110	-18	28.8
A 1300/950	860	1044	-7	19.7
A 1300/800	855	1045	-20	27.4
B 1150/920	865	1107	-14	31.3
B 1150/820	856	1118	-10	31.3
B 1200/950	858	1102	-28	32.7
B 1200/800	906	1100	-40	33.7
C 1150/950	760	970	+28	23.8
C 1150/800	765	991	-14	30.2
C 1200/950	896	1033	-18	29.1
C 1200/800	882	1042	-12	32.2
D 1150/850	923	1240	-24	34.8
D 1150/800	936	1218	-32	37.4
D 1300/950	892	1116	-	31.0
D 1300/800	939	1236	-	35.7

It can be seen that variations in properties as a function of the solution temperature and the finishing rolling temperature did not follow any simple trend, therefore these are briefly considered for each individual alloy.

Alloy A: after a 1200°C ST, with a decrease in FRT a progressive increase in σ_p with a corresponding increase in DBTT was found. However, following the 1300°C ST, σ_p was independent of FRT, but T_C decreased with FRT.

Alloy B: the mean free distance d appeared to be almost constant for all ST and FRT schedules, but as $d^{-\frac{1}{2}}$ increased, T_C decreased. The best combination of properties coincided with the 1200ST/800FRT treatment; viz: σ_p 906Nmm^{-2} and T_C-40°C.

Alloy C: the 1150°C ST treatment resulted in the lowest σ_p values of the whole series, but by raising the ST to 1200°C, increases in σ_p>100Nmm^{-2} were recorded, giving data similar to that obtained for Alloy B.

Alloy D: these specimens showed the smallest mfd figures, with σ_p > 900Nmm^{-2} and T_C < -20°C.

STRENGTHENING IN ACICULAR FERRITE STRUCTURES

Brozzo et al [1] considered that the mechanical strength and cleavage resistance of low C bainites appeared to be controlled by different structural parameters and not, as in the case of polygonal ferrite steels, by the same structural unit. This may also be the case for an acicular ferrite containing islands of an M-A phase.

Correlations between proof stress and austenite grain size, bainite packet size and martensite lath width have been derived, but the relationships between strength and structure for an acicular ferrite steel has received much less attention. In this preliminary attempt we have assumed that the small volume % of the M-A phase can be ignored and we have treated the material as consisting entirely of acicular ferrite. The mean free distance has been measured as previously described [8].

Using the data from Pickering [16], estimates of the strength levels expected from grain size and substantial solid solution strengthening can be made. By subtracting the solid solution strengthening, expressed for the present alloys as 15.4 (2.1%Mn-2%Cr), from the experimental value for the 0.2% proof stress, figures for the grain size strengthening are obtained. The data are shown in Fig. 1 where σ_M(Nmm^{-2})= $\sigma_{0.2PS}$-15.4 (2.1%Mn-2%Cr) is plotted against $d^{-\frac{1}{2}}$(mm$^{-\frac{1}{2}}$). The lower line, F shows strength levels which might be expected in solute-free ferrite due to grain size strengthening plus a friction stress, expressed as $\sigma_{0.2PS}$ = 54+17.4 $d^{-\frac{1}{2}}$.

Clearly, substantial solid solution and grain size effects do not fully account for the strength of Alloys A to D as shown by the line M in Fig. 1. The correlation coefficient, r for the graph is 0.83.

STRUCTURAL OBSERVATIONS

All the alloys which had undergone the 1200/800 treatment showed an acicular ferrite structure together with an M-A phase [15] which for Alloys B, C and D was ∿11% by volume, Fig. 2. (The M-A phase was not quantified for Alloy A). This phase was frequently identified by heavy microtwinning, Fig. 3, while the acicular ferrite was observed to contain a high dislocation density.

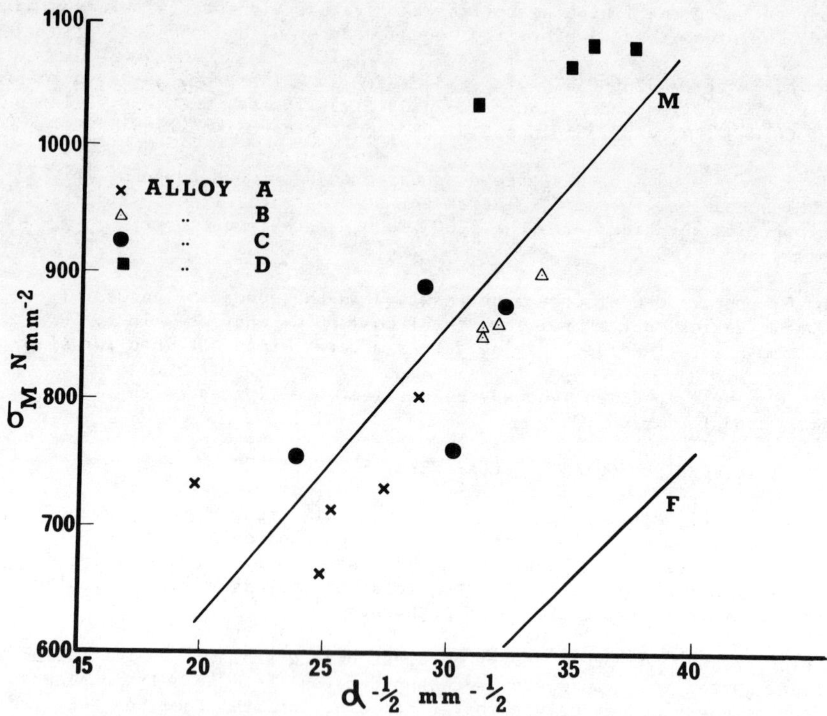

Fig. 1. A plot M of the modified proof stress, σ_M versus the mean free distance d, expressed as $d^{-\frac{1}{2}}$. F is the graph for solute free ferrite.

Some localized precipitation of cementite was found in all the alloys, but it was more pronounced in the niobium free Alloy C. For Alloys B, C and D, x-ray microanalysis showed the cementite to contain chromium.

Niobium carbide, which was identified by a combination of selected area electron diffraction and microanalysis, was however, the predominant precipitate in Alloys B and D. These particles also contained a small amount of chromium. There were noticeable differences in the size and distribution of the NbC precipitation in the two alloys. In Alloy B, the particles were 10-20nm in size, Fig. 4, whereas in Alloy D a finer particle size of <10nm was found, Fig. 5, together with a low density of larger particles, up to 50nm in size. The NbC in Alloy D was uniformly distributed in both the acicular ferrite and the M-A regions. However, in Alloy B these precipitates formed clearly defined networks, Fig. 4, with the individual cells being ~0.5 microns in size.

Since these networks cut across both the ferrite and the M-A phase boundaries, it is suggested that the NbC particles must have precipitated in austenite, probably at subgrain boundaries formed as a result of recovery after the controlled rolling schedule was complete.

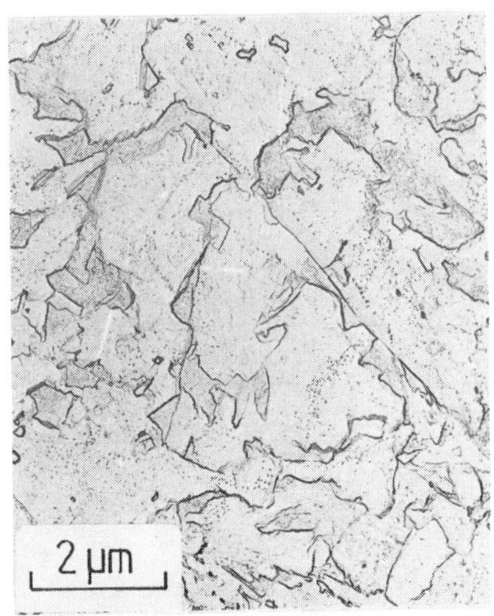

Fig. 2. B1150/820. TEM micrograph showing M-A islands dispersed in acicular ferrite (carbon replica).

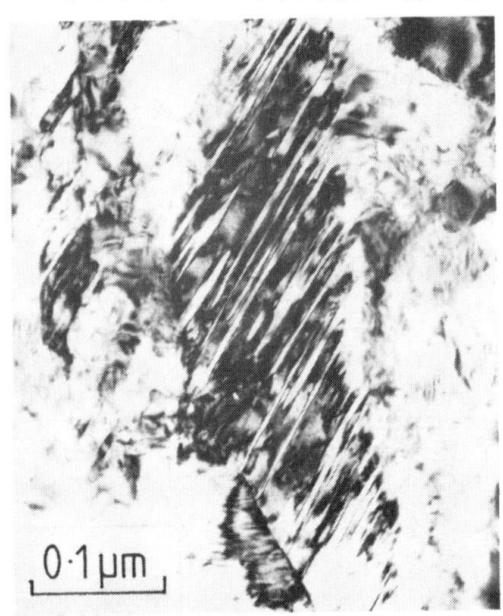

Fig. 3. B1150/820. TEM micrograph of a M-A island showing an internally twinned lath of martensite (thin foil).

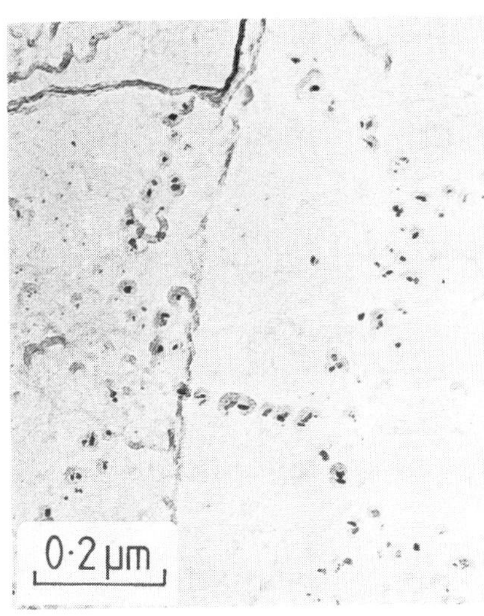

Fig. 4. B1150/820. TEM micrograph showing cellular distribution of niobium carbide particles (carbon replica).

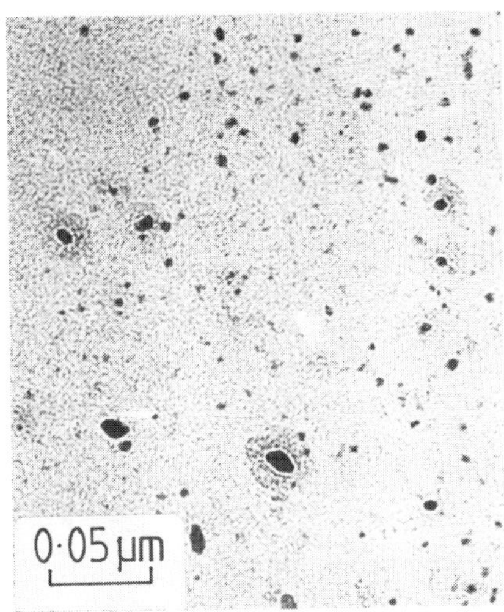

Fig. 5. D1150/800. TEM micrograph showing uniform distribution of niobium carbide particles.

DISCUSSION

The correlation between strength and grain size in Fig. 1 can explain most of the strengthening found experimentally as an 0.2% proof stress in Alloys A to D. However, other components of strengthening are present and are significant. The structural observations of the alloys rolled to 800°C have shown that an M-A phase is presented in quantities of around 10%. Also, electron microscopy has indicated that a fine particle dispersion exists in the Cr-Mn and Cr alloys and that larger particles outline an 0.5 micron size unit, probably a prior austenite sub-grain network. This latter observation suggests that following an 800°C finish rolling temperature, the austenite did not fully recrystallize. The samples also supported a substantial dislocation density. All these observations provide evidence for additional sources of strengthening. Work is in progress to quantify these contributions and an analyse of these results will lead to a fuller understanding of the mechanisms controlling the strength of acicular ferritic steels.

REFERENCES

1. P. Brozzo, G. Buzzichelli, A. Mascanzoni and M. Mirable, Met.Sci.11, 213, (1977).
2. J.P. Naylor, Metall. Trans A, 10A, 861, (1979).
3. M.J. Roberts, Metall.Trans. 1, 3287, (1970).
4. J.H. Woodhead and J.A. Whiteman in 'Processing and Properties of Low Carbon Steel' (edited by J.M. Gray), p145, Met.Soc.AIME, New York, (1973).
5. L.O. Norstrom, Scan.J.Met. 5, 159, (1976).
6. T.N. Baker and N.A. McPherson in Heat Treatment '76, p151, Metals Soc., London (1976).
7. T.N. Baker and R.L. Reuben, in Advances in the Physical Metallurgy and Applications of Steels, p213, Metals Soc., London, (1982).
8. R.L. Reuben and T.N. Baker, Met.Technol. 11, 6, (1984).
9. R.L. Reuben and T.N. Baker, Mat.Sci.Eng., 65, 199, (1984).
10. D.W. Smith and R.F. Hehemann, J.Iron Steel Inst., 209, 476, (1971).
11. G. Buzzichelli and A. Mascanzoni,,Mem.Sci.Rev.Metall., 73, 335, (1976).
12. H. Nevalainnen, V. Ollilainen and L. Vihavainen, Scan.J.Met., 5, 193, (1976).
13. J.Y. Koo and G. Thomas, Metall.Trans.A., 8A, 525, (1977).
14. R.L. Reuben, C.S. Wright andT.N. Baker - to be published
15. V. Biss and R.L. Cryderman, Metall.Trans. 2, 2267, (1971)
16. F.B. Pickering in Microalloying '75, p9, Union Carbide Corp., New York, (1977).

Relationship Between Microstructure and Notch Toughness Properties in 2.25 Cr-1Mo Steel Weld Metal

R. Chandel*, R. Orr*, J. Gianetto*, J. T. McGrath*, B. M. Patchett and C. Bicknell****

**CANMET, Energy, Mines and Resources Canada, Ottawa K1A 0G1, Canada*

***Department of Mineral Engineering, University of Alberta, Edmonton, Canada*

ABSTRACT

Experimental welds were prepared in 2.25 Cr - 1 Mo steel by the submerged arc and the gas metal arc narrow gap welding processes. Optimum notch toughness properties (>54 J at -40°C) were achieved in the post-weld heat treated condition in weld metals exhibiting a fine bainitic microstructure and a low inclusion content (oxygen + sulphur < 0.035%). The resistance to cleavage fracture was related to bainitic ferrite packet size while ductile fracture resistance was improved by reducing the inclusion volume fraction.

KEYWORDS

Narrow gap welding; 2.25 Cr - 1 Mo weld metal; notch toughness; strength; bainite; inclusions; reactor vessels.

INTRODUCTION

Thick-wall reactor vessels for the hydrotreating of heavy oil and tar sands bitumen have stringent mechanical property requirements for the 2.25 Cr - 1 Mo base material and weld metal. For example, the notch toughness requirement for heavy section welds specifies 54 J charpy impact energy at -40°C after post-weld heat treatment.

Narrow gap welding processes are being developed to provide the required weld properties as well as improved productivity in the joining of thick sections (1). The objective of the present program was to study the relationship between weld metal microstructure and notch toughness for a series of welds prepared in 2.25 Cr - 1 Mo steel by the submerged arc (SAW-NG) and the gas metal arc (GMAW-NG) narrow gap welding processes.

EXPERIMENTAL PROCEDURE

Materials and Weld Procedures

Narrow gap welds were made in SA-387 Grade 22, Class 2, 2.25 Cr - 1 Mo steel, 38 mm in thickness. The submerged arc narrow gap welds (SAW-NG) were prepared using a solid wire and three fluxes with a basicity index ranging from 1 to 3. A two pass per layer technique was used in joining a parallel-sided weld preparation with a gap width of 15 mm. The Miller SAW-NG system, with a square wave AC 1000 power source was employed operating at a heat input of 2 kJ/mm. A single pass per layer technique, with a gap width of 12.7 mm, was used for the narrow gap weld prepared by the gas metal arc narrow gap process (GMAW-NG) at a heat input of 3.7 kJ/mm. A Kobe twist wire (2) with an Ar + 20% CO_2 shielding gas mixture were the weld consumables. The welding consumables are listed in Table 1.

TABLE 1 Welding Consumables

Weld	Process	Wire	Flux*	Gas mixture
W_1	SAW	SD_2 Cr Mo	OP121TT	-
W_2	SAW	SD_2 Cr Mo	Linde 124	-
W_3	SAW	SD_2 Cr Mo	OP76	-
W_4	GMAW	Kobe twist wire	-	80 Ar + 20 CO_2

*basicity index (B.I.) OP121TT = 3.0, OP76 = 2.7, Linde 124 = 1.0

The compositions of the base material and welding wires are listed in Table 2. The welds were given a post-weld heat treatment of 10 h at 690°C.

TABLE 2 Composition of Weld Materials, wt %

Material	C	Mn	Si	S	P	Ni	Cr	Mo	Al	O
Weld W_1	0.10	0.55	0.22	0.009	0.013	0.21	2.68	1.05	0.02	0.026
Weld W_2	0.06	0.67	0.49	0.011	0.009	0.21	2.63	1.06	0.015	0.085
Weld W_3	0.10	0.53	0.19	0.007	0.007	0.20	2.70	1.03	0.015	0.022
Weld W_4	0.09	0.68	0.24	0.013	0.006	0.20	2.22	1.03	0.01	0.052
Base plate	0.10	0.49	0.20	0.028	0.008	0.33	2.27	1.03	0.01	-
SAW wire	0.12	0.48	0.14	0.007	0.007	0.19	2.83	0.82	0.045	-
GMAW wire	0.08	0.46	0.33	0.009	0.004	0.15	2.33	1.08	-	-

All weld metal tensile specimens were machined with their long axis parallel to the welding direction. Charpy impact specimens were taken from the top and bottom locations and were notched through thickness at the 1/4 W and 1/2 W positions (W = weld width). The weld metal microstructures were quantified by optical microscopy. Fractured surfaces were characterized by the SEM.

RESULTS

MECHANICAL PROPERTIES

All welds met the reactor vessel design strength requirements (UTS = 515-690 MPa, YS = 420 MPa min), with the GMAW-NG weld having the highest yield and ultimate strengths. Full Charpy transition curves for the welds are shown in Fig. 1. All welds, with the exception of weld W_2 met the notch toughness property requirement of 54 J at -40°C. Weld W_3 had the highest toughness over the full test temperature range.

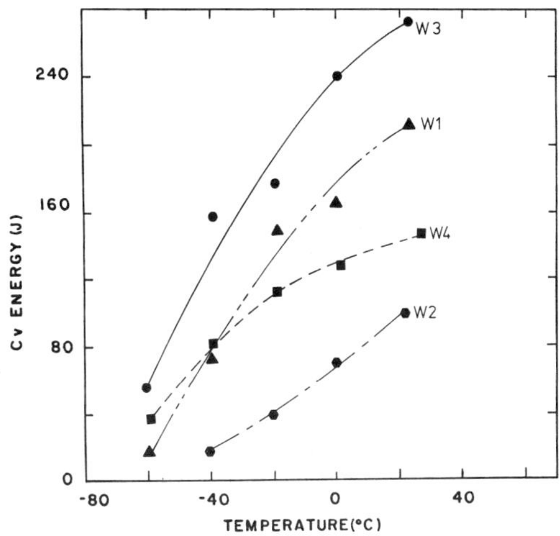

Fig. 1 - Charpy impact transition curves for narrow gap welds. Volume fraction of inclusions: W_1 = 0.48%, W_2 = 0.86%, W_3 = 0.40%, W_4 = 0.50%.

MICROSTRUCTURE

The microstructure of the welds was characterized in terms of as-deposited and reheated structure, the type and amount of transformation products and the volume fraction of inclusions. In the SAW-NG welds, the individual beads contained an as-deposited, columnar region and a reheated region in which the as-deposited structure had been subjected to the thermal effects of adjacent weld beads. In the reheated region, as shown in Fig. 2(a), the prior austenite grains varied from polygonal grains, 30 to 40 μm in size, to coarser elongated grains depending upon the austenitizing temperature that was reached prior to transformation. The as-deposited region contained much coarser (150 to 250 μm wide) elongated columnar grains [Fig. 2(b)]. Measurements made at the centre of the weld beads through full weld sections indicated that the reheated, recrystallized and transformed region occupied 72% of the total structure in weld W_3 and 65% in welds W_1 and W_2. The SAW-NG welds W_1, W_2 and W_3 featured a totally bainitic structure in both the as-deposited and the reheated regions. Using the terminology of Torronen (3),

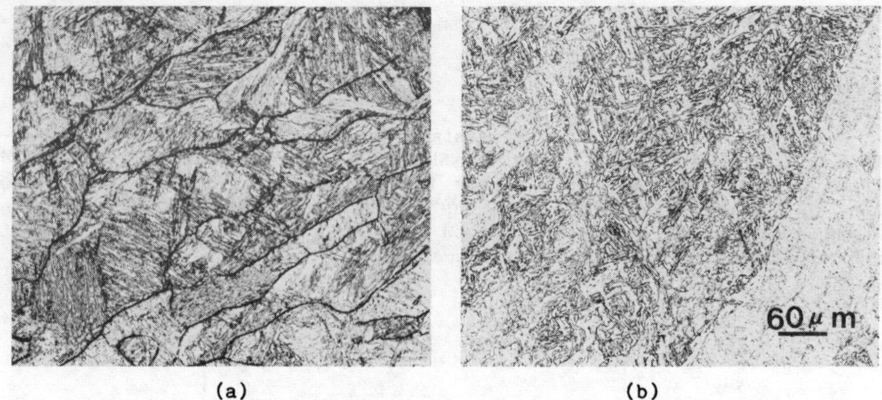

(a) (b)

Fig. 2 - Weld metal microstructure for SAW-NG weld W_3
(a) coarse-grain reheated region
(b) as-deposited columnar region

the bainitic structure within a prior austenite grain was composed of individual ferrite packets which formed bundles. These features are identified in Fig. 3(a). For welds W_1 and W_3, the packets were elongated in shape and had a width of approximately 3 μm. The packets were slightly coarser in weld W_2. Although transmission electron microscopy (TEM) remains to be done to fully characterize the bainitic structure, it was assumed from other work on similar bainitic structures in a quenched and tempered Cr-Mo-V alloy steel that there were high-angle boundaries between packets. Torronen (3) and Kotilainen (4) observed that the orientation relationship between neighbouring packets in the same bundle was mainly a twin orientation. This observation is important in defining the resistance of the bainitic structure to cleavage fracture.

The weld beads in the GMAW-NG weld W_4 contained 100% reheated structure. Although the individual beads were larger than the SAW-NG weld beads, the higher heat input per layer was sufficient to totally recrystallize the immediate underlying bead. As shown in Fig. 3(b), the bainitic ferrite packets were irregular in shape with less tendency to the elongated packets observed in the SAW-NG welds W_1 and W_3. Packet boundaries were also not easily distinguished.

The results of the measurement of inclusion volume fraction for all welds using the Quantimet 900 are indicated in Fig. 1. The particles were spherical in shape, and, although the chemical composition was not identified, they were assumed to be oxides and sulphides. The SAW-NG weld, W_2, prepared with the Linde 124 low-basicity flux had the highest volume fraction of inclusions. This observation corresponded with weld W_2 having the highest oxygen + sulphur content as listed in Table 2. The GMAW-NG weld W_4 had the second-highest volume fraction of inclusions.

Fractured surfaces of welds W_1, W_2, W_3 and W_4 impact tested at -40°C were examined in the SEM. At this temperature, all weld samples exhibited a combination of cleavage facets and ductile tearing ridges. The cleavage facets were generally <10 μm in size, suggesting that they could correspond to the bainitic ferrite packet size.

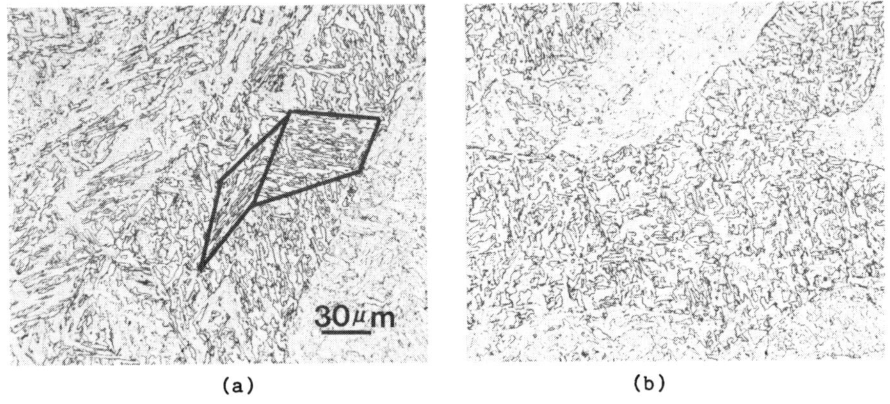

Fig. 3 - Bainitic structure in SAW-NG and GMAW-NG welds
(a) weld W_1 - bundles of packets are outlined
(b) weld W_4

DISCUSSION

The notch toughness tests conducted on narrow-gap welds in 2.25 Cr - 1 Mo steel in the post-weld heated condition produced by the SAW-NG and the GMAW-NG processes have shown that the notch toughness requirement of 54 J at -40°C was achieved in three of four welds tested. The notch toughness properties can be explained on the basis of weld microstructure.

In the upper shelf region of the Charpy transition curve, where the fracture mode is by ductile tearing, inclusions act as sites for microvoid coalescence. Thus, an increased volume fracture of inclusions should be associated with decreased resistance to ductile fracture and a lower Charpy impact energy (5). Such a correlation between inclusion content and Charpy impact energy in the ductile fracture regime was found for the four welds studied. Weld W_3 with the lowest oxygen + sulphur content and lowest volume fraction of inclusions had the highest toughness at 25°C, while weld W_2 with the highest inclusion content had the lowest toughness.

At lower temperatures, particularly in the transition region, there was a mixed mode of fracture with cleavage and ductile tearing taking place. The resistance to ductile tearing in the transition region would still be controlled by inclusion content. Various microstructural features including prior austenite grain size, bainite bundle size and bainite ferrite packet size can contribute potentially to the control of cleavage fracture resistance. The fractographic evidence has indicated that the unit fracture length for the four weld samples was related to ferrite packet dimensions. This implied that the cleavage fracture mechanism involved initiation and propagation across a ferrite packet. Continued fracture would require a reinitiation at the packet boundaries.

Confirmation that high-angle boundaries separated the bainitic ferrite packets remains to be shown by transmission electron microscopy. For the SAW-NG welds, the ferrite packet dimensions were similar with packets in weld W_2 being slightly coarser. Packet size and shape are a function of bainite transformation temperature and cooling rate (3,4). The ferrite packets were more irregular and less elongated in shape in the GMAW-NG weld W_4. The shape change may be a function of cooling rate. Weld W_4 was prepared with a higher heat input, 3.7 kJ/mm, and hence a slower cooling rate than the SAW-NG welds. It is not certain whether the irregular-shaped packets are more resistant to cleavage fracture than the elongated packets.

It is well known (6) that cleavage fracture resistance in bainitic structures in C-Mn steels is related to prior austenite grain size or bainite bundle size. In these steels the bainite structure contains parallel bainitic ferrite packets with a small misorientation across packet boundaries. Thus a crack can initiate and easily propagate across a complete bainite bundle. The unit crack length or cleavage facet size is therefore represented by the bainite bundle size. These observations contrast with the present study of 2.25 Cr - 1 Mo steel weld metal in which the packet size rather than the bainite bundle size controlled cleavage fracture initiation.

CONCLUSIONS

For the narrow-gap welding of thick section 2.25 Cr - 1 Mo steel by the SAW-NG and the GMAW-NG processes, weld metal notch toughness properties, i.e., 54 J at -40°C, can be obtained if the choice of welding consumables and operating parameters provide a fine bainitic microstructure with a low inclusion content ($O_2 + S < 0.35\%$). For the SAW-NG process, wires with a low impurity content ($S < 0.01\%$) and highly basic fluxes are the recommended consumables. The GMAW-NG process can produce welds with low inclusion content if the shielding gas contains $\leq 20\%$ CO_2. Heat inputs of <4 kJ/mm for narrow-gap welding will provide cooling rates resulting in fine, bainitic transformation products.

ACKNOWLEDGEMENTS

The authors wish to thank M.W. Letts for his assistance in the metallographic examination, Dr. M. Shehata and B. Casault for their work on inclusion characterization, and D. Lusk for NDT evaluation of the weldments.

REFERENCES

1. J.T. McGrath, B.M. Patchett, W.R. Tyson and R.F. Knight, Can Met Quart, 22, 274 (1983).

2. S. Kimura, I. Ishihara and Y. Nagai, Welding J., 58, 44 (1979).

3. K. Torronen, "Microstructural Parameters and Yielding in a Quenched and Tempered Cr-Mo-V Pressure Vessel Steel"; Tech. Res. Centre Finland Publ. (1979).

4. H. Kotilainen "The Micromechanisms of Cleavage Fracture and their Relationship to Fracture Toughness in a Bainitic Low Alloy Steel"; Tech. Res. Centre Finland Publ. (1980).

5. T. Gladman, B. Holmes and I.D. McIvor, Proc BSC/ISI Conf., The Effects of Second Phase Particles on the Mechanical Prop. of Steel, 68 (1971).

6. F.B. Pickering, Symp Transformation and Hardenability in Steels, Climax Moly. Co. of Mich, 109 (1967).

Fracture Behaviour of High-Strength Steels at Low Temperatures

M. Yao, R. P. Zhang, Y. Zhou and J. X. Zhan

Harbin Institute of Technology, Harbin, People's Republic of China

ABSTRACT

In this work, fracture behaviour of high-strength steels with tempered martensite structure are investigated over a temperature range from -196°C to 20°C. The cold-short phenomenon in high-strength steels is compared with that in mild steels. The influences of prolonged holding time during austenizing, temper embrittlement and carbon content are studied too.

KEYWORDS

High-strength steel; fracture toughness; cold-short phenomenon; quasi-cleavage fracture; temper embrittlement.

INTRODUCTION

Until now, investigations on cold-short phenomenon were carried out mainly in mild steels [1-4], but rarely in high-strength steels with tempered-martensite structure. In this work, low temperature fracture behaviour is investigated on a low-alloy steel 34SiMnCrNiMoV ("34" for short) heat-treated to high-strength level. The influence of prolonged holding time during austenizing and that of temper embrittlement are studied too. In order to study the effect of carbon content on low temperature fracture behaviour, another steel 46SiMnCrNiMoV ("46" for short) with the same alloying elements but higher carbon content is also used.

EXPERIMENTAL PROCEDURE

The chemical compositions and heat treatments of the used steels are given in TABLE 1 and TABLE 2.
Tension and fracture toughness tests were carried out over a temperature range from -196 ℃ to 20 ℃. Specimens used in this

investigation are 2.6mm thick. Fracture toughness tests were carried out in accordance with ASTM Method E740-80 using surface-precracked specimen. Fracture surfaces were observed by using SEM. The percentages of quasi-cleavage area below the crack tip in the flat part of the fracture surface at different temperatures were established.

TABLE 1 Chemical Composition of Steels

Steel	Elements, weight percent								
	C	Si	Mn	Cr	Mo	V	Ni	S	P
"34"	0.34	1.40	0.98	1.18	0.45	0.024	0.38	0.002	0.02
"46"	0.46	1.28	0.75	1.13	0.57	0.011	0.32	0.003	0.012

TABLE 2 Heat Treatment of Steels

Steel	Condition	Heat Treatment Procedure
"34"	1	15min at 930 °C, O.Q.; Tempered 2hr at 280 °C.
	2	5hr at 930 °C, O.Q.; Tempered 2hr at 280 °C.
	3	15min at 930 °C. O.Q.; Tempered 2hr at 400 °C.
"46"	4	15min at 930 °C, O.Q.; Tempered 2hr at 280 °C.

RESULTS AND DISCUSSIONS

Experimental data of yield strength σ_y, fracture toughness K_{1C}, and percentages of quasi-cleavage area in fracture surface at different temperatures are summarized in TABLE 3 and Fig. 1. As examples, SEM fractographs of steel "34", normally heat treated (condition 1) and fractured at 20 °C, -80 °C and -196 °C, are shown in Fig.

Features of Cold-short Phenomenon in High-strength Steels

It can be seen from experimental data that, with the lowering of test temperatures, the yield strength of steels increases, whereas the fracture toughness decreases. These results indicate that, as in mild steels, cold-short phenomenon occurs also in high-strength steels with tempered martensite structure. But there exist some important differences between them. It is well known [1-4] that, in mild steels, the ductile-brittle transition occurs in a rather narrow temperature range; the energy to fracture above this temperature range (upper shelf energy) is much higher than that below this range (lower shelf energy). This kind of cold-short behaviour is related to a change in the fracture mechanism: from microvoid coalescence at high temperatures to cleavage at low temperatures. Whereas, in high-strength steels, the fracture toughness decreases with temperature in a gradual manner; the difference between fracture toughness at room temperature and that at low temperatures is much smaller. This kind of cold-short behaviour is a result of a change of fracture mode from low energy microvoid coalescence to quasi-cleavage. As we know, at room temperature, the toughness of high-strength steels is much lower than that of mild steels. But at low temperatures, their difference is diminished; and at -196°C, the fracture toughness of high-strength steels is near or even higher than that of mild steels, which is about 80-130 kgf-mm$^{-3/2}$[5-6].

TABLE 3 Summary of Experimental Data

Steel	Condition	Testing Temperature °C	Fracture Toughness kgf-mm$^{-3/2}$	Yield Strength kgf/mm^2	Percentage of Quasi-cleavage Area*
		20	195.7	157	15
		-40	183.0	169	30
	1	-80	146.5	173	60
		-120	118.0	174	85
		-196	85.0		100
		20	247.0		5
		-40	212.5		20
"34"	2	-80	182.4		50
		-120	135.2		90
		-196	115.0		100
		20	196.2	145	20
		-40	152.5	151	60
	3	-80	116.0	159	80
		-120	86.5	162	100
		-196	76.5		100
		20	154.8	174	30
		-40	132.0	181	65
"46"	4	-80	107.0	186	90
		-120	91.0	190	100
		-196	79.0		100

*Established just below the crack tip in the flat part of the fracture surface.

Another important feature of low temperature fracture behaviour of high-strength steels is in the fracture morphology in the transition temperature range. In both mild and high-strength steels a mixed fracture can be observed in the transition temperature interval, but they are different in morphology. For mild steels, the mixed feature of fracture can be observed visually on the fracture surface: the cleavage (crystalline) facets gather in the middle of fracture surface, and is surrounded by fibred áreas, which gradually diminish when the temperature is lowered. So, the fracture is "macro-mixed" in the transition temperature interval. For high-strength steels, some changes in macroscopic morphology of fracture surface can also be observed: the percentage of shear lip diminishes while the flat area in the middle of fracture surface increases with the lowering of temperature. But it can be discovered under SEM that, even at the same position in the flat area of the

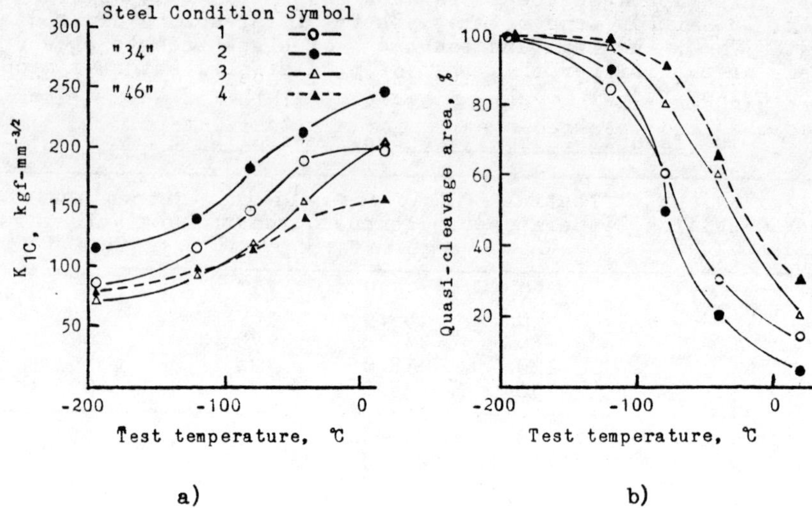

Fig. 1. Effect of test temperature on fracture behaviour of high-strength steels:
a) Fracture toughness, K_{1C}
b) Percentage of quasi-cleavage areas below crack tip in central part of fracture surface.

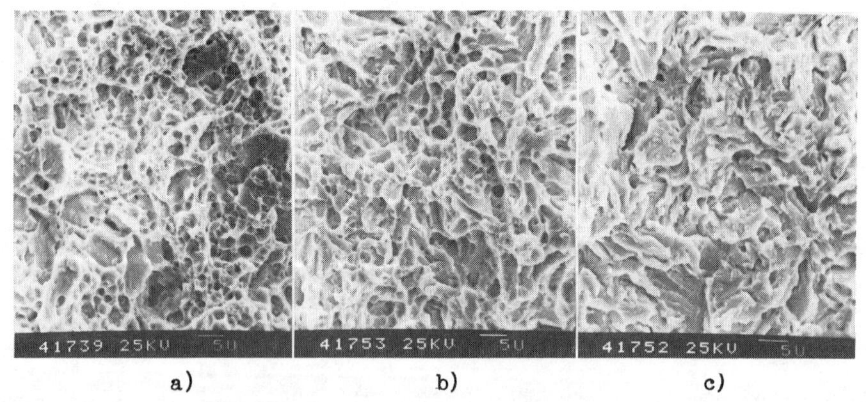

Fig. 2 SEM fractographs of normally heat-treated 34SiMnCrNiMoV steel at
a) 20 ℃; b) -80 ℃; c) -196 ℃

fracture surface, the fracture pattern is mixed and consists of randomly distributed dimples and quasi-cleavage facets, percentage of the latter increases with the lowering of temperature. So, in high-strength steels, the fracture is "micro-mixed" in the transistion temperature interval.

These features of low temperature fracture behaviour of high-strength steels may be related to their microstructure features. In quenched and tempered steels, martensites contain a lot of dislocations and precipitated particles, and their lattices are severely distorted; besides that, there may exist, during austenizing, a number of undissolved carbide particles. In comparison with annealed structure, the resistance to microvoid coalescence of high-strength steels is significantly reduced, while their resistance to cleavage/quasi-cleavage may differ little. So in high-strength steels difference between energy(toughness) to microvoid coalescence and that to quasi-cleavage is not so large as in mild steels.

Furthermore, it is believed [7] that in quenched structure the distribution of carbon is not uniform. Then, the micro-areas with higher carbon content and strength tend to fracture by quasi-cleavage mechanism, while the micro-areas with lower carbon content tend to fracture by microvoid coalescence mechanism. From these points, the features of low temperature fracture behaviour of high-strength steels can be explained satisfactorily.

Influence of Holding Time During Austenizing

It can be seen from Fig. 1 that, when the holding time during austenizing is prolonged, the fracture toughness of high-strength steels is improved and the percentage of quasi-cleavage micro-areas in fracture surface at the same temperatures is reduced. This can be related to the facts that carbide particles may dissolve more completely during prolonged austenizing and the structure obtained is more homogeneous.

Influence of Temper Embrittlement

In our experiment, temper-embrittle treatment has little effect on fracture toughness of high-strength steels at room temperature, when fracture occurs mainly by microvoid coalescence; but, at lower temperature, it reduces the toughness of steel and increases the percentage of quasi-cleavage area in fracture surface. Our results coincide with that in [8]. So,the temper embrittlement has little effect on the resistance to microvoid coalescence, but reduces the resistance to quasi-cleavage.

Influence of Carbon Content

By comparing the fracture behaviour of "34" with that of "46" at different temperatures, it can be seen that carbon reduces the toughness of steels at all temperatures. But the effect is more significant at higher timperatures when fracture occurs mainly by microcoid coalescence, and the effect is rather small at lower temperatures when fracture occurs mainly by quasi-cleavage.

CONCLUSIONS

1. As in mild steels, cold-short phenomenon occurs also in high-strength steels. But the effect of temperature on toughness is not so strong as in mild steels and the decreasing of fracture toughness with temperature is in a more gradual manner.

2. The cold-short phenomenon in high-strength steels is related to a change of fracture mode from low-energy microvoid coalescence to quasi-cleavage.

3. The morphology of fracture surface of high-strength steels in the transition temperature interval is "micro-mixed": SEM pattern at the same position in the fracture surface consists of randomly distributed dimples and quasi-cleavage facets.

4. Prolonged holding during austenizing improves the fracture toughness not only at room temperature but also at low temperatures.

5. Temper-embrittle treatment has little effect on fracture toughness of steels at room temperature, but reduces it at low temperatures.

6. Carbon reduces the toughness of high-strength steels. The effect is significant at high temperatures, and is small at low temperatures.

REFERENCES

1. J. F. Knott, Fundamentals of Fracture Mechanics, Butterworths, London (1973).
2. A. S. Tetelman and A. J. McEvily, Fracture of Structural Materials, John Wiley & Sons, Inc., New York (1967).
3. W. S. Pellini and P. P. Puzak, Trans. ASME, ser. A, 86, 429 (1964).
4. M. Yao, Y. K. He, D. M. Li and C. Z. Zhou, Proc. of ICF6, New Delhi, Vol. 2, p. 1423 (1984).
5. F. M. Bouldger et at, Proc. of ICF2, p. 180 (1969).
6. E.T. Wessel and W.G.Clark, Proc. of ICF2, p.825(1969).
7. B.V. Rao and G.Thomas, Proc. of Int. Conf. on Martensitic Transformations, p. 12(1979).
8. J.E. King et al, Proc. of ICF4, vol.2, p.299(1977).

Fracture Toughness Characteristics of Austenitic and Martensitic High-Chromium Cast Irons

S. B. Biner

Bradley University, Department of Manufacturing, Peoria, IL 61625, USA

ABSTRACT

The fracture behavior of a series of austenitic and martensitic high-chromium cast irons (15%Cr & 2.7%C) containing a constant volume percentage of M_7C_3 eutectic carbides in distributions that, varied from continuous network to isolated globules was studied.

In these alloys, fracture toughness was found to be controlled by attainment of a critical stress-state ahead of the crack tip despite the dimpled fracture surfaces between the failed carbides by cleavage mode. For austenitic alloys this stress-state is strongly influenced by the eutectic carbide distribution, whereas the martensitic alloys exhibited consistent fracture toughness values, independent from the eutectic carbide morphology, due to uniformity of the secondary carbide precipitation during hardening heattreatments.

KEYWORDS

High chromium cast iron, fracture toughness, eutectic carbide, abrasive-wear, molybdenum.

INTRODUCTION

Although high-chromium (high-cr) cast irons have excellent resistance to abrasive wear, their applications are limited, because of their poor fracture toughness characteristics. The metallurgy of these alloys has been studied extensively and overviewed in [1-3]. However, data relating to fracture toughness is scarce [4-6]. In this study, the normal continuous network of as-cast eutectic carbide structure was altered into isolated globules by additions of molybdenum and application of high temperature heattreatments. Failure mechanism in these alloys was studied by using specimens containing sharp cracks and blunt notches and correlated with the microstructural features.

EXPERIMENTAL STUDIES

The chemical composition of the alloys are given in Table 1 prepared by induction melting. The bars 15 mm x 20 mm x 300 mm in dimensions were cast in green sand molds from a temperature of about 1450°C. High temperature heattreatments were performed in a vacuum furnace at 1180°C. After certain holding periods cooling was expedited by an argon jet. In order to obtain martensitic matrix, specimens re-heated to an austenization temperature of 975°C and cooling was again in the vacuum furnace.

Table 1. Chemical Composition of Cast Irons wt%

Alloy No.	I	II	III	IV
C	2.79	2.79	2.60	2.60
Cr	16.2	16.0	15.7	15.2
Si	0.65	0.66	0.68	0.72
Mn	0.28	0.29	0.28	0.27
Mo	----	0.69	2.23	4.03

Fracture toughness tests specified by ASTM E-399 were performed in three point bending on 12 mm x 18 mm cross-section fatigue precracked specimens. The slow notch bending tests were carried out on charpy sized specimens without any fatigue pre-cracking. The notches in the specimens were 2 mm in depth, having 60 degree included angle and root radii of 0.25, 0.4 and 0.7 mm. The tensile stress acting ahead of the notch at fracture (σ_F)was determined from the model suggested by Wilsaw et. al. [7]. For a given austenitic alloy, σ_F values were almost identical for each notch geometry (maximum ∓ 3% difference) and the martensitic alloys exhibited a constant σ_F value for all cases, irrespective to high temperature heattreatment and to the notch type.

The quantitive data related to eutectic carbide distribution (i.e. volume, mean carbide particle size Co, and spacing, Lo) of the alloys for each holding period were determined in two perpendicular planes by linear intercept method.

RESULTS

The effects of additions of Mo as an extra alloying element to the base 2.7%C and 15%Cr iron composition were only considered for the carbide phase, since its effects on matrix properties of high-Cr cast irons are well established [1-3]. There was no visible effect of Mo on the as-cast structure of eutectic carbides. Soaking for various times up to 72 hours at 1180° C did not cause any noticeable variation in the eutectic carbide volume percentage and remained to be approximately 25.0%. However in the case of Mo containing alloys continuous network was broken up and a marked spheroidization and coarsening was observed. This is achieved at shorter heattreatment periods with increasing Mo content. (Fig. 1).

The variation of fracture toughness of the austenitic and martensitic alloys with holding periods at 1180°C are summarized in Figures 2a and 2b respectively. For austenitic alloys, as a optimum condition, it will be seen that, 2% Mo additions and 8h heattreatment at 1180°C, there is a 35% increase on the unheattreated alloy and 50% increase over that Mo free composition. Figure 2a also indicates that for each addition level there is a critical heattreatment period which gives the maximum fracture toughness value of that addition level. For the martensitic alloys as can be seen from Fig. 2b, neither alloy addition nor the application of high temperature heattreatments produced any significant effect. An almost constant level of fracture toughness values was observed.

The fracture surface of the austenitic alloys was dominated by cleavage cracking of the eutectic carbides and gave substantially higher area percentage than the measured quantitatively on the flat polished surfaces. In the martensitic alloys the failure of eutectic carbides occurred to a lower extent than in the austenitic alloys in agreement with the work of Gahr et.al.[8]. In both austenitic and martensitic alloys remaining fracture surfaces exhibited a dimpled failure of the matrix. Formation of voids occurred around the precipitated secondary carbides, but there was no apparent necking nor extensive void growth. Due to the large amount of secondary carbide precipitation resulting from hardening heattreatments, a finer and higher density of dimpled failure was observed in the martensitic alloys. (See Fig. 3).

MODELING OF FRACTURE MECHANISM AND DISCUSSION

For the austenitic alloys, the stress values obtained by using the dislocation pile-up mechanism [9] for nucleation of cracks in the eutectic carbides are very much lower than the failure stress (σ_F) values determined experimentally from notch bending specimens. Therefore the final event loading to fracture of the austenitic alloys could be described as, the breaking of the matrix ligaments between two or more adjacent cracked eutectic carbides. In the absence of a macro crack, the total work done for the failure of the austenitic matrix, as a result of micro-cracks within the eutectic carbides as an infinite body could be approximated as:

$$G_A = \frac{\sigma_F^2}{E} \; a \; (1-\nu^2) \tag{1}$$

where σ_F is the failure stress values obtained from notched fracture tests, a is the crack length in the eutectic carbides (equivalent to mean carbide size Co) E is the Young's modulus and ν is the Possion's ratio. In order above assumptions made to be valid, at fracture G_A values should be the same order of the magnitude for a wide range of eutectic carbide size, since matrix characteristics of the alloys are the same. In fact, by inserting the experimental σ_F values together corresponding Co values $G_A = 137.6 \mp 24$ J/m^2 was found. This enabled calculations to be made for the σ_F values of the alloys (which were not tested in notch bending test) by using their eutectic carbide size.

It is now well documented that, for the correlation of fracture toughness from a micromechanistic approach, the stress/strain state determined for the micro-scale should act over a microstructurally characteristic distance or process zone ahead of the macroscopic crack [10-12]. Thus fracture toughness could be expressed as:

$$K_{IC} = \sigma_F \sqrt{2\pi x_0} \tag{2}$$

The process zone (X_0) was taken to be 3.5 times the eutectic carbide spacing. By using eq. 2 and 3, the calculated K_{IC} values for all austenitic alloys were compared with experimentally obtained valid K_{IC} values in Fig. [5].

The model adopted here is analogous to that for cleavage or intergranular failure of steels in the lower shelf fracture region, although the austenitic matrix failed in a partial dimpled fracture mode. This contradictory behavior could be explained by considering the strong influence of stress tri-axiallity on void growth from the models developed for the ductile fracture [13-14]. In the case of austenitic high-Cr cast irons, the failure of large eutectic carbides could limit the amount of plastic deformation occurring in the austenite. However, local stress tri-axiallity which controls the final ductile rupture process is elevated by the microcracks in the eutectic carbides. Thus, the stress component becomes more dominating term than the strain component for void growth in these alloys, yielding to a stress-state dependent fracture mechanism. This can also be substantiated by observation of consistent fracture stresses σ_F independent from notch geometry. Therefore, fracture in these alloys controlled by size and distribution of large eutectic carbides as indicated by eq. (2) and (3) and as shown in Fig. (4a) with experimental values.

In the case of the martensitic alloys, the dislocation pile-up mechanism which is necessary event for the failure of the large eutectic carbides described above, could be largely inhibited by an entangling mechanism that results from the extensive precipitation of secondary carbides within the martensitic matrix (see Fig. 3). This, in here and also in Ref. 8 is substantiated by observation of lesser extend failure of the eutectic carbides on the fracture surfaces than the austenitic alloys. Therefore, during the fracture process, size and distribution of large eutectic carbides does not have any influence as indicated by Fig. (4b). By following the same arguments, the model developed for austenitic alloys could be extended to martensitic alloys by taking the secondary carbide parameters. For all the martensitic alloys tested, a constant failure stress was observed, resulting from the uniformity of the secondary carbide precipitation after some hardening heattreatment. Therefore the model developed would yield constant fracture toughness values as observed experimentally.

CONCLUSIONS

1. The failure mechanism in high-cr cast irons is stress-state controlled and for the austenitic alloys it is strongly influenced by the eutectic carbide distribution. Therefore, by carefully controlling the eutectic carbide distribution, the fracture toughness of the austenitic alloys could be improved considerably.
2. The fracture toughness of these alloys can be forecast from the microstructural parameters as described in the text.

REFERENCES

1. F. Maratray and U. Nanot, "Transformation Characteristics of Cr and Cr-Mo Cast Irons", Climax Molybdenum Brochure.
2. R. S. Jackson and J. Dodd, Metallurgia 76, 107, (1967).
3. T. E. Norman, A. Soloman and D. V. Doanes, AFS. Trans. 67, 242 (1959).
4. R. W. Durman, British Foundryman 69, 141 (1976).
5. D. E. Diesburg, ASTM-STP 559, 3 (1973).
6. D. E. Diesburg, "Toughness of As-cast and Heat Treated White Cast Irons", Climax Molybdenum Internal Report, L-212-113, January, 1973.
7. T. R. Wilsaw, C. A. Raw and A. S. Tetelman, Eng. Fracture Mech. 1, 191 (1968).

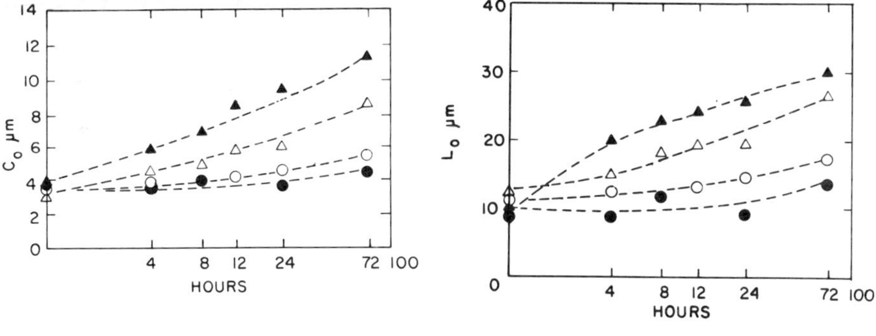

a b
Fig. 1 Variation of A-size and B-spacing of the eutectic carbides with holding periods at 1180°C.

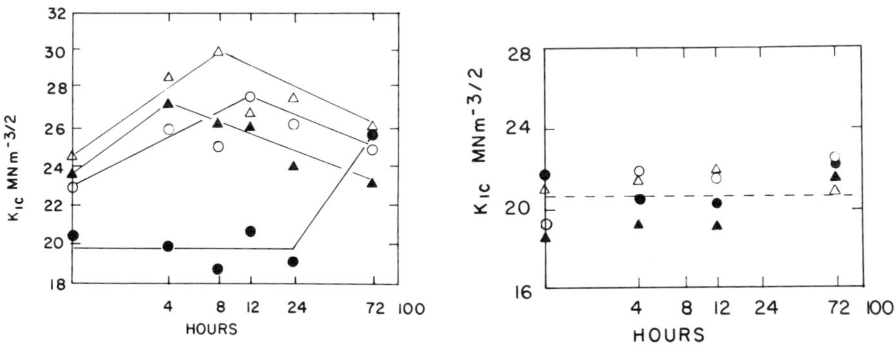

a b
Fig. 2 Variation of fracture toughness with the holding periods at 1180°C. (A-Austenitic alloys, B-Martensitic alloys)

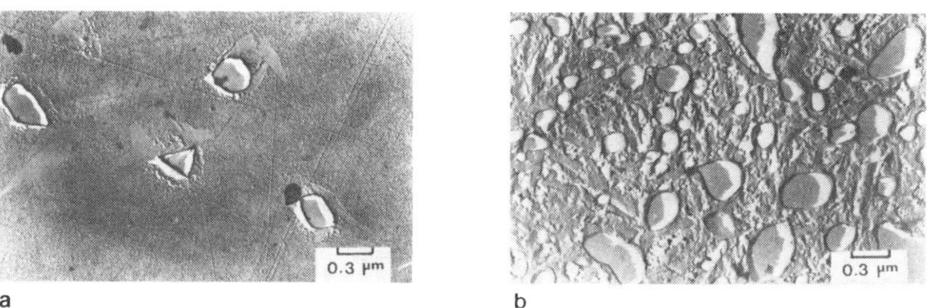

Fig. 3 Comparison of the secondary carbide morphologies of the austenitic and the martensitic High-Cr cast irons. (A-Austenitic alloys, B-Martensitic alloys)

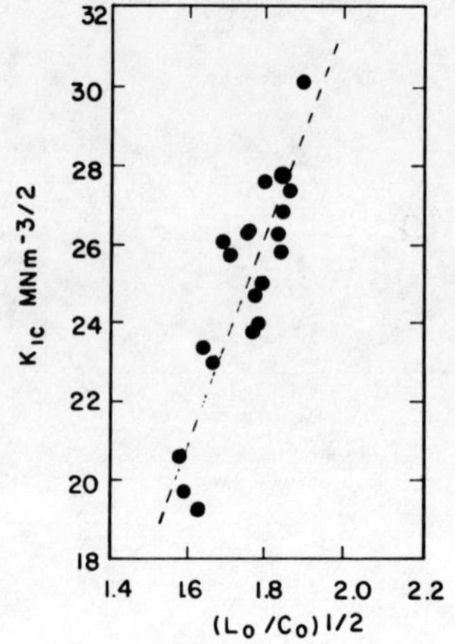

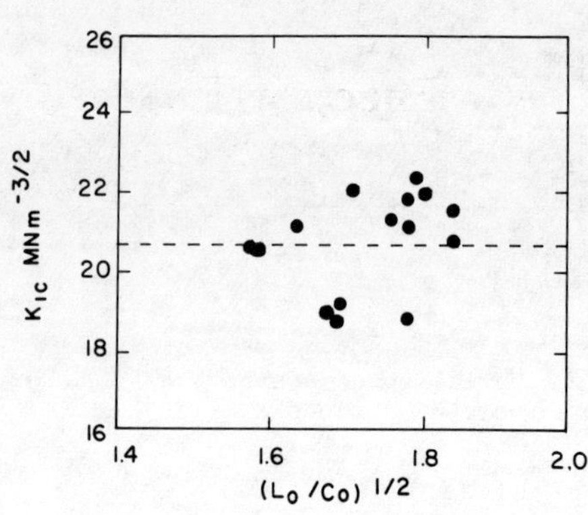

a b

Fig. 4 Variation of the fracture toughness with the eutectic carbide parameters (A-Austenitic alloys, B-Martensitic alloys)

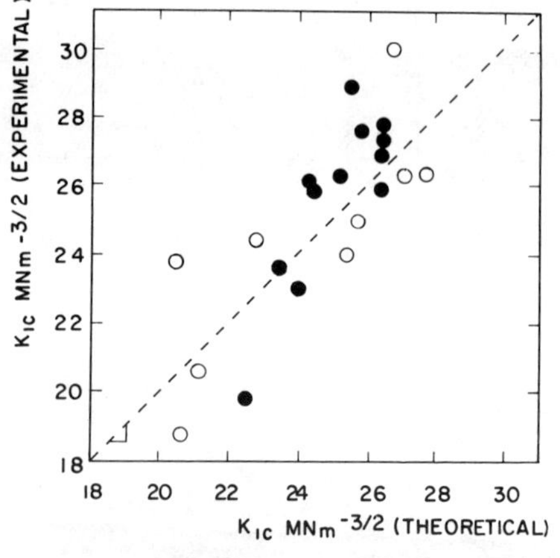

Fig. 5 Comparison of experimentally and theoretically obtained fracture toughness values for the austenitic alloys.

NOTE: In Figures: following signs for the O-alloy I, ●-alloy II, Δ-alloy III and ▲-alloy IV, respectively.

8. K. Z. Gahr and D. V. Doane, Met. Trans., 11A, 613 (1980).
9. J. T. Barnby and E. Smith, Met. Sci. J. 1, 56 (1967).
10. R. O. Ritchie, J. R. Knott and J. R. Rice, J. Mech. Phys. Solids 21, 395 (1973).
11. R. O. Ritchie, B. Francis and W. L. Server, Met. Trans. 21 834 (1976).
12. R. O. Ritchie, W. L. Server and R. A. Wullart, Met. Trans. 10A, 1557 (1979).
13. J. R. Rice and D. M. Tracey, J. Mech. and Phys. Solids, 17, 201 (1969).
14. J. W. Hancock and A. C. MacKenzie, J. Mech. Phys. Solids, 34, 147 (1976).

Effect of Tungsten and Molybdenum on the Strength and Toughness of Super-high Strength Steels

Wu Rengen and Du Lanku

**Department of Metallic Materials and Technology, Harbin Institute of Technology, Harbin, People's Republic of China*

ABSTRACT

After the quench and being tempered at 200°C three percent Cr medium carbon steel possesses more retained austenite , so that high strength and toughness can be obtained. The effect of W and Mo on tempered martensite embrittlement and temper embrittlement are studied. The temper embrittlement at 500°C is eliminated by W. The alloy containing W, after being tempered at 500°C, possesses higher strength and toughness, as well as higher resistance to stress corrosion cracking (SCC).

KEYWORDS

Tugnsten; molybdenum; chromium; low alloying super-high strength steel; tempered martensite embrittlement; temper embrittlement; stress corrosion cracking (SCC).

INTRODUCTION

Three percent Cr medium carbon steel, in the tempered martensite condition, possesses higher toughness as a result of containing more retained austenite after the quench. It is proved that the tempered martensite embrittlement is associated with the transformation of interlath retained austenite into carbides while being tempered over a range of 200-400°C [1,2] . This differs from that occurring near 500°C where embrittlement of these alloys is not dependent on structural changes, but arise from segregation of impurity elements to the prioraustenite grain boundaries [3,4] . This paper is an enquiry into the effect of tungsten and molybdenum on the strength, toughness and temper embrittlement of the super-high strength steels.

MATERIALS USED AND EXPERIMENTAL PROCESS ADOPTED

The three alloys A,B and C used in the experiments contain 0.32 pctC +2.86 pctCr +0.34 pctMo, 0.31 pctC +2.60 pctCr +0.44 pctMo +1.10 pctW, and 0.29 pctC +3.04 pctCr +0.68 pctMo with other alloying elements respectively. They were air induction melted in 150kg ingots and remelted in the electroslag process. After forming and annealing they were quenched in oil at 930°C, and then cooled down in air after being tempered at different temperatures for two hours. The specimens of the alloys were all tested in conventional ways, including tensile tests, Mesnager impact tests, bend tests for K1c plane strain fracture toughness and WOL specimen test for K1scc in 3% NaCl solution. They were also examined by optical and electron microscopy, scanning electron microscopy and Auger spectroscopy.

RESULTS AND ANALYSES OF THE EXPERIMENTS

A. The microstructure of the three alloys used in the experiments, packet martersite was typically present after being tempered at 200°C. The observations showed that these packets consisted of dislocated laths, retained interlath austenite films and tempered intralath carbides, mostly epsilon. The amount of the retained austenite in alloy A was larger than that in alloy B and C. At 300°C, a great amount of retained austenite remained in these alloys. After being tempered above 300°C, the retained austenite decomposed into strange interlath carbides, which led their microstructure to be similar to an upper bainite, and the interlath carbides grew larger. After being tempered above 400°C, the carbides were coarsened. Both interlath and intralath carbides coarsened more at higher temperatures, and the recovery of dislocation within the martensite laths becameresolvable. After being tempered at 550°C, recrystallization of the matrix occurred.

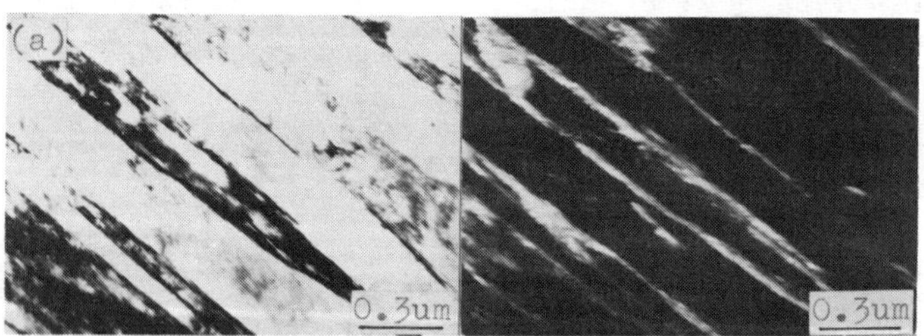

(a) Bright field (b) Dark field

Fig.1 Electron micrographs of alloy A at tempering temperature 200°C.

B. After being tempered at 200°C, alloy A possessed the highest strength and impact toughness, because, at this time, the martensite had not yet decomposed completely, and the effect of carbon on strengthening solid solution could play a considerable role. The carbon had transformed into carbides which could lead

TABLE 1 Mechanical Properties of The Alloys Used In Experiments

* - K_Q (value)

Alloys	Tempering T^0C	$\sigma_{0.2}$ MN/m^2	σ_b MN/m^2	ψ %	δ %	a_k $\frac{MN\cdot m}{m^2}$	K_{1c} $MN\cdot m^{-3/2}$	K_{1scc} $MN\cdot m^{-3/2}$
A	200	1548	1966	39.1	11.8	0.764	71.86	21.91
B	200	1515	1955	40.0	10.9	0.666	69.19	21.14
C	200	1441	1887	43.9	11.0	0.735	81.29	23.00
A	500	1329	1670	42.1	12.9	0.372	53.61	-----
B	500	1304	1544	55.5	13.6	0.588	52.68	-----
C	500	1331	1727	47.1	13.4	0.451	57.86	-----
A	550	1224	1398	49.2	15.5	0.470	100.62*	49.86
B	550	1322	1565	55.0	11.8	0.559	115.19*	47.52
C	550	1228	1500	57.7	15.1	0.617	115.72*	49.86

to the tempered martensite embrittlement. It follows that the steel surely possessed a high strength, high impact toughness, sufficient resistance to SCC: σ_b=1965.9MN/m², $\sigma_{0.2}$=1548.4MN/m², a_k=0.764MN·m/m², ψ=39.1%, δ=11.8%, K_{1c}=71.86MN/m$^{3/2}$, K_{1scc}=21.91 MN/m$^{3/2}$ (see TABLE 1).

At higher tempering temperatures, the interlath retained austenite had decomposed into carbides, in addition to the carbides precipitated from the martensite, and consequently the resistance to cracking and crack propagation were greatly reduced. This led to the decrease of the values a_k from 0.76 at 200°C to its minimum 0.37 at 500°C continuosly (Fig.2), and to the increase of quasi-cleavage fractures (Fig.3), so presented martensite embrittlement. The more was the content of retained austenite in alloy A, the lower the value a_k dropped down.

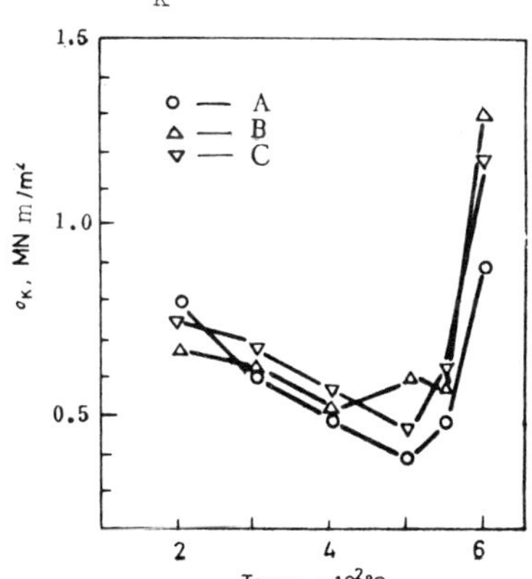

Fig.2 Effect of tempering temperatures on the impact toughness of alloy A,B and C.

At 500°C on the fracture surface appeared mainly intergranular fracture with partly transgranular fracture.
According to the Auger spectrum analysis, after being tempered at 500°C at the prior-austenite grain boundaries occured a segregation of the impurity elements—phosphorus, sulphor and carbon.

C. Alloy B containing tungsten, in comparison with alloy A, had a higher impact toughness (0.588MN/m²) at 500°C (Fig.2). This illustrated that its temper embrittlement at 500°C was obviously eliminated, and the fracture surface no intergranular rupture was found. The Auger spectrum analysis also showed that the segregation of the impurity elements at the prior-austenite grain boundaries were greatly reduced. This result apparently gave rise to the addition of alloying element tungsten. However, our experiments, as shown in Fig.4, demonstrated that the alloying element tungsten had no such an effect on the plane strain toughness k_{1c} at 500°C. It is quite evident that the value a_k is very sensitive to the weakening of the prior-austenite grain boundaries, whereas the value k_{1c} is not.
At 550°C, alloy B showed, to a small extent, a secondary hardness as indicated in Fig.5, whereas the value k_{1c} drammatically increased. This perhaps was attributed to the recrystallization of the alloy matrix. In this case, alloy B showed better strength and toughness: σ_b=1565 MN/m², $\sigma_{0.2}$=1322 MN/m², ψ=55%, δ=11.8%, a_k=0.559 MN·m/m², k_{1c}=115.19 MN·m$^{-3/2}$.

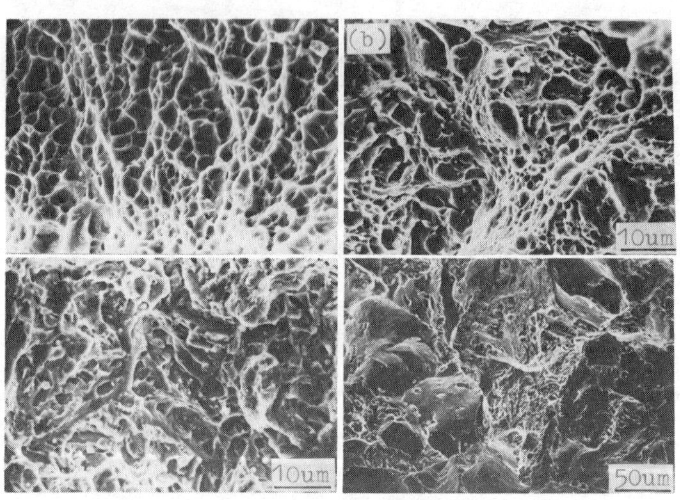

Fig.3 Fracture surfaces of the impact specimens of alloy A by SEM after being tempered at 200°C (a), 300°C (b), 400°C (c) and 500°C (d).

D. Alloy C with a low content of carbon had the highest level k_{1c} among the three alloys used in the experiments after being tempered below 500°C (Fig.4). Although the content of Mo in

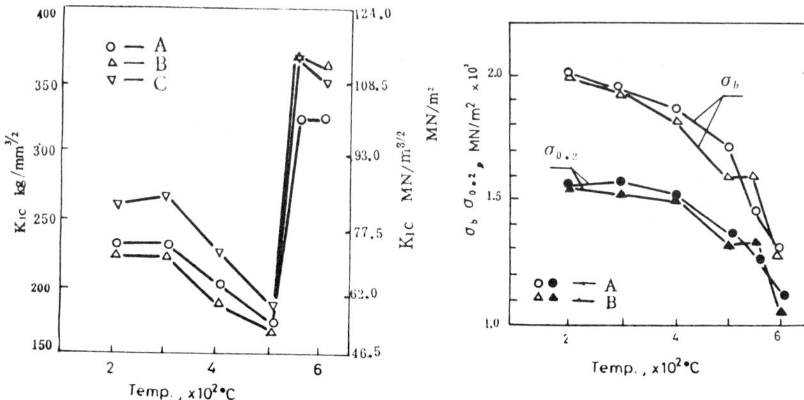

Fig.4 Effect of tempering temperatures on the plane strain fracture toughness k_{1c}.

Fig.5 Effect of tempering temperatures on σ_b and $\sigma_{0.2}$ of alloy A and alloy B.

alloy C (0.68%Mo) was more than that in alloy A or alloy B, no sign of effect on temper embrittlement had been found (Fig.2), and the fracture mode of specimens remained intergranular.

CONCLUSIONS

1. After the quench and in the state of being tempered at 200°C, 3pct Cr medium carbon steel (alloy A) contains more retained austenite, which are distributed in the martensite interlaths and will make a good effect on the toughness. At this time a part of carbon is kept in the martensits, and plays a role in strengthening the material, however the embrittlement of the tempered martensite has not yet developed. It follows that the material used will possess a higher toughness.

2. Tungsten can be used in alloys to reduce the segregation of impurity elements—phosphorus, sulphur and carbon to the prior-austenite grain boundaries, so as to eliminate the embrittlement after the alloy is tempered at 500°C. But in our experiments we have not found that molybdenum has the same behavior in the identical conditions. According to the facts described above, after being tempered at 550°C, alloy B with a content of tungsten will possess the most optimal strength and toughness in comparison with that of alloy A or C.

3. It should be pointed out that the inconsistant variation between the impact toughness a_k and the plane strain fracture toughness k_{1c} in the process of tempering is because the former is very sensitive to the weakening of the prior-austenite grain boundaries, whereas the later is not.

REFERANCE

1. M.Sarikaya, A.K.Jhingan, and G.Thomas: Metall. Trans. A, Vol. 14 A, pp. 1121-1133. (1983)
2. G.Thomas: Metall. Trans. A, Vol. 9A, pp.439-450, (1978).
3. S.K.Banerji, C.J.McMahon, Jr, and H.C.Feng: Metall. Trans. A, Vol. 9A, pp. 237-247, (1978).
4. G.R.Speich and W.C.Leslie: Metall. Trans., Vol. 3, pp. 1043-54, (1972).

Effects of Microstructure and Test Technique on the Toughness-Strength Relationship of Fully Pearlitic Eutectoid Steel

D. J. Alexander* and I. M. Bernstein

Department of Metallurgical Engineering and Materials Science, Carnegie-Mellon University, Pittsburgh, PA 15213, USA
**Now at University of Cambridge, Department of Metallurgy and Materials Science, Cambridge, UK*

ABSTRACT

The static fracture toughness of fully pearlitic eutectoid steel having combinations of extremes of prior austenite grain sizes and pearlite interlamellar spacings was measured. The coarse grained–coarse spacing microstructure had the highest toughness, while the other microstructures had similar but lower toughness levels. It is suggested that the toughness depends on crack deflection and ligament formation during the initial crack advance, factors which themselves are related to the microstructure by the cleavage facet size (controlled by the austenite grain size) and the plastic zone size (determined by the yield strength and hence the interlamellar spacing). Finally, the results are compared to previous measurements of the dynamic toughness.

KEY WORDS

fracture toughness; pearlite; cleavage; crack deflection; fracture facet size; plastic zone size.

INTRODUCTION

The nature of the pearlite transformation in ferrous eutectoid alloys allows the prior austenite grain size and the pearlite interlamellar spacing to be varied essentially independently, through selection of the austenitizing and isothermal transformation temperatures, respectively. Previous work (1–5) has shown that these are the two microstructural features which primarily control the mechanical properties. The present work (6) extends these to study the effects of experimentally obtainable extremes in prior austenite grain size and pearlite interlamellar spacing on the fracture toughness of fully pearlitic eutectoid steel. Two main aims were to understand the effects of varying grain size on toughness at constant yield strength, and to compare the fracture toughness values with dynamic fracture toughnesses measured from precracked instrumented Charpy impact specimens for similar microstructural variables (1).

EXPERIMENTAL PROCEDURE

All testing was conducted on specimens cut from a rail supplied by the Association of American Railroads. The composition was determined (4) to be, by weight percent, 0.80 carbon, 0.17 silicon, 0.84 manganese, 0.013 sulfur, 0.018 phosphorus, and the balance, iron, which conforms to the AISI 1080 designation. Oversize specimens taken from the rail head were austenitized under a slight positive pressure of argon, to minimize decarburization. The samples were transferred within 5 seconds to deoxidized, graphite–covered, stirred lead baths for the transformation to pearlite. Table 1 shows the heat treatments, and the resultant microstructures and mechanical properties (6).

After machining, the 25.4 mm thick compact specimens were precracked to a nominal a/W value of 0.5, at 20 Hz, R = 0.1, K_{max} ~ 20 MPa$\sqrt{m}$, on an MTS servohydraulic testing machine. The crack front was located by optical examination, and the nominal a/W value was determined using lines previously scribed on the front and back of the specimen. The specimens were fractured on the MTS in stroke control, with a constant crosshead velocity of 1.27 x 10^{-2} mm/s. The displacement was recorded using a clip gage extensometer mounted on integrally machined knife edges at the crack mouth. K_Q values of the toughness were calculated as per ASTM E399.78 (7).

TABLE 1: Heat Treatments, Microstructures, and Room Temperature Mechanical Properties

SPECIMEN	AUSTENITIZATION TEMPERATURE AND TIME	TRANSFORMATION TEMPERATURE	AUSTENITE GRAIN SIZE (μm)	INTERLAMELLAR SPACING (μm)	YIELD STRENGTH (MPa)	FRACTURE TOUGHNESS (MPa$\sqrt{m}$)
FF	800°C/1hr	550°C/10min	30	0.13	662	51
FC	800°C/1hr	690°C/1hr	30	0.32	331	48
CF	1000°C/3hr	550°C/10min	190	0.12	690	47
CC	1000°C/3hr	690°C/1hr	190	0.30	352	61

RESULTS

The results of the mechanical testing, as well as the fractography, are presented below. The various microstructures are identified with a two letter code, the first letter referring to the prior austenite grain size, and the second to the pearlite interlamellar spacing. The letters F and C are used for fine and coarse, respectively.

The microstructures resulting from the heat treatments were fully pearlitic. Occasionally, small areas of proeutectoid ferrite were observed, but these were very limited and should not affect the mechanical properties. These properties and the microstructural parameters are presented in Table 1 (6).

Analysis of the toughness of the microstructures was complicated by the frequent occurrence of pop-ins during testing. Often, the pop-ins were large enough to exceed the 2% crack growth allowed (7), although final fracture did not occur until much greater loads. For such cases the specification definition for K severely underestimates the toughness of the specimen, which still had substantially more load carrying ability than the low K value would suggest. Therefore, the maximum load carried by the specimen and the initial fatigue precrack length were used to calculate an apparent but physically more realistic toughness, called K_{max}, and these values were used to compare the relative toughnesses of the various microstructures.

The data (see Table 1) indicates that of the microstructures examined in this study, the Coarse grain size - Coarse pearlite spacing combination (CC) is significantly tougher. The other three microstructures have very similar toughnesses of about 48 MPa$\sqrt{m}$ (44 ksi$\sqrt{in}$) while the CC toughness is about 61 MPa$\sqrt{m}$ (56 ksi$\sqrt{in}$). The similarity in the toughness of the other three microstructures is surprising, as their microstructures differ drastically.

Examination of the fractured specimens showed increasingly rough fractures, in the order FF, CF, FC and CC, with CC clearly the roughest. The fracture mode was all cleavage, although some ductile tearing was found at the edges of the fracture facets, which were in turn similar in size to the respective prior austenite grain sizes. The cleavage river markings indicated cleavage began in numerous locations at the tip of the fatigue precrack, which showed some signs of crack blunting. No inclusions or other particles were found at the cleavage initiation sites.

DISCUSSION

The data shows that the effects of microstructure on the fracture toughness are not simply predictable. The high strength material shows no grain size effect, while the low strength material shows an increase in toughness with increasing grain size. These results are rather surprising, particularly in light of a previous study which showed an increase in the DBTT as the grain size was increased (1). In addition, there is a well known metallurgical rule of thumb that reducing the grain size results in an increase in toughness. However, there is little evidence in the literature which shows that grain refinement will improve the static lower shelf fracture toughness, where fracture occurs by cleavage, or even quasi-cleavage. While increases in toughness with decreases in grain size on the lower shelf have been reported (8,9), most investigations show essentially no increase in toughness as the grain size is reduced (10-12). Often at temperatures in the transition range the toughness of the fine grained microstructures is greater than for the coarse grained material (13-16), but even in these cases, there is no grain size effect on the lower shelf; in fact, the toughness may even decrease slightly (13-15). Some investigations show substantial decreases in toughness with finer grains (17,18). These results suggest that grain size effects are complex, and that the present results do have precedents.

The current results appear explainable with the aid of a simple model which accounts for both grain size and interlamellar spacing effects, through their respective control of the fracture facet size and yield strength (6). The model can also be used to help rationalize results from previous studies.

We envisage that for 'static' loading the toughness is controlled by the deflection of the crack front and the fracture of numerous ligaments, both of which are created by the initial growth of the crack front. Although the fatigue precrack is quite straight, and fairly flat and planar, subsequent cleavage growth from the crack front is distinctly nonplanar. The crack front will tilt, twist, split and bow as cleavage begins at different portions along the crack front (Fig. 1). Not only will the local stress intensity at the crack tip thus be reduced (19-22), but adjacent regions of the crack front will become separated by ligaments which must fracture before crack advance can continue (23-26). The applied load must be increased to raise the local stress intensity and allow ligament fracture, with a resultant increase in the toughness values.

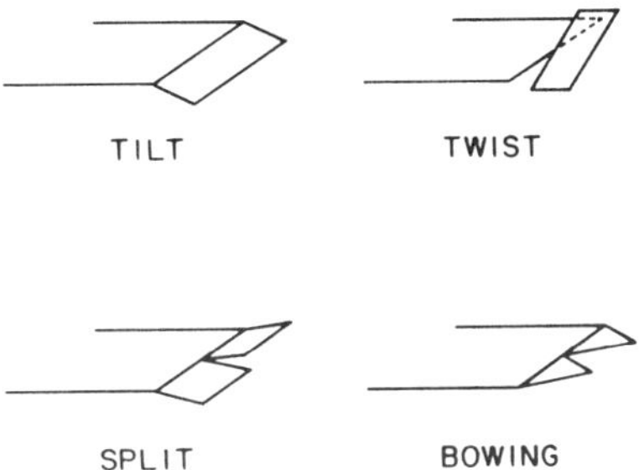

Fig. 1. Types of crack advance.

Both the grain size and the interlamellar spacing will affect crack deflections and ligament formation. Considering first the effect of the interlamellar spacing, the resultant yield strength will control the size of the plastic zone, and this will determine the amount of material that is sampled, i.e. the process zone (see Fig. 2). Finely spaced pearlite will have a high yield strength and so only a small region ahead of the crack tip will be sampled, which will obviously limit the extent of initial microcracking and deflection. The interlamellar spacing will also determine the cleavage fracture stress (6), affecting the ease of cleavage as the crack tip is loaded. While the magnitude of the stresses ahead of the crack tip will scale with the yield stress, it is the ratio of the cleavage fracture stress to the yield stress which will determine the propensity for and ease of cleavage occurring. Previous work (6) has shown that this ratio decreases as the interlamellar spacing decreases. Thus, high strength material will have a greater tendency for cleavage, as the stresses ahead of the crack tip can more readily exceed the cleavage fracture stress.

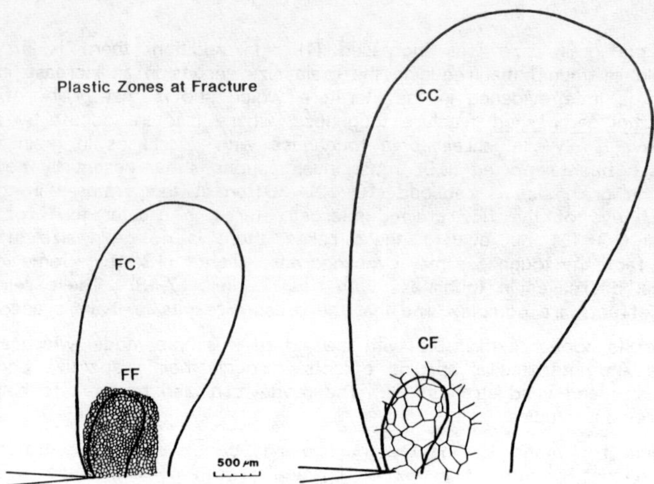

Fig. 2. Comparison of plastic zones at fracture to microstructure.

The role of the austenite grain size is reflected in the cleavage fracture facet size, which represents instances where the crack travels through regions of similarly oriented ferrite, with only minor deviations of the crack path (1–3). Thus, the size of the prior austenite grains will control the scale of the easily cracked areas of the microstructure. Once local cleavage has initiated, it will extend over a distance related to the austenite grain size, and this will determine the deflection of the crack. Larger fracture facets will create larger differences in regions along the crack front. This will produce an increasingly nonplanar crack front, which should increase the toughness. The austenite grain size also determines the size scale of regions of similarly oriented pearlitic ferrite. Thus, for a given plastic zone size, as shown in Fig. 2, more regions of various orientations will be sampled as the grain size is reduced. Therefore, initial cracking is more likely to occur closer to the crack plane as the grain size is reduced, and so a fine grained specimen will show less crack deflection and a flatter fracture surface, as was indeed observed.

Combining these two effects, the microstructure which should have an enhanced toughness is the coarse grain – coarse spacing combination, since it maximizes both the scale of the crack path jumps, and the size of the plastic zone. If the grain size is refined, then a large number of differently oriented regions will be encountered ahead of the crack tip. Initial cracking can develop close to the crack plane, without the need for large deviations. Even if a microcrack should be present away from the crack plane, the many more regions available in the pastic zone greatly increase the likelihood of finding another suitably oriented area closer to the initial crack plane. Thus, the tendency for crack deviations and crack front splitting will be lowered, as will the subsequent fracture toughness.

Similarly, decreasing the interlamellar spacing will reduce the toughness. The plastic zone size will be greatly reduced, due to the increased yield strength. This will limit the amount of material sampled, and so the chance of cracking occurring off the crack plane is reduced, as is the crack deflection. In addition, the stress ratio is lowered, favoring cleavage. The ligaments created will be smaller, ligament fracture will be more likely to occur by cleavage, and the toughness will be lower in this case also.

Results from a previous study (1) show a different microstructural dependency than that outlined above. In this case, the dynamic fracture toughness, K_{ID}, was measured. Based on limited evidence (27) where changes in temperature had similar effects on K_{ID} and K_{IC} for one pearlitic microstructural condition, with K_{ID} being consistently lower than K_{IC}, one might expect these different measures of toughness to show similar microstructural dependencies, as is usually observed (28); however, it is not clear that this is always the case. The previous data showed that the lower shelf dynamic fracture toughness increased as the grain size was refined, unlike the present results, where a coarse grain produces superior static fracture toughness values.

A possible explanation for the different behavior may be found in the widely different strain rates employed in the two test techniques. Dynamic impact tests involve crack tip strain rates on the order of 10^2 s^{-1}, while static tests have strain rates on the order of 10^{-2} s^{-1} (17). This difference will increase the yield stress, and greatly reduce the plastic zone size in the dynamic test. From our previous discussions a large plastic zone is essential in allowing out-of-plane cracking and subsequent crack deflection. If such crack splitting is not possible, crack deflection and ligament fracture may no longer control the toughness.

Since there is no available data on the effect of high strain rates on the yield stress of pearlite, single phase iron was used to estimate the increase in the yield stress; pearlite and iron have similar strain rate dependences at slower strain rates (29), so this procedure seems reasonable. Data from Leslie and co-workers (30) show that the yield stress would increase by about 200 MPa for strain rates increasing from 10^{-2} to 10^2 s^{-1}. From Hyzak's and Bernstein's data (1), a typical yield stress is about 500 MPa, which would then be increased at high strain rates to 700 MPa; K_{ID} is about 35 MPa$\sqrt{m}$, while the equivalent static toughness would be about 45 MPa$\sqrt{m}$. Therefore, from the square of the ratio of the respective toughnesses to the yield strengths the plastic zone in the dynamic test would be more than 3 times smaller than in the static test. This large difference may limit the crack splitting in the dynamic test, and so change the fracture controlling processes. In this case, the toughness depends primarily on the frequency of crack path changes, as proposed by Hyzak and Bernstein; thus the toughness should increase as the grain size is reduced, as was observed.

Further support for this explanation is found by comparing the present data for the two coarse grained microstructures. In this case, the yield stresses vary by about a factor of 2, so the plastic zones differ by a factor of 4. The result is a decrease in the toughness of the high strength material, in a manner similar to that observed in the dynamic case where the higher strain rate increased the yield stress.

The higher yield stress in the dynamic test will result in higher peak stresses at the crack tip, but the high strain rate should not change the cleavage fracture stress, which is insensitive to strain rate. Thus, the ratio of the cleavage fracture stress to the yield stress will be lowered, and so the initiation of cleavage should become easier as well. In this case, the frequency of initiation events may determine the toughness, so the finer grain size which would have more cleavage facets along the crack front would achieve higher toughnesses.

Unfortunately, a direct comparison of the most important case is not possible, for although microstructures similar to the present CF, FF, and FC conditions were tested, no samples in the coarse grain – coarse spacing (CC) condition were examined in the previous work. Since it is this microstructure which shows the highest static toughness, it would be very interesting if the value of the dynamic toughness was available for comparison.

Finally, the previous study also showed that the DBTT increased as the grain size increased (1), which can be interpreted as a reduction of toughness. The different microstructural response observed in the present study may reflect the differences in the variables measured in the different tests. The fracture toughness test measures the resistance to crack initiation, and only very limited crack growth, while the transition temperature test measures the total energy absorbed during the complete fracture of the test piece. Thus, the total crack propagation process is reflected in the energy obtained. In this latter case, the total number of the crack deflections may dominate, rather than the scale of only the first deflections, as is the case for the static fracture toughness tests.

In summary the difference observed in the microstructural control of the fracture toughness under static and dynamic conditions appears to be related to the different strain rates used and the different measures of toughness employed. The high strain rates increase the yield stress and thus reduce the plastic zone size, so crack splitting is reduced, and the microstructural dependency is altered. If toughness reflects crack propagation rather than initiation, the frequency of crack deflection rather than its initial magnitude becomes the controlling factor.

CONCLUSIONS

The microstructure in the present study which has the highest static fracture toughness is the coarse grain-coarse spacing material. Toughness depends on crack splitting and ligament formation during the initial crack growth from the fatigue precrack, which are in turn influenced by the pearlite spacing and the austenite grain size through their respective control of the yield strength and the fracture facet size. The large facet size and large plastic zone size for the CC microstructure promote large deflections and thus maximize the toughness. The much higher strain rate involved in impact testing alters the fracture processes, which results in a different microstructural dependency in dynamic testing.

ACKNOWLEDGEMENTS

We would like to thank the Association of American Railroads for their financial support, and Dr. A.W. Thompson for helpful discussions.

REFERENCES

1. J.M. Hyzak and I.M. Bernstein, *Met. Trans.*, **7A**, 1217 (1976).
2. Y.-J. Park and I.M. Bernstein, *Fracture 1977*, D.M.R. Taplin, ed., ICF4, University of Waterloo Press, Waterloo, Canada, 33 (1977).
3. Y.-J. Park and I.M. Bernstein, *Met. Trans.*, **10A**, 1653 (1979).
4. G.T. Gray III, Ph.D. Thesis, Carnegie-Mellon University, (1981).
5. J.J. Lewandowski, Ph.D. Thesis, Carnegie-Mellon University, (1983).
6. D.J. Alexander, Ph.D. Thesis, Carnegie-Mellon University, (1984).
7. *Plane-Strain Fracture Toughness of Metallic Materials - Annual Book of ASTM Standards*, ASTM E 399-78a.
8. N. Okumura, *Metal Science*, **17**, 581 (1983).
9. F.R. Stonesifer, *Eng. Fracture Mech.*, **10**, 305 (1978).
10. D.A. Curry and J.F. Knott, *Metal Science*, **10**, 1 (1976).
11. S.K. Chaudhuri and R. Brook, *Int. Jl. Fracture*, **12**, 101 (1976).
12. S. Ensha, Ph.D. Thesis, UCLA, (1974), DAHC-04-69-C-0008, U.S. Army Research Office, Durham.
13. W. Dahl and W.-B. Kretzschmann, *Arch. Eisen.*, **47**, 313 (1976).
14. W. Dahl and W.-B. Kretzchmann, *Fracture 1977*, D.M.R. Taplin, ed., ICF4, Univ. of Waterloo, Waterloo, Canada, 17 (1977).
15. T. Yokobori and S. Konosu, *Eng. Fracture Mech.*, **9**, 839 (1977).
16. W.L. Phillips, *Met. Trans.*, **4**, 388 (1973).
17. R.O. Ritchie, B. Francis and W.L. Server, *Met. Trans.*, **7A**, 831 (1976).
18. E.R. Parker and V.F. Zackay, *Eng. Fracture Mech.*, **7**, 371 (1975).
19. K.T. Faber and A.G. Evans, *Acta Met.*, **4**, 565 (1983).
20. B.A. Bilby, G.E. Cardew and I.C. Howard, *Fracture 1977*, D.M.R. Taplin, ed., ICF4, Univ. of Waterloo, Waterloo, Canada, 197 (1977).
21. S. Suresh, *Met. Trans.*, **14A**, 2375 (1983).
22. B. Cotterell and J.R. Rice, *Int. Jl. Fracture*, **16**, 155 (1980).
23. W.W. Geberich and N.R. Moody, *Fatigue Mechanisms - ASTM STP 675*, ASTM, 292 (1979).
24. J.M. Krafft, *Applied Matls. Research*, **1**, 88 (1964).
25. R.E. Stoltz, N.R. Moody and M.W. Perra, *Met. Trans.*, **14A**, 1528 (1983).
26. E. Smith, *Eng. Fracture Mech.*, **19**, 601 (1984).
27. A.S. Tetelman, S. Ensha, and J.N. Robinson, Report to Association of American Railroads, Chicago, Illinois, September 28, (1972).
28. T.J. Koppenaal, in *Instrumented Impact Testing - ASTM STP 563*, ASTM, 92 (1974).
29. T. Takahashi and M. Nagumo, *Trans. JIM*, **11**, 113 (1970).
30. W.C. Leslie, R.J. Sober, S.G. Babcock, and S.J. Green, *Trans. ASM*, **62**, 690 (1969).

Investigation of Microstructure in QLT-Treated Steel 12Ni4CrMo

S. Chengyi (C. Y. Soong), S. Yunhua, Z. Guangzhong and Wu Guangda

Central Iron and Steel Research Institute, Beijing, People's Republic of China

ABSTRACT

A microstructural investigation has been conducted on the martensitic steel 12Ni4CrMo which was subjected to QLT heat treatment (intercritical heat treatment). By means of TEM and SEM the morphology, phase constitution and crystallography of the QLT-treated steel were analysed in detail. It was shown that effective grain refinement, configuration of the zigzag prior austenite grain boundaries and suitable carbide dispersion are the important structural features in connection with the QLT toughening.

KEYWORDS

Intercritical heat treatment; QLT heat treatment; low temperature toughness; effective grain refinement; lath martensitic steel.

INTRODUCTION

In order to improve the low temperature toughness and reduce the temper embrittlement in low carbon martensitic steels, a number of intercritical heat treatments have been developed in recent years, and have been the subjects of numerous investigations.(1-7) These studies attributed the improved toughness to the complicated multi-phase structures, such as the refinement of the prior austenite grain size, the finely dispersed ferrite-tempered martensite aggregates, or the dense distribution of thermostable austenite precipitating along the boundaries of the original martensite laths, i.e. the reversion of austenite. Since the results reported varied considerably with the chemical composition, the microstructure and the heat treatment regime, it is essential to do the research on different grades of alloys and steels. In the present work a three-step intercritical heat treatment-QLT heat treatment was performed on a 4%Ni steel containing the carbide former Cr+Mo$\approx$2(wt%). The aim of this work is to gain a clear understanding of the microstructure in this QLT-treated steel so as to approach the structure-toughness relationship.

EXPERIMENTAL

The materials used in this work were electroslag-refined and hot-forged into 20mm sq bars and subsequently normalized at 900°C. Main chemical composition was 0.12C-4.3Ni-1.5Cr-0.5Mo(wt%). For comparison two kinds of experimental heat treatment were performed. Intercritical heat treatment (QLT): 860°C quenching (Q)+ Ac_1-Ac_3 region requenching (L) + 630°C tempering (T). Conventional heat treatment (QT): 860°C quenching (Q) + 650°C tempering (T). A variety of techniques was used in this work, including thin foil transmission electron microscopy, scanning electron metallography, x-ray diffraction and quantitative metallography. The impact tests were carried out by using a Impact Test Machine JB30A made in China. The Charpy V-notch test specimens were prepared according to the chinese national standard YB19-64.

RESULTS AND DISCUSSION

Correlation of Low Temperature Toughness, Phase Constitution with L-Treatment Quenching Temperature

Figure 1 shows the -196°C CVN toughness improved markly by QLT treatment. Ac_1-Ac_3 region quenching temperature (L-temperature) strongly affected the low temperature toughening effect. The optimum temperature has been established as 720°C (Ac_1+50°C). When the investigated steel was subjected to a QLT heat treatment with the L-temperature being 720°C, the -196°C CVN toughness was about 4.5 times the value of the conventionally treated steel at the similar yield strength levels.

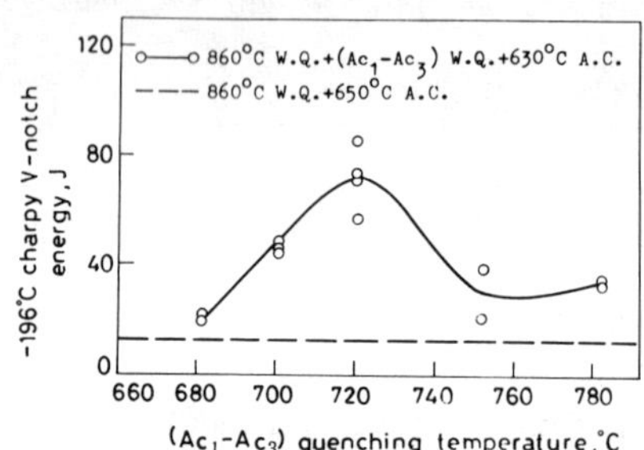

Fig. 1. Effect of Ac_1-Ac_3 region quenching temperature on the -196°C Charpy V-notch toughness of the QLT-treated steel. (Line of dashes showing the average value for the QT-treated steel)

A series of SEM micrographs (Fig. 2.a-e) shows the formation and growth of the newly formed austenite. When the steel heated in Ac_1-Ac_3 region, the globular austenite Ag and acicular austenite Aa appeared (Fig. 2.a). At 720°C the volume fraction of Aa reached the maximum, Aa checkered with the original martensite laths, while Ag fully distributed along the prior austenite grain boundaries and some original martensitic packet boundaries (Fig. 2.b). Above that temperature Aa turned into Ag, mainly for Ag expanded and swallowed up Aa rather than the coalescence of Aa itself as generally considered (Fig. 2.c-e). Just at 720°C the original lath martensite was well separated by these fresh phases, and Ag impelled the

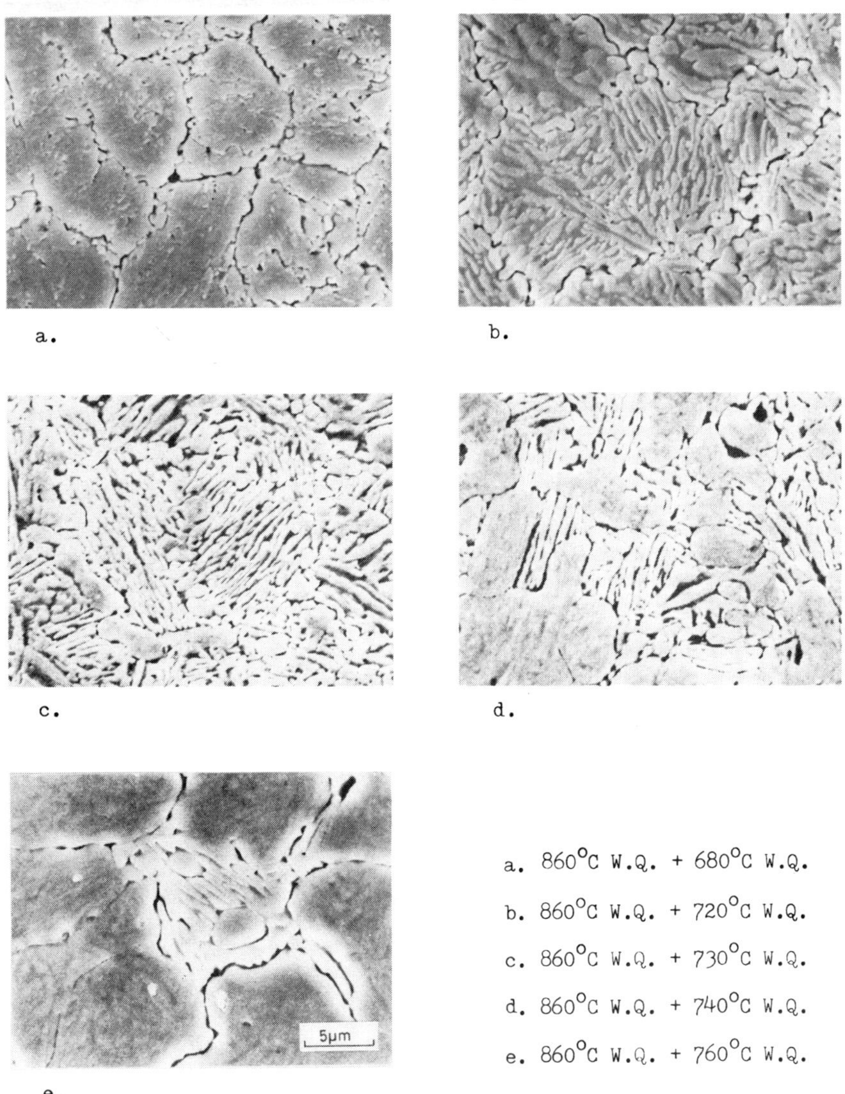

Fig. 2. SEM micrographs showing the formation and growth of the newly formed austenite in Ac_1-Ac_3 region.

mainly because γg expanded and swallowed up γa rather than through the coalescence of γa itself as commonly considered (Fig. 2.c-e). Precisely at 720°C the original lath martensite was well separated by these fresh phases, and γg forced the migration of the prior austenite grain boundaries into a "zigzag" form (Fig. 2.b). Quantitative metallography reveals that the area of the prior austenite grain boundaries in the QLT-treated steel (L-temperature 720°C) increased by 2-2.5 times as compared with the conventionally treated steel. The functions of zigzag grain boundaries might be: 1. to dilute the impurities segregated in grain boundaries; 2. to increase the chances for grain boundaries to prevent the microcrack propagation.

Microstructure Changes in QL Condition and in QLT Condition

Figure 3 shows the TEM microgragh of a QL-treated specimen, Ac_1-Ac_3 region quenching temperature being 720°C. In this condition globular austenite transformed into globular martensite Mg, meanwhile acicular austenite transformed into acicular martensite Ma. The majority of these two kinds of martensite were heavily dislocated. In this way a quenched martensite-residual ferrite dual-phase microstructure was produced, giving an intense internal stress. In Fig. 3 the obvious difference in the TEM image contrast between the two phases proves the essential distinction between them. In residual ferrite (R.F.) the appearance of the bend extinction coutours indicates that R.F. was a soft phase and was heavily strained.

Fig. 3. TEM micrograph showing the dual phase structure-newly formed martensite (dark) and residual ferrite (bright) after 860°C W.Q. + 720°C W.Q.

T-treatment introduced two processes into the steel: "tempering" for the quenched martensite; "ageing" for the residual ferrite (supersaturated ferrite). Both phases decomposed into ferrite plus carbide. The elimination of the difference in the TEM image contrast between the martensite and ferrite reveals the decrease of the essential distinction between them. Fig. 4 shows the morphology and crystallography of a specimen quenched at 860°C, requenched at 720°C and tempered at 630°C. Some granular ferrites in different crystallographic orientation surrounded the lath-like ferrites, most of which in similar crystallographic orientation (low angle misorientation). Some newly formed fine grains and cell-like subgrains were inbe-

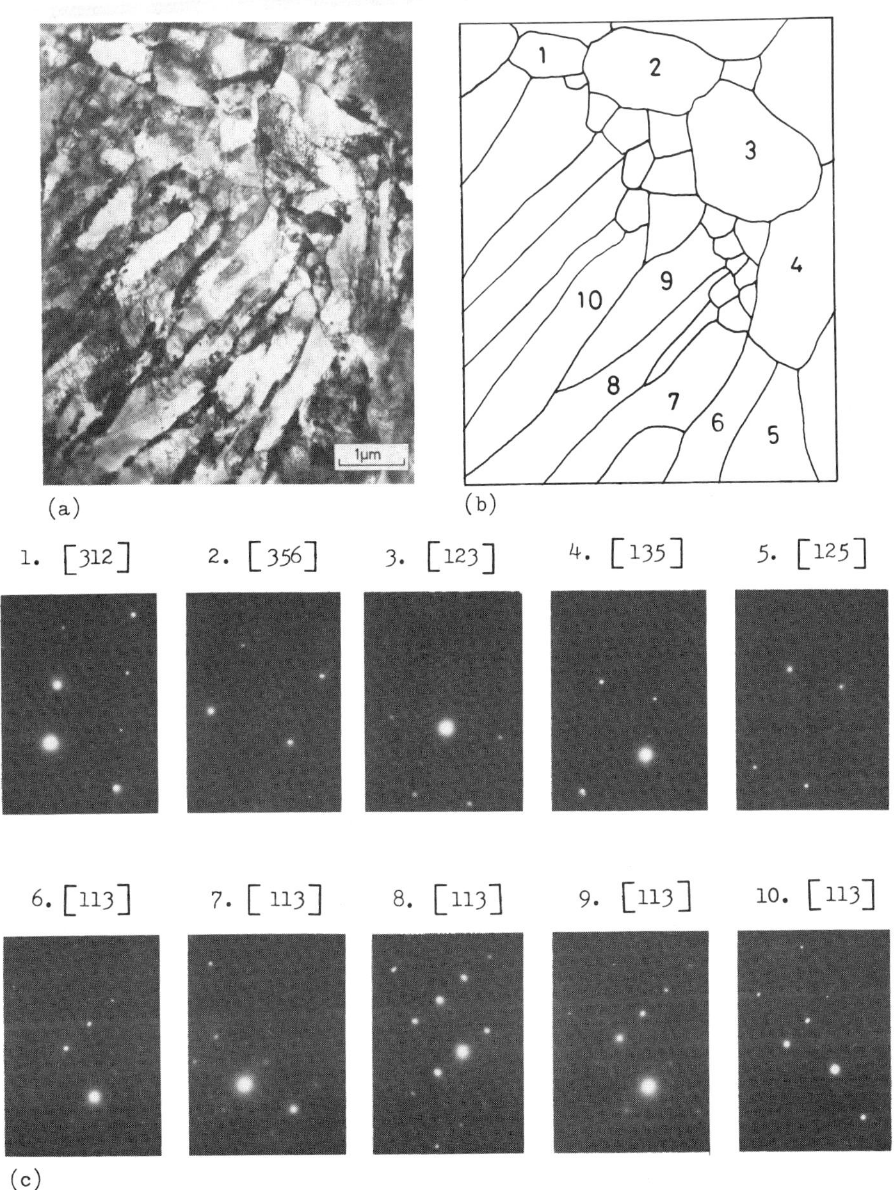

Fig. 4. (a) TEM micrograph of a QLT-treated specimen. 860°C W.Q. + 720°C W.Q. + 630°C A.C.
(b) Schematic drawing of (a).
(c) Electron diffraction pattern from the areas in the sequence showed by (b) and the indices of the zone axes.

tween. These above-mentioned phenomena caused "effective grain refinement" (packet size refinement). In addition, the residual ferrite was strengthened due to small carbides precipitating and pinning the dislocations. It is well known that the low temperature fracture of ferritic steel often takes place as a result of the cleavage along some certain crystallographic planes, e.g. (001). The critical cleavage stress increases with the refinement of grain size. Moreover, strengthening of the residual ferrite makes an important contribution to the good combination of strength and toughness.

In QLT condition, only few of austenite particles (reverted austenite particles) have been found by using selected area electron diffraction and dark field image technique on TEM. X-ray diffraction result shows that the volume fraction of austenite in the QLT-treated steel was less than 1.5%. It could be attributed to the lower Ni content and the higher Cr+Mo content in the experimental steel, making the austenite unstable and easily decompose into carbide plus ferrite during the tempering process.

SUMMARY

1. The low temperature toughness of the lath martensitic steel 12Ni4CrMo was improved markedly by QLT heat treatment. Ac_1-Ac_3 region quenching temperature strongly affected both the toughness and the microstructure.

2. The globular austenite γg and acicular austenite γa formed during heating in the temperature range Ac_1-Ac_3. At 720°C γa interspersed between the original martensite laths, while γg was fully distributed along the prior austenite grain boundaries and forced the migration of the boundaries into a "zigzag" form. Over 720°C γa turned into γg mainly because γg expanded and swallowed up γa rather than through the coalescence of γa itself.

3. After requenching at 720°C the newly formed austenite transformed into dislocation martensite and the residual ferrite was supersaturated and heavily strained. During tempering at 630°C two kinds of processes occured: "tempering" for the quenched martensite; "ageing" for the residual ferrite. Both phases decomposed into ferrite plus carbide. These phenomena caused "effective grain refinement". The residual ferrite was strengthened due to small carbides precipitating and pinning the dislocations during the ageing process.

4. Effective grain refinement, configuration of the zigzag prior austenite grain boundaries and suitable carbide dispersion were considered to be the important factors for the QLT treatment to toughen the 12Ni4CrMo steel.

REFERENCES

1. B.G.Sazonov, Metalloved. Obrab. Metall. USSR, 4, 30 (1957).
2. J.H.Roux, Mem. Sci. Rev. Met. 57, 507 (1960).
3. J.Leger, Ph. Maynirr, M.Toitot and P.Bastien, Mem. Sci. Rev. Met. 68, 313 (1970).
4. A.M.Polyakova and Sadovsky, Met. Sci. Heat. Treat. 1, 46 (1970).
5. S.Nagashima, T.Ooke, S.Sekino, H.Mimura, T.Fujishima, S.Yano and H.Sakurai, Trans. ISIJ. 11, 402 (1971).
6. J.Wada and D.V.Doane, Metall. Trans. 5, 231 (1964).
7. J.I.Kim, C.K.Syn and J.W.Morris,Jr., Metall. Trans. A. 14A, 93 (1983).

Correlation Between Microstructure and Fracture Toughness of Titanium Alloys

C. Müller, A. Gysler and G. Lütjering

Technische Universität Hamburg-Harburg 2100 Hamburg 90, Federal Republic of Germany

ABSTRACT

Fracture toughness tests were performed on Ti-6Al-4V with four different microstructures: fine lamellar, bi-modal with low and high volume fractions of primary α_p, and equiaxed structures. Some specimens were loaded to K-values within the stable crack extension regime, unloaded and sectioned, to determine the crack tip configurations in the specimen interior. Crack propagation for all microstructures occurred through the nucleation of secondary cracks and their connection with the main crack. The equiaxed and bi-modal structures exhibited a similar resistance against crack extension from the fatigue pre-crack (K_o) and almost identical crack tip opening displacements (CTOD). The fine lamellar microstructure showed much lower K_o and CTOD values. The resistance against unstable fracture decreased from bi-modal microstructures with 50 % α_p over those with volume fractions below 40% and the equiaxed structure, to lamellar specimens, showing the lowest resistance.

KEYWORDS

Fracture toughness; microstructure; secondary cracks; Ti-6Al-4V.

INTRODUCTION

The microstructure of (α+β) titanium alloys can be varied over a wide range, either by heat treatment or by thermomechanical treatments (1). Numerous investigations have shown that the fracture toughness behavior of titanium alloys depends strongly on microstructural parameters, such as geometrical arrangement of α and β phases (lamellar, equiaxed, bi-modal), dimensions of phases, texture of α-phase, and oxygen or hydrogen content (e.g. 2,3). With regard to the dependence of toughness on the geometrical arrangement of α and β phases in most cases an inverse relationship between yield stress and fracture toughness is reported in the literature (2,3). However the yield stress might not be a particular suitable parameter to describe the fracture behavior of a material. In a recent study (4) it was observed that the fracture toughness of Ti-6Al-4V with equiaxed and lamellar phase morphologies decreased with decreasing α-phase dimensions, although the yield stress exhibited the opposite ranking.

The purpose of the present investigation was to study the fracture behavior of well defined microstructures of the most widely used titanium alloy Ti-6Al-4V. Fracture toughness tests were performed on bi-modal microstructures with low and high volume fractions of primary α-phase, and on an equiaxed structure. In addition a fine lamellar microstructure was included in order to simulate the fracture behavior of the lamellar portions of the bi-modal structures.

EXPERIMENTAL PROCEDURE

The tests were performed on a Ti-6Al-4V alloy, supplied by RMI, Niles, OH. The chemical composition (in wt.%) was 6.5 Al, 4.0 V, 0.17 Fe, 0.17 O_2, 0.11 N_2. The equiaxed and bi-modal microstructures were prepared by thermomechanical processing. Blanks of the as-received bar were homogenized at 1050°C for 15 min, water quenched, and then cross-rolled at 800°C in multiple passes to a deformation degree of $\phi = -1.25$. The equiaxed microstructure was obtained by a recrystallization treatment at 800°C for 15 h. Bi-modal microstructures with different volume fractions of primary α-phase were produced by annealing for 1 h in the temperature range between 955 and 980°C (Table 1). The fine lamellar structure was prepared from blanks which were water quenched from 1050°C and annealed at 800°C for 1 h. The final quenching temperature was 800°C and the aging treatment 500°C for 24 h. The dimensions of the α-phase for bi-modal microstructures was about 7µm, while the equiaxed structure exhibited α-grain diameters of about 4µm. The width of the α-phase of the fine lamellar structure was about 2µm and the lengths up to several hundred micrometers. The texture of the α-phase for the equiaxed and bi-modal microstructures was a strong basal texture type (5), while the lamellar structure was essentially texture free.

Table 1 Heat Treatments and Tensile Properties of Ti-6Al-4V

Microstructure	Heat Treatment [1]	Volume % α_p	$\sigma_{0.2}$ (MPa)	σ_F (Mpa)	$\varepsilon_F = \ln(A_o/A_F$
Lamellar	1050°C/15min/WQ 800°C/ 1h/WQ	0 [2]	1056	1278	0.15
Bi-modal	980°C/ 1h/WQ 800°C/ 1h/WQ	30 - 40 [2]	1060	1510	0.59
Bi-modal	970°C/ 1h/WQ 800°C/ 1h/WQ	~ 55 [2]	1053	1500	0.64
Bi-modal	955°C/ 1h/WQ 800°C/ 1h/WQ	60 - 70 [2]	1038	1438	0.58
Equiaxed	800°C/15h/WQ	~ 85 [3]	1030	1404	0.59

[1] Aging: 24 h 500°C; [2] rest lamellar; [3] rest β

All mechanical tests were carried out at room temperature with the stress axis parallel to one of the rolling directions. Tensile tests were done on round specimens with gage dimensions of 4 mm diameter and 15 mm length, using an initial strain rate of $10^{-3}s^{-1}$. The tensile properties, together with the heat treatment and volume fraction of α_p are summarized in Table 1.

Fracture toughness tests were performed on CT-specimens with a thickness B of 8 mm and a width W of 32 mm. The fatigue pre-cracking (R=0.1, sine wave, 30Hz) and the fracture toughness tests were carried out in vacuum ($<10^{-4}$Pa) on a servohydraulic testing machine. The total notch length (chevron notch plus 3 mm fatigue crack) was about 16 mm (0.5 W). A constant piston displacement velocity of 1mm/min resulted in an initial rate of stress intensity factor increase of 1.7 $MPa\cdot m^{1/2}\cdot s^{-1}$. The load F was monitored as a function of crack opening displacement COD, measured with a clip-gage in front of the chevron notch.

The crack tip configuration in the center of specimens (plane strain region) was studied by loading specimens close to K_5 (6) a lower K-value to study the beginning of crack extension from the fatigue pre-crack. The unloaded specimens were then sliced in the mid-thickness (0.5B). Small samples containing the crack tips were polished, and slightly etched to reveal the microstructure.

EXPERIMENTAL RESULTS

The results of the fracture toughness tests are summarized as bar-graphs in Fig. 1. The lower line in each bar represents the stress intensity factor K_e where The F-COD curve deviated from the linear elastic line. The second line indicates K_5, calculated according to ASTM (6), while the upper line gives K_{max} where unstable fractur occurred. The top line of bars marked with an arrow shows the K-values where these specimens were unloaded for sectioning. The measured local crack extensions are indicated in Fig. 1. It was tried to unload specimens at about K_5 and at a lower value to determine K_0, where crack extension just started. For lamellar and equiaxed microstructures this was difficult since the stress intensity regime was very narrow between K_0 and K_{max}.

From Fig. 1 it can be seen that K_0 was very similar (about 70 $MPa\cdot m^{1/2}$) for the bi-modal microstructures, independently of primary α-phase volume fraction (within the range studied here). The corresponding value for the equiaxed structure was only slightly lower (about 65 $MPa\cdot m^{1/2}$), while the fine lamellar microstructure exhibited a considerably lower value of only 50 $MPa\cdot m^{1/2}$.

The highest resistance against unstable fracture was observed for bi-modal structures with α_p volume fractions above 55% (K_{max} > 80 $MPa\cdot m^{1/2}$). Bi-modal structures with volume fractions below 40% α_p and equiaxed microstructures (about 85% α_p) exhibited lower K_{max}-values of 70 $MPa\cdot m^{1/2}$ and between 65 and 70 $MPa\cdot m^{1/2}$, respectively. The lowest resistance was measured for the fine lamellar structure (52 $MPa\cdot m^{1/2}$).

The crack tip studies by SEM of unloaded specimens are summarized in Figs. 2 and 3. Examples of the surroundings of the fatigue pre-cracks at the beginning of crack extension are shown in Fig. 2. The bi-modal microstructures with low and high volume fractions of α_p exhibited the largest CTOD values of over 20µm (Figs. 2b and c). The equiaxed structure showed only a slightly lower value of about 20µm (Fig. 2d) but already crack extension of about 150µm, while the lamellar specimen revealed a considerably lower CTOD of below 10µm (Fig. 2a) and large crack extension. These micrographs also showed, that small secondary cracks are nucleated in front of the blunted pre-crack and that these microcracks are eventually linked together with the advancing crack front.

The crack tip areas of the specimens loaded to K_5 are shown in Fig. 3. Crack propagation occurred through nucleation of secondary cracks within the

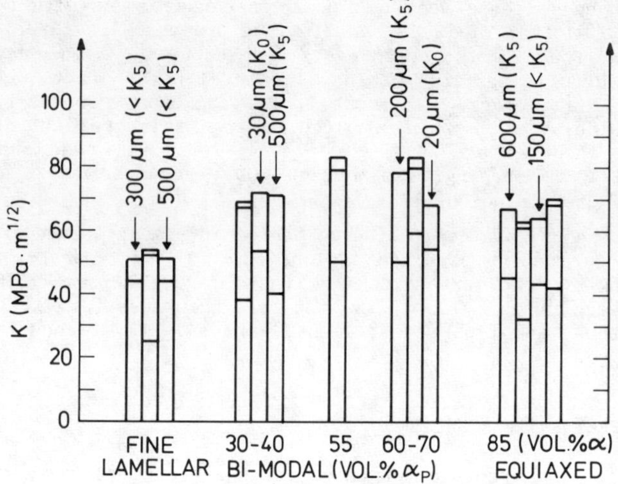

Fig. 1. Stress intensity factors for Ti-6Al-4V (Vacuum). Lower line: K_e; second line: K_5; top line: K_{max}. Arrows: sectioned specimens and crack extensions in specimen interior.

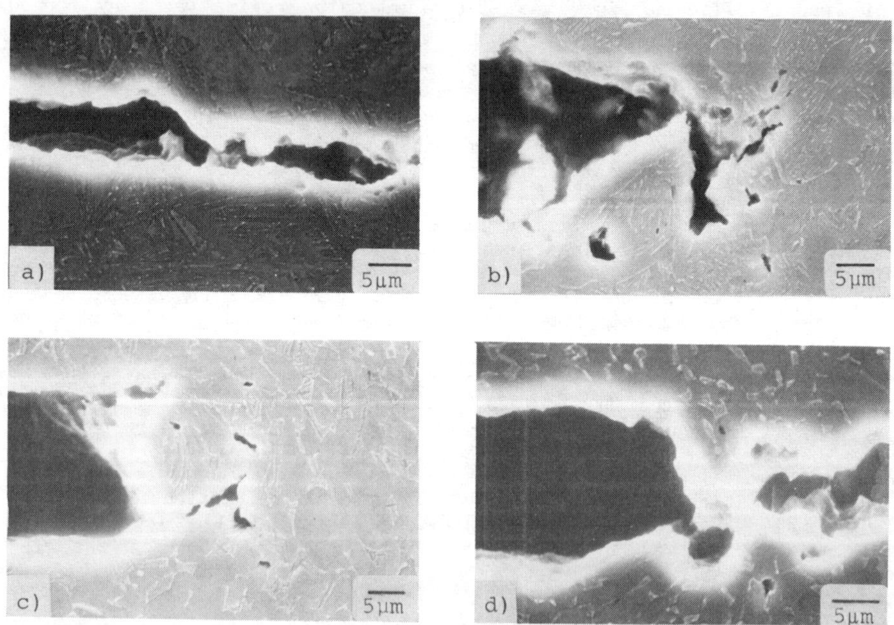

Fig. 2. Crack tip configurations at about K_0. a) lamellar; b) bi-modal, 35%α_p; c) bi-modal, 70%α_p; d) equiaxed, 85%α_p. (CPD →).

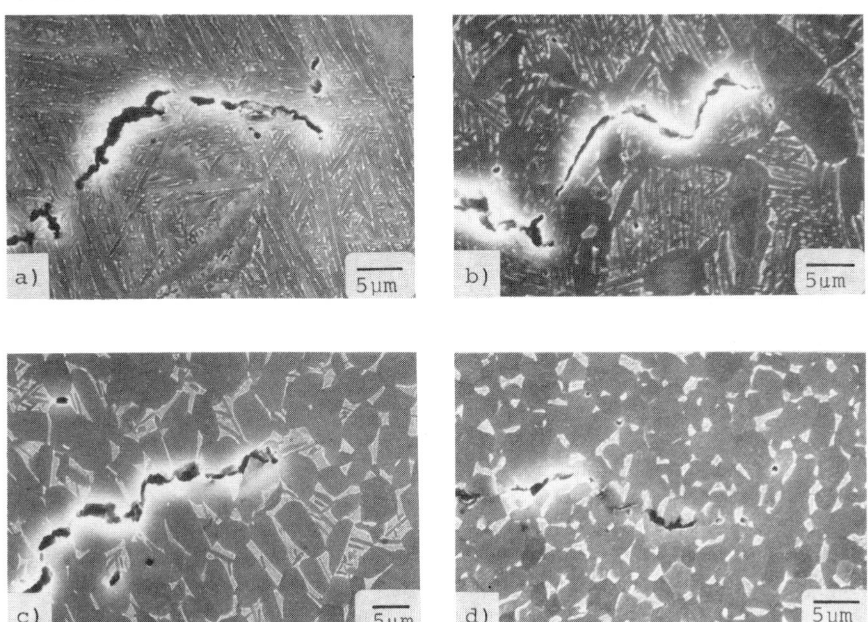

Fig. 3: Crack tip areas at K_5. a) lamellar; b) bi-modal, 35%α_p; c) bi-modal, 65%α_p, d) equiaxed, 85%α_p (CPD →).

plastic zone and their subsequent linkage with the crack front. These microcracks seemed to nucleate at phase boundaries, primarily at α/β interfaces. Crack propagation then occurred along α grain boundaries or α/β phase boundaries in equiaxed (Fig. 3d) and bi-modal structures with high α_p volume fractions (Fig. 3c), while the cracks seemed to advance predominantly within the lamellar matrix for the bi-modal microstructure with low α_p volume fraction (Fig. 3b).

DISCUSSION

The results of this study showed that the fracture toughness of Ti-6Al-4V can be varied considerably by changing the microstructure (Fig. 1), although the yield stresses of these different structures were nearly the same (Table 1). This result is in contrast to the inverse correlation between fracture toughness and yield stress usually reported in the literature (e.g. 2,3). Based on the SEM studies of unloaded and sectioned specimens it seems that the K_O-values where crack extension starts (Fig. 1), and the corresponding CTOD values (Fig. 2) are more dependent on those mechanical properties which are related to the fracture process, for example to fracture stress and ductility (Table 1). The K_O and CTOD values of bi-modal and equiaxed microstructures were observed to be similar, and this was also found for the σ_F and ε_F values. On the other hand, the lamellar structure showed much lower values for K_O and CTOD and also for the fracture stress and ductility in comparison to bi-modal and equiaxed microstructures (Table 1).

However it should be pointed out that the fracture toughness of different microstructures can not be determined unambiguously on the basis of tensile fracture stress and ductility alone. It was found in earlier investigations

(4,7) that the fracture toughness depends for example strongly on grain or phase dimensions. Those results showed that a coarse grain size material exhibited higher toughness values as compared to a fine grain size material as long as the fracture mechanism did not change. However, the tensile ductility and fracture stress of materials with different grain or phase dimensions showed the opposite ranking (4,7).

Crack propagation in a fracture toughness test was found to occur through the nucleation of secondary cracks within the plastic zone and their connection with the main crack (Figs. 2 and 3). The observed differences in resistance against crack propagation between K_o and unstable fracture for the different microstructures would have to be explained on the basis of this type of propagation mode.

The low resistance against crack extension of the fine lamellar microstructure is thought to be a result of the large number of potential crack nucleation sites due to the close spacings of the individual lamellae (2µm). The resulting high volume fraction of secondary cracks (4) might in turn promote the low crack propagation resistance. Bi-modal microstructures with α_p volume fractions below about 40% consist of a fine lamellar matrix in which the individual primary α grains are embedded. Since crack extension in such a bi-modal structure was found to occur primarily within the lamellar matrix (Fig. 3b) the low crack propagation resistance might then be explained in a similar way as for the lamellar microstructure. However to prove this assumption that fine lamellar structures have a higher volume fraction of secondary cracks than either equiaxed or bi-modal microstructures with high α_p volume fractions one would have to prepare a large number of sectioned specimens to obtain a better three dimensional microcrack distribution.

The bi-modal microstructures with primary α volume fractions of more than 50% showed the highest resistance against unstable fracture (Fig. 1). It seems therefore that the fine lamellar portions of bi-modal microstructures have the most beneficial effect on crack propagation resistance if they are evenly dispersed within the primary α grain structure as the matrix. If the β-phase was not transformed to a lamellar microstructure during quenching and annealing, as in the case of equiaxed microstructures, then the resistance against unstable fracture decreased markedly (Fig. 1). One reason for this behavior may result from the smaller α-phase dimensions of the equiaxed structure (~4µm) in comparison to the bi-modal microstructure (~ 7µm), since it is known that a grain size reduction decreased the resistance against unstable fracture (4).

ACKNOWLEDGEMENT

This work was supported by the EPRI, Palo Alto, CA.

REFERENCES

1. M. Peters and G. Lütjering, EPRI CS-2933 (1983).
2. N.E. Paton, J.C. Williams, J.C. Chesnutt and A.W. Thompson, in "Alloy Design for Fatigue and Fracture Resistance", AGARD-CP-185, 4-1 (1976)
3. C.A. Stubbington, in "Alloy Design for Fatigue and Fracture Resistance", AGARD-CP-185, 3-1 (1976).
4. A. Gysler and G. Lütjering, Proc. 5th Intern. Conf. on Titanium, Munich, DGM (1984).
5. M. Peters, A. Gysler and G. Lütjering, Met. Trans. 15A, 1597 (1984).
6. E399-81, Annal Book of STM Standards, ASTM, Philadelphia (1981).
7. A. Gysler, V. Bachmann and G. Lütjering, in "Fracture and the Role of Microstructure", p. 155, EMAS, Warley (1982).

Mechanisms of Dynamic Fracture in 1045 Steel

M. R. Bayoumi and M. N. Bassim

Metallurgical Sciences Laboratories, Department of Mechanical Engineering, University of Manitoba, Winnipeg, Manitoba R3T 2N2, Canada

ABSTRACT

Study of the fracture behavior of annealed 1045 steel under quasi-static conditions at strain rates corresponding to $\dot{\varepsilon} \cong 10^{-3}$ s^{-1} (equivalent to $\dot{K} \cong$ 1 MPa $\sqrt{m}/s$) and under dynamic loading rates corresponding to $\dot{\varepsilon}$ in excess of 10^3 s^{-1} ($\dot{K} = 10^6$ MPa $\sqrt{m}/s$) was undertaken. The dynamic fracture tests were performed using a Split Hopkinson Bar system. Furthermore, the fractured specimens were examined in a scanning electron microscope to measure the stretch zone width, stable crack growth and presence and density of dimples and inclusions as well as the size of facets. The transition from a predominantly ductile behavior at an almost cleavage fracture was identified as a function of strain rate. A quantitative relationship characterizing the variation of the stretch zone width with strain rate is given.

KEYWORDS

Fracture toughness, J_{Ic}, dynamic fracture, scanning electron microscopy, stretch zone, ductile fracture, cleavage fracture, loading rate, AISI 1045 steel.

INTRODUCTION

The analysis of dynamic fracture and crack propagation problems under higher rates of loading has become more important in materials research. Studies in dynamic fracture have been limited to analysis based on linear elastic fracture mechanics concepts. In many engineering materials such as low and medium strength steels, fracture is presently analyzed in terms of elastic-plastic behavior using J-integral. AISI 1045 steel, like many other materials, tends to exhibit an elastic-plastic fracture behavior at low rates of loading and a predominantly linear elastic fracture behavior when the loading rate is increased to the dynamic range. Recently [1] wedge loaded compact tension specimens (WLCT) of 1045 steel were used for quasi-static and dynamic determination of J_{Ic} using a special arrangement of the Split Hopkinson Pressure Bar. The analysis of the results was based on the J_{Ic} expression as:

$$J_{Ic} = kA/Bb \tag{1}$$

where k is a constant which was found to be about equal to 1.0, A is the energy under load-displacement curve up to the point of onset of crack propagation, b is the remaining ligament and B is the specimen thickness.

While in a continuum scale, the analysis of fracture concerns itself with the stability of a macroscopic crack as discussed, for example by Liebowitz [2], in the microscopic scale, however, the fracture process is one of nucleation, growth and coalescence of microscopic voids or cracks. The microvoid kinetics determine when a macroscopic, continuum scale crack forms in the first place and also determine the stability of the crack, since the growth of the large crack occurs as the result of microvoid kinetics in the process zone at the crack tip.

Void nucleation is generally classified as homogeneous and heterogeneous [3]. Homogeneous nucleation occurs within the microstructures at sites such as, dislocation and networks. On the other hand, heterogeneous nucleation takes place at inclusions and second-phase particles as well as fracture of grain boundaries [3-7]. The following heterogeneous nucleation sites at inclusions are; (i) fracture of the inclusions themselves, (ii) separation of the interfaces, when the particle is not strongly bounded to the matrix and (iii) fracture of the matrix, which occurs when the fracture strength of the matrix is less than that of the inclusion or interface.

Based on void nucleation mechanisms [6] followed by cavity growth and coalescence [8], Bayoumi and Bassim [9] developed a correlation between the fracture toughness parameter J_{Ic} and the strain rate $\dot{\varepsilon}$ of the following form:

(a) for low strain rates

$$J_{Ic}(\dot{\varepsilon}) = K_o \frac{m_o \sigma_o}{C} L_p^* \left[\bar{\varepsilon}_{i,\alpha,\beta} \left[\sigma_{rr}(\dot{\varepsilon}) - Y(\dot{\varepsilon},n)\right| + \bar{\varepsilon}_{c,\alpha,\beta}(V_f,n)\right| \tag{2}$$

(b) for high strain rates

$$J_{Ic}(\dot{\varepsilon}) = S\rho^* \left[\bar{\varepsilon}_{i,\alpha,\beta}[\sigma_{rr}(\dot{\varepsilon}) - Y(\dot{\varepsilon},n)] + \bar{\varepsilon}_{c,\alpha,\beta}(V_f,n)\right|^2 \cdot$$
$$f(E,K,n,\bar{\varepsilon}_y) + 2\,\sigma_f \pi L_e^*(1 - \nu^2)/E \tag{3}$$

where c, K_o are constants, m_o is a constraint factor which depends on the material and testing conditions, σ_o is the flow stress, L_p^* is a characteristic length corresponding to the distance from the crack tip to a point where the strain reaches a critical value at crack initiation, $\bar{\varepsilon}_{i,\alpha,\beta}$ is the strain for void nucleation under stress state $\alpha = \sigma_2/\sigma_1$ and $\beta = \sigma_3/\sigma_1$ (where σ_1, σ_2 and σ_3 are the principal stress components), $\sigma_{rr}(\dot{\varepsilon})$ is the interface stress normal to the inclusion-matrix interface as a function of strain rate, $Y(\dot{\varepsilon},n)$ is the flow stress as a function of strain rate $\dot{\varepsilon}$ and strain hardening exponent η, $\bar{\varepsilon}_{c,\alpha,\beta}(V_f,n)$ is the coalescence strain under stress state α, β, as a function of volume fraction of voids V_f and strain hardening exponent n, S is the shape factor characterizing the geometry of the plastic zone, ρ^* is the Neuber's microsupport effect constant, $\bar{\varepsilon}_y$ is the yield strain, L_e^* is a characteristic distance which depends on the microstructure of the material, E is the modulus of elasticity, K is the strength coefficient and ν is the Poisson's ratio.

The present study is devoted to examine the stretch zone width, stable crack growth and presence and distribution of dimples, facets and inclusion content when testing annealed AISI 1045 steel under both quasi-static conditions which have a strain rate corresponding to $\dot{\varepsilon} \cong 10^{-3}\ S^{-1}$ ($\dot{K} \cong$ MPa $\sqrt{m}/s$) and dynamic loading conditions with a strain rate $\dot{\varepsilon} \cong 10^{3} s^{-1}$ ($\dot{K} \cong 10^{6}$ MPa$\sqrt{m}/s$). A relationships of the dependence of the stretch zone width with strain rates for strain rate sensitive materials such as 1045 steel is developed.

EXPERIMENTAL PROCEDURE

Material: Annealed AISI 1045 steel was used. Optical microscopy of the microstructure for etched as well as polished only samples are shown in Figures 1 and 2 respectively.

Quasi-static and Dynamic Testing: Wedge loaded compact tension specimens were used for both quasi-static and dynamic tests. The procedure of using a Split Hopkinson Bar for dynamic testing of this steel was described [1].

Scanning Electron Microscopy: Halves of broken WLCT specimens were examined in a scanning electron microscopy (ISI-100B) using mostly a tilt angle β = 30°. Measurements of the stretch zone width, stable crack growth as well as counting the size and distribution of dimples and facets on the fracture surface were carried out.

RESULTS AND DISCUSSION

Scanning electron micrographs of the fractured surface of the specimens tested at strain rate range from $\dot{\varepsilon}$ = 0.38 x $10^{-3}s^{-1}$ to 2.5 x 10^{3} s^{-1} are shown in Fig. 3. On these micrographs are identified the stretch zone width as well as the size of stable crack growth if any. It is evident that both these parameters decrease as the strain rate is increased to the dynamic range.

The measurement of the stretch zone width provides an alternative method to determine the fracture toughness J_{Ic} [10 - 13] using the following equation:

$$J_{Ic} = \left[4\ \sigma_o d/(\cos\beta + \sin\beta)\right](1/G) \tag{4}$$

where, d is the measured length of the stretch zone on the micrographs, β is the incident angle of the beam (tilt angle) and G is the magnification. Thus using equations (2), (3) and (4), the relation between the stretch zone width d and the strain rate $\dot{\varepsilon}$ can be written as follows:

(a) at low strain rates

$$d(\dot{\varepsilon}) = \frac{G(\cos\beta + \sin\beta)}{4} \cdot \frac{K_o m_o}{C} L_p^* \left[\bar{\varepsilon}_{i,\alpha,\beta}[\sigma_{rr}(\dot{\varepsilon}) - Y(\dot{\varepsilon},n)] + \bar{\varepsilon}_{c,\alpha,\beta}(V_f,n)\right] \tag{5}$$

(b) at high strain rates

$$d(\dot{\varepsilon}) = \frac{G(\cos\beta + \sin\beta)}{4} \cdot S\rho^* \left[\bar{\varepsilon}_{i,\alpha,\beta}[\sigma_{rr}(\dot{\varepsilon}) - Y(\dot{\varepsilon},n)] + \bar{\varepsilon}_{c,\alpha,\beta}(V_f, n)\right|^2 \cdot$$

$$f(E,K,n,\bar{\varepsilon}_y) + 2\,\sigma_f \pi L_e^*(1 - \nu^2)/E \qquad (6)$$

Following the same approach as reported in [9], the correlation between the stretch zone width d and the strain rate $\dot{\varepsilon}$ for AISI 1045 steel can be presented as shown in Fig. 4. It is shown that d varies with $\eta\left(\text{Log}(\dot{\varepsilon}/10^{-3})\right)$ at low strain rates and with η^2 at high strain rates. The intersection of the straight line and the parabola in Fig. 4 will give a value of a strain rate where the transition in fracture behavior occurs. Below this value, fracture is predominantly ductile with significant plasticity at the crack tip. Above this value of strain rate, the fracture behavior is mostly cleavage and the corresponding stretch zone ahead of the crack tip is very limited.

Examples of the fractured surface of the specimens which reveal the size and distribution of dimples and facets under both quasi-static and dynamic loading conditions, are shown in Fig. 5. It is evident from these observations that ductile fracture is the main feature of the specimens tested quasi-statically while the cleavage fracture is obtained in dynamic tests.

The average void density (dimple density) for the ductile fracture features was determined to be in the range of 2.04×10^4 to 5.109×10^4 void/mm^2, while the size of the dominant dimples is in the range of 0.0133 mm to 0.0193 mm with an average interdisplacement of 0.06 mm. The cleavage fracture for dynamic loading mainly characterized by facets with an average size of 0.023 mm.

The density of inclusion content (see Fig. 2) is found to be in the range of 1.9×10^3 to 2.8×10^3 particles/mm^2, while the pearlite colony size is about 0.05 mm (see Fig. 1).

The above measurements relating the density of dimples to that of inclusions indicate that the void density is much greater than that of particle density by a factor of about 20. These conclusions are consistent with those found by other investigators [14] for AISI 304 stainless steels where the void density exceeded the particle density by a factor of 100. This suggests that the mechanisms of void formation in these steels are homogeneous, in the sense that voids occur within the grains and are related to dislocation mechanisms such as tangles, cells and low angle boundaries. Only few dimples seem to originate at second phase particles as evident from Fig. 5a.

For the specimens fractured in the dunamic range, the fracture surface is predominantly cleavaged with some indication of limited ductility. The facet size measured on the micrographs appears to be of the same order of magnitude as the pearlite colonies shown in Fig. 1. This results indicate that in the annealed 1045 steel, the pearlite colonies, which are stronger and more brittle than the ferritic matrix, the dynamic loading causes a brittle (cleavage) fracture of these colonies while the ferritic matrix undergoes limited ductile fracture as evidenced in Figs. 5b.

CONCLUSIONS

1 - A quantitative relationship which characterizes the variation of the stretch zone width with strain rate is given for low and high strain rates in equations (5) and (6).

2 - Fracture of annealed AISI 1045 steel at a low strain rate is predominantly ductile while the brittle fracture (cleavage) is the main feature at high strain rates.

3 - Ductile fracture at quasi-static condition give dimples density greater than inclusion density by a factor of 20.

4 - Cleavage fracture at high strain rates reveals a facet size with the same order of magnitude as the pearlite colonies with limited ductility due to the ferritic matrix.

ACKNOWLEDGEMENT

The support of the Natural Sciences and Engineering Research Council of Canada is acknowledged. Also the technical help of Mr. J. Van Dorp and Mr. R. Hartle is appreciated.

REFERENCES

1. M.R. Bayoumi, J.R. Klepaczko and M.N. Bassim, Journal of Testing and Evaluation, 12, 316 (1984).
2. H. Liebowitz (ed.), "Fracture", Vols. I - VII, Academic Press, New York (1968 - 1972).
3. A.L. Stevens, L.E. Pope, In Metallurgical Effects at High Strain Rades, R.W. Rohde, B.M. Butcher, J.R. Holland and C.H. Karnes, Eds., Plenum Press, New York (1973) p. 473.
4. S.H. Goods and L.M. Brown, Acta Met., 27, 1-15 (1979).
5. K. Tanaka, T. Mori and T. Nakamura, Phil. Mag., 21, 267 (1970).
6. A.S. Argon, J. Im and R. Safogler, Metall. Trans. A, 6, 825 (1975).
7. D.R. Curran, L. Seaman and D.A. Shockey, In Shock Waves and High-Strain-Rate Phenomena in Metals, M.A. Mayers and L.E. Murr, Eds., Plenum Press, New York, (1981) p. 129-167.
8. F.A. McClintock, J. Applied Mechanics, Transactions of the American Society of Mechanical Engineers, Series E, 35, 368 (1968).
9. M.R. Bayoumi and M.N. Bassim, In Modelling Problems in Crack Tip Mechanics, J.T. Pindera, Ed., Martinus Nijhoff Publishers, Dordrecht (1984) p. 173-182.
10. P. Nguyen-Duy and S. Bayard, Journal of Engineering Materials and Technology, Transactions of ASME, 103, 55-61 (1981).
11. K.F. Amouzouvi and M.N. Bassim, Materials Science Engineering, 55, 257-262 (1982).
12. M.R. Bayoumi and M.N. Bassim, In Proceedings of the International Conference of Materials, Vol. 2, ICM4, Stockholm, Sweden (1983) pp. 803-811.
13. A.J. Krasowsky and Vainshtok, International Journal of Fracture, Vol. 17, No. 6, 579-592, (1981).
14. M.A. Meyers and C.T. Aimone, Progress in Materials Science, Vol. 28, pp. 1-96, (1983).

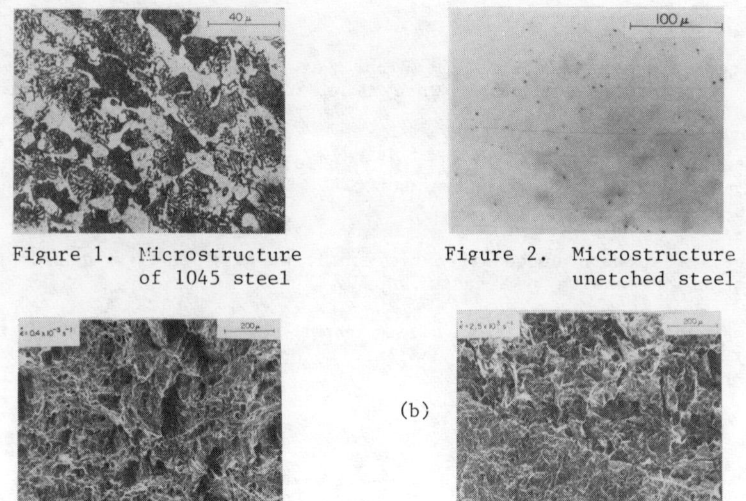

Figure 1. Microstructure of 1045 steel

Figure 2. Microstructure of unetched steel

Figure 3. Stretch zone width of a) quasistatic and b) dynamic specimens

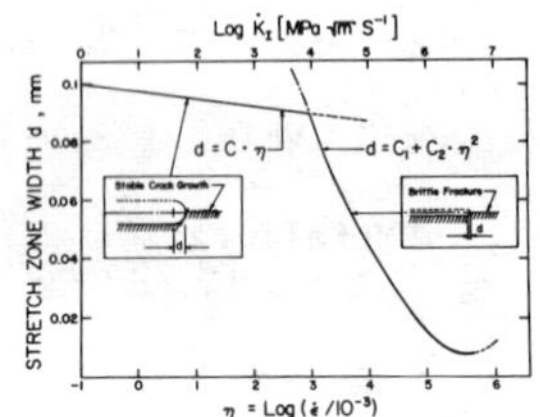

Figure 4. Variation of stretch zone with strain rate.

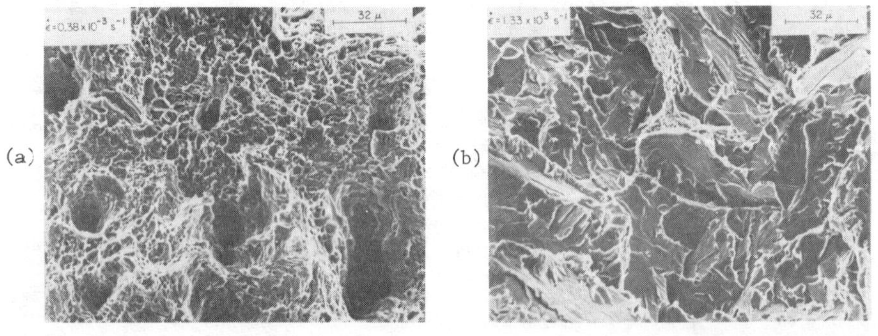

Figure 5. Fractured surface of a) quasistatic and b) dynamic specimen

On the Brittle-to-Ductile Transition of Silicon

G. Michot and A. George

Laboratoire de Physique du Solide, Associé au CNRS n° 155, Ecole des Mines, Parc de Saurupt, 54042 Nancy Cedex, France

ABSTRACT

Fracture experiments were carried out at various deflection rates on pre-cracked silicon single crystals between 900 and 1250 K. An abrupt transition from cleavage with limited plasticity to general yielding at the crack tip was observed. A model based on crack tip shielding through dislocation nucleation and glide is developed, which accounts for the deflection rate dependance of the measured temperature of transition and for the different behaviour of floating zone and Czochralski grown silicon crystals.

KEYWORDS

Silicon, fracture, brittle-to-ductile transition.

INTRODUCTION

At room temperature silicon single crystals are fully brittle and fail by cleavage on {111} planes, which is readily explained by the strong lattice friction that opposes dislocation motion, because of covalent bonding. At sufficiently high temperature (T > 900 K in conventional conditions) silicon can be plastically deformed since dislocations have gained by thermal activation a sufficient mobility to glide and multiply. As the critical stress for cleavage is not -or very little- temperature dependant, a brittle-to-ductile transition is expected to result from the decrease of the flow stress with raising the temperature. In fact this brittle-to-ductile transition is more complex since, in a crystal which already contains a sharp crack, deformation will in most cases take place first at and around the crack tip because of the stress concentrations arising there. Plastic deformation can blunt or shield the crack tip which modifies the conditions required for further crack propagation, making it more difficult.

St John (1) first studied the brittle-to-ductile transition in silicon, using pre-cracked samples and determing the transition temperature as a function of the imposed strain rate. A similar study was resumed but with two different grades of silicon. The results reported here give a new basis to relate the brittle-to-ductile transition to microscopic parameters such as dislocation mobilities or the ease of dislocation multiplication.

EXPERIMENTAL

Samples were obtained from dislocation free, n type ($\rho > 5\ \Omega$.cm) silicon single crystals grown by Wacker. Two grades were used : floating zoned (FZ) Si (Waso quality) and Czochralski (CZ) pulled Si containing 3.5×10^{17} Oxygen atoms per cm^3. Tapered double cantilever beam samples were machined in (211) wafers of $\sim$ 0.8 mm thickness. For pre-cracking at room temperature, crack initiation along a {111} plane was done by forcing an alumina wedge in the notch and crack arrest obtained by a compression stress applied on side faces, following the technique of St John (1). Details have been given elsewhere (2,3). During high temperature experiments, samples were kept in a reducing atmosphere (10 % H_2, 90 % N_2). Pre-cracked samples were loaded in opening mode using two pin holes, at temperatures ranging from 900 K to 1250 K. The displacement of pins was controlled in order to give a constant value of the deflection rate $\dot{\delta}$ comprised between 1.6 and 500 µm/mm. The recorded tensile load P was converted into stress intensity factor K_I using the calibration function calculated by Srawley and Gross (4) :

$$K_I = P\ f\left(\frac{a}{w}\right) / B\sqrt{w} \tag{1}$$

where B is the sample thickness, w the sample width, a the crack length (accurately measured by X-ray topography after pre-cracking (2,3)) and f the tabulated function.

RESULTS

The brittle-to-ductile transition can be depicted as follows :
(i) Below the transition temperature T_c, the load-deflection curve is strictly linear. The value of K_{IC}, corresponding to the critical load for fracture P_c through formula (1) is constant (K_{IC} = 0.94 MPa$\sqrt{m}$) in a very wide range of temperature from room temperature up to a few K below T_c.
(ii) Above T_c, the load-deflection curve departs from linearity for $P > P_\ell$ and could exhibit a maximum when the test is continued. We have also determined corresponding values K_ℓ from formula (1). In this domain, a general plastic deformation is observed leading to a macroscopic bending of the two beams of the sample and avoiding immediate failure.
(iii) The transition takes place in a very narrow range of temperature ($\sim$2 K). In the immediate vicinity of T_c, on the low temperature side, the load-deflection curve remains linear up to fracture which occurs for a much higher stress intensity factor, up to $K_c \sim 3\ K_{IC}$.

The temperature dependence of K_c/K_{Ic} and $K\ /K_{Ic}$ for different deflection rates has been represented in Fig.1 for both FZ and CZ silicon samples. The values of T_c measured in FZ Si are in close agreement to those obtained by St John (1). On the other hand, in CZ Si, the transition occurs at higher temperature for a fixed $\dot{\delta}$: T_c is shifted up by 41 K at $\dot{\delta}$ = 5 µm/mm and by 31 K at $\dot{\delta}$ = 50 µm/mm. T_c could not be determined in CZ Si at $\dot{\delta}$ = 500 µm/mm because of a very large scatter in the data.

K_ℓ values could not be determined very accurately. K_ℓ is a decreasing function of temperature both at a given $\dot{\delta}$ when $T-T_c$ increases and at a given $T-T_c$ when $\dot{\delta}$ increases.

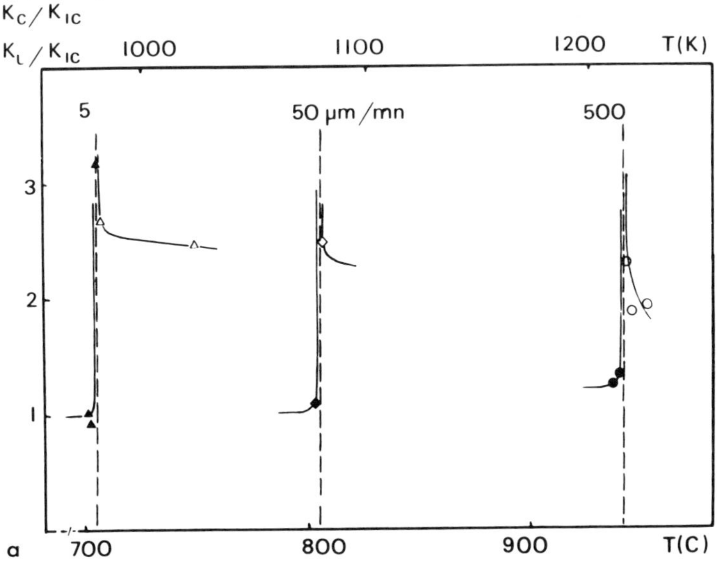

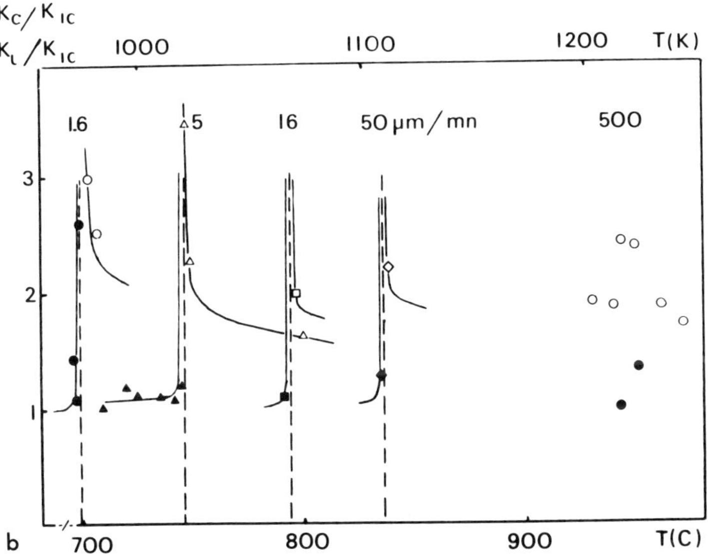

Fig.1 : Critical stress-intensity factors at fracture or flow versus temperature for different deflection rates. Full symbols : fracture, open symbols : flow. (a) FZ Silicon, (b) CZ Silicon.

DISCUSSION

In the fracture process of a ductile material some plastic work must be supplied in addition to the energy required to break atomic bonds. From the microscopic point of view, plastic strain around the crack tip can modify the situation in two ways : First, crack tip blunting occurs when emission or absorption of dislocations with a component of their Burgers vector normal to the crack plane and suitable sign occurs, leading to an increase of the crack tip root radius and then to stress concentrations decrease. A second possibility is simply that the stress tensor around the tip is changed by shielding : the emitted dislocations exert long range back stresses which oppose crack opening. In this latter case the interaction between crack tip stress field and dislocation stress fields leads to a decrease of the effective stress intensity factor felt at the tip for the same externally applied load (5).

St John (1) and Haasen (6) proposed a model based on crack tip blunting to explain the Arrhénius relation between $\dot{\delta}$ and T_c. The corresponding activation energy of $\sim$ 1.9 eV found in (1) was identified to that of dislocation glide. In this case it would correspond to shear stresses of $\sim$ 250 MPa (7) which might be not unrealistic very close to the tip. Fig.2 is an Arrhénius plot of present results. It can be observed that apparent activation energies are different for FZ Si (2.0 eV) and CZ (2.4 eV). Such a difference is difficult to explain if, as postulated in (1,6), dislocation velocity alone is rate controlling since it is the same in FZ and CZ Si at high stresses (8).

In our opinion, crack tip blunting is negligible in the loading conditions used. This assertion relies on the following argument : as the active slip planes cut the crack tip at some angle, blunting can only be very localized (Fig.3). Roughly for one emitted dislocation, the tip radius could increase from b to 2b at the intersecting point of the glide plane with the crack tip and the effect extends over a length of $\sim$ 2b. Just doubling the crack tip

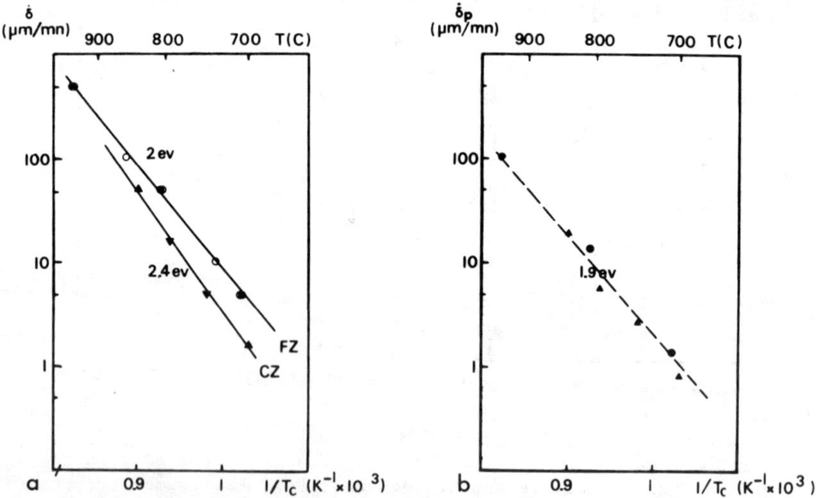

Fig.2 : (a) Arrhénius plot of deflection rate, $\dot{\delta}$ versus critical temperature, T_c (b) Arrhénius plot of plastic opening rate $\dot{\delta}_p$ versus critical temperature, T_c.

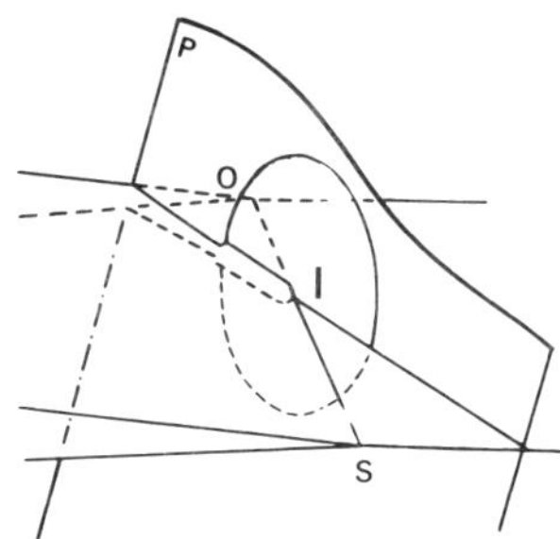

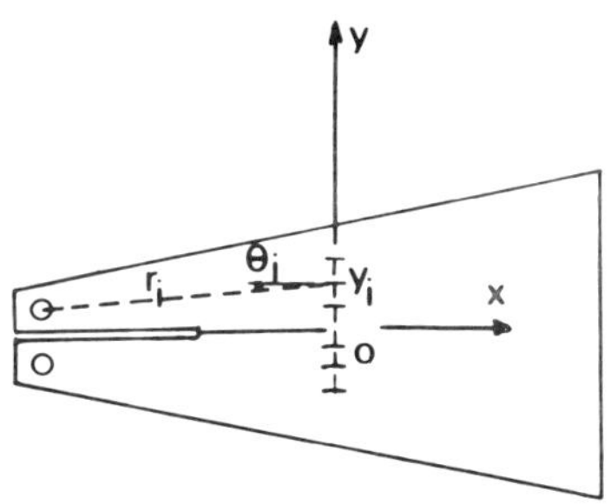

Fig.3 : Localized blunting at I of the crack tip OS by a dislocation emitted in plane P.

Fig.4 : Sample and model of plastic bending (see annex 1).

radius would require a very high dislocation density, much higher than revealed by etch pit counts (3). In addition ∿ 50 % of the total number of dislocations have no blunting effect (9,10). It was shown also that the increase of toughness at 300 K of samples previously relaxed at high temperature could be explained by the shielding model (9).

At 300 K, the elastic opening of the sample heads δ_e is related to the applied load P through the compliance C_e. At high temperature, the opening δ is related to P through an apparent compliance C_a and is the sum of δ_e and the plastic component δ_p. In dynamical loading at a deflection rate $\dot{\delta}$, the plastic opening rate is :

$$\dot{\delta}_p = (1 - C_e/C_a)\ \dot{\delta} \tag{2}$$

It is assumed that δ_p comes from bending of the two beams of the sample and the observed plastic zone, described in details in (2,3,10), is depicted as a wall of N edge dislocations normal to the crack plane, passing through the tip (Fig.4). δ_p is proportional to N (see annex 1) that is $\dot{\delta}_p$ to $\dot{N}$, the nucleation rate of dislocation at the tip.

At fixed $\dot{\delta}$, the crack tip stress field $\bar{\sigma}_e$ increases, but emitted dislocations create a stress field $\dot{\sigma}_p$ of opposite sign. This shielding depends on the ductility through $\dot{N}$. If the rate of increase of shielding $\dot{\bar{\sigma}}_p$ is smaller than $\dot{\bar{\sigma}}_e$ brittle fracture occurs. If $\dot{\bar{\sigma}}_p > \dot{\bar{\sigma}}_e$ on the contrary, general yielding is observed. The critical temperature is characterized by the exact compensation :

$$\dot{\bar{\sigma}}_e + \dot{\bar{\sigma}}_p = 0 \tag{3}$$

and a steady state stress is established. In this condition, it seems reasonable to assume that $\dot{N}$, controlled by the dislocation velocity, is constant :

$$\dot{N}\ \alpha\ \exp\ (-\ Q/k\ T_c) \tag{4}$$

where Q is the activation energy for dislocation glide and, from (2) to (4)

$$\text{Log}\ \dot{\delta}\ (1 - C_e/C_a)\ \alpha\ (-Q/k\ T_c) \tag{5}$$

Using the measured values of C_e and C_a, a single curve is obtained for both CZ and FZ crystals (Fig.2b), with Q = 1.9 eV (the scatter is due to difficulties in measuring C). In our view, the difference of T_c between CZ and FZ Si results not from a difference of dislocation mobility but from an easier multiplication of dislocations in FZ Si. This, which is consistent with the known mechanical behaviour of these two materials, can be ascribed to a different density of dislocation sources (microdefects) related to the presence of oxygen atoms. As temperature increases, oxygen precipitates are formed which can act as sources and the shift of T_c is smaller. Above 1673 K, any measurement of T_c becomes impossible because the development of deformation bands is no longer restricted to the vicinity of the crack tip.

REFERENCES

1. C.F. St John, Phulos. Mag. 32, 1193 (1975).
2. G. Michot, K. Badawi, A.R. Abd al Halim and A. George, Philos. Mag. 42A, 195 (1980).
3. G. Michot, Thèse d'Etat INPL Nancy (1982).
4. J.E. Srawley and B. Gross, NASA Techn. Note D. 3820; 12 (1967).
5. B.S. Majumdar and S.U. Burns, Acta Metall. 29, 579 (1981).
6. P. Haasen, in Atomistic of Fracture, NATO Adv. Res. Inst. Plenum, New-York (1981).
7. K.H. Küster and H. Alexander, Physica B 116, 594 (1983).
8. M. Imai and K. Sumino, Philos. Mag. 47A, 599 (1983)
9. G. Michot, A. George and G. Champier, Proceedings of the 4th European Conference on Fracture, Leoben, Austria, (1982).
10. G. Michot, A. George and G. Champier, Applications of X-ray Topographic Methods to Materials Science, p.377, Plenum, New-York (1984).

ANNEX 1

With the sample dimensions and notations of Fig.4 : $y_i \leqslant 4$ mm and $x \geqslant 14$ mm. Therefore tg $\theta_i \sim \theta_i$ and $r_i \cong x$. The displacement u_y at any point M due to the ith dislocation is :

$$u_y^i = \frac{-b}{4\pi(1-\nu)}\left[(1-2\nu)\ \mathrm{Log}\ r_i + \frac{1}{2}\cos 2\theta_i\right] = -\frac{b}{4\pi(1-\nu)}\left[\frac{1}{2} + (1+2\nu)\ \mathrm{Log}\ x - \left(\frac{y_i}{x}\right)^2\right]$$

For the N dislocations constituting the plastic zone, the total displacement is :

$$u_y = -\frac{b\,N}{4\pi(1-\nu)}\left[\frac{1}{2} + (1-2\nu)\ \mathrm{Log}\ x\right] + \frac{b}{4\pi(1-\nu)}\sum_{i=1}^{N}\left(\frac{y_i}{x}\right)^2$$

$$u_y = \frac{-b\,N}{4\pi(1-\nu)}\left[\frac{1}{2} + (1-2\nu)\ \mathrm{Log}\ x\right] + \frac{b}{4\pi(1-\nu)\,x^2}\int_0^{y_{max}} y^2\,\rho(y)\,dy \qquad \text{(A1)}$$

Etch pit counts have shown (3) that the distribution of dislocations is that of an inverted pile-up with an appreciable density for $y/x \ll 1$ only. In these conditions the second term in (A1) can be neglected and the plastic opening $\delta = 2\,u_y$ is proportional to the number of emitted dislocations. (This simple model of the plastic zone is convenient to calculate the plastic opening but should not be used to estimate the dislocation back stress exerted on the crack).

Microstructural Effects on Flow Localization in 7000 Series Aluminium Alloys (Al-Zn-Mg-Cu)

J. J. Lewandowski and J. F. Knott

Department of Metallurgy and Materials Science, University of Cambridge, UK

ABSTRACT

Yield strength and microstructure were independently varied in 7010 and 7475 aluminium alloys in an attempt to isolate microstructural effects on flow localization. Specimen geometries providing increasingly severe stress states (torsion, tension, notch-bend) were tested at room temperature to determine whether the attainment of a critical shear strain controls fracture in these alloys. The results indicate that a critical shear strain criterion alone is not sufficient to characterise failure. Finally, the microstructural effects on flow localization in these alloys were compared with effects on fracture toughness.

KEYWORDS

Aluminium alloys, flow localization, toughness.

INTRODUCTION

Continuum analyses (1) generally predict that flow localization will occur under conditions of zero, or very low work-hardening. The Lüders strain exhibited by low carbon steels in uniaxial tension is a classic example of macroscopic flow localization accompanying a limited region of zero work-hardening. Deformation beyond the Lüders strain, however, results in positive work-hardening and the flow becomes essentially uniform. The conditions which produce microscopic flow localization and fracture are less well understood, particularly since most materials exhibit finite rates of work-hardening throughout their stress-strain curves.

Both finite-element analysis (2) and experimental observations (3,4) indicate that increased rates of work-hardening promote a more diffuse spread of plasticity than that exhibited by zero work-hardening materials. There is, however, evidence that materials which exhibit positive work-hardening also exhibit failure by intense flow localization (5,6). It is not clear whether this is due to microstructural effects, such as cracking, or due to some local loss in work-hardening capacity.

The present work was conducted in an attempt to distinguish effects of microstructure from those of work-hardening on flow localization. 7000 series aluminium alloys (Al-Zn-Mg-Cu) were chosen as initial test materials because they may be heat-treated to identical initial yield strengths yet possess different microstructures, thereby eliminating comparison of behaviour between microstructures with different yield strengths. Different specimen geometries providing variations in stress state were tested to determine whether a critical value of shear strain would be a criterion for fracture in these alloys. These results are correlated with previously published values of fracture toughness for these alloys.

EXPERIMENTAL

Table 1 and Table 2 list the alloy compositions and heat treatments. Microstructural details are provided elsewhere (7). Cylindrical tensile specimens and torsion specimens with gauge diameter 5.4mm and gauge length 25.4mm were tested to failure at room temperature, as were blunt-notched bend specimens identical to the design of Griffiths and Owen (2). Both the tensile and torsion specimens were polished to a 1.0 μm diamond finish prior to testing. The torque-twist data obtained on the torsion tests were converted to τ-γ plots in the manner described elsewhere (8), while the strains at fracture in the notch-bend tests were calculated with an available finite-element stress analysis (2). The strain-rates $\dot{\varepsilon}$ and $\dot{\gamma}$ applied in the tension and torsion tests were 1×10^{-3}/sec, while the bend tests were loaded in four-point bending at a crosshead rate of 1.7×10^{-2} mm s^{-1}. These test speeds/strain rates were chosen to minimize environmental effects.

RESULTS

Table 3 lists the tensile properties obtained on the alloys tested. Underaged (UA) specimens exhibited serrated flow and macroscopically flat fracture surfaces which were aligned at approximately 45° to the tensile axis, Fig. 1. The fracture surfaces of the UA specimens contained roughly equal proportions of equiaxed dimples and shear dimples. Overaged (OA) tensile specimens did not exhibit serrated flow and exhibited "cup-cone" fractures, Fig. 2a, which contained a bimodal dimple population as shown in Fig. 2b.

No evidence of serrated flow was obtained for any of the specimens tested in torsion. UA specimens exhibited one macroscopic region of concentrated shear which resulted in immediate failure. The details of the surface deformation associated with this "instability band" (6) is shown in Fig. 3 for an UA specimen unloaded immediately prior to failure. OA specimens exhibited a number of individual instability bands, Fig. 4, each of which formed independently at roughly the same macroscopic shear strain. Subsequent flow was constrained to these regions of intense strain. Although failure eventually occurred in one of the instability bands, the failure strains in the OA alloys were considerably larger than those exhibited by the UA alloys. Table 4 summarises both the failure strain, γ_f, and strain at the onset of instability, $\gamma_{inst.}$, for each of the alloys tested. Fracture surfaces of the torsion specimens exhibited smeared features.

Table 5 summarizes the values of σ_{nom}/σ_y and notch root strain at fracture for the notch-bend specimens. Little difference between the two alloys was observed. Both UA and peak-aged (PA) specimens exhibited cracking at the notch prior to general yield, while no evidence of cracking was observed in OA alloys loaded past general yield. Cracks emanating from the notch for specimens loaded with the notch in tension, Fig. 5, followed the 'logarithmic spiral' (1) slip lines ahead of the notch. Specimens tested with the notch in compression and loaded to notch closing displacements in excess of that producing failure in specimens loaded in tension exhibited small indentations at the notch surface, but no signs of voiding or shear along macroscopic slip traces.

DISCUSSION

The tensile properties obtained in the present work were generally consistent with previous results on these alloys (7,9). The occurrence of serrated flow during tensile testing of the UA alloys suggests that dynamic strain-aging may be responsible for the low ductility slant fractures exhibited by the UA specimens. It has been suggested (9) that the constant pinning and unpinning of dislocations reduces work-hardening capacity while not significantly changing the work-hardening exponent, n. Flow localization is apparently facilitated under such conditions by the passage of dislocations which produce locally soft regions in the microstructure. The combination of non-shearable η ($MgZn_2$) precipitates and the lack of serrated flow in the OA alloys appears to prevent large strain concentrations, resulting in a more diffuse spread of plasticity and the 'cup-cone' type fractures in Fig. 2.

The absence of serrated flow in torsion testing of UA specimens is likely to be due to the differences in specimen geometry and strain gradient between the tensile and torsion specimens. Although serrated flow has been reported in torsion tests of an as-quenched aluminium alloy (10), the limited amount of serrated flow exhibited by the tensile specimens in the present work is probably insufficient to detect when testing is conducted in torsion. Despite the lack of serrated flow, failure of the UA specimens was coincident with the appearance of a macroscopic band of intense flow, consistent with the tensile results. Although OA alloys also exhibited instability bands at relatively low shear strains, Table 4 indicates that these instabilities were weak in that considerable deformation was accommodated in the bands prior to failure.

The trends exhibited by the fracture strains obtained on the notch-bend specimens were consistent with the resistance to localized flow exhibited by the tensile and torsion specimens. However, the numerical values at the onset of localization (i.e. Tables 3,4,5) indicate a strong dependence on imposed stress state. The onset of localized flow in torsion is simply $\gamma_{inst.}$, while in the UA tensile specimens it is appropriately chosen as the uniform true strain. The difficulty in locating and identifying instability bands in the notch-bend tests necessitates the choice of the notch root strain at fracture initiation. This will certainly be an upper bound for the strain to the onset of localized flow.

According to the von Mises yield criterion, the maximum tensile stress across a slip band increases in the proportion 0.557:1.0:1.15 on going from a state of pure shear (i.e. torsion) to uniaxial tension, and plane strain tension at the surface of the notch root, respectively. The instability strains and fracture strains, when compared in this context, illustrate a strong dependence of instability on the imposed stress state. Thus, it appears that the attainment simply of a critical shear strain is an insufficient criterion for localized flow and fracture in these alloys.

It might be expected that the fracture toughness values of these alloys would correlate with their resistance to localized flow, and that the toughness would therefore increase in the series PA:UA:OA. This is not always observed (7,11,12). Although PA alloys generally exhibit the lowest toughness of the three aging conditions, OA alloys often exhibit lower toughness than UA alloys of equivalent strength (7,11,12). The apparent discrepancy here may be resolved by examining the operative fracture mechanisms in sharp crack tests in comparison to those in the present tests. In related work (11,12), the lower toughness associated with OA alloys coincided with a fracture mode change from fast shear and dimpled rupture in UA specimens to an intergranular fracture mode. Thus, although the non-shearable aging precipitates reduce flow localization within the grains for OA alloys, grain boundary regions become preferred paths for deformation and fracture. The absence of grain boundary fracture in the present work is likely a result of the lower attainable stresses in the specimen geometries tested.

CONCLUSIONS

1. UA specimens of 7010 and 7475 aluminium alloys exhibit macroscopic instability bands in tension and torsion, resulting in failure at low strains. Low ductility in the UA condition is associated with the occurrence of dynamic strain aging.

2. OA alloys exhibited weaker flow instabilities in tension and torsion. The presence of non-shearable precipitates appears to delay the premature localization of flow exhibited by the UA specimens.

3. Fracture in these alloys does not appear to be controlled simply by a critical shear strain; but is affected also by the stress-state in test-pieces of different geometries.

ACKNOWLEDGEMENTS

The authors would like to thank Professor R.W.K. Honeycombe, F.R.S. and Professor D. Hull for provision of research facilities, and ALCAN for provision of material. One of the authors (JJL) would also like to acknowledge the support of a NATO Postdoctoral Fellowship in Science.

REFERENCES

1. R. Hill, The Mathematical Theory of Plasticity, O.U.P. (1950).
2. J.R. Griffiths and D.R.J. Owen, J. Mech. Phys. Solids, 19, 419 (1971).
3. W.W. Gerberich, Exp. Mech., 4, 335 (1964).
4. D.J. Alexander, J.J. Lewandowski and A.W. Thompson, "Work-hardening effects on notched bar yielding", submitted for publication.
5. J.Q. Clayton and J.F. Knott, Met. Sci., 10, 63 (1976).
6. C.A. Griffis and J.W. Spretnak, Trans. ISIJ, 9, 372 (1969).
7. C.Q. Chen and J.F. Knott, Met. Sci., 15, 357 (1981).
8. G. Dieter, Mechanical Metallurgy, McGraw-Hill, (1976).
9. J.E. King, C.P. You and J.F. Knott, Acta Met., 29, 1553 (1981).
10. P.G. McCormick, Acta Met., 30, 2079 (1982).
11. I. Kirman, Met. Trans., 2, 1761 (1971).
12. G.G. Garrett and J.F. Knott, Met. Trans. A, 9A, 1187 (1978).

TABLE 1 Composition of Aluminium Alloys Tested, wt.%

	Zn	Mg	Cu	Cr	Zr	Fe	Si	Mn	Ti
7010	6.2	2.35	1.7	<0.01	0.14	0.11	0.06	<0.01	0.03
7475	5.6	2.20	1.6	0.12	<0.02	0.09	0.05	<0.01	0.03

TABLE 2 Heat Treatments*

	Solid Solution	Quench	Aging
7010(U)			120°C/ 2 hr
7010(P)	470°C/1 hr	Cold water	120°C/24 hr
7010(O)			120°C/24 hr + 170°C/16 hr
7475(U)			120°C/ 2 hr
7475(P)	470°C/1 hr	Cold water	120°C/24 hr
7475(O)			120°C/24 hr + 170°C/12 hr

*U = Underaged, P = Peak-aged, O = Overaged.

TABLE 3 Tensile Properties of 7010 and 7475*

	0.2% σ_y (MPa)	UTS (MPa)	ε_f (in/in)	ε_u (in/in)	n	Fracture Surface
7010(U)	435	520	0.37	0.13	0.11	slant
7010(P) +	520	565	0.28	-	0.08	slant
7010(O)	460	525	0.54	0.11	0.09	cup-cone
7475(U)	410	510	0.28	0.11	0.15	slant
7475(P) +	515	570	0.24	-	0.09	slant
7475(O)	430	490	0.53	0.03	0.11	cup-cone

*U = Underaged, P = Peak-aged, O = Overaged. + From Ref. 7.

TABLE 4 Torsion Results

	$\gamma_{inst.}$	γ_{fail}	$\Delta\gamma$*
7010(U)	0.55	0.55	-
7475(U)	0.50	0.50	-
7010(O)	0.43	2.50	2.07
7475(O)	0.35	2.10	1.75

$*\Delta\gamma = \gamma_{fail} - \gamma_{inst.}$

TABLE 5 Notch-Bend Results

	σ_{nom}/σ_y	ε_{notch}
(U)	2.2	0.07
7010,7475(P)	2.0	0.045
(O)	>2.3	*

*Exceeds limit of stress analysis.

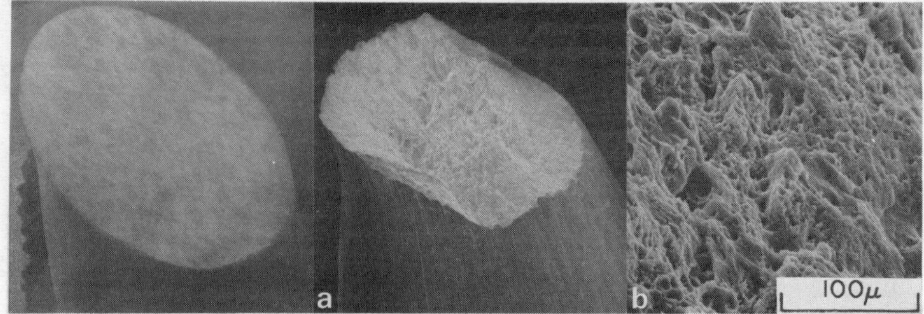

Fig.1 UA Tensile Specimen.

Fig.2 a) OA Tensile Specimen.
b) Fracture Surface of OA Specimen.

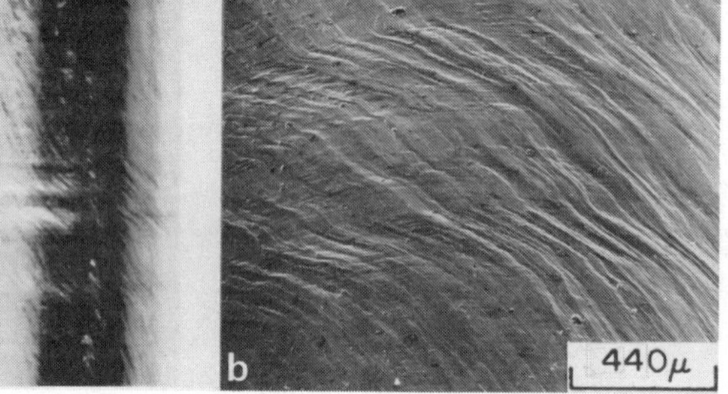

Fig.3 UA Torsion Specimen. Arrows locate 'instability band'.

Fig.4 a) OA Torsion Specimen. Arrows locate 'instability bands'.
b) High magnification view of a).

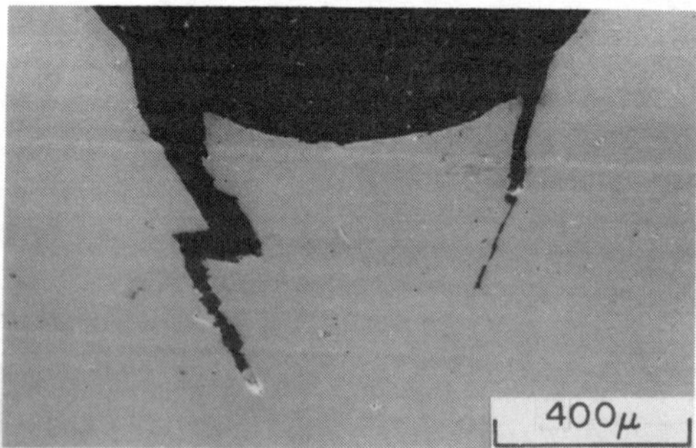

Fig.5 Fracture in notch-bend specimen follows 'logarithmic spiral' slip lines.

Experience with the Key Curve Method for Evaluation of Crack Growth and J Integral During Ductile Tearing

W. R. Tyson* and B. Marandet**

*PMRL, CANMET, Ottawa K1A 0G1, Canada
**IRSID, 78105 St. Germain en Laye, France

ABSTRACT

The "key curve" method introduced by Ernst and others has been used to estimate crack growth and J integral values during ductile tearing of 9% Ni steel tested in three-point bending at room temperature. The effect of ligament length and side grooving on the key curve was evaluated experimentally, and R-curves of fatigue precracked samples with and without side grooves were determined over a range of ligament lengths. The results support the usefulness of the "key curve" method.

KEYWORDS

Elastic/plastic crack growth; R curve; key curve; ductile tearing; J integral.

INTRODUCTION

For characterization of toughness during ductile tearing, measurement of the J integral as a function of crack extension Δa (the "R curve") is generally performed. Measurement of Δa using a single specimen is possible by a variety of techniques, including unloading compliance and potential drop. Recently, however, Ernst et al. (1) have proposed a method of analysis whereby the J-Δa curve may be extracted using data on load and load-point displacement only, in conjunction with a so-called "key curve". This is obviously an attractive possibility, as it enables R-curve determination from uninterrupted tests, which is especially useful at dynamic loading rates.

It is the objective of the present work to evaluate the "key curve" method for the determination of R curves for ductile tearing using three-point bend samples.

THE "KEY CURVE" METHOD

If J is expressed in the form $J = \eta A/b$ where $\eta = \eta(a/W)$ with a the crack length, W the width, and $b = W - a$, $A = \int_0^{\Delta}(P/B_N)d\Delta$ with P/B_N the load per unit net thickness and Δ the load-point displacement, then P/B_N is a function of a/W and Δ/W, and separability is automatically implied (1), with a convenient form for bending being:

$$\frac{P}{B_N} = \frac{b^2}{W} g(\frac{a}{W}) H(\frac{\Delta}{W}) \quad . \qquad \text{Eq 1}$$

Then Ernst et al. (1) have shown how values of crack growth da and J integral may be evaluated from a $P - \Delta$ record for a cracked specimen with a knowledge of g and H, in particular of the "key curve" function H'/H where $H' = dH/d(\Delta/W)$. For three-point bending (3PB), $g(a/W) \sim 1$, $\eta \sim 2$, and the equations take a particularly simple form:

$$\frac{da}{d\Delta} = \frac{b}{2W}(\frac{H'}{H} - \frac{dP}{Pd(\Delta/W)}) \qquad \text{Eq 2}$$

$$\text{and} \quad \frac{dJ}{d\Delta} = -\frac{J}{2W}(\frac{H'}{H} - \frac{dP}{Pd(\Delta/W)}) + \frac{2P}{B_N b} \quad . \qquad \text{Eq 3}$$

In the present work, the key curve was found experimentally using specimens with a blunt notch to delay the onset of tearing. The validity of the separability stated in Eq 1 was tested, and Eq 2 and 3 were used to evaluate R curves.

EXPERIMENTAL PROCEDURE

The material used was a 9% Ni steel in the double normalized and tempered condition (ASTM A353), supplied as 20 mm thick plate. Composition (wt %) was: 0.054 C, 9.10 Ni, 0.605 Mn, 0.077 Cr, 0.096 Mo, 0.073 Cu, 0.001 S, 0.005 P, 0.212 Si, 0.023 Al and 0.013 Ti. Mechanical properties at room temperature were: 0.2% yield strength 627 MPa, ultimate tensile strength 780 MPa, and work hardening index n = 7.7. Stress-strain curves were described well by power law hardening (Ramberg-Osgood equation), there being no yield plateau.

Three-point bend specimens were machined in transverse orientation, with dimensions 17.5 x 35 x 170 mm. A variety of crack preparations were used, with different crack root radii (ρ = 0, 0.1, 1 mm) and crack depths (a_o/W = 0.3 to 0.8), both with and without side grooves (ratio of net width to gross width B_N/B = 0.75 and 1.0). Specimens with ρ = 0 were prepared by fatigue pre-cracking, with observance of the precautions outlined in ASTM E813. Side grooves were made with a groove angle of 45° and root radius of 0.25 mm.

Specimens were tested in 3PB with span S = 4W at room temperature in a screw-driven universal testing machine at a crosshead travel speed of about 1 mm/min. Load and load-line displacement were recorded during the test, and the point of initiation of ductile tearing was determined from the minimum in AC potential drop across the crack mouth (2). Specimens were heat tinted after unloading to mark the position of the crack, and broken open at liquid nitrogen temperature to measure the initial (by 5-point average) and final (by 9-point average) crack lengths.

RESULTS

Figure 1 shows typical results for the variation of load with load-point displacement, for blunt-notched ($\rho = 1$) and fatigue precracked ($\rho = 0$) specimens of initial crack depth $a_o = 0.5$ W with side grooves ($B_N/B = 0.75$). The dashed curve is the extrapolated H function (Eq 4). The arrows on the curves mark the initiation of ductile tearing, detected by the AC potential drop technique; the blunt-notched specimen was unloaded at this point, while deformation was continued for the fatigue pre-cracked specimen up to $\Delta/W \sim 14 \times 10^{-2}$. Tearing began before maximum load was reached in all cases, and the minimum in AC potential corresponded well with the beginning of deviation from the P - Δ curve extrapolated to conform with power-law hardening (described by Eq 4 below).

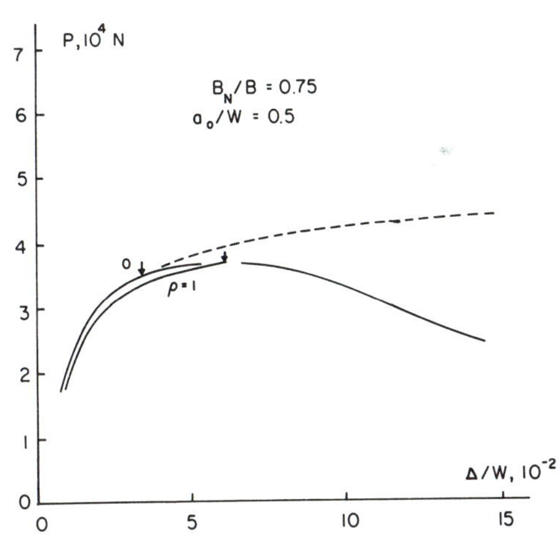

Fig. 1. Typical curves of load vs load-point displacement.

As seen in Fig. 1, the P - Δ curve is lowered by increasing the crack tip radius. Side grooving had the opposite effect, i.e., the P/B_N - Δ curve was raised by about 20%. These observations demonstrate that flow stresses at the notch tip are raised by the increased constraint accompanying increased notch sharpness and side grooves.

When curves of $(P/B_N)/(b^2/W)$ were plotted vs Δ/W, it was found that Eq 1 was obeyed within 10% over the crack depth range a/W = 0.3 to 0.8. There was a trend to lower curves within this range for a/W > 0.6 for side-grooved samples, but this is not general since a slight trend in the opposite direction was found for plane-sided samples.

Values of J_i at initiation were calculated from the usual formula $J = 2A/(B_N b)$ where A is the work done up to initiation for the fatigue pre-cracked samples, and are listed in the table. Reproducibility of J_i was ±~10%, with no significant effect of side-grooving or initial crack depth. Values of Δa and J (integrated values of da and dJ) were calculated using Eq 2 and 3, and plotted in Fig. 2 for (a) plane sided and (b) side-grooved specimens. H'/H was obtained from a power-law fit to the data for blunt-notched samples, viz:

$$\frac{\Delta}{W} = \frac{\Delta_{el}}{W} + \frac{\Delta_{pl}}{W} = \alpha H + \beta\left(\frac{H}{H_o}\right)^n \quad . \qquad \text{Eq 4}$$

Values of α, β, H_o, and n were found to be 6.3×10^{-5} MPa^{-1}, 3.10×10^{-2}, 300 MPa, and 8.8 for $B_N/B = 0.75$, and 8.0×10^{-5} MPa^{-1}, 3.34×10^{-2}, 260 MPa, and 8.8 for $B_N/B = 1$. Values of α contain a contribution of ~10 to 15% from deflection of the machine and load-point indentation, for which no correction was attempted. The form of Eq 4 is consistent with the finite element

TABLE

$\frac{B_N}{B}$	$\frac{a_o}{W}$	J_i, kJ/m^2	Final Δa, mm	
			Calculated	Measured*
1.0	0.5	264	1.49	1.77
	0.7	239	0.99	1.13
0.75	0.4	239	5.98	6.44
	0.5	276	4.41	4.99
	0.5	267	3.96	4.62
	0.6	229	1.97	2.14
	0.6	245	3.71	4.01
	0.7	212	2.16	2.40

*Includes stretch zone width ~0.1 mm

results of Kumar et al. (3), ignoring the "Irwin plastic zone" correction to the elastic term which was found to be negligible in the present case. Since the loads for both elastic and plastic deformation are increased by 25% side-grooving by a constant factor of ~20% (when loads are normalized by the net section as in Eq. 1), the key curves (H'/H) for B_N/B = 1 and 0.75 are practically identical.

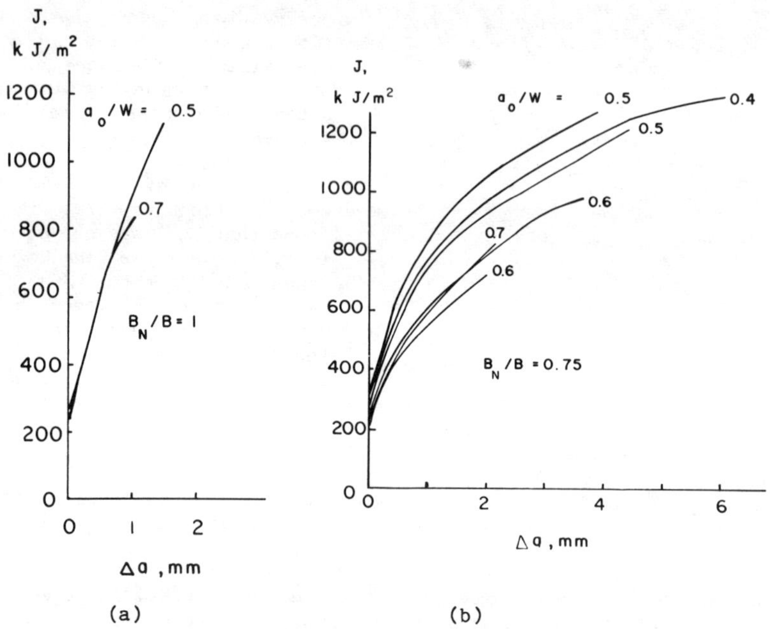

Fig. 2. R curves calculated by the "key curve" method.

Crack growth from blunting is not included in the data plotted in Fig. 2, i.e., Δa = 0 corresponds to the blunted crack tip position. Measured and calculated crack extensions are given in the table. The measured Δa includes the stretch zone (estimated by SEM examination of fracture surfaces to be ~0.1 mm at initiation), while the calculated Δa does not. Adding the stretch zone to the calculated Δa produces quite good agreement with measured Δa, although the former is systematically low by a small amount (~5%).

DISCUSSION

The "key curve method" has been demonstrated in this work to be straightforward to apply and to lead to acceptably reproducible results. Successful application depends on two important aspects: selection of a suitable key curve, and detection of the onset of crack growth.

In the present work, construction of $H(\Delta/W)$ was simplified by the absence of a yield plateau and the consequent ability to fit the stress-strain curve and load-load point displacement curve with a power law relationship. For most structural steels, a yield plateau is observed, and the $H(\Delta/W)$ function will be correspondingly more complex in the initial stages. However, H may be estimated accurately from experimental data on blunt-notched (or sub-sized) specimens. It has been verified in this work that, while $H(\Delta/W)$ may be shifted by alterations in constraint accompanying notch blunting or side grooving, H'/H (which is used in calculations, viz Eq 2 and 3) is insensitive to constraint since the shift corresponds to a multiplicative factor which cancels when a ratio is taken.

Reliable detection of the initiation of crack growth is essential. In principle, the key curve method is applicable from the start of deformation, including the elastic range; in practice, however, H'/H during elastic deformation is so large that very small errors in measurement of P or Δ cause large errors in crack growth estimation. In the present work, initiation was detected by the AC potential drop technique, which may be used without interrupting the test and thus is compatible with the key curve method; reproducibility of J_i values (see table) implies that initiation was accurately detected. In principle, initiation may also be identified with deviation of the load-load point displacement curve from the extrapolated H curve (dotted in Fig. 1); however, care must be taken in this case to adjust the curve representing zero crack growth (i.e., H) for constraint by assuring good fit with the initial part of the curve for the fatigue pre-cracked specimen (up to onset of crack growth), and to define quantitatively a criterion to detect the point of deviation.

The reliability of the method may best be judged from the accuracy of prediction of Δa, which may be seen from the table to be good, although systematically low by ~5%. The resistance curves (Fig. 2) show good reproducibility, although for side-grooved samples, the curves are significantly lower for $a_o/W > 0.5$. Also, in agreement with other work, the R curves for plane-sided specimens (Fig. 2a) have slightly higher slopes (tearing moduli) and continue to higher values of J, although initiation values (J_i) are similar for samples with and without side-grooves. It should be noted, however, that crack fronts became significantly curved during ductile tearing for plane-sided samples, while they remained quite straight for side-grooved specimens.

It is currently proposed (3) that R-curves remain valid up to $J \leq$ (min of B, b) $\sigma_y/25$ and $\Delta a \leq 0.06$ b, provided that $\omega = (b/J_i)(dJ/da) > 10$. These conditions are met in this work only for the initial parts of the R-curves (roughly, up to $J \sim 400$ kJ/m^2), and hence J-dominance is questionable for the latter part of the curves. This may explain the drop in R curves with increasing a_o/W noted in Fig. 2b, although not the fact that there is no dependence on a_o/W in Fig. 2a. This difference could be related to the observed lowering of the normalized $P - \Delta$ curves with increasing a_o/W for side-grooved samples which was not found for plane-sided samples; in effect, this would imply that g (Eq 1) becomes increasingly smaller than unity with increasing a_o/W for side-grooved samples. This would result, following the full development given by Ernst (1), in overestimation of Δa and underestimation of J using Eq 2 and 3 for the side-grooved samples for $a_o/W > 0.5$.

More fundamentally, the key curve method is based on the premise that normalized load is a unique function of a and Δ (Eq 1). In particular, in 3PB, g = 1 and so $(P/B_N)/(b^2/W)$ should be a unique function of Δ/W and independent of (a, Δ) history. In the present experiments, it was noted that the normalized load values, when calculated at the end of crack growth using measured values of b and P, were systematically higher by ~10% than values expected from extrapolation of H curves (remembering to take account of increased constraint for sharp notches). This would imply that g (Eq 1) should be an increasing function of a/W during tearing, which would lead to underestimation of Δa using Eq 2 by an amount close to that observed as the systematic error in the table. The physical origins of this dependence of $(P/B_N)/(b^2/W)$ on load history are not clear. However, according to the present results, the effect is small, and the resulting errors (~5-10%) in Δa and J are not serious. Moreover, the comparison has been made with extrapolated H values since measurements were possible only up to a maximum $\Delta/W \sim 10 \times 10^{-2}$ (limited by crack growth initiation in blunt-notched specimens). Verification of the validity of the extrapolation with no crack growth would require further measurements with sub-size specimens, although there is no reason to expect that loads would be higher than predicted by the power-law fit (Eq 4).

CONCLUSION

The "key curve" method has been found to give reproducible resistance curves, accurate to within 10% in Δa and J, from measurement of load and load-point displacement without the necessity of independent estimation of crack length.

ACKNOWLEDGEMENTS

This work was performed while one of the authors (WRT) was on a scientific exchange at IRSID. The authors are grateful to Creusot-Loire for the supply of material, and to the support staff at IRSID for technical assistance with specimen preparation and performance of the experiments. Discussions with Mlle. E. Maas were helpful.

WRT is grateful to CANMET and IRSID, and to the France/Canada exchange program, for the opportunity to do this work.

REFERENCES

1. H.A. Ernst, P.C. Paris and J.D. Landes, "Fracture Mechanics: Thirteenth Conference", ASTM STP 743, 476 (1981).
2. B. Marandet, G. Labbe, J. Pinard and M. Truchon, "Détection de l'amorçage et suivi de la propagation d'une fissure par variation du potentiel électrique en régime alternatif", Rapport IRSID RE 549 (1978).
3. V. Kumar, M.D. German and C.F. Shih, "An engineering approach for elastic-plastic fracture analysis", EPRI Report NP-1931 (1981).

Contribution of Local Deformation to Fracture Toughness Through Fractographic Measurements

A. W. Thompson

Department of Metallurgical Engineering and Materials Science, Carnegie-Mellon University, Pittsburgh, PA 15213, USA

ABSTRACT

Most efforts at modeling microstructural aspects of fracture toughness incorporate a measure of local deformation or strain. In general, there is no macroscopic means of measuring this strain. However, measurement by quantitative fractography of dimple depth and diameter (expressed as the ratio, *M*, of the two values) has been shown to be helpful in understanding microvoid coalescence (MVC) types of fracture. The use of *M* has been extended to quasi-cleavage and tearing topography (TTS) fractures, and to "blocky" fractures, for which a "regional" *M* is appropriate. For each of these cases, *M* is readily incorporated into modeling of fracture micromechanisms, and has the advantage of showing a direct relation to the physical process of fracture. Experimental work to date is sparse but appears to follow the expected magnitudes of *M* as well as the correlation to fracture ductility and toughness.

KEYWORDS

Fracture, micromechanisms, quantitative fractography, toughness, fracture mode.

INTRODUCTION

Attempts to construct models of the local fracture process at crack tips have a long history, going back at least to Krafft (1). Originally, such models incorporated the mechanical properties of tensile specimens, but it was realized that the states of both stress and strain at a crack tip were quite unlike those in tensile specimens. Subsequently, there have been other proposals, as has been reviewed (2,3). For example, the local fracture strain at a crack tip in plane strain can be approximated by the ductility of a plane-strain tensile specimen, such as a Clausing specimen (4). Crack tip strains have been an important focus in analyses of crack tip deformation and fracture, particularly when unit crack advances occur with substantial local strain or under strain control (3). The problem has been, however, that direct measurement of these strains is difficult, and that the combined plane strain and triaxial stress states at a crack tip clearly cannot correspond exactly to the behavior of any mechanically-simpler specimen. Accordingly, a need exists both to make such experimental measurements, and also to incorporate realistically both appropriate micromechanical properties and also microstructural variables into fracture mechanics analyses (3).

One approach to this problem is through quantitative fractography, not only in terms of the relatively macroscopic aspects such as "stretch zones" (2), but in terms of microscopic measurements of plastic features which occur during strain-controlled fracture processes such as microvoid coalescence (MVC), quasi-cleavage (QC) (5,6), and tearing topography (TTS) (7) fractures. One such proposal of this type, which is summarized here, is to measure the depth as well as the width of microvoid halves or dimples in MVC fractures (8-10). Any such work draws upon the still-developing field of quantitative fractography, the status of which has recently been summarized in several places (11-16). This paper outlines recent work and emphasizes the need for further development of fracture micromechanism modeling.

Most of the discussion here is about MVC fracture, partly because both the nuclei and the fracture process are well understood (17,18). For simplicity, it is assumed here that the nuclei for microvoids are second-phase particles, whether inclusions, dispersoids or precipitates. There is no longer any real dispute that particles are the dominant nuclei in most materials, although microvoid nucleation in slip bands, slip band intersections, or cell walls (17,19-22) presumably would lead to similar growth and coalescence processes to those associated with particles. The resulting microvoid halves on the fracture surface, or "dimples", may have irregular shapes, the mean dimensions are readily measured on conventional SEM (scanning electron microscope) fractographs.

There has been an interest for some time in interpreting changes in the MVC process through measurement of fracture surface dimples, and in a number of cases, pronounced changes have been found, as reviewed elsewhere (8,18,23,24). A variety of experimental variables, such as temperature, strain rate, specimen geometry (particularly as stress or strain state is affected), introduction of embrittling species such as hydrogen, and others, all can cause the size of MVC dimples to differ. Moreover, there have now been proposed models to account for what appear to be the limiting cases in such behavior (8,23).

The development of those models was complicated by the difficulty that the usual relationships of quantitative metallography (25) do not apply to fracture surfaces, since such surfaces are rarely either planar or random in orientation, and thus do not fulfull the requirements of a metallographic sectioning plane (8,11,23,26). In the case of MVC, it is evident that increasing depth of dimples increases the non-planarity or "micro-roughness" (8) of the fracture surface. The definition adopted (8,9) for micro-roughness is expressed by the parameter M, given by

$$M = h/w, \qquad [1]$$

where the dimple depth h and width w are depicted in Fig. 1. As discussed previously (8,9), as $M \to 0$, dimples become vanishingly shallow and the fracture surface tends toward local flatness and toward meeting the criteria for a metallographic section. As M increases, however, the fracture surface is increasingly unlike a metallographic section, and the "depth sampled" (23,26) below the mean fracture plane, by the fracture process, also increases. As discussed elsewhere (8,9,27), typical or expected values of M appear to lie in the range 0.5 to 1.

There are relatively few experimental data to which this prediction may be compared. However, in earlier work (8,9,27) it was pointed out that there do exist a number of results for fracture of pre-cracked specimens (28), in which it was shown that M values were in the range 0.5 to 1.0, as expected. However, experiments on copper (29) in notched bend specimens showed that M was a function of location in the specimen, varying from about 0.7 at the specimen center to values near unity along the specimen sides. This illustrates the importance of stress state in the magnitude of M, or conversely, the value of M in reflecting the stress state during the local fracture process. The same *may* be illustrated by tensile results on dimple width w only (30), which varied systematically from specimen center to surface, but it must be emphasized that no depth measurements were made, preventing M calculations.

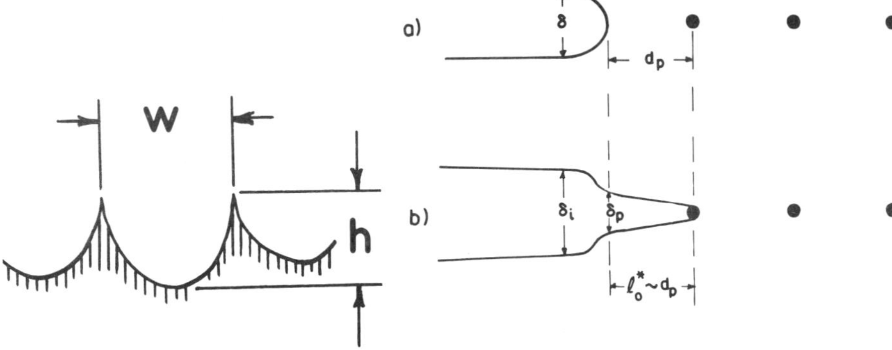

FIG. 1. Schematic illustration (8) of *M* definition, equation [1] in text, in terms of dimple depth h and width w. As discussed in the text, this schematic may also apply to TTS fracture, except that the fracture surface features in TTS are not rotationally-symmetric, as dimples approximately are, but are merely curved surfaces, often of complex curvature, which connect tear ridges (7).

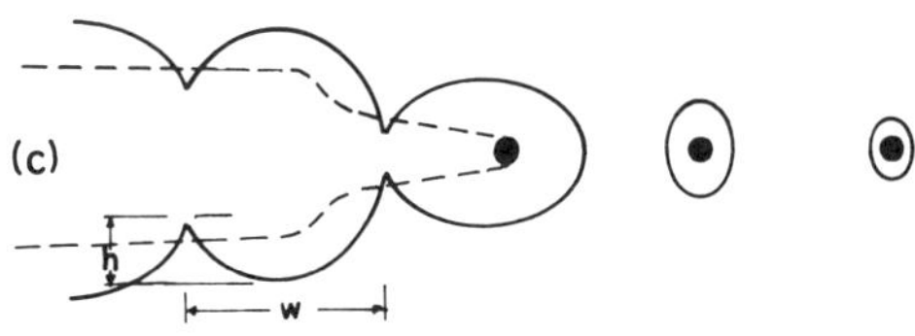

FIG. 2. Sketch of crack extension parameters under plastic conditions, as discussed in the text. From ref. 3.

ROLE OF LOCAL STRAIN IN TOUGHNESS

There are a variety of approaches which have been taken to modeling of toughness of reasonably ductile materials; a recent review is that of Firrao and Roberti (2). The most interesting of these approaches, in the present context, are those oriented to a critical distance. Such ideas are most familiar in the form of the proposal that a critical stress over a critical distance can be required for fracture, due to Ritchie, Knott and Rice (31). For fracture initiation (an appropriate criterion for fracture if K_{Ic} or J_{Ic} are to be used as descriptors) which occurs by MVC processes, however, it is likely more appropriate to use a local critical strain criterion for fracture, a concept which goes back at least as far as McClintock and Irwin (32). From the ideas of several other workers (3,32–35) has come the concept that the effective plastic strain ϵ_p must equal or exceed a critical plastic strain ϵ^* over some distance ahead of the crack tip, which is usually written as ℓ_0^*. Whether this ϵ^* is necessarily a *fracture* strain, since it is maintained some distance *ahead* of the crack tip, is at present an unresolved issue (27). This viewpoint has had some success (3) in describing existing data, such as upper shelf toughnesses (35).

When toughness is very high, it may be overly conservative to quantify toughness even by J_{Ic}, as has been argued elsewhere (3). In such cases, one may use what is often called "crack growth toughness", measured (3) by such parameters as dJ/da (a = crack length) or the dimensionless "tearing modulus" T_R. There are then additional points of interest regarding use of local fracture strain measures. Recent work in this area has tended to use local size scale parameters to characterize behavior. Some of these are shown in Fig. 2. Fig. 2(a) shows an idealized, blunted crack tip shape, with the distance between particles d_p and the crack tip opening δ indicated. In Fig. 2(b) an increment of crack extension has occurred, associated with a crack tip opening increment δ_p (here $\delta + \delta_p = \delta_i$); the extension is assumed to correspond approximately to the critical distance ℓ_0^*. Fig. 2(b) also depicts the assumption made in some treatments, that the extension about corresponds to d_p, although there is no requirement that this be so, and indeed, there are good reasons for doubting any necessary connection (3) between ℓ_0^* and d_p. In Fig. 2(c) is a sketch of a more realistic crack tip shape for plastic fracture by MVC, which also shows how h and

SMA2-N*

w, and thus M, can be fitted into this picture. An important point to be made here is that d_p refers to the population of particles which are effective in nucleating microvoids, not to all particles in the material.

The analyses leading to Fig. 2 make it clear that a direct experimental measure of parameters such as ℓ_0^* and d_p, as well as some method of measuring local strains themselves, particularly ϵ^*, would be of great value, both to test models for fracture processes, and also to provide better empirical constants in the relevant equations. One means of doing so is in terms of the fractographic information mentioned above, in M measures or related parameters.

As an example of an approach to strain measurement, it should be possible to use h and w data to measure the local plastic deformation (post nucleation), provided that the diameter of the nucleating particle, D_p, is known. In that case one can write the longitudinal and transverse deformations as the ratios of h and w/2, respectively, to D_p, so that the local strains are

$$\epsilon_L = \ell n\,(h/D_p) \qquad [2a]$$

and $$\epsilon_T = \ell n\,(w/2D_p), \qquad [2b]$$

respectively. The kernel of the second of these terms, $w/2D_p$, resembles a fracture parameter devised by Garrison (36), which he called R_v/R_i, with the two radius terms, R, referred to the *v*oid and *i*nclusion, respectively. Garrison's interest was to develop a relation among this transverse deformation parameter (which he seems to have regarded as true strain, while in fact it is the inverse log thereof), the initial inclusion spacing X_0, and the fracture toughness. However, the logarithm of R_v/R_i, as with the parameters above, measures strain directly.

The use of fracture surface micro-roughness information to obtain local strains appears to have been first proposed by Thompson and Ashby (9). They observed that one can regard the dimples as approximately circular but filling the fracture surface, so that the number of dimples per unit area, N_A, is given by $4/\pi w^2$. Combining this with conventional quantitative metallographic relations and recalling (23,26) that the fracture zone depth is of height 4h, one finds (9) that

$$h/D = (M^2/3f)^{1/3}\ . \qquad [3]$$

This is just half the longitudinal deformation parameter defined earlier. Again, its logarithm is the local strain, but now the strain is associated with fracture, since M is measured on the fracture surface. Strictly, it should be regarded as a measure of local (post-nucleation) strain, but if the nucleation strain is small, then the fracture strain ϵ_f is given by the logarithm of equation [3]. This relation shows that the local fracture strain can depend upon only the micro-roughness M and on the volume fraction of nucleating particles. (It is important to reiterate that **f** refers only to those particles which actually nucleate microvoids (23), not to the total particle population.) Since this makes available a measure of the local strain which actually occurred during the fracture process, it provides a way to evade the difficulties in estimating a crack-tip strain from strains measured in some other type of specimen.

The local strain expressed as the logarithm of equation [3] can then be inserted into relations for whichever fracture process is of interest. For example, fracture toughness J_{IC} can be notionally written (23) as $J_{IC} = \sigma_0 \cdot \epsilon_f^* \cdot \ell_0^*$, where the σ and ϵ terms are regarded as the mean effective values and ℓ_0^* is a (microstructural) length. Ordinarily, one takes the mean effective flow stress σ_0 to be the average of yield and ultimate strengths, estimates ϵ_f^* from some other fracture strain, such as that obtained from a Clausing specimen, and then calculates ℓ_0^* from the known value of J_{IC}. It should now be possible, however, to obtain the appropriate local fracture strain from M values, as described above, giving more confidence to the overall calculation and to resulting values of ℓ_0^*.

This use of *M* for local strain measurement as part of quantitative assessment of fracture toughness parameters (3) is not yet tested directly, yet there are already proposals (10) for extension of *M*-based concepts to several other fracture modes which are locally plastic, but which do not fit the MVC description. Among these are quasi-cleavage or QC (5,6,10). For the case of the tearing topography surface or TTS (7), it appears likely that the concept illustrated in Fig. 1 and equation [1] will apply, except that the fracture surface features are not rotationally-symmetric dimples, but are irregular curved surfaces between tear ridges (7).

Although the concept of *M* is still very new, it appears well worth exploration, experimental test, and extension. The prospect that all the locally plastic modes, MVC, QC and TTS, may be treatable by similar quantitative fractographic concepts, through measurement of parameters such as *M*, is most encouraging and certainly deserves additional work.

ACKNOWLEDGEMENTS

I appreciate many helpful discussions with R.O. Ritchie, M.F. Ashby and W.M. Garrison on the topic of this paper. Support was provided by the U.S. National Science Foundation under grant no. 81-19541 and by the Association of American Railroads through the Affiliated Laboratory at Carnegie-Mellon University.

REFERENCES

1. J.M. Krafft, *Appl. Mater. Res.* **3**, 88-101 (1964).
2. D. Firrao and R. Roberti, *Metall. Sci. and Technol.* **1**, 5-13 (1983).
3. R.O. Ritchie and A.W. Thompson, *Metall. Trans. A* **16A** (1985) in press.
4. D.P. Clausing, *Int. J. Fract. Mech.* **6**, 71-85 (1970).
5. C.D. Beachem and R.M.N. Pelloux, in *Fracture Toughness Testing and Its Applications* (ASTM STP 381), pp. 210-44. ASTM, Philadelphia (1965).
6. C.D. Beachem, *J. Basic Eng. (Trans. ASME, Series D)* **87**, 299-306 (1965).
7. A.W. Thompson and J.C. Chesnutt, *Metall. Trans. A* **10A**, 1193-96 (1979).
8. A.W. Thompson, *Acta Met.* **31**, 1517-23 (1983).
9. A.W. Thompson and M.F. Ashby, *Scripta Met.* **18** 127-30 (1984).
10. A.W. Thompson, in *Advances in Fracture Research*, Proc. 6th Int. Conf. on Fracture (edited by S.R. Valluri, D.M.R. Taplin, P.R. Rao, J.F. Knott and R. Dubey), Vol. 2, pp. 1393-99. Pergamon, Oxford (1984).
11. S.M. El-Soudani, *Metallography* **7**, 271-311 (1974).
12. J.L. Chermant and M. Coster, *J. Mater. Sci.* **14**, 509-34 (1979).
13. E.E. Underwood and S.B. Chakrabortty, in *Fractography and Materials Science*, ASTM STP 733 (edited by S.N. Gilbertson and R.D. Zipp), pp. 337-54. ASTM, Philadelphia (1981).
14. J.L. Chermant and M. Coster, *Int. Metals Reviews* **28**, 228-50 (1983).
15. S.M. El-Soudani, *Fundamentals of Quantitative Fractography*, Ph.D. dissertation, Univ. of Cambridge (1980).
16. K. Banerji and E.E. Underwood, in *Advances in Fracture Research*, Proc. 6th Int. Conf. on Fracture (edited by S.R. Valluri, D.M.R. Taplin, P.R. Rao, J.F. Knott and R. Dubey), Vol. 2, pp. 1371-78. Pergamon, Oxford (1984).
17. J.R. Low, *Prog. Mater. Sci.* **12**, 1-96 (1963).
18. A.W. Thompson, in *Effect of Hydrogen on Behavior of Materials* (edited by A.W. Thompson and I.M. Bernstein), pp. 467-77. TMS-AIME, New York (1976).
19. A.W. Thompson and J.C. Williams, in *Fracture 1977*, Proc. 4th Int. Conf. on Fracture (edited by D.M.R. Taplin), Vol. 2, pp. 343-48. Univ. of Waterloo Press, Waterloo, Ont. (1977).

20. R.N. Gardner, T.C. Pollock and H.G.F. Wilsdorf, *Mater. Sci. Eng.* **29**, 169–74 (1977).
21. S.H. Goods and L.M. Brown, *Acta Met.* **27** 1–15 (1979).
22. R.N. Gardner and H.G.F. Wilsdorf, *Metall. Trans. A* **11A**, 659–69 (1980).
23. A.W. Thompson, *Metall. Trans. A* **10A**, 727–31 (1979).
24. A.W. Thompson and I.M. Bernstein, in *Advanced Techniques for the Characterization of Hydrogen in Metals* (edited by N.F. Fiore and B.J. Berkowitz), pp. 43–60, TMS-AIME, Warrendale, PA (1982).
25. E.E. Underwood, *Quantitative Stereology.* Addison-Wesley, New York (1970).
26. D.J. Widgery and J.F. Knott, *Metal Sci.* **12**, 8–11 (1978).
27. A.W. Thompson, in *Fracture: Measurement of Localized Deformation by Novel Techniques* (edited by W.W. Gerberich and D.L. Davidson). TMS-AIME, Warrendale, PA, in press.
28. C.P. You, , Ph.D. dissertation, Univ. of Cambridge, 1984.
29. J.D.G. Groom, *Effects of Prestrain on Fracture*, Ph.D. dissertation, Univ. of Cambridge, 1971.
30. R.J. Coyle, J.A. Kargol, and N.F. Fiore, *Scripta Met.* **14**, 939–42 (1980).
31. R.O. Ritchie, J.F. Knott and J.R. Rice, *J. Mech. Phys. Solids* **21**, 395–410 (1973).
32. F.A. McClintock and G.R. Irwin, in *Fracture Toughness Testing and Its Applications* (ASTM STP 381), pp. 84–113. ASTM, Philadelphia (1965).
33. J.R. Rice and M.A. Johnson, in *Inelastic Behavior of Solids* (edited by M.F. Kanninen, W.F. Adler, A.R. Rosenfield and R.I. Jaffee), pp. 641–72. McGraw-Hill, New York (1970).
34. A.C. Mackenzie, J.W. Hancock and D.K. Brown, *Eng. Fract. Mech.* **9**, 167–88 (1977).
35. R.O. Ritchie, W.L. Server and R.A. Wullaert, *Metall. Trans. A* **10A**, 1557–70 (1979).
36. W.M. Garrison, *Scripta Met.* **18** 583–86 (1984).

SECTION 10

Cyclical Deformation and Fatigue

Déformation cyclique et fatigue

Effect of Surface Cracks induced by Hydrogen on the Fatigue Properties of AISI 304 Stainless Steel

P. S. Pierantoni*, P. E. V. de Miranda and R. Pascual***

**Instituto Militar de Engenharia, 22290 Rio de Janeiro, Brazil*
***COPPE-EE-Universidade Federal do Rio de Janeiro; 21910, Rio de Janeiro, Brazil*

ABSTRACT

The effect of hydrogen induced delayed cracks on the fatigue properties of AISI 304 stainless steel was investigated. Specimens were submitted to one or more hydrogenation-outgassing cycles, after which they were fatigued in reversed bending. SEM and optical observations of the lateral and fracture surfaces were also performed. It was found that fatigue life is reduced by a hydrogenation-outgassing cycle due to the introduction of surface cracks. This is particularly important for high cycle fatigue conditions. Repeated hydrogenation-outgassing cycles affect only to a minor extent the fatigue life.

KEYWORDS

Hydrogen embrittlement; fatigue; hydrogen induced cracks;surface cracks; austenitic stainless steels.

INTRODUCTION

Sufficient evidence has been already accumulated in the last years showing that austenitic stainless steels are susceptible to hydrogen embrittlement, although to a much lower extent that it is found for b.c.c. materials (1-3).
Hydrogen has a relatively high solubility in austenite, but a very low diffussivity,so that a very thin surface layer of high hydrogen concentration is produced in an austenitic stainless steel when placed in a hydrogen rich environment. This produces high compressive stresses in that surface layer and a partial transformation to the hexagonal ε martensite phase (4).Since the surface layer is very thin (usually about 20 μm) the bulk of the material remains unaffected and the effects observed on most of its properties are small, if the material is sufficiently thick. On the other hand, drastic fragilization effects have been reported for very thin samples (5,6).
If the material is outgassed, hydrogen scapes, generating surface tensile stresses which give rise to delayed cracks.Also, there is a partial transformation of the ε martensite into the b.c.c. α' phase as well as back into austenite.
Hydrogen effects in austenitic stainless steels are then a

surface effect and it is natural, in view of this fact, to believe that their fatigue properties, being highly dependent on the surface condition, can be affected.

Very little work has been done in this area, which has been recently reviewed by Schuster and Altstetter (7). In brief, the presence of a hydrogen atmosphere produces a decrease in the fatigue life of smooth specimens and an increase in the fatigue crack growth rate in notched precracked specimens. Both effects are enhanced in the more unstable stainless steels.

All these results are related to the effect of a hydrogen environment <u>during</u> the fatigue process. To the authors' knowledge, only some preliminary results have been reported (3) on the effect of the delayed cracks produced <u>after</u> outgassing a hydrogenated sample. Since this is a condition that can arise in practical applications, the objective of the present work was to study the influence of surface cracks induced by hydrogenation-outgassing cycles on the fatigue life of an AISI 304 stainless steel.

EXPERIMENTAL PROCEDURES

Flat hour-glass fatigue specimens with a 20 mm minimum gage were machined from a 4 mm thick AISI 304 stainless steel sheet, having the following chemical composition (in weight %): Cr: 18.85; Ni: 8.46; C: 0.07; Mn: 0.14; Si: 0.46; P: 0.026; S: 0.03; Fe: balance.

They were heat treated for 30 minutes at 1100 ^{o}C, quenched in water (yielding an average grain size of 80 μm) and electropolished. Fatigue tests were performed in reversed bending. Fatigue life was determined for three sets of specimens: 1) Non-hydrogenated; 2) Hydrogenated and outgassed; 3) Submitted to repeated cycles of hydrogenation and outgassing. SEM and optical observations of the lateral and fracture surfaces were also performed. Hydrogenation was obtained by cathodic charging of the specimens during 16 hours with a current density of 1000A/m^2 using a 1N H_2SO_4 electrolyte poissoned with As_2O_3. After that, the specimens were outgassed at room temperature for 8 hours and then heated at 110 ^{o}C for 18 hours. This procedure was used for the specimens of group 2 which were at that point ready for fatigue testing. For the specimens belonging to group 3 the sequence of cathodic charging and outgassing was repeated either two or five times before the fatigue test.

Vickers microhardness determinations of the annealed, hydrogenated and fatigued samples were also performed using a load of 25 g.

RESULTS

After the specimens were hydrogenated and outgassed a great number of small cracks could be observed on their surface. Groups of approximately parallel cracks appeared in each grain, but the distribution was highly inhomogeneous. This can be seen in Fig. 1. The presence of martensitic phases is apparent in region A indicated in Fig. 2.

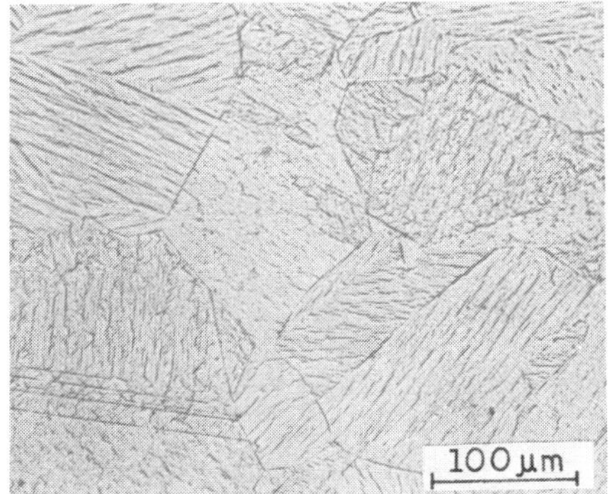

Fig. 1. Optical micrograph of a hydrogenated and outgassed sample showing hydrogen induced delayed cracks. Unetched.

Fig. 2. Other area of the same sample shown in Fig. 1. Note the presence of martensitic phases in region A. Unetched.

The average density of cracks was determined by semiquantitative measurements on a large number of areas and resulted in a value of 2.6×10^{10} cracks/m^2. These cracks are very shallow, with a depth of the order of 15 µm (3).
Fatigue life obtained for the three groups of specimens tested are represented in Fig. 3. The solid line represents the cyclic

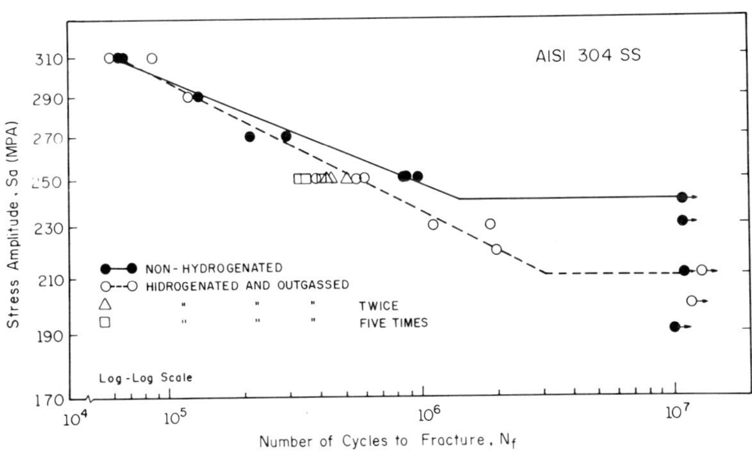

Fig. 3. Cyclic stress amplitude vs. number of cycles to failure for non-hydrogenated and hydrogenated samples.

stress amplitude vs. cycles to failure curve for the non-hydrogenated stainless steel, while the dashed line corresponds to the specimens with one cycle of hydrogenation and outgassing. A substantial decrease in fatigue life is observed in the high cycle fatigue regime (low cyclic stresses) but this effect tends to disappear as the stress amplitude increases. A maximum reduction of the fatigue life of about 40% was observed. The fatigue limit also decreased as a consequence of the hydrogenation-outgassing process.
For specimens on group 3 only tests at a cyclic stress amplitude of 250 MPa were performed. Only a slight decrease in fatigue life was observed for samples that were hydrogenated and outgassed more than once. Those submitted to five hydrogenation-outgassing cycles presented, as an average, shorter lives than specimens that suffered only two cycles. However, the experimental scatter makes any further analysis more difficult.
Figures 4 and 5 show micrographs obtained from the fatigue fracture surfaces of non-hydrogenated and hydrogenated and outgassed samples, respectively. Fracture surfaces were relatively flat for the non-hydrogenated samples while those corresponding to the hydrogenated and outgassed specimens were irregular, presenting well defined steps. However, no major differences were observed between the fatigue fracture modes in both types of specimens.

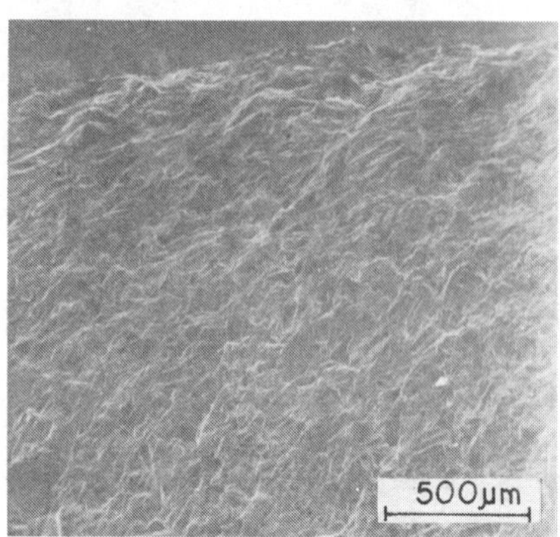

Fig. 4. SEM micrograph of the fatigue fracture surface of a non-hydrogenated sample.

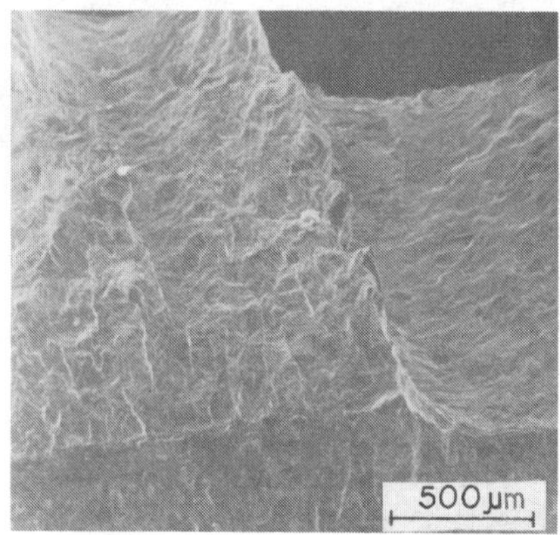

Fig. 5. SEM micrograph of a hydrogenated and outgassed sample showing an irregular fatigue fracture surface with steps.

The microhardness results, for several specimen conditions are shown in Table 1. It can be seen that the hydrogenated specimens present a higher value of hardness than the non-hydrogenated ones and also that a further increase in hardness is produced by the fatigue process in both cases. Only a moderate increase in microhardness is observed for specimens hydrogenated twice.

TABLE 1: Vickers microhardness results

Specimen Condition	Annealed	Annealed and fatigued	Hydrogenated	Hydrogenated and fatigued	Hydrogenated twice	Hydrogenated twice and fat.
Microhardness (H_V)	189	251	268	322	278	339

DISCUSSION

A fatigue failure process can be divided in two parts: initiation and propagation of cracks. The dashed line in Fig. 3 corresponds to hydrogenated specimens that already have small cracks at the beginning of the tests. The fraction of the fatigue life spent in the initiation and propagation stages depends on the level of applied cyclic stress (or strain). Propagation of cracks is dominating at high stresses (low cycle fatigue region) while the initiation process consumes a major part of the fatigue life as the stress or strain amplitude is lowered and the high cycle fatigue regime is reached. The presence of hydrogen induced delayed cracks should produce, then, a substantial reduction in the fatigue life in the high cycle fatigue region and should have a smaller effect in the low cycle fatigue regime, in agreement with the results shown in Fig. 2. As a first approximation, it could be assumed that the initiation stage has been eliminated for the hydrogenated specimens. In this case, the difference between the fatigue lives shown by the solid and dashed lines in Fig. 2 could represent the extent of the crack initiation stage for the non-hydrogenated samples. However, it should be born in mind that some differences may arise on the propagation of cracks produced by fatigue and those originated by the hydrogenation-outgassing process. Several factors could contribute to this behaviour: i) The dislocation structure developed by the fatigue process itself is expected to be different from the one that occurs in the surface region of hydrogenated samples. ii) Phase transformations occur in the hydrogenated and outgassed specimens prior to the fatigue test. iii) Surface hardening (Table 1) is produced by the hydrogenation process due to the presence of hydrogen induced martensitic phases and plastic deformation. iv) A fatigue induced partial reversion of the martensitic phases into austenite has also been observed (8). v) Substantial differences were observed between the fatigue fracture surfaces of non-hydrogenated and hidrogenated and outgassed specimens.
It is uncertain, as yet, which of the five items above described would play the main role and if and in which sense they would interact among them.
Repeated cycles of hydrogenation and outgassing affected only to a minor extent fatigue lives with respect to those observed after just one cycle. It should be emphasized, however, that the discussion of the above results is limited due to the small number of hydrogenation-outgassing cycles herein shown.
Hydrogenated and outgassed specimens will have, as mentioned before, a high density of surface cracks. If such a specimen is rehydrogenated, surface cracks will provide appropriate paths for hydrogen to diffuse into the material so that a thicker surface layer will be affected. Upon outgassing, deeper cracks

could form, leading to a further deterioration of the fatigue properties. However, it could also be argued that, in rehydrogenated material, the tensile stresses responsible for the generation of cracks can be, at least to a certain extent, elastically accomodated by the preexisting cracks so that no much further damage would be produced. The results described above apparently agree with this last hypotheses.
From the point of view of engineering applications of stainless steels, it can be concluded that, in structures that are submitted, during service life, to cycles of hydrogenation and outgassing, followed by fatigue conditions, a more conservative design criterium has to be used in order to account for the effect of surface cracks. This is specially true for high cycle fatigue conditions. This situation can arise in equipments which undergo intermittent operating routines and also due to service interruption for periodic maintenance.

CONCLUSIONS

- Fatigue life of AISI 304 stainless steel is reduced by a hydrogenation-outgassing cycle due to the introduction of surface cracks. This effect becomes more significant for high cycle fatigue conditions.
- For the conditions used in the present work, repeated hydrogenation-outgassing cycles further affect only to a minor extent the fatigue life.

ACKNOWLEDGEMENTS

The authors acknowledge financial support which made it possible to develop this work through FINEP, CNEN and CNPq (under contracts numbers 400361/83-EQ and 402289/84-MM).

REFERENCES

1. Y. Rosenthal, M. Mark-Markowitch, A. Stern and D. Eliezer, Mater. Sci. Eng., 67, 91(1984).
2. A.W. Thompson. Metal Progress, 110, 30(1976).
3. T.C.V. Silva, R. Pascual and P.E.V. Miranda, Proc. Int. Conf. on fracture prevention in energy and transport systems, Rio de Janeiro, 208(1983).
4. N. Narita, C.J. Altstetter and H.K. Birnbaum, Met. Trans., 13A, 1355(1982).
5. H. Hänninen and T. Hakarainen, Corrosion-NACE, 36, 47(1980).
6. M.V. Lelé and P.E.V. Miranda, Metalurgia-ABM, 40, 673(1984).
7. G. Schuster and C.J. Altstetter, Met. Trans., 14A,2085(1983).
8. P.S. Pierantoni, P.E.V. Miranda and R. Pascual, to be published.

Influence of Grain Size and Dispersion of Small Particles on Crack Initiation and Growth during Fatigue in Age Hardened AlMgSi Alloys

K. Detert, R. Scheffel and R. Stünkel

Institut für Werkstofftechnik, Universität GH Siegen, Paul-Bonatz-str. 9–11, 5900 Siegen, Federal Republic of Germany

ABSTRACT

Several age hardened AlMgSi alloys with different additions of Mn were investigated. They were submitted to different deformation and heat treatment processes to obtain differences in grain size and particle distribution. Fatigue of smooth samples at 0.25 and 0.7 per cent strain amplitude was studied until the first microcracks were observed. Hard inclusion particles of the size of 5 to 10µm present in all the alloys played a dominant role as nucleation sites. Grain boundaries at the surface acted also as nucleation sites when slip bands in the neighboring grains showed large differences of their orientation. Fatigue bands intersecting the surface of the sample played only a minor role as nucleation sites. The Mn addition and the lower grain size lead to a more uniform distribution of less observable fatigue bands, which in turn slowed down the occurence of microcrack nucleation at grain boundaries and glide bands. Those influences on microcrack initiation were however masked by the influence of the hard particles. Therefore no significant differences of the number of cycles unto failure in these alloys were observed. This number of cycles in the samples of 10mm diameter appeared almost in total to be associated with the necessary number of cycles to develop a microcrack of 10 to 20µm size into a crack of a few millimeter length. The apparent larger crack growth rate in the stage of microcracks compared to the crack growth rate of large cracks at the same low ΔK values can be explained by the absence of crack closure for the growth of microcracks, which governs the rate of crack growth of large cracks at low values of ΔK applied.

KEYWORDS

Fatigue, Precipitation-hardened alloys, Aluminium alloys with up to 1 per cent Si and up to 1 per cent Mg, Age hardened AlMgSi alloys, Crack initiation by fatigue, Crack growth by fatigue, Small crack growth, Microcracks.

ALLOYS INVESTIGATED

The precipitation hardened alloys of aluminium with main additions of about 1 per cent Mg and 1 per cent Si, which were investigated, are listed in table 1. The alloys 41, 42, 44, 45 were manufactured as small heats of 50kg weight with different Mn additions and hot rolled to slabs of a thickness of 25mm. The alloy 43 was commercially manufactured as a large heat and hot rolled into a sheet of a thickness of 25mm and then solution treated at 480°C for 1 hr. The grain size is listed as mean diameter in the cross section of the sheets, also in table 1. The micrograph of the longitudinal section is shown in fig. 1 - 3. It demonstrates how the grain size is affected by the Mn addition and by the different manufacturing procedures. In fig. 1b -3b the content of small particles present in the alloys is shown. The age hardening treatment was applied to rectangular rods of 25mm thickness. The rods were first annealed for 1 hr at 540°C then quenched in water and then aged for 2 to 6 hrs at 150°C in order to obtain a hardness value of 90 to 100 HB. The hardness values obtained are listed in table 2 together with the results of tensile tests on these alloys.

PROCEDURE FOR TESTING

Low cycle and high cycle fatigue were applied by an automatic controlled tensile testing machine in a push-pull cycle on uniaxial loaded specimens. The strain amplitude was kept constant for all tests with number of cycles to failure within 10000. The stress amplitude was controlled for all specimen tested in the lower strain range when the number of cycles to failure were higher than 10000 according to the required strain amplitude. Strain amplitude was measured by a clip-gage with a reference length of 25mm or by strain gage technique with a reference length of 5mm. Two different types of specimens were used to apply the cyclic load. The specimens with circular cross section of 10mm diameter had a length of 30mm of almost uniform cross section according to ASTM E 666-80. Those specimens were primarily used to study low cycle fatigue, as was reported elsewhere /1/. The surface of these specimens with a few exceptions was only machined to a roughness of 2 - 4µm, since roughness as smooth as 0,5µm obtained by polishing did not show any particular influence on cycles to failure. In order to better observe the initiation of microcracks, also samples with rectangular cross section were made. The surface of the rectangular section was electropolished. Additionally in some cases those electropolished surfaces were then submitted to anodic oxidation according to the method developed by Barker /2/, to make visible the faces of different grains due to differences of reflection of polarized light. The microcracks of fig. 1a - 3a were obtained by the same method.

The crack initiation during fatigue was particularly studied at two distinct amplitudes of strain. At the strain amplitude of $\pm$ 0.7 per cent with the ratio ε min/ ε max = R = -1 the number of cycles to failure was about 200 - 500 in all the alloys studied. At half this number a stabilized stress amplitude varied between 280 and 350N/mm² with a plastic strain amplitude of 0.2 to 0.5 per cent. At the strain amplitude of $\pm$ 0.25 per cent and a ratio R = -1 the number of cycles to failure was about $1 \cdot 10^5$ to $3 \cdot 10^5$. The stabilized stress amplitude at about half this number was between 170 to 180N/mm² with practically no measurable plastic strain amplitude.

RESULTS

At the strain amplitude of ± 0.7 per cent only a few cycles in the order of 10 are necessary to initiate a microcrack, that means below 5 per cent of the number of cycles until failure the first microcracks are present. At the strain amplitude of ± 0.25 per cent about 10000 to 15000 cycles, which represents 3 to 10 per cent of the number of cycles to failure, are required to produce the first microcracks. The size of the first detectable microcrack was about 20 to 50 micrometer in depth or length resp. After 40000 cycles the first small surface cracks of a depth of 200 to 500μm could be detected by dye penetration technique.
At both strain amplitudes inclusions of about 5 to 10 micrometer diameter at the surface of the samples acted as preferred sites for the origin of the microcracks as shown in fig. 4a. These inclusions represent hard particles, which do not undergo plastic deformation as does the surrounding Aluminium matrix. Those particles at the surface may seperate from the matrix and eventually break off as a whole as shown in fig. 4b. There was no fracturing of these particles observed, as found by Foth in the alloy AlCuMg (2024) /3/. At the high strain amplitude of ± 0.7 per cent microcracks apparently appeared also as grain boundary cracks or slip band cracks without the presence of a coarse particle as shown in fig. 5. Fatigue-glide bands were observed in certain grains favorably oriented with respect to the surface of the sample and the direction of alternating stress, when the number of cycles exceeded 20 to 40 percent of the number of cycles to failure. These glide bands could be detected as very faint parallel lines when strained at ± 0.25 per cent. Those bands appeared more pronounced at the higher strain amplitude as shown in fig. 6. In the alloy 41 with no Mn content those bands appeared to be most clearly visible with a wider distance than in all the other alloys studied pointing to the fact that in contrast to the other alloys coarse glide may occur only in alloy 41.
Distinct extrusions or intrusions at the polished surface could not be observed. A distinction of stage 1 from stage 2 of crack growth by fatigue was observed in a few cases yet only for a rather small distance of about 10 to 20μm from the surface as shown in fig. 7.
In general crack growth by fatigue however started as stage 2 perpendicular to the surface right from the beginning. It was also observed that microcracks situated next to each other may combine during growth to form a single crack of a larger size. In the samples tested this phenomena of combining small cracks during their growth appeared not to be as general as reported by Foth in his investigation /3/.

DISCUSSION

The microscopic examination can explain why the difference of microstructure did not show any significant differences of low cycle fatigue behavior /1/. One would have expexted an influence of the dispersion of fine α-MnSiAl particles on fatigue, which are present in all the alloys except of alloy 41. They have a size of 0.05 to 0.1μm. These particles suppressed indeed coarse glide which accellerates the microcrack formation at boundaries as one expects /4/. One would have expected also an influence of the difference in grain size. Since finer grains slows down the microcrack initiation as shown by Ruch /5/. In the alloy 43 was the smallest grain size present. Except for the difference in the cyclic stress strain curve no significant difference of the low cycle fatigue behavior dependent on strain amplitude was observed.

Apparently the presence of coarse particles of a size of 5 to 10µm dominate the occurence of microcracks during fatigue that all other influences were masked. These particles were present in all the alloys studied. In the alloy with higher Mn content the number of inclusion particles was somewhat higher than in the Mn free alloy. The highest content was present in the alloy 43 with the smallest grain size.The hard coarse particles are Mg_2Si particles which were not brought into solution during solution treatment and AlFeSi particles as shown by Evans /6/ and Edwards and Martin /7/. Almost the total number of cycles unto failure had been used to develop a macroscopic small crack of a few millimeter depth from a microcrack of a size of 10 to 20µm. Surprizingly the differences of grain size in those samples did not exert a distinguishable influence on the growth of those microcracks. One would expect such an influence from the growth behavior of large cracks at low values of the stress intensity range ΔK /8/. It had been shown that in age hardened aluminium alloys crack growth is influenced by the grain size in particular at low ΔK. This behavior had been explained by crack closure which is influenced by the grain size /9/, which reduced the applied $\Delta K = K_{ma} - K_{min}$ to an effective $\Delta K_{eff} = K_{max}\ K_{op}$ K_{op} is the value of stress intensity factor, when the crack opens up. According to Newman /10/ no such an effect of crack closure is present in the stage of growth of microcracks. Estimating the crack growth rate for those microcracks from the measurements of this investigation of strain amplitude of $\pm$ 0.25 per cent using the expression of small surface cracks $K=1.12\sigma\sqrt{\pi a}$ da/dN is faster for these microcracks as determined for large cracks at the same applied ΔK. See fig. 8. If one uses the positive stress amplitude of alternating stress $\Delta\sigma = \sigma_a$ as value for the stress range, the values for da/dN of microcracks correspond to those values of crack growth dependent on K_{eff} extrapolated to the low values of ΔK prevalent for those microcracks. It may support the hypothesis that growth of microcracks are not influenced by crack closure at low ΔK values, if an expression of ΔK is justifiable for microcracks of a size well within the grain diameter. In conclusion it can be determined from the measurements, that the size of the hard inclusion particles control the number of cycles unto failure of those small samples in all the alloys by determining the size of microcracks which then grow. When such inclusions of a size of 5 to 10µm are present other microscopic features of the structure of these alloys play only a minor role as crack nuclei in forming microcracks .

REFERENCES

1. K. Detert and R. Scheffel, Low cycle fatigue of AlMgSi alloys, to be published
2. L. J. Barker, Trans. Am. Soc. Met. 42, 347 (1950)
3. J. Foth, DFVLR-FB 84-29 (1984)
4. G. Lütjering, DFVLR-FB 74-70 (1974)
5. W. Ruch and V. Gerold, Proc. 4th ECF Leoben, Austria, 383 (1982)
6. D. W. Evans, Z. Aluminium 38, 219 (1962)
7. L. Edwards and J. W. Martin, Adv. Fract Research, Perg. Press, Proc. 5th ICF Cannes, France, 323 (1981)
8. R. Scheffel, R. Stünkel and K. Detert, Vortrag 16. Sitzung AK Bruchvorgänge Karlsruhe, Germany (DVM Berlin) 239 (1984)
9. R. Scheffel and K. Detert, Proc. 5th ECF Lissabon, Portugal, 805 (1984)
10. J. Newman, Proc. of AGARD Conf. Toronto, Can. AGARD-CP-3-28, Nr. 6 (1984)

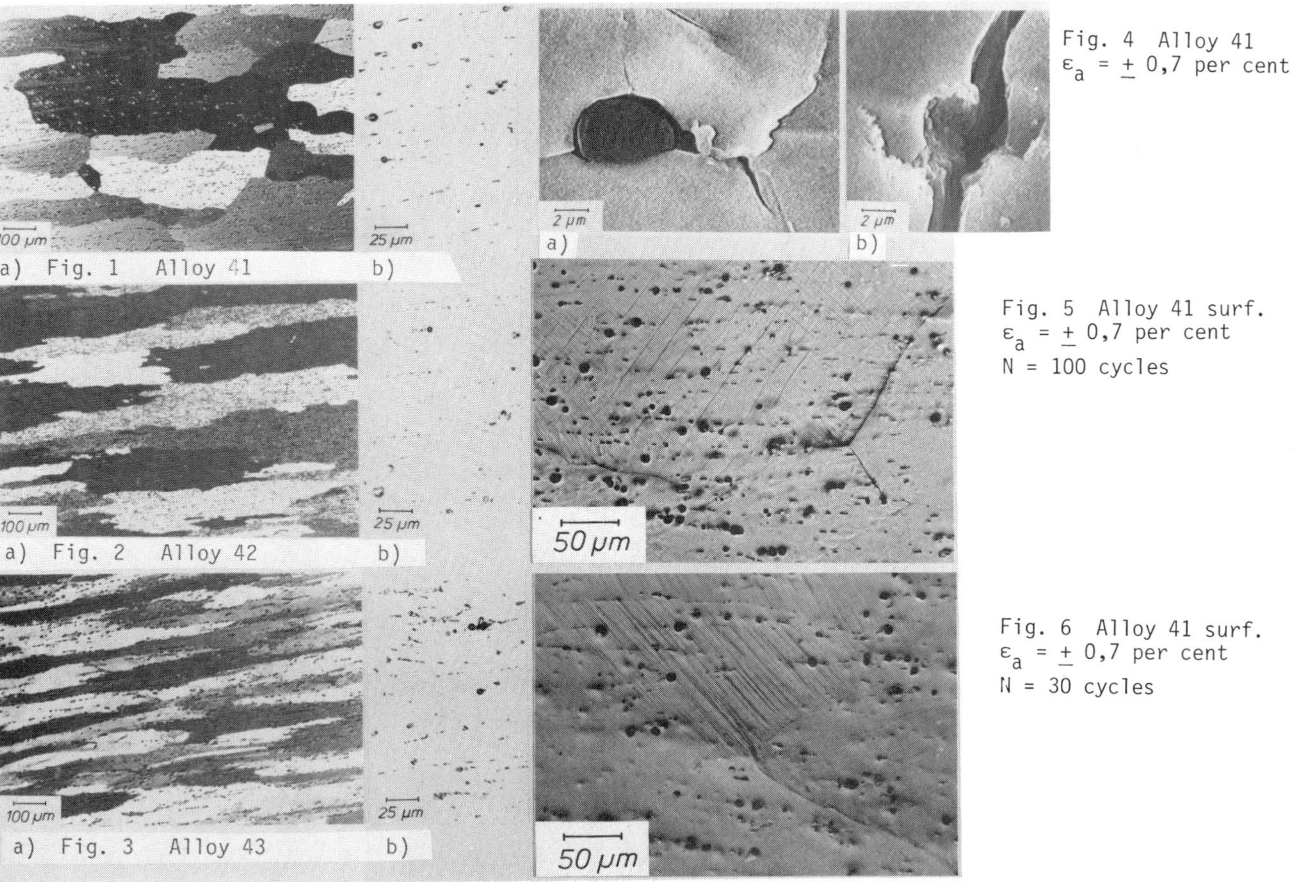

a) Fig. 1 Alloy 41 b)

a) Fig. 2 Alloy 42 b)

a) Fig. 3 Alloy 43 b)

a) b) Fig. 4 Alloy 41
$\varepsilon_a = \pm\ 0{,}7$ per cent

Fig. 5 Alloy 41 surf.
$\varepsilon_a = \pm\ 0{,}7$ per cent
N = 100 cycles

Fig. 6 Alloy 41 surf.
$\varepsilon_a = \pm\ 0{,}7$ per cent
N = 30 cycles

TABLE 1 Chemical Composition of Alloys Investigated (wt per cent)

Alloy Nr.	Si	Mg	Mn	Fe	grain size (diameter μm)
41	0.92	0.91	0.01	0.25	280
42	0.98	0.91	0.21	0.26	306
43	1.05	1.02	0.70	0.21	175
44	0.95	0.95	0.44	0.29	260
45	0.88	0.89	0.97	0.30	210

TABLE 2 Mechanical Properties of Alloys Investigated

Alloy Nr.	Hardness HB Nr.	Rp0.2 N/mm²	Rm N/mm²	A per cent	Z per cent
41	101	221	327	25	43
42	95	198	291	26	40
43	100	310	405	13	27
44	98	207	281	23	33
45	96	198	291	20	40

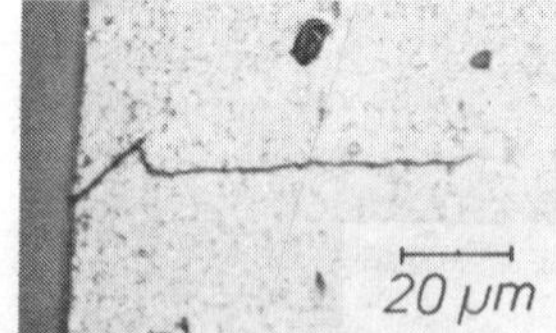

Fig. 7 Alloy 41
sect. with crack
$\varepsilon_a = \pm 0{,}7$ per cent
N = 100 cycles

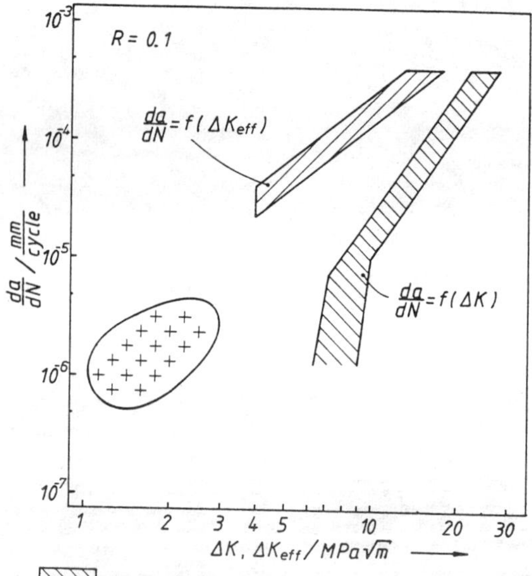

Fig. 8 : Comparison of crack growth rate for large and small cracks in AlMgSi alloys.

Fatigue Crack Initiation and Early Growth in an Austenitic Stainless Steel

Nam Soon Chang and W. L. Haworth*

Department of Metallurgical Engineering, Wayne State University, Detroit, Michigan, USA

ABSTRACT

Initiation and early growth of microcracks during fatigue were studied in an age hardening stainless steel (A286). The underaged material (UA) is more resistant to fatigue failure than overaged material (OA), particularly at low stress amplitude. Crack initiation begins very early in both materials and continues throughout the fatigue life; the microcrack density is significantly higher in OA material and is strongly stress-dependent in both UA and OA. Microcracks grow intermittently, but on the average they decelerate and stop growing at a critical length of about 50 microns. A few cracks continue to grow, and accelerate to cause failure. Fatigue lives are discussed in terms of crack initiation, early growth, and coalescence.

INTRODUCTION

Understanding the mechanisms of initiation and early growth of microcracks is essential to the development of improved predictions of fatigue life. In order to understand the fatigue fracture process fully in a material that is not pre-cracked, and ultimately to anticipate fatigue failure, it is necessary to investigate the behavior of microcracks randomly distributed on the unnotched smooth specimen, including their initiation, early growth and coalescence. The minimum data required for this analysis are the number of microcracks and their lengths throughout the fatigue life. However, very little quantitative information is available for the relationships between the number of cracks, microcrack growth rate, stress amplitude, and fatigue life in steels, although such studies have been made in Al alloys [1,2]. The present work has two objectives: first, to investigate the initiation and early growth of microcracks during fatigue in an age-hardening stainless steel with reference to slip character and microstructure; second, to establish experimentally how crack density and microcrack growth rates depend on stress amplitude and determine what effect these relationships have upon fatigue life.

*Current Address: Division of Materials Research, NSF, Washington, DC 20550

MATERIALS AND PROCEDURE

The material used in this study is a commercial age-hardening stainless steel (A286). Its chemical composition in weight percent is 24.71 Ni, 14.07 Cr, 2.26 Ti, 1.34 Mo, 0.26 V, 0.18 Al, 0.15 Si, 0.13 Mn, 0.049 C, 0.012 P, 0.006 S, 0.005 B, balance Fe. Specimens were machined from round rod in solution-treated condition. Flats were milled on either side of the gauge length for metallographic observation during the fatigue tests. The specimens were then solution treated again in air at 982°C for 1h and oil quenched. Subsequent aging was conducted in air, either at 650°C for 10h for the underaged (UA) condition, or at 780°C for 100h for the overaged (OA) condition, followed by air cooling. These two conditions show similar yield strengths, but generally, UA material is more ductile and stronger, as shown in Table 1. Both materials cyclically harden initially, and soften later in the fatigue life. The UA microstructure consists of fine, submicroscopic γ' precipitates (ordered $Ni_3(Ti,Al)$) in the matrix with TiC particles and abundant annealing twins, while in OA the lamellar phase η(Ni_3Ti) develops at grain boundaries in addition to overgrown, large γ'. The grain size is 17 microns for both materials.

Table 1. Tensile Properties of A286 Investigated

Alloy	0.2%YS,MN/m^2	UTS,MN/m^2	RA,%
UA	445	875	61
OA	450	799	49

All specimens were polished to 0.05 micron surface finish by hand. The fatigue tests were conducted in fully reversed axial loading (R = -1), using a sinusoidal load waveform at either 10 or 15 Hz. The tests were carred out in air at room temperature, and were interrupted at intervals to perform quantitative metallography _in situ_ on the flat surface of specimens, using Nomarski interference contrast in the optical microscope. Selected specimmens were removed for SEM examination.

RESULTS AND DISCUSSION

The fatigue life curves for both heat treatments are shown in Fig.1. The UA material has superior fatigue strength, particularly at low stress amplitudes: at 365MPa, for example, UA material fails after less than 4 x 10^5 cycles while OA specimens do not fail in tests extending beyond 5 x 10^6 cycles. The UA material exhibits relatively coarse, predominantly planar slip (Fig. 2a) which is characteristic of a deformation mechanism in which dislocations shear precipitate particles. The OA material shows a more homogeneous distribution of slip, particularly at low stress amplitudes (Fig. 2b): the slip lines are finer and less planar than in the UA material. Slip in the OA case is consistent with Orowan looping of dislocations around relatively large, incoherent γ' precipitate particles which are not sheared during the deformation process. In both UA and OA material, crack initiation is observed along slip bands and twin boundaries. Surface features which may be slip band extrusions or wear debris produced by crack-face rubbing are associated with many of the cracks (Fig.3). In OA material, cracks also begin at grain boundaries and along η phase lamellae as shown in Fig.4. SEM fractography confirms that after cracks initiate in slip bands, early crack growth is crystallographic.

The superficial crack density (number of microcracks exceeding 5 microns in length, per unit surface area) is shown as a function of elapsed fatigue cycles in Figs. 5a and 5b. Cracks are first observed after 1% or less of the fatigue life, but crack initiation continues throughout the life at all stress amplitudes. The crack density at any stage of life depends strongly on stress amplitude; its maximum value ranges from about $20/mm^2$ at runout in high cycle fatigue to more than $10^3/mm^2$ at failure in low cycle fatigue. The crack density at a given stress amplitude is significantly lower in UA than OA material at a fixed number of fatigue cycles. Thus, the more homogeneous slip distribution in OA material does not confer higher resistance to crack initiation as one might expect; the deleterious effect of the η phase may be a more important factor than the slip character itself in nucleating fatigue cracks. The higher initial rate of initiation in OA material (compared to UA) is consistent with the effectiveness of phase lamellae as crack initiation sites; these effective sites may be exhausted as cycling proceeds. In UA material, coarse slip bands and twin boundaries provide the primary initiation sites throughout the life so the relative initiation rate shows less change as the life fraction increases.

Average crack growth rates (da/dN) of individual surface microcracks ranging in length from about 10 to 1000 microns are shown in Figs 6a and 6b for UA and OA material, respectively, as a function of the nominal stress intensity factor range $\sigma_a a^{1/2}$ (where σ_a is the applied stress amplitude and a the half crack length). No correction has been made for crack shape or aspect ratio. The observed growth is intermittent, especially for very short cracks: thus the growth rates shown in Fig.6 represent a lower bound to the "real" rate at which the crack advances between intermittent arrests. For clarity, since the growth rate at constant length depends on stress amplitude, the data for a single amplitude are shown in Fig.6.

Crack growth in both UA and OA material follows a characteristic pattern shown by the V-shaped curve. Below a critical length (at the surface) of about two or three grain diameters (50 microns), the average growth rate of an individual microcrack tends to fall as its length increases, and most cracks that reach the critical length range then stop completely. The relatively few cracks that grow beyond about 50 microns in length then generally accelerate as they continue to grow. Wide variations in growth rate are observed for different cracks under the same (overall) conditions and often for the same crack as cycling proceeds. Both the variability of crack growth rate, and the overall tendency for short cracks to decelerate and stop, can be rationalized on the basis of crack closure effects. First, microplasticity produced with individual grains by slip band formation at the crack tip may contribute to crack closure if slip is blocked by a grain boundary or a second phase particle. The crack closure stress (σ_{cc}) will then rise (3), reducing the effective local stress intensity factor and slowing down the crack. If σ_{cc} is large enough to reduce ΔK_{eff} to zero, the crack stops completely. If the crack is able to advance into the next grain, however, ΔK_{eff} increases again and the crack accelerates. Another factor which may influence microcrack growth rates is the presence of both Mode I and Mode II displacements accompanying crystallographic crack growth, which will produce a faceted fracture surface. Irregular fracture morphologies, together with wear debris produced by crack face rubbing (Fig.3) will lead to crack closure by wedging open the crack at discrete contact points along the crack face. This roughness-induced crack closure raises σ_{cc} and reduces the growth rate; it will be minimal for extremely short cracks, so that the initial growth rate (Fig.6) is relatively high.

Finally, the data in Fig.6 show that at a given stress amplitude and crack length, the average growth rates of microcracks are significantly lower in UA than OA materials, particularly for the shortest cracks observed. This implies that short cracks in OA material reach the critical length of about 50 microns more rapidly than they do in UA material, which presumably contributes to the reduced fatigue resistance in the overaged condition. An important factor in the higher average crack growth rate in OA material may be crack coalescence: the crack density is relatively high in this condition, so cracks are more likely to coalesce and continue to grow with fewer intermittent arrests. The difference in crack growth rate is much smaller once the cracks exceed the critical length of approximately 50 microns; this is consistent with the greatly reduced likelihood of crack arrest as the crack length develops beyond two or three grain diameters. The rapid initiation and early coalescence of microcracks in OA material appear to be the major factors in reducing its fatigue strength.

CONCLUSIONS

1. UA material is more resistant to fatigue failure than OA, particularly at low stress amplitudes.

2. Microcrack densities are significantly lower in UA than OA at a given stress amplitude and number of fatigue cycles. Crack densities may exceed $1000/mm^2$, but are strongly stress-dependent; cracks are observed before 1% of fatigue life at all stresses.

3. Microcrack growth is intermittent. For very short cracks, the average growth rate decreases as the crack lengthens, and most cracks stop growing at a critical length of about 50 microns. These observations can be rationalized in terms of crack closure. A few cracks grow longer and accelerate to cause failure.

4. The relatively low fatigue strength of OA material appears to depend primarily on the early initiation and rapid coalescence of cracks in this aging condition. Crack initiation at η phase lamellae may be an important factor in this process.

REFERENCES

1. D. Sigler, M. Montpetit and W. L. Haworth, Met.Trans. 14A, 1983, 931

2. Lu Ling Tu and W. L. Haworth, Proc. 2nd Int. Conf. Fatigue and Fatigue Thresholds (C. J. Beevers, ed.), EMAS Ltd., England 1984, p.120.

3. W. L. Morris, Met. Trans. 11A, 1980, 1117.

ACKNOWLEDGEMENT: This work was supported by the US Army Research Office, Metallurgy and Materials Section, under contract No. DAA29-81-K0114.

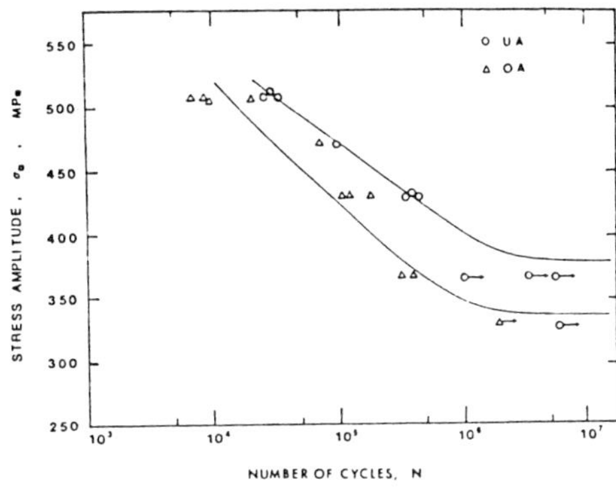

Fig. 1. Fatigue life data for underaged A286 stainless steel.

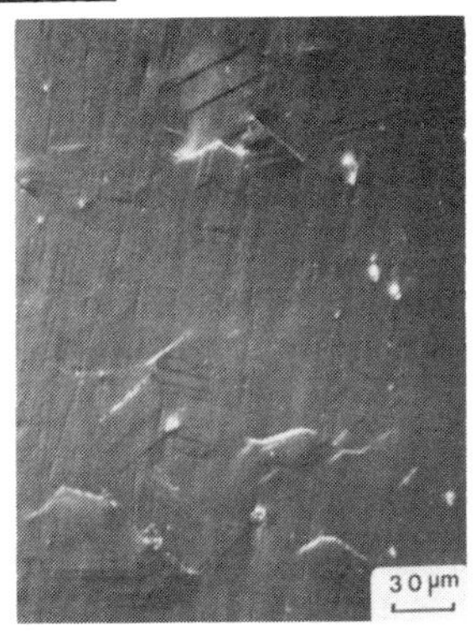

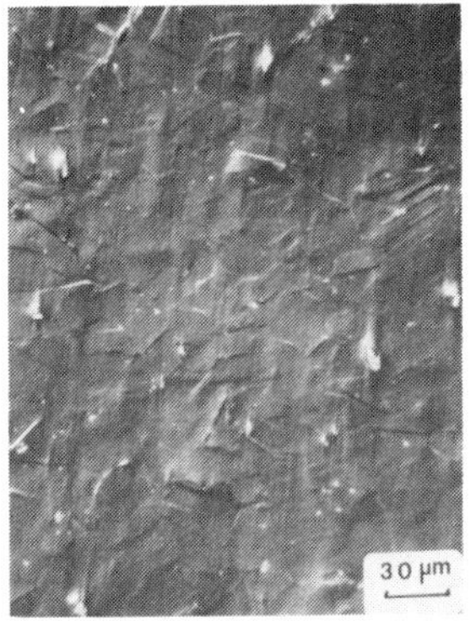

Fig. 2. a. UA material. Surface appearance after 3 x 10^6 cycles at 365 MPa. b. OA material. Surface appearance after 3 x 10^5 cycles at 369 MPa.

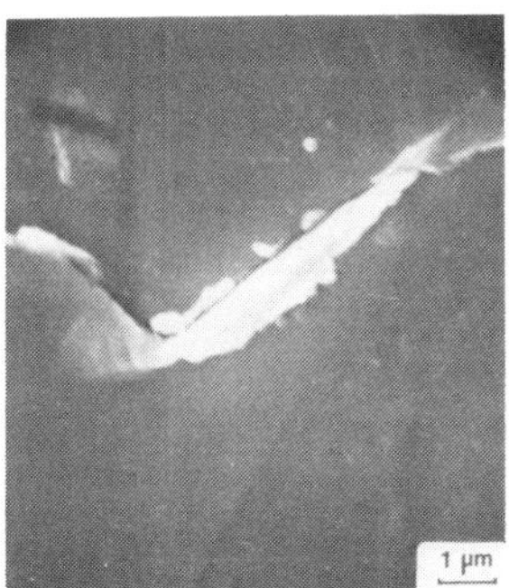

Fig. 3.

Slip band crack initiation in OA material at high stress amplitude (473 MPa, late in life).

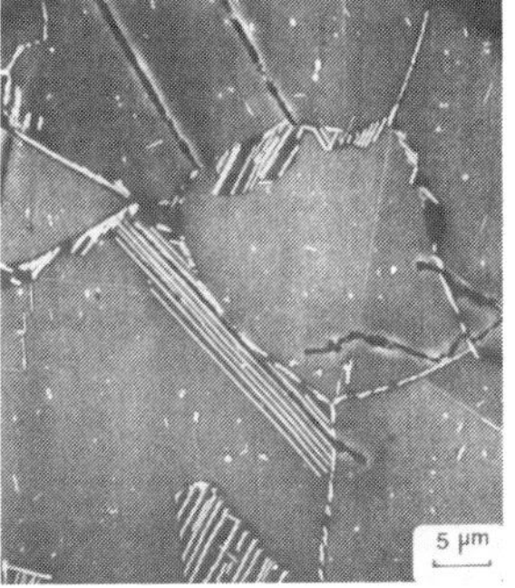

Fig. 4.

Crack initiation at η phase lamellae, twin boundaries and slip bands in OA material. 367 MPA, etched surface.

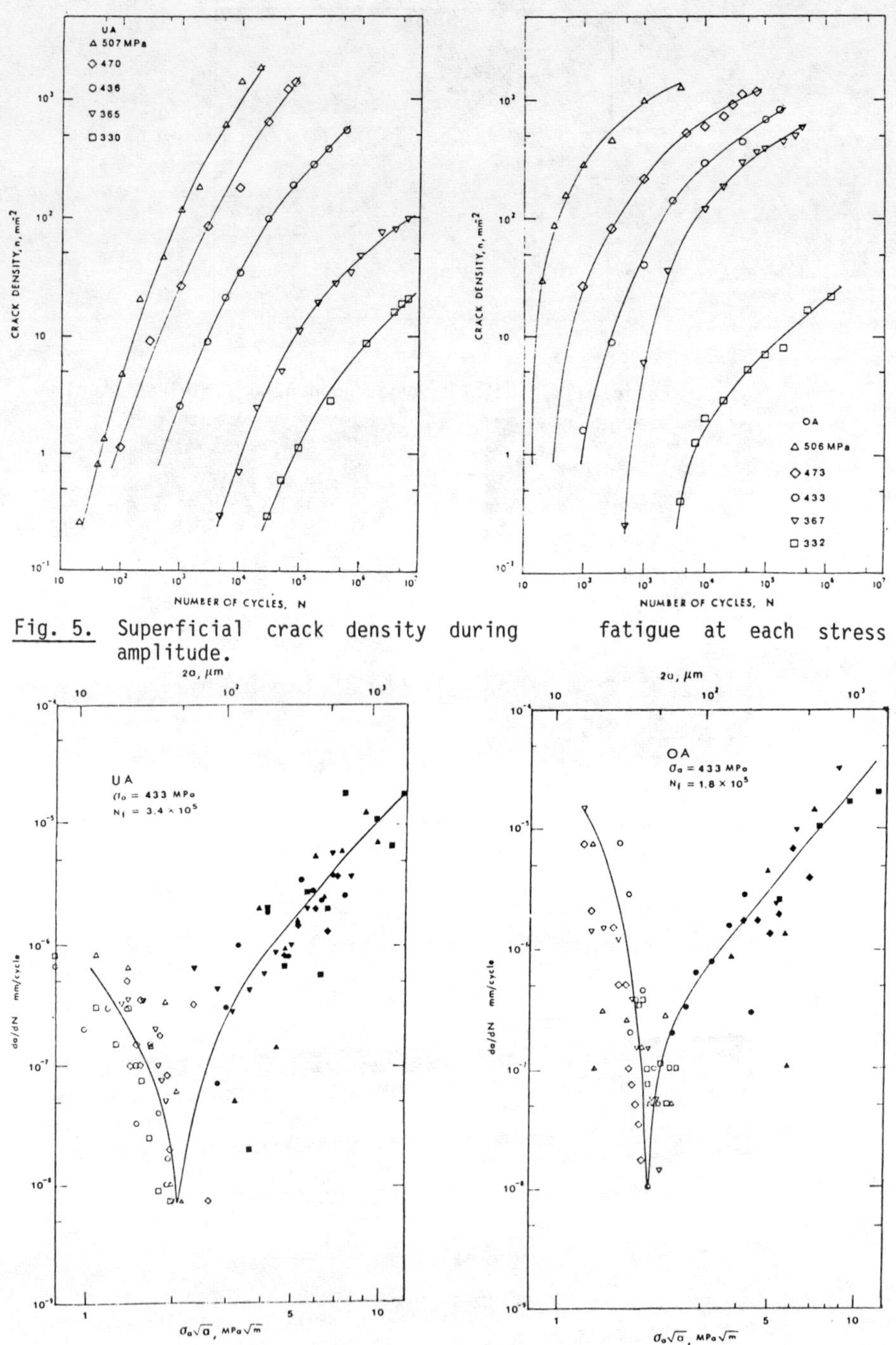

Fig. 5. Superficial crack density during fatigue at each stress amplitude.

Fig. 6. Average growth rates for indvidual microcracks at 433 MPa.

Amorcage en Fatigue a Partir d'une Entaille

A. Bignonnet et H. Buthod-Cuan

Institut de Recherches de la Sidérurgie Française (IRSID), 78105 Saint Germain en Laye Cedex, France

RESUME

Ce travail concerne les conditions d'applications de critères d'amorçage en fond d'entaille en état de plasticité localisée. Pour l'étude de ces critères, nous avons établi les lois d'écrouissage cyclique et de résistance à la fatigue oligocyclique d'un acier de construction. Des essais d'amorçage en fatigue sur des éprouvettes entaillées avec des K_t de 3 et 5 ont été réalisés. Avec ces résultats nous avons testé deux critères basés sur l'approche locale (Neuber et Molski-Glinka).

Le modèle de Molski-Glinka basé sur l'énergie de déformation locale donne une meilleure estimation que la règle de Neuber des contraintes et déformations locales. Les résultats montrent que si l'on s'intéresse à l'amorçage, le modèle de Molski-Glinka ne fait pas intervenir un coefficient de réduction de contrainte en fatigue, mais que, outre les caractéristiques cycliques du matériau, il suffit de connaître le K_t de l'entaille (le coefficient de concentration de contrainte K_t correspondant à l'entaille) pour formuler le critère d'amorçage.

INTRODUCTION

Pour l'étude de l'amorçage en fond d'entaille et en conditions de plasticité localisée, les critères les plus communément utilisés sont basés sur une approche locale, c'est-à-dire sur l'hypothèse que seules les contraintes et déformations locales gouvernent les mécanismes d'amorçage.

Nous avons repris ici cette approche et confronté deux modèles. Le premier critère développé par Topper, Wetzel et Morrow (1) étend la théorie de Neuber (2) à la fatigue en fond d'entaille. Le second critère est établi à partir d'une méthode énergétique de calcul élastoplastique des contraintes et déformations en fond d'entaille, développée par Molski et Glinka (3).

MATERIAU ET TECHNIQUES EXPERIMENTALES

Le matériau étudié est un acier de construction métallique du type E36 Z35 notamment utilisé pour la fabrication des structures métalliques soudées d'exploitation pétrolière en mer. La composition chimique et les caractéristiques mécaniques de cet acier sont données aux tableaux 1 et 2.

Tableau 1 : Composition chimique de l'acier E36 Z35 (% poids)

C	Mn	Si	S	P	Ni	Cr	Mo	Al	N_2
0,145	1,40	0,29	0,003	0,006	0,42	0,075	0,031	0,019	0,012

Tableau 2 : Caractéristiques mécaniques de l'acier E36 Z35

Re (MPa)	Rm (MPa)	A (%)	Z (%)
380	540	30	75

Les essais d'amorçage sur éprouvettes entaillées ont été réalisés en flexion 4 points sur une machine à résonance (fréquence $\sim$ 80 Hz) avec un rapport de charge R=0,1. Les entailles à fond semi-circulaire (rayon 0,6 mm, finition alésage fin) ont été choisies pour créer une concentration de contrainte telle que $K_t = 3$ pour une série et $K_t = 5$ pour une autre série.

La détection de l'amorçage a été faite à partir des mesures de variation de la complaisance de l'éprouvette. Pour détecter cette variation de complaisance, nous enregistrons pendant l'essai un signal δ' tel que : $\delta' = \delta - \alpha P$, après avoir ajusté au début de l'essai la valeur α de sorte que, quel que soit P, $\delta' = 0$; ceci permet d'amplifier considérablement le signal δ'. L'amorçage d'une fissure se traduit alors par une variation de δ'. Par cette technique, on détecte des fissures d'environ 0,2 mm de profondeur.

Les essais de fatigue plastique ont été réalisés sur des éprouvettes à partie utile cylindrique ($\emptyset$ = 15 mm) sollicitées en déformation totale imposée avec une vitesse de déformation constante ($\dot{\varepsilon} = 4 \times 10^{-3}\ s^{-1}$).

RESULTATS EXPERIMENTAUX

Lorsque l'on utilise l'approche locale pour bâtir un critère d'amorçage, on assimile le fond d'entaille à une éprouvette lisse qui subirait les mêmes contraintes et déformations que ce fond d'entaille. La ruine de l'éprouvette cylindrique ou de l'élément de volume en fond d'entaille se produira alors pour un même nombre de cycles de chargement.

Les résultats d'essais d'amorçage sur éprouvettes entaillées sont reportés sur la figure 1. La contrainte ΔS en ordonnée représente la variation de la contrainte nominale rapportée à la section nette sous l'entaille, le nombre de cycles à l'amorçage est en abscisse.

Les essais sur éprouvettes lisses nous ont permis de caractériser le comportement du matériau en fatigue plastique. L'acier E36 Z35 est un matériau qui se stabilise rapidement. La loi d'écrouissage cyclique à l'état stable est donnée à la figure 2. Cette loi peut s'écrire sous la forme :

$$\frac{\Delta\sigma}{2} = K' \left(\frac{\Delta\varepsilon_p}{2}\right)^{n'} = 770 \left(\frac{\Delta\varepsilon_p}{2}\right)^{0,13} \qquad [1]$$

La courbe de résistance à la fatigue plastique (ou courbe de Coffin-Manson) est donnée à la figure 3. La loi de Coffin-Manson peut s'écrire sous la forme d'une somme des termes élastique et plastique :

$$\frac{\Delta\varepsilon_t}{2} = \frac{\sigma'_f}{E}\ (2N)^b + \varepsilon'_f\ (2N)^c = 0,58\ (2N)^{-0,12} + 60\ (2N)^{-0,57} \qquad [2]$$

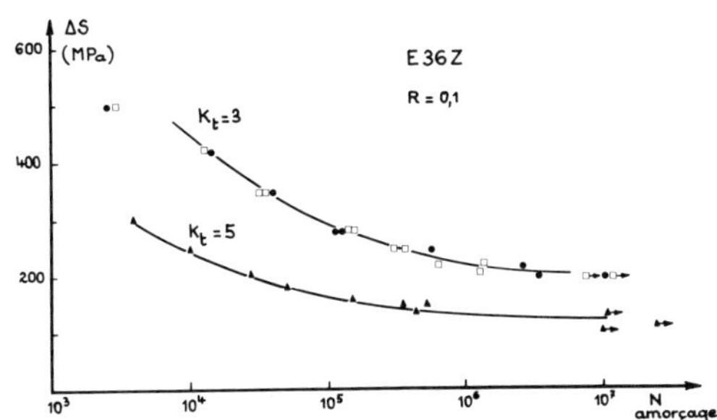

Fig. 1 - Courbes de résistance à l'amorçage sur éprouvettes entaillées en E36 Z35.

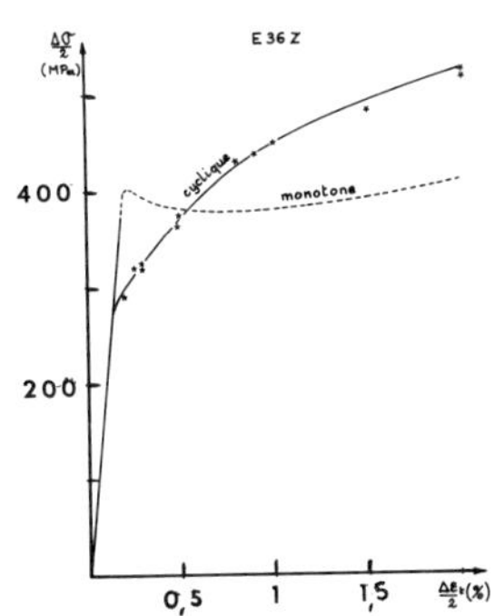

Fig. 2 - Courbe d'écrouissage cyclique pour l'acier E36 Z35.

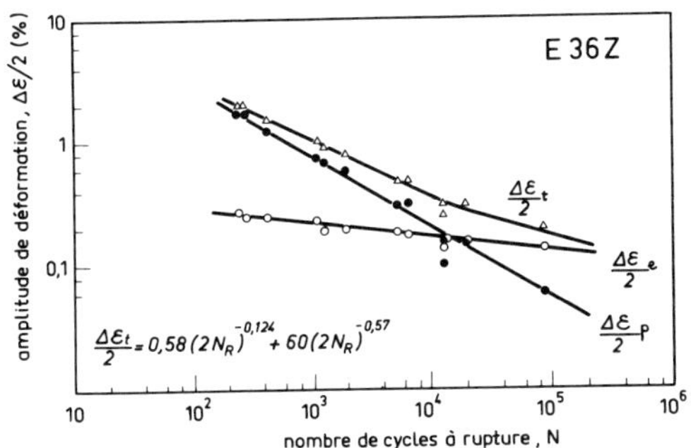

Fig. 3 - Courbe de Manson-Coffin pour l'acier E36 Z35.

CRITERES D'AMORCAGE

Utilisation de la règle de Neuber

La règle de Neuber donne les contraintes et déformations locales par la relation : $K_t^2 = K_\sigma . K_\varepsilon$ [3]

avec : $K_\sigma = \frac{\sigma}{S}$ et $K_\varepsilon = \frac{\varepsilon}{e}$

où S et e sont les contrainte et déformation nominales ; σ et ε sont les contrainte et déformation locales en fond d'entaille.

Il en résulte que pour un chargement nominal élastique : $\sigma.\varepsilon = \frac{K_t^2 . S^2}{E}$.

De nombreux auteurs (par exemple 4, 5, 6, 7, 8) ont appliqué la règle de Neuber au cas de la fatigue en remplaçant les contraintes et déformations locales, σ et ε ainsi que la contrainte nominale S par la variation de ces paramètres au cours d'un cycle de fatigue : Δσ, Δε et ΔS. Les auteurs précités remarquent que cette démarche conduit à une prévision des durées d'amorçage trop pessimiste. La substitution de K_f (coefficient de réduction de contrainte en fatigue) à K_t permet d'après ces auteurs d'obtenir des prévisions meilleures, comparées aux résultats de l'expérience.
Comme le remarque TRUCHON (9), à ce stade du raisonnement le problème a changé de nature : on a adjoint à la règle de Neuber une loi de résistance à la fatigue, et c'est cet ensemble qui constitue un critère d'amorçage. Rappelons que par définition K_f est égal au rapport de deux limites

d'endurance : $K_f = \frac{\sigma_D \text{ (lisse)}}{\sigma_D \text{ (entaillée)}}$. Il est toujours sous-entendu que les limites d'endurance ont été calculées sur la base de la rupture des éprouvettes et non de l'amorçage des fissures dans ces éprouvettes. Dans le cas des éprouvettes entaillées, l'existence d'un fort gradient de contrainte en fond d'entaille fait qu'une fissure, une fois amorçée est susceptible de ne pas se propager. Dès 1957 Frost et Dugdale (10) montrent que les valeurs de limites d'endurance basées sur <u>le seul amorçage</u> des fissures en fond d'entaille se rangent suivant une loi : $K_t\ \sigma_D$ (ent.) = Constante = σ_D (lisse).

Dowling (11), à partir de ces mêmes constatations, interprète "l'effet K_f" par la présence de fissures non propageantes dans les éprouvettes entaillées. L'utilisation du coefficient K_f pour la formulation d'un critère d'amorçage en fond d'entaille ne paraît donc pas justifiée et pourrait conduire à des estimations n'allant pas du côté de la sécurité. Dans le cas d'un critère basé sur la règle de Neuber, l'amélioration de la prédiction par l'utilisation de K_f tient en fait à ce que l'on vient ainsi contrebalancer un tout autre phénomène. En effet comme l'ont signalé certains auteurs (4, 6, 7, 8), la règle de Neuber surestime les contraintes et déformations locales et l'utilisation de K_f à la place de K_t permet, artificiellement, de réduire cette surestimation.

Dans ce qui suit nous allons vérifier que l'on peut bâtir un critère d'amorçage à partir de la règle de Neuber, à condition d'utiliser un facteur correctif pour tenir compte de la surestimation des contraintes et déformations locales. Cependant, en aucun cas, l'utilisation de K_f comme facteur correctif n'est justifiable.

S'agissant du critère d'amorçage on peut donc partir de la relation [3] en utilisant toutefois un facteur correctif sur les contraintes et déformations locales ; si le chargement nominal est élastique, on écrira alors :

$$\sqrt{E\ \Delta\sigma\ \Delta\varepsilon} = \alpha . K_t\ \Delta S \qquad [4]$$

ce qui permet de relier le chargement nominal aux sollicitations locales.

D'autre part, connaissant la loi de résistance à la fatigue du matériau $\Delta\varepsilon = f(N)$ (relation [2]) et la loi de comportement cyclique $\Delta\sigma = f(\Delta\varepsilon)$ (relation [1]), on en déduit la relation qui existe entre la fonction $\sqrt{E\Delta\sigma\ \Delta\varepsilon}$ et le nombre de cycles N soit :

$$\sqrt{E\ \Delta\sigma\ \Delta\varepsilon} = \sqrt{4K'E\ (\varepsilon'_f\ (2N)^c)^{n'+1} . \left[\frac{K'}{E}\ (\varepsilon'_f\ (2N)^c)^{n'-1} + 1\right]} \qquad [5]$$

ou encore

$$= \sqrt{4K'E\ (\varepsilon'_f\ (2N)^c)^{n'+1} . \left[\frac{\sigma f}{E\ \varepsilon'_f}\ (2N)^{b-c} + 1\right]}$$

Connaissant cette caractéristique du matériau, et le facteur correctif α, on peut prévoir de cette manière le nombre de cycles à l'amorçage correspondant à un chargement nominal donné en résolvant le système constitué par les équations [4] et [5].

Expérimentalement, à partir d'essais sur éprouvettes entaillées et sur éprouvettes lisses, on peut remonter à la valeur du facteur de correction. Pour nos conditions expérimentales (K_t = 3 et 5) on vérifie la stabilité du facteur correctif (figure 4) et on prendra $\alpha = 0{,}80$. A la figure 5 on montre la courbe reliant les conditions locales au nombre de cycles à l'amorçage en fond d'entaille (équation [5]).

<u>Critère selon la méthode Molski-Glinka</u>

La méthode de Molski-Glinka (3) est basée sur l'énergie de déformation absorbée localement. Pour un état de contrainte uniaxial en fond d'entaille ces auteurs proposent la règle suivante : $K_t^2 = \frac{W_\sigma}{W_S}$ [6]

où : $W_\sigma = \int_0^\varepsilon \sigma_{(\varepsilon)}\, d\varepsilon$, énergie de déformation par unité de volume au point le plus sollicité à fond d'entaille

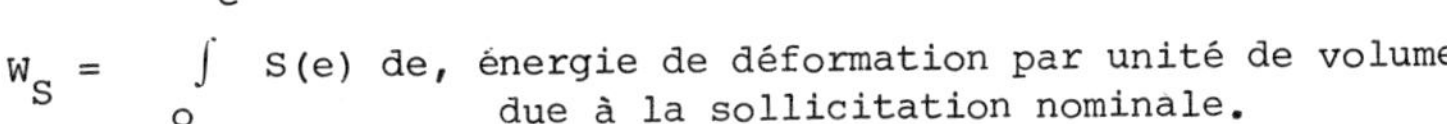

$$W_S = \int_0^e S(e)\,de,$$ énergie de déformation par unité de volume due à la sollicitation nominale.

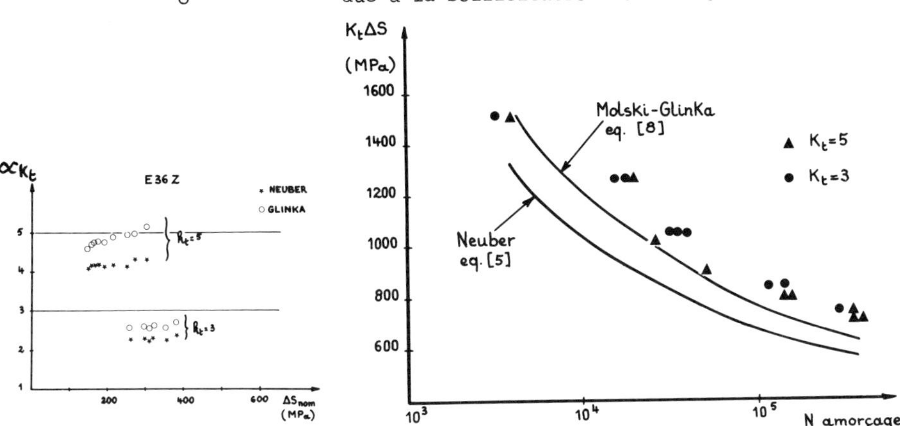

Fig. 4 - Vérification expérimentale de la stabilité du facteur correctif α pour l'application de la règle de Neuber.

Fig. 5 - Critère d'amorçage selon Neuber ou Molski Glinka pour l'acier E36 Z35.

Connaissant les courbes de traction, on calcule aisément W_σ et W_S. L'application à la fatigue de cette équation se fait simplement en remplaçant la contrainte (S, σ) par la variation des contraintes (ΔS, Δσ) et la courbe de traction par la courbe d'écrouissage cyclique : $\frac{\Delta\varepsilon}{2} = \frac{\Delta\sigma}{2E} + \left(\frac{\Delta\sigma}{2K'}\right)^{1/n'}$.

On obtient alors : $$W_{\Delta\sigma} = \frac{\Delta\sigma^2}{2E} + \frac{2\ \Delta\sigma}{n' + 1}\left(\frac{\Delta\sigma}{2K'}\right)^{1/n'} = 2E\left(\frac{\Delta\varepsilon_e}{2}\right)^2 + \frac{4K'}{n'+1}\left(\frac{\Delta\varepsilon_p}{2}\right)^{n'+1}$$

et $$W_{\Delta S} = \frac{\Delta S^2}{2E}$$ (pour un chargement nominal élastique).

La méthode proposée par Molski-Glinka (3) prévoit effectivement des contraintes et déformations locales plus faibles que celle obtenues en appliquant la règle de Neuber. De plus à l'aide de nombreuses mesures de déformations locales effectuées sur éprouvettes entaillées, les auteurs (3) ont constaté un bon accord entre l'expérience et la prévision par la méthode de l'énergie.

Pour le critère d'amorçage on partira de l'équation [6] que l'on peut écrire, pour un chargement nominal élastique :

$$K_t\ \Delta S = \sqrt{4E\left[\frac{\Delta\sigma^2}{4E} + \frac{\Delta\sigma}{n'+1}\left(\frac{\Delta\sigma}{2K'}\right)^{1/n'}\right]} = \sqrt{4E\left[E\left(\frac{\Delta\varepsilon_e}{2}\right)^2 + \frac{2K'}{n'+1}\left(\frac{\Delta\varepsilon_p}{2}\right)^{n'+1}\right]} \qquad [7]$$

Connaissant la loi de résistance à la fatigue du matériau [2] et son comportement cyclique [1] on en déduit la relation reliant les conditions locales et le nombre de cycles à l'amorçage en fond d'entaille (figure 5) :

$$\sqrt{4E\left[\frac{\Delta\sigma^2}{4E} + \frac{\Delta\sigma}{n'+1}\left(\frac{\Delta\sigma}{2K'}\right)^{1/n'}\right]} = \sqrt{4K'E\left(\varepsilon'_f(2N)^c\right)^{n'+1}\cdot\left[\frac{K'}{E}\left(\varepsilon'_f(2N)^c\right)^{n'-1} + \frac{2}{n'+1}\right]} \qquad [8]$$

ou encore $$= \sqrt{4K'E\left[\frac{E}{K'}\left(\frac{\sigma'_f}{E\varepsilon'_f}(2N)^b\right)^2 + \frac{2}{n'+1}\left(\varepsilon'_f\ (2N)^c\right)^{n'+1}\right]}$$

Pour résoudre ce système d'équation ([7] et [8]) il n'est plus nécessaire de connaître, comme dans le critère basé sur la règle de Neuber, le facteur

correctif qui nécessite une caractérisation longue et coûteuse La figure 4 montre la vérification expérimentale de ce dernier point.
La sensibilité de la méthode de détection de l'amorçage d'une fissure conditionne largement la conformité de l'expérience aux résultats des équations analytiques.

CONCLUSIONS

L'utilisation de l'approche locale pour prédire les durées de vie de composants mécaniques ou d'éléments de structure, est largement répandue. Si localement le régime des contraintes est élasto-plastique on peut estimer les durées d'amorçage en fond d'entaille à partir de données de fatigue obtenues sur éprouvettes lisses.
Pour pallier la surestimation des contraintes et déformations locales déduites de la règle de Neuber, il faut appliquer un facteur correctif afin d'obtenir des prévisions d'amorçage en accord avec l'expérience. Nous avons déterminé le facteur correctif dans le cas d'entailles avec un K_t de 3 et de 5. En revanche, la substitution de K_f à K_t dans la formulation du critère d'amorçage n'est pas justifiable. La détermination classique de K_f se faisant par des essais à rupture (amorçage + propagation d'une fissure) il sera d'autant plus dangereux d'utiliser un tel paramètre que la taille de la structure à dimensionner sera grande par rapport à la taille de l'éprouvette utilisée pour déterminer K_f. La méthode de Molski-Glinka basée sur l'énergie absorbée localement, donne plus précisément les contraintes et déformations locales. On en déduit un critère d'amorçage en fond d'entaille beaucoup plus fiable nécessitant seulement la connaissance des caractéristiques de fatigue oligocyclique du matériau et du coefficient de concentration de la pièce.

REMERCIEMENTS

Cette étude a été financée par l'Association de Recherche sur le comportement des Structures Métalliques en Mer (ARSEM), qui en a autorisé la publication.

BIBLIOGRAPHIE

1 - T.H. Topper, A.M. Wetzel and J.D. Morrow. "Neuber's rule applied to fatigue of notched specimens". J. Mater., 4, (1969), 200.

2 - H. Neuber. "Theory of stress concentration for shear strained prismatic bodies with arbitrary non-linear stress-strain law". J. Applied Mech., 28, (1961), 544.

3 - K. Molski and G. Glinka. "A method of elastic-plastic stress and strain calculation at a notch root". Mat. Sci. Eng., 50, (1981), 93.

4 - A. Baus, H.P. Lieurade, G. Sanz, M. Truchon. "Etude de l'amorçage des fissures de fatigue sur des éprouvettes en acier à très haute résistance possédant des défauts de formes et de dimensions différentes". CIT du CDS, (1978), n°1, pp. 161-190.

5 - R.M. Wetzel. "Smooth-Specimen Simulation of Fatigue Behavior of Notches". Journal of Materials, JMLSA, Vol. 3, n°3 (1968), pp. 646-657.

6 - J.D. Morrow, R.M. Wetzel, T.H. Topper. "Laboratory Simulation of Structural Fatigue Behavior". ASTM STP 462, (1970), pp. 74-91.

7 - A. Conle, H. Nowack. "Verification of a Neuber-based Notch Analysis by the Companion-Specimen Method". Experimental Mechanics, Feb. 1977, pp. 57-63.

8 - M. Truchon. "Application of Low-Cycle Fatigue Test Results to Crack Initiation from Notches. ASTM STP 770, (1982), pp. 254-268.

9 - M. Truchon. "L'amorçage des fissures de fatigue à partir d'entailles. Application aux joints soudés". Rapport IRSID RE 987 - 1982.

10 - N.E. Frost, D.S. Dugdale. "Fatigue Tests on Notched Mild Steel Plates with Measurements of Fatigue Cracks". J. Mech. Phys. Solids, (1957), Vol. 5, pp. 182-192.

11 - N.E. Dowling. "Fatigue of Notches and the Local Strain and Fracture Mechanics Approaches". ASTM STP 677, (1979), pp. 247-273.

Formation of Extrusions and Intrusions in Copper and in a Mild Steel

J. I. Dickson, S. Turenne, J. B. Vogt and J. P. Bailon

Département de génie métallurgique, Ecole Polytechnique de Montréal, Montreal, Quebec, Canada

ABSTRACT

The study of the characteristics of short extrusions and intrusions by scanning electron microscopy in copper and in a renitrogenized mild steel demonstrated that the shortest fatigue microcracks resolved were segmented and often unaccompanied by extrusions or accompanied by only very fine extrusions. Similar characteristics were often found at the extremities of somewhat longer intrusion-extrusion pairs. The observations generally indicated that the extrusions formed as a result of the prior presence of microcracks. This indication was particularly clear in the case of steel.

KEYWORDS

Extrusions, intrusions, microcracks, crack initiation, fatigue, copper, mild steel

INTRODUCTION

The occurrence during fatigue of extrusions and intrusions (or microcracks) on well-polished metal surfaces have been of considerable interest since Wood [1] first showed their existence. Early explanations for extrusions were generally based on dislocation theory and usually required improbable sequence of events to occur. In more recent years, two more attractive explanations have appeared. Neumann, Fuhlrott and Vehoff [2] explained extrusion-intrusion pairs by the difference in width of slip bands at a surface microcrack in tension and in compression. In tension, the width at the surface is narrow since the crack is open. A wider slip band is obtained in compression since crack closure permits shear stresses to act on the crack faces. The sum of the slip steps of different widths then results in an extrusion on the side of the crack which makes an obtuse angle with the free surface. Hunsche and Neumann [3] obtained experimental evidence on copper single crystals supporting this model. Extrusions formed on the predicted side of the crack. Intrusions were found without extrusions but not vice versa.

Essmann, Gösele and Mughrabi [4] have proposed a model based on vacancy concentrations and irreversible slip processes in persistent slip bands to explain the formation of extrusions. The re-entrant surface steps at the extrusions then act as stress concentrators giving rise to the formation of the intrusions at these steps. As the surfaces of the extrusions roughen as a result of irreversible glide processes, intrusions are also initiated within the width of the extrusions. While a considerable amount of experimental support for the details of this model has been cited [4,5], much of it is indirect, in that it not evident that it is associated with the initiation of very small (e.g., $\leqslant 5\ \mu m$) microcracks.

The objective of the present study was to experimentally verify by high-resolution scanning electron microscopy whether the extrusions gave rise to the intrusions or vice versa. Two polycrystalline materials were chosen, copper (grain size of 150 μm) and a renitrogenized mild steel (0.10 wt %C, 0.58% Mn, 0.001 % Al and 104 ppm N, grain size of 30 μm). The total strain amplitudes, $\Delta\varepsilon/2$, chosen were those which in previous work [6,7] on the same materials had produced persistent slip bands with ladder-like walls within the bulk.

A few samples were studied after fully-reversed push-pull testing. Some of the samples were subsequently bent and the portion submitted to tensile stresses in bending also observed.

The study focussed on characterizing the aspects of the smallest microcracks and extrusions observable, and comparing these characteristics to that of parallel longer microcracks and extrusions within the same grains.

OBSERVATIONS AND DISCUSSION

Copper : For the amplitudes employed, the copper samples showed a strong tendency for extrusions and intrusions to appear as pairs. The intrusions generally appeared to occur preferentially (Fig. 1) on the obtuse angle side of the extrusions, in agreement with previous observations [3]. Since this is the side which can be viewed more easily, considerable effort was spent verifying whether this was a true effect, by tilting the specimen and also by studying the extremities of extrusions and gaps in extrusions. This verification was not completely conclusive, but suggested that the effect was real. Often, intrusions were present (Fig. 2) on both sides of extrusions. Occasionally, individual extrusions were found for which an intrusion appeared only on the side where the extrusion makes an acute angle with the surface. On careful verification, some of these observations were shown to be the result of an extrusion curling and hiding the microcrack on the obtuse angle side.

The shorter microcracks and extrusions at the extremities of many longer microcracks-extrusions consisted of short segments, ≃ 0,5-3 μm separated by uncracked gaps (Fig. 3). The smallest isolated microcracks found (Fig. 1) also were of similar length to the shorter segments at extremities of longer microcracks. Such short microcrack segments unaccompanied by extrusions could often be found (Fig. 3) extending past the extremities of extrusion-microcrack pairs, including quite short pairs ≃ 20-30 μm in length. In other cases, extrusions were present on the microcrack segments at the extremities but decreased in height as the extremity was approached. In still other cases, regular height extrusions accompanied the microcracks right up to their extremities. For wide extrusions presenting parallel intrusions within the width, similar closely-spaced microcrack segments

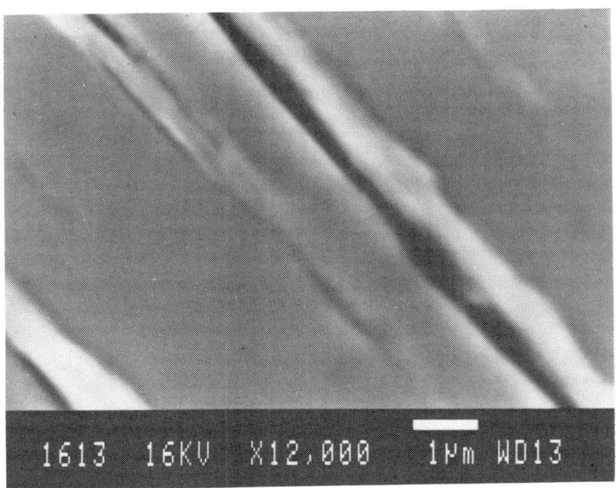

Fig. 1. Large and fine microcracks present at the obtuse angle corner of extrusions. Some of the smaller microcracks are accompanied by finer extrusions; others are not accompanied by extrusions (copper, $\Delta\varepsilon/2$ = 0.085%, end of fatigue life).

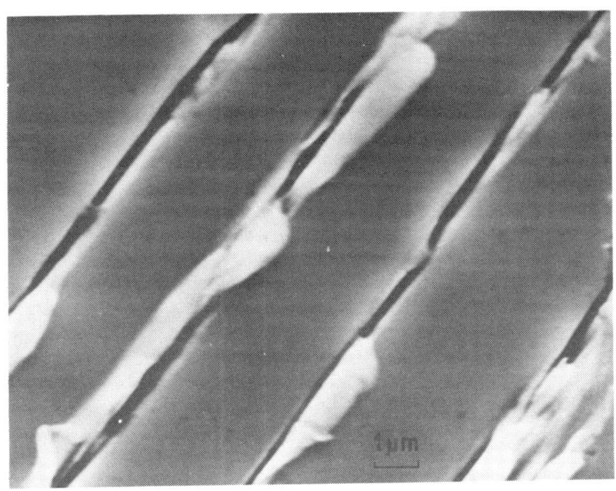

Fig. 2. Extrusions at times on one side, at times at the other side of microcracks. Some of the microcracks present are hidden by the extrusions (copper, $\Delta\varepsilon/2$ = 0.085%, ≃ 10% of fatigue life).

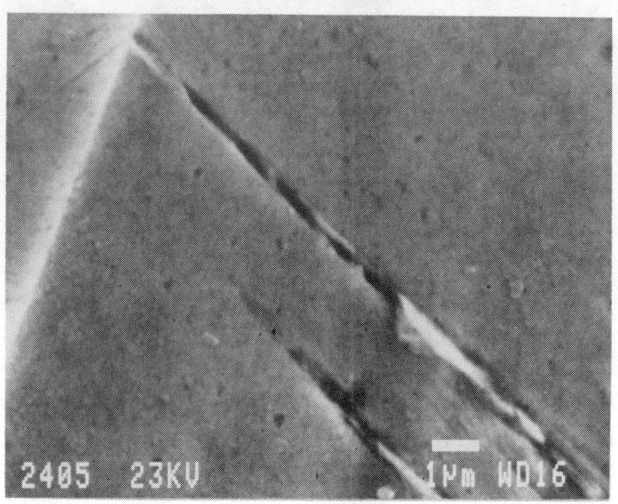

Fig. 3. Microcrack segments extending past the extremities of extrusion-intrusion pairs (copper, $\Delta\varepsilon/2$ = 0.075%, end of fatigue life).

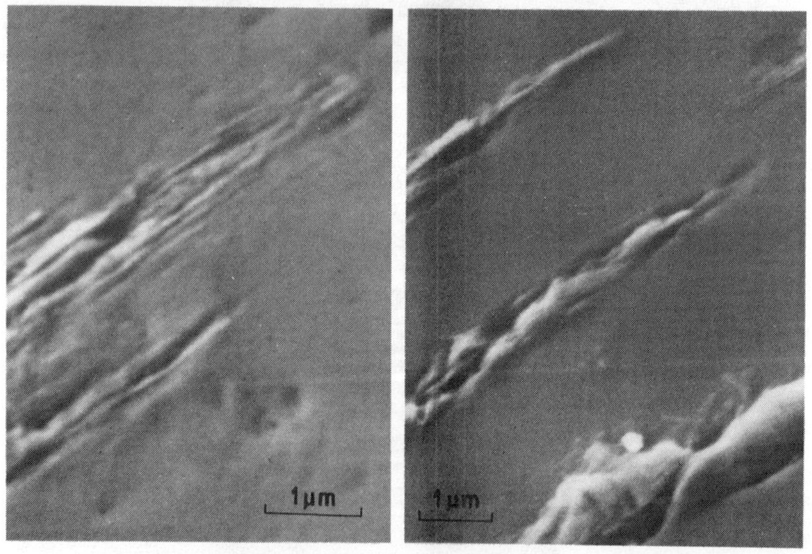

Fig. 4 and Fig. 5. Parallel microcrack segments extending past the extremities of wider bands of extrusions-intrusions (copper, $\Delta\varepsilon/2$ = 0.085%, ≃ 10% of fatigue).

could also be found (Figs. 4 and 5), extending past some extremities of extrusions.

The shortest isolated microcracks resolved were usually either not accompanied by extrusions (Fig. 1) or were accompanied by extrusions of much lower height than those accompanying longer microcracks. Individual isolated microcracks were rarely resolved. However, groups of 3-5 microcrack segments were more often resolved. Short isolated extrusions apparently not accompanied by intrusions were very rare. As well at extrusion extremities, only very rarely did the extrusions appear to extend beyond the intrusions. Numerous cases of extrusions hiding intrusions were demonstrated. It should also be noted that small extrusions could be resolved more easily than small fine intrusions. Essentially identical observations were obtained prior to and after bending the fatigued samples, although microcracks were easier to resolve on the bent samples.

The aspects of the shortest microcracks resolved, demonstrated that many formed prior to the formation of extrusions and also suggested that the extrusions formed as a result of the presence of the microcracks. The aspects of the extremities of longer intrusion-extrusion pairs showed that the intrusions in this region consisted of short segments of microcracks and also often suggested that the extrusions formed as a result of the prior presence of the microcracks. The aspects of the observations generally either agreed well or were at least compatible with the explanation proposed by Neumann, Fuhlrott and Vehoff [2] and only rarely agreed with that proposed by Essmann, Gösele and Mughrabi [4].

Renitrogenized mild steel : For the testing conditions employed, the renitrogenized mild steel presented a lesser tendency to have extrusions associated with microcracks. The microcracks were often curved in this b.c.c. metal. Both extrusions and intrusions were segmented (Figs. 6 and 7). No clear tendency for the intrusions to occur on one particular side of the extrusions was noticed. Extrusions, however, were often flattened and tended to hide microcrack portions and to make it difficult to identify the obtuse and acute angle corners. Short microcracks and microcrack extremities not associated with extrusions (Fig. 7) were observed much more frequently than in copper. Extrusions not accompanied, at least at their extremities, by microcracks were not found. The observations indicated, even more strongly than for copper, that the microcracks formed first and that their presence gave rise to the extrusions. From the present observations on this steel, however, good agreement with the model of Neumann, Fuhlrott and Vehoff [2] was not demonstrated.

CONCLUSIONS

From this study performed on two polycrystalline materials and for strain amplitudes giving rise to persistent slip bands with a ladder-like substructure, the experimental evidence indicated that microcracks generally preceded the formation of extrusions and that extrusions resulted from the presence of the microcracks.

ACKNOWLEDGMENTS

Financial support from the NSERC (Canada) and FCAC (Quebec) research support programs is gratefully acknowledged.

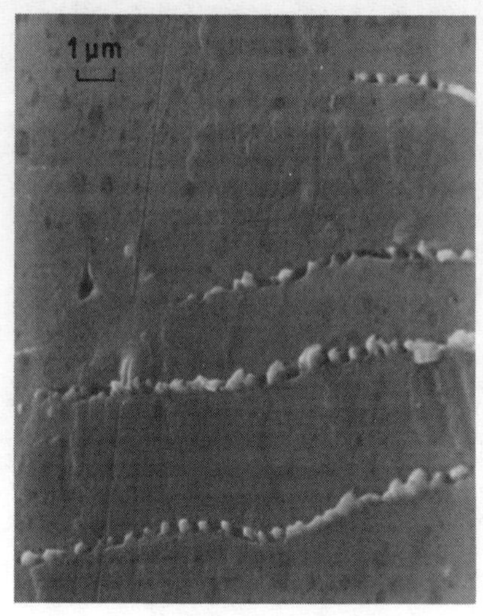

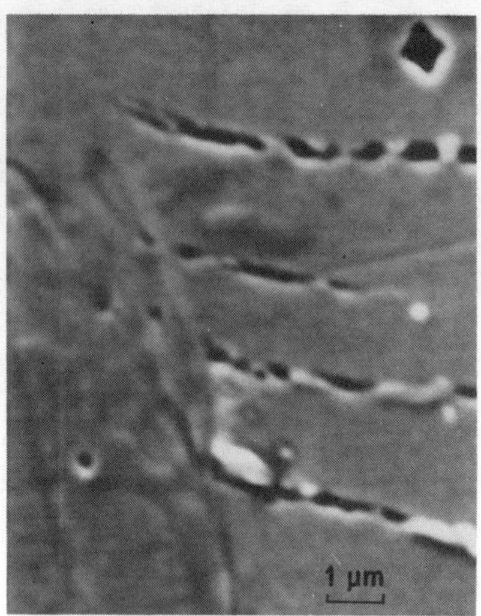

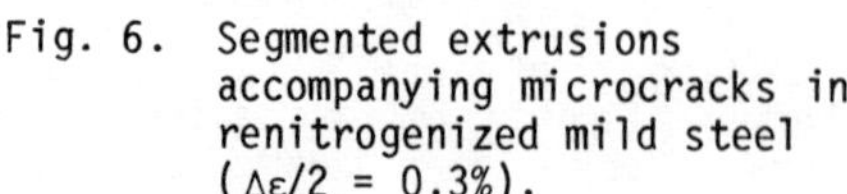
Fig. 6. Segmented extrusions accompanying microcracks in renitrogenized mild steel ($\Delta\varepsilon/2$ = 0.3%).

Fig. 7. Absence of extrusions along short microcracks and along microcrack segments past the extremities of extrusion-intrusion pairs, (renitrogenized mild steel $\Delta\varepsilon/2$ = 0.3%).

REFERENCES

1. W.A. Wood, Phil. Mag. 3, 692 (1958).
2. P. Neumann, H. Fuhlrott and H. Vehoff, in Fatigue Mechanisms, ASTM STP 675, edited by J.T. Fong, Am.Soc. for Testing and Materials, 371 (1979).
3. A. Hunsche and P. Neumann, in Advances in Fracture Research edited by D. François, Pergamon Press, Vol. 1, 273 (1981).
4. V. Essmann, V. Gösele and H. Mughrabi, Phil.Mag. 44, 405 (1981).
5. H. Mughrabi, R. Wang, K. Differt and V. Essmann, in Fatigue Mechanisms: Advances in Quantitative Measurement of Physical Damage, ASTM STP 811, edited by J. Lankford, D.L. Davidson, W.L. Morris and R.P. Wei, Am.Soc. for Testing and Materials, 5 (1983).
6. J. Boutin, M.A.Sc. Thesis, Ecole Polytechnique de Montréal (1983).
7. A.E. Geckinli and J.I. Dickson, unpublished research.

Etude par Emission Acoustique de la Deformation Cyclique de Polycristaux d'Aluminium de Haute Purete Sollicites en Traction Compression

A. Slimani, J. Chicois, R. Fougères et P. Fleischmann

Groupes d'Etudes de Métallurgie Physique et de Physique des Matériaux, UA CNRS 341, INSA de Lyon, 69621 Villeurbanne Cedex, France

RESUME

L'emission acoustique continue provoquée par sollicitation cyclique en traction compression d'Al 5N est étudiée en fonction du temps, du nombre de cycles, de la période et de l'amplitude de la déformation plastique. Les résultats sont interprétés à l'aide d'un modèle.

MOTS CLES

Aluminium, émission acoustique, dynamique des dislocations, fatigue.

INTRODUCTION

Des travaux récents ont montré que l'émission acoustique (EA) continue qui accompagne la déformation plastique de matériaux c.f.c. comme le cuivre ou l'aluminium contient des informations concernant la dynamique des dislocations (1)(2). Cette technique de mesure peut être appliquée au suivi, en temps réel, des mouvements des dislocations lors de la fatigue cyclique d'aluminium de haute pureté et fournir des informations complémentaires aux autres méthodes de caractérisation du comportement dynamique employées par ailleurs (3)(4).
L'objectif de ce travail s'inscrit dans la perspective de la modélisation des propriétés mécaniques de fatigue à partir de la théorie des dislocations : l'E.A. apporte une caractéristique du mouvement des dislocations au cours de la déformation plastique elle-même ; les caractéristiques de l'E.A. sont des valeurs moyennes comme le sont les grandeurs mécaniques mesurées.

DISPOSITIF EXPERIMENTAL

La figure 1 présente le schéma fonctionnel de l'installation expérimentale de sollicitation de l'éprouvette et de mesure de l'émission acoustique.
La machine de fatigue hydraulique a une cinématique entièrement élastique et elle est pilotée par un ordinateur PDP 11-23. La sollicitation cyclique est réalisée à vitesse de déformation plastique $\dot{\varepsilon}_p$ constante et à amplitude de déformation plastique ε_p imposée.

Les éprouvettes sont en aluminium de haute pureté (99,999 %) à l'état polycristallin avec une taille de grain de l'ordre de 2 mm.
La description complète des essais de fatigue, leurs caractéristiques ainsi que la préparation des éprouvettes sont définies en (8), qui figure au proceeding du présent Congrès.
L'émission acoustique est recueillie par un capteur large bande centrée autour de 1 Mhz. Après amplification de 60 dB le signal est mesuré par un millivoltmètre à valeur efficace VX 207 A Metrix dont la constante de temps est d'environ 0.4 s. Le résultat de la mesure est enregistré avec l'effort sur un enregistreur bicourbe et acquis par l'ordinateur, en même temps que la déformation et l'effort. Les informations sont enregistrées sur disquette et peuvent être exploitées en temps différé.

RESULTATS EXPERIMENTAUX

La procédure expérimentale utlisée consiste à solliciter une éprouvette de fatigue en traction-compression symétrique. Le nombre de cycle atteint est suffisant pour obtenir la saturation de l'EA et de la contrainte maximale dans chaque cycle. La vitesse de déformation plastique $\dot{\varepsilon}_p$ est déterminée à partir de la durée assignée au régime plastique dans chaque cycle. Pour simplifier, cette durée est appelée la période du cycle. En général la même éprouvette est réutilisée à des taux de déformation plastique maximum croissant d'un facteur 5 entre deux essais, la période du cycle étant maintenue constante.
La figure 2 présente, pour un cycle, un résultat typique de variation de la contrainte et de la tension efficace du signal d'EA en fonction du temps. Il faut noter que le signal d'EA est faible et qu'il est complètement noyé dans le bruit électronique lorsque la déformation plastique maximale est inférieure à $1.5\ 10^{-4}$.
En fonction du nombre de cycles, la figure 3 présente les variations de l'amplitude du pic d'EA associé au passage élastique-plastique et de la contrainte maximale dans chaque cycle. L'amplitude du pic d'EA en traction est la hauteur AB définie sur la figure 2, en compression la hauteur DE. La contrainte maximale σ_M augmente alors que l'EA diminue jusqu'à saturation.
La figure 4 présente la variation de l'EA en fonction de la vitesse de déformation plastique à déformation plastique maximale de $5\ 10^{-4}$ dans le cas d'un état saturé. La variation de la vitesse de déformation plastique est obtenue par modification de la période du cycle.
En fonction de la déformation plastique maximale, la figure 5 présente l'évolution de l'EA et de la contrainte maximale à saturation σ_S.

INTERPRETATION ET DISCUSSION

- Bases de l'interprétation des mesures d'E.A.

L'émission continue qui accompagne la déformation plastique des matériaux c.f.c. est liée aux mouvements des dislocations. La relation qui existe entre les signaux d'E.A. reçus et les caractéristiques microscopiques géométriques et dynamiques de ces mouvements a été décrite par Rouby et al. Avec des simplifications courantes, à savoir, mouvement d'une boucle de dislocation sur l'aire d'un rectangle, distance entre point d'observation et localisation de la source très grande vis-à-vis des grandeurs définissant ce rectangle, Rouby et al montrent que le déplacement ultrasonore de l'onde longitudinale qui atteint le point d'observation s'écrit :

$$u_L\ (d,t) = \frac{b\, C_T^2\, L}{4\, \pi\, C_L^3 d}\ v\, (t - d/C_L) \qquad /1/$$

b : vecteur de Burgers; C_T, C_L : vitesses des ondes transversales et longitudinales; L : largeur de l'aire balayée; d : distance entre source et

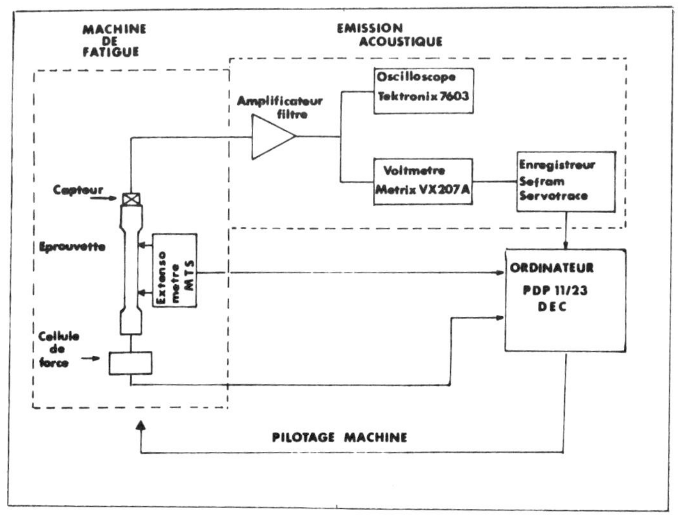

Figure 1. Schéma synoptique de l'appareillage

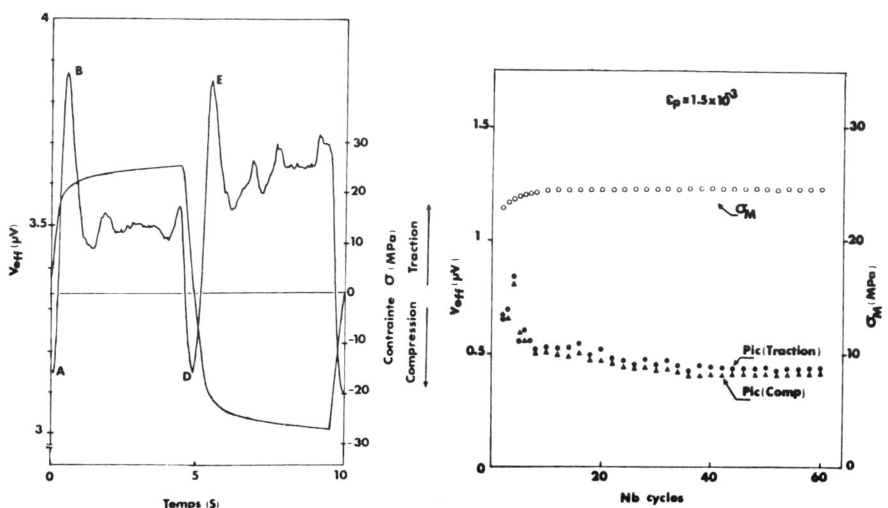

Fig 2.Contrainte et EA en fonction du temps dans un cycle (ε_p = 2 10^{-3}, période 10 s).

Fig 3. Contrainte maximale et hauteur du pic d'EA en fonction du nombre de cycles.

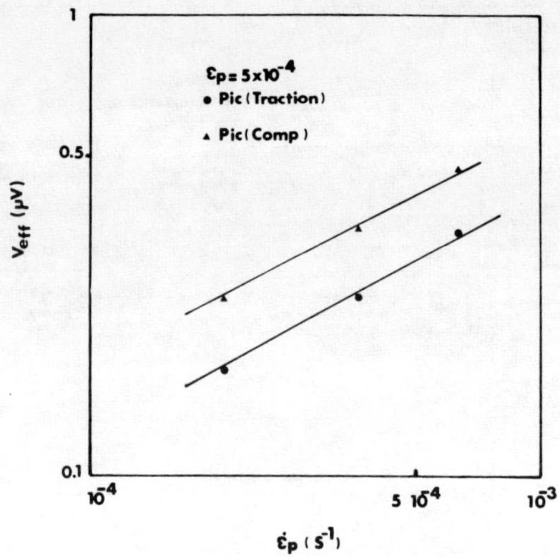

Fig. 4 : Influence de la vitesse de déformation plastique (état saturé mécaniquement à 5 10^{-4}).

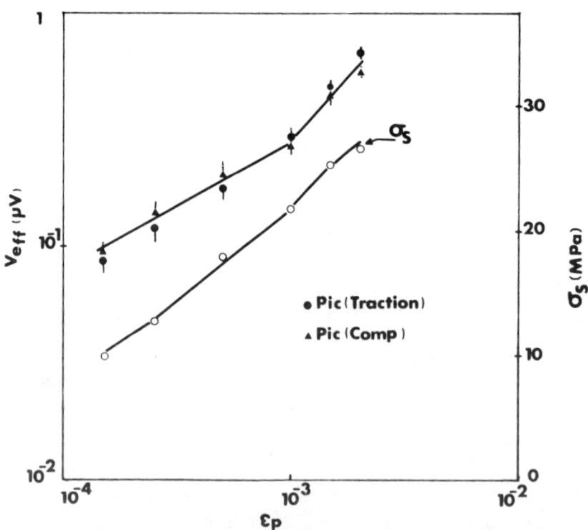

Fig. 5 : Influence de la déformation plastique sur le pic d'EA et sur la contrainte à saturation. Période du cycle 10 s.

point d'observation; v : vitesse de la source. La source se déplace selon la longueur du rectangle.
Cette expression est tout a fait similaire à celle décrite par Wadley et Merhabian (5) et est parfaitement compatible avec les expressions issues des travaux de Hsu et Ono (6).
Lorsqu'un grand nombre de dislocations sont actives simultanément sans cohérences spatiales ou temporelles entre elles, ce qui est le cas dans la déformation plastique homogène des matériaux c.f.c., ces déplacements arrivent au capteur de façon complètement , décorrélés entre eux, et le signal délivré par celui-ci a l'allure d'un bruit : c'est l'émission acoustique continue. Dans ces conditions, les énergies des signaux ultrasonores élémentaires sont sommables pour obtenir, au rendement du capteur près, l'énergie du signal délivré par celui-ci. En supposant que le capteur a une bande passante infinie, en posant v_o l'amplitude maximale de la vitesse de la source et en supposant que $\int_0^t v^2\,(t)\,dt \propto v_o^2$ on trouve que l'énergie du signal élémentaire est : $E = K\,L^2\,v_o^2$, K étant un facteur incluant tous les termes constants de l'expression /1/.
Le nombre de signaux de ce type qui atteignent le capteur par unité de temps est donné par la relation d'Orowan et Taylor dans le cas où toutes les sources sont identiques : $\dot{N} = \dot{\varepsilon}_p\,V/Ab$, $\dot{N}$: nombre de boucles de dislocations mobiles par unités de temps; $\dot{\varepsilon}_p$: vitesse de déformation plastique; V : volume plastifié; A : aire moyenne balayée par les dislocations; b : vecteur de Bürgers.
La puissance du signal s'écrit alors : $P_L = K\,(\,\dot{\varepsilon}_p\,V/Ab)\,L^2\,v_o^2$ /2/.
En supposant $A = L^2$ la tension efficace, qui est proportionnelle à la racine carrée de cette puissance, s'écrit : $v_{eff} = (K\;\dot{\varepsilon}_p\,V/b)^{\frac{1}{2}}\,v_o$ /3/.

- Interprétation des résultats

Lors d'une déformation plastique, seuls les paramètres $\dot{\varepsilon}_p$ et v_o varient, les résultats sont donc à interpréter en fonction de ces grandeurs.
La proportionnalité entre la tension efficace et la racine carrée de la vitesse de déformation plastique a déjà été signalée et observée de nombreuses fois en traction monotone sur l'aluminium (7). Nos propres mesures indiquent clairement cette proportionnalité lorsqu'on fait varier la période du cycle effort déformation à un taux de déformation plastique maximale imposé inférieur à 1.10^{-3} (fig. 4, pente = 0.5 en echelle log.). Cette loi d'évolution conduit également à une tension efficace du signal d'EA qui augmente comme la racine carrée de l'amplitude de déformation plastique imposée lorsque la période de l'essai reste constante (la figure 5, jusqu'à $1.\,10^{-3}$, présente une pente de l'ordre de 0.5). Ce type d'évolution a été observé quelque soit le paramètre d'EA exploité : amplitude du pic correspondant au passage élastique-plastique ou le niveau atteint en fin de déformation plastique. Physiquement cela signifie que les mêmes phénomènes doivent se produire dans un temps plus ou moins court selon la vitesse de déformation plastique imposée.
Au-delà de 10^{-3} la loi en racine carrée n'est plus vérfiée : l'EA croît plus vite que $\dot{\varepsilon}_p^{\frac{1}{2}}$. Par exemple à 1.5 10^{-3} le coefficient de proportionnalité entre les logarithmes de la vitesse et de la tension efficace est de 0.7.
On peut interpréter ces résultats en terme de nombre de boucles de dislocations mobiles disponibles : en-dessous de 10^{-3} un grand nombre de dislocations potentiellement mobiles existent et toute variation de la vitesse de déformation ne conduit qu'à des variations du nombre de boucles de dislocations activées sans que les caractéristiques moyennes du glissement changent. Par contre aux grandes vitesses de déformation une saturation dans le nombre de boucles de dislocations mobiles peut être atteinte et un nouveau phénomène physique doit intervenir pour permettre la déformation. Cela correspond à une augmentation des vitesses des

dislocations v_o selon la relation /3/ et donc une augmentation de l'EA plus rapide que prévue. Cette interprétation est d'ailleurs confortée par les mesures d'atténuation des US faites par Vincent et al. (8) dans les mêmes conditions expérimentales qui montrent une saturation de la variation d'atténuation des ondes ultrasonores à partir de 0.8×10^{-3}, susceptible d'être interprété en terme de saturation de la densité de dislocations. De la même manière I. Waille et al (9) ont montré dans les mêmes conditions expérimentales, que la fraction des éléments de volume déformés déduits des cycles de fatigue contrainte-déformation, atteingnent sensiblement la valeur unité pour des amplitudes de déformation voisines de 10^{-3}.
La forme de la courbe tension efficace en fonction de la déformation ou du temps (figure 2) dépend très peu du taux de déformation maximum. En régime purement élastique il n'y a pas d'émission acoustique car $\dot{\varepsilon}_p$ est nul. Au début du régime plastique, $\dot{\varepsilon}_p$ croît très rapidement à sa valeur nominale. Le fait que l'EA présente alors un pic important indique, selon la relation /3/, que les vitesses moyennes des dislocations sont élevées. Elles décroissent ensuite ce qui conduit à une diminution de l'EA.
Le terme v_o qui intervient dans l'expression /3/ de la tension efficace est la vitesse maximale des dislocations pour un saut balayant l'aire A ou parcourant le chemin L. L'EA mesurée dans le domaine de fréquence du Megahertz n'est sensible qu'à des phénomènes qui se produisent à une échelle de temps de l'ordre de la microseconde. Les observations au M.E.T. in situ faites par ailleurs (10) montrent que la durée du balayage des cellules par les dislocations est bien supérieure à la microseconde. Les sauts élémentaires auxquels l'EA est sensible s'effectuent donc sur des parcours beaucoup plus petits que la taille des cellules et correspondent sans doute à des parcours entre défauts du réseau notamment des dislocations. La diminution de vitesse des boucles de dislcoations envisagée ci-dessus peut provenir d'un accroissement des contraintes internes à grandes distances lorsque le taux de déformation plastique augmente.
Il est également possible d'interpréter la diminution de l'EA en fonction du nombre de cycles par l'accroissement des contraintes internes, l'EA restant stable dès que la saturation mécanique est atteinte. Enfin les dissymétries observées entre les niveaux d'EA en traction et compression, ne peuvent être expliquées qu'en termes de vitesse de dislocations plus ou moins importantes selon le sens de la déformation. Cela pourrait là encore être un effet lié aux contraintes internes mais cela demande de nouvelles expérimentations.

REFERENCES

1 C.B. Scruby, K.N.G. Wadley et J.E. Sinclair, Phil. Mag. A, 44, 2, 249, (1981).
2 D. Rouby, P. Fleischmann et C. Duvergier, Phil. Mag. A, 47, 5, 671, (1983).
3 J.T.H de Hosson, A. Huis, H Tamler et O. Kanert, Acta Metall, 32, 8, 1205, (1984).
4 A. Vincent et J. Perez, Phil. Mag. A, 40, 3, 377, (1979).
5 K.N.G. Wadley et R. Mehrabian, Mat. Sci. Eng., 65, 245, (1984).
6 S.Y.S. Hsu et K. Ono, Proc. 5th Int. Acoustic Emission Symp. Tokyo, 294, (1980).
7 M.A. Hamstad et A.K. Mukherjee, I.C.S.M.A.4 Proc., 2, 574, (1976).
8 A. Vincent, A. Hamel, J. Chicois et R. Fougères, proc. du présent Congrès.
9 I. Waille, J. Chicois, L. Vincent, A. Vincent et R. Fougeres, Fatigue 84, Proc. 2nd Int. Conf. on Fatigue, 61, (1984).
10 G. Guichon, J. Chicois, C. Esnouf et R. Fougeres, Fatigue 84, Proc 2nd. Int. Conf. on Fatigue, 31, (1984).

Evaluation of the Effect of Microstructure on Fatigue Crack Propagation using a Fractographic Technique

S. Nunomura*, Y. Higo*, M. Arai* and H. Hayakawa**

**Tokyo Institute of Techonology Research Laboratory of Precision Machinery & Electrometallurgy, Nagatsuta, Midori-ku, Yokohama, Japan*
***Nippon Steel Pipe Company, Central Research Laboratory, Watarida, Kawasaki-ku, Kawasaki, Japan*

ABSTRACT

To investigate the effect of micro structures on the fatigue crack propagation of metals, ß-brass, Al-1%Si alloy and Cu-Fe alloys were heat treated and fatigue tested with a special loading program consisted of two height pulses in accordance with a certain regulation. Fractography illustrated the interactions between the fatigue crack fronts and the micro structures; the precipitates, inclusions, grain boundaries, the phase boundaries and the voids. Some of the micro structures assisted the fatigue crack propagation. On the other hand some of them prevented the growth of it. The crack growth in opposite direction to the main crack growth from the voids have been confirmed.

KEYWORDS

Fatigue crack propagation; Fatigue striation; Micro structure; Fractography; Modulated pattern method; Void coarsence.

INTRODUCTION

Besides the tensile strength, the hardness and so on, the resistance against the fatigue crack propagation is an important strength parameter of structural materials. It is generally considered that the micro structure effects affect little on the fatigue crack propagation behaviour in medium stress intensity region, Paris' region or the stage II, while there are many counterevidences(1). Since there are too many micro structural parameters, it is impossible to identify their individual effect by macroscopic fatigue crack observation (da/dN). The effects of the parameters seems to compensate each other, and to result above mentioned uncertainty. It is required to clarify the individual effect of material micro structure on the fatigue crack propagation. Despite the fatigue striations gives us a lot of informations on microscopic propagation behaviours of the fatigue crack front, the interaction with grain boundaries, second phase particles or voids is still not well understood. Since striations appear only under appropriate combination of applied stress and orientation of the crystallographic planes, the striation areas are distributed locally and isolated each others. Although being in an area, the striations are discontinuous at the grain boundaries. The informations are limited to within a grain. A semi-analytical method of the fatigue crack propagation with a periodical overload during the fatigue test had been developped and it was successfully found that one striation corresponded to one load cycle while crack growth occured on the load rising half cycle[2], but the information from them was still in a grain. Authors developped a method to number the every striation lines on the fatigue fractured surface, Modulation pattern method[3], which is applicable to quantitative analysis of the interaction between the fatigue crack front and the micro structure of the structural alloys. The purpose of present study was to evaluate the individual effect of the micro structure parameters on the fatigue crack propagation using three type alloys which were represent the alloys having the coarse, the medium and the fine second phases, respectively.

EXPERIMENTAL PROCEDURE

Materials used were ß-brass (BB), Al-1%Si alloy (AS) and Cu-(1-3%)Fe alloy (CF). They were high purity alloys and have simple micro structures to simplify their effects on the fatigue crack propagation. Alloy AS was melted and cast using 99.99% Al and 99.9999999%Si, and other alloys were essentially of pure binary composition. The alloys were homogenized and then thermo-mechanically processed to produce desirable micro structure size and shape in each alloy, as indicated in Tables 1 and 2. 25mm (AS) or 6mm(BB and CF) thickness CT test-pieces were machined in LT orientation to ASTM specification E647. The macroscopic fatigue crack propagation rates da/dN vs ΔK curves were determined at a frequency of 20Hz and at an R of 0.1 in a servo-hydraulic machine. Crack detection was by a DC electric potential drop method, with a resolution of 0.01mm.

The simple periodical cyclic overload method cannot identify the simultaneous generated striations in the individual grains nor can count the number of loading cycles spent to advance the crack front through grain boundaries. The method used is that to modulate the striation spacings an electric signal generated in a computer system is supplied to the fatigue tester. Figure 1 shows a part of the modulated load spectrum which is consist of two height pulses in accordance with a certain regulation. With

TABLE 1 Heat Treatment and 2nd Phase Size (ß-Brass)

Specimen	Heat Treatment		2nd Phase	
	Temp (K)	Time(hr)	volume(%)	distance(μm)
BB - F	473	2	27.0	59.1
BB - M	673	2	39.2	71.1
BB - C	873	2	74.6	87.9

TABLE 2 Interparticle Distance, Dia and Proof Strength

Specimen	Temper(1) (K)-(hr)	Temper(2) (K)-(hr)	Precipitate $\bar{l}$(μm)	$\overline{\text{dia.}}$(μm)	(MPa)
AS - FS	293- 24	573- 10	0.39	1.4	31.7
AS - FR	293- 24	673- 10	0.52	1.9	31.0
AS - CS	77-	573- 10	1.60	3.4	25.0
AS - CR	77-	673- 10	1.20	2.8	26.4

this reference signal, fatigue tests are carried out the under constant ΔK or the constant load range control. The fatigue fracture surfaces show a combination of narrow and width striation spaces corresponding to the load spectrum, Fig. 2. An over load ratio of 112.5% was used, which corresponded to 160% in the spacing (n=4). Using the symbols "S" and "O" to represent the standard load and the overload cycles, respectively, the modulated pattern used is of the form: O-1,S-i,O-j,S-K (i=1~5,j=1~9 and k=1~10) with a period of 6,525, which corresponds to the crack length of 0.1-1mm in the stable crack growth range. This combination of the difference in spacings and the period is enough to identify the number of cycles for all striations of the fracture surface referring to the fatigue crack length data.

RESULTS AND DISCUSSION

(a) Coarse Second Phase: ß-Brass

The volume fraction and the mean interphase distance of the 2nd phase (α) with heat treatment condition of BB alloy are given in Table 1. In this alloy the volume fraction of the 2nd phase is so large that the crack cannot bypass it, then the fatigue crack propagation depends mainly on summation of those in each phase. According to the fractographic observation the ß phase more resists to the fatigue crack propagation than phase and from the da/dn vs ΔK curves it is deduced to the same conclusion (Fig.4). However, the rate was not the simple summation and there were a few interactions in regard to the phase orientation, the phase boundary and etc. It was found with the modulated pattern analysis that there were clear hesitate cycles on the boundary in ß to α crack propagation, but in α to ß

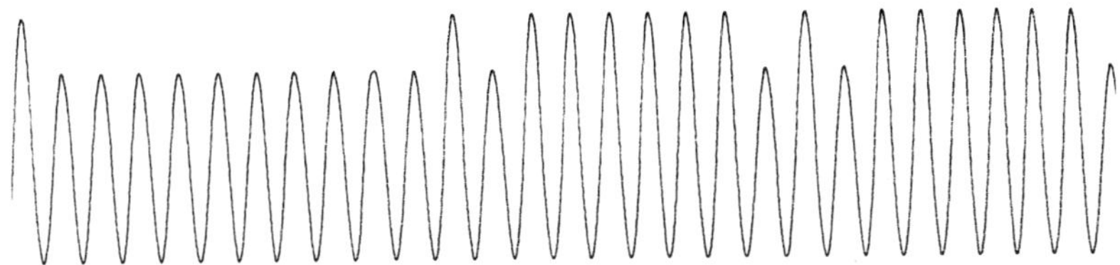

Fig.1. Modulated loading spectra for the fatigue test. Pattern form is -O1-S10,O1-S1-O6-S1,O1-S1-O6-S2,

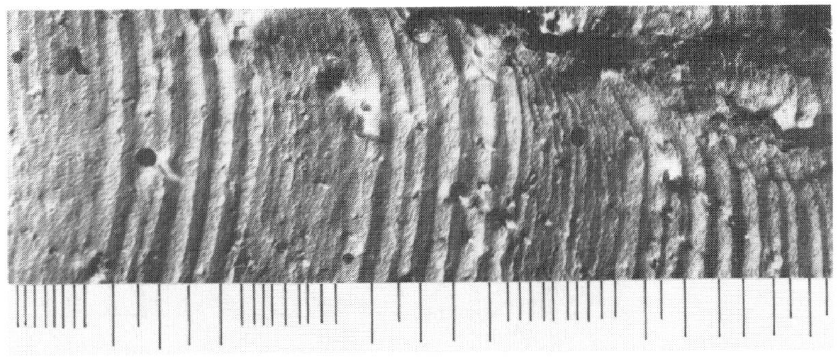

Fig.2. An example of modulated fatigue striation. Pattern format is -S8,O1-S1-O3-S9,O1-S1-O3-S10,O1-S1-O4-S1,

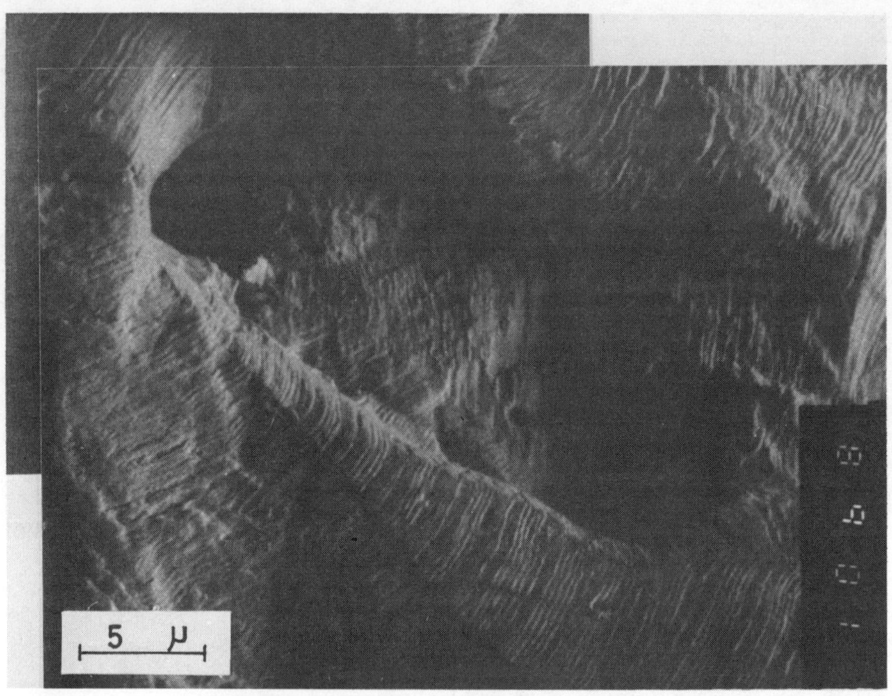

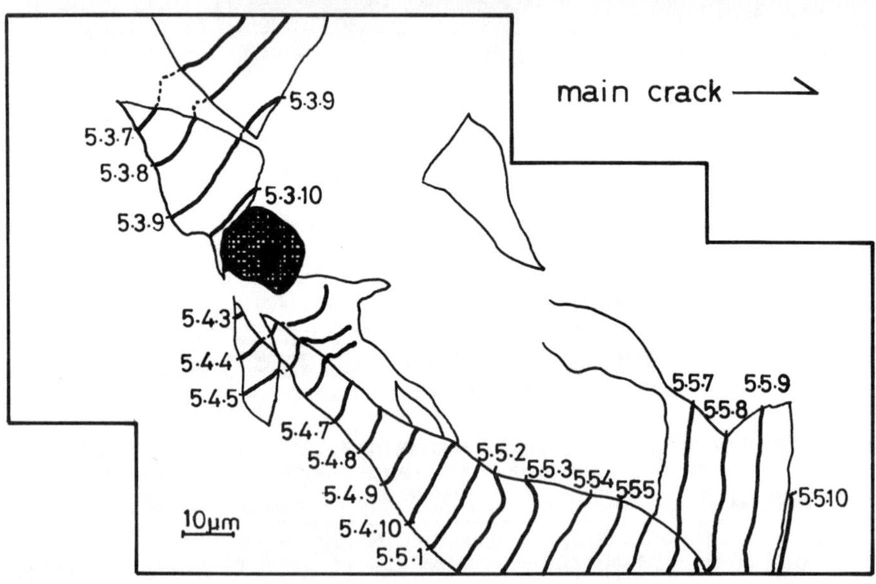

Fig.3. Interaction between the fatigue crack front and inclusion. A few hesitate cycles are found just before the inclusion. Striation in ß phase is approximately perpendicular to the orientation of the phase, but to macro crack growth direction.

crack propagation. Similar hesitate cycles were found just before the some inclusions as shown in Fig.3. Most of the striations in ß phase were not perpendicular to the macro cracking direction due to take minimum length in higher resistant zone (Fig. 3). These result suggest that there is best morphology of the 2nd phase for the fatigue crack propagation.

(b) Medium size precipitates: Al-1%Si Alloy

The interparticle distance and diameter and the proof stress $\sigma_{0.2}$ with aging condition of AS alloy are given in Table 2. The proof stress was in inverse proportion to the mean interparticle distance and followed on Orowan's law. da/dN vs ΔK curves of AS-alloy are given in Fig. 5. The effect of the micro structure was seen in the propagation rates of CS and CR at higher ΔK region where the former was 2.8 times faster than the later and this difference was statistically significant. The deviations in experimental values of FS and FR were greater than those of CS and CR, and then the effects of micro structure on the propagation rate shown in Fig. 5 was less significant. The facts that the rates of higher proof strength FS and FR were between those of lower proof strength CS and CR and that of lowest proof strength CS were slower than the others indicate that the propagation rate does not depend on the proof strength of the materials, while most of the fatigue crack propagation models are depend on the slip in the process zone. According to their fractographs, roughness of the fatigue fractured surface were depend on the mean precipitate size, namely, the precipitate works as an obstacle on the fatigue crack propagation. Since the precipitate in CS is the strongest, it shows the best fatigue property. It also indicates an usefulness of the proof strength on the da/dN since the rates of FS and FR is not less than that of CR. In this precipitate size range, the propagation rate depends on both strengths of precipitate and

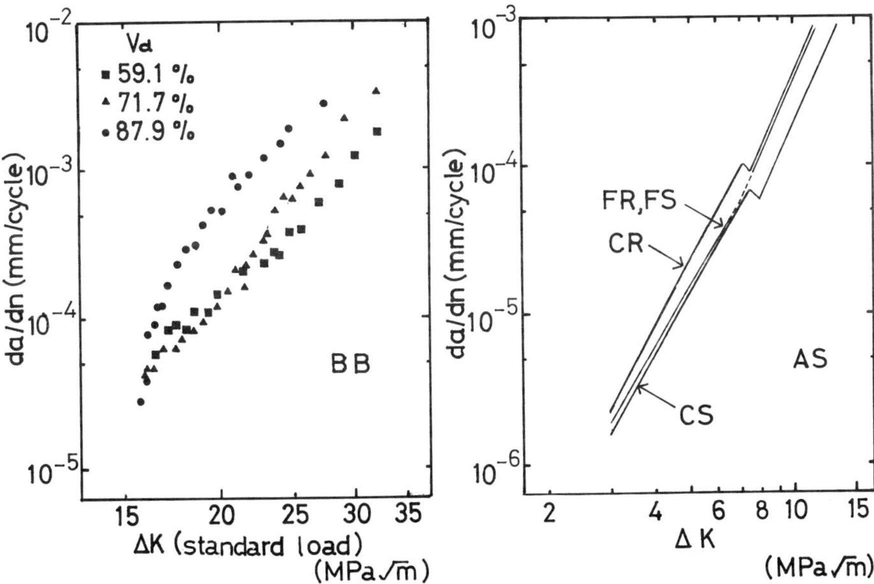

Fig.4. da/dN vs ΔK curves for micro structure with different volume fractions (Alloy BB).

Fig.5. da/dN vs ΔK curves for micro structure with different sizes of precipitation (AS)

matrix and whether the crack bypasses them or not.

(c) Fine Precipitates : Cu-Fe allyos

Forged Cu-Fe alloys which contained about 1%, 2% and 3% Fe were solution treated at 1223K, and aged at 973K for 3-24 hr in vacuum. The mean interpercipitate distance and the diameter of Cu-1.08%Fe alloy were ranged between 2400A-5000A and 360A-740A respectively[4]. Difference in da/dN vs K curves of fine precipitate Cu-Fe alloys where the interprecipitate distance and the diameter varied by factor of 4 was insignificant, while the proof strength varied by factor of 2 (figures abbreviated). It was found that the numbers of precipitates on the fatigue fracture surfaces observed by the SEM photographs were less than a tenth of those on the section observed by the metallographs. It is indicated that most of crack front bypasses the strong fine precipitates and then the effect of micro structure is not predominant. Figure 6 shows an interaction between the crack front and the precipitate where the crack growth was prevented by a precipitate. It suggests that a locally dense distributed precipitates works as an obstacle to the crack growth where the crack front does hardly bypass it. Some fatigue crack propagation mechanism consisted of the void coarsence[5], but there was not clear evidence whether the void grew before arrival of the crack or after arrival of it. Modulated striation patterns show the direction and the relative sequence of the crack growth. A void originated from a precipitate was grown by the striation before arrival of main crack front in opposite direction of it (Fig.7). It seems to accelarate the fatigue crack growth.

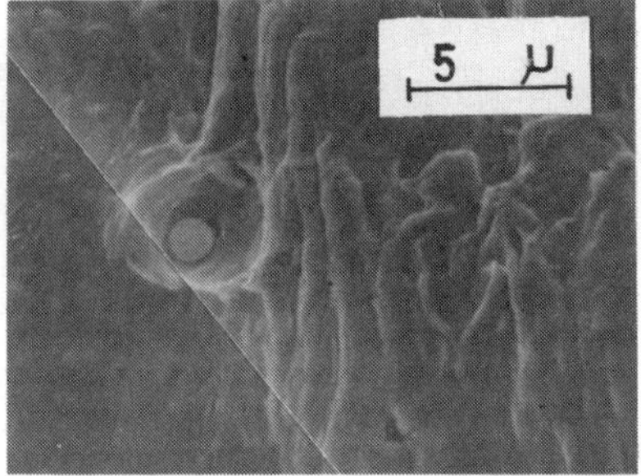

Fig.6. An obstacle for FCP (Cu-3%Fe)

CRACK →

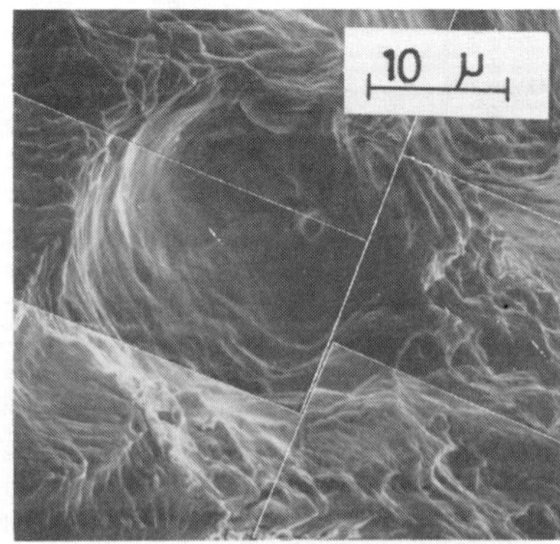

Fig.7. An opposite direction FCP

CONCLUSION

It has been shown that fatigue crack crack propagation test under program loads have yielded a better understanding of the effect of micro structure. Strong 2nd phase reduces the fatigue crack propagation rate if it is prevented to bypass the crack front on the phase. Crack propagation caused directional reversal due to secondary cracking ahead was observed.

REFERENCES

1. S.Nunomura and G.Kobayashi, J.Japan Light Metals,28,547(1878)
2. R.W.Herzberg, ASTM STP415, p202(1967)
3. S.Nunomura, Y.Higo, I.Tsubono and H.Hayakawa, Fract. Mech. Tech. Appl. to Mat. Evaluation & Design, p121, Martineus Nijoff,Hague(1983)
4. K.Monzen, Master Thesis of Tokyo Inst. of Technology (1979)
5. T.Yokobori, Phy. Strength & Plasticity, p.327(1969) MIT Press

Mecanismes de Propagation des Fissures de Fatigue d'Alliages d'Aluminium A9 et 7×49 Frittes (P/M) avec Oxydation Controlee

K. Benrefad, Ph. Bompard et D. Francois

Laboratoire Matériaux de l'Ecole Centrale des Arts et Manufactures, Châtenay Malabry, France

RESUME

Le présent travail porte sur la propagation des fissures de fatigue au voisinage du seuil dans des produits en aluminium frittés (A9 et 7x49) ayant subi avant compactage et filage différents degrés d'oxydation par traitement de la poudre sous atmosphère contrôlée. Pour pouvoir étudier l'influence propre de l'oxyde sur les propriétés mécaniques, des mesures ont d'abord été faites sur l'aluminium A9 élaboré par voie conventionnelle (I/M) ou par métallurgie des poudres (P/M). Ensuite l'étude est poursuivie sur les alliages 7x49 (I/M) ou (P/M) très chargés en éléments d'addition. Les résultats montrent que la résistance à la fissuration en fatigue dépend de la nature de l'alliage et du degré d'oxydation: l'oxyde peut avoir un rôle bénéfique dans le cas de l'aluminium A9, il est systématiquement néfaste dans l'alliage à haute résistance 7x49.

MOTS CLES

Alliages d'aluminium, Métallurgie des poudres, Oxyde, Endommagement, Fatigue, Fissuration, Seuil de propagation.

INTRODUCTION

L'utilisation de la métallurgie des poudres (P/M) dans l'élaboration des alliages d'aluminium à haute résistance mécanique a permis d'obtenir un gain important par rapport à la métallurgie conventionnelle "I/M" tant en résistance à la corrosion sous tension (1) qu'en résistance à l'amorçage en fatigue (2). Cependant la résistance à la propagation en fatigue reste insuffisante dans le domaine des faibles vitesses, ceci étant classiquement attribué à la finesse de la sous structure des alliages (P/M).

Par ailleurs, les alliages d'aluminium (P/M) contiennent de l'oxyde résiduel apparu lors de l'atomisation de la poudre. Le rôle de celui-ci sur la résistance à la propagation n'a jamais été étudié auparavant. Ce travail a pour but d'apporter un éclairage sur cette question à travers l'étude des mécanismes de fissuration d'alliages d'aluminium (P/M) contenant différents taux d'oxyde, obtenus par oxydation volontaire de la poudre.
Pour une meilleure compréhension des mécanismes de fissuration en fatigue et compte tenu de la structure très complexe des alliages d'aluminium (P/M), nous avons jugé utile d'étudier au préalable les effets de l'oxyde dans un aluminium pur A9 (P/M) avant de faire l'étude sur un alliage à haute résistance 7x49.
Les alliages étudiés ont été élaborés au Centre de Recherches de Cégédure Péchiney à Voreppe.

MATERIAUX ETUDIES ET LEURS CARACTERISTIQUES METALLURGIQUES

Les matériaux étudiés sont des alliages d'aluminium A9 et 7x49 élaborés par

les deux procédés (I/M) et (P/M). Leur composition chimique et leur degré d'oxydation sont données par les tableaux 1, 2 et 3
L'aluminium pur A9 (I/M) ou (P/M) est étudié à l'état recuit. Les dimensions des grains ainsi que les conditions de traitements sont donnés par le tableau 5.

Les alliages 7x49 (P/M) ou (I/M) ont subi un traitement thermique T6 : mise en solution à 470°C pendant 4h + trempe à l'eau froide + maturation à l'ambiante pendant trois jours + revenu à 120°C pendant 24h suivi d'un refroidissement à l'ambiante. Leur structure est dans l'état restauré.

L'étude par microscopies optique et électronique à transmission ainsi que par diffraction des rayons X de la nature et de la distribution des oxydes montrent que :

- les produits A9 (P/M) contiennent deux populations d'oxyde (Al_2O_3-γ) comme dans le SAP classique : une fine distribution de particules durcissantes et des amas alignés dans la direction du filage. Les paramètres structuraux de cet oxyde sont donnés par le tableau 4.

- les alliages 7x49 (P/M) contiennent une seule population d'oxyde (film fragmenté de MgO) localisé aux joints des grains; ces derniers coïncidant à une déformation près avec les grains de poudre initiaux.

Dans le cas de l'A9 comme dans le cas des alliages 7x49, les oxydes sont identifiés par diffraction des rayons X, selon la méthode de Seeman-Bohlin décrite dans la référence (5).

Tableau 1. Composition chimique en (µg/g) dans l'aluminium pur A9.

Aluminium pur A9

Procédé	éléments (µg/g)					
	Fe	Si	Cu	Mg	Zn	Ti
I/M	5	16	4	4	3	3
P/M	310	550	60	15	3	3

Tableau 2. Composition chimique des alliages 7x49 (I/M) et (P/M).
(XXX) : en (µg/g)

Alliages 7x49

Procédé	éléments % en poids								
	Fe	Si	Cu	Mn	Ti	Cr	Zr	Mg	Zn
I/M	0.19 (1900)	0.06 (600)	0.9	0.27	0.03	0.21	0.13	3.3	9.3
P/M	0.16 (1600)	0.06 (600)	0.9	0.27	0.03	0.21	0.13	3.3	9.3

Tableau 3. Degrés d'oxydation des alliages étudiés.

Alliages	Aluminium A9				Alliages 7x49			
degré d'oxydation	F témoin	A-0.4 P/M	A-0.5 P/M	A-1. P/M	G témoin	B-0.4 P/M	B-0.6 P/M	B-2 P/M
$\%O_2$ en poids	-	0.37	0.49	1.1	0.8 (µg/g)	0.39	0.61	1.94

Paramètres structuraux		$1.1\%O_2$	$0.49\%O_2$	$0.37\%O_2$
Fines	λ (µm)	1.05	0.82	0.82
particules	ϕ (µm)	0.45	0.33	0.33
Amas	λ (µm)	16	16	16
d'oxyde	ϕ max (µm)	10	4	4

Tableau 4. Paramètres structuraux de l'oxyde dans A9 (P/M)
λ: espacement, $\emptyset$: dimension.

désignations	A9 (P/M)- T=450°C, t=15mn			A9 I/M
	A-0.4 %0.37 O_2	A-0.5 0.49% O_2	A-1 1.1% O_2	F T : 335°C t = 5mn
Morphologie		Aplatie		Equiaxe
Dimension des grains(mm)	L : 1 TL: 0.2 TC: 0.1	L : 0.4 TL: 0.1 TC: 0.04	L : 0.2 TL: 0.08 TC: 0.03	0.28

Tableau.5 . Dimension et morphologie des grains de A9(P/M) et (I/M)
T : Température de traitement
t : Durée de maintien

PROCEDURE EXPERIMENTALE

Propagation des fissures de fatigue

Les éprouvettes utilisées sont du type C.T. et sont prélevées de façon à étudier la résistance à la fissuration en sens L.T et T.L.

Tous les essais de fatigue ont été faits à l'ambiante à l'aide d'une machine servohydraulique asservie, à une fréquence de 10 à 30 HZ.

La propagation de la fissure est suivie en surface au moyen d'un microscope, et sa longueur est mesurée avec un micromètre donnant des valeurs à ± 0.01mm.

La sollicitation en traction est sinusoïdale avec un rapport de charge minimale sur charge maximale constant , R = 0.1.

Le facteur d'intensité de contrainte K est donné par l'expression classique applicable aux éprouvettes standards C.T (6).

$$K = \frac{P}{B\sqrt{W}} \quad f\left(\frac{a}{W}\right)$$

La détermination des vitesses intermédiaires ou faibles est faite en réduisant progressivement l'amplitude du facteur d'intensité de contrainte ΔK (par paliers successifs) pour éviter les effets de surcharge. Nous avons entrepris des réductions de moins de 10% de ΔK à chaque palier, et ce lorsque la fissure s'était accrue d'une longueur supérieure à 30 fois la dimension de la zone plastique monotone provoquée par la charge antérieure. Soit $\Delta a \geqslant \frac{30}{3\pi}\left(\frac{Kmax}{\sigma Y}\right)^2$ où σY est la limite d'élasticité conventionnelle et Kmax est le facteur d'intensité de contrainte maximal atteint au cours d'un cycle.

RESULTATS EXPERIMENTAUX

Caractéristiques mécaniques

Les caractéristiques de traction sont consignées dans les tableaux 6 et 7. Ces résultats montrent que :

- l'oxyde (Al_2O_3) renforce l'aluminium pur en provoquant une augmentation de la contrainte d'écoulement en traction, comme cela a été constaté dans le SAP (7).
- l'oxyde (MgO) présent dans les alliages 7x49 (P/M) provoque, par rapport à l'alliage de référence, une dégradation de la résistance mécanique pour l'alliage le plus chargé en oxyde et une dégradation de l'allongement à rupture pour les alliages moyennement oxydés. La dégradation de la résistance est dûe à l'appauvrissement de la matrice en Mg qui est alors oxydé, Mg étant un élément indispensable au durcissement structurale de ces alliages. Quant à la dégradation de l'allongement à rupture, elle est dûe au film d'oxyde fragmenté aux joints de grains qui provoque une rupture intergranulaire à cupules.

L'influence de l'oxydation du Mg sur la diminution de la résistance est confirmée également par l'évolution de la microdureté pendant le vieillissement, figure 1.

Tableau 6.
Caractéristiques mécaniques de A9(P/M) et (I/M)

Propriétés \ Alliages	A - 0.4 (P/M)	A - 0.5 (P/M)	A - 1 (P/M)	A 9 (I/M)
R 0.2 (M Pa)	79	81	90	18
Rm (M Pa)	106	110	130	75
A (%)	27	25	19	58

Tableau 7.
Caractéristiques mécaniques des 7X49(P/M) et (I/M) ainsi que le gain de (P/M) par rapport à (I/M).

Propriétés \ Alliages	B témoin I/M	B- 0.4 P/M	Gain %	B- 0.6 P/M	Gain %	B-2 P/M	Gain %
R0.2 (M Pa)	642	599	-6.7	645	+0.5	301	-55.1
Rm(M Pa)	658	625	-5.1	655	-0.5	383	-41.8
A(%)	5.9	1.9	-67.8	2.0	-66.1	4.3	-27.1

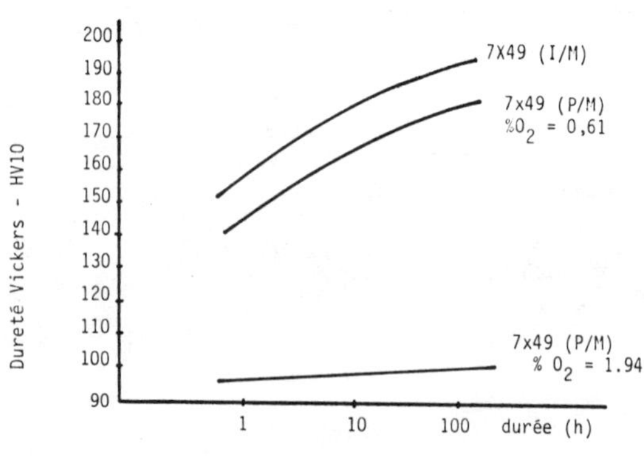

Figure 1
Dureté VICKERS pendant la maturation des alliages indiqués.

Propagation des fissures de fatigue

Les résultats expérimentaux de la résistance à la fissuration en fatigue montrent que la présence de l'oxyde a des conséquences favorables ou défavorables selon la nature de l'alliage et du degré d'oxydation préalable de la poudre.

Cas de l'aluminium A9, figure 2

L'effet positif de l'oxyde se manifeste uniquement dans le domaine des faibles vitesses. Nous attribuons ceci au durcissement provoqué par les fines particules d'oxyde, qui provoquent une diminution de le dimension de la zone plastique en fond de la fissure de fatigue. En conséquence, la vitesse de propagation dans A9 (P/M) se trouve diminuée, à ΔK égal, par rapport à A9 (I/M). Les amas d'oxyde qui sont non durcissants parce que trop gros, jouent au contraire un rôle défavorable. Leur effet endommageant apparait très nettement dans le cas de l'A9-1%O_2 sollicité dans le sens T.L, car ces amas sont alors grossiers et alignés dans la direction de propagation.

Pour les vitesses intermédiaires ou élevées ($\frac{da}{dN} > 10^{-5}$ mm/cycle), il y a compensation entre les effets durcissants des fines particules et endommageants des amas d'oxyde. Ceci n'a pas lieu dans le domaine des faibles vitesses en sens T.L. comme en sens L.T pour A9(P/M)-0.4/0.5%O_2 ou en sens L.T pour A9(P/M)1%O_2 car la dimension de la zone plastique cyclique devient alors du même ordre de grandeur que l'espacement entre les amas ou inférieure, fig.3.

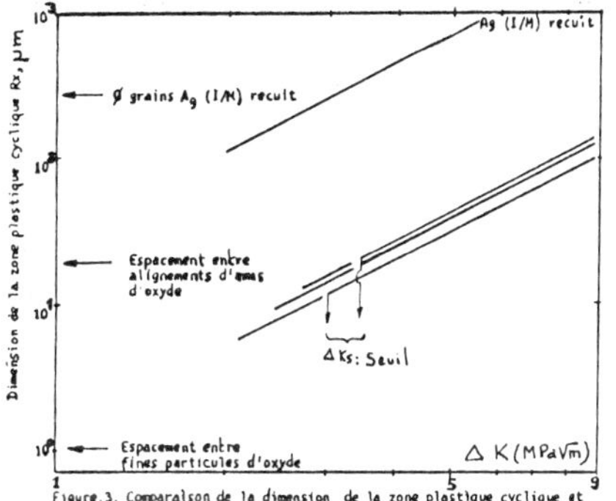

Figure.3. Comparaison de la dimension de la zone plastique cyclique et des paramètres structuraux des produits A9 (I/M) et (P/M).

Matériaux	A9 (I/M)	A9 (P/M)		
	σ_Y = 18MPa	0.4% O_2 σ_Y= 79MPa	0.5% O_2 σ_Y= 81 MPa	1.1% O_2 σ_Y= 90 MPa
Sens L.T K_s MPa$\sqrt{m}$	2	3.3	3.3	3.0
Gain	0	65	65	60
Sens T.L K_s MPa$\sqrt{m}$	2	3.3	3.3	1.8
Gain %	0	65	65	-10

Tableau. 8. Seuil de propagation de A9 (I/M) et de A9 (P/M). σ_Y limite d'élasticité conventionnelle à 0.2%.

Le seuil de propagation qui est défini pour une vitesse de l'ordre de 10^{-7} mm par cycle est amélioré par la présence des fines particules. Il est logique de corréler cet effet au rôle bénéfique de celles-ci sur la limite d'élasticité conventionnelle en remarquant que l'augmentation de cette dernière s'accompagne d'une amélioration du seuil (tableau 8), comme c'est le cas du cuivre et de ses alliages étudiés par HIGO, PICKARD et KNOTT (8).

Cas des alliages 7X49-T6, fig.4

Les alliages 7X49(P/M) moyennement oxydés sont très peu ductiles (tableau 6) et de ce fait, il est très difficile de faire des mesures de vitesses de propagation. L'alliage le plus chargé en oxyde présente une résistance à la fissuration dégradée par rapport à celle de l'alliage de référence. Ceci est dû à la présence de l'oxyde aux joints des grains et à la dégradation de la résistance de la matrice, liée à l'appauvrissement en Mg.
Les mesures faites sur l'alliage de référence (I/M), à l'état T6 non détensionné, sont incertaines, à cause de la courbure très prononcée du front de propagation freiné en surface par les contraintes résiduelles de compression.

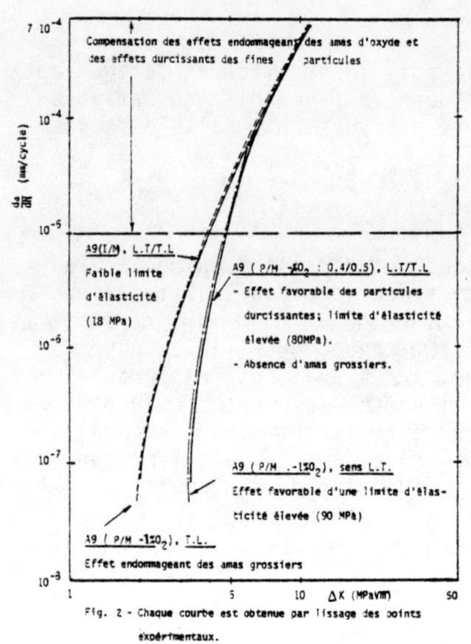

Fig. 2 - Chaque courbe est obtenue par lissage des points expérimentaux.

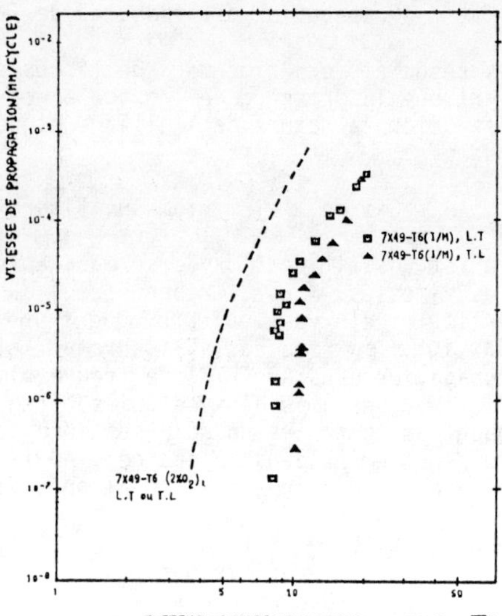

Figure.4. Comparaison de la résistance à la propagation des fissures de fatigue dans les alliages 7X49-T6(P/M) et (I/M).

CONCLUSION

Les résultats de cette étude permettent de conclure que :
Une population des fines particules d'oxyde peut jouer un rôle favorable sur la résistance à la fissuration si la matrice est susceptible d'un durcissement par une phase rigide dispersée, et défavorable pour les alliages à durcissement structural ne pouvant pas bénéficier d'un durcissement supplémentaire lié à la présence de cet oxyde.
Les amas grossiers d'oxyde ne sont pas durcissants, mais endommageants et, de ce fait, il sont défavorables à la résistance à la fissuration.
Les effets favorables des particules durcissantes (si le degré d'oxydation n'est pas élevé, ici $\%O_2 \leqslant 0.5$) se manifestent dans le domaine des faibles amplitudes de ΔK et au seuil de propagation. Pour les amplitudes de K plus élevées, la présence d'une double population d'oxyde fait qu'il y a compensation entre les effets favorables des particules durcissantes et endommagements des amas d'oxyde, la dimension de la zone plastique cyclique étant supérieure à l'espacement entre amas.

BIBLIOGRAPHIE

1 J.P. LYLE and W.S. CEBALAK, Met. Trans. A. vol.6, (1975), 685-699

2 J.P. LYLE, Jr and W.S. CEBULAK, Metals. Eng. Quarterly, vol.14, n°1, (1974)52

3 D.P. VOSS, European Office of Aerospace Research and Developpment, Report : EOARD-TR-80-1, (1979), London, England).

4 R.E. SANDERS, Jr. and W.L. OTTO, Jr. Profress in Powder Metallurgy vol. 34 (Proc. Conf.) Los Angeles, Calif. (1978), 294-311

5 A. GUINIER, Théorie et technique de la radiocristallographie, Ed. Dunod (1964-Paris).

6 D. FRANCOIS et L. JOLY, La rupture des Métaux, p.205 Masson et Cie 1970-Paris

7 P. GUYOT, Thèse d'Etat, Série A, N° 119, (Orsay-1965)

8 Y. HIGO, A.C. PICKARD and J.F. KNOTT Met. Sci., (1981), 233-240

Effect of Dispersoids on Fatigue Crack Propagation in Al-Zn-Mg Alloys

M. Harrison and J. W. Martin

Oxford University Department of Metallurgy and Science of Materials, Parks Road, Oxford, UK

ABSTRACT

The effect of Mn-bearing and of Zr-bearing dispersoids upon fatigue crack propagation in a peak-aged Al - 5.5wt% Zn - 1.7wt% Mg alloy has been studied. The alloys had similar grain-sizes, and comparable yield stresses.

The da/dN versus stress intensity range curves were measured. The effect of the Mn-dispersoids was to increase crack propagation rates, and to lower the threshold stress intensity range, as well as increasing the proportion of intergranular fracture. This could arise from cavitation at grain boundary dispersoid particles within the crack-tip plastic zone.

In contrast, the alloy with the higher volume fraction of Zr-dispersoid showed slower crack propagation rates and a higher threshold, with little intergranular fracture. This suggests that these dispersoids cause a homogenization of slip, thus reducing grain boundary stress concentrations.

The alloy with a low volume fraction of Zr dispersoids(which were also of much smaller particle size) showed the maximum intergranular fracture and maximum propagation rates. This indicates that these dispersoids are sheared, leading to increased planarity of slip.

KEYWORDS

Fatigue crack propagation; Aluminium alloys; Dispersoids

INTRODUCTION

Commercial heat-treatable aluminium alloys contain dispersoids, which are fine particles of intermetallic phases arising from the addition of a transition element such as Cr, Mn, Zr or V. They control grain growth during heat-treatment, and under conditions of monotonic straining they have been shown to promote homogenization of slip (1) which, together with the refined grain size, increases the tensile ductility and also the fracture toughness (2,3). Edwards and Martin have also shown that the addition of Mn-bearing dispersoids to peak-aged Al-Mg-Si alloys can result in improved properties with regard to the initiation and propagation of fatigue cracks under high cycle (4) and low cycle (5) conditions.

Little systematic work has been carried out in this area on the medium-strength Al-Zn-Mg alloys, and the object of the present work has been to compare the effects upon fatigue crack propagation of Mn-bearing and Zr-bearing dispersoids. The former consist of a uniform dispersion of non-coherent particles of intermetallic phase; the latter are distributed as clusters of coherent particles, and their slip-homogenizing effects are not well characterized. A ternary (dispersoid-free) alloy was studied for comparative purposes.

This information is needed for the defect-tolerant approach to design against fatigue failure, and this approach should give further insight into the role of microstructure in this context.

EXPERIMENTAL

The alloys were prepared from high purity (low Fe) material in order to minimize the occurrence of coarse residual particles, whose effects might swamp those of the dispersoids themselves. The composition of the alloys is given in Table 1:

Table 1: Composition of the alloys.

Alloy wt%	Cu	Fe	Mg	Mn	Si	Ti	Zn	Zr
CT	0.01	0.004	1.71	0.001	0.004	0.001	5.43	-
CMH	0.001	0.007	1.66	0.42	0.08	0.001	5.47	-
CZL	0.001	0.007	1.67	0.003	0.006	0.001	5.37	0.08
CZH	0.001	0.01	1.64	0.004	0.01	0.001	5.37	0.15

All the alloys were subjected to thermo-mechanical treatments such that the grain structures were equiaxed and recrystallized. They were given a common solution-treatment at 475°C, followed by water quenching and 24 hr ageing at room temperature. The alloys were then artificially aged at 130°C for 60 hr, which corresponded to peak hardness. The yield stresses and grain sizes of the alloys are given in Table 2.

Table 2: Grain sizes and Yield Stresses

Alloy	Grain Size (microns)	Yield Stress (MPa)
CT	105	411
CMH	101	405
CZL	105	378
CZH	80	355

Each alloy had similar dispersions of the ageing precipitate, both within the grains and at grain boundaries. The distribution of dispersoid particles in alloy CMH and CZH are shown in figs.1 and 2. The particle-size of the dispersoids is approximately 0.1 microns in each alloy; in CZL the particles were much finer, being of the order 0.01 microns

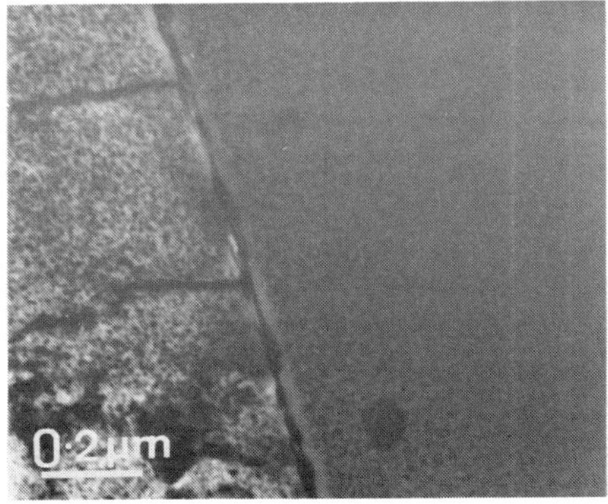

Fig.1. TEM of alloy CMH

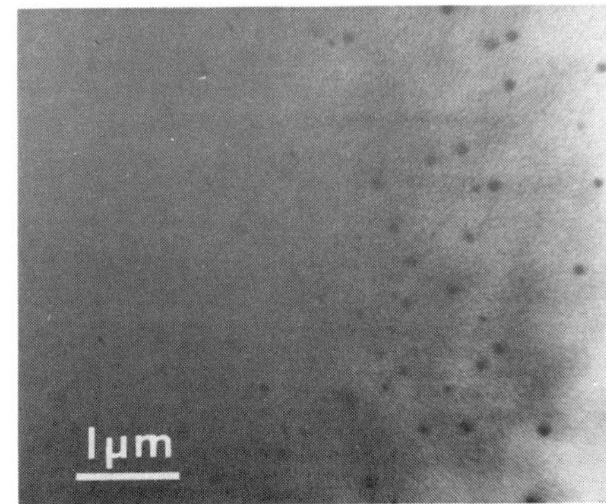

Fig.2. TEM of alloy CZH

Fatigue crack propagation tests were conducted on a servohydraulic machine operating in closed loop configuration under load control. Tests were performed on CT specimens of 6 mm thickness in laboratory air at constant stress intensity. Load-shedding was controlled by a microcomputer interfaced to the fatigue machine. Crack length was monitored by a DC potential drop technique.

Log. crack growth rate versus stress intensity range curves were plotted for the alloys, and the effects of the dispersoids upon the fatigue thresholds established.

EXPERIMENTAL RESULTS

Fig.3 shows the crack propagation behaviour of the four alloys, and it is seen that, in comparison with the dispersoid-free alloy CT, the manganese-dispersoid alloy CMH shows an increased crack propagation rate over the entire stress intensity range studied. It also shows a lower threshold stress intensity range than the ternary alloy.

It is recognized that crack closure may cause the effective stress intensity to be less than that recorded in fig.3. Closure was in fact monitored by means of a back-face strain gauge on the test-piece, and corrected curves plotted. It was found, however, that the relative positions of the curves for alloys CT and CMH were unchanged, and it is concluded that the differences recorded are a real effect.

With regard to the zirconium-containing alloys, it is apparent that a high volume fraction of this dispersoid decreases the crack propagation rate, and raises the threshold. The opposite appears to be true of the low-Zr alloy.

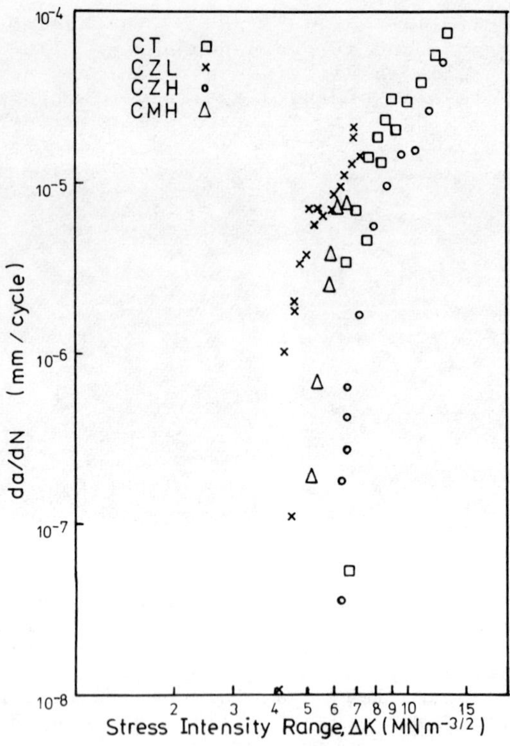

Fig.3. da/dN vs Stress Intensity Range data.

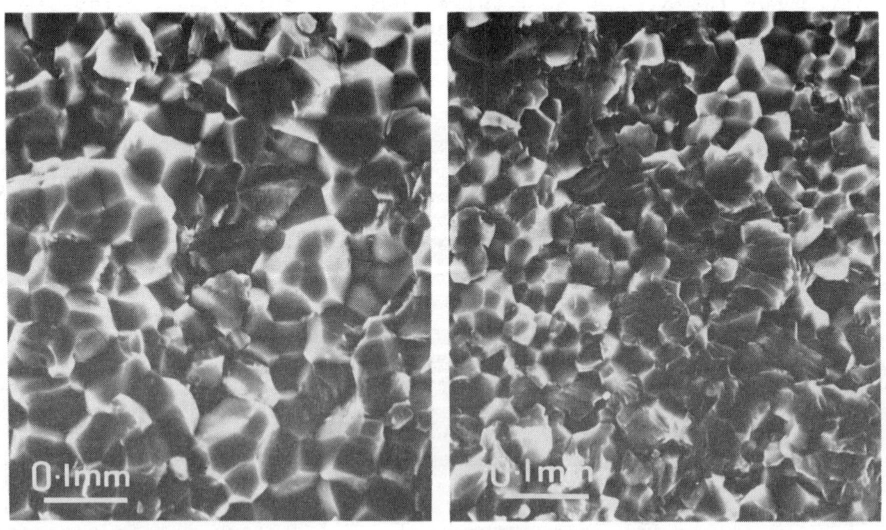

Fig.4. SEM of alloy CT

Fig.5. SEM of alloy CMH

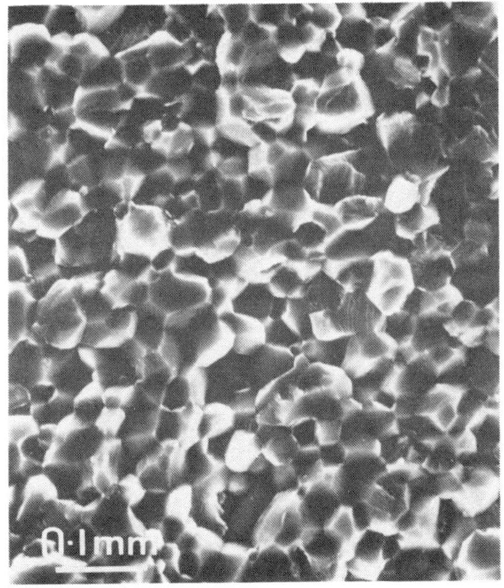

Fig.6. SEM of alloy CZL

Fig.7. SEM of alloy CZH

The fracture surfaces of all the alloys were examined in the SEM, and the results are shown in figs.4 -7. These all correspond to regions where the crack was propagating at stress intensity ranges close to the threshold value. The ternary alloy, CT, appears to have fractured in a predominantly intergranular manner. The proportion of intergranular failure appears to increase, however, on passing from CT to CMH and then to CZL, and this trend is accompanied by accelerating propagation rates. Alloy CZH, however, appears to exhibit mainly transgranular fracture.

DISCUSSION AND CONCLUSIONS

The present observations show interesting differences from those reported by Edwards (6) on the effect of Cr-bearing dispersoids on fatigue crack propagation in a high-strength Al-Zn-Mg-Cu alloy. The latter alloy showed little tendency for intergranular failure in fatigue, and at low growth rates the dispersoids lowered the stress intensity range threshold. It was suggested that the dispersoids tended to homogenise slip, since materials which exhibit inhomogeneous slip will show greater slip reversibility in the crack tip plastic zone, and thus have slower fatigue crack growth rates (7).

This process cannot be dominating the behaviour of alloys CMH and CZL, as no decrease in the proporation of intergranular fracture is observed; in fact the opposite trend is found. An alternative mechanism would be the cracking or decohesion of dispersoid particles ahead of the crack tip: this would accelerate crack propagation by the interlinking of those voids which happen to lie on grain boundaries.

The above process is probable in CMH, since the dispersoid phase is incoherent and little deformed by the plastic deformation ahead of the crack. In CZL, however, the fine particles of dispersoid are semi-coherent, and are able to undergo plastic deformation (8). This particle-shearing process is likely to increase the inhomogeneity of slip, thus increasing grain boundary stress concentrations within the crack-tip plastic zone. Such a process would account for the increased proportion of intergranular failure observed in CZL.

CZH, on the other hand, exhibits negligible intergranular fracture, and this implies that the increased particle size of the dispersoid in this alloy has prevented particle-shearing. Rather than cracking or decohering on deformation, the resulting homogenization of slip appears to suppress intergranular fracture (5).

Conclusions.

1. Incoherent Mn-bearing dispersoids accelerate crack propagation, and it is considered that this arises by the fracture or decohesion of the particle/matrix interface in the plastic zone ahead of the crack tip.

2. The semi-coherent Zr-bearing dispersoids appear to be able to act in two ways, depending on their size.

3. When finely dispersed (with particle sizes of the order 0.01 microns) the Zr-dispersoids are sheared by dislocations in the crack tip plastic zone. This further homogenizes slip and promotes intergranular fracture

4. When the dispersoids are coarser (0.1 microns diameter) they are evidently not sheared, but they homogenise the slip and suppress intergranular fracture.

ACKNOWLEDGEMENTS

The authors are grateful to Professor Sir Peter Hirsch, FRS, for the laboratory facltitities made available, and to the Science and Engineering Research Council and Messrs Alcan Ltd for financial and material support.

REFERENCES

1. J.M. Dowling and J.W. Martin, Acta Met. 24, 1147 (1976)
2. K.C. Prince and J.W. Martin, Acta Met. 27, 1401 (1979)
3. J.A. Blind and J.W. Martin, Mat.Sci.Eng 57, 49 (1983)
4. L. Edwards and J.W. Martin, Metal Science 17, 511 (1983)
5. L. Edwards and J.W. Martin, ICSMA 6, 873 (1982)
6. L. Edwards. Proc.4th Riso Int. Symp. on Metallurgy and Materials Science (1983)
7. E. Hornbogen and K-H. Zu Gahr, Acta Met. 24, 581 (1976)
8. M. Zedalis, L. Filler a M.E. Fine, Scripta Met.16,471 (1982)

Energy Required for Fatigue Crack Propagation

N. Ranganathan, J. Petit and J. de Fouquet

Laboratoire de Mécanique et de Physique des Matériaux, ENSMA, Rue Guillaume VII, 86034 Poitiers Cedex, France

ABSTRACT

The hysteric energy dissipated during fatigue crack propagation has been measured over a wide range of ΔK, and in three different environments on the high strength aluminum alloy, 2024 T351, using a compliance gauge. The results obtained show that for ΔK values greater than a critical level, ΔK_{cr} which depends on the R value and the environment, the specific energy U assumes a constant value of 2.6×10^5 J/m^2. For ΔK values lower than ΔK_{cr}, U increases as ΔK decreases. It is shown that this phenomenon can be explained by the fact that crack advances during each cycle for $\Delta K > \Delta K_{cr}$ and it advances after every N_f cycles for $\Delta K < \Delta K_{cr}$. A phenomenological relationship between U and N_f is proposed.

INTRODUCTION

Weertman /1/, modelising the crack tip plastic zone as a continuous array of dislocations and with the assumption that an element in this zone breaks when the cumulative displacement reaches a critical value, proposed the following relation for the crack growth rate :

$$da/dN = A \, \Delta K^4 / \mu \sigma_C^2 \, U \qquad (1)$$

where A is a constant, U the specific energy required to creat a unit surface, μ the shear modulus and σ_C the cyclic yield strength. Since then, a number of experimental techniques have been developed to measure the parameter U which is of significant importance /2,3,4/ and a compilation of most of the results is found in /2/.

In this paper a new technique is used to measure the hysteric energy dissipated during fatigue crack propagation in the 2024-T351 aluminum alloy in air, vacuum and high purity N_2 environments.
On the basis of the results obtained certain aspects of near threshold behaviour are discussed.

EXPERIMENTAL DETAILS

The results presented in this study pertain to the following tests :

1) Threshold tests at an R ratio of 0.5 in air, vacuum and high purity N_2 (3ppm H_2O + 1ppm O_2).

2) Threshold and constant ΔP tests at an R value of 0.1 in air and vacuum.

All tests were carried out using an Instron servo-hydraulic machine equipped with an environmental chamber /5/.
Test frequencies were 35 Hz during threshold tests and 5 Hz at high da/dN values during constant amplitude tests.
The specimen used is of the modified compact tension type of 10 mm thickness, with a C.O.D. gague mounted directly under the loading axis.

The crack growth was followed using an optical technique and at selected values of crack growth rates, the load displacement diagram, highly amplified to accentuate any hysterises, were obtained at a frequency of 0.02 Hz /6/. The area of the hysterises loops were measured using numerical integration techniques which gave directly the hysteric energy Q expressed in Joules/cycle. The specific energy U in Joules/m^2 is defined as :

$$U = Q/B\ 2\ da/dN \quad (2)$$

where da/dN is the crack growth rate and B the specimen thickness.

EXPERIMENTAL RESULTS

In Fig.1 the relationship between da/dN and ΔK for load ratios of 0.5 and 0.1 and for the three environments considered is presented. The curves presented here are similar to the ones obtained in several other aluminum alloys /7/ showing enhanced environmental effect at an R value of 0.5.

The crack growth behaviour in N_2 brings forth the typical transition behaviour which takes place at a near constant da/dN value, named as $(da/dN)_T$ of about 5.10^{-6} mm/cycle.

For $da/dN < (da/dN)_T$ the crack growth behaviour in N_2 is similar to that obtained in air and for $da/dN > (da/dN)_T$ the crack growth curve follows closely the one obtained in vacuum.This behaviour, which has been observed on other aluminum alloys as well has been attributed to the effect of water vapour adsorption and subsequent H_2 embrittlement at low crack growth rates where the crack remains stationary for several cycles before advancing /7,8/.

The relationship between the specific energy U and ΔK is presented in the neat two figures, treating separately the results obtained at R = 0.5 (Fig.2) and R = 0.1 (Fig.3).

At an R value of 0.5 for $\Delta K > 7$ MPa$\sqrt{m}$, value of U is essentially constant at about $2.5\ 10^5$ J/m^2. This value is called U_{cr} and ΔK value for which it is reached is called ΔK_{cr}. For $\Delta K < 7$ MPa$\sqrt{m}$, in air U increases as ΔK decreases and reaches a second plateau of about $2\ 10^6$ J/m^2 in the range $3 < \Delta K < 5.5$ MPa$\sqrt{m}$. For lower ΔK values near thershold U increases drastically and reaches a values as high as $5\ 10^8$ J/m^2.
In vacuum U increases gradually as ΔK decreases and at comparable ΔK values U is always higher than in air (about 10 times).
In N_2 in the range $8 < \Delta K < 12$ MPa$\sqrt{m}$, the evolution of U with respect to ΔK is similar to that obtained in vacuum. For $\Delta K < 8$ MPa$\sqrt{m}$ a plateau behaviour similar to the one observed in air obtained at a U value ofabout $3\ 10^6$ J/m^2 in the range $4 < \Delta K < 7.5$ MPa$\sqrt{m}$. For tests conductedat R = 0.1, it can be been that for $\Delta K > 12$ MPa$\sqrt{m}$ in air and $\Delta K > 19$ MPa$\sqrt{m}$ in vacuum the value of U is essentially constant at about $2.5\ 10^5$ J/m^2. For lower ΔK values

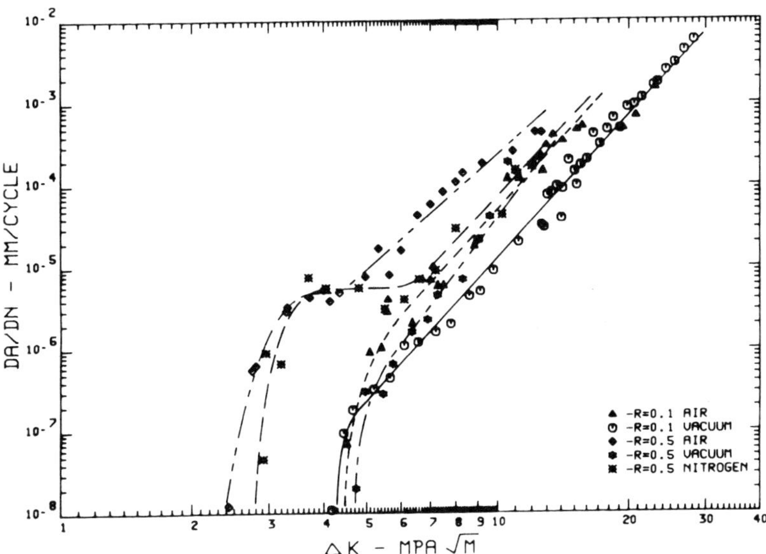

FIG.1 – Crack propagation curves s at R = 0.1 and R = 0.5 in Air, Vacuum and Nitrogen.

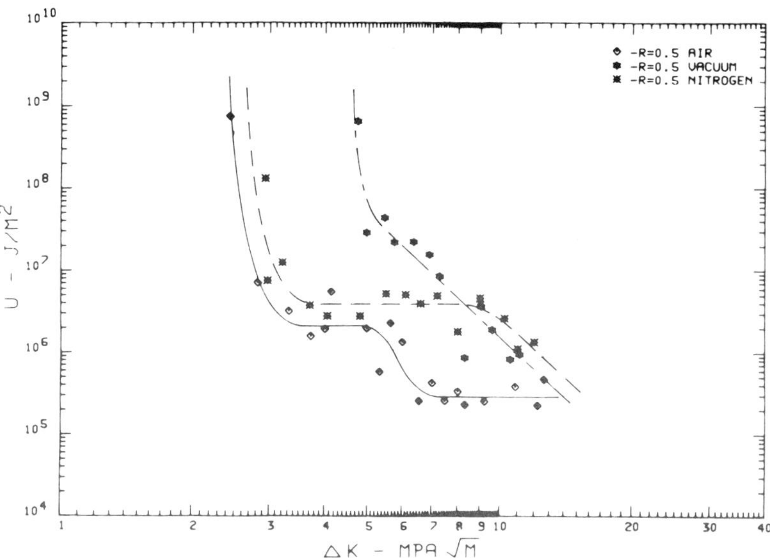

FIG.2 – Relationship between the specific energy U and ΔK at R = 0.5 in the three environments.

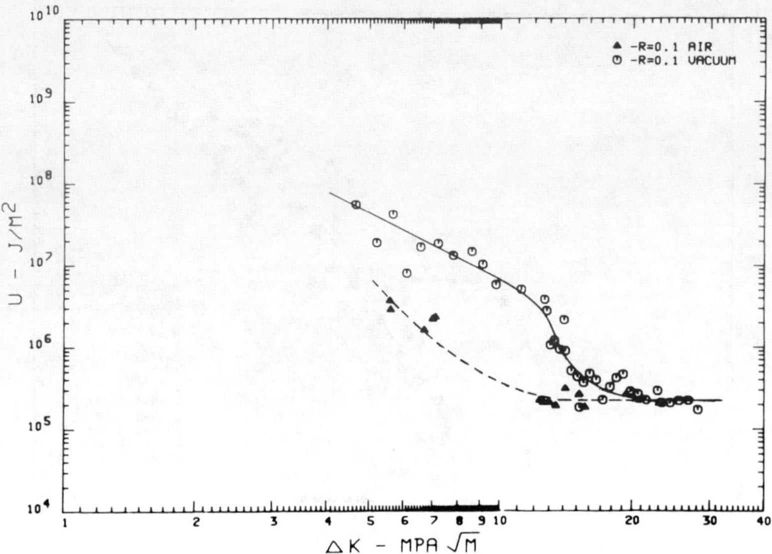

FIG. 3 – U – ΔK relationship at R = 0.1 in Air and Vacuum.

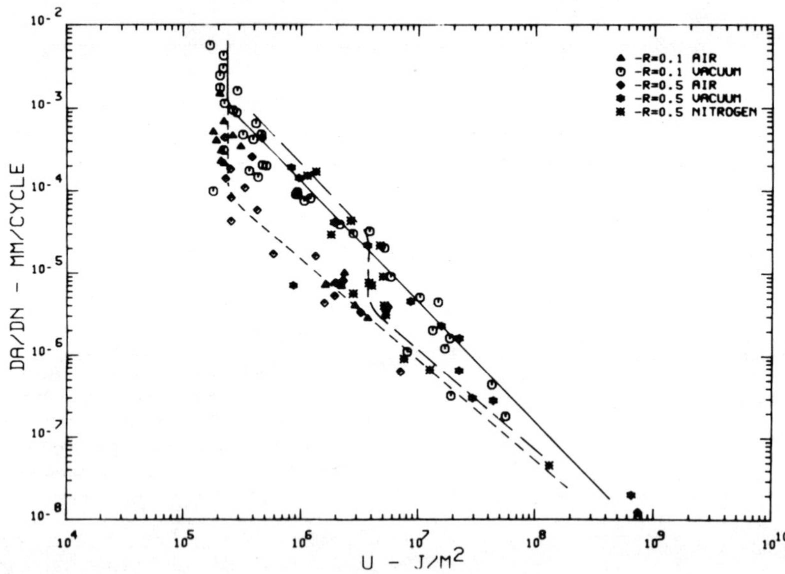

FIG.4 – Relationship between the crack growth rate and the specific energy for all the studied experimental conditions.

U increases as ΔK decreases in vacuum, while in air sufficient data points are not available.
Thus it is assumed that U_{cr} value previously defined is constant and independant of environment with an average value 2.6 10^5 J/m^2 and a standard deviation of 0.6 10^5 J/m^2. At the sametime ΔK_{cr} is dependant on environment and the R value.
Liaw et al. /2/ have measured U using the strain gague technique for the 2024 alloy in the T4 condition and they have reported values of 2.6 10^5 and 3.2 10^5 J/m^2 at ΔK values of 7.8 and 15.5 MPa $\sqrt{m}$ respectively in dry Argon atmosphere at an R value of 0.05. Eventhough the heat treatment and the atmosphere used are not exactly the same it is interesting to note that their values are within the scatter band obtained for U_{cr} in this study.
The relationship between da/dN and U is presented in Fig.4 for all the experimental condition studied. It can be seen that in vacuum U_{cr} value is reached for da/dN 8 10^{-4} mm/cycle. This value is called $|da/dN|_{cr}$. For da/dN less than this value the relationship between these two parameters is linear in the logarithmic scale (independant of R) suggesting a power law.

A similar result is obtained air while $|da/dN|_{cr}$ in air is much lower at about 10^{-4} mm/cycle.

In N_2 the initial data points fall on the curve obtained in vacuum and after the transition the data points tend to follow the curve obtained in air. Near the threshold all the data points tend towards the same value showing an asymptotic bahaviour.

DISCUSSION

a) U-ΔK relationships

A general observation made in this study is that the specific energy increases as near threshold condition are reached, however the relationship depends strongly upon the environment and the R value.

Davidson /9/ has proposed a qualitative model to explain the effect of environment on the specific energy. He predicts that the specific energy should reach a constant value at high ΔK values corresponding to fully developed stage II crack propagation. This result is confirmed in this study as for $\Delta K > \Delta K_{cr}$ U is independant of the environment and the R value and reaches U_{cr}. He has also suggested an agressive environment and a higher R value should lower the value of ΔK for which this limiting behaviour is reached. This phenomenon has been quantified in terms of the evolution of ΔK_{cr} values. The same author suggests an initial decrease of U as ΔK decreases which has not been observed in this study. It can be argued that results obtained in this study could be influenced by crack closure effects. However tests conducted at R = 0.5, where closure effects are minimal did not show any initial decrease.
Near the threshold U increases drastically probably due to the predominance of mode II crack opening as suggested by Davidson.

b) U-da/dN relationship

A critical fractographic study of the specimens used in this study and to be presented in another paper /10/ has shown that striation or striation like markings or coarse slip steps are observed from near threshold conditions upto a crack growth rate about 10^{-3} mm/cycle and is defined here as a microscopic step.

This step size coincides with the macroscopic growth rate for $da/dN > 2.10^{-4}$ mm/cycle in air and for $da/dN > 6\ 10^{-4}$ mm/cycle in vacuum. In this range it is assumed that crack advances during each cycle by one step. For da/dN value lower than the above mentioned values, the number of cycle N_f required to for the crack to advance by one step is obtained by :

$$N_F = p/da/dN \qquad (3)$$

In vacuum the value of p is constant at about $6\ 10^{-4}$ mm for $da/dN < |da/dN|_{cr}$ and in air the value of this parameter is on the order of $2\ 10^{-4}$ mm for $da/dN < |da/dN|_{cr}$ initially, and it increases near threshold.
Typically, N_F thus defined increases as threshold conditions are reached and attains a value higher than 20000 cycles for the last experimental point.
By using regression techniques the following relation is proposed between U and N_F for $U < U_{cr}$ for the results obtained in vacuum.

$$U = U_{cr}(N_F)^{0.69} \qquad (4)$$

In air as can be seen from Fig.4 a similar relationship can be assumed when p is constant and where H_2 embrittlement does not occurs.
This relation is similar to the one proposed by Halford et al. relating fatigue life to hysteric work in low cycle fatigue /11/.

CONCLUSIONS

1) The specific energy U associated with fatigue crack propagation has been measured in three different environments and over a large ΔK range for the 2024-T351 Aluminum alloy.

2) For ΔK greater than a critical value, ΔK_{cr} depending on R and environment U is equal to $2.6 \pm 0.6\ 10^5$ J/m^2.

3) For $\Delta K < \Delta K_{cr}$ U increases as ΔK decreases and a strong environmental effect is detected.

4) A phenomenological relation between U and the number of cycles N_F required for the crack to advance by a microscopic step is derived for $\Delta K < \Delta K_{cr}$ in vacuum.

REFERENCES

1. J.Weertman, Int. J. Fract., Mech. 1973, vol.9, p.125.
2. P.K. Liaw,S.I. Kwun and M.E. Fine Met. Trans. 1981, Vol.12A, p.49.
3. P.K. Liaw, M.E. Fine and D.L. Davidson. Fat. of Egng. Mat. and St. 1980, Vol.3, p.59.
4. T.S. Gross and J. Weertman, Met. Trans.,1982, Vol.13A, p.2165.
5. J.Petit et al. Rev. de phys.Appl. 1980, Vol.15, p.919.
6. A. Ohta and E. Sasaki, Int. J. of Fract., Vol. 11, 1975, p.1049.
7. J. Petit, TMS-AIME, Symposium, Philadelphia, oct. 1983.
8. J. Petit and A. Zeghloul, "Fatigue Thresholds" Proceedings of 1st . ISFT, Stockholm, J. Bäcklund et al. Eds. EMAS Ltd, U.K., 1981, p.563.
9. D.L. Davidson, Fat. Engng. Matls. and St., 1981, Vol.3, p. 229.
10. N. Ranganathan et al. (to be published).
11. G.R. Halford, J. of Matls, 1966, Vol.1, p. 3.

Influence of Microstructures and Residual Stress on Fatigue Crack Propagation Rate of High Carbon Steels

X. Z. Feng*, B. X. Liu*, W. J. Gao and S. S. Gao****

**Department of Metals and Technology, Harbin Institute of Technology, People's Republic of China*
***Xinlong Industrial Company, Harbin, People's Republic of China*

ABSTRACT

The cold-work tool steels T10A and GCr15 have been studied with regard to the influence of austenitization temperature, tempering temperature and the residual stress on the crack propagation rate and the critical crack length. Different hardening and tempering procedures have been examined. The crack propagation rate is dependent strongly on the austenitization temperature, the tempering temperature and the surface compressional residual stress in specimens.

KEYWORDS

Fatigue crack propagation rate; critical crack length; compressional internal stress; high carbon steels; heat treatment.

INTRODUCTION

Most common tool materials used in cold stamping and cold deforming are the high carbon steels. When heat treating cold-work tools with large dimensions, it is virtually impossible to form martensitic transformation products in central portion. The purpose of the investigation now reported has been to study the combined influence of the microstructures, hardness and the residual stress on the crack propagation rate in high carbon steels.

EXPERIMENTAL PROCEDURES

Specimen Preparation

The chemical composition of the steels used in this study is listed in Table 1.

TABLE 1 Composition of GCr15 and T10A (wt %)

Steel	C	Si	Mn	Cr	S	P
GCr15	0.99	0.20	0.33	1.49	0.02	0.027
T10A	1.00	0.20	0.30	-	0.02	0.030

Specimen size and geometries are shown in Figs. 1 and 2.

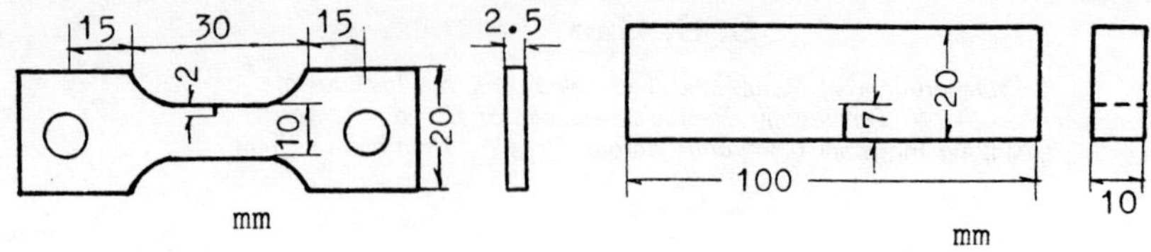

Fig.1 Tensile fatigue specimen

Fig.2 Bending fatigue specimen

Specimens were austenitized in molten salt bath according to the schedule in Table 2. The specimens were then single edge notched by spark machining using a 0.12mm diam molybdenum wire. The faces of the specimens were mechanically polished to facilitate observation of the crack.

Test Procedure

Fatigue testing was conducted at room temperature in air. A sinusoidal load was applied at a frequency of 7Hz. For any test, the maximum nominal stress was less than 10 per cent of the yield stress of the steels. The R ratio was equal to 0.22 for tensile fatigue test and 0.44 for bending fatigue test throughout this study. The growing fatigue crack was observed with a 30 magn. micrometer telemicroscope. Measurments of the crack length were made approximately every 5000 cycles.
The stress intensity factor for any crack length was calculated from following relationship (1):

For tensile fatigue test $\Delta K = \Delta\sigma . (a.\pi)^{\frac{1}{2}} . f(a/w)$ (1)

For bending fatigue test $\Delta K = \Delta P . f(a/w)/(B.W^{\frac{1}{2}})$ (2)

EXPERIMENTAL RESULTS

Fatigue Crack Propagation Cycles N_p and Critical Length a_c

Increasing the tempering temperature leads to a substantial increase of N_p and a_c, Table 3. Increasing the austenitization temperature leads to a decrease of N_p but there was no significant effect on the a_c.

TABLE 2 Heat Treatment Schedule of Steels

Symbol	Steel	Hardening (°C)	Grain Size (No)	Tempering (°C)	HRC Surf.	HRC Cent.	Microstructure* Surf.	Microstructure* Cent.
A	GCr15	840, Oil Quenched	9	150	63	63	TM	TM
B				300	60	60	TM	TM
C				470	47	47	TT	TT
D	GCr15	900, Oil Quenched	8	150	63	63	TM	TM
E				300	60	60	TM	TM
F				470	48	48	TT	TT
G	GCr15	1000, Oil Quenched	6	150	63	63	TM	TM
H				300	60	60	TM	TM
I				470	48	48	TT	TT
J	T10A	780, Through Water to Oil Quenched	8	250	60	40	TM	T
K				350	55	40	TM	T
L				400	50	40	TT	T
M				450	45	36	TT	S
N				500	40	31	TT	S
O				550	32	30	TS	S

*TM tempered martensite, TT tempered troosite, TS tempered sorbite, T troosite, S sorbite.

Fatigue Crack Propagation Rate da/dN

The da/dN increases continuously with the decrease of tempering temperature for the through hardened specimens, Fig.3, while the da/dN increases and then decreases significantly for non-through hardened, Fig.4. Increasing the austenitization temperature leads to a substantial increase of da/dN.

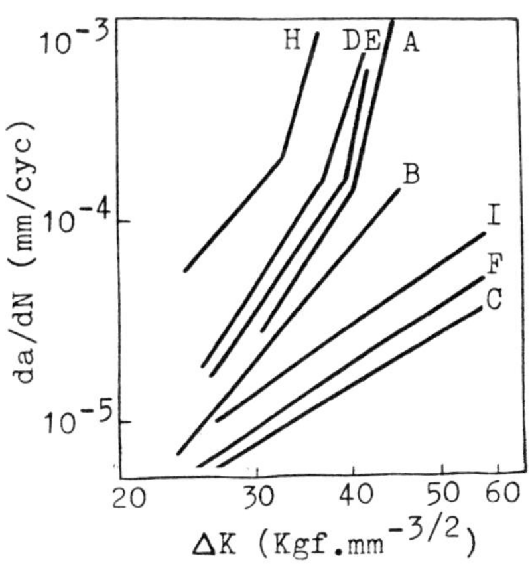

Fig.3 Fatigue crack propagation rate of the through hardened specimens of GCr15.

TABle 3 N_p, a_c and da/dN

Sym-bol	N_p (Cycle)	a_c (mm)	da/dN*
A	30000	3.5	1.00 10^{-3}
B	65000	3.9	2.00 10^{-4}
C	165000	6.2	2.50 10^{-5}
D	25000	3.3	2.00 10^{-3}
E	40000	3.8	8.50 10^{-4}
F	120000	5.9	4.00 10^{-5}
G	10000	3.2	—
H	25000	3.7	5.00 10^{-3}
I	100000	6.0	7.00 10^{-5}
J	289000	10.2	2.00 10^{-6}
K	211000	10.3	4.00 10^{-6}
L	171000	11.0	1.30 10^{-5}
M	179000	11.6	1.25 10^{-5}
N	190000	12.6	1.20 10^{-5}
O	232000	14.9	1.10 10^{-5}

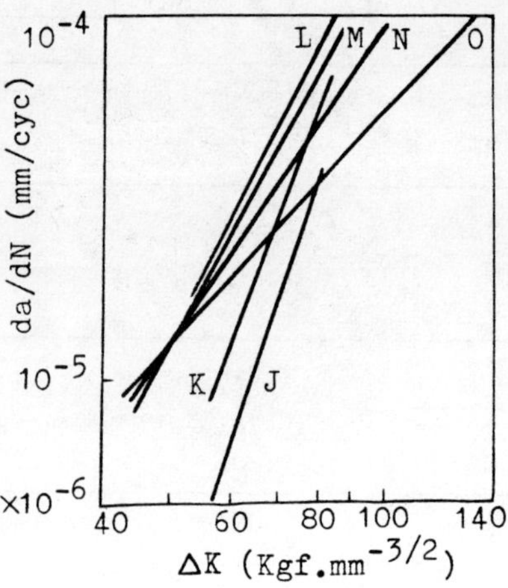

Fig.4 Fatigue crack propagation rate of the non-through hardened specimens of T10A.

*ΔK=40 Kgf·mm$^{-3/2}$

Residual Stress on surface of specimens

Measurement of residual stress was made on polished specimens using a X-ray diffractometer. The data of measurment given in Table 4.

TABLe 4 Residual Stress on Surface of Specimens

(non-through quenched, T10A)

Symbol	Hardening (°C)	Tempering (°C)	HRC Surface	HRC Center	Residual Stress (Kgf/mm^2)
J	780, Through Water	250	60	40	-53.90
N	to Oil	500	40	32	-24.70

Fractography

Considering the SEM results, the fracture surfaces of the specimens austenitized ≤ 900°C and tempered at all temperatures used were characterized by transgranular separation, Fig.5 a. The fracture surfaces of the specimens austenitized and tempered under 300°C were characterized by intergranular separation, Fig.5 b, while the specimens tempered above 300°C were characterized by transgranular separation.

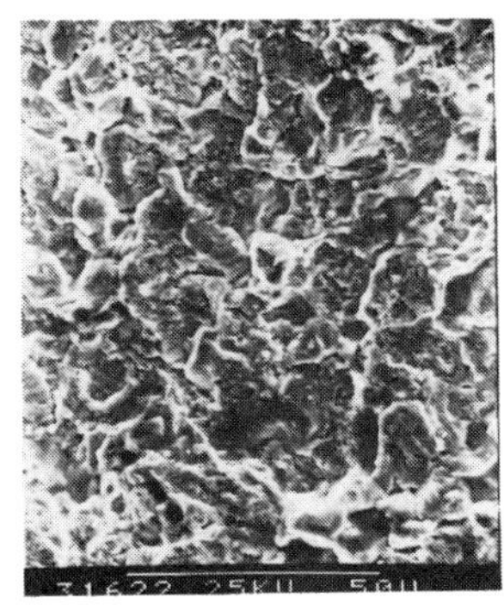

Fig. 5 a)

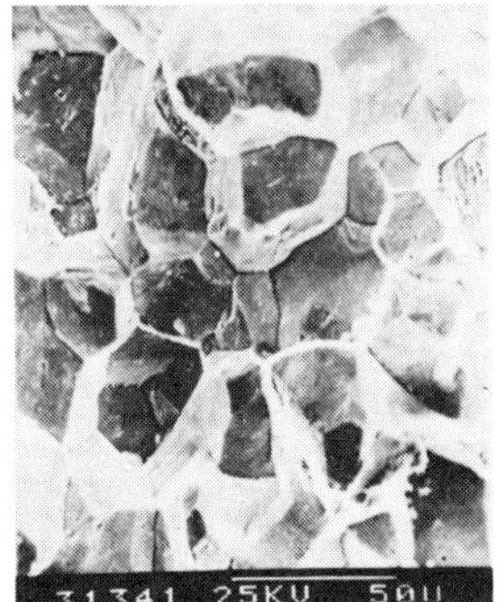

Fig. 5 b)

Fig.5 Fatigue fracture of specimens subjected to treating procedure: a) D, b) G.

DISCUSSION

Effect of Treating Temperature

The da/dN decreases with increasing tempering temperature at high ΔK but it increases at low ΔK by extrapolating. This is in agreement with results obtained on other materials (2). The da/dN increases with increasing austenitization temperature. This effect, we believe, is related to the factors such as austenitic grain size, fracture mechanism and microstructures (3). Future experimental research is need to clarify the causes of this influence.

Effect of Residual Stress

There is substantial different relationship between da/dN and the tempering temperature. The values of da/dN at intermediate ΔK levels given in Table 3. At the ΔK=40 Kgf.mm$^{-3/2}$ the non-through hardened specimens show lower crack propagation rate than through hardened. The difference is very large for the low temper. We believe that the dependence of da/dN on depth of hardened layer is due to the variation in internal stress in specimens. The greater part of compressional internal stress remains in outer layers of the non-through hardened specimens after low temper and counteract the part of load stress.

CONCLUSIONS

The principle findings from this work are summarized below:
The da/dN of high carbon steels are dependent on the microstructures and the residual stress. The surface compressional internal stress set up by non-through hardening leads to a very slow crack propagation rate.
High austenizing temperature with low temper of the steels results in intergranular fatigue fracture. It increases crack propagation rate but nearly has no effect on the critical crack length.

REFERENCES

1. W. F. Brown and J. E. Srawley, ASTM STP No 410 (1966).
2. P. L. Thielen and M. E. Fine, Met. Trans. Vol. 6A, 2133 (1975).
3. K. Ando, Proceedings of Intenatinal Conference on Mechanical Behaviours of Materials, Boston, 513, (1976).

Propagation de Fissures en Mode Mixte I + II par Utilisation du Disque Bresilien

D. Hiebel*, P. Poudou*, J. Ayel*, M. Louah et G. Pluvinage****

**I.F.P. Rueil Malmaison, France*
***Laboratoire de Fiabilité Mécanique, Université de Metz, Metz, France*

RESUME

Des expériences de fissuration en mode mixte I + II ont été réalisés par utilisation d'éprouvettes du type disque brésilien.

MOTS CLES

Fatigue ; mode mixte ; disque brésilien.

INTRODUCTION

Dans de nombreuses situations réelles, une fissure sollicitée cycliquement peut être soumise à un mode polymodal de fissuration, c'est-à-dire à la superposition de 2 ou 3 des modes élémentaires I, II et III. La simulation de ce type de fissuration en laboratoire peut se réaliser en utilisant un chargement uniaxial ou polyaxial. La fissuration polymodale sous chargement polyaxial utilise généralement des éprouvettes cruciformes sollicitées dans deux directions orthogonales par 2 vérins. Ce type de méthode permet d'accéder à tout trajet de chargement en phase où, déphasée, elle présente les inconvénients d'être coûteuse tant du point de vue du prix de l'éprouvette que de la machine d'essais ; de délicats problèmes d'axialité sont en outre à surmonter.

Les laboratoires d'essais préfèrent généralement utiliser des éprouvettes appropriées soumises à un chargement uniaxial. En restreignant notre description au mode mixte I + II de fissuration, on peut citer l'utilisation de panneaux à entaille centrale inclinée, d'éprouvettes de flexion quatre points excentrés. Dans la présente étude, nous avons utilisé une éprouvette habituellement d'emploi réservé à la rupture en mode mixte : le disque entaillé. Ce qui caractérise l'interprétation des essais de fatigue polymodal est une référence constante aux théories de la rupture en mode mixte.

Toutefois, le problème de la fissuration en fatigue présente une complexité supplémentaire. La fissure est généralement non linéaire, et il n'est souvent pas très réaliste de la considérer comme une fissure branchée. Le branchement est en outre supposé très petit. Compte tenu de la complexité de ce problème, de l'absence de solutions appropriées, on se contente dans la plupart des cas d'assimiler cette fissure curviligne à une fissure rectiligne équivalente mais d'inclinaison variable, solution que nous avons retenue.

La présente étude montre la réalisation d'essais de fissuration en mode I + II utilisant des disques entaillés en acier à haute résistance et soumis à un chargement uniaxial de compression. Leur interprétation au travers de différentes théories sera en outre examinée.

DESCRIPTION ET CALIBRATION D'EPROUVETTES DITES "DISQUE BRESILIEN"

Les éprouvettes en forme de disques fissurés soumis à un chargement de compression sont utilisées depuis plusieurs années pour étudier la rupture des roches et des bétons (1) (2). Ces éprouvettes ont été signalées pour la première fois par des brésiliens (3) chargés d'étudier la rupture de rouleaux de fontes emplis de béton utilisés pour déplacer une vieille église à Rio de Janeiro. Ces éprouvettes ont la particularité de présenter un champ de contrainte relativement uniforme (excepté sur le bord) sous l'effet du chargement qualifié de "traction indirecte". La géométrie est représentée sur la figure 1 (P est la charge appliquée, R le rayon du disque, β l'angle d'inclinaison de la fissure de longueur 2a). Le calcul des facteurs d'intensité de contraintes K_I et K_{II} en tête de la fissure sous l'effet de la charge de compression P a été réalisé par 3 auteurs KIM (4), AWAJI (5) et ANKINSON (6).

Ces différentes formules étant équivalentes, nous avons, par commodité, retenu celle d'ANKINSON :

$$K_I = N_I \sigma_o \sqrt{\pi a} \quad ; \quad K_{II} = N_{II} \sigma_o \sqrt{\pi a} \tag{1}$$

avec N_I et N_{II} des coefficients fonctions du rapport $(\frac{a}{R})$ et reportés dans (6)

$$\sigma_o = \frac{P}{\pi RB} \tag{2}$$

B est l'épaisseur du disque.

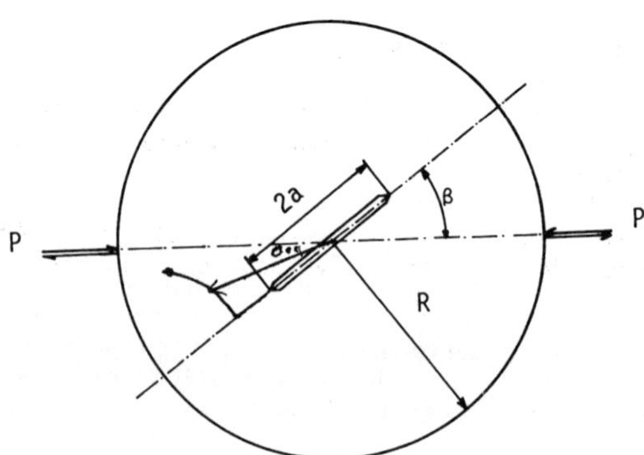

Fig. 1 Disque brésilien soumis à un chargement de mode I + II

MATERIAU ET CONDITIONS EXPERIMENTALES

Les disques ont été réalisés dans un acier à haute limite d'élasticité : l'acier 35NCD16, dont les caractéristiques mécaniques principales sont :

- limite d'élasticité : Re = 1 390 MPa
- résistance ultime : Rm : 1 870 MPa
- dureté : Hrc = 53

Le disque entaillé est préfissuré en mode I puis la fissure est inclinée d'un angle β afin de réaliser les conditions de fissuration en mode mixte I + II. Pour cela, la charge P est maintenue constante durant l'essai avec un rapport de charge R = 0,1 ; le signal du chargement a une fréquence de 40 Hz et une forme sinusoïdale.

Le trajet de la fissure est suivi à l'aide d'une loupe binoculaire selon deux directions orthogonales et dont l'une est est parallèle à la direction de chargement. Le relevé des extensions de la fissure selon les deux directions x et y permet de définir la longueur et la direction de la fissure équivalente selon la figure n°1.

L'utilisation de diverses valeurs de l'angle β a permis de faire varier le degré de polymodalité de la fissuration (défini par le rapport $(K_{II}/K_I)_o$ initial dans les conditions reportées dans le tableau 1.

TABLEAU 1

β_o	5°	9°5	14°5	25°
$(K_{II}/K_I)_o$	0,4	0,9	1,8	73,9
θ_o exp	30°	41°	59°	69°
θ_o th	35°	50°	62°	71°

Pour des angles supérieurs à 27°5 environ et pour des fissures relativement courtes (a/R < 0,3), K_I prend une valeur négative et la propagation de la fissure de fatigue ne s'effectue plus à l'extrémité de la préfissure. Pour cette raison, le domaine où β > 25° n'a pas été interprété.

RESULTATS ET INTERPRETATIONS

* La longueur et la direction de la fissure équivalente permet de calculer en tout point de la fissure le facteur d'intensité de contrainte en mode I et II selon la formule 1.

* La longueur projetée de la fissure a_y et le nombre de cycles de chargements permet de connaître la vitesse de fissuration

$$\frac{da_y}{dN}$$

Les résultats sont reportés dans un diagramme bilogarythmique

$(\frac{da_y}{dN}$ - paramètre de chargement).

Les paramètres de chargement étudiés sont ceux habituellement invoqués pour gouverner la fissuration en fatigue.

= l'amplitude du facteur d'intensité de contrainte en mode I (ΔK_I)

= l'amplitude du facteur d'intensité de contrainte équivalent (ΔK_{eq})

= l'amplitude de la variation de la densité d'énergie de déformation (ΔS).

Un certain nombre de faits expérimentaux semblent infirmer que le critère de propagation de la fissure en mode mixte soit $\Delta K_{II} = 0$. Ce critère indique que la direction de propagation est celle pour laquelle le mode II est nul et que le paramètre de chargement gouvernant le phénomène est ΔK_I.

* La fissure tend à se remettre le plus rapidement possible en mode I.

* L'angle initial de bifurcation Θ_o, aux erreurs de mesure près, est celui associé à ce critère (Tableau 1).

* Le dépouillement des résultats faisant intervenir les paramètres ΔK_{eq} et ΔS montrent que ceux-ci n'évoluent pas de façon biunivoque avec la longueur de la fissure projetée. Plus précisément, ces paramètres présentent préalablement une diminution, suivie d'une augmentation ; simultanément, la vitesse de fissuration est toujours croissante.

  Toutefois, un dépouillement des résultats selon le critère $\Delta K_{II} = 0$ fait apparaître une influence de l'angle d'inclinaison initial de la fissure β_o (figure 2) et donc du mode II. Un dépouillement utilisant une fissure branchée équivalente et la prise en considération de la longueur réelle de propagation réduit l'amplitude de ce phénomène.

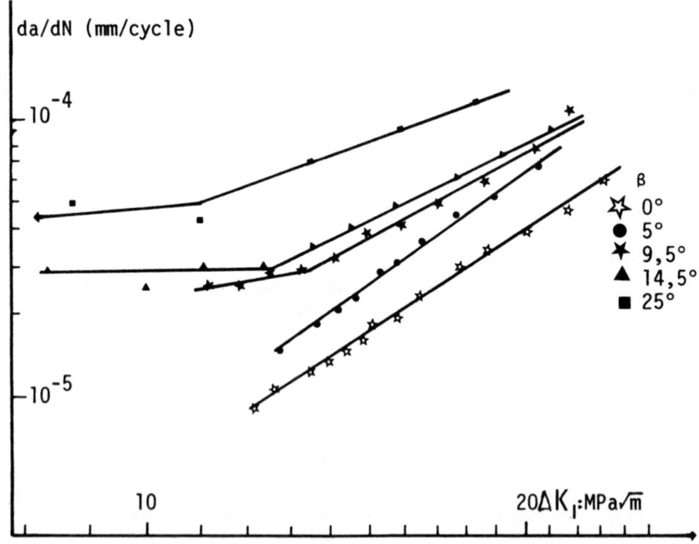

Fig. 2 Comparaison entre les courbes de vitesse en fonction de l'angle β

CONCLUSION

On peut réaliser aisément, par sollicitation uniaxiale, une fissuration polymodale I + II en utilisant des éprouvettes du type disque brésilien. Il apparaît que la fissuration en mode mixte est une étape transitoire avant le retour de la fissuration en mode I. Une interprétation correcte des résultats nécessite de tenir compte du chemin réel parcouru par la fissure et de sa forme géométrique.

BIBLIOGRAPHIE

1. G. HONDROS, The evaluation of Poisson's ratio and the modulus of materials of a low tensile resistance by the Brazilian (indirect tensile) test with a particular reference to concrete, *Australian Journal of Applied Science,* **vol. 10,** pp 243-268 (1959).

2. J.P. HENRY et J. PAQUET, Mécanique de la rupture appliquée aux dépouillements des essais brésiliens sur disques et anneaux de roches, *CRAS T 284,* **n°23,** pp 511-514 (Juin 1977).

3. F.L.L.B. CARNIERO et A. BARCELLOS, Concrete tensile strength, *Union of testing and research laboratory for materials and structures,* **n° 13** (1953).

4. S.C. KIM et H. KITAGAWA, A method of determination of mixed mode fracture toughness of little materials under compression, Proceeding of *Fracture Mechanics and Technology*, Hong-Kong (edited by G.C. SIH, C.L. CHOW), **vol II,** pp 1011-1019 (1977).

5. H. AWAJI et S. SATO, Combined mode fracture toughness measurement by the disk test, *Journal of engineering materials and technology,* **vol 100,** pp 175-182 (1978).

6. C. ATKINSON, On dislocation densities and stress singularities associated with cracks and pile-ups in inhomogeneous media, *Int. J. of Engnng Sciences,* **vol. 10,** pp 45-71 (1972).

Description Phenomenologique de la Viscoplasticite Cyclique avec Viellissement

C. el Ghali, K. Saanouni et C. Bathias

Groupe Mécanique, U.A. 849 du CNRS, Universite de Technologie de Compiegne, France

RESUME

Dans ce travail nous nous intéressons à la description macroscopique du vieillissement dynamique observé dans l'Udimet 500 à 700°C et son influence sur l'écrouissage isotrope.
En partant d'un modèle simple de viscoplasticité cyclique à écrouissage purement isotrope sans seuil, nous montrons qu'il est possible de bien rendre compte des essais de traction-compression par l'introduction d'une équation d'évolution supplémentaire associée au phénomène de vieillissement.

MOTS CLES

Viscoplasticité, vieillissement, écrouissage isotrope

1. INTRODUCTION

Les superalliages à base de Nickel, utilisés dans le domaine aéronautique ou nucléaire, sont généralement le siège d'une instabilité microstructurale du m oins pour une certaine plage de températures (1, 2).
Comme l'a déjà montré Coffin (3), l'Udimet 500 subit une forte instabilité microstructurale pour des températures inférieures à 800°C.
D'un point de vue physique, l'instabilité d'un milieu métallique est induite par la variation dans le temps de sa microstructure par un processus déterminé (précipitation, mode de déformation, ordre-désordre, etc...) ce qui entraine une évolution dans le temps de certaines propriétés rhéologiques du milieu. D'un point de vue mécanique, cette instabilité (vieillissement) se traduit soit par un adoucissement soit par un durcissement, soit par un adoucissement-durcissement alternatif dans quelques cas complexes.

2. ETUDE DU VIEILLISSEMENT DANS L'UDIMET 500

Une campagne d'essais de fatigue en traction-compression réalisée sur l'Udimet 500 à 700°C (cycle sinusoïdal) (4), avait montré un écrouissage cyclique d'autant plus prononcé que la période de sollicitation augmente ($2 < T < 100$) comme le montre la Fig. 1. On remarque clairement que plus T augmente, plus

la contrainte au cycle stabilisé augmente et plus l'amplitude de déformation irréversible diminue, traduisant ainsi une augmentation du domaine d'élasticité.
Une étude des processus physiques du vieillissement menée en parallèle (5) indique que le mécanisme de consolidation cyclique de l'Udimet 500 à 700°C serait dû à la remise en ordre des précipités γ' par diffusion entre les paires de dislocations. En effet, il a été montré que l'homogénéité du glissement augmente avec la période. Le ruban de défaut d'ordre étant réordonné par diffusion, la contrainte permettant à la deuxième dislocation de la paire de pénétrer dans γ' devra ainsi être augmentée. Ce phénomène doit se saturer pour les très grandes périodes et le phénomène de vieillissement-durcissement cèderait la place devant l'écoulement visqueux avec adoucissement.
Pour traduire l'influence du vieillissement sur l'écrouissage isotrope, nous avons choisi un modèle simple traduisant le comportement viscoplastique à écrouissage purement isotrope (6). Ce modèle utilise la déformation plastique cumulée :

$$\dot{p} = \sqrt{2/3}\sqrt{\dot{\epsilon}_p : \dot{\epsilon}_p} \qquad 1/$$

comme variable associée à l'écrouissage.
L'équation d'état est choisie sous la forme suivante :

$$F(\sigma, \epsilon_p, \dot{\epsilon}_p) = J_2(\sigma) - K_p^{1/m}\,\dot{p}^{1/n} = 0 \qquad 2/$$

où $J_2(\sigma)$ est le second invariant du tenseur déviateur s des contraintes, K, m, n sont des coefficients ne dépendant que du matériau et de la température.
L'expression 2/ pourrait s'écrire encore sous la forme :

$$\dot{p} = \left[J_2(\sigma)/K\,p^{1/m}\right]^n \qquad 3/$$

En utilisant la méthode globale avec le choix suivant du pseudo-potentiel de dissipation (7) :

$$\Omega = K/n+1\,\left[J_2(\sigma)/K\right]^{n+1}\,p^{-1/m} \qquad 4/$$

L'hypothèse de normalité généralisée, conduit à l'expression suivante du tenseur des vitesses de déformation viscoplastique :

$$\dot{\epsilon}_p = \partial\Omega/\partial\sigma = 3/2\,\dot{p}\,s/J_2(\sigma) \qquad 5/$$

où $\dot{p}$ est donné par la relation 3/.
Ce modèle rigoureusement valable pour des chargements simples et non cycliques exclue l'existence d'un seuil d'écoulement. L'effet du seuil est considéré implicitement dans la forte non linéarité des fonctions choisies. Son utilisation pour des chargements cycliques relève de l'approximation, mais son avantage pratique réside dans l'obtention de solutions analytiques simples. Le degré de non linéarité du modèle, étant gouverné par les constantes K, m, n ; pour traduire une évolution implicite du domaine d'élasticité (induite par le vieillissement), nous pouvons considérer que K, m, n sont des fonctions de la déformation viscoplastique et du temps. Dans le cas de l'Udimet 500 à 700°C, le dépouillement des résultats expérimentaux (4) a montré que les coefficients m, n restent sensiblement constants pour la plage de période explorée. Par contre K varie beaucoup en fonction de la période du cycle. L'effet du vieillissement pourrait donc être introduit par l'intermédiaire de la fonction K dépendant de p et de t. Ceci est compatible avec la forme multiplicative choisie pour introduire l'écrouissage . L'ensemble de ces constatations expérimentales suggère le choix suivant pour la loi d'évo-

lution de la nouvelle variable K :

$$dK = \Phi(K, p)\ dp + \Psi(K, p)\ dt \qquad 6/$$

Dans le cas de la traction-compression, les fonctions Φ et Ψ sont choisies telle que :

$$dK = a\ (Q_1 - K) \left| d\epsilon_p \right| + b\ (Q_2 - K)\ dt \qquad 7/$$

où a, b, Q_1 et Q_2 sont des constantes dépendant de la température. La variation de K en fonction du nombre de cycle s'écrit approximativement sous la forme suivante :

$$dK/dN = 2a\ (Q_1 - K)\Delta\epsilon_p + b\ (Q_2 - K)\ T \qquad 8/$$

où T est la période du cycle. Au cycle stabilisé, K névolue plus avec N. Ainsi nous obtenons sa valeur assymptotique K_s :

$$K_s = 2a\ Q_1\ \Delta\epsilon_p + b\ Q_2\ T/(2a\Delta\epsilon_p + b\ T) \qquad 9/$$

Dans ce cas, la loi de comportement élasto-viscoplastique au cycle stabilisé (traction-compression) s'écrit :

$$\begin{aligned} \epsilon &= \epsilon_e + \epsilon_p \\ \epsilon_e &= \sigma/E \\ \dot{\epsilon}_p &= \dot{p}\ \mathrm{Sgn}\ (\sigma) \\ \dot{p} &= \left[\sigma/K_s\ p^{1/m}\right]^n \\ K_s &= 2a\ Q_1\Delta\epsilon_p + b\ Q_2\ T/(2a\Delta\epsilon_p + bT) \end{aligned} \qquad 10/$$

La détermination de tous les paramètres de cette loi se fait par identification avec des résultats expérimentaux obtenus et en fluage pur et en écrouissage,relaxation sur des boucles d'hystérésis stabilisées à différentes périodes. Les valeurs suivantes ont été obtenues avec l'aide d'un programme d'identification graphique.

$$Q_1 = 800,\ Q_2 = 1500,\ a = 15,\ b = 0.2,\ n = 12,\ m = 4$$

Quelques exemples de comparaison entre les résultats expérimentaux et le modèle , sont présentés sur les figures 2 et 3. Ces résultats montrent une bonne description qualitative du phénomène de vieillissement observé.
La figure 4 montre une péridction par le modèle de cycles stabilisés pour des périodes allant de 1 s à 10^8 s. On remarque que le durcissement par le vieillissement est bien traduit par le modèle pour des périodes comprises entre 1 et 100 s conformément à l'expérience, mais pour des périodes supérieures, le modèle montre une évanescence totale du vieillissement traduisant une prédominance de l'écoulement visqueux. Il semble donc qu'il existe une période critique au delà de laquelle le phénomène de la viscosité l'emporte sur le vieillissement. La détermination expérimentale de cette période nécessite la réalisation d'essais à même déformation totale imposée et à des périodes supérieures à 100 s (travail en cours).

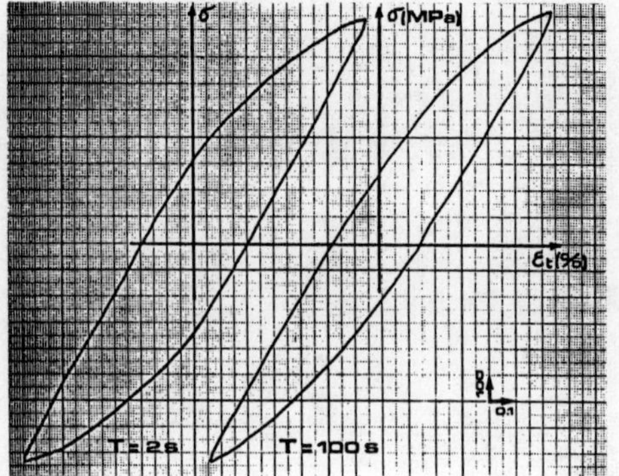

Fig.1 Boucles d'hystérésis expérimentales

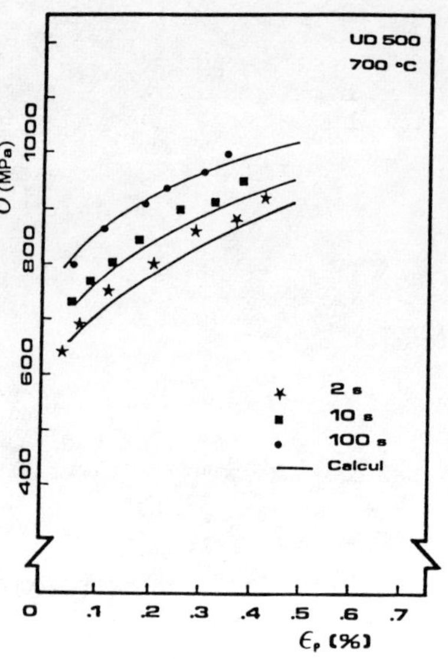

Fig.2 Courbes d'écrouissage cyclique

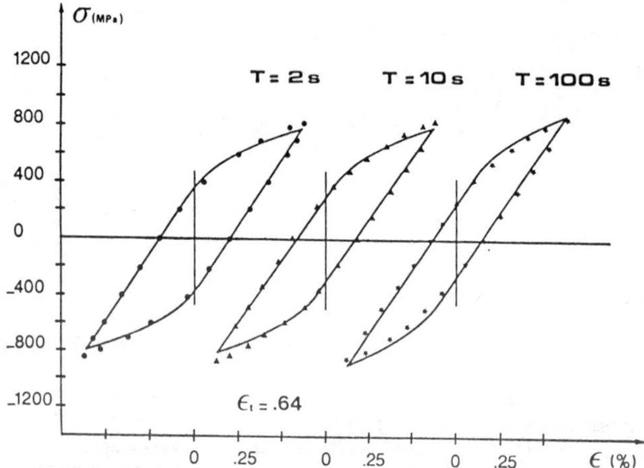

Fig.3 Comparaison modèle-expérience

3. DISCUSSION

La généralisation de nos résultats à des chargements plus complexes (fatigue multiaxiale, chargements hors phase, etc...) n'est pas possible avec ce modèle pour des raisons évidentes. Aussi pour obtenir une description plus rigoureuse du vieillissement d'un point de vue thermodynamique et généralisable à des trajets de chargements non simples, faut-il utiliser des lois à variables internes utilisant explicitement la notion de seuil d'écou-

lement pour décrire l'écrouissage.
Ces lois doivent permettre une description satisfaisante des phénomènes spécifiques aux chargements cycliques (adoucissement-durcissement, effet Bauschinger, restauration, écrouissage cinématique, etc... (7). Dans ce cadre l'équation d'état peut s'écrire :

$$J_2\ (\sigma - \mathcal{X}) - k - R - K\,\dot{p}^{1/n} = 0 \qquad 11/$$

où $\mathcal{X}$ est le tenseur de contraintes internes cinématique représentant le mouvement du centre du domaine d'élasticité dans l'espace des contraintes (écrouissage cinématique). k est la taille initiale du domaine d'élasticité. R représente la variation du rayon du domaine d'élasticité (écrouissage isotrope). $\sigma_V = K\,\dot{p}^{1/n}$ représente la contrainte visqueuse (fluage).
La figure 5 donne une schématisation de ces différentes quantités dans le cas de traction-compression.
Si l'on admet que l'écrouissage cinématique, n'est pas affecté par le vieillissement, ce dernier pourrait être introduit par l'intermédiaire de R et σ_V.

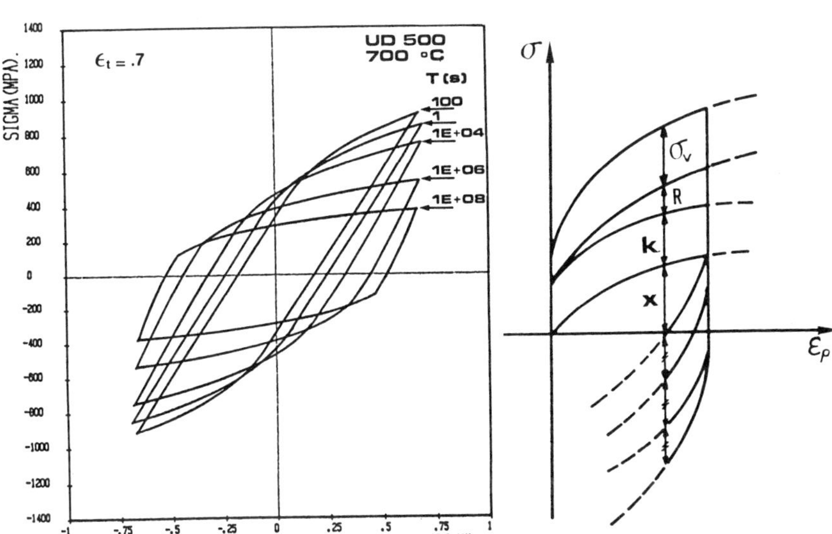

Fig.4 Possibilité du modèle

Fig.5 Décomposition de la contrainte en viscoplasticité

4. CONCLUSIONS

Le vieillissement des superalliages est un phénomène non négligeable qu'il faut considérer dans les lois de comportement, tout au moins dans le cas de l'Udimet 500.
Le modèle utilisé bien que simple, montre qu'il est possible de modéliser macroscopiquement le vieillissement. Il traduit assez correctement les résultats de traction-compression dans la plage de périodes étudiées. Par

contre, il ne s'apprête pas à une généralisation au cas des structures à chargement complexe. Il faut donc aborder ce problème dans un cadre thermodynamique plus rigoureux.

REFERENCES

1. Clavel M., Pineau A., Mater. Science and Engng, 34, 275 (1978)
2. Merrick H.F., Metall. Transactions, 5, 891, (1974)
3. Coffin L.F., Metall. Transactions, 2, 3105, (1971)
4. Saanouni K, Thèse 3ème cycle Université de Compiègne (1982)
5. Savoie J., Clavel M., à paraitre dans Scripta Metallurgica
6. Lemaitre J., Thèse Es Sciences, (1971)
7. Chaboche J.L., Thèse Es Sciences, (1978)

Effects of Temperature and Hold Times on Low Cycle Fatigue of Astroloy

S. J. Choe, N. S. Stoloff and D. J. Duquette

Materials Engineering Department, Rensselaer Polytechnic Institute, Troy, NY 12181, USA

ABSTRACT

Low cycle fatigue (LCF) and creep-fatigue-environment interactions of HIP Astroloy were studied at 650°C and 725°C. All tests were performed in high purity argon in an attempt to minimize environmental effects. Also, correlation between LCF and fatigue crack growth (FCG) rate for continuous cycling were tested utilizing data from experiments at each temperature. The results showed that the model proposed by Kaisand and Mowbray was successful in predicting the magnitude and trend of the fatigue crack growth rate from LCF data. Raising the temperature from 650°C to 725°C did not change the fracture mode, while employing tensile hold caused a change in fracture mode and was more damaging than raising the temperature by 75°C. All samples displayed multiple fracture origins, which is initiated transgranularly in continuous cycling tests and intergranularly in hold time tests. An examination of secondary crack showed no apparent creep damage. Oxidation in high purity argon appeared to be the major factor in LCF life degradation due to hold times.

KEYWORDS

Low cycle fatigue, high temperature, Astroloy, creep-fatigue-environment interactions, fatigue crack growth rate prediction, hot isostatic pressing

INTRODUCTION

New powder metallurgy superalloys using cost-effective techniques, such as hot isostatic pressing, offer great potential for gas turbine disk applications. Several PM nickel-base superalloys have been developed and evaluated for improved turbine performance[1].

Low cycle fatigue behavior and creep-fatigue interactions are the most important properties to be considered to define the materials' capabilities for service components.

Due to the simultaneous occurrence of creep and oxidation, high temperature fatigue is a complicated process. The relative importance of these effects are not well understood at the present time, especially for advanced materials.

The purpose of this investigation was to examine the effect of creep-fatigue interactions under controlled environmental conditions.

It has been generally observed that low cycle fatigue is predominantly a crack propagation process. The major fraction of total life (∿99%) is consumed in the propagation of small stage II crack [2,3]. Recently, Kaisand and Mowbray[4] developed a model which relates low cycle fatigue and fatigue crack growth rate, and observed the excellent correlation for steels and aluminum alloys.

In this study, this model was also evaluated for its ability to predict the crack growth rate from the low cycle fatigue data, using the data from experiments at each temperature on a hot isostatically pressed Astroloy.

EXPERIMENTAL PROCEDURE

A. Material

The alloy studied in this program was produced by hot isostatic pressing (HIP) the argon atomized pre-alloyed powder blend. The HIP cycle was performed at 1190°C and at 103MPa for 3 hrs. Prior to heat treatment, 25mm thick disks were cut from the as-HIP cylinder.

The alloy was solution treated at 1116°C for 2 hrs, followed by a complex precipitation and aging heat treatment as follows: 871°C/8hrs/AC (air cooled), 982°C/4hrs/AC, 649°C/24hrs/AC,760°C/8hrs/AC.

The composition and mechanical properties of the alloy are described in Table I.

B. LCF tests

Cylindrical specimens with a gage length 7.6 mm and gage diameter 3.3mm were machined from the peripheries of the disk. LCF tests were conducted in argon at 650°C and 725°C, using a closed loop servohydraulic testing machine.

All tests were performed under total axial strain control in a fully reversed mode (R=-1, A=∞). Continuous tests were performed at a frequency of 0.33Hz using a triangular wave form strain cycle. The total strain range was varied from 1.5% to 2.7%. Creep-fatigue tests were performed with the same loading and unloading rates except that 2 min. or 5 min. hold times were imposed at the maximum tensile strain.

Failed specimens were examined by scanning electron microscopy to identify fatigue crack origins and fracture modes, and then sectioned axially to examine secondary cracks and creep activities, if any.

TABLE I

Chemical composition: wt%

C	Cr	Co	Mo	Zr	Ti	Al	W	B	Ni
0.02	15	17	5	<0.01	3.5	4	<0.05	0.025	55.4

Mechanical properties

T	UTS (MPa)	0.2% σ_{ys} (MPa)	%El	%RA
R.T.	1393	936	26	31
538°C	1287	869	26	28

C. Prediction of FCG rate from LCF data

An interactive computer program was developed to obtain da/dN *vs* ΔK curves directly from the LCF data[5].

Specific information on the model and notation used are detailed in the original paper[4]. The predicted data were compared with the experimental data[6] that were generated in the same machine and on the same heat of material. Due to the absence of data for this PM alloy, the Young's modulus of U-700 at 538°C was assumed to be the same value with that of Astroloy at each temperature. The 538°C 0.2% σ_{ys} value shown in Table I was also used for calculation.

RESULTS

A. Microstructure

A typical microstructure of HIP Astroloy is shown in Fig. 1. The grain size of this alloy is 40∿60μm. Large undissolutioned γ' (0.3∿1μm) is located at the grain boundaries as large discrete particles. Intermediate sized γ' (0.1-0.3μm) and fine γ' (0.02-0.05μm) are evenly distributed throughout the material. The white particles on the grain boundaries are borides (A) and $M_{23}C_6$ carbides (B). There seems to be no size difference between these two phases (0.2-2μm). Relatively more borides were observed than $M_{23}C_6$ carbides. The total γ' volume fraction was measured as 0.42.

B. LCF behavior

LCF test results at 650°C and 725°C are summarized in Fig. 2. Both increasing temperature and tensile holds reduced LCF lives, however, the increase in temperature of 75°C did not affect the LCF lives as severely as did the hold times. The 5 min. tensile hold reduced the fatigue life more than the 2 min. hold.

The Coffin-Manson exponent remained approximately constant at a value of 0.9 with or without hold time.

All fracture surfaces were covered with oxide and displayed multiple fracture origins. A notable effect of hold time was seen in the

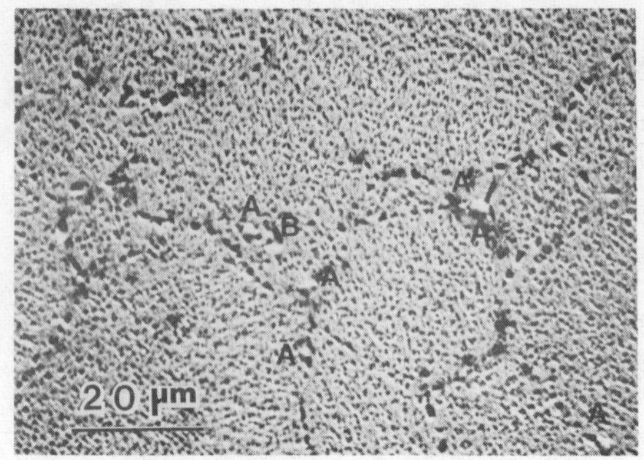

Fig. 1 Back scattered SEM micrograph of HIP Astroloy showing undissolved γ' (dark image on g.b.), (A) M_3B_2 carbides, (B) $M_{23}C_6$ carbides, fine γ' (inside grain)

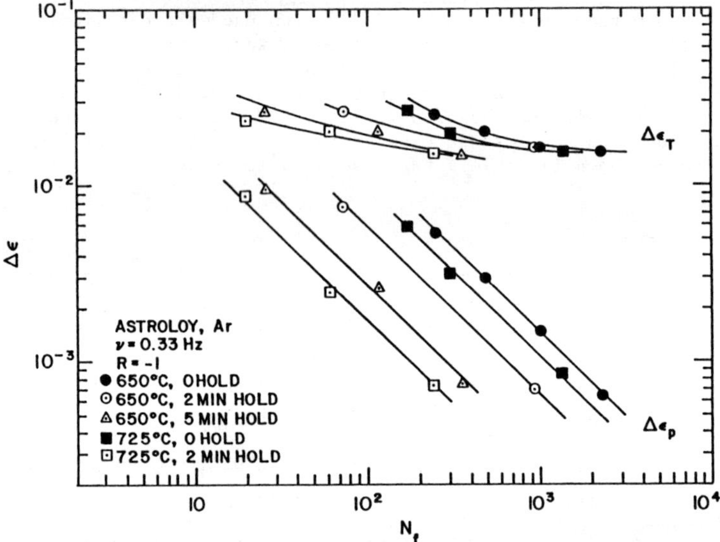

Fig. 2 LCF behavior of HIP Astroloy

initiation zone, which was always transgranular (TG) for continuous cycling and intergranular (IG) for hold time samples, as shown in Fig. 3(a) and (b). Pre-existing pores did not affect the crack initiation, however some inclusions and precipitates were usually observed in the origin.

Striations were noted in many continuously cycled samples, but generally were limited to the region close to the origin; however no striations were observed in hold time samples at 725°C or in 5 min. hold time samples at 650°C, perhaps due to the presence of substantial oxide films.

The most notable feature of the hold time effect was to produce

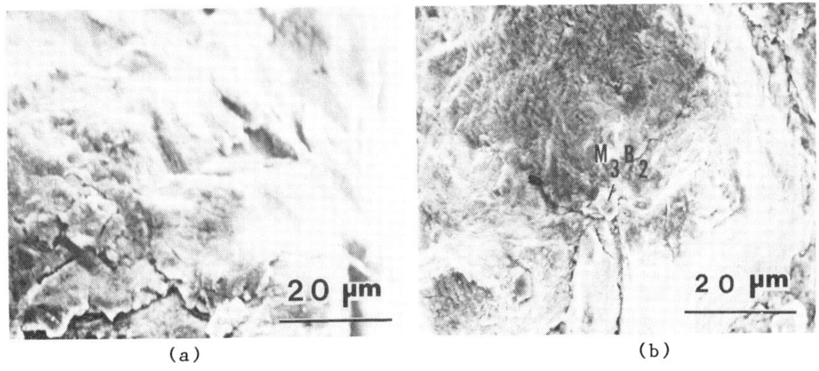

(a) (b)

Fig. 3 LCF crack origins showing (a) TG initiation in continuous cycling test, 650°C, $\Delta\varepsilon_t$ = 1.56% (b) IG initiation in 2 min. hold, 650°C, $\Delta\varepsilon_t$ = 2.63%.

a mixture of TG + IG cracking over most of fracture surface compared to only TG fracture in samples subjected to continuous cycling, as shown in Fig. 4(a) and (b).

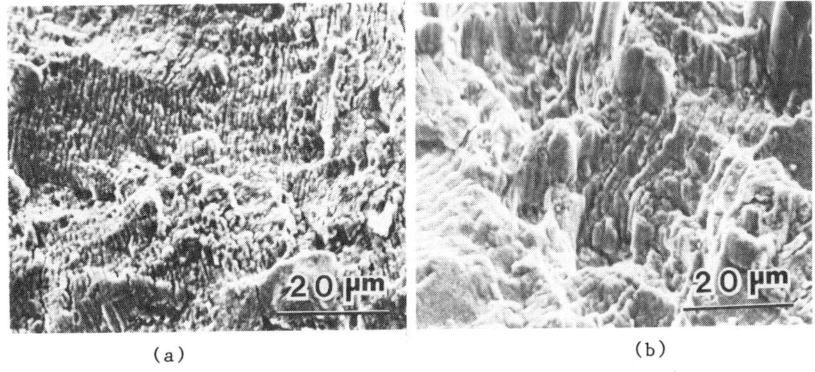

(a) (b)

Fig. 4 SEM fractographs showing (a) TG fracture surface in continuous cycling test, 650°C, $\Delta\varepsilon_t$ = 1.56%, N_f = 2321 (b) TG+IG fracture in 2 min. hold, 650°C, $\Delta\varepsilon_t$ = 1.67%, N_f = 945

The effect of temperature on the crack propagation mode is shown in Fig. 5(a) and (b). Raising the temperature from 650°C to 725°C did not change the TG crack propagation mode, but showed increased striation spacings.

Figure 6 shows a secondary crack in a longitudinal section of the specimen tested at 650°C, 2 min. hold, $\Delta\varepsilon_t$=2.05%. No apparent creep cavitation was observed in this and any other samples.

C. Comparison of experiment with prediction

Comparisons of predicted FCG rate from LCF data vs experimental data are shown in Fig. 7(a) and (b). Excellent correspondence between prediction and experiment was demonstrated in describing the magnitude and trend of the data.

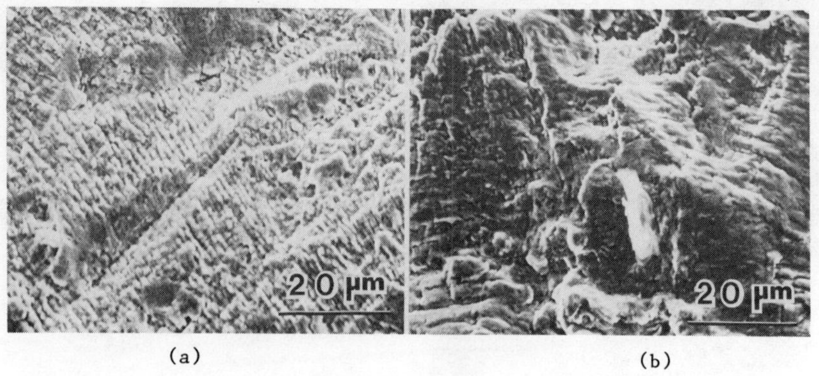

(a) (b)

Fig. 5 Effect of temperature on striation spacings
(a) 650°C, $\Delta\varepsilon_t$ = 2.03%, N_f = 484 (b) 725°C, $\Delta\varepsilon_t$ = 1.98%, N_f = 307

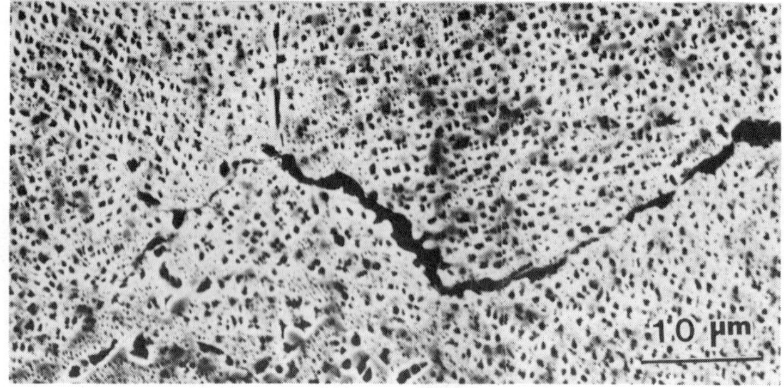

Fig. 6 Secondary crack formed in HIP Astroloy at $\Delta\varepsilon_t$ = 2.05%, 2 min. hold, 650°C

DISCUSSION

High temperature fatigue behavior is controlled by several competing factors. Creep processes at the crack tip may either enhance crack growth rate due to additional damage, or retard crack growth due to crack tip blunting or bifurcation. These processes are further complicated by oxidation at the crack tip.

The absence of creep deformation and the samll temperature dependence of LCF lives support the idea that the creep contributions are relatively insignificant for this alloy under these test conditions. A negligible creep rate was also observed at 650°C by Pelloux and Huang[8]. Therefore, the degradation of LCF lives by superposition of tensile hold was predominantly an environmental effect due to oxygen.

Elevated temperature, environmental effects are very complicated phenomena to understand. Even very minor amounts of oxygen can result in fatigue crack gorwth rates comparable to air[9]. In some cases, small amounts of oxygen have been shown to be more damaging than higher levels[10].

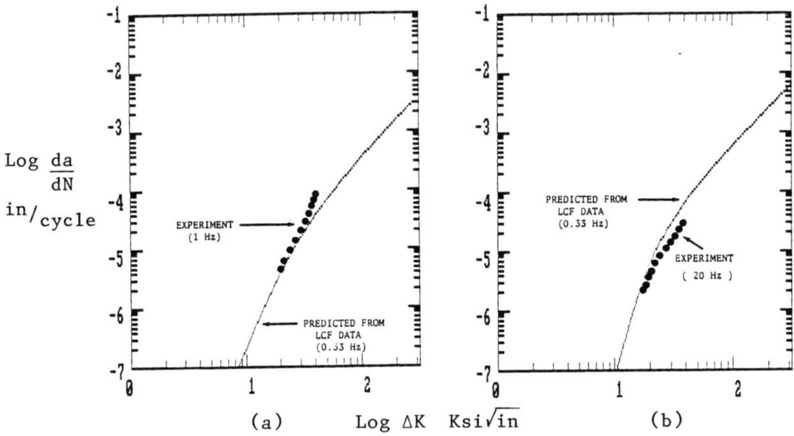

Fig. 7 Comparison between experimental and predicted fatigue crack growth rate for HIP Astroloy in continuous cycling test (a) 650°C (b) 725°C

It is known that mechanical loading can increase the oxidation rate by several orders of mangitude[11]. The measured oxide thickness on fracture surface was of the order of 50 to 200 nm, which was much thicker than the 10nm oxide film observed on the external polished surfaces in crack growth experiment[6]. Therefore, it is believed that, during the hold time, oxygen diffuses rapidly to the grain boundary to form an oxide wedge in the grain boundary, because grain boundaries are the more rapid diffusion paths. On subsequent cycling, cracking takes place along the oxide-matrix boundary or in oxide, resulting in a change in crack propagation mode from TG to mixed TG + IG mode. It is interesting to note that a 15 min. hold in air at 650°C caused IG crack propagation for the same alloy[7]. The same reasoning can be applied to the change in initiation mode due to hold time.

The reduced LCF life associated with hold time appears to be caused by Cr denuded weak zones adjacent to the grain boundaries, which result in high, localized crack propagation rates. The oxide film produced in argon in compact tension specimens was shown to be a chromium rich oxide by Auger electron analysis[6].

The excellent correlation between experiment and predicted fatigue crack growth rate from LCF data demonstrates that the model proposed by Kaisand and Mowbray can be successfully utilized for complex nickel based alloys.

CONCLUSIONS

1. LCF lives of Astroloy were drastically reduced by tensile hold in high purity argon at 650°C and 725°C, when compared to continuous cycling tests.

2. Tensile holds caused change in crack propagation and initiation mode, from TG to TG + IG and from TG to IG respectively.

3. The time dependent damage was found to be primarily controlled by environmental interaction, even in high purity argon.

4. Fatigue crack growth rate were predicted successfully from LCF data utilizing Kaisand and Mowbray's model.

ACKNOWLEDGMENT

This research was supported by NASA-Lewis Research Center under Grant No. NAG 3-22, under the direction of Dr. R.V. Miner. The authors would like to thank Dr. Miner for supplying material and for helpful discussions.

REFERENCES

1. R.V. Miner, J. Gayda and R.D. Maier, Met. Trans. A, **13A**, 1755 (1982).
2. J.C. Grosskreutz, Metal Fatigue Damage, ASTM STP 495, 5 (1971).
3. B. Tompkins, Phil. Mag., **18**, 1041 (1968).
4. L.R. Kaisand and D.F. Mowbray, J. Test. and Eval. **7**, 270 (1979).
5. S.J. Choe and D. Lee, term project in course CAE, RPI (1984).
6. S.V. Golwalkar, Ph.D. Thesis, RPI (1984).
7. B.A. Cowles, J.R. Warren and F.K. Haake, NASA NAS 3-20367 (quoted from P & WA Technical Report for CR-165123), **70** (1980).
8. R.M. Pelloux and J.S. Huang, Creep-Fatigue-Environment Interactions, AIME, 151 (1981).
9. S. Floreen and R.H. Kane, Met. Trans. A, **10A**, 1745 (1979).
10. Y. Hosoi and S. Abe, Met. Trans. A, **6A**, 1171 (1975).
11. R.P. Skelton and J.I. Bucklow, Metal Science, **12**, 64 (1978).

Fatigue and Creep Behaviour of Alloy 800H at Elevated Temperatures

B. A. Lerch*, B. Kempf*, D. Steiner and V. Gerold***

**Max-Planck-Institut für Metallforschung, Institut für Werkstoffwissenschaften, 7000 Stuttgart, Federal Republic of Germany*
***Now at IBM, 7030 Böblingen, Federal Republic of Germany*

ABSTRACT

The fatigue and creep properties of Alloy 800H were investigated at 800°C. Fracture was transgranular for LCF specimens with some intergranular cracking occurring at lower strain rates. For strain rates between $2x10^{-4}$ and $2x10^{-3}$ s^{-1} no effect of $\dot{\varepsilon}$ on N_f was observed. In contrast to this, slow/fast tests caused intergranular cracking and a slight reduction of N_f. Creep specimens always fractured transgranularly except for a zone of intergranular cracking at the surface. Grain boundary cavities were not observed in any of cracking behaviour.

KEYWORDS

Alloy 800H, high temperature fatigue, creep, slow/fast cycles, cracking mechanisms, oxidation.

INTRODUCTION

The austenitic stainless steel Alloy 800H is employed in areas involving high temperatures and aggressive environments. Such critical applications require a good knowledge of the material behaviour as well as the ability to predict the life of the component. These are often done through extrapolation of short time test data. This study was initiated to give reliable short time test data and to suggest limits for life time extrapolations.

EXPERIMENTAL PROCEDURE

The 19.5 mm diameter rods were received in a solutioned (1125°C/30 min.+WQ) state. The composition of the alloy is given in Table 1. Ageing treatments were performed at 800°C for 1,4,8,16,100 and 300 h, cooling afterwards in air.

TABLE 1 Composition of Alloy 800 (w/o)

Ni	Cr	Ti	Al	C	Si	Mn	Co
31.11	20.26	0.32	0.34	0.07	0.46	0.68	0.23

LCF Tests

Symmetrical fatigue tests were performed at 800°C on cylindrical specimens. The specimens were first heat treated for 16 h at 800°C + air cooled and subsequently mechanically and electrolytically polished. Fully reversed plastic strain tests ($R\varepsilon_p$=-1) were performed at constant total strain rates ε_t between $2x10^{-4}$ and $2x10^{-3}sec.^{-1}$. A constant plastic strain range ($\Delta\varepsilon_p$=0.3%) was maintained throughout testing.

Tests having an asymmetric cycle form were performed on identically prepared specimens. The plastic strain range (0.3%) was again kept constant, but different strain rates in tension ($\dot{\varepsilon}_T$) and compression ($\dot{\varepsilon}_C$) were employed. The strain rates varied from $2x10^{-5}$ to $1x10^{-2}sec.^{-1}$.

Creep Tests

Constant stress creep tests were performed on specimens with 5 mm diameters and 40 mm long gauges which were heat treated like the fatigue specimens. The temperature was measured by Ni-NiCr thermocouples at three positions on the gauge length. The test temperature was 800°±1°C.

RESULTS

Microstructure

The as-received material had an average equiaxed grain size of 70 μm and remained constant during further heat treatment at 800°C. It was particle free except for an occasional Ti(C,N) particle. After one hour of ageing at 800°C, discrete carbides of the type $Cr_{23}C_6$ (1) had already precipitated, covering the grain and twin boundaries. These carbides remained basically unchanged up to a maximum ageing time of 300 h. With increasing ageing, carbides (likewise $M_{23}C_6$) also precipitated in the matrix. Precipitation of γ' was not observed during either heat treatment or testing. The microstructure after 16 h at 800°C (as used for the test specimens) is shown in Fig.1.

Symmetrical LCF Tests

Specimens tested at a $\Delta\varepsilon_p$ = 0.30% hardened to a saturation stress early in the life. This stress was strain rate dependent as shown in Table 2. There was no detectable detrimental effect of the strain rate on the number of cycles to failure (N_f). At both strain rates cracking initiated at various places on the gauge surface. They propagated transgranularly at both strain

TABLE 2 Results of the LCF Tests

No	$\dot{\varepsilon}_t$ +) $10^{-4}s^{-1}$	σ_T MPa	σ_C MPa	N_f ++)	t_f ++) h
1	20	154	154	2830	3.9
2	20	152	152	3472	4.8
3	2	145	145	2369	32.9
4	2/20	127	159	1088	6.9
5	2/100	120	172	1234	6.5
6	0.2/100	94	162	671	32.3

T = tension C = compression

+) Slow/fast tests (nos. 4 to 6) are indicated as $\dot{\varepsilon}_T/\dot{\varepsilon}_C$

++) N_f and t_f are defined as a 20% load drop from the saturation load value.

rates (Fig. 2) and striations were clearly visible. At the slower strain rate, some intergranular cracking was observed. The fracture surfaces and gauges of all specimens were heavily oxidized, with the oxidation being more severe at the slower strain rates. Large globules of oxide were frequently seen at grain boundaries (Fig. 3).

Optical micrographs of secondary cracks taken from longitudinal cross-sections indicated the cracks were filled with an oxide layer, whose thickness increased with decreasing strain rate. Two different oxide structures could often be identified (Fig. 4). The lighter colored oxide next to the matrix was found to be a Cr-Mn rich oxide, while the inner oxide was Fe-rich.

Asymmetrical LCF Tests

A group of specimens were tested having the strain rates in tension slower than those in compression (slow/fast tests). The results are also listed in Table 2. The largest ratio employed was a factor of 500. The number of cycles to failure was reduced by at least a factor of two compared to the symmetrical tests. Fractographic examination indicated the fracture surfaces to be primarily intergranular (Fig. 5). These surfaces as well as the gauge lengths were heavily oxidized, especially in the grain boundary areas. As in the other experiments cavity formation was not observed.

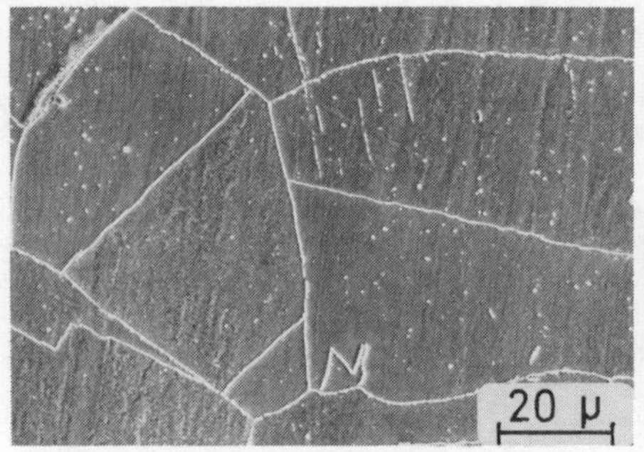

Fig. 1. Microstructure after 16h at 800°C. (SEM).

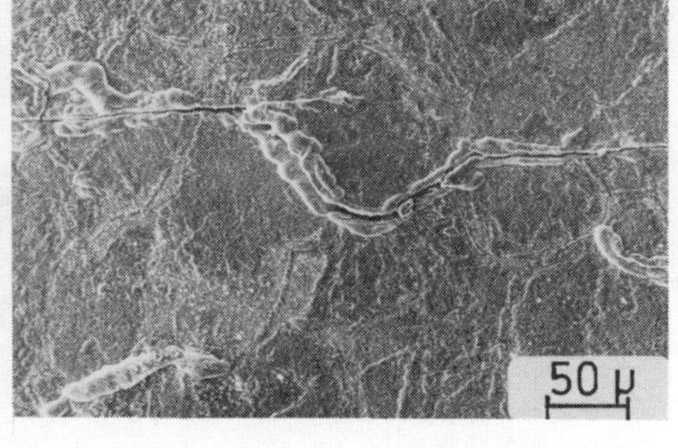

Fig. 2. Cracking and oxidation on the gauge surface of LCF test 3 (SEM).

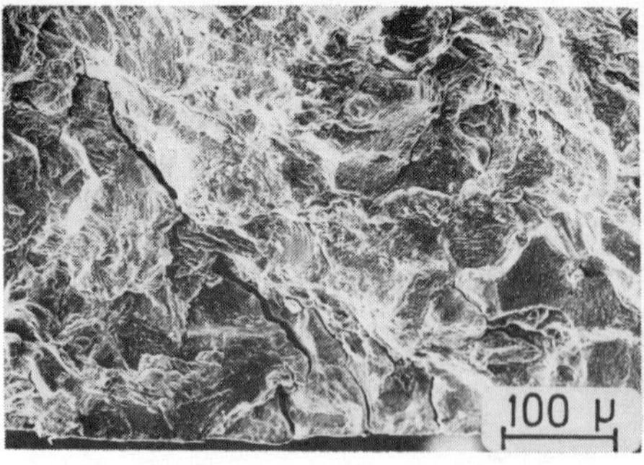

Fig. 3. Fracture at the surface of LCF test 3 (SEM).

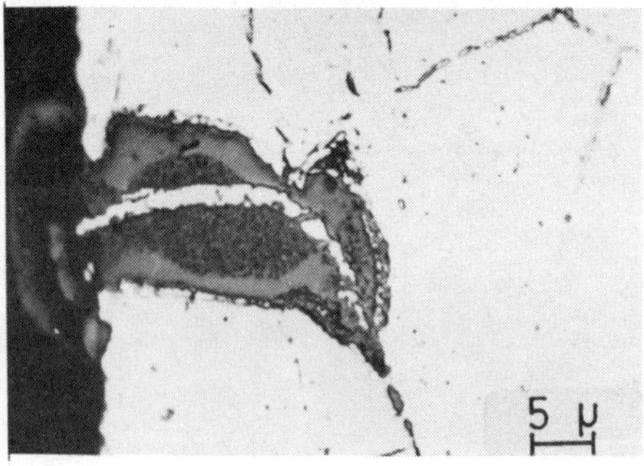

Fig. 4. Intercrystalline secondary crack of LCF test 3 (light micrograph). Longitudinal section.

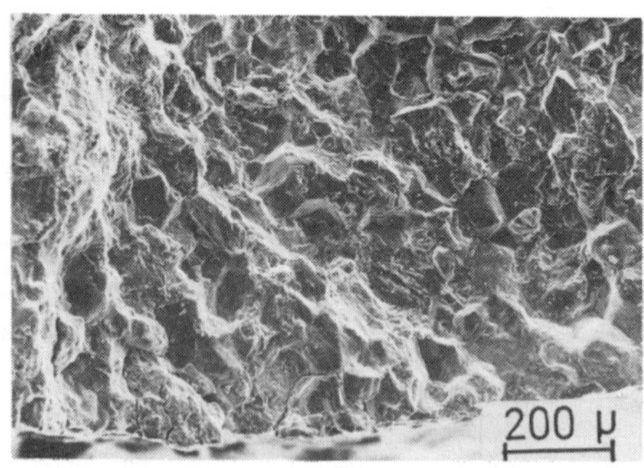

Fig. 5. Intercrystalline fracture surface of LCF test 5 (slow/fast). (SEM).

Fig. 7. Intercrystalline outer fracture zone of a low stress creep test (σ = 70 MPa) (SEM).

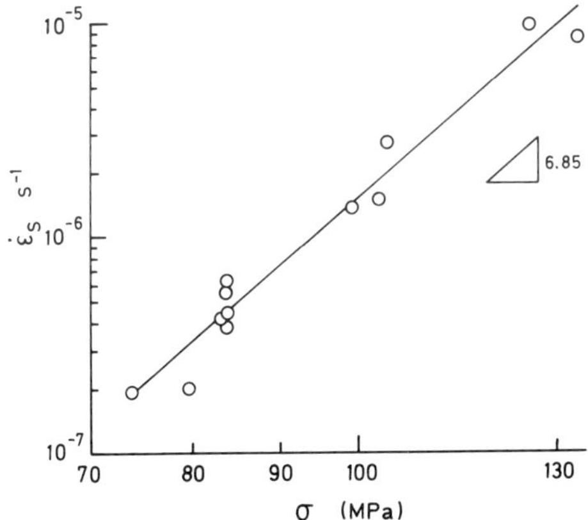

Fig. 6. Stress dependence of the stationary creep rate $\dot{\varepsilon}_s$ at 800°C.

Creep Tests

Twelve constant stress creep tests were performed between 70 and 130 MPa (The 0.2% yield stress is about 120 MPa). All specimens deformed in the classical three stage creep mode. The stationary creep rate $\dot{\varepsilon}_s$ was found to be stress dependent resulting in a power law relationship (Fig. 6)

$$\dot{\varepsilon}_s/s^{-1} = 3.1x10^{-20}(\sigma/MPa)^{6.85}$$

with a correlation coefficient r = 0.981. The creep rupture life time t_f decreased with increasing $\dot{\varepsilon}_s$ and yielded the following relationship:

$$t_f/s = 3.4x10^{-4}(\dot{\varepsilon}_s/s^{-1})^{-0.86}$$

with r = 0.976. The fracture ductility and the reduction in area decreased drastically with increasing fatigue life.

Fractographic analyses indicated an intergranular fracture mode in an outer zone near the gauge surface and a transgranular mode in the interior. The outer zone increased drastically with decreasing creep stress and covered 80% of the fracture area at σ = 70 MPa (Fig. 7). At 130 MPa the fraction was less than 25%. In no case could cavity formation at grain boundaries be found.

DISCUSSION

It can be observed from the creep tests that the main fracture had propagated from the gauge surface. The zone of intergranular cracking was restricted only to the outer grains, due to the severe oxidation at the grain boundaries. The longer tests had a much larger intergranular zone which can be explained by a preferred oxidation in the grain boundaries due to boundary diffusion of oxygen and stress-enhanced oxidation. If there is no other damage present the oxidation of grain boundaries seems to be the first damage to occur during a creep test. At still lower stresses a 100% intergranular fracture might be expected. The results of the symmetrical LCF tests indicate that the influence of oxidation on the fracture mode is less severe. This is not surprising since the life time t_f of the specimens do not cover the whole range observed in the creep experiments. Therefore, only some intercrystalline fracture can be found in the slower strain rate tests. The main damage results from transcrystalline fatigue crack propagation. It is possible that at a still lower strain rate the contribution from intergranular cracking could be larger reducing N_f significantly.

In the case of the slow/fast LCF tests a marked change to intercrystalline fracture and lower values of N_f and t_f was observed compared to the symmetrical tests. This can not be fully explained by oxidation effects but must have its origin in some other damage occurring in the grain boundaries due to asymmetric cycling. So far, no void formation (neither r nor w type) as it may be expected from results by Baik and Raj (2,3) and as it is discussed by Min and Raj (4) could be found during the present experiments.

CONCLUSIONS

1. During symmetrical LCF tests fracture was found to be transgranular independent of the strain rate of the test ($2x10^{-4}$ to $2x10^{-3}s^{-1}$). Only at the lower rate could some intergranular fracture be observed.
2. Slow/fast fatigue tests produced a shorter N_f and t_f than by symmetrical tests having an equivalent $\Delta\varepsilon_p$. Intergranular cracking was observed but no grain boundary cavitation was present.
3. Creep fracture occurred intergranularly by oxidation in an outer zone and transgranularly in an inner zone. The size of the outer zone increased with increasing life time.
4. Oxidation contributes to the damage in all cases. An additional mechanism must however be responsible for the intergranular failure in the slow/fast LCF tests.

ACKNOWLEDGEMENTS

This research is part of the European research project COST 501. The financial support by the Bundesministerium für Forschung und Technologie, Bonn, Germany is gratefully acknowledged. The alloy investigated was provided by the Vereinigte Edelstahlwerke, Kapfenberg, Austria.

REFERENCES

1. H. Aigner, O. Demel and H.P. Degischer, Z. für Werkstofftechnik 14, 24 (1983).
2. S. Baik and R. Raj, Met. Trans. 13A, 1215 (1982).
3. S. Baik and R. Raj, Met. Trans. 13A, 1207 (1982).
4. B.K. Min and R. Raj, Canad. Metall. Quart. 18, 171 (1979).

Effects of Segregation and Environment on Fatigue Crack Growth at Elevated Temperatures

P. Bowen and J. F. Knott

Department of Metallurgy and Materials Science, University of Cambridge, UK

ABSTRACT

For tests performed at 290°C, marked decreases in fatigue crack growth rates have been obtained in vacuum, compared with growth-rates in air. In addition, fractographic and metallographic observations indicate that the mechanism of crack growth is dramatically different in the two environments. In vacuum, crack growth is dominated by a transgranular "facetting" mechanism, which proceeds along lath-lath and packet-packet boundaries, and produces a very rough fracture surface. In air, crack growth is dominated by a "striation" type (continuum) mechanism, which largely ignores microstructural features except at both low and very high crack growth rates, and produces a less rough fracture surface. Significant effects of mean stress on crack growth rates are observed only in air.

KEYWORDS

segregation, environment, fatigue crack growth, high strength steel, mean stress, vacuum testing.

INTRODUCTION

In previous work [1], it was observed that fatigue crack growth in A533B steel, tested in air at 290°C over a limited stress-intensity range, $\Delta K = K_{max} - K_{min}$, was associated with the occurrence of a large number of intergranular facets on the fracture surface. The steel was in a coarse-grained martensitic condition and it was postulated that the facets were produced by the dynamic segregation of sulphur, already present within the steel in solid solution, to the crack-tip region. More usually, the presence of intergranular facets during fatigue crack growth is attributed to an effect of interaction with an external environment [2,3]. In order to attempt to decide between these alternative mechanisms, tests were performed at 290°C, over a wide range of ΔK values in air and in vacuum, at both low and high load ratios (characterised by the R ratio, where $R = K_{min}/K_{max}$).

EXPERIMENTAL

The material used was a commercial grade of A533B of the composition given in Table 1. A coarse-grained martensitic microstructure was produced by austenitising for 1h at 1250°C and quenching in oil to room temperature. The 0.2% proof stress measured at room temperature is 1300 MPa [4], and detailed microstructural information for this condition has been reported previously [1,4]. Fatigue crack growth tests were performed, using single-edge-notch (SEN) bend testpieces (TL orientation). These were fatigued in a Mand servo-hydraulic testing machine of 30 kN capacity, equipped with a high-temperature vacuum chamber, which could be operated under interactive microprocessor control. Heating was effected by quartz lamps, and temperature control to within ± 1°C was maintained at 290°C. The vacuum tests were carried out at a total pressure of 3.10^{-7} mbar.

Crack growth was detected by means of the D.C. potential-drop technique. Threshold tests were performed at R ratios of both 0.1 and 0.5, at a frequency of 40 Hz and the threshold value, ΔK_{th}, was taken to be that value of ΔK at which no crack growth could be detected in 8h. Constant load tests were employed to obtain crack growth rates at high values of ΔK, and such tests were carried out at a frequency of 20 Hz.

Metallographic observations were made on polished and etched nickel-plated sections. Fractographs were obtained using an ISI 100A scanning electron microscope, operating at 20 kV, 0° tilt.

TABLE 1 Composition of A533B plate, wt.%

C	Mn	Ni	Mo	P	S	Sn
0.25	1.51	0.63	0.53	0.005	0.005	0.005

TABLE 2 Summary of tests performed at 290°C

Environment	R ratio	ΔK_{th} (MPam$^{\frac{1}{2}}$)	m	A(da/dN in mm/cycle)
Lab. air	0.1	5.8	2.0	$1.5.10^{-7}$
Lab. air	0.5	3.7	1.7	$6.2.10^{-7}$
Vacuum	0.1	13.0	6.0	$3.5.10^{-13}$
Vacuum	0.5	11.0	5.7	$1.8.10^{-12}$

RESULTS

The fatigue crack growth-rate (da/dN) curves obtained at 290°C for low and high mean stress in air and in vacuum are shown in fig. 1. The generally accepted three stages of fatigue crack growth are also illustrated schematically in fig. 1. With respect to the experimental points, it can be seen that regions I (threshold) and II (Paris law) are present for the tests in air, but that in vacuum, region II appears to continue down to threshold. Crack growth rates obtained under vacuum conditions are lower than those obtained in air and this effect becomes more pronounced as the value of ΔK decreases. This leads to higher ΔK_{th} values under vacuum conditions, see Table 2. Also shown in Table 2 are the values of the constants A and m, derived from the Paris equation

$$da/dN = A\ \Delta K^{m} \tag{1}$$

The value of m is approximately 2 in air, and 6 in vacuum. From fig. 1 it can be seen that there is little effect of mean stress, on either da/dN or ΔK_{th} in vacuum. In air, however, the higher value of mean stress (R = 0.5) increases da/dN by a factor of two, and decreases ΔK_{th} from approximately 6 to 4 MPam$^{\frac{1}{2}}$.

For tests performed in vacuum, two distinct regions can be distinguished from both the optical sections and fracture surfaces: a transgranular "facetting" region which dominates at all values of ΔK, except at values close to ΔK_{th}. An optical micrograph of this mode of crack growth is shown in fig. 2(a) and a fractograph is shown in fig. 2(b). Crack growth appears to proceed along lath-lath interfaces - fig. 2(a). Little evidence of striation growth is observed; a very smooth, flat, featureless region is seen at ΔK values close to ΔK_{th} as shown in fig. 3(a) and fig. 3(b). No intergranular facets are observed in vacuum.

For tests performed in air, striation growth largely dominates the other modes of crack growth, see fig. 4, except at very high values of ΔK (i.e. when K_{max} approaches K_{IC}) where transgranular facetting is again observed, and at values of ΔK close to ΔK_{th} where intergranular facets are observed, in addition to the flat featureless regions observed in vacuum (fig. 5). In near threshold regions, there is some indication that more intergranular facets are observed at higher mean stress.

Therefore it appears that crack growth in air occurs largely by striation growth, whilst, in vacuum, crack growth occurs largely by transgranular facetting. The effect of environment on the mechanism of crack growth is illustrated in fig. 6 for a crack subjected to a constant stress intensity range.

DISCUSSION

Effects of environment

In the present study, da/dN values obtained at 290°C in vacuum are lower than those obtained in air. Such observations have been reported in the literature, and in the present work two mechanistic explanations may be suggested for the increases in da/dN in air compared with those in vacuum (the reference state):

(1) environmental interaction

Such interaction may be considered to derive from both physical adsorption [5] and/or environmental embrittlement [2,3]. Physical adsorption can be considered as preventing any possibility of full slip-reversibility i.e. complete rewelding of the newly-formed crack surface at the crack-tip during the "unloading" or compressive part of the loading cycle. Therefore, the increase in crack length per cycle will be greater in air than in vacuum. It is also possible (at low da/dN) for an external embrittlement mechanism to cause the value of da/dN in air (where intergranular facets are observed) to be higher than that in vacuum (where no intergranular facets are observed). It is, however, clear that significant increases in da/dN in air can be obtained for ΔK values at which no intergranular facets are seen. The total absence of intergranular facets in vacuum indicates that the mechanism of the dynamic segregation of sulphur from internal sources alone, as suggested previously [1], is not sufficient to produce intergranular facets during fatigue at 290°C;

(2) increased crack closure under vacuum conditions

In the present investigation the fracture surfaces produced in vacuum are clearly rougher than those produced in air - fig. 6. This rougher surface may lead to increased closure, and hence reduced effective ΔK, in vacuum, resulting in lower da/dN values. It is important, however, to note that any such increased closure in vacuum due to roughness, will be produced because of the transgranular facetting mechanism of crack extension which is essentially absent in air. Attention should therefore be directed towards the fundamental underlying processes that control the mechanisms of crack growth in air and in vacuum, as a consequence of which increased closure effects may be produced in vacuum. It is not clear how the dry-air environment suppresses the transgranular facetting crack growth, that appears to proceed along lath-lath interfaces in vacuum, although it is possible to suggest a schematic mechanism by which physical adsorption might influence the subsequent preferred crack plane [6].

Effects of mean stress

The increased da/dN value produced at higher mean stress (R = 0.5) in air may be attributed either to reduced closure effects (increased effective ΔK) and/or to the increased number of intergranular facets suggested by fractographic observations. Note that in vacuum no effect of high mean stress (R = 0.5) on da/dN is observed. Since in vacuum an extremely rough crack profile is obtained, and no effect of closure is seen, it is tempting to suggest that any effects of roughness induced closure in air will be minimal because the crack profile is less rough. It must be accepted, however, that the extremely rough crack profile produced in vacuum could still permit closure effects to occur even at the higher mean stress level (R = 0.5). It is unlikely that these would be as strong as those at lower mean stress, but, strictly, it is not possible at this stage, unequivocally, to establish the precise reason for the increased da/dN values at high mean stress in air.

CONCLUSIONS

(1) At 290°C not only are marked decreases in da/dN (and marked increases in ΔK_{th}) obtained in vacuum, compared with values in air, but also marked differences are observed in the crack growth mechanism. In vacuum, growth occurs largely by a transgranular facetting mechanism which proceeds along lath-lath interfaces and results in a very rough fracture surface. In air, growth occurs largely by a striation mechanism, and results in a less rough fracture surface.

(2) In air, intergranular facets are observed at low values of da/dN only. There is a total absence of intergranular facets in vacuum.

(3) Significant effects of increased mean stress (R = 0.5) are only observed in air.

REFERENCES

1. P. Bowen, C. A. Hippsley and J. F. Knott, *Acta metall.* 32, pp. 637-647, 1984.
2. R. O. Ritchie, *Int. Met. Reviews*, 20, pp. 205-230, 1979.
3. P. C. Paris et al., ASTM STP 513, pp. 141-176, 1972.
4. P. Bowen and J. F. Knott, *Met. Sci.*, 18, pp. 225-235, 1984.
5. R. M. N. Pelloux, ASM *Trans. Quart.*, 62, pp. 281-285, 1969.
6. P. Bowen, Ph.D. Thesis, "Effects of microstructure on toughness in pressure vessel steel", Univ. of Cambridge, October 1984.

ACKNOWLEDGEMENTS

One of the authors (PB) was supported during the course of this work by a CASE award from AERE, Harwell and latterly by a Goldsmiths' Junior Research Fellowship at Churchill College, Cambridge. Useful discussions with Dr. C. A. Hippsley at Harwell, and with Miss J. Kendall are gratefully acknowledged. Thanks are also due to Professor R. W. K. Honeycombe, F.R.S. and Professor D. Hull for provision of research facilities.

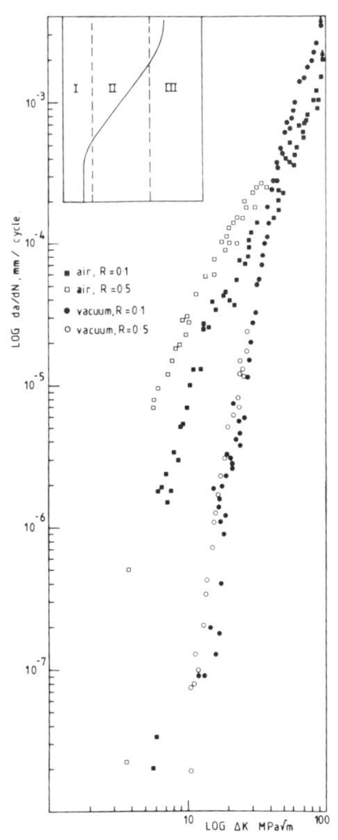

Fig. 1 Fatigue tests performed at 290°C.

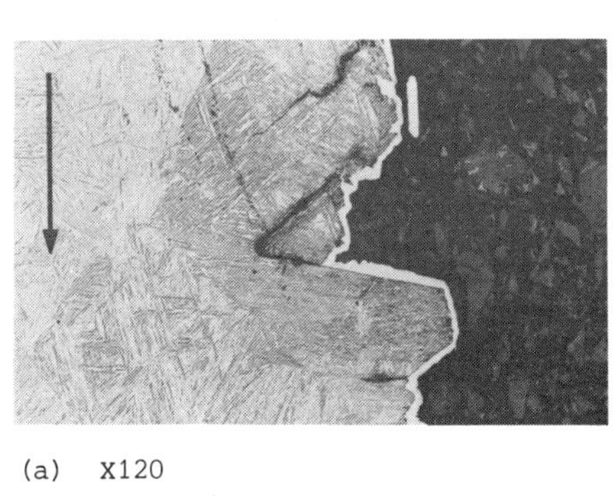

(a) X120

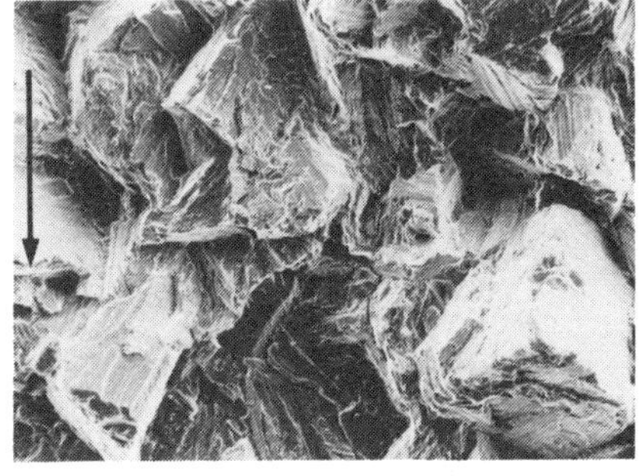

(b) X200

Fig. 2(a) and (b) Transgranular facetting observed in vacuum (a) ΔK = 16 MPa $m^{\frac{1}{2}}$, R = 0.5, (b) ΔK = 20 MPa $m^{\frac{1}{2}}$, R = 0.1. Direction of crack growth is arrowed.

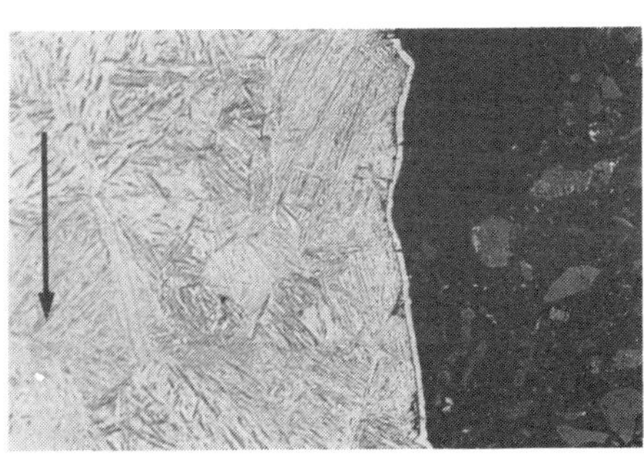

(a) X90

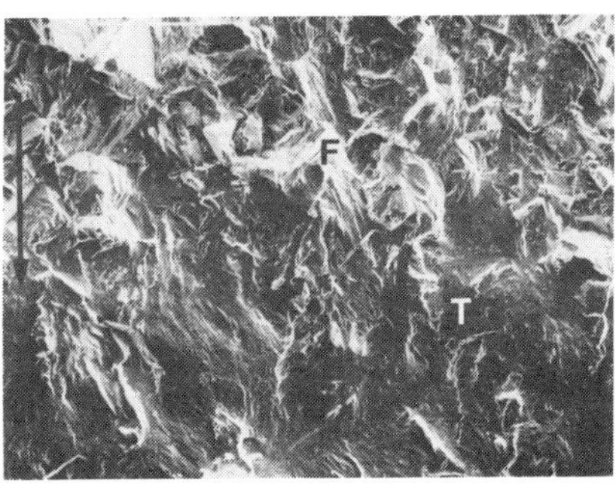

(b) X60

Fig. 3(a) and (b) Smooth crack growth in vacuum observed in near threshold regions, ΔK = 13 MPa $m^{\frac{1}{2}}$, R = 0.1. Note the change from facetted growth (F) to featureless growth (T) observed in (b).

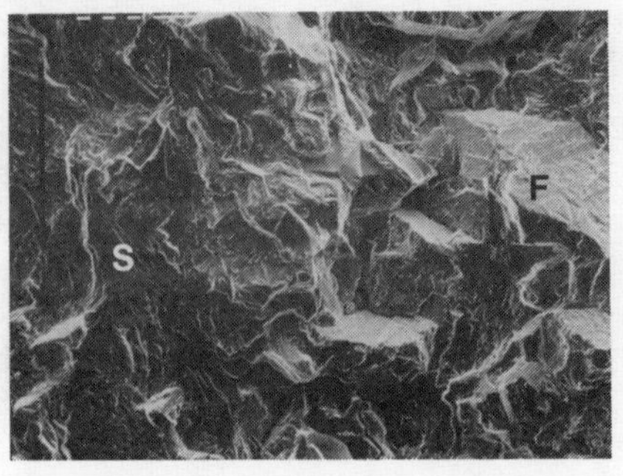

X90

Fig. 4 Crack growth in air showing striation controlled growth (S) with occasional facetting (F).
$\Delta K = 33$ MPa $m^{\frac{1}{2}}$, R = 0.1.

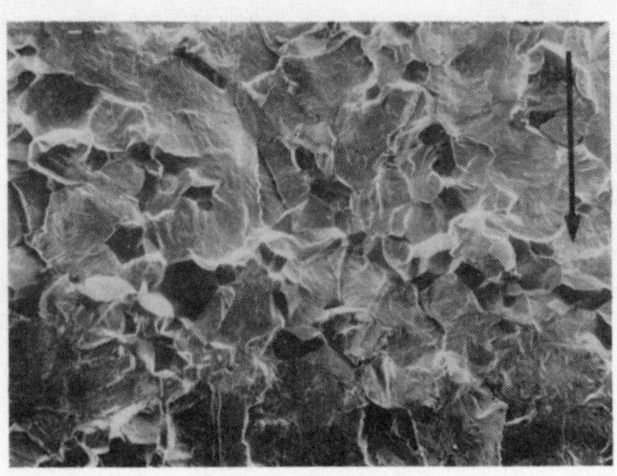

X40

Fig. 5 Crack growth in air in threshold regions. Large numbers of intergranular facets are seen.
$\Delta K = 4$ MPa $m^{\frac{1}{2}}$, R = 0.5.

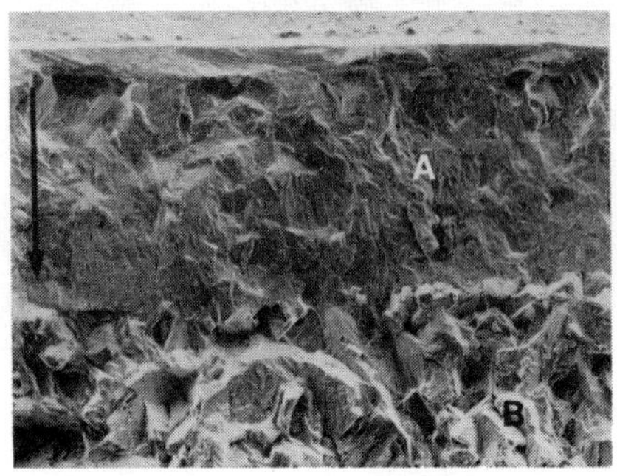

X40

Fig. 6 A comparison of crack growth in air (A) and in vacuum (B).
$\Delta K = 25$ MPa $m^{\frac{1}{2}}$, R = 0.1.

Evolution Structurale et Cavitation en Volume dans un Alliage Ni-Ge 6 at % Fatigue A O,5 Tf

B. Arnaud, R. Le Hazif et G. Martin

Centre d'Etudes Nucléaires de Saclay, I.R.D.I., Département de Technologie, Section de Recherces de Métallurgie Physique, 91191 Gif sur Yvette Cedex, France

RESUME

Nous étudions les microstructures développées par fatigue à chaud dans des solutions solides à base nickel. Nous présentons les résultats acquis sur un alliage Ni-Ge à 6% en atomes de Ge, fatigué entre 0,4 et 0,5 fois la température de fusion (Tf), sous une contrainte alternée de ± 100 MPa, à la fréquence de 2,5 Hz. Nous avons observé que :

- la microstructure de fatigue (distribution des dislocations) évolue continuellement pendant l'essai; nous n'avons pas noté de configurations d'équilibre.
- Des cavités se développent en volume; elles sont visibles après un nombre de cycles (seuil) qui dépend des conditions expérimentales (température et fréquence).
- Le domaine d'apparition des cavités en fatigue, est très voisin de celui qui correspond au gonflement du même alliage bombardé aux ions Ni^+.
- Quelques observations complémentaires montrent que les effets observés dans NiGe 6% at, existent aussi dans d'autres alliages (NiSi) et dans Ni pur.

MOTS-CLE

Fatigue à chaud ; solution solide; Ni; Ni-Ge; sous-structures; cavitation.

INTRODUCTION

Bien que les études des microstructures développées en cours de fatigue soient très nombreuses (2, 3), il existe très peu d'informations sur celles qui se forment dans les solutions solides sollicitées en fatigue. Rappelons que dans les métaux purs, les divers types de configuration de dislocations qui se développent en fatigue sont bien connus (veines, échelles, labyrinthes, cellules, etc...) et bien répertoriés dans les monocristaux (4-10) comme dans les polycristaux (11-16), ce qui a permis, entre autres, d'en dresser des cartes de domaine d'existence (15). Il est de plus connu que la déformation peut engendrer l'agglomération de défauts ponctuels, et par fatigue à chaud (17-26) il a souvent été observé une cavitation intergranulaire qui peut être rapide (10^2 cycles suffisent) et abondante (variation de densité proche de 10^{-2} (20)). Des

modèles de condensation de lacunes dans des zônes de concentration de contrainte avec diffusion le long des joints (17, 24) et/ou dans les grains (18, 21) existent. D'autres travaux signalent la possibilité d'une précipitation de lacunes en cavités à une échelle beaucoup plus fine (27-37) ou d'amas de défauts ponctuels non identifiés (38, 39). Ces agglomérats ont été pour la plupart détectés par des mesures d'analyse en volume : (annihilation des positrons (30-33), diffusion de neutrons aux petits angles (34-36) résistivité électrique (27-29, 38)) ce qui empêche de localiser ces amas.
En ce qui concerne les alliages simples, ce sont surtout les alliages biphasés qui ont été étudiés, l'intérêt étant porté sur la résistance des précipités au cisaillement par les dislocations (40-41).
Par contre à notre connaissance il n'existe pas d'étude systématique des microstructures de fatigue à chaud dans les solutions solides. Or en ajustant la composition d'alliage, on agit sur les propriétés des défauts ponctuels et sur la largeur de dissociation des dislocations du fait de la variation de l'énergie de faute. Répertorier ces effets et les comprendre présente un intérêt tant du point de vue fondamental, qu'appliqué à long terme. Du point de vue fondamental, on a signalé ailleurs (42) une analogie possible entre la phénoménologie de l'évolution de la microstructure dans des systèmes soumis à une sollicitation mécanique prolongée et dans des systèmes sous irradiation. Dans les deux cas le matériau est maintenu loin de sa configuration d'équilibre par un apport permanent d'énergie. Nous appuyant sur cette analogie, nous avons entrepris une étude comparative des microstructures développées au cours d'essais de traction-compression dans le nickel et les solutions solides NiGe et NiSi. Dans ces alliages, sous irradiation, suivant la température et la puissance injectée (flux instantané d'irradiation) des évolutions de microstructure qualitativement différentes sont observées (43). Nous pensons qu'une pareille situation se retrouve en fatigue à chaud, où la puissance injectée peut être modifiée en variant la fréquence et l'amplitude des cycles.
Nous présentons ici les résultats d'une étude faite sur du NiGe à 6%, qui montre, entre autres, l'existence d'un domaine de température à l'intérieur duquel une cavitation en volume se développe.

CONDITIONS EXPERIMENTALES

Elles ont été largement décrites par ailleurs (1), signalons simplement que nos essais sont effectués en traction-compression; la charge est imposée selon une loi triangulaire à une fréquence de 2,5 Hz. La température est comprise entre 400 et 600°C soit 0,4 à 0,5 Tf. Nous n'avons pas pu faire de mesure de la déformation par cycle, le montage ayant été optimisé afin de permettre une trempe rapide des échantillons en fin d'essai ou en cas de rupture prématurée. Les éprouvettes d'un diamètre de 3mm, sont ensuite découpées, perpendiculairement à l'axe, afin d'obtenir des disques qui sont amincis et polis en vue de leur examen par microscopie électronique.

RESULTATS EXPERIMENTAUX

Microstructure de Dislocations

Les résultats détaillés sont publiés dans (1) et nous rappelons ici les observations faites à température fixée (500°C) et à nombre de cycles croissant (Nc) (10^4 à 10^6 cycles) d'une part, à Nc fixé (5.10^5) et température croissante (400°C < T < 600°C) d'autre part. Le tableau 1 résume les résultats obtenus. Nous remarquons, que dans les 2 cas, le schéma d'évolution des structures est identique. La structure passe progressivement d'une répartition homogène des dislocations dans les grains, (la taille initiale des grains est de 70 μ) à une sous-structure cellulaire de taille moyenne 0,5 à 1 μ, dont les parois sont bien formées et les murs épais; par la suite les cellules semblent plus lâches, elles ne sont plus toutes fermées, et, la densité de dislocations dans

Tableau 1 Structures Observées

N \ T°C	400	450	500	600
10^6			cellules + cellules laches + polygonisation	
5.10^5	homogene	cellules	polygonisation + cellules laches	polygonisation quasi-totale
10^5			cellules laches	
5.10^4			cellules	
10^4			homogene	

les parois est plus faible. A un stade plus avancé la structure est plus hétérogène, elle est composée de cellules lâches et de plages où la recristallisation a commencé. Au stade final de l'évolution structurale la recristallisation semble totale, avec de nouveaux grains dont la taille est nettement inférieure à la taille initiale, et proche de la taille des premières cellules. A ce stade il semblerait que l'alliage se comporte comme un alliage neuf, mais avec des grains plus petits que l'alliage initial, un nouveau cycle d'évolution pourrait alors commencer; ce point reste à vérifier, il demande de très grands nombres de cycles qui sont rarement atteints, la rupture de l'éprouvette intervenant avant.

Cavitation en Volume

De nombreuses éprouvettes présentaient une cavitation en volume. La Fig. 1 montre les conditions de sollicitation donnant lieu au phénomène.

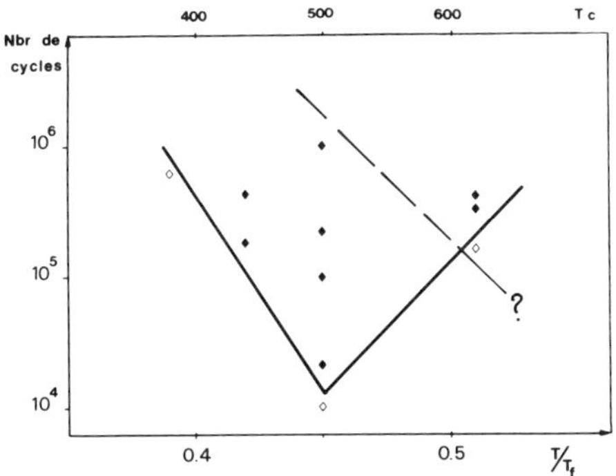

Fig. 1. Domaine d'apparition de la cavitation en volume. Alliage NiGe 6% at; fréquence 2,5 Hz contrainte ± 100 MPa.

La cavitation a lieu entre 0,4 et 0,5 Tf. La forme, la taille et la distribution des cavités sont fonction des conditions expérimentales (1). A 0,4 Tf, près du seuil d'apparition, les cavités sont très petites (< 20 nm), elles sont situées dans des plages représentant moins de 1% du volume. La densité dans ces plages est de l'ordre de $10^{15}/cm^3$. Vers 0,45 Tf, et 10^5 cycles environ, la taille des cavités est de 100 nm; le volume des plages est de quelques %; dans ces plages la densité est plus faible $10^{14}/cm^3$. A 0,4 Tf, 5.10^5 cycles nous étions dans un domaine proche de la structure homogène. A 0,45 Tf et 10^5 cycles nous commençons à quitter le domaine des cellules; on peut voir sur la Fig. 2 une répartition de cavités qui semblent être sur les sites d'une ancienne sous-structure cellulaire.

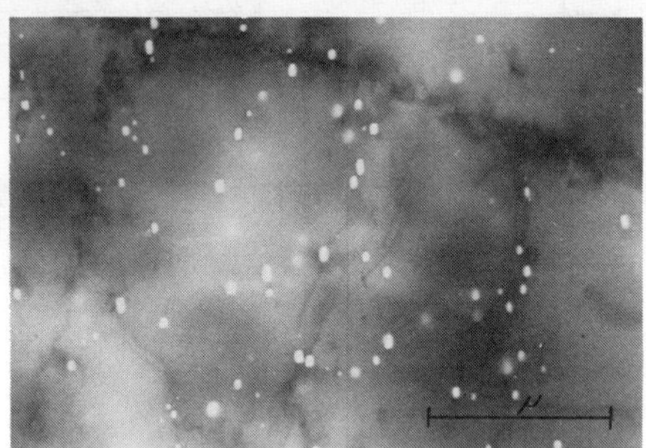

Fig. 2. Cavités en volume dans NiGe 6% at. Fatigué à 2,5 Hz, 10^5 cycles, contrainte ± 100 MPa

A plus haute température 0,5 Tf, et 5.10^5 cycles, dans le domaine de recristallisation, le volume des plages à cavités est inférieur à 0,1 % du volume; la densité de cavités est de $10^{13}/cm^3$ et seules quelques grosses cavités de plus de 100 nm sont observées.
La forme des cavités et la distribution de celles-ci dans les plages sont aussi fonction du domaine où elles sont formées. A 0,4 Tf et 5.10^5 cycles elles sont plutôt équiaxes et réparties au hasard; pour les plus fortes déformations et plus hautes températures, en plus de cavités équiaxes, on trouve des cavités allongées selon une direction <111>; elles peuvent être au hasard ou en bande. Un recuit de contrôle, a montré que les cavités de taille inférieure à 20 nm disparaissent en 3 minutes à 570°C; les plus grosses (100 nm) voient leur diamètre diminuer de moitié après quelques minutes à 620°C.

DISCUSSION ET CONCLUSION

La succession des microstructures à nombre de cycles croissant et température fixée rappelle pour les faibles nombres de cycles les résultats obtenus dans du laiton (11). D'autre part, extrapolant les cartes de formation des types de structures dues à Shiraï et Weertman (15) on constate que nos conditions expérimentales sont très compatibles, pour les grands nombres de cycles, avec celles qui conduisent à l'apparition des cellules et des sous-grains. L'évolution de la microstructure à nombre de cycles donné et à température croissante est très proche de l'évolution observée à température constante et nombre de cycles croissant. Ceci nous amène à suggérer que les mêmes successions de structures se produisent, et à nombre de cycles croissant elles seront d'autant plus rapide que la température augmente. La structure n'atteint pas de

configuration stationnaire, mais semble se détruire et se reconstruire continuellement.
La cavitation que nous observons montre les limites des modèles couramment admis pour la cavitation sous contrainte cyclique (21, 24, 25) : d'une part nous observons des cavités qui ne sont pas liées à des joints de grains, d'autre part ces cavités semblent vides de gaz, comme le suggère la rapidité du recuit dans le microscope. On admet en général que seules les concentrations de contrainte au niveau des joints (21, 24) ou une pression de gaz dans les cavités (25) peuvent expliquer la croissance de cavités en sollicitation cyclique. Lorsque les cavités ont été observées en volume (36, 37), il a été proposé que ce sont des cavités formées aux joints de grains qui auraient migré après coup, la migration des joints en fatigue étant bien connue (44, 45).
Les modèles classiques reposent sur l'idée que seul un transfert de lacunes des joints vers les cavités contribue à la croissance de celles-ci, le flux de lacunes provenant de la différence des concentrations d'équilibre de lacunes au joint sous contrainte et à la cavité. Comme la contrainte est alternée, cette différence change de signe et il faut soit une pression de gaz dans la cavité pour empêcher que celle-ci ne s'effondre totalement au cours du demi-cycle où la cavité alimente le joint en lacunes (25), soit tenir compte de termes du deuxième ordre en contrainte (qui ne changent pas de signe au cours du cycle) et des concentrations de contrainte aux rugosités (21) ou ondulations (24) du joint pour obtenir une croissance des cavités.
Une telle interprétation ne saurait être générale : nos observations montrent par exemple un cas de configuration où les cavités sont plutôt réparties en couronnes rappelant le contour de cellules observées à faible nombre de cycles. D'autre part des amas de défauts ponctuels (dont la nature lacunaire ou interstitielle n'a pas été déterminée) ont été signalés dans les murs des cellules (38, 39). De même on sait qu'une addition de Ge à Cu (31) ou Ni (31) favorise l'agglomération des lacunes formées par laminage à température ambiante.
Nos observations montrent donc que des cavités peuvent (sinon germer) au moins se développer sous contrainte cyclique, sans être dans un joint de grains, ni être stabilisées par une pression de gaz. Nous avons montré ailleurs (1) que la simple élimination aux cavités de la sursaturation de lacunes produites par déformation plastique pourrait rendre compte de la croissance de celles-ci.
La démarche proposée pour rendre compte de cette croissance, s'appuie sur le fait que le domaine de températures d'apparition des cavités en fatigue, est le même que celui observé pour le gonflement de cet alliage irradié aux ions Ni^+ (46). On compare les vitesses de production et d'élimination des défauts dans les 2 cas (irradiation et fatigue); et on montre que les forces motrices de la croissance des cavités peuvent être du même ordre de grandeur bien que les puissances injectées soient bien plus importantes sous irradiation. On montre alors que ce résultat n'est pas surprenant, car dans le domaine de température où nous travaillons une grande partie des lacunes d'irradiation s'éliminent par recombinaison mutuelle avec les interstitiels, alors qu'elle se fait par diffusion (donc plus lentement) vers les dislocations sous fatigue.
Terminons enfin, en signalant que nous avons aussi observé de la cavitation dans du NiSi[1] et du Ni, sollicités dans le même domaine de température.
Une étude sur l'effet de la fréquence est en cours, et les premiers résultats montrent qu'une augmentation de celle-ci, décale le domaine d'apparition des cavités vers les plus hautes températures. Le trait pointillé sur la Fig. 1, montre le domaine d'apparition des cavités pour NiGe fatigué à ± 100 MPa sous 10 Hz. Ici encore, on remarque une analogie entre les effets d'irradiation et la fatigue, sous irradiation l'augmentation du flux décale le pic de gonflement vers de plus hautes températures.

[1] (Ni-Si à 4% en at)

BIBLIOGRAPHIE

1 - B. Arnaud, R. Le Hazif, G. Martin - soumis à Acta Met. (1984)
2 - C. Laird - Plastic Def. Mat. 6 (1976) 101
3 - J.C. Grosskreutz, H. Mughrabi - Constitutive equation in plasticity (1975) 29, Ed. A. Argon, M.I.T. Editions
4 - P. Lukàs, M. Klesnil, J. Krejor - Phys. Stat. Sol. 27 (1968) 545
5 - P. Lukàs, M. Klesnil - Phys. Stat. Sol. 37 (1970) 833
6 - P.J. Woods - Phil. Mag. 28 (1973) 155
7 - A. Abel, M. Wilhelm, V. Gerold - Mat. Sci. Eng. 37 (1979) 187
8 - T. Tabata, H. Fujita, M. Hiraoka, K. Onishi - Phil. Mag. 47 (1983) 841
9 - N.Y. Jin - Phil. Mag. 48 (1983) L 33
10 - F. Ackerman, L.P. Kubin, J. Lepinoux, H. Mughrabi - à paraître
11 - J.C. Grosskreutz, W.H. Reimann, W.A. Wood - Acta Met. 14 (1966) 1549
12 - J.E. Pratt - Acta Met. 15 (1967) 319
13 - S.P. Bhat, C. Laird - Fat. of Eng. Mat. and Struc. 1 (1979) 59
14 - A.T. Winter, O.B. Pedersen, K.V. Rasmussen - Acta Met. 29 (1981) 735
15 - H. Shirai, J.R. Weertman - Scripta Met. 17 (1983) 1253
16 - J.O. Nilson - Scripta Met. 17 (1983) 593
17 - D. Hull, D.E. Rimmer - Phil. Mag. 4 (1959) 673
18 - R.P. Skelton - Phil. Mag. V14 (1966) 563
19 - M.V. Speight, J.E. Harris - Met. Sci. Jal 1 (1967) 83
20 - A. Gittins - Met. Sci. Jal 2 (1968) 51
21 - J. Weertman - Met. Trans. 5 (1974) 1743
22 - R. Raj, M.F. Ashby - Acta Met. 23 (1975) 653
23 - T. Saegusa, J.L. Weertman - Scripta Met. 12 (1978) 187
24 - J.R. Weertman - Can. Met. Quately 18 (1979) 73
25 - H. Trunkaus - Scripta Met. 15 (1981) 825
26 - K.U. Snowden, D.S. Hughes, P.A. Stathers - Met. Sci. 2 (1981) 73
27 - W. Kleinert, W. Schmidt - Phys. Stat. Sol. (a) 60 (1980) 69
28 - R. Franke, W. Kleinert, W. Schmidt - Phys. Stat. Sol. (a) 67 (1981) 469
29 - W. Kleinert, R. Franke - Phys. Stat. Sol. (a) 53 (1979) K 177
30 - P. Hautojärvi, A. Vehanen, V.S. Mikhalenkov - Sol. Sta. Can. 19 (1976) 309
31 - A. Vehanen, J.Y. Li Kauppila, P. Hautojärvi, V.S. Mikhalenkov - Phys. Stat. Sol. (a) 52 (1979) K 73
32 - G. Andrews - Phys. Stat. Sol. (a) 67 (1981) K 127
33 - T. Lepistö, J.Y. Li Kauppila, P. Kettunen, P. Hautojärvi - Phys. Stat. Sol. (a) 67 (1981) K 93
34 - P. Page, J.R. Weertman, M. Roth - Acta Met. 30 (1982) 1357
35 - P.O. Kettunen, T. Lepistö, G. Kostorz, G. Göltz - Acta Met. 29 (1981) 969
36 - M.H. Yoo, J.C. Ogle, B.S. Borie, E.H. Lee, R.W. Hendricks - Acta Met. 30 (1982) 1733
37 - R. Page, J.R. Weertman - Acta Met. 29 (1981) 527
38 - R. Gonzalez, J. Piqueras, L. Bru - Phys. Stat. Sol. (a) 34 (1976) K 9
39 - J. Piqueras, J.C. Grosskreutz, W. Frank - Phys. Stat. Sol. (a) 11 (1972) 567
40 - H. Mughrabi - RISO (1983) 65
41 - E.A. Starke, G. Lütjering - Fatigue et Microstructure ASM (1979)
42 - G. Martin - Ann. Chim. Fr 6 (1981) 46
43 - G. Martin, R. Cauvin, A. Barbu - Phase transf. during irradiation, Ch. 2, Ed. Nolfi, Applied Science Published
44 - P. Yavari, T.G. Langdon - Acta Met. 31 (1983) 1595
45 - T.G. Langdon, R.C. Gifkins - Scripta Met. 13 (1979) 1191
46 - A. Barbu - Communication personnelle

Effect of Waveshape and Hold-time on Fatigue Crack Propagation in Inconel X-750

F. Gabrielli* et R. M. Pelloux**

**CNR-ITM, via Induno 10, 20092 Cinisello B., Italy*
***Massachusetts Institute of Technology, Cambridge, MA 02139, USA*

ABSTRACT

Fatigue crack growth rates (FCGR) of Inconel X-750 were measured at 650°C with different loading waveforms. The waveshapes were: a) triangular, b) trapezoidal with different hold-times at maximum load and c) unsymmetrical with fast-slow and slow-fast loading-unloading stress rates. FCGR were compared with creep crack growth rates (CCGR) measured in single edge notch specimens at 650°C.
The inclusion of hold-times at K_{max} produces an increase of FCGR and, when hold-times are large, leads to fatigue crack growth mechanisms which are similar to those observed in high temperature creep cracking.
High temperature FCGR measured with unsymmetrical waveshape with slow loading and fast unloading are higher than those measured with fast-slow cycles.
The role of time dependent processes in accelerating high temperature fatigue crack propagation is discussed in detail.

Keywords: Fatigue, creep, crack growth, crack growth mechanisms, fatigue striations, creep cavities, oxidation.

INTRODUCTION

Fatigue fracture of many high temperature components often occurs by initiation and propagation of cracks from stress concentration regions due to thermal gradients or to preexisting flaws. The rational design approach for a structure, where preexisting flaws cannot be eliminated, is to determine the remaining life on the basis of crack growth. However, elevated temperature fatigue is complicated by the presence of time dependent phenomena. Time dependent damaging processes produce microstructural changes resulting in a marked decrease of fatigue resistance with subsequent reduction of the expected fatigue lives evaluated on the basis of time independent fatigue.

Crack growth by fatigue, creep or simultaneous processes of creep and fatigue involves a large number of parameters such as temperature, frequency, waveform, hold-time, microstructure, environment, etc.

The present investigation intends to clarify the effects of some of these variables in high temperature fatigue of Inconel X-750, a γ-γ' hardenable Ni base superalloy, which is used in several applications in the aerospace and nuclear industries. The effect of temperature, frequency and environment has already been reported [1,2]. The effect of environment was evaluated by comparing test data obtained in aggressive environment with data in inert environment. In air tests the environmental effects are principally due to oxidation at the crack tip. Oxidation damage at high temperature is mainly transgranular at high frequency and intergranular at low frequency.

In the present study the effects of different waveshapes and hold-times in a triangular waveform are analyzed. Fracture mechanisms are studied and correlated to the macroscopic fatigue crack growth data.

MATERIAL AND EXPERIMENTAL PROCEDURES

The Inconel alloy X-750 used in this investigation was supplied by International Nickel Co in the form of 12.7 mm thick plate of the following composition (wt%): Ni 73, Cr 15.5, Fe 7, Ti 2.5, Al 0.7, Nb 1, Mo 0.5, Si 0.2, C 0.04, after annealing (2h-1150°C/AC). The average grain size was 0.12 mm and the mechanical properties are summarized in Table I.

Single edge notch tensile specimens were used in this investigation. The test specimens, machined from the plate, had a rectangular cross section of 11.7x 4.4 mm^2 and a starter notch, 1 mm deep, was cut with a string saw.

Fatigue crack growth tests were performed in load control on a closed loop servo-controlled hydraulic testing machine with a 110 kN load capacity. A Radio Frequency Lepel unit was used to heat the specimen with a temperature control of ±4°C. Triangular loading was chosen because it is easy to insert a hold--time at maximum load while keeping a constant loading rate.

In this study, low frequency fatigue crack growth rates have been measured for different waveshapes. In the dual rates continuous cycling tests a positive loading rate, different from the negative loading rate, was employed, viz, fast -slow and slow-fast loading in a sawtooth wave with rise times of 0.5 and 80 s, respectively. The R ($=K_{min}/K_{max}$) ratio was kept equal to 0.05.

Crack length was determined with a travelling telescope with 0.01 mm accuracy. Fatigue crack growth rates (FCGR) were correlated to the stress intensity range ($\Delta K=K_{max}-K_{min}$). The stress intensity factor was calculated with an expression derived by Harris [3] for an edge notch crack in an edge notched bar not subjected to bending.

All the fracture surfaces were examined with a SEM in order to identify the micromechanisms of fracture.

RESULTS

The effect of the introduction of hold-times in a triangular waveform (1 Hz) on high temperature FCGR of Inconel X-750 is shown in Fig. 1. In general FCGR increase when the hold-time (HT) is increased. The comparison of the data obtained with no hold-time and with HT=20 s shows that at very low ΔK FCGR with HT=20 s tend to values measured with no hold-time. The $\log\frac{da}{dn}$ vs $\log \Delta K$ curve shows a rapidly increasing crack growth rate with increasing ΔK. At high ΔK, the introduction of the hold time produces an increase of the FCGR and a decrease of the critical stress intensity range ΔK_c.

Table I: Mechanical properties of Inconel X-750.

T °C	σ_{ys} MPa	σ_{uts} MPa	ε_p %
24	448	772	50
24	441	820	50
550	324	689	44
650	552	621	7
650	531	607	7
750	503	559	4

When hold-times increases, the increase of FCGR is more remarkable at low ΔK and, apparently, all the $\log\frac{da}{dn}$ vs $\log \Delta K$ curves

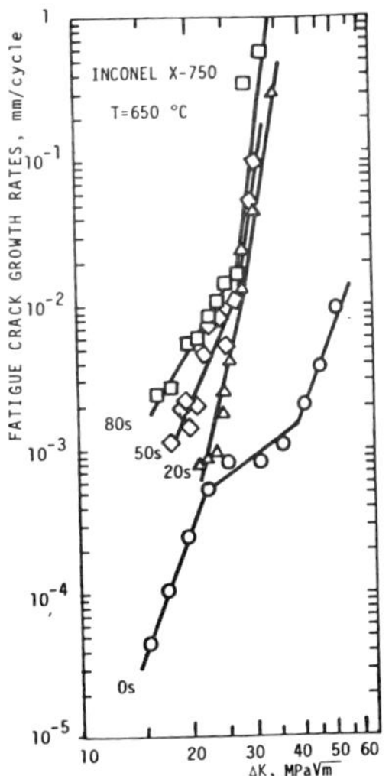

Fig. 1 - Hold-time effect on FCGR of Inconel X-750 at elevated temperature (Baseline data is for 1 Hz).

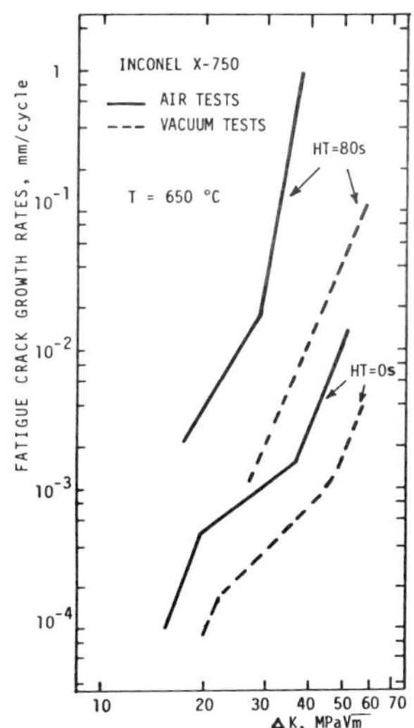

Fig. 2 - Oxidation effect on FCGR in Inconel X-750 tested with and without hold-time (1 Hz).

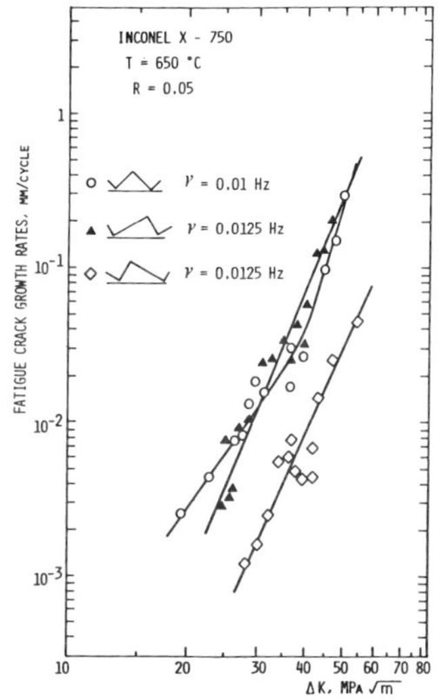

Fig. 3 - Effect of waveshape on high temperature FCGR of Inconel X-750.

with HT≥ 20 s tend to the same ΔK_c,Fig.1.

The effect of oxidation is illustrated in Fig. 2 which compares the fatigue crack growth tests in air and in vacuum ($p_{O_2}<10^{-5}$torr). It is evident that the aggressive nature of oxygen manifests itself more in cycles with hold-times, and it is apparent that this effect is not negligible at low FCGR.

The effects of different waveshapes with dual rate waveshapes continuous cycling is shown in Fig. 3. The FCGR obtained with unsymmetrical waves were compared with those obtained with symmetrical triangular waveshape with the same maximum load and nearly equal period. No significant difference is observed in the FCGR with symmetrical and slow-fast waveshape. FCGR, however, decrease by an order of magnitude in fast-slow tests. This effect can be attributed to oxidation and creep processes at the crack tip.

FRACTOGRAPHY

All the fatigue fracture surfaces were extensively examined for a better understanding of the different fracture mechanisms operative in the range of ΔK that have been investigated.

The specimens fatigued at 1 Hz, 650°C show a transgranular failure mode with different mechanisms depending on ΔK level: i) a crystallographic fracture mode at low ΔK, ii) a fracture mode by ductile striations at intermediate ΔK with very small regions of intergranular fracture and iii) void coalescence at high ΔK. A detailed analysis of fracture mechanisms in fatigue crack propagation of Inconel X-750 versus temperature and frequency has been already reported [4].

When a hold-time is inserted at maximum load in the triangular waveform at 650°C, the fatigue striations (Fig. 4) are replaced by intergranular fracture mode portions which become larger with increasing hold-times. The fracture surface of the specimen fatigued with HT = 20 s is largely intergranular with some limited transgranular regions which disappear when hold-time is increased to 50 and 80 s. With HT>20 s the fracture path is completely intergranular (Fig. 5) with secondary cracks at grain boudaries. The mechanical cyclic effects of crack opening and closing with rather high stress rates appear to be negligible and are unable to produce a transgranular part of fracture.

The fractographic analysis of the specimen fatigued in vacuum with HT = 80 s shows an intergranular fracture similarly to the specimens tested in air but without secondary cracks. Furthermore several creep cavities, that in the air tests were obscured by oxidation, were observed (Fig. 6); the cavity size increases on increasing of the ΔK.

The intergranular character of the fracture surfaces is found also in specimens fatigued with unsymmetrical waveshapes. However the fracture surfaces of the specimen tested with a fast-slow waveshape exhibit weaker oxidation effects than in specimens tested with slow-fast and symmetrical waveshapes.

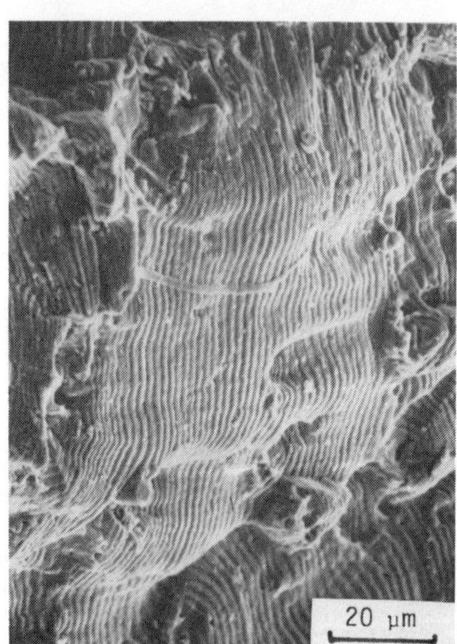

Fig. 4 - Fatigue fracture surface of Inconel X-750 air tested at 650°C with triangular waveshape (1 Hz, HT= 0 s).

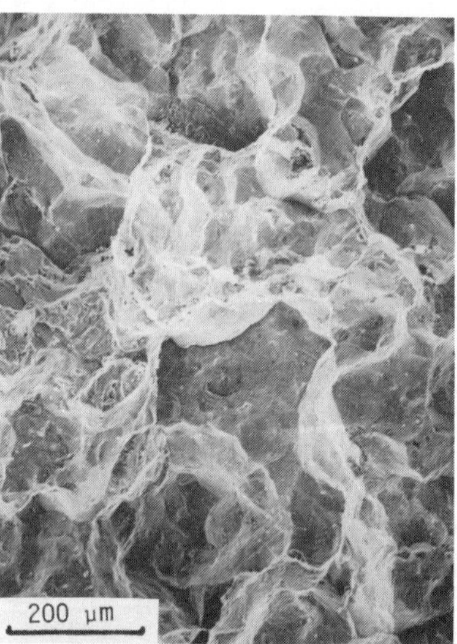

Fig. 5 - Intergranular fracture in Inconel X-750 fatigued in air at 650°C with HT= 80 s.

DISCUSSION

In cycle dependent fatigue, high temperature fatigue crack growth is essentially due to oxidation processes at the crack tip owing to the aggressive nature of air environment. Oxygen transport into the crack tip plastic zone is favored by cyclic loading [5,6]; it can diffuse along the persistent slip bands to enhance the stage I growth or in the bulk from the crack tip to enhance the stage II.

In time dependent fatigue, oxidation processes operate in conjunction with creep mechanisms to produce an intergranular crack growth mode. In general grain boundaries are characterized by the presence of cavities and/or by mechanisms of brittle decohesion depending on the relative contribution of creep and oxidation. Creep mechanisms at a crack tip include grain boundary sliding, growth and coalescence of cavities. Furthermore the oxygen diffusing along the boundaries produces chemical reactions with grain boundary phases and enhances the flux of vacancies promoting rapid nucleation and growth of cavities [7]. Grain boundary diffusion may also cause embrittlement, which contributes to the debonding and brittle decohesion of grain boundaries.

Intergranular damage is the typical mechanism found after fatigue testing of Inconel X-750 at high temperature with sufficient hold times (>20 s). These fracture mechanisms were also observed after fatigue tests with triangular symmetrical waveshape with the same period as the trapezoidal waveshape. The FCGR values are similar for both waveshapes in air and in vacuum (Fig. 7), indicating that the duration of the load cycles is very important in high temperature fatigue at very low frequencies.

With increasing hold-times at low ΔK creep-fatigue damaging mechanisms contribute in considerable way to the increase of the FCGR. As soon as the damage spreads uniformly in the material a fast fatigue crack growth, that seems independent on creep processes, occurs.

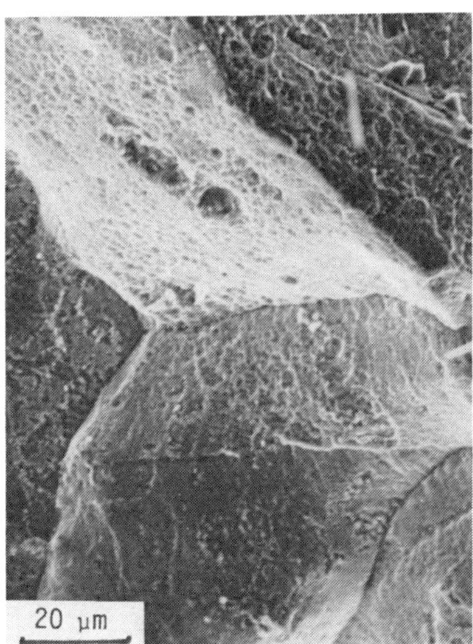

Fig. 6 - Cavities in fracture surface of Inconel X-750 vacuum fatigued with hold-time of 80 s.

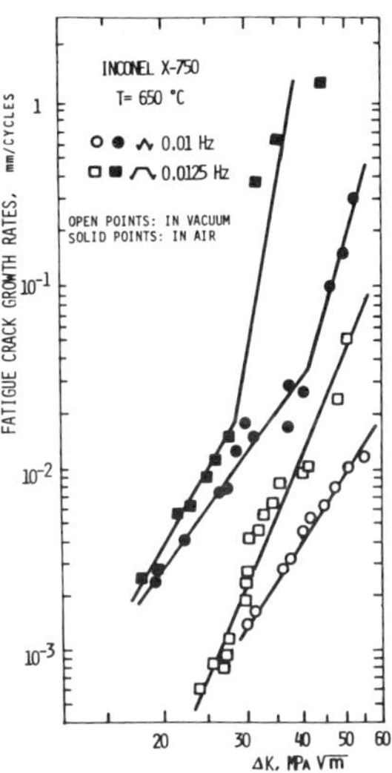

Fig. 7 - Comparison of data obtained with triangular and trapezoidal waveshapes with same maximum load and nearly equal period.

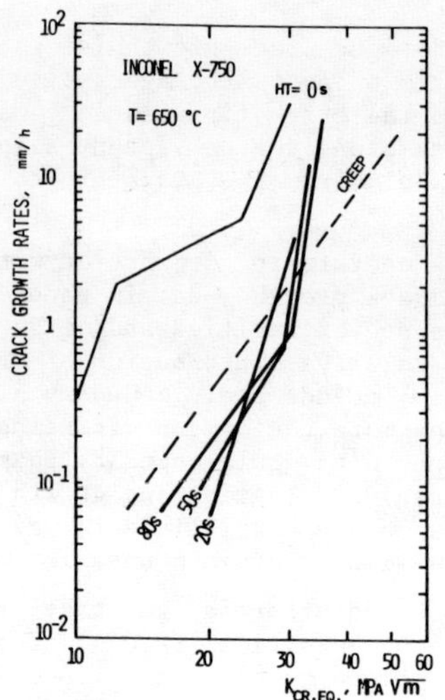

Fig. 8. Comparison of crack growth rates per unit time in Inconel X-750 creep tested and fatigue tested with different hold-times.

FCGR and creep crack growth rates were compared by transforming the $\log\frac{da}{dn}$ vs $\log \Delta K$ plots to $\log\frac{da}{dt}$ vs $\log K_{cr.eq.}$ with the mathematical transformations reported in a previous work [1] and, for trapezoidal waveshapes, assuming $K_{max}=K_{cr.eq.}$. The $\frac{da}{dt}$ data transformed from the FCGR data and the creep crack growth rates are plotted in Fig. 8. The comparison of fatigue data with creep data shows that FCGR per unit time with zero hold-time are significantly higher than creep rates and FCGR with HT $\geq$ 20 s. This leads to the conclusion that, for the triangular waveshape, cyclic mechanical damage prevails over the creep damage. This is confirmed also by the transgranular fracture mode with ductile fatigue striations.

The introduction of hold-times leads to FCGR per unit time which, within the limits of data dispersion are comparable with the creep crack growth rates. However, at low ΔK, a small systematic difference is obtained with trapezoidal waveshapes. This difference can be attributed to the closure effects taking place at the crack tip, during cyclic tests. Such phenomena prevent the exposure of fresh material, thus decelerating the crack growth per unit time because of the reduced contribution of oxidation mechanisms.

CONCLUSION

Fatigue crack growth data, obtained at 650°C with different waveshapes have shown that the introduction of hold-times in a triangular waveform produces an increase of the FCGR and a decrease of the critical stress intensity range ΔK_c. Sufficiently large hold-times lead to fatigue crack growth mechanisms similar to those observed in high temperature creep crack growth.

High temperature FCGR measured with unsymmetrical waveform with slow loading rate and fast unloading are higher than those measured with fast-slow cycles due to synergistic effects of creep and oxidation at the crack tip.

This research work was performed at MIT and sponsored by the Air Force Office of Scientific Research under Grant no. 76-2928E. One of the authors (F.G.) was formerly Visiting Scientist at MIT on leave of CNR-ITM that is gratefully acknowledged for the financial support.

REFERENCES

1. F. Gabrielli and R. M. Pelloux, *Metall.Trans.A*, 13A, 1083(1982).
2. F. Gabrielli and R. M. Pelloux, in *Proceed. 8th Congress on Material Testing*, OMIKK-Technoinform, Budapest, p.622(1982).
3. D. D. Harris, *J. Bas. Engn. Trans. ASME*, 89, 49(1967).
4. F. Gabrielli and R. M. Pelloux, *La Met. It.*, 10, 449(1981).
5. K. Sadananda and P. Shahinian, *J. Engng. Mat.Technol.*, 100, 381(1978).
6. J. W. Swanson and H. L. Marcus, *Metall. Trans. A*, 9A, 291(1978).
7. R. H. Cook and R. P. Skelton, *Int. Metall. Rev.*, 19, 199(1974).

Anisotropie de Deformation Plastique Cyclique et de Fissuration par Fatigue dans l'Acier 316 L a 600 °C

R. Roux, J. Charrier et C. Gasc

Laboratoire de Mécanique et de Physique des Matériaux, U.A 863 au C.N.R.S., ENSMA, Rue Guillaume VII, 86034 Poitiers, France

RESUME

Des éprouvettes prélevées dans une tôle fortement texturée en acier 316 L et sollicitées en fatigue oligocyclique à 600 °C présentent des directions de propagation des fissures reliées directement aux directions principales de la tôle. L'étude de la structure et des mécanismes de glissement a permis d'expliquer pourquoi la propagation se produit toujours suivant l'épaisseur de la tôle.

INTRODUCTION

Les modèles mécaniques d'endommagement et de propagation des fissures de fatigue /1,2,3,4/ ne permettent pas jusqu'à présent de tenir compte de l'anisotropie du matériau. Celle-ci, qui peut résulter de la texture créée par la mise en forme ou d'une structure orientée, est pourtant susceptible de modifier les mécanismes d'amorçage et de propagation des fissures.

Nous avons testé en fatigue oligocyclique à 600 °C des éprouvettes en acier inoxydable 316 L prélevées dans une tôle épaisse, soit dans le sens du laminage soit suivant le sens travers long. Cette tôle présente une orientation préférentielle très marquée due au laminage. Des résultats antérieurs sur un matériau analogue faisaient état d'un comportement différent dans les directions L et T.L /5/. Nous avons cherché dans cette étude à préciser et expliquer les différences éventuelles concernant l'amorçage et la propagation des fissures pour ces deux directions de sollicitation.

STRUCTURE ET TEXTURE

La tôle est hypertrempée à partir de 1150 °C après laminage. Les grains d'austénite sont légèrement allongés (45 à 90 μm dans le sens long, 35 à 75μm dans les sens travers). On trouve environ 2 % de ferrite résiduelle sous forme de plaquettes ($e = 4\ \mu m$, $d = 40\ \mu m$) parallèles au plan (L, T.L.), situées au joints de grains de l'austénite.

La texture a été déterminée par enregistrement des figures de poles des plans (111) et (200). Elle est fortement marquée puisque dans 85 % des grains le réseau est désorienté de moins de 10 degrés par rapport à l'orientation principale (fig. 1) :

Plan de la tôle : (100) - Direction de laminage L : [110]
Direction travers long T.L.: [$\bar{1}$10] - Direction travers court T.C : [100]

La figure 2 résume les observations relatives à la structure et à la texture, et permet de situer ces éléments par rapport aux éprouvettes prélevées dans les directions L ou T.L.

COMPORTEMENT EN FATIGUE

Les éprouvettes ont été sollicitées à 600 °C en fatigue oligocyclique jusqu'à rupture en traction compression à déformation totale imposée $\Delta\varepsilon_t$ = 0,7 %. Les essais ont été effectués soit avec un extensomètre longitudinal, soit avec un extensomètre axial ; les durées de vie obtenues sont indépendantes du type d'extensomètrie, sauf localisation particulière de la fissure principale.

Les durées de vie sont pratiquement les mêmes pour les éprouvettes prélevées dans le sens long et dans le sens travers long ($N_R \sim 3000$). De même l'évolution de la contrainte en haut de cycle au cours du cyclage est identique pour les deux orientations.

L'amorçage des fissures de fatigue est peu sensible à l'effet d'anisotropie. On observe en effet pour les deux orientations des sites d'amorçage multiples tout autour du fût de l'éprouvette (fig.3). Cependant les mécanismes de plastification sont sensibles à l'anisotropie puisque par fluage à un niveau de contrainte équivalent à celui de l'essai de fatigue (250 MPa), on constate dès la fin du fluage primaire une forte anisotropie du coefficient de Poisson plastique /6/. Ceci a été retrouvé au cours du cyclage avec l'extensomètrie diamétrale /7/.

La propagation des fissures est, elle aussi, sensible à l'effet d'anisotropie puisqu'elle se produit toujours dans la direction T.C., c'est-à-dire suivant l'épaisseur de la tôle, aussi bien pour une sollicitation suivant L que suivant T.L. Cette conclusion résulte de la localisation de la section de rupture finale. Elle est confirmée par l'observation au M.E.B. des stries de fatigue.

DISCUSSION ET INTERPRETATION

Si on applique une force suivant la direction [011](direction L de la tôle), le calcul des coefficients de Schmidt montre que le plan de glissement actif est ($\bar{1}\bar{1}$1) et que, dans ce plan, deux directions de glissement son possibles : [110]et [101]. L'amplitude des déplacements dans les trois directions principales de la tôle provoqués par ces deux systèmes de glissement est proportionelles aux valeurs suivantes :

L : 0,5 T.L. : 0,5 T.C. : 0,7

Les déplacements induits par le glissement dans le plan de la tôle (L ou T.L.), et dans la direction perpendiculaire T.C., sont donc d'amplitude différente. De plus, puisque les deux systèmes de glissement sont également probables, et compte tenu de la dispersion autour du pic de texture, on peut admettre qu'ils interviennent chacun dans la moitié des grains. On observe alors que les déplacements suivant T.L. dus à ces deux systèmes sont de sens opposés du fait de la symétrie par rapport au plan ($01\bar{1}$) (fig.4). Leur effet s'annule donc pour l'ensemble des grains. Par contre, il n'en est pas de même dans les direction T.C. et L où les déplacements sont de mêmes sens dans tous les grains quel que soit le glissement considéré.

Si on applique une force suivant la direction ($0\bar{1}1$) (direction T.L. de la tôle), on parvient à des conclusions identiques du fait de la symétrie de la texture.

Dans le cas de l'éprouvette prélevée dans le sens long, si on compare la sollicitation imposée à deux amorces de fissure situées sur le fût de l'éprouvette, l'une du côté T.C. et l'autre du côté T.L., la première est sollicitée en modes I + II et la seconde en mode I + III. En effet, le mouvement en fond de fissure dans le sens long correspond au mode I, alors que le mouvement dans le sens T.C., correspond respectivement aux modes III et II (fig.5). Pour les deux amorces de fissures, on a donc un mode I d'ouverture qui correspond à la fissuration par stries. A celui-ci se superpose pour l'amorce T.C. un mode II de forte amplitude et de même sens dans tous les grains (qui intervient également dans les modèles de stries),ainsi qu'un mode III, mais de plus faible amplitude et dont l'effet s'annule sur l'ensemble des grains.

Pour l'amorce T.L., le mode I est accompagné au contraire d'un mode III de forte amplitude, l'effet du mode II s'annulant entre les grains voisins.

Dans le cas des éprouvettes prélevées en Travers-Long, on retrouve les mêmes sollicitations pour les amorces situées respectivement du côté T.C. et du côté L.

Puisqu'on a observé que les fissures se propagent toujours à partir des amorces situées du côté T.C., il apparaît que la superposition des modes I + II contribue plus efficacement à la propagation que celle des modes I + III, ce qui est conforme aux modèles classiques de propagation par stries /8/. Ces stries, qui sont présentes sur toutes les surfaces fissurées, sont donc bien caractéristiques du mécanisme de fissuration par fatigue dans ce matériau à 600 °C, et ne peuvent se former en l'absence de sollicitation en mode II.

Il faut également envisager un effet possible des plaquettes de ferrite sur la propagation des fissures. Les directions L et T.L. jouent un rôle identique. Une fissure se propageant dans l'une de ces direction coupe les plaquettes suivant un grand diamètre. Par contre, la propagation observée suivant T.C. correspond à une fissure coupant les plaquettes suivant leur épaisseur, et celles-ci paraissent donc offrir une moindre résistance dans ces conditions. On observe d'ailleurs que les stries ne sont pratiquement pas modifiées au passage de ces plaquettes dont l'épaisseur est voisine de la valeur de l'interstrie.

CONCLUSION

La texture très marquée de la tôle provoque, pour les deux directions de sollicitation L et T.L., des déplacements en fond de fissure correspondant à une sollicitation en mode mixte : I + II pour une fissure se propageant dans la direction T.C., I + III pour une fissure se propageant dans une direction perpendiculaire.

La disposition des plaquettes de ferrite résiduelle est telle qu'une fissure se propageant selon T.C. les coupe dans leur épaisseur, et selon un grand diamètre dans les autres cas.

La propagation de la fissure principale ayant toujours lieu dans la direction T.C., malgré des amorçages multiples, on doit conclure que ces deux éléments on un effet déterminant sur les conditions de propagation, ce qui est cohérent avec les modèles de propagation par stries.

REMERCIEMENTS

Ce travail a été effectué au sein du GRECO "Grandes déformation et Endommagement" du CNRS, et grâce à l'aide financière du Groupement d'Intérêt Scientifique "Rupture à chaud".

REFERENCES

1. L. Kaechele, The Rand Corporation RM 36500-PRR (1963).
2. J. Lemaitre, J.L. Chaboche, J.M.P. V2 n°3, 317 (1978).
3. J.L. Chaboche, Thèse Etat, Paris (1978).
4. C. Caileteau, H. Policella, G. Baudin, C.R. Ac. Sc. Paris, 292 série II-1103.
5. G. Brun, J.P. Gauthier,P. Petrequin Mem. Sc. Revue Met., 461 (1976).
6. R. Roux, J. Charrier, C. Gasc, ECF5 815 (1984).
7. J. Petit, Mémoire CNAM, Poitiers (1983).
8. C. Laird, A.S.T.M. - S.T.P. 415 (1967).

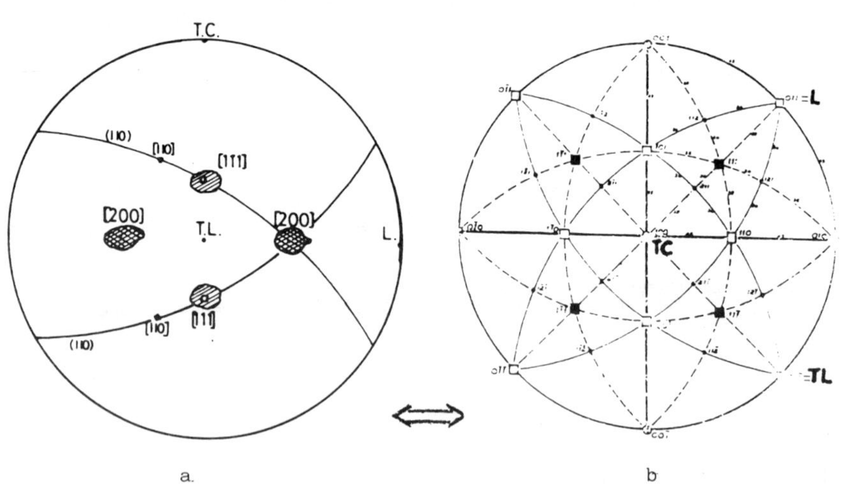

FIG1 : a) Texture obtenue
b) Texture rapportée à la projection stéréographique standard.

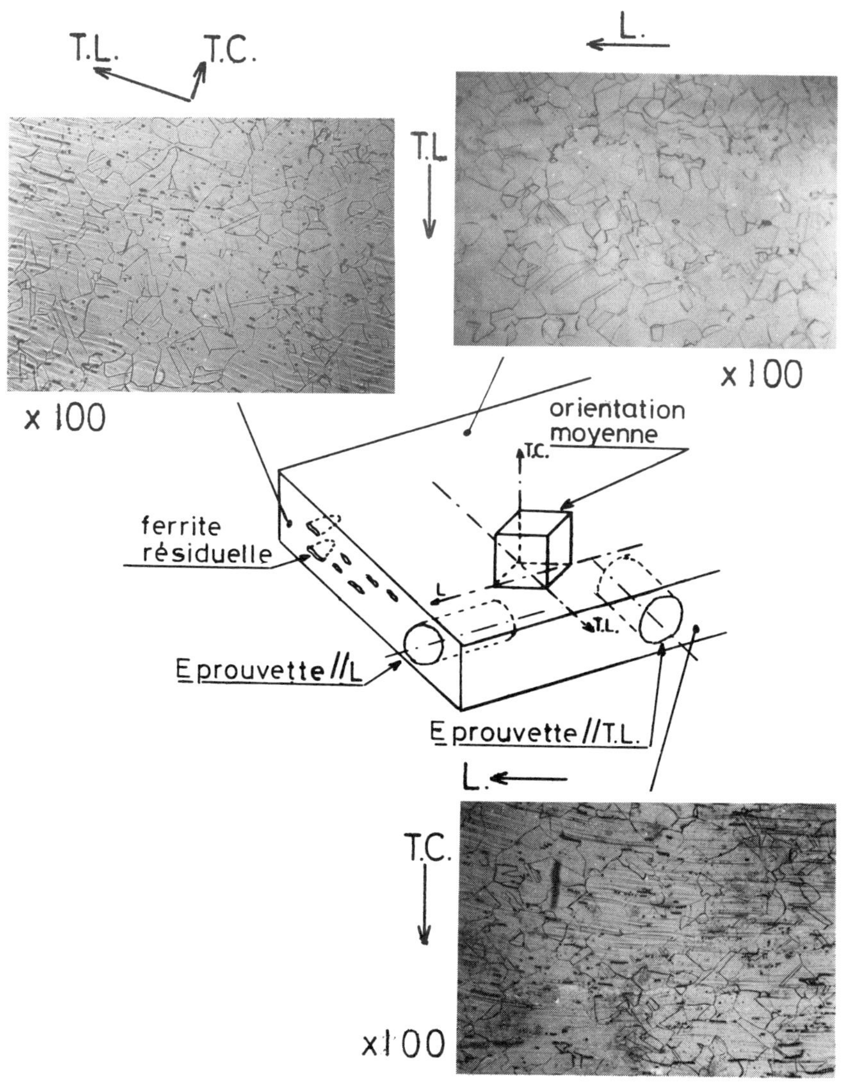

FIG.2 : Configuration micrographique - Texture -
Prélèvement des éprouvettes.

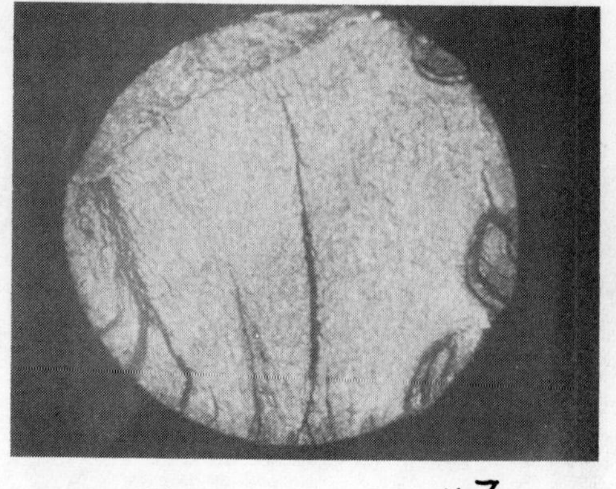

FIG.3 : Amorce de fissure et fissure principale

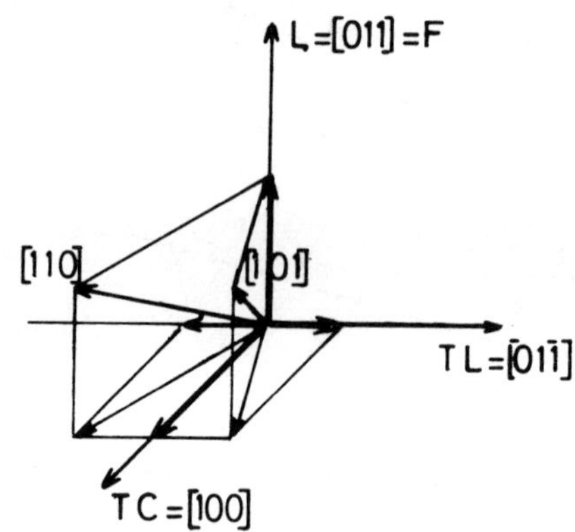

FIG.4 : Déplacements dus au glissement - cas où F//L.

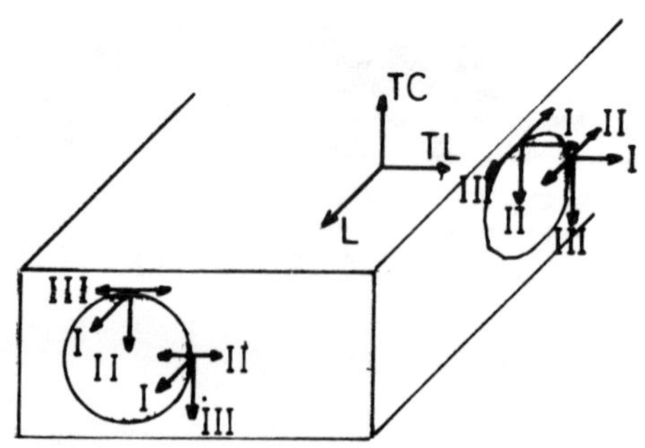

FIG.5 : Différents modes suivant l'emplacement des amorces de fissure

Influence of Filtration on Fatigue and Fracture Properties of Alloy LC4

Jia Jun*, Jiang Lieguang*, Li Qingchun*, Jiang Xiangquan, Liu Wanjiang**, Yuan Xiuxia** and Li Manliang****

**Harbin Institute of Technology, People's Republic of China*
***Northeast Light Alloy Fabrication Plant, People's Republic of China*

ABSTRACT

The study proved that filtration technology increases not only normal mechanical properties of alloy LC4, but also its fatigue property of both casting and wrought temper. The filtration can make the fatigue life of its casting temper increased up to 20%, and the fatigue cycle of wrought temper with fatigue cracking expanded to 1 mm has been increased up to 17%. The fitration technology changed the shape of curve S--N of wroght material, slowed the cracking germ and expansion rate of preliminary cracking down. The filtration technology can increase fracture toughness of alloy LC4 both in casting and wrought temper up to 12% and 16% respectively. The study has also proved that the filtration technology has influence over the solidification process and solid phase changing process.

KEYWORDS

Casting; Aluminum base alloys; Filtration; Mechanical properties; Fatigue properties; Fracture toughness.

I. INTRODUCTION

In these years, many works about the improvement of application properties of aluminum alloys have been published (1--7). The alloy LC4 is the same alloy as alloy 7075 in U.S; as alloy B95 which is now called alloy 1950 in Soviet Union (8). The influence of filtration upon the plasticity, fracture toughness and fatigue property of alloy LC4 has mainly been surveyed in this paper.

II. EXPERIMENT METHOD

The ingots used for experiment are semi-continuously cast both with and without filtration in production conditions. The ingot is 360 mm diameter and 5000 mm long. During the first half of the ingot casting, the molten metel was going through ceramic filtrating pipes with 14 mesh, and then went into the die. The second half of the ingot was cast without filtration.

The samples for mechanical property testing were cut from the homologous positions of the filtrated and non-filtrated part of ingot. All the samples were heat treated and aged by the following systems: 470^{+5}_{-1} °C, hold for 1 hour, quench in water; aged at 135^{+5} °C, hold for 16 hours.

After ingot hand-forged in multi-direction, the samples were cut from the homologous positions of the blocks. The heat treatment system of the samples was the same as that for casting samples.

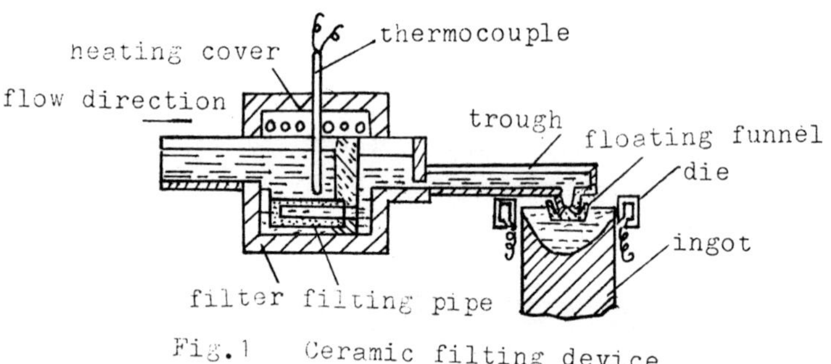

Fig.1 Ceramic filting device

III. THE EXPERIMENT RESULT

1. Analysis of composition and aluminum oxide and structure survey of samples.

The compositions and the aluminum oxide were checked by bromomethyl alcohol solving method combined with X-ray powder method. The content of Al_2O_3 in alloy LC4 was 44 PPM before filtration, and 30 PPM after filtration. The composition was not changed before and after the filtration.

The macro-structure of casting state of alloy LC4 shows that the structure became a little bit grosser after filtration.

2. The influence of filtration technology on mechanical property

Fig.2 shows the average variations of common mechanic properties of alloy LC4. From fig. 2 it can be seen, σ_b, $\sigma_{0.2}$ and δ_5 are increased both in casting and wrought state after filtration. It should be noted that the filtration technology have increased the plasticity of alloy LC4 in wrought state obviously, the elongation has been increased 33.8% in average, and the fluctuating scope has been reduced from 2.00-10.3% down to 8.33-10.00%.

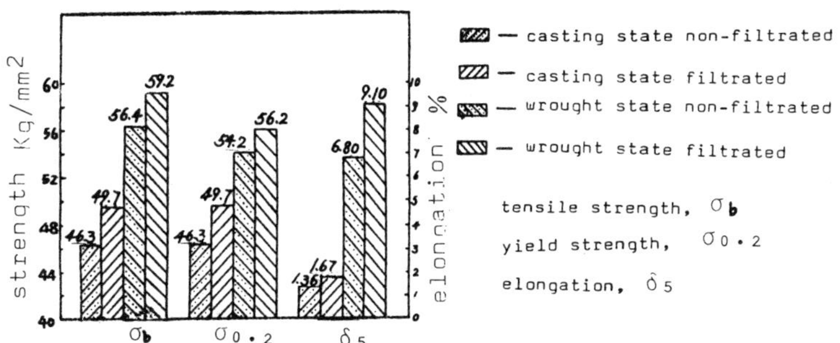

Fig.2 The influence of filtrating purification on normal mechanical properties of alloy LC4

3. The influence of filtration on fatigue property

The LC4 samples used for fatigue testing are made in two commercial common ways: rotating bending fatigue sample and three-point bending fatigue sample.

Under the load of 11.0 Kg/mm^2, the rotating fatigue life of alloy LC4 in casting state without filtration is 6.16×10^5 cycles in average, for the wrought sample with filtration the life is 7.376×10^5 cycles in average. The filtration technology can increase fatigue life of alloy LC4 in casting temper up to 19.5%. Figure 3 shows the curve S--N of alloy LC4 in wrought temper. It is clear from the curves showed on fig. 3 that under various stress the endurance of the samples that had been filtrated is longer than that of samples that had not been filtrated.

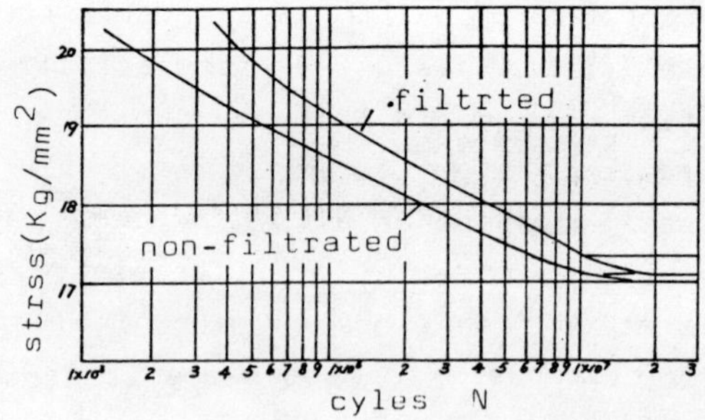

Fig. 3 S-N curve of alloy LC4 (wrought state)

During observation in the experiment we found the cracking germ duration in filtrated material is longer than that in material without filtration. Under the same experiment condition, the fatigue crack in samples with filtration expands to 1.0 mm after 453,000 cycles in average, but for the samples without filtration the crack expands to 1.0 mm after 386,000 cycles. The cycle number of filtrated samples is 17.39% higher than that of non-filtrated. It can be noted from the two regression curves in figure 4, that the filtration technology can obviously increase the resistant capacity to fatigue crack expanding when the difference of stress density factor ΔK is low.

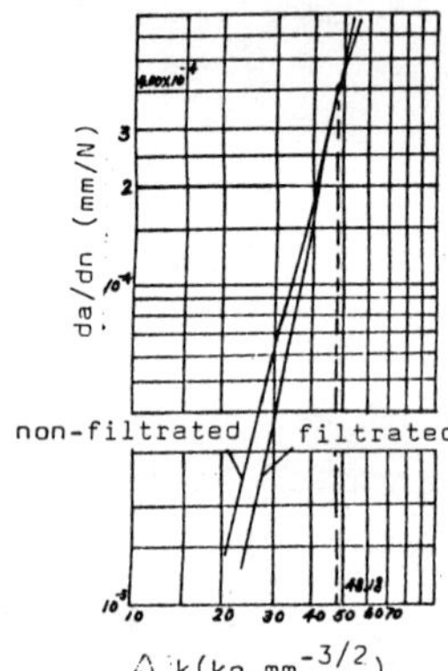

filtrated: $da/dn=1.89\times10^{-12}(\ k)^{4.95}$

non-filtrated: $da/dn=5.28\times10^{-11}(\ k)^{4.07}$

Fig. 4 da/dN-ΔK relationship of wrought alloy LC4

4. influence of filtration technology on fracture toughness of alloy LC4

During testing we have noted that the K_{1c} value of wrought material without filtration fluctuates in a large scope, from 58.7 to 109.4 $Kg/mm^{3/2}$; but the K_{1c} of wrought material with filtration fluctuates in a much smaller scope, from 89.1 to 111.1 $Kg/mm^{3/2}$. So we can say that the filtration technology make the property

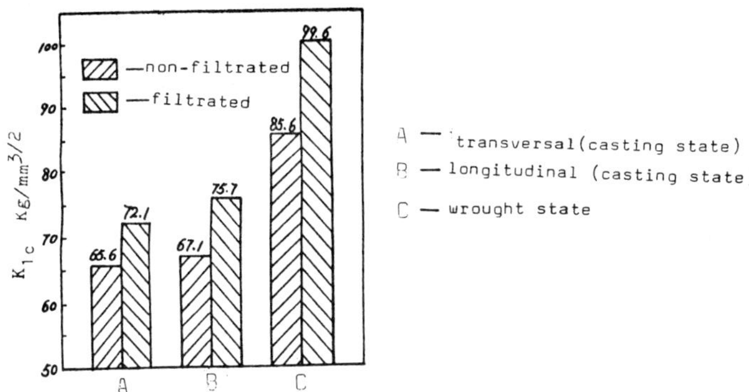

Fig.5 The influence of filtrating purification on fracture toughness of alloy LC4

of this alloy more stable.

5. The influence of filtration technology on the density of alloy LC4

Figure 6 shows the density of alloy LC4 checked by analytical balance with accuracy of 1/10,000.

6. The influence of filtration technology on age hardening of alloy LC4

Figure 7 shows the age hardening curves of alloy LC4 before and after filtration . that means, filtration influences on the alloy's solid phase changing process--age hardening process.

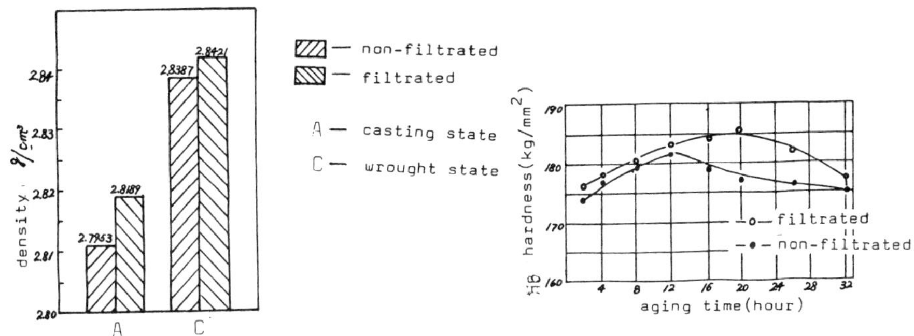

Fig.6 The influence of filtrating purification on density of alloy LC4

Fig.7 The influence of filtrating purification on age harden curve of alloy LC4

IV. CONCLUSION

1. The filtrating purification has the influence on the solidification, increases the density and reduces porosity; The

filtration has the influence also on solid phase variation process of aluminum alloy-aging. After filtration the aging precipitates distribute evenly, the peak value of age hardning is increased, over-age time is elongated.

2. The filtration purification can increase nirmal mechanic properties of cast and wroufht alloy,especially the plasticity of wrought metal has been increased obviously,hte fracture toughness of cast and wrought aluminum alloy has been increased, consequently the quality stability of product is increased .

3. Filtration can increase fatigue property of cast and wrought alloy LC4,change the shape of S-N curve of the material , increase the resistance capacity to over load of the material, slower the crack germ and crack expanding rate during the start of the crack.

REFERENCES

1. M. Khohain, Corrosion.35,285(1981).
2. Y.W.Kim,J.of Metals.32,No.12,12(1980).
3. B.M.Markochef, ZavodskayaLabolatorya.No,3345(1965).
4. M.E.Drits,Fracture of Aluminum Alloy.p.136,Metallei press, Moskva(1980).
5. Japan Society of Foundry,Influence of inclusion in aluminum alloy foundry on fatigue property,Reserch report 28(1981).
6. Jia Jun, Journal of Harbin Institute of Technology.43,70 (1981).
7. Jia Jun, Foundry.99,25,Edited by F.I.C.M.E,China(1983).
8. Marshall J.Wahll,Handbook of Soviet Alloy Compositions.12-7 Metals and Ceramics Infoemation Center(1975).

Fatigue of Ordered Alloys

A. K. Kuruvilla and N. S. Stoloff

Materials Engineering Department, Rensselaer Polytechnic Institute, Troy, NY 12180, USA

ABSTRACT

The influence of order on the fatigue behavior of two alloys, $(Fe,Ni)_3V$ and FeCo-V has been investigated. The former forms an Ll_2 superlattice while the latter forms a B2 superlattice. Even though both alloys systems display enhanced fatigue lives subsequent to ordering, the rate of crack propagation is found to increase with order in the B2 superlattice and decrease with order in the Ll_2 superlattice. This behavior is explained on the basis of the characteristic dislocation structure in the ordered lattice.

KEYWORDS

Ordered alloys, fatigue, crack propagation, threshold stress intensity, super-dislocations.

INTRODUCTION

The suppression of cross slip and the reduction in the number of available slip systems observed with long range order suggests that crack nucleation is less likely under cyclic loading in ordered alloys. There have also been investigations which suggest that the rate of crack propagation decreases with increasing planarity of slip.[1] The onset of order would therefore be expected to enhance fatigue properties.

The relatively few investigations carried out to date of the influence of long range order on fatigue processes indicate that fatigue lives are generally enhanced by ordering. In Cu_3Au single crystals[2] and in Ni_3Mn and FeCo-V[3], the fully ordered alloy displayed longer lives as compared to the disordered alloy at identical stresses. In a study of the fatigue behavior of Mg_3Cd, Whitehead and Noble[4] showed that ordering did not promote any enhancement in fatigue properties in this alloy. This behavior was attributed to the unique nature of slip in this alloy. While the disordered material deforms primarily by basal slip, the ordered material deforms equally easily by both basal and prismatic slip.

Most fatigue studies in ordered alloys have focussed on the effect of order on fatigue life. To date, there has been only one study that deals with the influence of order on fatigue crack propagation.[5] In that study, ordering resulted in a decrease in the rate of crack propagation and an increase in the threshold stress intensity for crack growth in an $(Fe,Ni)_3V$ alloy. The superior crack growth resistance of the ordered alloy was attributed to the unique dislocation structure in the ordered lattice. This study, which is an extension of the previous work, compared the influence of order on the crack growth characteristics of two different superlattices, namely $L1_2$ $(Fe,Ni)_3V$ and B2 FeCo-V.

EXPERIMENTAL PROCEDURE

The alloys used in this study were of the $(Fe,Ni)_3V$ and FeCo-V type. The compositions of the alloys are shown in Table 1. The $(Fe,Ni)_3V$ alloy designated by LRO-42 and LRO-60 (the difference in these being the amount of rare earth additions) were tested in the fully disordered and fully ordered conditions. The FeCo-V alloy was tested in the fully disordered and partially ordered conditions, S = 0.4 and S = 0.7. The heat treatments involved in producing these ordered states are shown in Table 2.

TABLE 1 Composition of Alloys, by Weight

LRO-42:	39.5% Ni, 37% Fe, 22.4% V, 0.4% Ti, 1000 ppm Ce
LRO-60:	39.5% Ni, 37% Fe, 22.4% V, 0.4% Ti, 400 ppm Ce
FeCo-V:	48.75% Fe, 49.10% Co, 2.03% V.

TABLE 2 Heat Treatment

Alloy	Treatment	Conditions
LRO alloys:	Disordering	- Anneal at 1100°C for 20 mins. and quench
	Ordering	- 650°C/24 hrs, 600°C/24 hrs., 500°C/48 hrs., furnace cool.
FeCo-V :	Disordering	- Anneal at 840°C for 1 hr. and iced brine quench
	Ordering	- S = 0.4: furnace cool to 690°C and quench S = 0.7: furnace cool to 650°C and quench S = 1.0: 500°C/5 hrs., furnace cool.

High cycle fatigue tests were performed on LRO-42, in both the disordered and fully ordered conditions using a Satec mechanical fatigue machine at a frequency of 30 Hz. Crack propagation experiments were conducted on LRO-60, in both the disordered and fully ordered conditions, using a closed loop servo-hydraulic fatigue machine. Compact tension specimens (31.8 mm x 30.5 mm x 2.5 mm) were used and crack growth was monitored using a traveling microscope. The test frequency was 20 Hz at a stress ratio, $R = \sigma_{min}/\sigma_{max}$ = 0.1 to 0.2. High cycle fatigue tests were performed in air while the crack growth tests were conducted in argon.

Crack propagation in the FeCo-V alloy was studied under exactly the same experimental conditions as described previously, except that the tests were run in a vacuum of about 7×10^{-8} torr. All of the experiments were conducted at room temperature.

The fracture surfaces were examined in a scanning electron microscope. Transmission electron microscopy was performed on samples from test specimens to obtain information regarding the dislocation structure in these alloys. The LRO alloys were electropolished using a solution of 25% H_2SO_4 in methanol at 28-32V in the temperature range -15°C to -25°C. For FeCo-V alloys, the electrolyte was a solution of 10% perchloric acid in acetic acid. The temperature range for jet polishing was 10°C to 12°C and the voltage employed was about 30V.

RESULTS

The results of high cycle fatigue tests on LRO-42 revealed that ordering enhanced the fatigue resistance of this alloy by way of improved fatigue life. In both the ordered and disordered conditions, the alloy displayed transgranular cracking.

The effect of order on crack propagation in the closely related alloy LRO-60 is shown in Fig. 1. Ordering promotes a decrease in the rate of crack propagation. It is interesting to note that the "threshold" stress intensity for crack propagation (arbitrairly selected at $da/dN = 10^{-9}$ m/cycle) increases by a factor of about 1.4. With increasing stress intensity, the crack growth curves approach each other, thus suggesting that the beneficial effects of order may be not sustained over the entire regime of crack growth. The mode of fracture was similar in both conditions. The initial stage of crack growth was highly crystallographic (see Fig. 2a) and striations were observed in the later stages of crack growth (Fig. 2b). The effect of order on the rate of crack propagation in FeCo-V is shown in Fig. 3. In this alloy, the rate of crack propagation is <u>accelerated</u> with an increase in the degree of order. Crack growth could not be monitored in the fully ordered condition, $S = 1$ because the crack grew extremely rapidly. The fracture surfaces had some interesting features. In the disordered condition crack growth was transgranular and appeared ductile (Fig. 4a) except in the overload region which displayed brittle cleavage facets (Fig. 4b). With an increase in the degree of order to $S = 0.4$, there were a few cleavage facets on an otherwise ductile transgranular fracture surface (Fig. 4c) but the overload region displayed brittle cleavage facets (Fig. 4d) similar to that observed in the disordered specimen.

DISCUSSION

The movement of superdislocations and the resultant decrease in cross slip is thought to be responsible for the superior fatigue properties of ordered alloys.[3,7] In high cycle fatigue, LRO-42 in the ordered condition was superior to the same alloy in the disordered condition (Fig. 1). Surface slip observations revealed that slip was planar in both the ordered and disordered condition and therefore the only difference was in the presence of characteristic superdislocations in the ordered lattice, see Fig. 5. The enhanced fatigue resistance resulting from ordering can therefore be attributed to decreased cross slip which, in turn, could enhance resistance to both crack nucleation and/or propagation.

Crack growth experiments revealed that ordering enhanced the resistance of LRO-60 to fatigue crack propagation, especially in the near threshold region, see Fig. 1. As suggested in a previous publication[5], the influence of long range order on crack propagation, especially in the near threshold region, would be expected to be reflected in changes in crack tip opening

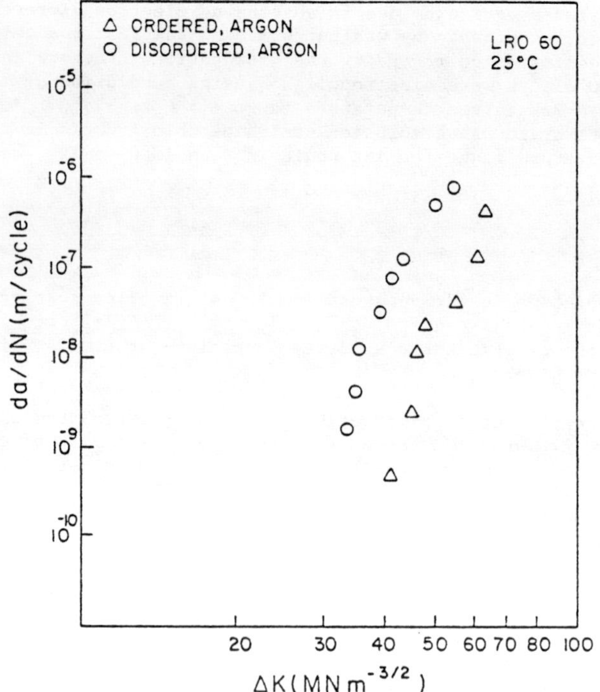

Fig. 1 Influence of order on crack propagation in LRO-60.

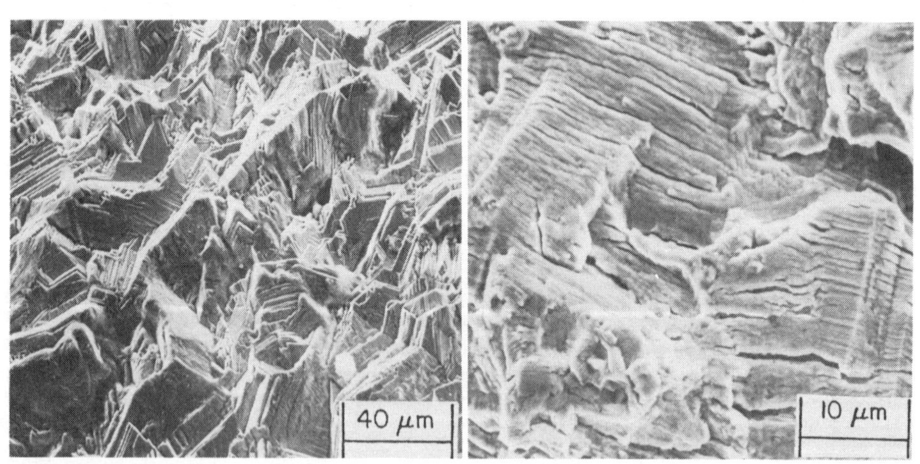

Fig. 2 Fracture surface of ordered LRO-60.

a) Crystallographic cracking. b) striations.

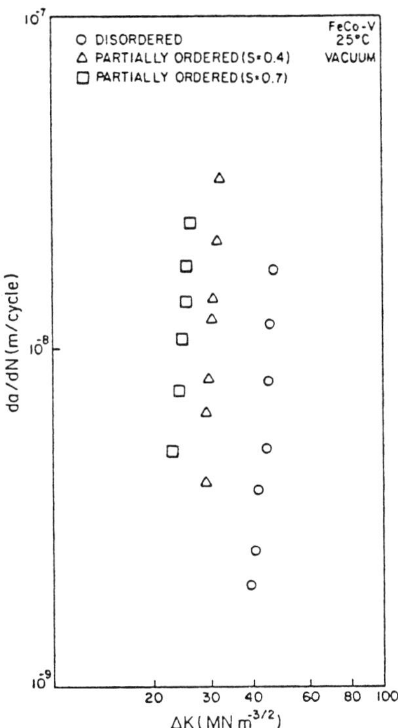

Fig. 3 Influence of order on crack propagation in FeCo-V.

displacement, CTOD. On one model[6], ΔK_{Th} might be governed by the least possible CTOD which should be of the order of one Burgers vector length, b, in disordered material. For tension-tension loading, R=0 and $\Delta K_{Th} = K_{max}$. The crack tip opening can then be represented as:

$$CTOD = b = \frac{\Delta K^2_{Th}}{E\sigma_y} \quad (1)$$

This leads to the expression

$$K_{Th} = \alpha\sqrt{E\sigma_y b} \quad (2)$$

It is not known how E is influenced by order in LRO-60, but it may be assumed that the effect is small, as in other ordered systems.[7] Yield stress is decreased by about 20% with order in this alloy. However, the most significant variable to be affected by long range order is the Burgers vector associated with dislocation nucleation at a crack tip. Superlattice dislocations, Burgers vector - 2b, are responsible for plastic deformation in $(Fe,Ni)_3V$ and other $L1_2$-type ordered alloys as can be seen in Fig. 5. The effective value of b in equation 2 is thus doubled, leading to an increase of $\sqrt{2}$ in predicted values of ΔK_{Th} on either model. This is of the same magnitude as the increase in ΔK_{Th} for da/dN = 10^{-9}m/cycle, (as close to threshold as could be approached in these experiments) between disordered and ordered LRO-60 as can be seen in Fig. 1. The advantages of ordering

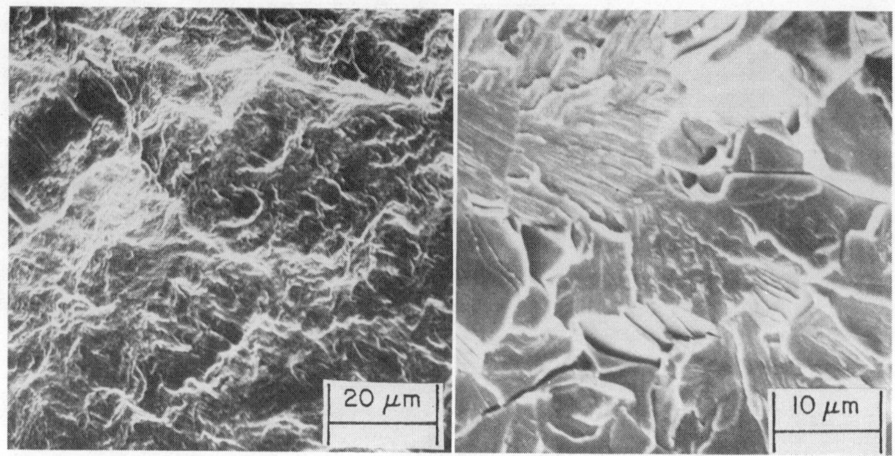

Fig. 4 a and b. Fracture surface of disordered FeCo-V.

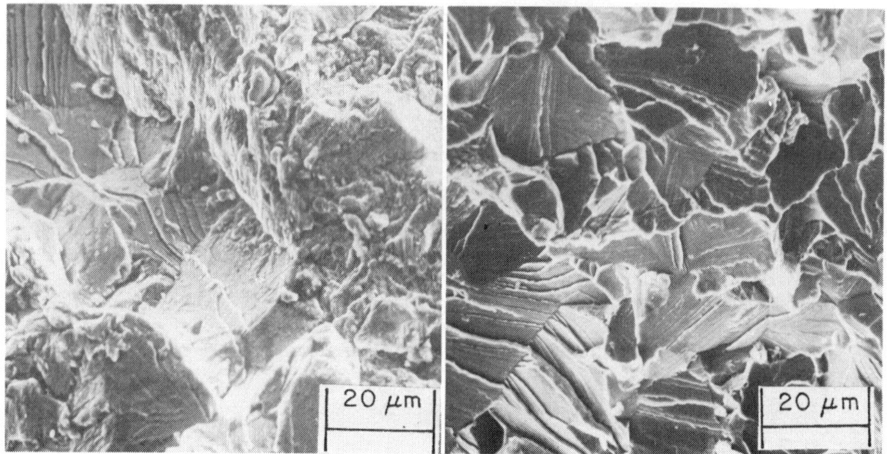

Fig. 4 c and d. Fracture surface of partially ordered (S=0.4) FeCo-V.

is retained even in the mid-range of crack growth. At higher ΔK values, possible effects of strain-induced localized disordering remove much of the benefits of the ordered structure.

This analysis cannot be applied to the crack growth behavior of the FeCo-V alloy. An increase in the degree of order leads to a decrease in the resistance to fatigue crack propagation. In tensile experiments, it has been observed that there is a drastic decrease in ductility with ordering in this alloy.[7] As the alloy orders fully into a B2 superlattice, the tensile elongation decreases to about 4%. This significant decrease in ductility is accompanied by a drastic change in the mode of fracture. While the disordered

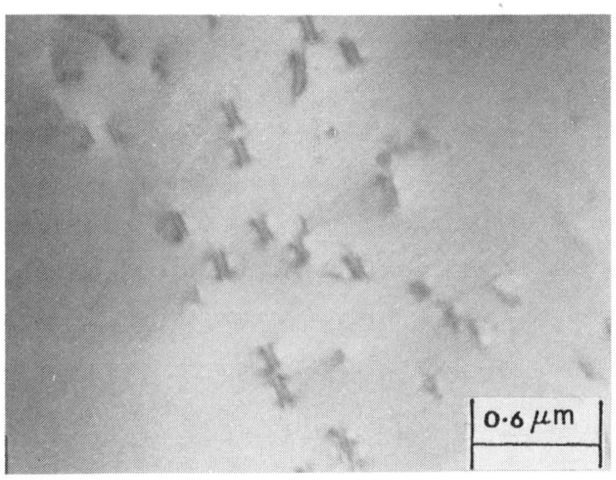

Fig. 5. TEM micrograph of LRO-60 showing paired dislocations.

FeCo-V alloy displays a dimpled fracture surface when tested in tension, the ordered alloy displays an absolutely brittle cleavage type of fracture. The change in fracture mode results from the drastic decrease in slip relaxation processes that accompany ordering either by a reduction in the number of slip systems or a decrease in cross slip or both.[7] Another important factor to consider is that of slip compatibility. Poor compatibility of slip would result in catastrophic propagation of cracks, once they are nucleated. During fatigue crack propagation of FeCo-V, therefore, it seems that an increase in the degree of order leads to a reduction in accommodation of stresses concentrated at the crack tip. This results in enhanced crack propagation in the ordered conditions. In the partially ordered condition, a growing crack is likely to encounter ordered and disordered regions. The presence of ordered regions would be expected to promote a higher rate of crack propagation, based on the previous discussion. The presence of some cleavage facets on an otherwise ductile transgranular fracture surface in partially ordered (S=0.4) FeCo-V, see Fig. 4c, suggests that the cleavage arises probably from the crack growing in grains that are unfavorably oriented for slip. Since even ordered FeCo-V has many possible slip systems (even with reduced ease of cross slip), stress concentration can be accommodated in most grains. It is, in any case, clear that the superlattice dislocations in ordered FeCo-V are not beneficial in increasing the resistance to fatigue crack propagation in this alloy.

In considering fatigue behavior of ordered alloys, there seems to be a few important factors that have to be considered. Reduced cross slip, improved homogeneity and fineness of slip enhance the resistance of the ordered alloy to fatigue crack nucleation. Even in the FeCo-V alloy, a previous study[3] has shown that the fully ordered condition of Fe-Co-V is superior to the disordered condition in high cycle fatigue. While ordering may, therefore, be advantageous in resisting crack nucleation, its effect on crack propagation could be quite different. If there is sufficient ductility in the ordered alloy ahead of a crack tip to accommodate the stress concentration, ordering could be expected to enhance the resistance of the alloy to fatigue crack propagation. In the ordered condition, LRO-60 displays a tensile elongation of about 25% whereas FeCo-V displays a ductility of about 4%. If ordering leads to a decrease in ductility ahead of a crack tip (by virtue of reduced cross slip, poor slip compatibility, etc.), the alloy would show decreasing fatigue crack growth resistance with ordering. The key to alloy design for superior fatigue properties may, therefore, by the development of ductile long range ordered alloys. It is interesting to note here that the crack growth resistance of ductile long range ordered alloys LRO-42 and LRO-60 compares favorably with that of other high temperature alloys in the temperature range 25°C - 600°C.[5] The most significant observation is the superior threshold stress intensity values of these ordered alloys when compared to other structural alloys. At high ΔK values, the advantages of ordering disappear, probably due to strain-induced localized disordering.

CONCLUSIONS

1) Crack growth resistance of ductile LRO alloys is improved by long range order, while crack growth rates are accelerated by ordering in FeCo-V.

2) Superlattice dislocations may enhance or decrease crack growth resistance depending upon the ability of the alloy to support high stresses at a crack tip without unstable fracture.

3) High cycle fatigue resistance of $(Fe,Ni)_3V$ and FeCo-V alloys is improved by long range order, suggesting that crack nucleation is suppressed by ordering in both superlattices.

ACKNOWLEDGEMENTS

The authors are grateful to the Allegheny Ludlum Steel Corp. for providing the FeCo-V alloy. This research was supported by the U.S. Dept. of Energy under the Energy Conversion and Utilization Technologies (E-Cut) Program, and by the National Science Foundation under Grant No. DMR84-09593.

REFERENCES

1) A. J. McEvily and R. C. Boettner, Acta Met., 11, 725, 1963.

2) G. Rudolph, P. Haasen, B. L. Mordike and P. Neumann, Proc. First International Conf. on Fracture (Sendai, Japan), 1965, 2, 501.

3) R. C. Boettner, N. S. Stoloff, and R. G. Davies, Trans. Met. Soc. of AIME, Vol. 236, (1966), p. 131.

4) R. S. Whitehead and F. W. Noble, Journal of Matls. Sci. 5 (1970) 851-861.

5) A. K. Kuruvilla and N. S. Stoloff, Metall. Trans. to be published.

6) E. Tschegg and S. Stanzl, Acta Met. 1981, 29, pp. 33-40 (quoting a private communication of F. McClintock).

7) N. S. Stoloff and R. G. Davies, Prog. in Mat. Sci., 1966, 13, 1-84, New York, Pergamon Press.

Failure Prediction of Fatigued Aluminum Alloys by Quantitative X-ray Determination of Microplasticity

S. Weissmann* and T. Takemoto**

**Department of Mechanics and Materials Science, College of Engineering Rutgers University, Piscataway, NJ, USA*
***Shunan R & D Laboratory, Nishin Steel Corporation, Shinnanyo, Yamaguchi, Japan*

ABSTRACT

Fundamental X-ray studies were carried out to characterize and measure the distribution of residual elastic strains and microplasticity in tensile-deformed silicon crystals which functioned as a model material. By applying the concepts, the methodology of investigation and novel X-ray instrumentation which emerged from these single crystal studies to a commercial Aℓ2024 alloy, cycled in air and in corrosive NaCℓ solution, the accrued fatigue damage of the alloy was determined and on this basis nondestructive failure predictions were made. The principal guideline for the analysis of the accrued fatigue damage was the direct measurement of the strainhardened microplastic regions by X-ray double-crystal diffractometry using a computer-aided rocking curve analyzer (CARCA).

KEYWORDS

Failure prediction; fatigued aluminum alloys; corrosion fatigue; microplasticity determination; strain distribution determination; CARCA; Pendellösung; X-ray topography.

INTRODUCTION

It is well known that the fracture of most materials, with the exception of those which exhibit extreme brittleness, is preceded by plastic deformation which has a profound effect on the stability of residual elastic stresses [1] and the generation and distribution of plasticity induced by cyclic deformation and their relationship to the residual elastic strains has been the concern of many X-ray investigations [2-8]. The studies revealed complicated variations of microstrains with cycling on the surface and in depth. To clarify experimentally the relationship between induced microplasticity and residual elastic strains, a fundamental characterization study of the strain distribution was first carried out on tensile-deformed silicon crystals which functioned as a model material, and the conceptional guidelines derived from the single crystal studies lead to the determination of the accrued prefracture damage and to failure prediction of cycled Aℓ2024 alloys.

To characterize the relationship between induced microplasticity and residual elastic strains it is necessary that the material should contain none of it prior to deformation. Silicon crystals fulfill this requirement because they can be obtained free of dislocations and of elastic strains. Above 60% of the absolute melting temperature the crystals become ductile and behave like metal crystals [9,10]. Because of their initial crystal perfection advantage is taken of the sensitivity to disturbances of the dynamical interaction of wavefields by lattice defects. By employing Pendellösung Fringe Topography (PFT) [11] microplastic zones become characterized by the destruction of the fringe pattern and residual elastic strains by bending and increased narrowing of the fringe spacings as the strain field increases [12]. Systematic time and temperature dependent recovery studies of plastic zones, constraining residual elastic strains, were undertaken. The PFT was combined with quantitative plastic strain measurements based on X-ray double-crystal diffractometry (DCD) [12,13] and with quantitative, point-to-point measurements of elastic strain gradients, based on measurements of local variations of reflected intensities by microfluorescent densitometry [13,14,15]. These studies revealed the sites of maximum elastic strain concentration because such sites, being stress-aided, exhibited the earliest recovery characteristics.

The following results were obtained: (1) Residual elastic strains induced by deformation are enclaved by constraining workhardened plastic zones. (2) The degree of workhardening in the plastic zone governs the magnitude of the residual elastic strains at the elastic-plastic boundary. The plastic zone functions as a stress raiser and the steepest elastic strain gradients are associated with workhardened plastic zones at notches or cracks. (3) Compared to the bulk the surface layers of ductile materials have a special propensity for dislocation generation and workhardening. (4) Large lattice misalignment is closely linked to the fracture of the material and may outline the macroscopic future fracture path [17].

ANALYSIS OF ACCRUED DAMAGE AND FAILURE PREDICTION OF Aℓ2024 ALLOY CYCLED IN AIR

The principal guideline for the analysis of the accured damage is the direct measurement of the strainhardened microplastic regions. The latter are determined by measuring accumulations of excess dislocation densities which can be sensitively assessed from rocking curve measurements. Each grain is viewed to function as the test crystal in a double-crystal diffractometer arrangement. A computer aided rocking curve analyzer (CARCA) was developed [18] which analyzes the lattice defects of a very large grain population swiftly and precisely. The objective of CARCA is to characterize the deformation stage of a material in terms of grains that have undergone the largest lattice distortions. The criterion of failure predictability is the critical dislocation density, ρ^*, obtained in terms of a critical value β^* of the rocking curve halfwidth since the results of single crystal studies have shown that regions of β^* maxima are associated with regions of largest strain concentrations [12,13,17].

Figure 1 shows the dependence of the average rocking curve halfwidth, $\bar{\beta}$, of a large sampled grain population on the fraction of fatigue life n/n_f of a cycled Aℓ2024-T4 allow. A straight line relationship was obtained for low cycle fatigue when the surface was investigated using CrKα radiation [19]. The results of high cycle fatigue performed on the same alloy by Pangborn et al. [20] using the DCD method were verified showing a straight line relationship when the fatigued grains in the bulk were analyzed by MoKα radiation which is capable of penetrating beyond the surface layer. The

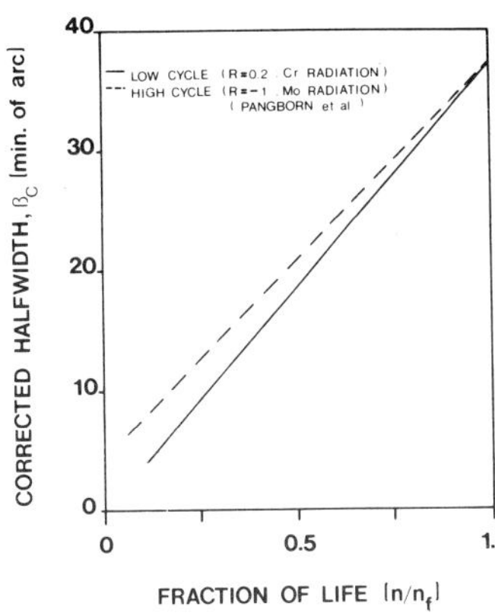

Fig. 1. Dependence of $\bar{\beta}$ on n/n_f for low cycle fatigue and bulk properties of high cycle fatigue of Aℓ2024.

CARCA study also corroborated the results of the analysis of Pangborn et al. for surface grains showing that in high cycle fatigue a characteristic plateau dependence exists in the $\bar{\beta}$ vs. n/n_f curve. This plateau is associated with the special propensity of surface grains to workharden, compared to the bulk grains, and thus exert an impediment to the egression of dislocations in the bulk. It was shown that when the $\bar{\beta}$ of the grain population in the bulk approached that of the surface layer failure was imminent. Similar results were obtained subsequently by Weiss and Oshida [21] who studied the X-ray line broadening of an Aℓ7050-7 7651 alloy, subjected to reverse bending fatigue.

On the basis of a calibration curve such as that shown in Fig. 2, the independence of failure prediction on the loading history was demonstrated by spectral loading studies. The direct estimate of remaining life from the X-ray data was accurate to 10% of the fraction determined by cycling the alloy to failure [19,20].

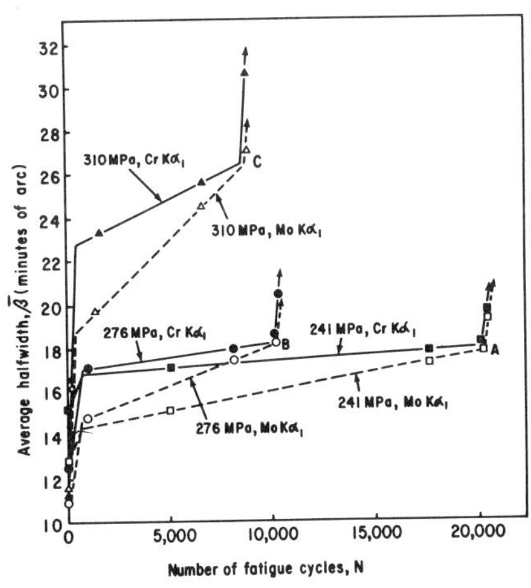

Fig. 2. Dependence of $\bar{\beta}$ on number of cycles N at various stress levels of Aℓ-2024-T4 cycled in 3.5% NaCℓ solution.

FAILURE PREDICTABILITY IN LOW CYCLE CORROSION FATIGUE

The same principle of defect structure analysis and failure prediction which were applied to the alloy cycled in air were applied to the Aℓ2024-T4 alloy cycled in tension ($R = 0.1$) in corrosive 3.5% NaCℓ solution. Figure 2 shows the dependence of $\bar{\beta}$ on the number of fatigue cycles. The solid lines of the curves pertain to the measurements with $CrK\alpha_2$ radiation, while the dotted lines refer to those performed with $MoK\alpha_1$ radiation. The absorption of the specimen permitted only surface grains to be analyzed with chromium radiation. By contrast, the short wavelength of molybdenum radiation permitted analysis of grains exceeding a 50 μm depth distance. At the points where the curves pertaining to surface and bulk grain intersect (points A, B and C), which corresponds to the attainment of the β^* value, catastrophic failure set in [22]. With the exception of very low cycle fatigue (310 MPa) the plateau dependence of β values for surface grains was again observed. When large strain amplitudes are applied, the barrier effect of the surface layer to dislocation egression becomes ineffective, resulting in early nucleation of surface cracks and most of the life is taken up by crack propagation.

DEPENDENCE OF CORROSION FATIGUE LIFE ON CYCLIC STRESS

To investigate the dependence of life in high cycle corrosion fatigue (HCCF) on stress, a Goodman's diagram was constructed in Fig. 3. The cyclic stress and the mean stress were calculated by the following relations: $\sigma_{cyc} = 1/2\ \sigma_{max}(1-R)$, $\sigma_{mean} = 1/2\ \sigma_{max}(1+R)$. As shown in Fig. 3 the fatigue strength in corrosive medium was found to be 40 MPa. Inspection of Fig. 3 shows that the corrosion fatigue life depended strongly on the cyclic stress and not on the mean stress. The increases of β values, shown in Fig. 4, show also the dependence on maximum cyclic stress. To establish a basis of comparison between corrosion fatigue and stress corrosion behavior a static load test in 3.5% NaCℓ solution was performed. Although a stress of 350 MPa was applied which was higher than the yield stress (Y.S. = 276 MPa) no failure took place after 150 hours. The $\bar{\beta}$ values were about 30' of arc.

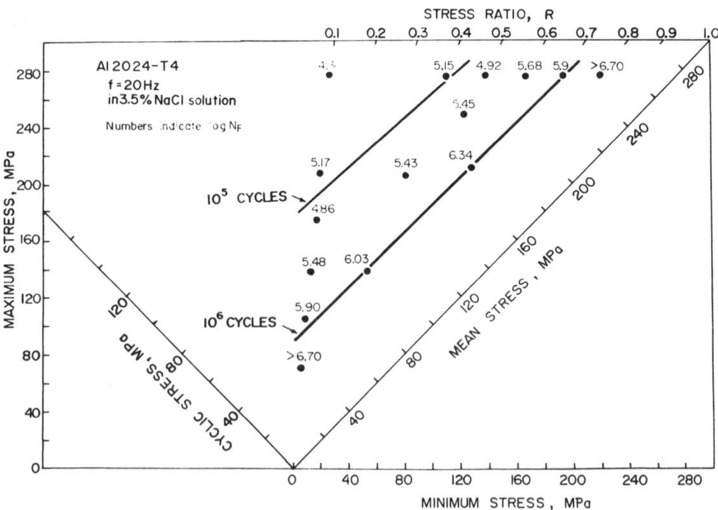

Fig.3. Goodman's diagram of Aℓ-2024-T4 cycled in 3.5% NaCℓ solution, f=20 Hz.

The critical halfwidth, β*, at which catastrophic failure occurred in HCCF was also determined. The dependence of β* on applied stress for different R values is shown in Fig. 5. It will be noted that β* did not increase appreciably until the maximum stress was higher than 160 MPa. This value corresponds to about 60% of the static yield stress. At maximum stress, the static yield stress, β* was only about 12' of arc. Evidently cycling in the corrosive environment lead not only to failure but also to reduced microplasticity.

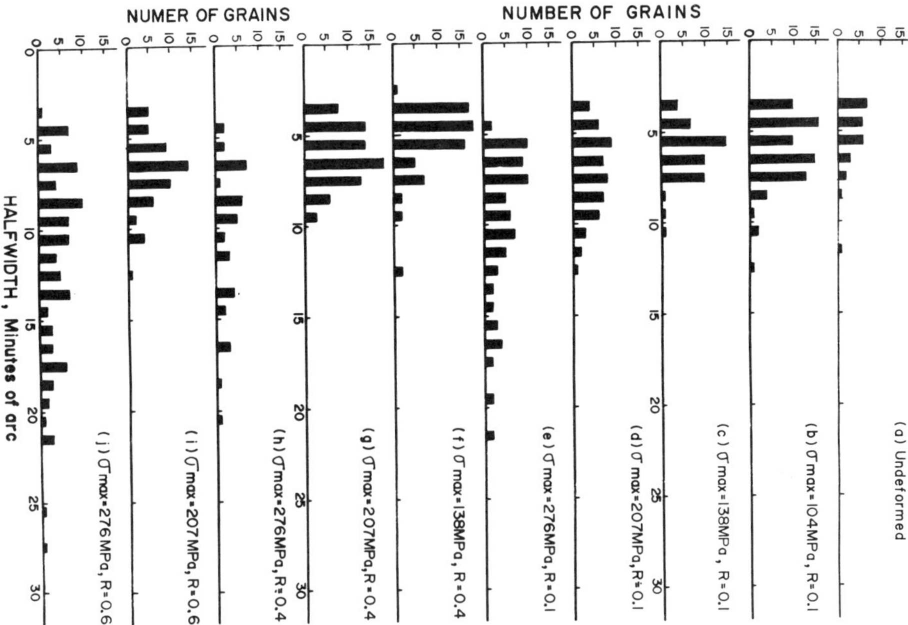

Fig. 4. Distribution of rocking curve halfwidths at various stress levels and R values for corrosion fatigued Aℓ-2024-T4.

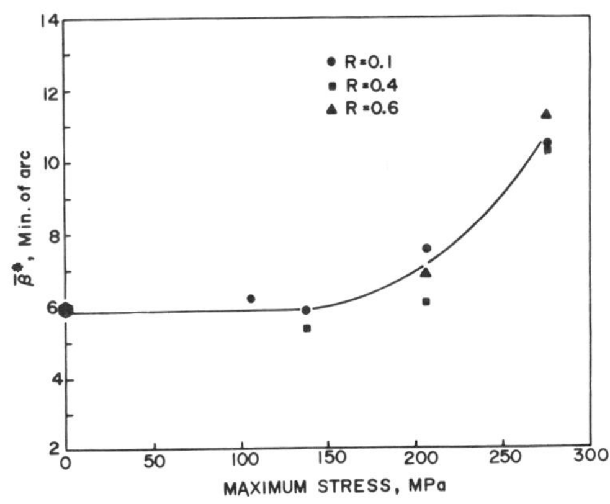

Fig. 5. Dependence of critical halfwidth, b*, on applied maximum stress for corrosion fatigued Aℓ-2024-T4.
Aℓ-2024-T4.

REFERENCES

1. E. Macherauch in *Eigenspannungen* (edited by Deutsche Gesellschaft fur Metallkunde), 41 (1980).
2. S. Taira and K. Hayashi, Proc. 7th Jpn. Congr. on Testing Materials, Japan Soc. of Testing Materials, Kyoto, 1 (1964).
3. S. Taira and K. Hayashgi, Bull. JSME **9**, 627 (1966).
4. S. Taira,T. Gotto and Y. Nakomo, Proc. 12th Jpn. Congr. on Materials Research, Soc.Mat. Sci., Kyoto, 8 (1969).
5. S. Taira, K. Tanaka, T. Tanabe, Proc. 13th Jap. Congr. on Materials Research, 14 (1970).
6. M. Nago and V. Weiss, Trans. ASME, Ser. H. **99**, 110 (1977).
7. S. Kodama, H. Misawa and Kurebayashi in Proc. of Fatigue-84, Birmingham,1171 (1984).
8. H. K. Kuo and J. B. Cohen, *Mat. Sci. and Eng.* **61**, 127 (1983).
9. H. Alexander, *Phys. Status Solidi* **26**, 725 (1968).
10. H. Alexander, *Phys. Status Solidi* **27**, 391 (1968).
11. N. Kato and A. R. Lang, *Acta Cryst.* **12**, 787 (1959).
12. Y. Tsunekawa and S. Weissmann, *Mat. Sci. and Eng.* **17**, 511 (1975).
13. H. Y. Liu, G. J. Weng and S. Weissmann, *J. Appl. Cryst.* **15** 594 (1982).
14. Z. H. Kalman,J. Chaudhuri, G. J. Weng and S. Weissmann, *J. Appl. Cryst.* **13**, 290 (1980).
15. J. Chaudhuri, Z. H. Kalman, G. J. Weng and S. Weissmann, *J. Appl. Cryst.* **15**, 423 (1982).
16. S. Weissmann, V. A. Greenhut, J. Chaudhuri and Z. H. Kalman, *J. Appl. Cryst.* **16**, 606 (1983).
17. H. Y. Liu, W. E. Mayo and S. Weissmann, *Mat. Sci. and Eng.* **63**, 81 (1984).
18. R. Yazici, W. E. Mayo, T. Takemoto and S. Weissmann, *J. Appl. Cryst.* **51**, 89 (1983).
19. W. E. Mayo and S. Weissmann in Applications of X-ray Topographic Method to Materials Science (edited by S. Weissmann, F. Balibar and J. F. Petroff) p. 311. Plenum Press, New York (1984).
20. R. N. Pangborn, S. Weissmann and I. R. Kramer, *Met. Trans.* **12A**, 109 (1981).
21. V. Weiss and Y. Oshida in Proc. of Conf. on Fatigue-84 (edited by C.J. Beevers), p.1151, Birmingham, England (1984).
22. T. Takemoto, S. Weissmann and I. R. Kramer in Fatigue, Environment and Temperature Effects (edited by J. Burke and V. Weiss), p. 71. Plenum Press, New York (1983).

Material Properties for Life Assessment of Valves

J. Zdarek, K. Bohatec, C. Budac and M. Mikolasek

SIGMA Research Institute, Prague, Czechoslovakia

ABSTRACT

Some important results of experiments performed on material samples from the valve bodies used in the classical and nuclear power stations are summarized. Also results of the stress state, strength and life calculations are evaluated and some differences in the life prediction methods are compared.

KEYWORDS

Low cycle and thermal fatigue, creep-fatigue, strength and life calculations, evaluation methods, valves.

INTRODUCTION

High requirements on the properties of materials used for the valve construction proposed for classical and nuclear power stations are derived from the requirements on the long term life under elevated temperatures. The valve bodies are working under severe loading conditions which also includes the nonstationary thermal regimes. The stress-deformation conditions of the low cycle thermal fatigue occures in the critical places of the valve bodies.

The stress-deformation state in the valve body is determined by the three dimensional final element method /FEM/ under all loading conditions. A complex material research including isothermal low cycle fatigue, thermal fatigue with positive loading cycle and combined thermal fatigue with hold times was simultaneously performed. The obtained results were used for the life assessment of the valve body. Some differences in the life assessment methods according to the reference stress method, the ASME Code and the soviet code are compared in the conclusion.

MATERIAL RESEARCH

Most frequently used construction material for valve bodies in classical and nuclear power stations is low alloyed feritic perlitic steel 1/2 Cr 1/4 Mo 1/4 V. A series of fundamental and special tests were performed in the framework of the complex assessment of this material. The samples for the tests were taken from the forging of the valve nozzle. Since this material was tested mainly with regard to the long term service above 500°C, tests were partialy made with the thermally aged material. The aging regime was 3 000 hr/550°C/air. It is evident from the tensile tests that the aged material shows lower values of stress resistance against nonelastic deformation and higher plastic properties at the service temperature. From the comparison of the notch toughness values results it is seen that aged material has higher transition temperature and higher notch toughness at the service temperature.

The low cycle fatigue curves are shown on fig. 1. The material in as received condition is a typical thermally and mechanically strengthened material, that slowly softens during testing and his Manson-Coffin curve has relatively shallow slope. The average deformation rate was kept constant during testing at a value of $\dot{\varepsilon} = 6 \cdot 10^{-3}\ s^{-1}$.

The aged material, as it is seen from fig. 1, has higher resistance against the low cycle fatigue, particularly in the range of higher deformations. It corresponds to better plastic properties at the temperature 550°C.

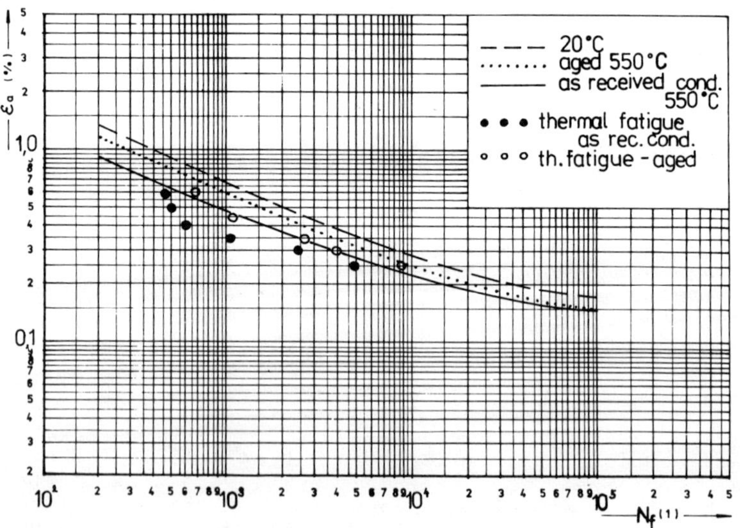

Fig. 1. The low cycle fatigue and thermal fatigue test results.

For the assessment of the effect of temperature changes on the low cycle fatigue life several thermal fatigue tests were made on the as received and aged material with the temperature changes of 550°C ⇄ 450°C.

Only positive type of cycle was applied, which means that the maximum tensile loading was synchronised with the maximum temperature. The data from the literature also identify this type of the thermal fatigue as the most damaging one.

From our presented results is possible to conclude that despite the fact that the temperature range is not large the effect on the life reduction is relatively large especially in the vicinity of the deformation amplitude ε_a = 0,4%. For the higher and lower deformation levels the thermal fatigue results are gradually closer to the low cycle fatigue curve for the highest temperature, i.e. 550°C. The thermal fatigue results of the aged material are again better that those of the material in the as received condition.

The aim of the next experimental work was to assess the influence of the creep portion in the low cycle fatigue on life and service performance of the material. One of the experiments was performed simply with lower average deformation rate while keeping the deformation control. The deformation rate was about fifty times lower than it is in typical low cycle fatigue tests. This type of test is very convenient to be described by the modified Manson-Coffin equation.

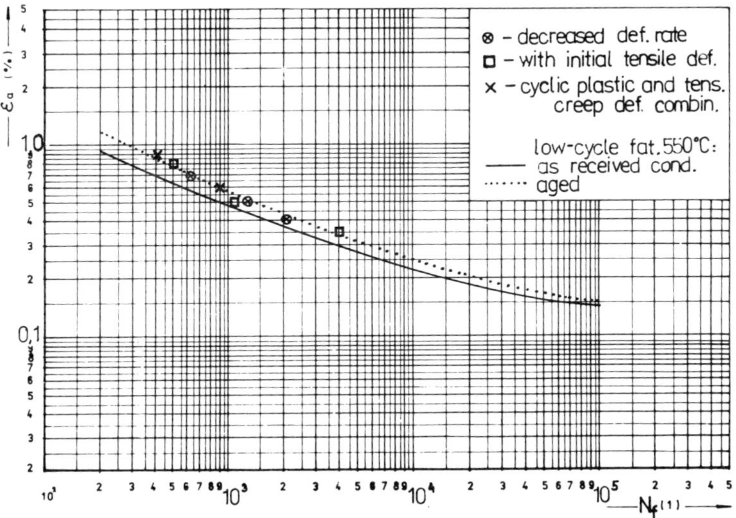

Fig. 2. The results of the low cycle fatigue tests with the creep effect.

The results shown on fig. 2, by open circles nevertheless indicate that cyclic life still increases compared with basic curve, most probably due to the improvement of the plastic properties during the long term exposition to the temperature and are between the results of the as received and aged material. The agresive influence of the creep is not noticeable. Similar results show further two types of tests.

The experimental results marked by full squares are those of the low cycle fatigue /550°C/ tests of the preliminary crept material on the stress level σ = 200 MPa until the controled longitudinal creep deformation ε_c = 3%. The results shown by cross means evaluation by the strain-range partitioning method.

STRENGTH CALCULATION AND LIFE ASSESSMENT

The analytical computational methods and the finite element method are used for the stress state calculation. Because it is not possible to evaluate each type of the valve body by the three dimensional finite element model, we do use the simplified analytical methods, and for the typical and characteristic valve shapes we use the three dimensional FEM model. At the present time we have available several automatic mesh generatores. By use of them it is possible to generate three dimensional FEM model with the input data including several specific valve body dimensions. The stress states due to internal pressure, external forces and moments acting on connecting nozzles and due to non-stationary thermal regimes are determined. The typical boundary conditions are shown on fig. 3.

The evaluation of strength and long term life in all critical places and sections is based on the calculated stress state verified by experimental tests. All stress categories were evaluated in detail according to the soviet code /1/ and the reference stress method /2/. Partial assessment was carried out according to the ASME Code N-47 /3/.

Let us look closer on the cyclic strength evaluation under combined creep-fatigue conditions. In the above chapter we have shown in detail the experimental results used for the comparison with the calculational formulas. The reference stress method uses for the cyclic strength evaluation the ASME Code method. If we consider elastic solution according to /3/, there are several fatigue curves for different temperatures available. The creep-fatigue interaction is covered by the safety factors.

For the cyclic strength evaluation by calculation we have used the calculational formulas which include the creep effect in such a way that an expected change of the plastic properties due to long term influence of the creep can be substituted. This property is the value of the reduction of the area which is also the most important property for the low cycle or thermal fatigue. The boundary value of the reduction of the area was derived from the extrapolated creep tests during that this value was measured

/4/. This evaluation procedure of the cyclic strength with the influence of the creep is proposed to be included in the prepared soviet code for calculation at elevated temperatures.

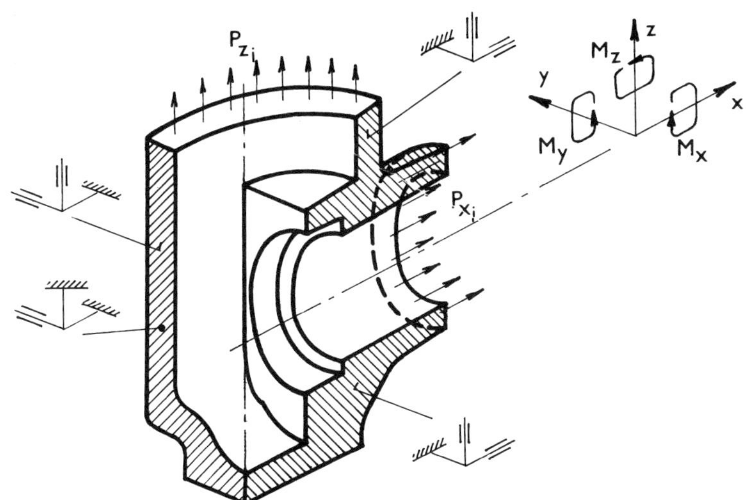

Fig. 3. Stress calculation boundry conditions for mechanical loading.

CONCLUSIONS

The obtained results from the material research confirmed very good material properties of the 1/2 Cr 1/4 Mo 1/4 V steel, namely that the improvement of the plastic properties and toughness due to long term exposure to the higher temperatures about 550°C is more pronounced than the deterioration by the creep. This conclusion confirmes the proper use of the calculational formulas for the cyclic strength assessment under the influence of the creep by using the specific material properties and particularly the boundary values of the reduction of the area determined by the long term creep tests.

The complex material research and the detailed stress evaluation by modern calculational methods enables us to evaluate the conditions of the long term strength and service life of the valves for classical and nuclear power plants. Based on the general analysis if is possible to guarantee extended strength requirements for all loading conditions and long term service life up to 200 000 hours.

REFERENCES

1. The life assessment for the low cycle fatigue and creep. Technical material RTM 108.031.105-77, /1977/.

2. ASME Boiler and Pressure Vessel Code, Code Case N-47, /1979/.

3. R.A. Ainsworth, I.W. Goodall, High - temperature structural design methods in Gas Cooled Reactors Today, BNES, London, /1982/.

4. V. Foldyna, private communication, /1984/.

Quantitative Fractographic Examination of Aircraft Components Tested Under a Fatigue Spectrum Loading

E. Abramovici, M. Burak, K. C. Overbury and D. R. Turner

Canadair Ltd. Montreal, Quebec, Canada

ABSTRACT

The objectives and techniques of fractographic examination as a part of an aircraft certification program were described. The safe life of a main landing gear lug fabricated from a high strength aluminum alloy forging was then estimated based on the results of quantitative fractographic examination.

KEY WORDS

Fractography, striation count, safe-life, landing-gear, 7175 high strength-aluminum alloy.

INTRODUCTION

As a part of a business-jet certification program, critical components are subjected to extensive testing to a flight representative loading spectrum. Some of these components are tested to fatigue (i.e. safe life) requirements while others are tested for damage tolerance and residual strength (i.e. fail safe) requirements. [1]

The loading spectrum used in testing these components includes a complete Ground-Air-Ground (G-A-G) flight sequence of aircraft loads, repeated many times for the required number of flights. For specific components this aircraft G-A-G spectrum may be conservatively condensed to simplify the test program.

During the test program, cracks may appear or be artificially introduced on the component. Fractographic examination of the crack area, along with additional supporting strength, metallurgical and analytical data, will be required to ultimately determine the life of the components or minimum inspection periods for safe aircraft usage.

This paper describes the fractographic techniques used to examine a crack which developed on a component during testing, and briefly illustrates the use of the information developed by this examination towards determination of the components safe life for use on the aircraft.

OBJECTIVES OF FRACTOGRAPHIC EXAMINATION

Fractographic examination of cracks developed during component testing has several major objectives:

1) To determine the exact origin of the crack and the direction of crack propagation.
2) To identify, on a microscopic scale, the applied loading spectrum and to establish the influence of certain spectrum loads on the advancement or retardation of the crack propagation.
3) To determine the number of spectrum loading cycles from the start of crack initiation to the end of testing, in order to substantiate the analytical crack-growth curves for the component under test.
4) To examine the crack propagation front and provide information regarding inspectable crack size, crack tunnelling, etc.
5) To provide information about possible material or manufacturing defects which could be a source of crack nucleation or which could be associated with acceleration of crack propagation.

The importance of each objective will, of course, depend upon the component being examined and, more specifically, upon the design concept and the requirements imposed on the component (i.e. safe life or fail-safe).
The results of the fractographic examination are then reviewed and used in the final determination of the component safe life or safe inspection periods.

TECHNIQUES

In order to provide answers for the above objectives, qualitative and quantitative fractographic techniques have been developed over a number of years. These techniques make use of optical microscopes, as well as of modern high-technology, scanning electron microscopes.

Qualitative fractographic examination will provide answers to the objectives regarding the initiation site, the general direction of propagation, material defects, etc. This examination employs low and high power optical microscopes, different light sources, cleaning and etching solutions, etc. For the purpose of this paper, these techniques will not be described, since these techniques are extensively covered in failure analysis hand books and reports.
Quantitative fractographic examination can answer all objectives regarding crack growth rate, retardation effect, propagation front, inspectable crack size, etc. This examination employs electron microscopy techniques. This paper will attempt to describe some of our results using quantitative fractographic examination.
The basic principle of quantitative fractographic examination lies in the measurement of fatigue striation spacing. Basically, as has been demonstrated by many researchers, "each load excursion was responsible for one striation and that the size (that is, distance between markings) depended strongly on the stress amplitude". [2] Based on this principle of striation counting, two techniques are employed:

1) In the case of a simple, constant amplitude, loading spectrum, a number of fields located along an imaginary straight line which passes through the origin, are photographed. Subsequently, each photograph is examined, the striation spacing is measured, then plotted versus the distance from the origin.

2) In the case of multistep loading spectrum (or block spectrum), an identification of an entire loading block is made first. After establishing the appearance and the succession of the fatigue striations within the block, a number of fields containing blocks are photographed and the block lengths are measured. The results are plotted versus the distance from the origin. During identification of the loading spectrum with the microscopic fracture imprint, an analysis of the retardation effect is performed, as well.

EXPERIMENTAL RESULTS

For this paper, a main landing gear lug was selected. The part was fabricated from a 7175-T736 high strength aluminum alloy die forging per AMS 4149. The landing gear had been declared a 'safe life' component, therefore the fractographic examination concentrated on the objectives considered most important for safe life requirements.

For test purposes, the aircraft spectrum was conservatively simplified into three distinct test cycles of landing gear ground loads. Each cycle containing landing, taxiing, braking, turning and take-off loads. These test loading cycles were applied to the gear in blocks representing 20 aircraft flights, with each block containing 1A, 3B* and 13C cycles. A description of the spectrum is given in Fig. I. The loadings were repeated for a total of 3001 blocks, at which time the test was halted due to a failure in the main fitting upper lug. Following an analytical review,it was determined that the critical stresses on the lug resulted from secondary effects due to reactive loads at the gear/aircraft interface. The reactive loads were calculated for each test cycle and are reproduced in Fig. I.

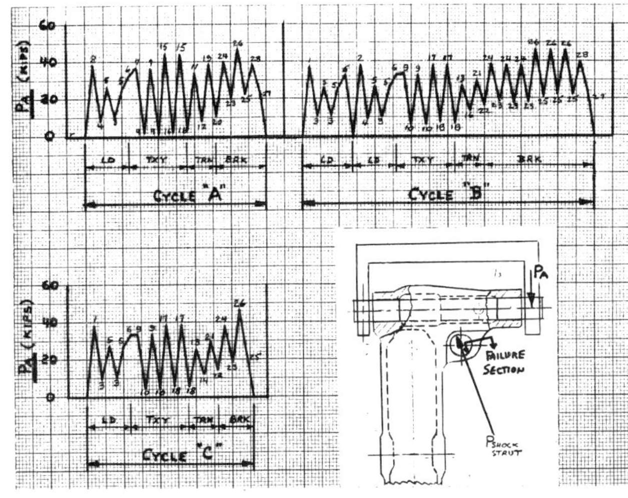

LD ... Landing
TXY .. Taxiing
TRN .. Turning
BRK .. Braking

Fig. I

*Note: each B cycle simulated 2 aircraft flights.

QUALITATIVE FRACTOGRAPHIC EXAMINATION

A general view of the fractured lug, as received in the Laboratory, is shown in Fig. 2. The fracture followed a radial path, was relatively smooth and flat, except for the last stage, which had an inclination of approximately 45°.

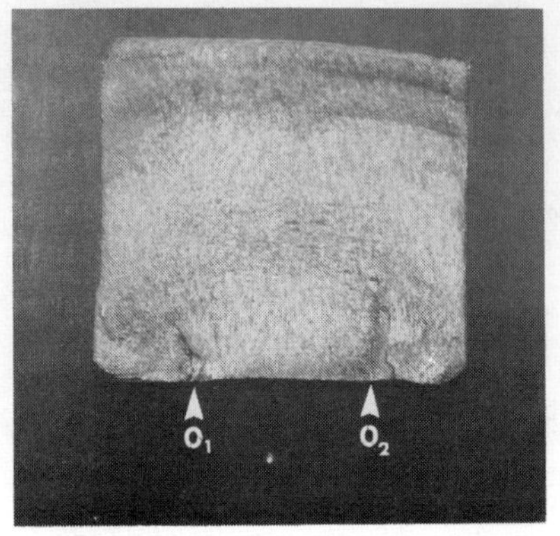

Fig. 2

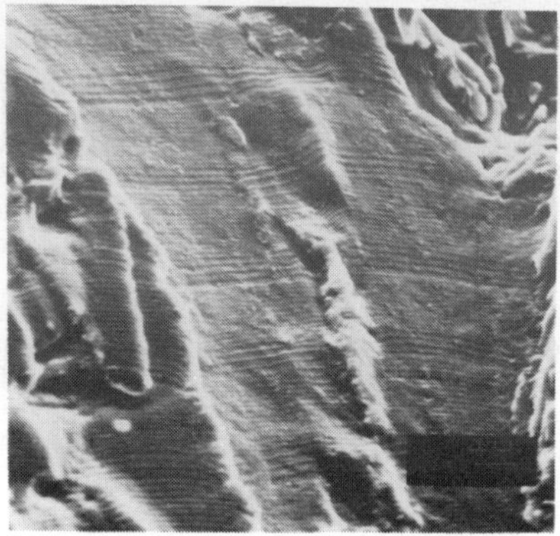

Fig. 3

Macro examination revealed the presence of two origins, both situated on the inner circumference of the fitting lug: origin O_1 located approximately 7.1mm (0.28") from the reworked side of the fitting and origin O_2 located approximately 5.3mm (0.21") from the opposite side of the fitting, as shown in Fig. 2. No mechanical damage was associated with the initiation sites.

QUANTITATIVE FRACTOGRAPHIC EXAMINATION

Scanning electron microscope examination was carried out to reveal the micro-morphology of the fracture. A consecutive pattern, or blocks of fatigue striations was found, which could be matched to the applied loading spectrum, Fig. 3. This identification is shown in Fig. 4. On the small SEM micrograph on the right (original magnification: 2050x) a group of 17 striations could be matched with a block of cycles as indicated by the spectrum loading. On the enlarged SEM micrograph on the left (original magnification: 8200x) the fine striations could be matched to significant changes in loads applied during the test. The numbers represent the corresponding loading cycle as indicated in the spectrum description. (BR represents the braking sequence). It is evident that "braking" under a spectrum of type "B", produced a deeper crack advance than the "braking" under a spectrum of type "A" or "C". This identification revealed a perfect match, on the microscopic scale, to the spectrum of loads and therefore each of the three cycles (identified as cycle "A", "B" and "C") had produced a characteristic imprint on the fracture micromorphology.

A striation count was carried out on the crack field which originated at O_1. Thirty micro fields were selected, the striations were photographed and measured in each field. All thirty fields selected were representative of cycles of type "C" only.

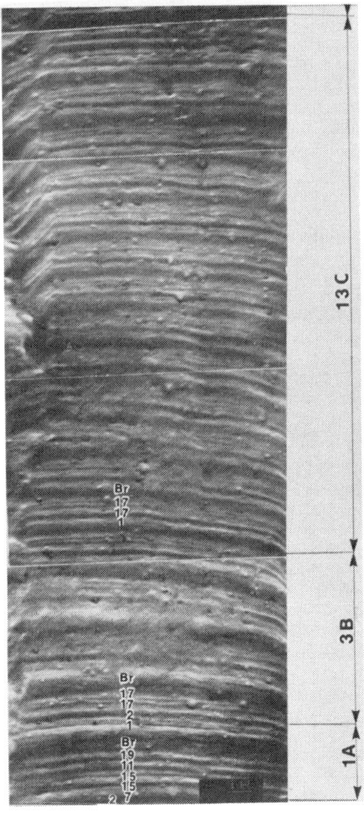

Fig. 4

Fig. 5

The results of the striation count revealed that a minimum of 21,000 "C" cycles (32312 flights) were necessary to propagate the crack from the origin to a point located approximately 17.6mm (0.7") from the origin, at which time testing was halted.

An estimate of the total number of cycles necessary to propagate a crack long enough to be detectable* from the lateral side of the fitting, was calculated using the area from Fig. 5, where a macroscopic crack arrest mark is visible around origin O_2. A striation count was carried out to estimate the number of cycles necessary to propagate the crack from the initiation to this crack arrest mark. The result indicated 5660 cycles (type C"). Assuming that the crack propagated in a parallel path, beyond the crack arrest mark, a total of 2540 cycles "C" were estimated as required to propagate the crack further to a depth which would produce a crack of 1.3mm (0.050") on the lateral side of the fitting. Therefore it was estimated that, after initiation, a total of 8200 cycles "C" would produce a detectable crack on the lateral side of the part. The crack growth curve versus the number of flights as derived from the fractographic examination, is shown in Fig. 6.

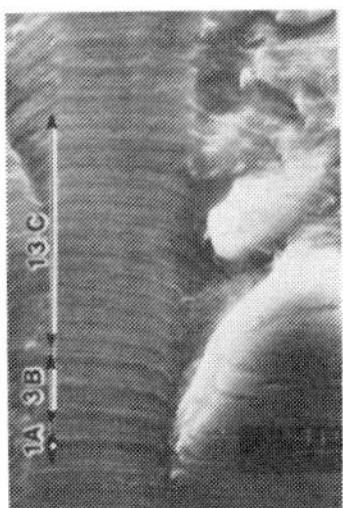

* A length of minimum 1.3mm (0.050") is considered to be detectable.

SUBJECT: CRACK GROWTH CURVE DATA

POINT NUMBER	CRACK LENGTH (MM)	ESTIMATED NUMBER OF FLIGHTS
1	2.43	12115
2	2.51	12633
3	2.70	13524
4	2.78	13903
5	3.07	15232
6	3.54	16732
7	3.84	17501
8	4.09	18053
9	4.47	18866
10	4.68	19512
11	5.00	20061
12	5.32	20676
13	5.94	22223
14	6.74	23376
15	7.16	23784
16	7.73	24109
17	8.09	24435
18	8.72	24918
19	9.35	25523
20	9.67	25770
21	10.25	26107
22	11.01	26575
23	11.72	27121
24	12.53	27575
25	13.66	28449
26	14.10	28783
27	14.57	29227
28	15.06	29720
29	16.51	31304
30	17.56	32312

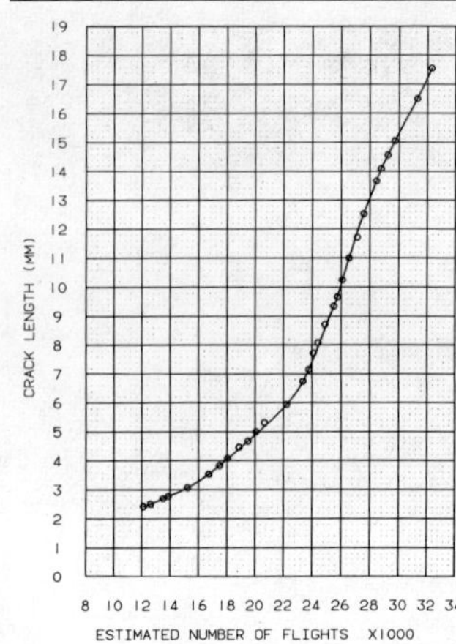

Fig. 6

ESTIMATION OF THE SAFE LIFE OF THE COMPONENT

Using a safe life fatigue scatter factor of 4.0, a preliminary safe life for this component was then determined to be

$$(3001\text{x}20 - (21000\text{-}8200))\text{x}\,\frac{20}{13}\,\text{x}\ 1/4 = 10{,}005 \text{ safe aircraft flights}$$

where: 3001 - total number of blocks applied during the test
21000 - the estimated number of cycles "C" necessary for propagation
8200 - the estimated number of cycles "C" necessary to produce a "detectable" crack
20/13 - transformation parameter

As mentioned earlier, the test loading spectrum was derived conservatively. Using the results of the fractographic examination, a more exhaustive analytical review of the failure is in progress which will take advantage of the conservatism in the test spectrum, to derive a higher, more realistic safe life for this component. The analytical review is however beyond the scope of this present paper.

CONCLUSIONS

Fractographic examination of components subjected to fatigue and damage tolerance testing is demonstrated to be a useful method in the estimation of the safe life of these components.
The techniques used in fractographic analysis allowed the loading spectrum applied during the testing to be identified on the microscopic fracture imprint and therefore estimation of the number of cycles necessary to propagate the crack could be accurately determined.

REFERENCES

1. FAR Part 25.571, Amendment 45
2. J.C. McMillan, R.W. Hertzberg, ASTM STP 436, 89-121

A New Fatigue-Creep Interaction Map for a Big ESR-cast-to-shape Gas Turbine Disc (ECD)

G. L. Chen, Q. F. He, L. Gao, X. S. Xie, S. H. Zhang*, F. Y. Chang and Y. C. Zhao*****

**Beijing University of Iron and Steel Technology, Beijing, People's Republic of China*
***Red Flag Machine Works, Xian, People's Republic of China*
****Shanghai Fifth Steel Plant, Shanghai, People's Republic of China*

ABSTRACT

A new fatigue-creep interaction map for a big ECD (930mm dia.) is constructed to show the full view under a wide range of combinations of fatigue and creep stresses on the basis of fatigue-creep interaction curves. Comparison of this map with that for wrought disc of the same superalloy indicates that the ECD is more suitable for long time stationary gas turbine survice. These maps also provide the essential information on which types of deformation and fracture mechanisms are operating under certain service condition. On the basis of these maps, several features of fatigue-creep interaction are discussed. In order to improve the fatigue and creep properties of ECD, a suitable solution treatment or a HIP treatment is necessary.

KEYWORDS

Fatigue-creep interaction map; Superalloy; ESR-Cast-to-Shape disc.

INTRODUCTION

The ECD has successfully passed 800 hr engine test (1) and 800 hr test flight on a aircraft gas turbine. However, the fatigue property of ECD under service stress condition is doubtful due to the coarse columnar grain configuration. For the successful application of ECD, a fundamental understanding of its fatigue-creep interaction behaviors is utmost important. This paper tries to construct a new fatigue-creep interaction map to show the full view under a wide range of combinations of fatigue and creep stresses and to discuss the characteristic mechanical properties of ECD and the interaction mechanisms.

MATERIAL AND PROCEDURES

The chemical composition of ECD is similar to that of V57. Heat treatment schedule used is 1050°C/4h o.c.+700°C/32h a.c. Since the fatigue and creep properties of ECD is strongly related to the orientation of the specimen (2), all specimens used in the present investigation have an orientation of 45 degrees with the axis of columnar grains. The fatigue-creep interaction tests are load control tests at a frequency of 1 HZ. Each test is conducted at a constant maximum stress (3). Since the maximum stress (σ_{max}) is constant, both fatigue stress (stress amplitude σ_a) and creep stress (or mean stress σ_m) can be varied simultaneously in the opposite directions by changing the minimum stress. All tests are performed at 650°c in air.

RESULTS

The σ_a-N_f curve evaluated at constant σ_{max} is shown in Fig. 1. This fatigue -creep interaction curve is different from the normal S-N curve and creep curve. Figure 1 indicates that the interaction curve can be envisaged as a combination of pure fatigue curve and pure creep curve. On the upper part of the curve the fracture life is related mainly to the fatigue stress (σ_a). On the bottom part of the curve the rupture life is determined by the creep stress (σ_m). At the nose of the curve both σ_a and σ_m have a great influence on the fracture life. Figure 2 shows the first type of fatigue-creep interaction map which includes five curves : pure fatigue curve, pure creep curve, and three interaction curves for different σ_{max}. It can be seen from Fig. 2 that all interactions cause the fracture life curve to move to the shorter life region. Also, as the maximum stress increases the interaction curves move to the shorter life region, and the nose of curve moves to the area with high σ_a. With the different stress dependences of fracture lives at constant temperature, the full map can be classified into three regions (Fig. 2) : i.e. interaction with predominant fatigue fracture (F region), interaction with predominant creep fracture (C region), and interaction with mixed fracture (FC region).

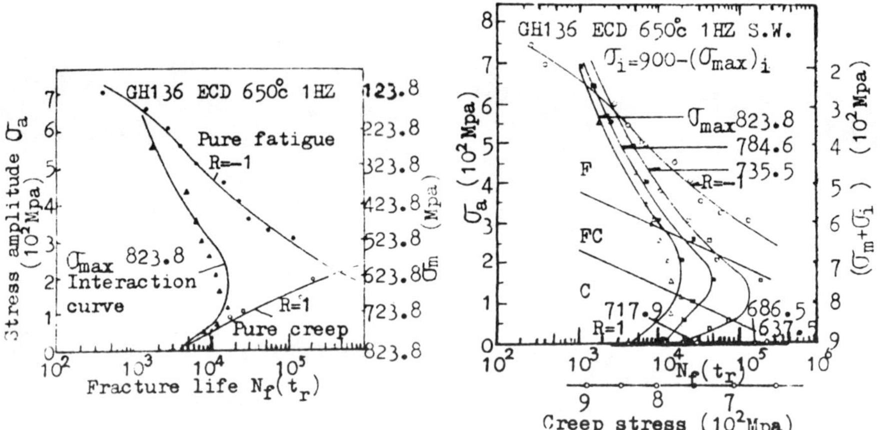

Fig. 1 Fatigue-creep interaction curve Fig. 2 Fatigue-creep interaction map

Life equations are listed in Table 1.

Same treatments have been done for the effect of such combinations on dynamic secondary creep rates, and another type of interaction map is constructed on this basis (Fig. 3). From this map the $\dot{\varepsilon}_s$ is a function of σ_{max}, σ_a and σ_m. In F region $\dot{\varepsilon}_s$ increases with increasing σ_a and σ_{max}. In C region $\dot{\varepsilon}_s$ increases with increasing σ_{max} and σ_m. All $\dot{\varepsilon}_s$ equations are listed in Table 1.

The interaction map can be analized, just like Hales' work (4), on the basis of crack initiation and propagation. The crack initiation and propagation can be schematically expressed in Fig. 4 which includes four curves : fatigue crack initiation curve and fracture curve (f_0f_0 and FF), creep crack initiation and fracture curve (C_0C_0 and CC). If point A locates above point D (Fig. 4), the

TABLE 1 Regression Equations

σ_{max} Mpa	F region		C region	
735.5	$\sigma_a=7.23\cdot10^3 N_f^{-0.32}$,	r=-0.98	$\sigma_m\cdot t_r^{0.053}=1.23\cdot10^3$,	r=-0.99
	$\dot{\varepsilon}_s=2.83\cdot10^{-14}\sigma_a^{3.43}$,	r=0.99	$\dot{\varepsilon}_s=2.255\cdot10^{-66}\sigma_m^{21.5}$,	r=0.99
784.6	$\sigma_a=7.43\cdot10^3 N_f^{-0.34}$,	r=-0.99	$\sigma_m\cdot t_r^{0.154}=3.28\cdot10^3$,	r=-0.97
	$\dot{\varepsilon}_s=1.483\cdot10^{-13}\sigma_a^{3.28}$,	r=0.99	$\dot{\varepsilon}_s=9.416\cdot10^{-41}\sigma_m^{12.61}$,	r=0.98
823.8	$\sigma_a=8.48\cdot10^3 N_f^{-0.365}$,	r=-0.98	$\sigma_m\cdot t_r^{0.083}=1.64\cdot10^3$,	r=-0.98
	$\dot{\varepsilon}_s=1.351\cdot10^{-11}\sigma_a^{2.67}$,	r=0.99	$\dot{\varepsilon}_s=1.666\cdot10^{-37}\sigma_m^{11.55}$,	r=0.99
R=-1	$\sigma_a=2.50\cdot10^3 N_f^{-0.18}$,	r=-0.99		
	$\dot{\varepsilon}_s=5.071\cdot10^{-20}\sigma_a^{5.164}$,	r=0.98		
R=1	$\sigma\cdot t_r^{0.057}=1.31\cdot10^3$,	r=-0.97		
	$\dot{\varepsilon}_s=1.969\cdot10^{-53}\sigma^{17.026}$,	r=0.98		

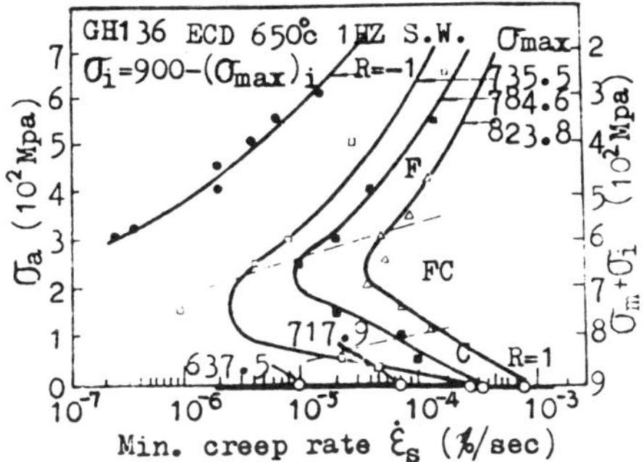

Fig. 3 Interaction map for dynamic secondary creep rates.

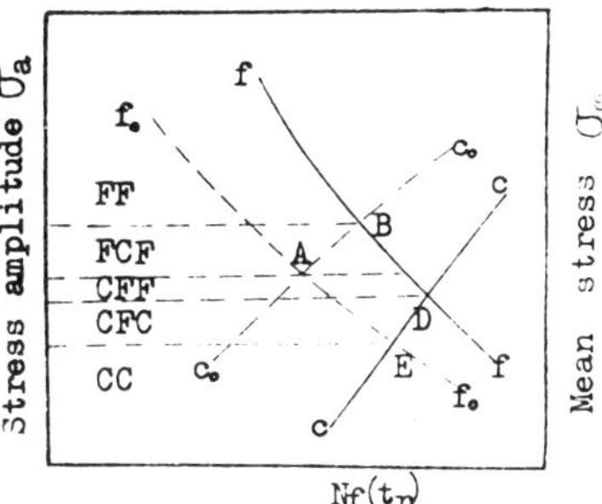

Fig. 4 Schematic view of the crack initiation and propagation.

following five regions must exist : i.e. fatigue crack nucleation and propagation to fracture (FF region); fatigue and creep crack nucleation and predominant fatigue crack propagation to fracture (FCF region); creep crack initiation followed by fatigue crack nucleation and competition with creep crack and propagation to fracture (CFF region); creep crack initiation followed by fatigue crack nucleation and final creep crack propagation to fracture (CFC region); creep crack initiation and propagation to fracture (CC region). If the point A is below the point D, the CFF region will change to FCC region in which fatigue crack initiation is followed by creep crack initiation and propagation to fracture. The experimental observations on crack initiation and propagation illustrate the existance of FF. FCF. CFF. CFC and CC regions. The appearance fracture surface of specimens tested in FF region is similar to the normal fatigue fracture surface. The appearance of fracture surface of CC specimen is similar to the normal creep fracture surface characterized by creep voids. The typical appearance of fracture surfaces of CFC and FCF specimens are shown in Fig. 5a and 5b respectively. The basic characteristics of the fracture surface for CFC specimens is the creep voids with seperated fatigue fracture areas near the surface of the specimen (Fig. 5a). The fatigue striations with creep voids appear to be the typical fracture appearance for FCF specimens (Fig. 5b). The appearance for CFF specimens can be characterized by completely mixed distrubution of fatigue striation and creep voids (Fig. 5c), but the ratio of striation to void areas is varied with the ratio of σ_a to σ_m. According to the results of our observation, a complete interaction fracture mechanism map has been constructed (Fig. 6).

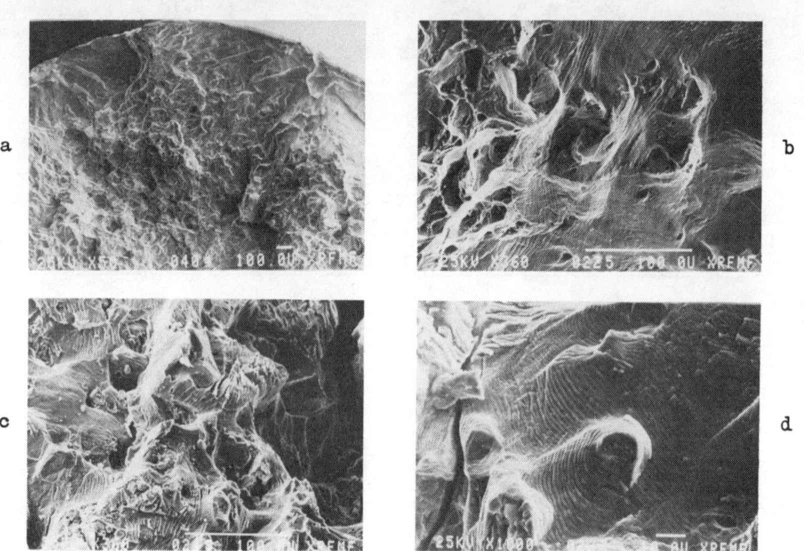

Fig. 5 SEM of fracture appearance. Specimens tested in a. CFC region b. FCF region c. CFF region d. fracture sources

DISCUSSION

Figure 6 indicates that the area of F region is relatively larger than that of C region, and the CFC region does not locate in the C region but in the FC region. These facts imply that the creep resistance of this disc is relatively high and the fatigue resistance relatively low. The comparison of the map for ECD with the map for wrought disc of the same superalloy (Fig. 7) indicates that the fracture life in the C region for ECD is longer than that of wrought disc. If the service condition is similar to such stress condition, like that for the long time stationary gas turbine survice, the application of ECD may be reasonable. These maps also provide the essential information on which types of deformation and fracture mechanisms are operating under certain service condition for the ECD. Figure 5d shows that the primary phases formed due to dendritic segregation and casting defects play as the sources of fatigue crack. This gives a explanation why the fatigue resistance of ECD is relatively low. A suitable solution heat treatment or a HIP treatment should result in improving fatigue and creep properties of ECD.

On the basis of the interaction map, the fatigue-creep interaction must be in nature distinguished among F. C. and FC regions. In F region the interaction can be expressed as the influence of creep stress and strain on the fatigue life. In C region the interaction can be indicated by the influence of fatigue stress and cyclic strain on the creep rupture life. In FC region the interaction is a mixed interaction. The effect of σ_m on the fatigue life is shown in Fig. 8a. Figure 8b shows the influence of σ_a on the creep rupture life. All interactions for ECD result in reduction of fracture life due to the following features of interaction : (a) the increased detrimental effect of primary phases or casting defects which promotes a tendency of initiation of subsurface crack (Fig. 9a). (b) the promotion of fatigue crack initiation and propagation by the creep voids acting as the sources of fatigue crack (Fig. 9b).

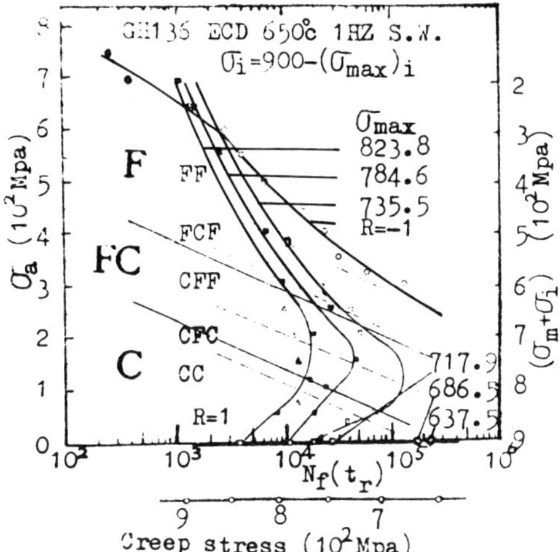

Fig. 6 Interaction map for fracture mechanism

Fig. 7 Interaction map for wrought disc

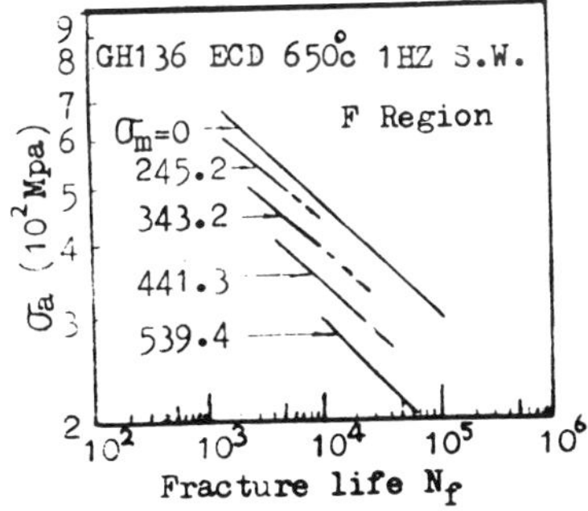

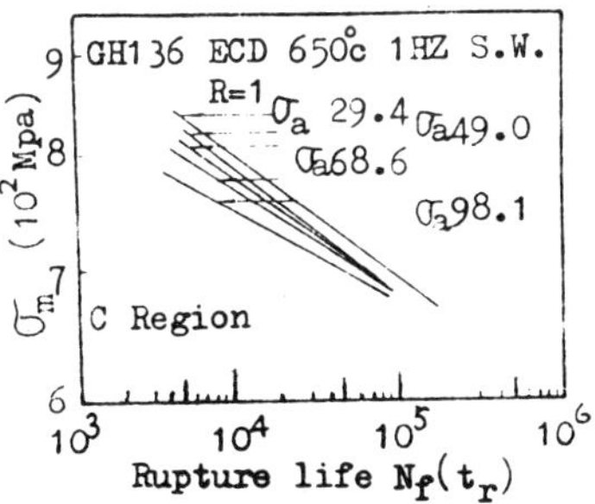

Fig. 8 The effect of σ_a and σ_m on fracture life

(c) the appearance in the stage I fracture plane of slip traces and shallow creep voids (Fig. 9c). (d) increase in fatigue striation spacings and the appearance of small creep voids between fatigue striations (Fig. 9d). (e) increase in the number of secondary cracks on creep fracture surface. All these interactions result in promoting fatigue crack initiation and increasing the extent of plastic zone at crack tip and crack growth rate.

CONCLUSIONS

(1) Fatigue-creep interaction map shows clearly the full view under a wide range of the combination of fatigue and creep stresses. This is the way to show the interaction characteristics of materials under combined stresses.
(2) The creep resistance of ECD is relatively high, but the fatigue resistance is relatively low. The ECD is more suitable for stationary gas turbine service.
(3) A suitable solution treatment or HIP treatment is necessary for increasing the fatigue and creep properties of ECD.
(4) Interactions must be in nature distinguoshed among F. C. and FC regions. Various interaction mechanisms can act under different combined stresses.

REFERENCES

1. J.F.Chang, G.P.Hu, L.Gao, in SUPERALLOY 1980 (edited by J.K.Tien etc.), p.245 ASM Metals Park Ohio 44073.
2. Q.F.He Ph.D. Thesis B.U.I.S.T. (in press).
3. G.L.Chen, J. of Aeronautical Materials, 3, 47 (1983).
4. R.Hales, in FATIGUE OF ENG. MATER.AND STRUCTURE (Pergamon Press Ltd.), 3, p.339 (1980).

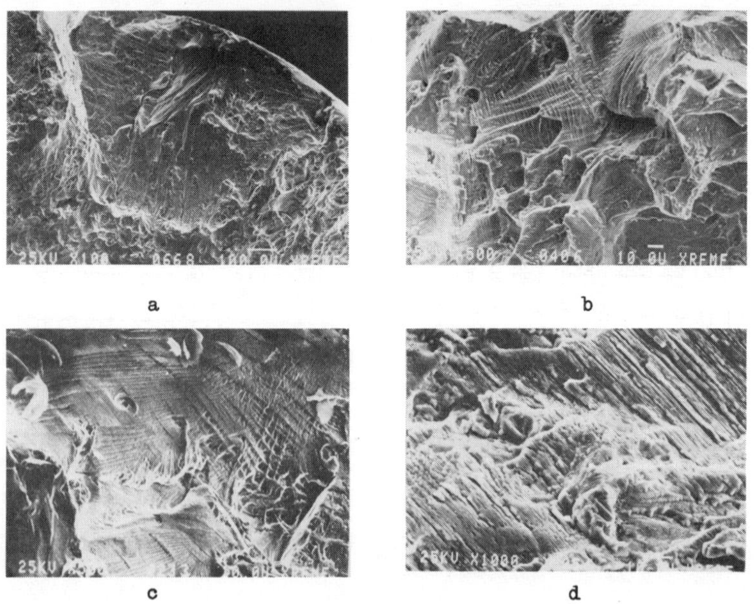

Fig. 9 Features of interactions

A Correlation of Threshold Stress and Stress Intensity Condition for Fatigue Cracks

D. Kujawski[1] and F. Ellyin

Department of Mechanical Engineering, University of Alberta, Edmonton, Alberta T6G 2G8, Canada

ABSTRACT

One of the current topics of research in fatigue failure is the quantitative determination of the transition from the endurance limit concept to that of the threshold stress intensity one. It is postulated that a critical crack size a_0 can be defined for which these two concepts are equally contributing to the threshold condition. Based on this premise, a relationship is proposed which can be used to determine the threshold stress and stress intensity condition for the transition range of crack sizes, i.e. short cracks.

KEYWORDS

Threshold stress; threshold stress intensity; fatigue; crack propagation; endurance limit; short cracks; long cracks; critical crack size.

INTRODUCTION

Traditionally the problem of threshold condition in fatigue has been approached by considering the smooth and precracked specimens. For smooth specimens, the stress range is termed the endurance limit, $\Delta\sigma_{e0}$, which could be applied without causing fatigue failure for a large number of cycles, usually 10^7-10^9 cycles. The endurance limit of machine parts with small flaws depends mainly on the crack length 'a', and load ratio $R = P_{min}/P_{max}$. Frost [1], Kobayashi and Nakazawa [2] found the parameter $(\Delta\sigma/2)^n \cdot a$ = const. useful in analyzing dependence of the endurance limit for short crack data.

The application of fracture mechanics to fatigue threshold condition is based on defining a threshold stress intensity range, ΔK_{TH0}, below which a 'long'

---

[1]On leave from Institute of Machine Design Fundamentals, Warsaw Technical University, Warsaw, Narbutta 84, Poland.

fatigue crack does not detectibly propagate. The usual definition for ΔK_{THO} corresponds to a crack growth rate from 10^{-11} to 10^{-12} m/cycle. Harrison [3], Frost et al.[4] and Ando [5] proposed the following relations for the threshold condition for fatigue cracks, $\Delta K = \Delta K_{TH}$ for $R \geq 0$, and $K_{max} = K_{TH}$ when $R \leq 0$.

For a given mean stress, specimen geometry, environment, and loading frequency, the endurance limit and stress intensity threshold are considered to be material properties. In this paper, the interrelation between threshold stress and stress intensity is discussed.

THRESHOLD CONDITIONS FOR FATIGUE CRACKS

The condition for which a detectible crack does not propagate is defined by ΔK_{THO} and $\Delta\sigma_{e0}$ for 'long' and 'short' cracks, respectively. These two approaches overlap for short cracks and have caused a conceptual difficulty.

The effect of crack size on the threshold condition for fatigue cracks is shown schematically in Fig. 1. In it the stress range $\Delta\sigma$ (solid line) necessary to cause fatigue crack growth is plotted against an equivalent crack length, a_{eq}, defined as:

$$\Delta K = \alpha\Delta\sigma(\pi a)^{1/2} = \Delta\sigma\,(\pi a_{eq})^{1/2}, \tag{1}$$

where α is a function of specimen geometry and crack shape. It can be seen from Fig. 1 that the solid curve lies between two straight dashed lines representing the constant smooth specimen fatigue limit and constant stress intensity range condition. This curve approaches the smooth specimen fatigue limit, $\Delta\sigma_{e0}$, for 'very short' cracks, and the constant stress intensity range, ΔK_{TH0}, for 'long' cracks. The ΔK_{TH} = const. region establishes a set of crack length "a_{eq}" and applied stress range,

$$\Delta\sigma \leq \Delta\sigma_{TH} = \Delta K_{TH}(\pi a_{eq})^{-1/2}, \tag{2}$$

for which further crack extension does not occur. One can conclude from Eq. (2).

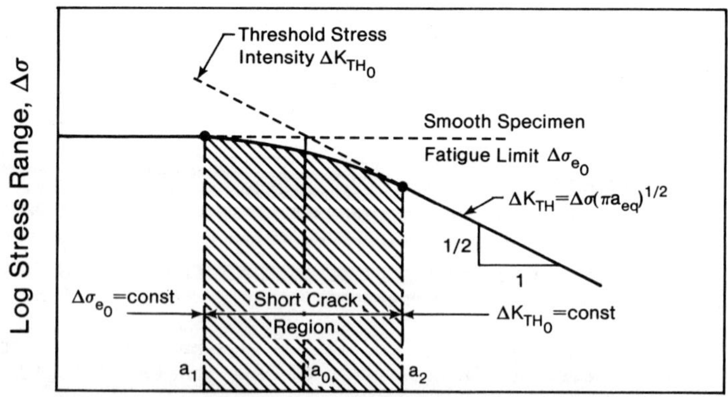

Fig. 1. Effect of equivalent crack length on threshold condition

that as the crack size becomes exceedingly small, the required threshold stress should increase and could exceed the smooth specimen fatigue limit. A set of crack sizes $a_1 \leq a_{eq} \leq a_2$ could therefore exist (shaded area in Fig. 1), for which $\Delta K_{TH} \leq \Delta K_{TH0}$ as well as $\Delta\sigma_{TH} \leq \Delta\sigma_{e0}$ define the condition that crack is blocked or arrested.

The value of a_2 represents the limit above which constant stress intensity range approach can be used and a_1 represents the limit below which initial crack length has no effect on the fatigue life. Thus, for cracks whose lengths lie between the limits a_1 and a_2, one would expect that each of the conditions $\Delta\sigma_{TH} = \Delta\sigma_{e0}$ and $\Delta K_{TH} = \Delta K_{TH0}$ will become non-conservative. Therefore, a critical crack size a_0 can be defined as,

$$\Delta K_{TH0} = \Delta\sigma_{e0}(\pi a_0)^{1/2}, \tag{3}$$

where these two concept ie. constant stress and constant stress intensity range determine simultaneously the threshold condition with equal rank. The idea of critical crack size a_0 was introduced by some investigators for explaining the dependence of ΔK_{TH} and $\Delta\sigma_{TH}$ on crack size [6,7,8]. Smith [6] suggested that $\sigma_{TH} = \sigma_{e0}$ when the crack length is shorter than a_0, and above this length it is expressed by $\Delta K_{TH} = \Delta K_{TH0}$ for long cracks.

Usami and Shida [9] proposed that the threshold condition could be expressed by the critical size of the plastic zone ahead of the crack tip. Tanaka et al. [10] assumed that the threshold condition for crack growth is determined by the condition whether the slip band near the crack tip propagates into an adjacent grain or not. Both models show the crack length dependence on the ΔK_{TH} and $\Delta\sigma_{TH}$, and a comparison of the experimental data was provided for various materials found in the literature [10]. The materials used were low carbon steels [1,8,9,11,12], high strength steel [13], 13 Cr cast steel [9], copper [14], and aluminum [15].

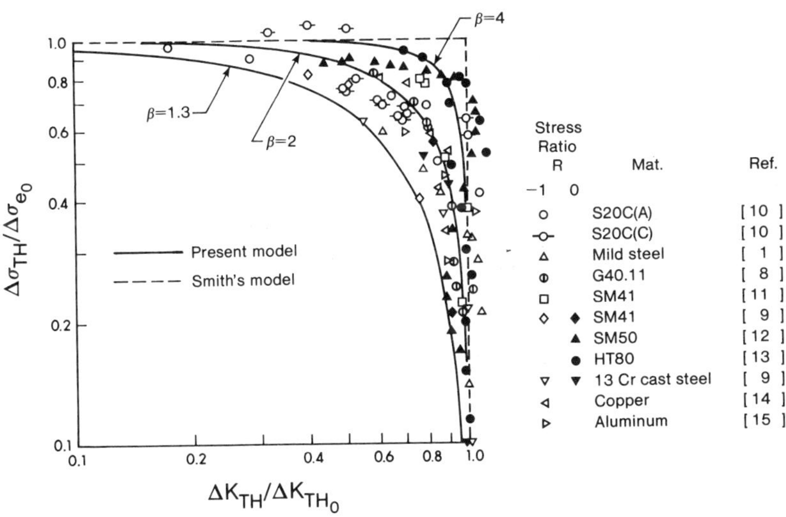

Fig. 2. Normalized relationship between threshold stress and threshold stress intensity

Normalized relation between threshold stress $\Delta\sigma_{TH}/\Delta\sigma_{e0}$ and corresponding threshold stress intensity $\Delta K_{TH}/\Delta K_{TH0}$ reported by Tanaka et al. [10] are plotted in Fig. 2. Considering the scatter of experimental data, it can be said that the data points lie symmetrically with respect to the line $(\Delta\sigma_{TH}/\Delta\sigma_{e0})=(\Delta K_{TH}/\Delta K_{TH0})$. Therefore, it may be postulated that the normalized relation between threshold stress and threshold stress intensity can be described by the following equation:

$$(\Delta\sigma_{TH}/\Delta\sigma_{e0})^{\beta} + (\Delta K_{TH}/\Delta K_{TH0})^{\beta} = 1 . \qquad (4)$$

The solid lines in Fig. 2 show the qualitative predictions of Eq.(4) for different values of exponent β. The dashed lines in Fig. 2 represents the Smith's model [6]. It is seen from Fig. 2 that the data points fall between the two curves corresponding to β=1.3 and β=4, and lie closely to the curve β=2. Using Eq. (2 and 3), Eq. (4) can be written in the form of,

$$\Delta\sigma_{TH} = \Delta\sigma_{e0}\left[1 + (a_{eq}/a_0)^{\beta/2}\right]^{-1/\beta}, \text{ or } \Delta K_{TH} = \Delta K_{TH0}\left[1 + (a_0/a_{eq})^{\beta/2}\right]^{-1/\beta} \qquad (5)$$

It is of interest to note that Usami and Shida [3] and Tanaka et al. [10] models give similar results to that of Eq. (5) if β=2. Also Haddad et al.'s [8] expression for the threshold stress is equivalent to the first of Eq.(5) with β=2.

DISCUSSION

It is well known that the threshold stress intensity increases and the fatigue limit decreases with the increasing grain size. It is also known that both of these threshold values decrease with the increasing R ratio. Since values for a_0 given by Eq.(2) are obtained based on the ΔK_{TH0} and $\Delta\sigma_{e0}$ values for a particular R ratio, it follows that a relationship among the critical crack length, grain size and R exists. Haddad et al. [8] have shown that a_0 varies linearly with the grain size, however, it did not relate directly to the material's microstructural characteristic length of the low strength steels investigated. Recently, Morris et al. [16] have shown that larger grains at the surface of 2219 - T851 aluminum accumulate tensile plastic strain at a stress amplitude approximately equal to the fatigue limit. These surface layer phenomena, ie. the reduced flow resistance and effect of R ratio and grain size which cause a progressive increase in local microplasticity, are responsible for the crack initiation and fatique strength reduction. Ellyin and Kujawski have shown [17] that the cumulative tensile deformation can significantly decrease the absorbed energy to fracture. The microstructure influence on the threshold condition can be described through a relationship between a_0 and β. When a crack size $a_{eq} = a_0$, the constant stress and constant stress intensity range contribute equally to the threshold condition, i.e. in Eq. (4) both terms are equal.

The value of β can therefore be determined by matching Eq. (4) with the experimental values for $\Delta\sigma_{TH}/\Delta\sigma_{e0}$ or $\Delta K_{TH}/\Delta K_{TH0}$ at $a_{eq} = a_0$. The values of β thus calculated are plotted vs a_0 in Fig. 3 for six different materials. Figure 3 suggest that β varies linearly with a_0 in a log-log space for the materials investigated. It should be noted that a relation between β and a_0 for 13 Cr cast steel does not follow the trend shown in Fig. 3. This divergence is probably due to existence of natural defects such as micro shrinkage cavities [9] which affect the fatigue strength.

Figures 2 and 3 suggest that for a material with the value of a_0 smaller than those shown in Fig. 3, a higher value of β can be expected and consequently experimental points of the threshold condition for such a material will lie closer to Smith's model [8].

The theoretical relation of $\Delta\sigma_{TH}/\Delta\sigma_{e0} = \alpha_1$ and $\Delta K_{TH}/\Delta K_{TH0} = \alpha_2$ obtained from Eq. (5), are plotted in Fig. 4. The curves with $\alpha_1 = \alpha_2$ equal 0.95 and 0.9 represent 5 and 10 percent deviation from the threshold stress or threshold stress intensity range of constant stress or constant stress intensity range condition, respectively. Each of these curves established a set of relative crack size a_{eq}/a_0 and the corresponding value of β. Based on experimental data Taylor and Knott [19] have attempted to correlate the value of a_2 with the critical crack size a_0, (see Fig. 1). They found that most data lie between $a_2 = 10a_0$ and $a_2 = a_0$, and they suggested that $a_2 = 4a_0$ is a reasonable approximation. It is interesting to point out that the relation between a_2 and a_0 noted by Taylor and Knott from the experimental observation agreed with the proposed model for low and medium strength metals for which $\beta \approx 2$, see Fig.4.

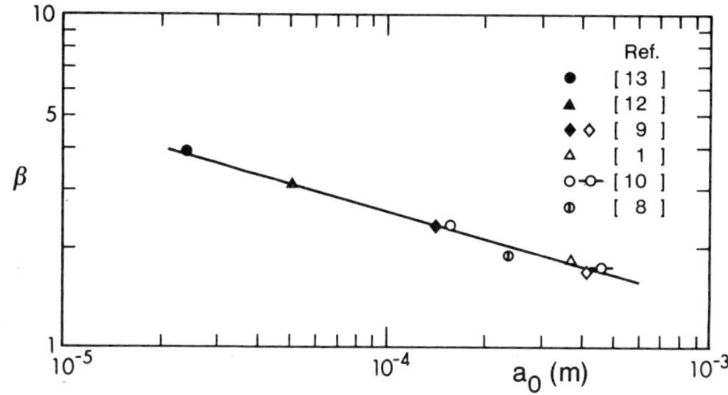

Fig. 3. Relation between exponent β and critical crack size a_0.

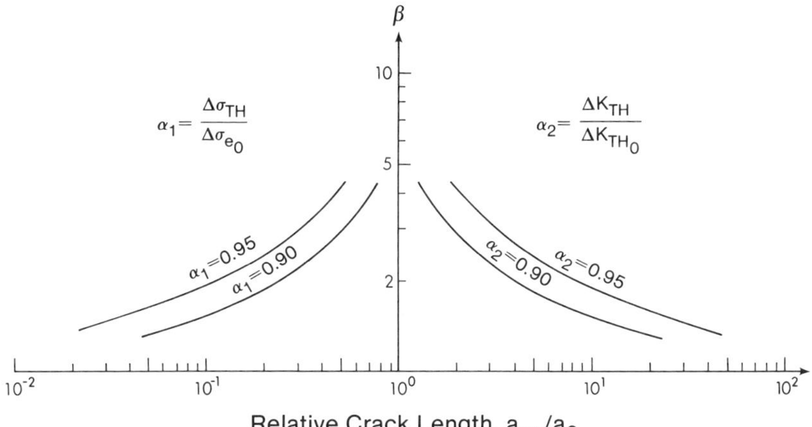

Fig. 4. Curves of constant relative threshold stress, and relative threshold stress intensity in terms of exponent β and relative crack length.

CONCLUSIONS

A number of proposed models relating the threshold stress and stress intensity are shown to be particular case of the one developed herein. Predictions of the proposed relation herein are in agreement with the experimental data. The interrelation between threshold stress and stress intensity is given by Eq. (4), and in terms of the critical crack size by Eq. (5).

ACKNOWLEDGEMENT

The financial support of the Natural Sciences and Engineering Research Council of Canada (NSERC Grant No. A-3808) and the Izaak Killam Memorial Foundation is greatfully acknowledged.

REFERENCES

1. N.E. Frost, Proc. Inst. Mech. Engrs. 173, 811 (1959).
2. H. Kobayashi and H. Nakazawa, Trans. Japan Soc. Mech. Engrs. 36, 1789 (1970).
3. J.D. Harrison, Metal Construct. Br. Weld. J. 17, 93 (1970).
4. N.E. Frost, L.P. Pook and K. Danton, Engng. Fract. Mech. 3, 109 (1971).
5. S. Ando, Mech. Behavior of Mat., Proc. 1971 Inter. Conf. on Mech. Behavior of Materials, 1, 331 (1972).
6. R.A. Smith, Inter. J. Fracture, 13, R.717 (1977).
7. P. Lukaŝ and M. Klesnil, Mater. Sci. Engng. 39, 61 (1978).
8. M.H. El Haddad, K.S. Smith and T.H. Topper, J. Engng. Mater. Technol. 101, 92 (1979).
9. S. Usami and S. Shida, Fatigue of Engng. Mater. & Struct., 1, 471 (1979).
10. K. Tanaka, Y. Nakai and M. Yamashita, Intern. J. Fracture 17, 519 (1981).
11. H. Ohuchida, S. Usami, and A. Nishioka, Trans. Japan Soc. Mech. Engrs. 41, 703 (1975).
12. H. Kitagawa and S. Takahashi, Proc. 2nd Inter. Conf. on Mech. Behavior of Materials, Boston, p. 627 (1976).
13. H. Kitagawa and S. Takahashi. Trans. Japan Cong. on Material Research, p. 106 (1979).
14. N.E. Frost, J. Mech. Engng Science, 5, 15 (1963).
15. N.E. Frost, J. Mech. Engng Science, 6, 203 (1964).
16. W.L. Morris, R.V. Inman, M.R. James, J. Mater. Science 17, 1413 (1982).
17. F. Ellyin and D. Kujawski, J. Press. Vessel Technol. 106, 342 (1984).
18. J. Lankford, Inter. J. Fracture, 16, R7 (1980).
19. D. Taylor and J.F. Knott, Fatigue of Engng Mater. & Struct., 4, 147 (1981).

Influence of Microstructure on Fatigue Limit Range $\triangle\delta\omega$ and Threshold $\triangle\varkappa_{th}$

Du Bai-Ping and Li Nian

Research Institute for Strength of Metals, Xi'an Jiaotong University, Xi'an, Shaanxi Province, People's Republic of China

ABSTRACT

According to the measurements of ΔK_{th} and $\Delta\sigma w$ values of various microstructure in 12CrNi3A steel, the equation $\Delta K_{th}=1.12\Delta\sigma w\sqrt{\pi a}$ is shown to be valid, where "a" is a material parameter, named fracture unit, which relates to the microstructure.

Another formula showing influence of the fracture unit on ΔK_{th} value is established:

$$\Delta K_{th}=K_c^m + 0.7\,\sigma_i\sqrt{d}$$

Here, d is the fracture unit size. σ_i is the frictional stress in dislocation moving. K_c^m is the critical value of microscopic ΔK.

Reasons for the lowest ΔK_{th} value of the martensitic structure are explained.

KEYWORDS

Threshold; fracture unit; fatigue limit.

INTRODUCTION

ΔK_{th} data of various kinds of steel, collected by Tanaka[1] show that the martensitic structure exhibits lower ΔK_{th} value than the others. Generally, data presented by different authors were produced under different experimental conditions, if the specimen size, the stress ratio R etc. are different. It would not be appreciable to put them together in comparison.

The object of this paper is to compare the ΔK_{th} value of two kinds of steel with different microstructures in the same experimental condition. The other purpose is to study the size and the properties of the fracture unit and to explain the reasons of the lowest ΔK_{th} value for martensitic structure.

MATERIALS AND EXPERIMENTAL METHODS

Steel 12CrNi3A, 40Cr are employed in this study. Chemical compositions are shown in Table 1.

TABLE 1 Chemical Compositions of 12CrNi3A and 40Cr Steel in Weight %

	C	Si	Mn	P	S	Cr	Ni
12CrNi3A	0.13	0.30	0.49	0.006	0.0053	0.47	2.83
40Cr	0.40	0.39	0.69	0.014	0.011	0.96	

Heat treatments and the mechanical properties are in Table 2.

TABLE 2 Heat Treatments, Microstructures and Mechanical Properties

	Grop	Quenching	Tempering	Size A.G μm	Size M.P μm	$\Delta\sigma_w$ MPa	σ_i MPa	ΔK_{th} mpa$\sqrt{m}$ Exp	CAL
12CrNi3A	A1	1200.OQ	680	109*				8.9	
	A2	1200.OQ	600	109*		542	644	8.23	9.2
	A3	1200.OQ	400		(2-7)*	794	870	4.75	4.79
	A4	1200.OQ	200		(2-7)*	823	564**	3.5	4.3
	B	1050.OQ	600	27*		639		7.3	
	C1	870.O.W.Q	600	18*		639	629	6	6.6
	C2	870.O.W.Q	400		(2-7)*	776	995	5	4.97
	C3	870.O.W.Q 850.O.W.Q	200		(2-7)*	852	746**	4.0	4.6
40Cr	E	870.O.W.Q	200		(1-3)*		695**	4.1	4.0

A.G.: Austenitic grain. M.P. Martensitic packet

* Fracture unit. ** Limit of proportionality

Specimen is ectangular bar of 120x25x13 mm size. The stress ratio R used in three point bending fatigue experiment is 1/3. The threshould is determined by decreasing K value test.

EXPERIMENTAL RESULTS AND DISCUSSIONS

On the Relationship Among $\Delta\sigma_w$, ΔK_{th} and Microstructure

The relationship among $\Delta\sigma_w$, ΔK_{th} and microstructure can be expressed by formula[2].

$$\Delta K_{th} = 1.12 \Delta \sigma_w \sqrt{\pi a} \quad (1)$$

Here, a is material parameter named as "fracture unit", which is related to the microstructure. The austenite grain size can be taken as the fracture unit for the high temperature tempered microstructure while the width of the martensite packet for the low temperature tempered that.

Influence of Fracture Unit on ΔK_{th}

Form Table 2, it is clear that ΔK_{th} of martensite is the lowest one among all microstructures.

The structure of the high temperature tempered steel is characterized by the austenitec grain size Fig. 1 shows the influence of the austenitec grain size d on ΔK_{th}, and can be expressed as follows:

$$\Delta K_{th} = 4.6 + 400\sqrt{d} \ \text{MPa}\sqrt{m} \quad (2)$$

The propagation of a fatigue crack in the first stage is noncontinuous. When the crack tip slip bands are blocked by the grain or packet boundary shown in Fig. 2;3, the crack stops its propagation. Once a crack stopped by the grain boundary, dislocations pile up, and create large stress in its tip. It may produce a new dislocation source or dirive out the pin dislocations in the next grain. Thus, plastic deformation would take place there, and a new crack nucleus under the cyclic stress. If shear stress τ_c can drive a new dislocation source at a distance, r, from the grain boundary, which shown in Fig. 4. The following Hall-Petch relation can be used:

$$\tau_c = (\tau - \tau_i)\left(\frac{W_o}{r}\right)^{\frac{1}{2}}$$

where, τ is the shear stress of the slip plane, τ_i is the frictional shear stress in dislocation moving. W_o is the size of the blocked slip band zone, which equals to the distance from crack tip to grain boundary.

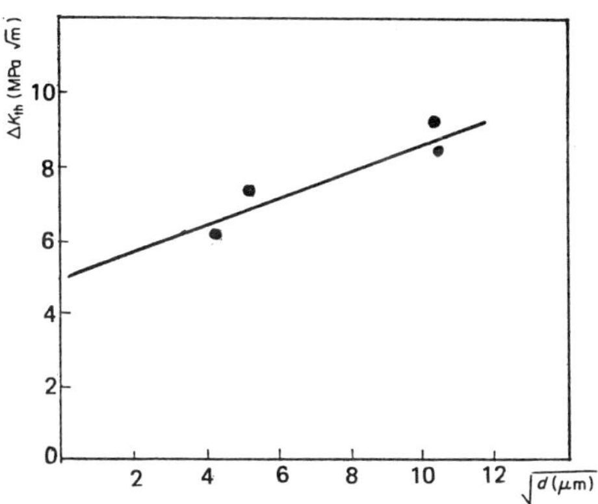

Fig. 1 ΔK_{th} vs Austenitic grain, $\sqrt{d}$

Fig. 2 Fatigue crack growth in the threshold region showing a crack tip stopped near a grain boundary

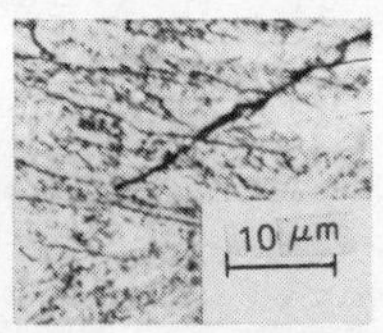

Fig. 3 Fatigue crack growth in the threshold region showing a crack tip stopped near a martensite plate or packet

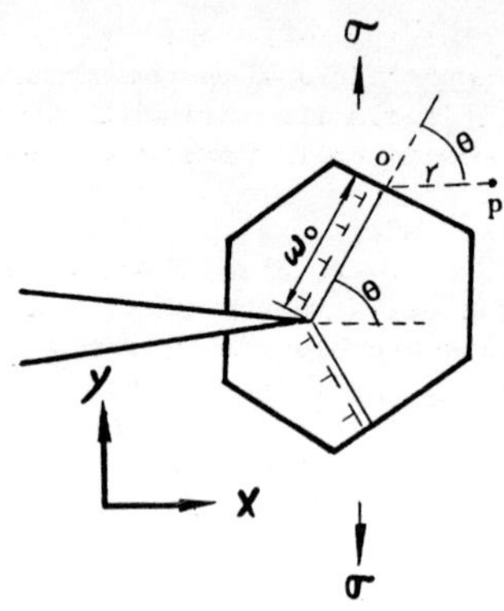

Fig. 4 A model of crack-tip slip band blocked by grain boundary

If it is expressed in normal stress, then

$$\sigma c = (\sigma - \sigma_i)\left(\frac{W_o}{r}\right)^{\frac{1}{2}} \qquad (3)$$

where, σc, σ and σi are normal stresses corresponding to τ_c, τ and τ_i, respectively. When σ takes a right angle to the slip plane at "O" point of Fig. 4, then

$$\sigma = \frac{1}{2}(\sigma y + \sigma x) + \frac{1}{2}(\sigma Y - \sigma x)\cos\frac{\theta}{2} + \tau xy \sin 2\theta$$

where, θ is the angle of crack plane to the slip plane. When $\theta=70.5^o$, τ is τ_{max}, the normal stress σ is

$$\sigma = 0.416K_1 / \sqrt{2\pi w} \qquad (4)$$

where, w is the distance of the spot that σ acts on to crack tip. The mean normal stress σm, acts on the slip plane of a unit thickness, is

$$\sigma m = \int_0^{\omega_o} \frac{0.416K_1}{\sqrt{2\pi w}}\, dw \times 1 / W_o \times 1 = 0.83K_1 / \sqrt{2\pi W_o} \qquad (5)$$

From formula (3) and (5), ΔK_{th} is given by

$$\Delta K_{th} = 1.2\sigma c\sqrt{2\pi r} + 1.2\,\sigma i\sqrt{2\pi W_o} \qquad (6)$$

According to experimental observation shown in Fig. 2;3. W_o is in the following range: $W_o = (0.03 - 0.07)$ d

If W_o = 0.05d, then formula (6) can be change into

$$\Delta K_{th} = K^m_c + 0.7\ \sigma i \sqrt{d} \qquad (7)$$

Suppose K^m_c = $K\sigma c$, it is a critical value of ΔK, which initiates a new crack nucleus under noncontinuous crack propagation. When the crack tip slip bands are blocked by the grain boundary, for the same microstructure the σc is approximate the same[3]. Therefore, K^m_c is nearly the same too, it can be obtained through the experiment from formula (7). If the formula (2) is compared with (7), the physical meaning of each term in formula (2) will be clear. When the fracture unit in 12CrNi3A steel is the sustenite grain size, then

$$\Delta K_{th} = 4.6 + 0.7\sigma i \sqrt{d} \qquad (8)$$

From TABLE 2, it is shown that the varying range of σi is about 3 percent, thus, σi is reasonably to be considered as a constant, from formula (2) and (8). σi = 570 MPa. This is the experimental average value of frictional resistance on dislocation moving in different austenite grain size of high temperature tempered steel.

Next, we will discuss that if the martenite packet size can be employed as a fracture unit, it is still available to use formula (7). Thompson suggested[4] that, Hall-Petch formula can be applied to d > 1um, TEM observation showed that the studied steel after low or medium temperature tempering, the martensite packet exhibits clear cut boundaries. Its width is large than 1um. We assume the width of the martensite packet is standing for the fracture unit d[2]. Formula (7), therefore, can be applied. K^m_c of 400°C tempered 12CrNi3A steel was obtained from the experiment, when 1200°C quenching, K^m_c = 3.45, when 870°C quenching, K^m_c =3.53, its average value is 3.5. Formula (7) is then expressed as

$$\Delta K_{th} = 3.5 + 0.7\ \sigma i \sqrt{d} \qquad (9)$$

The martensite packet width is 2 to 7 um. From formula (1) d=4.5um[2], this size is used for formula (9), then ΔK_{th} is given and listed on TABLE 2. Since the martensite packet size is more or less the same in both low and medium tempered specimens, K^m_c = 3.5 should be kept in constant. Thus, formula (9) can be applied in low temperature tempering range, the calculated value is listed in TABLE 2. Measured ΔK_{th} value is similar to the calculated one. Small deviation is thought to be resulted in the fact that, a few parts of martensite packet boundary have been smeared out after raising the tempering temperature to 400°C. Therefore, the mean size of the fracture unit is slightly larger than that of 200°C tempered.

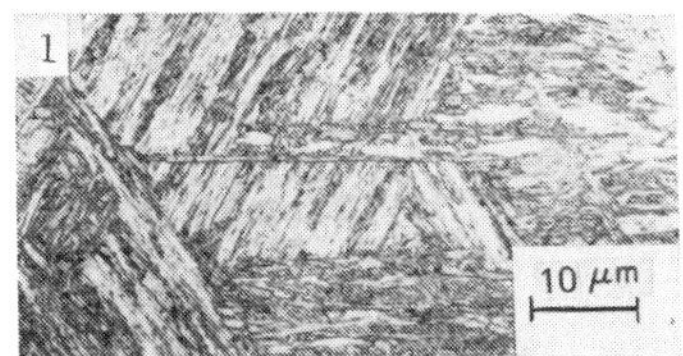

Fig. 5 Microstructure in 12CrNi3A Steel quenched at 1200°C and tempered at 200°C (Fig. 5-1) and 400°C (Fig. 5-2)

If, formula (9), which is obtained from 12CrNi3A steel, is applied to 200°C tempered 40Cr steel. The measured width of the martensite packet is 0.5 to 3.5 µm, let d=2 µm, the calculated value is just the same as the experimental one. If K_c^m in formula (8) and in (9) are compared, it is noticed that K_c^m in the austenite is larger than that in the martensite. The former is 4.6, while the latter is 3.5. There are several problems need to be studied further, for instance, whether the frictional stress σ_i could be replaced by limit of proportionality σ_p and the role of carbon content on ΔK_{th}.

It is true that ΔK_{th} should be changed to ΔK_{theff} in calculation. However, at high stress ratio, say R=1/3 in our experiment, ΔK_{th} seems to be valid.

CONCLUSIONS

For low and medium tempered 12CrNi3A and 40Cr steel, in our experiment, the following conclusions can be made:

1. The relationship between ΔK_{th} and $\Delta\sigma_w$ can be expressed in the formula

$$\Delta K_{th} = 1.12\ \sigma_w\sqrt{\pi a}$$

   a is the size of the fracture unit related to the microstructure.

2. ΔK_{th} is dependent on K_c^m, σ_i and d. The relationship among them can be expressed as follows

$$\Delta K_{th} = K_c^m + 0.7\ \sigma_i\sqrt{d}$$

$$K_c^m = 4.6 \quad \text{for austenite grain}$$

$$K_c^m = 3.5 \quad \text{for martensite packet}$$

3. ΔK_{th} for martensite is low, this is because K_c^m of martensite packet is smaller than K_c^m of austenite grain. The size of the fracture unit in martensite is much smaller than that in austenite grain. Therefore, the acting distance of dislocation frictional stress σ_i is small, and the function of σ_i is decreased.

REFERENCES

1. K. Tanaka, J. ISI., Japan, vol.67, No.2, Feb. 1981, p.245-261.
2. Li Nian, Du Bai-Ping and Zhou Hui_Jiu, Int. J. Fatigue vol.6. No.2, April 1984, P.89-94.
3. Cracknell A, Petch, N.J., Acta Met., 3, 186 (1955).
4. Thompson A.W., Scripta Met., 8, 145 (1974), Acta Met., 23, 1337 (1975).

Influence de l'Etat de Revenu de l'Alliage d'Aluminium 7075 sur la Fermeture des Fissures de Fatigue au Seuil de Non Propagation Sous Vide

M. C. Lafarie-Frenot et C. Gasc*

**Laboratoire de Mécanique et de Physique des Matériaux, ENSMA, Rue Guillaume VII, 86034 Poitiers Cedex, France*

RESUME

Quatre états microstructuraux d'un alliage d'aluminium à haute résistance (7075 Al) ont été comparés dans le domaine des basses vitesses de fissuration en fatigue sous vide. Les auteurs montrent que l'état microstructural de l'alliage influe de façon importante sur les conditions de fermeture de la fissure, du fait des grandes différences de rugosité. Cependant, après correction des effets de fermeture, les valeurs de $\Delta K_{eff\ seuil}$ dépendent encore sensiblement du mode de rupture plus ou moins cristallographique.

MOTS CLEFS

Fatigue - Alliage d'Aluminium 7075 - Seuil de non propagation - Vide - Fermeture.

INTRODUCTION

Au cours de ces dernières années, de nombreux auteurs ont montré l'influence significative de la rugosité des surfaces de rupture sur la fermeture d'une fissure de fatigue, en particulier lorsque la taille des aspérités est du même ordre de grandeur que les déplacements des lèvres de la fissure (1,2). Par ailleurs, il est connu que les alliages d'aluminium à précipitation durcissante présentent des faciès de rupture par fatigue plus ou moins cristallographiques selon le degré de cisaillabilité des précipités, lorsqu'on s'approche du seuil de non fissuration de ces alliages (3,4). Dans une éprouvette polycristalline, lorsqu'un mécanisme de rupture cristallographique est mis en jeu, de grandes déviations locales du chemin suivi par la fissure sont observées, dues à l'orientation respective des grains par rapport à la direction du chargement (5). Lorsque de tels alliages sont sollicités dans l'air du laboratoire, on superpose à la fermeture induite par la rugosité des surfaces, une fermeture due à la présence de couches d'oxyde d'autant plus épaisses que l'on s'approche du seuil de non fissuration (6,7) et qui peut masquer l'effet particulier de la rugosité. C'est pourquoi nous avons essayé, au cours de ce travail, de mettre en évidence l'influence de la rugosité des surfaces de rupture, elle-même reliée à l'état de revenu de l'alliage (7075 Al), sur la fermeture d'une fissure de fatigue, en effectuant des essais de seuil de non propagation sous vide.

CONDITIONS EXPERIMENTALES

Quatre états microstructuraux de l'alliage d'aluminium à haute résistance 7075 Al, obtenus par différents revenus après trempe ont été comparés dans le domaine des basses vitesses de fissuration en fatigue : sous revenu T351, T651, surrevenu T7351 et surrevenu hors normes 24 h à 200 °C. Dans le tableau I, sont notées les caractéristiques mécaniques de ces alliages, ainsi que les caractéristiques de la précipitation durcissante.

TABLEAU I

	0,2 % σ_y (MPa)	0,2 % σ_{yc} (Mpa)	Précipitation	
T351	458	584	GP+M'	≃ 10 Å cisaillable
T651	527	566	M'+(M+T')	≃ 60 Å cisaillable
T7351	470	448	M +T'	≃100 Å Partiellement cisaillable
SR	234	227	M	≃200 Å non cisaillable

Il faut noter que la limite d'élasticité de l'état surrevenu est très faible par rapport aux trois précédentes, ce qui entraînera, pour une même valeur du facteur d'intensité des contraintes, une taille de zone plastifiée plus grande. Des essais de seuil de non fissuration, avec un facteur d'intensité des contraintes décroissant, ont été effectués dans le vide ($p \leq 10^{-3}$ Pa) à 40 Hz, avec des éprouvettes compactes (type CT) d'épaisseur 12 mm et pour une valeur du rapport de charge R = 0,1. L'ouverture et la fermeture de la fissure à chaque cycle ont été étudiées à la fréquence d'essai (f = 40 Hz) à partir des variations de complaisance de l'éprouvette, mesurées à l'aide d'une jauge de déformation collée sur la face arrière de l'éprouvette. Lorsqu'une vitesse de fissuration de quelques 10^{-8} mm/cycle est obtenue, un dernier enregistrement est effectué puis l'éprouvette est démontée et observée avec un microscope optique sur lequel un dispositif original de mise en charge a été adapté, qui permet de simuler le dernier cycle de chargement.

RESULTATS EXPERIMENTAUX - DISCUSSION

Dans le tableau II sont données les valeurs caractéristiques obtenues au seuil de non fissuration pour les quatre alliages étudiés (Pour la détermination de $P_{ouverture}$, voir réf.4).

TABLEAU II

	P_{Max} (daN)	P_{min} (daN)	da/dN (mm/c)	$P_{ouverture}$ (daN)	ΔK_{seuil} (MPa$\sqrt{m}$)	$\Delta K_{eff\ seuil} = K_{Max} - K_{ouv}$ (MPa$\sqrt{m}$)
T351	300	30	6 10^{-8}	145	7	3,7 ± 0,2
T651	150	15	2 10^{-8}	60	5	3,6 ± 0,2
T7351	120	12	3 10^{-8}	33	3,2	2,7 ± 0,2
SR	110	11	1,5 10^{-8}	45	3,2	2 ± 0,2

L'observation de la fissure sur la face polie de l'éprouvette, ainsi que celle des faciès de rupture par microscopie à balayage, fait apparaître de grandes différences de rugosité des surfaces rompues avec l'état microstructural de l'alliage : ces différences ont été précédemment décrites (4,8) et reliées à la nature de la précipitation durcissante.

Dans l'état T351, la présence de zones GP parfaitement cisaillables conduit à un mécanisme de rupture cristallographique qui entraîne un faciès de plus en plus rugueux lorsque la vitesse diminue. L'état T651 présente une transition lorsque la taille de la zone plastifiée devient du même ordre de grandeur que la taille du grain et on observe une rugosité qui augmente brutalement lors de cette transition pour devenir du même ordre de grandeur que celle du T351 lorsqu'on arrive au seuil de non propagation. Les états surrevenus, quant à eux, présentent un faciès très plat, l'incohérence et la grosse taille des précipités favorisant l'activation d'un plus grand nombre de plans de glissement ainsi que du glissement dévié ; ceci conduit à un chemin de fissuration très rectiligne, peu sensible à l'orientation des grains, et à une rugosité très faible qui diminue avec la vitesse.

Un système original de mise en charge de l'éprouvette a permis d'observer au microscope optique et de mesurer l'évolution du déplacement des deux lèvres de la fissure pour différentes valeurs de la charge appliquée, ce qui permet de simuler le dernier cycle de chargement.

L'écart entre les lèvres de la fissure a été mesuré là où la direction locale de la fissure est perpendiculaire à l'axe de chargement, et avec une précision de 1 µm. Ces mesures permettent de tracer pour les quatre alliages étudiés les profils moyens de la fissure pour différentes charges au cours du dernier cycle de fatigue (Voir fig.1). Il faut noter que de nombreux points de contact entre les lèvres de la fissure existent tout au long de celle-ci, là où localement le chemin suivi par la fissure se rapproche de la direction du chargement. Lorsque les surfaces de rupture sont très accidentées (T351 et T651 aux basses vitesses de propagation), ces points de contact sont très nombreux et existent encore même pour la charge maximale du cycle de chargement imposé. La présence de telles aspérités qui frottent l'une contre l'autre au cours du cycle empêche le rapprochement des lèvres de la fissure lorsque la charge diminue, ce qui conduit à une fissure qui apparaît "ouverte" à la charge nulle (T351 et T651). Par contre, lorsque les surfaces de rupture sont peu accidentées, la fissure apparaît fermée sur toute sa longueur à la charge minimale du cycle (T7351 et surrevenu).

Sur la figure 2 sont représentées les variations de l'écartement des lèvres en fonction de la charge, mesurées en différents points de la fissure. A une distance donnée de la pointe de la fissure, et quel que soit l'alliage, cet écartement ne croît qu'au delà d'une charge que l'on peut définir comme la charge d'ouverture en ce point.

* Au cours d'un cycle de chargement, cette ouverture s'effectue progressivement de l'entaille vers la pointe de la fissure pour les états rugueux (T351 et T651). La charge pour laquelle la fissure est totalement ouverte correspond à celle obtenue par les mesures de complaisance.

	T351	T651
Mesure optique	160 daN	70 daN
P_o complaisance	145 daN	60 daN

* Par contre, l'ouverture se produit simultanément sur toute la longueur de la fissure et pour une charge plus faible correspondant exactement à celle obtenue par les mesures de complaisance dans l'alliage T7351. (P_o = 33 daN). Les valeurs mesurées sur l'état surrevenu étant très faibles, leur relative imprécision ne permet pas de conclure avec certitude ; cependant il semble que la partie de la fissure proche del'entaille s'ouvrirait la dernière, et pour une charge correspondant à celle des mesures de complaisance (P_o = 45 daN) (Fig.1 et 2).

Dans tous les cas , la fermeture tout le long de la fissure au cours d'un cycle évolue comme la rugosité des surfaces de rupture. Ainsi, pour le T351, la fissure s'ouvre d'abord vers l'entaille, et pour le surrevenu d'abord vers la pointe, là où la rugosité est la plus faible. De même, la charge d'ouverture est d'autant plus basse que la rugosité est faible.Puisque sous vide et avec un rapport de charge R = 0,1 la fermeture concerne toute la longueur de fissure (4), c'est l'évolution de la rugosité des surfaces de rupture tout le long de la fissure qui influe sur la valeur de la charge pour laquelle la fissure est totalement ouverte. Il apparaît donc que l'ouverture de fissure est sensible à l'histoire de la rugosité tout autant qu'à celle de la zone plastifiée au cours de l'essai. Le paramètre rugosité est donc essentiel, et lié directement à l'état structural de l'alliage, et aux mécanismes de fissuration.

Cependant, il faut rappeler que la prise en compte de la fermeture pour le calcul d'une valeur de ΔK_{eff} au seuil de non propagation de ces alliages ne conduit pas à une valeur unique du seuil (Tableau II).Celui-ci res- donc sensible aux mécanismes de rupture déterminés par l'état structural.

CONCLUSIONS

Dans le vide, en l'absence d'oxydation des surfaces de rupture, l'état microstructural influe de façon importante sur les conditions de fermeture de la fissure, du fait des grandes différences de rugosité. De plus, pour un même état microstructural, la rugosité évolue avec la vitesse de propagation de la fissure, ce qui conditionne la cinématique de la fermeture. Il apparait donc que l'histoire du chargement dynamique détermine les conditions de fermeture à travers l'évolution de la rugosité, au même titre qu'à travers l'évolution de la zone plastifiée.

REFERENCES

1. I.C. Mayes and J. Baker, Fatigue Engng Mater.Struct., 4, 79 (1981)
2. K. Minakawa and A.J. Mc Evily, Scripta Metall, 15, 633 (1981).
3. J. Petit, P. Renaud and P. Violan Proceedings of the 4th Eur.Conf. Fracture, 426 (1982).
4. M.C. Lafarie-Frenot and C. Gasc, Fatigue Engng Mater.Struct., 6, 4,329 (1983).
5. M.C. Lafarie-Frenot and C. Gasc, Proceedings. of the 6th, Int.Conf. on Fracture, 3 (1984).
6. R.O. Ritchie,S. Suresh and C.M. Moss J. Engng. Mater. Techn. Trans. ASME 102,293 (1980).
7. J. Petit, "Fatigue crack growth threshold concepts", TMS. AIME pub. 3,24 (1984).
8. M.C. Lafarie-Frenot and C. Gasc, Proc. of Fatigue 84,2,687 (1984).

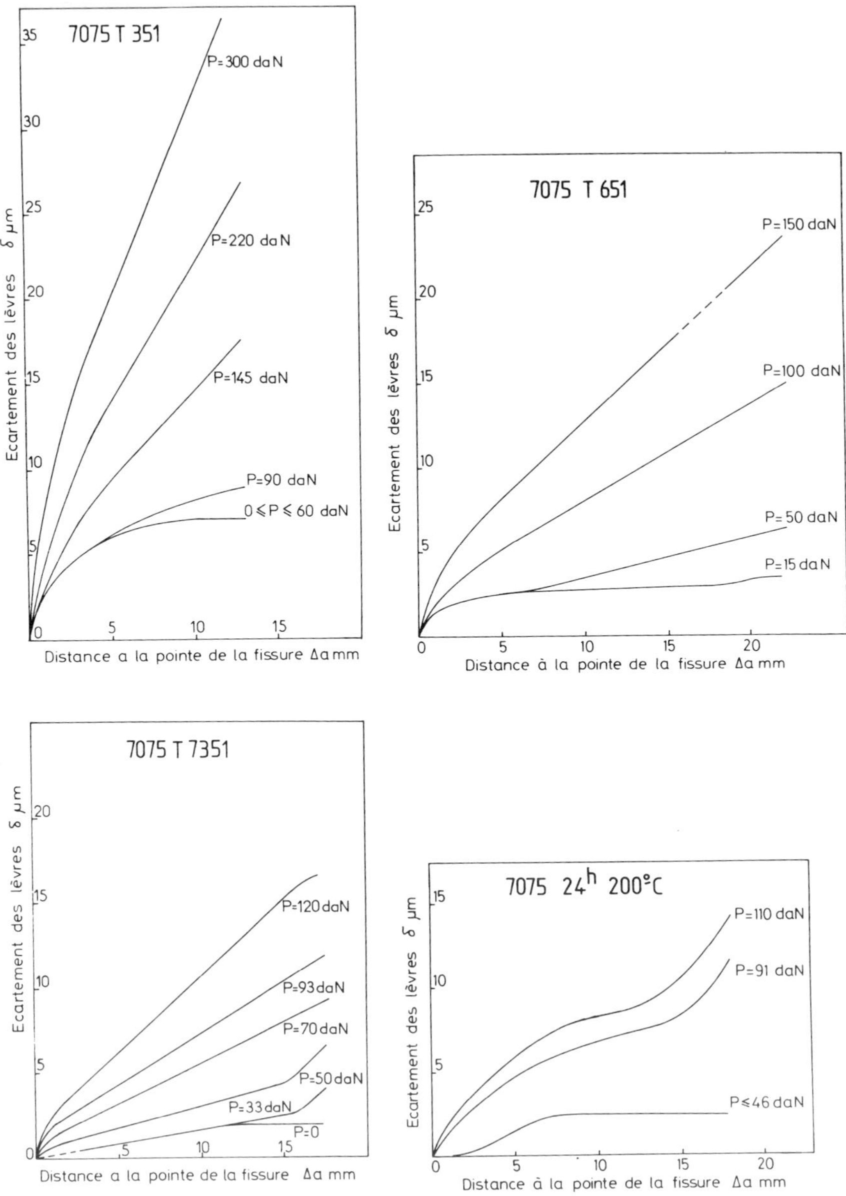

FIG.1 - Profils moyens de la fissure pour différentes charges du dernier cycle de fatigue et pour les quatre états de revenu.

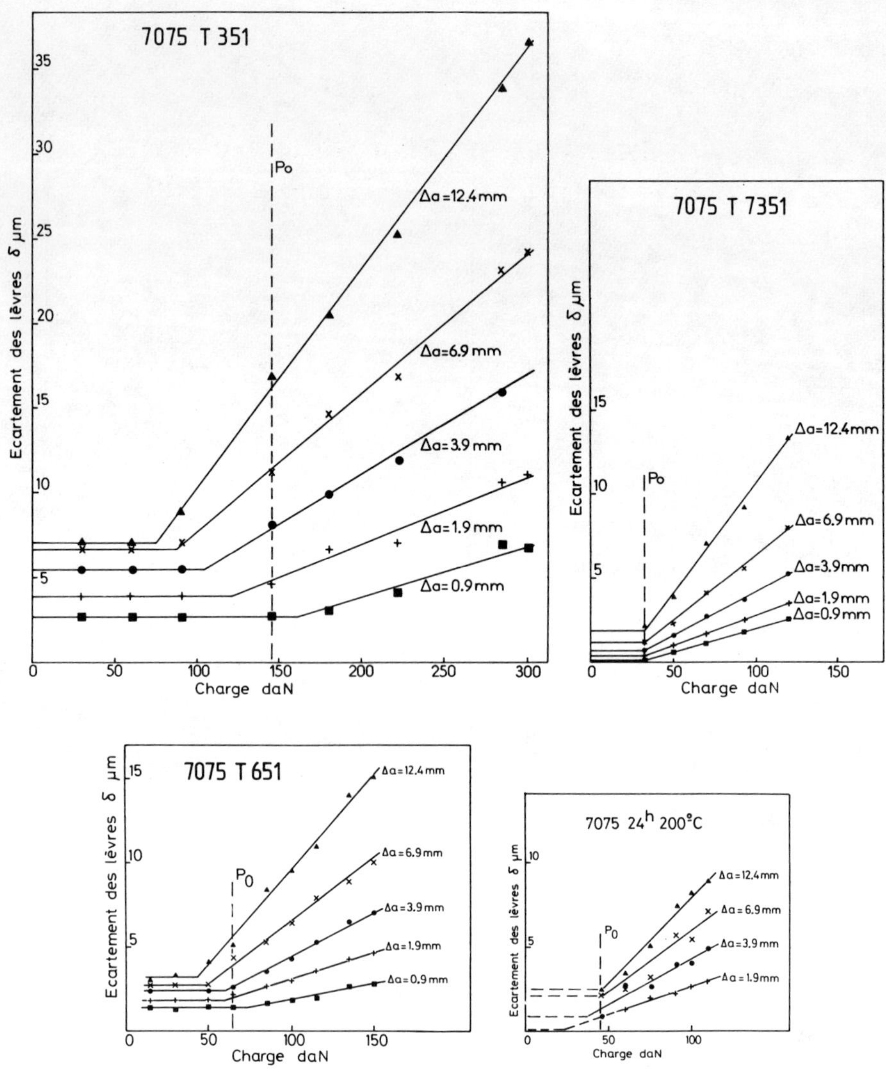

FIG.2 - Evolution de l'écartement entre les deux lèvres de la fissure avec la charge appliquée, à différentes distances (Δ a) de la pointe de la fissure, et pour les quatre états de revenu. P_o indique la valeur de la charge d'ouverture déterminée par les mesures de complaisance.

On the Effect of Microstructure on Crack Closure in the Near-threshold Region

A. J. McEvily and K. Minakawa

Metallurgy Department, University of Connecticut, Storrs, CT 06268, USA

ABSTRACT

The factors governing crack closure, particularly in the near-threshold region, are reviewed. Plasticity-induced closure in the continuum-sense is found not to be responsible for closure in plane strain. Rather roughness-induced closure, in the absence of environmental effects, is the dominant form of closure. This form of closure is strongly related to the microstructure, and examples of this interplay are given.

KEYWORDS

Crack closure; near-threshold fatigue crack growth; roughness; microstructure.

INTRODUCTION

Crack closure is now recognized to have an important influence on fatigue crack growth in the near-threshold region, and it is therefore of interest to understand the origin of the closure process. Crack closure was first discussed by Elber (1) in terms of the residual plastic stretch left in the wake of an advancing fatigue crack. The residual stretch approach is basically a continuum approach and has been treated theoretically as such for plane stress conditions (2,3). Recent experimental evidence however favors a process of closure, particularly in plane strain, which reflects microstructural characteristics such as grain size and planarity of glide, with closure occurring as the result of point-to-point contacts rather than in a uniform manner. For example, it has been observed by Bowles and Schijve (4) that in the case of high-strength aluminum alloys closure in the plain strain region is not uniform across the fracture surfaces immediately behind the crack tip, but instead occurs at discrete contact points associated with stretched ligaments left in the wake of the crack. On the other hand more-or-less continuous closure was observed in the plane stress regions along the specimen surfaces. Further, in a number of alloys there is evidence that plane-strain closure is due to a mismatch of topographical features across the mating fracture surfaces (5), a process known as roughness-induced closure (6). Additionally, oxidation (7) can contribute to closure particularly in the near-threshold region where plane strain conditions essentially prevail. The purpose of the present paper is to review the various characteristics of the plain-strain closure processes in the near-threshold region, with particular attention given to the influence of microstructure on closure.

CLOSURE IN THE NEAR-THRESHOLD REGION

Crack closure behavior in the near threshold depends strongly on microstructure, as can be judged from Fig. 1 wherein for purposes of comparison several different types of closure behavior are shown. In Fig. 1 the stress intensity factor level at the opening load, Kop, is plotted as a function of ΔK for R=0 loading. It is seen that the value of Kop may be almost zero or remain constant with ΔK, Fig. 1a (8). In other cases Kop may rise as the threshold level is approached and then remain constant, Fig. 1b (8,9). In a third case the Kop behavior depends on the test frequency, Fig. 1c (10), and in the final example the maximum value of Kop is observed above rather than at the threshold level, Fig. 1d (11). In each of these cases the maximum value of Kop is less than Kmax, i.e., the crack tip is open over some range even at threshold. It is also noted that these closure characteristics were determined with clip gauges mounted at the mouth of the specimens so that the average closure behavior across the thickness of the specimen is obtained. We next consider the reasons for these various types of closure behavior.

With respect to plasticity-induced closure quantitative theoretical treatments for plane strain as in the case of plane stress do not appear to be available. However, the effect, if any, should be dependent upon yield stress, but in comparing aluminum alloys of similar yield strength one can find a wide variation in closure behavior, even for alloys of similar grain sizes. It seems particularly significant that in the case of the IN9021 aluminum alloy, Fig. 1a, which has a yield strength similar to that of 7075-T6 that no closure at all was observed (8). The conclusion reached from this finding is that residual plastic stretch is not a significant contribution to the closure process in the plane strain region. We look elsewhere therefore for the causes of the observed levels of closure.

In the absence of environmental effects it is our contention that the primary type of closure is roughness-induced not only in the near threshold region but also above as well. This roughness arises as the result of the nature of the crack advance process. Above the near threshold region the process of crack advance is relatively insensitive to microstructure and Young's modulus is the dominant material parameter. In this range the stress intensity at the crack tip is sufficiently high to promote slip on at least two intersecting slip planes, each at an angle to the nominal direction of propagation, and Mode I deformation results. As the crack advances the path over many cycles changes from one plane to the other giving rise to a serrated type of roughness in the direction of propagation if the crack surface is viewed in profile. In the near-threshold range where the stress intensity is much less only one plane may be active and over a much longer distance before deviation onto a secondary plane occurs. Where only a single plane is active over a large distance the deformation is of the Mode II type. After a relatively small advance on the secondary

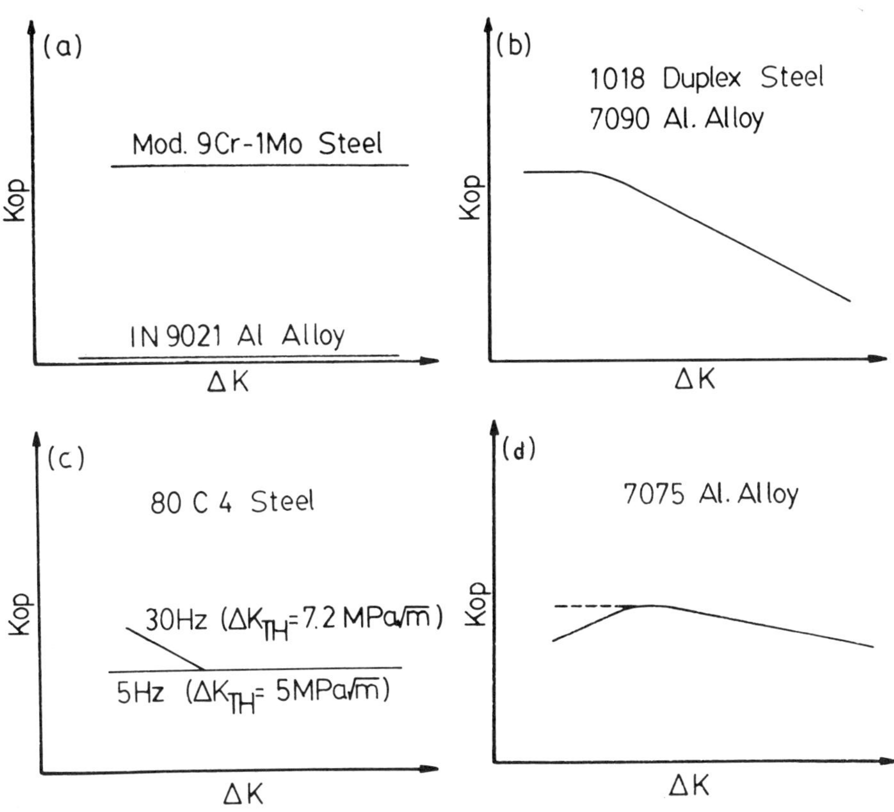

Fig. 1. Various types of Kop behavior as a function of ΔK in the near-threshold region of fatigue crack growth.

plane the crack may return to the primary system. The net result is that the overall size of facet developed can be larger in this region than at higher growth rate regions, but this will depend on the nature of the microstructure. Where larger facets can develop there will be an increase in the closure level associated with the mismatch of the larger facets on unloading.

We next examine some specific instances to illustrate these concepts. For example, the modified 9Cr-1Mo steel (8) of Fig. 1a has a tempered martensitic microstructure. The prior austenitic grain size was 20 μm, and the complexity of the martensitic microstructure formed within these grains imposes a limit on the size of facets which can be developed. Although a few isolated facets were observed which exhibited the characteristics of Mode II growth the overall scale of roughness did not change on moving from the intermediate into the near-threshold growth region. As a consequence the closure level is independent of ΔK. Similarly, when the scale of roughness is extremely small as in the case of the IN9021 alloy, a circumstance resulting from an extremely small grain size (0.1-1.0 μm) the closure level will be independent of ΔK, but in this case at virtually at the zero level. Since the R-dependency of the threshold level is due to the R-dependency of the closure effect, where no closure is observed there is no variation in threshold level with R, as observed for IN9021 (8). Furthermore the threshold for R=0 conditions will be relatively low, being only 0.8 MPa$\sqrt{m}$ for the IN9021 alloy.

In those cases where there is an increase in closure as the threshold level is approached there is a corresponding increase in the average facet size and the scale of roughness. This was particularly evident in the case of a duplex 1018 steel which consisted of 30 μm ferrite grains encapsulated by martensite (9). The presence of the strong martensite inhibited general slip in the soft ferrite and near threshold crack advance occurred primarily along one slip system giving rise to a high closure level. Above the near-threshold region roughness persisted, but on a much finer scale. The threshold level for this alloy was also high, at 14 MPa$\sqrt{m}$ for R=0.05. The same steel in the normalized condition had a ferrite grain size of 10 μm. It also exhibited an increase in closure near threshold, but not to the same extent as in the case of the duplex condition. The threshold level for the normalized condition at R=0.05 was 8.3 MPa$\sqrt{m}$. The high strength aluminum alloys have grain sizes in the range 1-15 μm, and they also show an increase in closure level which may reach a saturation value, Fig. 1b, as the threshold is approached, again because of an increase in the scale of roughness. The scale of roughness in influencing closure has also been noted for a pearlitic steel heat treated to develop either a coarse or fine pearlitic microstructure, with the coarser microstructure developing more roughness and a higher closure level (12).

The increase of closure with frequency, Fig. 1c, is due to oxidation (10). The observed frequency effect may be due to the development of a higher temperature at the crack tip as the result of a lower rate of dissipation per cycle of the heat carried by fretting. Oxide-induced closure has been studied by Ritchie and co-workers (7) and Stewart (13) in the case of steels. It appears to be an effect more pronounced in certain steels than others (14,15). Further, the effect appears to decrease in ferritic steels with an increase in chromium content (16). There is also evidence that prolonged cycling of steel at the threshold level can lead to an increase in oxide and fretting debris and a resultant decrease in ΔK_{eff} (17). The effect has also been observed in an aluminum alloy, but only in an overaged condition for which it was markedly pronounced (18). Furthermore, it is noted that in a certain aluminum alloy fretting can lead to the reduction of K_{op} rather than an increase as with oxidation induced closure by reducing the scale of roughness.

The significance of Mode II on closure has been indicated above. The Mode II process of crack growth near threshold was first discussed by Ohtsuka (19), and the implications for closure were discussed by McEvily (20). However, it is interesting to note that not all Mode II crack advance processes lead to increased closure, as is evident from the work of Matsuoka et al. (21) and Ohta et al. (22) with 304 stainless steel. This austenitic alloy is characterized by a low stacking fault energy giving rise to planar slip. As shown in Fig. 2a, the threshold level for this alloy is low and independent of R, in contrast to the more usual behavior indicated for the A553 low alloy steel. The absence of an R-dependency is an indication of a low closure level at threshold. The closure level increases as shown in Fig. 2b as the ΔK increases and Mode I deformation occurs. The fractographic features published by Matsuoka et al. show that the type of roughness

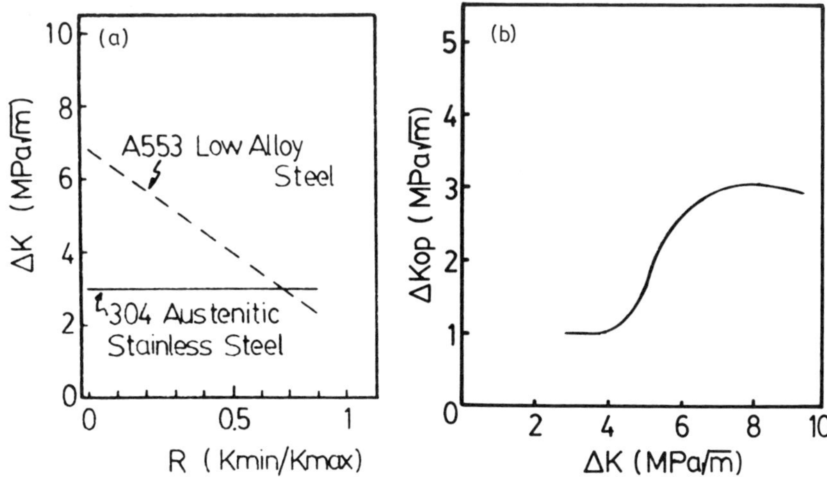

Fig. 2(a) Variation of ΔK_{TH} as a function of R for 304 stainless steel and A553 low alloy steel.
2(b) ΔKop variation for 304 stainless stell as a function of ΔK. (R=0) Based on comparison of da/dN vs. ΔK plots for R=0.1 and R=0.5-0.9 (21).

commonly found in other alloys is absent in this alloy although clear-cut indications of Mode II growth are present, but that the nature of the roughness differs from that observed in other alloys. To illustrate this difference two different types of roughness are schematically shown in Fig. 3. The zig-zag character to the roughness in the direction of crack advance, Fig. 3a, gives rise to the mismatch that leads to roughness induced closure in steels and aluminum alloys. This feature is absent in the stainless steel as indicated in Fig. 3b so that the mating surfaces are able to nestle together on unloading without developing a closure effect. The facets extend completely across the grains in the 304, and tilt at grain boundaries. This tilting will also give rise to a zig-zag pattern, but of a relatively long wavelength, with a minimal contribution to closure. The facets must be inclined to the tensile direction in order to develop a shear stress along them. Any preferred orientation may also serve to reduce the closure level. The absence of a zig-zag pattern within a grain may be due to a high rate of cyclic hardening on other slip planes in this alloy. Planarity of glide in itself is not sufficient to account for the absence of the zig-zag pattern since this type of pattern has been observed in nickel-base superalloys. These alloys develop the usual roughness and have threshold values well above those for 304 stainless (23,24). Since these alloys are much stronger to begin with the potential for hardening secondary systems will be less pronounced allowing the zig-zag pattern to develop so as to keep the average crack plane normal to the driving tensile stress.

To complete the listing of factors which affect closure we can add crack path meandering on a much coarser scale, a process observed in titanium alloys (25,26), aluminum alloys (27) and steels (9,28). As discussed by

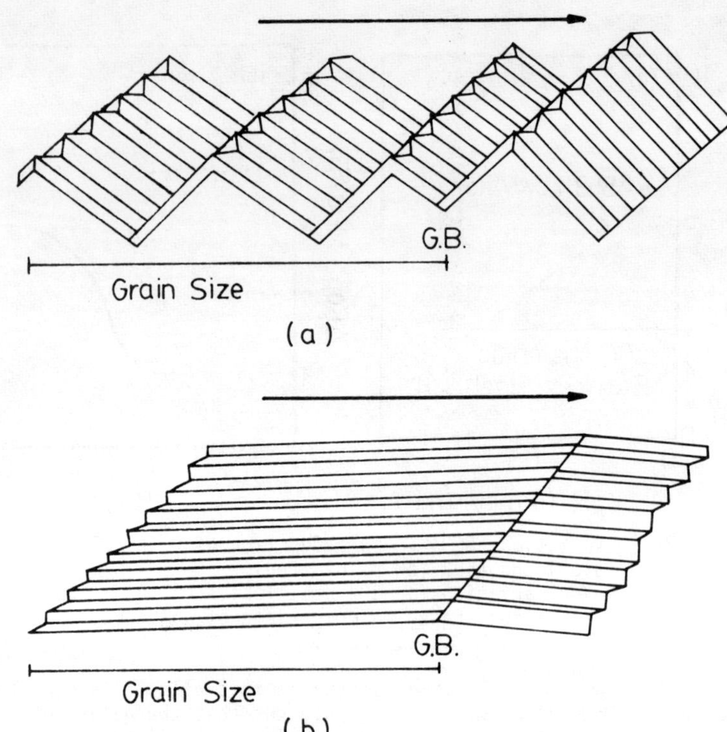

Fig. 3. Illustration of two types of roughness: (a) Zig-zag mode II facet. Typical of aluminum alloys and ferritic steels. (b) Planar mode II facet. Typical of 304 stainless steel.

Beevers and co-workers (29) this type of crack growth can lead to contact being made at isolated points well behind the crack tip, a type of roughness-induced closure. As a result the tip itself is propped open, a process referred to as non-closure (30). For this type of contact during the loading portion of the cycle separation may occur at contact points well behind the crack tip and lead immediately to deformation and opening at the crack tip. In contrast where meandering does not occur there is an unzipping process behind the crack tip and the opening point progressively moves to the crack tip during the opening process.

CONCLUSIONS

1. A continuum type of plasticity-induced closure does not occur during plain-strain fatigue crack growth.

2. In the absence of environmental influences the primary source of closure is roughness-induced.

3. Both the scale of the roughness as well as wave-length of the roughness affect the level of closure that can be developed.

REFERENCES

1. W. Elber, in "Damage Tolerance in Aircraft Structures", ASTM STP 486, p. 230, American Society for Testing and Materials, Philadelphia, PA (1971).
2. B. Budiansky and J. W. Hutchinson, J. Appl. Mech., Trans. ASME, Series E, 45, p. 267 (1978).
3. J. C. Newman, Jr., in "Methods and Models for Predicting Fatigue Crack Growth under Random Loading", ASTM STP 748, p. 53, American Society for Testing and Materials, Philadelphia, PA (1981).
4. C. Q. Bowles and J. Schijve, in "Fatigue Mechanisms: Advances in Quantitative Measurement of Physical Damage", ASTM STP 811, p. 400, American Society for Testing and Materials, Philadelphia, PA (1983).
5. K. Minakawa and A. J. McEvily, Scripta Met., 15, p. 633 (1981).
6. R. O. Ritchie and S. Suresh, Met. Trans. A, 13A, p. 937 (1982).
7. R. O. Ritchie, S. Suresh and C. M. Moss, J. Eng. Mat. Tech., Trans. ASME, Series H, 102, p. 293 (1980).
8. A. J. McEvily, K. Minakawa and H. Nakamura, to be in "Fracture: Interactions of Microstructure, Mechanisms and Mechanics, Proc. of a symposium, Los Angeles, February, 1984, AIME, Warrendale, PA.
9. K. Minakawa, Y. Matsuo and A. J. McEvily, Met. Trans. A, 13A, p. 439 (1982).
10. A. Bignonnet, A. Dias and H. P. Lieurade, in "Advances in Fracture Mechanics", Proc. of ICF6, 3, p. 1861, Pergamon Press, Oxford (1984).
11. A. J. McEvily, Annual Report, AFOSR-81-0046 (January, 1982).
12. G. T. Gray, III, J. C. Williams and A. W. Thompson, Met. Trans. A, 14A, p. 421 (1983).
13. A. T. Stewart, Eng. Fract. Mech., 13, p. 463 (1980).
14. R. O. Ritchie, in "Analytical and Experimental Fracture Mechanics, p. 81, Sijthoff and Noordhoff (1981).
15. S. Suresh and R. O. Ritchie, in "Fatigue Crack Growth Threshold Concepts", p. 227, AIME, Warrendale, PA (1984).
16. W. D. Zhu, K. Minakawa and A. J. McEvily, unpublished research, University of Connecticut.
17. M. Kikukana, M. Jono and K. Tanaka, in "Proc. of 2nd. Intl. Conf. Mechanical Behavior of Materials", Special Vol., p. 254, ASM (1976).
18. S. Suresh, A. K. Vasudévan and P. E. Bretz, Met. Trans. A, 15A, p. 369 (1984).
19. A. Otsuka, K. Mori and T. Miyata, Eng. Fract. Mech., 7, p. 429 (1975).
20. A. J. McEvily, Metal Science, 11, p. 274 (1977).
21. S. Matsuoka, S. Nishijima, C. Masuda and S. Ohtsubo, in "Advances in Fracture Research", Proc. of ICF6, 3, p. 1561, Pergamon Press, Oxford (1984).
22. A. Ohta, E. Sasaki and M. Kosuge, Trans. National Research Institute for Metals, Japan, 19, p. 183 (1977).
23. J. N. Vincent and L. Rémy, in "Advances in Fracture Research", Proc. of ICF5, 3, p. 1357, Pergamon Press, Oxford (1981).
24. R. A. Venables, M. A. Hicks and J. E. King, in "Fatigue Crack Growth Threshold Concepts", p. 341, AIME, Warrendale, PA (1984).
25. N. J. Walker and C. J. Beevers, Fatigue of Engng. Mater. Struct. 1, p. 135 (1979).
26. J. A. Ruppen and A. J. McEvily, Fatigue of Engng. Mater. Struct. 2, p. 63 (1979).

27. A. K. Vasudévan, P. E. Bretz, A. C. Miller and S. Suresh, Mater. Sci. and Engng., 64, 113 (1984).
28. V. B. Dutta, S. Suresh and R. O. Ritchie, Met. Trans. A, 15A, p. 1193 Threshold Concepts", p. 327 (1984).
29. C. J. Beevers, K. Bell and R. L. Carlson, in "Fatigue Crack Growth Threshold Concepts", p. 327 (1984).
30. C. J. Beevers, in "Advances in Fracture Research", Proc. of ICF5, 3, p. 1335, Pergamon Press, Oxford (1981).

Crack Closure and the Stress Ratio Dependence of the Fatigue Threshold in a Nickel-base Alloy

J. Byrne*, T. V. Duggan* and C. J. Beevers**

**Engineering Faculty, Portsmouth Polytechnic, UK*
***Department of Metallurgy and Materials, University of Birmingham, UK*

ABSTRACT

The influence of crack closure and stress ratio on the fatigue threshold stress intensity range ΔK_{th}, is studied for different crack configurations in the nickel base alloy Nimonic 105. It is shown that the stress ratio dependence of ΔK_{th} cannot be entirely explained by crack closure and the use of an effective stress intensity range, $\Delta K_{eff.th}$. An additional mean stress effect on the fatigue threshold level is indicated.

KEYWORDS

Fatigue threshold; Nimonic 105; stress ratio dependance; crack closure; mean stress.

INTRODUCTION

Near-threshold fatigue crack growth (fcg) and the fatigue threshold stress intensity range (ΔK_{th}) have been found generally to be extremely sensitive to cyclic stress ratio $R(= \sigma_{min}/\sigma_{max})$(1). Studies of a wide range of steels and non-ferrous alloys tested in room temperature air have indicated a marked reduction in ΔK_{th} and an increase in fcg rates when increasing R from 0 to 0.9. This is of practical importance under mean tensile stress conditions due to standing loads or residual stress.

A series of empirical relationships between ΔK_{th} and R have been proposed e.g. (2).

$$\Delta K_{th} = \Delta K_{th,o}(1-R)^{\gamma} \qquad \text{..... (1)}$$

where $\Delta K_{th,o}$ is the threshold value at R = 0 and γ is material dependant.

Even under all tensile stress cycling, some closure of crack faces may occur. This can be caused by the plastic zone "wake" of stretched material formed when growing a crack during threshold determination(3), by wedging ("non-closure") of the fracture surfaces at asperities(4) or by blocking with oxidation/corrosion product(5).

It has been proposed that the R effect on threshold might be entirely due to crack closure(6). Also a reduced or negligible R dependence has been found at negative R values(7), with increased strength(8), in inert atmospheres (e.g. (9)) and in thicker sections(10). The last three of these effects may be explained by reduced crack closure giving similar values of effective threshold stress intensity range ($\Delta K_{eff,th}$). Use of the effective range of K, implies that crack growth is controlled by the range of stress intensity factor during which the crack tip is open, and therefore by the range of crack tip opening displacement.

EXPERIMENTAL DETAILS

The material studied is a nickel base alloy, Nimonic 105, in the age hardened condition, full details of which have been published earlier(11). Three types of crack configuration were produced in 4-point bend specimens:-

(i) through the thickness, from sharp vee notches;

(ii) corner cracks, by machining down pre-cracked notched specimens; and

(iii) contained and non-contained semi-elliptical cracks, from finned specimens, by machining off a fin from pre-cracked specimens.

The machined-down specimens were vacuum annealed prior to threshold determination as described in (11). Testing was carried out in room temperature air on an Amsler Vibrophore at frequencies between 80 and 100 Hz, utilising the d.c. potential difference crack monitoring technique. Fatigue crack growth rates were calculated using the 3-point secant method, and standard stress intensity factor solutions for each crack configuration were used(11). Crack closure was monitored using compliance change measured using both a conventional COD clip-on gauge and a back face strain (BFS) gauge.

A load shedding procedure was adopted to determine the fatigue threshold condition, this being defined as no detectable crack growth being evident for at least 10^7 cycles. This corresponds to an fcg rate of less than 5 x 10^{-9} mm/cycle for a crack length increase detection sensitivity of 0.05 mm. Thresholds were determined for stress ratios between 0.1 and 0.7.

R DEPENDENCE OF ΔK_{th} FOR THROUGH CRACKS

The nominal value of ΔK_{th} was significantly reduced by increasing R, from a value of 8.8 ± 0.6 MPa√m at R = 0.1 to 6.1 ± 0.8 MPa√m at R = 0.5. A linear regression giving: $\Delta K_{th} = 9.2 - 5.9$ R.

This predicts a ΔK_{th} value of 5.1 MPa√m at R = 0.7. However, experimental results for semi-elliptical cracks at R = 0.7, had an average value of 4.0 ± 0.25 MPa√m. Taking this into account, the R dependence of ΔK_{th} is better represented (within ± 2.8%) by equation (1) i.e. $\Delta K_{th} = 9.4(1-R)^{0.64}$.

A significant difference observed between the through cracked results and the contained semi-elliptic results, was the absence in the latter of any detectable crack closure, i.e. $\Delta K = \Delta K_{eff}$.

CRACK CLOSURE AND THE THRESHOLD CONDITION

The fact that crack closure can occur at values of $K > K_{min}$ in all tension fatigue cycles has been clearly indicated by a wide variety of experimental techniques. Direct physical evidence of fracture surface contact at specimen surfaces has been observed using surface replicas (e.g. 4,11,) and from direct microscopic observation of specimen surfaces (12). Furthermore, there is now a strong body of physical evidence for closure indicated by the observed presence of fretting product on fracture surfaces for room temperature air tests(13).

In applying ΔK_{eff} to the threshold condition to obtain $\Delta K_{eff,th}$, the original definition of Elber(3) has been commonly used:

$$\Delta K_{eff} = K_{max} - K_{op} \qquad \text{..... (2)}$$

where K_{op} has been estimated from the load at which the crack is indicated to be fully open, e.g. on a COD/load plot (Fig. 1).

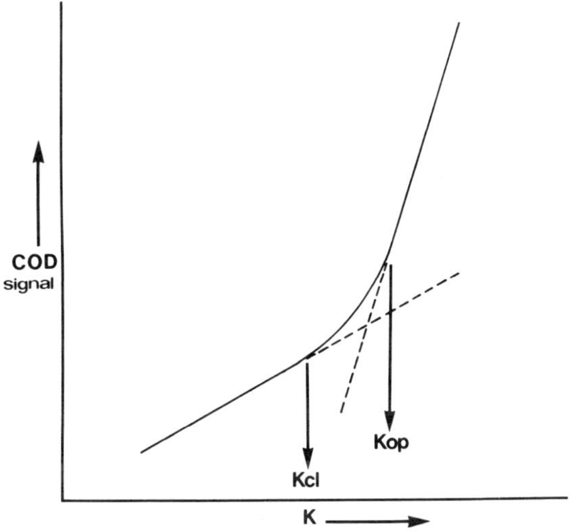

Fig. 1. Schematic illustration of variation in COD with increasing K, showing the definitions of K_{op} and K_{cl}.

However it is evident(11) that near-threshold "faceted" fcg may proceed by a shear mode of fracture (i.e. locally Mode II in nature), implying that crack growth could occur below K_{op}. Thus an alternative definition of ΔK_{eff} was employed in assessing the R dependence of $\Delta K_{eff,th}$:-

$$\Delta K_{eff} = K_{max} - K_{cl} \qquad \text{..... (3)}$$

where K_{cl} is the value of K where the crack is estimated to be fully closed.

Whilst use of $\Delta K_{eff,th}$ as defined by equation (2) significantly reduces the R dependence of threshold, a detectable reduction in $\Delta K_{eff.th}$ still occurs with increasing R, particularly for contained semi-elliptic cracks, showing no detectable closure, for which $\Delta K_{eff} = \Delta K$ (Fig. 2). Application of equation (1) to the closure-free results gives $\Delta K_{th} = 6.3\ (1-R)^{0.29}$.

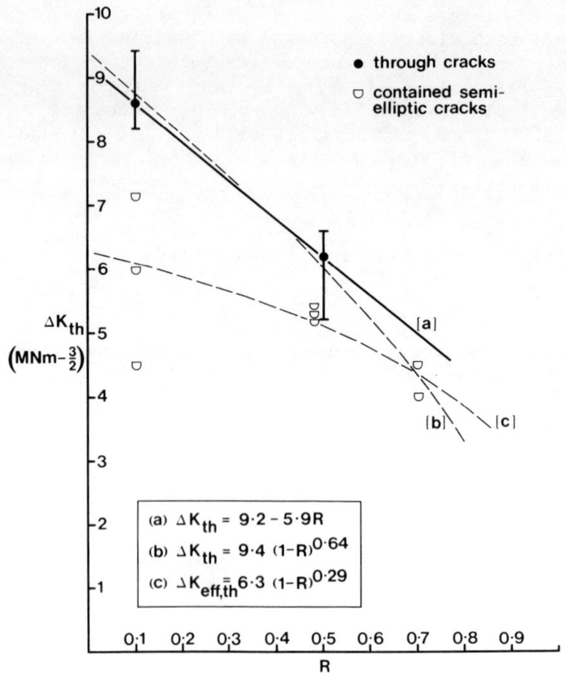

Fig. 2. Stress ratio dependence of nominal ΔK_{th}, showing the fit of three possible mathematical models.

This reduction in γ from 0.64 (for nominal ΔK_{th}) to 0.29 (for $\Delta K_{eff.th}$) indicates that γ is only partially dependent on crack closure, since a value of zero is necessary for a constant $\Delta K_{eff,th}$ with increasing R. Though it is notable that the value of 0.29 for γ is lower than any previously quoted, apart from results obtained under vacuum.

It is considered that it may be more appropriate, where closure occurs at greater than K_{min}, to relate $\Delta K_{eff.th}$ to the <u>effective</u> stress intensity ratio:

$$R_{eff} = K_{cl}/K_{max} \qquad \text{..... (4)}$$

Use of ΔK_{eff} rather than K_{max} to correlate the crack tip stress field with threshold and fcg rate, assumes that there is more significance in the reversed stress intensity range rather than the maximum value. Therefore in using $\Delta K_{eff.th}$ to describe the threshold condition, it seems reasonable to use R_{eff} to characterise both the mean stress and the ratio of effective minimum to maximum stress intensity factor in the fatigue cycle. Of course for a closure free cycle $R = R_{eff}$. However, even using this approach, the global $\Delta K_{eff.th}$ data shows a decrease with increasing R_{eff} (Fig. 3), such that when linearly regressed: $\Delta K_{eff.th} = 6.7 - 3R_{eff}$.

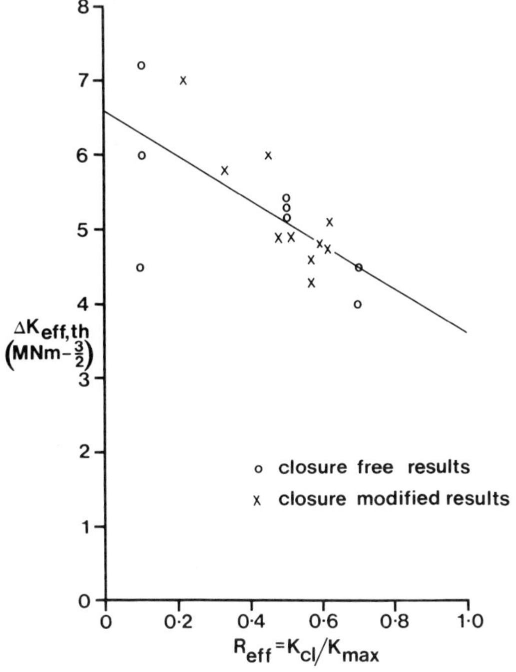

Fig. 3. Variation of $\Delta K_{eff,th}$ with R_{eff} where $\Delta K_{eff} = K_{max} - K_{cl}$ (regressed linear slope = -3).

Also applying equation (1) gives a γ value of 0.29, further supporting the conclusion that both crack closure and increasing mean stress contribute to the stress ratio dependence of the threshold condition, at least in air for the alloy studied. A similar reduction in ΔK_{th}, under assumed closure-free conditions, at stress ratios greater than 0.5 has been observed by Beevers et al(14) for the Ni-base alloy IN901, and by Powell(15) for Ti-6Al-4V and Ti5331S alloys. However, for fine grained high strength Ni-base alloy Astroloy(16) and the ultra-fine grained high strength Al-alloy IN9021-T4 (17), little or no R dependence of ΔK_{th} was found up to R = 0.8.

The precise mechanism by which a higher mean stress intensity causes a reduction of ΔK_{th} in the absence of closure is not clear, and this effect is clearly not a general one in view of some of the results obtained for other alloys. A possible mechanism is an increased tearing of ligaments between advanced facets on near-threshold crack fronts.

CONCLUSIONS

1. Decrease in the fatigue threshold stress intensity range (ΔK_{th}) with increasing stress ratio (R) is not entirely explained by crack closure.

2. Significant reduction in effective threshold ($\Delta K_{eff.th}$) occurs with increasing stress ratio for closure-free and closure-modified results.

3. Even when stress intensity ratio is modified to allow for closure, as $R_{eff} = K_{cl}/K_{max}$, a reduced $\Delta K_{eff.th}$ is obtained with increasing R_{eff}.

4. The R dependence of ΔK_{th} is indicated to be due to a combination of crack closure and mean stress effects.

REFERENCES

1. R.O. Ritchie, Int. Met. Rev., 20, 205 (1979).

2. P.E. Irving and C.J. Beevers, "Microstructural Influences on Fatigue Crack Growth in Ti-6Al-4V", Mat. Sc. and Eng., 14, 229 (1974).

3. W. Elber, Eng. Fract. Mech., 2, 37 (1970).

4. N. Walker and C.J. Beevers, Fatigue of Eng. Mat. and Struct., 1, 135 (1978).

5. A.T. Stewart, Eng. Fract. Mech., Vol. 13, 463 (1980).

6. R.A. Schmidt and P.C. Paris, ASTM STP 536, 79 (1973).

7. A. Ohta and E. Sasaki, Eng. Fract. Mech., Vol. 9, 655 (1977).

8. R.O. Ritchie, J. Eng. Mat. and Tech., Trans. ASME., Series H, Vol. 99, 195 (1977).

9. R.J. Cooke, P.E. Irving, G.S. Booth and C.J. Beevers, Eng. Fract. Mech., Vol. 7, 69 (1975).

10. J.K. Musuva and J.C. Radon, ICF5 Cannes, France, 1365 (1981).

11. J. Byrne and T.V. Duggan, in "Fatigue Thresholds - Fundamentals and Engineering Applications", ed. C.J. Beevers, EMAS, 759 (1982).

12. H.U. Staal and J.D. Elen, Eng. Fract. Mech., 11, 275 (1979).

13. S. Suresh and R.O. Ritchie, ICF5, Cannes, France, 1873 (1981).

14. C.J. Beevers, K. Bell and R.L. Carlson, in "Fatigue Crack Growth Threshold Concepts", ed. D. Davidson and S. Suresh, Met. Soc. A.I.M.E., 327 (1984).

15. B.E. Powell, PhD Thesis, Portsmouth Polytechnic (1985).

16. R.A. Venables, M.A. Hicks and J.E. King, in (14), p. 341.

17. A.J. McEvily and K. Minakawa, Ibid, p. 517

Crack Growth Behaviour of 2¼ Cr-1 Mo Steel at Elevated Temperatures

W. K. Lee and H. J. Westwood

Ontario Hydro Research Division, Toronto, Ontario, Canada

ABSTRACT

This paper appraises two predictive models for creep-fatigue crack growth data correlation. The hyperbolic sine equation and the linear damage concept could both model the crack growth behaviour for some of the available experimental data. In particular, the former could account for the effect of frequency and loading ratio. The latter provided a conservative upper bound value for hold time tests in which creep-fatigue interactions were important.

KEYWORDS

Creep-fatigue, crack growth, stress intensity range, models, data extrapolation, power plant components.

INTRODUCTION

Creep-fatigue crack growth data on 2-1/4 Cr-1 Mo steel are being generated as input to a multidisciplinary project to optimize the service performance of critical components in thermal power boilers [1,2]. Since crack growth testing is time-consuming and expensive, extrapolation of limited data to operational conditions is required. This paper examines two methods for modelling and predicting crack growth rates. Although ΔK is used as the correlating parameter, the methodology is equally applicable if other parameters are used.

FATIGUE CRACK GROWTH BEHAVIOUR

The Paris-Erdogan [3] equation of the form $da/dN = a(\Delta K)^b$ is often used to relate crack growth rate da/dN to stress intensity range ΔK in the linear (Stage II) portion of the typical da/dN vs ΔK plot and has frequently been applied for life prediction and defect assessments. The equation is not applicable, however, to the increasing growth rate (Stage III) situation nor to the very low, near threshold (Stage I) rates. Furthermore, the overall shape of the da/dN vs ΔK curves is dependent on the specific test conditions. For these reasons, attention has been directed at mathematical models which can analytically represent a wide range of crack growth rates [4].

This paper describes application of the hyperbolic sine model and also a linear damage summation approach to the data currently available from the high temperature tests on 2-1/4 Cr-1 Mo. Details concerning material and test method have been given [5].

HYPERBOLIC SINE MODEL [6]

Background

In this interpolative method, frequency (ν), load ratio (R) and temperature (T) are fundamental parameters in a crack growth model described by the following equation:

$$\log \left(\frac{da}{dN}\right) = C_1 \sinh [C_2(\log \Delta K + C_3)] + C_4$$

where C_1 is material constant; C_2 is a function of R, ν, T; C_3 is a function of C_4, ν, R; C_4 is a function of ν, R, T.

The model can examine the effect of frequency, loading ratio and temperature on crack growth rates as shown schematically by Figure 1. For example, along the frequency line, a change in frequency will result in different C_2, C_3, C_4 values and hence the crack growth rates can be predicted from the new hyperbolic sine equation.

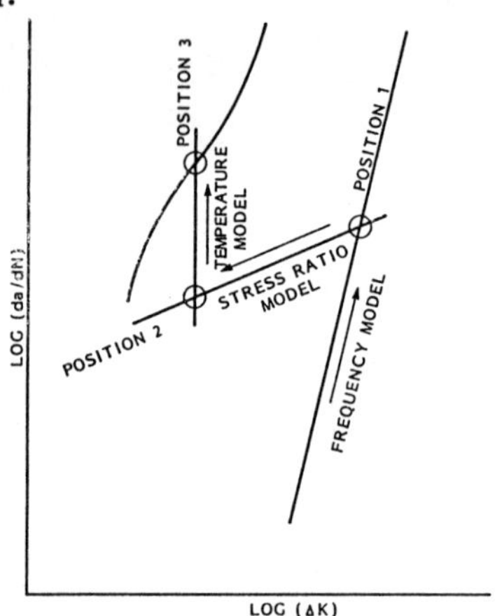

FIG. 1 SCHEMATIC DIAGRAM FOR THE HYPERBOLIC SINE MODEL/6/

Application to Present Data on 2-1/4 Cr-1 Mo

The data were first plotted on log (ΔK) vs log (da/dN) graphs and then C_3, C_4 were estimated from the inflection point for each curve. With frequency as the variable, three sets of data were grouped together and both C_3 and C_4 were tabulated as a function of the frequency. C_3 was found to be constant and independent of the test frequency, whereas C_4 showed a power law relation with the test frequency. C_1 and C_2 were then determined for each test frequency by solving the hyperbolic sine equation using two pairs of experimental data. Upon correlation with other C_1 and C_2 values from other test frequencies, a unique expression to describe these parameters was obtained.

For the crack growth tests under triangular loading with R = 0.05 and frequencies of 0.001 Hz, 0.0067 Hz and 0.067 Hz, the data and the corresponding hyperbolic sine curves are plotted in Figure 2. The governing relationships are:

$\log C_1 = 0.6 \log \nu + 0.6;\ 1/\nu = 0.8\,(C_2)^2;\ C_3 = \log 32;\ C_4 = -6.5\,(\nu).$

In general, the curves fit the data very well for the second stage.

The same methodology was applied to the 0.001 Hz results for R = 0.5 and 0.05. As shown in Figure 3, reasonable fit was again achieved with the following coefficients for the hyperbolic sine function:

R = 0.05	R = 0.5
$C_1 = 0.063;\ C_2 = 35;$	$C_1 = 0.21;\ C_2 = 21.35$
$C_3 = \log 32;\ C_4 = -5.1$	$C_3 = -\log 16.37;\ C_4 = -5.1$

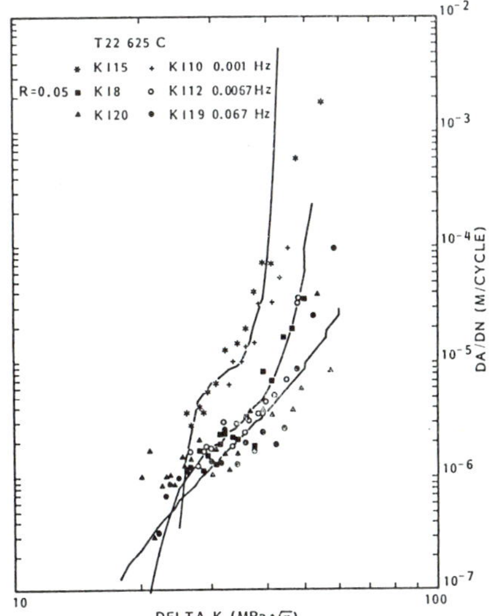

FIG. 2 EFFECT OF FREQUENCY ON CRACK GROWTH, USING THE HYPERBOLIC SINE MODEL

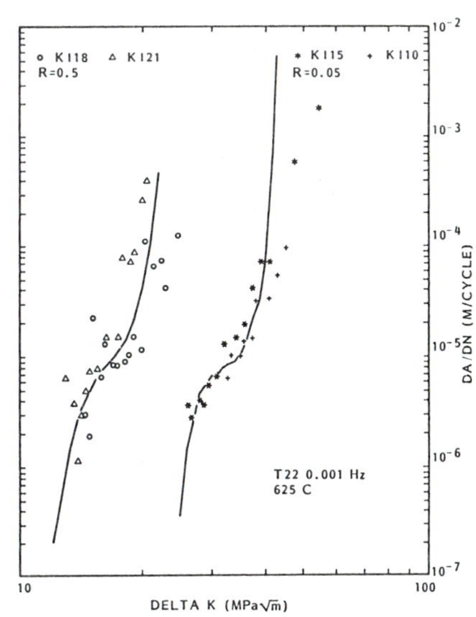

FIG. 3 EFFECT OF R RATIO ON CRACK GROWTH USING THE HYPERBOLIC SINE MODEL

With further data from forthcoming tests with R = 0.3, definition of any R ratio dependency of the coefficients will be facilitated.

An attempt to predict the crack growth rate for a different test frequency using the current hyperbolic sine equation is shown in Figure 4. The predicted crack growth rates for 0.0003 Hz are higher than the experimental values obtained for 0.0067 Hz and 0.001 Hz. This behaviour is expected since lower cyclic frequency tends to produce more creep-fatigue damage in the material.

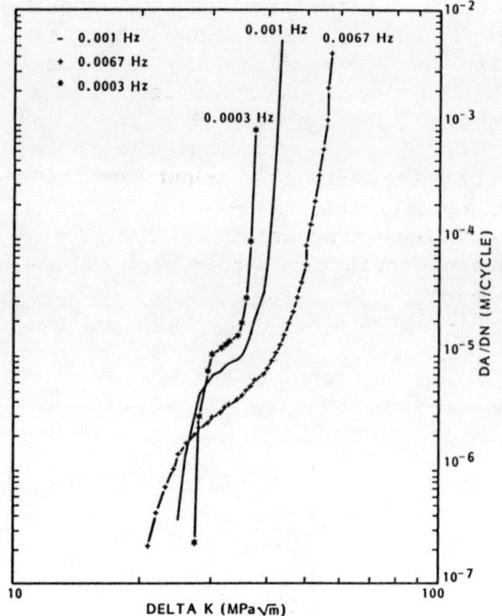

FIG. 4 PREDICTION FOR CRACK GROWTH RATES AT 0.0003 Hz USING EXPERIMENTAL RESULTS FROM 0.0067 Hz AND 0.001 Hz

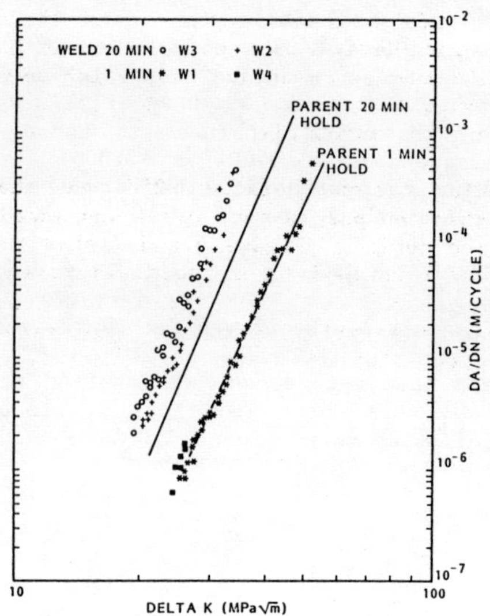

FIG. 5 CRACK GROWTH DATA FOR PARENT AND WELD MATERIALS TESTED AT 625°C WITH 0.0067 Hz RAMP RATE

It has been suggested that, being symmetric about an inflection point, the sine function cannot describe all three crack growth stages since experimental data usually lack such symmetry [7]. The correlation level for the Stage II cracking can be considered excellent but, for Stage I and III, lack of data precludes a valid comparison. For the Stage I region, the threshold ΔK values are higher for the higher frequency data (Figures 2 and 4) which would not be the expected behaviour. With more data points in the threshold region, the equation coefficients might be adjusted to achieve better curve fitting in this region.

LINEAR DAMAGE MODEL [8]

As reported [2,5], in tests at 0.0067 Hz, addition of 1 min and 20 min hold times increased the cyclic crack growth rate to higher values than were observed in tests at 0.001 Hz without hold periods. The hyperbolic sine equation cannot address the effect of hold time so that an alternative approach is required for creep-fatigue crack growth rate data. It was felt worthwhile, to explore a linear damage treatment of the hold time test data.

For the linear damage approach, crack growth data generated under both fatigue and creep loading conditions are required. Creep crack growth data are available elsewhere, from tests on 2-1/4 Cr-1 Mo weld metal in the quenched and tempered condition [9,10]. Results from in-house work, see Fig. 5, show little difference in crack growth behaviour between parent and weld metal specimens except in 20 min hold time tests. Accordingly, it was felt acceptable to use these external data to permit some appraisal of the linear damage approach.

The quenched and tempered data [9,10] from tests at 565°C were utilized to estimate the creep crack growth contribution at 625°C for the parent material. It was felt that the use of 565°C data was permissible since the in-house results had shown little dependence of crack growth on temperature in this range [2].

Using fatigue data at 625°C for the parent material, the amount of crack extension per fatigue cycle was added to the creep crack growth for the hold time period. Figure 6 and 7 show the theoretical prediction and the actual experimental data expressed in terms of crack growth per second for the one minute and 20 minute hold time tests, respectively. For the former, the parent material test data were very similar to predicted values; for the latter, the prediction produced higher crack growth rates.

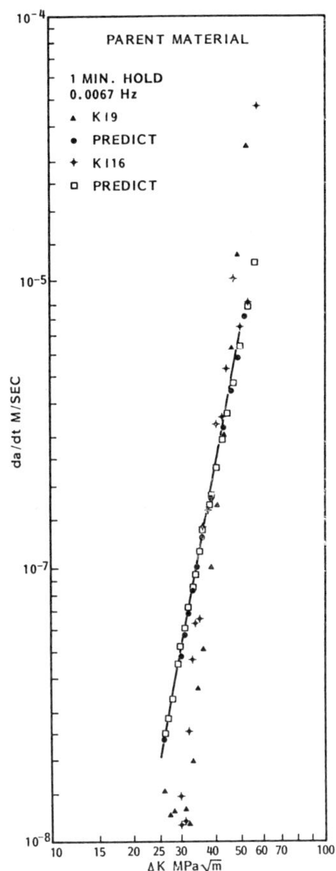

FIG. 6 APPLICATION OF LINEAR DAMAGE METHOD TO 1 MIN HOLD 625° DATA

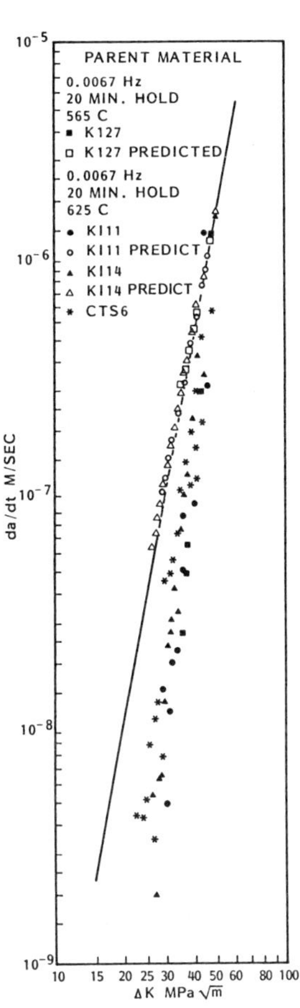

FIG. 7 APPLICATION OF LINEAR DAMAGE METHOD TO 20 MIN HOLD DATA

The similarity between the prediction and the experimental data suggested that the contribution from creep was small and the 565°C creep data were sufficient to simulate the 625°C behaviour. Another possibility is that for 2-1/4 Cr-1 Mo steel, creep-fatigue interaction is not additive and the actual growth is less than the sum of creep and fatigue components.

The linear summation method at least provides an upper bound for the crack growth rates. For the 20 minute hold time results, the prediction is above the actual data for both 565 and 625°C tests. The use of weld material data gave a conservative upper bound prediction for the crack growth behaviour in parent material. These observations suggest that the contribution from creep can become significant when the hold time increases and the microstructure changes to the coarse grained weld heat affected zone material. The linear damage prediction, even when using creep data for weld microstructure at lower temperature, can provide a conservative upper bound for the parent material at 625°C. With more appropriate data for the parent material, a better prediction would be achieved.

CONCLUSIONS

1. High temperature crack growth data of the form da/dN vs ΔK have been modelled in terms of two available approaches.

2. The hyperbolic sine model showed good data correlation for different test frequencies and loading ratios though correlation in the near-threshold region was compromised by shortage of data.

3. The linear damage approach to data from tests incorporating hold times provided an upper bound value for creep-fatigue damage prediction; more accurate prediction would require better creep crack growth data than was available.

REFERENCES

1. R.B. Dooley, R.L. McQueen and D. Sidey, Proc. American Power Conference, p 208, (1982).
2. W.K. Lee, Ontario Hydro Research Division Report No 83-430-K, 1983.
3. P.C. Paris and F. Erdogan, Trans. ASME Series D., p 528, (1963).

M.S. Miller and J.P. Gallagher, In Fatigue Crack Growth Measurement and Data Analysis (Edited by S.J. Hudak, Jr and R.J. Bucci), ASTM, STP738, p 205, (1981).

5. P.M. Sooley, H.J. Westwood, W.K. Lee, D. Sidey and D.W. Hoeppner, ASME PVP, Vol 71, "Thermal and Environmental Effects in Fatigue: Research Design Interface", ed C.E. Jaske, S.J. Hudak, Jr. and M.E. Mayfield, June 1983.
6. D.L. Sims, C.G. Annis, Jr and R.M. Wallace, Air Force Materials Laboratory Report #AFML-TR-76-176, Part III, (1977).
7. A. Saxena, S.J. Hudak Jr., and G.M. Jouris, Engineering Fracture Mechanics, 12, 103, (1979).
8. ASME Code Case N47, Appendix T, Section III, Class 1.
9. M.J. Siverns and A.T. Price, Int Journal of Fracture, 9, 199, (1973).
10. D.J. Gooch, Effects of Geometric Constraint on Analysis Methods for Creep Crack Growth in 2-1/4 Cr 1 Mo Weld Metal, CEGB Report No TPRD/L/2384/N82, (1983).

Fatigued Copper Single Crystals Studied by Small-angle Neutron Scattering

T. Lepistö*, V.-T. Kuokkala*, P. Kettunen*, G. Kostorz and R. Schmelczer**

**Institute of Materials Science, Tampere University of Technology, SF-33101 Tampere 10, Finland*
***Institute für Angewandte Physik, ETH Hönggerberg, GH-8093 Zürich, Switzerland*

ABSTRACT

The structure of single and multiple slip oriented copper single crystals fatigued at small plastic shear strain amplitudes was studied using small angle neutron scattering (SANS) and transmission electron microscopy (TEM). The increase of the small angle scattering intensity of the fatigued samples was found to be primarily due to dislocations, including dipoles, but also a volume effect of voids with a diameter of about 2 nm and a volume fraction of about 10^{-5} was detected. Transmission electron microscopy using the weak-beam technique was applied to get additional information about the distribution of the point defects in the structure.

KEYWORDS

Fatigue; copper single crystals; small angle neutron scattering.

INTRODUCTION

In f.c.c. metal single crystals oriented for single slip, cyclic straining at low constant plastic strain amplitudes results in the formation of persistent slip bands (PSBs). This behavior is well studied in copper showing an almost linear increase in the volume fraction of PSBs from 0 to 100 % with plastic shear strain amplitudes from $6x10^{-5}$ to $7.5x10^{-3}$, respectively /1/.

The dislocation arrangements in the matrix and in the PSBs have been studied intensively by transmission electron microscopy (TEM). These determinations have shown the matrix to be composed of the so-called loop patches or vein structure, dense dislocation arrangements separated by channels of low dislocations density /2/. The PSBs consist of dislocation walls and almost dislocation-free channels between them. The wall spacings have been found to exhibit a unique value of the order of 1.3 µm. The walls of PSBs lie perpendicular to the primary Burgers vector $1/2[\bar{1}01]$ and are reported to consist of dense dislocation arrangements composed mainly of edge dislocation dipoles and multipoles /3/.

Single crystals orientated for multiple slip have not been studied in detail /4/. Although matrix structures similar to those in single slip orientated crystals have been reported, the formation and structure of PSBs are still unclear. According to published results, arrangements resembling PSBs appear also in these crystals /4/. Much more work is, however, needed to clarify the situation and especially to get valuable information to correlate the behavior of single crystals and polycrystals and the corresponding dislocation arrangements.

The present study on the dislocation arrangements in copper single crystals orientated for single and multiple slip gives new experimental information to clarify the situation. These crystals show also some evidence about the existence of dipoles and other structural features in them. The small-angle neutron scattering (SANS) method is shown to be a potential method to study certain details in the deformed structures.

EXPERIMENTAL

High-purity copper (99.999+%) was used to grow single crystals with axes $[\bar{1}11]$ and 5° from [011] towards $[\bar{1}12]$ by a modified Bridgman method in an atmosphere of purified argon. The as-grown crystals were shaped by spark erosion to have a gauge section of 25 mm in length and 10x12 mm^2 or 5x7 mm^2 in cross-section, and subsequently annealed in an atmosphere of argon +3 % hydrogen at 1073°C for a period of four days. This time was sufficient to reduce the dislocation content to about 10^4 cm^{-2}. Cooling to room temperature was done very slowly to allow hydrogen to diffuse out of the samples. After this, special steel-made ends were soldered into each crystal allowing proper fixing into the testing machine. In the following, single slip oriented crystals are denoted by 5/* and multiple slip oriented crystals by 6/*.

Specimens were fatigued in a universal servohydraulic testing machine controlled by a minicomputer at true plastic shear strain amplitudes of 0.1 % and 0.15 %. The detailed test procedures are given in an earlier paper /4/.

The SANS measurements were carried out at the Institute Laue-Langevin (ILL) in Grenoble, France, with the instrument D11. Details on the experimental procedure are given in paper /4/.

RESULTS AND DISCUSSION

The SANS measurements with fatigued copper single crystals yielded scattered intensity (differential cross-sections) as a function of Q (= $4\pi \sin\theta/\lambda$ where θ is half the scattering angle and λ is the wavelength of the incident neutrons). These scattering curves may contain contributions from different types of defects (voids, individual dislocation, dipoles). The total scattered intensity can be expressed by an equation

$$I_{tot} = K_{dip}Q^{-1} + K_{dis}Q^{-3} + K_{voids}e^{-\frac{Q^2R_G^2}{3}} + K \qquad (1)$$

where the first two terms represent the scattering from dipoles and dislocations, the third term takes into account the scattered intensity of voids and K is a material and test dependent constant. The Q range where the first three terms may be expressed as stated is not necessarily the same, and special evaluations are necessary in all cases.

Figure 1 shows the measured SANS intensities (I vs Q) of single slip oriented copper single crystals fatigued to three different stages of cyclic hardening, and of a well-annealed undeformed reference sample. With increasing number of fatiguing cycles the scattered intensity increases, in particular at smaller Q values, where the extra scattering follows a Q^{-3} dependence as figure 2 shows, and can thus be attributed to dislocations.

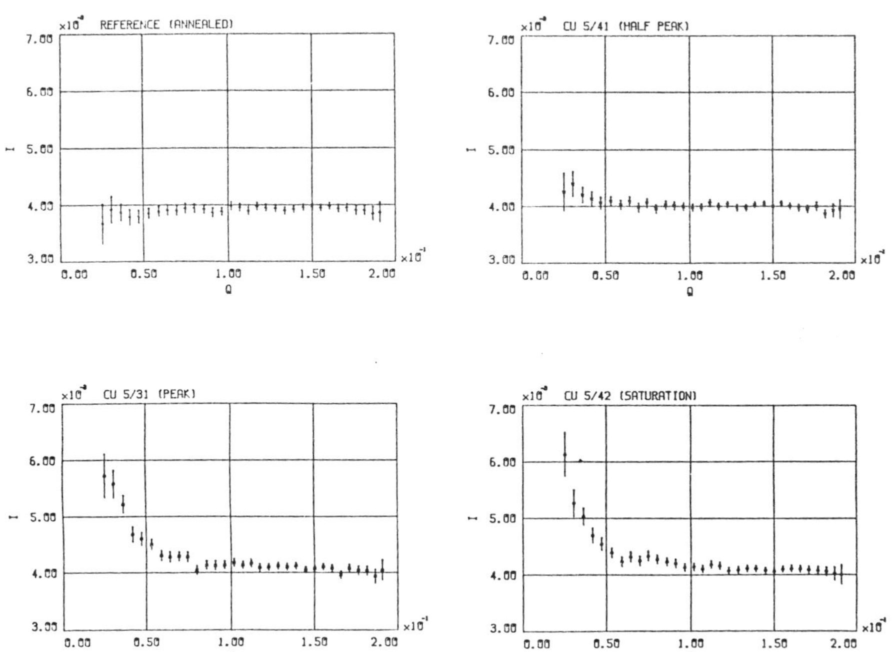

Fig. 1. The measured small-angle neutron scattering intensities for single slip oriented copper single crystals fatigued to different stages of cyclic hardening.

At higher Q values ($Q \gtrsim 0.7$ Å^{-1}) the inclination of the I vs Q curve increases with proceeding fatigue cycling. Applying the Guinier approximation to the difference intensity curve of the sample fatigued to the peak value of the cyclic hardening (c.f. ref. 6) and the reference sample, the radius of gyration of the scattering items was determined, as shown in figure 3. The average diameter of these scattering particles, which are believed to be vacancy clusters or microvoids, is about 2 nm with the assumption of a spherical shape. The estimated volume fraction of voids is of the order of 10^{-5}.

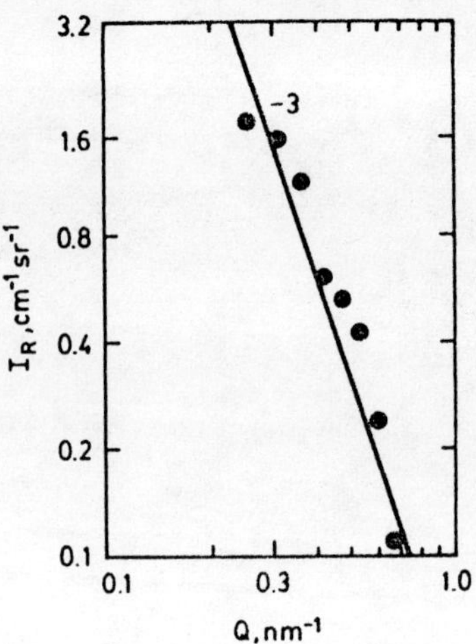

Fig. 2. The scattering of 5/31-Ref. at small Q values /5/.

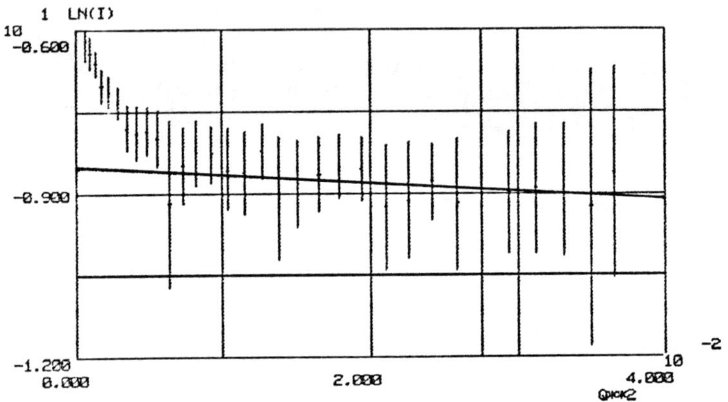

Fig. 3. Guinier plot of the intensity difference 5/31-Ref.

With a multiple slip oriented single crystal fatigued at a constant plastic shear strain amplitude of 0.1 % to 40,000 cycles the neutron small-angle scattering showed a deviation from the Q^{-3} dependence normally found for $Q \lesssim 0.7$ nm^{-1}. The exponent of the pover law fit was about -2.3 indicating also scattering due to dislocation dipoles /5/.

Transmission electron microscopy using the bright-field, dark-field and weak-beam techniques verified the existence of small point defects in the crystals. The point defects tend to concentrate into the channel/wall boundaries and into the walls both in single and multiple slip oriented crystals as figure 4 shows. The existence of vacancy-type loops in the channels was also verified by the detailed TEM-studies.

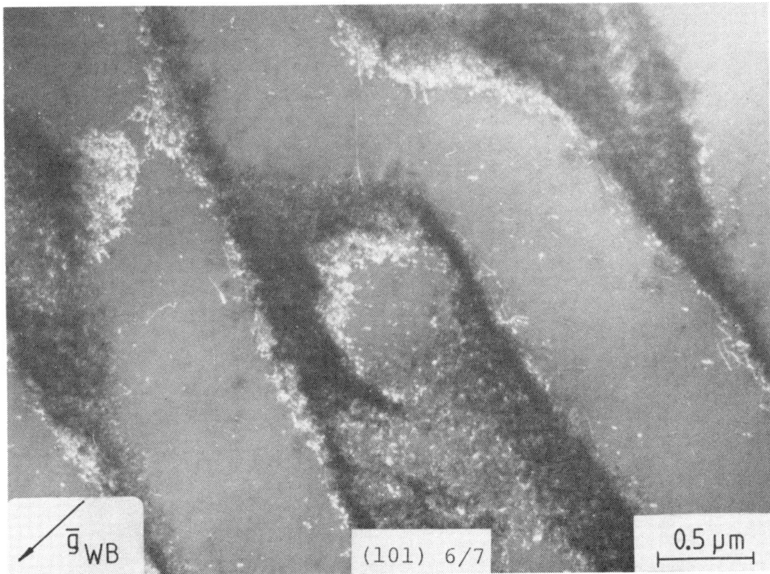

Fig. 4. The point defect distribution in a multiple slip oriented copper single crystal fatigued at a constant plastic shear strain amplitude of 0.1 %, revealed by the weak-beam TEM technique.

ACKNOWLEDGEMENTS

The authors want ot express their sincere gratitude for the Institute Max v. Laue-Paul Langevin for the possibility to use the instrument D 11 for the present measurements. Sincere thanks are directed also to Mr. Cerri and Mr. Petry for the skilful assistance during the measurements. The financial support provided by the IAEA, Vienna, is also gratefully acknowledged.

REFERENCES

1. H. Mughrabi, *Continuum Models of Discrete Systems 4*, O. Brulin and R.K.T. Hsich, North Holland, Amsterdam. p. 241 (1981).
2. Z.S. Basinski, A.S. Korbel and S.J. Basinski, *Acta Metall.* **28**, 191 (1980).
3. J.G. Antonopoulos, L.M. Brown and A.T. Winter, *Phil. Mag.* **34**, 549 (1976).
4. P. O. Kettunen, T. Lepistö, G. Kostorz and G. Göltz, *Acta Metall.* **29**, 96 (1981).
5. G. Kostorz, MRS Fall Meeting, Boston (1984).
6. T.K. Lepistö, V.-T. Kuokkala, Proc. 5th Risø Int. Symp. Metall. Mat. Sci., Risø Nat. Lab., 371 (1984).

Dislocation Behaviour and Slip/ Twinning Interactions During the Cyclic Plastic Deformation of bcc-Fe-Cr-Ni Alloys

T. Magnin*, A. Fourdeux*, J. Lepinoux, L. P. Kubin**, P. Chomel***, M. Fagot*** et J. P. Cottu*****

**Ecole des Mines, 158 cours Fauriel, 42023 Saint-Etienne Cédex, France*
***Faculté des Sciences, L. A. 131 C.N.R.S., 86022 Poitiers, Cédex, France*
****INSA, Avenue de Rangueil, 31077 Toulouse Cédex, France*

ABSTRACT

This paper examines the role of twinning on the cyclic plastic deformation behaviour of a bcc stainless steel which deforms both by slip and twinning at 300K. Well defined macroscopic tests and "in situ" experiments in transmission electron microscopy are used. The mechanical behaviour, particularly the influence of twinning on the cyclic hardening stage and the crack initiation processes, is then discussed in terms of slip/twinning interactions in the bcc lattice.

KEYWORDS

Cyclic deformation, fatigue, dislocation, twinning, slip, bcc alloys, "in situ" experiments.

INTRODUCTION

Deformation twinning in the bcc lattice is generally favoured by conditions of low temperature, high strain rate, and by the presence of some substitutional elements [1]. In these conditions, the mobility of screw dislocations $\frac{1}{2}$ <111> is very reduced, as a consequence of their original structure. The movement of partials is then favoured. Under an applied shear stress a multilayer fault on {112} planes can occur and this evolves into a twin [2]. The formation of twins in the bcc lattice is related to the interrelationship between slip and twinning as classically schematized by the following reaction :

$$\text{(A)}\quad \frac{1}{2}\langle 111\rangle_{\text{screw}} \underset{\text{slip}}{\overset{\text{twinning}}{\rightleftharpoons}} 3 \times \frac{1}{6}\langle 111\rangle \ [3].$$

Nevertheless, the influence of a cyclic shear stress on the evolution of this reaction has never been systematically studied. This is the main objective of this paper which concerns the cyclic deformation behaviour of a bcc Fe-Cr-Ni alloy which deforms both by slip and twinning at room temperature [4,5]. The aim of this work is then to explain the macroscopic mechanical behaviour of alloy in terms of slip/twinning interactions, using a combination of will defined mechanical tests and in situ experiments by transmission electron microscopy (TEM).

EXPERIMENTAL PROCEDURE

The ferritic stainless steel examined in this study contains, in wt, [25-26] Cr, 1 % Mo, 5 % Ni and less than 100 ppm C + N. It was waterquenched from 1523K. The slip/twinning interactions are observed and quantified : (1) at a macroscopic scale, using carefully controlled tension-compression tests, performed on a servohydraulic machine at imposed plastic deformation ($10^{-4} < \frac{\Delta\varepsilon_p}{2} < 10^{-2}$) and imposed strain rate ($10^{-5}\,s^{-1} < \dot{\varepsilon} < 10^{-2}\,s^{-1}$) on fatigue specimen (10 mm gauge length and 5 mm diameter). (2) at a microscopic scale, using both classical observations in TEM and "in situ" experiments performed on thin foils with two kind of tests : (i) alternating shear at room temperature [6] and (ii) tension-compression tests in the temperature range 150-380K [7].

DISLOCATION BEHAVIOUR

The introduction of Ni to the bcc Fe-26 Cr matrix induces a very different deformation mode compared to the alloy without Ni. In situ experiments were achieved on Fe-26Cr-1Mo-5Ni specimens precycled on the servohydraulic machine during 3 cycles at $\frac{\Delta\varepsilon_p}{2} = 4.10^{-3}$. Fig. 1 shows the extension of a dislocation loop during the tensile phase and clearly indicates that the deformation regime of the alloy is typical of that of bcc metals at low temperature. The mobility of screw dislocations is very reduced compared to the mobility of edge segments. Large friction stresses take place. An increase in the temperature to 380K, or a variation of the strain rate ($10^{-5}\,s^{-1} < \dot{\varepsilon} < 10^{-2}\,s^{-1}$), does not induce any change in the dislocation behaviour, contrarily to what happens in the Fe-26Cr-1Mo alloy [8]. Moreover, because of the local stresses due to the waterquench, the displacement of screw dislocations is very discontinuous and induces the formation of sharp slip bands. Finally, it has been found that screw dislocations can exhibit an asymmetrical behaviour between tension and compression as for bcc alloys deformed in the low temperature regime [8,9].
Thus, the introduction of nickel in the bcc Fe-26Cr matrix induces a very pronounced increase of the athermal temperature, which favors the nucleation of twins at room temperature. Fig. 2 shows a fragmented twin created during the "in situ" deformation of a thin foil and which is classical in the bcc lattice [10]Fig. 3 shows an intersection of twins at grain boundary and the resulting accommodation by slip and twinning. It can be noticed that : (1) the twins are very often to be found in pairs, (2) dislocation piles-up are very profuse at twin matrix interfaces.
The observation of such a behaviour for a bcc metal at $T \simeq 0.2$ T melting is quite original.

SLIP/TWINNING INTERACTIONS DURING CYCLING

Fig. 4 clearly indicates that twinning occurs during the first cycles and is associated with pronounced strain recovery and Baushinger effects. These phenomena have been explained in terms of restored elastic energy [5]. During the first cycles, the twin interfaces which act at barriers to the gliding lattice dislocations induce a rapid hardening. Then, the reaction (A) is displaced to favour slip when the density of mobile dislocations is higher. In the same way, it has been found that the cumulative plastic deformation by twinning $\Sigma\,\varepsilon^{1}_{pt}$ (Fig. 4) is constant whatever the strain amplitude. When twinning is not operating after the first cycles, the hardening rate decreases and the saturation regime is rapidly reached. It has been mentioned elsewhere [11] that the saturation regime can be modelled

as for other bcc metals in the low temperature regime.
Crack initiation processes are also related to the slip/twinning interactions. Fig. 5 shows the Coffin-Manson curves for alloys with and without Ni. In the low plastic strain range ($\frac{\Delta\varepsilon_p}{2} < 10^{-3}$), the fatigue life is very reduced when twinning occurs (by a factor 15 at $\frac{\Delta\varepsilon_p}{2} = 10^{-4}$). Crack initiation takes place in the twins or at twin interactions (Fig. 6). Because of twinning, a fatigue limit corresponding to the quasi-reversible motion of edge dislocations as for many bcc alloys at low temperature and low strain amplitude [9] does not exist. In the high plastic strain range ($\frac{\Delta\varepsilon_p}{2} \geq 2.10^{-3}$), twinning induces a less pronounced reduction of the fatigue life (by a factor 3 at $\frac{\Delta\varepsilon_p}{2} = 4.10^{-3}$). Crack initiation is intergranular, sometimes at twin/grain boundary intersections but generally due to the strong shape change effect of the grains induced by the asymmetrical behaviour of screw dislocations between tension and compression, as for the Fe-26Cr-1Mo alloy at low temperature [8].
In conclusion, it can be said that the cyclic mechanical properties of the Fe-Cr-Ni alloy are directly related to the evolution of the slip/twinning interactions. This study also clearly shows the influence of a cyclic shear on the evolution of the interrelationship between slip and twinning in the bcc lattice.

REFERENCES

1 - S. Mahajan and D.F. Williams, Int. Met. Rev., 18, 43 (1973)
2 - M.S. Duesbery, V. Vitek and D.K. Bowen, Proc. Roy. Soc., Ser. A, 332, 85 (1973).
3 - S. Mahajan, Acta Met., 23, 671 (1975).
4 - T. Magnin and F. Moret, Scripta metall., 16, 1225 (1982)
5 - T. Magnin, L. Coudreuse and A. Fourdeux, Mat. Sci. Eng., 63, L5 (1984).
6 - J. Lepinoux, L.P. Kubin, Phil. Mag. (1985) (in the press).
7 - P. Chomel, M. Fagot and J.P. Cottu, Scripta metall., 19, 2 (1985).
8 - T. Magnin, A. Fourdeux and J.H. Driver, Phys. St. Sol. A, 65, 301 (1981)
9 - H. Mughrabi, H. Herz and X. Stark, Int. Jr. Fract., 17, 193 (1981)
10 - J. Levasseur, Thesis, Paris (1972).
11 - L. Coudreuse and T. Magnin, Proc. Journées Int. de Printemps, SFM editor, Paris, 159 (1984).

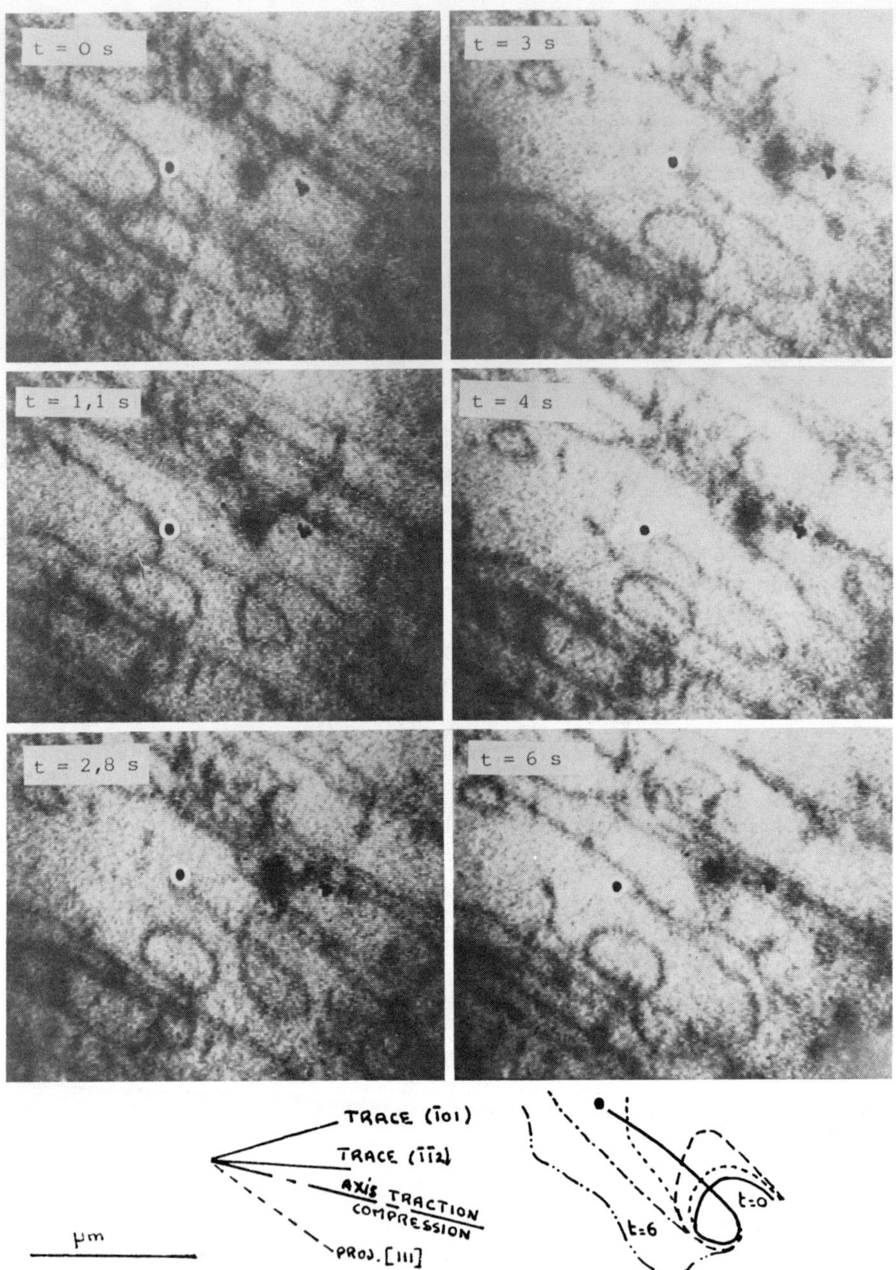

Fig. 1. "In situ" experiments on a Fe-26Cr-1Mo-5Ni thin foil obtained from a specimen precycled during 3 cycles at $\frac{\Delta\varepsilon_p}{2} = 4.10^{-3}$. Observation of the evolution of a dislocation loop during the tensile part of a cycle.

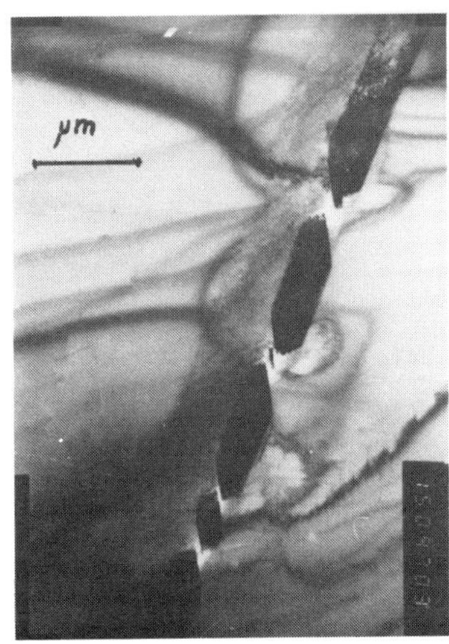

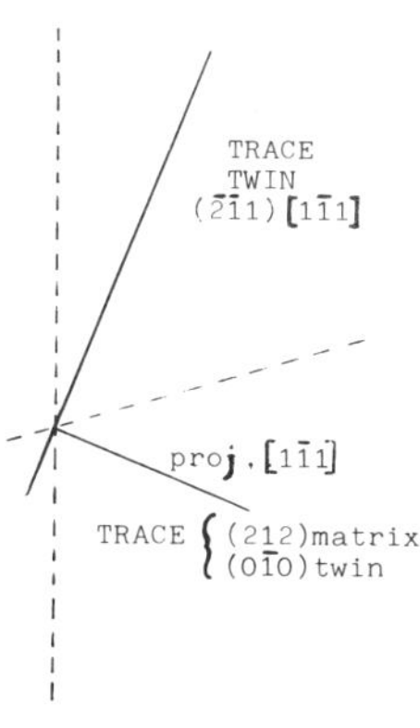

Fig. 2. Fragmented twin observed after an "in situ" deformation of a Fe-26Cr-1Mo-5Ni thin foil.

Fig. 3. Intersection of twins at grain boundary and plastic accommodation by slip and twinning (G.B. = grain boundary).

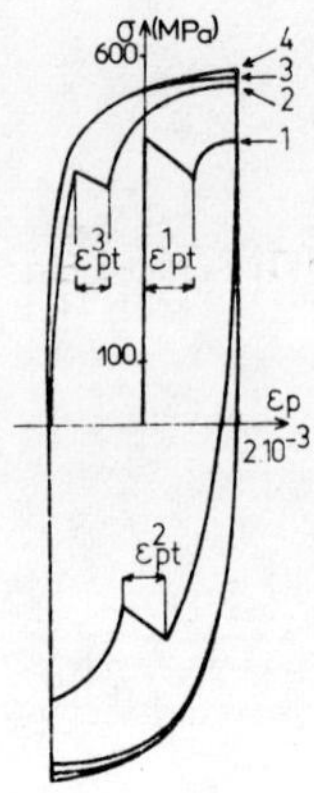

Fig. 4. Stress-strain curves of the Fe-26Cr-1Mo-5Ni alloy during the first cycles at $\frac{\Delta\varepsilon_p}{2} = 2.10^{-3}$ and $\dot{\varepsilon} = 10^{-3}s^{-1}$.

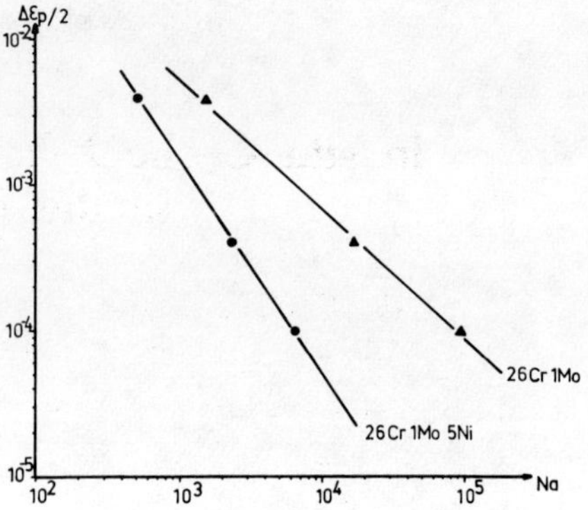

Fig. 5. Coffin-Manson curves of Fe-26Cr-1Mo and Fe-26Cr-1Mo-5Ni alloys at $\dot{\varepsilon}=10^{-3}s^{-1}$

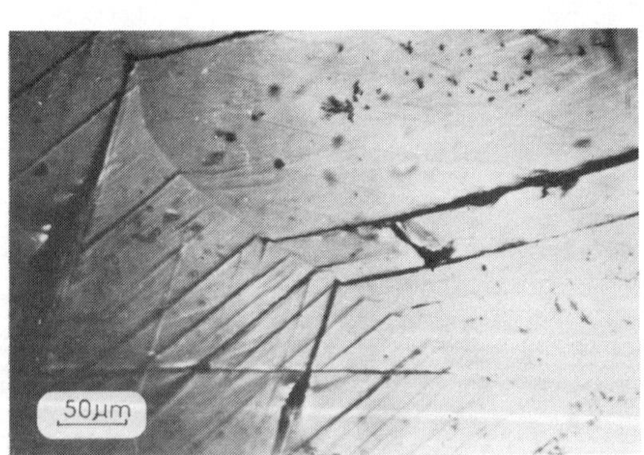

Fig. 6. Crack initiation in twins at low plastic strain amplitude ($\frac{\Delta\varepsilon_p}{2} = 10^{-4}$, $\dot{\varepsilon} = 10^{-2}s^{-1}$, 6500 cycles)

In situ Cyclic X-Y Mechanical Tests in the HVEM

A. W. Sleeswyk*, M. E. Kassner and G. J. Kemerink***

**Laboratory Algemene Natuurkunde, Materials Science Centre, Westersingel 34, 9718 CM Groningen, The Netherlands*
***Naval Postgraduate School, Department of Mechanical Engineering, Code 69, Monterey, CA 93940, USA*

ABSTRACT

Mechanical analysis shows that dislocation movement is reversed by tensile stressing in two mutually perpendicular directions. Buckling of foil specimens liable to occur in other cyclic tests can thus be avoided. A design for performing such x-y EM tests *in situ* is presented, together with experimental observations which tend to show mutual influence of dislocations over distances of the order of 10^4b.

KEYWORDS

In situ mechanical tests in EM, cyclic testing, reverse dislocation motion, foil buckling.

In a recent article, Kubin and Lépinoux (1) review the various known devices for cyclic *in situ* mechanical testing in the electron microscope. The purpose of such tests is to make the mobile dislocations in the foil move back and forth by varying the load conditions on the specimen. Reversal of the load is, of course, the most obvious way of achieving this, but a drawback of that method is that buckling is liable to take place in at least one of the two modes of deformation.

The magnitude of the stress σ_B at which buckling of a square thin foil of thickness t, length of the edges d, occurs, is given by the expression (2):

$$\sigma_B = K{\cdot}E\,(t/d)^2, \qquad (1)$$

in which E is Young's modulus and K a factor which depends on the boundary conditions and the manner of loading. K ranges from ∿3 to ∿20. The influence of Poisson's ratio, which may raise the buckling stress by ∿10 % is neglected in our estimate. Buckling may occur during compressive loading or shear loading.

Compressive loading occurs in the method of Imura et al. (3) and in the experiments of Cottu and Landowski (1). In the first, the specimen is glued over a hole in a thick foil which is alternately bent upwards and downwards, and the four edges of the specimen may be considered as clamped, in the second, simple compression occurs in one mode of loading, two edges are clamped, the other two are free. If we consider a specimen measuring 1 x 1 mm, thickness 2.10^{-4} mm, in the first case the value of K is 7.7, and eq. (1) yields: $\sigma_B/E = 0.31 \cdot 10^{-6}$ for the buckling stress. For the second case, with K = 3.3, $\sigma_B/E = 0.13 \cdot 10^{-6}$ results.

When a shear is applied along the edges, which are simply supported, the situation is as in the shear device of Kubin and Lépinoux (4). Then, K = 7.7 and $\sigma_B/E = 0.31 \cdot 10^{-6}$.

These findings imply that in these devices elastic buckling precedes plastic deformation, which, even in very soft crystals, does not take place at stresses lower than $\sim 10^{-4} \cdot E$. The buckled foil will undulate, and as a consequence, the mechanical conditions at which dislocation movement takes place will depend on the location in the specimen, and these conditions will generally differ from those which would be produced in a flat specimen. Because of the great depth of field of the electron microscope, the elastic buckling does not manifest itself in the electron microscopic images of the defect structure.

A way to avoid buckling is to apply only tensile loads to the specimen. In a two-dimensional solid, a tensile stress alternating between two mutually perpendicular stress axes sa_1 and sa_2 would result in a reversal of the shear stresses between each mode of deformation, hence in a reversal of the dislocation movement, as illustrated in the accompanying diagram (Fig. 1).

In three dimensions the situation is more complicated. The stress axis sa_2 lies in a plane perpendicular to sa_1, and whether or not a tensile stress σ_2 along sa_2 reverses the shear stresses caused by σ_1 depends on the position of sa_2 relative to the glide plane, gp, and the glide direction, gd (Fig. 2).

The glide direction gd and sa_1 enclose an angle λ_1, and these two directions define a plane which is enclined at an angle α_1 to the glide plane gp. The angle between the normal to gp and sa_1 is ϕ_1. Spherical goniometry gives the relation: $\cos \phi_1 = \sin \alpha_1 \cdot \sin \lambda_1$, which, substituted in the customary

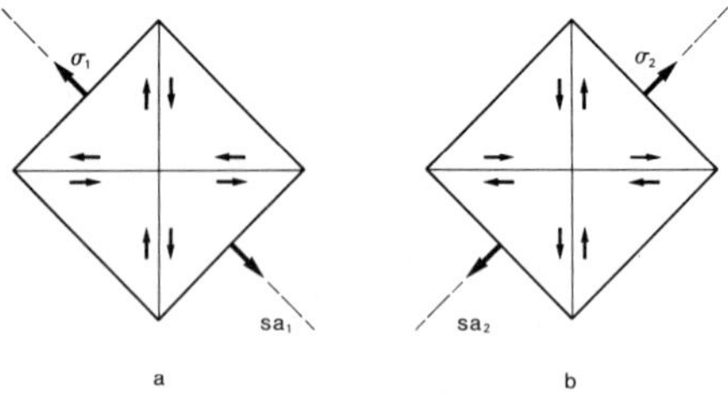

Fig. 1. Reversal of shear stress in x-y tensile tests on a two-dimensional solid.

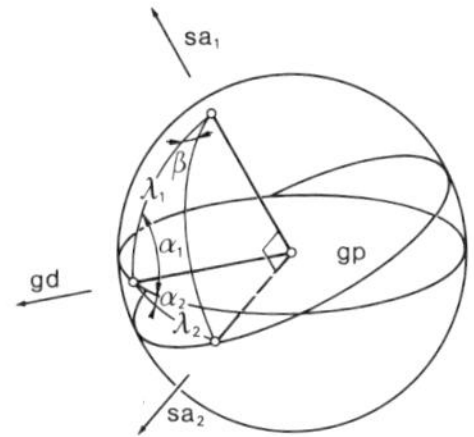

Fig. 2. Geometry of x-y tensile tests in relation to glide in a three-dimensional solid.

expression for Schmid's law, $\tau_1/\sigma_1 = \cos\phi_1 \cdot \cos\lambda_1$, results in:

$$\tau_1/\sigma_1 = 1/2 \cdot \sin\alpha_1 \cdot \sin 2\lambda_1 \qquad (2)$$

As shown in the diagram, the angle β between the plane containing sa_1 and gd, and the plane containing sa_1 and sa_2, determines the position of sa_2 in the plane perpendicular to sa_1. Now, an expression can be derived for the quotient τ_2/σ_2 in terms of α_1, λ_1 and β. In the expression for Schmid's law: $\tau_2/\sigma_2 = -1/2 \sin\alpha_2 \cdot \sin 2\lambda_2$ the equalities $\cos\lambda_2 = \cos\beta \cdot \sin\lambda_1$ and $\cos\lambda_1 = -\tan\beta \cdot \cot(\alpha_1 + \alpha_1)$ are substituted, which results in:

$$\tau_2/\sigma_2 = -1/2\ (\sin 2\beta \cdot \cos\alpha_1 \cdot \sin\lambda_1 + \cos^2\beta \cdot \sin\alpha_1 \cdot \sin 2\lambda_1) \qquad (3)$$

In the accompanying set of diagrams (Fig. 3) both τ_1/σ_1 and τ_2/σ_2 have been

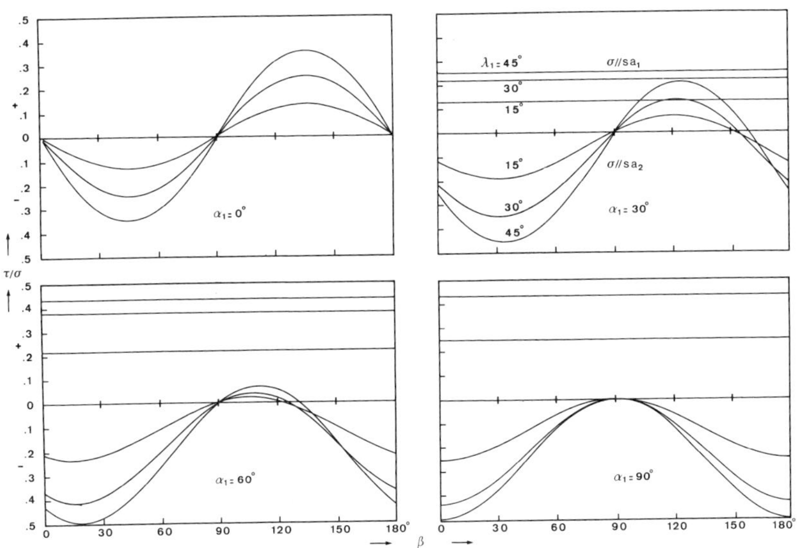

Fig. 3. τ/σ vs. β diagrams for σ_1 and σ_2.

plotted for 4 values of α_1 and 3 values of λ_1, while for τ_2/σ_2 the variation with β is given. The main conclusion that can be drawn is the following: for the orientations for which τ_1/σ_1 is large, i.e. $60^o < \alpha_1 < 90^o$, application of σ_2 along sa_2 will result in shear stresses of opposite sign but of the same magnitude for $\beta \approx 0$ and $\beta \approx 180^o$, and of much smaller magnitude for $\beta \approx 90^o$. For the orientations for which τ_1/σ_1 is small, i.e. $0 < \alpha_1 < 60^o$, application of σ_2 along sa_2 may result in shear stresses of the same magnitude, both positive and negative.

In practice, this implies that the majority of the dislocations moving forward when a tensile stress was applied along sa_1 will either not move or move backward when the tensile stress is applied along sa_2. Only rarely will a dislocation then continue to move forward.

In the holder built for performing *in situ* tensile tests in mutually perpendicular directions in the electron microscope, the specimen foil bridges a gap between the two movable platens on which it is mounted. The connecting pin (see Fig. 4) activates the movement of each platen. When it moves from the tip of the holder, it pushes the x-platen in the direction of the longitudinal axis of the holder, when it moves towards the tip, it causes the y-platen to rotate around the fulcrum; the specimen is then pulled in a direction which is approximately perpendicular to the longitudinal axis. Windows have been provided in the holder frame and in the y-platen for letting the electron beam pass through.

Results obtained on polycrystalline high-purity aluminium are illustrated in the paired photomicrographs, A, B and C (fig. 5). In Figs. B and C the tensile axis is perpendicular to the one in A. The largest displacement is made by the dislocation in the upper half, which in Fig. A glides from NE to SW, shortening itself considerably in the process, and which reverses its path in Fig. B and C.

This dislocation appears to interact with the dislocation oriented NW-SE, of which the lower part annihilates itself at the free surface (Fig. C): the upper part is presumably prevented from doing so by the first dislocation. Near the bottom of the image, one of the two closely parallel dislocations in Fig. A, appears to have annihilated itself in Fig. B. The other dislocation has moved a little distance between Figs. A and B, but in Fig. C it

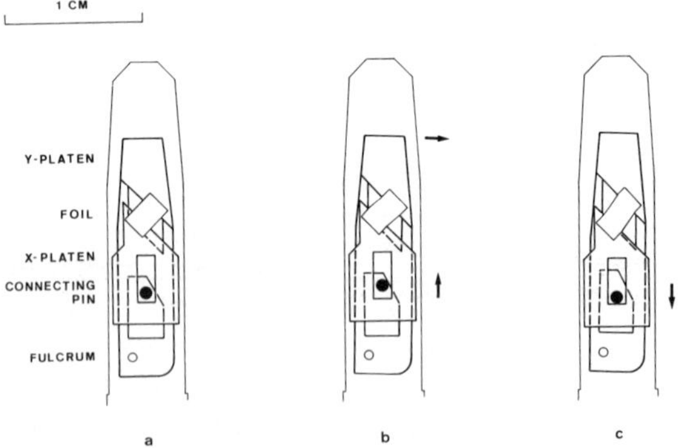

Fig. 4. Schematic of x-y tensile holder.

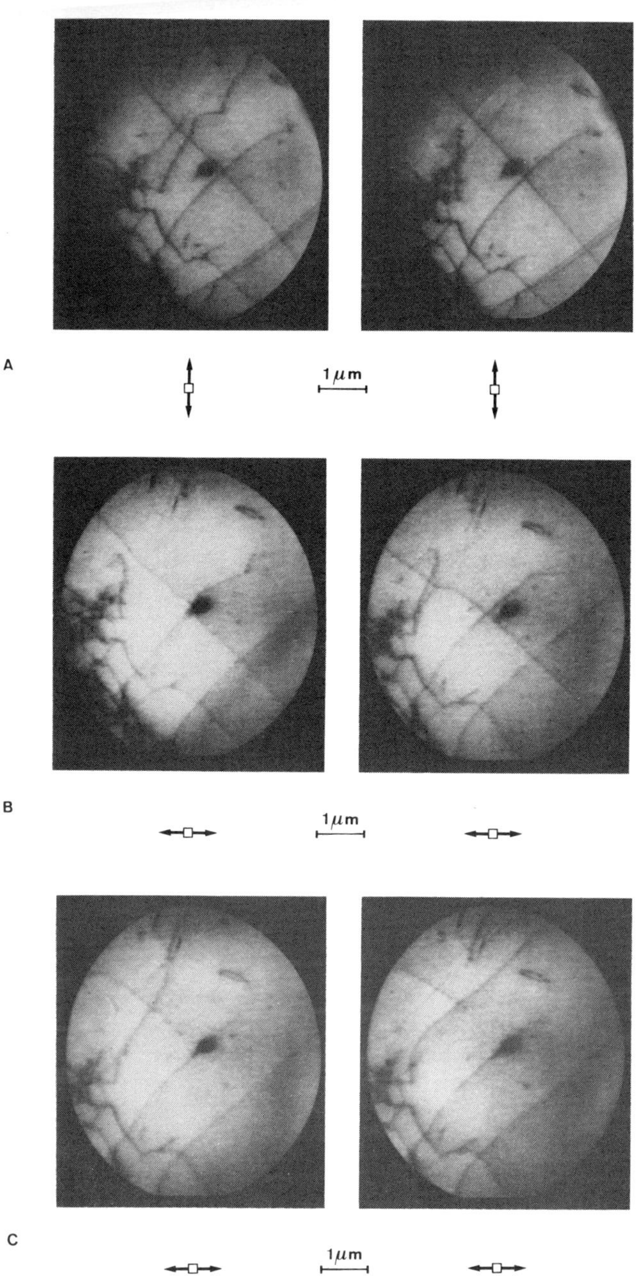

Fig. 5 (A, B, C). Successive dislocation configurations under x-y tensile stressing.

has not moved at all. The most plausible interpretation would seem that of the original three parallel dislocations, the one which was annihilated had a Burgers vector of opposite sign to the one close-by, while the Burgers vector of the other dislocation was of the same sign. Consequently, when the dislocation had been annihilated, the other two, attracting each other, could move to a closer distance. They may have been prevented from moving further by the presence of the most mobile dislocation which reversed its path.

It would seem then, that not only do dislocations react to cyclic x-y loading as predicted by the mechanical analysis, but also that they induce each other to move at distances some μm apart, i.e. at a distance of $\sim 10^4$ Burgers vectors.

ACKNOWLEDGEMENTS

We are much indebted to Messrs. N. Drost and J. Jansen for their able assistance during the experiments in the 1 Megavolt electronmicroscope, and to Mr. M. Mulder who machined the special holder. This research was sponsored by the Stichting Technische Wetenschappen, the Metaalinstituut T.N.O. and the Rijksuniversiteit Groningen in the Netherlands and the Lawrence Livermore National Laboratory in the U.S.A. The experiments were performed while one of us (M.E.K.) held a Fulbright Senior Scholarship.

REFERENCES

1. L. Kubin and J. Lépinoux, *J. Microsc. Spectrosc. Electron.* 9, 319 (1984).
2. R.J. Roark, *Formulas for Stress and Strain.* p. 302 ff. McGraw-Hill, New York (1954).
3. T. Imura, T. Ishikawa, Y. Sakitani, T. Tono, C. Morita, H. Saka, K. Koda, A. Yamamoto and S. Saimoto, *Proc. 8th Int. Congr. on EM, Canberra.* 1, p. 168 (1974).
4. J. Lépinoux, *La déformation en sollicitation cyclique de monocristeaux de cuivre, étude par microscopie électronique "in situ".* Thesis, Univ. Poitiers (Superv. L. Kubin), 1983.

Orientations of Dipolar Walls in Cyclically Deformed 316 Stainless Steel

G. L'Espérance, J. B. Vogt and J. I. Dickson

Département de génie métallurgique, École Polytechnique de Montréal, Montreal, Quebec, Canada

ABSTRACT

The study of a region of a grain of cyclically deformed 316 steel showed four different orientations of dipolar walls agreeing with a {210}, {111}, and with two different {211} or {311} orientations. All of these orientations have been predicted by a recent model for dipolar walls containing dislocations of two Burgers vectors.

KEYWORDS

Cyclic deformation, fatigue, dipolar walls, dislocation substructures, f.c.c. metals.

INTRODUCTION

In the last few years, considerable attention [e.g., 1-6] has been focused on dipolar walls, such as "labyrinth" walls, which form in f.c.c. metals during cyclic deformation involving dislocations of more than one Burgers vector. Dickson, Boutin and L'Espérance [6] have recently explained the two labyrinth wall orientations, namely perpendicular {210} and {100} unambiguously identified in copper [1] and in 316 stainless steel [5], in the case of [$\bar{1}$01] and [101] Burgers vectors (referred to here only by their direction). This explanation is based on the construction of a Taylor's network of dipole loops essentially edge in character, with alternating columns of dislocations of each Burgers vector. Fig. 1 schematizes this Taylor's network in the plane containing both orientations of edge dislocations corresponding to the long segments of the dipole loops. In this plane, the effective stacking directions are those which bisect the acute and obtuse angles between the edge dislocation orientations. The third effective stacking direction is that normal to this plane. A wall is obtained by limited stacking in one of these stacking directions. It is further argued [6] that the dipole loops are swept into walls by edge dislocations parallel to the long loop segments and that, therefore, stacking in the direction bisecting the acute angle between edge dislocation orientations is mechanistically favoured and the wall normal to this

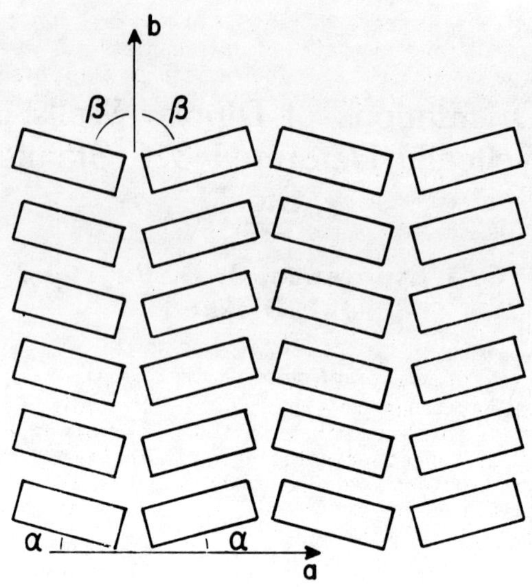

Fig. 1. Schematic representation of possible stacking sequence of dipolar loops of two Burgers vectors and slip systems in the plane containing the two edge dislocation orientations. The effective stacking directions are the two directions which bisect the acute and obtuse angles between the orientations of the long segments of dipole loops and the direction normal to the plane containing these two edge dislocation orientations.

TABLE 1

Burgers Vector 1	Slip Plane 1	Burgers Vector 2	Slip Plane 2	Wall 1	Wall 2	Wall 3
$[\bar{1}01]$	(111)	$[101]$	$(\bar{1}11)$	(100)	(012)	$(0\bar{2}1)$
$[\bar{1}01]$	$(1\bar{1}1)$	$[101]$	$(\bar{1}\bar{1}1)$	(100)	$(0\bar{1}2)$	(021)
$[\bar{1}01]$	(111)	$[101]$	$(\bar{1}\bar{1}1)$	(001)	(210)	$(\bar{1}20)$
$[\bar{1}01]$	$(1\bar{1}1)$	$[101]$	$(\bar{1}11)$	(001)	$(2\bar{1}0)$	(120)
$[\bar{1}01]$	$(1\bar{1}1)$	$[011]$	$(1\bar{1}1)$	$(\bar{1}12)$	$(1\bar{1}1)$	(110)
$[\bar{1}01]$	$(1\bar{1}1)$	$[011]$	$(\bar{1}\bar{1}1)$	$(\bar{1}30)$	$(\bar{3}\bar{1}5)$	(312)
$[\bar{1}01]$	(111)	$[011]$	$(\bar{1}\bar{1}1)$	(110)	$(\bar{1}13)$	$(3\bar{3}2)$
$[\bar{1}01]$	(111)	$[011]$	$(1\bar{1}1)$	$(\bar{3}10)$	(135)	$(\bar{1}\bar{3}2)$

Wall 2 contains the two edge dislocation orientation
Wall 3 should be mechanistically disfavoured

direction mechanistically disfavoured.

The walls geometrically predicted in f.c.c. metals by this model can be summarized by considering a pair of perpendicular and a pair of non-perpendicular Burgers vectors and the possible slip planes for the long loop segments [6]. These different predictions are summarized in Table 1.

From Table 1, it can be seen that labyrinth walls of both $\{100\}$ and $\{210\}$ orientations can be predicted for a pair of mutually perpendicular Burgers vectors. As discussed elsewhere [6], the actual wall orientations which should form will be influenced by such factors as their stability, their ability to accommodate excess dislocations of one Burgers vector or slip system and their versatility in accommodating different pairs of Burgers vectors. The prediction of this model appears to be in good agreement [6] with the available experimental evidence.

The present investigation was carried out in a region of a grain containing different wall orientations observed in a thin foil from 316 stainless steel deformed at a total strain amplitude of 0.5% at 22°C. This grain contained several different orientations of dipolar walls and dislocation of several Burgers vectors and the objective was to identify as many of these walls as possible and to compare them with the predictions obtained by constructing a Taylor's network of the general type indicated in Fig.1.

RESULTS AND DISCUSSION

The approximate orientations of the walls lettered in Fig. 2 were identified by micrographs taken for different beam directions. This orientation was then determined with greater precision by employing controlled tilting experiments (e.g., Fig. 3) about a beam direction when possible normal to the pole of this wall plane and noting the change in the projected wall thickness. Of the five sets of walls lettered, wall B as well as D were found to have a $(\bar{1}02)$ orientation; wall D' had an orientation that deviated somewhat from $(\bar{1}02)$. This $(\bar{1}02)$ wall orientation is that which is dominant in this region and a number of other walls were shown to be parallel. Wall A, near a grain boundary and parallel to two neighbouring walls, was found to have a $(\bar{1}11)$ orientation. Wall C was identified as either (112) or (113); wall E as either (121) or (131).

The pairs of slip systems which can produce these particular wall orientations, according to the previous proposals [6], were then deduced. The $(\bar{1}02)$ should thus involve [011] dislocations (with long segments) on $(\bar{1}\bar{1}1)$ and $[0\bar{1}1]$ dislocations on $(\bar{1}11]$. The $(\bar{1}11)$ wall could involve any pair combination of [110], $[0\bar{1}1]$ and [101] Burgers vectors on the $(\bar{1}11)$ slip plane. The (121) can be predicted from the Taylor's network involving [011] and [110] dislocations both on the $(1\bar{1}1)$ slip plane; the (131) wall from the same pairs of Burgers vectors but on the respective cross-slip planes $(\bar{1}\bar{1}1)$ and $(\bar{1}11)$. The (112) or (113) wall would both involve [011] and [101] dislocations. The slip plane would be $(\bar{1}\bar{1}1)$ for the (112) wall; the respective slip planes would be $(1\bar{1}1)$ and $(\bar{1}11)$ for the (113) wall. According to these predictions, the four different orientations of walls found in the same region of a grain would require only four Burgers vectors taken in the proper pair combinations, and three of these wall orientations, including the most dominant, involve the same [011] Burgers vector in the pair combinations. The only identified wall orientation which does not involve this Burgers vector consisted of wall A and two essentially parallel walls present in the immediate vicinity of a grain boundary. This

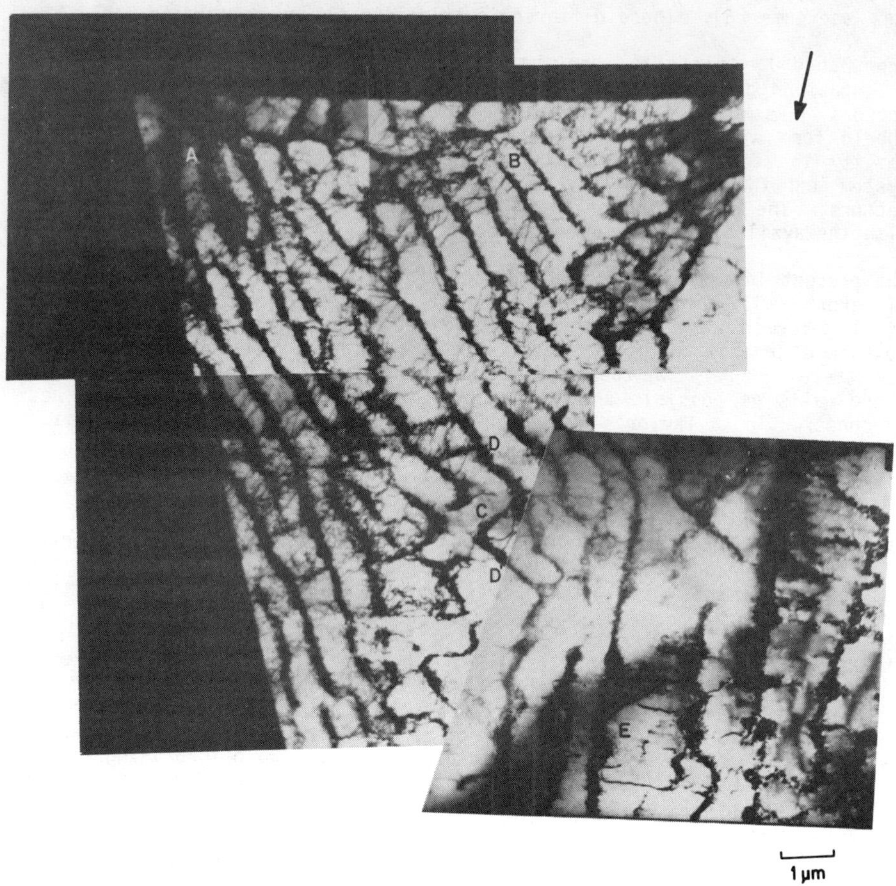

Fig. 2. Mosaic of grain containing walls A, B, C, D and E characterized in this study. The electron beam direction, B, was close to $[2\bar{1}1]$; g = $(11\bar{1})$.

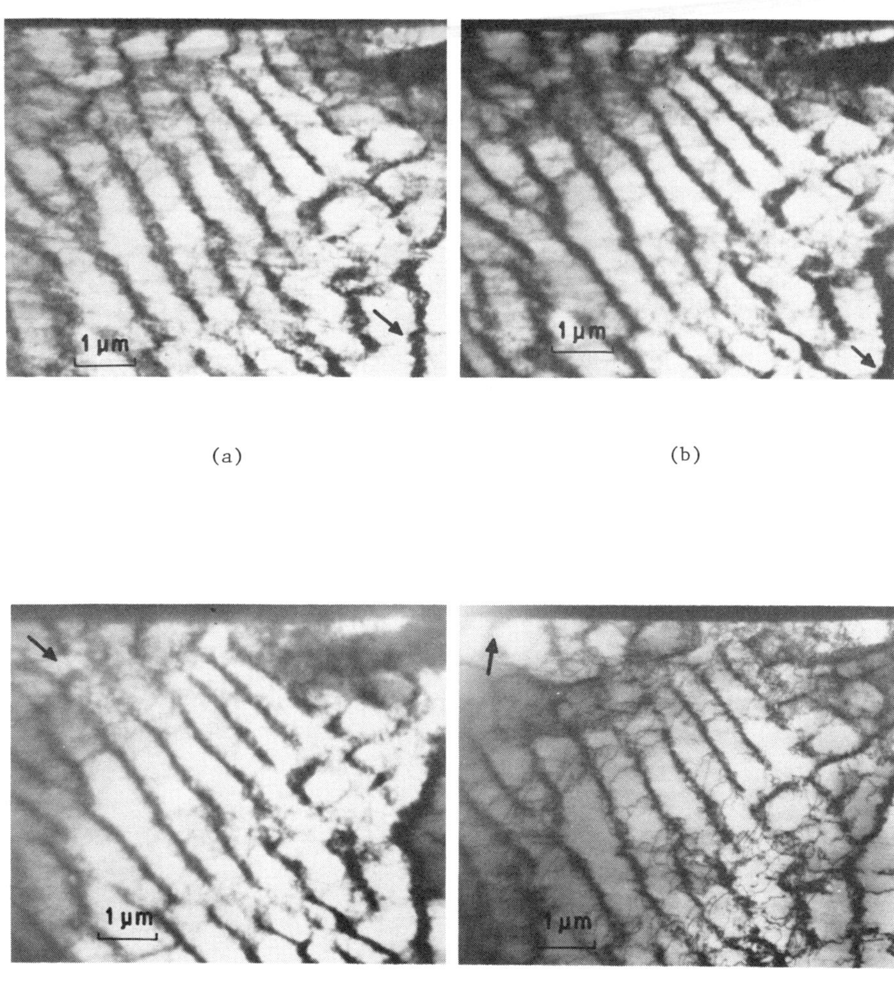

(c) (d)

Fig. 3. Observation of walls B at different electron beam directions to determine the orientation of the walls by considering changes in their projected widths : a) $\underline{B} \simeq [101]$ situated $\simeq 74°$ from $[\bar{1}02]$; $\underline{g} = (020)$; b) $\underline{B} = [201]$ situated $\simeq 92°$ from $[\bar{1}02]$; $\underline{g} = (0\bar{2}0)$; c) $\underline{B} \simeq [301]$ situated $\simeq 100°$ from $[\bar{1}02]$; $\underline{g} = (020)$; d) $\underline{B} \simeq [2\bar{1}\bar{1}]$ situated $\simeq 88°$ from $[\bar{1}02]$; $\underline{g} = (\bar{1}\bar{1}1)$.

aspect of the observations is logical, since only a limited number of Burgers vectors would normally be expected to operate within this small region.

Although the Burgers vectors of dislocations in this area were not completely analyzed, dislocations presumably screw in character were found parallel to the traces of the required Burgers vectors. Dislocations having a [011] Burgers vector were unambiguously identified and very frequent.

The slip planes involved in these predictions are of secondary importance [6], since it is generally considered that extensive cross-slip to annihilate the screw dislocations is a necessary condition for the formation of dipolar walls. The slip planes involved, however, should be important in determining the wall orientation when an excess number of dislocations of one slip system must be accommodated. The present observations thus suggest that [011] $(\bar{1}\bar{1}1)$ should be the dominant slip system in this region.

All of the wall orientations indicated from the present observations can be predicted from the construction of a Taylor's network of dipolar loops of two Burgers vectors and slip systems. The $\{210\}$ orientation has been previously reported for labyrinth walls [1]. The present observation demonstrates its occurrence in a non-labyrinth configuration.

CONCLUSIONS

The present observations thus provide evidence that dipolar walls of $\{210\}$, $\{111\}$ as well as $\{112\}$ or $\{113\}$ orientations do form during the cyclic deformation of this f.c.c. metal and thus provide further experimental support for the proposals of Dickson, Boutin and L'Espérance [6] concerning the dipolar walls expected for pairs of Burgers vectors.

ACKNOWLEDGMENTS

Financial support from the NSERC (Canada) and FCAC (Quebec) programs is gratefully acknowledged. The sample studied was provided by Marie Bernard and Bui-Quoc Thang.

REFERENCES

1. F. Ackermann, L.P. Kubin, J. Lepinoux and H. Mughrabi, Acta metall., 32, 715 (1984).
2. P. Charsley, Mater.Sci.Eng. 47, 181 (1981).
3. N.Y. Yin and A.T. Winter, Acta metall., 32, 1173 (1984).
4. K. Mecke and C. Blochwitz, Crystal Res. and Technol., 17, 743 (1982).
5. G. L'Espérance, J.B. Vogt and J.I. Dickson, submitted to Mater.Sci. Eng.
6. J.I. Dickson, J. Boutin and G. L'Espérance, submitted to Acta.metall.

A DSC and TEM Study of the Cyclic and Monotonic Hardening of Al-5% Mg

J. H. Driver* and J. M. Papazian**

**Department de Materiaux, Ecole Nationale Superieure des Mines, 158 Cours Fauriel, 42023 Saint Etienne, Cedex, France*
***Research and Development Center, Grumman Aerospace Corporation, Bethpage, NY 11714, USA*

ABSTRACT

Transmission electron microscopy (TEM) and differential scanning calorimetry (DSC) have been used to characterize the cyclic and monotonic hardening mechanisms in Al-5Mg. Cyclic deformation to saturation produced a high density of dislocation loops which gave rise to an endothermic reaction in the DSC scans. This reaction occured at ~150°C with an enthalpy ΔH_r = 0.2-0.5 J/g and was shown to be due to annihilation of the dislocation loops. Additional experiments showed that the loops were not present in cold rolled or cold drawn material, but could be generated by fatigue subsequent to the cold work.

KEY WORDS

Fatigue; aluminum alloys; TEM; DSC; cyclic hardening, dislocation loops; Al-5%Mg; Al-5%Mg-0.5%Ag.

INTRODUCTION

Al-Mg alloys are known to harden substantially by plastic deformation, either by conventional cold working practices or by fatigue. Transmission electron microscopy (TEM) shows that monotonic hardening occurs by the buildup of a high density of tangled dislocations whereas cyclic hardening is associated with the formation of a high dislocation loop density [1,2]. It has also been shown that differential scanning calorimetry (DSC) can be successfully used for precipitate characterization in fatigued commerical aluminum alloys [3]. The aim of this study is to characterize the strenghtening mechanisms in cold-worked and fatigued Al-Mg by DSC and TEM and, in particular, to show that the DSC thermograms can be correlated with the dislocation substructures developed by different deformation modes.

EXPERIMENTAL

The high purity Al-5 Mg alloy containing 5.05 wt% Mg, 0.04 wt% Fe and 0.04 wt% Si used in this work was either in the annealed state (recrystallization anneal of 1 h at 460°C and water quenched), or cold worked as follows: (a) cold drawn 30% reduction in area to a true strain ε = 0.26 or (b) cold rolled 87% to a true strain ε = 2. The cylindrical fatigue specimens of 14 mm gage length and 6 mm diameter were cycled in symmetrical tension-compression at constant plastic strain amplitudes, $\Delta\varepsilon_p/2$, in the range $5x10^{-5}$ to $2x10^{-2}$ [1]. Tests were conducted to either macroscopic crack intiation (N_f), or to some life fraction, and the specimens sectionned perpendicular to the load axis for either TEM (disc technique) or DSC. The DSC measurements were made on 6 mm diameter discs at a heating rate of 10°C/min using pure aluminum samples for reference material as previously described [3].

RESULTS

Cyclic Saturation of Annealed Al-5Mg

The saturation cyclic stress-strain curve, $\sigma_s(\varepsilon_p)$, of the Al-5Mg alloy in the annealed state exhibits a well defined plateau at ~270 MPa for plastic strain amplitudes between 10^{-4} and $2x10^{-3}$ [1]. As discussed previously [1], the high fatigue hardening capacity and the saturation plateau behavior can both be attributed to the formation of a high density of homogeneously distributed fine dislocation loops. Figure 1(a) shows a typical example of the dislocation microstructure in a specimen cycled to failure at a plastic strain of 10^{-4}. The observed loops have dimensions ranging from 5 to 100 nm in length (typically 25 nm) and from 10 to 30 nm in height with a density of the order of $10^{15}/cm^3$. At higher plastic strain amplitudes the matrix loop density does not appear to increase signficantly but intense slip bands are developed in the bulk to accomodate the higher strains.

Figure 2 shows the DSC plots of Al-5Mg in different conditions: (a) before fatigue, (b) fatigued at $\Delta\varepsilon_p/2 = \pm1.1x10^{-4}$ to failure (same specimen as in

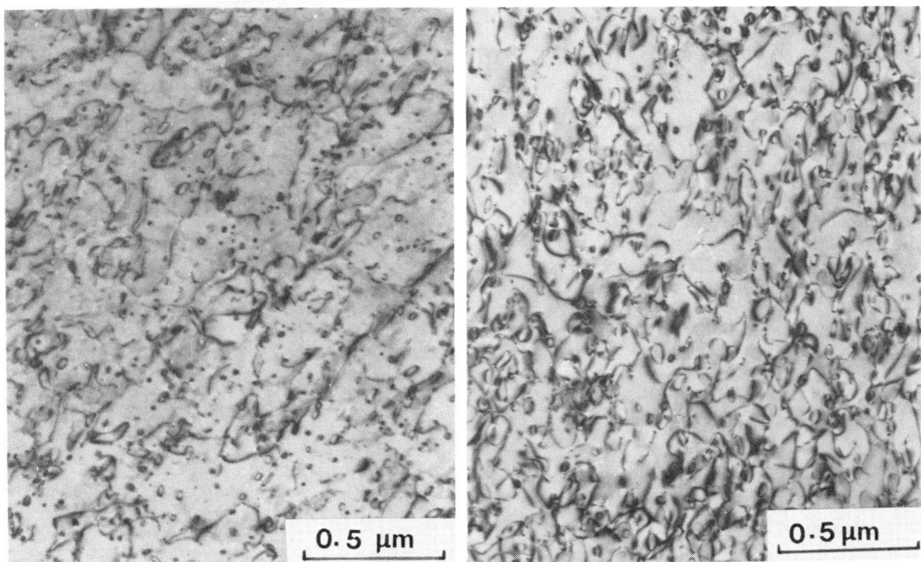

(a) as fatigued (b) fatigued plus 15 min at $145^{o}C$

Fig. 1. Dislocation structures after fatigue at $\Delta\varepsilon_p/2=\pm1x10^{-4}$ to failure.

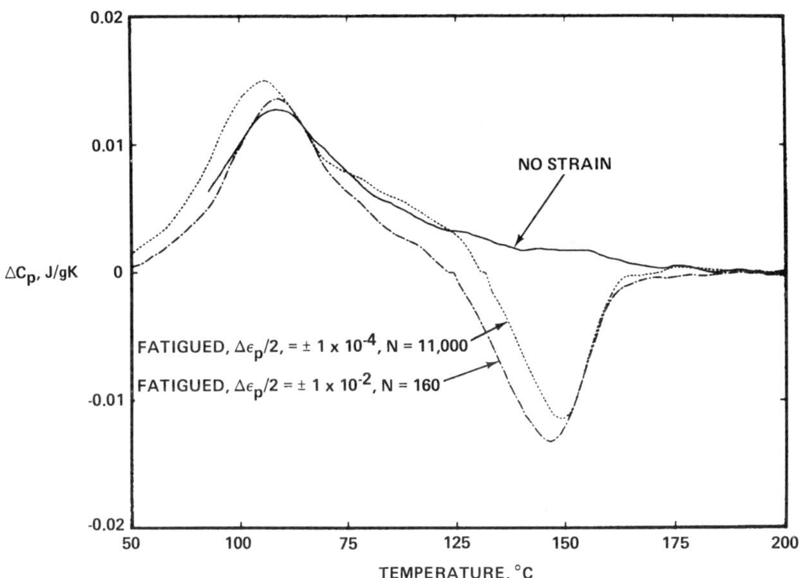

Fig. 2 DSC results from Al-5Mg: before fatigue, fatigued at $\Delta\varepsilon_p/2$ = $\pm1x10^{-4}$ for 11,000 cycles and fatigued at $\Delta\varepsilon_p/2$ = $\pm1x10^{-2}$ for 160 cycles.

Fig. 2), and (c) fatigued at $\Delta\varepsilon_p/2 = \pm 1.1x10^{-2}$ to failure. All these conditions give rise to low temperature endotherms at ~80^oC with reaction enthalpies $\Delta H_r \simeq 0.5$ J/g, which are thought to be due to GP zone disolution [4,5].

Figure 2 also shows the presence of a hitherto unreported exotherm at ~150^oC ($\Delta H_r \simeq 0.5$ J/g) in the fatigued samples. The 150^oC exotherm was found to be characteristic of the cyclically hardened state, at saturation, for all plastic strain amplitudes, and was therefore studied in some detail. A 150^oC exothermic reaction could be due to a variety of causes such as precipitation on fatigue-generated dislocations, or some dislocation recovery process. To determine to origin of the exotherm, TEM was carried out on the sample fatigued at $\Delta\varepsilon_p/2 = \pm 1.1x10^{-4}$ and then aged 15 min at 145^oC; the resulting microstructure is shown in Fig. 1(b). There is no sign of precipitation but clear evidence for a substantial decrease in the density of the fine loops. Only a relatively low density of larger loops, together with the original dislocation debris, remains. A further feature is the rounding off, at 145^oC, of the jogged dislocations which thread through the fine loops matrice structure. It appears therefore that the 150^oC peak is due to the coalescence and annihilation of fatigue generated loops. This interpretation is also consistent with microhardness results; after 15 min at 150^oC the Vickers microhardness of the fatigued samples decreases from 105 (as fatigued) to 85. The annealed value was 70.

Cyclic Hardening

To correlate the DSC results with the development of the dislocation substructure during the cyclic hardening phase, Al-5Mg specimens were fatigued at $\Delta\varepsilon_p/2 = \pm 3x10^{-4}$ for different numbers of cycles N (or equivalent different cummulative plastic strains ε_p cum = 2 $\Delta\varepsilon_p$ N). This material showed slow cyclic hardening behavior; saturation requires about 2000 cycles. Figure 3 gives the DSC plots of specimens cycled 15, 150, and 1500 cycles (ε_p cum values of 0.018, 0.18, and 1.8 respectively). During hardening, there is a tendency for the peak temperature of the low temperature endotherm to shift to slightly lower temperatures indicating that fatigue may reduce the zone size. Previous work in Al-Cu has shown that smaller GP zones dissolve at lower temperatures [6]. The important feature of the thermograms is the complete absence of the 150^oC exotherm in the two specimens fatigued for relatively low numbers of cycles, and the

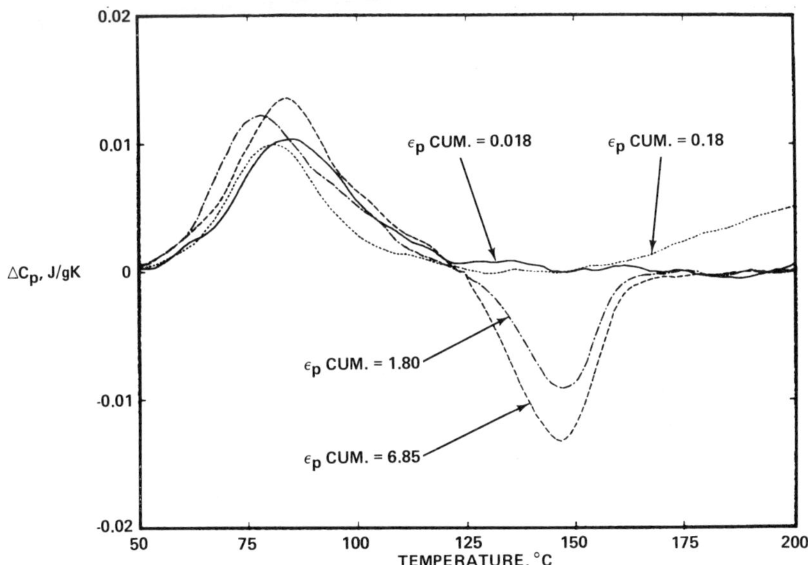

Fig. 3. DSC results from Al-5Mg fatigued at $\Delta\varepsilon_p/2 = \pm 3 \times 10^{-4}$ to 15, 150 and 1500 cycles or fatigured at $\Delta\varepsilon_p/2 = \pm 10^{-2}$ to failure.

appearance of the exotherm in the specimen cycled close to saturation. The dislocation microstructures were examined by TEM. After 15 cycles long cusped dislocation are formed together with some loops, after 150 cycles a higher dislocation density is developed but the loop density remains relatively low. Only after 1500 cycles can one find the high loop density microstructure typical of cyclic saturation. These results confirm the suggestion that the 150°C exotherm is due to loop coalescence, and also indicate that DSC can be a sensitive technique for diagnosing the degree of fatigue hardening in Al-Mg alloys.

Cold-Worked Material

Additional DSC experiments were performed on cold worked material. The 150°C exotherm was not present after monotonic plastic deformation to equivalent total strains, but appeared when these samples were fatigued after cold work. This indicates that fatigue hardening of the cold-worked alloy is also due to fine loop generation, a result difficult to obtain by conventional TEM because of the highly dislocated microstructure.

DISCUSSION AND CONCLUSIONS

The results on the fatigued Al-5Mg alloy show that at cyclic saturation a fine dislocation loop structure is developed which is associated with a 150°C exotherm in the DSC scans. The sharpness and reproducibility of the exotherm is quite remarkable; previous calorimetric studies of dislocation recovery after monotonic deformation tend to show slow heat evolutions over a wide temperature range [7]. The present results, using plastic strain controlled fatigue tests on simple alloys, are due to the development of a remarkably uniform loop distribution which then appears to be annihilated by coalescence within a relatively narrow temperature range.

There is surprisingly little recent data on the stored energy in fatigued metals. Some old data on low stress fatigue of copper [8] gives ~0.13 J/g released in the temperature range 300 to 400°C (0.42-0.5 T_m). Expressed in J/mole, this value (8 J/mole) is very close to that of the 150°C (0.45 T_m) exotherm in fatigued Al-Mg (0.3-0.5 J/g or ~8-13 J/mole). It is well known that fatigue in copper develops a loop matrix structure, so the correlation is reasonable.

The essential conclusions of this study can be summarized as follows:

(i) DSC scans of Al-5Mg fatigued to cyclic saturation reveal a 150°C exotherm which is shown by TEM to be related to the annhilation of fine dislocation loops.

(ii) This exotherm occurs only after fatigue of these alloys in different metallurgical conditions but not after monotonic deformation to equivalent strains.

REFERENCES

1. J.H. Driver and P. Rieux, Mat. Sci. Eng., 68, 35 (1984).
2. S.I. Kwun and M.E. Fine, Scripta Met. 18 981 (1984)
3. J.M. Papazian, R.J. Delasi and P.N. Adler, Metall. Trans. 11A, 135 (1980).
4. T. Sato, Y. Kojima and T. Takahashi, Metall. Trans. 13A, 1373 (1982).
5. K. Osamura and T. Ogura, Metall. Trans. 15A, 835 (1984).
6. J.M. Papazian, Metall. Trans. 13A (1982).
7. M.B. Bever, D.B. Holt and A.L. Tichener in Progress in Materials Science, 17, Ed. B. Chalmers, J.W. Christian. T.B. Masalski, (1973).
8. L.M. Clarebrough, M.E. Hargreaves, G.W. West and A.K. Head, Proc. Roy. Soc., 242A, 160 (1957).

Etude du Comportement Dynamique des Dislocations au Cours de la Sollicitation de Fatigue par Mesures d'Attenuation d'Ondes Ultrasonores

A. Vincent*, A. Hamel, J. Chicois** et R. Fougères****

**Laboratoire de Traitement du Signal et Ultrasons, Bât. 502, INSA de Lyon, 69621 Villeurbanne Cédex, France*
***Groupes d'Etudes de Métallurgie Physique et Physique de Matériaux, U.A. C.N.R.S. N° 341, Bât 502, INSA de Lyon, 69621 Villeurbanne, Cédex, France*

RESUME

Dans le cas de l'aluminium pur polycristallin, nous avons mesuré les variations d'atténuation d'ondes ultrasonores au cours de sollicitations cycliques. Les résultats obtenus sont interprétés en terme de comportement dynamique des dislocations. Les résultats obtenus montrent l'intérêt d'une telle approche.

MOTS-CLES

Fatigue ; atténuation ultrasonore ; dynamique des dislocations.

INTRODUCTION

Ces dernières années, de nombreux travaux ont été réalisés dans le domaine de la fatigue des métaux à l'état mono ou polycristallin. Des articles de revue successifs font périodiquement le point sur ces travaux /1/,/2/,/3/,/4/,/5/. Les études effectuées présentent deux types d'approche. La première cherche à définir sur les boucles de fatigue contrainte-déformation des grandeurs mécaniques caractéristiques : contrainte maximale du cycle, limite élastique et, récemment, l'énergie relative dissipée /6/. La deuxième caractérise par MET classique les structures de dislocations obtenues : bandes de glissement permanentes, veines et canaux, structures labyrinthes ou encore cellules de dislocations. A partir des résultats expérimentaux fournis par ces deux approches, on tente, à l'aide de modèles de dislocations, de rendre compte des propriétés mécaniques définies sur les cycles. Une des limites d'une telle approche réside dans le fait que ces propriétés mécaniques sont obtenues au cours de la fatigue elle-même alors que les structures de dislocations sont observées sur des états fatigués, mais non contraints. Pour palier à cette difficulté, des observations ont été réalisées sur des états fatigués irradiés aux neutrons rapides et pour lesquels les dislocations restent bloquées dans la position atteinte au cours de la sollicitation /7/. Des expériences de fatigue in situ ont aussi cherché à obtenir des informations sur la dynamique des dislocations /8/. Ces expériences sont irremplaçables pour identifier les mécanismes microscopiques de la fatigue. Elles ne permettent cependant pas d'obtenir des grandeurs caractéristiques du comportement moyen des dislocations pendant la sollicitation de fatigue.

L'objectif du présent travail est de chercher à obtenir de telles données par des mesures d'atténuation d'ondes ultrasonores réalisées au cours des cycles de fatigue. En effet, il a été montré que, dans l'aluminium pur, l'atténuation ultrasonore était liée à l'environnement des dislocations en défauts ponctuels, ainsi qu'à la densité de dislocations mobiles /9/.

I. CONDITIONS EXPERIMENTALES

Les essais de fatigue ont été réalisés sur un aluminium pur (99,999 %) à l'état polycristallin. Les échantillons sont prélevés dans des barres fortement écrouies par étirage. Le taux de déformation, défini par la réduction de section des barres avant et après étirage, est d'environ 70 %. L'usinage conduit à des éprouvettes de fatigue de longueur 70 mm, à extrémités filetées et dont la partie utile est un cylindre de 40 mm de long et de 9 mm de diamètre. Après étirage et usinage, les éprouvettes sont traitées pendant 2 heures à 450°C, puis refroidies lentement au four. La taille de grain ainsi obtenue est de l'ordre de 2 mm.

Les essais de fatigue sont effectués en traction-compression sur une machine d'essai spécifique à cinématique entièrement élastique et permettant ainsi d'obtenir des niveaux de déformation très faibles. La déformation ε de l'éprouvette est mesurée par des extensomètres placés sur sa partie utile : aux basses amplitudes ($10^{-6}<\varepsilon<10^{-4}$), on utilise quatre jauges à fil résistant de 1000 Ω chacune et montées en pont complet ; aux fortes amplitudes ($10^{-4}<\varepsilon<10^{-2}$), nous avons utilisé un extensomètre MTS dont la base de mesure est de 20 mm. La machine sollicite l'éprouvette à déformation plastique imposée $\Delta\varepsilon_p$ grâce au pilotage d'un mini-ordinateur DIGITAL PDP11. La fréquence de l'essai est de 0,1 Hz. En fonction du temps, la déformation varie linéairement. Tous les essais ont été effectués à la température ambiante. Une étude au microscope électronique en transmission a montré qu'une structure cellulaire de dislocations était formée dès les premiers cycles /10/.

Par ailleurs, les ondes utilisées pour les mesures ultrasonores ont une fréquence égale à 15 MHz. Elles se propagent dans une direction parallèle à l'axe de l'échantillon selon un mode de transmission de type echo /11/. On mesure la variation d'atténuation $\Delta\alpha$ entre l'état sollicité en fatigue et l'état initial de l'échantillon.

Les propriétés mécaniques et les mesures ultrasonores sont déterminées pour chaque cycle. Nous avons étudié l'effet du nombre de cycles et de l'amplitude de déformation plastique imposée. Dans ce dernier cas, nous utilisons une même éprouvette pour étudier des valeurs d'amplitude croissante, le rapport entre deux déformations successives étant au moins égal à 5. Nous avons vérifié que, en-dehors des premiers cycles, les résultats obtenus sont sensiblement les mêmes que ceux d'une éprouvette vierge.

II. RESULTATS EXPERIMENTAUX

Nous avons représenté pour chaque cycle l'évolution de la variation d'atténuation en fonction de la déformation ε. En outre, sur chaque figure, nous avons également reporté, pour l'état saturé mécaniquement*, le cycle contrainte-déformation $\sigma=f(\varepsilon)$ correspondant. Les figures 1, 2 et 3 représentent respectivement les résultats obtenus pour des amplitudes de déformations plastiques égales à 3, 200 et 1500.10^{-6} qui illustrent l'ensemble des types de cycles $\Delta\alpha=f(\varepsilon)$.

---

*Nous considérons que la saturation mécanique est atteinte pour un nombre de cycles N lorsque, pour le (N+50)^{e} cycle, la variation de contrainte maximale entre N et N+50 est inférieure au 1/100e de l'augmentation de contrainte maximale obtenue pour les N premiers cycles.

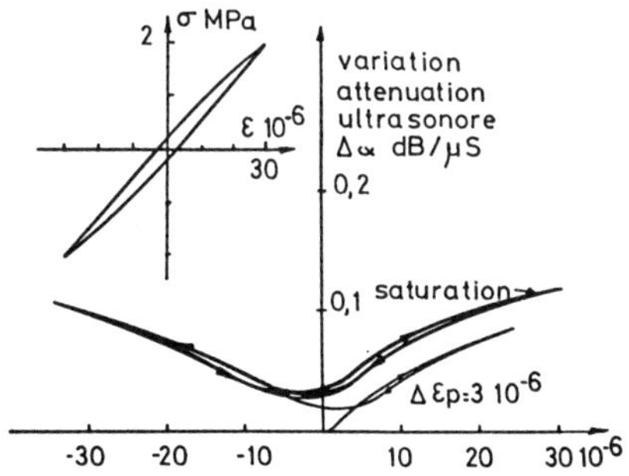

Fig. 1. Variation de l'atténuation (Δα) avec la défomation de fatigue ε pendant le durcissement cyclique. Cycle contrainte-déformation (σ=f(ε)) à saturation. $\Delta\varepsilon_p=3\times10^{-6}$.

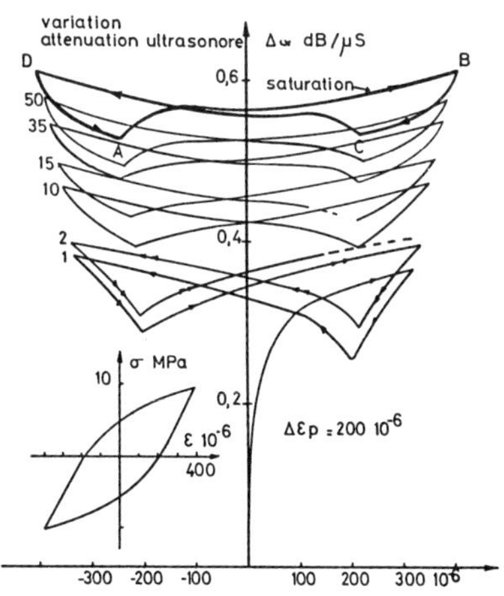

Fig. 2. Idem Fig. 1. $\Delta\varepsilon_p=200\times10^{-6}$.

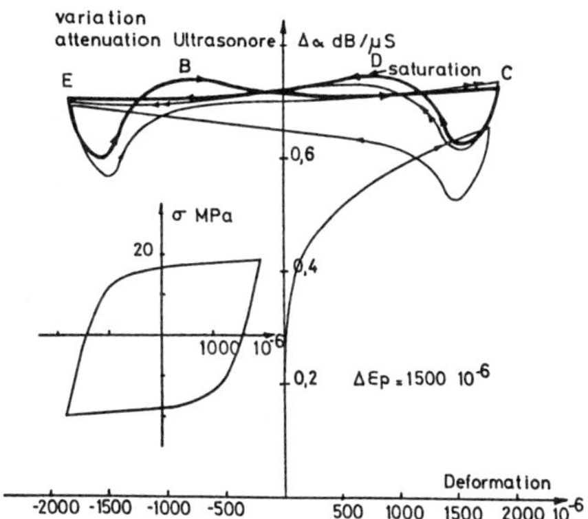

Fig. 3. Idem Fig. 1. $\Delta\varepsilon_p=1500\times10^{-6}$.

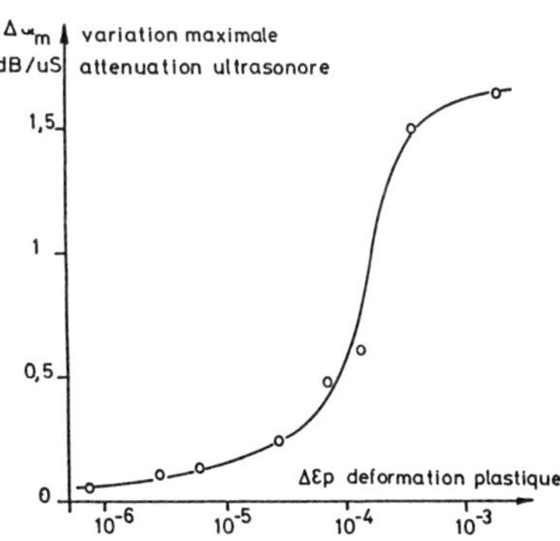

Fig. 4. Variation de l'atténuation maximale ($\Delta\alpha_{max}$) avec l'amplitude de déformation plastique imposée $\Delta\varepsilon_p$. Cas de la saturation.

Pour les très faibles amplitudes (fig. 1), on constate que les variations d'atténuation avec la déformation présentent un comportement quasi réversible, la valeur de l'atténuation étant à peu près la même à la charge ou à la décharge pour une déformation donnée.

Aux amplitudes plus fortes, le comportement atténuation-déformation devient irréversible.

Dans le domaine $5\times10^{-6}<\Delta\varepsilon_p<800\times10^{-6}$, les cycles $\Delta\alpha=f(\varepsilon)$ présentent une augmentation continue de l'atténuation dans le domaine plastique (partie AB en traction et CD en compression de la figure 2 pour laquelle $\Delta\varepsilon_p=200\times10^{-6}$). En outre, le niveau de $\Delta\alpha$ croît en fonction du nombre de cycles et se stabilise pour un nombre de cycles N (de l'ordre de quelques dizaines) sensiblement égal à celui de la saturation mécanique.

Au contraire, pour $\Delta\varepsilon_p>800\times10^{-6}$ (figure 3 pour laquelle $\Delta\varepsilon_p=1500\times10^{-6}$), il apparaît sur les cycles stabilisés mécaniquement une saturation de $\Delta\alpha$ particulièrement marquée (partie BC en traction et DE en compression).

Enfin, pour l'état saturé mécaniquement, la courbe de la figure 4 représente l'évolution de la variation d'atténuation maximale $\Delta\alpha_{max}$, en fonction de l'amplitude de déformation plastique imposée. $\Delta\alpha_{max}$ est définie sur un cycle comme la différence entre le niveau maximal de l'atténuation au cours du cycle saturé et le niveau observé sur l'état vierge de l'échantillon.

III. DISCUSSION DES RESULTATS

III.1. Bases de l'interprétation des mesures ultrasonores

Des travaux antérieurs /9/, effectués sur l'aluminium soumis à des sollicitations statiques, ont montré que l'atténuation ultrasonore α dépend de la densité de dislocations Λ et de leur mobilité. Cette dernière dépend notamment de l'environnement des dislocations en défauts ponctuels et, par conséquent, de la longueur libre ℓ des segments de dislocations qui vibrent sous l'effet des ondes ultrasonores. En première approximation, on peut considérer le modèle de la corde vibrante /12/, qui prévoit une atténuation α proportionnelle à $\Lambda\ell^4$ (1). Ainsi, une variation d'atténuation $\Delta\alpha$ correspond à une variation de ℓ et/ou de Λ. Plusieurs cas sont à considérer selon le caractère de réversibilité plus ou moins prononcé des cycles $\Delta\alpha=f(\varepsilon)$.

. Cycles $\Delta\alpha=f(\varepsilon)$ quasi réversibles :

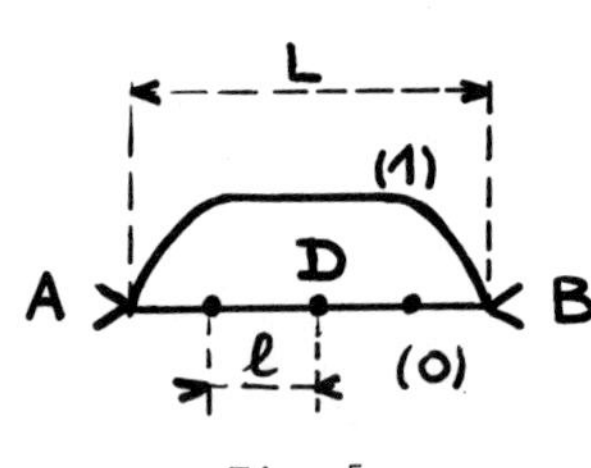

Fig. 5.

$\Delta\alpha$ provient d'un dépiégeage des dislocations, à partir des défauts ponctuels ayant migré vers ces dislocations avant le début de l'expérience, selon un schéma représenté par la figure 5. On a représenté un segment de dislocation compris entre deux ancrages "durs" A et B et piégé par des défauts ponctuels D (situation 0). Sous l'effet de la contrainte appliquée, par exemple en traction, la dislocation peut passer de la position (0) à la position (1). Ceci correspond à une augmentation de ℓ ($\ell\rightarrow L$ (fig. 5)), qui entraîne une augmentation d'atténuation $\Delta\alpha$ selon la relation (1). Un retour de la déformation à valeur nulle entraîne un repiégeage de la dislocation (passage de la position 1 à la position 0).

Si l'environnement en défauts ponctuels reste le même lorsque la déformation croît et si les segments de dislocations présentent la même longueur L, la variation d'atténuation $\Delta\alpha$ est proportionnelle à la densité de dislocations mobiles Λ, c'est-à-dire à la densité de dislocations désancrées.

. Cycles $\Delta\alpha=f(\varepsilon)$ nettement irréversibles

L'irréversibilité indique que les déplacements (0)→(1)→(0) du schéma de la figure 5 ne se produisent plus. Les variations de α doivent être interprétées essentiellement en terme de variation de la densité des dislocations. A environnement en défauts ponctuels identique, une augmentation de α correspond à une création de dislocations, au contraire une diminution doit être interprétée comme une annihilation. Cette évolution n'est cependant valable que lorsque la densité de dislocations Λ n'est pas trop élevée, de telle sorte que la réduction de L ne l'emporte pas sur l'augmentation de Λ.

III.2. Interprétation des résultats

A partir de ces considérations, il est possible d'interpréter les résultats expérimentaux présentés ci-dessus de la manière suivante :

. aux très basses amplitudes de déformation (fig. 1), les cycles $\Delta\alpha=f(\varepsilon)$ sont quasi réversibles. Un tel comportement peut être décrit par le schéma de la figure 5, c'est-à-dire un dépiégeage des dislocations à partir des défauts ponctuels. On peut penser que ces défauts ponctuels sont constitués par les éléments résiduels dans l'aluminium. C'est le cas, par exemple, du magnésium qui présente une forte énergie d'interaction avec les dislocations. Compte tenu de la faible valeur de leur énergie d'interaction, les lacunes ne doivent pas jouer un rôle important.

Si l'on suppose les ancrages durs fixes, la composante non élastique de la déformation (fig. 1) pourrait être due à la courbure des dislocations entre ces obstacles. L'ouverture du cycle $\sigma=f(\varepsilon)$ à $\sigma=0$ s'expliquerait alors par un retour retardé de la dislocation vers sa position d'équilibre (situation (0) fig. 5) sous l'effet d'un frottement visqueux dû aux défauts ponctuels. La déformation anélastique ε_a, calculée pour la contrainte maximale du cycle, représenterait alors une borne supérieure de l'ouverture du cycle.

Avec les hypothèses prises habituellement sur la courbure des dislocations (parabolique), la valeur de la tension de ligne ($1/2Gb^2$, avec G le module d'élasticité, b le vecteur de Burgers) et la longueur des arcs de dislocations ($L=1/\sqrt{\Lambda}$), on montre aisément que $\varepsilon_a=(f^2.\sigma)/6G$ (2), avec f le facteur de Schmidt moyen. Si, de plus, on pose $\sigma=E\varepsilon$ (avec E module d'Young), on a $\varepsilon_a=(f^2.\varepsilon)/2$ (3). Une déformation $\Delta\varepsilon_p=3\times10^{-6}$ correspond à une déformation totale $\varepsilon=30\times10^{-6}$, soit un rapport $\Delta\varepsilon_p/\varepsilon=1/10$. En prenant $\varepsilon=30\times10^{-6}$ et $f=1/3$, on trouve, avec la relation (3), une valeur $\varepsilon_a/\varepsilon\simeq1/20$, soit un rapport deux fois plus petit. Ceci signifie que les ancrages durs ne sont pas fixes et qu'au cours de la fatigue à $\Delta\varepsilon_p=3\times10^{-6}$, il existe effectivement une déformation microplastique correspondant à un mouvement de va et vient de ces ancrages. En outre, lors du passage à déformation nulle, le piégeage des dislocations par les défauts ponctuels redevient le même, ce qui explique le minimum d'atténuation observé pour $\varepsilon=0$ (fig. 1).

. aux amplitudes de déformation intermédiaires (fig. 2), l'irréversibilité des cycles $\Delta\alpha=f(\varepsilon)$ doit être interprétée en terme de création de nouvelles dislocations. Dans ce domaine, en effet, on peut supposer que les dislocations sont toutes dépiégées, qu'elles parcourent de grandes distances et ne reviennent plus à leur position initiale pour $\varepsilon=0$: dans ces conditions, l'effet des défauts ponctuels est négligeable. Ainsi, il y a augmentation de la densité de dislocations sur un cycle donné, mais aussi augmentation de cette densité en fonction du nombre de cycles jusqu'à stabilisation.

. aux fortes amplitudes (fig. 3), la saturation de l'atténuation en fonction de la déformation du cycle peut correspondre soit à une densité de dislocations constante, soit à une augmentation de Λ, dont les effets seraient contrebalancée par une réduction de L. Cette dernière hypothèse semble peu probable, car, selon la relation (1), L devrait varier comme $\Lambda^{-1/4}$. En fait, L varie généralement comme $\Lambda^{-1/2}$. De plus, la première hypothèse d'une densité de dislocations sensiblement constante pour $\Delta\varepsilon_p > 800 \times 10^{-6}$ est tout à fait en accord avec les interprétations des résultats d'émission acoustique obtenues dans les mêmes conditions expérimentales /13/.

CONCLUSION

Cette étude a montré que les variations d'atténuation d'ondes ultrasonores mesurées au cours de cycles de fatigue étaient susceptibles d'apporter des renseignements qualitatifs intéressants concernant le comportement dynamique des dislocations. L'obtention de résultats quantitatifs demande à ce que soient mesurées également les variations de vitesse ultrasonore, comme l'ont montré les travaux antérieurs en sollicitation statique /9/. De plus, pour l'interprétation des variations d'atténuation, il faudrait prendre en compte l'effet des désorientations cristallines dues à la présence de la structure cellulaire des dislocations.

BIBLIOGRAPHIE

1. D. Kuhlmann-Wilsdorf and C. Laird, *Mat. Sci. Engng.* 27, 137 (1977), Ibid 37, 111 (1979).
2. D. Kuhlmann-Wilsdorf, *Mat. Sci. Engng.* 39, 127 (1979), Ibid 39, 231 (1979).
3. H. Mughrabi, *Strength of Metals and Alloys, Proc. of ICSMA5*, Aachen (1979) (edited by P. Haasen, V. Gerold and G. Kostorz), 1615, Pergamon Press, Oxford (1979).
4. H. Mughrabi and R. Wang, *Proc. of 2nd RISØ Inter. Symp. on Metal. and Mat. Sci.*, 87 (1981).
5. T. Magnin, J. Driver, J. Lepinoux et L. P. Kubin, *Rev. Phys. Appl.* 19, 467 (1984), Ibid 19, 483 (1984).
6. J. Chicois, C. Esnouf, G. Fantozzi, A. Vincent et R. Fougères, *Journal de Physique*, Supplément n° 12, Tome 44, C9.777 (1983).
7. H. Mughrabi, *Proc. Fourth Int. Conf. on Continum Models of Discrete Systems*, Stockholm (edited by O. Burtin and R. K. T. Hsieh), 241, North Holland, Amsterdam (1981).
8. J. Lepinoux, Thèse 3e Cycle, Université de Poitiers, n° 884 (1983).
9. A. Vincent et J. Perez, *Phil. Mag.* A40, 3, 377 (1979).
10. G. Guichon, J. Chicois, C. Esnouf and R. Fougères, *Fatigue 84, Proceedings 2nd Int. Conf. on Fatigue and Fatigue Thresholds*, Fatigue, Birmingham, U. K., Sept. (J. C. Beevers editor), 31 (1984).
11. A. Vincent, J.-L. Bouvier-Volaille et P. Fleischmann, *J. Phys.* E15, 765 (1982).
12. A. Granato and K. Lucke, *J. Appl. Phys.*, 27, 6, 583 (1956).
13. A. Slimani, J. Chicois, R. Fougères et P. Fleischmann, *Proceeding* du présent congrès.

On the Cyclic Deformation Behavior of Notched Specimens

H. J. Böhmer, D. Eifler and E. Macherauch

Institut für Werkstoffkunde I, Universität (TH) Karlsruhe, Federal Republic of Germany

ABSTRACT

Cyclic deformation curves were determined both in companion specimen tests as well as by the application of Neuber's rule to direct measured nominal stress - total strain - hysteresisloops. Different cyclic deformation behavior was found for notched and companion specimens. The consequences of multiaxiality and inhomogeneity of the stress state as well as of the influence of residual stresses on the cyclic response of the notch root can be assessed.

KEYWORDS

Fatigue, smooth and notched specimens, multiaxiality, cyclic deformation, Neuber's formula

INTRODUCTION

The cyclic deformation behavior of notched parts is of great significance both in fundamental as well as in applied respects. Whereas notch strains very often can be measured by strain gages, great difficulties occur in the exact determination of notch stresses. Notch effects usually are described by stress or strain concentration factors K_σ and K_ε, which represent the ratio between effective and nominal stress or strain values. If elastic deformation conditions are fulfilled, K_σ and K_ε are equal to the theoretical elastic stress concentration factor K_t. However, elastic-plastic notch root deformations require the application of approximation formulas for the calculation of stress resp. strain concentration factors [1]-[3]. The actual relationship between notch stress and appertaining notch strain is influenced by fatigue phenomena like cyclic softening or hardening and notch effects due to the multiaxiality of the stress state and the development of residual stresses in the net section of the notched specimens. Proceeding from Neuber's formula the present paper describes a procedure by which information about the effective stress-strain-history at the notch root can be obtained from measured nominal stresses and total notch strains. Furthermore, the limits for the simulation of the cyclic deformation response of the notch root area by smooth companion specimens will be discussed.

MATERIAL AND TEST CONDITIONS

Load controlled push-pull tests with a frequency of 5 Hz have been performed in servohydraulic testing machines with smooth and notched specimens of normalized SAE 1045. Additionally strain controlled tests with smooth companion specimens, submitted to the course of the measured notch strain, have been carried out. Smooth and two types of notched specimens with different notch root radii and K_t-values of 2.2 and 3.0 have been investigated. The specimens were annealed after machining for releasing residual stresses. For selected cycles hysteresisloops were registrated. To measure notch strains in loading direction four strain gages with a grid lengths of 0.2 mm were applicated at the notch root as shown in Fig. 1. In the smooth specimens strains were determined by a capacitive extensometer. However, in the case of companion specimen tests strain gages with grid lengths of 6 mm were used. From the measured data the development of the cyclic deformation curves were evaluated. The measured data were sufficient to determine the local resp. the integral cyclic deformation response of the smooth resp. notched specimens.

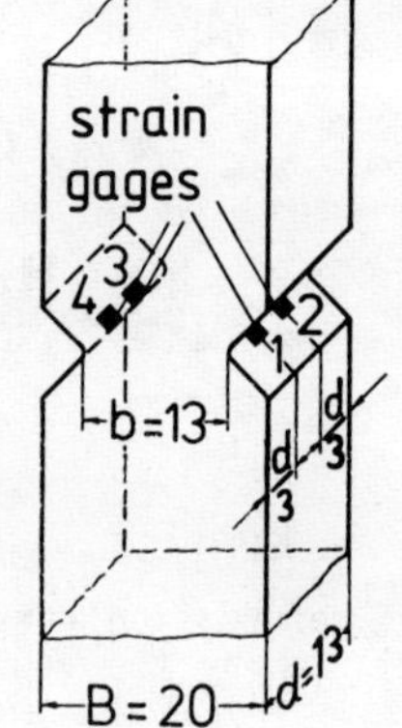

Fig. 1

EFFECTIVE STRESS-STRAIN-BEHAVIOR OF THE NOTCH ROOT

During cycling of notched specimens the nominal stress in the net section and the total notch strain parallel to the loading stress determine the shape of the hysteresisloop, as shown in the upper part of Fig. 2. As a consequence of the notch root elastic-plastic deformations during tensile loading the first principal notch stress reaches a value σ_{max} less than $K_t S_{max}$ where S_{max} is the maximum nominal stress. Unloading the specimen to zero nominal stress ($S = S_m = 0$) at the notch root a compressive residual stress σ_c^{res} will be produced. This is illustrated in the lower part of Fig. 2. In the compression half-cycle the notch stress σ_{min} will be less than $K_t S_{min}$ and after unloading to $S = S_m = 0$ a tensile residual stress σ_t^{res} will remain at the notch root. Consequently, the width of the nominal stress - notch strain - hysteresisloop (upper part of Fig. 2) differs considerably from the width of the notch stress - notch strain - hysteresisloop (lower part of Fig. 2) and the apparent plastic notch strain amplitude at the mean value of nominal stress $\varepsilon_{a,p}(S = S_m = 0)$ and the effective plastic notch strain amplitude at the mean value of notch stress $\varepsilon_{a,p}(\sigma = \sigma_m)$ are not equal.

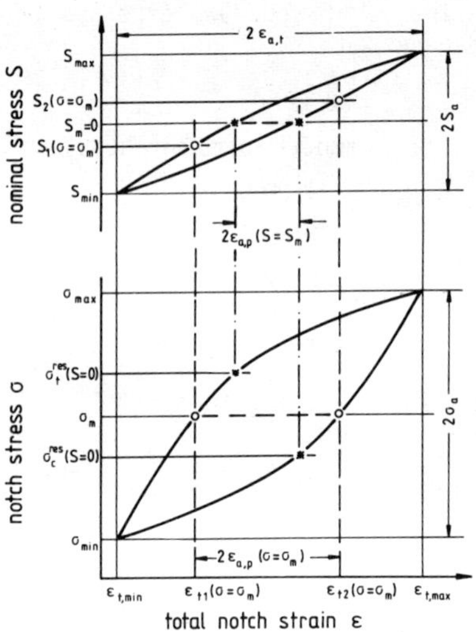

Fig. 2 Nominal stress - total notch strain - and notch stress - total notch strain - hysteresisloops

Thus, the cyclic deformation behavior of notched specimens can only be described correctly if the dependance of notch stress amplitude, total notch strain amplitude and effective plastic notch strain amplitude on the number of cycles will be known. Whereas the total notch strain amplitude can be determined directly from strain gage measurements, it is more difficult to get correct values of the effective notch stress amplitude and the effective plastic notch strain amplitude. However, using Neuber's formula [1],[4] at least an approximate determination of both values is possible [5]. Assuming a nominal elastic strain range Δe_N, Neuber's formula applied to cyclic loading yields

$$\Delta\sigma\Delta\varepsilon = \frac{K_t^2\,\Delta S^2}{E} \tag{1}$$

With $\Delta\sigma = 2\sigma_a$, $\Delta\varepsilon = 2\varepsilon_{a,t}$ and $\Delta S = 2S_a$ one obtains

$$\sigma_a = \frac{K_t^2 S_a^2}{E\varepsilon_{a,t}} \tag{2}$$

which can be used to calculate the effective notch stress amplitudes σ_a from the measured nominal stress - notch strain - hysteresisloops. Inserting $\Delta\sigma = \sigma_a$, $\Delta\varepsilon = \varepsilon_{t1} - \varepsilon_{t,min}$ and $\Delta S = S_1 - S_{min}$ in eq. 1, the relationship

$$\varepsilon_{t1}-\varepsilon_{t,min} = \frac{K_t^2}{\sigma_a E}(S_1-S_{min})^2 = \frac{\varepsilon_{a,t}}{S_a^2}(S_1-S_{min})^2 \tag{3}$$

will be obtained, by which the unknown coordinates of the point S_1, ε_{t1} in the upper part of Fig. 2 can be determined. As shown in Fig. 3, the unknown values ε_{t1} and S_1 lie on a parabola, the intersection of which with the increasing part of the nominal stress - notch strain - hysteresisloop leads to the point wanted. A similar procedure for the second unknown point S_2, ε_{t2} results in

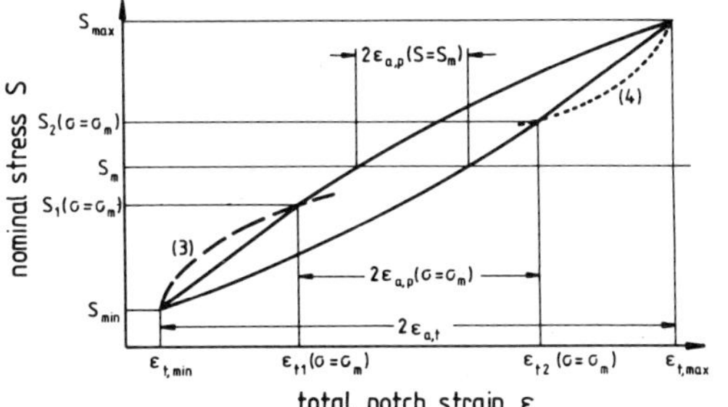

Fig. 3 Determination of effective plastic notch strain amplitude from measured nominal stress - total notch strain hysteresisloop

$$\varepsilon_{t2}-\varepsilon_{t,max} = \frac{K_t^2}{-\sigma_a E}(S_2-S_{max})^2 = -\frac{\varepsilon_{a,t}}{S_a^2}(S_2-S_{max})^2 \tag{4}$$

which, as also shown in Fig. 3, again describes a parabola intersecting the decreasing part of the nominal stress - notch strain - hysteresisloop at the point wanted. With the knowledge of ε_{t1} and ε_{t2}, however, the effective plastic notch strain amplitude is determined.

Contrary to the usual application of approximation formulas for the calculation of notch stresses and strains, based on stabilized cyclic stress-strain-relationships derived from smooth specimen behavior, the presented method is free of errors due to inadequate stress-strain-laws since the relationship measured between nominal stress and appertaining total notch strain is influenced by fatigue softening and hardening as well as by notch effects. Only the approximation formula itself limits the quality of the estimation.

EXPERIMENTAL RESULTS

In the following effective plastic notch strain amplitudes $\varepsilon_{a,p}(\sigma = \sigma_m)$ are taken for the determination of cyclic deformation curves of the notched specimens. In all cases the real curves end if the strain gages at the notch root were destroyed by crack initiation. The remaining specimen life till fracture is indicated by the dotted parts of the curves.

In Fig. 4 cyclic deformation curves of notched specimens with K_t = 3.0 are shown fatigued at different nominal stress amplitudes. As can be seen from the left part of Fig. 4, low nominal stress loadings first lead to pure elastic deformations followed by softening effects after a certain number of cycles and finished by a saturation in the development of the plastic strain amplitudes. At higher nominal stress amplitudes, the notch root areas deform elastic-plastically from the beginning and reach no saturation till crack initiation. Since load-controlled tests were performed, the effective notch stress amplitudes change during the specimen life depending on the amount of plastic deformation. This is shown by the curves in the right part of Fig. 4. Only in the case of pure elastic notch root loading the notch stress amplitude is equal to K_tS_a.With increasing plastic notch strain amplitude the effective notch stress amplitude decreases.

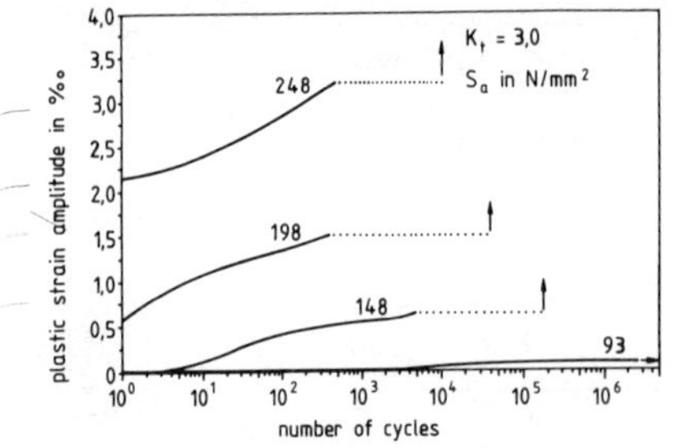

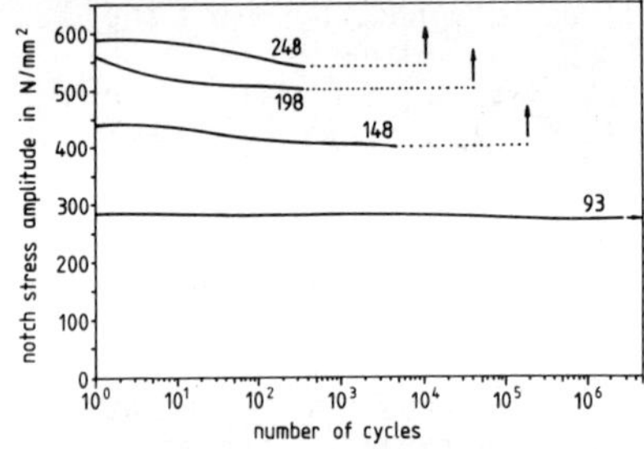

Fig. 4 Development of the effective plastic notch strain amplitudes and the notch stress amplitudes

Figure 5 illustrates the differences in the cyclic deformation behavior of smooth and notched specimens with K_t = 2.2. In both cases plastic strain amplitudes are plotted against the number of cycles for different stress amplitudes, nominal stress amplitudes in the case of smooth specimens and effective notch root stress amplitudes in the case of the notched specimens. Since the effective stress amplitudes of the notched specimens are not constant, the respective amplitudes at crack initiation have been chosen as parameters. As can be seen, in both types of specimens under comparable stress amplitudes plastic deformations start after similar numbers of cycles but reach markedly larger values in the case of smooth than in the case of notched specimens. The accumulated plastic strains up to crack initiation are smaller in notched than in smooth specimens. Furthermore, smooth specimens work-hardened after an initial stage of cyclic softening, whereas notched specimens either work-softened up to saturation at lower stress amplitudes or work-softened without saturation up to crack initiation at higher stress amplitudes.

Figure 6 illustrates the influence of the stress concentration factor K_t on the cyclic deformation curves for different effective notch stress amplitudes. In all cases cyclic softening occurs. The extend of the softening decreases with increasing K_t. In the medium life range at comparable effective stress amplitudes K_t = 3.0 seems to result in smaller plastic strain amplitudes than K_t = 2.2. However, these differences disappear at effective stress amplitudes leading to cycle numbers up to crack initiation larger than $5 \cdot 10^4$.

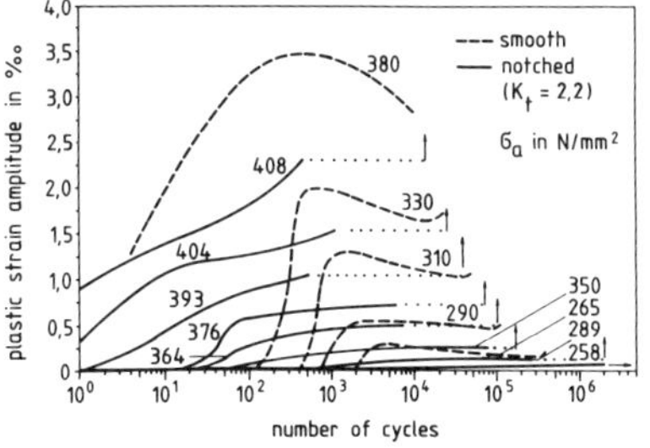

Fig. 5 Comparison of the cyclic deformation curves for smooth and notched specimens

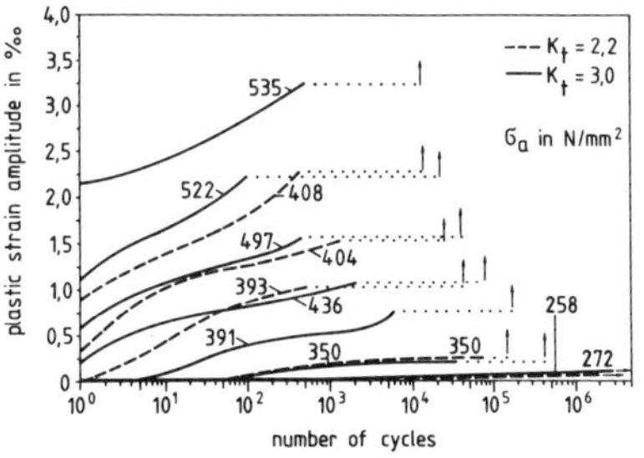

Fig. 6 Comparison of the cyclic deformation curves for two differently notched specimens

In Fig. 7 results from companion specimen experiments are compared with directly determined plastic notch strain amplitudes. It shows cyclic deformation curves for K_t = 2.2 and nominal stress amplitudes of 130 and 151 N/mm². The curves belonging to the smooth companion specimens are dashed. In the case of S_a = 151 N/mm² the strain gage at the notch root failed before

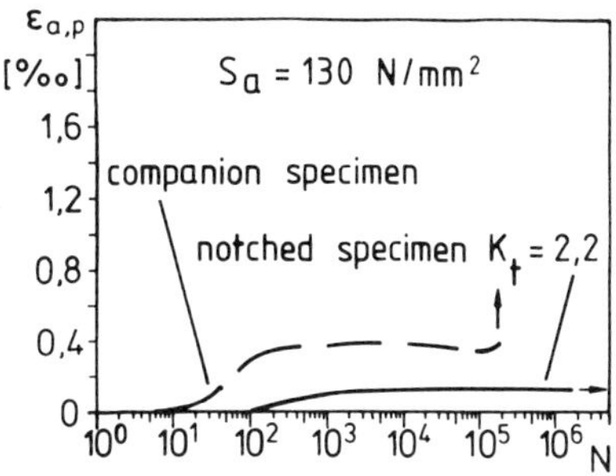

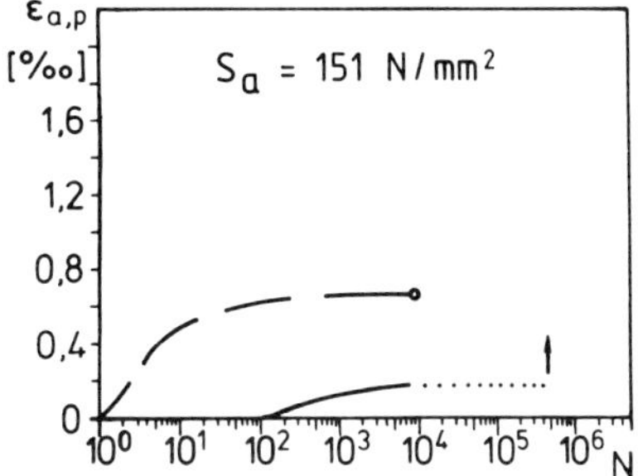

Fig. 7 Cyclic deformation curves of notched (K_t=2.2) and companion specimens

the companion specimen was broken. In all experiments the companion specimens soften much earlier than the notched ones. In the case of S_a = 130 the companion specimen fractured even before cracks initiated at the notch root of the notched specimen.

DISCUSSION

The method presented to assess the deformation behavior of notched specimens proceeding from Neuber's formula and basing on nominal stresses and direct notch strain measurements considers the influence of both, the multiaxiality of the loading stress state and the residual stress state on the stress-strain-relationship. Consequently, no necessity exists to approximate the stress-strain-relationship of the notch root area by any stress-strain-law derived from experiments with smooth specimens.

At comparable stress amplitudes the start of plastic deformations due to cyclic softening seems to be insensitive to notch effects, whereas the amounts of plastic strain amplitudes are smaller in the notch roots of notched specimens than with smooth specimens. The plastic notch strain amplitudes decrease with increasing K_t. This may be explained by the influence of the multiaxial notch stress state and the stress gradient on the cyclic plastic deformations. In no case during the fatigue experiments with notched specimens work-hardening was observed. Obviously the plastic deformations in the notch root lead to a redistribution of the stress state in the net section in such a way that highly stressed regions will be relieved partially whereas neighbour regions will be stressed more and starts to soften. Consequently, no work-hardening in the very surface of the plastic deformed zone at the notch root can be observed up to crack initiation.

Under comparable stress amplitudes the amounts of accumulated plastic strains up to crack initiation are smaller in notched specimens than in smooth ones and decrease with increasing K_t. In this connection it should be emphasized that in smooth specimens very early small fatigue cracks develop but a relatively long time passes before crack propagation can be observed. On the contrary, as a consequence of stress concentration in the notch root of notched specimens, small fatigue cracks start quickly to propagate and form macroscopic cracks. Thus, in the case of smooth and notched specimens differences between accumulated plastic strain values up to crack initiation depend strongly on the definition of crack initiation. On the other hand, the differences measured also could be caused by inhomogeneous deformation processes at the notch root. The fact is that at the notch roots cracks always developed at some very exposed regions with relatively deep roughness valleys or near inclusions. It can be supposed that the strain gages were mostly not sticked at these regions. Therefore it can be expected that the plastic strain amplitudes at these places are larger than the recorded values. Recent TEM-observations seem to confirm this statement.

The discrepancies in the cyclic deformation behavior of notched and smooth companion specimens are strongly influenced first by the multiaxiality of the notch stress state and second by the cyclic changes of the residual stress state due to macroscopic inhomogeneous plastic deformations of the notched specimens. Thus, from simulations of the notch root behavior with smooth samples as companion specimen experiments [6] or Neuber-control experiments [7] correct results can only be expected if the mentioned notch effects are not very pronounced as in the case of large notch root radii and specimens of relatively small thickness. On the other hand, in the case of sharp notches larger errors may occur if the stress-strain relationships of smooth specimens are considered as equivalent to the relationship between the largest effective notch stress component and the appertaining strain.

ACKNOWLEDGEMENT

The financial support of these investigations by the Deutsche Forschungsgemeinschaft is gratefully acknowledged.

LITERATURE

[1] H.Neuber, J. Appl. Mech. (Trans. ASME), 544 (1961)
[2] P.Heuler, T.Seeger, Proc. Int. Symp. on Low-cycle fatigue strength and Elasto-plastic behavior of Materials, DVM, 319 (1979)
[3] T.Seeger, A.Beste, Fortschrittsberichte der VDI-Zeitschriften Reihe 18 Bd. 2, (1977)
[4] T.H.Topper, R.M.Wetzel, J.D.Morrow, J. of Mater., JMLSA, 200 (1969)
[5] H.J.Böhmer, D.Eifler, E.Macherauch, Z. Metallkunde to be publ. (1985)
[6] J.H.Crews, Jr. and H.F.Hardrath, Exp. Mech. 313 (1966)
[7] R.M.Wetzel, J. of Mater., JMLSA, 3, 646 (1968)

Equilibrium Shapes of Dislocation Lines and Activation of Dislocation Sources Under the Influence of Internal Stresses

H. G. Schmid* and U. Essmann**

**Institut de Génie Atomique, EPF-Lausanne, Switzerland*
***Max-Planck-Institut für Metallforschung, Institut für Physik, 7000 Stuttgart 80, Federal Republic of Germany*

ABSTRACT

Equilibrium shapes of dislocation segments in stress gradients are calculated. The results are discussed with respect to determinations of internal stress fields by direct observations of bent dislocation segments.

Internal stresses facilitate the operation of dislocation sources. It is shown that not only the Orowan condition has to be fulfilled for the escaping of a bowed out dislocation but that also the gradients of the internal stresses have to be taken into account. Within a stress gradient the escaping dislocation may arrive at a stable position which can be overcome with the help of the applied stress.

KEYWORDS

Orowan stress , persistent slip bands, dislocation shapes, internal stresses, dislocation sources, fatigue.

INTRODUCTION

Internal stresses in deformed crystals can be investigated by measuring the radii of curved dislocation segments with the help of transmission electron microscopy (TEM). For an observation of dislocation shapes which are representative of the bulk crystal it is necessary to "freeze-in" the as deformed dislocation arrangement and prevent relaxation during unloading or subsequent thinning for electron microscopy examination. A number of techniques have been reported in the literature [1, 2].

A systematic study of the stress distribution by TEM has been published for copper crystals fatigued by alternating tension and compression [2].

The subject of the following discussion are the equilibrium shapes of dislocation loops in stress gradients and which conclusions can be drawn with respect to the internal stress τ_i. In general it is assumed that a dislocation bowing out between two pinning points is stable as long as τ is smaller than the Orowan stress $\tau_{or} = 2T/(bd_o)$, where d_o denotes the distance between the pinning points. If $\tau > \tau_{or}$ the dislocation loop escapes and spreads out in its slip plane. This, however, supposes that the stress τ is locally constant, which does not hold with respect to the fluctuating internal stresses as will be shown in the following discussion.

BENT DISLOCATION SEGMENTS AND INTERNAL STRESSES IN PSBs.

Cyclic slip localization in so-called persistent slip bands (PSBs) is a characteristic feature of wavy-slip materials fatigued at low to intermediate strain amplitudes[3]. PSBs are lamallae of high local glide activity which lie parallel to the active (primary) glide plane. In pure copper the lamellae have thicknesses of about 1 µm and exhibit a characteristic dislocation arrangement. A periodic structure of long dislocation walls subdivides the PSB lamella into channels of about 1.2µm in width. The walls are oriented perpendicular to the effective Burgers vector and consist of edge dislocation dipoles. A low density of dislocations having mainly screw character is observed in the channels. The stress distribution in the stress applied state has been determined by measuring the curvature radii of bent dislocations in the channels [2] . Drawings of the observed dislocation arrangement and of the determined stresses are shown in fig. 1 and 2, respectively.

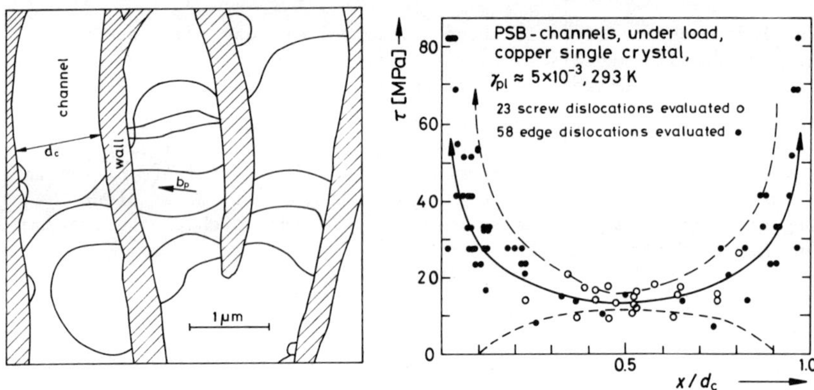

Fig. 1 Dislocation shapes from a TEM micrograph from a section parallel to the primary glide plane of a copper single crystal [2]

Fig. 2 Measured local stresses in channels between PSB-walls in the stress applied state τ_a=28MPa [2]. d_c channel width, —— mean value of the measured stress values, ---- calculated limit of suggested stress profiles.

One may interprete the measured local stresses in fig. 2 as a stress profile which does not markedly change along the walls of the PSBs. One then obtains the stress distribution (a) in fig. 3. A test of this assumption is possible by comparing observed and calculated equilibrium shapes of dislocations. Our calculations made use of the DeWit -Koehler definition [4] $\tau = T(\theta)/\rho b$, a numerical procedure [5] and the isotropic approximation of the line tension $T(\theta) = T_e + (T_s - T_e)\cos^2\theta$ [4] , T is the line tension, ρ the bending radius, θ denotes the character of dislocation; T_s and T_e are the line tensions of the screw and the edge dislocations, respectively. For the calculations we use the values T_s = 2.5 nN and T_e = 1 nN as determined experimentally [2] Thereby we avoid inconsistences between the measured stress profile which is

based on these experimental line tension values and the dislocation shapes calculated in suggested stress profile (see also [7]).
Some calculated shapes are shown in fig. 4. Experimentally one finds essentially two types of dislocation lines : Those which bow out from the walls and those which traverse the channels in the following called "edge" and "screw" dislocations respectively. An "edge" dislocation segment becomes unstable when the dislocation acquires screw character at the two pinning points. The shape then corresponds to the Orowan condition. The half loop(a1) in Fig. 4 has been calculated supposing the stress distribution (a) in Fig.3. Its shape agrees with observed configurations.

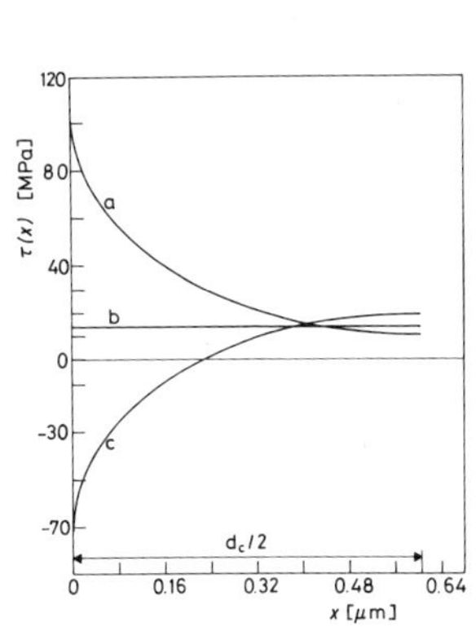

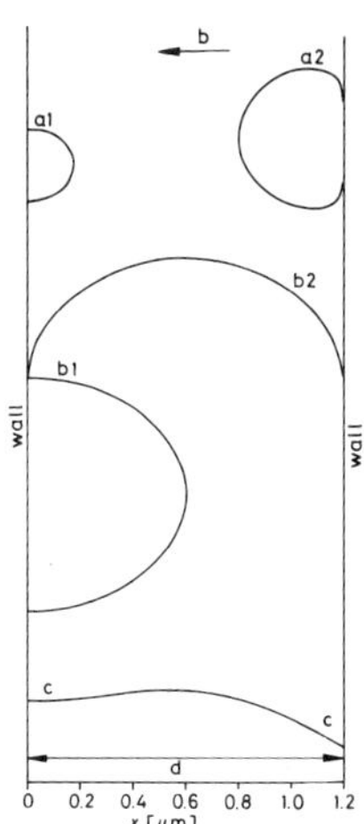

Fig. 3 Assumed stress distributions $\tau_{(x)}$
a) suggested stress profile according to [2]
b) constant stress
c) stress profile which changes the sign near the wall.

Fig. 4 Equilibrium positions of dislocations calculated using the stress profiles a, b, and c
The index 1 and 2 correspond to "edge" and "screw" dislocations, respectively.

If a "screw" dislocation bows out it becomes unstable once it acquires edge character where it touches the walls. The calculation of the dislocation line starts at one of the walls with a segment having edge character. The calculated dislocation line is in equilibrium position when the dislocation line can reach the other wall where it becomes pinned. With the above mentioned stress profile one obtains the dislocations line a2 in fig. 4 which cannot reach the opposite wall and which hence is not stable. Therefore we conclude that under the supposed stress conditions equilibrium shapes which link the walls cannot exist in contrast to the observations. This discrepancy suggests that the supposed stress distribution is not able to give a complete description of the experimental observations.

In order to solve this contradiction we refer to the fluctuations of τ_i along the walls which can be inferred from the observations [2]. For an estimate of the magnitude of these fluctuations we consider a number of different stress profiles and calculate the corresponding equilibrium shapes of dislocations which then are compared with the experimental findings.

The most simple assumption is that τ is constant across the channel (line b in fig. 3). The calculation shows that under this assumption equilibrium positions of "screw" dislocations are only stable for stresses $\tau < 13$ MPa (line b2 in fig. 4). This implies that in general larger stresses in the channel cannot be detected by "screw" dislocations linking the walls since then no equilibrium configurations do exist.

Occasionally larger stresses in the middle of the channel have been determined. In our terms this can only be understood if one supposes a stress profile which falls off at the walls. A corresponding stress profile is given in fig. 3 and denoted by c. One then obtains the equilibrium configuration c in fig. 4. Note that the sign of the stress changes in this case. This leads to a point of inflection in the dislocation line. Configurations of this type can be found in micrographs like fig. 1. Accordingly the internal stresses along the walls exhibit rather large fluctuations.

ORIGIN OF FLUCTUATING INTERNAL STRESSES IN PSBs.

The fluctuactions can be understood in terms of the mechanism by which the walls are maintained. As shown elsewhere the gliding screw dislocations draw out edge dislocations along the walls [2,6]. Once the screw dislocations annihilate with dislocations of opposite sign gliding on adjacent glide planes edge dislocation segments are deposited in the walls. The segments are connected via the cross slip plane. Their length is typically 2 µm in copper[6]. The plastic deformation of the walls implies that edge dislocations of opposite sign enter the walls from both sides. In steady state the edge dislocations annihilate mutually on the average. However, locally a small surplus of dislocations of one sign may exist temporarily within a wall. Furthermore, keeping in mind that the wall is about 1µm high (height or PSB lamella) dislocation dipoles having a width of 1µm may exist within the wall. Hence we conclude that also the dipole strength along the wall fluctuates. The fluctuation length of the internal stresses along the walls is suggested to be typically 1µm.

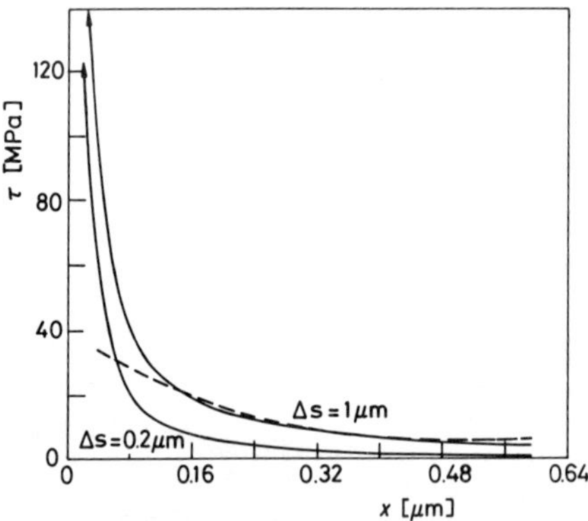

Fig. 5 The influence of the resolved shear stresses τ induced by straight edge dislocation segments of the length Δs.

Fig. 5 shows the stress τ produced in the channel by surplus edge dislocation segments which are deposited on the border of a wall in the glide plane according to Brown's theorem [8]. The dashed line in Fig. 5 denotes half the scattering width of the experimental points in fig 2. The coincidence between the full and dashed curve in fig. 5 supports the assumption that surplus dislocations in the wall are responsible for the observed stress fluctuations. The agreement is rather good in the middle of the channel. The curves deviate from each other in the immediate neighbourhood of the wall (within about 0,15µm).

The calculated influence of the surplus dislocations can be added or substracted from the average stress profile in fig. 2. This leads to the dashed borderlines. In the region of $x/d < 0.15$, however the measured small edge dislocation shapes occur only under high stresses. In order to obtain a better description of the stress distribution, one has also to consider the screw dislocations which enter the wall. With a spectrum of profiles lying within the borderlines, the observed dislocation shapes can be explained by our calculations.

OPERATION OF DISLOCATION SOURCES IN STRESS GRADIENTS

Local peaks of the stress distribution activate dislocation sources having high Orowan stresses. The loops then have to spread in stress gradients. In the following it will be shown that the escaping dislocation loops may arrive at stable positions in the stress gradients. In order to demonstrate this effect, we suppose a cylinder symmetrical stress peak. For our qualitative discussion we consider only the behaviour of the dislocation along the coordinate x and do not determine the exact shape of the loop. In the centre of the stress peak the Orowan condition must be fulfilled. The loop expands as long as (see introduction)

$$x\tau(x) > \frac{T}{b} \qquad (x > d_o/2)$$

where we suppose that the radius of the loop $\rho = x$. With increasing x the stress $\tau(x)$ may, however, decrease such that

$$x\tau(x) = \frac{T}{b} \qquad (x > d_o/2)$$

In this case the dislocation stops in an equilibrium position.

This holds for all stress fields $\tau(x) \sim x^{-n}$ with $n > 1$. As shown in fig. 5 and discussed above such stress fields are expected to prevail in PSB lamellae.

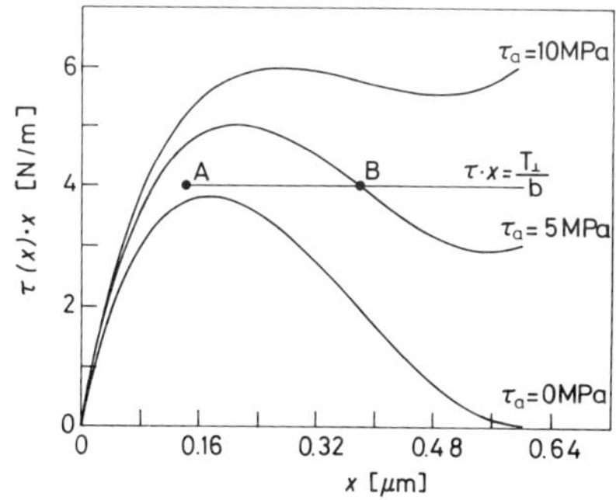

Fig. 6 The distribution of $\tau(x) \cdot x$ using the stress profile a. in Fig. 3 according to [2]

In order to illustrate this behaviour we suppose along x the stress distribution $\tau(x)$ denoted by a in fig. 3. According to fig. 4 this distribution leads to an Orowan radius of about $d_o/2 \approx 0.12\mu m$ in copper ($T_e/b = 4N/m$). The behaviour of the dislocation can best be seen by plotting $x\tau(x)$ over x (see fig. 6). In order to demonstrate the influence of an applied stress τ_a we write $\tau(x) = \tau_i(x) + \tau_a$. We consider three cases $\tau_a = 0$, 5 and 10 MPa (We neglect the corresponding decrease of d_o). For $\tau_a = 0$ MPa the dislocation bends but can not fulfill the Orowan condition A in fig. 6. For $\tau_a = 5$ MPa the dislocation fulfills the Orowan condition but comes to rest in the next stable position B its shape being comparable to loop a2 in fig. 4. For $\tau_a = 10$ MPa the dislocation line escapes without being trapped in a stable position.

CONCLUSIONS

In order to obtain the stress distribution of internal stresses one has to compare the observed dislocation shapes with calculated ones. In case of PSBs afair agreement is obtained between observed and calculated shapes as one supposes rather large stress fluctuations.

ACKNOWLEDGEMENTS

The authors wish to express their thanks to Dr. R. Gotthardt, Dr. S. Gorjachev and Dr. M. Morris for many hours of fruitful discussions and reading the manuscript and also to Mrs. Fahrni for typing the manuscript.

REFERENCES

1. U. Essmann, Phys. Stat. Sol. 3, 32 (1963)
2. H. Mughrabi, Cyclic Plasticity of Matrix and Persistent Slip Bands in Fatigued Metals, in Continuum Models of Discrete Systems 4, eds. O. Brulin and R.K.T. Hsieh, North Holland, Amsterdam, 241 (1981)
3. H. Mughrabi, Mater. Sci. Enging., 33, 207 (1978)
4. G. DeWit, J.S. Koehler, Phys. Rev. 116, 1113 (1959)
5. H. Schmid "Plastizität von Kalium", phil. thesis, Univ. of Vienna (1978)
6. U. Essmann, phil. mag. A, 45, 171 (1982)
7. J. Lepinoux, L.P. Kubin accepted for phil. mag.(1984)
8. L.M. Brown, Phil. Mag.15,363(1967)

Back Stress Variation Along the Hysteresis Loop: the Effect of Microstructure

Y. S. Chung and A. Abel

School of Civil and Mining Engineering, University of Sydney, N.S.W. 2006, Australia

ABSTRACT

Fully reverse low cycle fatigue tests were conducted on five aluminium alloys which can be divided into three groups on the basis of strengthening mechanisms. They are the age hardening alloys 6061, 6351 and 7005, the solid solution strengthened 5083 and dislocation and subgrain strengthened 1200 alloys.

During cycling a stable hysteresis loop emerged typically after about 10% of the fatigue life. Using this stable state the back stress variation was studied by analysing the hysteresis behaviour when unloading and reverse loading from different stress states along the loop.

The results indicate a rapid release of back stresses, typically in the 2 to 15% range of the applied prestrain, followed by a rapid build up of new back stresses the rate of which gradually diminishes as the end of the cycle is approached. These two processes, depending on the strain at which load reversal is initiated, can lead to three distinct hysteresis response: the operation of the Bauschinger effect, the non-existence of that effect and what appears to be the operation of a negative Bauschinger effect.

These results are discussed with reference to the strengthening mechanism associated with the type of microstructure.

KEYWORDS

Low-cycle fatigue; aluminium alloys; hysteresis loop; back stresses; Bauschinger effect; microstructure.

INTRODUCTION

It is generally accepted that there is a reversal in the back stress and stored energy along half of a hysteresis loop (1-5). The actual position along the loop, where the back stress is zero, is, however, not well established. Moreover, very little information is available on the rate of dissipation of the back stresses after stress reversal and its relation to the microstructure.

The experimental data of Wilson (2) shows that the reverse strain which reduces the average value of the long range back stresses in the matrix to zero is about 3% after a prestrain of 9%. A lower value of 1.6% is obtained for a low carbon steel with a lower volume fraction of precipitates. More generally, reverse straining reduces the internal stresses to zero at a strain which is usually between one third and one half of the previously applied forward strain (1,6). The assumption of a linear decrease of back stress after reverse loading (7) implies that the back stress is zero mid-point along the hysteresis loop. A similar linear variation of back stress is assumed by Tanaka and Matsuoka (5) on the basis of a hardening theory (8). In that approach the plastic strain at which the back stress is zero varies from a constant value of half the total applied strain at small applied strain ranges, to a smaller value at higher strain ranges. The transition occurring at a critical plastic strain range where the accumulation of back stresses have become saturated due to stress-relieving mechanisms such as cross-slip of dislocations and/or plastic deformation of the particles. Experimental result on stored energy change along the hysteresis loop (4) shows that the ratio of the plastic strain where energy release occurs to the applied plastic strain range is about 0.3 and is independent of the plastic strain range.

As an extension of previous work (9) the present paper explores these ideas on five aluminium alloys covering age hardening, solution strengthening and subgrain strengthening microstructures.

EXPERIMENTAL RESULTS

The strengthening microstructures ranged from subgrain and dislocation strengthening in the 1200 alloy; solid solution strengthening by Magnesium in 5083, and the high density of GP zones and finely dispersed precipitates in the precipitation hardened alloys, 7005, 6351 and 6061. A TEM investigation (10) has confirmed the three types of microstructures associated with these alloys. The five alloys were plant-fabricated with commercial tempers. The low cycle fatigue tests, with fully reverse strain-controls, were performed on a 250 kN Instron, TT-K model, using the automatic transfer points as in (9). A relatively stable cyclic state was obtained typically after about 10% of the fatigue life, that is, approximately the same hysteresis loop has been traced over and over again. This stable state was used for the stress reversal tests conducted at different stress states along the hysteresis loop.

To improve the accuracy of the measurements, a compulog was used to collect stress-strain values. Data collection commenced along the hysteresis loop preceding the stress reversal and using a 0.2 second interval between successive readings. Typically about 150 to 200 points were scanned for each stress reversal test. Analysis of the data on the computer involved the determination of the back stresses, σ_b (11-12) and the Bauschinger parameters, β_σ in terms of stress, β_ε strain and β_E energy parameters (13).

The hysteresis behaviour at the stress states 1,2,3 ... along the loop, were analysed in terms of the Bauschinger parameters (Fig. 1) and the back stresses were calculated by using Cottrell's model, the result of which is shown in Fig. 2. A very rapid back stress dissipation takes place upon load reversal in all of the alloys and this is shown for three representative alloys in Fig. 3. The strain value at which the back stress, σ_b, equals zero is a function of the microstructure as well as the applied cyclic strain range, $\Delta\varepsilon_p$, as shown in Figs. 4 and 5. The loop shape variation of the three representative alloys are shown in Figs. 6 to 8 where the Bauschinger energy parameter is plotted against the plastic strain at which stress reversal took place from a stable hysteresis state. Any one curve in these Figures is made up from the results obtained from a number of stress reversal experiments similarly to those outlined in relation to Fig. 1. The terminating lines at the end of any curve simply indicate the total applied cyclic strain amplitude for that particular load reversal test.

DISCUSSION

Load reversals from various stress states provided by a relatively stable hysteresis loop indicated three distinct cases. Using the results illustrated by Fig. 1 the operation of the Bauschinger effect, B.E., is observed between points 1 and 7 and load reversal from point 8 produced what may be termed as a negative Bauschinger effect, that is, the reverse yield stress appears to be higher than the peak stress achieved during the pre-stressing. Somewhere between points 7 and 8 there is a stress state from which load reversal would have resulted in a zero Bauschinger effect. These three hysteresis responses will be discussed below.

During most of the "forward" deformation cycle the deformation processes generate enough back stresses which in turn contribute to the operation of a Bauschinger effect. This effect, however, shows an optimum manifestation, as measured through the energy parameter β_E, which is sensitive to the prevailing microstructure. Taking the experimental values presented in Fig. 1, it seems that the deformation processes increase the Bauschinger effect approximately up to 6×10^{-3} forward strain when the applied strain amplitude is $\sim 10 \times 10^{-3}$. Similar conclusion is reached when one considers the details presented in Fig. 2 where the relative size of the back stress value tapers off at approximately the same prestrain value.

Looking at the results presented in Fig. 7 the significance of 6×10^{-3} prestrain is more evident for this precipitation hardened alloy. Cycling up to 6×10^{-3} prestrain shows increasing tendency in the value of β_E and beyond that prestrain the Bauschinger effect looses in effectiveness. That particular optimum prestrain value is of the order of 3×10^{-3} and 8×10^{-3} for the 1200 and 5083 alloys respectively, as shown in Figs. 6 and 8. One other strong trend displayed by Figs. 6-8 is that the smaller the applied strain amplitude is, the higher is the Bauschinger energy parameter.

If one makes a strong association between back stresses and Bauschinger effect then these results suggest that back stress generation is strongly dependent on both the applied strain and the microstructure. This view is further strengthened by the results shown in Fig. 3.

The measured values of strain at which $\sigma_b = 0$ appears to be small and as shown in Fig. 5 it amounts to only 2 to 15 per cent of the applied cyclic strain amplitude. At that deformation state there is no detectable Bauschinger effect and on Fig. 1 the load reversal position giving this state is somewhere between points 7 and 8. More premature load reversal leads to a negative Bauschinger effect. The existence of the customary as well as the negative Bauschinger effect suggest that simultaneously back stresses with opposite signs may exist. A negative back stress, σ_{b2}, accumulated between points R and 11 on Fig. 1 assists deformation between points 11 and 7,6 and a positive one, σ_{b1}, being generated during the current deformation cycle which opposes the deformation in the present direction. As σ_{b2} diminishes in magnitude while σ_{b1} grows between the points F and 7,6 a negative B.E. changes to zero B.E. before the customary B.E. emerges.

Fatigue with random loading and cumulative damage may represent areas where the consideration of the negative B.E. should be relevant. It should be noted that the zero B.E. stress state is more than 50% of the maximum stress applied in that particular cycling depicted by Fig. 1.

How long after stress reversal do the negative back stresses disappear completely is a question more difficult to answer. As loading from point F to 8 to 7 etc will create back stresses immediately, of the σ_{b1} type, it is convenient to merge the broken line with the solid σ_b at around the points 5 and 6, simply on the basis of curve fitting. Accepting this would automatically lead to the broken line representing the decay of σ_{b_2} which curve terminates exactly at the same value at which σ_{b1} and σ_b merges. The sudden change in the B.E. stress parameter at that point, Fig. 1, gives some backing to this proposal.

SUMMARY

The back stress variation along the stable hysteresis loop, formed typically after about 10% of the fatigue life, was obtained by analysing the unloading and reverse loading response at different stress states along the loop. The back stress generation and dissipation was found to be strongly dependent on both the applied strain and microstructure.

Load reversals from various stress states along the stable hysteresis loop indicated three distinct cases: the operation and non existence of the Bauschinger effect and the existence of what may be termed as a negative Bauschinger effect. This latter may play a part in assessing damage production in a random loading fatigue case.

ACKNOWLEDGEMENT

This research was sponsored by the Australian Aluminium Development Council and this support is gratefully acknowledged.

REFERENCES

1. J.D. Atkinson, L.M. Brown and W.M. Stobbs, Phil. Mag. 30, 1247 (1974).
2. D.V. Wilson, Acta Met. 13, 807 (1965).
3. C.E. Feltner and C. Laird, Acta Met. 15, 1633 (1967).
4. G.R. Halford, Ph.D. Thesis, University of Illinois, Urbana (1966).
5. K. Tanaka and S. Matsuoka, J. of Mat.Sci., 11, 656 (1976).
6. W. Stobbs, Work Hardening in Tension and Fatigue, A.W. Thompson (editor) (1977).
7. P.R. Strutt, J. of Aust. Inst. of Met. 8, 2, 115 (1963).
8. K. Tanaka and T. Mori, Acta Met. 18, 931 (1970).
9. Y.S. Chung and A. Abel, ICSMA6, 825 (1982).
10. Y.S. Chung, Ph.D. Thesis, University of Sydney (1981).
11. A.H. Cottrell, Dislocations and Plastic Flow in Crystals, Oxford University Press, London (1953).
12. D. Kuhlmann-Wilsdorf and C. Laird, Mat. Sci. Eng. 37, 111 (1979).
13. A. Abel and H. Muir, Phil. Mag. 26, 489 (1972).

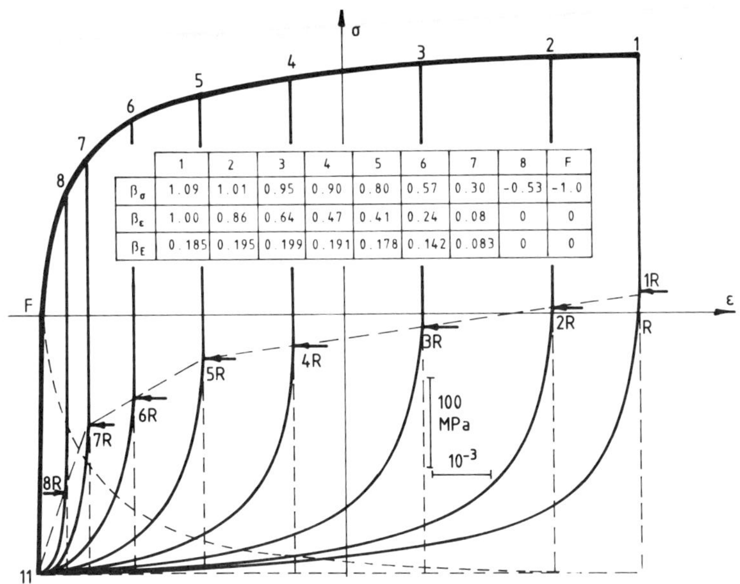

	1	2	3	4	5	6	7	8	F
β_σ	1.09	1.01	0.95	0.90	0.80	0.57	0.30	-0.53	-1.0
β_ε	1.00	0.86	0.64	0.47	0.41	0.24	0.08	0	0
β_E	0.185	0.195	0.199	0.191	0.178	0.142	0.083	0	0

FIG. 1 LOAD REVERSALS FROM VARIOUS STRESS STATES

$\Delta\varepsilon_p = 9.8\times10^{-3}$; Alloy 6061

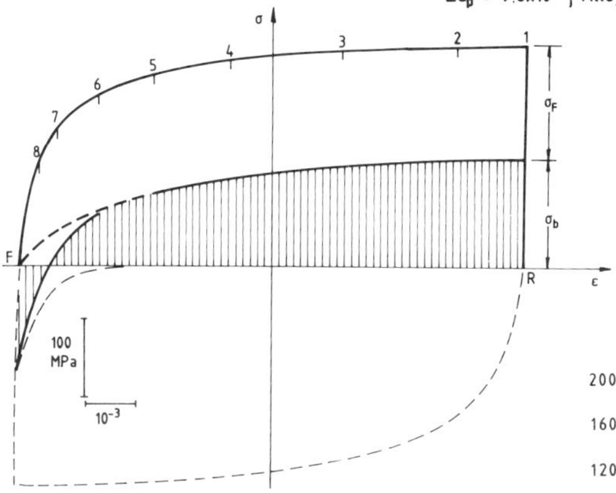

FIG. 2 COMPONENTS OF THE FLOW STRESS ALONG THE HYSTERESIS LOOP

$\Delta\varepsilon_p = 9.8\times10^{-3}$; Alloy 6061

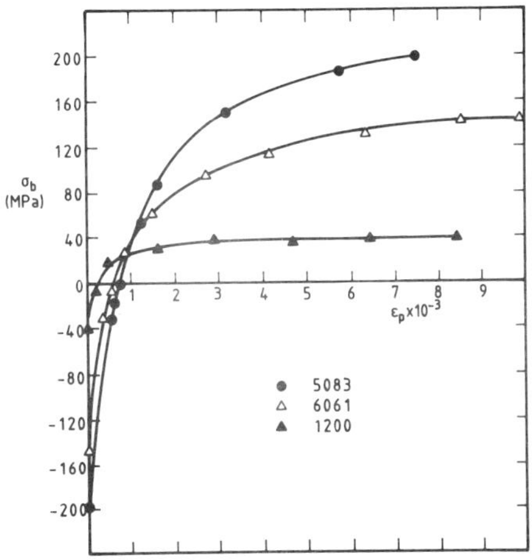

FIG. 3 BACK STRESS VARIATIONS ALONG THE HYSTERESIS LOOP

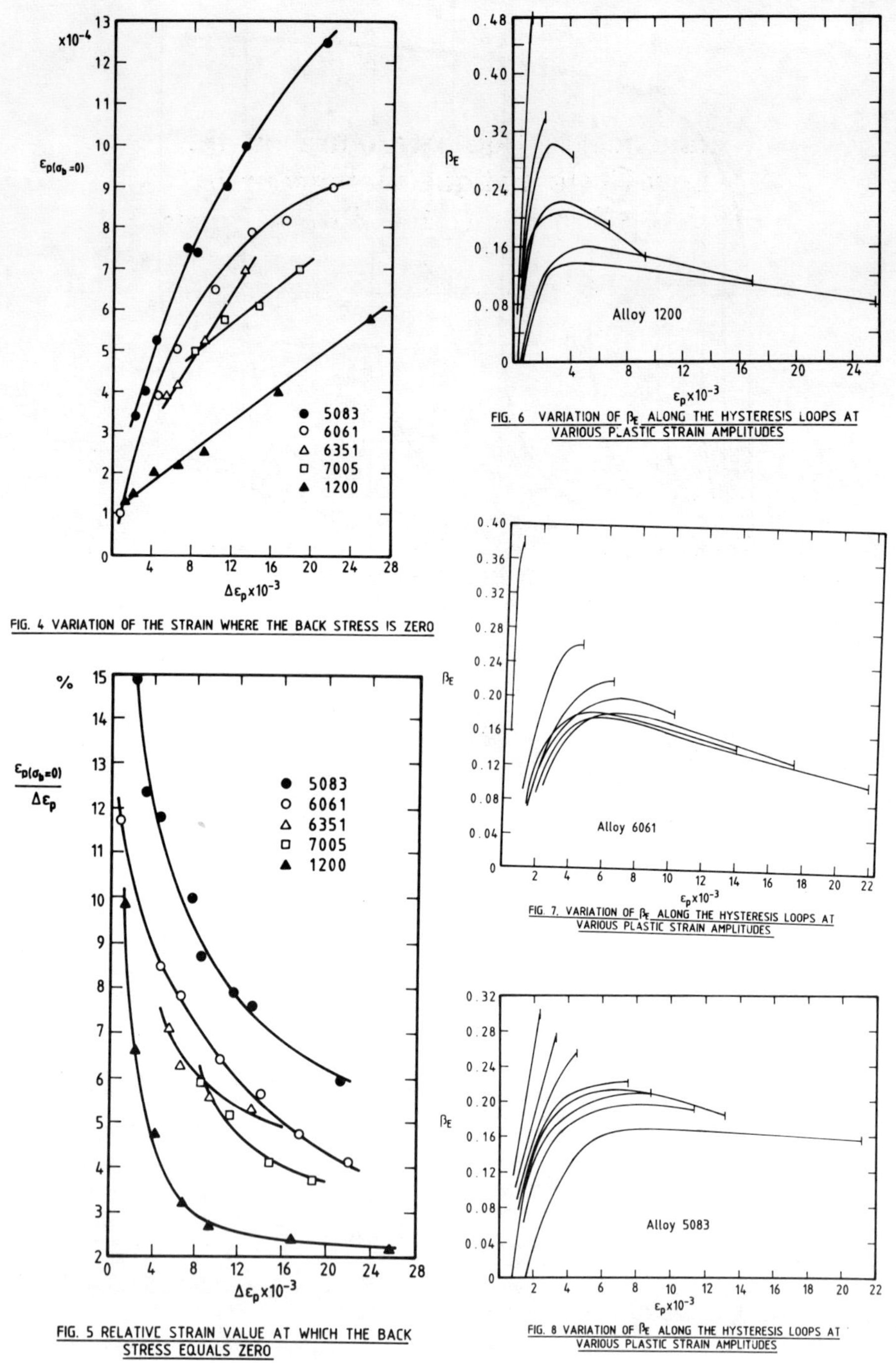

FIG. 4 VARIATION OF THE STRAIN WHERE THE BACK STRESS IS ZERO

FIG. 5 RELATIVE STRAIN VALUE AT WHICH THE BACK STRESS EQUALS ZERO

FIG. 6 VARIATION OF β_E ALONG THE HYSTERESIS LOOPS AT VARIOUS PLASTIC STRAIN AMPLITUDES

FIG. 7. VARIATION OF β_E ALONG THE HYSTERESIS LOOPS AT VARIOUS PLASTIC STRAIN AMPLITUDES

FIG. 8 VARIATION OF β_E ALONG THE HYSTERESIS LOOPS AT VARIOUS PLASTIC STRAIN AMPLITUDES

Influence of Microstructure on the Low-Cycle Fatigue Behaviour in 11.5% Chromium Dual-Phase Steels

F. Blum, N. R. Comins and M. P. Shaw

National Institute for Materials Research, C.S.I.R., P.O. Box 395, Pretoria 0001, South Africa

ABSTRACT

Two chromium containing dual-phase steels, containing different proportions of martensite and ferrite, have been examined under low-cycle fatigue conditions. Differences between monotonic and cyclic stress-strain behaviour are discussed with respect to phase proportions present and dislocation configurations observed by transmission electron microscopy. Prestrain effects are considered in terms of dislocation rearragement and are compared with the effect of pre-existing transformation induced dislocations on the monotonic stress-strain response.

KEYWORDS

Dual-phase steels, ferrite, martensite, fatigue, TEM, dislocation structures, pre-strain.

INTRODUCTION

The low alloy dual-phase steels, characterized by a martensite-ferrite microstructure, are a relatively new class of steels which show an attractive combination of strength and ductility. Furthermore, they display continuous yielding behaviour, a high initial work-hardening rate and a low yield to ultimate tensile strength ratio [1,2,3]. The mechanical properties are generally sensitive to the volume fraction of martensite and the morphology of the phases [4]. For most dual-phase materials studied, the tensile properties have been found to vary approximately linearly with the martensite volume fraction [2,3,4], unlike the low cycle fatigue behaviour which does not follow a rule of mixtures. Typically, dual-phase steels show cyclic hardening up to about 30% martensite and beyond this level cyclic softening occurs [5].

Steels based upon the addition of approximately 12% chromium have been studied extensively in the past, largely with a view to optimizing the corrosion resistance and mechanical properties of fully ferritic microstructures [6]. However, more recently, the properties of similar steels with dual-phase (ferrite-martensite) microstructures, made available by suitable thermomechanical processing and heat treatment, have aroused increasing interest [7]. It is therefore of value to study the monotonic and cyclic deformation behaviour of these alloys in comparison with results obtained for low alloy dual-phase steels.

An earlier investigation [8] showed that the alloys displayed continuous yielding and a three stage work-hardening index under tensile loading, similar to that reported in vanadium-bearing low alloy dual-phase steel [2]. The transformation-induced dislocation distributions in the ferrite and their subsequent development during deformation was also studied.

The aim of the present work is to investigate the low-cycle fatigue properties of an 11.5% chromium steel (designated 3CR12) in the as-processed form for two alloy compositions (viz. 0.68% and 1.21% Ni, see Table 1), and with a more isotropic microstructure of the first alloy produced by further laboratory heat treatment. The specific objectives were to study the response to cyclic straining, the effect of prestrain on this behaviour and the microstructural developments induced by cyclic plastic deformation.

MATERIAL AND EXPERIMENTAL PROCEDURES

The chemical compositions of the investigated steels are summarized in Table 1, and examples of the microstructure are given in Fig. 1. In the as-processed condition alloys 1 and 2 were characterized by a heavily banded microstructure with martensite volume fractions of approximately 30% and 70% respectively. The heat treatment of alloy 1 produced a more isotropic globular structure with increased martensite content of about 60%.

TABLE 1 Chemical Composition of 3CR12

Designation	C	S	Mn	Si	Ti	Cr	Ni	N	Fe
Alloy 1	0.024	0.013	1.32	0.43	0.35	11.5	0.68	0.014	balance
Alloy 2	0.027	0.010	1.24	0.52	0.24	11.54	1.21	0.012	balance

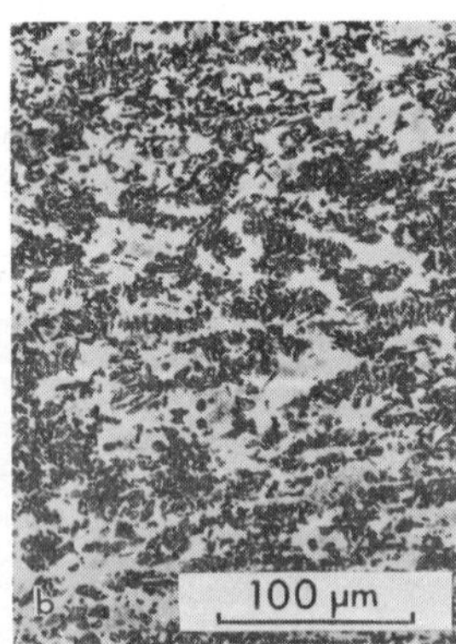

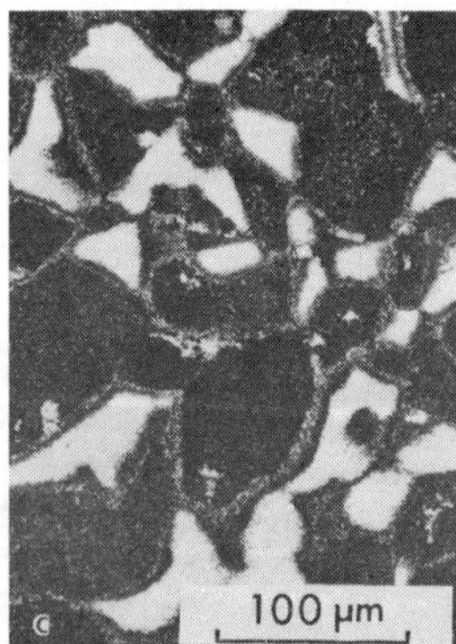

Fig. 1. Optical micrographs of the microstructures of alloy 1.
(a) as-processed, longitudinal (b) as-processed, transverse
(c) heat-treated. Martensite appears dark and ferrite light

Incremental step tests were performed for the determination of cyclic stress-strain (CSS) curves using a single specimen [9]. A limited number of companion specimens were also cycled under constant total strain amplitude to failure and the stable hysteresis loops obtained from these tests were used to verify the application of the incremental step test procedure for the dual-phase steels under investigation. An Instron servohydraulic testing system was used and all tests were performed under total strain control at room temperature.

For the microstructural investigations, samples were cycled at preset total strain amplitudes until stable hysteresis loops were obtained. Transverse sections from the gauge lengths were prepared for transmission electron microscopy using standard techniques.

RESULTS

Mechanical Behaviour

The initial cyclic response for tension-compression (R=-1) cycling with a strain amplitude of 0.6% for alloys 1 and 2 is shown in Fig. 2. For the as-processed material with the banded microstructure, samples from both the longitudinal and long transverse directions (with respect to the rolling direction) were tested and displayed similar trends. In order to investigate the effects of prior tensile deformation on cyclic response, samples were cycled about a mean tensile strain level of 2.5% with a strain amplitude of 0.6% as shown in Fig. 3, where both the change in the stress range and in the mean stress level with cycle number is indicated. There is clear evidence of cyclic stress relaxation in both alloys.

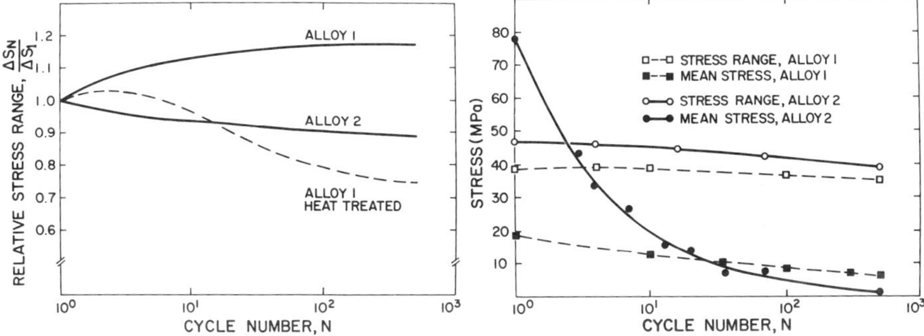

Fig. 2. Initial response to tension-compression cycling

Fig. 3. Initial response to tension-tension cycling

The corresponding cyclic stress strain curves are compared to the monotonic behaviour in Figs 4 and 5. The increase in strength of alloy 1 in the heat treated condition is consistent with the increased martensite volume fraction, but little difference in the corresponding cyclic stress strain curves is evident. Alloy 2, with the higher martensite content, shows cyclic softening.

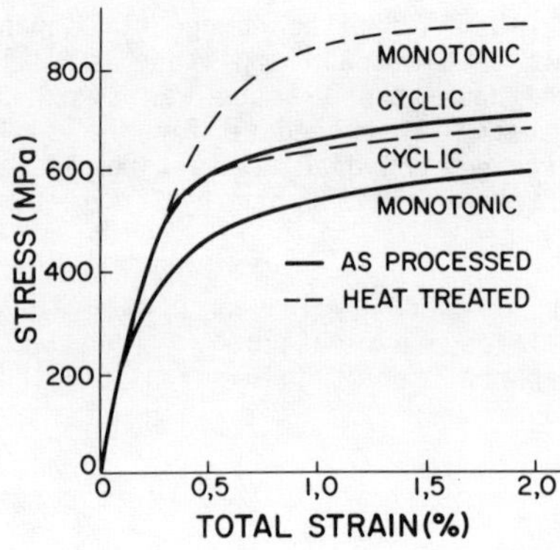

Fig. 4. Monotonic and cyclic stress strain curves for alloy 1 in the as-processed and heat treated condition

Fig. 5. Monotonic and cyclic stress strain curves for alloy 2

In order to assess the dislocation distribution development, the structures in undeformed and cyclically deformed samples were studied. Figure 6 shows the typical dislocation structure observed near the ferrite/martensite interfaces in undeformed heat treated material, in comparison with that encountered near the centre of the corresponding ferrite grain. Figures 7(a) and (b) show the distribution in samples cyclically stabilized under R=-1 conditions with total strain amplitudes of 0.5 and 1.0% respectively.

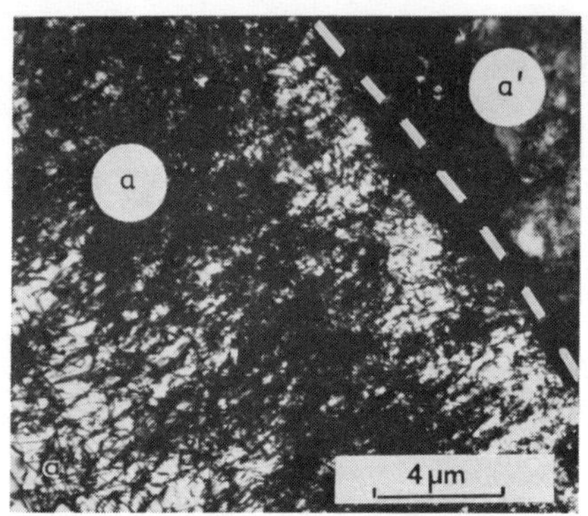

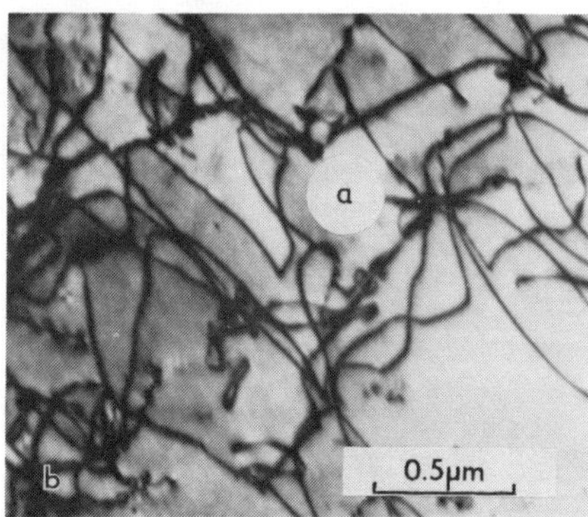

Fig. 6. Bright field transmission electron micrographs (a) showing the dislocation structure adjacent to the ferrite (α)/martensite (α') interface in an as-transformed specimen, (b) showing the dislocation density at the centre of a 20 μm diameter ferrite grain.

DISCUSSION

Cyclic Behaviour

The cyclic deformation response of the alloys investigated follows the trends observed in low-alloy dual-phase steels [5]. At lower martensite contents,

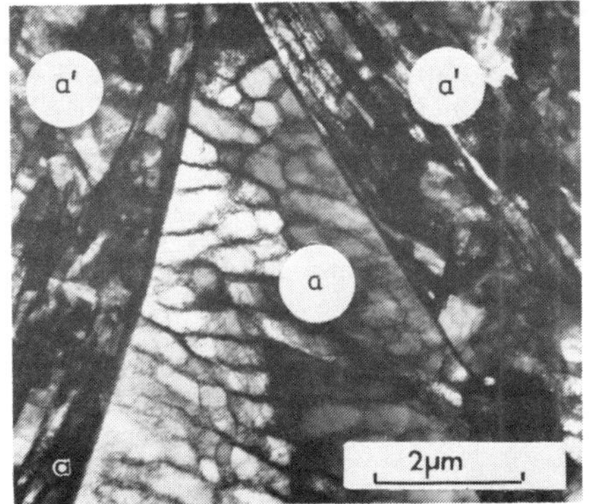

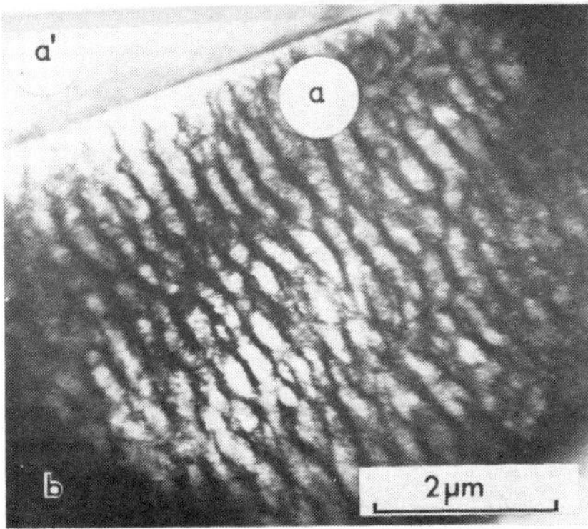

Fig. 7. Bright field transmission electron micrographs (a) showing dislocation cells in ferrite (α) produced by cyclic stabilization at a total strain amplitude of 0.5%, (b) dislocation structure in ferrite (α) after cyclic stabilization at a total strain amplitude of 1%

around 30%, cyclic hardening behaviour is observed (see Fig. 4). Although the tensile properties are enhanced by higher martensite fractions, cyclic softening behaviour was observed in the heat-treated specimens. This was apparent with considerably different martensite morphologies.

It was found further that, for both alloys 1 and 2 cyclic softening occurred in the tension-tension cycling condition. Cyclic softening for dual-phase steels in the prestrained condition has been reported by Sherman and Davies [10]. They found that as-processed vanadium-containing dual-phase steel cyclically hardened and that it softened after prestraining between two and eight percent. The cyclic stress strain curves for all conditions, however, fell within a relatively narrow band.

It can also be seen from Fig. 3 that the mean stress decreases with the number of cycles, and diminishes after about 50 cycles for both types of material cycled in the prestrained condition. This decrease in mean stress has been referred to as cycle dependent stress relaxation, or cyclic creep, and has been explained in terms of accumulation of plastic strain with each cycle of repeated straining associated with a rearrangement of unstable high density dislocation substructures [11,12]. Cycle-dependent stress relaxation can play an important role when high stresses are produced as the result of single large overloads.

Microstructure

Unlike the dual-phase steels studied by Sherman et al. [13], in the alloys investigated here there is convincing evidence of a high density of transformation-induced dislocations in the ferrite-martensite interface, with a decreasing density on progressing into a ferrite grain (see Fig. 6). Previous work has shown that on monotonic deformation, the high density near the interface is retained while the overall density increases within the ferrite phase [8]. The work-hardening behaviour, however, supports strain partitioning between the phases at the typical strain levels used in the cyclic work.

The dislocation structures observed in the cyclically deformed material show some significant changes. At both levels of strain ($\Delta\varepsilon$ = 0.5 and 1%), a clear fatigue cell structure is in evidence, uniformly distributed across the ferrite grains. It is immediately apparent that there is no evidence of transformation-induced dislocations near the interfaces, with the cell structures progressing right up to the boundaries, similar to that observed by Sherman et al. [13]. The cell structures, also, are not as equiaxed as those reported by these authors, with the present case showing distinct wall alignment. The cell size is on a scale of about 0.5 μm and decreases with increased cyclic strain amplitude. There appears to be a trend for the cell walls to thicken at these higher strain levels. It is interesting to note, in contrast to Sherman et al. [13], that all ferrite grains observed contained dislocation cells, even when surrounded by martensite, and no evidence could be found in the present case to suggest 'shielding' of some fraction of the ferrite.

The changes in dislocation structure observed indicate that cyclic straining is capable of significantly rearranging transformation-induced dislocations, and therefore it is reasonable that distributions produced by monotonic prestraining would react similarly. Thus the decreasing mean stress level on cyclic deformation is explicable in these terms.

The lath martensite, seen in all the alloys, is highly dislocated even in the undeformed condition. Detailed study of strain partitioning to this phase is therefore difficult using transmission electron microscopy, although some rearrangement of dislocations might be expected.

REFERENCES

1. R.G. Davies, Met. Trans. **9A,** 671 (1978).
2. W.R. Cribb and J.M. Rigsbee, in Structure and Properties of Dual-Phase Steels (edited by R.A. Kot and J.W. Morris), p.91. The Metallurgical Society of AIME, New Orleans (1979).
3. A.R. Marder, Met. Trans. **13A,** (1982).
4. N.J. Kim and G. Thomas, Met. Trans. **12A,** 483 (1981).
5. A.M. Sherman and R.G. Davies, Int. J. Fatigue, January 1981, 36 (1981).
6. J.Z. Briggs and T.D. Parker, in The Super 12% Cr Steels, Pub. Climax Molybdenum Co., Ann Arbor, Michigan (1965).
7. A. Ball and J.P. Hoffman, Met. Tech. September 1981, 329 (1981).
8. M.P. Shaw, F. Blum and N.R. Comins, Proc. Inaug. Int. 3CR12 Conference, p. 193, Johannesburg 1984.
9. R.W. Landgraf, JoDean Morrow and T. Endo, J. Materials, JMLSA, **4(1),** 176 (1969).
10. A.M. Sherman and R.G. Davies, Met. Trans. **10A,** 929 (1979).
11. JoDean Morrow and G.M. Sinclair, in Symposium on the Basic Mechanisms of Fatigue, ASTM STP 237, p. 83. American Society for Testing and Materials (1959).
12. P. Parikh and E. Shaprio, in Recent Developments in Mechanical Testing, ASTM STP 608, p. 106. American Society for Testing and Materials (1976).
13. A.M. Sherman, R.G. Davies and W.T. Donlon, in Fundamentals of Dual-Phase Steels (edited by R.A. Kot and B.L. Bramfitt), p. 85. The Metallurgical Society of AIME, Chicago (1981).

Cumulated Transversal Dilatation During Cyclic Loading of FCC Symmetrical Single Crystals

P. Franciosi

Lab. P.M.T.M., C.N.R.S., Univ. Paris Nord, Av. J. B. Clément, 93430 Villetaneuse, France

ABSTRACT

Tests on Silver and Copper FCC single crystals show that during cyclic loading along symmetrical crystallographic axes a cumulated transversal dilatation is observable and can be interpreted as a lattice rotation contribution to the effective hardening, which is different in tension and in compression, and what allows different systems combinations to be preferably active for the same orientation according to the applied load sign.

KEYWORDS

Single crystals, cyclic loading, axisymmetric loading.

INTRODUCTION

When single crystals are loaded in symmetrical orientations, more slip systems than required by the boundary conditions are simultaneously potential. It has been previously shown /1/ that a good prediction of the preferred active slip modes according to observed behaviour can be obtained from a stability criterion on the general form

$$W^{*} = \dot{\sigma}^{*}_{\alpha\beta}\, \dot{\varepsilon}^{T}_{\alpha\beta} - \dot{\sigma}_{\alpha'\beta'}\, \dot{\varepsilon}^{T*}_{\alpha'\beta'} \qquad \text{minimum} \tag{1}$$

where $\dot{\varepsilon}^{T}$ and $\dot{\sigma}$ are respectively the total strain rate and stress rate tensor and where $\alpha\beta$ and $\alpha'\beta'$ respectively denotes their assigned terms while (*) means any admissible solution.

When the elastic strain is neglected ($\varepsilon^{T} = \varepsilon^{P}$) one can write :

$$W^{*} = \sum_{g=1}^{m} \sum_{l=1}^{m} H^{gl} \dot{\gamma}^{*l} \dot{\gamma}^{*g} - 2 \sum_{g=1}^{m} \dot{\sigma}_{\alpha'\beta'}\, R^{g}_{\alpha'\beta'}\, \dot{\gamma}^{*g} \quad \text{minimum} \tag{2}$$

where $\dot{\gamma}^{*g}$ denotes any slip rate of a distribution which satisfies to the boundary conditions :

$$\dot{\varepsilon}^{P}_{\alpha\beta} = \sum_{g=1}^{m} R^{g}_{\alpha\beta}\, \dot{\gamma}^{*g} \tag{3}$$

with $R^g_{\alpha\beta}$ geometrical coefficients associated with the actual crystallographic loading orientation.

In(2) the $|H|$ matrix is called the effective hardening matrix and contains both the physical hardening $|h|$ from the usual phenomenological hardening law

$$\dot{\tau}_c^{*g} = \sum_{l=1}^{m} h^{gl} \dot{\gamma}^{*l} \tag{4}$$

and a lattice rotation effect which, from the Schmid plastic flow criterion

$$\dot{\tau}^{*g} = \sigma_{ij} \dot{R}_{ij}^{*g} + \dot{\sigma}_{ij}^{*} R_{ij}^{g} \leq \dot{\tau}_c^{*g} \tag{5}$$

for any g slip system or

$$\dot{\sigma}_{ij}^{*} R_{ij}^{\;g} \leq \dot{\tau}_{ij}^{*g} - \sigma_{ij} \dot{R}_{ij}^{*g} = \sum_{l=1}^{m} (h^{gl} - \sigma_{ij} \alpha_{ij}^{gl}) \dot{\gamma}^{*l} = \sum_{l=1}^{m} H^{gl} \dot{\gamma}^{*l} \tag{6}$$

takes the form of a matrix $\sigma_{ij}\alpha_{ij}$ where the α_{ij} terms are given by

$$\dot{R}_{ij}^{*g} = \sum_{l=1}^{m} \alpha_{ij}^{gl} \, \dot{\gamma}^{*l}$$

Since the α_{ij} matrices only depend on the loading orientation, for a given situation this rotation effect is only sensitive to the stress terms sign, and for an axial loading, is opposite in tension and in compression. In a previous theoretical analysis of the preferred slip combinations in axial loading along symmetrical axes /2/, one has shown that preferred slip modes in tension can differ from preferred ones in compression and vice versa. Then, the non assigned terms of the strain rate tensor are allowed to be different in tension and in compression, what can lead during cyclic loading to a cumulated transversal dilatation as follows : if 3 refers to the load axis, one has in tension :

$$\Delta\varepsilon_{33}^{T} \approx \Delta\varepsilon_{33}^{P} = \sum_{g=1}^{m} R_{33}^{g} \; \Delta\gamma^{g} \tag{7}$$

and in the simplest case of equilibrated slip on the m active systems :

$$\Delta\gamma^{g} = \Delta\gamma = \Delta\varepsilon_{33}^{P} \Big/ \sum_{g=1}^{m} R_{33}^{g} \tag{8}$$

while along the transverse direction 1 (or 2) one has :

$$\Delta\varepsilon_{11}^{P} = \Delta\gamma \sum_{g=1}^{m} R_{11}^{g} = \Delta\varepsilon_{33}^{P} \sum_{g=1}^{m} R_{11}^{g} \Big/ \sum_{g=1}^{m} R_{33}^{g} \tag{9}$$

Now, if in compression a m* systems combination is active, it comes :

$$\Delta\varepsilon_{11}^{P(c)} = \Delta\varepsilon_{33}^{P(c)} \sum_{g=1}^{m^*} R_{11}^{g} \Big/ \sum_{g=1}^{m^*} R_{33}^{g} \tag{10}$$

Then, during a complete cycle where $|\Delta\varepsilon^P_{33}| = |\Delta\varepsilon^{P(c)}_{33}|$ one obtains the cumulated plastic dilatation along 1 axis :

$$\text{Cum } \Delta\varepsilon^P_{1i} = |\Delta\varepsilon^P_{33}| \left(\sum_{g=1}^{m} R^g_{11} \Big/ \sum_{g=1}^{m} R^g_{33} - \sum_{g=1}^{m^*} R^g_{11} \Big/ \sum_{g=1}^{m^*} R^g_{33} \right) \qquad (11)$$

where both elastic strain and rotation are neglected.
This cumulated dilatation is expected from (11) proportional to both $|\Delta\varepsilon^P_{33}|$ and the cycles number N.

The now reported experimental results from tests on Ag and Cu symmetrical crystals allow a qualitative confirmation of the existence of this lattice rotation effect which leads to microscopic shape changes.

EXPERIMENTAL INVESTIGATION

Cyclic loading of Silver and Copper axisymmetric single crystals have been performed at room T° at strain amplitudes up to 1 % and at small strain rate ($\dot{\varepsilon}$ 10^{-4} s^{-1}). Cycling is monitored by an extensometer and strain gages are sticked in parallel to the transversal axes of the specimens to measure the cumulated transversal dilatation. The specimens are rigidly fixed on an Instron type machine so that a zero total rotation is a convenient assumption to calculate the $|\alpha_{ij}|$ matrices. From the whole results, the theoretically expected cumulated transversal dilatation do significantly exist and is much more important for Ag than for Cu. The here reported results are concerned with a <110> axis (for Ag and Cu) and with a <122> axis (for Cu).

- for a <110> axis, figure 1, the observed slip mode in tension is always

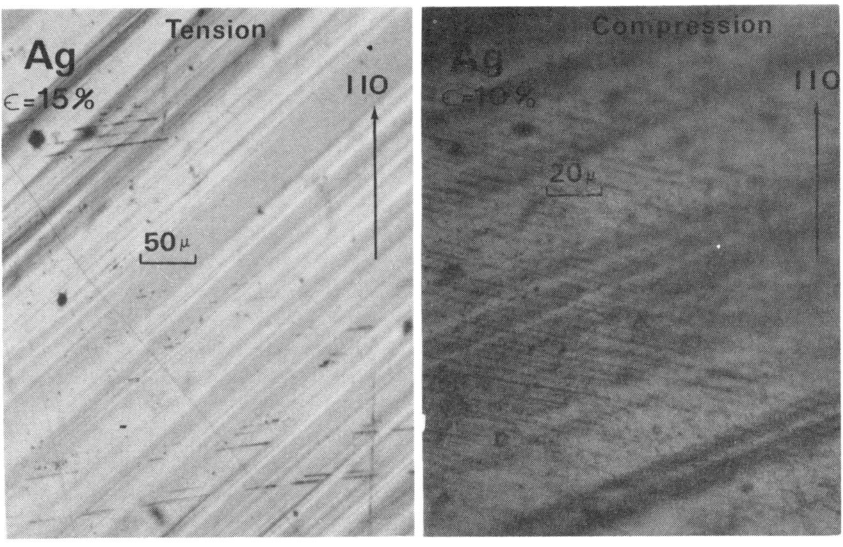

Fig. 1 : Micrographies of a <110> Ag axis : a) tension ; b) compression.

single glide, while in compression, non coplanar double slip is observed at small strain for anisotropic metals such as Ag or Cu (low stacking fault energy).

In tension, the stress strain curve shows an initial stage I of weak hardening while in compression one observes the parabolic shape characteristic of multislip /3/. During cyclic loading along <110>, the hardening slope, as shown for Ag on Fig. 2, is much higher in compression than in tension, in accord with the slope differences observed between monotonic tensile and compressive loading, what reflects the change from single slip mode to multislip.

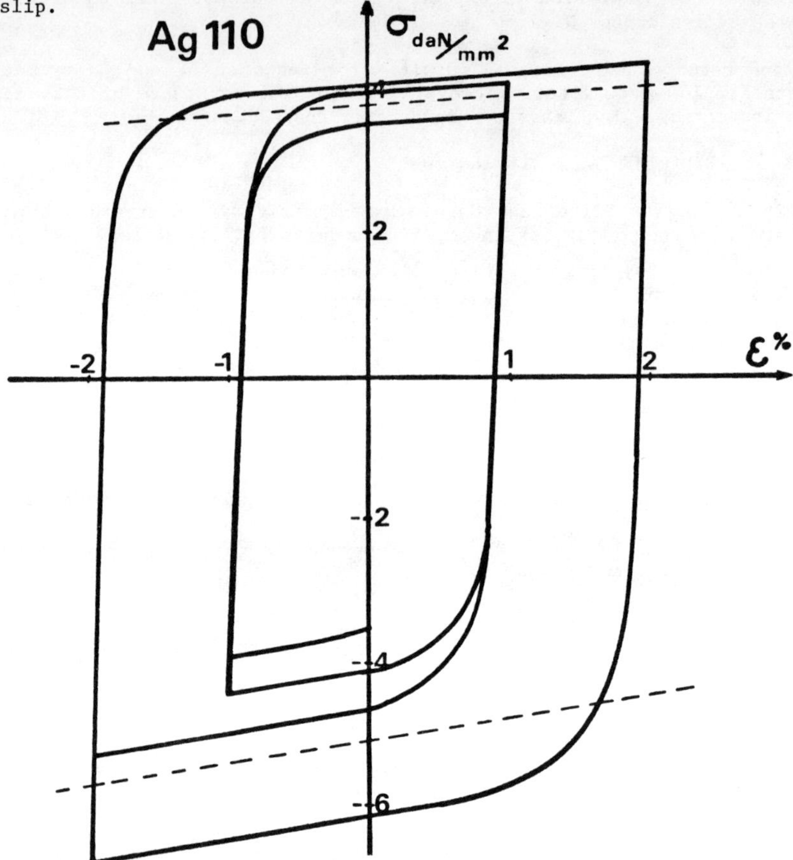

Fig. 2 : Cyclic σ/ε curve for Ag <110> axis.

Measured cumulated transversal dilatations for a Ag <110> axis are plotted fig. 3 : A and B curves correspond with two identical specimens : on one transversal direction the gage signal was negligible (unfavourable face geometry) and not reported here. On the other direction, the cumulated dilatation is about 1,5 % (per cycle) of the $\Delta\varepsilon_{33}^{P}$ cycling strain amplitude. According to the simplifying assumptions ($\varepsilon^{e} = 0$, $\Delta\gamma^{g} = \Delta\gamma$, $\omega^{e} = 0$) the linearity with both $\Delta\varepsilon_{33}^{P}$ and N appears consistent with the theoretical interpretation from (7) and (11).

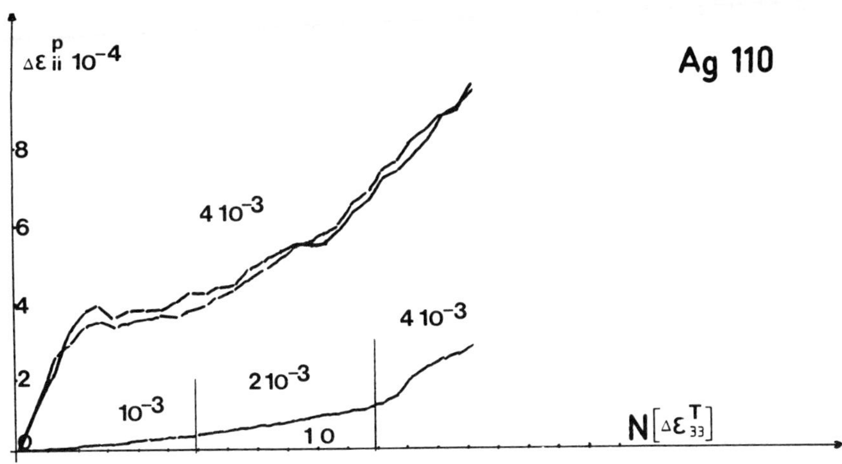

Fig. 3 : $\Delta\varepsilon^{P}_{ii}$ for Ag <110> axis

The results for Cu <110> and <112> axes are plotted fig. 4.

- For a <112> axis, the preferred slip mode is single glide in tension and coplanar double glide in compression. As for Ag the linear relation with both Δε 33 P and N is satisfactorily verified according to the simplifications.

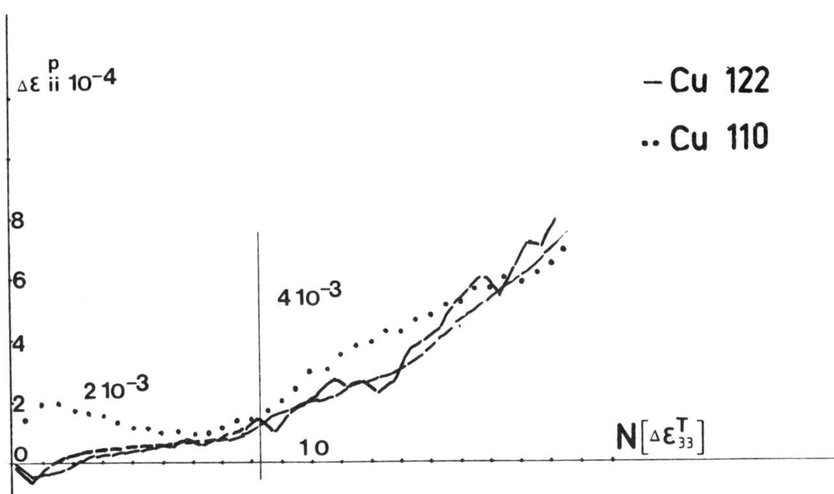

Fig 4 : $\Delta\varepsilon^{P}_{ii}$ for Cu<110> and Cu<122> axes

CONCLUSION

While for BCC crystals such an experimentaly observed transversal cumulated dilatation /4/ - even for non symmetrical axis orientations - is interpreted in terms of a <111> <112> slip systems asymmetry, i.e. of a physical hardening asymettry (an asymmetric $|h|$ matrix in our approach) , for the FCC single crystals, it can be interpreted in terms of a lattice rotation effect included in the effective hardening ($|H|$ matrix in (2)) but is only expected for few symmetrical axis orientations.

Now, although this cumulative strain effect in FCC crystals (what finally means microscopic shape changes) is smaller than for the BCC structure, as for BCC /4/, it can be expected to contribute in the initiation of fatigue cracks at FCC polycrystals grain boundaries.

REFERENCES

/1/ Franciosi P. Thesis (1984) Paris.

/2/ Franciosi P., Zaoui A. Acta Metall, 30 (1982) p. 1627.

/3/ Nakada Y., Kocks V.F., Chalmers B. Trans. Metall Soc. Aime. 230 (1964) 7 607.

/4/ Mughrabi H., Wuthrich Ch., Phil. Mag. 33 (1976), p. 963.

Cyclic Deformation of Molybdenum Single Crystals

J. A. Planell* and F. Guiu

Department of Materials, Queen Mary College, University of London, Mile End Road, London E1 4NS, UK
**At present: Departamento de Metalurgia, E.T.S. Ingenieros Industriales de Barcelona, Universidad Politécnica de Cataluña, Diagonal 647, 08020-Barcelona, Spain*

ABSTRACT

Molybdenum single crystals with single glide orientation were cyclically deformed in tension-compression at four different temperatures and at a constant strain-rate of 6.0×10^{-4} s^{-1}. The cyclic stress-strain curves were determined and that obtained at 400 K was studied in detail. It shows four stages like other b.c.c. crystals deformed near their transition temperature. The dislocation substructures have been analysed by T.E.M. in foils prepared from specimens deformed in each region.

KEYWORDS

Cyclic deformation, Molybdenum single crystals, Transition temperature, Dislocation substructures.

INTRODUCTION

It is now well established that b.c.c. metals exhibit a low temperature behaviour the main characteristics of which are the strong temperature and strain-rate dependence of the yield stress and the slip and stress asymmetry with respect to tension and compression when deformed at and below their transition temperature (for a review, see 1,2,3). Although the cyclic deformation properties and the dislocation substructures developed have been far less explored, a general pattern of behaviour can be drawn from recent studies by different authors working with different crystals and it can be summarized as follows: At temperatures below the transition temperature, the c.s.s. curves have a quasi-parabolic shape, and multiple glide and stress asymmetry are observed (4,5,6). A dislocation substructure consisting of a network of long screw dislocations with primary and secondary Burgers vectors is developed (6). Around the transition temperature, the c.s.s. curves display four regions (7,8,9). Different cyclic hardening behaviour is observed in each region, deformation takes place by single glide, stress and glide asymmetry are observed and changes in shape can be measured (8,9,10). Large numbers of both edge and screw primary dislocations are involved in the dislocation substructures reported (6,11,12). At a higher temperature regime, achieved using slower strain-rates, the different stages of the c.s.s. curve, the stress and glide asymmetry and the changes in shape all disappear (7,9,11). Screw dislocations are rarely seen and it seems that dislocations of the primary system are predominant (11,12).

EXPERIMENTAL METHOD

Molybdenum single crystals with a total interstitial content of less than 75 p.p.m. by mass were grown with single glide orientation. Specimens were cyclically deformed in tension-compression in a servohydraulic testing machine under strain control and at different temperatures. The c.s.s. curves were determined by plotting axial saturation stress against axial saturation plastic strain using the "incremental step method" (i.s.m.). The total strain-rate used in all tests was $6.0 \times 10^{-4}\ s^{-1}$. The specimens cyclically deformed directly at different constant plastic strain amplitudes were sliced parallel to both primary and conjugate glide planes and thin foils for T.E.M. observa tions were prepared. The "image-absence" technique was used to identify the dislocation Burgers vectors. Dislocation densities in the different substructures were measured.

EXPERIMENTAL RESULTS

Cyclic Stress-Strain Curves at Different Temperatures

C.s.s. curves at 350, 375, 400 and 450 K were determined and are shown in Fig. 1. At 350 K the c.s.s. curve is quasi - parabolic like at room temperature (5), although the level of stresses is lower. At 375 K a plateau may exist but was missed in the i.s.m. determination. Such suggestion is based on the softening observed in the cyclic hardening curve obtained at $\varepsilon_T = \pm 0.48 \times 10^{-3}$. The c.s.s. curves at 400 and 450 K respectively show four different regions, one of them being a distinctive plateau. As it can be noticed, the stress asymmetry decreases with increasing temperature.
These results confirm that Mo follows the same general pattern of behaviour as other

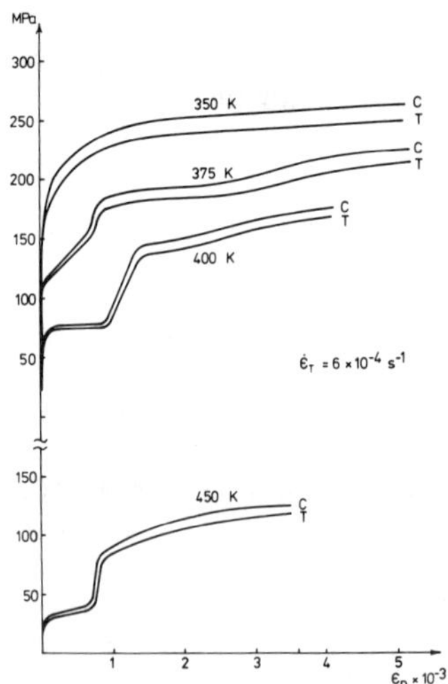

Fig. 1. C.s.s. curves at 350,375, 400 and 450 K obtained by the i.s.m. (T) and (C), are saturation stresses in tension and compression.

b.c.c. crystals. When the metal is deformed around its transition temperature, the c.s.s. curves display four clear stages, while below the transition temperature, the c.s.s. curves are quasi-parabolic. In order to observe the flattening of the c.s.s. curves, temperatures higher than 450 K would probably be needed.

Dislocation Substructures at 400 K

The temperature of 400 K was chosen for the present investigation due to the clear and extensive plateau exhibited by the c.s.s. curve. Four regions can be differentiated and they correspond to region 1 or the microstrain region, $\varepsilon_p < 0.2 \times 10^{-3}$; region 2 or the plateau region, $0.2 \times 10^{-3} < \varepsilon_p < 0.9 \times 10^{-3}$; region 3 or the rapid hardening region, $0.9 \times 10^{-3} < \varepsilon_p < 1.2 \times 10^{-3}$; and region 4 or the high strain amplitude region, $\varepsilon_p > 1.2 \times 10^{-3}$. Several virgin specimens were cycled at different plastic strain amplitudes typical of the various regions and at different cumulative plastic strains. Thin foils were then prepared and dislocation substructures studied.

The general features of the dislocation substructures observed in foils parallel to the primary glide plane $(\bar{1}01)$ from specimens deformed in the plateau region can be described as a random distribution of poorly developed dislocation tangles variable in sizes and shapes and with large free areas between them. Small dislocation loops, dipoles and debris with primary Burgers vector $[111]$ were mainly accumulated near the tangles. The detailed analysis of the tangles showed that they consisted mainly of primary dislocations attached to a backbone constituted by a secondary dislocation or a group of short non-primary segments. Fig 2 shows a small tangle under different diffracting conditions, and the presence of primary and non-primary dislocations is clearly revealed. Table 1 shows the measurements of the average local dislocation density in the tangles, $\bar{\rho}_L$, and the percentages of primary dislocations present in them, P%, in foils prepared from specimens deformed at different plastic strain amplitudes and cumulative plastic strains. It can be noticed that the percentage of primary dislocations in the tangles increases with both the plastic strain amplitude and the cumulative plastic strain. In the conjugate plane the results can be correlated to those found in the primary plane.

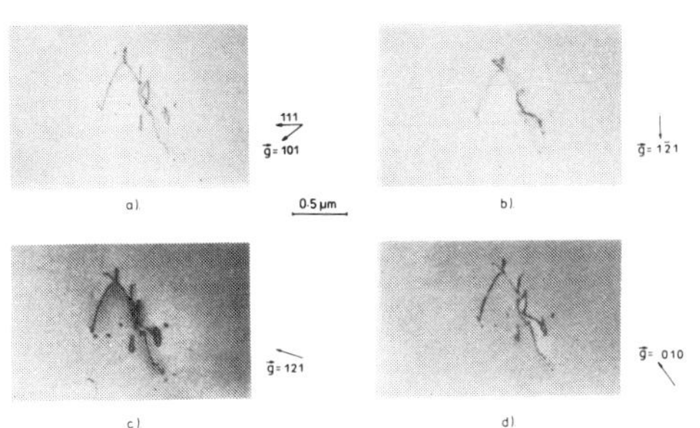

Fig. 2. Burgers vector extinction analysis of a tangle in a foil parallel to the primary glide plane from a specimen deformed in the plateau region.

The dislocation substructures observed in foils parallel to the primary glide plane prepared from specimens deformed in region 3 were very inhomogeneous. Two main types of dislocation arrangements were present: a greater number of tangles

Table 1. Dislocation Densities in Regions 2,3 and 4, Measured in the Primary Glide Plane

			$\bar{\rho}_L$/cm^{-2} tangles	P % tangles	$\bar{\rho}_L$/cm^{-2} walls	P % walls	$\bar{\rho}_L$/cm^{-2} bundles	P % bundles
Region 2	$\varepsilon_p = \pm 0.35 \times 10^{-3}$	$\varepsilon_c = 0.85$	1.1×10^9	67%				
		$\varepsilon_c = 2$	2×10^9	71%				
	$\varepsilon_p = \pm 0.77 \times 10^{-3}$	$\varepsilon_c = 0.85$	1.3×10^9	73%				
		$\varepsilon_c = 2$	2×10^9	79%				
Region 3	$\varepsilon_p = \pm 1.04 \times 10^{-3}$	$\varepsilon_c = 5$	2×10^9	70%	9×10^9	55%		
	$\varepsilon_p = \pm 1.13 \times 10^{-3}$	$\varepsilon_c = 4.5$	2×10^9	60%	2.5×10^{10}	46%		
Region 4	$\varepsilon_p = \pm 1.21 \times 10^{-3}$	$\varepsilon_c = 0.07$					1.3×10^{10}	52%
		$\varepsilon_c = 2$					2.2×10^{10}	50%
	$\varepsilon_T = \pm 1.95 \times 10^{-3}$	$\varepsilon_c = 0.34$					2×10^{10}	51%

with the same morphology as those observed in the plateau, and dislocation walls. Such different substructures could be present in different areas of a single foil. The average local dislocation density in the tangles and walls, $\bar{\rho}_L$, and the percentages of primary dislocations, P%, are shown in Table 1. The percentages of primary dislocations in the tangles tend to decrease with plastic strain amplitude, and in the walls such percentages are even lower. This suggests an increased activity of secondary slip systems.

Observations made in foils parallel to the conjugate glide plane revealed two main dislocation arrangements: tangles similar to those found in the plateau, in greater numbers, and a ragged dislocation cell structure. A detail of a wall is shown in Fig. 3. The percentages of secondary dislocations were as in the primary plane, higher than 50%; dipoles, loops and debris with conjugate Burgers vector 111 could be observed in them which suggests the activity of conjugate dislocations sources in agreement with observations made in the primary plane. The dislocation substructures observed in foils parallel to either the primary or the conjugate planes from specimens deformed at the lower plastic strain amplitudes of region 4 can be described as a random distribution of unconnected dislocation bundles of variable sizes as shown in Fig. 4. The bundles are dislocation arrangements more developed in size than the tangles and containing dislocation loops, dipoles and short segments of both primary and secondary dislocations. At higher strain amplitudes the bundles seem to merge and form an elongated cell structure. Table 1 shows the average local dislocation density in the bundles, $\bar{\rho}_L$, and the percentages of primary dislocations, P%. One of the tests was carried out under total strain amplitude control. It can be noticed that the dislocation density in the bundles is approximately a factor of 10 higher than in the tangles observed in the plateau. The percentages of primary dislocations in the bundles are similar to those found in the walls of region 3.

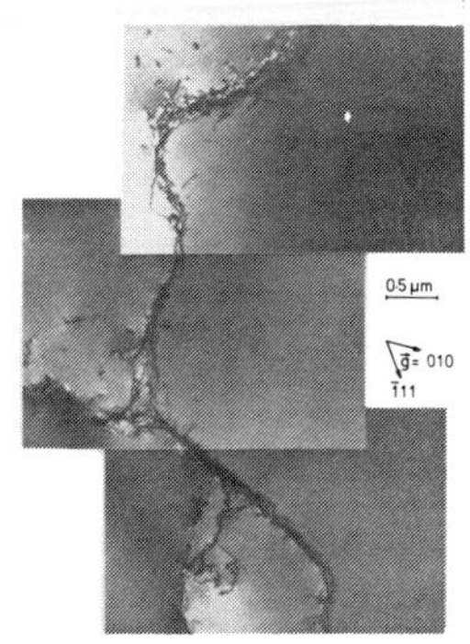

Fig. 3. Detail of a wall in the cell structure observed in a foil parallel to the conjugate glide plane from a specimen deformed in region 3.

DISCUSSION

The percentages of primary dislocations measured in the different substructures observed in the different regions of the c.s.s. curve at 400 K reveal that glide seems to be accommodated mainly by the primary slip system, although secondary slip systems become active at plastic strain amplitudes typical of region 4, in agreement with other T.E.M. observations in Nb (6), α-Fe (10), and Fe-26Cr-1Mo (11) single crystals and with slip line observations and changes in shape measurements in the same metals (8,9,13). The activity and mobility of primary screw dislocations in the plateau region is confirmed by the presence of prismatic loops, dipoles and debris with primary Burgers vector near the tangles, and which must have been formed by double cross-slip of primary screw dislocations. Consequently, multiplication of primary dislocations must take place in the plateau region in agreement with the mechanical results interpretation obtained in Nb (9).

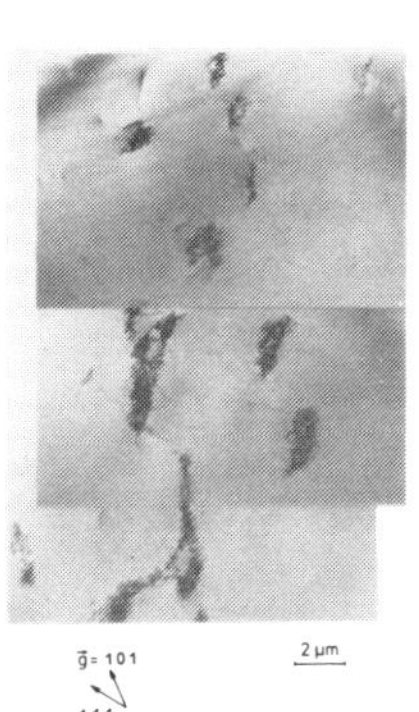

Fig. 4. Elongated bundles in a foil parallel to the primary glide plane from a specimen deformed in region 4.

When plastic strain amplitude increases to levels typical of region 3, the percentages of primary dislocations in the tangles decrease in comparison to those found in the plateau. However, the study of the cyclic hardening behaviour in this region is needed in order to give an interpretarion of the wall structure observed and the increased activity of secondary slip systems (13).

In region 4, large numbers of both primary and secondary dislocations are found in the bundles from the beginning of deformation as the test carried out at $\varepsilon_c = 0.07$ reveals. Therefore, although primary dislocations account for more than 50% of the average local dislocation density in the bundles, secondary sources are readily activated from the first cycles of deformation.

The influence of temperature on the cyclic deformation behaviour can be now explained in terms of the effect of temperature on the mobility of screw dislocations. The interpretation of the tensile deformation behaviour of Mo single crystals at different temperatures given by Luft and Kaun (14) can be extended to the cyclic deformation properties. At low temperatures, edge dislocations on secondary systems become mobile before primary screw dislocations due to their low mobility. Therefore, secondary dislocations are responsible for the high rate of hardening and the quasi-parabolic shape of the c.s.s. curves. The results on high purity Nb crystals cyclically deformed at 77 and 195 K (6) provide a good evidence. At temperatures around the transition temperature, primary screw dislocations are more mobile and the activation of secondary dislocations is delayed, which explains the presence of the plateau in the c.s.s. curves. At higher plastic strain amplitudes, secondary sources are activated (6,9,10,11), and cyclic hardening in regions 3 and 4 takes place. At temperatures above the transition temperature, both edge and screw dislocations have an equally high mobility and consequently, the primary slip system will be predominant as evidence suggests (10,11). The low activity of the secondary glide systems will be responsible for the low rate of hardening exhibited by the c.s.s. curves.

ACKNOWLEDGEMENT

One of the authors (J.A. Planell) would like to thank the CIRIT from the Generalitat of Catalunya for financial support during part of this work.

REFERENCES

1. J.W. Christian, 2nd Int. Conf. on the Strength of Metals and Alloys, Vol. 1, Asilomar, American Society for Metals, p. 31, (1970).
2. L.P. Kubin, Rev. Def. Behaviour of Metals, 1, 244, (1977).
3. B. Sestak, 5th Int. Conf. on the Strength of Metals and Alloys, Vol. 3, Pergamon Press, p. 1461, (1979).
4. B. Etemad and F. Guiu, Scripta Met., 8, 931, (1974).
5. M. Anglada, B. Etemad, J.A. Planell and F. Guiu, Scripta Met., 14, 1319, (1980).
6. F. Ackermann, Dr. Rer. Nat. Thesis, Institut für Physik, University of Stuttgart, (1981).
7. H. Mughrabi, K. Herz and X. Stark, Acta Met., 24, 659, (1976).
8. H. Mughrabi and Ch. Wüthrich, Phil. Mag., 33, 963, (1976).
9. M. Anglada and F. Guiu, Phil. Mag., 44, 499 and 523, (1981).
10. H. Mughrabi, K. Herz and X. Stark, Int. J. of Fracture, 17, 193, (1981).
11. T. Magnin, A. Fourdeaux and J.H. Driver, Phys. Stat. Sol. (a), 65, 301, (1981).
12. H. Mughrabi, F. Ackermann and K. Herz, ASTM Special Technical Publication, nº 675, p. 69, (1979).
13. J.A. Planell and F. Guiu, to be published.
14. A. Luft and L. Kaun, Phys. Stat. Sol., 37, 781, (1970).

Etude Comparative du Comportement en Fatigue Oligocyclique d'un Alliage Austenitique a l'Etat Moule et a l'Etat Corroye

M. Gerland, J. Charrier, J. de Fouquet et J. P. Amirault

Laboratoire de Mécanique et de Physique des Matériaux U.A.863 au C.N.R.S., ENSMA Rue Guillaume VII-86034 Poitiers Cedex, France

RESUME

Le comportement en fatigue oligocyclique d'un alliage austénitique Nicral C35, à l'état moulé et à l'état corroyé, a été étudié en fonction de la température, de 20 à 1000 °C, et de l'amplitude de sollicitation totale. Les mécanismes d'endommagement et de propagation des fissures, dans les deux états, associés à une consolidation cyclique importante au dessous de 700 °C et à un adoucissement continu au dessus de 700 °C ont été déterminés par microscopie en transmission et à balayage.

MOTS-CLES

Fatigue oligocyclique, alliage austénitique, structure moulée ou corroyée, haute température, microfractographie (MEB), dislocations (MET), déformation Bauschinger.

INTRODUCTION

Les structures coulées présentent de façon générale une moindre résistance en fatigue, liée à la dimension des cristaux dendritiques, à l'anisotropie et aux ségrégations au niveau des interfaces entre dendrites /1-3/. Cette moindre résistance est un obstacle à l'utilisation des pièces coulées qui par ailleurs présentent des avantages de faisabilité soit dans le cas de pièces de forme complexe, soit dans le cas de grandes séries. La présente étude a pour objet la comparaison, dans le cas de l'alliage Nicral C35, du comportement en fatigue oligocyclique à chaud du matériau à l'état moulé et du matériau à l'état corroyé. Les courbes de consolidation ou d'adoucissement cyclique, les boucles d'hystérésis, l'amplitude de la déformation Bauschinger et les évolutions microstructurales observées par microscopie électronique sur lames minces ont permis de caractériser les mécanismes d'endommagement cyclique des deux matériaux et d'expliquer l'amorçage et les modes de propagation des fissures entre l'ambiante et 1000 °C.

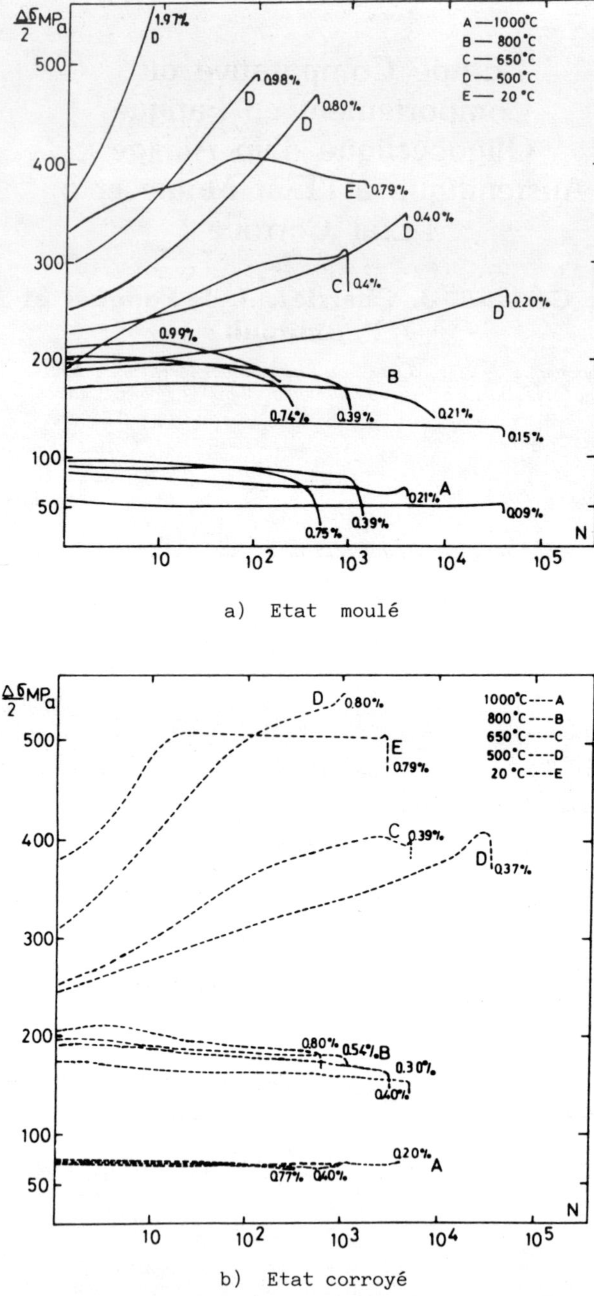

FIG.1 - Evolution des contraintes en fonction du nombre de cycles.

CONDITIONS EXPERIMENTALES

Les conditions expérimentales sont décrites de façon détaillée dans la référence /4/. Les essais de fatigue ont été réalisés à amplitude de déformation cyclique totale $\Delta\varepsilon_t$ imposée de $\pm$ 10^{-3} à $\pm$ 10^{-2}, à vitesse de déformation constante de 5.10^{-4} s^{-1}.

RESULTATS EXPERIMENTAUX

* <u>Comportement cyclique</u>

Sur la Fig. 1a,b sont représentées les courbes d'évolution de la contrainte cyclique maximale à $\Delta\varepsilon_t/2$ fixée en fonction du nombre de cycles aux différentes températures respectivement pour l'état moulé (a) et pour l'état corroyé (b). On observe d'une part l'influence importante de la température sur l'écrouissage cyclique de part et d'autre de 700 °C, d'autre part la sensibilité beaucoup plus grande de la structure coulée à l'effet d'amplitude de sollicitation aux températures élevées. Les courbes de Manson-Coffin sont représentées sur la Fig.2a,b respectivement pour l'état moulé (a) et l'état corroyé (b), pour les températures de 500, 800 et 1000 °C.

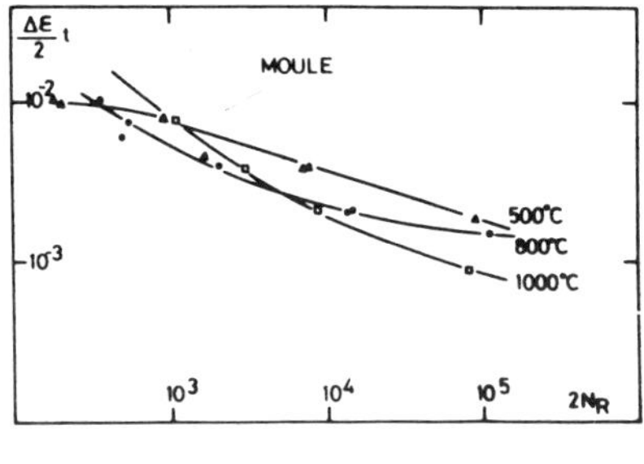

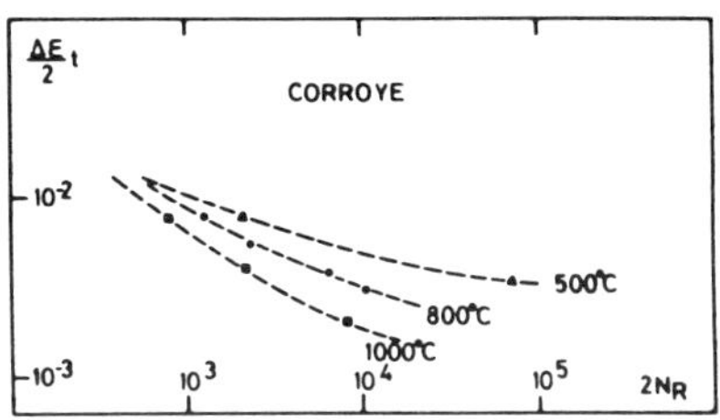

a) état moulé b) état corroyé

FIG.2 - Courbes de Manson-Coffin

* <u>Etude microfractographique</u>

. Etat moulé

Sauf à 800 °C et quelle que soit l'amplitude de sollicitation cyclique, l'amorçage et la propagation sont à prédominance interdendritique, en particulier à forte amplitude. Les stries de fatigue sont rares et mal définies (Fig.3a). De très nombreux précipités sont visibles au niveau des faciès de propagation, surtout dans les joints (Fig.3b). La rupture finale est totalement interdendritique.

. Etat corroyé

Sauf à 650 °C l'amorçage et la propagation sont à prédominance transgranulaire. Des stries de fatigue nettement visibles et recouvrant des plages étendues apparaissent après rupture à 500 et 650 °C (Fig.3c). Comme dans

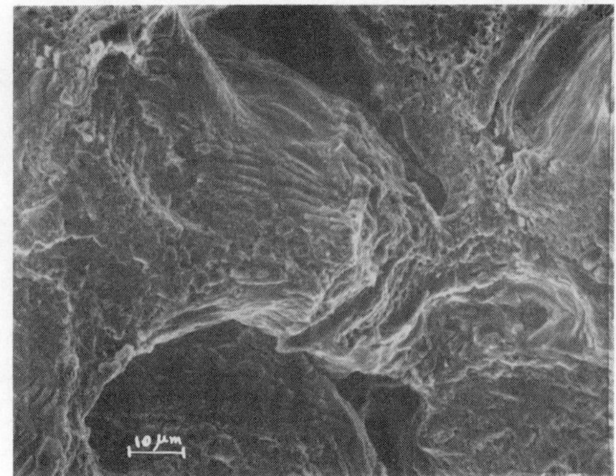

FIG.3a - Etat moulé 500 °C

$$\frac{\Delta\varepsilon_t}{2} = 0{,}99\ \%$$

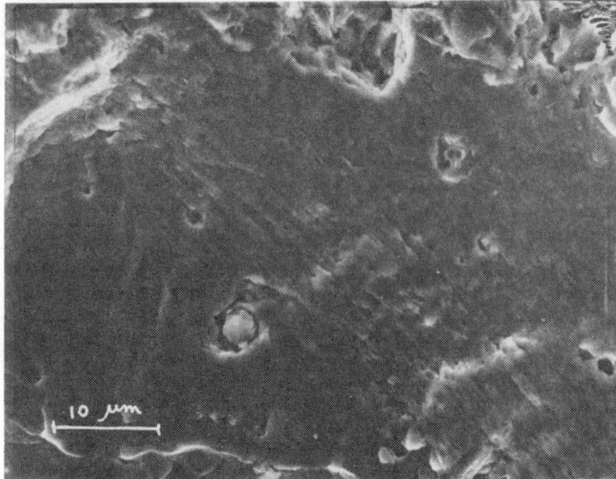

FIG.3b - Etat moulé 800 °C

$$\frac{\Delta\varepsilon_t}{2} = 0{,}15\ \%$$

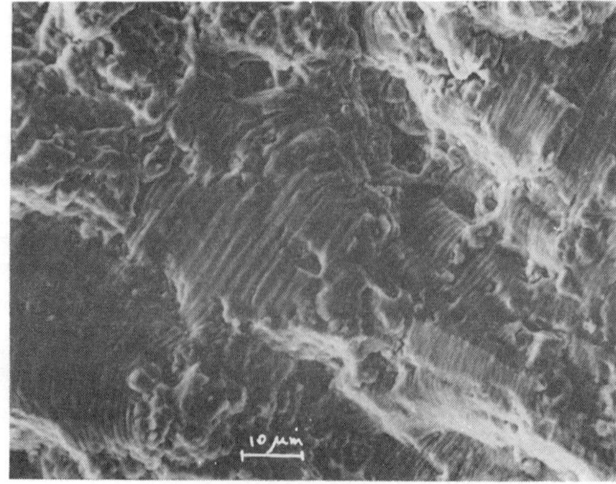

FIG.3c - Etat corroyé 650°C

$$\frac{\Delta\varepsilon_t}{2} = 0{,}39\ \%$$

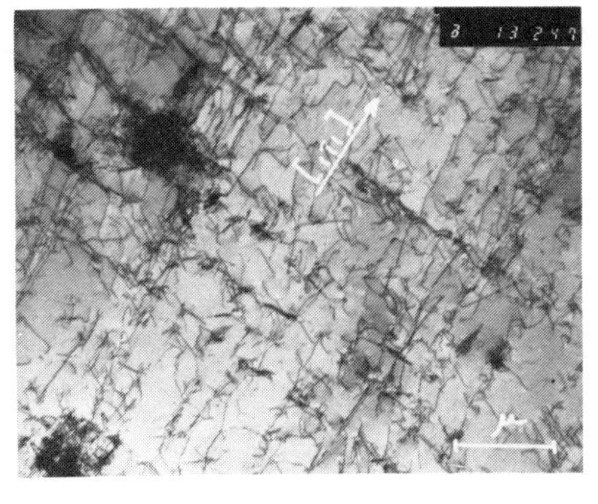

a) 20 °C

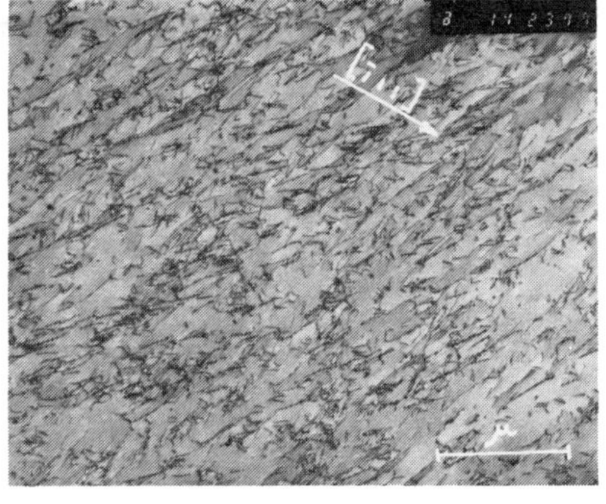

b) 500 °C

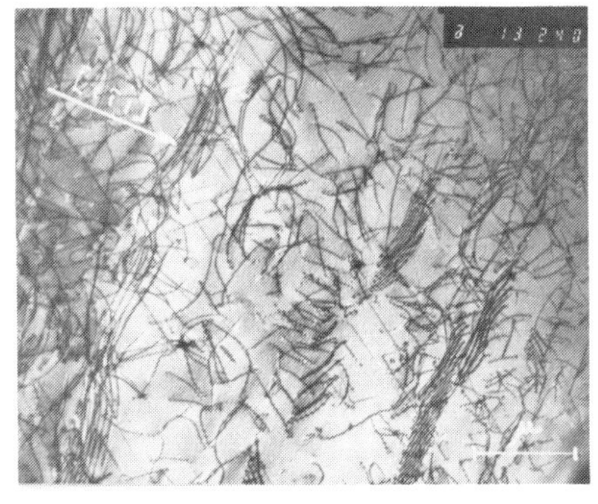

c) 650 °C

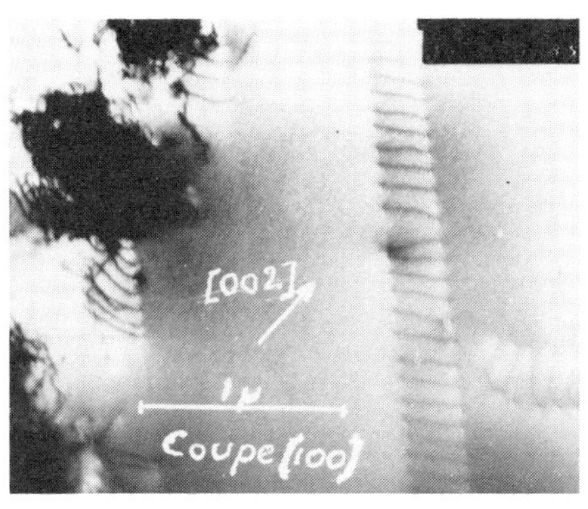

d) 800 °C

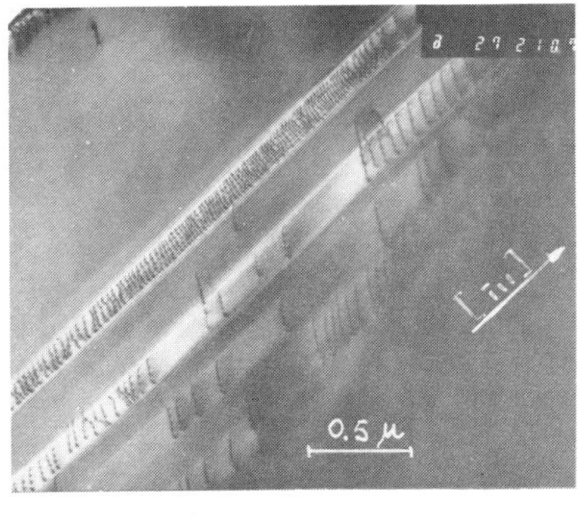

e) 1000 °C

FIG.4 - Eprouvettes fatiguées 10 cycles

$$\frac{\Delta\varepsilon_t}{2} = 0,40\ \%$$

l'état moulé, les traces de nombreux précipités arrachés au cours de la propagation apparaissent sur les faciès.

* Arrangement des dislocations

La configuration des dislocations après fatigue, d'autant moins homogène que la température est élevée, est assez semblable dans les deux états. L'arrangement est plutôt cellulaire à basse température et sous forme d'empilements à haute température (Fig.4). Au cours de la durée de vie, leur densité augmente considérablement à 500 °C, et de façon assez importante à 20 et 650 °C. A plus haute température, la structure évolue sous forme de sous-joints, surtout à l'état corroyé /4/.

ANALYSE ET CONCLUSION

Les différences de durée de vie ne sont pas considérables. Toutefois les durées de vie sont systématiquement plus longues à l'état corroyé qu'à l'état moulé, sauf à 1000 °C. Cette augmentation s'explique principalement par la propagation plus lente et plus transgranulaire des fissures dans l'alliage corroyé, alors que les joints interdendritiques jouent un rôle prépondérant dans l'alliage moulé.

Au dessous de 700 °C, l'effet le plus marquant est la très forte consolidation cyclique qui provoque la décohésion des interfaces précipités-matrice responsable de l'endommagement des joints à l'état moulé. La consolidation maximale et continue à 500 °C a pour conséquence un stade de propagation très court. Par contre à 20 °C et à 650 °C, la consolidation moins intense et plus rapide explique un stade de propagation qui occupe une fraction importante de la durée de vie.

Au dessus de 700 °C, le comportement cyclique se décompose en trois stades :
- un régime transitoire correspondant à l'établissement d'une structure des dislocations quasi stationnaire qui résulte à la fois de leur interaction mutuelle et de leur interaction avec les précipités de la matrice, fonction de la température.
- un régime quasi stabilisé qui occupe la majeure partie de la durée de vie et au cours duquel se forment les microfissures.
- une déconsolidation finale rapide associée au développement d'une fissure macroscopique.

L'évolution de la déformation Bauschinger au cours de la vie de l'éprouvette a montré que cette déformation est maximale au moment de l'amorçage des premières microfissures, et au contraire minimale lorsque la propagation est transgranulaire par stries.

REFERENCES.

1. C. Amzallag, P. Rabbe, G. Gallet, H.P. Lieurade, Mém. Sci. Rev. Mét. 75, 161 (1978).
2. J. Davidson, Rapport Creusot-Loire, E.CR. 763 (1977).
3. W.D. Cao, X.P. Lu, J.J. Liu, A.D. Yang, Mat. Sci. Eng. 63, 165, (1984).
4. M. Gerland, Thèse Doctorat d'Etat, Poitiers (1984).

Fatigue Deformation Near Regions of Surface Damage in FCC Crystals

P. Charsley and J. D. M. White

Physics Department, University of Surrey, Guildford, UK

ABSTRACT

Surface deformation features have been studied near indentations on single crystals of Cu after low amplitude fatigue. Persistent slips bands (PSB's) have been observed on localized regions although PSB's did not form elsewhere. This effect persists, with some modification, after annealing. Consideration of the various contributions suggests that the detailed dislocation microstructure is the most significant.

KEYWORDS

Fatigue deformation, persistent slip bands, dislocations, indentation, Cu monocrystals.

INTRODUCTION

Materials used in an engineering context invariably have surface imperfections which are avoided in laboratory specimens used for fundamental research. This work is an attempt to determine the effect on the fatigue behaviour of damage due to single indentations on otherwise perfect monocrystalline copper surfaces. Since the development of microcracks from PSB's is the most likely form of failure, we have investigated their formation near indentations for several different crystal orientations, both for as-indented specimens and for specimens which have been annealed after indentation.

EXPERIMENTAL TECHNIQUES

Crystals 2 mm thick were grown, by the Bridgeman technique, from 99.99% Cu of the shape and orientations shown in Figs.1(a) and 1(b). The central points on Fig.1(b) correspond to the specimen axes and the extreme points to the directions of the specimen edges. Specimens were mechanically and

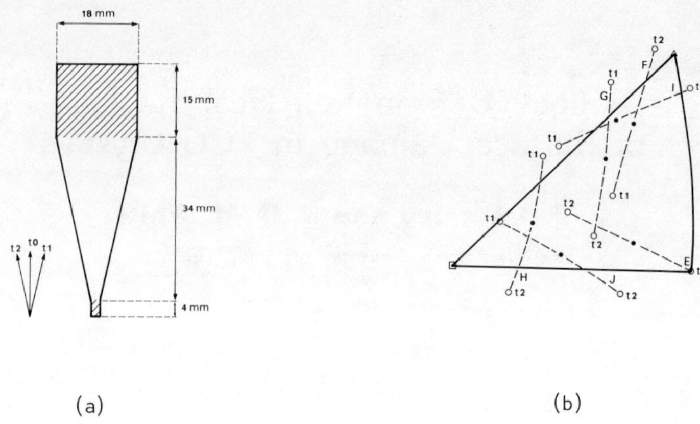

(a) (b)

Fig.1. Details of Cu specimens

(a) Specimen shapes. (b) Specimen orientations; the central point is the direction of the central axis and the extreme points correspond to the specimen edges.

electrolytically polished, annealed in vacuo at 700°C and then indented at four points along the axes using a Vickers diamond pyramid indenter. Indentations were ∿ 12 μm deep and oriented with the side parallel or perpendicular to the tensile axis. After a light electropolish specimens were fatigued in bending, at room temperature, at a constant surface total strain amplitude of 8.10^{-4}. The specimen geometry ensured that this amplitude was constant along the axis away from the grips. PSB formation was observed using a Reichart microscope. Deformation produced by indentation was observed with Nomarski contrast. In some experiments a specimen was indented in two places, annealed in vacuo at 950°C, and then indented at two additional points. This allowed a direct comparison of two kinds of damage on a single specimen.

EXPERIMENTAL RESULTS

For all orientations PSB's developed near indentations but at no other points along the specimen axes; each orientation had a characteristic pattern of bands. Figs.2.(a) and (b) show the PSB's on specimen E after $1.2 \cdot 10^4$ and $4.5 \cdot 10^5$ cycles. PSB's were first seen after $7 \cdot 10^3$ cycles; testing was continued for $7 \cdot 10^5$ cycles at which stage PSB's were still not visible along the axis except near the indentations. That the bands were indeed PSB's was supported by electropolishing experiments and by SEM results which showed pronounced intrusions and extrusions. In Fig.2(b) some bands are very short whereas several others are long; the long bands continued to extend, at a decreasing rate, throughout the test. In orientations E and H the short bands were more pronounced than for other orientations whereas for orientations F and G the greatest lengths of single PSB's were achieved (up to ∿ 1 mm after ≳ 10^5 cycles). When the indentation pit was completely removed by electropolishing

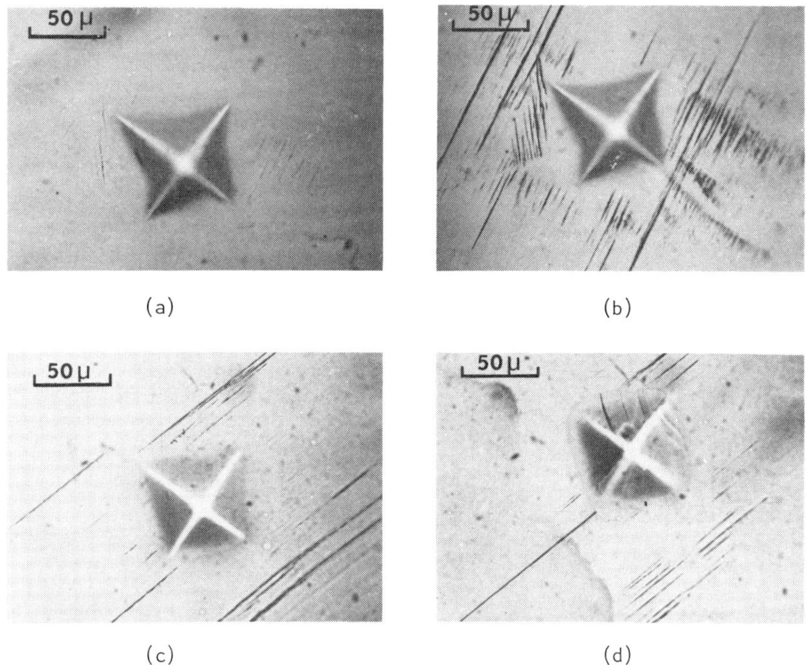

Fig.2. PSB's formed near indentations
(a) Orientations E, N = 1.2 · 10^4 cycles. (b) Orientation E, N = 4.5 · 10^5 cycles. (c) Orientation F, N = 3.6 · 10^4 cycles. (d) Orientation F, annealed at 950°C after indentation, N = 3.6 · 10^4 cycles.

(orientation F) the pattern of PSB's produced by fatigue was unchanged; however the total PSB length was reduced by ∿ 30%.

Figures 2(c) and 2(d) show the effect on PSB formation of annealing a specimen of orientation F, at 950°C for 1½ hours, after indentation. After annealing there remains a strong localization of PSB formation with the additional effect of PSB's being initiated at the edges of the pit and within. In the case of unnealed indentations PSB's are not observed within the pits nor at the edges and corners, for any orientation. As a result of annealing an indentation the total PSB length was reduced by ∿ 20% after 4 · 10^4 cycles but initiation occurred after a smaller number of cycles.

DISCUSSION

The enhancement of PSB formation near indentations is expected to arise from one or more of three effects; there is a geometrical stress-raising effect due to the indentation, there are expected to be long-range internal stresses due to the deformation around the indenter and a complex dislocation microstructure will be produced. The stress-raising effect has

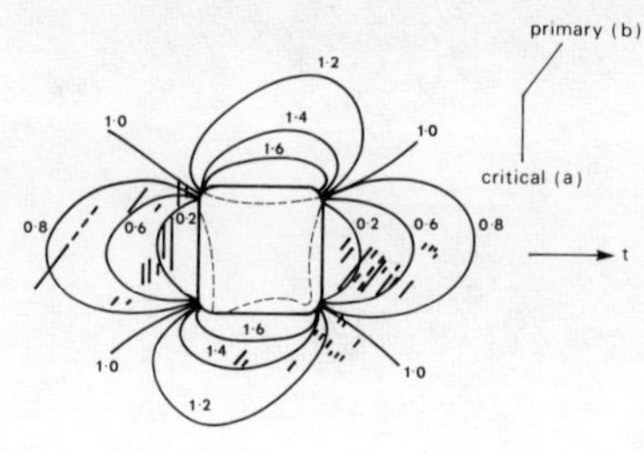

Fig.3. Stress enhancement factors, calculated for a square hole for orientation E, resolved onto the primary system. The PSB's formed after $2.5 \cdot 10^4$ cycles are shown schematically for comparison.

been calculated, for orientations E and F, on the assumption that the surface stresses are of the greatest significance and that the indentation can be treated as a square hole. The method of Greenspan (1) was used and the stress-enhancement factors were resolved onto each of the {111}/<110> slip systems. Figure 3 shows the stress-enhancement contours for the system B IV, for orientation E, with the positions of PSB's formed after $2.5 \cdot 10^4$ cycles superimposed. There is no significant correlation. However PSB's which initiated at the edges of hemispherical pits, formed electrolytically on specimen G, were accurately predicted by a similar calculation for a circular hole in a plate. This proves that surface stress-enhancement due to surface geometry can be of importance but that it is not significant in the case of un-annealed indentations. The same type of calculation was applied to the case of annealed indentations for orientation F but poor correlation was found. Since annealing would remove all internal stresses the PSB initiation sites must be determined by a dislocation microstructure (cell structure) which remains after annealing. It is therefore probable that it is the dislocation microstructure which is the most important in determining the PSB initiation sites near un-annealed indentations.

Several models have been used to calculate the internal stresses around indentations (2, 3). In general terms they predict compressive stresses at the surface very close to the indentation, which may account for the presence of PSB's at the edges and corners in Fig.2(d) while being absent in Fig.2(c), on the assumption that compressive stresses inhibit PSB initiation. However none of the calculations of internal stress have given satisfactory predictions of the distribution of the sites of PSB initiation. The micrograph of the indentation slip for orientation E shown in Fig.4(a) suggests one reason for this, namely the extensive plastic relaxation when the indenter is removed which is not allowed for in the models used. The slip occurring as a result of indentation is shown schematically in Fig.4(b) superimposed on the active slip planes deduced by modifying the Dyer (4) model for indentation on the {100} faces of Cu. By comparing this diagram with Fig.2(b) it can be seen that PSB initiation is well correlated with regions in which slip on plane B intersects slip on planes A or C. A more detailed analysis suggests that the expected slip systems are capable of forming Lomer-Cottrell barriers. Observations reported previously (5) support this idea where PSB's were observed to form near the specimen edge for orientation E. In this region intersecting fatigue slip on planes B and A provided sites for the initiation of PSB's on the primary system. This is being investigated by TEM.

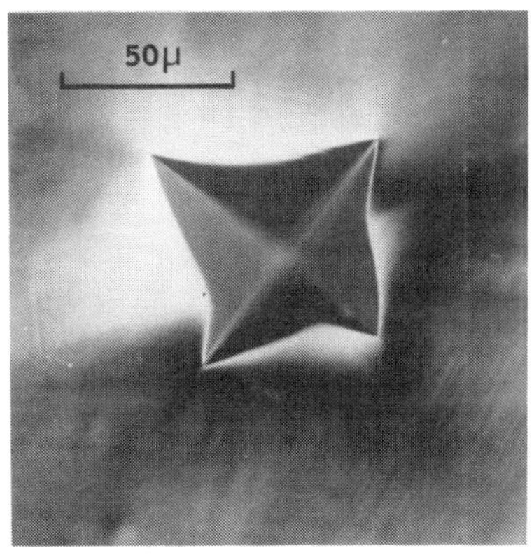

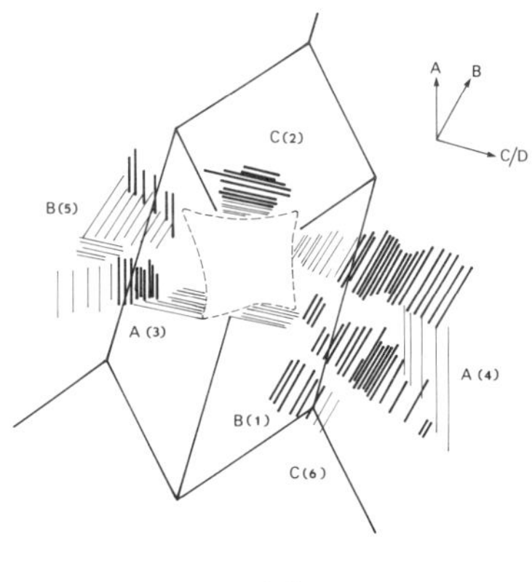

(a) (b)

Fig.4. Indentation slip for orientation E
(a) Micrograph using Nomarski contrast. (b) Diagram of active slip planes predicted on modified Dyer model with observed indentation slip shown schematically. This figure should be compared with Fig.2(b).

It should be noted that the long PSB's propagate well away from the indentations into regions where it can be assumed that there is no direct effect due to the indentations. Thus surface damage can cause the early initiation of PSB's some of which can propagate extensively. This effect can be explained on the assumption that the end of a PSB provides a region of stress concentration and indicates the likely reduction in fatigue life due to surface damage.

ACKNOWLEDGEMENTS

The authors would like to thank Dr. P.E.J. Flewitt and Professor K.E.Puttick for valuable discussions and Mrs. Sheila Rudman for assistance with the diagrams.

REFERENCES

1. M. Greenspan, Q.Appl.Math 2, 60 (1944).
2. R. Hill, "The Mathematical Theory of Plasticity", Clarendon Press, Oxford (1950).
3. C.M. Perrott, Wear 45, 293 (1977).
4. L.D. Dyer, Trans. A.S.M. 58, 620 (1965).
5. P. Charsley, K.E. Puttick and J.D.M. White, Scripta Met. 15, 1201 (1981).

SECTION 11

Polymers, Ceramics, Composites

Polymères, céramiques, composites

Deformation a Haute Temperature d'un Composite Lamellaire: Aspects Microstructuraux et Effets des Interfaces

M. Ignat

Institut National Polytechnique de Grenoble; Laboratoire de Thermodynamique et Physico-Chimie Métallurgiques, E.N.S.E.E.G., Domaine universitaire B.P. 75, 38402 St Martin d'Hères, France

RESUME

Les points les plus marquants des résultats obtenus lors de l'analyse de la déformation d'un composite lamellaire déformé en fluage et en relaxation de la contrainte sont rappelés. A partir de ces points, nous présentons ici un modèle unitaire qui permet de reproduire et interpréter de façon cohérente le comportement du composite lors des essais mentionnés.

MOTS-CLES
Interfaces, fluage, relaxation de la contrainte, modèle unitaire, évolution microstructurale, essais.in-situ.

INTRODUCTION
Dans les nombreuses études qui interprétent le comportement des matériaux composites, on utilise le concept de phase renforçante. Le but de cet article est de synthétiser la déformation d'un composite lamellaire, et de montrer que à haute température la déformation peut s'interpréter en tenant compte principalement du rôle des interfaces.

METHODES EXPERIMENTALES
L'alliage que nous avons choisi est un composite lamellaire obtenu à partir du mélange eutectique $Al-Al_2Cu$ [1]. Les lamelles des deux phases ont une épaisseur moyenne de 2μm. Les caractéristiques structurales du composite, ainsi que les éprouvettes et dispositifs utilisés lors des expériences sont décrits dans les références 2 et 3.

RESULTATS EXPERIMENTAUX
Comportement macroscopique du composite. Nous allons rappeler brièvement les points les plus marquants concernant le comportement en fluage et en relaxation de la contrainte du composite lamellaire :
- Les essais de fluage, montrent en fonction de la température, un stade quasi-stationnaire dont les limites dans le domaine contrainte appliquée et température T ont été déterminées. (Figure 1). Pour atteindre un fluage quasi-stationnaire, il faut que la contrainte appliquée dépasse une limite qui est la contrainte seuil (σ_s). Celle-ci décroît avec la température.

- Les essais de relaxation de la contrainte mettent en évidence l'éxistance d'une contrainte asymptotique (σ_∞, Figure 1), indépendante de l'état de traction initial et prenant les mêmes valeurs que la contrainte seuil σ_s, dans toute la gamme des températures communes aux deux types d'essai : ce point sera utilisé pour bâtir le modèle unitaire de déformation du composite.

OBSERVATIONS MICROSTRUCTURALES : MECANISMES DE DEFORMATION

Les observations au niveau de la microstructure de déformation ont été faites principalement en microscopie électronique classique et à très haute tension (THT, essais in-situ). Nous résumons premièrement les principaux processus qui se développent dans les phases, pour ensuite insister sur les mécanismes actifs aux joints. La phase Al commence à se déformer par des dislocations émises aux interfaces et se déplaçant sur des plans [111] et [110] ayant un facteur de SCHMIDT pratiquement nul. Ceci montre que l'aluminium se déforme surtout par un cisaillement parallèlement aux joints de phases. Aux températures les plus basses de nos essais, la phase Al_2Cu est initialement exempte de dislocations en fluage. Ceci n'est pas le cas pour les essais de relaxation de la contrainte, pour lesquels apparaissent des petits amas de dislocations répartis de façon très hétérogène. Au plus hautes températures de nos essais, les plans de glissement : identifiés dans la phase Al_2Cu sont du type {101}, {112}, {001}. Certaines de ces observations ont été détaillées dans les références 4 et 5. En ce qui concerne le joint de phases certains phénomènes observés en cours de déformation indiquent très clairement que sa participation est décisive. Entre autres : les sources de dislocations (Figure 2a et b) ainsi que le mouvement et réarrangement de dislocations interfaciales (Figure 2c). Un calcul récent [15] montre que les champs de contrainte associés à ces arrangements sont importants, et peuvent être à l'origine de l'apparition de microfissures dans le composite. le joint doit donc accommoder la déformation entre des phases plastiquement différentes, en assurant, à l'échelle microscopique la continuité de la déformation. Celle-ci peut être interprétée en écrivant la continuité des vecteurs de BURGERS des dislocations qui intéragissent au joint, ainsi que l'ont fait FORWOOD et CLAREBROUGH pour un joint de grains. En effet, nous constatons que les plans G_1 (011) et G_2(111) dans la lamelle Al de la Figure 3 (observation in-situ) contiennent respectivement les vecteurs de BURGERS a/2 [011] et a/2 [110] dont la somme correspond à un vecteur a/2 [121]. Le modèle et la direction de ce dernier sont voisins du vecteur 1/2 [111] de la lamelle Al_2Cu adjacente, et contenu dans le plan {011} Al_2Cu à partir duquel sont émises des dislocations qui cisaillent le joint de phases.

DISCUSSION

- Description mécanique du comportement du composite : schéma rhéologique

Le comportement macroscopique en fluage et en relaxation de la contrainte du composite peut être décrit à travers d'un simple schéma rhéologique tenant compte des effets suivants : la contrainte appliquée en fluage ou imposée initialement en relaxation, doit dépasser une valeur limite qui rhéologiquement peut être schématisée par un frottement solide. Les comportements des phases et interfaces peuvent être alors schématisés par des éléments type MAXWELL (ressort-amortisseur). Leur disposition est guidée par la contribution prédominante du joint à haute température et par l'absence de déformation de la phase Al_2Cu à basse température. La disposition des éléments rhéologiques conduit alors au schéma de SCHWEDOF ([7], Figure 4, §R). Il reste à fournir une explication physique de la contrainte seuil ou asymptotique, et de la viscosité globale η, représentative du mécanisme de déformation prédominant.

Pour la contrainte seuil ($\sigma_{S,\infty}$), nos observations expérimentales ainsi que nos mesures nous conduisent à admettre que c'est une grandeur qui est liée aux propriétés des interfaces dans le composite. Elle est interprétée comme une contrainte limite à partir de laquelle les cissions interfaciales sont suffisament importantes pour contribuer à la déformation irréversible. Pour la viscosité (η) caractéristique du composite nous supposons premièrement que les mécanismes de diffusion sont prédominants à haute température, sans préjuger du mécanisme de diffusion précis (volume, joints....), la vitesse de déformation correspondante est alors semblable à celle proposée par FROST et ASHBY [8] :

$$\dot{\varepsilon} = A \frac{\mu b D_0}{RT} \left(\frac{\sigma}{\mu}\right)^n \exp - \frac{Q}{RT} \qquad (1)$$

Chaque terme ayant sa signification habituelle. L'énergie d'activationQ va caractériser le mécanisme de déformation prépondérant. Regroupant les termes constants, et en tenant compte de la contrainte seuil ou asymptotique, pour le fluage on obtient :

$$\dot{\varepsilon} = \frac{A}{T} \left(\sigma - \sigma_{S,\infty}\right)^n \exp - \frac{Q}{RT} \qquad (2)$$

En relaxation de la contrainte, par intégration des équations d'état avec la relation (2) pour vitesse de déformation plastique on obtient :

$$\sigma = \left[\frac{K A(\sigma_0, T)}{T}\left(\exp - \frac{Q}{RT}\right) t\,(n-1) + \frac{1}{(\sigma_0 - \sigma_S)^{n-1}}\right]^{\frac{1}{1-n}} + \sigma_{S,\infty} \qquad (3)$$

σ_0 étant la contrainte initiale en relaxation, K une rigidité équivalente du système, t le temps... les autres termes ayant leur signifiation habituelle. Par la suite, nous avons fait le traitement numérique de nos résultats à l'aide de la méthode d'analyse en composantes principales [9]. Les écarts moyens entre valeurs calculées et mesurées rapportées aux valeurs mesurées sont de 23% pour le fluage et de 7.2% pour la relaxation de la contrainte. Ils sont reportés sur la Figure 4 : (F et R respectivement). La valeur des paramètres ajustables communs aux deux essais sont pour l'exposant de la contrainte n2,13 et l'énergie Q103 KJ/mole.

CONCLUSION

Les valeurs de n et Q confirment le fait qu'il s'agit bien d'un phénomène de transport de matière aux joints, séparant les deux phases du composite. En effet la valeur de n=2 est mentionnée dans plusieurs modèles de déformation contrôlés par le mouvement de dislocations aux joints de grains [10 à 13]. La valeur de l'énergie d'activation obtenue, égale à 103 kJ mole^{-1} est pratiquement la même que celle obtenue par HO et WEATHERLY [14] Q 97 kJ mole^{-1} qui ont étudié dans le même composite la migration de noeuds triples $Al_2Cu/Al/Al$, activée par la diffusion du cuivre aux interfaces. Le modèle unitaire proposé basé d'une part sur la quasi-identité des contraintes seuil et asymptotique et d'autre part sur l'effet prépondérant des interfaces, nous permet de relier deux types d'essais très différents. Les résultats obtenus, bien meilleurs que ceux obtenus par des lois de déformation classiques démontrent que à haute température, les mécanismes contrôlant la déformation d'un composite lamellaire sont liés aux propriétés de diffusion aux interfaces.

REFERENCES

1. J.P. RIQUET, F. DURAND, Mat.Res.Bull., 10, 399 (1975).
2. M. IGNAT, F. DURAND, Phys.Stat.Sal. (a), 51, 567 (1979).
3. M. IGNAT, D. PROULX, R. BONNET, F. DURAND, Acta Met., 29, 1607 (1981).
4. M. IGNAT, R. BONNET, D. CAILLARD, J.L. MARTIN, Phys.Stat.Sal. (a) 49, 675, (1978).
5. M. IGNAT, D. CAILLARD, Proceedings on "In Situ composites IV" M.R.S. Symposium Vol. 123, Elsevier Science Publishing 6, North-Holland (1982).
6. C.T. FORWOOD, L.M. CLAREBROUGH, Phil. Mag. 1, 44, 31 (1981).
7. B. PERSOZ, Introduction à l'étude de la rhéologie, Dunod Paris, 1960.
8. H.J. FROST, M.F. ASHBY, "Deformation Mechanisms Maps" CUED/C/MATS/TR90, Cambridge University, (1981).
9. G. SAPORTA "Théories et méthodes de la statistique" Publications J.F.P. Editions Technip, Paris, (1978).
10. A.K. MOKHERJEE, Mater.Sc.Eng., 8, 83, (1971).
11. B. BURTON Mater.Sc.Eng., 10, 9, (1972).
12. R.C. GIFKINS, Met.Trans., 8A, 1507, (1977).
13. J.R. SPINGARN, W.D. Nix, Acta Met. 27, 921, (1979)
14. E. HO, W.C. WEATHERLY, Acta Met., 23, 1451, (1975).
15. M. IGNAT, M. DUPEUX, "Progress in Science and Engineering of composites" Ed., ICCM-IV, Tokyo, p. 1361, (1982).

FIGURES

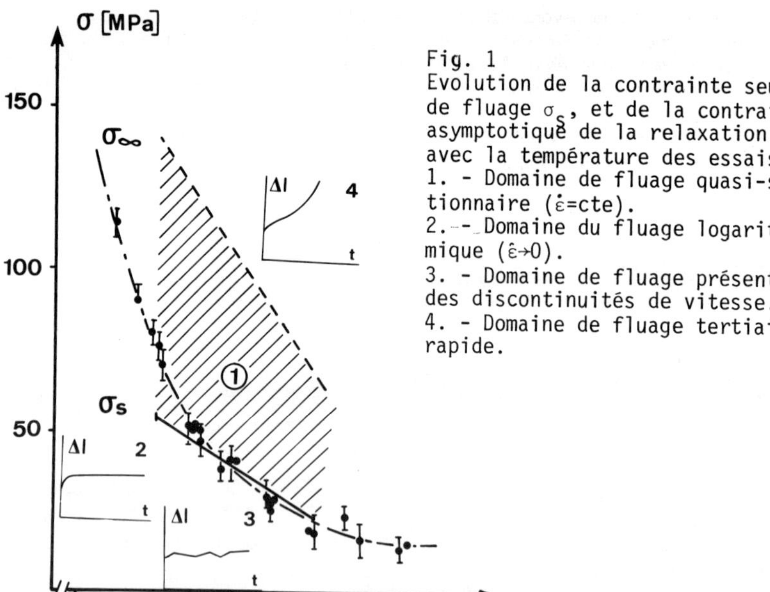

Fig. 1
Evolution de la contrainte seuil de fluage σ_S, et de la contrainte asymptotique de la relaxation σ_∞ avec la température des essais.
1. - Domaine de fluage quasi-stationnaire ($\dot{\varepsilon}$=cte).
2. - Domaine du fluage logarithmique ($\dot{\varepsilon}\to 0$).
3. - Domaine de fluage présentant des discontinuités de vitesse.
4. - Domaine de fluage tertiaire rapide.

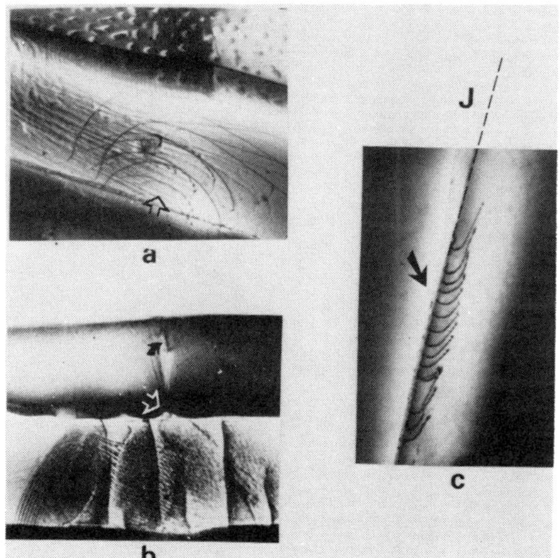

Fig. 2
(a,b), Sources de dislocations aux 1000 kV, joints de phases (l'épaisseur des lamelles est d'environ 2 µm),
a. - Emission dans une lamelle Al (essai in situ, 573 K, 1000 kV).
b. - Emission dans une lamelle Al_2Cu (essai in situ, 623 K, 1000 kV).
c. - Défauts aux joints de phases (J).

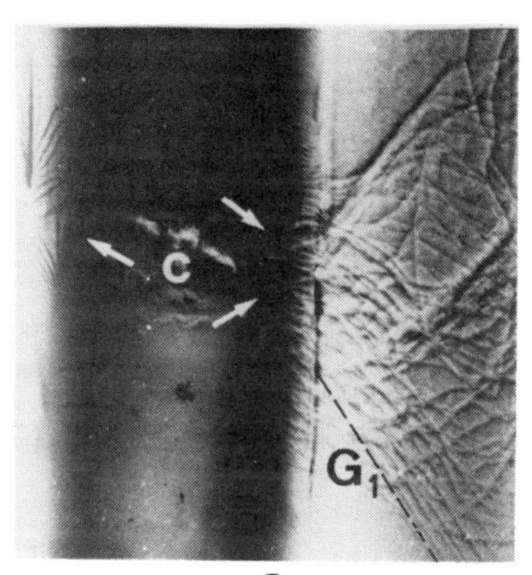

a

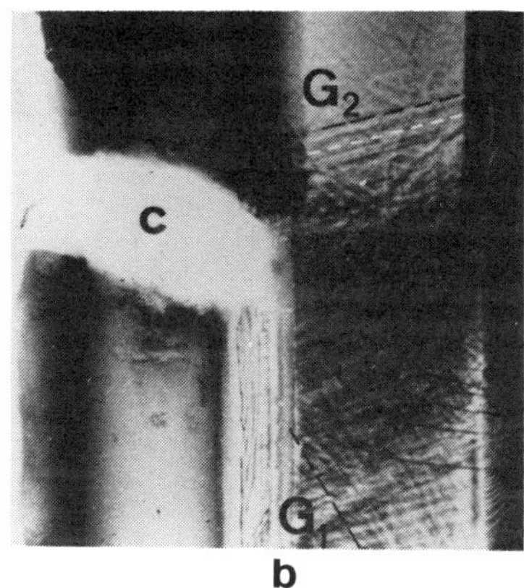

b

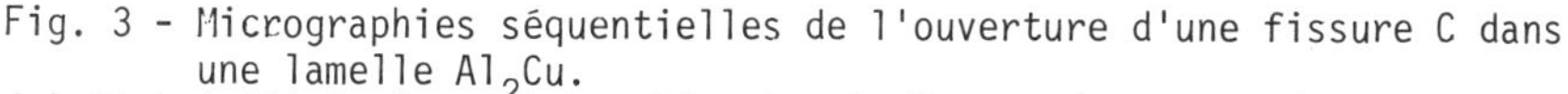

Fig. 3 - Micrographies séquentielles de l'ouverture d'une fissure C dans une lamelle Al_2Cu.
(a) Etat initial, les flèches blanches indiquant des zones du joint de phases à l'intérieur desquelles les défauts interfaciaux sont en mouvement.
(b) Activation des plans G_1 et G_2 dans la lamelle Al adjacente en cours de relaxation de la contrainte. (Essai in situ 623 K, 1000 kV).

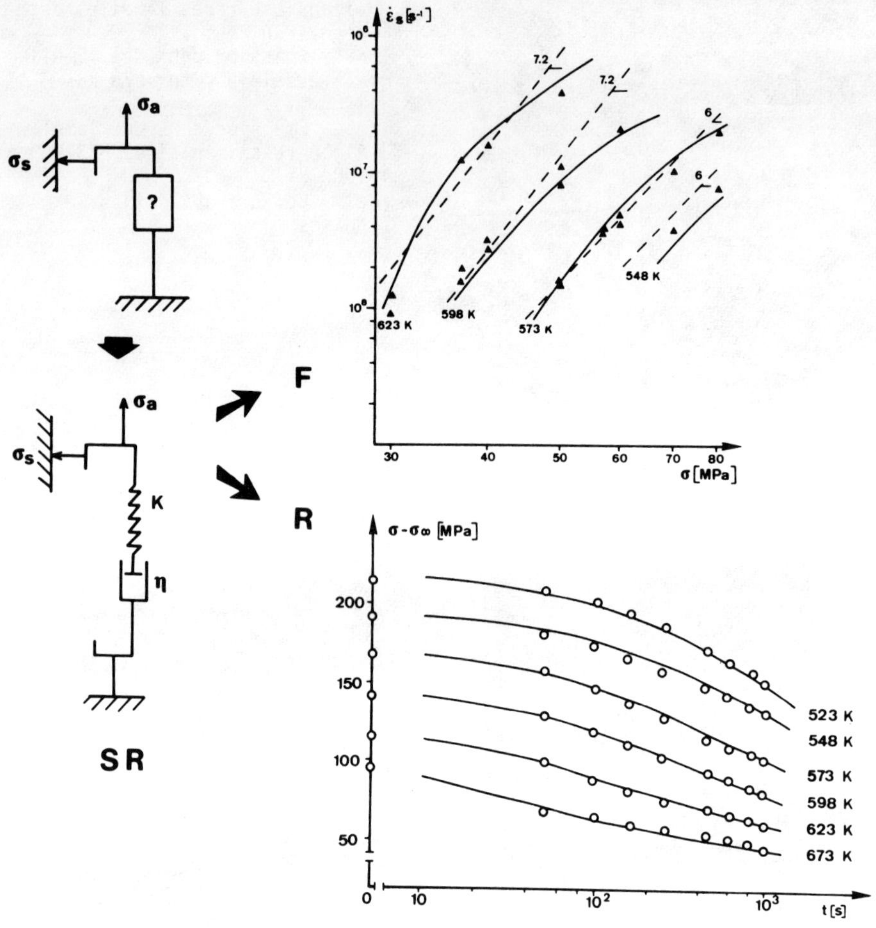

Fig. 4
Schéma rhéologique représentant la déformation du composite.(SR)
F : diagramme logarithmique représentant la vitesse de fluage quasi-stationnaire en fonction de la contrainte appliquée. Les points noirs correspondent aux mesures expérimentales, les traits pleins aux représentations classiques et les traits pleins au modèle unitaire.
R : Essais de relaxation de la contrainte. Prédictions du modèle : traits pleins, par rapport aux mesures expérimentales, points ronds.

Thermally Induced Residual Stress and its Relaxation in Al-Al_3Ni Eutectic Composites

O. Yanagisawa and T. Yano

Faculty of Engineering, Hiroshima University, Saijo-Shitami, Higashi-Hiroshima 724, Japan

ABSTRACT

The residual stress of Al-Al_3Ni eutectic composites caused by several kinds of thermal history and the relaxation behavior of it at constant temperature were analized.

KEYWORDS

Al-Al_3Ni eutectic composites; residual stress; thermal expansion-cotraction; stress relaxation; threshold stress.

INTRODUCTION

The residual stress remarkably influences the tensile and compressive deformation behaviors, especially the yield stresses in eutectic composites. Also, the thermal fatigue behaviors in the fiber-reinforced composites are directly affected by the residual stress. In this study, the residual stress of Al-Al_3Ni eutectic composites caused by several kinds of thermal history were determined from the tensile and compressive stress-strain curves. In addition, the relationship between the thermal expansion-contraction curve and the residual stress was examined, and then the relaxation behaviors of the residual stress were analized at constant temperatures.

EXPERIMENTAL METHOD

Al-Al_3Ni eutectic composites were prepared by unidirectional solidification of Al-6.18 mass%Ni master alloy at the solidificaton rate of 1.03×10^{-5} m/s and temperature gradient of about 6 K/mm. Dimensions of tensile and compressive test specimens were 18mm and 12mm gauge length, respectively, and 6mm diameter. These specimens, which had been annealed at 500Kx18ks, were tested in tension and compression after keeping them at 373K, 473K, 573K or 673K for 3.6ks in order to investigate the influence of the residual stress on the deformation at elevated temperatures. Strain was measured with strain-gauge (below 473K) and extensometer(above 573K), which was attached to the extruded portion of the gauge section of the specimen. Dimensions of the test specimens for thermal expansion measurement were 5mm diameter and 15mm length. Changes in length were measured by a dilatometer in a heating-cooling process ($|dT/dt|$ =0.25K/s) and at constant temperatures. Fluctuation of temperature was ± 0.05K at constant temperatures, which corresponds to that in the strain of ± 1×10^{-6}.

RESULTS

Figure 1 shows the schematical method of determining the residual stress σ_m^R in the matrix from the tensile and compressive stress-strain curve. It has been confirmed by the experiments that the stress-strain curve shows a linear work-hardening in Stage II and a hysteresis loop under the loading-unloading process for tensile and compressive deformation [1]. A dot-dash-line in Fig. 1 represents the stress component in the fibers and it passes through point P or Q, the center of the tensile or compressive hysteresis loop, respectively. The slope of this line is equal to V_fE_f, the product of volume fraction and Young's modulus of the fiber. The stress value at point O'in Fig. 1, the intersection of this line and the stress axis, is equal to $-V_m\sigma_m^R$ (= $V_f\sigma_f^R$, where σ_f^R is the residual stress in the fiber). As shown in Fig. 1, the value of σ_m^R, obtained schematically from the tensile hysteresis loop, is nealy equal to that of the compressive one. Therefore, it may be considered that the Bauschinger effect in the matrix, induced in the thermal history, does not remarkably influence the measured value of σ_m^R.

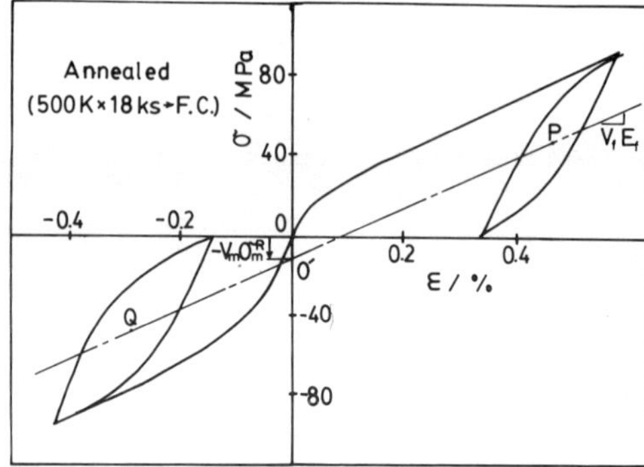

Fig. 1. Tensile and compressive stress-strain curves in annealed specimens showing the method of determining the residual stress in the matrix σ_m^R.

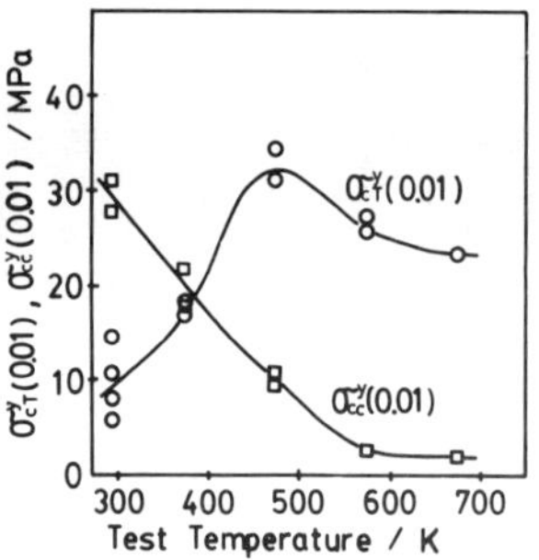

Fig. 2. Variation of the tensile and compressive 0.01 % proof stresses with test temperature.

The compressive proof stress is slightly bigger than the tensile one at room temperature, while the tensile proof stresses are bigger than the compressive ones at temperatures above 473K. Variation of the tensile and compressive 0.01% proof stress $\sigma_{cT}^y(0.01)$, $\sigma_{cC}^y(0.01)$ of the composite with test temperature is shown in Fig. 2.. This is much different from the usual temperature dependency of the proof stress. It suggests that the proof stress of this composite is considerably influenced by the residual stress.

The tensile and compressive proportional limit σ_{mT}^P, σ_{mC}^P, 0.01% proof stress $\sigma_{mT}^y(0.01)$, $\sigma_{mC}^y(0.01)$ of the matrix without the residual stress and the residual stress σ_m^R in the matrix are shown in Fig. 3 as a function of test temperature. The proportional limit and proof stress of the matrix without the residual stress were calculated by rule of mixtures and the mean value of measured residual stress. When the sign of the residual stress is plus, that is, in the tensile stress state, the tensile plastic deformation has occured in the matrix in previous thermal history and then the Bauschinger effect is found in the initial stage of the compressive deformation. Therefore, the proportional limit or proof stress at a relative small strain cannot be defined without distinguishing tensile or compressive state. Then, the tensile proportional limit and proof stress are shown to compare with the residual stress, when the residual stress in the matrix is tensile. The compressive ones are shown, when the residual stress is compressive. The residual stress in the matrix is in the tensile stress state at room temperature. This residual stress gradually decreases and then changes into the compressive state at temperatures above about 473K. Such a change of the residual stress during the heating process may be well understood, after considering the difference in the thermal expansion coefficient of both phases [2,3]. Also, this result suggests that the residual stress is stable even after keeping them at 673K for 3.6ks.

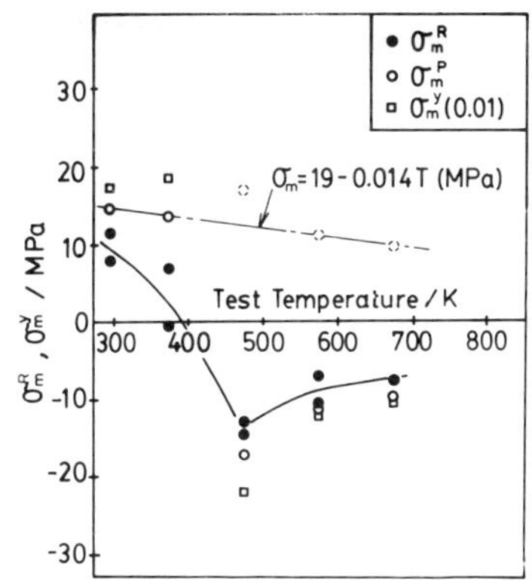

Fig. 3. Proportional limit σ_m^P and 0.01% proof stress $\sigma_m^y(0.01)$ of matrix without the residual stress, and residual stress σ_m^R as a function of test temperature.

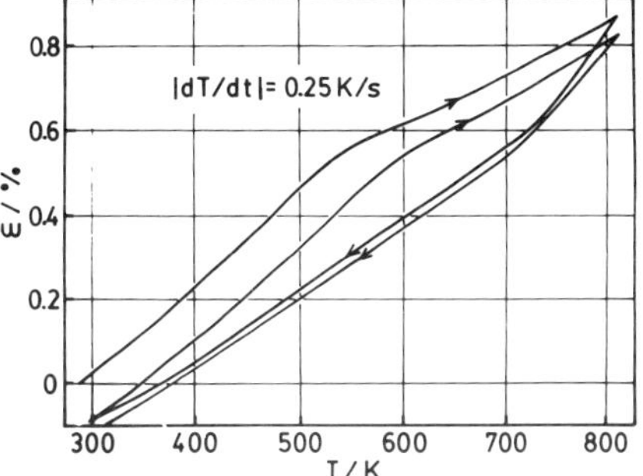

Fig. 4. Thermal expansion-contraction curves of Al-Al_3Ni eutectic composite.

Thermal expansion-contraction curve at temperatures between 293K and 823K is shown in Fig. 4. The slope of the expansion and contraction curve changes at about 573K in the first heating process and at about 720K in the first cooling process, respectively. These changes correspond to the compressive yielding in the matrix during the heating process and the tensile yielding in the matrix during the cooling process [2,3]. Expansion and contraction curves at various constant temperatures, after heating or cooling, are shown in Fig. 5. Slight expansion are found in keeping at 423K, after heating from 293K. When the sample is heated to 523K, after keeping it at 423K for 10.8ks, contraction is found inversely. When the constant-temperature keeping experiment was carried out after heating by 100K, the contraction strains at constant temperatures tend to increase with the temperature rising to 723K (Fig. 5(a)). On the other hand, the expansion of a specimen is observed at a constant temperature of 723K, when it has been heated from 293K to 823K and then cooled to 723K after keeping it at 823K for 10.8 ks. The expansion strains tend to increase with the temperature decreasing to 523K after cooling by 100K.

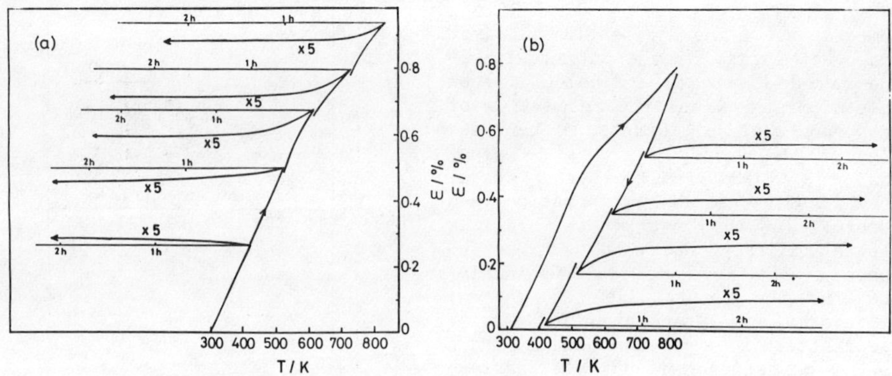

Fig. 5. Variation of strain of the composite with time at constant temperatures of 423K, 523K, 623K, 723K and 823K, (a) after heating process, (b) after cooling process.

DISCUSSION

The equilibrium equation of stress, the compatibility assumption of strain and the constitutive equation in each phase of this composite are assumed as follows, based on the rule of mixtures [3],

$$V_f\sigma_f + V_m\sigma_m = 0 \tag{1}$$

$$\Delta\varepsilon_c = \alpha_f\Delta T + \Delta\varepsilon_f = \alpha_m\Delta T + \Delta\varepsilon_m,\ \Delta\varepsilon_m = \Delta\varepsilon_{mel} + \Delta\varepsilon_{mpl} \tag{2}$$

$$\sigma_f = E_f\varepsilon_f,\ \sigma_m = E_m\varepsilon_{mel},\ \sigma_m = \sigma_m(T, \varepsilon_{mpl}, \dot{\varepsilon}_{mpl}) \tag{3}$$

where, V, E and α are volume fraction, Young's modulous and thermal expansion coefficient, respectively. The characters, f and m, refer to fiber and matrix, respectively. Then, strain of this composite is given as eq. (4) by using eq. (1)~(3),

$$\varepsilon_c = \alpha_f(T - T_M) - V_m\sigma_m^R/V_fE_f \tag{4}$$

where, T_M is melting point, σ_m^R is residual stress in the matrix and it is assumed that $\varepsilon_c = \sigma_m^R = 0$ at $T = T_M$. The curve, predicted from eq. (1)~(4) and $\sigma_m = 19 - 0.014T$ (MPa), was relatively well consistent with the measured one. When the matrix is in the compressive stress state in the heating process at temperatures higher than about 460K, the composite contracts at these constant temperatures. On the contrary, when the matrix is in the tensile stress state in the cooling process at temperatures below 720K, the composite expands at these constant temperatures (Fig. 5). It may be reasonable to regard this behavior as the relaxation of the residual stress by creeping in the matrix.

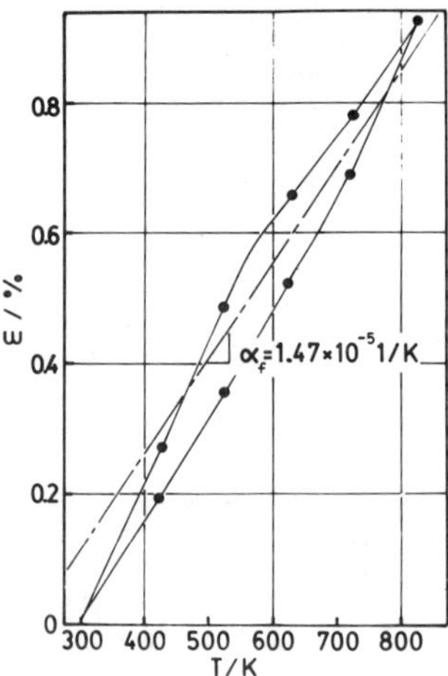

Fig. 6. Strain of the composite after keeping for 7.2ks at each constant temperature in Fig. 5 as a function of the keeping temperature.

Strain of the composite after keeping at each

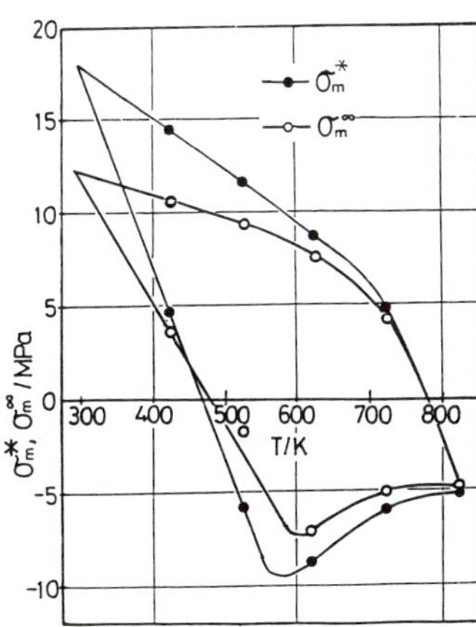

Fig. 7. Residual stress in matrix σ_m^* after keeping for 7.2ks and threshold stress σ_m^∞ as a function of temperature.

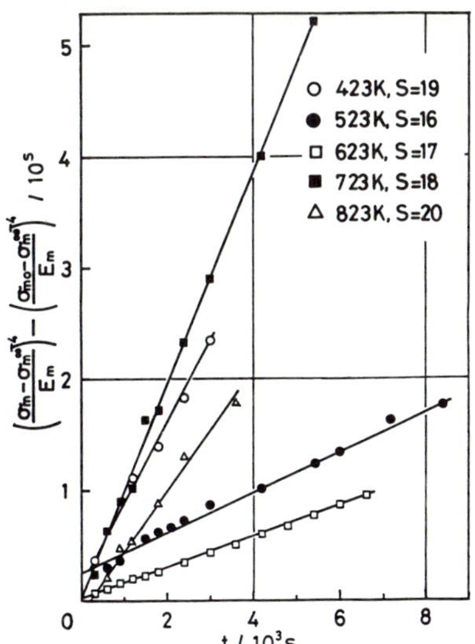

Fig. 8. The value of the left hand side in eq.(8) as a function of time, t.

constant temperature for 7.2 ks is shown in Fig. 6 as a function of the keeping temperature. Because the results in Fig. 5(a) and (b) were obtained from the different specimens and then both strains at 823K were not the same, strains after sufficient relaxation were made to coincide in Fig. 6. The hysteresis loop in Fig. 6 is regarded as that of a very small heating and cooling rate. The residual stress in the matrix can be obtained from the strain of the composite, ε_c, and the strain $\varepsilon_{co}=\alpha_f(T-T_M)$, in which the residual stress is 0 at the same temperature, with eq. (4).

$$\sigma_m^R=-V_fE_f(\varepsilon_c-\varepsilon_{co})/V_m \qquad (5)$$

Here, ε_{co} is given by $\varepsilon_{co}=(\varepsilon_{ch}-\varepsilon_{cc})/2$. ε_{ch} and ε_{cc} are strains of the composite after keeping at constant temperatures between 623K and 723K, at which the matrix plastically deforms, for 7.2 ks in the heating and cooling process, respectively. Also, it is assumed that the absolute values of the residual stresses corresponding to ε_{ch} and ε_{cc} are the same. (ε_{co} is represented by the dot-dash-line in Fig. 6. However, the strain at an origin does not influence eq. (5).)

It was assumed that the fiber is elastic and the matrix creeps. Constitutive equation for creep deformation in the matrix is assumed as,

$$\dot{\varepsilon}_{mpl}=A[(\sigma_m-\sigma_m^\infty)/E_m]^n \exp(-Q/RT). \qquad (6)$$

Stress relaxation rate at a constant temperature is given as eq. (7), with eq. (1)~(3) and eq. (6).

$$\dot{\sigma}_m=-V_fE_fE_m/E_c\cdot A[(\sigma_m-\sigma_m^\infty)/E_m]^n\cdot \exp(-Q/RT) \qquad (7)$$

By integrating eq. (7) with time, t, and assuming that $\sigma_m=\sigma_{mo}$ at t=0, eq. (8) is obtained.

$$[(\sigma_m-\sigma_m^{\infty})/E_m]^{1-n} - [(\sigma_{mo}-\sigma_m^{\infty})/E_m]^{1-n}$$

$$=\{[(n-1)V_f E_f/E_c]\cdot \exp(-Q/RT)\}\cdot t = H(T)\cdot t \qquad (8)$$

When the relationship between the residual stress in the matrix, which is determined from eq. (5) and Fig. 5 and 6, and time, t, is analized with eq. (8), it is necessary to assume the threshold stress σ_m^{∞} and the stress exponent n. Then, it was assumed that the stress exponent n=5 for pure aluminum [4] and the threshold stress σ_m^{∞} is 0 or σ_m^{*}, which is the residual stress after keeping for 7.2 ks at each constant temperature in Fig. 6. However, in these cases, the relationship between the value of the left hand side of eq. (8) and time is not proportional, which was expected from eq. (8). Then, the threshold stress σ_m^{∞} was determined so as to satisfy the proportional relation of eq. (8) for n=5. The threshold stress σ_m^{∞} determined by this method is shown in Fig. 7 and the referential relationship between the value of the left hand side of eq. (8) and time is shown in Fig. 8, which satisfies the proportional relation. Furthermore, the activation energy for eq. (8) was 126 kJ/mol, which was obtained from the slope in Fig. 8. This value of activation energy is approximately equal to that of the self diffusion for pure aluminum [4].

CONCLUSION

(1) The residual stress in the matrix is in the tensile stress state at room temperature and changes into the compressive one at temperatures higher than about 400K, when the specimen is subsequently heated.
(2) The thermal expansion-contraction curve of this composite shows the hysteresis loop during the heating and cooling process at temperatures between 293K and 823K. This loop relates to the residual stresses in both phases.
(3) The changes in the strain of this composite at constant temperatures after heating and cooling processes are influenced by the sign of the residual stress. When the residual stress in the matrix is in the compressive, the composite contracts. When the residual stress in the matrix is in the tensile state, the composite expands.
(4) The relaxation behaviors of the residual stress of this composite at constant temperatures can be well described by the form of the steady-state creep rate law for pure aluminum with the threshold stress.

REFERENCES

1. O.Yanagisawa and T.Yano,*J.Japan Inst.Metals* 6,583(1984).
2. K.Wakashima,T.Kawakubo and S.Umekawa,*Metall.Trans.* 6,1755(1975).
3. G.Garmong,*Metall.Trans.* 5,2183(1974).
4. O.D.Sherby,*Acta Metall.* 10,135(1962).

Interphase Boundary Sliding and Fracture of Lamellar Al-$CuAl_2$ Composite at High Temperatures

T. Watanabe, Y. Sato, H. Kanda and S. Karashima

Department of Materials Science, Faculty of Engineering, Tohoku University, Sendai, Japan

ABSTRACT

Interphase boundary sliding and fracture, and the effect of different types of growth faults on the processes have been studied microscopically on the lamellar Al-$CuAl_2$ composite. It has been shown that sliding can occur at evry two interphase boundaries, probably originating from a special structural feature of interphase boundaries in the eutectic alloy composite. It has also been demonstrated that extensive interphase boundary sliding can occur being associated with the presence of growth faults such as lamellar termination and ripple or misfit boundary. We may consider that the localized sliding very likely introduces the bending stress which facilitates the fracture of brittle $CuAl_2$ (θ) layer controlling the ductility and fracture of the composite. It is concluded that the presence of growth faults is an important key factor controlling interphase boundary sliding and fracture of the lamellar Al-$CuAl_2$ eutectic composite at high temperatures.

KEYWORDS

Interphase boundary sliding; High-temperature fracture; Lamellar eutectic composite; Al-$CuAl_2$ composite; Effect of growth faults.

INTRODUCTION

Eutectic composites have drawn much attention of many workers in the area of high-temperature alloy design because interphase boundaries contained are very thermally stable and have a greater strengthening effect in addition to composite stengthening effect of two-phase alloy. However, it has been pointed out that in the case of eutectic composites, the rule of mixtures does not provide a general rationalization of creep strength because the density of growth faults becomes more important as the volume fraction increases, in particular in the case of lamellar eutectic alloy composites whose volume fraction exceeds 30% (1). Moreover, interphase boundary sliding can take place at high temperatures and it may play some important roles in creep deformation and fracture, as grain boundaries do (2). However, unfortunately, interphase boundary sliding and fracture, and the effect of growth faults such as termination, ripple fault or lamellar misfit boundary (3,4), in particular, lamellar uetectic composites have been a little studied so far (5-7). The present paper reports microscopic observations on interphase boundary sliding and fracture, and the effect of different types of growth fault in $Al-CuAl_2$ eutectic alloy which is widely known as a model lamellar eutectic composite with the volume fraction of 48%.

EXPERIMENTAL PROCEDURE

The lamellar $Al-CuAl_2$ composite was grown from the melt of an Aluminium-34mass % copper alloy prepared by vacuum melting of high purity aluminium (99.999%) and copper (99.998%). The lamellar spacing was about 8 µm (interplanar spacing ≃ 4 µm). Tensile creep specimens which contained interphase boundaries (lamellar structure) at 45° to the specimen axis were cut by a spark machine in order to facilitate sliding at the boundaries. Creep tests were carried out at high temperatures ranging from 670K to 740K at creep stress from 10 to 40 MPa in argon. The test was interrupted several times for each specimen for the observations on the boundary sliding and fracture with a scanning electron microscope.

RESULTS AND DISCUSSION

(1) Characteristics of Interphase Boundary Sliding

Creep deformation occurred mainly due to interphase boundary sliding and crystal deformation of aluminium rich α phase layers. It was found that interphase boundary sliding does not take place at all boundaries but certain boundaries more extensively. Fig.1(a) and (b) clearly demonstrate this. Upon SEM observation, the boundaries which showed an extensive sliding had much brighter contrast than others. The occurrence of localized sliding was also observed on the side surface of the specimen arranged at 45° to the growth direction. Sliding occurred to different extents at different positions being associated with the presence of ripple faults or lamellar misfit boundaries,as shown in Fig.1(c). Higher magnification observations revealed that extensive sliding at certain interphase boundaries is associated with the presence of lamellar termination near those boundaries. Fig.2 shows that extensive sliding occurred at an interphase boundary from the position of A to B near a termination indicated by the arrow, as schematically shown by the attached diagram. It has become evident that localized interphase boundary sliding is always associatted with growth faults such as termination and ripple fault.

(2) "Ad-hoc" Sliding at Interphase Boundaries for the Lamellar Eutectic Alloy

Another important finding in the present investigation is that interphase boundary sliding can take place at every two boundaries, not at all boundaries as shown in Fig.3 although there were some exceptions especially at large creep strain. Pairs of α (slightly bright contrast) and θ (dark contrast) phases slide to each other. Similar sliding behaviour of interphase boundaries in the lamellar $Al-CuAl_2$ composite has been observed recently by Gaymard and Bonnet (8) in high-temperature cyclic shear deformation at the temperatures from 633K to 733K. They have explained the interesting sliding feature by assuming the presence of microscopic staircase structure at interphase boundaries. Similar sliding behaviour can be seen from micrographes in the paper which reported interphase boundary sliding in lamellar Pb-Sn eutectic composite, although there is no description about it by the authors (9).

The observed "ad-hoc" sliding at interphase boundaries is considered to be associated wich the difference in the fine-structure between two neighbouring interphase boundaries bordering θ or α phase, probably introduced during growth of the lamellar composite. Therefore,we may consider that this "ad-hoc" sliding must be an intrinsic nature of interphase boundary sliding particularly for eutectic composites. So far, the mechanism of interphase boundary sliding has been discussed in terms of a dislocation model involving lattice dislocations from the grain interior (10), quite similarly as the case of grain boundary sliding (11).

(3) The Effect of Growth Faults on Interphase Boundary Sliding

The present investigation has revealed that interphase boundary sliding occurs very locally in the specimen, especially being associated with the presence of lamellar termination and ripple fault or misfit boundary. The localization of interphase boundary sliding appears to play an important role in fracture of Θ phase which is considered to support the strength of the composite. Interphase boundary sliding could continuously occur during creep deformation if the occurrence of fracture of Θ phase would not finalize the deformation by the propagation of fracture to beighbouring α and Θ layers.

(4) Comparison between Interphase and Grain Boundary Sliding

In eutectic alloy composites, grain boundaries also exist although their density is much lower than that of interphase boundaries. It is interesting to study whether there exists any difference in sliding behaviour between the two types of boundaries in the same test condition, since the structure of

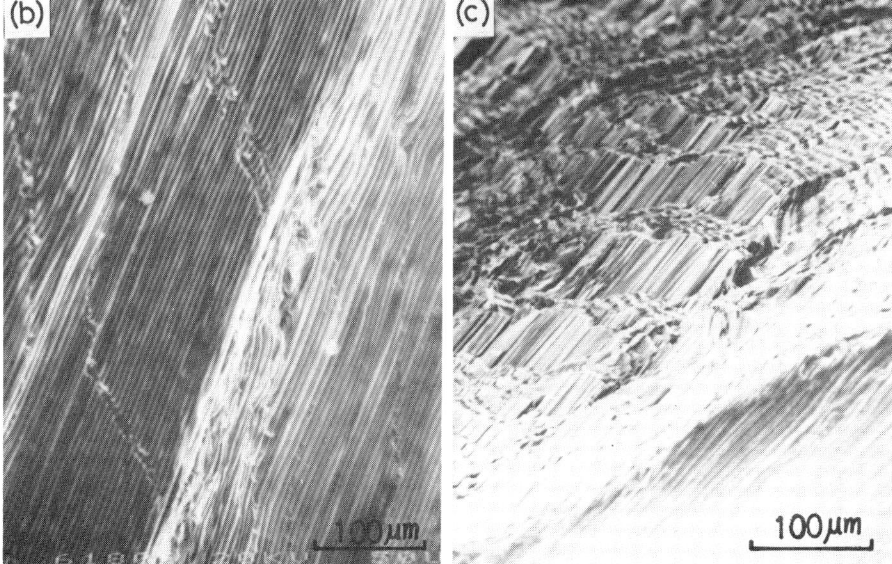

Fig.1 SEM observations on deformation and interphase boundary sliding.
(a) low magnification
(b) high magnification
(c) oblique view of the side and front surfaces.

Note extensive sliding associated with ripple faults on the side surface.

Fig.2

Extensive interphase boundary sliding near a lamellar termination.

The characteristic feature of sliding is shown schematically.

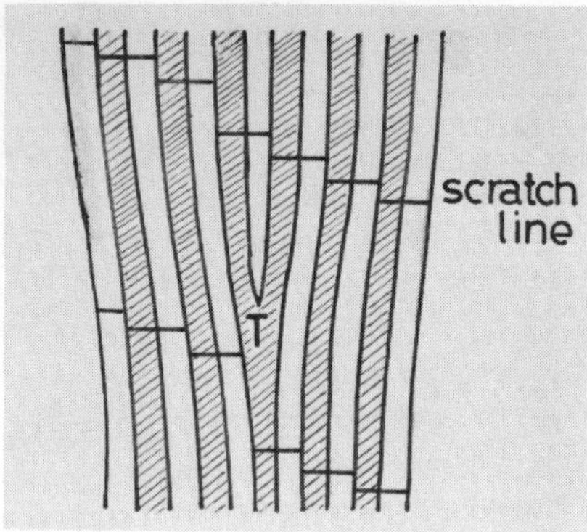

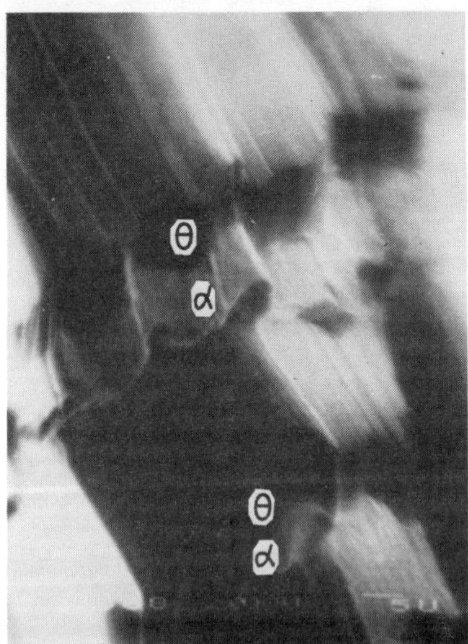

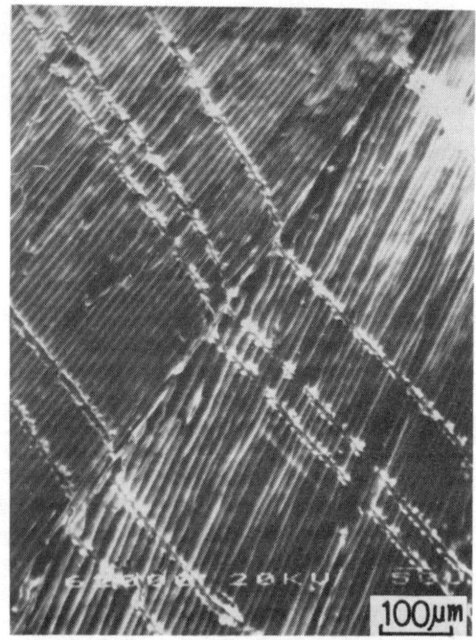

Fig.3 Interphase boundary sliding occurring at every two boundaries. Dark layers are Θ phase and bright ones α phase.

Fig.4 Grain boundary sliding occurring in the lamellar Al-$CuAl_2$ composite (The boundary tilt angle =10°)

interphase boundaries is known to be of semi-coherent narure and rather more ordered than general grain boundaries. Fig.4 shows that a grain boundary whose misorientation angle was about 10°, could slide more easily than interphase boundaries. Different tendencies for sliding between grain and interphase boundaries have also been reported on creep deformation in several two-phase alloys (12-14).

(5) Fracture Behaviour of the Lamellar Al-$CuAl_2$ Composite

Fig.5 demonstrates that the fracture of Θ phase layer takes place in the vicinity of interphase boundary which exhibited an extensive sliding (remember that interphase boundaries involving extensive sliding look much brighter than others with less sliding upon SEM observation), and a crack propagates involving fracture in neighbouring Θ phase layers through shear or localized deformation of α phase layers in front of the crack. Fracture of Θ phase layer occurs in very brittle manner even at high teperature with little plastic deformation as clearly seen on the fracture surface (Fig.6). Further observations on brittle nature of the fracture of Θ phase layer are shown in Fig.7(a) and 7(b) which show the appearance of the Θ phase layer after large deformation and slipping-off of the α phase layer. It is very likely that the localization of interphase boundary sliding associated with growth faults will introduce the bending stress and facilitate the brittle fracture of hard Θ phase layer which controls the ductility and fracture of the lamellar eutectic composite. Therefore, the result of the present investigation suggests that the elimination of growth faults from the lamellar composite would avoid the localization of interphase boundary sliding and reduce the introduction of the bending stress, to result in the enhancement of the ductility of the composite at high temperatures.

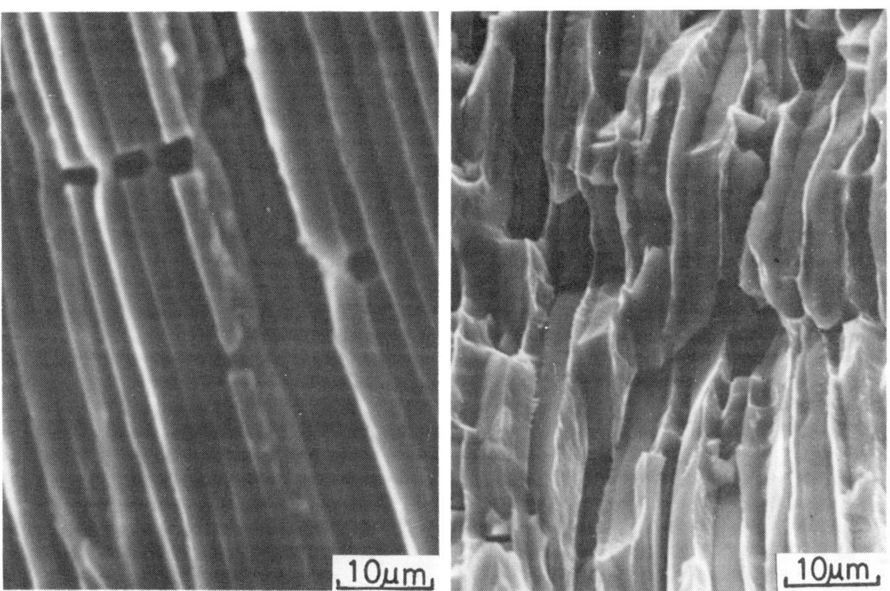

Fig.5 The initiation and propagation of cracks in the Θ phase layers intervened by the α phase layers.

Fig.6 SEM micrograph of fracture surface.

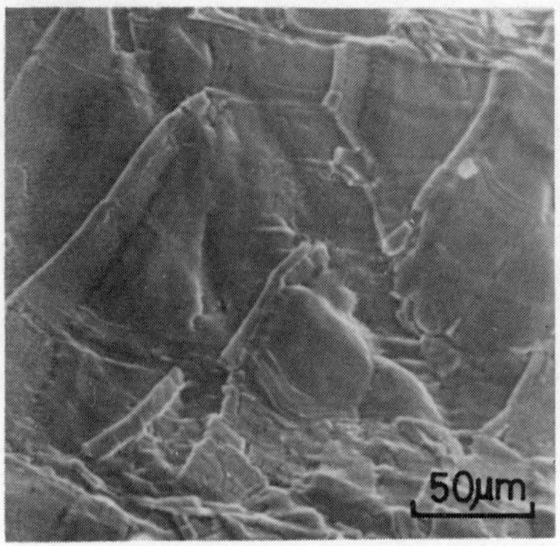

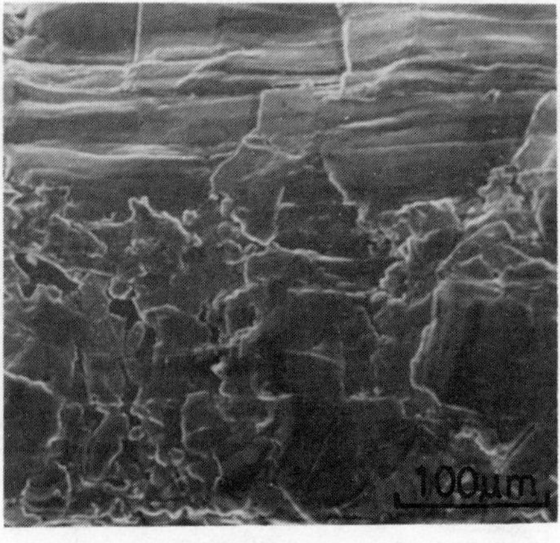

(a) (b)

Fig.7 SEM micrographs demonstrating fragile nature of Θ phase layer at high temperature.

Detrimental effects of growth faults on high temperature strength and fracture of the lamellar eutectic composites should be more recognized and studied in order to develop more ductile and strong composite materials.

CONCLUSION

It was found that interphase boundary sliding can occur at every two boundaries and that the presence of growth faults such as termination and ripple fault enhances the sliding during creep deformation in the Al-$CuAl_2$ lamellar eutectic composite. The localization of interphase boundary sliding at certain boundaries likely introduces the bending stress which will facilitate the fracture of the hard Θ phase layer. It is concluded that the presence of growth faults is an important key factor controlling deformation and fracture induced by interphase boundary sliding in the lamellar Al-$CuAl_2$ eutectic composite.

Acknowledgements

This work was supported by a Grant-in-Aid for Scientic Research from the Ministry of Education,Science and Culture. One of the present authors (T.W.) appreciates the provision of visiting professorship by Ecole Nationale Superieure des Mines de Saint-Etienne where this paper was written.

REFERENCES

1. N.S.Stoloff: Alloy and Microstructure Design, ed.by J.K.Tien and G.S.Ansell, Academic Press, New York, p.65(1976).
2. T.Watanabe, Met.Trans.,A14, 531(1983).
3. L.M.Hogan,R.W.Kraft and F.D.Lemkey, Advances in Materials Researches, ed.by H.Herman, Wiley-Interscience,New York,Vol.5, p.83 (1971).
4 D.D.Double, Mater.Sci.Engr.,11,325(1973),5.M.Ignat,R.Bonnet,D.Caillaerd and J.L.Martin, Phys.Stat.Sol.,(a)49,675(1978).
6 M.Ignat,and M.Depeux,Progress in Science and Engineering of Composites, ICCM-IV, Tokyo, p.1361 (1982), 7. M.Ignat and R.Bonnet, Acta metall.,31 1991(1983), 8. P.Gaymard and R.Bonnet, Revue Phys.Appl.,16,145(1981).
9. A.Eberhardt and B.Baudlet, J.Mater.Sci.,9,865(1974), 10. H.Suzuki,T. Takasugi and O.Izumi, Acta metall.,30,1647(1982).
11. T.Watanabe,M.Yamada,S.Shima and S.Karashima, Phil.Mag.,A40,667(1979).
12. T.Chandra,J.J.Jonas and D.M.R.Taplin,J.Mater.Sci.,13,2380(1978).
13. R.B.Vastava and T.G.Langdon, Acta metall., 27,251(1979).
14. R.Shariat,R.B.Vastava and T.G.Langdon, Acta metall.,30,285(1982).

Stress Rupture of Al-Ni Eutectic Composites

M. H. Abdel Latif, and M. A. Rizk

Department of Design and Production Engineering, Ain Shams University, Cairo, Egypt

ABSTRACT

Al-Ni eutectic alloys were prepared from commercial purity Al and Al-Ni master alloy by melting under controlled Argon atmosphere. Eutectic composites with different microstructures were then obtained by unidirectional solidification at predetermined growth rates, higher and lower than those required for lamellar/rod-like transition. Stress rupture tests were carried out on the directionally solidified (DS) eutectic composites as well as DS Al and as cast eutectic alloy. The stress rupture properties were determined for the prepared specimens and related to their microstructures. The DS rod-like Al-Ni eutectic composite exhibited superior rupture properties to the lamellar composite, as cast Al-Ni and DS Al.

KEYWORDS

Directional solidification (DS); eutectic composites; lamellar structure; rod-like structure; rupture stress; rupture life.

INTRODUCTION

When directionally solidified with different rates, the Al-Ni eutectic composite has been shown to develop a rod-like structure (1,2). However, directional solidification at very low growth rates below the range for rod-like/lamellar transition leads to the development of a lamellar structure(1).

Stress rupture and creep behavior of eutectic composites is intimately related to their structures. Finer microstructures lead to the improvement of stress rupture life and a decrease in the minimum strain rate (3,4). The occurance of a cellular structure was found to be detrimental to stress rupture and creep properties(4).

In this work, the effect of microstructure and intermetallic compound morphology of the Al-Ni eutectic alloy on the stress rupture behavior is studied. Specimens representing the different structures namely, as cast, DS Rod-like and DS lamellar eutectic composites were prepared from commercial purity materials and stress ruptured. Parallel tests were also performed on the DS Al representing the matrix material of the composites.

EXPERIMENTAL PROCEDURE

The Al-Ni eutectic alloys were prepared from commercial purity materials with Ni content of 6.2 wt%. Melting and solidification were carried out under a controlled Argon atmosphere. Rods from microstructurally accepted ingots were homogenized at 723°K for 3 days and represented the as cast alloy. Other rods were DS under a purified dynamic Argon atmosphere in a specially constructed apparatus. Two growth rates were applied, namely 33.8μm/sec to produce a rod-like structure and 1.7μm/sec to produce a lamellar structure. All DS specimens were preheated at 1173°K for one hour and the temperature gradient was kept constant at 10°K/mm.

Microstructure examination was performed on cross and longitudinal sections of as cast Al-Ni, DS Al, and DS rod-like and lamellar composite specimens. Stress rupture specimens 5 mm in diameter and 20 mm in gage length were machined and polished from the central part of the solidified rods. Stress rupture tests were carried out under constant load in a specially constructed m/c. Temperature was kept constant during the tests at 573°K, Using a temperature controller with an accuracy of $\pm$ 2°K. Specimens were placed in the uniform temperature zone of a specially designed electric resistance furnace. The temperature difference along the gage length was less than 2°K. One half of each ruptured DS eutectic composite specimen was longitudinally sectioned and optically examined for possible rod or lamellae fracture at and below the fracture surface. Fracture morphology was studied in the other half using scanning electron microscopy. The percentage elongation of the ruptured specimens was measured according to ASTM standards No. E 139-66 T.

RESULTS AND DISCUSSION

The microstructure of the as cast Al-Ni eutectic alloy and DS eutectic composite grown at 33.8 and 1.7 μm/sec are shown in Fig. 1. All materials exhibited an intermetallic Al_3Ni phase dispersed in a solid solution of Al matrix. However, the nature of the dispersed phase was different in each material. In the as cast, the Al_3Ni phase was randomly oriented in both the cross and the longitudinal sections, Figs. 1 a and b. In the DS composite grown at 33.8 μm/sec, the Al_3Ni phase was in the form of rods aligned parallel to the longitudinal growth direction, Fig. 1c. A colony structure was observed for this composite, Fig. 1d, which is attributed to the commercial purity of the materials used(5). The rods at the colony boundary was observed to be coarser and slightly bent towards the boundary, Fig. 1c. In the DS composite grown at 1.7 μm/sec, the Al_3Ni phase was in the form of lamellae rather than rods as is evident from the cross section, Fig. 1e. This lamellae exhibited a uniform thickness and were perfectly aligned in the longitudinal growth direction, Fig. 1f. No colony structure was observed for this composite, Fig. 1e, the growth rate R was reasonably low leading to a high G/R value. This observation seems to be supported by Kraft and Albright finding that, with a known amount of impurity, colony formation can be suppressed, provided that the G/R value is high enough(6).

The stress rupture results of the DS Al-Ni eutectic composite for both rod-like and lamellar structures, the as cast Al-Ni eutectic and DS commercial purity Al is shown in Fig. 2. A Larson-Miller plot of these results is shown in Fig. 3.

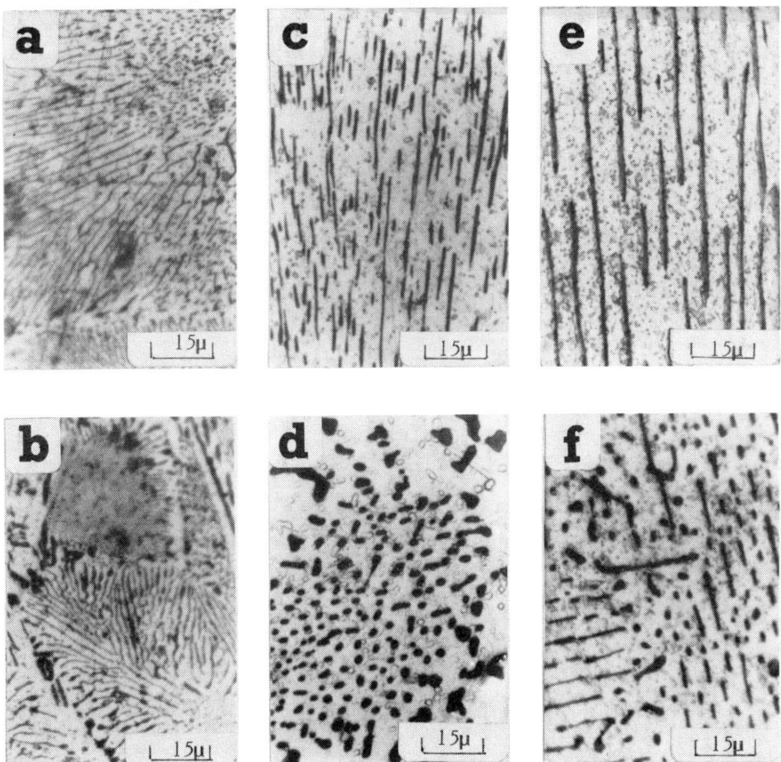

Fig. 1. Microstructure of Al-Ni eutectic alloys.
As cast (a) longitudinal, (b) cross.
DS, R=33.8 μm/sec (c) longitudinal (d) cross.
DS, R=1.7 μm/sec (e) longitudinal, (f) cross.

Figure. 4, is deduced from Fig. 1, and gives the rupture stress for the different alloy structures corresponding to certain rupture times. These data are important for the alloy design to sustain a certain stress for a required service life. Figure. 5, shows a similar plot of the rupture stress divided by the UTS (stress ratio) for the different materials versus rupture time.

It is depicted from the figures that the DS Al-Ni eutectic composites are superior to the as cast Al-Ni eutectic and DS commercial purity Al as indicated by a higher rupture stress for the same rupture life. This is due to the strengthening effect of the Al_3Ni phase which is dispersed in the Al matrix of the DS composites in the form of rods or lamellae aligned in the longitudinal stress direction. However, if this strengthening effect is compensated by plotting the stress ratio versus rupture time, Fig. 5, the DS commercial Purity Al is shown to be superior. Thus apart from the strengthening effect, the Al_3Ni can be detrimental to the rupture life of the alloy and suggests that rupture of the alloy may start by early Al_3Ni phase fracture.

Although the DS rod-like composites showed a colony structure with coarse and misaligned rods at the colony boundary, they exhibited higher rupture stresses

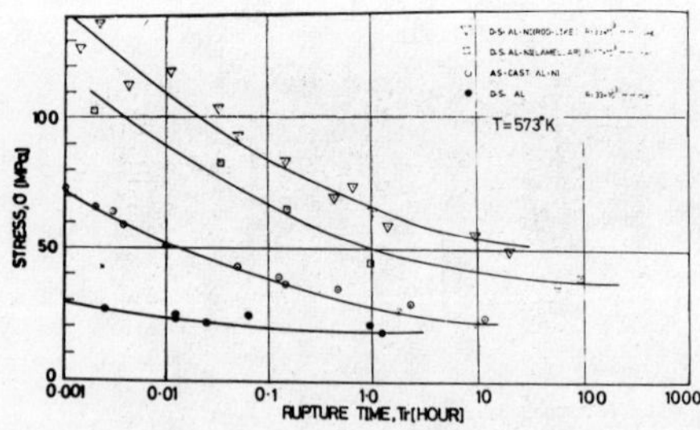

Fig. 2. Log rupture time versus rupture stress for the different structures.

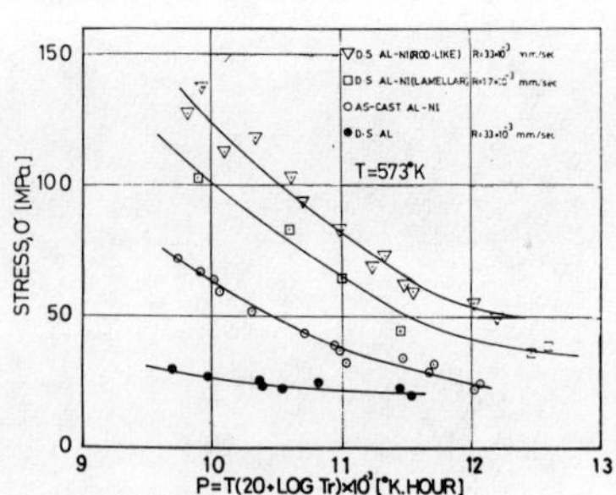

Fig. 3. Larson-Miller plot for the stress rupture results.

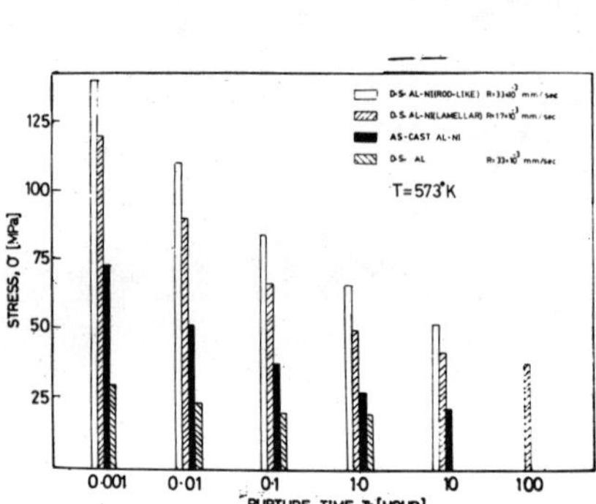

Fig. 4. Rupture stress corresponding to certain rupture times for the different structures.

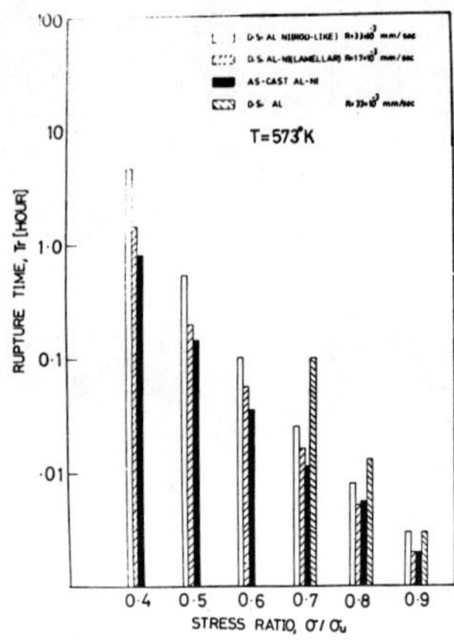

Fig. 5. Rupture stress ratio versus log rupture time for the different structures.

than the lamellar composites which were colony free. This may be explained by the increasing strengthening effect caused by a higher aspect ratio of the Al_3Ni rods compared to the lamellae, given the same volume fraction.

The percentage elongation to rupture, E_l, was determined and plotted versus the rupture time for the as cast Al-Ni, DS Al and DS rod-like Al-Ni composite in Fig. 6. E_l values for the lamellar composite were very low and excluded from the figure. Due to the constraint imposed by the aligned Al_3Ni rods or lamellae, the DS composites showed much lower E_l values than the as cast Al-Ni and DS Al. The macrofracture appearance is shown in Fig. 7, and indicates that the as cast Al-Ni and DS Al (left) exhibited more ductile fracture than the DS composites (right).

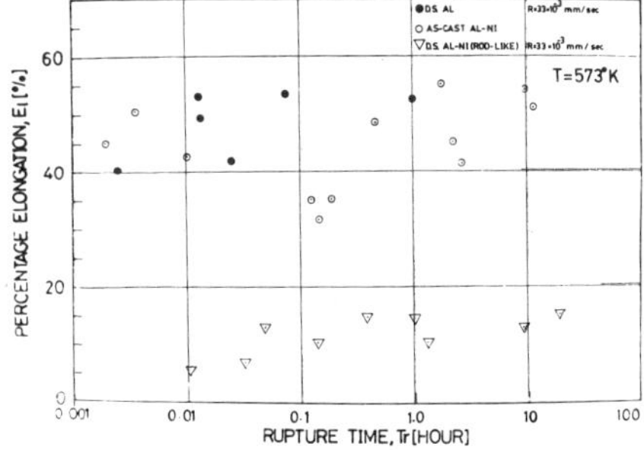

Fig. 6. Percentage elongation versus log rupture time for the as cast Al-Ni, DS Al and DS rod-like composite.

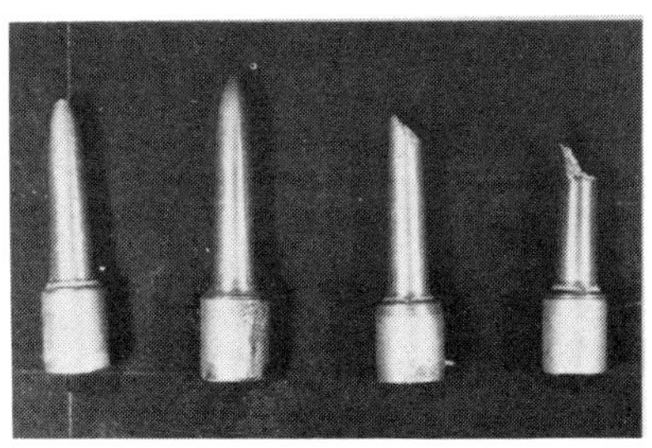

Fig. 7. Macrofracture appearance of as cast Al-Ni, DS Al, rod-like, and lamellar DS composites. (left to right).

Microscopic examination of Al_3Ni rod fracture at and below the fracture surface of the rod-like composite is shown in Fig. 8. Considerable rod fractures were observed perpendicular, parallel and inclined to the rod axis for a specimen stress ruptured at a high stress level, Fig. 8a. This was associated with void formation and suggested that fracture occurs by rod fracture, void formation, growth and coalescence. Considerable rod fracture was also observed at a distance of 1 mm below the fracture surface of the same specimen, Fig. 8b. At low stress levels, microscopic examination revealed lower rod fracture density at 1 mm from the fracture surface, associated with considerable void growth mainly at rod/matrix interface and rod ends, Fig. 8c. On the other hand, fewer lamellae fractures were observed in the lamellar composite at 1 mm from fracture surface at both high and low stress levels, Figs. 9 a and b, respectively. No void growth was observed below the fracture surface. This suggests that rupture of the lamellar

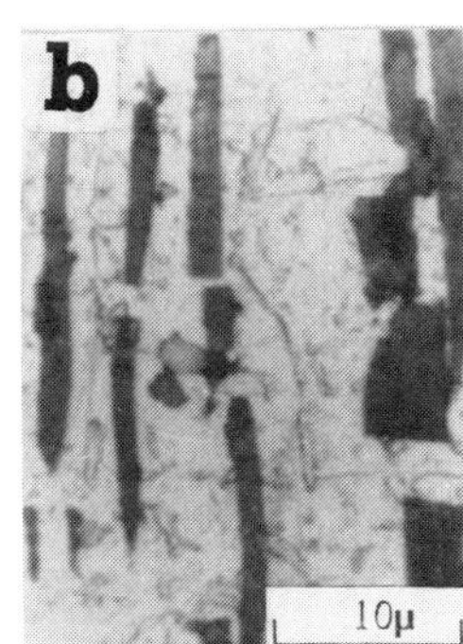

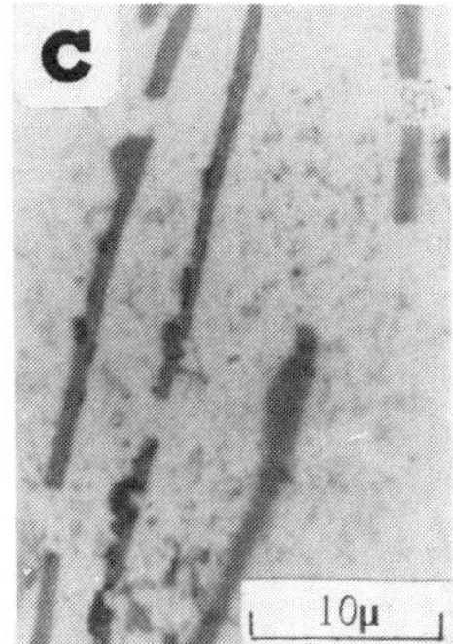

Fig. 8. Al_3Ni rod fractures at and below the fracture surface.
(a) Fracture surface, high stress level.
(b) 1 mm below fracture surface, high stress level.
(c) 1 mm below fracture surface, low stress level.

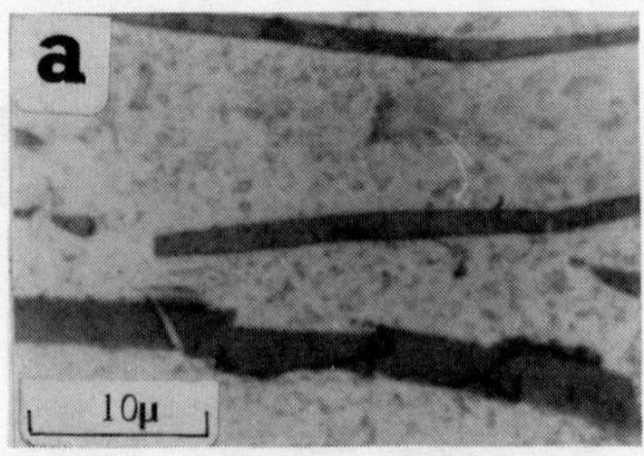

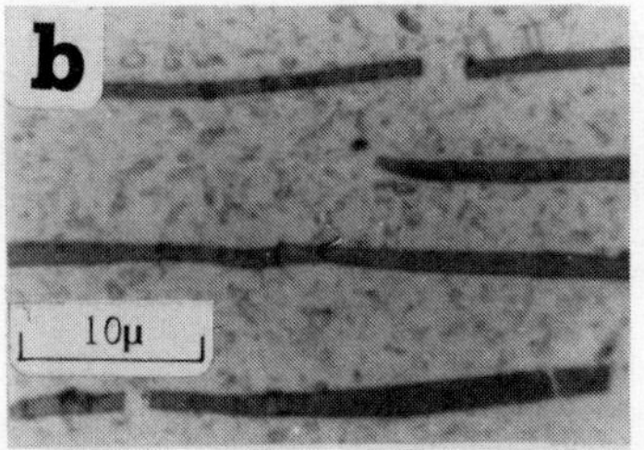

Fig. 9. Al_3Ni lamellae fractures 1 mm below fracture surface.
(a) High stress level. (b) Low stress level.

composite may readily occur after few lamellae fractures. This is certainly not the case for the rod-like composite which can tolerate multiple rod fractures before rupture thus showing superior rupture properties.

CONCLUSIONS

Due to their higher aspect ratio for the same volume fraction, the Al_3Ni rods in rod-like composites can undergo multiple fracture. Therefore, rod-like composites exhibited higher rupture stresses than lamellar composites for the same rupture lives. Moreover, percentage elongations to rupture for rod-like composites were higher than for lamellar. This is accounted by the less constraint imposed by Al_3Ni rods on matrix flow and the existance of a colony structure in the rod-like composites.

REFERENCES

1. F.D. Lemkey, R. W. Hertzberg, and J. A. Ford, Trans. Met. Soc. AIME 233, 334 (1965).
2. E. M. Breinan, E. R. Thompson, G. P. McCarthy, and W. J. Herman. Met. Trans. 3, 221 (1972).
3. E. M. Breinan, E. R. Thompson, and W. K. Tice, Met. Trans. 3, 211 (1972).
4. P. N. Quested, E. Bullock, and M. Mc Lean, Proceedings of Sheffield international conference on solidification and casting, Ranmoor House Sheffield Univ. P 71 (1977).
5. H. W. Weart, and D. J. Mach, Trans. Met. Soc. AIME. 664 (1958).
6. R. W. Kraft, and D. L. Albright, Trans. Met. Soc. AIME. 221, 95 (1961).

Fracture Toughness of SiC/Al Metal Matrix Composites

D. F. Hasson* and C. R. Crowe**

**Department of Mechanical Engineering, U.S. Naval Academy, Annapolis, MD 21402, USA*
***Material Science and Technology Division, Naval Research Laboratory, Washington, D.C. 20375–5000, USA*

ABSTRACT

The fracture toughness of discontinuous SiC/6061-T6 aluminum metal matrix composites has been measured and correlated with notch acuity, microstructure, and fracture morphology. Results show a simple relation between notch acuity and fracture toughness, and data correlates with a model derived from small-scale yielding mechanics.

KEYWORDS

Metal matrix composites; silicon carbide-aluminum; fracture toughness, microstructure.

INTRODUCTION

Discontinuous silicon carbide/aluminum alloy (SiC/Al) metal matrix composites (MMC's) have exhibited improved physical and mechanical properties as compared to the properties of the wrought matrix alloy[1,2]. The tensile ductility and fracture properties of the composite, however, are less than those of the wrought alloy[3,4]. Although the materials show limited ductility on a macroscopic scale, SEM fractography indicates that fracture occurs by a ductile mechanism with plastic deformation localized adjacent to the crack tip. This produces a fracture surface consisting of fine dimples, the size of which are on the order of the subgrain size[5].

Illuminating the factors which control fracture toughness is the object of the present study.

EXPERIMENTAL

The materials used in this study were nominal T6 heat treated 20v/o SiC_w/6061 aluminum extruded tube, 20v/o SiC_w/6061 aluminum plate, and 25v/o SiC_p/6061 aluminum plate. Details of these materials are described elsewhere[3,4]. Fracture toughness specimens from the tube were in standard

(1/2)T compact tension configuration. The fracture test specimens taken from the plate materials had the geometry of a (1/2)T compact tension specimen except the thickness was 6.35 mm rather than 12.70 mm. The notches had various root radii. Fracture toughness tests were performed in duplicate and the testing procedure and data analysis conformed to ASTM E399 procedures where appropriate. Stereo pair scanning electron microscope (SEM) was performed on representative fracture surfaces. Dimple height, h, measurements were made from the stereo pairs.

RESULTS AND DISCUSSION

Mechanical Properties

The tensile strength properties for the various materials and orientations are given in Table 1. Also included in Table 1 are reference values for wrought Al 6061 in the T6 condition. Comparison of 6061 -T6 Al and SiC/Al 6061 -T6 properties shows significant increases in modulus, yield stress and ultimate stress due to the addition of the SiC. Whisker additions are seen to be somewhat more effective in strengthening than are particulate additions. Ductility, however, is less in the SiC/Al 6061 -T6 materials than in the unreinforced alloy. Orientation effects are more pronounced in the whisker material than in the particulate material. Figures 1(a) and (b)

TABLE 1: MECHANICAL PROPERTIES

	MATERIAL	E(GPa)	σ_y (MPa)	σ_{uts} (MPa)	%ELONG	K_{IC} (MPa $\sqrt{m}$)
PLATE	20v/o SiC_w/6061-T6 (L)	122	443	584	1.8	7.1 (L-T)
	20v/o SiC_w/6061-T6 (T)	91	409	480	3.8	6.3 (T-L)
	25v/o SiC_p/6061-T6 (L)	99	345	410	4.4	15.8 (L-T)
	25v/o SiC_p/6061-T6 (T)	98	350	409	4.0	14.5 (T-L)
	6061-T6 Al*	69	276	310	17	~37
TUBE	20v/o SiC_w/6061-T6 (L)	107	374	521	2.7	22.4 (L-C)
		—	—	—	—	14.0 (R-L)
		—	—	—	—	17.6 (C-L)

LEGEND: E = YOUNG's MODULUS; σ_y = 0.2% FLOW STRESS; σ_{uts} = ULTIMATE TENSILE STRENGTH; % ELONG = THE PLASTIC STRAIN (ϵ_p) AT FAILURE; AND K_{IC} = PLANE STRAIN FRACTURE TOUGHNESS

*DATA FROM ASM METALS HANDBOOK

are examples of the effect of notch acuity on the fracture toughness of SiC/Al. Figure 1(a) shows a linear relationship between the crack tip opening displacement and the notch root radius. This linearity implies that the fracture is a strain controlled process[6]. The data in Fig. 1(b) are shown correlated with the function:

$$K_{Ic}(\rho) = K_{Ic} \frac{(1 + \rho/21)^{3/2}}{(1 + \rho/1)} \quad (1)$$

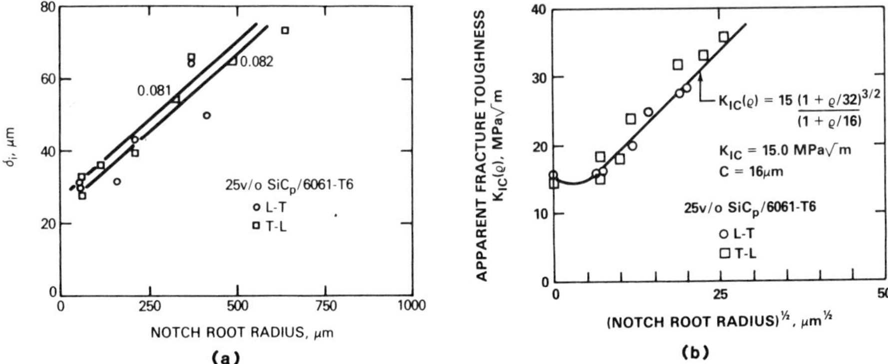

Fig.1. Effect of notch acuity on a) crack opening displacement, and b) K_{1c}

In this relation, ρ is the notch root radius and l is a characteristic length in front of the crack tip. All fracture toughness data available for SiC/Al MMC's fits equation (1). Plane strain fracture toughness values for the various materials and orientations are also listed in Table 1. The level of all K_{Ic} values for the composite are less than 65 percent of the K_{Ic} value of 36.82 MPa √m of wrought Al 6061 alloy in the T6 condition[8,9].

Scanning Electron Microscopy

Observations of the fracture paths of cracked, but not separated compact tension specimens were made. Figures 2 (a) & (b) show representative process zones at the crack tips for fractures in the whisker and particulate material, respectively. In both cases, the crack is observed to propagate in the matrix material by opening of voids at the ends of SiC particles. Crack process zone voids are highly localized and were not observed more than about two times the mean particle spacing from the plane of the fracture except in a region of high damage near the root of the notch. These observations are in accord with those of Mukherji and Chellman[10].

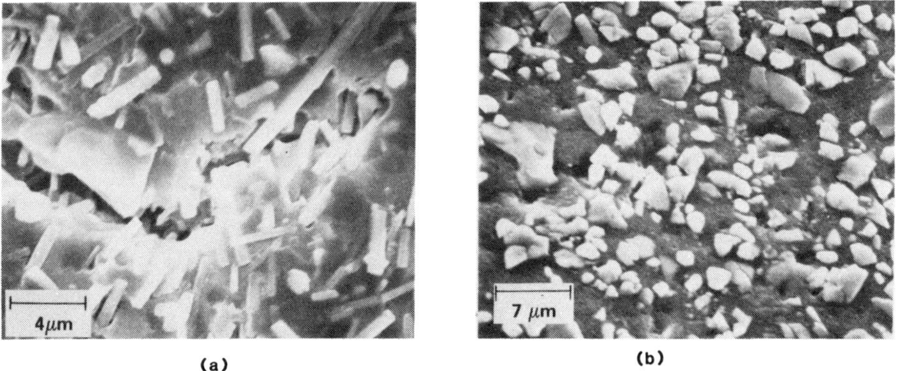

Fig.2. SEM micrographs of fracture process zone in a) SiC/Al and b)SiC /Al

Observations of fracture surfaces in the SEM revealed five distinct morphological features. In decreasing frequency of observation, these features are dimples, inclusions, non-bonded regions, non-infiltrated regions, and decohension at aluminum grain boundaries. Most of the fracture surface consisted of fine, equiaxed dimples of uniform size. Embedded in the base of approximately 50% of the dimples is a SiC particle tip which is covered with a coating of aluminum matrix[11]. Quantitative measurements of the microstructural data are given in Table 2.

TABLE 2: MICROSTRUCTURAL PARAMETERS

	MATERIAL	ORIENT	MFD(μm)	Sp(μm)	d_0(μm)	D_d(μm)	H_d(μm)
PLATE	25 v/o SiC_p/6061 T6	L	9.7	13.0	3.3	3.9	1.37
	25 v/o SiC_p/6061 T6	T	8.7	11.5	2.8	3.7	1.24
	20 v/o SiC_w/6061 T6	L	2.2	2.8	0.6*	1.2	1.26
	20 v/o SiC_w/6061 T6	T	3.3	4.1	0.8*	1.1	0.55
TUBE	20 v/o SiC_w/6061 T6	L	3.9	4.5	0.9*	1.8	1.60
	20 v/o SiC_w/6061 T6	R	3.5	4.3	0.8*	1.6	0.88
	20 v/o SiC_w/6061 T6	C	3.4	4.5	1.1*	1.3	0.75

MFD = MEAN FREE DISTANCE   Sp = PARTICLE SPACING d_0 = MEAN PARTICLE DIAMETER
D_d = MEAN DIMPLE DIAMETER   *(CALCULATED AS Sp-MFD) H_d = MEAN DIMPLE HEIGHT

Micromechanical Fracture Processes

The micromechanical processes are envisioned as shown in Fig. 3. First, crack tip blunting occurs with the development of a small-scale plastic zone in front of the crack tip. If δ_t is the crack tip opening displacement, theoretically the region of intense plastic strain extends a distance of $\sim 1.9\delta_t$ ahead of the crack[12-14]. The theory also indicates that for a stationary crack,

$$\delta_t = \frac{\alpha K_I^2}{\sigma_y E} \tag{2}$$

where α is a non-dimensional numerical constant which depends on the work hardening exponent, n of the material, K_I is the stress intensity at the crack tip, E is Young's modulus, and σ_y is the yield stress. As the plastic zone develops, it interacts with the microstructure. Plastic flow becomes concentrated around the embedded SiC particles and a process zone is established in which voids are opened at the ends of SiC particles. Fractographic evidence (Fig. 2) indicates that the voids open at the ends of SiC particles adjacent to the crack plane. The evidence also indicates that the decohesion occurs in the matrix leaving a thin coating of matrix material to cover the SiC particle which has initiated the void. It is not known whether the void is nucleated homogeneously in the aluminum or whether nucleation occurs at some defect or other matrix phase such as Mg_2Si precipitates in the aluminum; the major point being that the nucleation is associated with stress/strain localization caused by the SiC particles. Void growth occurs by stretching matrix in the ligaments. As the surfaces move apart and the matrix material at the crack tip reaches tensile instability, a chisel edge is formed and the leading void coalesces with the crack tip. The crack tip opening displacement at this stage is limited to the void height, 2h, since the opening deformation is limited to the growth and coalescence of the next void. Since the work hardening exponent is

~ 0.1, $\alpha \sim 0.44$ for the SiC/6061-Al material[13] and the plane strain fracture toughness, K_{Ic}, should be related to the mean dimple height, H_d, by,

$$K_{Ic} \sim 2.1 \sqrt{E\sigma_y H_d} \quad (3)$$

A comparison of measured toughness data with the predictions of Eq. 3 show agreement within 13% for the particulate plate material, within 65% for the plate whisker material, and within 28% for the tube whisker material.

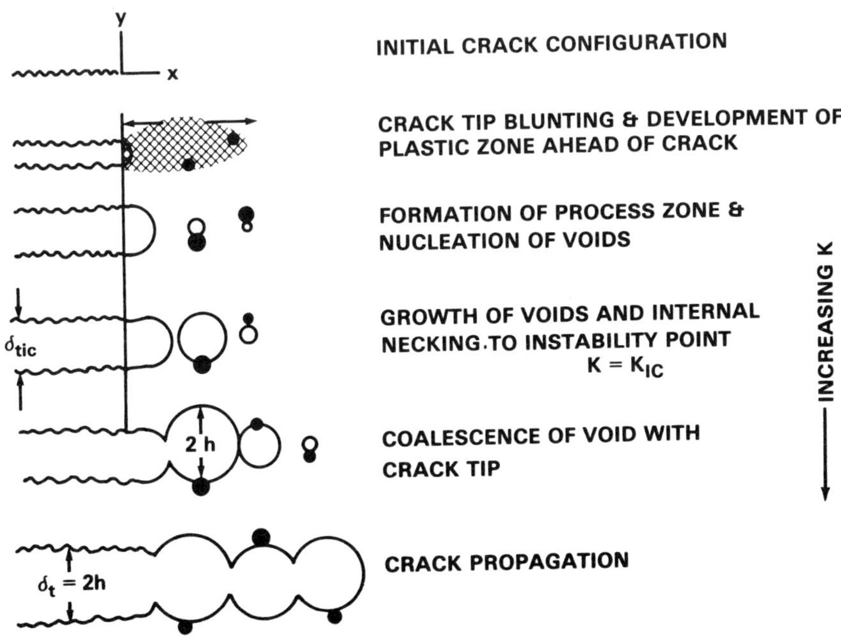

Fig.3. Schematic of processes leading to fracture in SiC/Al

CONCLUSIONS

Results indicate that fracture of SiC/Al-6061 MMC's is a strain controlled process in which the fracture toughness and notch acuity are connected through a simple mathematical relation. The fracture is observed to occur by a small-scale yielding process and the results correlated well with small-scale yielding mechanics models which appear in the literature.

ACKNOWLEDGEMENTS

The authors wish to express their appreciation to Messrs. W. Uhmladt, R. A. Gray, and S. Hoover and to Ms. N. Nutter for assistance with various aspects of this work. Financial support was provided by the Naval Sea Systems Command and the Office of Naval Research.

REFERENCES

1. A.P. DiVecha, C.R. Crowe, and S.G. Fishman, in Failure Modes in Composites (edited by J.A. Cornie and F.W. Grossman, AIME, Warrendale, PA., p. 406 (1978).
2. A.P. DiVecha, S.G. Fishman,and S.G. Karmarkar, J. Metals, 33, 12 (1981).
3. C.R. Crowe and R.A. Gray, in Failure Mechanisms in High Performance Materials (edited by J.G. Early and R. Shives) Cambridge U. Press, N.Y., 1984.
4. D.F. Hasson, S.M. Hoover, and C.R. Crowe, submitted to J. Mat. Science.
5. J.S. Ahearn and D.C. Cooke, "TEM and SEM of Fatigued SiC/Al Metal Matrix Composites," Martin Marietta Laboratories Report MML TR-3SC, January 1984.
6. J.F. Knott, Fundamentals of Fracture Mechanics, Butterworths, London, 1972.
7. A.J. Kinloch and J.G. Williams, J. Mat. Sci., 15, 987 (1980).
8. R.P. Kambour and S. Miller, J. Mat. Sci., 12, 2281, (1977).
9. J.G. Kaufman, in Fracture Prevention and Control (edited by D.W. Hoeppner), ASM, Metals Park, Ohio, p. 121, (1974).
10. T.K. Mukherji and D. Chellman, in Proc. Sixth Annual Discontinuous Reinforced Aluminum Composites Working Group Meeting, Metal Matrix Information Analysis Center, Report MMCIAC No. 000479, p. 305, April 1984.
11. R.J. Arsenault, Ibid, p. 1.
12. J.R. Rice and M.A. Johnson, in Inelastic Behavior of Solids (edited by M.F. Kanninen, W.F. Adler, A.R. Rosenfield, and R.I. Jaffee), p. 641, McGraw-Hill, N.Y. (1970).
13. R.M. McMeeking, J. Mech. Phys. Solids, 25, 357, (1977).
14. J.R. Rice and E. Sorenson, J. Mech. Phys. Solids, 26, 163, (1978).

Influence of Microstructure on the Toughness of Carbon Fibre/Plastic Composites

C. Y. Barlow*, M. V. Ward and A. H. Windle***

**University of Cambridge, Department of Metallurgy and Materials Science, Cambridge, UK*
***ICI New Materials Group, Wilton, Middlesbrough, Cleveland, UK*

ABSTRACT

A preliminary study is presented of the relationship between the microstructural aspects of failure and the fracture energy G_{1C} for cracking parallel to the fibres in long-fibre/thermoplastic matrix composites. Fracture energies are measured by a new technique, and fracture surfaces generated by the test are examined by scanning electron microscopy.

KEYWORDS

Carbon fibre composite; Thermoplastic; Fracture Surface; Fracture Energy.

INTRODUCTION

The aim of the present study is to identify the characteristic features of fracture surfaces in a particular class of composite material, and to establish their relevance to the energy absorbed to failure. The material used is based on the Aromatic Polymer Composite (APC) range manufactured by Imperial Chemical Industries, and the current investigation uses only uniaxially aligned long fibre material. A range of materials was generated by modifications both to the fibres and to the polymer matrix (polyether etherketone, PEEK), but only the latter series is discussed here. The samples are in the form of compression moulded plates, thickness 1-2mm, produced from stacked prepreg layers by a standardised procedure [1].

The first stage in the project was to design a fracture test which would both allow measurement of fracture energies from the samples available (of which supplies were in some cases very limited); and also produce well-defined fracture surfaces which were characteristic of the material. The fracture surfaces (which are observed using scanning electron microscopy, SEM) are reproducible, and provide a good basis for understanding the influence of composite formulations on fracture energy. Fracture surfaces are examined using scanning electron microscopy (SEM). It is important to be able to examine both fracture energy and fracture mode, as either alone may be misleading. For example, it is possible to obtain impressively high fracture energy values, but the accompanying fracture mode (e.g. excessive fibre-matrix debonding) may be undesirable for other reasons.

Fracture energy determination

The experimental procedure followed for the generation of fracture surfaces and the calculation of fracture energies is described in more detail elsewhere [2] but the salient points are outlined here. The failure mode chosen is fracture normal to the fibre axes and also normal to the prepreg layers. A crack is introduced by inserting a razor blade into the block so as to split off a section of thickness between 0.3 and 1.5mm. A crack extendss into the material ahead of the sharp edge of the blade and in the plane of the blade. The crack attains a steady-state length which remains constant as the blade is driven into the block. The region of the section being split off lying between the point of contact with the razor blade shoulder and the tip of the crack may be regarded as a cantilever beam, loaded at the razor blade and supported at the crack tip. Measurement of the crack length c; section thickness d; blade thickness h; and longitudinal Young's modulus E; allows the fracture energy for normal failure G_{IC} to be calculated from the equation

$$G_{IC} = \frac{3E' d^3 h^2}{8(c')^4}$$

based on the equation due to Obreimoff [3]. In this modified form E' is the bending modulus of the material, and is related to E by the equation

$$E' = E \left[\frac{c^2}{c^2 + 3d^2} \right]$$

[ref. 4]

The measured crack length c is related to the effective crack length c' by a graphically determined factor c, giving $c' = c + \Delta c$. This factor Δc is material dependent and may be positive or negative. It incorporates several different functions, and obviates both some experimental and theoretical difficulties. The theory does not take into account such features as end effects around the bent beam, and a constant factor may be used to reduce the error. The experimental difficulty arises from uncertainty in the position of the effective position of the crack tip. This is the load-bearing point, and it may not exactly coincide with the apparent position of the end of the crack. This implies the existence of a plastic zone ahead of the crack, and this feature is discussed below.

FRACTURE SURFACES

The fracture surfaces show the path through the material which the crack has followed, and the crack will follow the most favourable route available to it. The final fracture surface therefore carries some information on the relative strengths of the different regions of the material corresponding to the amount of fracture energy absorbed. Thus for example the fracture surface of one material may show clean fibres, but reducing the strength of the matrix results in fibres which are resin-coated as the crack now runs through the matrix rather than along the fibre-matrix interface. The same effect may be produced by increasing the strength of the interface, but different fracture energy values would be expected in these two cases.

A number of features of the fracture surfaces have been identified as being of relevance to the fracture energy of the material, with the microstructure of failure of the matrix being particularly significant. Figures 1 and 2 demonstrate the effect of using optimised and non-optimised matrices, while keeping other parameters constant. The matrix in fig. 2 shows a rough fracture surface with a reasonable degree of "macroductility". This term is used to describe plastic deformation of the matrix on the scale of the fibre diameter, and is seen as the matrix drawing up into peaks. The presence of macroductility seems to be essential to the attainment of fracture energies of $> 2kJm^{-2}$ in these systems. Where it is absent, as in fig. 1, extensive matrix cracking is seen. The fracture energies of these two materials were found to be:

Non-optimised matrix,	figure 1	$G_{IC} = 0.46 \pm 0.08\ kJm^{-2}$
Optimised matrix,	figure 2	$G_{IC} = 3.38 \pm 0.15\ kJm^{-2}$

For comparison, a commercial epoxy matrix composite (in which macroductility is absent) also gave a fracture energy of $0.46 \pm 0.06 kJm^{-2}$. The fracture surface does not however resemble that of figure 1, and it should be remembered that a similarity in fracture energy does not necessarily imply that the fracture strengths or fracture energies are comparable.

ACKNOWLEDGEMENTS

We are indebted to ICI for funding of this project and for provision of material. Thanks are due to Dr D. Briggs and Dr E. Nield for assistance in the programme.

REFERENCES

1. ICI New Materials Group Provisional Data Sheet APC PD.

2. Barlow, C.Y. and Windle, A.H., J. Mater. Sci. Letters, in press.

3. Obreimoff, J.W., Proc. Roy. Soc. A127, 290 (1930).

4. Farebrother, T.H. and Raymond, J.A., 1977, in "Polymer Engineering Composites" (Applied Science Publishers Ltd., London, England) ed. M.O.W. Richardson.

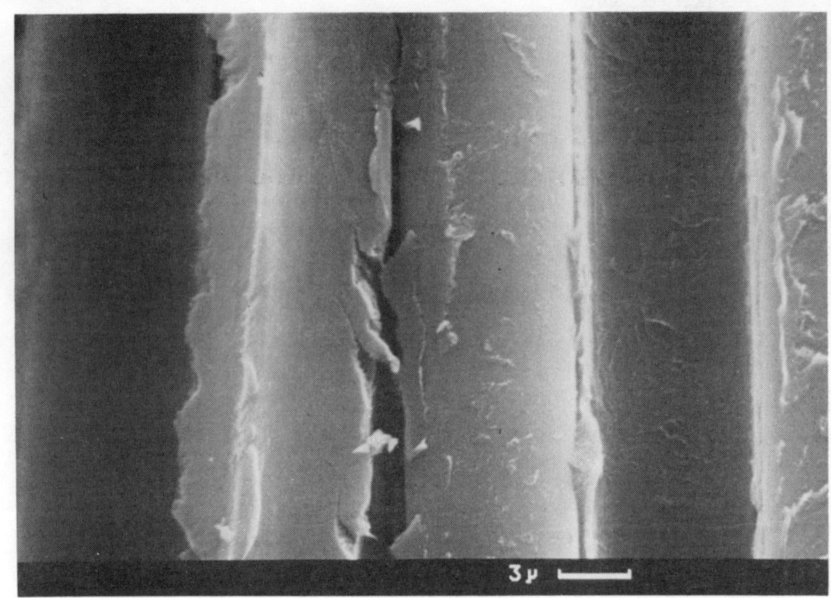

Fig.1 Non-optimised matrix

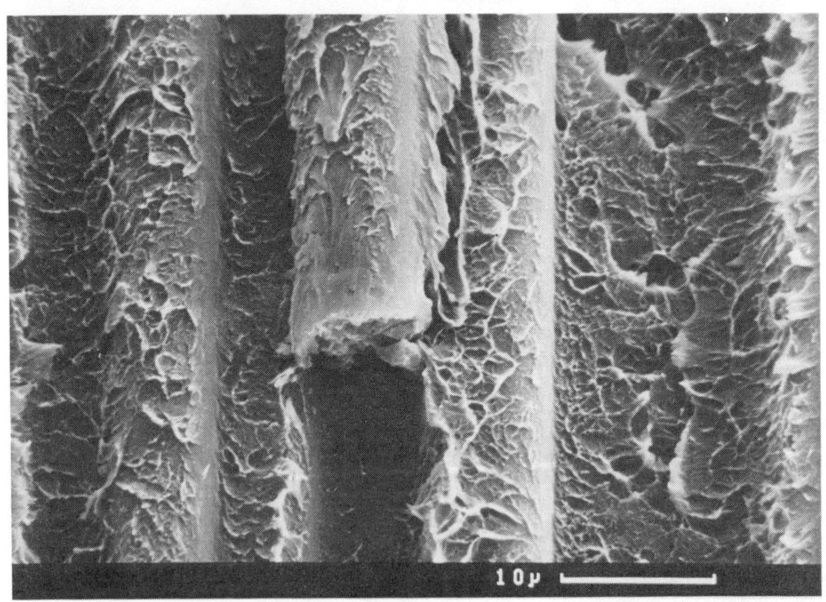

Fig.2 Optimised matrix

Static and Cyclic Behaviour of Reinforced Polymer Sandwich Beams

W. A. Lucier, U. H. Mohaupt and A. Plumtree

University of Waterloo, Department of Mechanical Engineering, Waterloo, Ontario N2L 3G1, Canada

ABSTRACT

The design of sandwich beams to flexural fatigue with a low span to depth ratio has been considered. The stiffness and stress levels at critical sites in the beam were found to be highly dependent on the relative thicknesses of the facings and the core.

Knowing the static and fatigue behaviour of the facing, core and interfacial bond materials, the designer will be able to determine the relative facing thickness required for a structurally optimal beam. In the case of beams with glass fibre/polyester cores, the relative thickness of carbon fibre/epoxy facings is expected to decrease 50% for optimal fatigue behaviour compared with optimal static behaviour.

KEYWORDS

Composites, Sandwich Beams, Composite Laminates, Carbon Fibre, Fatigue.

INTRODUCTION

The static design of sandwich beams was investigated by Kuenzi (1). His approach yielded the optimal ratio of the core to the combined weight of the faces for minimal overall beam weight. This ratio was 2:1 for a specified stiffness, 1:1 for a specified normal bending strength, and 1:3 if it were known that facing failure would occur due to local instability (buckling). Allen (2) developed a graphic representation of sandwich design that encompassed several physical limitations (facing stress, buckling, shear stress, deflection, etc.) and outlined a permissible design zone. Both these approaches were dedicated to lightweight core materials and very thin facings, hence are not generally suitable to low span to depth ratios. Lightweight core materials have insufficient strength for components with a low span to depth ratio and localized loading. Materials with higher strength and, consequently, higher density, such as random or chopped fibre reinforced plastics, are appropriate for these applications which will be considered in the present work.

THEORETICAL ANALYSIS

The flexural strength and stiffness of sandwich beams of constant weight tested in simple three point bending have been investigated. Typical facing and core materials were selected for analysis at particular spans to demonstrate the importance of facing thickness, material selection, and loading conditions in the design of composite sandwich beams.

In this study, a lightweight aluminum honeycomb (Aℓ-Hc) core was compared with one of glass fibre reinforced polyester (GF/P). For facings, carbon fibre reinforced epoxy (CF/E) and glass fibre reinforced epoxy (GF/E) were considered since they are representative of this group of materials. For the same reason, an epoxy interfacial adhesive has been included in our analysis.

A satisfactory bond between the facing and the core is essential to transfer shear and maintain structural integrity. Shear failure can occur at either the neutral axis or the interface of the laminate. In general, if the core material has a high shear strength, interfacial delamination will occur. Alternatively, for a weak core material, shear failure will take place near the neutral axis. Hence, only one mode of shear failure will dominate.

Lucier, Mohaupt and Plumtree (3) have developed equations to analyse the static behaviour of sandwich beams of these materials at spans of 0.3 m and 1.5 m. Bending stresses, shearing stresses and deflection for a unit (1N) load were determined using their analysis. The influence of percent facing material (representing the combined volume of facings as a percentage of the total beam volume) on facing bending stress is represented by Fig. 1. This bending stress drops markedly and continuously with increasing percent facing material for beams with GF/P cores. In contrast, beams with Aℓ-Hc cores exhibit a rapid decrease, then a significant increase in

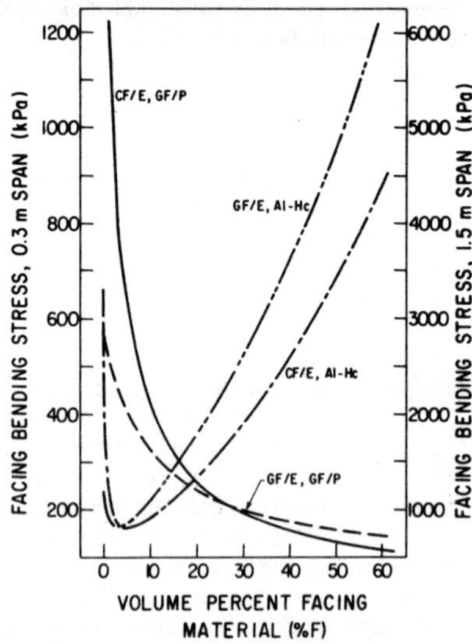

Fig. 1 Normal facing bending stress plotted against volume percentage facing material.

bending stress with percent facing material. In this latter case, minima occur at relatively thin facings (about 5%F).

The core and interface shear stresses for Aℓ-Hc beams were found to significantly increase with percent facing material. All shear stresses were minimal at very thin facings (< 2.5%F). The interfacial shear stress increased asymptotically toward that at the neutral axis with increasing percent facing. For beams with GF/P cores, the core shear stress was far less dependent on percent facing than the beams with the lightweight cores. Minima occurred for CF/E and GF/E at 20%F and 24%F, respectively. The interfacial shear stress showed a marked dependence on percent facing and also approached that at the core with increasing facing thickness.

Figure 2 shows the relationship between deflection and percent facing for a span of 0.3 m. Percent facing was found to have a strong influence on the deflection (or conversely stiffness) for beams with Aℓ-Hc cores. For these beams, deflection decreased to a minimum at approximately 2.5%F, then gradually increased as percent facing increased. The optimum stiffness corresponding to minimal deflection for beams of constant weight confirms the proposal of Kuenzi (1).

In contrast to the beams with lightweight Aℓ-Hc cores, those constructed with GF/P show monotonically decreasing deflection with increasing percent facing. Since the density of the GF/P core exceeds that of the facings, it is reasonable to deduce that stiffness of a beam of fixed weight would reach a maximum at 100% facing material.

Although the magnitude was about two orders greater, the deflection per unit load for a 1.5 m span varied with %F in a similar manner to that for the 0.3 m span given in Fig. 2.

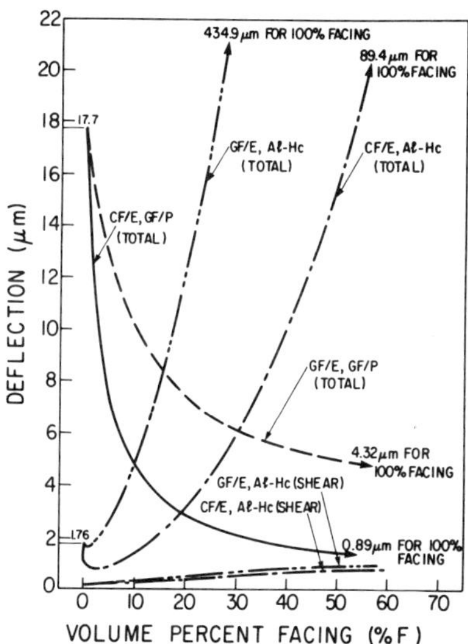

Fig. 2 Deflection plotted against volume percentage facing material for 0.3 m span.

Shear deflection was found to be negligible ($\leq$ 3.5% of total) in all cases, except for the Aℓ-Hc beams at a span of 0.3 m. Shear deflection contributed as much as 8% of the total at 2.5%F for GF/E, Aℓ-Hc beams and nearly 20% at 5%F for CF/E, Aℓ-Hc beams.

OPTIMIZATION PROCEDURE

Static Behaviour

An optimization criterion to fully utilize the strengths of the facing, interface adhesive, and core materials of a sandwich beam was developed by the authors (3). This was achieved by providing a factor of safety for normal bending equal to that for shear failure. This implies that the ratio of the facing bending stress to the core (or interface) shear stress must equal the ratio of their corresponding strengths.

For the materials selected, the ratio of these strengths were:

Facing	Core	Facing/ Core shear	Facing/ Interfacial shear
CF/E	GF/P	13	106
CF/E	Aℓ-Hc	377	106
GF/E	GF/P	7	63
GF/E	Aℓ-Hc	222	63

The largest strength ratio will generally dictate the mode of shear failure that will dominate. In the case of the Aℓ-Hc beams, core shear will occur before interfacial shear for the adhesive selected. The opposite is true for the GF/P core beams.

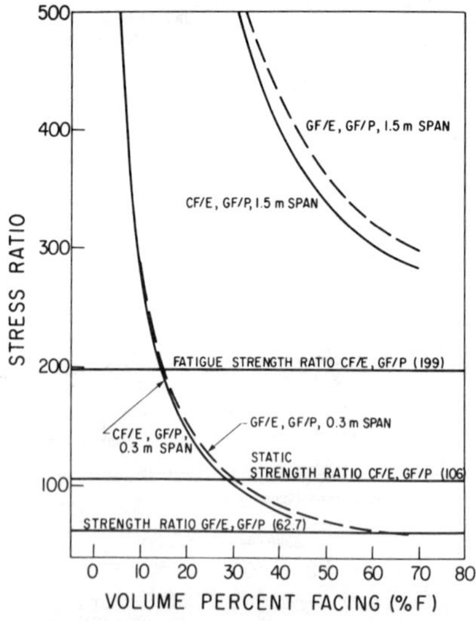

Fig. 3 Stress ratio plotted against volume percentage facing material for glass fibre/polyester core beams.

The ratio of bending stress in the facing to shear stress in the core for beams of constant weight is a function of the span to depth ratio, the material properties, and the relative facing thickness. Fig. 3 gives the relationship between stress ratio and percent facing for GF/P beams. An optimal percent facing occurs at the intersection of the stress ratio and the strength ratio. For stress ratios larger than the strength ratio, facing failures will occur. Conversely, below the strength ratio, shear failures will occur. For the 1.5 m span, no optimal percent facing exists as the ratio of facing to interfacial shear stress always exceeds the strength ratio. At the 0.3 m span, the optimal stress ratio corresponds to 28%F for the CF/E, GF/P beams, and 64%F for the GF/E, GF/P beams.

It was determined earlier (3) that for Aℓ-Hc core beams at the 1.5 m span, the ratio of facing to core shear stress was also excessively large, thereby negating the proposed optimization. At the 0.3 m span, optimal stress ratios occurred at 4.5%F for CF/E, Aℓ-Hc beams and at 3%F for GF/E, Aℓ-Hc beams. It is interesting to note that for these cases, the optimization based on stress ratio is in close agreement with the optimal stiffness.

To verify this optimization for static conditions, seven CF/E, GF/P beams were constructed with the optimal relative facing thickness of 28%F. Theoretical analysis predicted a load-bearing capacity of 6250 N and a stiffness of 506.4 kN/m for three point bending at a 0.3 m span. The mean failure load was found to be 6685 N, and the mean stiffness was determined as 438 kN/m. Four of the beams suffered interfacial delamination, while the remaining three failed by compressive facing fracture. The failure modes indicated that the underlying principles for the optimization were valid.

Cyclic Behaviour

The facing, core and interfacial bond materials may exhibit widely varying degrees of damage during fatigue. A sandwich beam design based on static properties would not necessarily be optimal for cyclic loading conditions. Bader and Johnson (4) found that unidirectional CF/E has a fatigue limit of approximately 75% of its static strength when subjected to three point flexure. The epoxy adhesive is also subject to fatigue. Wake (5) found that the shear fatigue limit for epoxies was in the range of 40% of their static shear strength. Thus, the strength ratio for the CF/E, GF/P beam would increase about twofold from 106 for static to 199 for cyclic endurance conditions. In accordance with Fig. 3, the optimal relative facing thickness for a 0.3 m span would be reduced to 14%F for fatigue from 28%F which was considered the optimum for static loading.

The effect of damage accumulation on the modulus of fibre reinforced polymers may have a secondary influence on this analysis. Nevadunsky, et al. (6) have monitored stiffness during fatigue testing and detected permanent changes at less than 1% of the total life to fracture. As the stresses at various sites in a sandwich beam are functions of the moduli of the core and facings, the stress ratio should therefore be expected to change during the life of the specimen. Further investigation would be required to accommodate this effect.

CONCLUSIONS

1. Sandwich beams with high strength, high density cores can be optimized at a constant weight by equalizing the ratio of normal bending and shear stresses to the ratio of the respective strengths. Best overall static performance and effective material utilization are achieved at relatively thick facings (>20% by volume) depending on the properties of the facing, core, and interface adhesive. These beams are most applicable to conditions with a low span to depth ratio and/or localized loading.

2. Optimal sandwich beam design for cyclic loading applications should be based on the fatigue strength ratio at the specified design life. This is particularly important if there is a significant difference in the fatigue ratios for the beam constituents. In general, the optimal thickness for unidirectional composite facings will decrease when considering cyclic rather than static loading conditions.

ACKNOWLEDGMENTS

The authors would like to express gratitude to the Hexcel Corporation for providing materials for testing and to the Natural Sciences and Engineering Research Council Canada for funding this research. Thanks are also extended to Mr. E. Huber for his assistance in preparing specimens and to Mrs. B. Hultin and Mrs. R. Hamel for typing this manuscript.

REFERENCES

1. E.W. Kuenzi, U.S. Forest Service Research Note FPL Q86 (1965).
2. H.G. Allen, Analysis and Design of Structural Sandwich Panels, Pergamon Press, Oxford (1969).
3. W.A. Lucier, U.H. Mohaupt, A. Plumtree, Mechanical Behaviour Prediction of Composite Laminate Beams, presented at the Second IAVD Congress on Vehicle Design, Geneva, March 1985.
4. M.G. Bader, M. Johnson, Composites **5** , 2 (1974).
5. W.C. Wake, Adhesion and the Formulation of Adhesives, 167, Applied Science Publ. (1982).
6. J.J. Nevadunsky, J.J. Lucas, M.J. Salkind, J. of Comp. Mat. **9** (1975).

Review on Fatigue Behavior of Graphite/Epoxy Composites

S. V. Hoa, H. McQueen and D. Doan

Department of Mechanical Engineering, Concordia University, Montreal, Quebec H3G 1M8, Canada

ABSTRACT

Increasing use of composite materials, especially graphite/epoxy composites, has raised a question of the reliability of these materials under cyclic loading. As a result, their fatigue behavior has been extensively studied.

This paper presents a review on fatigue behavior of graphite/epoxy composites that has been reported by numerous researchers in the literature. Laminate types, S-N curves, and fatigue mechanisms of notched and unnotched specimens are the aspects in which this review work is treated. Several fatigue damage types and fatigue failure modes have been found. However, the current understanding of fatigue behavior of graphite/epoxy composites is still insufficient.

KEYWORDS

Composite materials, graphite/epoxy composites, fatigue behavior, fiber, matrix, debonding, delamination, laminates, fatigue mechanism.

INTRODUCTION

Fatigue and fracture of composites have been extensively studied in the last decade due to their increasing use in the aerospace industry. Nevertheless, understanding composite fatigue behavior is a long-term research and study because of the diverse kinds of fibers, matrix materials, fiber orientations and stacking sequences, and the variety of fatigue damage types. As a result, fatigue failure of composites is generally a combination of matrix cracking, debonding, delamination, void growth, and fiber breakage. Therefore, a criterion for fatigue failure of composites cannot be defined in terms of a single mode, but it must be a very complex formula containing many variables.

The objective of this paper is to review technical literature on fatigue behavior of graphite/epoxy composites in the aspects of laminate types, S-N curves, and fatigue mechanisms.

LITERATURE REVIEW

Laminate Types

Researchers have studied and reported their results of the following specific graphite/epoxy laminates:

1. On and off-axis: $[\theta]_n$, θ = 0,10,15,30,45,60 and 90 deg. [1].
2. Angle-ply: $[\pm\theta]_{nS}$, θ = 15,30,45,60, and 75 deg. [2,3].
3. Symmetrically balanced: $[0/\pm\theta/0]_S$, θ = 15,30,45,60,75 and 90 deg or $[0/\pm 30/90]_S$ [4]. Orthotropic laminates: $[0/90]_{nS}$ or $[90/0]_{nS}$ are classified in this group.
4. Quasi-isotropic $[0/\pm 45/90]_S$, $[90/\pm 45/0]_S$, $[45/90/-45/0]_S$ [5].

S-N Curves

A quantitative method to understand the fatigue behavior of a material is the classical S-N curve. Whitney [8] reported his studies on the S-N curve characterization of graphite/epoxy laminates based on a work done by Hahn and Kim [6]. This approach involves two basic assumptions:

1. A power law representation S-N curve, and;
2. A two-parameter Weibull distribution of time-to-failure.

More details are as explained below.

The power law of S-N curve can be written in the mathematical form of

$$CNS^b = 1 \tag{1}$$

where S is the stress range, N is the number of cycles to failures, and b anc C are material constants.

Two-parameter Weibull distribution is given by

$$R(N) = \exp\left[-\left(\frac{N}{No}\right)^{\alpha_f}\right] \tag{2}$$

where R(N) denotes the reliability (probability of survival), No is the characteristic time-to-failure (location parameter), and α_f is the fatigue shape parameter.

Rewriting eq. (1)

$$N = C^{-1}S^{-b} \tag{3}$$

Substituting eq. (3) into eq. (2) and solving for S,

$$S = K\{[-\ln R(N)]^{-1/\alpha_f b}\}\ No^{-1/b} \tag{4}$$

where

$$K = C^{-1/b} \tag{5}$$

when N = No, $-\ln R(No) = 1$ and eq. (4) reduces to

$$S(No) = KNo^{-1/b} \tag{6}$$

From a practical viewpoint, one is more interested in obtaining an S-N curve for any R(N) rather than an S-No curve. Writing eq. (6) in the form

$$No = C^{-1}S^{-b} \tag{7}$$

and substituting into eq. (2) gives

$$R(N) = \exp[-(\frac{N}{C^{-1}S^{-b}})^{\alpha_f}] \tag{8}$$

Sovling for S leads to the following S-N relationship for any desired level of reliability R(N)

$$S = K\{[-\ln R(N)]^{-1/\alpha_f b}\}N^{-1/b} \tag{9}$$

Whitney collected fatigue data of the following T300/934 graphite-epoxy laminates:

1. $[0/45/0_2/-45/0_2/+45/0_2]_S$ - Laminate 1.
2. $[45/90/-45/90/-45/0/45/0]_S$ - Laminate 2.
3. $[\pm 45]_{2S}$ - Laminate 3.

and obtained the S-N curves of these three laminates for tension-tension fatigue loading with R(N) = 0.95 as shown in Fig. 1.

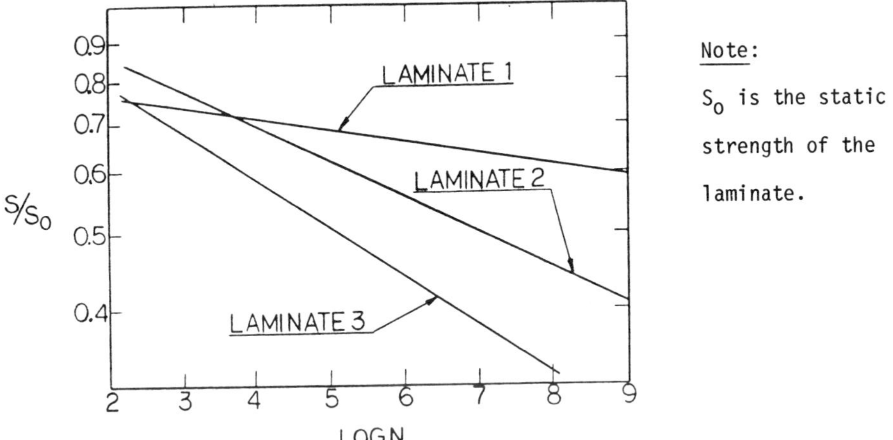

Fig. 1. S-N curves for tension-tension loading with R(N) = 0.95 [Ref. 8].

Fatigue Mechanisms

Another important aspect of fatigue of composite materials that has been extensively studied is fatigue mechanisms in these materials. A review of fatigue mechanisms of graphite/epoxy composites is divided into two major categories - unnotched laminates and notched laminates. Within each category, the influences of parameters like fiber orientations, stacking sequence and environmental effects will be discussed.

Unnotched Fatigue Mechanism

A fundamental study on fatigue behavior of graphite/epoxy composites was done by Awerbuch and Hahn [1]. In this work, they reported on fatigue failure modes of seven off-axis angle AS-3501A laminates: 0,10,20,30,45,60 and 90 deg. Using scanning electron microscope (SEM) and taking three-dimensional photographs of the fracture surface, they have found the following interesting points regarding failure modes:

1. Fracture surfaces include a combination of several interdependent failure modes such as fracture of individual fibers and fiber tows, matrix serration (shear failure), matrix cleavage, and matrix/interface cracking parallel to fibers.
2. Matrix serrations increase with longitudinal shear stress. In the absence of the shear stress, a cleavage type of failure prevails.
3. Matrix serrations are larger on the fatigure fracture surfaces and smaller on the static fracture surfaces.
4. Fracture surfaces are not planar; they consist of matrix-dominated failure regions, interface-dominated failure regions, and combinations thereof.

A notable work by Ramani and Williams [3] on fatigue behavior of angle-ply graphite/epoxy composites has drawn a few remarkable points:

1. Angle-ply laminates $[0/\pm 30]_{3S}$ are, in general, susceptible to tensile fatigue failures. An apparent fatigue limit is about 60% of the static strength in tension-tension fatigue.
2. Fatigue tests on $[0/\pm\theta/0]_{2S}$ show that the fatigue limit increases with increasing percentage of 0-deg. ply in the laminate and that the fatigue mechanism appears to be one of general failure along $\pm\theta$-deg. directions.
3. The dominant failure mechanism for the $[0/\pm 45/90]_S$ laminate is edge delamination propagating across the specimen to effect failure.

Masters and Reifsnider [9] investigated on cumulative damage development in two types of quasi-isotripic graphite/epoxy laminates and obtained the following results by an experimental method called the surface replication technique.

1. Both of Type I $[0/+45/-45/90]_S$ and Type II $[0/90/\pm 45]_S$ laminates developed transverse cracks in 90 deg. laminae. These transverse cracks were arrested initially at the lamina interfaces and did not propagate to adjacent plies.
2. In Type I laminates, the number of transverse cracks in 90 deg. lamina increased until a uniform pattern of periodically spaced cracks was formed over the entire length of the specimen. Coincident to this, transverse extension into the -45-deg. laminae, longitudinal cracks grew from transverse cracks in the 90-deg. lamina. They eventually joined together, linking all 90-deg. lamina cracks, and formed one continuous edge delamination along the specimen length.
3. Damage in Type II laminates also began with transverse cracking that initially spanned only the 90-deg. lamina. These cracks were much more numerous compared with that of Type I. As load increased, transverse cracks appeared in all of the remaining off-axis laminae.
4. Delaminations caused by fatigue loading were evident at the +45/-45-deg lamina interfaces in both types of specimens.

Tension-compression and compression-compression fatigue behavior of graphite-epoxy composites have been studied by a few researchers. A work by

Ryder and Walker [7] has given some insight in this area. Their findings form a large number of quasi-isotropic T300/934 graphite/epoxy coupons $[0/45/90/-45-45/90/45/0]_S$ are:

1. Tension-compression testing showed when delamination was noted prior to failure, normally 90% or more of the coupon life was exceeded. Therefore, delamination may possibly be a more useful definition of failure in T-C fatigue.
2. Coupons tested in T-C fatigue failed similar to those tested in T-T, but large out-of-plane buckling of the outer plies also occured, reducing the fatigue life after the onset of delamination. All coupons tested in T-C fatigue failed during compression load excursion.

The effects of environmental parameters, such as moisture and temperature, on fatigue behavior of graphite/epoxy composites have recently received more attention. Rotem and Nelson [4] have investigated the fatigue behavior of T300/5208 graphite/epoxy laminates at elevated temperatures. They showed that the experimental results are in good agreement with theoretical predictions. Using the temperature "shifting-factors" enable one to predict long-term behavior at some temperature from short-time testing at elevated temperatures.

Kriz and Stinchcomb [10] have studied the effects of moisture absorbed in $[0/\pm45/90]_S$ and $[0/90/\pm45]_S$ T300/5028 graphite/epoxy laminates during static tension and tension-tension cyclic loading. Their useful results are:

1. In general, moisture tends to reduce moduli, strength, and scatter of strengths for specimens tested.
2. Moisture can alter the chronology of damage development, but the damage in each laminate observed in prior to fracture appeared to be independent of moisture.

Notched Fatigue Mechanisms

The term "notch" is defined as cutouts which may include holes, slits,slots and cracks. Many structural components of composite materials contain those notch types due to flaws in materials or requirements of connecting components. The local stress state around a notch is extremely complex, and consequently the investigation of fatigue damage mechanisms of notched components is a formidable task.

Ramani and Williams [3], as mentioned above, have studied two types of notch, double-edge notched DEN and center circular hole CCN, on $[0/\pm\theta]_{3S}$, $[0/\pm\theta/0]_{2S}$ and $[0/\pm45/90]_{2S}$ AS/3501 graphite/epoxy laminates and have reported the following conclusions:

1. The $[0/\pm\theta]_{3S}$ and $[0/\pm\theta/0]_{2S}$ laminates demonstrate relative insentivity to T-T fatigue in the notched condition. The $[0/\pm45/90]_{2S}$, on the other hand, is susceptible to T-T fatigue failures in both the notched and unnotched conditions at very low stress levels through edge delamination effects.
2. Damage propagation from notched and holes in T-T fatigue occurs most exclusively along the loading direction. Damage is well contained between axial splits at hole edges or between specimen edge and notch root split for edge notches. Transverse spread of damage from notches and holes is very slight.

3. Progressive damage accumulation from edge notches into the specimen interior occurs under T-C fatigue.

One work reported by Whitcomb [5], has given more insight on experimental and analytical study of fatigue damage in notches graphite/epoxy laminates. Whitcomb has studied notched specimens containing a center hole of orthotropic, $[0/\pm45/0]_S$ and $[45/0/-45/0]_S$, and quasi-isotropic, $[90/\pm45/0]_S$ and $[45/90/-45/0]_S$, laminates. Below are his findings:

1. Fatigue damage was primarily delamination and ply cracking parallel to the fibers. In general, ply cracks did not propagate into adjacent plies of different fiber orientation. In all cases, fatigue cycling of notched specimens resulted in residual strengths greater than or equal to the virgin strength. Fatigue loading generally caused only small stiffness changes.
2. The location of delamination was sensitive to the fiber orientations, stacking sequence, and sign of loading. Finite element stress analysis indicated that both interlaminar normal stress and shear stress must be considered to explain the observed delamination. Furthermore, the altered stress distribution, concomittant with fatigue damage growth, can change the direction of delamination propagation.

CONCLUSION

Throughout the paper, the fatigue behavior of graphite/epoxy composite laminates has been reviewed from various notable works on this subject. As mentioned earlier, understanding the fatigue behavior of composite materials is a long-term reseach and study. All investigations which have been carried out just show us a rather general picture of fatigue damage types and fatigue failure modes occured in graphite/epoxy laminates. However, to determine which fatigue mechanism is the most important and critical to graphite/epoxy composites, further work is required. At this state of understanding, delamination seems to have a major role in the fatigue failure of composites.

REFERENCES

1. J. Awerbuch and H.T. Hahn, Fatigue of Fibrous Composite Materials, ASTM STP 723, p. 243. Am. Soc. for Testing and Materials (1981).
2. A. Rotem and Z. Hanshin, AIAA J. 14, 868 (1976).
3. S.V. Ramani and D.P. Williams, Fatigue of Filamentary Composite Materials, ASTM STP 636, p. 27. Am. Soc. for Testing and Materials (1977).
4. A. Rotem and H.G. Nelson, Fatigue of Fibrous Composite Materials, ASTM STP 723, p. 152. Am. Soc. for Testing and Materials (1981).
5. J.D. Whitcomb, Fatigue of Fibrous Composite Materials, ASTM STP 723, p. 48. Am. Soc. for Testing and Materials (1981).
6. H.J. Hahn and R.Y. Kim, J. of Composite Materials, 9, 297 (1975).
7. J.T. Ryder and E.K. Walker, Fatigue of Filamentary Composite Materials, ASTM STP 636, p. 3. Am. Soc. for Testing and Materials (1977).
8. J.M. Whitney, Fatigue of Fibrous Composite Materials, ASTM STP 723, p. 133. Am. Soc. for Testing and Materials (1981).
9. E.J. Master and K.L. Reifsnider, Damage in Composite Materials, ASTM STP 775, p. 40. Am. Soc. for Testing and Materials (1982).
10. R.D. Kriz and W.W. Stinchcomb, Damage in Composite Materials, ASTM STP 775, p. 63. Am. Soc. for Testing and Materials (1982).

Behavior of Polypropylene-Aluminum Laminates Under Tensile Loading

M. M. Farag and A. A. Hanna

Department of Engineering, American University in Cairo, Egypt

ABSTRACT

The present work is an experimental study of the tensile properties of two types of polypropylene-aluminum laminates of the kind in soft packaging. The behavior of the laminates under different strain rates and test temperatures was analysed in terms of the properties of their individual components, aluminum and polypropylene. At higher strain rates and lower temperatures the laminates failed by interlayer destruction and at lower strain rates and higher temperatures interlayer reinforcement was observed. Aluminum foils exhibited more ductility when supported by the polypropylene in the laminate and this resulted in an increase in the ultimate tensile strength of the laminates beyond that anticipated by the simple rule of mixtures. The strength of laminates σ as well as their individual components increased with increasing the strain rate $\dot{\varepsilon}$ according to the relationship:

$$\sigma = K_1 + K_2 \ln \dot{\varepsilon}$$

The values of the constants K_1 and K_2 for the laminates were found to fall between their values for aluminum foils and polypropylene films.

KEYWORDS

Multi-layer laminates; polypropylene; aluminum; tensile strength; failure modes; strain rate; temperature; rule of mixtures.

INTRODUCTION

In laminated composite materials the individual layers behave differently than if they were isolated. This is generally attributed to the mechanical interaction between the various layers. Two extreme cases of such interaction have been identified as mutual interlayer reinforcement and mutual interlayer destruction (1,2). In the first case a relatively brittle layer is protected against fracture by its neighbouring ductile layers and the composite undergoes a large ductile deformation. In the second case a relatively ductile layer fractures at relatively low elongation. The two extreme cases can be encountered in the cases where strong interfacial adhesion is present. A variety of intermediate behavior can be obtained in the cases of weaker adhesion where delamination can occur.

Bhateja and Alfrey (3) indicated that in the case of polymer-metal laminates complete interlayer reinforcement is favored for relatively thick polymer and thin metal layers with strong bonding. Complete interlayer destruction was, on the other hand, considered to be favored by relatively thin polymer and thick metal layers with strong bonding.

The present work is an experimental study of the tensile properties of some multi-layer laminates of the type used in soft packaging. The behavior of the laminates under different tensile strain rates and test temperatures will be analysed in terms of the behavior of their constituents.

MATERIALS AND EXPERIMENTAL TECHNIQUES

Two types of laminates which were made up of layers of polypropylene films (pp) and aluminum foils (Al) were used in the present work. The structure of laminate 1 is : "0.075 mm pp-primer-0.02 mm Al-adhesive-0.02 mm Al-primer-0.075 mm pp" and the structure of laminate 2 is : "0.150 mm pp-0.06 mm Al-primer-0.075 mm pp". The laminates were supplied by Hanco Coatings Incorporated, USA. The thickness fractions of aluminum in laminates 1 and 2 were 0.18 and 0.182 respectively. The as-received samples were tested in the longitudinal direction. The constituents of the laminates were also tested separately. To obtain the polypropylene film alone, the aluminum layers were dissolved by soaking the laminates in sodium hydroxide solution for 48 hrs at room temperature. The aluminum foils were obtained by dissolving the polypropylene layers in molten wax at 130°C for 10 min. The surfaces of the separated constituents did not show any signs of attack by the chemicals used to dissolve the rest of the laminate.

Tensile samples of the laminates, aluminum foil and polypropylene films were prepared by cutting using a razor blade and a template. This technique was found to yield consistant results when testing aluminum foils (5). The gage length of the tensile samples was 32 mm and the width was 6 mm. The tensile tests were carried out at constant cross-head speeds corresponding to initial strain rates of 0.1, 0.79 and 7.9 min^{-1}. Tests were carried out at room temperature (25°C) and 1°C. The latter temperature was achieved by testing the samples while submerged in a mixture of ice and water. The experimental results presented here are the average of at least five valid tensile tests for each condition.

RESULTS AND DISCUSSION

Ductility and Modes of Failure

Figure (1) shows some typical stress-strain diagrams for the laminates tested in the present work. Initially the slopes of the diagrams are high and then gradually decrease with increasing strain up to the ultimate tensile strength. The slopes are generally higher for higher strain rates and lower temperature. Fig. (1) shows two different types of laminate behavior. In the first type all the layers break together as shown in Fig. (2). This behavior is similar to that described as interlayer destruction in ref. (1-3) and was observed for the two types of laminates at all strain rates when tested at 1°C. It was also observed in laminate 1 at a strain rate of 7.9 min^{-1} and 25°C and in laminate 2 at strain rates of 7.9 and 0.79 min^{-1} at 25°C. In the second type of laminate behavior the aluminum layers broke first while the polypropylene layers continued to stretch at a reduced load. A typical fractured sample is shown in Fig. (3)

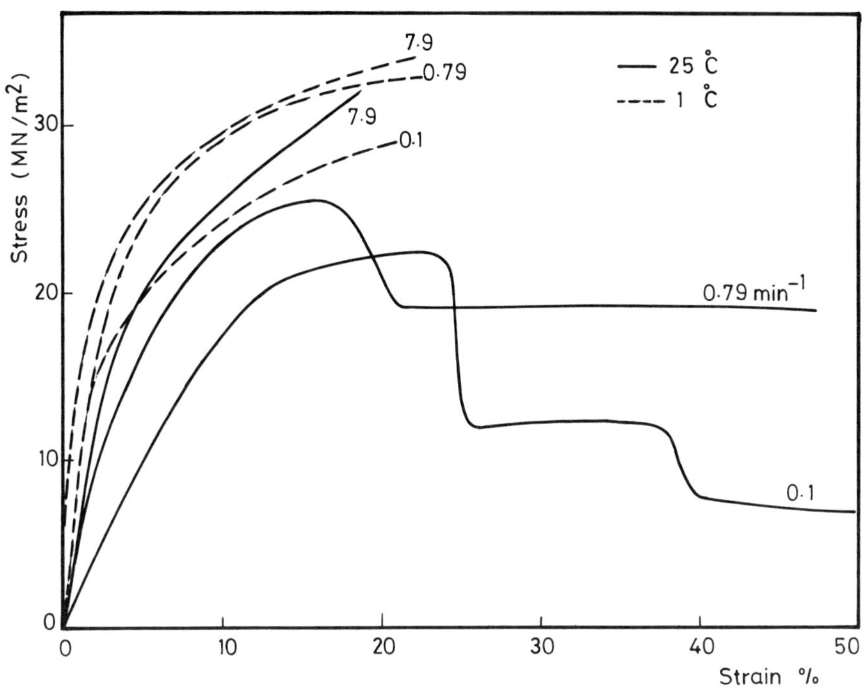

Fig. 1 Stress strain curves for laminate 1 at different strain rates and temperatures. The strain rates in min^{-1} are indicated on the curves.

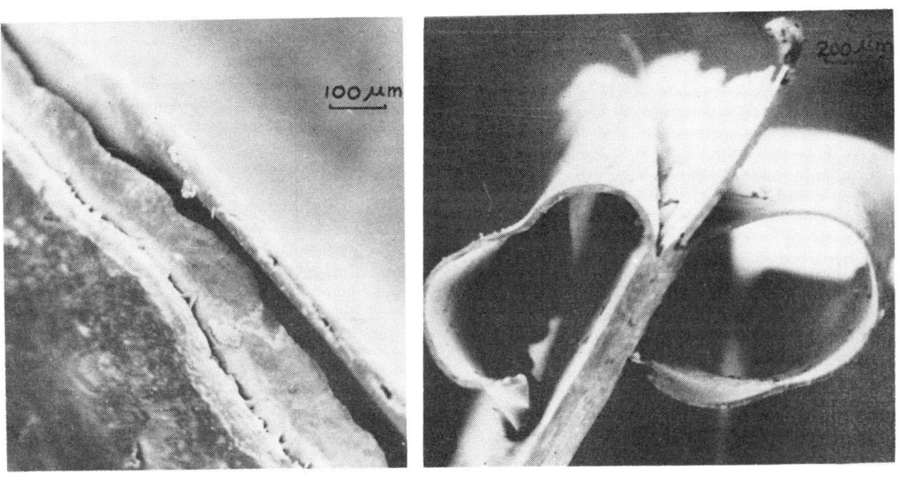

Fig. 2 Fracture surface of laminate 2 strain rate = 7.9 min^{-1}, $1^{o}C$.

Fig. 3 Fracture area of laminate 1 strain rate = 0.1 min^{-1}, $25^{o}C$.

and shows similar features to interlayer reinforcement described in ref. (1-3).

The values of strain at which fracture occurred in the aluminum layers while still in the laminates are compared with the fracture strains of the aluminum foils which have been separated from the polypropylene sheets, Fig. (4). The fracture strains of the separated aluminum foils are seen to be considerably lower than those of the aluminum in the laminates. The difference is attributed to the support given to the aluminum by the neighboring polypropylene layers. Similar behavior was observed by Bhateja and Alfrey (3). The present results show that the supporting action of the polypropylene is more effective at 1°C than at 25°C. Another interesting feature of the results in Fig. (3) is that the thicker aluminum foils show higher ductility than the thinner foils when separated out of the laminates. Similar behavior was observed by Eady and Gifkins (6) and was attributed to the strengthening effect the grains have on neighboring grains. For thinner foils the probability of one or two grains occupying the total thickness is higher. The presence of the foils in the laminates reduces the effect of thickness and the two types of aluminum foils exhibit similar ductility values as seen in Fig. (4).

Tensile Strength

Table (1) gives the experimentally measured average UTS values of the laminates and the separated constituents at different strain rates and temperatures. Increasing the strain rate caused the strength of the

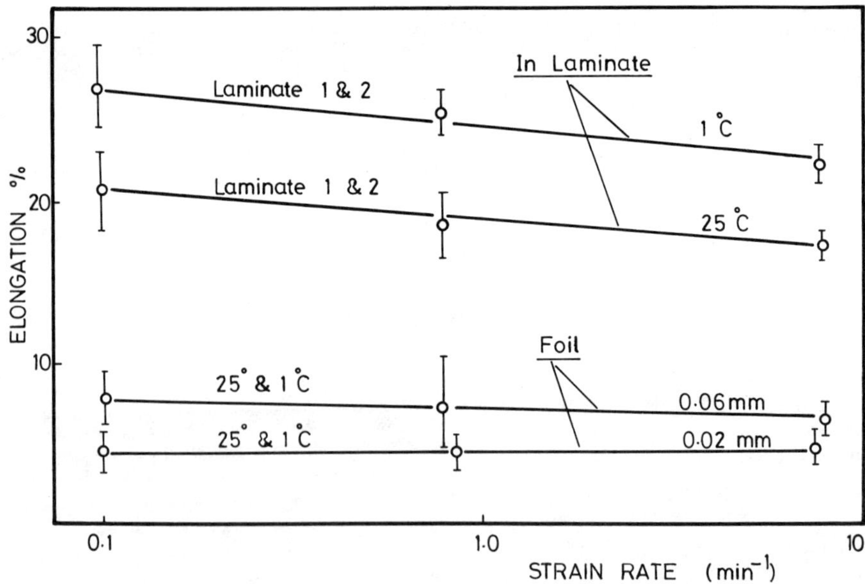

Fig. 4 Effect of strain rate on the elongation % of aluminum in the laminate and as a separate foil.

TABLE 1 Tensile Test Results for Laminates, Aluminum Foils and Polypropylene Films

Laminate	Temperature °C	Strain rate (min^{-1})	UTS Al (MN/m^2)	Flow stress pp (MN/m^2)	UTS laminate σ_ℓ (MN/m^2)	σ_c (MN/m^2)	σ_A (MN/m^2)	σ'_c (MN/m^2)
1	25	0.1	32	8.1	21.9	13	98	42.2
		0.79	35	12.5	24.8	17.1	83	25.2
		7.9	40	15.7	28.5	20.7	90	29.1
	1	0.1	32	17	29	20.1	98	31.3
		0.79	35	20	31.9	23.1	104	35.2
		7.9	40	21.9	34.8	25.6	104	36.8
2	25	0.1	37	8.1	20.5	14.3	81	21.0
		0.79	39.8	12.5	22.9	18.3	80	24.2
		7.9	42.1	15.7	26.7	21.3	80	26.7
	1	0.1	37	17	24.8	21.3	81.5	28.1
		0.79	39.8	20	26.7	24.2	37.7	29.0
		7.9	42.1	21.9	29.1	26.2	81.5	31.9

laminates, the aluminum and polypropylene to increase while decreasing the temperature caused the strength of the laminates and polypropylene to increase but did not affect the strength of aluminum. An attempt was made to calculate the UTS of the laminates using the simple rule of mixtures:

$$\sigma_c = \sigma_a t_a + \sigma_p t_p \qquad (1)$$

where σ_c is the calculated UTS of laminate. σ_a and t_a are the UTS and thickness fraction of aluminum foil and σ_p and t_p are the flow strength at the fracture strain of laminate and thickness fraction of polypropylene films. $t_a + t_p = 1$.

The values of σ_c are given in table (1) and are seen to be considerably lower than the experimentally measured UTS values for the laminates σ_ℓ. It is estimated that the main reason for the discrepancy between the measured and calculated values lies in the use of the UTS of the separated aluminum foils in the calculation. The ultimate tensile strains of the unsupported foils are considerably lower than the fracture strains of aluminum in the laminates, fig. (4). The strength of aluminum at its fracture strains when supported by polypropylene (σ_A) was calculated using the relationship:

$$\sigma_A = K \varepsilon^n \qquad (2)$$

where K is the strength coefficient and n is the strain-hardening coefficient. The values of K and n were determined from the experimental results of the aluminum foils and were found to be : $K = 275\ MN/m^2$ and $n = 0.7$ for the 0.02 mm - thick foil; $K = 158\ MN/m^2$ and $n = 0.55$ for the 0.06 mm - thick foil. The values of σ_A were inserted in the rule of mixtures, Eq. (1), and calculated laminate strength values σ'_c are given in table (1) and are seen to agree closely with the experimental UTS values σ_ℓ. The experimentally determined UTS results σ for the laminates, the aluminum foils and polypropylene were found to follow the relationship:

$$\sigma = K_1 + K_2 \ \ell n \ \dot{\varepsilon} \qquad (3)$$

where $\dot{\varepsilon}$ is the strain rate and K_1 and K_2 are constants. Similar relationships were observed for aluminum (7) and polypropylene (8). Table (2) gives the values of K_1 and K_2 for aluminum foils, polypropylene and laminates. The results show that the values of K_1 and K_2 for the laminates fall between the values of aluminum and polypropylene.

TABLE 2 Values of Constants K_1 and K_2 in Eq. (3)

Temperature	Constant	Al foil 0.02 mm	Al foil 0.06 mm	Polypropylene	Laminate 1	Laminate 2
25^oC	K_1	35.2	40	12.8	25	23.3
	K_2	1.7	1.7	1.3	1.6	1.65
1^oC	K_1	35.2	40	20.3	32.2	27
	K_2	1.7	1.7	0.83	1.3	1.04

CONCLUSIONS

The present results show that the behavior of laminates in tension is influenced by the loading conditions, the structure of the laminate as well as the relative properties of the components. Interlayer reinforcement is favored by slower strain rates and higher temperatures.

The support given to the aluminum foils by the polypropylene increased its ductility and enhanced the tensile strength of the laminate. Designing the laminate structure to enhance the ductility of the more brittle phase would also have beneficial influence on the strength.

REFERENCES

1. W.J. Schrenk and T. Alfrey, Polymer Engineering and Sci., 9. p 393 (1969).
2. W.J. Schrenk, Applied Polymer Symposia, No. 24, p. 9 (1974).
3. S.K. Bhateja and T. Alfrey, J. Composite Materials, 14, p. 42 (1980).
4. J.G. Beese and G.M. Bram, J. Engr. Mat. and Tech. Jan., p. 10 (1975).
5. J.A. Eady, G.D. Bennett, and N.F. Herbst, Aluminium, 58, p. 423 (1982).
6. J.A. Eady and R.C. Gifkins, Aluminium, 58, p. 593 (1982).
7. G.E. Dieter, Mechanical Metallurgy, 2nd Ed., McGraw-Hill (1976).
8. I.M. Ward, J. Mat. Sci., 6, p. 1397 (1971).

Tenacite en Dynamique d'une Alumine

B. Tolba*, P. Becker et G. Pluvinage****

**Laboratoire de Fiabilité Mécanique, Faculté des Sciences, Ile du Saulcy, 57045 Metz Cedex, France*
***Laboratoire de Métallurgie Structurale, Faculté des Sciences, Ile du Saulcy, 57045 Metz Cedex, France*

RESUME

La ténacité est déterminée en statique et en dynamique. Les résultats sont examinés en utilisant le modèle statistique de Weibull à deux et à trois paramètres ; ainsi, la dispersion dans les résultats du facteur d'intensité de contrainte critique K_{Ic} a pu être calculée pour les deux types d'essais. L'examen fractographique a confirmé le changement du processus de rupture.

MOTS CLES

Ténacité dynamique ; céramique.

INTRODUCTION

Les céramiques présentent une grande fragilité au choc. Habituellement, cette fragilité est caractérisée par la ténacité statique K_{Ic}. La tenue au choc est étudiée en utilisant une méthode de chargement dynamique. La dispersion des résultats dans la ténacité K_{Ic} est analysée à l'aide du modèle de Weibull à 3 paramètres.

MODELE STATISTIQUE DE WEIBULL

C'est un modèle obéissant à la théorie du maillon le plus faible. Il est purement empirique et s'applique à une quantité de problèmes.

Les propriétés mécaniques des composites, dans notre cas les céramiques, sont généralement étudiées en utilisant le modèle de Weibull où la probabilité de rupture pour une contrainte appliquée s'écrit :

$$P_f(V) = 1 - \exp\left[- V \left[\frac{\sigma - \sigma_u}{\sigma_o} \right]^m \right] \qquad (1)$$

avec σ : contrainte appliquée, V : volume contraint

σ_o : facteur de normalisation

σ_u : contrainte au-dessous de laquelle la probabilité de rupture est zéro

m : facteur décrivant l'inhomogénéité des tailles des défauts

Le modèle de Weibull peut s'écrire sous une autre forme (1) :

$$P_f = 1 - \exp\left[- \left[\frac{K_{Ic} - K_s}{K_o - K_s}\right]^{m_1}\right] \tag{2}$$

où K_{Ic} : facteur d'intensité de contrainte critique

K_s : facteur d'intensité de contrainte seuil

K_o : facteur de normalisation

m_1 : facteur décrivant le degré de dispersion

ETUDE EXPERIMENTALE

Matériau et Eprouvettes

Les éprouvettes sont usinées à partir de l'alumine DESMARQUEST AF99 de propriétés :

- composition chimique : 99,7 % Al_2O_3 α polycristalline
- masse volumique apparente : 3,9 g/cm³
- porosité ouverte : 0 %
- obtention par frittage sous pression isostatique

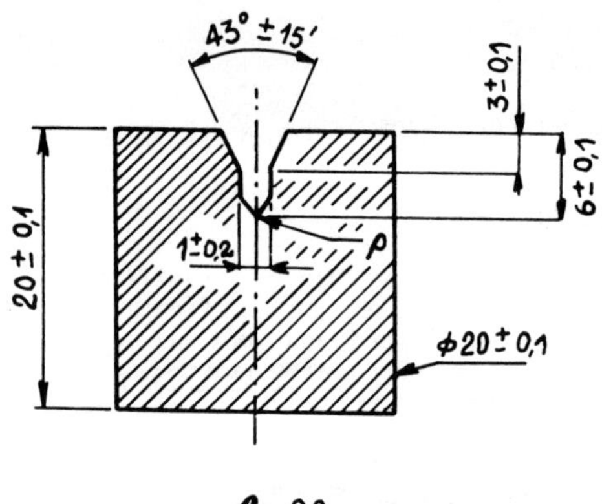

Fig. 1 Eprouvette d'essai WLSR

RESULTATS EXPERIMENTAUX

Ténacité à faible Vitesse de Chargement $\dot{K}_I \simeq 0,25$ MPa$\sqrt{m}$/s

Les essais sont réalisés sur une machine de traction-compression du type SERVOTEST. Le facteur d'intensité de contrainte critique statique K_{Ic} est obtenu par la relation suivante :

$$K_{Ic} = \frac{F}{D(W)^{1/2}} Y\ (a/W) \qquad (3)$$

avec F : force de rupture
D : diamètre de l'éprouvette
W : longueur de l'éprouvette
Y(a/W) : fonction de complaisance

La fonction de calibration est tirée de celle de l'éprouvette WLCT* décrite dans la référence (3). En tenant compte du frottement entre le coin et l'éprouvette, la force de rupture est calculée de la relation

$$F = \frac{P}{2\ \mathrm{tg}\ (\frac{\alpha}{2} + \mathrm{arc\ tg}\ \mu)} \qquad (4)$$

où P : force appliquée
μ : coefficient de frottement acier-alumine pris égal à environ 0,25
α : angle d'ouverture du coin

Les résultats statiques et dynamiques sont regroupés dans le tableau ci-dessous.

Ténacité à grande Vitesse de Chargement $\dot{K}_I \simeq 4.10^5$ MPa$\sqrt{m}$/s

Les essais sous chargement à grande vitesse sont obtenus en utilisant une méthode de chargement par train d'ondes et barres d'Hopkinson (Split Hopkinson Pressure Bar) présentées schématiquement en figure 2.

TABLEAU 1 Résultats des essais statiques et dynamiques

PARAMETRES DETERMINES	DYNAMIQUE	STATIQUE
Vitesse de chargement MPa $\sqrt{m}$/s	$\dot{K}_I \simeq 4.10^5$	$\dot{K}_I \simeq 0,25$
Facteur d'intensité de contrainte critique	K_{Ic} = 5,4 MPa$\sqrt{m}$	K_{Ic} = 4,7 MPa$\sqrt{m}$
Module de Weibull	m_1 = 3,5	m_1 = 10
Facteur de normalisation	K_o = 5,81 MPa$\sqrt{m}$	K_o = 4,68 MPa$\sqrt{m}$
Facteur d'intensité de contrainte seuil	K_s = 0	K_s = 0
Coefficient de corrélation	r = 0,947	r = 0,979
On obtient K_s = 2,75 MPa $\sqrt{m}$ pour m_1 = 1,64 et r = 0,977		

* Wedge Loading Compact Tension

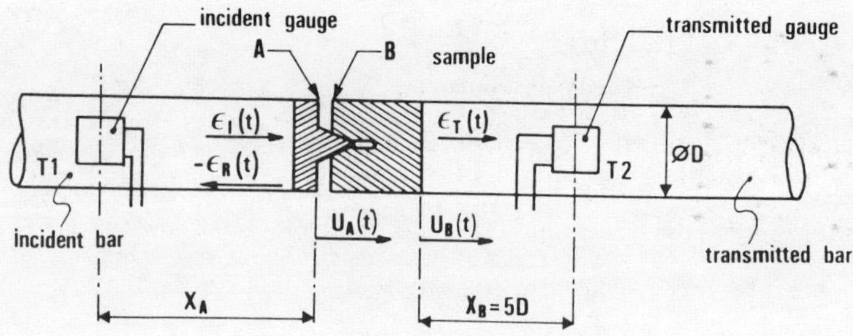

Fig. 2 Barres d'Hopkinson

Le facteur d'intensité de contrainte critique dynamique K_{Ic} est calculé à partir de la même relation qu'en statique (3).

Le modèle statistique de Weibull à trois paramètres permet de déterminer le degré de dispersion dans les résultats de K_{Ic} (m_1), le facteur d'intensité de contrainte seuil (K_s) et le facteur de normalisation (K_o) en traçant $\mathrm{LnLn}\left(\frac{1}{1-P_f}\right)$ en fonction de $\mathrm{Ln}K_{Ic}$ (Figure 3a, 3b).

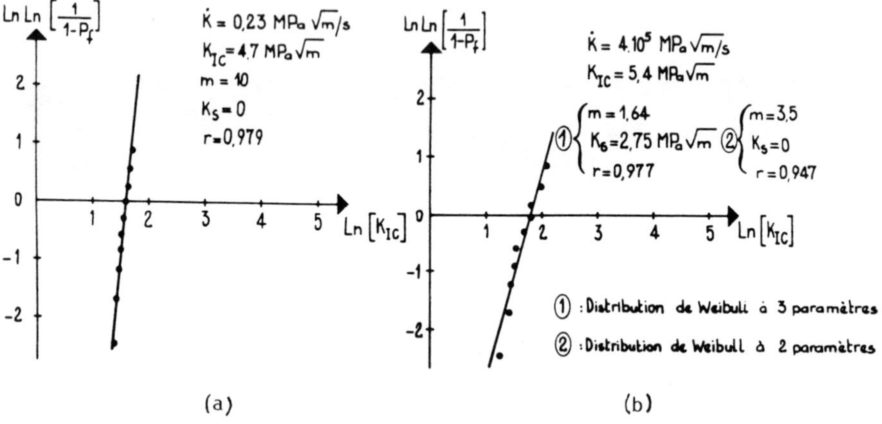

(a) (b)

Fig. 3 Diagramme de Weibull : (a) en statique - (b) en dynamique

Examens fractographiques

L'examen au microscope électronique à balayage des faciès de rupture montre une nette différence dans le processus de rupture entre les deux types d'essais. Nous constatons que la rupture en statique est intergranulaire (figure 4a), tandis qu'en dynamique, elle est mixte (inter et trans-granulaire) (figure 4b).

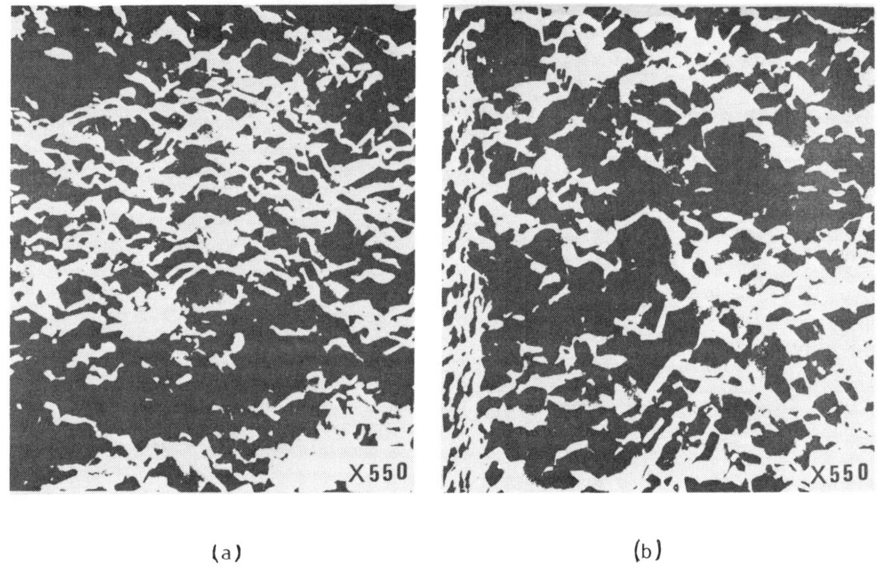

(a) (b)

Fig. 4 Faciès de rupture : (a) en statique - (b) en dynamique

DISCUSSION ET CONCLUSION

Le module de Weibull à faible vitesse de chargement est en bon accord avec les résultats donnés par d'autres auteurs ayant travaillé sur des alumines semblables (DAVIDGE (4) m = 13,2, FONTAINE (5) m = 12).

La ténacité en dynamique est plus grande que celle en statique, 5,4 MPa $\sqrt{m}$ et 4,7 MPa $\sqrt{m}$.

Le fait que le module m_1 est plus petit pour les conditions dynamiques (m_1 = 3,5) que pour les conditions en statique (m_1 = 10) montre l'écart dans les résultats des essais. Ceci peut être attribué au fait que des défauts de différentes tailles peuvent être activés simultanément. Ce phénomène est confirmé par les théories de KIPP (6) et KALTHOFF (7). Afin de vérifier ce processus de rupture, nous avons fait des examens fractographiques au microscope électronique à balayage. La surface de rupture d'une éprouvette sous chargement statique montre que la rupture est entièrement intergranulaire (figure 4a), tandis que pour le chargement dynamique, la surface de rupture présente un caractère mixte (inter et transgranulaire) (figure 4b).

BIBLIOGRAPHIE

1. K. WALLIM, The Scatter in K_{Ic} results, *Eng. Fract. Mech.*, (1984).
2. R.T. BUBSEY, Compliance calibration of the short rod chevron notch specimen for fracture toughness testing of brittle materials, *Int. J. of Fract.*, **18**, 2 (1982).
3. B. DAMBRINE, Thèse, Université de Technologie de Compiègne (1982).

4. R.W. DAVIDGE *et al.*, *Trans. Brit. Ceram. Soc.*, **66**, 8, 405 (1967).
5. J.C. FONTAINE, Thèse, Université de Technologie de Compiègne (1982).
6. M.E. KIPP, D.E. GRADY, E.P. CHEN, Strain rate dependent fracture initiation, *Int. J. of Fract.*, **16**, 5, 471 (1980).
7. J.K. KALTHOFF, S. WINKLER, W. BOHME, P.A. SCHOCKEY, Mechanical response of crack to impact loading, *Proc. Int. Conf. on Dynamical Properties and Fract. Dynamics of Eng. Materials* (1983).

SECTION 12

Wear Resistance

Résistance à l'usure

On the Abrasive Wear of AISI 4340 Steel

T. B. Wu and R. J. Young

Department of Materials Science and Engineering, National Tsing Hua University, Hsinchu, Taiwan, Republic of China

ABSTRACT

In order to clarify the role of carbide morphology on the wear property of AISI 4340 steel, abrasion tests were emploied on specimens having a lower bainite or tempered martensite microstructure but retained austenite. From transmission electron microscopy it has been observed that the bainite consists fine carbides uniformly distributed in ferrite matrix; however, the tempered martensite has long, coarse carbides present on the boundaries of martensite laths which also consists of fine carbides uniformly distributed in ferrite matrix. The bainite shows a better wear resistance than that of the tempered martensite. A slight improvement of this property also can be obtained on the latter by reducing the as-quenched retained austenite which eventually decomposed to ferrite and coarse carbides after tempering. Since there are no or little austenite retained in the tested specimens, the previous results is attributed to the difference of carbide morphologies among them.

KEYWARDS

Abrasive wear; 4340 steel.

INTRODUCTION

The role of microstructures on wear resistance of metals has been studied [1-5], and it is generally recognized that a microstructure with high toughness improves the wear resistance of metal. The improvement has been linked to flow stress and work hardening rates in simple systems. However, for metals with multiphase microstructures, simple correlations are less clear. Recently, investigations have been carried out on alloy steels [6,7]. In these works, the role of retained austenite was emphasized. It was found that the tempered bainitic microstructures containing retained austenite exhibited superior wear resistance to the tempered martensites containing little or no retained austenite. This beneficial effect was attributed to the ductile nature of retained austenite in bainitic microstructure. Nevertheless in these works, the role of carbide morphologies, which may also be an important factor controlling the wear behavior of alloy steel, has been ignored. Since this factor is not an invariant in the above works, the correlation between wear resistance and retained austenite is still not clear. It is thus interesting to study the role of carbide morphologies on wear behavior. In order to clarify this role, microstructures having the same hardness and containing little or no retained austenite are employed in this study.

EXPERIMENTAL

A commercial grade AISI 4340 alloy steel was selected for this study. The steel was investigated employing the following heat treatments in salt pots:

(a) austenitize at 845°C, 10 minutes → oil quench → temper at 400°C, 2 hours air cool,

(b) similar to (a) but quench in liquid nitrogen before tempering,

(c) austenitize at 845°C, 10 minutes → isothermally transform at 330°C, 1 hour → air cool.

All specimens were treated to have the same hardness, Rc 44.3±0.3.

A Shimadzu x-ray diffractometer with graphite monochromator was employed to estimate the retained austenite content in the various specimens. Copper K_α radiation was used and the scanning range was from 72° to 96° to include the $(311)_\gamma$, $(211)_\alpha$ and $(220)_\gamma$ peaks. Thin foils were prepared from the specimens and examined in a JEOL 100B transmission electron microscope to identify the various microstructures.

To compare the toughness from each treatment, the Charpy V-Notch impact tests were carried out. Average result was taken from six specimens for each heat treatment condition. The wear resistances were determined by two-body abrasive wear tests with a pin-on-disc tester, as shown in Figure 1. Rod specimens of 12.5 mm diameter were used in the tests. The rotating speed of disc was 1 cycle/sec, and the load was 0.913Kg. The abrasive paper used was SiC, 180 grit. Each specimen was run for 50 cycles before the weight loss was measured with a balance having precision of ±0.1 mg. A new abrasive paper was used for each measurement. Twenty measurements were performed for each specimen and three specimens for each treatment condition. The average weight loss was then determined and used to compute the wear ratio as:

$$\text{wear ratio} = \frac{\text{Average weight loss of each treatment}}{\text{Average weight loss of treatment (a)}}\ .$$

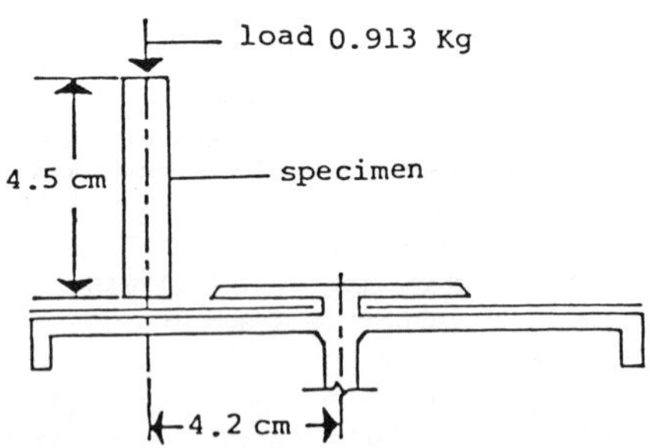

Fig. 1. Schematic diagram of the abrasive wear tester.

After wear testing, the scanning electron microscope was employed to examine the worn morphology of the specimen.

RESULTS AND DISCUSSION

The contents of retained austenite in specimens with different treatments are shown in Table I.

TABLE I Contents of Retained Austenite in Volume Percent

Treatments	Before Tempering	After Tempering	Isothermal
(a)	4.5	~ 0	—
(b)	3	~ 0	—
(c)	—	—	~ 0

It is found that there is a slight amount of retained austenite in the asquenched specimens. The content of this as-quenched retained austenite can be reduced by liquid nitrogen quenching in treatment (b). However after tempering at 400°C, all the austenite disappeared. Also, there is no retained austenite detected in the isothermal transformed specimens.

From transmission electron microscopy, a typical tempered martensitic microstructure resulting from treatment (a) is observed, as shown in Figure 2.

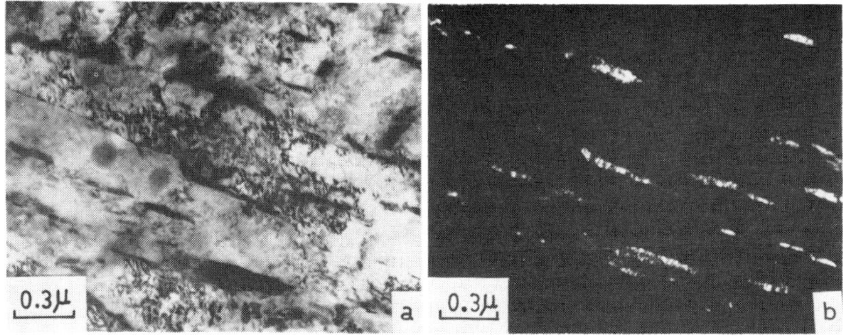

Fig. 2. Transmission electron microscopy of tempered martensite microstructure: (a) bright field image (b) dark field image of carbides.

Fine carbides are uniformly precipitated in the martensite matrix but with long, coarse carbides present on the boundaries of martensite laths. A similar result can be observed in the specimen of treatment (b). It is well known [8] that the as-quenched retained austenite in steel has the form of thin films separating the martensite laths, and it generally contains a high concentration of carbon. This as-quenched retained austenite becomes unstable when tempered, and decomposes to ferrite and carbides. Thus, from the position of the observed long, coarse carbides and previous results of x-ray diffraction, it is clear that these carbides have come from the decomposition of as-quenched retained austenite during tempering. On the other hand, a different carbide morphology can be observed in the specimen of treatment (c), as shown in Figure 3. It is a lower bainitic microstructure with carbides uniformly distributed in the ferrite matrix.

The results of Charpy impact and abrasive wear tests are shown in Table II. It can be seen that slightly better results can be achieved with treatment

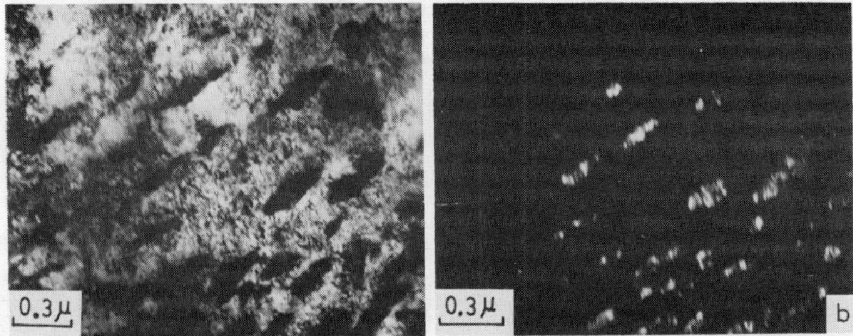

Fig. 3. Transmission electron microscopy of bainite microstructure: (a) bright field image (b) dark field image of carbides.

(b) than treatment (a). However, superior results can be obtained with treatment (c). From the results of scanning electron microscopy, as shown in Figures 4 and 5, spalling and cracking have been frequently observed around the worn traces. However, overall the situation for the specimen of treatment (c) is not as serious as for treatments (a) or (b). This is consistent with the results of Charpy and wear tests.

TABLE II Results of Charpy Impact and Wear Tests

	Treatment (a)	Treatment (b)	Treatment (c)
Charpy (m-Kg)	3.71 ± 0.07	3.81 ± 0.24	4.91 ± 0.37
Wear Ratio	1	0.951 ± 0.020	0.805 ± 0.018

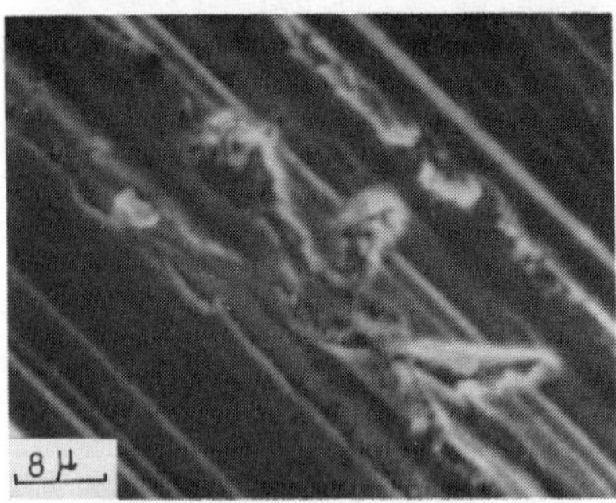

Fig. 4. Scanning electron microscopy of worn surface from specimen with tempered martensite microstructure.

Fig. 5. Scanning electron microscopy of worn surface from specimen with bainite microstructure.

In previous studies [6,7], the superior wear resistance exhibited by a bainitic microstructure was attributed to the substantial amount of retained austenite. However from the results in this work, containing little or no retained austenite, the bainite still shows a better wear resistance than the tempered martensite does. It is thus clear that the retained austenite may be an important but not necessarily the major factor controlling the wear behavior in alloy steels. The carbide morphology can be an important factor. This can be understood from the following arguments. The tempered martensite has a microstructure with long, coarse carbides on the boundaries of laths of martensite, which can prevent the slip of dislocations across the martensite laths during wear. Microcracks can then be initiated around these carbides due to the dislocation pile-ups, which account for the serious spalling and cracking around the worn traces on specimens with a tempered martensite microstructure. This is detrimental to the wear resistance. Consequently, a reduction of these detrimental carbides by reducing the as-quenched retained austenite can then improve the wear resistance, as well as toughness, of tempered martensite. This explains the result observed in specimens of treatment (b). On the other hand, in the bainitic microstructure, the carbides are distributed uniformly in ferrite laths without such detrimental carbide barriers between ferrite laths. A superior result can be thus obtained.

ACKNOWLEDGMENT

The authors would like to express their gratitude to National Science Council for the support of this research under contract No. NSC73-0405-E007-004. Thanks are also due to Drs. C.T. Hu and J.G. Duh for their assistances on this research.

REFERENCES

1. P.L. Hurricks, Wear, 26, 285, (1973).
2. E. Hornbogen, Wear, 33, 251, (1975).
3. N.P. Suh, N. Saka, and S. Jahanmir, Wear, 44, 127, (1977).
4. P.J. Mutton and J.D. Watson, Wear, 48, 385, (1979).
5. N. Prasad and S.D. Kulkarni, Wear, 63, 329, (1980).
6. N.J. Kar, Wear of Materials 1981, P. 415, American Society of Mechanical Engineers, New York (1981).
7. W.J. Salesky and G. Thomas, Wear of Materials, P. 298, American Society of Mechanical Engineers, New York (1981).
8. E.R. Parker, Met. Trans., 8A, 1025, (1977).

Analysis of Elastic and Plastic Deformation in a Layered Structure Subjected to Rolling-Sliding Contact

N. Tunca and E. E. Laufer

Technical University of Nova Scotia, P.O. Box 1000, Halifax, Nova Scotia B3J 2X4, Canada

ABSTRACT

A layered structure has been observed to develop on rollers in a high speed roller bearing, operated under conditions of insufficient lubrication. This consists of a hard outer layer (1150 HK), going abruptly to a soft layer (600 HK) and shading gradually to the normal material (750 HK). Finite element studies have been done on this system to study the effects of this layered structure on cyclic crack propagation. The results of these studies are compared with studies on homogeneous material, and with experimental observations.

KEY WORDS

Layered structure, fatigue, stresses, strains, fractures, FEM, crack propagation, Hertzian contact.

INTRODUCTION

In earlier work, a layered structure was observed to have formed on the surfaces of rollers of a failed bearing taken from an aircraft engine (1). The present work describes a similar observation on a bearing run in a laboratory test rig.

There are two theories for the origin of microcracks leading to rolling contact fatigue:

1. subsurface, or 2. surface originated fatigue.

Generally, one or the other of these mechanisms will predominate, depending on variables such as temperature, load, lubricant, microtopography, sliding, etc. Suh and Sin (2) investigated the propagation of a horizontal subsurface crack in a fully plastic solid, using the crack tip sliding displacement technique with FEM (Finite Element Method). The purpose of the present work is to present observations of a layered structure which developed at the surface of rollers in a bearing tested in a laboratory test rig, and to discuss the results of FEM calculations to determine the effect of this layered structure on crack propagation.

EXPERIMENTAL OBSERVATIONS

High speed roller bearings with rollers made from type 52100 steel were run in a laboratory test rig under conditions of excessive load and insufficient lubrication. Certain regions of the rollers developed a layered microstructure as shown in Fig. 1a. Three and occasionally four layers were observed, viz., a white etching region (WER), light etching region (LER) and a dark etching region (DER) shading smoothly into the normal etching (NER) or unaffected material. Figure 2 shows the corresponding microhardness profile. The layers are identified as martensite (LER), over-tempered martensite (DER) and the normal tempered martensite microstructure of the bearing (NER). The WER may be cementite, but was not positively identified. The layers are presumed to form due to occasional contact of the roller with the inner or outer raceway, leading to momentary surface temperature excursions into an austenite temperature range, followed by quenching by the relatively cold mass of the bearing. A similar mechanism has been described by Vingsbo and Hogmark (3). This explanation is supported by the occasional observation of double and even triple layers, resulting from a second or a third contact as shown in Fig. 1b and 1c. It was considered of interest to investigate the effect of this structure on crack initiation and propagation.

FINITE ELEMENT ANALYSIS OF FATIGUE CRACK PROPAGATION

An elastic-plastic finite element analysis, in conjunction with a crack growth criterion, was used to study crack growth behavior in a layered structure and in a homogeneous structure, under plane strain conditions. The finite element model was composed of two-dimensional quadrilateral elements of unit thickness. At the crack tip small elements of size 1 x 1 μm were used, the size was increased gradually away from the crack tip. A classical bilinear kinematic hardening theory was used in the analysis. The yield surface was assumed to have a form similar to the von Mises yield criterion. The load was applied to a roller of radius 3.175 mm by a counterformal inner ring of radius 14.75 mm, and consisted of a normal load, Fn, and a tangential load calculated from an assumed coefficient of friction as $F_t = \mu F_N$. The layered structure was considered to consist of LER of thickness 12 μm, followed by DER of thickness 16 μm, then NER. The material properties assumed are shown in Fig. 3, and are based on material properties estimated from microhardness measurements (4).

In LER the crack growth criterion used was based on the crack tip fracture stress. The crack tip stress was the nodal average stress computed from the stresses at the crack tip. Whenever the nodal average tensile stress equalled or exceeded a preset critical stress the crack-tip node was considered to break and the crack advanced to the next node. An initial deep pit (or crack) of length 6 μm was assumed. The length of the initial crack size was chosen from the average depth of pits observed on the surface of the rollers.

The stress distributions were calculated in both layered and homogeneous structures under 2250 N normal load and a tangential load (sliding friction coefficient, $\mu = 0.33$). Figures 4a and b show the normal (σ_x) and orthagonal shear stress distributions in the layered structure. Figure 5 shows the extent of plastic deformation in the layered and homogeneous media. The calculations showed that the normal tensile stress (σ_x) near to the leading edge of the contact was much greater in the layered structure (σ_x = 1042 MPa) than in the homogeneous structure (σ_x = 591 MPa). In both structures orthagonal shear stresses were in the same direction, especially close to the contact surface.

As a result of these observations a crack was introduced in three zones: I) At maximum tensile stress, II) At maximum shear stress, III) At low compressive stress (σ_y) as shown in Fig. 5.

In LER, the LEFM approach was valid, because the plastic deformation area was localized at the crack tip, but in DER and original structure, the plastic zone size was very big, then the crack tip shear displacement approach was applicable (2).

Table 1 shows the tensile, normal and orthagonal shear stresses at the crack tip in layered and homogeneous structures. In Zone I of the LER the normal stresses were in tension but the tensile stress was less than the fracture stress. If the local conditions change slightly, e.g., loading, μ, then the crack may propagate in Mode I with the contribution of Mode II. In zone III the hydrostatic stresses were not as high as in Zone II. A crack propagates mostly in zone III with Mode II (shear fracture mode).

Figures 6 and 7 show the plastic deformation area size and stress distribution in the layered structure when crack tip is at the DER. Especially in zone III the crack propagates with crack tip shear displacement mode. The same mechanism occurred in the original structure.

CONCLUSION

The load carrying capacity of LER in layered structure is higher than the homogeneous structure. Hard LER is therefore a protective layer of the component if it covers the whole surface and delays the initiation of cracks, but if the initial crack size reaches its critical value, then it propagates very quickly in the LER. Its propagation slows down or stops in DER or in homogeneous structure as a result of work hardening and high permanent deformation at the crack tip (5, 6).

In pure rolling, the surface of each cylinder is progressively displaced in the forward direction of rotation relative to the core by plastic shearing (7, 8, 9). If the tangential traction is in the same direction as that of rolling (so called negative sliding), the surface material displaces more in forward direction than in pure rolling condition.

Under plastic shearing with the contribution of a tangential traction, the crack is deviated in a forward direction in both the DER and in the homogeneous structure. If the applied load overcomes the work hardening effect the crack propagates with crack tip shearing displacement mode to cause spalling, but in the homogeneous structure the depth of spalling will be shallow. In the layered structure the depth of the spall will be slightly greater than the thickness of the LER.

Also, the boundaries of localized layered structures within a homogeneous structure are critical regions and act as stress raisers for crack initiation and propagation.

TABLE 1 The Stress Level at the Crack Tip in Layered and Homogeneous structures

REGION	ZONE	Tensile Stress at Crack Tip	σ_x at Crack Tip	σ_y at Crack Tip	σ_{xy} at Crack Tip
LAYERED STRUCTURE					
LER	I	1775	+2350	+1560	+150
	II	2400	-2600	-2550	+250
	III	2400	-2950	-1650	+300
DER	I	1370	+2075	+1575	+ 80
	II	1800	-3400	-3050	+300
	III	1750	-4250	-3150	+400
HOMOGENEOUS STRUCTURE					
NER	I	1450	+1400	+ 860	+150
	II	2050	-3800	-2750	+275
	III	1750	-2130	-1100	+375

REFERENCES

1. E.E. Laufer and T. Savaskan, Met. Tech. II, 530 (1984)
2. N.P. Suh and H.C. Sin, ASLE Trans., 26, 3, 360 (1983).
3. D. Vingsbo and S. Hogmark, Fundamentals of Friction and Wear, ASM, Metals Park, Ohio (1980).
4. R.J. Kar, R.M. Horn and V.F. Zackay, Met. Trans., 10A, 1711 (1979).
5. H.D. Dill and C.R. Saff, ASTM, STP No. 637, 141 (1977)
6. J.C. Newman, ASTM, STP No 637, 57 (1977).
7. A.W. Crook, Proc. Instn. Mech. Engrs., London, 171, 187, (1957).
8. G.M. Hamilton, Proc. Instn. Mech. Engrs., London, 177, 667 (1963).
9. J.E. Merwin and K.L. Johnson, Proc. Inst. Mech. Engrs., London, 177, 676 (1963).

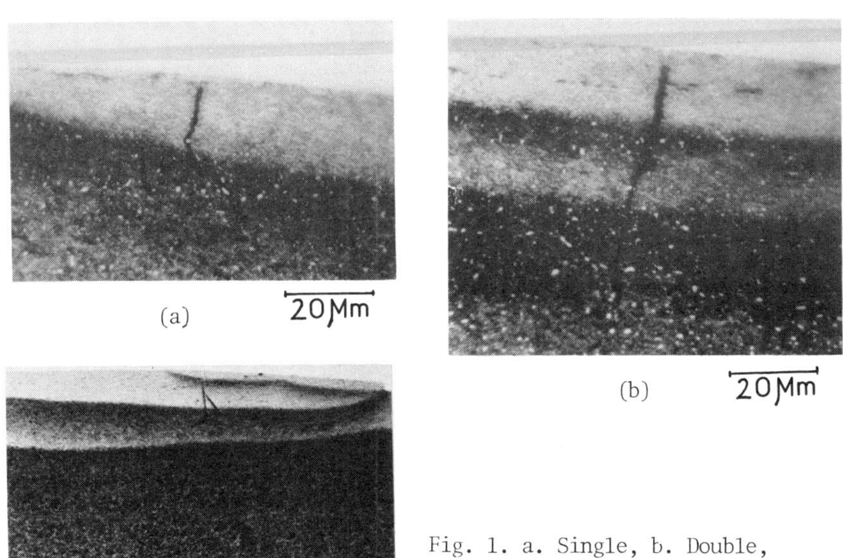

Fig. 1. a. Single, b. Double, c. Triple layered structure.

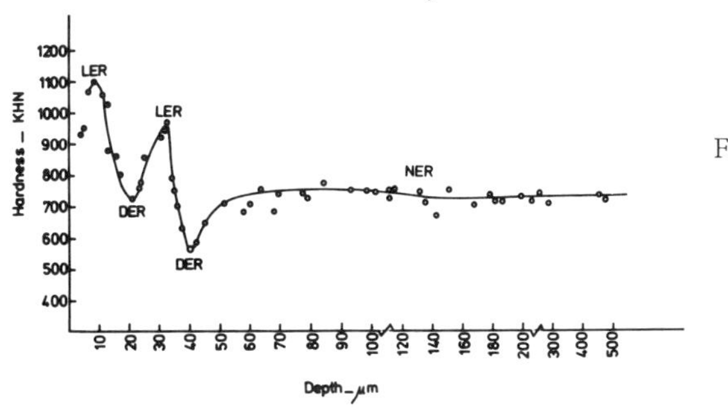

Fig. 2. Microhardness profile corresponding to Fig. 1b.

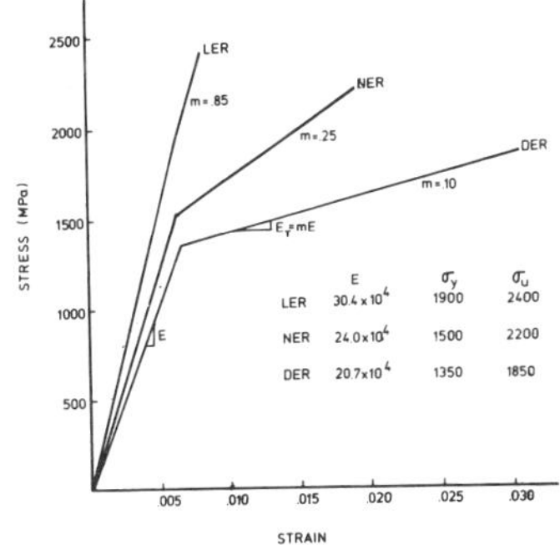

Fig. 3. Assumed stress-strain properties in layers.

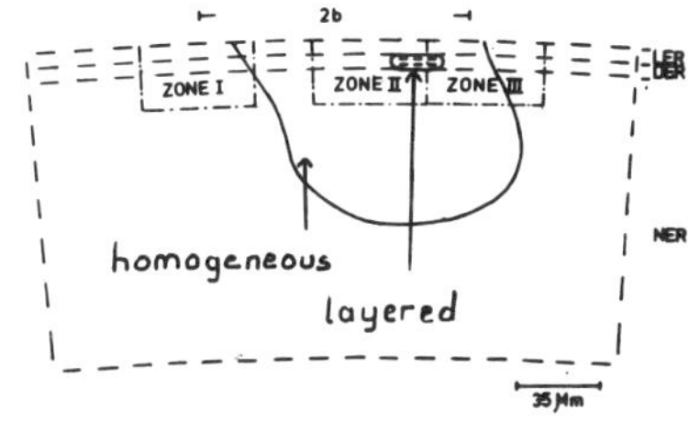

Fig. 5. Plastically deformed regions in layered and homogeneous structures, and location of zones (-·- zone)

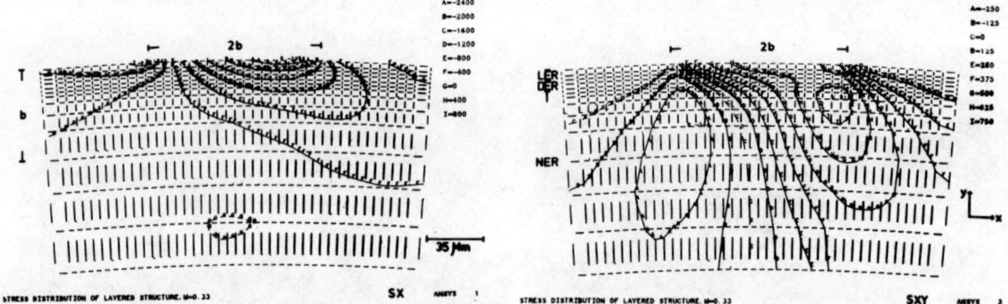

Fig. 4. Stress distribution of layered structure. b = 105 μm, F_N = 2250N. a. σ_x, b. σ_{xy}

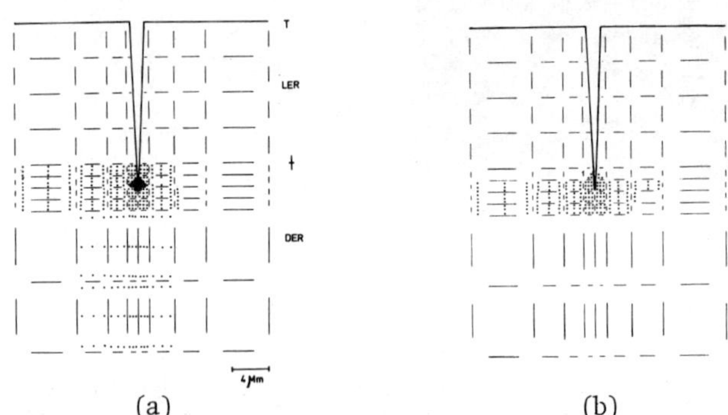

(a) (b)

Fig. 6. Plastic zone at crack tip integration points indicated. a. Zone I and II (♠ zone I) b. Zone III.

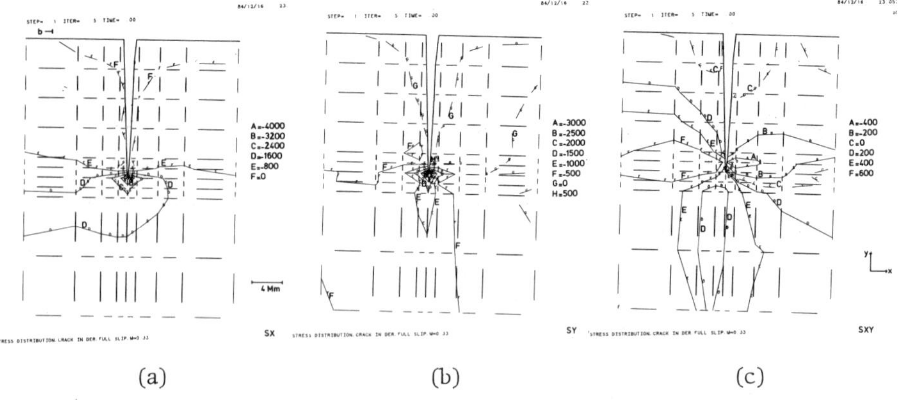

(a) (b) (c)

Fig. 7. Stress distribution at crack tip in zone III of layered structure. a. σ_x, b. σ_y, c. σ_{xy}

Effect of Inclusion Type and Control on the Wear of an Alloy Rail Steel

D. M. Fegredo*, M. T. Shehata*, A. Palmer* and J. Kalousek**

**PMRL, CANMET, Energy, Mines and Resources Canada, Ottawa K1A 0G1, Canada*
***Western Laboratory, National Research Council, Vancouver, B.C., Canada*

ABSTRACT

Six vacuum and induction melts of a Cr-Mo alloy rail steel were prepared with varying sulphide and oxide contents and hot-rolled to plate. Wear rates were determined, after heat treatment, in a dual disk-on-disk machine specially designed to simulate rail/wheel interactions on tangent and curved tracks. Dry wear rates and one fully lubricated (pure grease) rate were obtained. A marked, apparently linear dependence of dry wear rate on sulphide content is obtained with oxides having a lesser effect. The sulphide effect is related to both sulphide quantity and deformability since rare earth additions reduce the wear rate. Lubrication lowers wear rate very significantly, but similar inclusion characteristics are observed at the deformed surface.

KEY WORDS

Rails, wear, inclusions, sulphides, oxides, dry, grease.

INTRODUCTION

In spite of continuing improvements, rail wear currently costs Canadian railways $300 million annually (1). Increasing rail hardness generally increases rail wear resistance, however recent full-scale (2) and laboratory results (3,4) show that wear resistance increases very slowly or not at all above ~HV 370. Obviously, some route other than the indefinite increase in hardness is necessary to improve wear resistance.

Steel cleanliness is cited as one of the contributing factors (5) to railhead gauge face wear, and inclusions are known crack nucleators (6). Very thin wear flakes in combined mild-severe wear were attributed to highly deformed MnS (7) and extensive evidence of MnS "inclusion-cracks" have been observed in simulated rail/wheel laboratory tests (4). The work reported below was done primarily to quantify the effects of controlled sulphide and oxide distributions on dry wear rate. A fully-lubricated test (pure grease) is also reported.

EXPERIMENTAL PROCEDURE

One 50-kg and one 140-kg heat of alloy rail steel of basic composition 0.7% C, 0.8% Mn, 0.3% Si, 0.8% Cr, 0.20% Mo were air-melted in induction furnaces. An addition of 0.035% S was made to the former and one ingot (steel A) was cast. Five ingots were cast from the 140 kg heat (which contained 0.005% S) and then individually re-melted in a vacuum induction unit and cast as 20-kg ingots. Deoxidizers and other elements were added and pouring generally done in a reduced pressure of argon to produce various types of inclusions as follows:

(a) No additions; steels B and C. Steel C poured in air to produce more inclusions.
(b) Al deoxidation followed by a short hold to increase Al_2O_3; steel D.
(c) As (b), but with 0.025% S addition; steel E.
(d) As (c), but with additions of ~0.035% Ce and 0.025% La to produce dispersed rare earth sulphides; steel F.

All ingots were hot-rolled. Wear blanks were machined and transformed to pearlite of hardness HV 350-370. Disks with scaled-down rail profiles were finish-machined and tested (dry or lubricated) in a dual disk-on-disk rail/wheel wear testing machine designed to simulate rail/wheel interactions on straight or curved track (8). The tests simulated the loading on high (outer) rails in curved track and therefore the disks suffered gauge face and gauge corner attack in addition to running surface wear. Wear was measured by disk weight loss to 0.05% accuracy, as described elsewhere (9).

Using a Quantimet 900 automatic image analysis system, measurements were made of sulphide and oxide (Al_2O_3 or silicates) inclusions in all hot-rolled plates. The number and length (>10 μm) of inclusion-cracks were also obtained within ~200 μm from the worn surface on transverse sections cut radially from worn (dry) disks. Enough fields were examined to include all surface and sub-surface features.

EXPERIMENTAL RESULTS

The measurements of oxide and sulphide inclusions are given in Table 1.

TABLE 1
Inclusion characteristics and wear rates. Lengths >0.5 μm counted.

	Oxides			Sulphides			
Steel	Vol fraction %	Avg length μm	Total length $\mu m/mm^2$	Vol fraction %	Avg length μm	Total length $\mu m/mm^2$	Dry wear rate* μg/cycle
A	0.069	5.76	155	0.218	5.63	1574	24.9
B	0.0022	2.19	26	0.027	3.42	221	8.4
C	0.0050	2.79	49	0.033	3.49	251	8.5
D	0.0091	4.33	84	0.026	2.96	253	6.9
E	0.0212	3.42	262	0.152	5.29	1475	22.3
F	0.0108	2.09	95	0.177	4.08	776	17.5
D	Lubricated (pure grease) wear test*: 0.196 μg/cycle						

* wheelset load = 7000 N for all tests
Dry wear: 80 000 cycles per test. Lubricated wear: 2.04×10^6 cycles.

Figure 1 plots dry wear rate against the parameters shown.

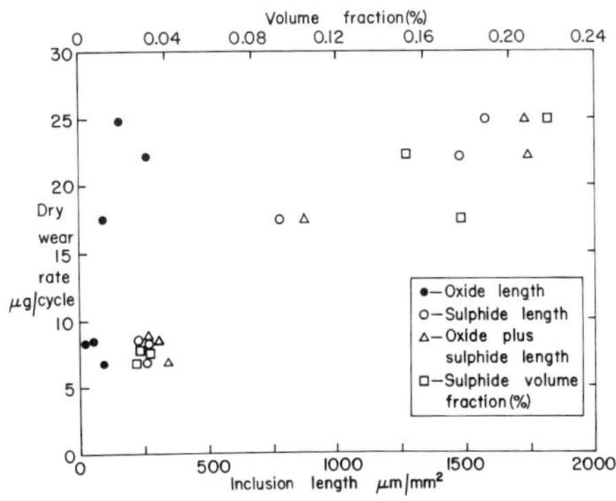

Figure 1. Dry wear rate vs the parameters shown, taken from Table 1.

Scanning electron micrographs of Ni-plated transverse sections from worn disks are given in Fig. 2. Figure 2(a) shows the gauge face, gauge corner and gauge corner/running surface junction of steel D; these regions undergo the severest wear in service. The shear lip (arrow) is typical. The sheared surface layer of the gauge corner is shown in Fig. 2(b). Much more surface flaking and sub-surface cracking are seen in the gauge corner of steel A, Fig. 2(c). All sub-surface cracks analyzed were associated with MnS, Fig. 2(d). Steel F contains R.E. sulphides, Fig. 2(e) which are much less plastic than MnS. Thus the gauge corner, Fig. 2(f), shows many relatively undeformed inclusions. But MnS associated with the R.E. sulphides will deform, Fig. 2(g). The lower friction coefficient in grease produces a much thinner deformed surface layer, Fig. 2(h). The ends of both surface cracks in Fig. 2(h) pass through MnS inclusions, Fig. 2(i).

Frequency distributions of the length of surface and sub-surface cracks are given in Fig. 3(a,b) for steels A, D and steels E, F respectively. Microhardness values vs depth from the worn surface of steel D (dry wear) are given for different locations in Fig. 4. These values are typical.

DISCUSSION

A description of rail/wheel contact stresses is outside the scope of this paper (see Ref 10). Qualitatively, the outer rail is mainly subjected to compressive and lateral forces. The latter arise from lateral creep which is most severe in the gauge face and gauge corner region. Thus, metal flows towards the gauge corner and gauge face forming the types of deformed, sheared surface layer and shear lip seen in Fig. 2(a).

Figure 1 indicates that dry wear rate is principally dependent on total sulphide length which depends on sulphide content and deformability. The relationship appears linear. Oxides will undoubtedly contribute to wear, but the main dependence on MnS for the higher sulphur concentrations is clear from Fig. 2(a) (steel D) and 2(c) (steel A). Energy dispersive analysis showed all sub-surface cracks analyzed to be associated in varying degrees with MnS inclusions whose very weak interfaces with the steel matrix readily part and are quite visible in the SEM, Fig. 2(d). Quantimet measurements of surface and sub-surface features in steels A and D, Fig. 3(a), show that many more short inclusion cracks and longer cracks occur in the deformed surface layer of steel A because of its higher sulphide content.

The importance of sulphide deformability is seen (i) from Fig. 1 in which the anomalous result occurs that steel E with 0.152 vol% sulphide has a higher wear rate than steel F with 0.177 vol% and (ii) from Fig. 2(c) and Fig. 2(f) which correlates the undeformed R.E. sulphides with a lower wear rate. Thus

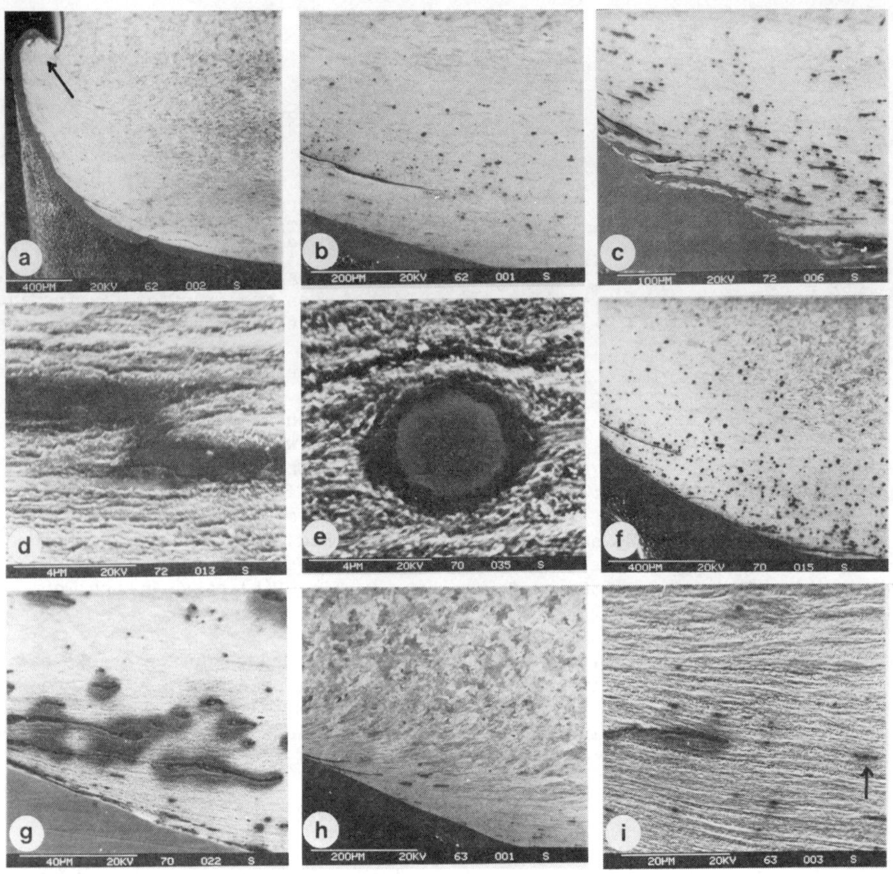

Figure 2. Scanning electron micrographs of Ni-plated transverse sections from worn disks, (a - g) dry wear, (h,i) lubricated wear (GF-gauge face; GC-gauge corner; RS-running surface; FC-field corner).
(a) Steel D (GF, GC and GC/RS junction); arrow points to shear lip (b,c) GC, Steels D and A respectively. (d) Junction of two MnS inclusion-cracks. Note MnS/steel interface cracks, (e) R.E. sulphide with associated, peripheral MnS, (f) GC of Steel F, (g) Pancaked MnS associated with relatively undeformed RE sulphides, (h) GC, Steel D, (i) Very thin end of surface crack ending in MnS inclusion (arrow), Steel D.

steel F has more short cracks and significantly fewer long cracks than Steel E, Fig. 3(b).

Since dry wear rate appears to be mainly related to sulphide content and deformability, rail/wheel contact stresses are important. High microhardness values are obtained at and near the worn surface in the vicinity of the gauge corner, Fig. 4. As-rolled sulphides commence to deform soon after entry into the sheared layer and continue elongating as the surface is approached. The

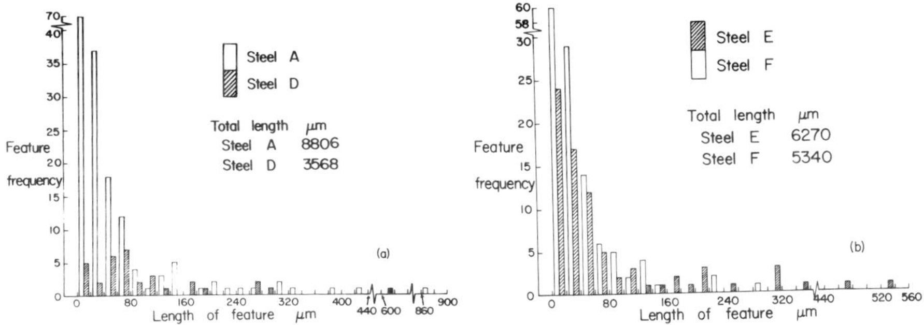

Figure 3 (a,b). Bar diagrams of feature (surface and sub-surface cracks) frequency vs feature length of Steels A, D and Steels E, F respectively.

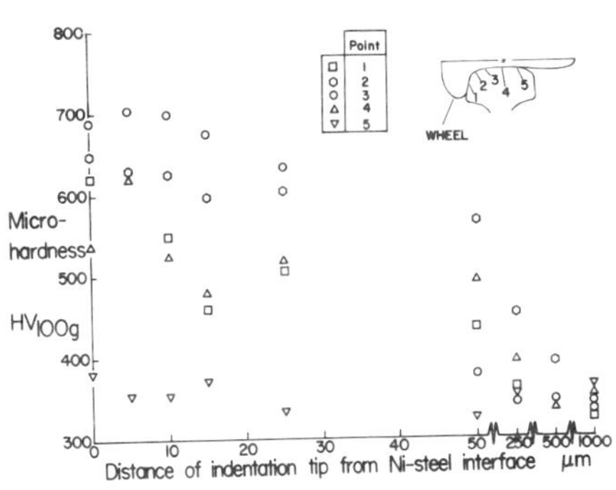

Figure 4. Microhardness vs depth from dry worn surface of Steel D for locations: 1-centre of GF; 2-centre of GC; 3-GC/RS junction, 4-centre of RS; 5-RS/FC junction.

inclusion envelope, which is the crack, increases in length and aspect ratio. However, many SEM observations indicate that isolated sub-surface inclusion cracks do not unequivocally appear to extend by steel matrix cracking at their tips even when fairly close to the worn surface. Matrix cracking, by fatigue, is a possibility either when sub-surface inclusion-cracks, which are basically internal notches, are closely spaced or when they penetrate through to the worn surface. But the ends of surface cracks, which are external notches, probably extend during repeated cycling by matrix cracking towards sulphides (and oxides), Fig. 2(i), because of the higher stress concentration of the surface crack.

A limiting friction coefficient of 0.09-0.12 is obtained in pure grease lubricated wear (9). This results in reduced lateral stresses and thinner deformed surface layers, Fig. 2(h). The overall thinner deformed surface layer results in shorter MnS inclusion-cracks near the surface. These, together with the much smaller lateral stresses will retard the production of surface cracks. The subsequent slower rate of surface crack growth by fatigue, compared to dry wear, leads to a significantly lower wear rate.

CONCLUSIONS

Under simulated conditions of rail/wheel contact on curved track, it is found that:

1. Dry wear rates principally depend on sulphide content and deformability for the sulphide and oxide concentrations used.

2. A linear relationship appears to exist between total sulphide length and dry wear rate.

3. Pure grease lubrication significantly lowers the wear rate.

ACKNOWLEDGEMENTS

We acknowledge with thanks the assistance of E. Connors, F. Lam, B. Casault, V. Chartrand and L. Clements.

REFERENCES

1. Roney, M.D. (1983). Contact mechanics and wear of rail/wheel systems Univ. of Waterloo Press, 271-291.
2. Steele, R.K. (1983). Can. Met. Quart., 22, 353-367.
3. Hodgson, W.H., R.R. Preston and J.K. Yates (1982). Proceedings of the Second International Heavy Haul Railways Conference, Colorado Springs, Colorado. Paper 82-HH-22. ISBN No. 0-9611854-0-6.
4. Kalousek, J., D.M. Fegredo and E.E. Laufer (1985). To be published in the Proceedings of the International Conference on Wear of Materials, Vancouver, B.C.
5. Steele, R.K. and R.P. Reiff (1982). Proceedings of the Second International Heavy Haul Railways Conference, Colorado Springs, Colorado. Paper 82-HH-24.
6. Marich, S, J.W. Cottam and P. Curcio (1978). Proceedings of the First International Heavy Haul Railways Conference, Perth, Australia, The Institution of Engineers Australia. Paper I.1, Session 303.
7. Bolton, P.J. and P. Clayton (1984). Wear, 93, 145-165.
8. Kalousek, J. (1982). Can. Met. Quart., 21, 67-72.
9. Kalousek, J., H. Ghonem and R.K. Steele (1984). Proceedings of the 11th Leeds-Lyon Symposium on Tribology, Burlington Press, Cambridge.
10. Ghonem, H., J. Kalousek, D.H. Stone and E.E. Laufer (1982). Proceedings of the Second International Heavy Haul Railways Conference, Colorado Springs, Colorado. Paper 82-HH-31.

Etude de l'Endommagement sous Sollicitations Multiples de Superalliages Recharges: Interpretation Structurale de la Resistance a l'Usure a Chaud

S. Tawfiq*, P. Fluzin*, G. Beranger* et A. Boucher**

**Division "Matériaux" ERA 910 du CNRS, Université de Technologie de Compiègne, France*
***UNIREC, Centre de Recherche d'UNIEUX France*

ABSTRACT

The surface degradation of hot work tools (hot working rolls) is often due to the action of synergy of different factors which can be either, thermal, mechanical, physico-chemical or tribological. This paper presents the main results of high temperature wear using a copper frotting piece. Some hard facing layers, (2 nickel base alloys and 2 cobalt base alloys, obtained by plasma semi-transfered arc proceeding), were tested. Weight loss, wear facies, surface roughness and the influence of the structure are explained. We illustrate the non compatibility of nickel base alloys towards copper and the good behaviour of cobalt base alloys.

KEYWORDS

Wear resistance; hard facing; hot rolling; nickel base alloys; cobalt base alloys; copper; surface roughness; influence of structure.

INTRODUCTION

Le phénomène d'usure est étroitement associé aux facteurs thermique, mécanique et physicochimique. Dans le cas du laminage à chaud, l'usure est la conséquence du glissement relatif entre le produit laminé et les cylindres de laminoir. Il semble qu'elle s'effectue par adhésion et/ou par abrasion. Elle est fortement influencée par l'état de surface du matériau. Le glissement entre les deux surfaces s'effectue en présence de films d'oxydes qui joueront, selon leur nature, un rôle abrasif ou lubrifiant [2]. Ce type de dégradation est généralement prépondérant pour les cages finisseuses des laminoirs. En effet, les sollicitations mécaniques sont importantes augmentant ainsi la charge appliquée à l'usure. Par ailleurs, l'altération de la surface résultant des facteurs physicochimique et tribologique augmente le taux de particules abrasives. Cette détérioration est de plus favorisée par une vitesse de rotation élevée des cylindres.

Compte tenu des nombreux facteurs intervenant de façon simultanée et/ou conjuguée dans le processus de ruine, le choix de la nuance métallique de compromis devient très difficile et dans certains cas impossible. Ce sont les raisons qui nous ont conduits à nous intéresser aux revêtements multicouches. La résistance à l'endommagement de plusieurs nuances d'aciers rechargés par le procédé plasma à arc semi-transféré (alliages A, B, C et D. Tableau I) a donc été étudiée.

METHODE EXPERIMENTALE

L'étude de la résistance à l'usure à chaud a été réalisée à l'aide d'un dispositif d'essai permettant d'intégrer la plus part des facteurs d'endommagement évoqués dans notre introduction (Fig. 1). La pièce (1) représentant le galet de laminage est en contact avec un cylindre de cuivre (2) simulant le métal à laminer (piste de frottement). La température de la piste, de l'ordre de 500°C, correspond à la température du métal laminé lors des dernières "passes" de laminage (finisseur). La rotation de l'éprouvette (1) sur cylindre de cuivre (2) fixe, accentue le frottement et accélère par conséquent le processus d'usure. Une charge constante de 2,1 kg est appliquée à l'ensemble. La force tangentielle entre les deux surfaces en contact et mesurée à l'aide d'un capteur de force afin de déterminer le coefficient de frottement. Le taux d'usure est appréhendé par la perte en poids qui est mesurée par double pesée avant et après chaque essai. Le coefficient d'usure est déterminé à partir de l'équation suivante [9] :

$$K = \frac{VP}{Ld}$$

V : usure volumique (mm^3)
L : charge (kg)
d : longueur de glissement pour 20000 cycles (mm)
P : dureté du matériau (kg/mm^2)

Les paramètres suivants ont été fixés pour tous nos essais :
- Température de la piste (cuivre) T = 500°C,
- Vitesse de rotation du galet 10 tours/mn,
- L'éprouvette est refroidie à l'eau déminéralisée à un débit constant de 20 cm^3/mn,
- Les éprouvettes ont la forme d'un disque de 70 mm de diamètre et de 5 mm d'épaisseur. Leur profil est légèrement arrondi. Chaque pièce est préalablement polie avant essai de telle sorte à posséder un Ra (la moyenne arithmétique des écarts de profil par rapport à la ligne moyenne) de 0.045 µm.
- Les pesées sont réalisées après un nettoyage adéquat tous les 2500 cycles de fonctionnement sur une balance de précision.

RESULTATS ET INTERPRETATIONS

Nous avons utilisé comme référence pour nos essais, la nuance Z38CDV5 classiquement employée dans l'outillage à chaud.

- Acier à outils à 5 % de chrome (Z38CDV5)

Cet alliage possède une perte de poids régulière et on n'observe pas de transition "usure douce/usure sévère" (Fig. 2). La perte de poids totale pour 20000 cycles est de 221 mg. Comme l'illustre la figure (3b), il présente un faciès d'usure essentiellement abrasif. Le profil de rugosimétrie indique un Ra de 0.121 µm pour un coefficient d'usure de 1.44 . 10^{-3} dans nos conditions d'essai. Cette nuance est facilement oxydable mais n'est pas sensible au collage avec le cuivre.

- Rechargements de base nickel

Deux rechargements de base nickel ont été étudiés ; il s'agit des alliages A et B. Leurs principales caractéristiques sont résumées dans le tableau I.

On peut observer à l'aide de la figure (2) une usure moindre du rechargement B qui possède par ailleurs une perte de poids assez régulière. La perte de poids totale pour 20000 cycles est de 176 mg et donc légèrement inférieure à celle de l'acier Z38CDV5. Le coefficient d'usure est de 1.10^{-3}. Le rechargement A est par contre moins performant ; l'usure qui est moins importante que celle du Z38CDV5 en deçà de 10000 cycles s'accélère brusquement au delà. La perte de poids totale est de 254 mg et le coefficient d'usure de $1.5 \cdot 10^{-3}$. L'examen des faciès d'usure (Fig. 3c) indique une détérioration importante de la surface par abrasion et par adhésion. On observe en effet un phénomène de collage particulièrement significatif avec ces nuances. Celui-ci, attesté par des analyses de surface à la microsonde électronique, accroît considérablement l'endommagement. Les paramètres rugosimétriques reflètent bien entendu cet état de fait, et le Ra de l'alliage A passe de 0.045 µm à 0.22 µm.

Le phénomène de collage peut en partie s'expliquer par l'affinité des matériaux l'un pour l'autre. On considére en effet [5] que la compatibilité tribologique de deux alliages exige que les métaux soient insolubles l'un dans l'autre, et ne forment pas de solution solide. Le cuivre et le nickel ne remplissent pas de telles conditions, et ils ont par ailleurs le même réseau cristallin (C.F.C.) et des caractéristiques très voisines.

La différence de comportement entre les deux alliages considérés et notamment leur propension à l'usure adhésive peut être interprétée en termes microstructural. En ce qui concerne l'alliage A, la matrice et les phases durcissantes [Ni_3 Nb, Ni_3 (Al,Ti)] sont sensibles au collage. Par contre, dans le cas de la nuance B, durcie par un mélange complexe de carbures et de borures [4], seule la matrice (80 % de la fraction volumique) est affectée par l'usure adhésive.

- Rechargements de base cobalt

Les deux alliages étudiés (alliage C et alliage D : tableau I) présentent les meilleures caractéristiques tribologiques (Fig. 2). Les pertes de poids sont très régulières et restent assez faibles ; 91 mg pour l'alliage D et 97 mg pour l'alliage C. Le coefficient d'usure est de ($0.567 \cdot 10^{-3}$) pour l'alliage D et de ($0.47 \cdot 10^{-3}$) pour l'alliage C.

Les faciès d'usure observés (Fig. 3d) sont peu perturbés et traduisent une usure essentiellement abrasive. Le paramètre rugosimétrique Ra est de 0.121 µm pour l'alliage C et de 0.078 µm pour l'alliage D, ce qui indique une légère différence de comportement. Celle-ci se matérialise par des rayures plus larges, plus profondes, mais moins denses dans l'alliage C que dans l'alliage D. Cette différence peut s'interpréter en considérant la nature, la quantité et la distribution des carbures : l'alliage C possède des carbures intergranulaires (du type $M_{23}C_6$) en proportion plus faible que ceux présents dans l'alliage D. Ces derniers de type M_7C_3 forment un mélange eutectique avec la matrice et représentent 12.6 % en teneur pondérale [6]. Ils confèrent à l'alliage D une plus grande dureté initiale (475-500 HV 0.1). L'alliage C est toutefois écrouissable à chaud et a subi dans nos conditions d'essai un durcissement superficiel de l'ordre de 20 à 30 HV 0.1.

CONCLUSION

Nous retenons des essais entrepris les points suivants :

- On observe, dans nos conditions de sollicitation, une forte sensibilité à l'usure adhésive des superalliages base nickel rechargés qui se manifeste dès les premières heures de fonctionnement. Ce phénomène se justifie en partie par une certaine incompatibilité tribologique du cuivre et du nickel.
- Les meilleures performances ont été obtenues avec les rechargements de base cobalt qui se détériorent essentiellement par usure abrasive. La différence de comportement entre les deux nuances testées reste quantitativement faible malgré des niveaux de dureté initiale très différents (tableau I).
- On notera cependant que les conditions particulières de solidification dues au procédé employé (plasma à arc semi-transféré) confèrent à la structure dendritique du métal (alliage C et alliage D) une orientation préférentielle radiale qui pourra d'une part, augmenter la résistance à l'usure à chaud mais d'autre part, faciliter la propagation de fissure, notamment par fatigue thermique. Notons que ces alliages sont testés en fatigue et choc thermique avant d'être testés en usure à chaud.

L'aspect de la sollicitation thermique est particulièrement important dans la mesure où la détérioration de l'état de surface d'un outillage pour travail à chaud est souvent due à la synergie de plusieurs facteurs d'endommagement [1,2].

REFERENCES

1. P. Fluzin, Thèse de Doctorat d'Ingénieur, U.T.C., 1983.
2. A. Magnée, G. Gaspard et D. Coutsouradis, 15e journée des aciers spéciaux, St Etienne, 13-14 Mai 1976.
3. C. Gayard, S. Hamar Thibault et C.H. Alibert, J. materials Science, N° 17, 1982.
4. CABOT CORPORATION, HAYNES N6, publication N° W.202 U.S.A. 1980.
5. A. Blouet et R. Gras, cours de tribologie, U.T.C. automne 1982.
6. Silence, Trans. ASM, Vol. 100, July 1978.
7. C. Sims and W.C. Hagel, the superalloys New York, 1972.
8. T. Kilner, R.M. Pilliard, G.C. Weatherly and C. Alibert, J. biomedical materials research, Vol. 16, 1979.
9. CABOT CORPORATION, publication N° W.207 U.S.A. 1982.

ALLIAGES	composition chimique %														dureté initiale
	C	Mn	Si	Ni	Cr	Mo	W	V	Co	B	Fe	Ti	Al	Nb	HV 0.1
Z38CDV5	0.38	0.3	1	-	5	1.25	-	0.5	-	-	*	-	-	-	500-550
alliage A	0.1	-	-	*	18	3	-	-	-	-	20	0.9	0.5	5.2	430-460
alliage B	1.1	<1	-	*	29	5.5	2	-	-	0.6	-	-	-	-	430-460
alliage C	0.25	<1	<2	3	27	5.5	-	-	*	-	-	-	-	-	370-460
alliage D	1.1	<1	1	<3	28	<1	4	-	*	-	-	-	-	-	475-500

* base

Tableau I

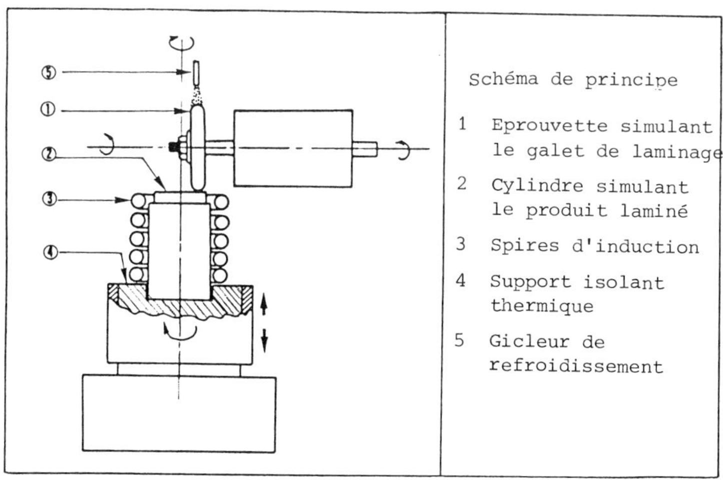

Fig. 1. Machine d'essai

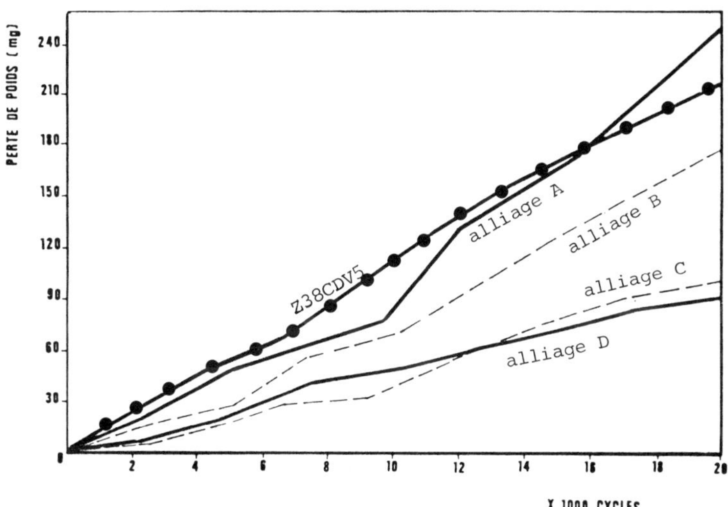

Fig. 2. Résistance à l'usure à chaud des alliages comparée avec celle du Z 38 CDV5

Fig. 3. Faciès et profils d'usure.

a. Surface initiale (profil.x100000)

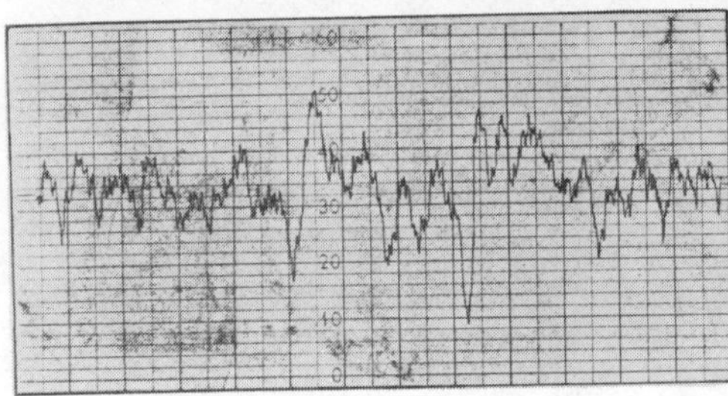

b. Faciès d'usure du Z38CDV5
(profil.x50000)

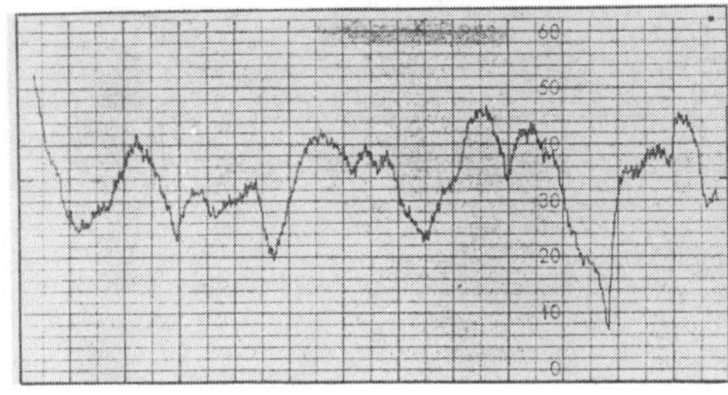

c. Faciès d'usure de l'alliage A
(profil.x20000)

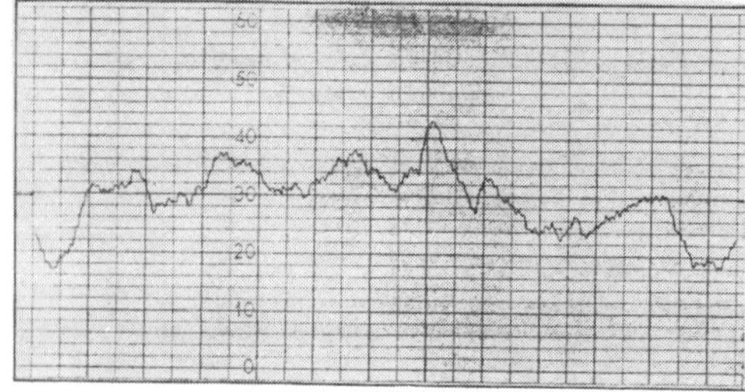

d. Faciès d'usure de l'alliage C
(profil.x50000)

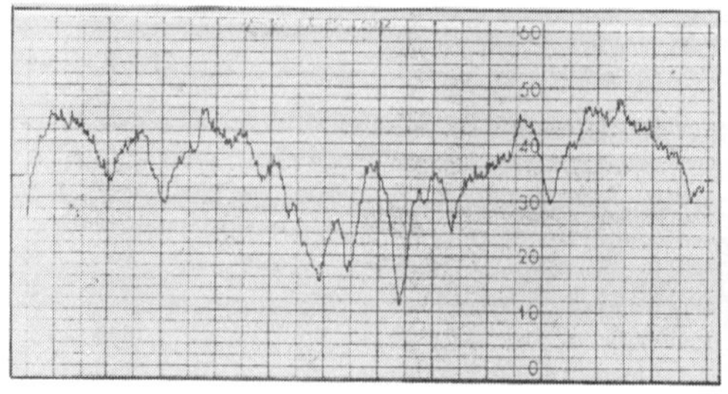

Comportement en Usure des Coussinets d'Alliages de Zinc-Aluminium

J. Masounave*, R. J. Banhurst and J. C. Farge****

**Institut de Genie des Materiaux, Conseil national de recherches Canada, 75 de Mortagne, Boucherville, Quebec J4B 6Y4, Canada*
***Centre de Recherche Noranda, 240 boul. Hymus, Pointe Claire, Quebec H9R 1G5, Canada*

RESUME

A l'aide des courbes de Stribeck, les mécanismes d'usure des alliages de ZA12 et ZA27, testés avec des arbres de grande et faible dureté, ont été précisés. Il est montré que le bronze 660 a un comportement très semblable aux alliages ZA, principalement au ZA27/arbre dur.

Les alliages de ZA ne présentent pas, en condition limite d'usure, de phénomène de bloquage, ce phénomène qui est expliqué par la formation d'une couche superficielle, riche en zinc et fer, qui se reforme au fur et à mesure qu'elle s'évacue dans les débris. De plus, la rugosité initiale a une grande influence sur l'usure des coussinets.

INTRODUCTION

Des travaux de laboratoires, commandités par l'Ilzro et réalisés par Battelle (1) ont montré que les alliages de zinc-aluminium (ZA), obtenus par coulée en sable, pouvaient être de bons concurrents aux alliages de bronze (2, 3) couramment utilisés pour les coussinets.(4)

Les alliages utilisés dans le présent travail ont été coulés en continu, par le centre de Recherche de Noranda (5,6), qui est impliqué dans le développement des alliages ZA depuis plusieurs années. (5, 6, 7)

Matériaux utilisés

Les compositions des alliages respectent la norme ASTM B669. Ils contiennent respectivement 12% (ZA12) et 27% d'Al (ZA27). Ils n'ont pas subi de traitement thermique subséquent.

Les arbres sont en acier 1144, couramment employé pour cette application. Un traitement thermique après trempe a permis d'en faire varier la dureté. L'usinage au tour a permis d'obtenir deux différentes rugosités (.2 ou .4 µm CLA).

La graisse (Sunoco Prestige 741EP) a été choisie pour sa grande variété d'application. Elle est proche de celle utilisée par Battelle pour ses expériences.

Montage expérimental

Le montage expérimental simule les conditions rencontrées en service. C'est un montage relativement classique (10) dans ce type de travaux (fig. 1). La charge est appliquée à l'aide de ressorts suffisamment longs afin que les variations causées par l'usure du coussinet n'entrainent pas de variations importantes de la charge. La température est mesurée par des thermocouples situés à trois millimètres de la surface interne du coussinet. Le coefficient de friction est mesuré par l'intermédiaire d'une lame de ressort instrumentée avec un pont de jauges. Ces trois lectures sont enregistrées périodiquement par un ordinateur de bureau. Une quantité de 0,1 mg de graisse est injectée automatiquement toutes les 20 minutes.(fig. 1) La vitesse du moteur (10 HP) est maintenue constante pendant tout l'essai. Afin d'accélerer l'obtention des résultats, une machine d'essais à 8 stations de travail a été construite.

Résultats expérimentaux

Chaque expérience était précédée d'une période de rodage de 6 h, sous une charge de 3.45 MPa avec la vitesse de rotation considérée ultérieurement. La figure 2 est un bon exemple de fonctionnement dans le régime mixte. Après une période initiale de transition où les valeurs de la température et du coefficient de friction restent basses, on assiste à de fortes variations de ses paramètres. Pendant la période initiale, il y a probablement accumulation de débris entre les deux surfaces. Ces débris finissent par s'évacuer, laissant de nouvelles surfaces fraiches en contact. Le régime d'usure sévère ($T \geqslant 120°C$, $f \geqslant .1$) apparait au bout de 100 h. Cette expérience montre clairement le danger présenté par des expériences de courte durée (1).

1. Charge critique

La charge est augmentée de 3.45 MPa à chaque 8 h, jusqu'à ce que l'usure sévère soit atteinte (fig. 3). La différence de comportement entre les différents alliages n'est pas significative. Rappelons que les alliages utilisés par Battelle ont été obtenus par coulée en sable.

2. Courbes de Stribeck (fig.4)

Le coefficient de friction f est mesuré en fonction du rapport $Z=\mu N/P$ où μ représente la viscosité (Poise), N la vitesse (rpm), et P la contrainte (Pa). Le coussinet n° 1 a fonctionné dans des conditions d'usure hydrodynamique, tandis que le n° 2 et 3 étaient dans des conditions d'usure mixte. Pour chaque matériau, testé avec un arbre donné, on a additionné chaque courbe individuelle (fig. 5 à 9) pour obtenir des courbes cumulées. Ces figures montrent la barre d'erreur expérimentale, importante en régime d'usure hydrodynamique. En tenant compte des imprécisions inhérentes à ce type d'expérience, il semble que les couples ZA27/arbre dur et bronze/arbre dur aient les comportements les plus favorables.

Discussion:

a- Elévation de température

A partir de la dissipation de la chaleur générée par le frottement, des modèles théoriques calculent l'élévation de température ΔT (3,8 à 17). En accord avec certains d'entre eux (3,8 à 11), trouve que ΔT est proportionnelle au produit f.N.P. (fig.10 et 11). Aux basses valeurs de fNP, la génération d'énergie est insuffisante pour échauffer le coussinet, tandis qu'aux hautes valeurs l'évacuation de chaleur par l'arbre devient importante.

Les modèles (8, 9, 13, 16) qui postulent une déformation de la surface (13), adsorption (16), déformation plastique (14), concluent à une variation ΔT proportionnelle à fN P, tandis que (13) prévoit une proportionalité avec fP N, surtout à haute vitesse. Ces prédictions ne semblent pas être vérifiées.

En superposant les résultats d'un grand nombre d'expériences (fig. 11), le mode d'usure (hydrodynamique, moyenne et sévère) devient apparent.

b- Courbe de Stribeck

Leloup (18, 19) a défini, à partir du calcul classique de la théorie de l'hydrodynamisme et de nombreuses approximations expérimentales, un paramètre L qui permet, de façon analogue à Stribeck, de décrire l'usure d'un coussinet (fig. 12: chaque point représente une expérience et a été calculé à partir de la valeur maximale de f.). Dans l'équation de Leloup V ou N représente la vitesse, d le diamètre et l la longueur du coussinet. Certains points de la courbe f(L), correspondants à l'usure mixte, se trouvent situés dans la région hydrodynamique. La description de Stribeck semble donc plus juste et plus simple, surtout dans la région proche du minimum.

c-Mécanisme d'usure

L'usure des coussinets peut être décrite par la déformation plastique et les transformations métallurgiques engendrées par l'élévation de température et la pression dans la couche qui se trouve sous la surface.

La figure 13 représente la surface d'un coussinet de ZA27 ayant travaillé dans le régime d'usure moyenne. L'usure se localise dans les marques des stries d'usinage de l'arbre.

Au fur et à mesure que les conditions d'opérations se rapprochent de l'usure sévère, cette couche superficielle augmente et recouvre toute la surface de frottement (fig. 14). Cette figure montre la transition entre la région usée et la région vierge du coussinet. La région usée est située en bas sur la figure. On retrouve les stries d'usinage de l'arbre sur la surface d'usure. On peut observer, au centre, la formation d'un débris qui s'évacuera avec la graisse.Cette couche est riche en zinc avec des quantités de fer variables de 10 à 40%, avec de faibles quantités d'aluminium, de 0 à 13%. Cette couche superficielle se retrouve sur les coussinets de ZA12 et ZA27.

Quand l'usure devient sévère, la couche superficielle riche en Zn-Fe augmente, et la formation des débris devient plus rapide (fig. 15).

Cette morphologie de l'usure correspond à un mécanisme d'adhésion. Contrairement à d'autres mécanismes d'usure, tel que l'abrasion par exemple, l'usure par adhésion engendre des variations brutales du comportement (cf. fig. 11). La transition entre l'usure mixte et sévère correspond au moment où les surfaces de l'arbre et du coussinet sont recouvertes de la couche riche en zinc. L'usure est d'abord concentrée dans les marques d'usinage de l'arbre (fig. 13), avant de s'étendre à toute la surface (fig.15). Il est donc normal que la rugosité de l'arbre ait une grande influence sur l'usure. (fig.16). Cette figure représente, pour une vitesse de 80 rpm, la contrainte qui délimite la région hydrodynamique de la région d'usure mixte.

Au fur et à mesure que la couche superficielle s'élimine sous forme de débris, une nouvelle couche est formée. Ce processus explique une grande différence de comportement avec les alliages de bronze: sous des charges excessives, les alliages de ZA n'entrainent pas de bloquage entre l'arbre et le coussinet, mais la formation continue de débris (fig.15).

Le mécanisme d'usure peut encore être précisé, en observant les surfaces d'usure selon des sections perpendiculaires à la surface d'usure. A la figure 17 (usure mixte), la surface d'usure est située vers le haut. Au centre de la figure, on remarque la microstructure typique d'un alliage obtenue par coulée continue. En se rapprochant de la surface d'usure, les dendrites sont déformées sur une profondeur de 60 à 80 μm environ. Sur la surface de frottement, la couche riche en zinc est apparente et, en certain endroit est proche de se détacher.

Dans la zone d'usure sévère (condition limite), la zone de transition, où la microstructure est déformée, a disparu (fig. 18). Il est probable que l'élévation de température associée à des contraintes élevées engendrent la formation d'une couche superficielle très épaisse, qui absorbe une grande partie de la chaleur créée par le frottement. Sur cette figure, on peut apercevoir un grand débris en voie de se détacher de la surface (en bas à gauche).

La même couche superficielle riche en fer et zinc se retrouve sur la surface de l'arbre. Des essais infructueux n'ont pas permis de mesurer la rugosité de l'arbre après usure, car il est difficile de décoller la couche superficielle causée par l'usure.

Le mécanisme responsable de la formation de cette couche superficielle reste encore à préciser.

Il est probable que tout mécanisme qui tendrait à augmenter la résistance au fluage de la matrice, aurait un effet bénéfique sur l'usure.

Conclusions:

1- Les courbes de Stribeck pour 2 alliages de ZA, testés avec deux arbres de dureté différente, ont été obtenues et comparées à celle du bronze 660. Le comportement du bronze 660 est comparable à celui des alliages ZA. L'alliage ZA27/arbre dur, semble au moins aussi bon que le 660/arbre dur.

2- Les mesures d'élévation de température au cours de l'expérience ont permis de vérifier certains modèles (ΔT proportionnel à fNP).

3- Une couche superficielle, riche en fer et zinc et pauvre en aluminium, se forme sur la surface du coussinet et de l'arbre. Les débris sont créés à partir de cette couche. Ils se reforment au fur et à mesure de leur évacuation. Ce mécanisme empêche un grippage (bloquage) entre l'arbre et le coussinet.

4- La rugosité initiale de l'arbre a une grande influence sur l'usure.

Références

1. Battelle-Columbus, (OH), Ilzro Project ZM 298. Progress Report n°5 (Dec. 1981)
2. The Design of Boundary Lubricated Cast Bronze Bearings, Handbook, Battelle, Columbus. Cast Bronze Bearing Institute Inc (1978)
3. Cast Bronze Bearing Design Manuel, H..C. Rippal, publié par Cast Bronze Bearing Institute Inc. (July 1971).
4. T. Calayag, D. Farres, SAE Technical Paper N° 820643, (Apr. 19, 1982).
5. E. Gervais, H. Levert, M. Bess, 84th Casting Congress, publié dans Trans of A.F.S., St Louis (Missouri) (April 1980).
6. E. Gervais, H. Levert, A.Y. Kandeil, 10th Int. Pressure Die Casting, Madrid, Spain, (May 10-14, 1981)
7. R. Banhurst, J. Masounave, CIM (Août 1983).
8. H. Block, Interdisciplinary Approach to the Lubrications of Concentrated Contacts, NASA SP-237, Washington (1970).
9. R.S. Fein, ASLE Trans. 3,34 (1960).
10. J.K. Lancaster, Source Book on Wear Control Technology, ASM, 384-422, (1978).
11. S.F. Murray, M.B. Petterson, D.F. Kennedy, Annual Report, INCRA Project 210, .(1975).
12. M.V. Korovchinski, Trans ASME, J. Basic Eng. D., 811-817, (1965).
13. E. Rabinowicz, Ed Wiley, N.Y., (1965).
14. F.P. Bowden and D. Tabor, Oxford Univ. Press, Clarendon Press, Oxford (1964).
15. J.C. Jaeger, J. Proc Royal Soc. NSW 76,203, (1942).
16. A.B. Crease, Tribology, 15-20, (Feb. 1973).
17. A. Beerbower, ASLE Trans. 14, 2, 90-104 (Apr. 1971).
18. L. Leloup, Mémoire de la Société Royale Belge des Ingénieurs et des Industriels, n° 3 et 4, 1950 (Bruxelles).
19. J. Bozet, Revue de l'institut français du pétrole, vol. 35,5, 885-913, (sept-oct 1980).

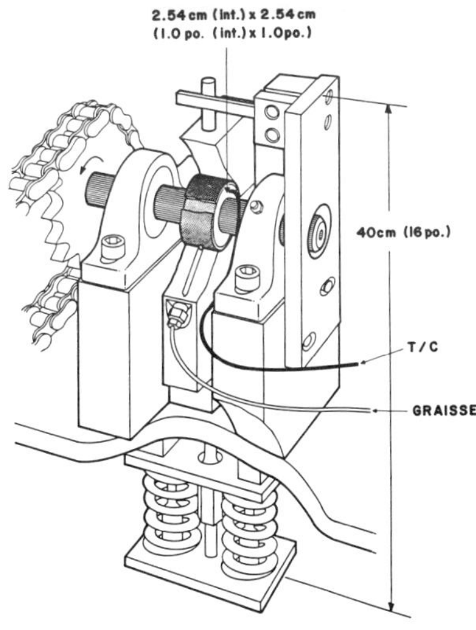

Fig. 1. Schema de principe du montage expérimental

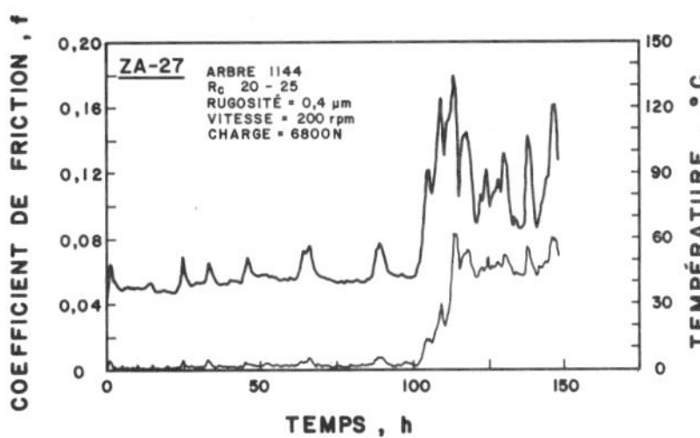

Fig. 2. Courbe d'usure typique (régime mixte)

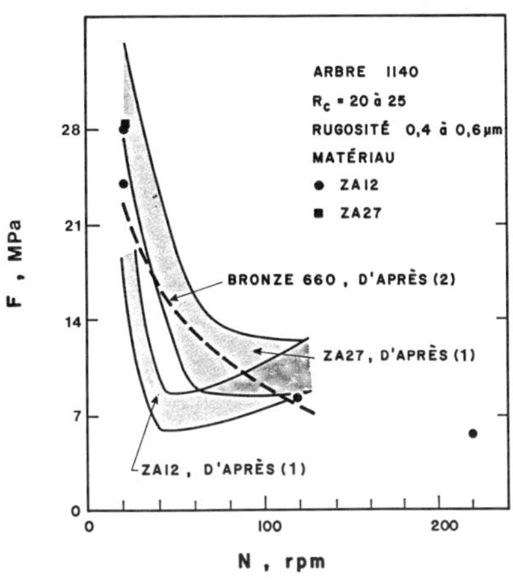

Fig. 3. Variation de la charge jusqu'à usure sévère vs. vitesse de rotation

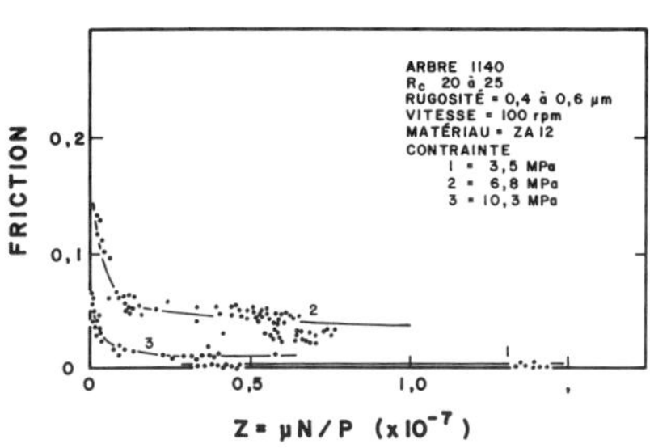

Fig. 4 Courbe de Stribeck

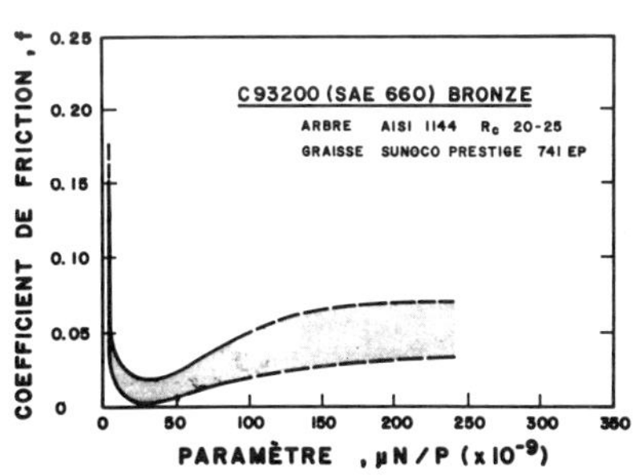

Fig. 5. Courbe de Stribeck

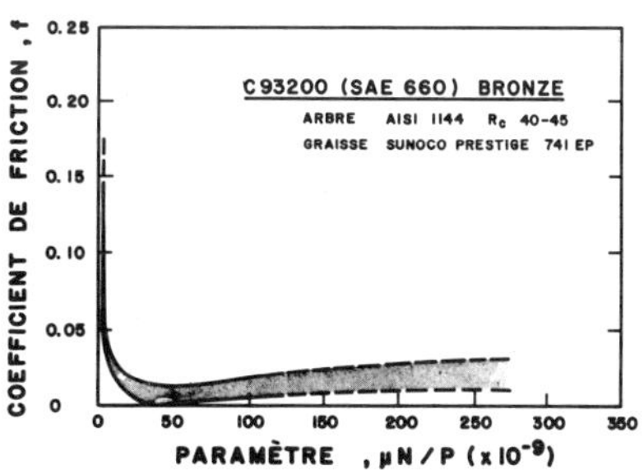

Fig. 6 Courbe de Stribeck

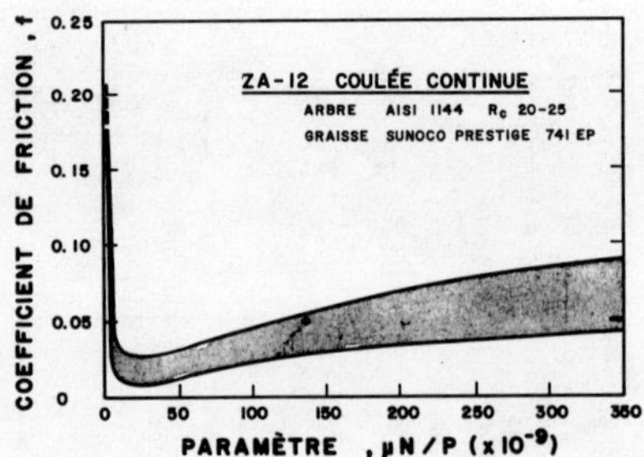

Fig. 7 Courbe de Stribeck

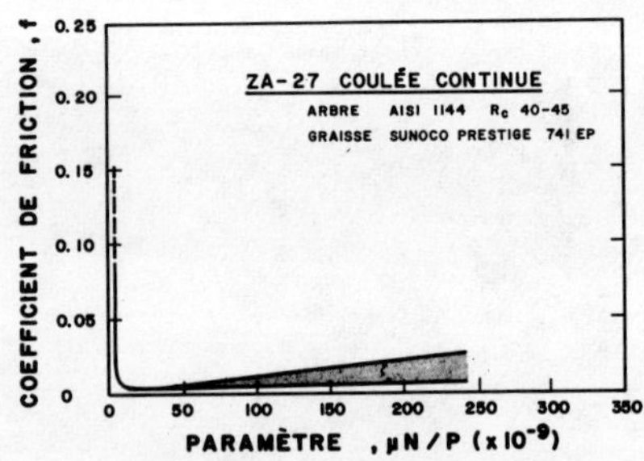

Fig. 8 Courbe de Stribeck

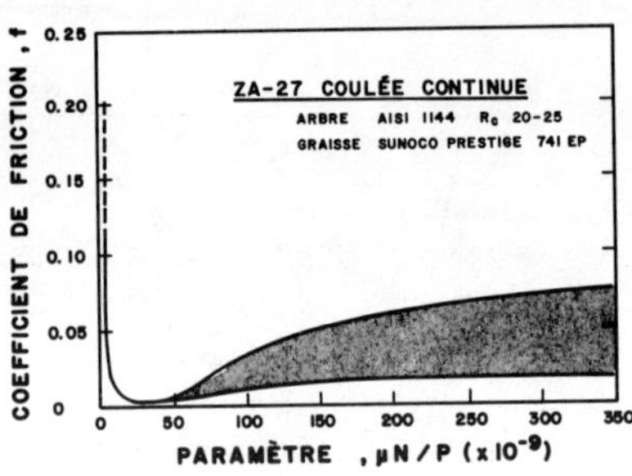

Fig. 9 Courbe de Stribeck

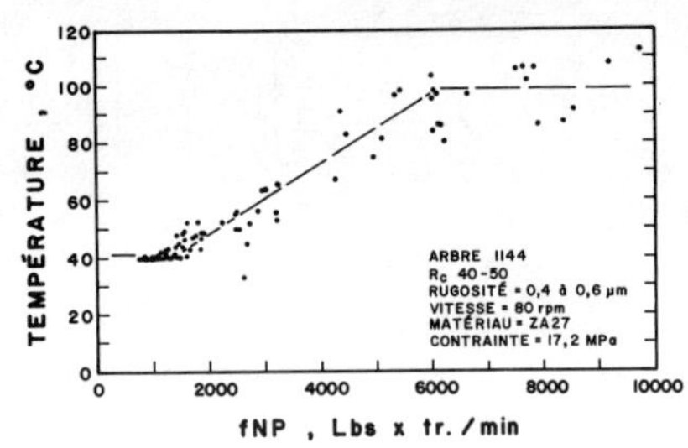

Fig. 10 Courbe d'énergie

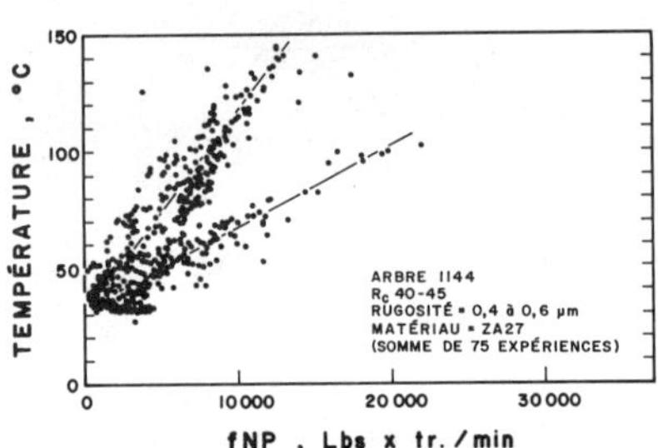

Fig. 11 Courbe d'énergie

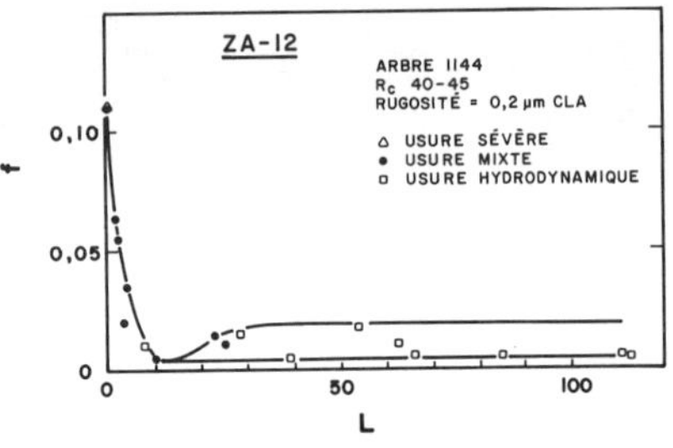

ZA-12

ARBRE 1144

Rc 40-45

RUGOSITÉ = 0,2 μm CLA

USURE SÉVÈRE

USURE MIXTE

USURE HYDRODYNAMIQUE

Z (x10⁻⁹)

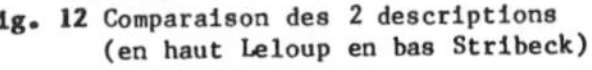

Fig. 12 Comparaison des 2 descriptions (en haut Leloup en bas Stribeck)

X1000 1970 10.0U IGM83

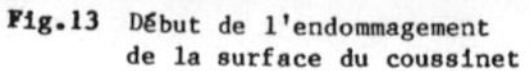

Fig.13 Début de l'endommagement de la surface du coussinet

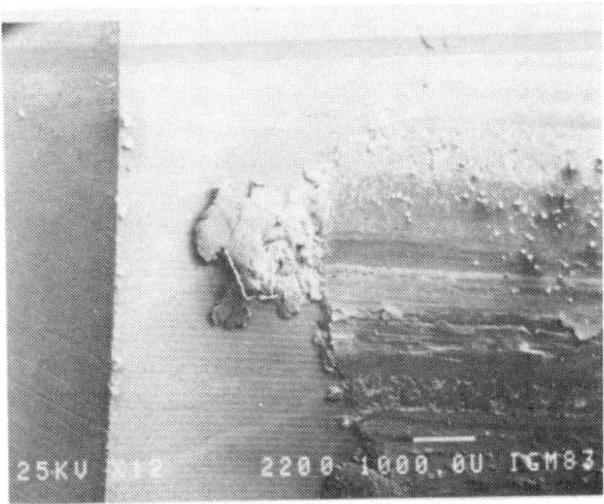

Fig. 14 Transition entre la zone usée (à droite) et la surface initiale du coussinet (à gauche)

Fig. 15 Formation d'un débris

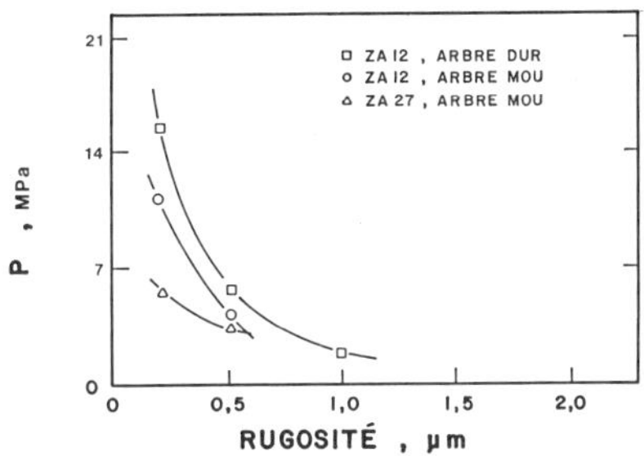

Fig. 16 Variation de la charge de transition entre l'usure hydrodynamique et l'usure moyenne en fonction de la rugosité (à 80 rpm)

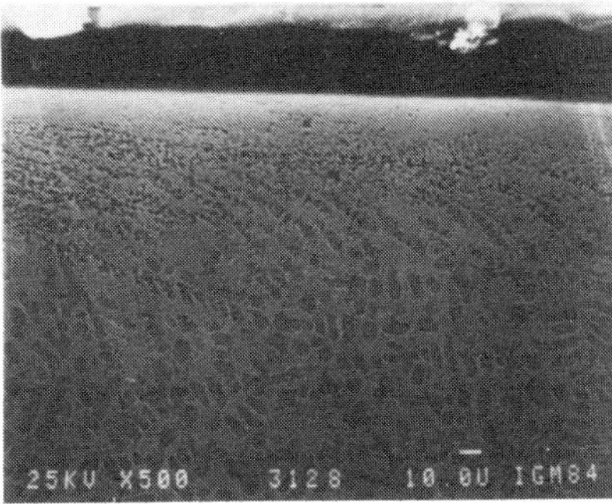

Fig. 17 Vue en section d'un coussinet de ZA27 usé en régime mixte. La surface d'usure est située en haut de la photo

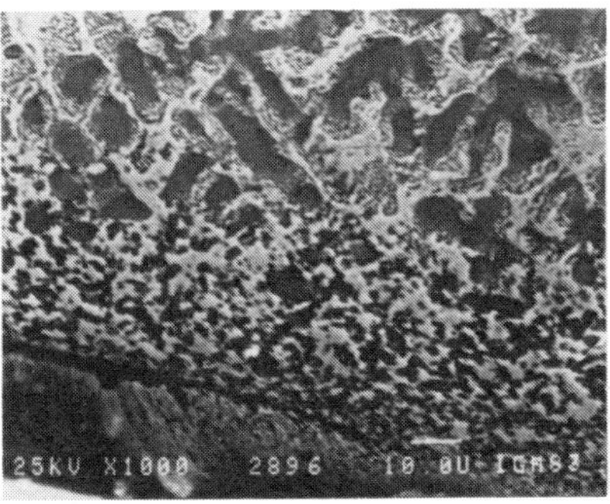

Fig. 18 Vue en section d'un coussinet de ZA27 usé en régime limite. La surface d'usure est située en bas de la photo.

Relation Entre la Resistance a l'Usure et la Structure de Fontes Blanches Riches en Chrome (20%)

J. M. Schissler*,, J. Saverna*, P. Dupin* et T. Mathia*****

**Laboratoire de Métallurgie associé au CNRS, U.A. 159, Ecole des Mines Parc de Saurupt, 54042 Nancy Cedex, France*
***Université de Nancy I, 54506 Vandoeuvre Cedex, France*
****Laboratoire de Technologie des Surfaces, Ecole Centrale de Lyon, 69130 Ecully, France*

RESUME

Cette étude a permis de mettre en évidence l'influence de la microstructure sur la résistance à l'enlèvement de matière simulé par essai sclérométrique. Le réseau des carbures eutectiques formé de carbures M_7C_3 et sa répartition entraînent l'existence d'une hétérogénéité de composition au sein de la matrice. Cette variable se répercute au niveau des caractéristiques morphologiques du sillon provoqué par l'essai sclérométrique. Le traitement thermique constitue également un facteur non négligeable puisqu'à dureté identique mais à phases en présence différentes, la structure n'offre pas une résistance similaire à l'enlèvement de matière. Cette étude montre également que la dureté ne peut être un paramètre unique garant d'une bonne résistance à l'abrasion.

MOTS CLES

Fonte blanche à haut chrome ; traitement thermique ; abrasion ; sclérométrie ; revenu ; ségrégations.

INTRODUCTION

Les fontes blanches à haut chrome constituent une famille d'alliages de plus en plus employés dans l'industrie du broyage. Durant l'emploi de ces matériaux, l'enlèvement de matière est tributaire de la nature de l'abrasif (composition, granulométrie, etc...) ainsi que l'aptitude de l'alliage à résister à l'action de l'abrasif. Le but de cette étude a été de montrer l'influence de l'évolution structurale obtenue par traitement thermique préliminaire sur la morphologie d'une rainure provoquée par un essai sclérométrique. Le choix de cet essai découle du fait que le sillon créé peut être considéré comme l'action élémentaire du processus aboutissant à la perte macroscopique de matière.

TECHNIQUES EXPERIMENTALES

Choix de l'alliage, élaboration
L'alliage à la composition suivante : (en poids) : 20 Cr ; 2,6 C ; 0,8 Si ; 0,8 Mn ; 0,03 P ; 0,02 S. Cette composition situe cette fonte comme étant légérement hypo-eutectique.
Après élaboration dans un four à induction, l'alliage a été coulé sous forme de cube de 100 x 100 x 100 mm dans un moule en sable silico-argileux. Tous les échantillons ont été prélevés au centre de ce cube par électro-érosion, dans un volume de 40 x 40 x 40 mm.

Traitements thermiques
Les traitements thermiques qui ont été appliqués ont été les suivants :

Type A : température d'austénitisation : θ_{γ} = 1150°C ; temps de maintien t_{γ} = 2 heures.

Type B : température d'austénitisation : θ_{γ} = 1000°C ; temps de maintien t_{γ} = 2 heures.

Les refroidissements ultérieurs ont été différents suivant le type :

Type A : refroidissement jusqu'à 20°C par trempe à l'eau

Type B : refroidissement identique à celui du type A suivi d'une immersion dans l'azote liquide pendant 30 minutes puis retour à la température ambiante.

Une partie des échantillons a ensuite subi un revenu à 700°C pendant 1 heure suivi d'un refroidissement à l'air jusqu'à la température ambiante. Les échantillons ainsi traités ont été dénommés A7 et B7.

Techniques d'observation
Quelle que soit la technique employée, la préparation des surfaces ou des divers spécimens a toujours été effectuée sans apport de chaleur ou refroidissement et sans apport extérieur de contraintes.

Pour examen par microscopie optique, la surface des échantillons a été polie puis attaquée avec le réactif de Valenta (1). Les examens par microscopie électronique ont été effectués sur des lames minces. La surface rainurée a été observée par microscopie à balayage. La répartition des éléments au sein de la structure a été mesurée grâce à l'emploi de la microsonde de Castaing. Les essais sclérométriques ont été réalisés grâce à un scléromètre entièrement informatisé (2) dont le microduromètre appliquait une charge nominale de 0,3 N.

RESULTATS ET DISCUSSION

A l'état brut de coulée, on a la structure représentée sur la Fig. 1. On retrouve bien l'aspect classique de ce type de fonte (3, 4, 5).
Le squelette eutectique est bien développé et la matrice inter-carbures eutectiques est plus ou moins transformée. La microscopie optique ne permet pas de discerner quels sont les divers produits qui sont apparus, dans cette matrice, le long de l'interface des carbures eutectiques. Si ces derniers sont très rapprochés, la matrice située entre eux est entièrement transformée. Par contre, si l'écart devient important, la zone centrale de cette matrice n'est pas perturbée. Cette matrice est austénitique et de ce fait, les produits de décomposition peuvent être multiples. C'est pour cette raison qu'un examen par microscopie électronique par transmission (TEM) s'impose (Fig. 2).

La structure eutectique est parfaitement bien définie et on note plusieurs points importants.
Premièrement, le carbure eutectique est du type M_7C_3. Il présente des fautes d'empilement et son interface n'est pas perturbé par une précipitation qui serait survenue pendant le refroidissement dans l'état solide.

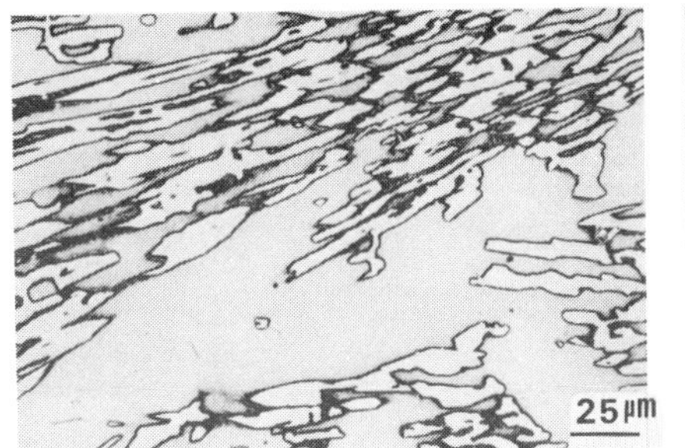

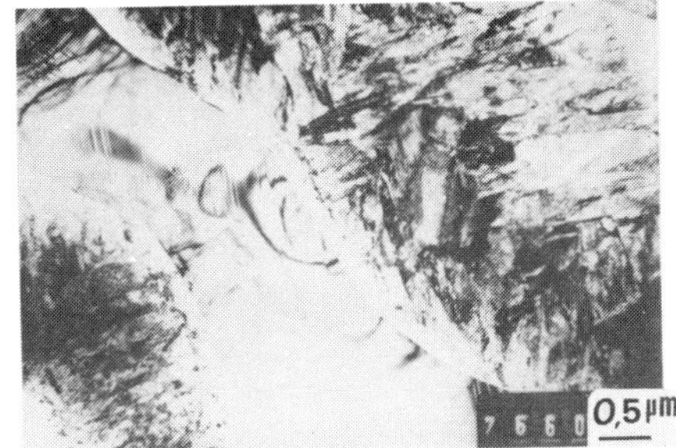

Fig. 1 - Aspect structural de la fonte à l'état brut de coulée.

Fig. 2 - Morphologie de l'eutectique à l'état brut de coulée.

Deuxièmement, la matrice ex-austénitique de l'eutectique est hétérogène. Le long de l'interface du carbure M_7C_3, elle est composée de bainite inférieure et cette dernière est remplacée par de la martensite si on examine la matrice plus éloignée de cette interface.

Afin d'expliquer ces variations structurales on a étudié les profils apparents de la ségrégation des divers éléments répartis dans une dendrite (Fig.3).

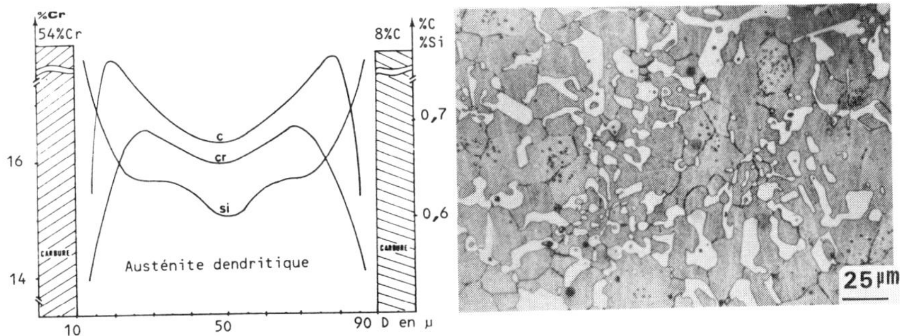

Fig. 3 - Profils apparents de répartition des éléments C, Cr, Si dans une dendrite.

Fig. 4 - Aspect structural de l'alliage A.

La variation du profil de concentration en chrome est assez importante et elle explique l'existence des diverses morphologies observées à l'état brut de coulée. A ce profil sont associés ceux du carbone et du silicium et on comprend aisément que la variation locale de composition entraîne une évolution de la trempabilité donc une évolution structurale. Au sein de l'ex-austénite eutectique, le taux de chrome est moins élevé que celui relevé dans une dendrite. Il en est de même pour le taux de carbone qui est également en équilibre local avec le silicium et c'est pourquoi on n'observe plus d'austénite non transformée dans ces zones mais, au contraire, un agglomérat bainito-martensitique. Toutes ces variations de concentrations en éléments sont

tributaires des divers coefficients de partition propre à chaque élément donc la taille du carbure, liée au choc thermique, joue un rôle. Le phosphore se trouve également rejeté et réparti le long de l'interface M_7C_3.
L'obtention d'une structure à l'état brut de coulée, reproductible, est donc aléatoire, d'où la nécessité d'un traitement thermique. La Fig. 4 illustre la structure de l'alliage après le traitement A. Les carbures eutectiques ont maintenant une forme plus globulaire. Le maintien à 1150°C a provoqué une remise en solution presque totale des carbures précipités en cours de réchauffage. Il reste toujours un amas de globules au centre des ex-dendrites. Ces carbures ont pris naissance au sein de l'austénite durant les premières phases du revenu. L'hétérogénéité de composition en silicium de cette austénite entraîne une cinétique d'évolution et de croissance des carbures très variable. Cela se répercute au niveau de la remise en solution à haute température. Tous les carbures ont été remis en solution dans les zones eutectiques situées entre les carbures M_7C_3. La matrice est austénitique. La dureté de l'alliage après traitement A est de 460 Hv30.
Si on effectue le traitement B, on obtient une matrice martensitique avec des traces d'austénite non transformée. La dureté de l'alliage est de 830 Hv30.
Un revenu à 700°C entraîne une évolution morphologique très nette de la matrice. La durée de 1 heure a été choisie car ce bref temps de maintien permet de bien mettre en relief l'influence des structures de départ sur les cinétiques de revenu.
Dans le cas de l'alliage A7 (Fig. 5), l'état adouci n'a pas été atteint. Cela est lié à la faible valeur du temps de revenu. On note une précipitation assez fine, en faible quantité, autour des amas globulaires pré-existants et le long des carbures M_7C_3. La structure austénitique et ses défauts ont participé à une orientation des précipités. On a apparition d'une structure de trempe secondaire provoquée par une transformation martensitique de l'austénite hypertrempée destabilisée par un revenu de courte durée à 700°C. La dureté atteint la valeur de 600 Hv30.
Dans le cas de l'alliage B7 (Fig. 6), le revenu a provoqué l'apparition de l'état adouci. La dureté a une valeur de 430 Hv30. Le revenu a mis en valeur l'influence des ségrégations préalablement signalées. Dans la matrice ex-martensitique de morphologie dendritique, on distingue très bien trois zones:

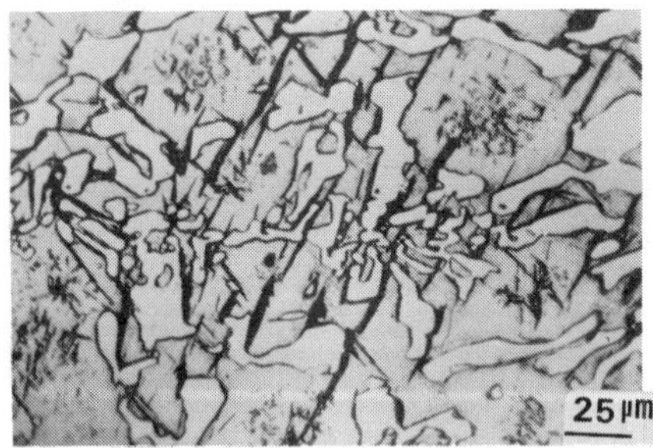

Fig. 5 - Aspect structural de l'alliage A7 après revenu à 700°C (1 heure)

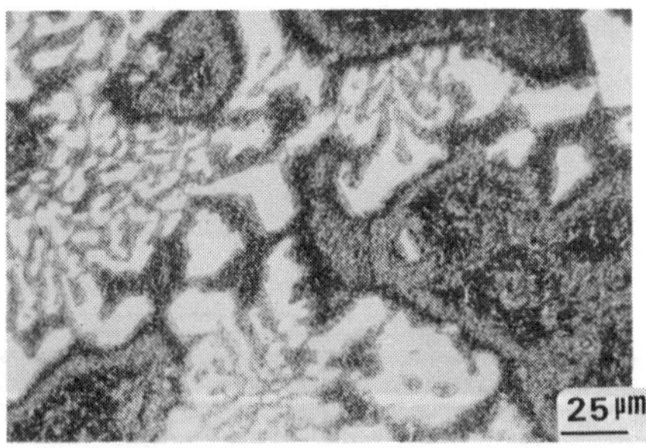

Fig. 6 - Aspect structural de l'alliage B7 après revenu à 700°C (1 heure)

le liseré, le centre et la face intermédiaire. La densité de précipités varie d'une zone à l'autre. La matrice inter-carbures eutectiques possède une

une précipitation très fine.
A ce stade de l'étude, il était donc intéressant d'observer la variation d'aspect de la rainure provoquée par l'essai sclérométrique, en fonction du traitement thermique.
A la suite du traitement A (dureté 460 hv 30), l'opération de rainurage provoque un sillon d'une largeur moyenne de 40 à 42 μ (Fig. 7). Celui-ci est rectiligne et ses bords sont assez déchiquetés. On remarque une microfissuretion plus ou moins régulièrement espacée. L'action conjuguée des variations de composition des diverses austénites et de la contrainte appliquée a provoqué une transformation martensitique dont l'amplitude varie en fonction de la température M_s locale, le taux de martensite formée sera d'autant plus élevé qu'on se rapprochera de l'interface M_7C_3-matrice. Ce phénomène créant une expansion volumique, l'interface sera donc une zone particulièrement sensible si on y associe l'action fragilisante du phosphore qui y est ségrégé. La formation martensitique engendrant un produit dur, cela peut initier une surface de type martensitique. Toutes ces remarques expliquent l'aspect déchiqueté des lèvres de la rainure et les décohésions localisées au niveau des amas eutectiques.
La rainure observée sur l'échantillon B (dureté 830 Hv 30) est moins large puisque la valeur moyenne mesurée entre les lèvres n'est que de 35 μ (Fig.8). Le fond de la rainure est rectiligne mais les bords sont moins déchiquetés. Il y a un développement ultérieur de martensite sous l'effet de la contrainte.

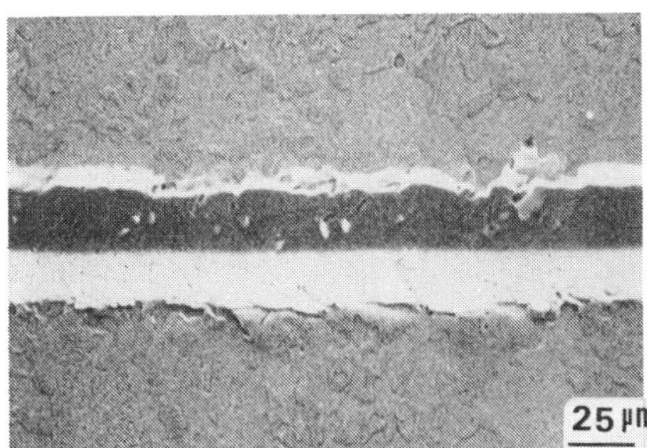

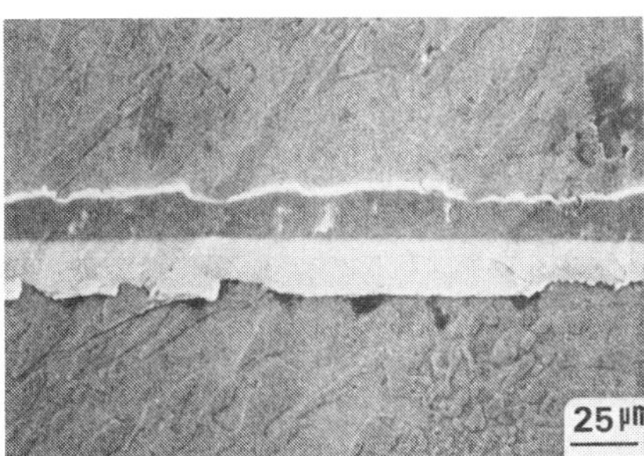

Fig. 7 - Aspect du sillon - Alliage A   Fig. 8 - Aspect du sillon - Alliabe B

On note l'influence de la phase M_7C_3, celle de la zone hautement perturbée de l'interface et celle de la zone centrale de la dendrite. La décohésion au niveau de l'interface explique le phénomène de "micro-spalling".

Le revenu effectué sur l'échantillon A7 affecte la largeur du sillon (Fig. 9). Malgré la dureté élevée (600 Hv 30) la largeur atteint une valeur moyenne de 45 μ, ce qui est proche de celle de l'échantillon hypertrempe A. On ne peut retenir le critère de la dureté comme paramètre sélectif de résistance à l'usure. Un examen des bords du sillon ne permet pas de rattacher leur aspect ni à celui de l'alliage A ni à celui de l'alliage B. Les microfissures associées à des variations de relief indiquent qu'il y a eu transformation martensitique. Le revenu effectué sur l'échantillon B7 (dureté 430 Hv 30) entraîne la formation d'une nouvelle morphologie des faciès de la rainure (Fig. 10), or l'alliage a une dureté à peu près identique à celle de l'alliage A. A dureté semblable, la résistance à l'usure peut être totalement différente, du moins dans le stade de son initiation. La rainure a une valeur moyenne de 60 μ. Par rapport à l'échantillon A, la variation est de l'ordre

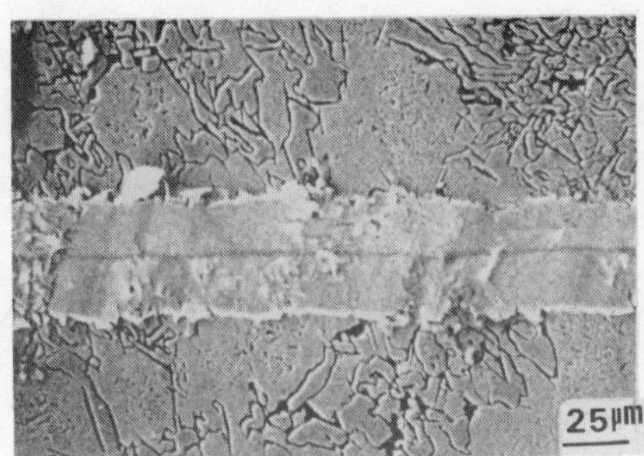

Fig. 9 - Aspect du sillon - Alliage A7

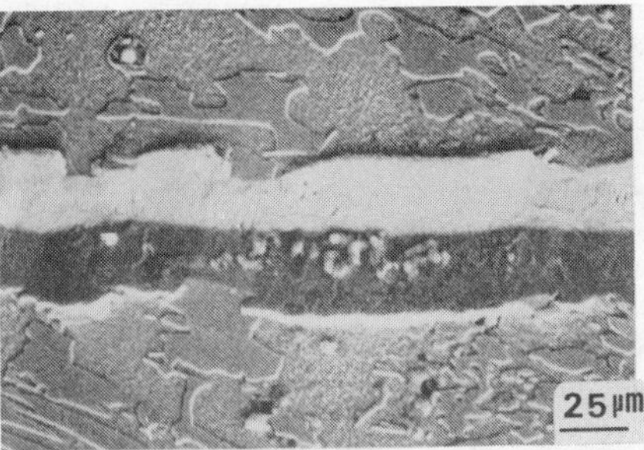

Fig. 10 - Aspect du sillon - Alliage B7

de 50 %. Cela ne peut être imputable qu'au changement de phases. Les bords déchiquetés évoluent en fonction de l'émergence des amas eutectiques. A fort grandissement, on note des successions de microfractographies en "vagues". Cela peut être relié à une transformation martensitique de l'austénite créée par échauffement local lié au passage de l'outil suivi d'un refroidissement brutal. Les amas eutectiques jouent un rôle d'autant plus important qu'ils sont répartis dans une matrice adoucie. L'échantillon B7 met en relief l'influence de l'hétérogénéité de structure puisque les bords de la rainure s'écartent d'autant plus que les carbures de revenu sont grossiers, donc au centre de l'ex-dendrite.

CONCLUSION

Cette étude a permis de mettre en évidence plusieurs points importants :

- la résistance à l'enlèvement de matière ne peut pas être directement relié à la dureté de l'alliage.
- à 20°C, la matrice martensitique est celle qui résiste le mieux. A dureté identique, il est nécessaire de connaître la nature des phases en présence. L'enlèvement de matière est plus faible après hypertrempe de la matrice qu'après adoucissement d'une structure ex-martensitique.
- la régularité de l'opération d'enlèvement de matière doit prendre en compte la répartition des carbures eutectiques et leur morphologie. La nature et la répartition des phases en présence dans la matrice sont des facteurs importants car en plus de l'hétérogénéité de phase intervient l'hétérogénéité chimique initiale, hétérogénéité qui module le taux local de phases.

BIBLIOGRAPHIE

1. C. Valenta, Iron and Steel Inst. 19, 179 (1930)
2. T. Mathia, Abrasion of Brittle Solids (à paraître)
3. K. Ogi, Y. Matsubara, K. Matsuda, AFS Trans <u>89</u>, 197 (1981)
4. P. Dupin, J. Saverna, J.M. Schissler, AFS Trans <u>90</u>, 711 (1982)
5. Jiantong-King, Wenhua-Lu, Kiatong-Wang, Wear of Mat Casme, 45 (1983).

Abrasive Wear of Cast Ni-Cr-C Alloys

E. Blank and E. Luchsinger

Département des matériaux, Ecole Polytechnique Fédérale de Lausanne, Switzerland

ABSTRACT

The effects of M_7C_3 carbides on abrasive wear of cast Ni-Cr-C alloys were investigated. The carbide volume fraction was varied from 0 to 30 per cent. In the case of coarse SiC abrasive grains, the wear resistance of carbide containing alloys was always inferior to the pure solid solution matrix. Carbides had a beneficial effect in the case of fine grained abrasive papers. Carbide spacings variations exerted only a small effect on wear resistance. Under all testing conditions, the wear behaviour of dendritic alloys followed a rule of mixtures between solid solution and eutectic. Wear mechanisms were discussed in terms of stress criteria and energy consumption.

KEYWORDS

Abrasive wear, cast two-phase alloys, carbides

INTRODUCTION

Cast alloys with dendritic microstructures very frequently exhibit good wear properties due to the presence of large amounts of hard and brittle interdendritic eutectic. Industrially important examples are cast non-ferrous metals (Al-Si, tin bronzes), high chromium cast irons, cast steels and high temperature alloys (heat resisting steels HC, HH, etc.). Contradictory informations about microstructural effects on abrasive wear properties are found in the literature. On the one hand, the wear rates of heterogeneous materials are reported to behave simply like the weighted sums of wear rates of the constituents.(1) On the other hand, microstructural parameters like interface strengths or size and morphology of second phase particles are invoked to account for deviations from the simply additive behaviour(2).

It is the purpose of this paper to improve the understanding of the effects of interdendritic carbides on abrasive wear of cast Ni-Cr-C alloys. The morphology of the interdendritic M_7C_3 carbides in this system is the same as in chromium cast irons but the matrix consisting of a very ductile (Ni,Cr,C) solid solution, is entirely different. Compared to alloy systems with a wear resistant matrix, this alloy should emphasize the effects of carbides precipitated from the melt. M_7C_3 carbides have a fibrous morphology which might exhibit an orientation effect. In order to obtain microstructures as uniform as possible, all test specimens were prepared by directional solidification. The wear tests were performed in a plane perpendicular to the growth axis in solidification. Thus the carbide orientation was the same in all tests.

MICROSTRUCTURE AND WEAR BEHAVIOUR

Ni-Cr-C alloys with compositions reaching from a pure solid solution to the hypereutectic alloy with 30 vol. % M_7C_3 particles were directionally solidified with two different cooling rates, 0.3°C/sec and 3°C/sec. Transverse sections of a dendritic alloy and a eutectic alloy, both grown at low cooling rate, are shown in Figs 1, 2.

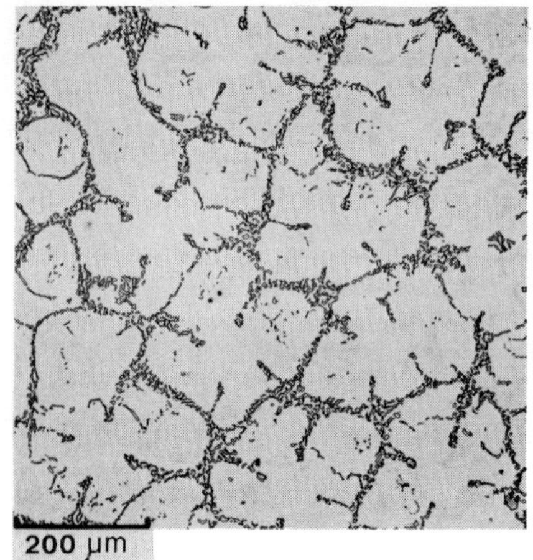

Fig. 1. Transverse section of a dendritic Ni-Cr-C alloy solidified at 0.3°C/sec.

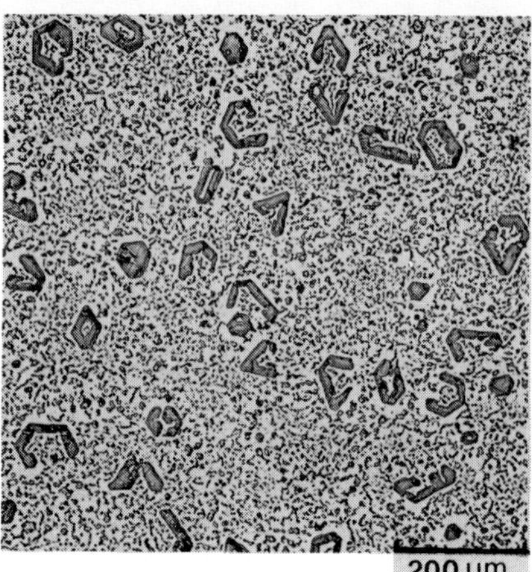

Fig. 2. Transverse section of a hypereutectic Ni-Cr-C alloy solidified at 0.3°C/sec.

Longitudinal sections of the same alloys but grown at 3°C/sec are given by Figs. 3, 4. From both figures, the fibrous M_7C_3 structure becomes obvious. Spacings of the fine fibres are very similar in both alloy types. However the eutectic material exhibits primary carbides whereas the dendritic one contains fibres perpendicular to the plane of the figure, separating the dendrite arms.

Wear tests were performed on a belt grinder using SiC grinding papers with 150, 240, 400 and 600 grit sizes. Weight losses and friction forces were measured as functions of the applied loads (5, 10 and 20 N, specimen diameter 5.8 mm). Part of the results are presented in Fig. 5. The most remarkable feature is the influence of load and grit size on wear. Only for grit sizes 400 and 600, the increase of the carbide volume fraction has a beneficial effect on wear properties. At high loads and small grit sizes, the wear resistance of the eutectic alloys is inferior to the value of the matrix.

In dendritic alloys solidified at the higher cooling rate, wear rates were generally somewhat smaller (< 10 %). There was no such general tendancy in eutectic alloys.

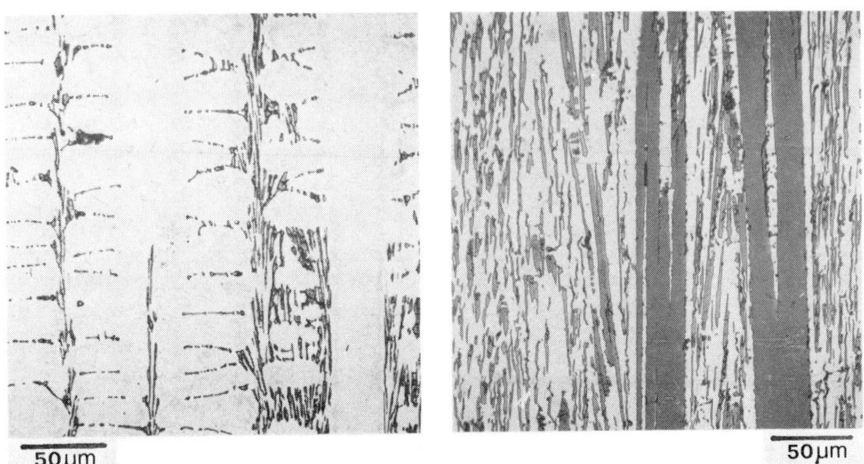

Fig. 3. Longitudinal section of dendritic Ni-Cr-C alloy solidified at $3^{o}C/sec$.

Fig. 4. Longitudinal section of a hypereutectic Ni-Cr-C alloy solidified at $3^{o}C/sec$.

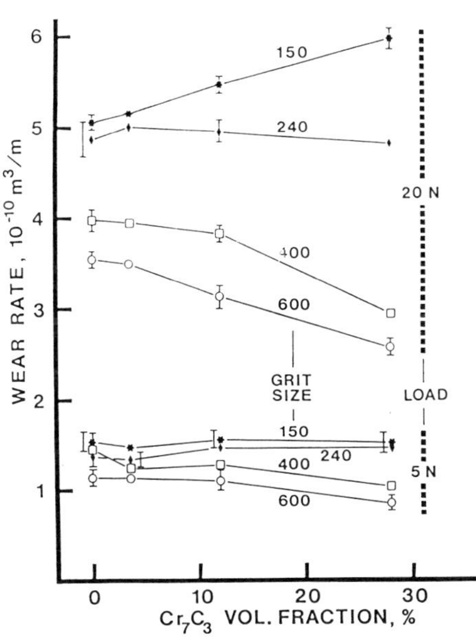

Fig. 5. Wear rate as a function of carbide volume fraction of alloys solidified at $0,3^{o}C/sec$.

Microscopic observations revealed that the matrix was worn by chip formation under all testing conditions. Carbide removal occured by a brittle fracture mechanism. In wear with fine-grained abrasives the carbides stood out from the wear surface whereas in the case of coarse-grained abrasives, heavy microcracking of the carbides in the subsurface range was observed.

DISCUSSION

The tests with fine grained papers have demonstrated that carbides precipitated from the melt can have a beneficial effect on the abrasive wear behaviour. This is clearly demonstrated by Fig. 6 where the experimental wear rates have been extrapolated towards 100 per cent carbides. According to this graph representing a rule of mixtures, the wear resistance of the carbides must be very high compared to the matrix. This is not obvious because the SiC abrasive grains are harder than the M_7C_3 carbides. Moreover, carbide distribution and morphology should influence the wear behaviour only to a minor degree since Fig. 6 expresses a simple mass effect.

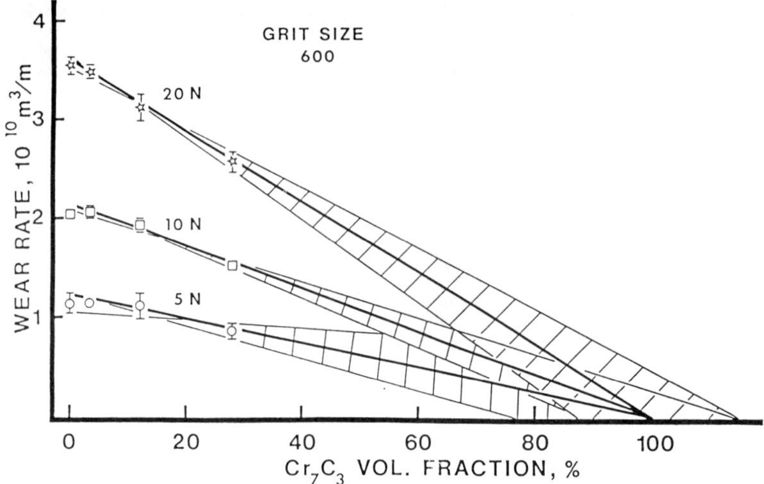

Fig. 6. Extrapolation of measured wear rates to 100 % carbides.

The basic difference between the two observed wear types becomes obvious if the removed volume per unit energy is plotted versus the carbide volume fraction where the grit size and applied loads appear as parameters (Fig. 7). For coarse abrasive grains, this specific volume increases linearily with the carbide volume fraction. It does not depend on the applied load. The opposite is true for fine grained papers. One explanation for the behaviour at small grit size (150) would be to assume that the carbides in the subsurface layer fail according to a stress criterion with very little energy consumption, The increase of the specific volume can then be considered as a measure for weakening of the microstructure: As the eutectic, carbide rich areas cannot support any more the load on the specimen, the abrasive grains penetrate more profoundly into the matrix. Thus the removed volume increases.

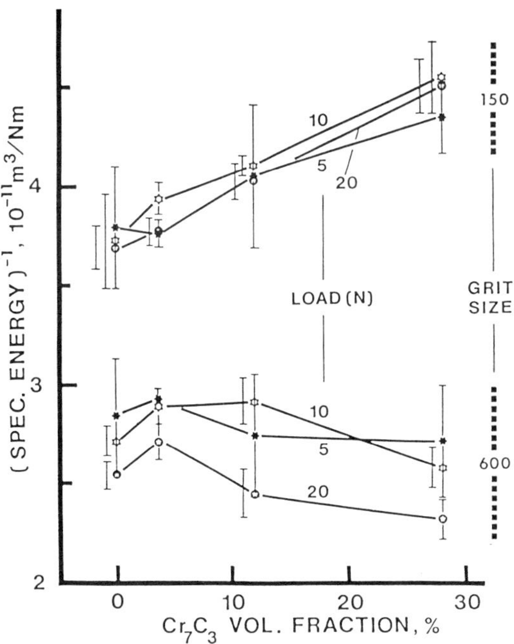

Fig. 7. Specific wear rate as a function of the carbide volume fraction.

At high grit sizes (600), there is no carbide cracking beneath the surface. On the contrary, the eutectic areas must share unproportionally big parts of the specimen load since these areas stand out from the wear surface. In order to explain the load dependance of the specific wear rate one might think of two different contributions to the energy consumption. First, there may simply exist an increased friction force due to complete filling of the free space between the abrasive grains at high load. Such a contribution would be expected to be most prominent in the solid solution matrix. It should decrease with increasing carbide volume fraction because carbides don't undergo plastic deformation. Fig. 7 however indicates a increase in friction energy rather than a decrease. Thus an additional energy consuming process must exist. The most likely mechanism seems to be energy consumption by "composite-type" work-hardening of eutectic areas (4). Thereby plastic energy would be dissipated and elastic energy would be stored. But this would also mean that wear depends on the microstructure because the elastically stored energy is a function of volume fraction and morphology of the brittle phase.

Carbide failure according to a stress criterion as mentioned above, has been related to the M_7C_3 Vickers hardness. It can be shown from the wear diagrams that as soon as the load per abrasive grain transfered to the alloy becomes larger than the product of carbide hardness times carbide section, carbides have a detrimental effect on wear resistance. This behaviour is encountered in hypereutectic materials as well as in interdendritic eutectic areas. Also in this case of a breakdown of the eutectic microstructure, dendritic alloys very strictly follow a linear rule of mixtures between the wear rates of the matrix and eutectic areas.

CONCLUSIONS

The investigations have shown that Khrushovs' rule of mixtures is appropriate for the description of wear rates of dendritic-type heterogeneous materials. In the case of catastrophic microstructure breakdown, however, the interdendritic eutectic as a whole must be considered as one of the constituent phases, rather than the carbides alone. The Vickers hardness of carbides is a critical parameter determining the critical load before microstructure breakdown occurs. Variation of carbide spacings and carbide size within the limits of cooling rates easily available in foundry practice, do not influence the wear rate to a considerable extent. Microstructure breakdown is controlled by a stress criterion. Below the critical stress, wear resistance seems to depend on the volume fraction of carbides and on the work-hardening capability of the eutectic regions.

ACKNOWLEDGMENT

This research is part of the Swiss National Project on Materials and Materials Resources which is supported by the Swiss National Foundation for Scientific Research.

REFERENCES

1. M.M. Khrushov, *Wear 28*, 69-88 (1974).
2. N. Saka, *in Fundamentals of Tribology*, eds. N.P. Suh & N. Saka, p. 135, MIT Press, (1980).
3. J.K. Fulcher, T.H. Kossel and N.F. Fiore, *Wear 84*, 313 (1983).
4. K. Tanaka and T. Mori, *Acta metall.* 18, 931 (1970)

Wear and Fatigue Properties of Laser Melted Cast Iron

H. W. Bergmann, W. Henning and B. L. Mordike

Institut für Werkstoffkunde und Werkstofftechnik, Technische Universität Clausthal, Federal Republic of Germany

ABSTRACT

This paper gives some examples of laser melted cast irons including microstructures and resultant properties. Possible applications of laser melted components are also discussed.

KEYWORDS

Laser melting, surface engineering, wear and fatigue behaviour.

INTRODUCTION

Various techniques have been used to modify the surface structure of grey cast iron in order to achieve a white, ledeburitic microstructure whilst retaining the grey core. The most prominent techniques are chilled surface casting and Tungsten Inert Gas Welding (TIG). Today, both methods are widely accepted in industry, for instance in camshaft production. The advantage of such treatments is the combination of the core ductility, due to stable Fe-C solidification resulting in an iron-graphite microstructure, and the wear resistance of the surface due to metastable solidification giving a $Fe-Fe_3C$ layer. Recently, the availability of high-powered lasers has enabled high speed surface melting and large scale automation. This paper describes the laser melting process, the microstructures obtained and a variety of resultant properties.

EXPERIMENTAL

SG iron with ferritic, pearlitic and bainitic matrix was surface melted using a 500 W CO_2 laser focussed beam with TEM 00 mode. CO_2 or He was used to increase the amount of incident energy absorbed by forming a plasma. The melted layer was between 0.3 and 0.5 mm thick depending on the feed rate. The surface was perfectly uniform, crack free and untarnished. The surface roughness was less then $\pm$ 8 μm in the lasered condition. The laser treatment caused negligible distortion. A typical picture is given in Fig.1.

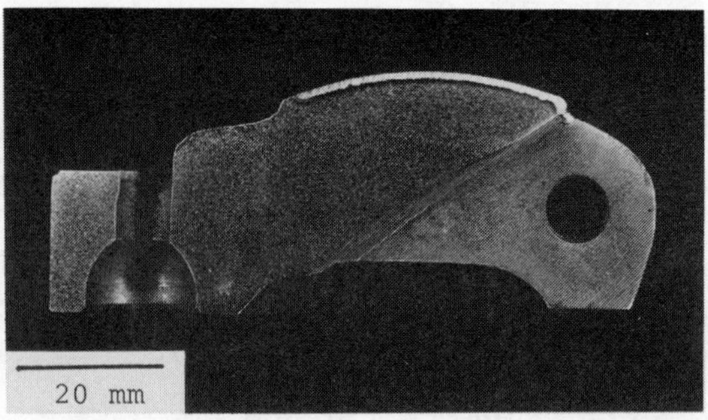

Fig. 1. Laser melted SG iron rocker arm

RESULTS AND DISCUSSION

Laser melting of cast irons may lead to surface hardness improvements by directional rapid solidification. Rapid solidification results in grain refinement, metastable phase formation, e.g., conversion of graphite to Fe_3C, and supersaturation which offers the possibilities of further transformation and/or precipitation hardening. A typical microstructure is given in Fig. 2 showing the ledeburitic case and the ferritic SG iron core. There are further improvements possible by laser alloying which are discussed elsewhere (1).

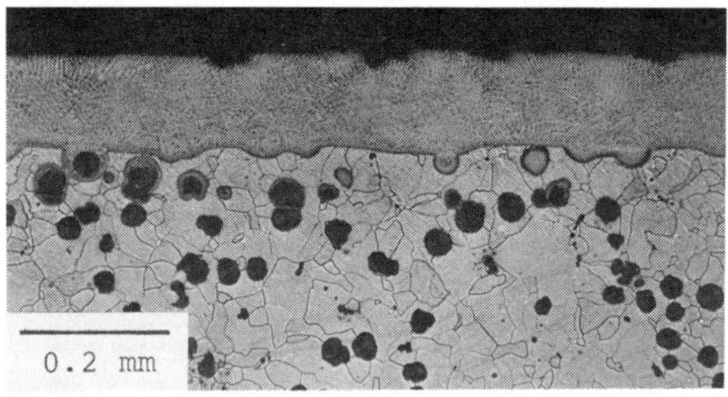

Fig. 2. Microstructure of laser melted SG iron

The mechanical properties of laser melted S.G. irons depend on both the area ratio of melt depth to substrate and HAZ to substrate. For S.G. iron the HAZ is small when compared to the melt depth and may be neglected in a first approximation. Fig. 3 shows a typical stress- strain curve for a laser melted test sample.

The stress-strain curves for untreated and laser melted test samples are similar, however, above the yield point, the case cracks circumferentially giving a feature on the curve not dissimilar to discontinuous yielding and this is accompanied by an acoustic emission. At least 1 % elongation occurs before the case cracks. Fractographs show a ductile fracture of the core and the more brittle fracture of the outer layer. Cracks propagate through the case mainly in the interdendritic eutectic

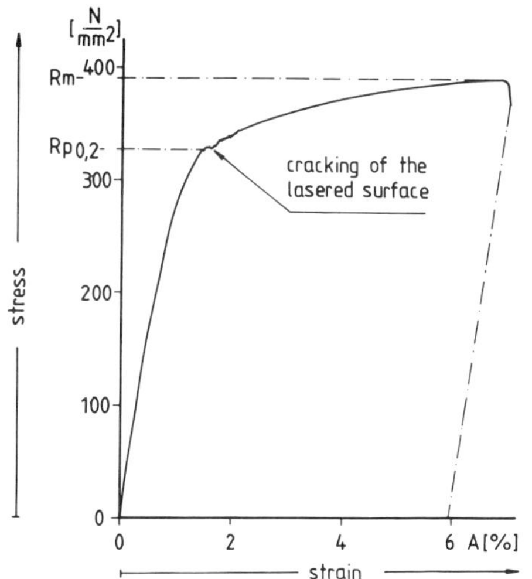

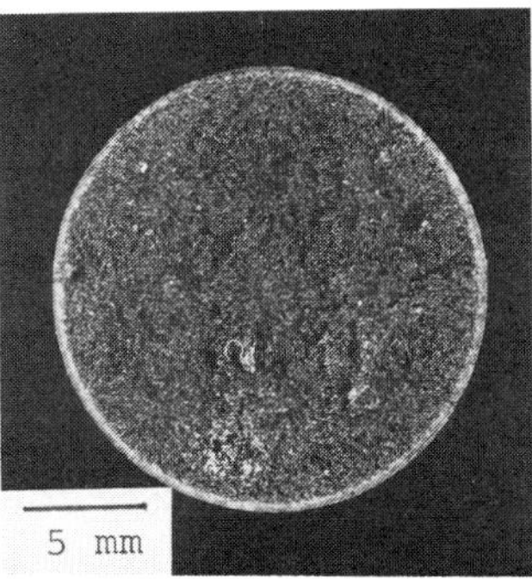

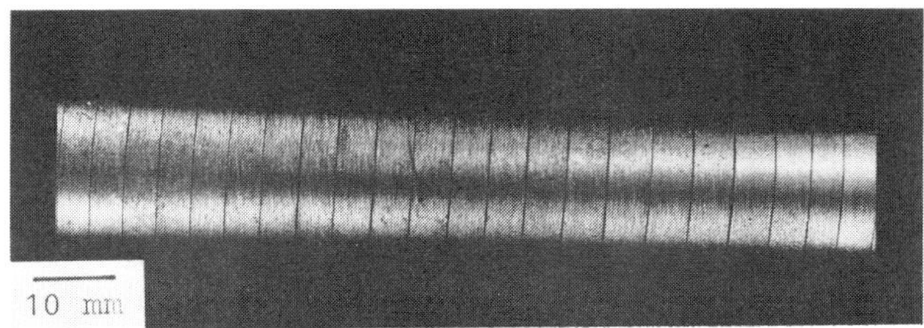

Fig. 3. Stress-strain curve of laser melted SG iron and fractured tensile test specimens

phase, where microporosity may also be present. A unique feature of this material after tensile testing is the occurrence of equidistant circumferential cracks along the gauge length. These cracks do not penetrate further than the case core interface because of workhardening of the ductile core. This effect is not influenced by metallurgical inhomogeneities between laser tracks as there are none and it occurs with longitudinal tracks as well as for spiral melting. This behaviour is observed for ferritic, pearlitic and bainitic core structures which have a higher toughness than the case. When a martensitic matrix is used, the initial case crack continues through the brittle core.
Laser surface melting results in internal stresses on solidification. Several factors contribute to the final stress situations, for example, the volume difference on solidification, differences in shrinkage in the solid state for core and surface due to both temperature differences and different coefficients of thermal expansion. However, the various transformations in both case and core have an effect. For cast iron with pearlitic matrix and flaky graphite (2) it was found that in the surface,

compressive stresses were present which change to tensile stresses at some distance below the surface. Depending on the case depth, the sign of the stress changes in either the core or the surface layer.

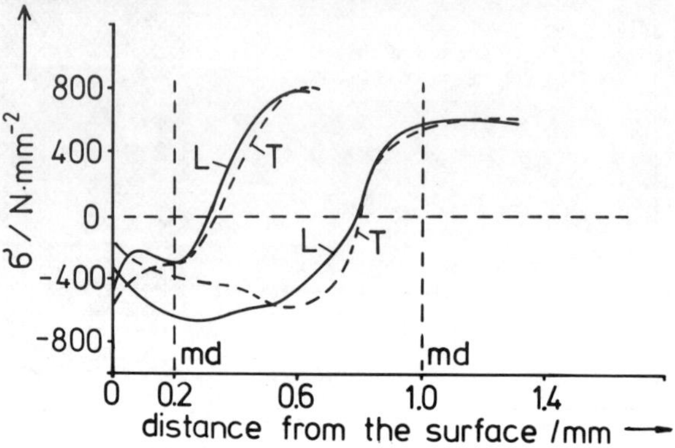

σ / N·mm⁻²
800
400
0
-400
-800
md
md
L
T
L
T
0 0.2 0.6 1.0 1.4
distance from the surface /mm

Fig. 4. Internal stresses in laser melted SG iron (md = case depth, L = stresses in longitudinal direction, T = stresses in transversal direction)

The fatigue properties of ferritic S.G. iron are given in Fig. 5. Untreated specimens were ground after machining. As no relevant data are available in the literature, optimised processing parameters for laser treatment had to be defined. This was done by keeping the case depth constant at about 10 % of the cross-section. Compared to the untreated S.G. iron, a decrease in fatigue limit was found in the as lasered condition. This was more pronounced when N_2, CO_2 or Ar was used as the protective gas, but less than with the He.

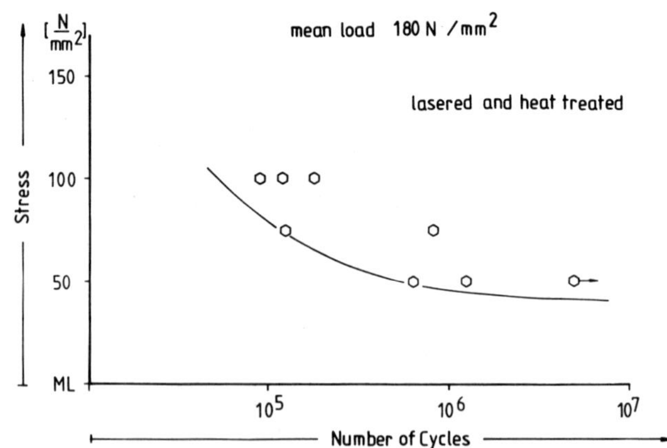

Fig. 5. Fatigue Properties of laser melted SG iron
a) 15 mm diameter, 0.3 mm case depth

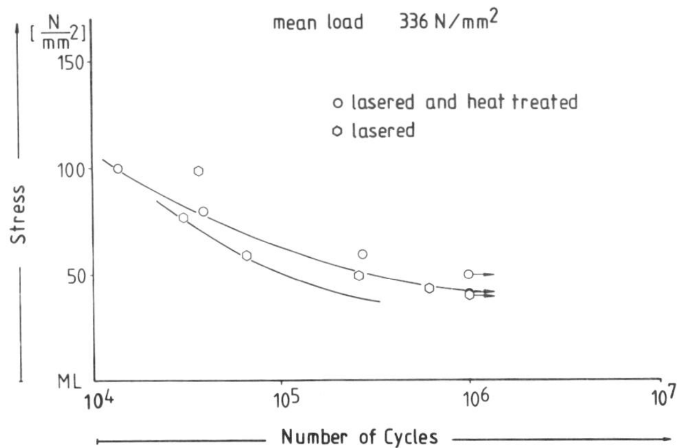

b) 8 mm diameter, 1 mm case depth

In addition, a spiral laser track gives better results than a longitudinal series of tracks. Grinding the surface, i.e., smoothing small amounts of roughness produced by the laser treatment leads to a small decrease in fatigue limit as compared with untreated value. The most favourable values were found when the specimens were annealed at 240 °C for 2 hrs after laser treatment with helium and subsequently ground.

The excellent wear properties of ledeburitic surface layers on cast iron has been demonstrated over recent years. Most of the work has been carried out on TIG melted surfaces, while for the new technique of laser melting, only a limited amount of data is available (3,4). For lasered grey iron with pearlitic matrix and flaky graphite, Bell and co-workers demonstrated that the wear properties in dry pin on disc tests are increased with laser melting by one order of magnitude compared to the untreated situation. There was still a significant increase compared to fully martensitic hardened microstructures.

The advantage of laser melted S.G. iron was demonstrated for rolls which run dry against each other with a definite relative slip, one wheel being driven and the other partially braked (Fig. 6). Two loads were applied (500 and 1000 Nmm^{-2}) and the humidity was controlled. For a slip of approximately 3 - 5 % and the higher load, melting occurred after a couple of minutes when steel rolls were used, but the cast irons showed no evidence of any deformation. For comparison, 1 % slip was used. It is obvious that the wear properties of lasered cast iron are superior to those of hardened, conventional steels. This improvement is less pronounced if the steels have been hard faced or nitrided. Combinations of lasered irons and steels must be avoided as massive deformation of the steel occurs. When laser treatments were carried out under He, better results were found than with other gases. Excellent results were obtained when laser melted S.G. irons were used in combination with TiN or TiC coated hardened steels, the TiN/TiC surface suffering the wear. After the tests, no deformation is visible on the ledeburitic surface.

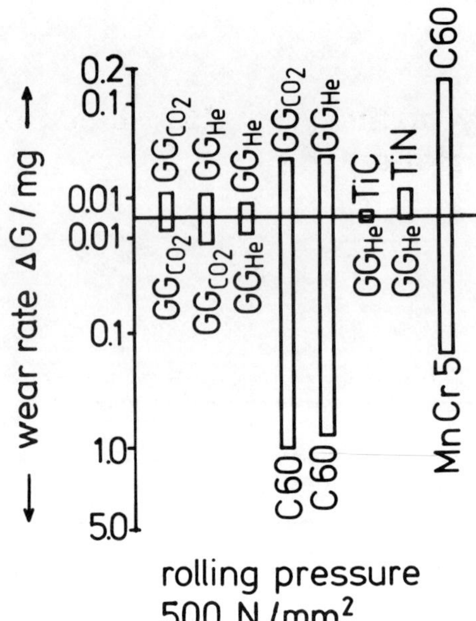

Fig. 6. Wear properties of laser melted SG iron

GENERAL COMMENTS AND POSSIBLE APPLICATIONS

Apart from the fact that laser melting enhances certain desirable properties mentioned above, there are additional beneficial features obtained with this technique. The melting process can be carried out on an almost finished component as the distortion associated with the process is negligible and surface roughness is in the order of 10 µm, which can easily be improved by grinding if necessary. There is no need for a dedicated handling system as a uniform and homogeneous case can be achieved if lasered within the focal distance, i.e., ± 1.5 cm for a 25.4 cm focal length lens. Laser melting is a hardening process which can be fully automated. It is also a fast and cheap process.

ACKNOWLEDGEMENTS

This work was supported by the VW-Stiftung.

REFERENCES

1. H.W. Bergmann and B.L. Mordike, Z. Werkstofftechnik 14, 228, (1983).
2. H.W. Bergmann, B.L. Mordike and T. Bell, Laser 83, München, Springer Verlag Berlin, (1985).
3. D.N.H. Trafford, T. Bell, J.H.P.C. Megaw and A.S. Bransden, Proc. Conf. Heat Treatment, (1981), The Metals Society, London.
4. H.W. Bergmann and W. Witzel, Textures in Rapidly Solidified Metals, Proc. ICOTOM 7, Holland, (1984).

Study of the Wear Resistance of Hot-dip Borized Steels

A. Salas, H. Balmori*, C. Godinez* and F. Moreno***

**Graduate School and Department of Metallurgical Engineering, ESIQIE-Instituto Politécnico Nacional, Apdo. Postal 75–874, México D.F. 07300*
***Central Laboratory of National Railways of Mexico*

ABSTRACT

The abrasion resistance of 1045, 4140 and 8620 steels borided by inmersion in a fused borax bath has been determined and compared with that of the same steels in the quenched and normalized conditions. The results show that the resistance to abrasive wear is superior for borided steels.

KEYWORDS

Wear; abrasion resistance; borizing; quenched; normalized; hot dip; steel.

INTRODUCTION

Abrasive wear is one of the most common problems found in industry. It is deleterious because it increases the frequency of maintenance and reposition of components (1). Thus, it is necessary to develop new wear resistant materials or processes that produce increasing resistance to abrasive wear.

Abrasive wear might be defined as the damage on a surface caused by a particle of higher hardness (2). These particles can be introduced between two surfaces from external sources or be formed between them by a chemical process. So, it is very difficult to design an experiment which reflects the actual conditions in which a material will work and that, at the same time, gives information suitable for a rigorous analisis (3,4). This is the reason why many tests to evaluate abrasion resistance only relate the weight loss suffered by a material which is loaded against a rotating disk to the total 'length' of swept surface.

Some of our previous work (5,6) has dealt mainly with the development of steels of very high surface hardness (ranging from 1300 to 1500 HVN) by difusion of boron into the surface of the steel to form an outer layer of FeB followed by a Fe_2B layer in contact with the substrate. The borizing method consists in the inmersion of the steel in a fused salt bath containing borax as the boron providing agent. This method is economical and of easy application because it does not require special facilities.

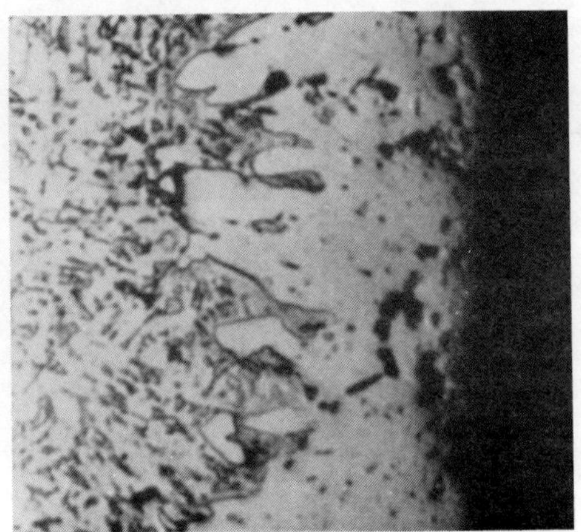

Fig. 1. Original microstructure of steel 8620 borided. 950 X

EXPERIMENTAL METHOD

Square bars of AISI 1045, 4140 and 8620 steels (12.7 x 12.7 x 40mm) were borided by inmersion in a fused salt bath composed of borax, NH_4Cl and NaCl (as activating salts) and Al powder (as reducing agent). The treatment was carried out at 1000°C for 5 hr. Prior to inmersion in the salt bath, specimens were ground down to 600 grit and degreased with acetone. Another group of specimens was kept for 3 hr at 900°C in a liquid salt bath and quenched in oil with severe agitation. Normalizing was carried out in a controlled atmosphere furnace at 900°C for 3 hr followed by cooling in still air.

Abrasion tests were performed by pressing the specimen, with a force of approximately 20 kg, against a rotating disk (169 mm diameter x 12.7 mm wide) at 90 RPM. The disk dragged a suspension of alumina of 40# Tyler mesh size in water (4 g:1ml). Every 1000 revolutions and up to 10,000 revolutions (approximately a swept distance of 5.3 km) the specimen weight was determined to the nearest 0.0001 g. The suspension was renewed every 3000 revolutions. More details on the specifications and apparatus employed might be foun in ASTM B 611 standard.

Wear resistance, A, in cm^3/rev, was calculated as

$$A = \frac{L}{D\ N} \times 10^5 \qquad (1)$$

where L is the weight loss (g), D is the density of the specimen (g/cm^3) and N is the number of revolutions of the disk. Densities were determined by the method of weight differences of the specimen in air and in distilled water.

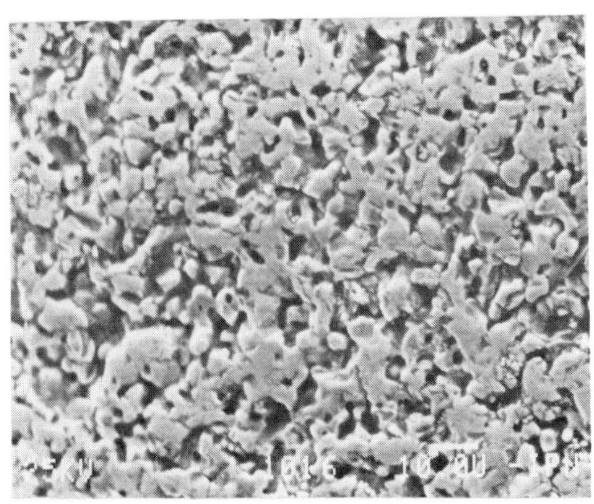

Fig. 2. Original surface of steel 8620 borided.

RESULTS AND DISCUSSION

Heat Treatments

The microstructures found in the as-treated specimens were as follws: normalized specimens were composed of a ferrite-pearlite mixture consistent with the carbon content of the steels; a fully martensitic structure was found in the quenched steels, and in the borided specimens there was a boride layer (Fig. 1) characterized (6) as FeB at the surface followed by a softer Fe_2B between the former and the substrate. On the original surfaces of the normalized and quenched specimens there were only some scratches and pits; on the borided steels there were many protuberances (Fig. 2). Surface hardnesses are given in Table 1.

TABLE 1 Surface Hardnesses* of the Steels

Steel	Normalized	Quenched	Borided
1045	72.5 (276)	83 (443)	86 (550)
4140	72 (270)	80 (394)	88 (626)
8620	68 (230)	82 (420)	85 (513)

*Hardnes numbers are in Super Rockwell 15N and (HVN).

Weight Losses

Weight loss as a function of the number of revolutions is shown in Fig. 3. It is clear that weight loss for the borided specimens is less than in quenched and normalized conditions. Quenched specimens lose weight somewhat slightly slower than the normalized ones. In all cases, weight loss is higher for 8620 steel, with small differences between 4140 and 1045 steels. In general, the higher the surface hardness, the slower the weight loss

In Fig. 3, some tests do not finish at 10,000 revolutions, although all of

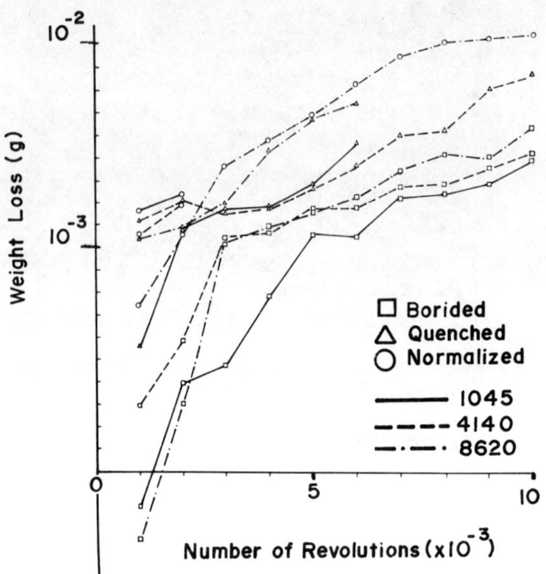

Fig. 3. Weight losses as a function of the number of revolutions of the disk.

them were given that number of revolutions. The reason for this is that after the last datum point reported for each specimen in Fig. 3, specimens begun gaining weight due to the building up of material at the specimen surface, as shown in Fig. 4. This obviously changed the original wear conditions. This happening might be important in some cases since upon gaining weight the specimens did not wear off for 3000 to 6000 revolutions, as in the case of normalized specimens 1045 and 4140; under these conditions it was not considered adequate to evaluate abrasion resistance since the original wear conditions were completely changed.

Wear resistance A is given in Fig. 5 (the lower the value of A, the higher the abrasion resistance). As a reflection of the behaviour shown in Fig. 3, the borizing treatments provided a higher abrasion resistance which was apparently reduced after 5000 revolutions to a level still higher than that found for the other heat-treating conditions. The resistance of the non-borided specimens kept constant or with only small variations. Up to now, the reasons for this behaviour are unclear, but as a hypothesis it might be thought that since the borided surface layer is composed mainly of FeB, which is hard, the initial abrasion resistance is high, but after some wear Fe_2B appears and the abrasion resistance diminishes because of the lower hardness of the last component.

The analysis of the abraded surfaces revealed the existence of scratches and pits in all cases (Fig. 6). In some instances, alumina particles were found in the specimen surface. This observation points out the occurrence of cutting abrasion processes (7). In this case, surface hardness is an important

Fig. 4. Abraded surface of steel 1045 after 8000 revolutions.

parameter since it represents the resistance of the surface against the penetrating particle; therefore, the higher the hardness the higher the abrasion resistance will be, as it is with the borided surfaces.

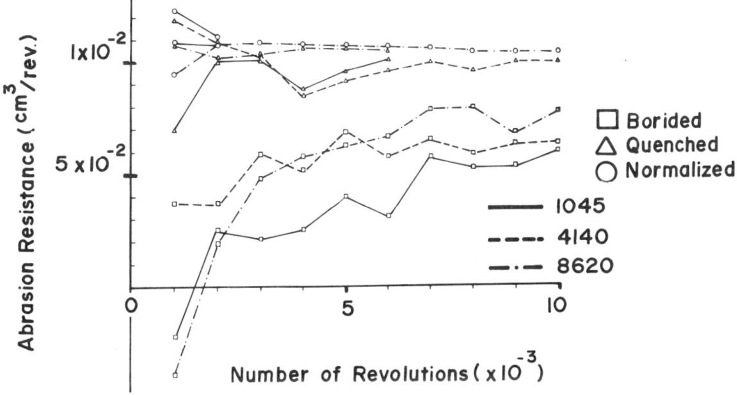

Fig. 5. Abrasion resistance, A, as a function of the number of revolutions of the disk.

Finally, to confirm the behaviour observed, the same kind of experiment was carried out under more severe abrasion conditions by using alundum of 60 mesh size. The weight loss increased dramatically, although the same trend was kept as shown in Table 2 for 1045 steel after 1000 revolutions.

CONCLUSIONS

1. Borided surfaces were more resistant to abrasion than steels quenched and normalized.

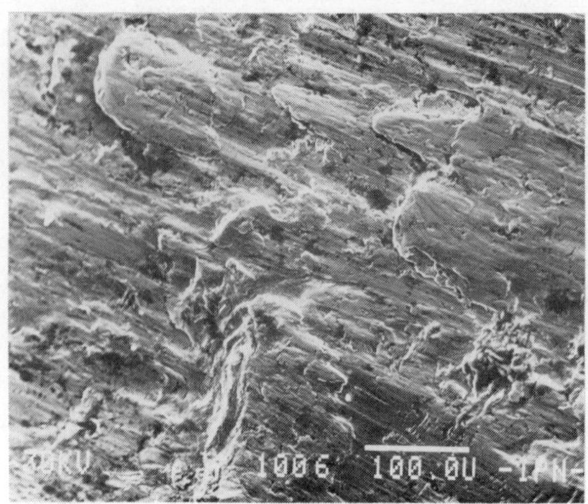

Fig. 6. Abraded surface of steel 4140 after 10,000 revolutions.

TABLE 2 Abrasion Resistance Against Alundum*

Condition	L (g)	A (cm /rev)
Normalized	0.6384	8.1849
Quenched	0.5016	6.4313
Borided	0.4422	5.6691

* Data are for steel 1045 after 1000 rev.
L, weight loss; A, abrasion resistance

2. Abrasion resistance increased with increasing surface hardness.

ACKNOWLEDGEMENT

The authors would like to express their gratitude to Dr. J. G. Cabañas for his advice and helpful discussions.

REFERENCES

1. T. S. Eyre, in Source Book on Wear Control Tehnology (edited by ASM), p. 1, Metals Park, Ohio (1978).
2. F. T. Barnwell, in Fundamentals of Tribology (edited by N. P. Suh and N. Saka), p. 401, MIT Press, Cambridge, MA (1980).
3. D. Larsen-Badse, in Source Book on Wear Control Technology (edited by ASM) p. 120, Metals Park, Ohio (1978).
4. R. B. Gundlach and J. L. Park, in Source Book on Wear Control Technology (edited by ASM), p. 126, Metals Park, Ohio (1978).
5. A. Salas and co-workers, in Proc. of VIII Inter-American Conf. on Mat. Tech. (edited by AEO), p. 29-7, San Juan, PR (1984).
6. A. Salas and co-workers, in Proc. of V Enc. Inv. Metalúrgica (edited by ITS), p. 20, Saltillo, MEX (1983).
7. N. P. Suh and co-workers, in Fundamentals of Tribology (edited by N. P. Suh and N. Saka), p. 493, MIT Press, Cambridge, MA (1980).

Microstructure of Deformation Relative to Cavitation Erosion

A. Karimi

Institut de Génie Atomique, Ecole Polytechnique Fédérale de Lausanne, CH - 1015 Lausanne, Switzerland

ABSTRACT

An austenitic stainless steel has been submitted to the vortex cavitation erosion. The damaged specimens have been examined in the Scanning Electron Microscope (SEM) to determine the surface deformation distribution and the weak points of the alloy against cavitation attack. Transmission Electron Microscopy (TEM) observation was made to characterize the material response and the deformation microstructures. The shock wave character of the applied stress due to the collapse of cavities and the heterogeneity of deformation in austenite, leads to the appearance of the coarse slip bands in the grain. These transform into macro-cracks from which erosion originates. Deformation twinning and strain induced α' martensite appear from the beginning of the test and their volume fraction increases with exposure time.

KEYWORDS

Erosion, Cavitation, austenitic stainless steel, micro-structure, deformation twinning, martensitic transformation.

INTRODUCTION

In the hydrodynamic systems, when a machine part is exposed to cavitation attack, it supports two kinds of dynamic stress fields. One is due to the working conditions, applied into the volume of the parts. The amplitude of this stress remains generally below the elastic limit. Another one is the pulsatory stress due to the collapse of vapour cavities, which acts essentially on the superficial layer the thickness of which is about 100μm (1). The amplitude of this pulsatory stress can reach 2 or 3 times the elastic limit of usual alloys. This is high enough to induce a plastic deformation with a rate of about $10^5 - 10^6 s^{-1}$ (2). These specific conditions of material loading, produce a microstructure which can be different from that observed after the conventional testing of materials i.e. creep, tensile tests,etc. For many years cavitation erosion investigations were devoted to qualitative data and comparative tests relative to erosion rates in an attempt to find higher strength materials. Only, in the last ten years, the strain induced sub-structures and their relation with erosion resistance, have recieved more attention. The appearance of deformation twinning and strain induced martensitic transformation have been evidenced in the superalloys (3,4), stainless steels (5,6) and Cu based alloys (7), and considered to be a benefic factor for erosion resistance. High erosion resistance of twinning alloys is attributed by Mahajan et al (8) to more efficient means of strain energy accommodation, ease of strain relaxation by glide at blocked twins and to the imporvement of ductility and work hardening capacity of the matrix. However the role of strain-induced martensitic transformation is correlated to the absorbtion of a fraction of the incident cavitation impact energy and to a higher strain energy requirement to fracture the martensite than the matrix (9). In the present work, the sub-substructure of an eroded austenitic stainless steel has been studied at different states of erosion, in an attempt correlate it with the erosion resistance.

EXPERIMENTAL PROCEDURE

The investigated alloy was an austenitic stainless steel with chemical composition of 0.032 (%wt) C, 17.7 %Cr, 13.5 %Ni, 2.70 %Mo, 0.95%Si, 1.20%Mm and $\sigma_{0.2}$=238 Mpa, σ r = 555 Mpa.

Cavitation erosion tests were carried out in a "vortex cavitation generator" This device is described in ref. (10) and consists in a Conic cavitation chamber, in which a rotating flow is created by tangential inlet and axial outlet of a flowing liquid. The vapour cavities are produced by sudden closure of the tangential inlet by a rotating valve (ω=0-2000 r.p.m.). They collapse immediately thus producing a hydrodinamic stress directed towards the specimen. In this device, controllable cavitation impacts can be produced so as to study the effects of a single collapse and the multiple impacts due to repeated collapses.

For TEM observations, foils of 3mm diameter and 0.2-0.3mm thickness were cut by spark-machining from the damaged surface and thinned electro-chemically in a solution of 20% perchloric acid + 80% methanol, at T=-10°C and V = 15v.

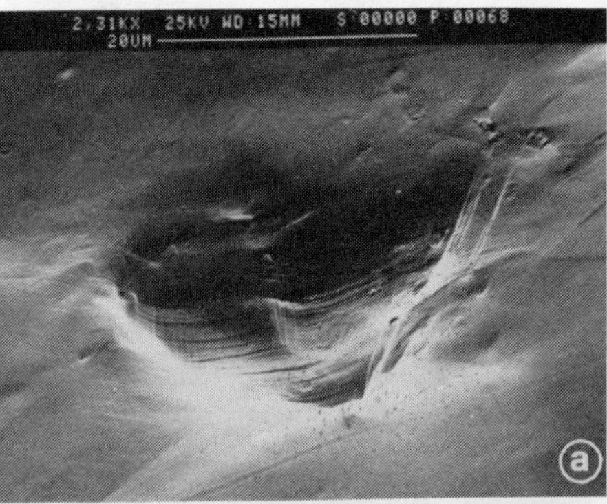

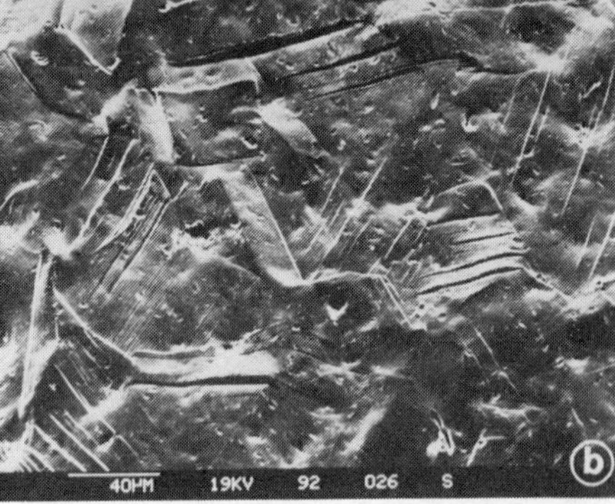

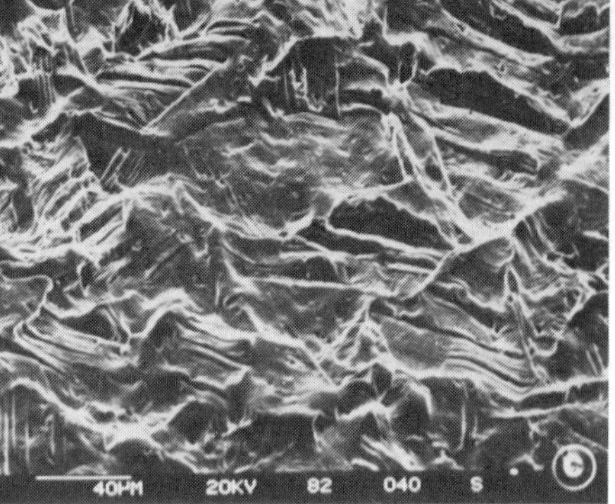

Fig. 1. Developpment of surface deformation with cavitation attack exposure time
a) single impact
b) multiple impact, 1,2 impacts / 100μm^2
c) erosion nucleation, 3,2 impacts / 100μm^2

EXPERIMENTAL RESULTS

a) Single impact

Each impact thus generated in the cavitation chamber can lead to the formation of a crater in stainless steel specimens. These craters have then been observed by Scanning electron microscopy. They have a diameter of d=5-25µm and are surronded by a plastic zone of upto 60µm (Fig. 1a). The maximum height of craters measured by SEM stereophotommetry (11) was h=2-10µm. Therefore, the diameter of the impacting hadrodynamic jet due to collapse can be estimated from $2R=(d^2/4 + h^2)/h$, where 2R=20-100µm. Tridimensional observation of the craters permits also to determine the angle of the impact direction with respect to the specimen surface. This angle was estimated generally in the range $\alpha=85\pm5^{o}$.

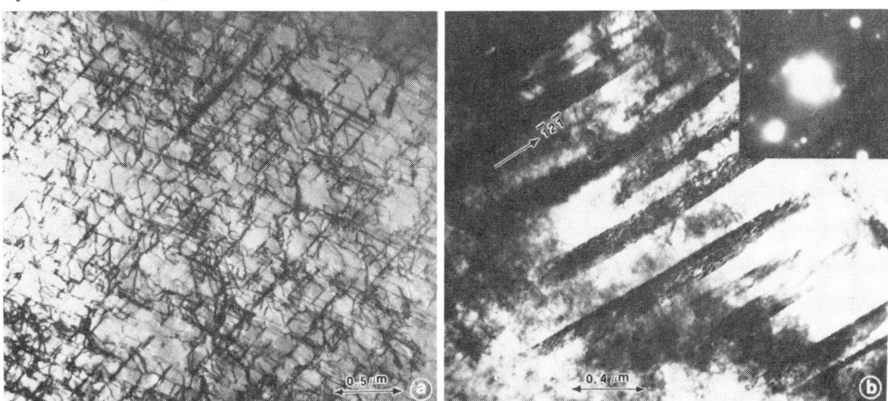

Fig. 2. Sub-surface microstructure of an eroded specimen at various depths a) h ≃ 60µm, b ≃ 5µm

b) Multiple impacts and erosion mechanisms

Surface deformation due to single impacts is essentially characterized by the appearance of relatively short slip bands which do not cross through the whole grains. But, repeated impacts broaden the damaged area with the development of slip bands which become coarser and larger and cross through the grains. The heterogeneous deformation of austenite induced by a repeated pulsatory stress seems to be similar to that observed after conventional mechanical tests such as tensile tests. It consists of relatively coarse slip bands, distant of a few tenth of a µm to a few µm. In spite of this, one can note frequently coarser slip bands which steps of about 1µm bound to the very localized character of impact deformation, as shown in Fig. 1b. These slip bands then develop and widen during the erosion tests, to convey to macrocracks from which erosion originates, and causing a weight loss of the alloy. An example of such a surface evolution is presented in Fig. 1c. In contrast with this, finer slips and mor uniform deformed zones, present a longer time for erosion initiation and a higher resistance to cavitation damages. This is probably due to a cushioning effect on several gliding planes and to the division of the applied stress into several glissile planes which reduce the impact intensity.

c) Deformation sub-structures

It is evident that, when the impact stress is high enough to induce a plastic deformation in the specimen, the amount of strain and the rate of hardening in the plastic tone decrease quickly as a function of the distance from the impact point. Whereas the cavitation erosion sub-structure depends on the one hand on the impact density i.e. the exposure time and on the other hand on the depth from the surface. The influence of these parameters i.e. time and depth, were studied using the TEM technique. The foils for TEM observations were thinned electro.chemically using an automatic twin jet polishing unit. By temporary interruption of one of the jet, the thinned zone was displaced at various depth from the damaged surface (H=0-200μm) so as to investigate the evolution of the microstructure. Two examples of microstructures of specimens submitted to the same erosion conditions are presented in Fig. 2. The exposure time corresponds to an average impact density of about 8.5 impacts/100μm^2. Fig. 2a shows the aspect of dislocation arrangements at a depth of ∿60μm There are generally straight and parallel dislocations denoting a high stress concentration in the slip planes. On the other hand, Fig. 2b shows the near surface (h≃5μm) microstructure of similar damaged specimen including wide and dense deformation twin bands. As shown also in Fig. 3, the configuration of cavitation erosion twins consists in wide bands crossing the grains. This is probably due to the local character of a repeated compressive stress of relatively limited amplitude which is only able to develop an activated twin rather than to initiate another less favorable system.

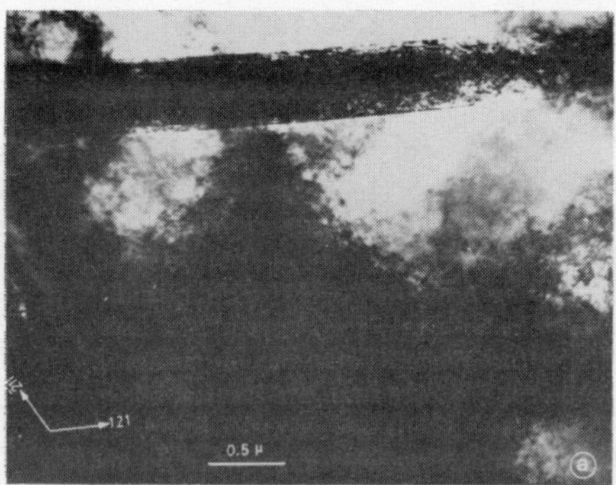

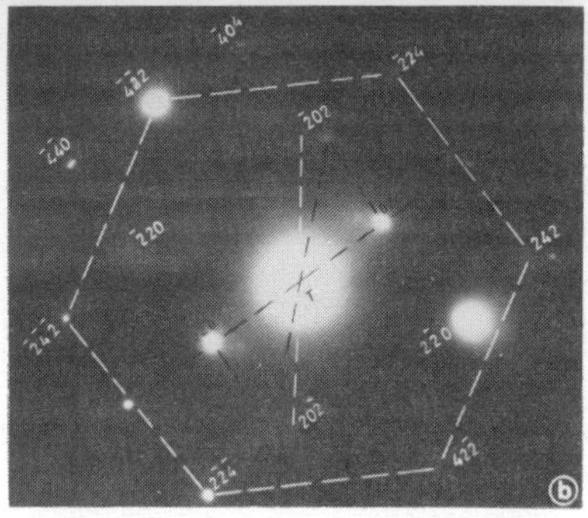

Fig. 3. Twin in a strongly strained region of the specimen
a) BF image
b) indexed SAD pattern

Strain induced α' martensite has also been frequently observed in eroded specimens. Fig. 4 shows an example of such a transformation in a specimen exposed to cavitation attack for 4.2 impacts/100μm^2. The crystallographic orientation relations between the matrix and the martensite have been determined by indexing the corresponding SAD patterns. Fig. 4a shows a bright field image of a cross section of the grain with a SAD pattern corresponding to the $[1\ \bar{2}\ \bar{3}]$ zone axis (fig. 4b). This diffraction pattern corresponds to the middle of the grain without transformation. However Fig. 4c shows the SAD pattern of the transformed area indicated by a rectangle in Fig. 4a. This transformed area has a$[11\bar{1}]$orientation for the martensite. One can not the following orientation relations : $[1\bar{1}1]$A || [011] M and $[\bar{1}\bar{2}1]$A || $[\bar{1}0\bar{1}]$ M. A dark field image on 101 M shows the width of the transformed area (∿2μm).

DISCUSSION

There is not a detailed information in the litterature concerning the morphology and the crystallographic orientation relations between deformation twins and martensitic transformations induced by cavitation erosion. However, hte (111)A || (011)M relationship observed in eroded specimen containing 4.2 impacts/ $100 \mu m^2$, is a common austenite-martensite orientation relationship found in quenched stainless steels (12). Murr et al (13) studying defect generation in shock loading deformation reported that shock-induced microstructures are governed by the same basic features which govern all deformation microstructures. The crystallographic and morphological features of shock induced twins do not differ substantially from the conventionally formed ones. Considering this and the similariry between shock loading and cavitation erosion conditions, one can expect a comparable nucleation mechanism between cavitation erosion microstructures and other conventional features. The limited amplitude and the pulsatory character of the cavitation collapse stress, develop the activated system rather than nucleate another less favorable one, giving raise to the wide bands of twin or martensite.

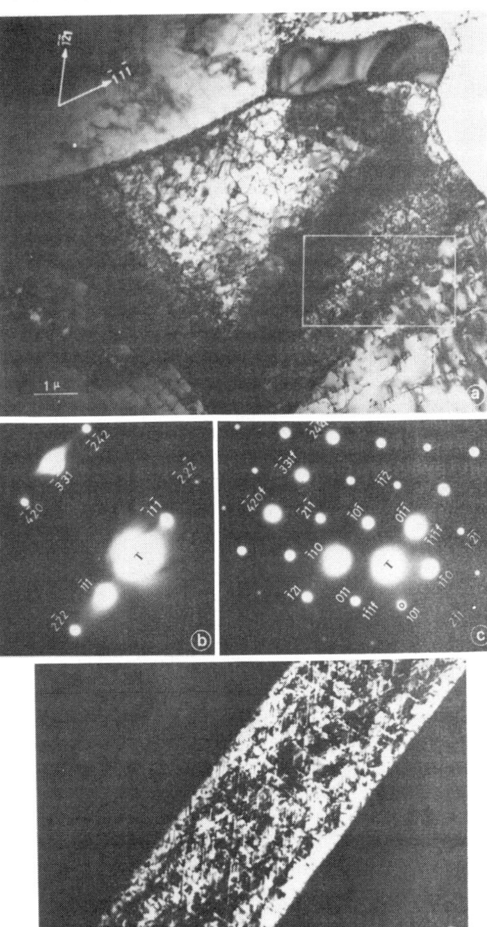

Fig. 4. TEM micrographs of α' martensitic transformation
a) BF image of a [1$\bar{2}\bar{3}$] cross section
b) SAD pattern of a non transformed area
c) SAD patterns of a combined martensite-matrix area
d) DF image of the transformed area (101) orientation

In spite of this, many attempts were made to determine the influence of twin and martensite appearance on the erosion resistance of materials. The appearance of martensite is generally considered as a benefical factor : either it absorbs a significant fraction of the collapse impact energy or a higher strain energy is required to fracture the martensite rather than the matrix (9). But, twinning effects have been interpreted in various manners such as, high expenditure of strain energy in twin boundaries (14), less heterogeneous distribution of stress concentration in twinned area (15), ease of strain relaxation by twin-glide combination and more efficient strain energy accommodations (8). However, Heathcock et al (6), associated the work hardening rate improved by the presence of twins and martensites to the erosion resistance.

A relatively greater hardening induced by twin and phase transformation has also been confirmed by Graham et al (16) in "MP" type alloys and by Murr et al (17). They observed that the hardening of metals and alloys can be at least 3-4 times greater under shock loading conditions as compared to hardening in a tensile test, for the same amount of strain.

ACKNOWLEDGEMENTS

I would like to thank Prof. J.L. Martin for his contribution to several discussions and reading of the manuscript. The Swiss Commission for the encouragement to scientific research (CERS) is acknowledged for the financial support of this program.

REFERENCES

1. M.S. Pellest, R.E. Devine, Trans. ASME, Ser. D, J. Basic Eng. 88, 691-700 (1966)
2. S. Fujikawa, T. Akamatsu, J. Fluid Mech. 97, 3, 481-512 (1980)
3. D.A. Woodford, Met. Trans. 3, 1137 1145 (1972)
4. C.J. Heathcock, A. Ball, Wear, 74, 11-26 (1981-82)
5. T.F. Pederson, S. Pederson, I. Hansson, Proc. 6th int. conf. on Erosion by liquid an solid impact, 4-8 sept. 1983, Cambridge, U.K.
6. C.J. Heathcock, B.E. Protheroe, A. Ball, Wear, 80, 311-327 (1982)
7. S. Dakshinamoorthy, M.S. thesis, Dept. of Mat. Sci. State Univ. of New York (1975)
8. S. Vaidya, S. Mahajan, C.M. Preece Met. Trans., 11A, 1139-1149 (1980)
9. K.A. Antony, W.L. Silence, 5th Proc. int. Conf. on Erosion by solid and liquid impact, 67-1, Cambridge, U.K. (1979)
10. A. Karimi, Proc. 3rd int. conf. on Mechanical Properties of materials at high rates of strain, J. Harding ed. 9-12 April, Oxford, U.K. (1984)
11. A. Boyde, Scanning Electron Microscopy, O.C. Wells ed. Mc Graw-Hill, New York (1974)
12. B.P.J. Sandvick, C.M. Waymann, Metallography 16, 199-227 (1983)
13. M. Meyers, L.E. Murr, Shock waves and high-strain rate phenomena in metals, plenum press, New-York, (1981)
14. R.J. Wasilenski, Scripta Met. 1, 45 (1967)
15. R. Priestnarm "Deformation twinning", R.E. Reed Hill, H.C. Rogers eds. Gorden and breack (1964)
16. A.H. Graham, J.L. Youngblood, Met. Trans. 1, 423-430 (1970)
17. L.E. Murr, D. Kulhman-Wilsdorf, Acta Met. 26, 847-857 (1978)

SECTION 13

Rapid Solidification

Solidification rapide et verres métalliques

Microstructure and Mechanical Properties of Rapidly Quenched Shaped Memory Alloys

S. Eucken and E. Hornbogen

Institut für Werkstoffe, Ruhr-Universität Bochum, Federal Republic of Germany

ABSTRACT

The microstructure and mechanical properties of shape-memory (SM) alloys produced by meltspinning were studied. Homogeneous structures of the β-phases were aspired in the following alloy systems Cu-Sn, Cu-Al, Cu-Al-Ni, Cu-Zn-Al, and Ni-Ti. Besides the undesired heterogeneous microstructures, 1. single columnar, 2. double columnar, 3. equiaxed, and 4. mixed, layered homogeneous grain structures could be obtained. The grain sizes are one to two orders of magnitude smaller than in conventionally solidified alloys. Tensile testing revealed that the single columnar microstructure provides the most favorable SM-behavior (pseudoelastic deformation up to 7 %). In addition, the strength of rapidly cooled materials was higher than that of the same conventionally solidified alloys.

KEYWORDS

Shape memory; pseudoelasticity; true yield stress; rapid quenching; meltspinning; type of grain structure.

INTRODUCTION

Prerequisite for the occurrence of the shape memory effect is a reversible martensitic transformation. It may originate from stress induced selection of $\beta \rightarrow \alpha_M$ transformation shear, or from shear inside the martensite structure $\alpha_M^+ \rightarrow \alpha_M^-$ [1,2,3]. Shape memory alloys can facilitate complicated motions which elsewise would require involved mechanical joints. SM-alloys are often needed as thin wires or ribbons. Usual production of these semi-finished products by wire drawing or rolling is tedious and expensive. Conventional solidification and powder compaction are the established methods for the production of these materials. The purpose of the present investigation is threefold:

A. formation of a ribbon in an efficient one-step process;
B. extension of the compositional range of the homogeneous β-phase and consequently of the M_s-temperature [4];
C. production of certain types of grain structure, which should favor large shape changes.

EXPERIMENTAL PROCEDURES

The alloys were molten by induction heating, ribbons with a thickness of 5 to 100 μm and 0.5 to 3 mm in width were prepared by meltspinning on a copper wheel with a diameter of 200 mm and wheel velocities between 10 and 52 m/s in a 0.8 bar He atmosphere (0.3 bar for Ni-Ti). The microstructure was investigated by light microscopy of cross and longitudinal sections, by scanning electron microscopy of mercury embrittled specimen [5], and by transmission electron microscopy. The martensite temperatures were determined by observation of the temperatures at which a shape memory effect occurs. All tensile tests were made at ambient temperature. Portions of the ribbon with good edge quality were selected to avoid considerable notch effects during the tensile tests.

RESULTS AND DISCUSSION

Microstructure

A homogeneous β-phase is always aspired for the SM-alloys. Therefore, out of the three possible crystallization reactions: primary, eutectic, and massive (or polymorphous) [6] only the latter

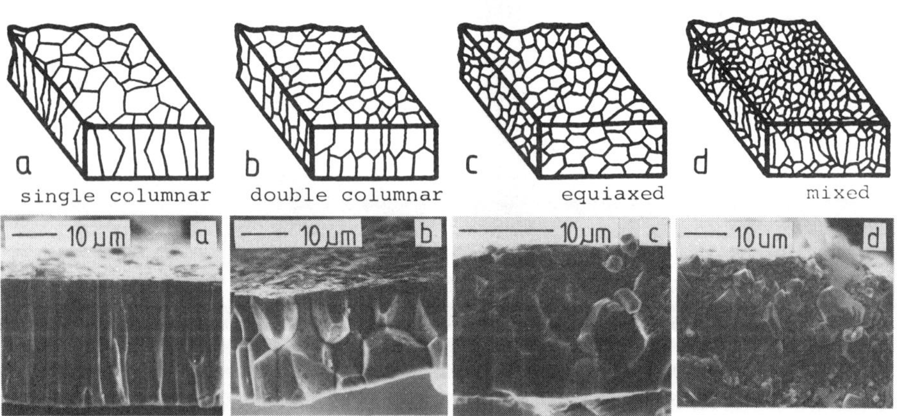

Fig. 1. Types of homogeneous β-phase grain structures (schematic and micrographs): a) Cu 26 wt.-% Zn, 3.8 wt.-% Al, wheel speed v = 50 m/s, nozzle diameter d = 0.65 mm, temperature of the melt T = 920 °C; b) Cu 11,7% Al, 3.5% Ni, v = 30 m/s, d = 0.75 mm, T = 1200°C; c) Cu 24.4% Sn, v = 25 m/s, d = 0.4 mm, T = 900° C; d) Cu 19.5% Sn, v = 20 m/s, d = 0.65 mm, T = 920° C.

will lead to the desired structure. Figure 1 shows schematic representations and SEM-micrographs of the observed microstructures. The single columnar grain structure forms at high wheel speeds, i.e. in a high temperature gradient. Crystallization starts at the interface with the wheel, and the nuclei grow to the free surface (Fig. 2a). This microstructure is also obtained by overheating of the melt and smaller cooling rates. Fig. 3 shows another single columnar structure. This time a ribbon had coiled several times around the wheel. Favorable conditions for heterogeneous nucleation at the identical type of material may explain this result.

The double columnar microstructure (Fig. 1b) originates at smaller cooling rates, respectively overheating, if crystallisation can start in addition in the temperature gradient at the free surface (Fig. 2b). At still lower cooling rates the equiaxed microstructure is found (Figs. 1c, 2c), if individual (homogeneous?) nucleation in the interior of the liquid cannot be suppressed. Another way to form this microstructure could be by heterogeneous nucleation at dispersed indissolved particles. A special case of an equiaxed grain structure is obtained by a low temperature of the melt and a high cooling rate. Grains originate of a diameter about equivalent to the thickness of the ribbon (Fig. 4).

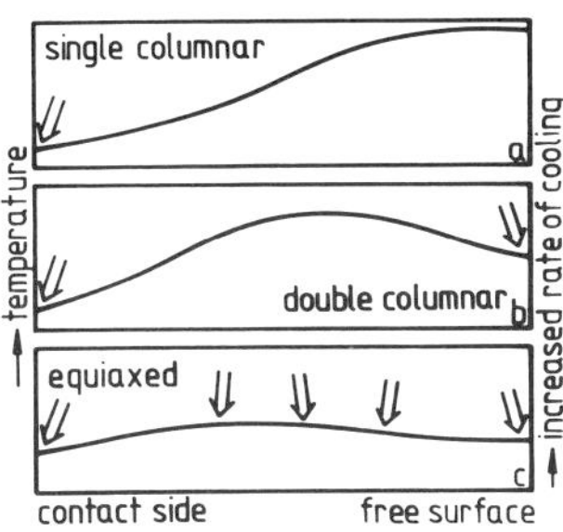

Fig. 2. Temperature gradients in ribbons; nucleation sites indicated as ⇒

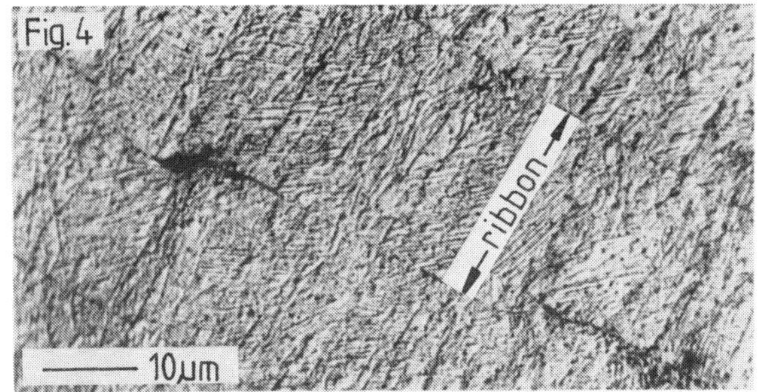

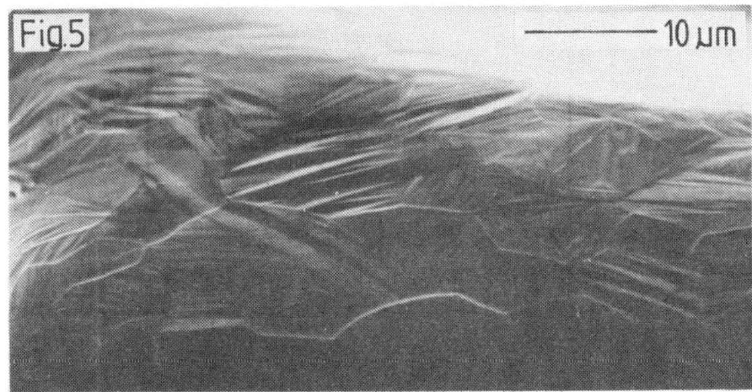

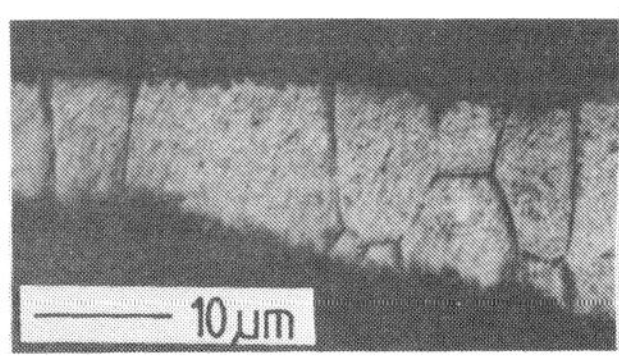

Fig. 3. Equiaxed grain structure with a grain size in the range of ribbon thickness (Cu 14.2 wt.-% Al, v = 52 m/s, d = 0.4 mm, T = 1050° C).

Fig. 4. Single columnar structure in a ribbon coiled several times around the wheel (Cu 21.6 wt.-% Sn, v = 52 m/s, d = 0.4 mm).

Fig. 5. Surface relief by $\beta \rightarrow \alpha_M$ transformation (CuZnAl).

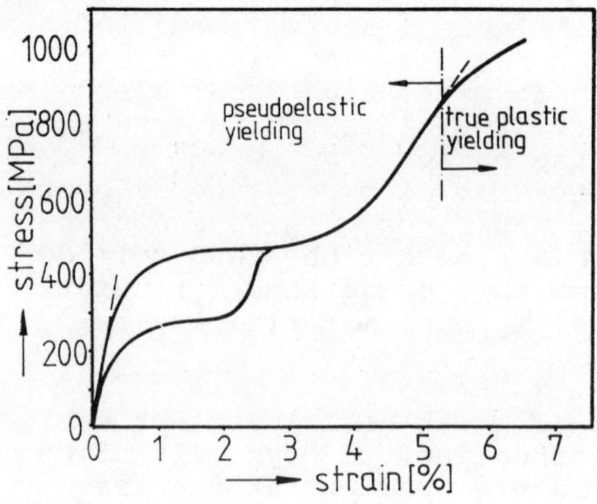

Fig. 6. Stress-strain curve of a Ni-Ti ribbon (M_s < 20°C).

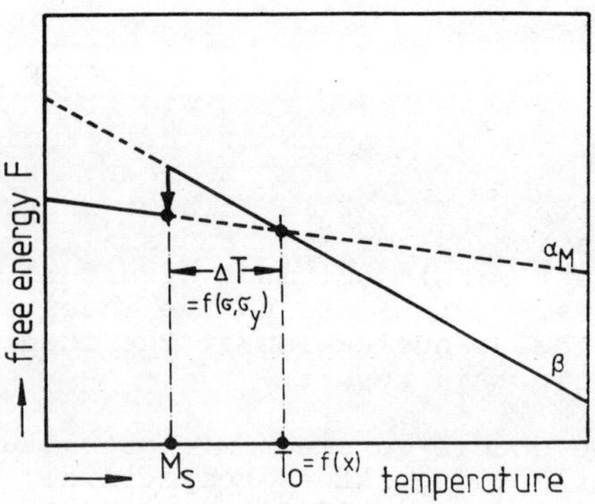

Fig. 7. Variation of M_s caused by chemical composition x, stress σ, and yield strength σ_y.

At intermediate cooling rates a structure originates which is characterized by layer of very small grains at the two surfaces and larger ones in the interior (Fig. 1d) [5].

Besides the massive reaction, eutectic crystallization may occur at compositions in the α + β-field of the phase diagram and at low cooling rates. Precipitation of a second phase from the as-crystallized β-phase is a second type of heterogeneous microstructure which may be observed [4]. Such structures are less suitable for our purpose. A desired case of secondary reaction is the martensitic transformation of the crystallized β-phase (Fig. 5).

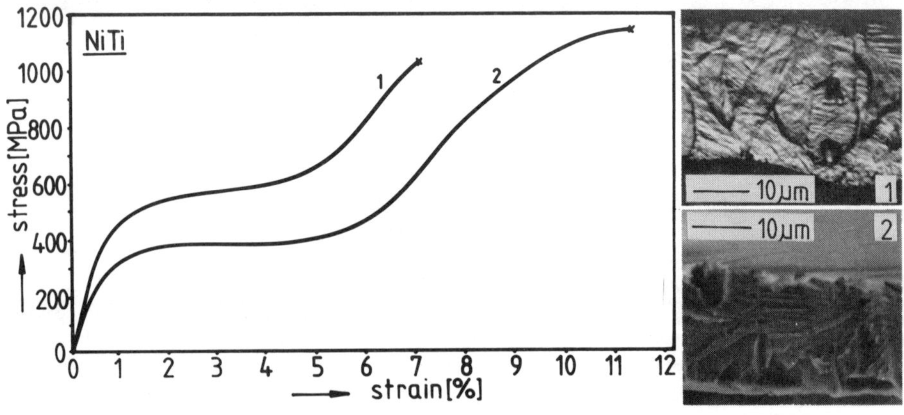

Fig. 8. Stress-stain curves of Ni-Ti ribbons and respective microstructures (M_s > 20° C) (v = 30 m/s, d = 0.4 mm, 0.2 bar He atmosphere!); fracture surface through martensite of specimen 2 (single columnar) and micrograph of specimen1: a multi grained microstructure prevails.

Mechanical Properties

A typical stress strain curve of a rapidly cooled shape memory alloy with M_s far below ambient temperature is shown in fig. 6. The large, pseudoelastic strain is due to oriented stress induced transformation. The effects of alloy composition x, external stress σ, and yield strength σ_y on the martensitic transformation are schematically shown in fig. 7. The undercooling ΔT required for the onset of the transformation is decreased by σ, but increased by hardening of the β-phase σ_y. T_o is affected by the chemical composition x, and also the degree of order in the β-phase [7,8]. (In β-CuZn, for example, M_s is raised about 30° C by σ = 100 MPa [9]). This produces two types of yielding of the ribbons under increasing tensile loading (Fig. 6). While the first is reversed during cooling (pseudoelastic), the second represents the start of true plastic deformation, i.e. the true yield stress by irreversible motion of dislocations and the limit above which shape memory material should not be loaded.

Two crystallographic mechanisms are responsible for the large reversible strains (Figs. 8,9, and 10). The selection of those orientational varieties of the $\beta \rightarrow \alpha_M$-transformation for which the external stress σ provides the larger component of the resolved transformation shear stress $\tau_{\beta \rightarrow \alpha}$. If the $\beta \rightarrow \alpha_M$-transformation has occured stress-free during cooling for $M_s > 20^o$ C, reorientation of the internal defect (twin) structure of the martensite $\alpha_M^+ \rightarrow \alpha_M^-$ (Figs. 8,9, and 10), produces a shape change which also is removed by the reverse transformation $\alpha_M \rightarrow \beta$, during reheating above A_s.

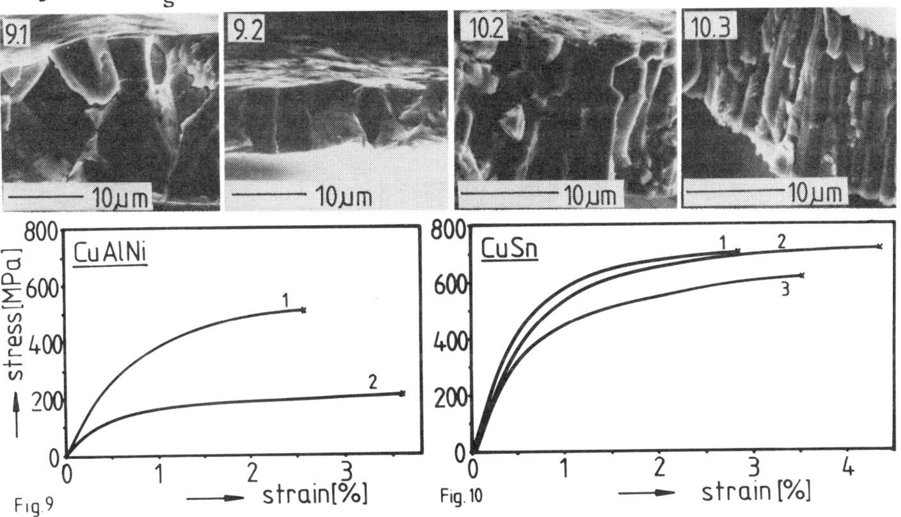

Stress-strain curves and respective microstructures:
Fig. 9. Cu 11.7% Al, 3.5% Ni, v = 30 m/s, d = 0,4 mm, T = 1200° C, $M_s > 20^o$ C; mercury embrittled specimen 1 (close to fracture surface) and fracture surface of spec. 2.
Fig. 10. Cu 23.5% Sn, v = 20 m/s, d = 0.65 mm, T = 1050° C; mercury embrittled spec.2 and 3(structure of spec.1 equivalent to spec.2).

The shape memory alloys produced by rapid cooling show all the features known from the materials produced by conventional methods [10,11,12,13]. The particular properties can be explained by the following factors which are all related to the special method of its production:

1. the shape of a thin ribbon;
2. a grain size which extends to the two free surfaces;
3. hardening of the β-phase by quenched-in dislocations, anti-phase domain boundaries, and vacancies.

The combination of the factors 1 and 2 explain the large shape changes which were measured. The maximum strain as-limited only by the crystallographic conditions of the $\beta \rightarrow \alpha_M$-transformation can be realized because of the lack of compatibility problems at grain boundaries especially in the single columnar grain structures (in bulk material a grain size/thickness ratio of 1 was found to be optimal, too [14]).

The factors indicated as 3, all contribute to raise the true yield stress of the alloys, and thus the range of loading conditions which they may be exposed to in use as a shape memory material.

ACKNOWLEDGEMENT

The support by the German Science Foundation (DFG project Ho 325/15-5) is gratefully acknowledged.

REFERENCES

1. L. Delaey and H. Warlimont, Progr. Mater. Sci. 18 (1974).
2. J. Perkins (editor), Shape Memory Effects in Alloys, Plenum Press, New York (1975).
3. C. M. Wayman, in: Physical Metallurgy (edited by R. W. Cahn and P. Haasen), Elsevier Science Publisher BV p. 1031 (1983).
4. S. Eucken and E. Hornbogen, in: Proc. 5th Int. Conf. Rap. Quenched Metals (editor: H. Warlimont and S. Steeb), North Holland Publishing, Amsterdam, in print (1984).
5. M. Blank, C. Caesar and U. Köster, ref. 4.
6. E. Hornbogen, ref. 4.
7. E. Hornbogen, Z. Metallkde. 75, 741 (1984).
8. E. Hornbogen, Acta Met. 33, in print (1985).
9. E. Hornbogen and G. Wassermann, Z. Metallkde. 47, 427 (1956).
10. J. V. Wood and P. H. Shingu, Met. Trans 15A, 471 (1984).
11. J. Perkins, Met. Trans 14A, 2229 (1983).
12. R. Oshima et al., in: Proc. 4th Int. Conf. Martensitic Transformation (edited by L. Delaey and M. Chandrasekaran) 749, Les Editions de Physique, Les Ulis, France (1982).
13. Q.-G. Rong, D.-S. Li and Y. Zhu, ref. 4.
14. I. Dvorak and E. B. Hawbolt, Met. Trans 6A, 95 (1975).

Microstructure and Mechanical Properties of Rapidly Solidified CuNi10 Alloys

C. Caesar and U. Köster

Department of Chemical Engineering, University of Dortmund
D-4600 Dortmund 50, Federal Republic of Germany

ABSTRACT

Rapid solidification results in a refinement of the grain size, elimination of segregation, extension of solid solubility above the equilibrium limit and the possibility to form novel non-equilibrium crystalline or amorphous phases. Since Cu-Ni is a system with complete miscibility in the solid state, it can be chosen as a model system in order to investigate the influence of grain size, shape and anisotropy on mechanical properties and deformation behaviour of rapidly solidified materials.
The microstructure of melt-spun Cu-Ni ribbons has been found to be very anisotropic and depend strongly on the solidification parameters: Columnar grains at high surface velocities of the quenching wheel (40 m/s), equiaxed grains at 10 m/s and a three-layered structure at intermediate velocities. Ribbons with the layered microstructure exhibit extremely high elongations for such a thin material. Yield and tensile strength have been found to increase with the surface speed of the quenching wheel, i.e. with decreasing grain dimensions. However, at very high velocities, i.e. even smaller widths of the columnar grains, a significant drop has been observed. Models will be discussed in order to explain why deformation in these ultrafine-grain materials is eased.

KEYWORDS

Microcrystalline materials; ultrafine-grain size; rapidly solidified materials; microstructure; Hall-Petch relationship.

INTRODUCTION

Rapid solidification of metals and alloys has become a field of extensive attention, since this method offers prospect of new or improved engineering properties /1/. The well known beneficial effects are refinement of the grain size, elimination of segregation, extension of solid solubility above the equilibrium limit and the possibility to form novel non-equilibrium crystalline phases or even metallic glasses. Usually all these effects superpose each other. Since Cu-Ni is a system with complete miscibility in the solid state, it can be chosen as a model system to study the influence of grain size as well as anisotropy and grain shape on mechanical properties and deformation behaviour of rapidly solidified materials. The aim of this paper is to characterize the fine-scale microstructure of melt-spun CuNi10 ribbons and to present first results on the microstructure/mechanical property relationship for these rapidly solidified materials.

METHODS AND MATERIALS

CuNi10 and CuNi30 alloys were prepared from Cu (Balzers 99.99) and Ni (Baker 99.99) in an RF furnace under an argon atmosphere of 400 hPa. Ribbons of these alloys were cast in an argon atmosphere of 400 hPa by melt-spinning onto a copper-wheel of 25 cm in diameter with surface velocities ranging from 5 to 40 m/s; all other solidification parameters were held constant.

The microstructure of these ribbons has been investigated by means of light and transmission electron microscopy (TEM); in addition, a preparation technique /2/ was applied using mercury embrittlement to reveal the grain structure by means of scanning electron microscopy (SEM). Surface and fracture morphologies have been examined by scanning electron microscopy.

Vickers microhardness was measured with loads of 15 to 25 g with regard to the limits for maximal load, which depends on the actual hardness and the thickness of the ribbon. Cold rolling was performed with a precision rolling mill specially designed for thin ribbons. Tensile tests have been performed at ambient temperature with gage lengths of 15 to 20 mm and a strain rate $\dot{\varepsilon} = 4 \cdot 10^{-4}\ s^{-1}$; care was taken to ensure a constant thickness/width ratio of the tensile specimens.

RESULTS

As shown schematically in fig. 1 the microstructure of rapidly solidified CuNi10 alloys depends strongly on the surface velocity of the quenching wheel and cannot be described simply by a grain diameter. At low velocities (about 5 to 10 m/s) equiaxed or only slightly elongated crystals are formed;

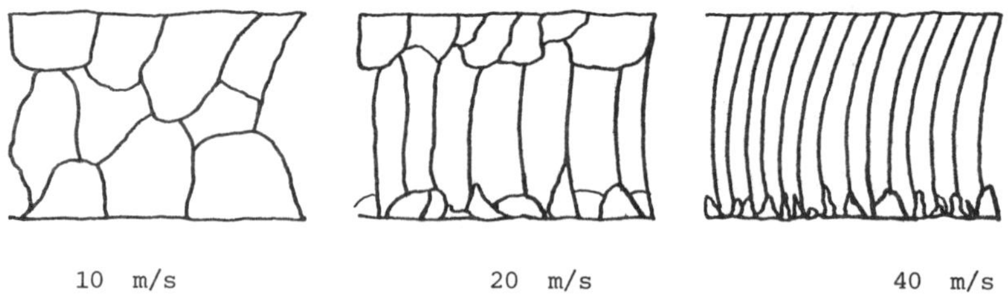

10 m/s 20 m/s 40 m/s

Fig. 1: Influence of surface velocity of the quenching wheel on the microstructure of melt-spun CuNi10 (schematically).

the grain size decreases with increasing velocity, e.g. 12 µm at 10 m/s. Surface velocities of 30 m/s and more result in columnar crystals extending from the contact to the free surface of the ribbon. The width of the columnar grains has been found to decrease from 2.5 µm at a velocity of 30 m/s to about 1.0 µm at 40 m/s. At surface speeds of about 15 to 25 m/s three regions can be distinguished in longitudinal sections of melt-spun ribbons: small equiaxed chill crystals form at the contact side from which a zone of slightly curved columnar crystals (about 5 µm in width) develops by growth selection; near the free surface a layer of equiaxed crystal has been observed which diminishes as the surface velocity increases.

The influence of the wheel velocity on microhardness HV, yield strength $R_{p0.2}$ and tensile strength R_m of melt-spun CuNi10 alloys is shown in fig. 2: The microhardness does not depend significantly on the microstructures formed by rapid solidification. $R_{p0.2}$ has been found to increase from about 60 MPa in ribbons with equiaxed grains with 12 µm in diameter to about 150 MPa in ribbons melt-spun at 30 m/s, but decreases significantly in materials with a smaller width of the columnar grains produced at 40 m/s. Taking into account the differences in strain R_m shows a very similar behaviour.

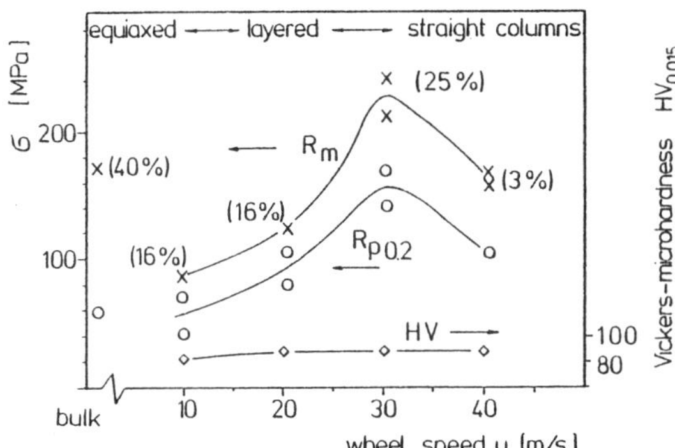

Fig. 2: Influence of the wheel speed on microhardness HV, yield strength $R_{p0.2}$ and tensile strength R_m of melt-spun CuNi10 alloys.

Ribbons with the three-layered microstructure exhibit extremely large (especially for such thin material) elongations of more than 20 %. The highest elongation (40 %) was observed in ribbons cast in vacuum at 20 m/s. Under these conditions the cooling rate at the free surface is reduced resulting in a thicker layer of equiaxed crystals. We assume that high elongation can occur only if the outer layers protect the intermediate one against the influence of microcracks from the surface.

Deformation of bulk samples of the ingot CuNi10 material is characterized by the formation of coarse slip bands with steps of 10 to 30 µm. Fig. 3 shows fracture morphologies and corresponding microstructures of the melt-spun ribbons: The equiaxed crystals in the ribbons melt-spun at 10 m/s elongate significantly until final rupture; slip bands are much finer than in the

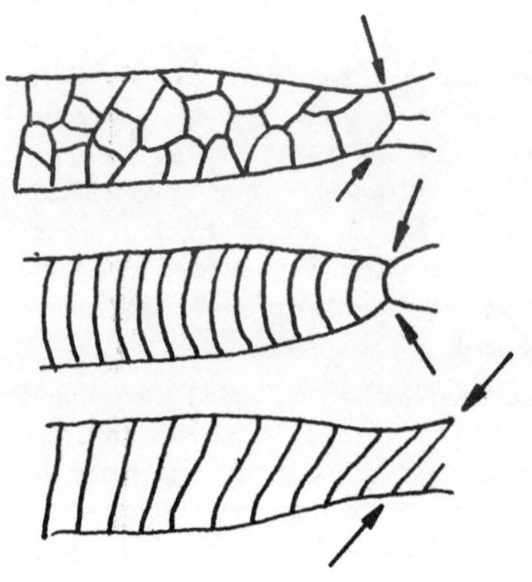

(a) Schematic sketch of fracture morphology and microstructure

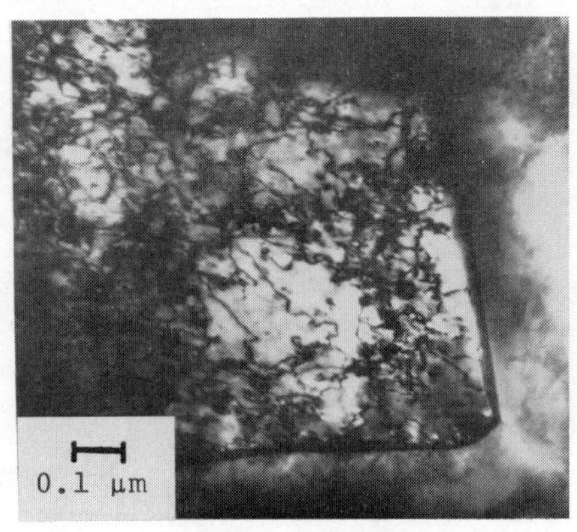

(b) CuNi10 (40 m/s): TEM after cold rolling (40% reduction in thickness)

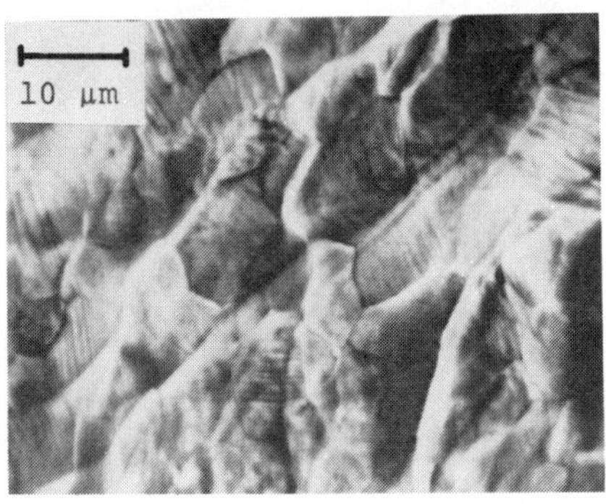

(c) CuNi10 (10 m/s): SEM of the fracture surface

(d) CuNi10 (10 m/s): Ribbon surface near the crack rim

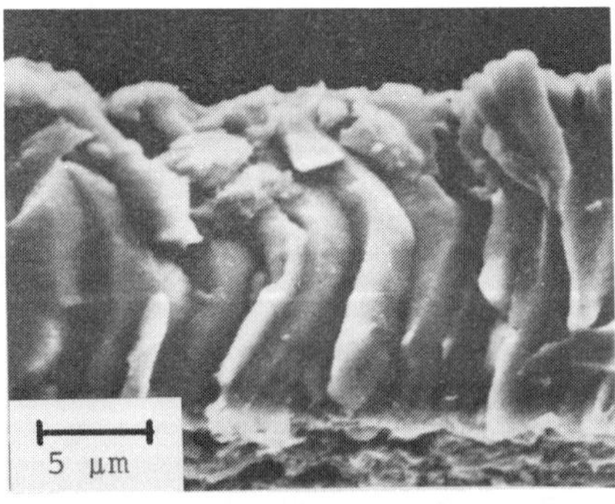

(e) CuNi10 (30 m/s): longitudinal section after cold rolling (50%)

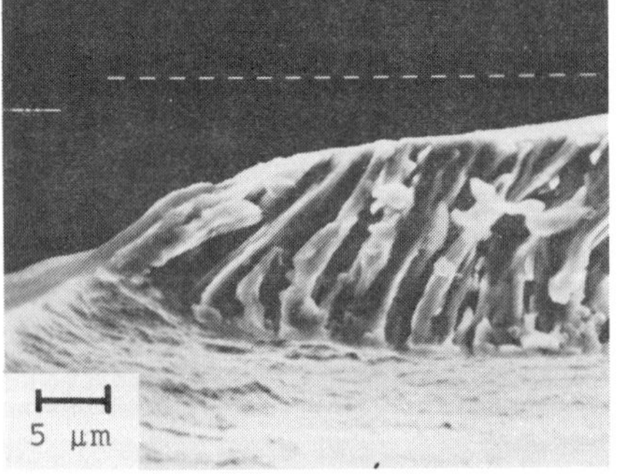

(f) CuNi30 (36 m/s): longitudinal section near crack

Fig. 3: Fracture morphology and microstructure of melt-spun Cu-Ni alloys

ingot material with about 5 mm grain diameter. In ribbons with columnar structure failure has been observed to occur by two modes depending probably on local stress conditions and the nickel content: deformation proceeds by bending of the columar grains which shorten and thicken in the final shear zone. In the other mode observed more often in CuNi30 deformation proceeds by a tilting mechanism of the columnar grains until they finally slide off under an angle between about 30 and 45°.

Cold rolling has been found to produce very similar microstructures. TEM studies after cold rolling of about 30% reduction in thickness indicate the formation of dislocation walls, but the dislocation density observed in ribbons cast at 40 m/s seems to be relatively low compared to specimens prepared at lower velocities of the quenching wheel. Cold rolling increases $R_{p0.2}$, but does not change R_m significantly.

DISCUSSION

Grain size has a measurable effect on most mechanical properties. At room temperature hardness HV, yield strength $R_{p0.2}$ and tensile strength R_m increase with decreasing grain size. Usually the yield strength is related to the grain size by the Hall-Petch equation:

$$R_{p0.2} = \sigma_o + K \cdot L^{-1/2}$$

where L is the average grain size; a similar equation is valid for the tensile stress at constant strain.

Whereas microhardness of the rapidly solidified microcrystalline CuNi10 ribbons has been found to be about equal to the value measured in the bulk ingot material with a grain size of about 5 mm, $R_{p0.2}$ increases as expected with increasing surface velocity of the quenching-wheel up to about 30 m/s, i.e. with decreasing grain dimensions, but shows a surprising drop at higher velocities. To some extent this unexpected behaviour might be related to the shape of the tensile specimens cut from the very thin melt-spun ribbons. Since Vickers microhardness measurements which are not as sensitive to the ribbon thickness do not show any increase in microhardness with decreasing grain dimensions, it seems likely that there are other mechanisms reducing strength in ultrafine-grain materials produced by melt-spinning.

In ultrafine nickel prepared by electroplating lesser strength values than those described by the Hall-Petch relation have been reported by Armstrong /3/, but not a reduction in strength as observed in our melt-spun ribbons. Armstrong assumes that the presence of micro-cracks may be responsible. From investigations on mechanical properties of Fe-Ni-B metallic glasses /4/ it is known that very fast surface velocities result in the formation of surface irregularities and serration of the edges thus reducing $R_{p0.2} \cong R_m$ significantly, thus reducing tensile strength, but not microhardness.

A sharp texture may also decrease the effectiveness of grain boundaries as slip barriers due to the alignment of slip systems in adjacent grains and hence the ease of propagation of slip.

From creep experiments von 316 stainless steel /5/ at about 650°C it is known that the single-phase alloy resists creep much better with a 65 µm grain size (ingot material) than with a 5 µm grain size (RSP-material). For CuNi10 ambient temperature equals to about 0.25 T_m, which is in reasonable agreement with the lower limit of homologous temperature for grain boundary sliding. Grain boundary sliding can occur in specimens which consist of ultrafine-grain material; the contribution of grain boundary sliding to the total creep strain is larger the smaller is the grain size. This effect may be increased due to the grain aspect ratio in the columnar microstructure of melt-spun ribbons with most grain boundaries perpendicular to the main stress axis. The failure mechanism observed in ribbons solidified at higher surface velocities and the relative low density of dislocations found by TEM in cold rolled ribbons with a very fine columnar structure probably indicate such a strong influence of grain boundary sliding by diffusional processes. In addition, due to the rapid solidification process, crystals in melt-spun ribbons (especially near both surfaces which exhibit the highest cooling rate) contain a very high vacancy concentration thus increasing diffusivity and dislocation climb.

For a better understanding of the mechanical behaviour of rapidly solidified CuNi10 ribbons further informations are urgent; therefore, in addition to a systematic cross sectional TEM analysis not finished so far investigations on creep behaviour as well as on the influence of temperature and strain rate are underway.

ACKNOWLEDGEMENT

The authors like to thank Prof. Dr. F. Haeßner (Lehrstuhl für Werkstoffkunde und Herstellungsverfahren, TU Braunschweig) for the opportunity to use the rolling mill specially designed for extrem thin ribbons and Dipl.-Ing. Speitling for the assistance in rolling the CuNi10 ribbons. The financial help of the DFG (Ko 668/5) is gratefully acknowledged.

REFERENCES

1. M. Cohen, B.H. Kear, R. Mehrabian, "Rapid Solidification Processing - An Outlook", in Rapid Solidication Processing, Principles and Technologies Vol.II, (Baton Rouge, 1980), p.1

2. M. Blank, C. Caesar, U. Köster, Proceedings Rapidly Quenched Materials (RQ 5), Würzburg 1984, in press

3. R.W. Armstrong, Proc. ICSMA 5 (Aachen 1979), p.795

4. U. Köster, U. Herold, H.-G. Hillenbrand, Scripta Met. 17, 867 (1983)

5. N.J. Grant, J.Metals 35, 20 (1983).

Aging Response of a Rapidly Solidified beta-Eutectoid Ti-9 wt% Co Alloy

S. Krishnamurthy*, I. Weiss, D. Eylon*** and F. H. Froes***

**Air Force Wright Aeronautical Laboratories, Materials Laboratory, AFWAL/MLLS, Wright-Patterson AFB, OH 45433, USA*
***Wright State University, School of Engineering, Dayton, OH 45435, USA*
****Metcut-Materials Research Group, P.O. Box 33511, Wright-Patterson AFB, OH 45433, USA*

ABSTRACT

Rapidly solidified ribbons of a hypoeutectoid Ti-9 wt% Co alloy were aged at three sub-eutectoid temperatures in the range of 525-650°C for times varying from 10 s to 24 hr. For comparison, bulk specimens of this composition were prepared by conventional button melting and swaging followed by a beta solution treatment and water quenching, and were subjected to the same aging treatments. Microstructural characterization of the alloy samples was carried out by optical microscopy, SEM, and STEM techniques. The aging behavior was studied by measuring microhardness values. The as-produced ribbons and quenched bulk specimens exhibited a retained beta microstructure with a grain size of 25 µm and 400 µm, respectively. Aging below the eutectoid temperature resulted in the precipitation of fine pro-eutectoid alpha phase at very short aging times (∿10 s) which was accompanied by a hardness peak. This was followed by the formation of Ti_2Co intermetallic compound and a decrease in hardness. The precipitation on aging occurred more rapidly and more uniformly in the ribbons than in the bulk specimens presumably due to a higher vacancy concentration and finer grain size in the ribbons. Because of this difference in precipitation, the ribbons exhibited higher hardness than bulk specimens at short aging times. Prolonged aging caused a significant drop in hardness due to microstructural coarsening.

KEYWORDS

Rapid solidification; ribbons; titanium-cobalt; eutectoid-former; precipitation; age hardening; alpha phase; beta phase; intermetallic.

INTRODUCTION

Rapid solidification (RS) processing has the advantages of increased material homogeneity, microstructural refinement, extension of solid solubility, and formation of unique structures. These benefits have been demonstrated in many conventional titanium alloys such as Ti-6Al-4V [1], Ti-5Al-2.5Sn [2], Ti-11.5Mo-6Zr-4.5Sn [1], and Ti-15V-3Cr-3Al-3Sn [3] as well as in novel alloy compositions containing rare earth elements [4], C and B [5]. Currently, the feasibility of developing high strength RS titanium alloys containing elements like Fe, Cr, Co, Cu, and Ni that lead to eutectoid decomposition of the beta phase is being investigated [6]. Past efforts on the development of eutectoid forming titanium alloys using a conventional ingot metallurgy (IM) approach were hampered due to alloy segregation and consequent workability problems. Additionally, the precipitation of compound particles at grain boundaries led to intergranular fracture and alloy embrittlement [7-9]. Thus, compound formation is usually suppressed by further alloying as in the case of commercial alloys Ti-13V-11Cr-3Al and Ti-10V-2Fe-3Al that contain eutectoid forming additions. The present study on a hypoeutectoid Ti-9 wt% Co alloy investigates the feasibility of using rapid solidification processing to obtain microstructural homogeneity and modification of precipitation characteristics for achieving a combination of high strength and ductility.

Titanium-rich Ti-Co alloys undergo a eutectoid decomposition of the beta phase into alpha+Ti_2Co at 685°C and 9.5 wt% Co as shown in Fig. 1 [10]. Quenching of Ti-rich alloys containing up to ∿6 wt% Co from the beta phase field to room temperature leads to the formation of martensite [11]. At higher Co contents, the metastable beta phase can be fully retained at room temperature. Previous work on Ti-Co alloys containing 8-10 wt% Co showed that good mechanical properties (1170MPa tensile strength and 10-15% elongation) result in the as beta-quenched condition, but tempering to produce alpha+Ti_2Co leads to embrittlement [9]. The alloy brittleness was observed in the form of intergranular fracture and was associated with grain boundary precipitation of the decomposition products alpha+Ti_2Co in the initial stages of tempering. With continued tempering, precipitation progressed from grain boundaries towards grain interiors. This was accompanied by a decrease in strength, a moderate improvement in ductility with a transition from intergranular to transgranular fracture [9]. The objective of the present work was to investigate the feasibility of producing a high strength Ti-Co alloy via rapid solidification processing and determine the advantages of this method over the conventional IM approach. In this communication, the precipitation behavior of a rapidly solidified Ti-9 wt% Co alloy is reported and compared with that of conventional IM Ti-9 wt% Co alloy samples.

EXPERIMENTAL PROCEDURE

The Ti-9 wt% Co alloy was prepared in the form of a cigar-shaped button by tungsten arc melting high grade elemental components. The button was remelted and solidified several times to increase homogeneity. The alloy was then swaged at 1040°C through successive dies to a final diameter of 10 mm. A portion of this swaged material was used for remelting and rapid solidification by the chill block melt spinning technique [12, 13]. This process consisted of arc melting the alloy in a water cooled copper crucible and forcing the molten alloy through an orifice at the bottom of the crucible on to a spinning heat extracting copper wheel. Thus, the molten jet was spun into long ribbons ∿2 mm wide and 50 μm thick. The remaining portion of the swaged rod was cut into 5 mm thick pieces. These IM bulk samples were encapsulated in evacuated vycor tubing and beta solution treated at 800°C for 30 min and quenched in ice water mixture while breaking the capsules. The solution treated bulk samples and melt spun ribbon samples were encapsulated in evacuated vycor for further aging heat treatments. Aging was carried out at three different sub-eutectoid temperatures (525°C, 575°C, and 650°C) for times varying from 10 s to 8.64×10^4 s (24 hr).

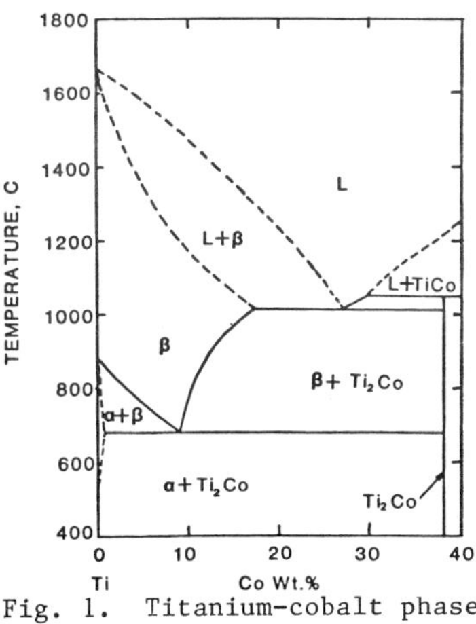

Fig. 1. Titanium-cobalt phase diagram [10].

Microstructural characterization of the alloys was carried out by optical microscopy, SEM, and TEM techniques. The compositions of precipitates and matrix were investigated in a scanning transmission electron microscope equipped with x-ray energy dispersive spectroscopy attachment. The age hardening response was studied by measuring microhardness values using a 136° apex square-based diamond pyramid indentor at a test load of 25 g. Readings were taken from 20 indentations on each sample to calculate the average hardness values.

RESULTS AND DISCUSSION

The aging response of bulk samples and RS ribbons of Ti-9 wt% Co alloy is shown in Fig. 2 where the microhardness is plotted as a function of aging time at 525°C and 575°C. Data for 650°C aging which showed a continuous decrease in microhardness with aging time are not indicated in this figure.

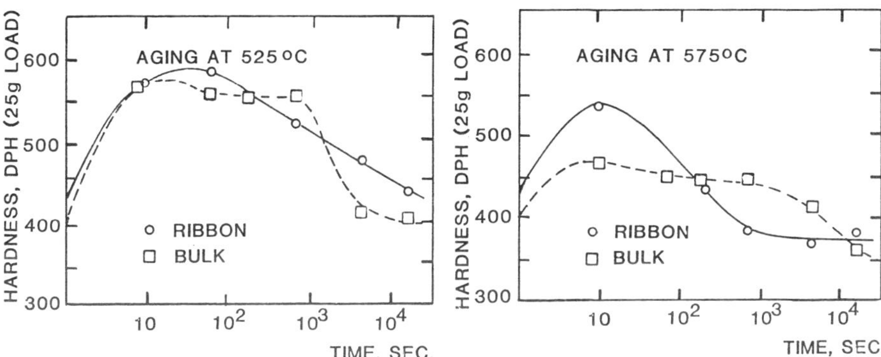

Fig. 2. Variation of microhardness with aging time for Ti-9 wt% Co alloy ribbons and bulk specimens aged at 525°C and 575°C.

Initially, the rapidly solidified ribbon exhibits a higher hardness than the solution treated bulk alloy sample. On aging at 525°C and 575°C, hardness peaks are observed at very short aging times (∿10 s). Beyond these hardness maxima, the aging response is significantly different for ribbon and bulk samples. In the case of RS ribbons, the aging at 525°C causes a gradual decrease in hardness following the peak value. At 575°C, a similar drop in the hardness of ribbons is observed in the aging interval $10\text{-}10^3$ s, with no appreciable decrease at longer aging times. In contrast, the hardness values for bulk samples reach a plateau in the $10\text{-}10^3$ s interval, and show a decrease at longer times. No marked difference is noticed in the hardness of ribbons and bulk samples after prolonged aging. The hardness values decreased further when aging was continued up to 8.64×10^4 s (24 hr).

Microstructural characterization of as-quenched ribbons and solution treated bulk specimens revealed a retained beta structure in both cases. The beta grain size was over 400 μm in the bulk specimens as compared to about 25 μm in the ribbons. The examination of the aged samples showed that similar microstructural changes occurred at each aging temperature. The microstructures obtained by the aging of bulk specimens for various periods at 575°C are presented in Fig. 3. In this coarse-grained material, the early stage of aging consists of precipitation of fine alpha phase predominantly near the original beta grain boundaries (Fig. 3a). Precipitation occurs to a lesser extent at the grain interiors (Fig. 3b). Further aging leads to an increase in the volume fraction of alpha phase (Fig. 3c) followed by some coarsening of alpha structure. The Ti_2Co intermetallic compound was not observed until after 15 min of aging. Fig. 3d shows the coarse alpha+Ti_2Co microstructure after 4 hr of aging at 575°C. The retained beta microstructure of as-produced RS ribbons is shown in Fig. 4a. When this material is aged at 575°C, the alpha+Ti_2Co structure evolves at very short aging times (<60 s) as indicated in Fig. 4b. This stage is followed by coarsening (Figs. 4c and 4d).

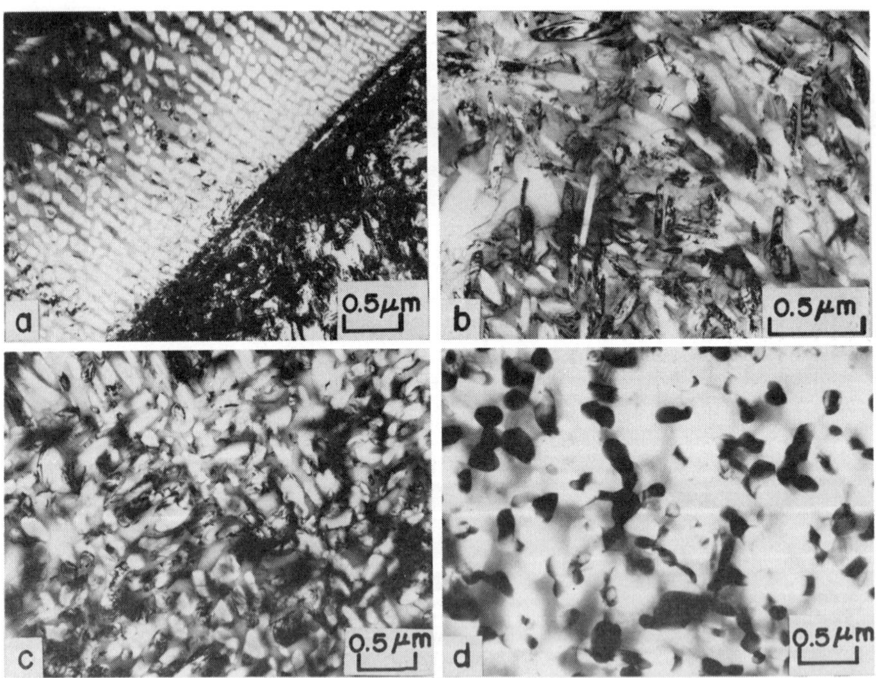

Fig. 3. TEM micrographs of Ti-9 wt% Co IM bulk specimens solution treated at 800°C for 30 min and water quenched, followed by aging at 575°C: (a), (b) aged 10 s, (c) aged 180 s, and (d) aged 14,400 s.

The microhardness data may now be rationalized in terms of the microstructural changes observed during aging. The higher initial hardness of the rapidly solidified alloy as compared to bulk material may be attributed to a higher vacancy concentration in the RS ribbon. In the initial stage of aging, the precipitation of alpha phase occurs on a fine scale within the beta matrix. This alpha precipitation can account for the rise in hardness of both IM bulk material and RS ribbons. In the case of bulk material, further aging leads to an increase in the volume fraction of the alpha phase followed by some coarsening. This correlates with the hardness level being maintained without any appreciable decrease during the aging interval $10-10^3$ s. After this aging period, the Ti_2Co intermetallic compound precipitates at the alpha/beta interphase boundaries and the alpha regions coarsen considerably. This stage is accompanied by a decrease in the hardness of the bulk specimens. In contrast, the precipitation of Ti_2Co compound occurs at shorter aging times in the RS ribbons. Consequently, the hardness drop is observed earlier in the case of ribbons.

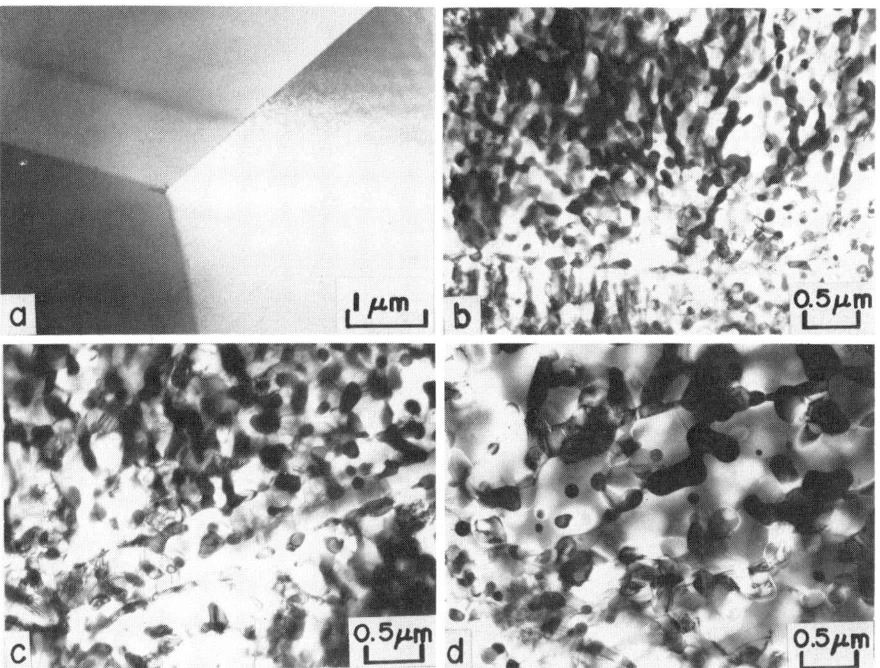

Fig. 4. TEM micrographs of Ti-9 wt% Co RS ribbons: (a) as-produced, (b), (c), and (d) represent microstructures obtained by aging at 575°C for 60 s, 600 s, and 14,400 s, respectively.

The microhardness results obtained in this work are consistent with those reported previously for a Ti-10 at% Co alloy aged in the temperature range 450-525°C [14]. In this earlier study, a hardness peak was observed after a few minutes of aging followed by a sharp drop in hardness corresponding to the precipitation of Ti_2Co [14].

CONCLUSIONS

Rapid solidification of Ti-9 wt% Co alloy results in a considerable refinement of microstructure. This is demonstrated by a comparison of the microstructures of bulk specimens and RS ribbons which showed a beta grain size of 400 μm and 25 μm, respectively. Aging of the bulk samples and RS ribbons at sub-eutectoid temperatures resulted in hardness peaks at very short times. The increase in hardness is associated with the precipitation of fine pro-eutectoid alpha phase accompanied by solute enrichment of the beta matrix. This initial stage of aging is followed by the formation of Ti_2Co intermetallic compound at the alpha/beta interphase boundaries leading to a decrease in hardness. The precipitation of alpha+Ti_2Co structure occurs more uniformly in the RS ribbons, but the kinetics of this reaction are very rapid presumably due to a higher vacancy concentration and a finer grain size in the RS material. A drop in hardness is noted after the appearance of Ti_2Co compound in the aged alloy microstructures. This may be attributed to the loss of solid solution strengthening due to the depletion of beta phase, and the coarsening of alpha+Ti_2Co structure.

ACKNOWLEDGMENTS

S. Krishnamurthy and D. Eylon gratefully acknowledge the financial support provided under AFWAL, Materials Laboratory Contract Nos. F33615-82-C-5001 and F33615-82-C-5078, respectively. The experimental assistance of P. E. Blosser during the course of this study is recognized. The assistance of W. A. Houston in optical metallography, R. D. Brodecki in SEM work, and R. E. Omlor in STEM analysis is also appreciated.

REFERENCES

1. T. F. Broderick, F. H. Froes, and A. G. Jackson, in Rapidly Solidified Metastable Materials (edited by B. H. Kear and B. C. Giessen), 28, p. 345. Elsevier Science Publishing Co., Inc., New York (1984).
2. A. G. Jackson, T. F. Broderick, F. H. Froes, and J. Moteff, in Rapid Solidification Processing Principles and Technologies, III (edited by R. Mehrabian), p. 585. National Bureau of Standards, Washington, DC (1982).
3. T. C. Peng, S. M. L. Sastry, J. E. O'Neal, and J. F. Tesson, Fall AIME Meeting, Louisville, KY (1981).
4. S. M. L. Sastry, P. J. Meschter, and J. E. O'Neal, Met. Trans. A 15A, No. 7, 1451 (1984).
5. S. M. L. Sastry, T. C. Peng, P. J. Meschter, and J. E. O'Neal, J. Metals 35, No. 9, 21 (1983).
6. S. Krishnamurthy, R. G. Vogt, D. Eylon, and F. H. Froes, in Rapidly Solidified Metastable Materials (edited by B. H. Kear and B. C. Giessen), 28, p. 361. Elsevier Science Publishing Co., Inc., New York (1984).
7. H. W. Worner, J. Inst. Metals 79, No. 3, 173 (1951).
8. F. C. Holden, H. R. Ogden, and R. I. Jaffee, Trans. AIME 206, 521 (1956).
9. F. W. Yakymyshyn, G. R. Purdy, R. Taggart, and J. Gordon Parr, Trans. ASM 53, 283 (1961).
10. F. L. Orrell and M. G. Fontana, Trans. ASM 47, 554 (1955).
11. Y. C. Huang, S. Suzuki, H. Kaneko, and T. Sato, in The Science, Technology and Application of Titanium (edited by R. I. Jaffee and N. E. Promisel), p. 691. Pergamon Press, London (1970).
12. S. J. Savage and F. H. Froes, J. Metals 36, No. 4, 20 (1984).
13. S. H. Whang and B. C. Giessen, in Rapid Solidification Processing Principles and Technologies, III (edited by R. Mehrabian), p. 439. National Bureau of Standards, Washington, DC (1982).
14. P. R. Swann and J. G. Parr, Trans. TMS-AIME 212, 276 (1958).

Rapid Solidification and Aging of a Near-eutectoid Titanium-Nickel Alloy

W. A. Baeslack III*, S. Krishnamurthy and F. H. Froes****

**Department of Welding Engineering, The Ohio State University, Columbus, OH 43210, USA*
***AFWAL Materials Laboratory, Wright-Patterson AFB, OH 45433, USA*

ABSTRACT

Rapidly-solidified microstructures were produced in a near-eutectoid Ti-5.5 wt% Ni alloy by laser surface melting and melt spinning, and subsequently heat-treated at sub-eutectoid temperatures for 3, 10 and 60 minutes. Light and scanning-electron microscopy found distinct morphological differences in the β grain morphologies and transformed-β microstructures of the RS products to result from differences in β grain nucleation and in the cooling rates experienced through the solidification and solid-state transformation temperature ranges. Heat treatment produced a decrease in the hardness of all as-solidified structures due to Ti_2Ni precipitation and/or coarsening.

KEYWORDS

Titanium-nickel; rapid solidification; laser melting; melt spinning; eutectoid; precipitation.

INTRODUCTION

The application of rapid-solidification (RS) processing to titanium alloys has become a subject of increased interest and study [1]. The distinct advantage of RS processing in producing unique microstructures and mechanical properties in titanium alloys which are not obtainable via conventional ingot metallurgy has perhaps best been demonstrated by the generation of alloys exhibiting eutectoid or near-eutectoid compositions. In conventional ingot metallurgy, the segregation of eutectoid-former elements such as iron, copper and nickel during initial alloy solidification and during processing has greatly limited their allowable levels. The recent application of RS processing to these alloy systems however, has shown that extremely fine-grained homogeneous microstructures can be produced, with the specific microstructure constituents and morphologies dependent on both the actual alloy chemistry and the nature of the rapid solidification processes [2]. Recent studies of an electron-beam melted and splat-quenched Ti-3 wt%Ni alloy have, for example, shown that rapid solidification and cooling rates can produce a fine-grained martensitic structure which, on aging, transforms to fine α plus a uniform distribution of Ti_2Ni precipitates. It is anticipated that such microstructures would be very conducive to high strength.

This paper presents the results of a continued investigation into the RS behavior of the Ti-Ni system. Specifically, the microstructures and heat treatment response of a near-eutectoid Ti-5.5 wt% Ni alloy produced by the laser melting and melt-spinning processes are compared and discussed.

EXPERIMENTAL

A Ti-5.5 wt% Ni alloy was prepared by the tungsten-arc melting of elemental components and subsequent drop casting into a cylindrical shape. Homogenization of the cast ingot prior to RS conversion included heat treatment at 1038°C for 24 hours followed by a two-step swaging operation which provided a total reduction in area of 50%. Laser surface melts were generated on wafer specimens cut transversely from the 10 mm diameter rod using a 600 watt enhanced-pulse CO_2 laser equipped with a high-speed traversing system and argon shielding. Cooling rates of about 10^5 and 10^3°C/s in the solidification range were obtained with continuous laser power of 550 watts and traversing rates of about 40 mm/s (designated as Type A) and 0.4 mm/s (designated as Type B), respectively. Conversion of the alloy to ribbon was accomplished by a melt-spinning process described previously [2]. Both the laser melts and melt-spun ribbons were heat treated at sub-eutectoid temperatures (T_{eutd} = 775°C) of 500 and 625°C for times of 3, 10, and 60 minutes. Characterization of both the as-solidified and heat-treated specimens included light and scanning-electron microscopy and Vicker's hardness testing (50 gram load).

RESULTS AND DISCUSSION

The as-swaged Ti-5.5 wt% Ni alloy microstructure illustrated in Fig. 1 exhibited a coarse, nearly-equiaxed prior-β grain microstructure. Slow cooling subsequent to swaging from temperatures exceeding the β transus promoted β decomposition to a "bainitic," nonlamellar-eutectoid microstructure comprised of grain boundary allotriomorphs and intragranular proeutectoid α plates separated by Ti_2Ni compound particles. This coarse α

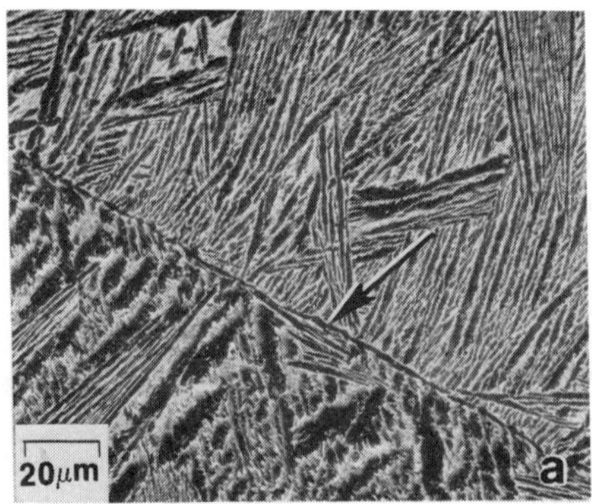

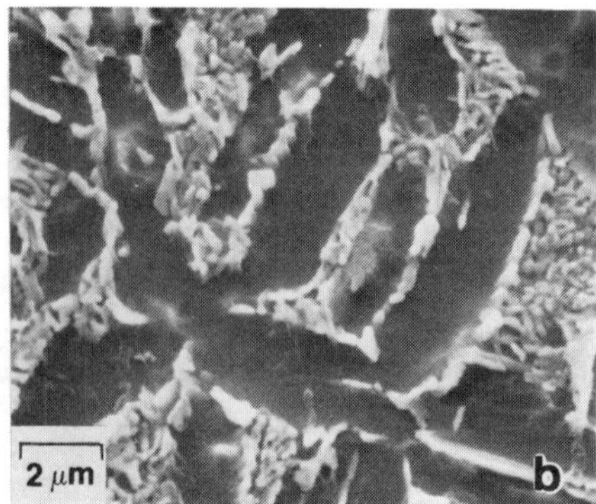

Fig. 1. SEM micrographs of as-swaged Ti-5.5 wt% Ni alloy. Arrow in (a) indicates prior-β grain boundary.

plate structure and ineffective strengthening by the incoherent Ti_2Ni particles contributed to a relatively low hardness (270 DPH).

Figure 2 illustrates the macrostructural characteristics of the Type A laser melt. A relatively uniform melt width and a typical Gaussian heat source, conduction-limited transverse cross-section shape can be observed. Solidification of the melt occurred epitaxially from the coarse β grain substrate by a columnar dendritic growth mode. As these grains grew, their substructure became increasingly finer due to increased cooling rates nearer the melt surface and ultimately transitioned to an extremely fine cellular product (approximately 1 μm cell spacing). Also evident near the melt zone surface were β grains which appeared columnar at the melt surface and nearly equiaxed in the transverse section, and which exhibited very fine celullar solidification substructures. Considering that the entire melt zones were typically contained within one or two β grains, and that growth occurred epitaxially, such a fine β grain structure would not be expected.

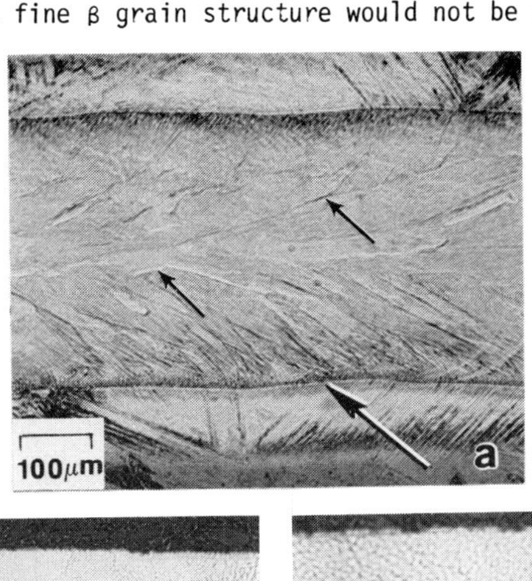

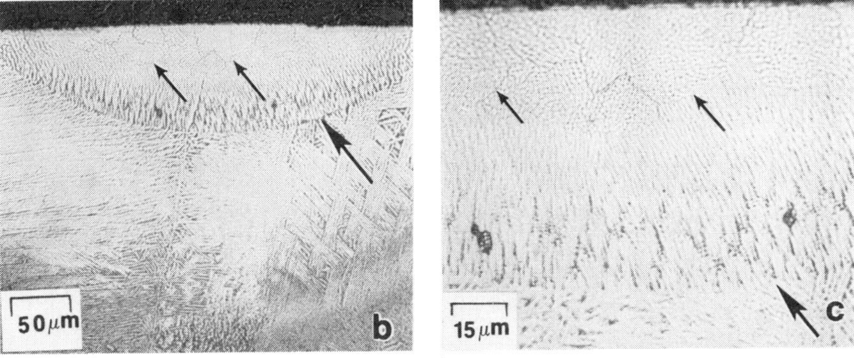

Fig. 2. Light micrographs of Type A laser melt showing (a) top surface and (b, c) transverse section. Large arrows indicate fusion boundary, small arrows indicate β subgrain boundaries.

Figure 3 illustrates the larger Type B laser melt zone, which also solidified epitaxially from the base metal substrate. Observation of the as-solidified surface showed a coarse, cellular solidification substructure (8-10 μm cell spacing). The SEM micrograph in Fig. 3b shows the fine transformed-β microstructure in the melt zone which prevented the clear delineation of β subgrains and the solidification substructure. Based on continuous-cooling transformation work of Kaneko and Huang [4], a completely martensitic transformed β structure would be expected. However, observation at high magnification indicated the presence of very fine Ti_2Ni particles at plate boundaries, suggesting at least a partial diffusional transformation. It is important to recognize that the delineation of HCP transformation products produced by a diffusionless-shear (martensitic) process versus a diffusional process is often not clear in titanium alloys, particularly in continuously-cooled structures. Several investigators have shown that transitions in HCP plate growth from a diffusionless/martensitic process to a diffusional process can be essentially continuous [5-7].

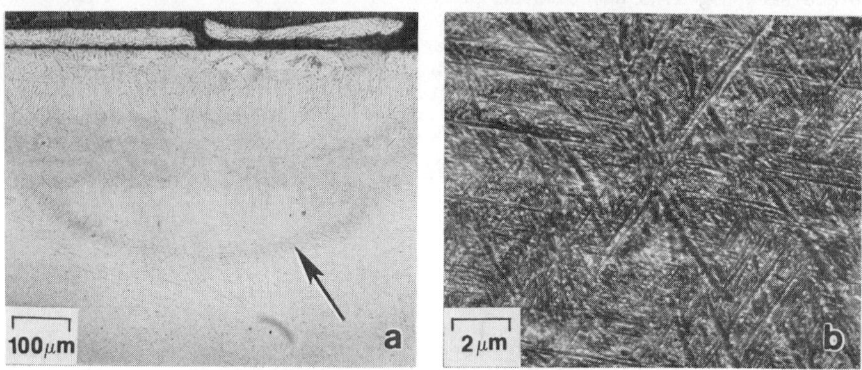

Fig. 3. Light and SEM micrographs of transverse section in Type B laser melt. Large arrow in (a) indicates fusion boundary.

Heat treatment of the Type A laser melt at 500°C promoted the formation of extremely fine Ti_2Ni precipitates both homogeneously (as characterized by SEM) and heterogeneously at solidification substructure and β subgrain boundaries. At the higher heat treatment temperatures precipitation occurred in a similar manner, but with a much coarser precipitate size. Figure 4 illustrates the Ti_2Ni precipitates in the Type A laser melt heat

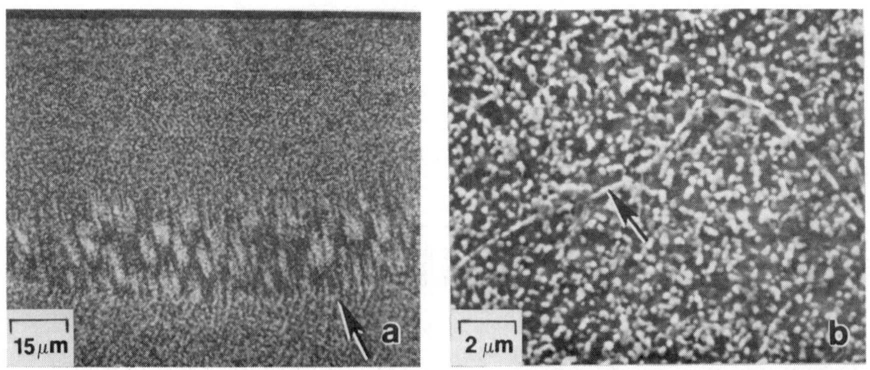

Fig. 4. Light and SEM micrographs showing microstructure of Type A laser melt aged at 625°C for 10 minutes, arrow in (a) indicates fusion boundary, arrow in (b) indicates β subgrain boundary.

treated at 625°C for 10 minutes. Note the heterogeneous precipitation at β subgrain boundaries (Fig. 4b). Heat treatment of the Type B laser melt resulted in a coarsening of the Ti_2Ni precipitate structure, which becomes particularly evident at 625°C. As expected, the precipitate coarsening at both temperatures increased with time at temperature.

In contrast to the elongated β grains in the laser melts, the melt-spun ribbons exhibited nearly equiaxed β grains approximately 10 to 20 μm in diameter. Evidence of a nonlamellar eutectoid microstructure comprised of extremely fine α plates and Ti_2Ni particles, rather than martensite, suggests that beta decomposition occurred after the ribbon separated from the quench wheel. Heat treatment of the structure promoted a coarsening of the intragranular plates, with the degree of coarsening increasing with both temperature and time at temperature. As shown in Fig. 5d, higher temperatures also promoted a coarsening and spheroidization of Ti_2Ni second phase particles.

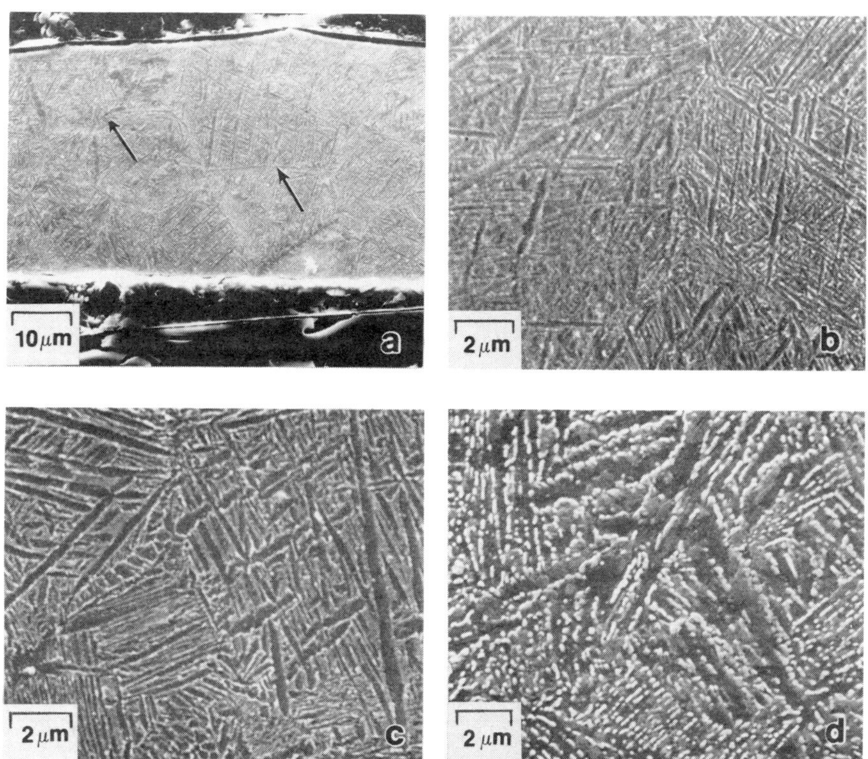

Fig. 5. SEM micrographs of melt-spun ribbon, (a) transverse section, arrows indicate prior-β grain boundaries, (b-d) transformed-β microstructures in as-solidified, aged at 500°C/10 min. and aged at 625°C/10 min. conditions, respectively.

Figure 6 shows the results of hardness testing on the as-solidified and heat-treated laser and melt-spun microstructures. The extremely fine as-solidified Type A laser melt microstructure exhibited the highest hardness.

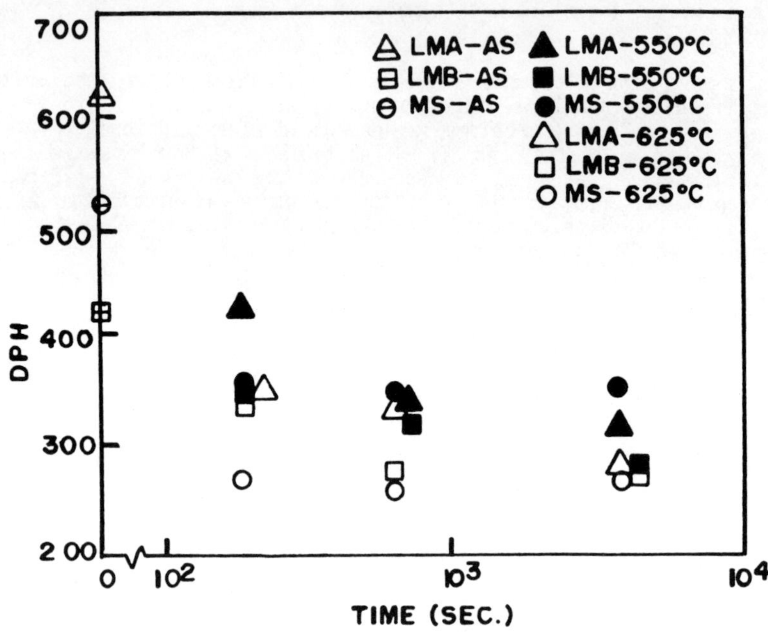

Fig. 6. Microhardness versus time curves for laser melt and melt-spun RS microstructures at two sub-eutectoid temperatures.

An extremely fine martensitic microstructure would be expected to contribute to such a high hardness, although the influence of fine Ti_2Ni particles is not as clear. Somewhat unexpectedly, the coarser melt-spun bainite structure was harder than the finer martensite/bainite structure in the as-solidified Type B laser melt. Heat treatment of all microstructures promoted a decrease in hardness, with the softening response being the most rapid in the melt-spun material. Earlier work on an ingot Ti-6 wt% Ni alloy [8] suggests that the drop in hardness resulted from the precipitation and/or coarsening of incoherent Ti_2Ni, which results in a depletion of Ni from the HCP matrix. A coarsening of the HCP plate structure concurrent with this Ti_2Ni precipitation and coarsening also likely contributed to the observed reductions in hardness.

REFERENCES

1. S. M. L. Sastry, T. C. Peng, P. J. Meschter and J. E. O'Neal, J. Metals 21, 35 (1983).
2. S. J. Savage and F. H. Froes, J. Metals 20, 36 (1984).
3. J. E. O'Neal, S. M. L. Sastry, T. C. Peng and J. F. Tesson, Microstructural Science, 11, 143 (1983).
4. H. Kaneko and Y. C. Huang, J. Jpn. Inst. Met. 27, 387 (1963).
5. H. M. Flower, S. D. Henry and D. R. West, J. Mater. Sci. 9, 57 (1974).
6. W. A. Baeslack III and F. D. Mullins, J. Mater. Sci. Let. 2, 715 (1983).
7. W. A. Baeslack III and F. D. Mullins, J. Mater. Sci. Let. 3, 333 (1984).
8. D. H. Polonis and J. Gordon Parr, Trans. AIME 206, 531 (1956).

Microstructure of Rapidly Solidified IN-100 Superalloy

Zhang Shouhua, Cheng Tianyi, Hu Benfu and Li Huiying

Beijing University of Iron and Steel Technology, Beijing, People's Republic of China

ABSTRACT

Thin ribbons of melt spun IN-100 superalloy are studied with TEM, SEM, EDS, electron and X-ray diffractions. The cast structures of these rapidly solidified ribbons are found to consist of microcrystalline, cellular and dendritic grains, while their microstructures are composed of γ solid solution, γ' precipitates and MC carbides. The microstructures are related to the cooling rates during solidification of the ribbons. The lattice parameters of γ solid solution increase with the cooling rates. Evolution of γ' precipitates follows a change in cooling rates. It appears in several stages: the formation of GP zones and their ordering, the formation of γ' precipitates and a morphological change from the spherical to cubic. MC carbide particles dwell at the subcell and simple cell boundaries, also in the interdendritic regions, dependent on the distance away from the bottom surface of a ribbon. Extracted MC carbides exhibit a daisy-like morphology and their chemical compositions are determined. The interdendritic segregations in the alloy are also studied.

KEYWORDS

Rapid solidification; IN-100 superalloy; melt spinning; cast structures; microstructure; microsegregation; tensile strength.

INTRODUCTION

Rapid solidification technique has received special attention in the development of superalloys in recent years. An experimental helium atomizer has been set up to produce superalloy powders [1]. Microstructural studies of rapidly solidified IN-100 superalloys have appeared in the literature[2,3,4,5], but the quantitative relationships between parameters of the solidification process and the resulting alloy microstructures are meager up to the present. This paper intends to correlate the microstructures with the cooling rates during solidification, aiming at the control of the structures of the alloy by regulating the cooling rates, obtained by whatever means.

MATERIALS AND METHODS

The chemical composition of the Ni-based IN-100 superalloy used is as follows: 0.16% C, 10.01% Cr, 3.10% Mo, 5.77% Al, 5.70% Ti, 14.93% Co, 0.94% V, 0.014% B, 0.026% Zr, balance Ni. The alloy has been vacuum-melted and cast into rods, which are then melt spun in air into thin ribbons 10–80 μm thick on a copper wheel 180 mm in diameter, rotating at 5200, 4400, 3500, 2000 or 800 rpm. The corresponding cooling rates, calculated from the secondary dendritic arm spacings (DAS) measured on the thin ribbons by Holiday's equation $DAS = 40(\dot{T})^{-0.33}$ [1], where $\dot{T}$ is the mean cooling rate from the melting point to the half melting point temperature of the alloy, are listed in Table 1. Microstructures of the thin ribbons are studied with TEM, SEM, EDS, electron and X-ray diffractions. The tensile properties are measured with a hand manipulated miniature testing device.

TABLE 1 The Measured Average Secondary DAS and the Calculated Cooling Rates in Ribbons

Rotating speeds, rpm:	4400	3500	2000	800
Ave. thickness of ribbons, μm:	10.4	17.1	30.5	83.3
Ave. DAS, μm:	0.228	0.297	0.608	0.874
Ave. cooling rate, $\dot{T}$, K/s:	6.3×10^6	2.8×10^6	3.2×10^5	1.1×10^5

RESULTS AND DISCUSSIONS

The cast structures of the rapidly solidified ribbons are observed to arrange themselves from bottom to top in the order of microcrystalline, cellular and dendritic grains, Fig. 1. The thicknesses of these zones vary with the cooling rates. When the cooling rate is increased, the dendritic zone becomes thinner and the cellular zone thicker. The microcrystalline zone is always very thin, less than 1 μm.

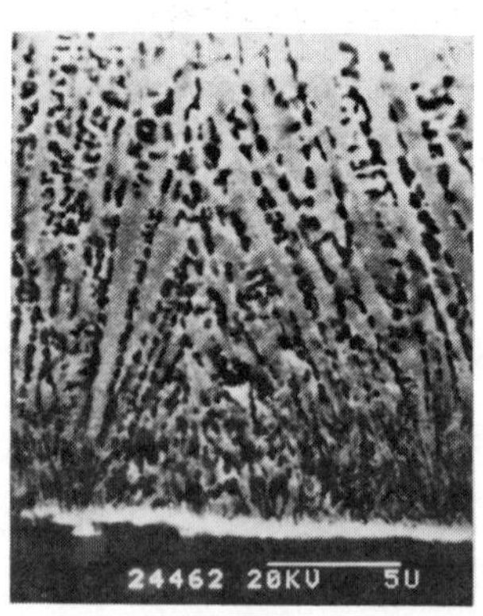

Fig. 1. Cast structure of a rapidly solidified ribbon. SEM.

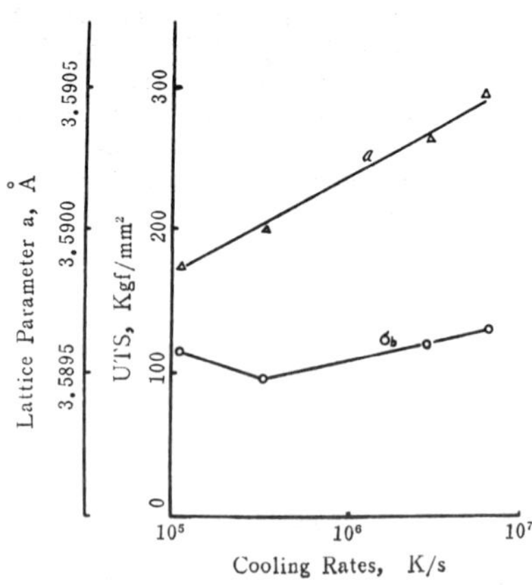

Fig. 2. Change of lattice parameters of γ solid solution and UTS of ribbons with cooling rates.

Since the flow of heat during solidification in a ribbon is principally unidirectional, the cooling rate varies along the thickness of a ribbon. The cooling rates at the points of cellular-dendritic transition and of microcrystalline-cellular transition are found to be in the order of 10^6 and 10^7 K/s respectively.

Only three microconstituents are observed in the rapidly quenched IN-100 alloy: γ solid solution, ordered γ' precipitates and MC type carbides. No M_3B_2 phase and $\gamma + \gamma'$ eutectics, both usually present in conventionally cast alloys, are observed. The matrices are composed of supersaturated γ solid solution, of which the lattice parameters, hence the degree of supersaturation of solute elements, are found to vary linearly with the logarithm of cooling rates, Fig. 2.

The evolution of γ' phase from γ solid solution is observed to follow a change in cooling rates, Fig. 3. A mottled appearance in the matrices occurs in ribbons spun at speeds 5200, 4400 and 3500 rpm, Fig. 3a, b. Electron diffraction patterns obtained from the matrices reveal the fact that the formation of GP zones and their subsequent ordering process have occurred in the alloy even when the cooling rate is as high as 10^7 K/s. The size of the mottled regions and the roughness of the appearance increase with the decrease in cooling

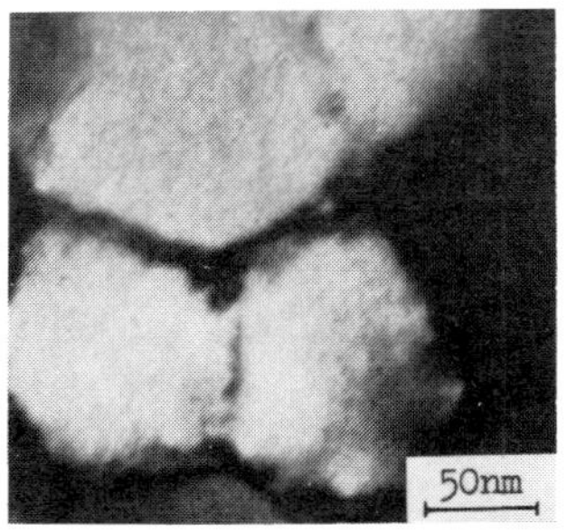

a. Mottled appearance. 5200 rpm.

b. Mottled appearance. 4400 rpm.

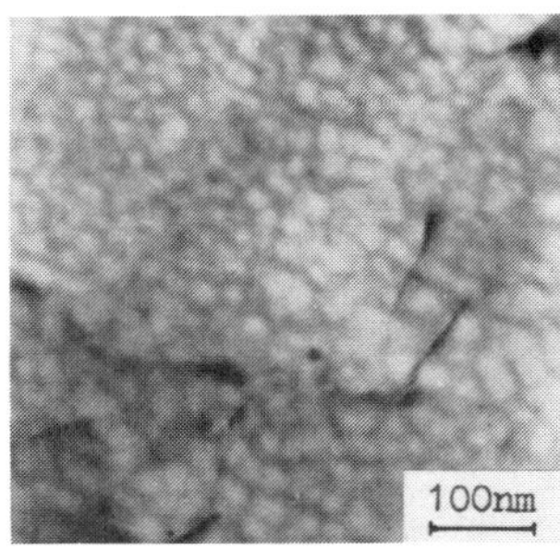

c. Spherical γ' precipitates. 2000 rpm.

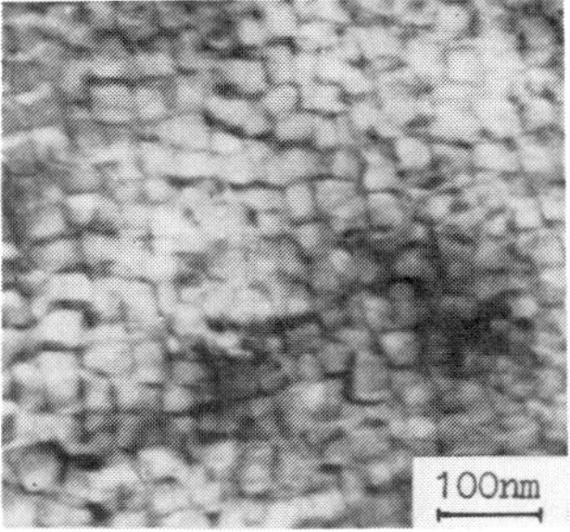

d. Cubic γ' precipitates. 800 rpm.

Fig. 3 Progress in the formation of γ' phase from γ solid solution shown in ribbons spun at different speeds. TEM.

rate. The mottled regions become clearly delineated in ribbons spun at 2000 rpm, Fig. 3c. This means that the formation of γ' precipitates is accomplished when the cooling rate is about 10^5 K/s. In ribbons spun at 800 rpm, the morphology of γ' precipitates has changed from spherical to cubic, Fig. 3d. In our previous investigations of René 95 superalloy, the formation of mottled regions can be suppressed at cooling rates around 10^7 K/s and no clearly defined γ' precipitates are observed at 10^5 K/s [6]. That IN-100 alloy contains a sum of 11.5% Al and Ti, while René 95 alloy only 6.2% explains the differences in kinetics for the evolution of γ' phase. However, the morphological change which occurs when the γ' precipitates have grown to a size of about 0.2 μm, is similar to the observations with René 95 [6] and Nimonic alloys [7].

During solidification, carbide particles have developed to decorate the grain boundaries. Few MC carbide particles are observed on the microcrystalline grain boundaries, Fig. 4, a fact indicating that little microsegregation has taken place. A cellular zone comes next to the microcrystalline zone. It is interesting to note that cellular grains at positions closer to the microcrystalline region bear subcell structures, Fig. 5, while farther ones are simple cell grains, Fig. 6. Carbide particles develop on both type of cell

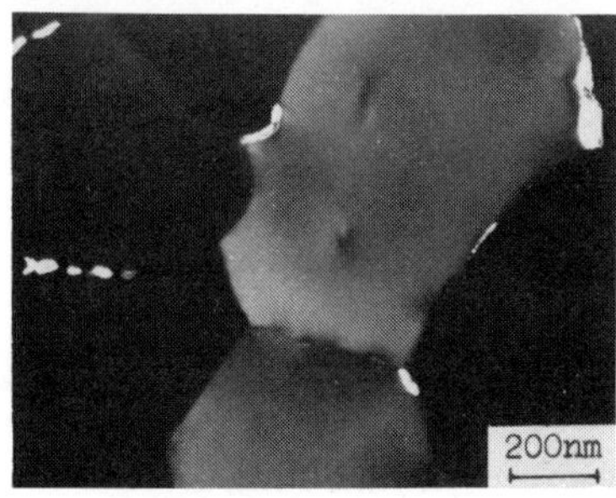

Fig. 4. Carbide particles on γ grain boundaries in microcrystalline zone. 2000 rpm. TEM, dark field.

Fig. 5. Carbide particles on subcell and cell boundaries near the mid-portion of a ribbon. 3500 rpm. TEM.

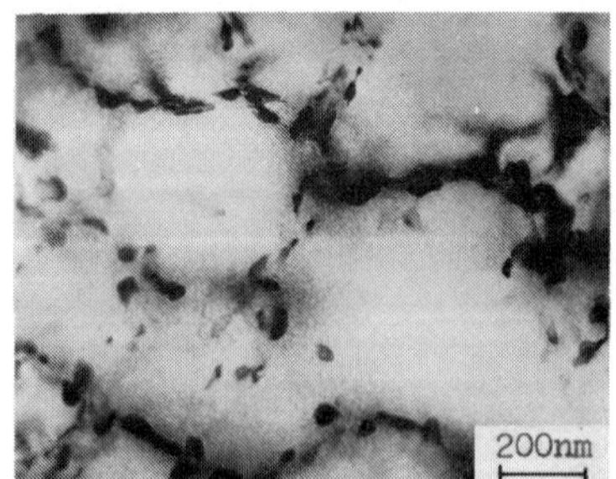

Fig. 6. Carbide particles on simple cell boundaries near the top-portion of a ribbon. 3500 rpm. TEM.

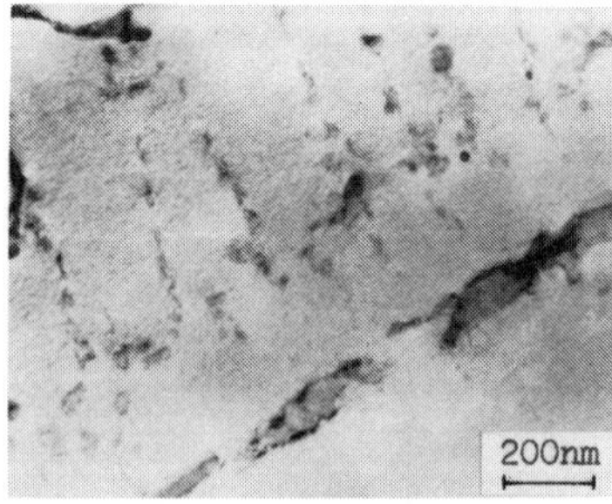

Fig. 7. Interdendritic carbide particles in the top-portion of a ribbon. 3500 rpm. TEM.

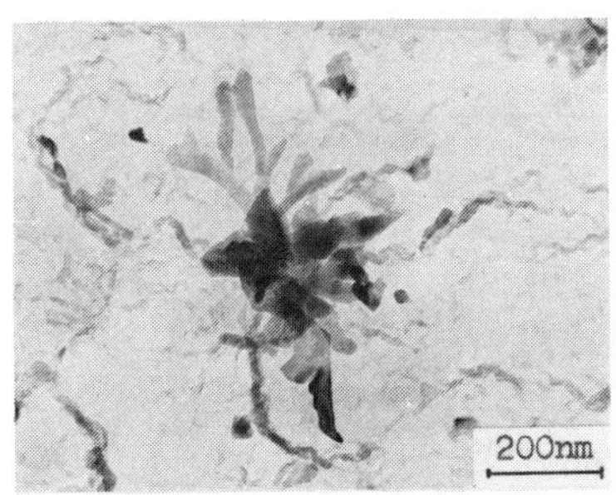

Fig. 8. Daisy-like carbides extracted from a ribbon spun at 4400 rpm. TEM.

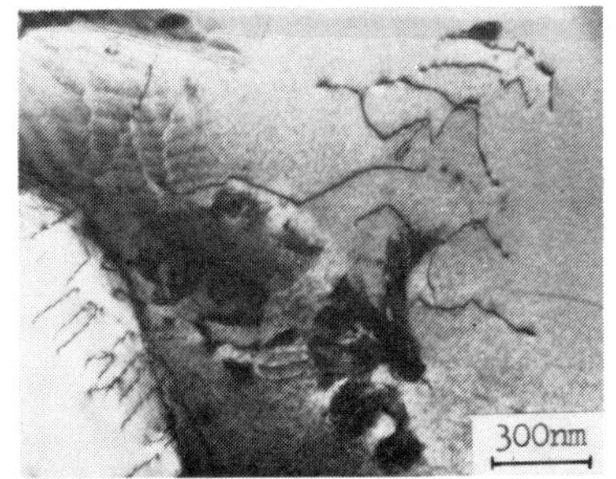

Fig. 9. Daisy-like carbides and dislocations inside a grain. 2000 rpm. TEM.

boundaries. Numerous large MC carbide particles are found in the interdendritic regions, Fig. 7. The distribution of carbides indicates that, when the position of a particular microstructure is located farther away from the bottom surface of a ribbon, i. e. when the cooling rate at the position is lower, the segregation distance of the elements and the extent of their segregation become greater and the size and amount of carbide particles become larger. The extracted MC carbide particles exhibit daisy-like morphologies, Fig. 8. Their chemical composition are shown in Table 2.

TABLE 2 Comparison of Chemical Compositions between Daisy-like and Conventionally Cast Carbides. EDS

Carbides	Al	Mo	Ti	V	Cr	Co	Ni
Conventionally cast	-	33.2	55.0	5.4	3.5	1.2	4.1
Daisy-like	5.8	33.2	40.7	6.6	9.9	1.4	2.2

We see that the chemical composition of daisy-like carbides deviates considerably from that found in conventionally cast alloys. The strong carbide-forming elements Ti, Mo and V concentrate themselves in both types MC carbides. While Ti, which is the strongest of these three elements, has not got enough chance to concentrate in the rapidly solidified carbides, Cr, which has a limited solubility in MC carbides, acquires no sufficient time to migrate away from them. Obviously, the growth rates of the daisy-like carbides are so high that a significant amount of non-carbide-forming element Al is entrapped in them. The daisy-like morphology itself suggests their high growth rates. As the cooling rate decreases, the carbides still keep the same morphology but the petals are getting thicker. Figure 9 shows daisy-like carbides engulfed by a growing γ grain and the dislocations generated in the matrix lattice in their immediate neighborhood [8]. Carbides also precipitate on the dislocations inside the γ grains, Fig. 3b.

The dendritic segregations of elements in both conventionally cast and rapidly solidified IN-100 alloys have been studied with EDS, and their segregation index values C_{max}/C_{min} are shown in Table 3. For the alloy in the rapidly solidified states, the segregations are greatly reduced. The segregation index of Ti is highest both for the conventionally cast and for the rapidly solidified states. The segregation index of an element decreases with its carbide-forming tendency. It is consistent with the known fact that MC carbides are concentrated in the interdendritic regions.

TABLE 3 Comparison of Dendritic Segregation Indices of Elements between the Conventionally Cast and the Rapidly Solidified Alloys. EDS

Alloy condition	Segregation indices					
	Al	Mo	Ti	Cr	Co	Ni
Conventionally cast	−1.18	+1.35	+1.61	+1.17	+1.23	−1.11
Rapidly solidified	+1.03	+1.06	+1.09	+1.01	−1.04	+1.01

Ultimate tensile strengths (UTS) of the ribbons change with the cooling rates, as shown in Fig. 2. They are 132.6, 119.9, 96.3 and 113.8 kgf/mm^2 at 4400, 3500, 2000 and 800 rpm respectively. They run parallel to the changes in lattice parameters of γ matrices till 2000 rpm, but when the spinning speed goes down further to 800 rpm, UTS increases again. This implies the predominance of the solution strengthening effect above 2000 rpm and the manifestation of the precipitation strengthening at lower speeds.

In conclusion, we have learned that, for a superalloy of a specific composition, its cast structures and microstructures after rapid solidification are regulated by the cooling rates. A similar structure is obtained at two points so long as their cooling rates are the same, with no regard to the shape ofthe material and the cooling medium.

REFERENCES

1. P. R. Holiday, A. R. Cox znd R. J. Patterson II, in Proc. Int. Conf. on Rapid Solidification Processing: Principles and Technologies (edited by R. Mehrabian, B. H. Kear and M. Cohen), p. 246, Claitors Publ. Div. (1977).
2. N. Shohoji, H. A. Davies H. Jones and D. H. Warrington, in Proc. 4th Int. Conf. on Rapidly Quenched Metals (edited by T. Masumoto and T. Suzuki), p. 1529, Jap. Inst. Met. (1982).
3. J. V. Wood, P. F. Bee, J.Mater. Sci. 15, 2709 (1980).
4. R. J. Patterson II, A. R. Cox and E. C. van Reuth, J. Metals, 39, No. 9, 34 (1980).
5. B. H. Kear, P. R. Holiday and A. R. Cox, Met. Trans., 10A, 191 (1979).
6. Zhang Shouhua, Wang Naiyi and Hu Benfu, to be published.
7. P. K. Footner and B. P. Richards, J. Mater. Sci. 17, 2141 (1982).
8. Mutaftschiev, in Dislocations in Solids (edited by F. R. N. Nabarro), vol. 5, p. 57, North-Holland Publ. Co. (1980).

Consolidation of Rapidly Solidified High Carbon Iron Base Alloys

D. Burchards, K. U. Kainer and B. L. Mordike

Institut für Werkstoffkunde und Werkstofftechnik, Technische Universität Clausthal, D 3392 Clausthal-Zellerfeld, Federal Republic of Germany

ABSTRACT

High carbon iron base alloys with additions of Cr, Si, Ni, Mn and Ti were melted and rapidly quenched by water atomising to produce powder. The rapid cooling (10^4 K/s) enables the retention of super- saturated metastable austenite with carbides. The alloy composition and particle size, fixed by the production parameters, determine the microstructure of the particle. The microcrystalline powder was consolidated using conventional powder metallurgical techniques. The mechanical properties of such material was determined and correlated with particle microstructure and the consolidated microstructure.

KEYWORDS

Rapid solidification, high carbon iron alloy, powder metallurgy, mechanical properties, microstructure.

INTRODUCTION

Powder produced by rapidly quenched conventional high carbon white iron can be consolidated using standard powder metallurgical techniques to produce hard but ductile materials (1,2). Rapid quenching enables microstructures to be obtained, which are characterized by the following (1): - fine-disperse microstructure, high degree of supersaturation, metastable phases and homogeneous distribution of phases and elements.

In consolidation and working an attempt must be made to retain the beneficial effects of rapid solidification. When this is successful technologically interesting conditions of properties can be achieved. Rapidly quenched powders can be consolidated in various ways (3,4): uniaxial warm compaction and rolling, warm isostatic pressing, warm pressing. Such consolidation Fe- 3% C and Fe- 1.5 Cr 3% C alloys exhibit high tensile strength at room temperature and superplastic behaviour at temperatures between 600 and 850°C (3,4). In the above work the powder of 55 µm size was used and consolidated at 650 - 670°C. In this work rapidly quenched powders of various particle sizes were consolidated by extrusion or forging. The aim was to obtain the same properties with less effort.

EXPERIMENTIAL DETAILS

The preparation and consolidation of the microcrystalline powder follows the scheme outlined in Fig.1. The alloy compositions listed in Table I were melted, water atomized and divided into into fractions of predetermined particle size. For each particle size billets were prepared by cold compaction and hot extruded or forged at a temperature between 850 and 950°C depending on the composition.

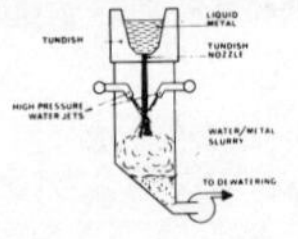

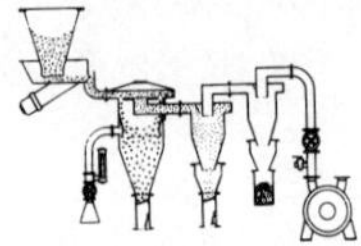

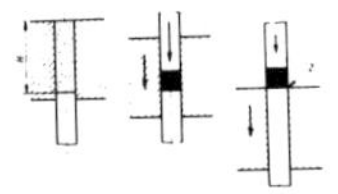

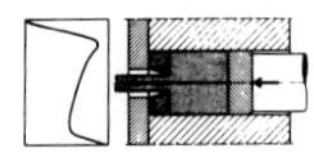

Fig.1 Flow chart

No.	C	Si	Mn	P	S	Cr	Ni	Ti	V
1	2.91	0.47	0.57	0.026	0.142	0.042	0.034	-	-
2	3.60	0.07	0.03	0.006	0.004	0.020	0.030	-	-
3	3.60	0.17	0.05	0.007	0.004	1.650	0.020	-	-
4	3.60	0.17	0.05	0.007	0.004	3.060	0.040	-	-
5	2.60	2.03	0.12	0.011	0.008	-	0.030	-	-
6	3.40	1.98	0.08	0.016	0.007	0.003	0.040	-	-
7	4.60	2.20	0.08	0.010	0.010	0.040	0.030	-	-
8	2.76	0.64	0.48	0.017	0.100	6.840	0.130	0.005	0.027
9	2.84	0.64	0.56	0.019	0.158	3.240	0.070	0.004	0.008
10	2.65	0.60	0.57	0.014	0.163	2.410	6.900	0.128	-
11	3.20	0.89	0.59	0.035	0.035	0.260	3.440	2.040	0.080

TABLE I Chemical analysis of the used alloys

RESULTS AND DISCUSSION

On the rapid solidification high carbon iron particles solidify either dendritically or cellularly with an alloy dependent residual eutectic (Fig.2) and supersaturated austenite.
The solidification rate increases with decreasing particle size. It can be estimated from the secondary dendrite arm separation.

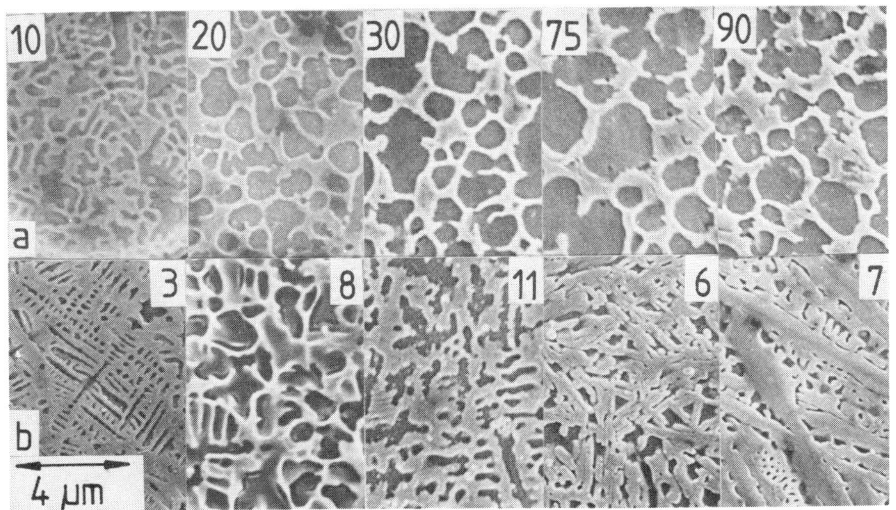

Fig.2 a) Alloy 10, microstructure as a function of particle size (10 - 90 µm)
b) Alloys 3,6-8,11, microstructures of 20 µm particles for different compositions.

In the 20 µm powders shown above the rate was 2 x 10^5 K/s. The dependence of solidification rate on particle size determines the microstructure of the particles. This is evident from Fig.2a for alloy 10. In this example very fine particles (d ∿ 10 µm) solidify dendritically as the temperature gradient is low and the growth rate is high (5). Large particles (d ∿ 20 µm) with medium temperature gradients and growth rates solidify cellularly. In change particles size (d > 30 µm) no further influence of cooling rate is apparent. The cell diameter remains virtually constant. This alloy composition was an exception. Most others investigated solidified dendritically over a large range of particle size. The alloy composition is also important in determining the microstructure. For constant particle size d = 20 µm) this effect is shown in Fig. 2b. How Cr-contents lead to fine dendritic solidification (specimen No. 3). On increasing the Cr-content the carbides became coarser and a change from dendritic to cellular solidification is observed (specimen No. 8). The eutectic fraction is reduced by increasing the Ni-content of Cr free alloys (for same carbon content, specimen No. 11). Nickel expands the γ-region and favours undercooling of austenite. Hypoeutectic Fe-C-Si alloys solidify dendritically (Fig. 4). At the eutectic composition (Fig. 2b, No 6) a cellular microstructure with a fine lamellar eutectic is obtained. The hypereutectic silicon alloy (spec. No 7) contains large needle like primary carbides with eutectic.
The dependence of hardness on particle size shows the influence of the microstructure on the mechanical properties of rapidly solidified white iron powders (Fig. 3).
The hardness increases with decreasing particle size. In addition the influence of hardness on alloy composition is marked.

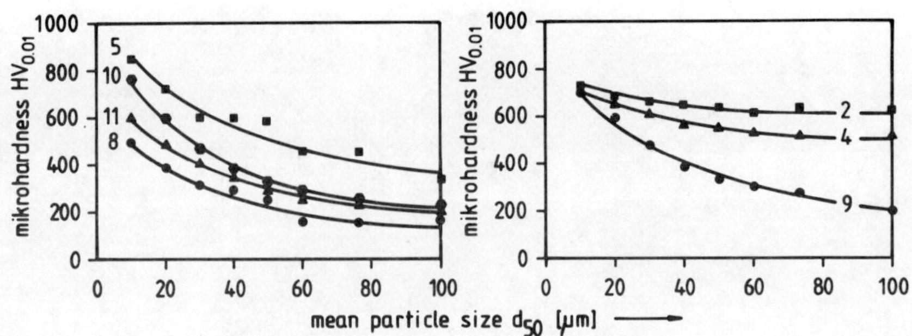

Fig.3 Hardness dependence of RST powder in the as quenched state

The highest hardness values are attained by the Si-containing alloy. In the serious Ni, Ti and Cr-containing powder lower values are obtained in the as quenched state. High carbon content unalloyed white cast iron powders (spec. No 2) show high hardness values with only weak influence of the particle size. If the RST powder is heat treated the supersaturated austenite decomposes. Depending on the alloying additions either graphite or carbide precipitates. Silicon decreases the solubility of carbon in austenite and leads to graphite precipitates (Fig. 4 e,f). After heat treatment the material is soft and the dependence on particles size disappears (Fig. 5, spec. No 5). Chromium a carbide former leads on heat treatment to the formation of ferrite and spheroidal mixed carbide (Fig. 4a,d). The material 8 and 9 are then very hard.

The structure changes on heat treatment much less than for Si-containing alloy powder (Fig. 4). Cr acts as a grain size stabilizer. A difference in the behaviour between the initially cellular and dendritic structures is not apparent. Ni increases the solubility of carbon in austenite and the volume fraction of carbide is thus increases marginally on heat treatment. The high hardness of heat treated specimens (Fig. 5, No 10) derives from martensite which can form on air-cooling this composition.

An attempt was made to extrude the RST powders. In some cases this was possible - in others the material can be considered as hot pressed or forged (density 97%). The extruded material is 100% dense apart from occasional pores.

The tensile properties and hardness of the extruded rods as well as the hardness of the forged material were determined as a function of the initial particle size (Table 2). The forged Cr alloys showed a high hardness for the coarse powders (d_{50} = 75 μm), low hardness for the medium sized powders (d_{50}=38 μm). The hardness increases again somewhat for the finer powders. The microstructure for the medium sized powders consists of a relatively coarse carbide net in contrast to the structures for the fine and coarse particles.

The extruded silicon containing alloys are very soft compared with chromium containing alloys and the hardness exhibits no dependence on particle size. The heat required for extrusion is sufficient to promote precipitations of graphite (Fig. 6b, No 5). Compared with conventional cast iron of this composition the powder product has a higher tensile strength (Table 2, No 5)

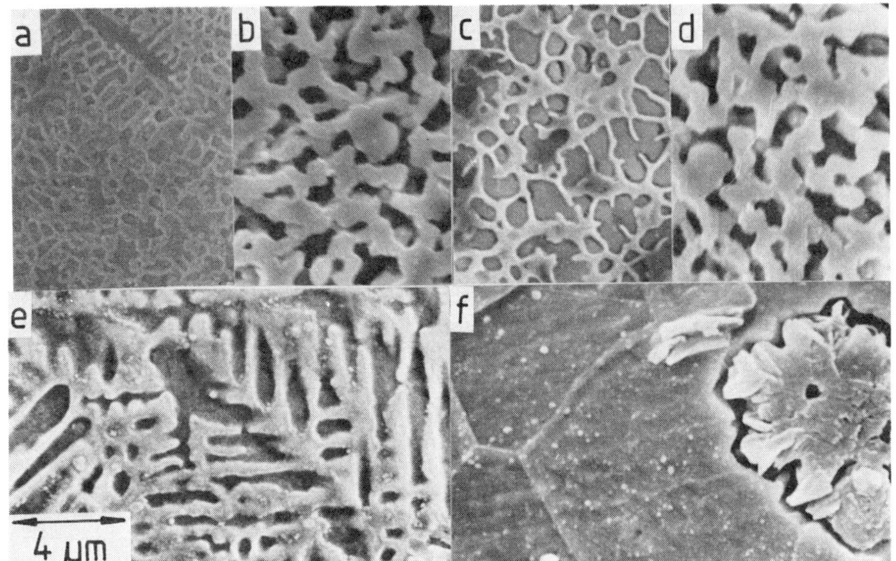

Fig. 4 Microstructure of as quenched and tempered RST iron base powder
a) No 9 as quenched, b) No 9 tempered
c) No 10 as quenched, d) No 10 tempered
e) No 5 as quenched, f) No 5 tempered

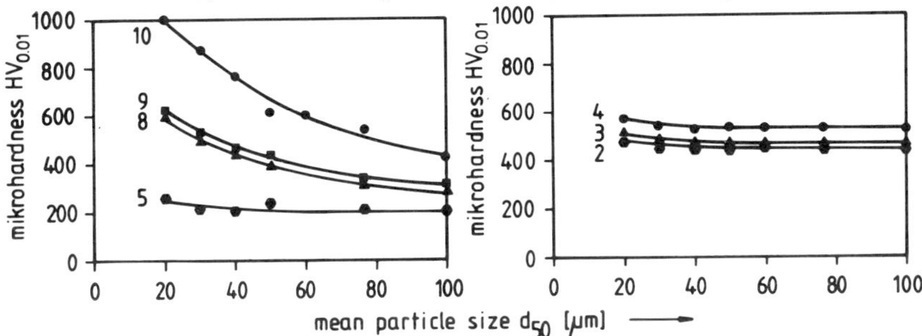

Fig.5 Hardness dependence of heat treated RST powder ($870^{\circ}C$, 30 min)

with good ductility (12 - 18 %). The hardness of Ni-containing forged specimens (No 10) increases continuously with decreasing particle size. The high hardness indicates the presence of martensite (Fig. 6b, No 10).
The effect on the microstructure and properties of using Tellurium as a grain refiner was investigated. 0.009 % was added to the melt. Te has a 10^3 greater effect on the supercooling on the melt compared with Cr (Fig. 6b, No 1) (6).
The superplastic behavior of high carbon content RST powder was investigated using the extruded specimen 4. The microstructure (Fig.6b, No 4) consists of a carbide network which ensures

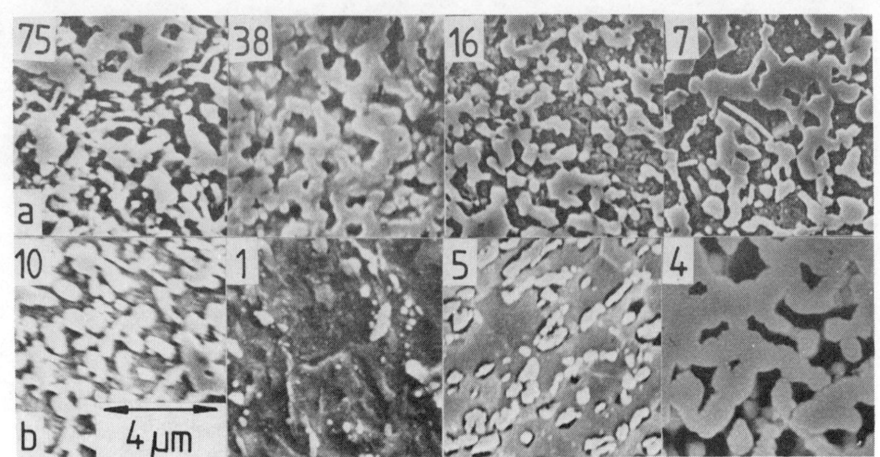

Fig.6 Microstructure of RST-powder a) as a function of the particle size (75 µm - 7 µm) No 8; b) No 1,4, 5,10 for d_{50} = 20µm

strength at room temperature. At temperatures of T ∿ 0.6 T_m (T=670°C)(1,2,3) the material is very plastic ε_p ∿ 60% which suggests that superplastic behaviour may be observed for three alloys.

No	temp. °C	d_{50} (µm)	HV 10	UTS N/mm²	No	temp. °C	d_{50} (µm)	HV 10	UTS N/mm²
1 +	RT	25.0	575	1098	5 +	RT	81.5	200	568
2 +	RT	17.0	543	859	5 +	RT	49.0	280	581
3 +	RT	40.0	460	1032	5 +	RT	25.0	290	583
4 +	RT	25.0	447	1145	10 *	RT	74.2	661	-
4 +	670	25.0	-	44	10 *	RT	14.8	824	-
4 +	670	17.0	-	92	10 *	RT	7.5	840	-
8 *	RT	77.5	692	-	11 +	RT	75.0	379	873
8 *	RT	38.0	477	-	11 +	RT	36.0	363	655
8 *	RT	16.2	523	-	11 +	RT	15.0	380	855
8 *	RT	6.9	529	-	11 +	RT	8.0	381	951

TABLE 2 Properties of extruded (+) or forged (*) RST powder

REFERENCES

1. H. Jones, Rapid Solidification of Metalls and Alloys, Institution of Metallurgists, London (1980).
2. B. Walser and O.D. Sherby, Metall. Trans. A. <u>10</u>, 1461 (1979).
3. O.A. Ruanu et al., Metall. Trans. A. <u>13</u>, 1785 (1982).
4. A.A. Bochvar et al.,Doklady Academia Nauk, USSR, 230 (1976).
5. R. Mehrabian, Met. Rev. <u>27</u>, 185 (1982).
6. A.A. Zhukov and I.A. Vashukov, Proc. of Solidification Technology in Foundary and Casthouse, Met. Society, London, (1980).

Study of Mechanical Properties of Al-Cu Alloys in Semi-solid State During Solidification

Li Qingchun*, Zeng Songyan* and Liu Chi*

**Department of Metallic Material and Technology, Harbin Institute of Technology, Harbin, People's Republic of China*

ABSTRACT

In this paper the influence of strain rate and trace additions on the strength, failure strain, rate of strength increase in semi-solid state and temperature range of the semi-solid of Al-Cu alloys during solidification is investigated with special apparatus.

KEYWORDS

strength, failure strain, rate of strength increase, semi-solid state, temperature range of the semi-sloid state, semi-solidus, solidification.

INTRODUCTION

The data of the alloy mechanical properties during solidification are indispensable to the casting process. Alloys in the semi-solid state possess the strength as a solid does. The mechanical properties in the temperature range of the semi-solid state are very low and increase as temperature of alloys decreases. So the temperature range of the semi-solid state and rate of strength increase, increase of tensile strength unit decrease of temperature, are also important parameters of alloys in the semi-solid state. That is the reason why the mechanical properties, temperature range of the semi-solid state and rate of strength increase of Al-Cu alloys are investigated.

It is difficult to measure the properties in the semi-solid state because they are very low. Previous studies[1,2] did not take the nonuniformity of the temperature distribution in the specimen into consideration. Especially there is little information about the influence of trace additions on the mechanical properties. In this paper the strength and failure strain of

Al-Cu alloys at constant temperatures in the temperature range of the semi-solid state during solidification were measured at a strain rate of 0.3 1/min with special apparatus. Using this apparatus the temperature deviation in the specimen is less than 0.5 ^{0}C and minimum strength which can be measured is 0.01 MPa.

EXPERAMENTAL RESULTS

The strength of Al-4%Cu alloy in the temperature range of the semi-solid state is shown in Fig. 1. And the temperature range of the semi-solid state(Rss), rate of strength increase(Rsi) and solidus(Ts) can be determined. Semi-solidus(Ty) can be defined as a line, below which the alloys begin to get strength. Failure strain is the ratio of failure elongation to the gauge length of the specimen.

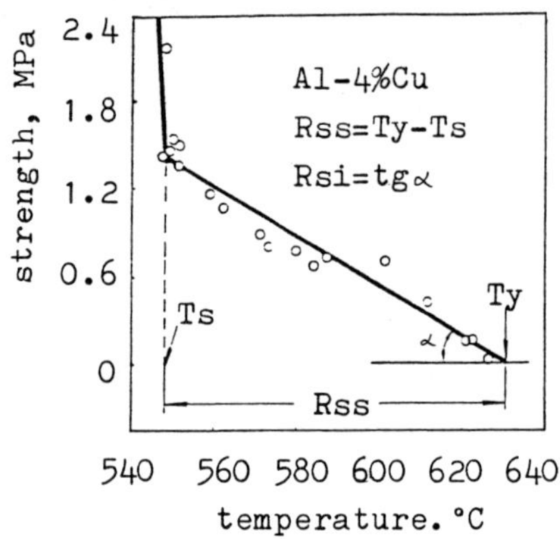

Fig. 1. The strength in the semi-solid state versus temperature of Al-4%Cu alloy.

Influence of Strain Rate on the Mechanical Properties

Figure 2 shows the influence of strain rates on the strength and failure strain of Al-4%Cu alloy in the temperature range of the semi-solid state.

Figure 2 shows the strength and failure strain increase as the strain rate increase, since the creep and stress relaxation do not occure if the strain rate is sufficiently high.

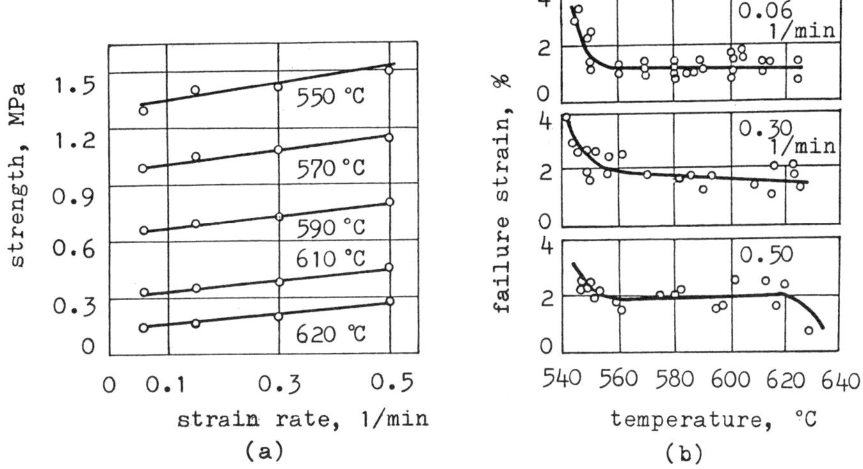

Fig. 2. Influence of strain rates on the strength and failure strain of Al-4%Cu alloy.

Influence of Additions of Rare Earths on the Mechanical Properties

Influence of additions of mischmetal(RE), La and Y on the strength and failure strain of Al-5%Cu alloy in the temperature range of the semi-solid state is shown in Fig. 3 and Fig. 4.

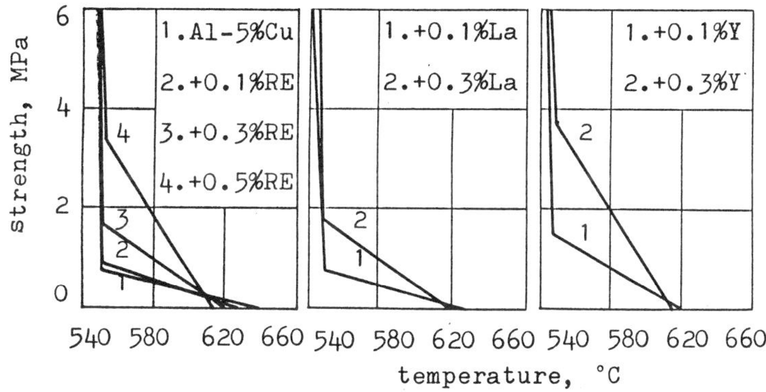

Fig. 3. Influence of additions of RE, La and Y on the strength of Al-5%Cu alloy.

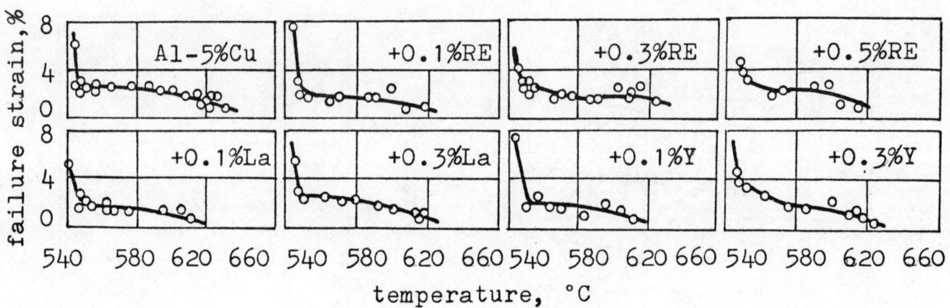

Fig. 4. Influence of additions of RE, La and Y on the failure strain of Al-5%Cu alloy.

The rare earths are able to lower the semi-solidus and increase the strength in later stage of the semi-solid state. When the additions of the rare earths are more than 0.3%, the failure strain in the later stage of the semi-solid state increases only slightly. The temperature range of the semi-solid state can be reduced and the rate of strength increase can be increased by the rare earth additions.

Redistribution of the rare earths in the alloy makes the rare earths accumulate mainly at grain boundaries. It will refine the grains and strengthen the boundary in the later stage of the semi-solid state, and change the mechanical properties.

Influence of additions of Mn and Ti on the Mechanical Properties

Influence of additions of Mn and Ti on the strength and failure strain of Al-5%Cu alloy in the temperature range of the semi-solid state is shown in Fig. 5.

By the addition of 0.3%Ti grains are refined and the influence of Ti addition on the mechanical properties is the same as that of the rare earth additions. By the addition of 0.9%Mn the grains are coarsened and the semi-solidus arises. At the same time the rate of strength increase increases. Influence of the additions of 0.3%Ti and 0.9%Mn will reduce the failure strain.

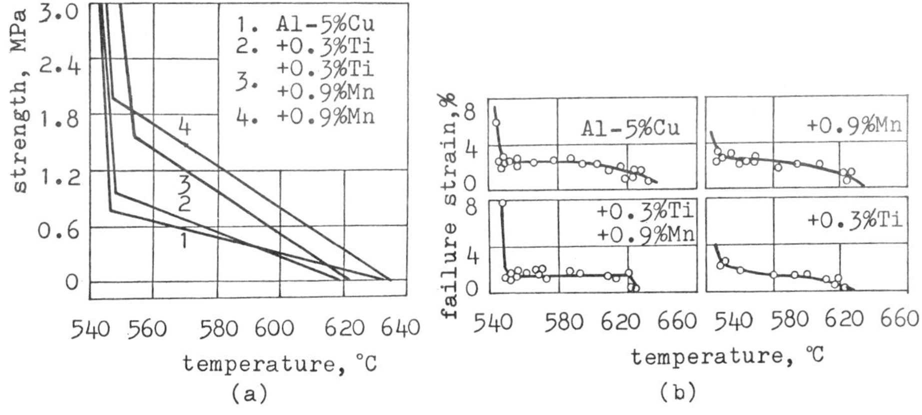

Fig. 5. Influence of additions of Mn and Ti on the strength and failure strain of Al-5%Cu alloy.

CONCLUSION

The mechanical properties of Al alloys in the semi-solid state during solidification are related to the strain rate, solidification process and grain structure. As strain rate increases from 0.06 1/min to 0.5 1/min, the tensile strength and failure strain increase. The intergranular liquid is the main factor influencing the mechanical properties. As temperature decreases the intergranular liquid decreases and the strength and failure strain increase. Additions of RE, La , Y, Ti and Mn will change the quantity and distribution of the intergranular liquid and thus influence the mechanical properties.

REFERENCES

1. Toshio Isobe, Masao Kubota and Sanji Kitaoka, J. Japan Foundrymen's Soc., 160, 53(1981).
2. Ching-Tsan Lin and Takeshi Nakata, J. Japan Foundrymen's Soc., 28, 56(1984).

Effect of Microstructure on Compressive Properties of Stir Cast (Rheocast) Zinc Alloys

H. LeHuy, J. Masounave and J. Blain

Industrial Materials Research Institute, National Research Council Canada, 75 de Mortagne Blvd., Boucherville, Quebec J4B 6Y4, Canada

ABSTRACT

Properties in compression of semisolid Zn-Al alloys with non dendritic structure produced by a stir-casting technique were studied. Small cylindrical specimens with different volume fraction of primary solid were deformed between two parallel plates at constant strain rate of 0.72 mm mm^{-1} s^{-1}. The testing temperature, ranging from 390 to 430°C, covered the first remelting stage of the alloys. The maximum measured compressive strength was found to be lowered by an increase in the amount of primary solid particles present in the alloys. These experiments illustrate the advantages of forming operations of metal alloys in their semisolid state.

INTRODUCTION

There are many recent reports in the literature describing studies[1-3] on sheared partially solid metal alloys and the application of the findings to innovate and improve casting processes. The fundamental point of this emerging solidification processing technique is a vigorous agitation of a solidifying alloy from the time solidification starts until the metal is cast. The stirring action breaks up the dendrites and creates a unique structure[4,5], where spherical or rosette shaped particles are suspended in the remaining liquid. The viscosity of sheared alloys can be maintained low[6], so that slurries with up to 0.60 fraction of solid can flow smoothly.

The stir-cast metal techniques are not limited to foundry applications. Recent studies of the mechanical processing of engineering materials in their sheared semisolid state showed much promise. In die-casting techniques[7,8], the die life may increase because of the low temperatures involved; in squeeze casting[9,10], the advantageous features of both casting and forging were combined; in metal forming[11-13], mechanical processing may reduce the force applied during forging or extrusion. In an earlier paper[14], it was mentioned the different deformation behavior of semisolid alloys which depends on their dendritic or non dendritic structure. For stir-cast alloys, new deformation mechanisms were observed. The compressive resistance is low and the microstructure is quasi-unchanged after deformation. In the same way, Taha et al[15] showed that extrusion can be performed advantageously in the semi-solid state with non dendritic alloys obtained by mechanical stirring during solidification. However, a better comprehension of the phenomena occuring in the related processes is needed. As opposed to other testing methods where complex deformation mechanisms occur congruently, the compression test between two parallel plates produces a simple deformation pattern which can be easily analyzed.

The aim of this paper is to report some experimental results concerning the effects of the microstructure on the compressive properties of a common foundry alloy. A zinc-aluminum alloy has been used in this investigation because of its casting interests and its relatively large solidification range.

EXPERIMENTAL DETAILS

A commercial zinc alloy ZA-27 with 27 wt% aluminum and 2 wt% copper was used for this study. Its nominal composition and the liquidus and solidus temperatures are given in Table 1. The stir-cast materials were prepared in batches of 10 kg of base alloy, continuously cooled and sheared from the fully liquid phase to the desired fraction of solid, held isothermally during 5 minutes then poured in graphite molds. Experimental casting details were already described in an earlier paper[6]. Four stir-cast alloys were selected for this study, with their solid fraction varying from .15 to .35. The temperature of the alloys was controlled to within $\pm$ 1°C by chromel-alumel thermocouples inserted in the apparatus. The stirring velocity was identical for all tests, i.e. v = 600 rpm, corresponding to a maximum shearing rate, $\dot{\gamma}_{max}$, of 500 s^{-1}. The cooling rate, G, measuring the rate of solidification, ranged from 6 to 10°C/min.

Table 1. Characteristics of the Zn-Al alloys studied.

Base Alloy				
Composition of ZA-27:	25-28 wt% Al; 2-2.5 wt% Cu; 0.15 wt% Mg; 0.002 wt% Pb; balance, zinc.			
Liquidus temperature:	492°C			
Solidus temperature :	437°C			
Eutectic temperature:	382°C			
Stir-cast specimens				
- Stirring temperature	480°C	472°C	465°C	460°C
- Initial fraction solid	.15	.23	.30	.35
- Density (g/cm^3)	4.76 $\pm$ .02	4.78	4.84	4.90
- Primary particles size	nucleous 30-50 μm	nucleous 30-50 μm	rosettes 100 μm	rosettes clusters 100-300 μm

The compression tests were conducted according to specifications put forward by the ASTM Standard methods of compression testing of metallic alloys as in Ref. 16. The testing specimens, machined from cast ingots, were cylindrical with a 20.05 mm diameter, and a height of 12.70 mm. They were deformed between two parallel plates of high hardened steel, heated to the desired temperature with resistive elements. Graphite lubricant was used to minimize the friction between the specimen and the plates. Chromel-alumel thermocouples embedded in the bottom and top plates were used to control temperature uniformity of the compressive specimens during the tests. The deformation was performed at a constant strain rate of 0.72 $mm.mm^{-1}$ s^{-1} (deformation velocity, 9 mm s^{-1}) using a standard 10 tons MTS testing machine. A minimum of three tests were performed on each alloy at different temperatures, ranging from 390°C to 430°C, covering the solid phase and the beginning of semisolid phase. The average results of these tests are presented in Table 2.

Table 2. Data of the compressive resistance of stir-cast alloys.

	Stir-Cast Temperature (°C)	480	472	465	460
(1)	Testing Temperature: 390°C				
	- max. compr. strength (MPa)	29 ±5	24.5 ±5	15.4 ±2	12.5 ±2
	- strain ε	.170	.145	.090	.065 ±.005
	- strain rate ε	.81	.78	.77	.74 ±.02
(2)	Testing Temperature: 410°C				
	- max. compr. strength (MPa)	7.25 ±2	6.25 ±2	4.9 ±2	2.7 ±1.5
	- strain ε	.040	.050	.030	.030
	- strain rate ε	.71	.71	.70	.70
(3)	Testing Temperature: 430°C				
	- max. compr. strength (MPa)	1.2 ±.80	.60 ±.40	-	-
	- strain ε	.060	.060	-	-
	- strain rate ε	.71	.71	-	-

CHARACTERISTICS OF STIR-CAST ALLOYS

Figure 1 shows typical micrographs of dendritic and non dendritic ZA-27 alloys. The structure of dendritic specimens (Fig. 1a) consists of aluminum-rich dendrites and an interdendritic eutectic rich in copper and zinc. In contrast, the stir-cast structure (Fig. 1b) shows degenerated dendrites embedded in a remaining liquid phase. The primary particles are fairly round in shape and homogeneously distributed in the cast ingots. The amount of primary particles, defined as fraction solid f_s, is a function of the stirring temperature. The measured values, obtained by metallographic measurements, are in reasonable agreement with the thermodynamic calculated ones, assuming thermodynamic equilibrium conditions. A numerical fit based on the phase diagram[17] of Zinc-Al alloys gives the following relation:

$$f_s = \frac{C_o - 1980.15 + 13.44\,T - 0.03\,T^2 + 2.10^{-5}\,T^3}{3736.80 - 20.24\,T + 0.04\,T^2 - 2.10^{-5}\,T^3} \qquad (1)$$

where C_o is the nominal concentration of the alloy.
T, the temperature in °C.

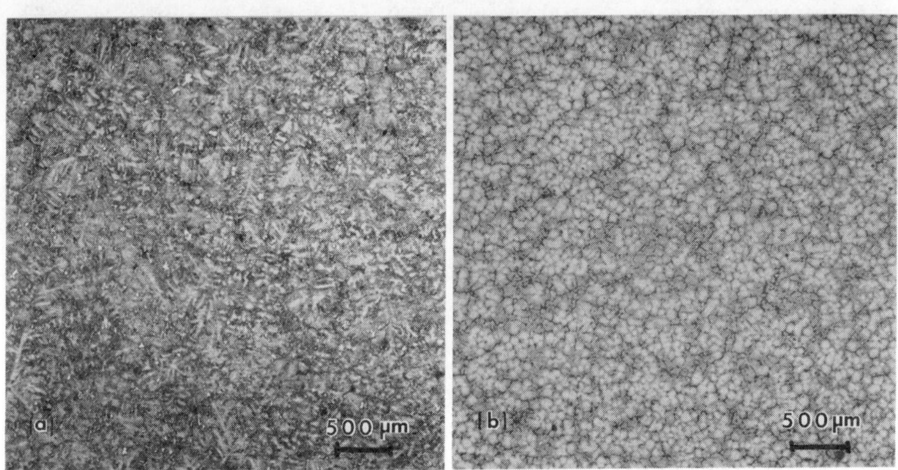

Fig. 1 Microstructure of the dendritic and non dendritic ZA-27 alloys.
a) Conventional mold casting; b) Stir-cast at 460°C

It has been previously reported[7] that non dendritic materials obey a power-law fluid model:

$$\eta_a = k\dot{\gamma}^n \qquad (2)$$

where η_a is the viscosity
and $\dot{\gamma}$ the shear rate.

At low shear rates (less than 600 sec^{-1}), the slurries display solid-liquid mixture with viscosity up to 10 poises. By decreasing the temperature, the slurries are thickened. Statistical measurements of particle size and their distribution show that the increase of solid fraction is principally due to enhanced nucleation[6].

The application of mechanical stirring produces a vortex which can lead to gases being trapped in the slurry. The measured bulk density of slurries, reported in Table 1, indicates an increase of density with the decrease of temperature up to a fraction solid of approximately .35 after which the density starts to decrease. This porosity, finely dispersed throughout the ingots, is clearly visible in Fig. 1b.

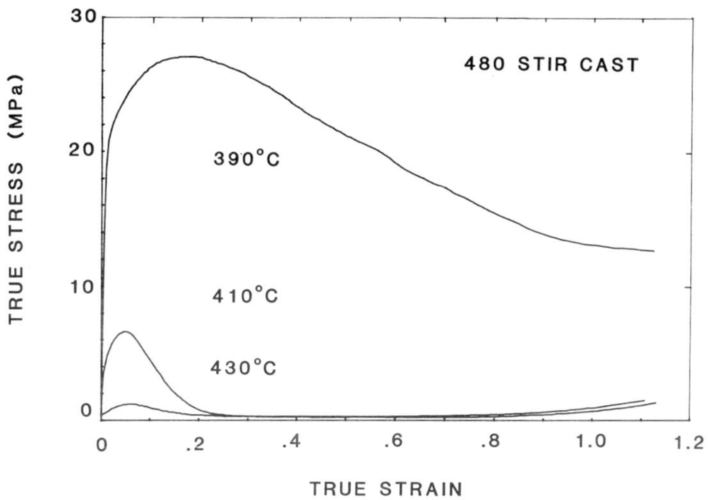

Fig. 2 True stress vs true strain of ZA-27 alloys deformed at 0.72 mm mm^{-1} s^{-1} and at various temperatures (see Table 1 for treatment).

Fig. 2 shows compressive stress versus true strain curves for stir-cast materials poured at 480°C, which were deformed at a constant strain rate of 0.72 mm mm^{-1} s^{-1}. By increasing the testing temperature, the amount of liquid present in the specimen during the compression increased, and therefore the measured stress decreased. A maximum stress is seen in the curves at a strain of about 0.06 to 0.20 depending on the testing temperature, followed by a decrease and finally an increase at higher strains. The first maximum in strength, particularly interesting for forming applications, is related to the resistance of the semisolid structure under deformation and is discussed in details in the next section. The following decrease of stress was attributed(13) to the breakdown of the structure whereas the increase of stress at higher strain (i.e. ε more than 0.8) was principally due to the friction of materials with the compression plates. Compared to the dendritic ones, the compressive resistance of stir-cast alloys is ten times lower. Furthermore, segregation in compressed specimens is less severe in stir-cast alloys than in dendritic ones. Studying the extrusion of Pb-Sn stir-cast alloys, Taha et al(15) found that non dendritic alloys behave like a viscous fluid. A possible explanation for both deformations is that the primary particles are relatively isolated from each other so they can move freely under strain(14).

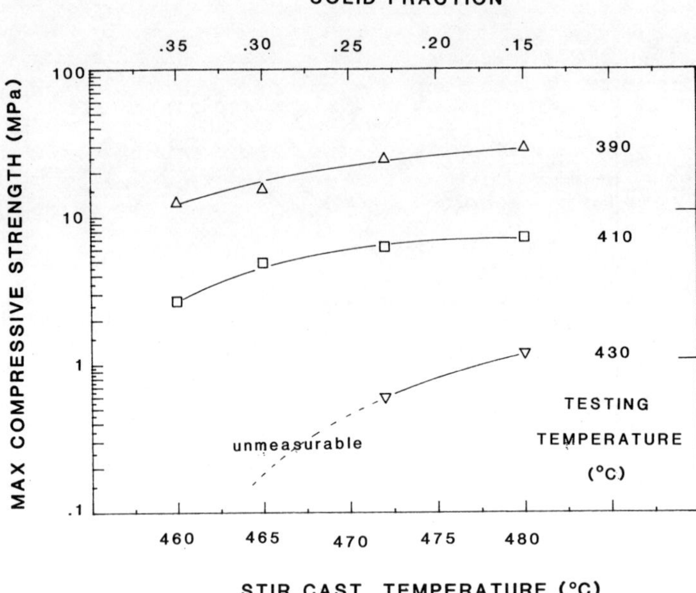

Fig. 3 Compressive resistance of sheared zinc-aluminum alloys in function of the initial solid fraction and at different testing temperatures. (See Table 1 for treatment)

Figure 3 shows the variation of the compressive resistance of stir-cast materials versus the initial solid fraction and at different testing temperatures ranging from the eutectic temperature (T_{eut} = 382°C) up to the solidus temperature (T_{sol} = 437°C). The experimental data shown in Table 2 for the four alloys with primary particle fraction of 0.15 to 0.35 indicate that the measured forces decrease rapidly when the amount of solid particles present in the slurries increases. For stir-cast alloys poured at 460°C and 465°C, the compressive resistance becomes negligible. In this case, the measured forces, which were less than 0.1 KN (corresponding to σ = .3 MPa) are very weak and near the measurement limit of the testing machine. The friction resistance between the sample and the plates is not negligible. It should also be noted that, when the compressive stress decreases, either by increasing the testing temperature or by increasing the initial fraction solid, the maximum stress shifts toward lower strains. When examining the deformation of semisolid Sn-Pb alloys, Suery et al(12) have already reported this phenomenon, but no consistent explanations were given. It is tempting to associate the breakdown of the structure with a decrease of the viscosity as reported by Laxmanan[13]. It is, however, clearly shown that the pseudo-plastic nature of the deformation of semisolid stir-cast alloys is due both to the initial primary solid phase (f_s) and to the fraction of liquid (g_L) formed during the tests. This unique property appears to be readily adaptable to practical forming process.

CONCLUSION

The study in compression of semisolid stir-cast zinc-aluminum alloys was conducted by varying the initial fraction solid of the slurries from .15 to .35. Results show that deformation is controlled both by the primary solid phase and by the fraction of liquid formed during the compression test. The decrease of compressive strength is more pronounced for large volume fraction of solid phase. By reheating the slurries in the semisolid state, the interparticle metal is remelted, leading the particles to move freely under strain. When the compressive stress decreases, the maximum shifts toward lower strains. No consistent explanations are yet given for this phenomenon.

ACKNOWLEDGEMENT

The authors gratefully acknowledge G. St-Amand for his assistance in the preparation of the alloys and M. Thibodeau for his technical assistance.

REFERENCES

1. N.A. El-Mahallawy, N. Fat-Halla and M.A. Taha, in Proceedings of the Fourth International Conference on Mechanical Behavior of Materials, (edited by J. Carlsson and N.G. Ohlson), p. 695. ICM4, Stockholm, Sweden (1983).
2. R.D. Doherty, H.I. Lee and E.A. Feest, Mat. Sci. and Eng., 65, 181-189 (1984).
3. T.Z. Kattamis, Proc. of the Int. Symp. Delft, 189 (1977), (Delft University of Technology 1977).
4. A. Vogel, R.D. Doherty and B. Cantor, "Solidification and Casting of Metals" (The Metals Society, London), 518, (1979).
5. Rheocasting, Proceedings of a workshop held at the AMMRC (1977), MCIC Report, Columbus, Ohio.
6. H. LeHuy, J. Masounave and J. Blain, J. of Mat. Sci., in press.
7. K.P. Young, R.G. Riek and M.C. Flemings, Proc. Conf. on the Solidification and Casting of Metals, Sheffield, 1977, (The Metals Society, London), 510 (1979).
8. C.C. Law, J.D. Hostetler and L.F. Schulmeister, Mat. Sci. Eng., 38, 123 (1979).
9. M. Kiuchi, S. Sugiyama and K. Arai, 20th Int. Machine Tool Design and Research Conference, Birmingham, 1979, London, McMillan 71-86, (1980).
10. D. Gelderloos and K. Karasek, J. Mat. Sci. Lett., 3, 232 (1984).
11. D.A. Pinsky, P.O. Charreyron and M.C. Flemings, Metall. Trans. B, 15, 173 (1984).
12. M. Suery and M.C. Flemings, Metall. Trans. A, 13, 1809 (1982).
13. V. Laxmanan and M.C. Flemings, Metall. Trans. A, 11, 1927 (1980).
14. H. LeHuy, J. Blain, G.L. Bata and J. Masounave, Mat. Sci. and Eng., 67, 229-232, (1984).
15. M.A. Taha and M. Suery, Metal Technology, 11, 226-230 (1984).
16. ASTM Stand. E9-81, in 1983 Annual Book of ASTM Standards, Vol. 03.01 ASTM, Philadelphia, PA, 140-151 (1981).
    ASTM Stand. E209-65, in 1983 Annual Book of ASTM Standards, Vol. 03.01, ASTM Philadelphia, PA, 365-373 (1981).
17. Metals Handbook, Vol. 8, 8th Edition, Am. Soc. for Metals, Metals Park, Ohio, U.S.A. ,(1972).